THE MATHEMATICAL PAPERS OF ISAAC NEWTON

VOLUME VII

1691-1695

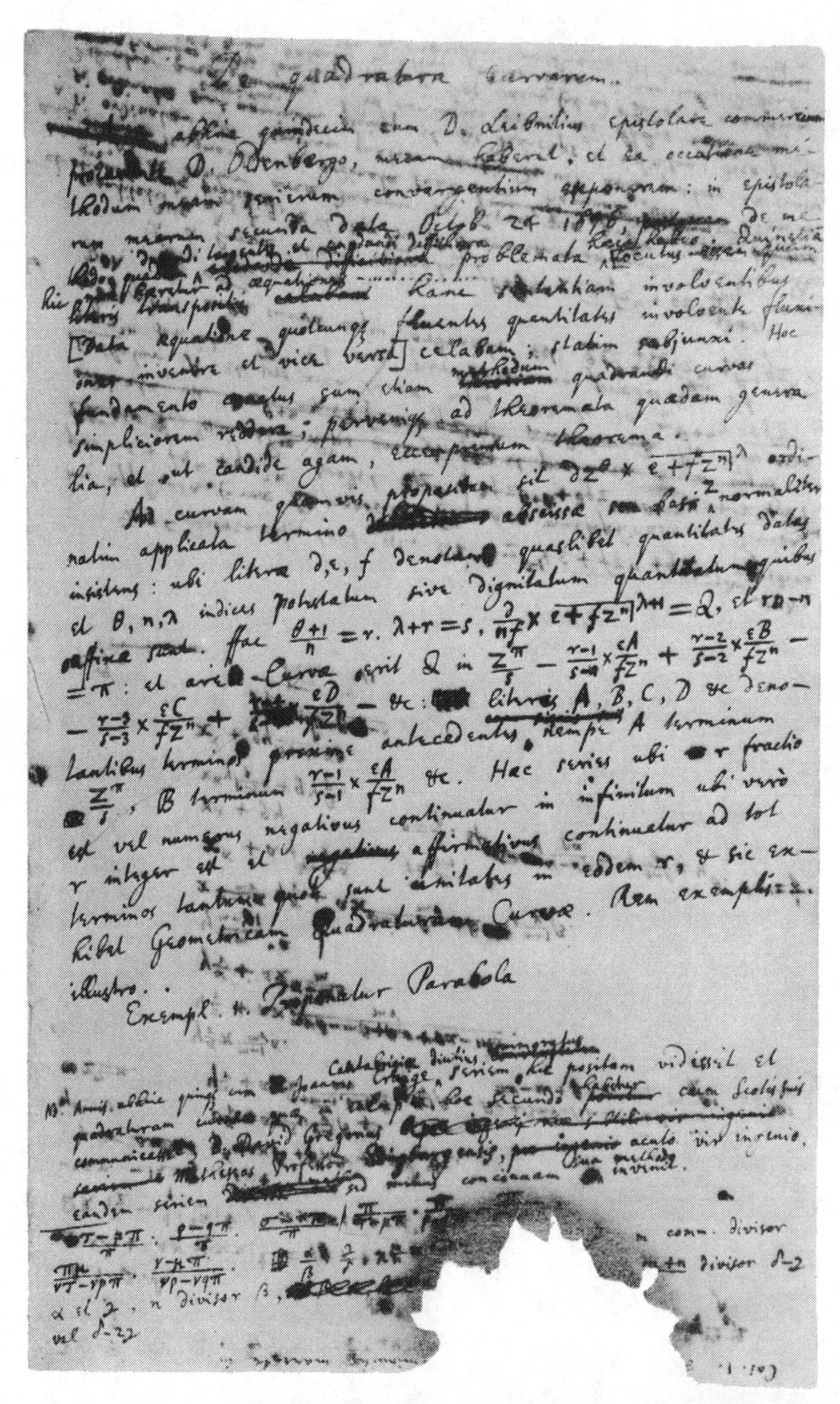

The opening page of Newton's 1691 tract 'De quadratura Curvarum' (**1**, §1).

THE MATHEMATICAL PAPERS OF ISAAC NEWTON

VOLUME VII

1691-1695

EDITED BY
D. T. WHITESIDE

WITH THE ASSISTANCE IN PUBLICATION OF
A. PRAG

CAMBRIDGE
AT THE UNIVERSITY PRESS
1976

CAMBRIDGE UNIVERSITY PRESS
Cambridge, New York, Melbourne, Madrid, Cape Town, Singapore, São Paulo

Cambridge University Press
The Edinburgh Building, Cambridge CB2 8RU, UK

Published in the United States of America by Cambridge University Press, New York

www.cambridge.org
Information on this title: www.cambridge.org/9780521087209

Notes, commentary and transcriptions

First published 1976
This digitally printed version 2008

A catalogue record for this publication is available from the British Library

Library of Congress Catalogue Card Number: 65–11203

ISBN 978-0-521-08720-9 hardback
ISBN 978-0-521-04589-6 paperback
ISBN 978-0-521-72054-0 paperback set (8 volumes)

TO MICHAEL HOSKIN
FOR THE UNSTINTED WARMTH OF HIS
ENCOURAGEMENT OVER SO MANY YEARS

PREFACE

The sixth volume of this edition of his extant mathematical papers closed with Newton, freshly returned to professorial routine in the sluggish backwater of Cambridge after a year's heady racing to and fro in the mainstream of political life in London, making a start on his self-imposed task of recasting and refining the text of the mighty *Principia* which he had given out publicly a little while before. In this sequel we draw as ever upon the vast wealth of his surviving worksheets and drafts chronologically to retrace his mathematical output over the four years from November 1691, terminating in the autumn of 1695 a few weeks before he took (in March the next year) his fateful decision to accept the Wardenship of the London Mint and so, after nearly thirty-five years, effectively sever his ties with a University which had intellectually mothered him and for so long quietly sheltered him from the harsh and earthy realities of the outside world.

These last years of his active tenure of the Lucasian chair at Cambridge were, I need not say, in large part passed in consolidating and retrenching familiar ground which Newton had long since gained, but they also saw fundamental advances into such novel terrains as those of the general Taylor expansion of an algebraic function and of the projective classification of curves, notably the component species of the cubic whose earlier equivalent Cartesian enumeration is here, too, given its final polish; and above all they reveal him taking a hard look back at the higher geometry of the 'ancients' as its traces have endured in the works of its exponents Euclid, Archimedes and Apollonius and of their 'synthesizer' Pappus. While those portions of Newton's present writings on the quadrature of curves and the cataloguing of cubics which were afterwards appended by him, as his 'Tractatus De Quadratura Curvarum' and 'Enumeratio Linearum Tertii Ordinis', to the *princeps* edition of his *Opticks* in 1704 will be well known in their essence to the *cognoscenti*, the hitherto unpublished later propositions of his 1691 'De quadratura' and the massive bulk of his ensuing geometrical researches, whose very existence—not to mention their intrinsic excellence—has gone publicly all but unrecorded in the nearly three centuries since Newton penned their script, will here come mint-fresh to all but the favoured few (if indeed there are any) who have previously gained access to the totality of his jumbled mathematical *Nachlass*. The other small pieces, on conchoidal cubics and foliate quartics and on a variety of contrivances *ad hoc* of rules for interpolation and approximate quadrature, which fill the remaining pages are not of momentous importance, but hold some surprises for those interested to pursue their detail. Let it be enough that the autograph manuscripts now reproduced are no mere resurrected historical curiosities fit only once more to gather dust in some forgotten corner, but will require the rewriting of more than one page in the historical textbook.

As ever, for permission to print the texts of papers in their custody, I stand principally indebted to the Librarian and Syndics of the University Library, Cambridge, but my gratitude is again owed to the same private owner who has, desiring only that his anonymity be preserved, once more allowed me freely to publish the complementary folios in his possession which exactly fill the main gaps in the former's collection of Newton's mathematical papers (and which were removed therefrom in Newton's own day by William Jones). While the University of Cambridge has now increasingly taken over the burden of financing the long hours of carrying out preparatory research, collating manuscript and printed material, and producing the finished editings of texts which are here presented, I must continue to acknowledge the monetary support provided, now as for more than ten years past, by the Sloan Foundation of Philadelphia, the Leverhulme Trust and the Master and Fellows of Newton's own Trinity College, Cambridge. What can I find to say in communally thanking them all which time has not unworthily stiffened into cliché? Perhaps that a monument such as this to an intellectual giant of the past is not raised on the shoulders, however broad in the flesh or willing in the spirit, of its editor alone.

It will be seen that my accredited coadjutors are reduced to be but one. This is, as always, to ignore the smaller efforts taken by others in affording me piecemeal factual information and corrections of minor detail: this volume, in particular, wears a kinder and more balanced face to David Gregory through the gentle persistence of Miss Christina Eagles. Adolf Prag, who remains to bedeck the title-page, has long since earned my supreme accolade of being accepted as ever-available and near-omniscient helpmate and prime critic in guiding my volumes through into their published form; so many of their finer nuances of mathematical, verbal and historical detail are his. To my other past helper Michael Hoskin, who has vanished from the title to be properly (I may hope) and worthily reincarnated in my dedication of this volume to him, I convey my warmest thanks for all he has done for me, as mentor and friend, these last twenty years.

To the Syndics of the Cambridge University Press, lastly, let me express my acknowledgement of all the hard work and expertise expended and applied by the staff of the Publishing and Printing Divisions in clarifying and beautifying the disorders and uglinesses of my submitted handwritten copy to be the object of surpassing printer's vertu which it has here become.

D.T.W.

26 February 1976

EDITORIAL NOTE

Once more we need say little to smooth the reader's way into a volume which in its style, layout and conventions closely models itself upon its predecessors. Our aim, as ever, is to be as faithful to the autograph papers which we reproduce as the confines imposed by the linearities of the printed page will permit. To fill out and clarify their meaning—such, at least, as we conceive it to be—we have again not fought shy of making editorial intercalation in Newton's often roughly, hastily drafted and ill-organized texts; particularly so when we present for modern inspection our versions of what in the original are mere calculations, set out without verbal connectives to indicate the sense and logical sequence which Newton himself understood. To the same end we have trivially liberated his words from grammatical illogicalities of gender, case-ending, plurality and mood; but at the same time have taken care to preserve all the significant idiosyncrasies, contractions, superscripts and archaic spellings which, if nothing more, add an authentic touch of the period to the clean, straight and marginally adjusted lines of type which are here the modern facsimiles of Newton's ink-blobbed, much-cancelled and often rudely scrawled manuscripts. Within the proprieties of modern idiom we have kept our facing English renderings of the principal Latin texts deliberately literal, designing our phrases to be but a prop to the fuller understanding of Newton's Latin words, rather than to be read in their own right as polished paraphrases of his intended meaning. In the case of secondary texts of minor importance, we have not hesitated to set these in the many untranslated appendices which here appear in yet greater profusion than before, along with the substance of those cancelled passages which are too long or complicated to attach in a pertinent footnote to their parent text. (As with anyone else, we may add in justification, the older Newton came to be, the more 'wasteful' are his preserved papers in their relative bulk. Is it sacrilegious to suggest that there is no point in making full and exact reproduction of every last one of his increasingly numerous and individually often minimally variant extant preliminary worksheets for, and posterior revises of, an item which itself is of but minor importance?) In both Latin original and our parallel English rendering *en face* a twin vertical rule in the margin alongside remains our signal that the line or passage so distinguished has been cancelled by Newton in the manuscript. Other conventions and notations here used ape those employed in previous volumes: in particular 'fl$(R\hat{B})$' (see page 452, note (21) below) is set by us to denote the 'fl[uxio]' (fluxion) of the arc subtended by the angle $\hat{B}$ in a circle of radius R, and the same is done *mutatis mutandis* elsewhere. In the interests of economy we have slightly standardized the considerable variety of subtly (but not, we think, intentionally) varying sizes and shapes in which Newton portrayed his multi-dotted, multiply accented and boxed symbols; in so doing we have nowhere deviated from the printed equivalents which Newton himself chose in delivering these to the world in his 'Tractatus de Quadratura Curvarum' in 1704. May we therefore be forgiven, if not excused. We would finally remind that forwards and backwards reference within this volume is often made by the convention **2**, 3, §2' (by which understand '[Part] **2**, [Section] 3, [Subsection] 2'), while we point to citations in previous volumes by the code 'IV: 219–20' (understand '[Volume] IV: [pages] 219–20').

GENERAL INTRODUCTION

The backdrop to the four years of fresh researches and reshapings of past discoveries here reproduced has already been set in our previous volume.[1] In the autumn of 1691 where we take up this new episode in the continuing saga of his mathematical development Isaac Newton, now late in his forty-ninth year, lay intellectually still very much in the shadow of the magisterial *Philosophiæ Naturalis Principia Mathematica* which he had wrought—imperfectly as that may be—half a dozen years before, caged by its very brilliance and originality, unable to transcend its mental confines. In between times, after a year spent in the glare and noisy bustle of political and social life in London, he had returned gloomily in the mid-winter of early 1690 to the slow, dull pace of college and professorial life in Cambridge, rather helplessly to flit moth-like round the bright-burning candle of his master creation, now and then[2] fanning its flame to a yet purer blaze. He must have felt the deep sadness of realizing that whatever else it was still in him to do would not stand comparison with what he had achieved in the crowded months of his *biennium mirabilissimum* in the middle 1680's.

Though his year amid the dazzle and event of the metropolis had (we may see in hindsight) left him for evermore dissatisfied with the puny material rewards of dedicated scholarship and the petty round of day-to-day academic life, Newton had outwardly appeared to slip back smoothly enough into being once more the remote Lucasian Professor denied by statute[3] the opportunity to play an active part in the governing of the University or the teaching of his own Trinity College, the resident Cambridge expert in all things mathematical, scientific and indeed (with his fellow Lincolnshireman Henry More now dead) even theological.[4] In his little lean-to 'elaboratory' tucked under the wall of Trinity's chapel he took up again, where he had left them off in 1688, his unending train of smoky, bubbling chemical and metallurgical experiments[5]

---

(1) See VI: xxiii-xxiv.

(2) Compare the remodellings of the *Principia*'s 'Liber primus' at this time which are reproduced in VI: 538–608.

(3) As confirmed by Charles II on 18 January 1664; see III: xxvii.

(4) We have primarily in mind the lengthy open letters on 'notable corruptions' of Scriptural texts which Newton passed to John Locke in November 1690 (see *The Correspondence of Isaac Newton*, **3**, 1961: 83–122 and 129–42).

(5) The loose sheets in ULC. Add. 3973.7–9 on which he penned the surviving record of these new experiments evidence his undaunted efforts on various dates between March 1691 and February 1696 to fuse new alloys of tin, copper, lead, bismuth and zinc, to attack these compounds with *aqua fortis* and other acids, and to sublimate them with antimony and *sal armoniacum* (on which see V: xiv, note (17)); compare G. D. Liveing's accurate and informative

—as ever to no real profit, it would seem. Though (so far as we know) he made no further deposit in the University Library of any of his new Lucasian lectures, and what their content may have been is anyone's educated guess, he presumably began again from time to time—if no longer with weekly frequency—to address his largely uncaring undergraduate audience from his professorial podium in the schools. After his previous experiences of so often reading thereon for 'want of Hearers' to 'ye Walls' of his lecture room(6) he did not, we may suppose, repeat his earlier mistake of trying to instil an undiluted version of his *Principia* into his listeners, whoever these might have been,(7) but aimed at the simpler goal of applying the principles of mechanics to the heavens with a lighter, more common touch.(8) To more senior acquaintances he showed himself helpful in smoothing their way into comprehending the fundamentals and basic lessons of his published *œuvre maîtresse*. When the young Richard Bentley, floundering badly under the weight of a long list of 'necessary' preparatory reading,(9) applied to Newton himself in late 1691 for

listing of their content in *A Catalogue of the Portsmouth Collection of Books and Papers written by or belonging to Sir Isaac Newton* (Cambridge, 1888): 19–20.

(6) As Humphrey Newton was long afterwards to observe to John Conduitt; see VI: xii, note (3).

(7) We have previously (again see VI: xxii, note (3)) cited William Whiston's testimony that it was with 'no Assistance' that he set himself 'with the utmost Zeal' to study the *Principia* in Cambridge at this time.

(8) There do exist in ULC. Add. 4005 several English fragments which may in some way or other have had to do, if not with Newton's public lectures, then at least with the private instruction in his chamber of students interested enough to seek him out there for which Lucasian statute also provided (see III: xxii–xxiii). These comprehended, notably, (ff. 23r–24r/25r) an outline of 'The Elements of Mechanics' and 'The Mechanical Frame of the world'; and (ff. 21r–22r) drafts of the first chapter 'Of the Sun & fixt Starrs' and part of the second one 'Of the Earth & Planets' of a tract on 'Cosmography', which was seemingly broken off to be reordered (on ff. 45r–50r) as a similar account of fifteen astronomical 'Phænomena'. (The uncancelled portions of the texts of these are printed in A. R. and M. B. Hall, *Unpublished Scientific Papers of Isaac Newton* (Cambridge, 1962): 165–9, 374–7 and 378–85 respectively.) We may add that in a contemporary memorandum 'Of educating Youth in the Universities' (ULC. Add. 4005.5: 14r–15v, first published by W. W. Rouse Ball in *The Cambridge Review*, **31**, No. 763 [for 21 October 1909]: 29–30 [= *Cambridge Papers* (London, 1918): 244–5] as 'A Seventeenth-Century Flysheet', and more recently refurbished by A. R. and M. B. Hall, *Unpublished Papers:* 369–73) Newton laid down the *desiderata* which he felt an introduction to the elements of exact science should strive for, requiring (f. 14r) 'The Mathematick Lecturer to read first some easy & useful practical things, then Euclid, Sphericks [solid geometry], the projections of the Sphere, the construction of Mapps, Trigonometry, Astronomy, Opticks, Musick, Algebra, &c. Also to examin &...instruct in the principles of Chronology & Geography'. (On the last compare IV: 10, note (28).)

(9) This list, written out in June 1691 by John Craige for William Wotton (acting on Bentley's behalf), is given in full by David Brewster in his *Memoirs of the Life, Writings and Discoveries of Sir Isaac Newton*, **1** (Edinburgh, 1855): 456–9. Substantial excerpts from it, more faithfully transcribed from the original (in Trinity College, Cambridge. R.16.38), are reproduced in *Correspondence*, **3**: 150–1.

a less formidably technical *entrée* to the book, he was rewarded with a much more concise programme of preliminary reading together with the sound advice that 'At ye first perusal of my Book it's enough if you understand ye Propositions wth some of ye Demonstrations wch are easier then the rest. For when you understand ye easier they will afterwards give you light into ye harder.'(10) And when the next year Bentley came to deliver the first series of commemorative Boyle lectures at St Martin's in London, taking as his theme the natural evidence for the existence of God, he again sought instruction from Newton and again was awarded it: 'When I wrote my treatise about our System', the latter returned on 10 December 1692, 'I had an eye upon such Principles as might work wth considering men for the beleife of a Deity & nothing can rejoyce me more then to find it usefull for that purpose',(11) and over the two ensuing months he went on to elaborate in fine detail his views and beliefs regarding the formation of the visible world and its underlying structure.(12)

During these last four Cambridge years Newton stayed close to home, leaving his University town only on a few brief visits to his native Lincolnshire or to London. If we combine the complementary records of his college buttery accounts and (incomplete) list of signings out of and back into Trinity,(13) and interlard these with what else we can extract from his contemporary letters (few as these are) we may establish more precisely that he passed three weeks in London from the last day or so of December 1691;(14) a fortnight away each in early June and again in early July 1693, both seemingly in trips north to comfort his half-sister Hannah as her husband Robert Barton lay dying from

(10) Joseph Edleston, *Correspondence of Sir Isaac Newton and Professor Cotes* (London, 1850): 274–5 [= *Correspondence of Isaac Newton*, 3: 156]. Newton added that 'When you have read y^e first 60 pages, pass on to y^e 3^d Book & when you see the design of that you may turn back to such Propositions as you shall have a desire to know, or peruse the whole in order if you think fit'.

(11) *Correspondence*, 3: 233. We have already cited in I: 9, note (24) Newton's equally often quoted following sentence: 'But if I have done ye publick any service this way 'tis due to nothing but industry & a patient thought.'

(12) *Correspondence*, 3: 233–40, 244, 246–56; see also A. Koyré, *From the Closed World to the Infinite Universe* (Baltimore, 1957): 179–89 and his 'Newton, Galilée et Platon', *Actes du IXe Congrès International d'Histoire des Sciences* (Barcelona/Madrid, 1959): 165–87 [Englished as 'Newton, Galileo and Plato' in the gathering of his *Newtonian Studies* (London, 1965): 201–20]. We have already sketched in VI: 56–7, note (73) the dynamical substratum of the discerning *critique* of Galileo's 'Platonic' conjecture as to the formation of the solar system which Newton set down for Bentley in his two last letters of 17 January and 25 February 1692/3 (*Correspondence*, 3: 240, 255–6).

(13) These are collected (from the scattered originals in Trinity College's Muniments Room) on pages lxxxix/xc and lxxxv respectively of Edleston's *Correspondence of Sir Isaac Newton and Professor Cotes* (note (10) above).

(14) See our more detailed comment on this London visit on page 11 below.

some wasting disease at Brigstock in Northamptonshire,[15] and a further fortnight in London in mid-September following; visited London again briefly at the beginning of September 1694;[16] and made two separate trips to Lincolnshire in September 1695.[17] Except for the troubled late summer of 1693 (to which we shall return in a moment), during these years Newton's life went calmly and uneventfully on, unafflicted by serious illness and disturbed only by the rare arrival of a visitor to talk over his work or an occasional flurry of correspondence. His stay in London in January 1692, probably inspired in the main by his determination to reveal to David Gregory just how far his pretensions to the 'prime' theorem on series-quadrature had been outdistanced by his own newly composed (and yet uncompleted) treatise on the quadrature of curves,[18] led *inter alia* to his renewing personal contact with the Swiss mystic and mathematician Fatio de Duillier,[19] and it is to the latter's vivid ensuing accounts of what in manuscript he was then shown that we know externally so much of Newton's mathematical interests at this time.[20] After Fatio visited Cambridge early the following November their intimacy

(15) In subsequently writing to 'Camebrig' on 24 August seeking Newton's 'advise' on the future settlement of his estate, Hannah observed that 'My Dear Husband ever since his return to Brigstock has been very ill....I find noe hopes of Cure but that hee lossis his flesh and strength very fast' (*Correspondence*, **3**: 278).

(16) See note (56) below.

(17) 'I am newly returned from a journey I lately took into Lincolnshire & am going another journey', Newton wrote to John Flamsteed on 14 September (*Correspondence*, **4**, 1967: 169), adding that he would have no 'time to think of ye theory of the Moon...this month or above'. The only hint we can trace of the pressing business which twice within a couple of weeks took him north from Cambridge is a letter from his half-brother Benjamin Smith in Colsterworth which, in acknowledging the success of a 'plaster' supplied by Newton to ease the pregnancy pains of Benjamin's wife, spoke 'our thankes for all your trouble and cost' (*ibid.*: 187).

(18) See pages 7–11 below for the context of this. The two principal versions of the treatise 'De quadratura Curvarum' which Newton put together over the early winter of 1691/2 are reproduced in **1**, §§1/2 following.

(19) See VI: xxiii, note (46). In amendment of what we there too vaguely adduced, Newton had met Fatio at least as early as 12 June 1689 when both attended the meeting of the Royal Society at which (see *ibid.*: note (45)) Christiaan Huygens, then on a visit to London, gave an account of his forthcoming *Traité de la Lumière*. Their names are found several times together in the Society's Journal Book over the ensuing months (compare *Correspondence*, **3**: 69, note (1)). In late August following, both Fatio and Huygens (as the latter recorded in his private diary) travelled by barge up the Thames with Newton to hear him plead his case before William for varying the statutes of King's College so that he might be made its Provost (on the failure of which attempted preferment see VI: xxiv, note (48)).

(20) Compare pages 12–13 below. When Newton came to incorporate in his 'De quadratura Curvarum' (as Case 4 of its Proposition XI) a rule whereby, in certain simple cases, a given first-order fluxional equation may be reduced to an exactly quadrable form by means of an appropriate multiplying factor, he paid Fatio the rare honour of explicitly crediting him with its invention (see page 78, note (68) following).

grew apace, despite Fatio's long-lasting moan thereafter about 'a grievous cold, which is fallen upon my lungs' contracted by him on his journey back to London:[(21)] Newton revealed quite remarkable restraint in pandering to Fatio's hypochondria, prescribing a variety of 'Imperial powders' to cure his malady,[(22)] and early in 1693 warmed to Fatio's thoughts of continuing to reside in England 'some years, chiefly at Cambridge',[(23)] but he remained ever cool regarding his *protégé*'s extremist views on the subject of the 'biblical prophesys'.[(24)] From our present viewpoint it is a pity that what there is of scientific interest in their considerable correspondence with each other at this time has to do with alchemical matters.[(25)] When in the spring of 1693 Fatio left London for Switzerland to claim a family estate 'such...as will keep me as long as I live, provided I go there again',[(26)] his close relationship with Newton lapsed and upon his return to England a year or so afterwards, there to spend the remainder of his long life, was resumed only in infrequent, casual encounter.[(27)]

The year before, to go back, Newton had not allowed his brief *contretemps* with David Gregory over publication of his 1676 series expansion of the area of a curve defined by a binomial equation to spoil the welcome which he

(21) Fatio to Newton, 17 November 1692 (*Correspondence*, **3**: 230, with '9^ber^' there mistakenly read as 'September', however).

(22) See Fatio's responses to Newton on 17 and 22 November (*Correspondence*, **3**: 230, 231–2).

(23) So Fatio wrote to Newton on 30 January 1692/3 (*Correspondence*, **3**: 242). Newton responded on 14 February that 'When I invited you hither [the previous November] I was contriving how you might subsist here a year or two' (*ibid.*: 245), and added a month later: 'The chamber next me is disposed of; but that which I was contriving was, that since your want of health would not give you leave to undertake your design for a subsistence at London, to make you such an allowance as might make your subsistence here easy to you' (*ibid.*: 263).

(24) To a long paragraph by Fatio on this topic on 30 January 1692/3 (*Correspondence*, **3**: 242) Newton replied succinctly on 14 February that he was 'glad you have taken ye prophesies into consideration & I believe there is much in what you say about them, but I fear you indulge too much in fansy in some things' (*ibid.*: 245). Fatio was afterwards, of course, to be pilloried at Charing Cross more than once for over-enthusiastically propagating his beliefs along with the fanatical Prophets from the Cévennes.

(25) See especially Fatio's letters to Newton on 4 and 18 May 1693 (*Correspondence*, **3**: 265–6, 268).

(26) So he wrote to Newton on 11 April (*Correspondence*, **3**: 391).

(27) It is rare, indeed, that the contemporary record registers their meeting at all. (But see note (70) below.) In his last letter to Newton on 18 May Fatio made a parting request that 'If it was in your way to come to town I should be very glad to see You and to confer with You' (*Correspondence*, **3**: 270). The Trinity buttery accounts (see note (13) above) record an absence by Newton from Cambridge for ten days in late May, and it is natural to suppose that he acceded to Fatio's wish to visit him in London. What happened so abruptly to terminate their intimacy can only be conjectured, but we need not go to such extremes of speculation as F. E. Manuel presents in psycho-sexual explanation thereof in Chapter 9 of his *A Portrait of Isaac Newton* (Cambridge, Massachusetts, 1968), where he dubs Fatio 'Newton's ape'.

afforded Gregory's friend (and first publisher in 1688 of the quadrature series in dispute(28)), the Edinburgh physician Archibald Pitcairne, when the latter visited Cambridge at the beginning of March 1692 *en route* to take up the chair of medicine at Leyden. On that occasion, as is well known, Pitcairne persuaded out of Newton an autograph 'De natura Acidorum' which, by way of a transcript of its content made by Gregory, was eventually published eighteen years later (in its original Latin and a somewhat differing English version) in the second volume of John Harris' *Lexicon Technicum*.(29) In the same vein Newton wrote a long letter to John Locke the following August, passing comment on two (al)chemical 'recipes' which Locke had transcribed for him out of the papers of the lately deceased Robert Boyle.(30) Locke had been a frequent correspondent of Newton's since December the previous year,(31) and on 3 May the latter had written: 'Now the churlish weather is almost over I was thinking w^th^in a Post or two to put you in mind of my desire to see you here where you shall be as welcome as I can make you....You may lodge conveniently either at ye Rose Tavern or Queen's Arms Inn'.(32) But, so far as we know, nothing came of Newton's proposal and there was a lull in their exchange of letters during a whole year from August 1692. That same month there arrived by post from Oxford a suggestion of much more lasting import: a plea from John Wallis that Newton should contribute something to the news expanded Latin edition of his *Algebra* which Wallis was then about to put to press in the second volume—but first to appear—of his collected *Opera Mathematica*. Wallis' letter itself is lost, but in the draft of his reply on 27 August returning his 'hearty thanks for giving me opportunity of adding or altering what may concern me in your book' Newton not only gave permission, should Wallis think it 'of any moment', for his infinite series for $\pi/2\sqrt{2}$(33) to be inserted 'where you speak of that [for $\pi/4$] of Mr Leibnitz', along with the

(28) See page 6, note (15) below.

(29) There is a facsimile reprint in (ed.) I. B. Cohen and R. E. Schofield, *Isaac Newton's Papers and Letters on Natural Philosophy* (Cambridge, 1958): 256–8; the original autograph of the 'De natura Acidorum', completed by a passage in David Gregory's transcription, is reproduced in *Correspondence*, **3**: 205–9.

(30) See *Correspondence*, **3**: 217–19; Locke's transcription of the two (al)chemical recipes had been included with his letter to Newton on 26 July (*ibid.*: 216–17). An unpublished variant draft of the latter's reply exists on ULC. Add. 3965.13: 469^r^. On 7 July Newton had written to Locke that he 'should be glad to assist...all I can, having a liberty of communication allowed me by M^r^ B[oyle]' (*ibid.*: 215).

(31) See his letters of 13 December 1691 and 26 January/16 February 1691/2 (*Correspondence*, **3**: 185–6, 192–3, 195).

(32) *Correspondence*, **3**: 214.

(33) The pairwise alternating one which he had discovered in 1676 (see IV: 208, 211–12) and soon after transmitted to Leibniz in his *epistola posterior* of 24 October of that year (see *Correspondence*, **2**, 1960: 120).

decoding of the anagram in which he had in his *epistola posterior* of 1676 to Leibniz scrambled the bare enunciation of his dual method for extracting the fluent 'root' out of a given fluxional equation,(34) but added that 'The two methods you desire depend upon a third mentioned neare the beginning of the Letter [of 24 October 1676] wch ought therefore to be first explained'.(35) Newton then went on to expound this primary method of squaring algebraic curves by series, instancing it not merely in the case of the *primum Theorema* which had been its sole exemplification in 1676, but also by a yet more general theorem of quadrature by expansion into series which he would appear to have newly found.(36) In a second (lost) letter of 17 September he adjoined a lengthy worked example of his method of extracting the fluent 'root' of a given fluxional equation term by term as an ascending or descending infinite series. Wallis' minimal reshaping of Newton's *ipsissima verba* in Chapter XCV of his Latin *Algebra* the next year (37) has its deserved niche in history as the first public account of the intertwined methods of fluxions and infinite series.

(34) See *Correspondence*, **2**: 129 and IV: 673. We have in II: 190–1, note (25) already quoted the unravelling of this anagram as Newton set it down at the time in his Waste Book (ULC. Add. 4004: 81^{v}).

(35) *Correspondence*, **3**: 219.

(36) See pages 70–1, note (49) following.

(37) In his *Opera Mathematica*, **2** (Oxford, 1693): 390–6, reproduced in **1**, Appendix 3 on pages 170–80 below, there incorporating the few small corrections which Newton entered in his library copy of the volume (now Whipple Science Museum, Cambridge. WS 1305). There exists, we would add, the yet unpublished draft (ULC. Add. 3977.7) of a letter from Newton to Wallis in about January of 1693, communicating his 'hearty thanks for the sheets of your book [which] is so neare finishing' and communicating these 'amendments'—too late for them to be set right in the published volume, even in its *errata*. We may be grateful that a printer's deadline prevented Wallis from taking note of Newton's letter (if indeed it was ever sent), since it also acted to stop the broadcasting of an unfortunate failure of memory on the latter's part when he went on to insist that 'The plague was in Cambridge in both ye years 1665 & 1666 but it was in 1666 y^{t} I was absent from Cambridge & therefore I have set down [at *Opera*, **2** 'pag. 368 lin. ult.' where he had a few lines above directed '*pro* 1665 *lege* 1666'] an amendm^{t} of ye year'. 'I wrote to you lately', he continued—evidently referring to one or other of his lost letters of 27 August and 17 September 1692—'that I found ye method of converging series in the winter between ye years 1665 & 1666. For that was ye earliest mention of it I could find then amongst my papers. But meeting since wth the notes [ULC. Add. 4000: 2^{r}–20^{v}, reproduced in I: 47–121] wch in y^{e} year 1664 upon my first reading of Vieta's works Schooten's Miscelanies & yor Arithmetica Infinitorum I took out of those books & finding among these notes [ff. 18^{r}–19^{v} = I: 104–11] my deduction of the series for the circle out of yours in yor Arithmetica Infinitorum: I collect y^{t} it was in y^{e} year 1664 that I deduced th[is] out of yo^{rs}. There is also among these notes [ff. 20^{r}/20^{v} = I: 112–15] Mercators series for squaring the Hyperbola found by y^{e} same method wth some others. But I cannot find y^{t} I understood y^{e} invention of these series by division & extraction of roots [as in ULC. Add. 3958.3: 72^{r}/70^{r}–71^{r}, reproduced in I: 122–34] or made any further progress in this business before the winter wch was between ye years 1665 & 1666. But in the winter & ye spring following by ye use of Division & extraction of roots I brought the method to be general, & then the plague made

And so to Newton's 'black year' of 1693, as his most recent general biographer[38] has dubbed it. What fresh is there here to say? Certainly, where scholars have, from the pedestals of their own stand-points, bickered ceaselessly this past century and a half over the possible causes and long-term after-effects of Newton's undeniable breakdown in health in the late summer of that year, we would be foolish to attempt any definitive assessments when the extant record offers but a blurred glimpse of the past reality. Between the abandoned draft of a business reply to Otto Mencke, publisher of the *Acta Eruditorum*, at the end of May[39]—a letter from Leibniz[40] (the only one he ever wrote directly to Newton) went similarly unanswered till the autumn—and the taut, melancholic, self-deprecating outbursts to Pepys and Locke nearly four months later to which we shall come in a moment, his extant correspondence (whatever might have perished) consists solely of the already mentioned plea addressed to him by his half-sister Hannah Barton for 'advise' on settling the estate of her dying husband Robert.[41] Fresh from spending much of the two preceding months away (or so we may assume) in Brigstock comforting his sister, Newton suddenly reappears to our modern eyes on 13 September, 'extremely troubled at the embroilment I am in' and having 'neither ate nor slept well this past twelve month, nor...my former consistency of mind',[42] temporarily resident in London 'At the Bull [Inn] in Shoreditch' whither the proffered bait of new possibilities of employment in the metropolis had lured him. But the mood to be gone from Cambridge had left him as rapidly as it had taken hold. To Samuel Pepys, who (through John Millington) had 'pressed' him to come to town, he sent off on that day a sorry, hang-tail,

---

me leave Cambridge. But I do not think it requisite that you should make a particular mention of these things. I believe you have said enough in ye beginning of ye 91th Chapter'—where, that is, Wallis had made extract of the pertinent portion of Newton's 1676 *epistola posterior* (*Correspondence*, **2**: 111–14).

(38) F. E. Manuel in his *Portrait of Isaac Newton* (note (27) above), where he so titles his Chapter 10 (pages 213–25).

(39) See *Correspondence*, **3**: 270–1; Mencke's letter of 1 February to which Newton began here to respond is lost. He did not come to review his unfinished first reply till some six months afterwards, and then on 22 November sent off to Mencke a considerably reorganized letter (*ibid.*: 291–2).

(40) *Correspondence*, **3**: 257–8. While Leibniz put this in the post in early March, it probably did not arrive in Cambridge—if we may go by the transit times between England and Germany of other letters at this time—till some weeks (or even months) later. We will examine in note (46) below the detailed answer Newton at length rendered in October to Leibniz' 'great expectation' of him to apply his utmost skill 'tum ut problemata quæ ex data tangentium proprietate quærunt lineas, reducantur optimè ad quadraturas; tum ut quadraturæ ipsæ (quod valde vellem) reducantur ad curvarum rectificationes' (*ibid.*: 257).

(41) See note (15) above.

(42) As he excused himself in his letter to Pepys on that day (see *Correspondence*, **3**: 279).

withdrawn refusal of some unspecified offer of a (Government?) position: 'I never designed to get anything by your interest, nor by King James's[!] favour, but am now sensible that I must...see neither you nor the rest of my friends any more, if I may but leave them quietly.'(43) And on the morrow he penned a celebrated outburst to Locke, one whose detail he could not recall a month later, or so he claimed:

'Being of opinion that you endeavoured to embroil me wth woemen & by other means I was so much affected with it as that when one told me you were sickly & would not live I answered twere better if you were dead. I desire you to forgive me this uncharitableness. For I am now satisfied that what you have done is just & I beg your pardon for my having hard thoughts of you for it &...also for saying or thinking that there was a designe to sell me an office, or to embroile me.'(44)

When his more balanced sanity returned, Newton excused himself to Locke by remarking that 'The last winter by sleeping too often by my fire I got an ill habit of sleeping & a distemper wch this summer has been epidemical put me further out of order, so that when I wrote to you I had not slept an hour a night for a fortnight together & for 5 nights together not a wink'.(45)

What 'office' it was that Newton here refused to purchase—if it was on sale at all—and what feminine entanglements he sought desperately to evade—did they exist outside of his own imagination—we shall probably never know. The exact rôles in conditioning his reactions to these played severally by his repressed deep-seated puritanism, his less well documented psycho-sexual frustration (coupled maybe with the totally putative onset of male menopause), his recent close brush with death within his immediate family and the inevitably attendant thoughts of his own mortality, his wider neuroses and quirks of personality, the physical effects of his recent 'epidemical distemper' and the mental strain of its debilities and of being deprived of sleep for so long, even the poison of the noxious fumes from his fire (that in his chamber or of the furnaces in his chemical elaboratory?): these will be argued for ever more. What we would here insist firmly upon is that we can trace no enduring influence of this short-lived breakdown in altering the course of his contem-

(43) Again cited from *Correspondence*, **3**: 279. Pepys was more than a little taken aback at receiving a letter which was, he wrote to Millington on 26 September, 'so surprising to me for the inconsistency of every part of it, as to be put into great disorder by it,...lest it should arise from...a discomposure in head, or mind, or both' (*ibid.*: 281).

(44) *Correspondence*, **3**: 280.

(45) Newton to Locke, 15 October 1693 (*Correspondence*, **3**: 284). He had given a similar account of the onset of his malady to Millington when he met him at Huntingdon on 28 September, there excusing himself, as Millington made haste to inform Pepys, 'upon his own accord, and before I had time to ask him any question' for 'a distemper that much seized his head, and that kept him awake for above five nights together' and asked so to be pardoned by Pepys, 'he being very much ashamed he should be so rude to a person from whom he hath so great an honour' (*ibid.*: 282).

porary scientific pursuits and practices, or more than momentarily affecting their rate of development. Having speedily made his peace with both Pepys and Locke, Newton passed in mid-October 1693 to make a long-delayed reply to Leibniz' letter of the previous spring, cheerfully unknotting for him the fluxional anagram which he had transmitted seventeen years before, and also answering Leibniz' request to explain how (by means of a curve's evolute) a reduction of the problem of rectification is made to one of the quadrature of curves.(46) And he was equally quick to oblige when in late November there arrived in Cambridge John Smith, Writing Master of Christ's Hospital at London, bearing a letter from Pepys(47) requesting that he test Smith's solution

(46) Newton to Leibniz, 16 October 1693 (*Correspondence*, **3**: 285). The general mode which he there presents for relating the problems of quadrature and rectification is essentially that, discovered by him long before in May 1665 (see I: 263) and afterwards elaborated in Proposition 9 of his October 1666 tract (see I: 432–4), whereby the length of an evolute curve is determined as the difference in length between the radii of curvature drawn to touch it at its two end-points from any of the family of involutes which the 'unwrapping' of its tangent generates; except that this tangent is now referred to the semi-intrinsic coordinates compounded of the distance x of its instantaneous meet with an arbitrarily fixed straight line measured from some given point in the latter, and of the difference z between its length at that point and its length when $x = 0$. At once $\dot{z}/\dot{x} = y/a$ is the cosine of the angle which the evolving tangent makes with the fixed line; whence, once the function $y \equiv f_x$ is determined from the given property of the curve whose arc-length is to be found, the latter's rectification is yielded by the quadrature of the area $az = \int_{,}^{x} y.dx$. Conversely, given the relationship between x and y and thence the cosine $y/a = \dot{z}/\dot{x}$ of the angle of slope of the corresponding evolute tangent, the family of tangents so constructed will envelop a curve whose related general arc-length is $z-b$. It does not follow, however, as Newton would have it, that such a rectification 'semper fieri potest Geometrice ubi fluxionum $\dot{x}$ et [$\dot{z}$] relatio geometrica est' since all but a few functions z having 'geometrical' fluxional derivatives will not themselves be algebraic. (Here we would warn, incidentally, that several slips of Newton's pen in writing y and $\dot{y}$ in place of z and $\dot{z}$ stand mostly uncorrected in all the several printings which the letter has received since the middle nineteenth century.) There Newton leaves it for Leibniz, but of course where in particular the evolving tangent is normal to the fixed line when $x = 0$—there having length b, say—and we take the evolute curve to have the general point (r, s) in the system of perpendicular Cartesian coordinates in which the fixed line is the abscissa $s = 0$ and its given point is the origin $(0, 0)$, then at once $x = s.(dr/ds)-r$ and likewise

$$z = s\sqrt{[1+(dr/ds)^2]}-\int_s^b \sqrt{[1+(dr/ds)^2]}.ds+b.$$

When—in apparent ignorance of Newton's earlier investigation—Jakob Hermann independently proposed the problem 'Invenire curvam vel curvas algebraicas, quarum rectificatio indefinita dependeat a quadratura cujusvis curvæ algebraicæ' in the *Acta Eruditorum* in August 1719, a solution was given in this form four years later by Hermann himself (*Acta* (April 1723): 174–9) and further extended the next year by Johann Bernoulli (*Acta* (August 1724): 356 ff. = *Opera Omnia*, **2** (Lausanne/Geneva, 1742): 582–92).

(47) Who commended 'y^e Bearer...noe less for what I personally know of his general Ingenuity...than for y^e general Reputation he has in this Towne (inferiour to none, but superiour to most) for his Maistery in the two Points of his Profession, namely, Faire–Writeing and Arithmetick, soe farr (principally) as is subservient to Accountantship' (*Correspondence*, **3**:

to a problem of dice currently the topic of conversation 'in this Towne, ...among Men of Numbers', namely: 'How much more or lesse Expectation *A* may (wth equal Lucke) reasonably have; of throwing at one or every Throw one Sice at least with Six Dyes, than *B* two Sices with Twelve, or *C* three with Eighteen Dyes?'(48) In reply on 16 December Newton laid out for Pepys a series of 'progressions of numbers' from which he drew, for 1, 2, 3, ..., 6 dice in succession the 'number of chances without sixes', the 'chances for one six & no more' and the 'chances for two sixes & no more',(49) and thence deduced that the probability, 31031/46656, of there ensuing at least one six with 6 throws of dice is greater than that, 1346704211/2176782336, of there falling at least two sixes with 12 throws, 'And so by producing the progressions to the number of eighteen dice...you will have the proportion of [the players'] stakes upon equal advantage'.(50) That there should be an odds-on chance of throwing at least one six (or any other face) in 6 throws of dice still appeared paradoxical to Pepys, and Newton patiently spelled out in yet a further letter to him on 22 December 1693 how it is that such 'chances' are not mutually

293). It would seem that Smith was responsible at least for organizing the 'blue-coat boys' usually allotted the task of drawing the winning tickets in London lotteries of this period, and may well have been himself paid for administering their running, if not helping to frame their rules and gauge their expectations of profit.

(48) Or so Pepys rephrased it for Newton on 9 December 1693 (*Correspondence*, **3**: 297). In his introductory letter on 22 December Pepys had queried more vaguely whether it is 'as easy a Taske' to 'fling a 6' with 6 dice, as '2 Sixes' with 12 dice, or as '3 Sixes' with 18 dice (*ibid.*: 294); and Newton rightly responded four days later that 'ye Question...seemed to me at first to be ill stated & in examining Mr Smith about ye meaning of some phrases in it he put the case of ye Question ye same as if *A* plaid with six dyes till he threw a six & then *B* threw as often wth 12 & *C* wth 18[,] the one for twice as many sixes [&] the other for thrice as many...' (*ibid.*: 295).

(49) See *Correspondence*, **3**: 299. In modern abridged notation Newton there tabulates in succession the chances, namely $\binom{j}{i}5^{j-i}$ in 6^j, of turning up i sixes, $i = 0, 1, 2$, in j dice, $j = 1, 2, 3, \ldots, 6$, understanding that the full number of throws are allowed to each player even after another has thrown his pertinent number of sixes in fewer 'flings' than he would (on average) need. He had, of course, imbibed the simple notion of equi-probability of chances to which he here appeals long before when, as an undergraduate at Trinity, he had made careful study of Huygens' tract *De Ratiociniis in Ludo Aleæ* (Leyden, 1657); see I: 58–62.

(50) *Correspondence*, **3**: 300. It will be clear from the previous note that the probability 'upon equal advantage' (as Newton put it) of throwing at least n sixes in $6n$ 'flings' of the dice will be $P(n) = \sum_{n \leqslant i \leqslant 6n} \binom{6n}{i} 5^{6n-i}/6^{6n}$. Since the (normal) distribution of the expansion of $\left(\frac{5}{6}+\frac{1}{6}\right)^{6n}$ is spread with uniform deviation round its largest term $\binom{6n}{n}\left(\frac{5}{6}\right)^{5n}\left(\frac{1}{6}\right)^{n}$, for sufficiently large n the probability $P(n)$ will come to differ minimally from $\frac{1}{2}$. Explicit proof that $P(n)$ indeed tends monotonically to $\frac{1}{2}$ as $n \to \infty$ is given by T. W. Chaundy and J. E. Bullard in their analysis of 'John Smith's problem' in *Mathematical Gazette*, **44**, 1960: 253–60. A more comprehensive survey of the historical background—with the suggestion that the problem owed more

exclusive.[51] The next year in May he went to an immense amount of trouble, no doubt again at Pepys' initial wish, in casting an alternative scheme of instruction for the 'blue-coat boys' of the Mathematical School at Christ's Hospital, but it would take us too far from our purpose to go into its detail.[52]

Above all, the considerable bulk of the mathematical papers reproduced in the pages here following which (so we assert and can many times solidly prove) derive from the two years after the autumn of 1693 will demonstrate that there was no sudden drop, in quantity or in quality, in Newton's technical output. As we have more than once observed in the previous volume[53] and as may many times more be seen below,[54] when David Gregory paid an extended visit to Cambridge early in May 1694 and was allowed virtually free run of Newton's private scientific papers, he found there a veritable treasure-house

---

to Pepys than to Smith in its origin—is presented by F. N. David in 'Mr Newton, Mr Pepys & dyse: A historical note', *Annals of Science*, **13**, 1957: 137–47 (summarized in her *Games, Gods and Gambling* (London, 1962): 125–9).

(51) *Correspondence*, **3**: 302–3, answering Pepys' letter of the previous day on the point (*ibid.*: 301–2).

(52) Much of the documentary material which supports this statement remains unpublished. The Mathematical Master at this time, Newton's former Trinity *confrère* Edward Paget (on whom see VI: xviii–xx), had drawn up a revised, more forward-looking curriculum for his School; and the Treasurer of Christ's Hospital, Nathaniel Hawes, was at a Committee meeting on 2 May 'desired when he goes to Cambridge on Friday next [4 May] to take with him a copy of the old and new schemes, and advise with the Professor and other Mathematicians in the University concerning them, and get their opinions in writing which of the two schemes they judge best' (see *Correspondence*, **3**: 366, note (2)). The secretary copy of Paget's scheme sent to Newton (ULC. Add. 4005.16: 85r–86r) and the draft (*ibid.*: 89v–88v) of his preliminary response to Hawes a few days later, raising 'a few Questions about ye new scheme of Learning proposed for yor foundation', both remain in manuscript. Likewise unprinted is the covering letter to Paget *and Professor Cotes* (the draft exists on ULC. Add. 3965.12: 330r) which Newton sent to London on 25 May along with his fuller reply to Hawes (Edleston, *Correspondence of Sir Isaac Newton* (note (10) above): Appendix: 280–92 = *Correspondence*, **3**: 357–65) and his own preferred 'New Scheme of Learning proposed for the Mathematical Boys in Christ's Hospital' (Edleston: 292–4 = *Correspondence*, **3**: 365–6; a number of variant preliminary castings of this in Newton's own hand exist at [in order of the sequence of their composition] ULC. Add. 4005.16: 100r/100v, 91r, 88r, 90r/90v, 87r/87v and Trinity College, Cambridge. R.5.4[22], the last a revision of the version first published by Edleston from the secretary copy in Christ's Hospital Court Book of that sent to Hawes). There followed two further letters of Newton to Hawes on 26 May (Edleston: 294–5 = *Correspondence*, **3**: 367–8) and 14 June (Edleston: 296–7), but yet again unprinted is a final letter to him on the topic in (?) July 1694 (the draft is ULC. Add. 4005.16: 93r) where Newton urges the advantage gained 'If Mr Stones Foundation [that is, the Mathematical School] be conjoyned wth ye Kings in a subservient way so that as often as any of the King's places become vacant by death they may be filled up not out of the grammar school but out of Mr Stones children of like standing in Mathematical learning'.

(53) See especially VI: 568–9, note (1); 578–9, note (21); 583, note (34); and 601, note (2).

(54) Particularly pages 208, note (28); 221, note (1) and 222, note (10); 269–70, note (55); 508–9, note (2).

of all things mathematical, some of which Newton was even then in the course of further elaborating and refining.(55) While the true depth and extent of these predominantly geometrical researches has never hitherto been made public, and we may accordingly forgive those who have previously seen only derivative sterility in all but a few isolated mathematical sorties made by him after 1693, or the rare resparking of his youthful fire in response to an occasional submitted problem, let us henceforth (if the tenacity of received opinion permits) correct this blindness to the past truth. All which is, conversely, not to deny that in Newton's scientific papers and correspondence of the early 1690's we may trace a slow but accelerating decrease in his elasticity to absorb fresh findings and his hitherto matchless capacity to attack novel problems and evolve new techniques of solution. Though we should not exaggerate what is at first a barely perceptible trend, his mathematical writings from 1693 onwards do indeed come more and more to look back to the glories of yester-year, their explicitly announced purpose less to create anew than to finish and polish earlier investigations left rough and incomplete. But this is the inevitable relentless attrition of old age, not the sudden and permanent debility of a mental storm or physical breakdown in health in the summer of that year.

(55) On pages 196–7 below we cite the detailed impression which Gregory took away with him from Cambridge of the grand mathematical project on which Newton was working at this time, his multi-volume treatise of *Geometria*, cast in both ancient analytical and modern fluxional moulds, whose surviving drafts are reproduced in our present Part 2 following. Upon his return to Oxford—having in July (see *Correspondence*, **3**: 380–2; and VI: 470–7) drawn from Newton a simplified step-by-step demonstration of the form of the 'figure wch feels ye least resistance in ye Schol. of Prop. XXXV Lib II' of his *Principia*—David Gregory began to organize what else he had gleaned during his Cambridge visit, first in individual mathematical memoranda and then in a full-blown elaboration, in 47 propositions, of a treatise on 'Isaaci Newtoni Methodus Fluxionum; ubi Calculus Differentialis Leibnitij, et Methodus Tangentium Barrovij explicantur, et exemplis plurimis omnis generis illustrantur. Auctore Davide Gregorio M.D. Astronomiæ Professore Saviliano Oxoniæ'. (Gregory's preliminary scheme, now Royal Society. Gregory MS: 64: 'Describenda et Chartis consignanda Mense Septembri MDCXCIV'—listed in his later catalogue of his papers as C79: 'Adumbratio nostræ [!] de fluxionibus methodi'—is reproduced in *Correspondence*, **4**, 1967: 15–16. He went on to draft the full compendium in late October, as several dates entered by him in his original manuscript, now St Andrews. QA 33G8/D12, establish beyond surmise.) While this loose collection of calculus problems was fairly widely circulated in its day—apart from Gregory's fair copy of its text (now in Christ Church, Oxford) there exist transcripts of it in the hand of William Jones and of John Keill, the latter (now ULC. Lucasian Papers [Res. 1894]: Packet No. 13) once in Newton's possession it would seem—it has never been printed, justly so since its content is all but wholly derivative from the researches of Newton and the published articles of such creative mathematicians as Leibniz and Jakob Bernoulli. For all that we may, here as elsewhere, praise Gregory's sincere aim of opening up Newton's close-held mathematical findings to the world at large, our reaction to such feebly wrought endeavours to do so must ever be one of sorrow that Newton could attract no more able and gifted a disciple to widen and extend the deep inroads into future mathematical discovery which he had himself wrought over the past three decades.

In yet one more change of scene, to pursue our brief chronological conspectus' we find Newton on 1 September 1694 conversing with John Flamsteed at Greenwich, optimistically claiming the moon's theory to be 'in his power' and stating that he needed only '5 or 6 equations' of its motion fully to capture it.(56) But though Flamsteed over the next few months sent him a hundred and fifty and more unprecedentedly accurate observations of the moon's passage which he himself had made over nearly twenty years from his small but efficient observatory atop Greenwich Park, Newton's confident initial expectation of being able to use these to determine the numerical parameters in an improved, dynamically based lunar theory—one whose basic structure was derived theoretically by assessing the disturbing action of the sun's gravitational pull on the simple Keplerian motion of a 'planetary' moon orbiting in an ellipse round the earth at a focus—came slowly to be eroded during the course of an extensive correspondence which he maintained almost without break with Flamsteed over the next year till it petered out in the late summer of 1695. Let us here forbear to cite any details.(57) For all his continuing show of hope that he had it in his grasp to achieve 'this Theory so very intricate & the Theory of Gravity so necessary to it, that I am satisfied it will never be perfected but by somebody who understands ye Theory of gravity as well or better then I do',(58) and despite some success in afterwards mocking up a modified Horrocksian kinematic model of lunar orbit both in a scholium appended to David Gregory's *Astronomia* in 1702(59) and in a recast scholium to Proposition XXXV of the *Principia*'s third book in its second edition in 1713,(60) when Halley qualified his published propositions on the moon as

(56) Or so David Gregory noted in a contemporary memorandum (C58, now Royal Society. Gregory MS: 26, reproduced in *Correspondence*, **4**: 7) where he wrote: 'D. Newtonus primo Septembris die 1694 Grenovici Flamstedium adiit, ubi locutus est de nova editione suorum *Principiorum*. Credit Theoriam lunæ esse in potestate: ad illius locum inveniendum opus erit 5 vel 6 æquationibus'.

(57) For an up-to-date introduction to the tangled mass of primary and secondary literature on this knotty technical topic we may refer to our own survey of 'Newton's Lunar Theory: From High Hope to Disenchantment' (*Vistas in Astronomy*, **19**, 1976: 317–28).

(58) As he described it to Flamsteed on 16 February 1694/5 (*Correspondence*, **4**: 87).

(59) *Astronomiæ Physicæ & Geometricæ Elementa. Auctore Davide Gregorio* (Oxford, 1702): 332–6, especially 333. Newton's preliminary English drafts in ULC. Add. 3966.10: [in sequence] 76, 77, 74, 88–9 and 82–3 (the last of which—except for the omission of its title 'The Theory of the Moon'—is reproduced in *Correspondence*, **4**: 322–6) do not differ in essence from Gregory's published Latin version (now available in facsimile in I. B. Cohen, *Isaac Newton's 'Theory of the Moon's Motion' (1702). With a bibliographical and historical introduction* (London, 1975): 123–8, along with reproductions of its three separate contemporary retranslations back into English, *Newtono inscio*).

(60) *Philosophiæ Naturalis Principia Mathematica* (Cambridge, $_{2}$1713): 421-5; Newton's drafts of this exist in three successive states in ULC. Add. 3966.9: 61, 67/68 and 65/63 respectively. The scholium reappeared in the *editio ultima* (London, $_{3}$1726: 459–65) unchanged except for one major excision.

'all sagacity' he wore, as we have previously remarked,(61) a thin smile, and when anyone threatened to utter any printed claim that he had indeed mastered the moon's motion he was downright upset.(62)

What effect this failure to construct a viable theory of the moon's path had on his wider confidence in his ability to go on making a fruitful contribution to scientific knowledge, we can only guess. One might almost entertain the hypothesis that if he had succeeded in 1695 in contriving an accurate 'Theory of ye Moon' then Newton would have gained strength to stay on in Cambridge, making sustained effort further to widen and deepen the mechanical and astronomical researches which he had begun to shape so magnificently in dynamical form in the 1680's and continued more recently to promote. As it was, however, from the early months of 1695 onwards he became more and more undecided and unsettled at Cambridge. In April John Wallis had, in sending along the newly printed *Volumen primum* of his collected works, renewed his entreaty that Newton should publish more of his hoard of mathematical and scientific papers—not least because, he wrote, he had just 'had intimation from Holland [that] your Notions (of *Fluxions*) pass there with great applause, by the name of *Leibnitz's Calculus Differentialis*'(63)—but Newton

(61) See VI: 27.

(62) If indeed not downright, tetchily angry. The most celebrated outburst of such wrath on Newton's part came in December 1698 when Flamsteed thought to add a paragraph to his forthcoming account of 'ye parallax of ye Pole star' which he (mistakenly) thought he had observed—this John Wallis, ever sharp-eyed to the saleability of a 'newsy' discovery, had at once collared for the third volume of his *Opera Mathematica* (then in press)—where he mildly reminded that it was he who had 'accommodated' Newton with accurate observations of the moon 'in ordine ad emendationem Theoriæ lunaris Horroccianæ[,] qua in re spero eum successus consecuturum expectationi suæ pares' (so he quoted his phrasing to Newton on 2 January 1698/9 after Gregory had informed the latter of its general content; see *Correspondence*, 4: 293). Newton blazed back four days later: 'Upon hearing occasionally that you had sent a letter to Dr Wallis about ye parallax of ye fixt starrs to be printed & that you had mentioned me therein with respect to ye Theory of ye Moon I was concerned to be publickly brought upon ye stage about what perhaps will never be fitted for ye publick & thereby the world put into an expectation of what perhaps they are never like to have. I do not love to be printed upon every occasion much less to be dunned & teazed by forreigners about Mathematical things or to be thought by our own people to be trifling away my time about them when I should be about ye Kings business. And therefore I desired Dr Gregory to write to Dr Wallis against printing that clause wch related to that Theory & mentioned me about it' (*Correspondence*, 4: 296; as is so often done, we have earlier quoted the penultimate sentence *hors de contexte* in v: xiv, note (14)).

(63) Wallis to Newton, 10 April 1695 (Edleston, *Correspondence of Newton and Cotes* (note (10)): Appendix: 300 = *Correspondence*, 4: 100). Newton's reply on 21 April is lost, but its gist is readily gatherable from Wallis' further letter on 30 April where he kept up his pressure to publish: 'Consider, that 'tis now about Thirty years since you were master of those notions about *Fluxions* and *Infinite Series*; but you have never published ought of it to this day....' Tis true, I have endeavoured to do you right in that point. But if I had published the same or like

dithered, even when Wallis went on to propose that he himself be permitted to have printed at Oxford (where 'we have most of the Cutts allready, & furniture fit for it'(64)) the full texts of the two letters of 13 June and 24 October 1676 to Leibniz whose content he had earlier summarized in his *Algebra*.(65) In the outcome, full publication of the *epistolæ prior et posterior* had to await the appearance of Wallis' third volume of *Opera Mathematica*(66) four years later, although maybe his persistence was the necessary goad which in June stimulated Newton privately to begin honing and augmenting the prior drafts of his researches into the species and properties of cubic curves,(67) to be effectively the 'Enumeratio Linearum tertij Ordinis' which he was in in 1704 to append to the *editio princeps* of his *Opticks*. A visit from Edmond Halley in August, 'about a designe of determining the Orbs of some Comets for me'(68) produced little harvest comparable to that of his momentous first trip to Cambridge eleven years before, though he was—a rare privilege!—permitted to carry Newton's 'Quadratures of Curves' back with him to London for transcription,(69) having doubtless urged (as he was again later

notions, without naming you; & the world possessed of anothers *Calculus differentialis*, instead of your *fluxions:* How should this, or the next Age, know of your share therein?' (*Correspondence*, **4**: 117). Wallis could not have foreseen, of course, the mighty industrial complex of recent Newtonian scholarship.

(64) Wallis to Newton, 30 May 1695 (*Correspondence*, **4**: 129). 'Mr Caswell or I', he added, 'will see to the correcting of the Press'.

(65) See IV: 672, note (54). A prod from Wallis on 3 July to correct a 'Transcript [now ULC. Add. 3977.1] of your two letters' which he had earlier sent, if only that 'so corrected... I might at lest leave them reposited in the Savilian Library amongst other Manuscript Papers; which will...confirm to you the reputation of your having discovered these notions so long ago' (*Correspondence*, **4**: 139), would appear to have gone unanswered by Newton, though there does exist the unfinished draft of such a response (ULC. Add. 3977.3, reproduced in *Correspondence*, **4**: 140–1) where he began by thanking Wallis for 'your pains in transcribing my two Letters of 1676' and then passing once more (compare note (37) above) faultily to recall that it was 'in ye beginning of the year 1666...I retired from the University into Lincolnshire to avoyd the plague'. The following November Wallis informed Halley that Newton still did not seem 'forward' for having his two 1676 letters printed in Oxford (see *Coreespondence*, **4**: 186).

(66) Where the full Latin texts of Newton's two letters are set in prime place in its appended 'Epistolarum Collectio' (*Opera Mathematica*, **3**, Oxford, 1699: 622–9, 634–45).

(67) Those reproduced on II: 10–88 and IV: 354–404. With their algebraic reductions by linear transformation recast into a rather less readily graspable appeal to the geometrical lie of the diametral (conic) hyperbola which shares a real asymptote with the general cubic, their content is subsumed to be the backbone of the final 1695 enumeration of the species of cubic curve which is set out on pages 588–644 below.

(68) So Newton reported to Flamsteed on 14 September 1695 (*Correspondence*, **4**: 169). 'He has', Newton went on—quoting from Halley's letter to him of a week earlier (on which see the next note), 'since determined ye orb of ye Comet of 1683 by my Theory & finds by an exact calculus that it answers all your Observations & his own to a minute'.

(69) As we have previously remarked in III: 12, note (29), some weeks after his return to London Halley wrote apologizing to Newton that 'I have not yett returned you your Quadra-

to do[70] that the treatise should be published. He did, it is true, momentarily revive Newton's interest in cometary orbits, the accurate construction of whose conical paths he had only roughly and readily achieved nine years before,[71] and there are in fact to be found among the latter's astronomical papers of this period[72] several new computations of elements of the orbit of the 'great' comet of 1680/1. In a lost letter of 1 October Newton communicated to Halley (his own?) observations of the comet of 1682 so that, by constructing their separate orbits, he might test—as a rough consideration suggested—'if it were not the same with that of 1607'.[73] But he could not match Halley's enthusiasm for such tedious numerical computations as were necessary to 'limitt the Orbs of all the Comets that have been hitherto observed'[74] and their corre-

---

tures of Curves, having not yet transcribed them, but no one has seen them, nor shall, but by your directions; and in a few days I will send you them' (E. F. MacPike, *Correspondence and Papers of Edmond Halley* (Oxford, 1932): 91 = *Correspondence*, **4**: 165). 'Since I left you', Halley had begun his letter, 'I have been desirous to make triall how I could obtain the position of the Orb of the Comet of 1683, and after having gotten some little direction from a course [coarse] Construction, I took the pains to examine and verifie it by an accurate Calculus, wherin I have exceeded my expectation, finding that a parabolick orb limited according to your Theory will most exactly answer all the Observations Mr Flamsteed and my self formerly made of that Comett, even within the compass of one minute'.

(70) Some seven years afterwards David Gregory entered in his private diary that 'On Sunday 15 Nov. 1702 He [Newton] promised Mr [Francis] Robarts, Mr Fatio, Capt. Halley & me to publish his Quadratures, his treatise of Light [*sc. Opticks*], & his treatise of the Curves of the 2d Genre' (W. G. Hiscock, *David Gregory, Isaac Newton and their Circle. Extracts from David Gregory's Memoranda, 1677–1708* (Oxford, 1937): 14). He had noted rather differently but two days before that 'Mr Newton is to republish his [*Principia*]; & therein give us his methode of Quadratures' (*ibid.*: 13).

(71) See vi: 481–507; and compare A. N. Kriloff, 'On Sir Isaac Newton's Method of Determining the Parabolic Orbit of a Comet' (*Monthly Notices of the Royal Astronomical Society*, **85**, 1925: 640–56).

(72) Notably those now in ULC. Add. 3965.11/14/18. An unfinished 'Constructio orbis Cometæ qui annis 1680 & 1681 apparuit...ex Observationibus...tr[ibus] quas Flamsteedius habuit Dec 21, Jan 5 & Jan 25' (Add. 3965.11: 170r) is reproduced at *Correspondence*, **4**: 167. On our pages 682–8 below we present the edited text of two complementary worksheets (Add. 3965.14: 586r and 3965.11: 165r) where Newton computes in two separate ways the slope to the meridian of the same comet's apparent path on 30 December.

(73) As Halley phrased it in his reply on 7 October (*Correspondence*, **4**: 173), adding his entreaty to Newton, 'when your more important business [in Lincolnshire] is over', that he should 'consider how far a Comets motion may be disturbed by the Centers of Saturn and Jupiter, particularly in its ascent from the Sun, and what difference they may cause in the time of the Revolution of a Comett in its very Elliptick Orb'.

(74) Halley to Newton, 21 October 1695 (*Correspondence*, **4**: 182). 'I have', [he there wrote,] 'almost finished the Comet of 1682 and the next you shall know whether that of 1607 were not the same, which I see more and more reason to suspect. I am now become so ready at the finding a Cometts orb by Calculation, that...I think I can make a shift without [rulers].'

spondence tailed off in late autumn as Halley lost his own interest in checking the identity of the two most recent apparitions of 'his' comet.(75)

A last letter from Flamsteed on 11 January 1695/6 querying the truth of a rumour he had heard that Newton had 'finished ye Theory of ye Moon *on uncontestable principles*'(76)—this went unanswered, of course—and then Newton wrote to Halley on 14 March to stop a 'report...sometime spreading among ye Fellows of ye Royal Society as if I was about ye Longitude at Sea' and likewise to obviate a 'rumour of preferment for me in the Mint'.(77) But the latter was true, despite his denial: Newton had been secretly arranging with Charles Montague to put in for just such a post, and within a week the latter wrote that the newly vacated post of Warden of the Mint, with its salary of 'five or six hundred pounds per An' and 'not too much bus'nesse to require more attendance than you can spare',(78) was his for the asking. Within another month the Royal Warrant confirming the appointment(79) was through and Newton, papers and belongings packed away in his trunks, was off on the road to London, never again to return to Cambridge except on the briefest of visits.

And there, since our concern is here only with Newton's last university years, we must leave him. Forgive our prejudiced sigh for the passing of a uniquely talented man from the environs, restrictive as in many ways they were, of the small Fenland town where he had passed the prime of his age for creative invention amid his books and the smoke of his laboratory fire. Many have it in them to be hard-headed businessmen, successful politicians, able organizers of people and administrators of government, even efficient Masters

(75) A final, undated letter from Halley on the topic relates that he 'could not get time to finish the account of the two Comets I promised you' (*Correspondence*, **4**: 190). We would add that the draft (ULC. Add. 3965.14: 605^r) of a following letter from Newton to Halley, reproduced in *Correspondence*, **4**: 184–5 under the date 'late October 1695', is in fact—as its revised version (Add. 3982.7, printed in E. F. MacPike, *Correspondence of Halley* (note (69) above): 199) makes yet clearer—of about January 1725.

(76) *Correspondence*, **4**: 192. Whoever Flamsteed's informant was (may be Halley?), he was plausible—and accurately informed?—enough to credit Newton with having discovered '6 severall Inæqualitys [of the moon's motion] &...nevertheless ye Calculation will not be much more troublesome or difficult then formerly'. A few years later Newton was to set seven such inequalities down in the 'Lunæ Theoria Newtoniana' which he allowed Gregory to publish in 1702 (see note (59) above).

(77) *Correspondence*, **4**: 193. The undercurrents of well-founded rumour which began in the winter of 1695/6 to put Newton in the running for a place in the Mint may have had their source in the report on the deterioration of the nation's silver coinage (now in London. Goldsmith's Company MS 62 'Recoinage of 1696') which Newton had prepared some while before, where *inter alia* he 'proposed a Price Control Board...to reduce prices...or at least limit their increase [which] was to operate...on the Chartered Companies of London' (J. Craig, *Newton at the Mint* (Cambridge, 1946): 9).

(78) Montague to Newton, 19 March 1695/6 (*Correspondence*, **4**: 195).

(79) The Mint's record of this, dated 13 April 1696, is reproduced in *Correspondence*, **4**: 200.

of the Mint and forceful Presidents of the Royal Society; only too few have ever possessed the intellectual genius and surpassing capacity to stamp their image upon the thought of their age and that of centuries to follow. Watching over the minting of a nation's coin, catching a few counterfeiters, increasing an already respectably sized personal fortune, being a political figure, even dictating to one's fellow scientists: it should all seem a crass and empty ambition once you have written a *Principia*.... But it did not to Newton. So quickly on to the final mathematical texts which he penned at Cambridge, and to our exegeses thereof in introduction and in footnote.

ANALYTICAL TABLE OF CONTENTS

PART 1

THE FIRST TRACT 'DE QUADRATURA CURVARUM'

(early winter 1691–2)

* NB. Unless otherwise specified, citations here and below are of manuscripts in the University Library, Cambridge.

'prime' theorem for the series-quadrature of binomials, 36. And to quadrinomials: notes upon these, 38. Parallel reductions of the quadrature of yet more complicated multinomials to simpler (and, if possible, straightforwardly quadrable) form, 40. Case 1: by lowering the degree of the multiplying power z^θ, 42. Case 2: by lowering the degree of the 'nomial', 44. Cases 3/4 conjoin these part-reductions and make suggestion for their extension, 46. Whence to find the simplest figures with which curves needing to be squared can be compared, 46. Further analogous theorems are mapped out, 46.

§2 (Add. 3962.3: 55^r, 56^v–57^v/3960.10: 167–71/3960.7: 121–35/3960.10: 173–9/3960.11: 181–95/ 3965.6: 38^r–39^r [in sequence] +private). The revised, augmented treatise of quadrature. The opening is much as before, 48. But Newton now quotes his 'prime' proposition on series-quadrature *verbatim* from his 1676 *epistola posterior*, along with its worked examples, 50. And attendant rectification of the cissoid, 52. The backwards extension of the prime theorem is given refined expression, 54. The earlier 'Reg. 1' becomes 'Prop. II'; and 'Reg. 2', on the transmuted quadrature of trinomials, is emended to be 'Prop. III', 56. Further transmutation of the primary reduced forms to be six secondary ones, 60. The 'simplest' such reduced forms, 62. Proposition IV: the fundamental rule for deriving the fluxion of a given equation *ab initio*, 62. Case 1: when the equation is surd-free, and Case 2: when it is 'irrational', 64. Proposition V: 'to find curves which can be squared' (by setting the abscissa as base variable and ordinate as fluxion, and determining the area as the fluent), 66. Proposition VI: 'to find a curve equal [in area] to a given one' (by equating the products of their ordinate and the fluxion of their abscissa), 66. Propositions VII and VIII: the ordinates of curves whose areas are defined by multinomial equations (obtained by finding their fluxions), 68. Proposition IX: the series-quadrature of a curve whose ordinate is defined by a compound multinomial equation, 70. Proposition X: the yet more general case where the ordinate is doubly multinomial, 70. Proposition XI: given a fluxional equation, to determine the corresponding fluent relationship, 70. 'The problem is the most useful of those commonly propounded in mathematics', 72. Case 1: where the fluxional equation lacks one or other of the fluents (and so is in directly integrable form); and Case 2: where the fluents can readily be separated, 72. An example invoking the area of a hyperbola (reduced to the equivalent logarithmic series), 74. Case 3: where by distinguishing related groups of terms in it the equation is seen to be directly integrable, 74. An example in a first-order fluxional equation, 76. Or can be made so by multiplying or dividing through its terms by some mononomial quantity: 'This rule or one like it was communicated some while ago by Mr Fatio', 78. Case 4: possibilities of determining the form of the fluent relationship by inspection of the fluxional equation, 80. 'This is worthy to be treated in more detail', 82. Case 5: picking out possible single terms in the fluent by inspection of the fluxional equation, 82. An example where the latter is of second order, 84. Another where it is of third, 86. A further one, of second order and involving surds, is abandoned by Newton (when he saw the complications in obtaining more than a particular solution?), 88. Case 6: a technique of reduction where the base variable is not present in the fluxional equation, 88. An example in which reduction is made to a hyperbola-area, 90. A variant mode of operation 'sometimes convenient', 90. Two more examples (constants of integration are ignored in both), 92. Case 7: the possibility of obtaining the fluent 'root' in exact form by first extracting it as an infinite series (by Proposition XII following), 92. Proposition XII: the equivalent fluent relationship obtained from a given fluxional equation as an 'unterminated converging series'. Case 1: the mode of extraction where the base variable is 'very small', 92. Case 2: the modified algorithm where the latter is very large, 94. Case 3: reductions to Cases 1/2 by adding or subtracting a constant from the base variable; in corollary Newton enunciates the particular 'Maclaurin' expansion of a quantity, 96. And also its general 'Taylor' series, 98. But is misled in thinking to be able to apply them here by his notation (which

PART 2

RESEARCHES IN PURE GEOMETRY AND QUADRATURE OF CURVES

(*c.* 1693)

segments of a triangle, 452. Further such theorems; 'but it is enough to have disclosed the method of investigation', 454. Similar fluxional relationships in more general figures allowed to 'flow' in specified ways: instanced where a given line-length slides in a given angle to intercept a given straight line, 456. Finding the (instantaneous) direction of motion in a curve by compounding the fluxions of related defining motions: when the latter are in given lines, 456. And when they are through fixed poles, 458. Representing such related fluxions: the 'goal-point' towards which the resultant tangential motion tends, 458. A first example of its construction in a curve defined Cartesian-wise by a given relationship between the 'angled' distances of its general point from two base lines, 460. The instance where these coordinates are the straight lines drawn from a point on a given circle through two fixed poles to their meet with a second given circle, 462. And where they are correlate arcs of the two circles; and yet more generally where the circles are replaced by other curves, 464. A second example in curves defined by bipolar coordinates: instanced in the (four) Cartesian ovals, 464. A third example in Descartes' 'instrument' for generating higher-order curves from the intersection(s) of a translating given curve and of a transversal constrained to rotate round a stationary pole and also to pass through a point fixed in the translating plane, 466. Instanced in Descartes' cases where the translating curve is an Apollonian parabola (yielding the Cartesian trident) and where it is a circle (yielding the conchoid), 468.

APPENDIX. Miscellaneous preliminary drafts for the final 'Liber 1'. [1] (Add. 4004: $129^r/129^v$). Geometrical equivalents of the arithmetical operations of multiplying, dividing and extracting roots (on taking some unit line as 'universal measure'), 470. But we should not take this congruence to mirror an essential identity: geometry should be free of arithmetical computation as far as possible, 471. [2] (Add. 3963.2: 11^r). A further exhortation to keep geometry uncontaminated by 'exotic terms': but 'if anyone feels otherwise, the matter is not important enough further to dispute about', 472. [3] (Add. 3963.2: $4^r/5^r$). A further warning not to 'confuse' arithmetic and geometry, despite their many analogies, 473. [4] (Add. 3963.3: $13^r/13^v$). What particular 'conditions' of curves are preserved under planar optical projection, and what altered: rules (suppressed by Newton from his main text as too digressive) listing what is maintained in the perspective figure, 475. Replacements allowable where intersections pass under projection to be 'absurd' (at infinity), 476. [5] (Add. 3963.3: 13^r–14^r). The preceding recast as six 'porisms', 476. General properties of curves 'induced' from simple cases, 477. Instanced in the general Newtonian diameter and his 'parallelepiped' rule of the constant ratio of products of intercepts, and so on, 478. Infinite points, asymptotes, parabolic and hyperbolic branches ('to be instanced in conics'), 478. [6] (Add. 3963.3: 14^v/ 963.10: 107^r). An enlarged scheme treating general properties of curves, 478. First kind: the general Newtonian diameter, 479. Second kind: Newton's 'parallelepiped' rule, 480. The ensuing construction of the tangent at a point on a given curve: instanced in the conic hyperbola (where it yields the pole-polar property), 481. [7] (Add. 3963.3: 21^r–26^r). An augmented revise of the 'first kind of properties', 482. The rectilinear locus whose ordinate is the aggregate of the intercepts made by a given line n-ple in a transversal of fixed direction; its higher-order (quadratic, cubic, ...) parabolic generalizations where the squares, cubes, ...of the intercepts are summed, 482. And where the products of the intercepts two, three, ...at a time are summed, 483. Problem 1: to construct the general Newtonian diameter of a curve with respect to a given ordinate direction, 483. Newton's terminology for these diameters and their related 'centres of ordination', 484. Extension to asymptotal intercepts, 485. Problem 2: to construct the analogous quadratic, cubic, ...diameters, 485. A Problem 3 'to find a parabola passing through given points and tending in an ordained direction' is merely enunciated, 486. Parabolic points (double/at infinity); other possibilities of multiple points in a curve, and their use in distinguishing curves into

PART 3

CARTESIAN ANALYSIS OF HIGHER PLANE CURVES AND FINITE-DIFFERENCE APPROXIMATIONS

(*c.* summer 1695)

LIST OF PLATES

PART 1

THE FIRST TRACT 'DE QUADRATURA CURVARUM'

(early winter 1691–2)

INTRODUCTION

When Newton in 1685 made the Scots mathematician John Craige welcome during an extended visit to Cambridge, showing him in the privacy of his Trinity rooms a selection of his yet unpublished papers on calculus and infinite series, he could not have foreseen the tangled consequences which that innocent and generous act would have over the next half dozen years. Craige was then on the point of publishing a short tract wherein he expounded a systematic (if far from general) 'Method of determining the quadrature of figures comprehended by straight lines and curves,[(1)] there gathering a variety of techniques of rational algebraic quadrature and arc-rectification developed over the preceding quarter of a century by John Wallis, Nicolaus Mercator, Isaac Barrow, Leibniz and most recently David Gregory 'in his very fine treatise *On the Dimension of Figures*'.[(2)] Greatly dissatisfied with a newly appeared article by Walther von Tschirnhaus which sought—none too successfully—to lay down criteria for such rational quadrature in the case of conic, cubic and quartic curves,[(3)] and hearing that Wallis was about to make public a summary of Newton's method for squaring curves,[(4)] Craige had at some unknown previous date written to Cambridge for enlightenment and was now in consequence granted the rare privilege of a private view of Newton's papers as they related to his interests. Craige himself specifies that he was then shown 'manuscripts', doubtless including the 'De Analysi',[(5)] in which quadratures were attained by reducing

(1) Our English rendering of the title of his *Methodus Figurarum Lineis Rectis & Curvis comprehensis Quadraturas determinandi* (London, 1685).

(2) 'clarissimus Nostras D. *David Gregorius* in pulcherrimo suo Tractatu, [*Exercitatio Geometrica*] *de Dimensione Figurarum*' (*Methodus Figurarum*: 12). On Gregory's 1684 *Exercitatio* see IV: 414–16; for the other contemporary mathematicians named by Craige see his *Methodus*: 3, 12, 16, 26–7 and 30–1.

(3) 'Methodus datæ figuræ rectis lineis et curva geometrica terminatæ aut quadraturam aut impossibilitatem ejusdem quadraturæ determinandi' (*Acta Eruditorum* (October 1683): 433–7). Tschirnhaus' title is deliberately reflected, of course, in that of Craige's 1685 work, in whose appended 'Animadversio in Methodum Figuras dimetiendi, A clarissimo Quodam Germano editam in Actis Eruditorum Lipsiæ publicatis' (*Methodus Figurarum*: 38–43) he was deeply critical of Tschirnhaus' pretensions.

(4) Doubtless from the *Proposal about Printing a Treatise of Algebra, Historical and Practical*... which Wallis circulated in 1683 and where (see IV: 413, note (20)) mention is made of his plan to treat 'lastly, the method of infinite *Series*...invented by Mr. *Isaac Newton*,...which is of great use for the...squaring of *Curve-lined Figures*...'. The quadrature series which Newton had communicated to Leibniz in October 1676 did not in fact appear in the 1685 edition of Wallis' *Algebra*, and this omission occasioned the sequence of events which we summarize in sequel.

(5) See II: 206–46.

the 'root' of the integrand to an infinite series(6) and it was presumably on this occasion that he was permitted to make the abstract from the 1671 fluxional tract to which we have earlier referred;(7) but, though to Tschirnhaus' asserted criteria for quadrability he was given two counter-instances—notably that of the curve of ordinate $kx(a^2+x^2)^{\frac{1}{2}}$ which he himself was quick to publish(8)—and while Problem 9 of the 1671 tract(9) would have afforded him some insight into Newton's several techniques of exact quadrature, he was not, it would appear, allowed more than a glimpse of the 'prime' theorem for squaring curves of

(6) In speaking in his 1685 tract of his *desideratum* of a general 'Methodus...Figurarum Quadraturas determinandi [quæ] ad omnes figuras extendatur (exceptis iis quæ a Curvis transcendentibus terminantur, quas nulla hactenus vulgata Methodus comprehendit)' he remarked that 'cum figuram aliquam Quadrando necesse sit & Radicem ex æquatione affecta (& supra Quadraticas ascendente) extrahere,...unicum remedium mihi cognitum est radicem istius æquationis in seriem infinitam (juxta Methodum clarissimi viri D. Isaaci Newtoni Geometræ non minus quam Analystæ præstantissimi) resolvere, quam prælo commissam esse à clariss. Wallisio [see IV: 672, note (54)] audimus, quamque insignis ipse D. Newtonus mihi in Manuscriptis pro summâ suâ humanitate communicavit' (*Methodus Figurarum*: 26–7).

(7) See III: 354–5, note (1).

(8) In the 'Animadversio' on Tschirnhaus' 1683 'general' method of algebraic quadrature which he appended to his *Methodus Figurarum...Quadraturas determinandi* (see note (3) above); it is there (p. 43) cited—without notice of its Newtonian provenance—as 'Æquatio naturam curvæ...exprimens $z^2 = \frac{\overline{m^2+x^2}[\times]x^2}{p^2}$ in qua x denotat abscissas...& z ordinatas, m & p quantitates datas & [d]eterminatas'. More than thirty years afterwards Craige recalled in the 'Præfatio' to his *De Calculo Fluentium Libri Duo* (London, 1718) that 'Calcul[i] fluentium...prima Elementa, cum Juvenis essem, circa Annum 1685 excogitavi: Quo tempore *Cantabrigiæ* commoratus D. *Newtonum* rogavi, ut eadem, priusquam prælo committerentur, perlegere dignaretur: Quodque Ille pro summa sua humanitate fecit: Nec-non ut Objectiones in Schedulis meis contra D. *D[e] T[schirnhausio]* allatas corroboraret, duarum Figurarum Quadraturas sponte mihi obtulit; erant autem harum Curvarum Æquationes $m^2y^2 = x^4+a^2x^2$ & $m^2y^2 = x^3+ax^2$' (signature [b2r]). In his article 'On the Early History of Infinitesimals in England' (*Philosophical Magazine* (4) **4**, No. 26 (November 1852): 321–30) Augustus De Morgan set forth (pp. 326–7) the curious argument that this passage relates to the amended 'Responsio ad Literas Domini D. T. *Lipsiam* missas Feb. 20. 1686' which Craige seven years later appended to his much reworked *Tractatus Mathematicus de Figurarum Curvilinearum Quadraturis et Locis Geometricis* (London, 1693): 55–61, and he was led consequently to speculate that 'it was this very tract of Craig's which immediately suggested to Newton the progress which the views of Leibnitz were [by then] making, and induced him to forward to Wallis [in August 1692!] the extracts from the *Quadr. Curv.* [which appeared in Wallis' *Opera*, **2**, 1693: 390–6]': both these unfounded suppositions were soon afterwards accurately demolished by H. Sloman in his augmented examination of *The Claim of Leibnitz to the Invention of the Differential Calculus* (Cambridge, 1860): 111–17.

(9) See III: 210–90, especially 236–64. We there remarked (*ibid.*: 237, note (540)) that the quadrature series which Newton communicated to Leibniz in 1676 (see next note) is but an easy generalization of the *Ordo secundus* of the 'Catalogus' of algebraic curves having exact quadrature which is there set out. No portion of the 1671 tract's 'Prob: 9' is, however, included in the 'Tractatus de Seriebus infinitis et Convergentibus' (reproduced in III: 354–72) as Gregory transcribed it from Craige's notes upon the tract after the latter's return to Scotland.

general ordinate $dz^{\theta}(e+fz^{\eta})^{\lambda}$ which Newton had communicated to Leibniz in his *epistola posterior* nine years before[10] and which he now described to Craige as 'able to exhibit innumerable quadratures of this sort by means of an infinite series which shall in given circumstances, breaking off, determine the geometrical quadrature of a propounded figure'.[11]

Upon his return to Scotland shortly afterwards, Craige soon became close friends at Edinburgh with both David Gregory and the physician (and learned amateur scientist) Archibald Pitcairne, informing them of the quality of Newton's quadrature series 'which both confessed to be completely unknown to them'[12] and citing the two instances of it which he had earlier been given in Cambridge. By building upon these and accurately divining their sequence Gregory found little difficulty in recovering Newton's general theorem for himself and thereupon, careless of first inventor's rights, immediately assumed ownership of the algorithm, telling Pitcairne of his 'discovery' and writing to a common friend Colin Campbell on 2 October 1686 that 'As for my Methode of quadratur I resolve godwilling to print it shortly'.[13] When Pitcairne in turn informed Craige of this intention, the latter was at first uncertain whether Gregory's generalization was identical with the original from whose instances it had been drawn and in the summer of 1688 he wrote to Cambridge requesting Newton to send along his full theorem so that comparison between the two might be made. Newton did so in a lost letter to Craige on 19 September, and when the two expansions were set down side by side their identity, superficial

(10) See *The Correspondence of Isaac Newton*, **2**, 1960: 110–29, especially 115–17.

(11) So Craige afterwards expressed it when he later wrote in the 'Præfatio' to his *De Calculo Fluentium* (see note (8)) that Newton, having given him particular counter-instances to disprove Tschirnhaus' method of quadratures, 'me interim certiorem fecit se posse hujusmodi [Curvas] innumeras exhibere per *Seriem Infinitam*, quæ in datis conditionibus abrumpens Figuræ propositæ Quadraturam Geometricam determinaret' (signature [b2^r]).

(12) 'In Patriam postea redeunti magna mihi intercedebat familiaritas cum Eruditissimo Medico D. *Pitcairnio* & D. *D. Gregorio*; quibus significavi qualem pro Quadraturis Seriem haberet D. *Newtonus*, quam penitùs ipsis ignotam uterq; fatebatur' (Craige, *De Calculo Fluentium*; Præfatio: [b2^r]),

(13) *Correspondence*, **2**: 451. Having 'in y^e meane time' urged Campbell to take as 'instances' of his quadrature method both the Newtonian curves, of ordinates $\sqrt{[(m^2+x^2)\,x^2/p^2]}$ and $\sqrt{[(ax^2+x^3)/n]}$, which (see note (8)) Craige had earlier passed onto him, Gregory added that 'I could give you millions of such but I choose y^e most easy and simple. You see these at y^e first brew seem not quadrable but by a Series but my present methode does them universally.' We may well wonder if Gregory would have been half as confident without Newton's indirect tutoring: when some time before May 1694 his attempt to duplicate Newton's result in Corollary 2 to Proposition XCI of Book 1 of the *Principia* (London, $_1$1687: 220–1, reproduced on VI : 225–6) led him to seek the integral of the 'formula' $ex(b^3x-c^2x^2-d^4)^{-\frac{1}{2}}$ he was at a loss to do other than weakly jot down in a memorandum (ULE. Gregory C 60) 'Sed hæc...cum proprie ad trinomium revocetur nequit per nostram methodum quadrari. tentandum an per methodum Newtoni.'

discrepancies in notation apart, was evident[14]—and so Craige no doubt hastened to tell Gregory. But it was too late to prevent Pitcairne publishing[15] the quadrature series as Gregory's invention without mention, to Craige's barely concealed disgust,[16] of Newton's prior discovery and private circulation of it a dozen years before. Newton himself, it would seem, was not informed at once of the plagiarism—if he had been, we may imagine the anger which the incident would have roused in him so soon after his upset at the appearance of Gregory's *Exercitatio* four years earlier[17]—and the onrush of political event in England carried him off almost at once into a momentarily crowded existence, far from academic squabble, as a University member of the Convention Parliament.[18]

The affair was all but forgotten, certainly, when more than two years afterwards in the summer of 1691 the Savilian Professorship of Astronomy at

(14) 'Post aliquot verò menses narrabat mihi D. *Pitcairnius* D. *Gregorium* Seriem similiter abrumpentem invenisse. Ego nullus dubitans, quin eandem ex duabus prædictis Quadraturis ipsi à me communicatis deduxerit, per Literas D. *Newtonum* rogavi, ut Seriem suam mihi transmittere vellet, ut an eadem esset cum *Gregoriana* perspicerem: Rogatui meo annuit Vir illustrissimus per Literas 19 Sept. 1688 datas: Nec mirum si parva esset inter utramque Seriem discrepantia, cum *Gregorius*, ex duobus illis Exemplis & indicatâ a me Seriei Newtonianæ indole, suam facile deducere potuisset' (*De Calculo Fluentium* (note (8)): Præfatio: [62r/62v]).

(15) In his *Solutio Problematis de Historicis seu Inventoribus* (Edinburgh, 1688). No copy of this rare tract today exists in any of the university and college libraries in Cambridge and it would appear highly unlikely that Newton saw a copy till at least three years later when he met Gregory for the first time on a visit to London (see below).

(16) Certainly, in sending to Colin Campbell on 30 January 1688/9 a 'general Method for finding the Curvature of any given curve...copied out of Mr Newton's manuscript'—that is, Problem 5 of the 1671 tract (see III: 354, note (1))—and promising 'for your further satisfaction...Mr Newton's Series for Quadratures, which he was pleased to send me, in a letter [*viz.* of the previous 19 September; see note (14) above] not long since' (*Correspondence*, **3**, 1961: 8), he was quick to add in parenthesis to the latter the confidence that 'I must tell you b[y] the by that I saw this Series at Cambridge & acquainted Dr Pit. & Mr Greg: with it, & told them the cheife propertie of it, *sc*: that it breaks of when the figur's Quadrable. At which time they were altogether ignorant of such a series as I can let you see by letters of the Dr: written to me at Cambridge; which astonished me to find no mention made of Mr Newton by the dr: but keep this to your selfe' (*ibid.*: 8–9). In 1718, ten years after Gregory's death, Craige was prepared to speak out publicly in giving his 'historiola' of the episode when it could matter only to him and to Newton: 'Hanc...Lectoribus impertire æquum videbatur, ut soli *Newtono* Seriem illam tribuendam esse cognoscerent. Satius quidem multo fuisset, si ipse (dum vivus esset) *Gregorius* eandem Orbi Mathematico communicasset, quodqȝ se facturum promisit per Literas dat. *Londini* 10, Oct. 1691. Me interim in iis hortatus est, ut, si quid haberem ad Memoriam ejus in hoc negotio juvandam, id ego quam citissime ad illum transmitterem; quod sine mora a me rem omnem fideliter ab initio narrante factum erat. Opus enim erat mihi facillimum, utpote qui omnes ejus & *Pitcairnii* Literas hanc materiam spectantes tum apud me habuerim, & adhuc habeo' (*De Calculo Fluentium* (note (8)): Præfatio: [b2v]).

(17) See IV: 413–17. As we observe below, when the full implications of Gregory's 'independent' discovery of the quadrature series were brought home to Newton in November 1691 his reaction then followed a similar pattern to that which it had seven years earlier.

(18) See VI: xxii–xxiv.

Oxford fell vacant and Gregory—then, as it chanced, in London on business for his Scots university[19]—applied to Newton to support his candidature for it. Though thereupon, 'having known him by his printed Mathematical performances & by discoursing wth travellers from Scotland & of late by conversing with him', Newton was quick to commend him to the electors to the Chair as 'prudent sober industrious modest & judicious & in Mathematiques a great Artist...very well skilled in Analysis & Geometry both new & old',[20] as the autumn of 1691 drew on Gregory came more and more to despair of his prospect of gaining the Professorship against the strong competition of his rivals Edmond Halley and, even more, John Caswell.[21] By early October, with still 'nothing done in the affair of Oxford', he was reluctantly making ready to journey back to Edinburgh to resume his professorial duties there when, evidently in a late move to promote his sagging chances of the Savilian Chair, he decided to

(19) Gregory wrote to Newton on 7 November 1691 to excuse himself for having 'been diverted long from writing to you as ye allowed me, by some affairs of our Coledge [at Edinburgh] that I was oblidged to manage with our Scots secretaries here' (*Correspondence*, **3**: 170). The 'affairs' had no doubt in part to do with Gregory's refusal the year before to take the newly required oath of allegiance to the English—rather than, as previously, the Scottish—throne (see *ibid.*: 171, note (1)); while his act of abjuration did not immediately cost him his professorship at Edinburgh, it left him with an uncertain prospect of continuing permanently in office there, and he was very happy to escape to the academic peace of an English university. When he first met Gregory in London in late July 1691 Newton evidently showed considerable interest in the institutions and methods of teaching in Scotland's 'Coledges', for—if only as a token of appreciation for supporting his candidature at Oxford—Gregory shortly afterwards wrote to Newton a long letter, dated '8 Agust', on the subject (reproduced in *Correspondence*, **3**: 157–62.)

(20) Or so Newton wrote from 'London. July ye 27th 1691' in the draft (ULC. Add. 4005.5: 16^{r}, given in *Correspondence*, **3**: 154–5) which he retained of his recommendation; the undated letter he sent to Arthur Charlett, secretary to the Professorship Electors, presumably on the same day (now in Bodleian MS Ballard 24, and first printed by John Nichols in his *Illustrations of the Literary History of the Eighteenth Century. Consisting of Authentic Memoirs and Original Letters of Eminent Persons*, **4** (London, 1822): 49–50) only trivially emends and amplifies this. In continuation he further asserted that Gregory 'has been conversant in the best writers about Astronomy & understands that science very well. He is not only acquainted wth books but his invention in mathematical things is also good....I take him to be an ornament to his Country...'. So much praise after only a few hours' meeting with the man! At this first encounter the topic of Gregory's purportedly independent discovery of Newton's quadrature series would seem to have been broached, since in his letter on 10 October following (quoted below) Gregory assumes that its prime author is broadly familiar with 'my series for quadratures', at least in the bare form in which Pitcairne had published it three years earlier (see note (15) above).

(21) On 27 August Gregory wrote to Newton that 'Ther hath been no further discourse about the Savilian profession of Astronomie, and I beleeve that without more noise Mr Casswell will gett in ther' (*Correspondence*, **3**: 166). Three months later on 26 November, only days before his election to the Savilian Chair, he added in the same vein that 'the electors for the Astronomy professor have not yet met, but so farr as I can perceive Mr Casswell will defeat Mr Hally and me in that affair' (*ibid.*: 181).

publish a revised version of his quadrature series. On the 10th, accordingly, he wrote to Newton

to give you account that Mr Hally and others have told me that it will be fitt to illustrate my series for quadratures with some examples and publish it in the [*Philosophical*] *transactions*. I beg then Sir that ye will give me your opinion of it and allow that I transmitt it to you to be revised, and published in form of a letter to you. And since, by what you have told me, I know that ye have such a series long agoe I entreat ye'l tell me so much of the historie of it as ye think fitt I should know and publish in this paper. Sir since I am resolved to doe in this according to your opinion, I hope ye will freely allow me it on the whole matter.(22)

When Gregory's augmented account of the quadrature series, formally addressed 'to the finest of [natural] philosophers, Isaac Newton',(23) duly arrived in Cambridge a month later accompanied by his further plea that 'besides your answer to it which ye [will] allow to be publick..., ye will be pleased to advertise me if ther be any mistake in it, for I wrote it when I was not much at leisure',(24) Newton obligingly began at once to draft an equally measured response directed 'to the very celebrated David Gregory, Professor of Mathematics at Edinburgh University':(25)

I have read through your letter, and I compliment you on the series set out in it and likewise on your plan to publish it afresh,(26) illustrated with examples. The method whereby you fell upon it is very elegant and neat, and will beyond doubt prove agreeable to your readers. But when with your usual politeness you request my own series from me in return, it is necessary first to point out several things in its regard. For instance, when the eminent Mr G. W. Leibniz exchanged correspondence with me fifteen years ago through the agency of Mr Oldenburg and I took the occasion to disclose my method of infinite series, I described the present expansion in the second, dated 24 October 1676, of my letters—the two, that is, out of which the eminent geometer Dr John Wallis expounded my method of infinite series in his *Algebra* [in 1685], there testifying that he

(22) *Correspondence*, **3**: 170. In his lost reply to Gregory's plea Newton evidently acquiesced, since in a following letter on 26 November Gregory had cause to remind him that 'you wrote that you would allow [yours] to be published with [my Series which Dr pitcairn published]' (*ibid.*: 181).

(23) 'Philosophorum optimo Isaaco Neutono apud Cantabrigenses matheseos professori...'. Gregory's submitted account of his series (now ULC. Add. 3980.6) is reproduced by H. W. Turnbull in *Correspondence*, **3**: 172–6 with pertinent notes thereon.

(24) *Correspondence*, **3**: 171.

(25) 'Viro Clarissimo Davido Gregorio in Academia Edinburgensi Matheseos Prof.' In sequel we render Newton's Latin draft (ULC. Add. 3980.11) into modern English; to stress the manner in which Newton at once reshaped it to be the opening paragraph of his ensuing tract 'De quadratura Curvarum', we reproduce its original text (first printed in *Correspondence*, **3**: 181–2) in appendix to this introduction.

(26) Pitcairne had, of course, published its bare enunciation in 1688; see note (15) above.

(27) In Wallis' own words 'Mr. *Isaac Newton*, the worthy Professor of Mathematicks in *Cambridge*,... about the Year 1664, or 1665,...did with great sagacity apply himself to that

had passed over much of what I had related in them.[27] Again, when your countryman Craige was with us six years ago on an extended stay, he perused my manuscripts (as he himself owns in the book of his he then published[28]) and at that time transmitted to you my quadrature of the curve[29] about which dispute had arisen in correspondence; it was then that you attacked the quadrature of curves afresh and hit upon your series. You are, of course, aware that after his subsequent return to Scotland Craige confirmed that he had seen my series; at that period I myself had no inkling either that he was stirring up contention about the matter or that you had found the series, nor had you, as I believe, previously learnt that I had hit on a closely similar series. But, that I may now satisfy your demand, I need to quote the text of my letter. And, seeing that what I shall perhaps communicate to you at some future date on the quadrature of curves depends on a certain analytical method which I there touched upon somewhat obscurely, I shall describe also what regards this method.

After a curt and none too well informed summary of its historical origins—'Fermat devised the elements of this method and Sluse advanced them'—Newton then began to transcribe his quadrature method as he had set it down for Leibniz in October 1676, stressing that it was taken over from 'a tract on infinite series and the method of fluent quantities' which he had composed five years earlier still.[30] But almost at once he broke off his formal epistle, never to complete it. What answer he returned to Gregory's anxious letter three weeks later on 26 November inquiring if his previous one had miscarried[31] we do not

Speculation [of Infinite Series]. This I find by Two Letters of his (which I have seen,) written to Mr. *Oldenburg*, on that Subject, (dated *June* 13, and *Octob*. 24. 1676,) full of very ingenious discoveries, and well deserving to be made more publick....He doth therein, not only give us many such Approximations fitted to particular cases; but lays down general Rules and Methods, easily applicable to cases innumerable; from whence such Infinite Series or Progressions may be deduced at pleasure; and those in great varieties for the same particular case....' (*A Treatise of Algebra, Both Historical and Practical*, London, 1685: 330). He concludes his ensuing excerpts from Newton's *epistolæ prior et posterior* for Leibniz (see *ibid.*: 330–46) with the observation that 'There is a great deal more (in these Papers) of like nature'.

(28) That is, Craige's 1685 *Methodus Figurarum*... *Quadraturas determinandi*; see note (1) above.

(29) Namely, that of Cartesian equation $pz = x\sqrt{[m^2+x^2]}$, whose exact 'square' was a crushing counter-instance to the would-be universal validity of Tschirnhaus' 1693 criterion for such quadrature; see notes (3) and (8) above.

(30) Slightly to paraphrase Newton's own remark 'De quodam...Tractatu quem annis abhinc viginti de seriebus infinitis et methodo fluentium quantitatum conscripseram.' His reference is, we need scarcely say, to his 1671 treatise on infinite series and fluxions; see III: 32–328, especially 237, note (540). In his *epistola posterior* itself he had spoken more vaguely of a 'Tractatu[s] de his seriebus [quem] conscripsi '(*Correspondence*, 2: 114).

(31) 'About three weeks agoe I wrote to you a letter wherin I gave some examples of Curves quadrable by my Series [and] entreated ye would let me know yours which was not unlike to it.... I am affraid this letter be lost and hath not come to your hands. But if it hath, I beg ye'l give me your opinion of it freely, and whither it ought to come abroad' (*Correspondence*, 3: 181). H. W. Turnbull's affirmation (*ibid.*) that Newton responded with the unfinished draft letter from which we have just now heavily quoted cannot be right, nor is there any reason for thinking that Gregory received a written reply of any sort.

know—perhaps it was given only by word of mouth when Newton met him in London at the end of December.(32) By then Gregory had been elected to the Savilian Professorship and the pressure upon him to make rapid English publication of his variant of the quadrature series was removed; it appeared only a year and a half later in the Latin edition of Wallis' *Algebra*, and without mention of Gregory's earlier move to have it printed along with Newton's equivalent prior expansion.(33)

For Newton himself that was far from the end of it. As his hitherto unpublished papers of the period reveal, the pattern of his private reaction to this new stimulus from Gregory to review his earlier researches into the quadrature of curves almost exactly parallels that of his earlier response in June 1684 to

(32) See the extracts from Gregory's 'Varia Astronomica et Philosophica 1691 Londini. 28 Dec^r' (ULE. Gregory C85) reproduced in *Correspondence*, **3**: 191, where for instance is cited Newton's 'opinion' on 'a good design of a publick speech' for him to give as the newly appointed Savilian Professor of Astronomy at Oxford: namely one 'to shew that the most simple laws of nature are observed in the structure of a great part of the Universe, that the philosophy ought ther to begin, and that Cosmical Qualities are as much easier as they are more Universall than particular ones'. (Gregory did indeed take heed of Newton's advice in the inaugural which he delivered 'in Auditorio Astronomico Oxoniæ' on the following 21 April; see P. D. Lawrence and A. G. Molland, 'David Gregory's Inaugural Lecture at Oxford', *Notes and Records of the Royal Society of London*, **25**, 1970: 143–78, particularly 159–65). It is true that Newton recorded his exit from Trinity College in December 1691 as being not till New Year's Eve (see Joseph Edleston, *Correspondence of Sir Isaac Newton and Professor Cotes* (London, 1850): lxxxv), but this is a readily explained slip.

(33) See John Wallis, *Opera Mathematica*, **2** (Oxford, 1693): 377–80. Gregory's revised and illustrated account of 'Methodus mea Quadraturarum' was there, much as before, presented as a formal letter, dated 'Oxoniæ 21 Julii 1692', to his fellow Savilian Professor and introduced with the non-committal observation (*ibid.*: 377–8) that 'Anno 1688 edidit *Archibaldus Pitcarnius* M.D.... Schedulam quam *Solutionem Problematis de Inventoribus* dixit, in qua methodum tradit à me inventam, cujus ope series deteguntur, quibus infinitæ numero Curvæ & spatia iis & rectis contenta nullius methodi hactenus cognitæ legibus subjecta mensurantur'. Wallis himself was well aware that Newton's *epistola posterior* gave the lie to this implied assertion of priority, and speedily wrote to Cambridge asking permission of the expansion's prime discoverer to publish the quadrature series as he had communicated it to Leibniz in October 1676; in lost letters to Wallis of 27 August and 17 September Newton not only gave his approval to such publication but also sent along, extracted from his newly written treatise 'De quadratura Curvarum', the generalization which duly appeared in print with his 1676 series the next year (*ibid.*: 390–1, reproduced in §2, Appendix 3 below). Meanwhile David Gregory during a visit to Holland in May/June 1693 (see *Correspondence*, **3**: 274, note (1)) had met Huygens and discussed with him his own variant 'Methodus...prout a D: Pitcarnio tibi tradita erat'; upon his subsequent return to Oxford he wrote to Huygens in mid-August 1693, enclosing a further summary of his method 'omnia prout a Wallisio in *Algebræ* suæ nova editione brevi in lucem mittendâ explicantur' (*Correspondence*, **3**: 275–6) and also an 'extrait'—'une grande feuille d'écriture', Huygens described it to L'Hospital on 16 June 1694 (*Œuvres*, **10**, 1905: 622) —taken from the companion 'Methodu[s] D. Newtoni qualem...illam descripsit Wallisius' which Huygens in turn eventually passed on to Leibniz in late August 1694 'puisque vous le souhaitez' (see *ibid.*: 669). The reaction of neither Huygens nor Leibniz to this attempt by Gregory to defend the independence of his discovery of the quadrature series is recorded.

Gregory's transmission of a copy of his *Exercitatio Geometrica*:[34] once provoked, his interest in the matter passed almost immediately from specifying and clarifying the historical context of their discovery to refining and reshaping their verbal exposition and then to developing their content and enlarging their applications. Straightaway, Newton in November 1691 minimally reworked the unfinished draft of his intended reply to Gregory's missive to be the opening of a short tract 'De quadratura Curvarum' wherein an expansion of the 'first Theorem' on general algebraic quadrature which he had communicated to Leibniz in 1676 was joined to a revision of his researches at the same period into the 'Quadrature of all curves whose æquations consist of but three termes'[35] to form a preliminary scheme[36] for determining the 'simplest figures with which exhibited curves can be compared'. And within days this was enlarged to be a developed treatise[37] on the same broad theme of the quadrature of algebraic curves, but now further embracing general methods for resolving fluxional equations into equivalent fluent ones either by exact techniques or, in default, by expanding their roots as infinite series, and thereafter applying these methods to solving problems in the curvilinear motion of geometrical points and in measuring the central forces induced in given dynamical orbits.

By late December Newton's new tract was already far enough advanced to completion for him to tell Gregory at the beginning of a three-week visit he made to London that he planned to publish it as the second of 'two parts of Geometry, the first the Geometry of the Ancients, the second of Quadratures'.[38] In the course of the same visit he likewise informed Fatio de Duillier more vaguely that his 'treatise of curves would soon see the light'.[39] A month afterwards, having

(34) Again see IV: 413–17, and compare note (17) above.

(35) See III: 373–82; compare §1: note (33) below.

(36) Reproduced in §1 following.

(37) The revised 'De quadratura Curvarum' given in §2 below.

(38) As Gregory noted down on 28 December among his 'Varia Astronomica et Philosophica' (see *Correspondence*, **3**: 191 and compare note (32) above.) The surviving manuscript drafts in which Newton afterwards variously implemented this plan for a dual-purpose 'Geometriæ Libri Duo' are reproduced in Part 2 following; see especially **2**, **3**, §§1/2.

(39) Or so Huygens repeated it back to Fatio in his letter of 29 February 1692 (N.S.) with the words 'Son Traité des Lignes a ce que mon frere [Constantijn] me mande (qui le tenoit de vous...) devoit bientost voir le jour' (*Notes and Records of the Royal Society*, **6**, 1949: 159 = *Correspondence*, **3**: 196) and commenting thereon that 'j[e l] 'attens avec impatience, esperant d'y apprendre toutes ces belles choses dont vous faites mention.... J'ay depuis peu...songé à cette affaire des quadratures,...et j'ay encore remarqué quelques regles et tentatives pour trouver la quadrature quand l'Equation d'une ligne quadrable est donnée, mais cela ne va pas bien loin lors que les Equations sont un peu composées, et je voudrois bien sçavoir si Monsieur Newton a des regles generales pour cela, et s'il peut connoitre quand la quadrature est impossible, ou quand elle depend de celles de l'Hyperbole ou du Cercle....Au reste je n'entens pas ce que signifie la fluxion de la fluxion...' (*ibid.*). How serious Newton's own intention of publishing his 'De quadratura' was at this time is difficult to determine, and any

looked through the text of the 'De quadratura' for himself, Fatio could report to Christiaan Huygens that it compared with Leibniz' letters and publications on 'Calculus differentialis' like 'a finished original' to 'a lame and very imperfect copy', adding that 'what he has on quadratures is infinitely more general than all that has been had before, and it is very simple and of wonderful use in all parts of geometry'.(40) Four weeks later still, early in March, he added vividly if with exaggeration that 'I have been chilled to see what Mr Newton has done, and have reproached him for rendering my own hard work useless and not wanting to leave anything for his friends who are come after him to do'.(41) But by then, as Fatio sadly admitted in the same letter to Huygens, the heat of Newton's interest in seeing his treatise into print was past(42) and his attention had

attempt to assess it must also, if we are to trust Joseph Raphson's later memory of an event then more than twenty years old (see III: 11), take into consideration his parallel plan at this period to revise his 1671 tract for publication. What connection (if any) there was between the two—or whether Raphson in his 1715 *History of Fluxions* merely confused Newton's 1671 'Treatise reviv'd' with his newly composed tract on quadrature—would now seem impossible to substantiate.

(40) Fatio to Huygens, 15 February 1692 (N.S.): 'Les Lettres que Monsieur Newton ecrivit à Monsieur Leibnitz il y a 15. ou 16. ans parlent bien...positivement.... Ce n'est que bien long temps aprez qu'il [Leibniz] a donné au Public les Regles de son Calculus Differentialis.... Et la manière dont il s'en est acquité est si eloignée de ce que Monsieur Newton a la dessus que je ne puis m'empecher en comparant ces choses ensemble de sentir bien fortement leur difference comme d'un original achevé, et d'une copie estropiée et tres imparfaite.... Monsieur Newton a tout ce que Monsieur Leibnitz paroit avoir, et tout ce que j'avois moi-même et que Monsieur Leibnitz n'avoit pas[!]. Mais il est encore allé infiniment plus loin que nous, soit pour ce qui regarde les quadratures, soit pour ce qui regarde la proprieté de la courbe quand il la faut trouver par la proprieté de la Tangente....Ce qu'il a sur les Quadratures est infiniment plus general que tout ce que l'on avoit auparavant, et il est tres simple et d'un usage merveilleux dans toutes les parties de la Geometrie' (*Œuvres complètes de Christiaan Huygens*, **10** (The Hague, 1905): 257–8 = *Correspondence*, **3**: 193–4). While Fatio here beyond doubt greatly over-estimates the depth and sophistication of his mathematical expertise in comparison with that of Leibniz, he had nonetheless since 1687 acquired by his own self-tutored efforts a working proficiency with the basic methods and procedures of elementary calculus, and so was not wholly incompetent here to pass judgement; as we shall see (§2: note (68) below) he partially repaid Newton for the privilege of being allowed a sight of his 'De quadratura' by bringing to his attention the technique of multiplying a given 'fluxional' equation through by a determinable factor in order to reduce it (where this is possible) to a directly integrable form.

(41) Fatio to Huygens, 17 March 1692 (N.S.): 'j'ai été glacé en voiant ce qu' a fait Monsieur Newton, et je lui ai reproché qu'il rendoit inutile mon travail et qu'il ne voulait rien laisser à faire à ses Amis qui sont venus aprez lui. Je croi que dans la suite il ne faudra pas que j'entreprenne d'étude un peu difficile et de longue haleine sans etre assuré de sa part que l'envie ne lui prendra pas de traitter le même sujet' (*Œuvres complètes*, **10**: 272, quoted in *Correspondence*, **3**: 198, note (5)).

(42) 'Monsieur Newton se relache déjà sur l'impression de son Traitté des lignes Courbes. Sa premiere chaleur est passée, et je croi qu'il s'accoutume peu à peu à juger qu'il n'est pas fort necessaire qu'il s'engage dans les embarras que l'impression d'un Traitté comme celui là traine necessairement aprez elle. Nous y perdrions beaucoup si ce Traitté ne paroissoit

already strayed to other matters.(43) Though Huygens continued to pump Fatio for further news of Newton's 'curious tract',(44) he had during his remaining years of life to rest content with the short extracts from the 'De quadratura' which Wallis included soon afterwards in his Latin *Algebra*.(45) And the ripple of excitement which had been briefly stirred in London and in Holland by the sudden prospect of Newton publishing his researches into quadrature faded as quickly as it had arisen.

To this external record of the circumstances in which Newton came to put together his first treatise 'De quadratura Curvarum' the surviving states of its text add little. In his opening paragraph he briefly remarks that 'Fifteen years ago, when Mr Leibniz held an exchange of letters with me through the agency of Mr Oldenburg..., I spoke in the second of my letters (dated 24 October 1676) of a certain method of drawing tangents and unknotting more difficult problems which I concealed in an anagram...'(46) and thereafter passes at once to cite *verbatim* from his *epistola posterior* his prime quadrature theorem for squaring a general algebraic binomial by a 'converging' series. In regard to his priority in its discovery he is content to adjoin a subdued note that, 'Five years ago, after John Craige had, during an extended stay in Cambridge, seen the series here

point' (*Œuvres complètes*, **10**: 271). 'Je ne sçai si je Vous ai dit', Fatio continued to Huygens, 'que Monsieur Newton y donne une Methode bien étendue de trouver la Courbe la plus simple dont depend la Quadrature d'une Courbe proposée. Il est certain que jusques à present il n'a encore rien paru de si beau dans la Geometrie abstraite que cet ecrit qui n'est que de quelques feuilles et qui ne seroit point trop long pour entrer dans une transaction'.

(43) And in particular to chemistry. When Archibald Pitcairne visited Newton in Cambridge on 2/3 March their talk was of 'acids', 'fermentations' and 'fiery spirits'; see the manuscript 'De natura Acidorum. written by Mr Newton's hand' (reproduced in *Correspondence*, **3**: 205–9) which Pitcairne then took away with him.

(44) 'Vous me faites de plus en plus envieux', he wrote to Fatio on 5 April following, 'pour ce curieux Traité de Mr Newton *de Lineis Curvis*. Estant achevé, et fort court, comment peut il s'excuser de ne le point publier sur l'embaras de l'impression? Ou bien vous Monsieur, que ne luy offrez vous vostre secours?' (*Œuvres complètes*, **10**: 279). Six months later he was more cautious in his enthusiasm when on 22 October he imparted to L'Hospital the news that 'Mr Fatio m'a fait esperer la publication d'un traité de Mr Newton sur ce sujet [des quadratures des courbes et du probleme renversé des Tangentes], qui, à son avis, en sçait bien plus que luy et Mr Leibnitz' (*ibid.*: 327).

(45) *Opera Mathematica*, **2** (Oxford, 1693): 1–482; the excerpts from the 'De quadratura' (*ibid.*: 390–6), communicated by Newton to Wallis the previous summer in letters of 27 August and 17 September (see note (33) above) whose texts have not themselves survived, are reproduced below in §2, Appendix 3, where in our commentary (*ibid.*: note (26)) we also record Leibniz' disappointed reaction to what he regarded, when a copy of Wallis' book finally reached him in September 1694, as an over-curt and insufficiently informative exposition by Newton of his methods 'pour l'inverse des tangentes'. No fuller version of the 'De quadratura Curvarum' appeared in print till 1704, nine years after Huygens' death; see note (61).

(46) Our translation of Newton's Latin on page 24 below.

set out and communicated the quadrature of the curve presented in this second example to his fellow Scots, Mr David Gregory, Professor of Mathematics at Edinburgh and a man of keen intelligence, came upon the same series, but in a less neat form, by a method of his own'.[47] That Newton did indeed pen his ensuing treatise in the early winter of 1691/2 is amply confirmed by the fact that a revise of Case 2 of Proposition XI of its second state is drafted over and around a letter written to him on 19 December 1691 by his former amanuensis Humphrey Newton.[48] And a brief reminder of the degree to which Fatio de Duillier at this period gained access to Newton's confidence occurs in the latter's acknowledgement—in itself unusual—regarding the employment of a multiplying factor to reduce a given first-order fluxional equation to directly quadrable form that 'This rule, or one like it, was communicated to me some while ago by Mr Fatio'.[49] But otherwise Newton's manuscript bears no impress of the sequence of events which brought it into being, and afterwards led him to leave it off unfinished.

In neither of its versions, of course, is the work conceived as a beginner's introduction to the calculus of fluxions. It began its existence[50] purely as a technical tract on the quadrature of algebraic multinomials, rationally so, if possible, or in terms of the areas of the 'simplest' curves in terms of which they may be squared or, failing that, as an infinite, convergent series of rational terms: in that initial form no concession is made to the ignorance of the untutored reader. In the broader-based revised treatise which Newton composed in its wake,[51] his now familiar method for attaining from first principles the fluxional derivative of a 'given equation involving any number of fluent quantities' makes a welcome reappearance, but the only definition given by him of the fundamental notion of the fluxion of a quantity remains the intuitive, loosely verbalized one that it is the instantaneous 'speed' of increase or decrease of its fluent at a given 'time' in its growth, and hence (when multiplied by the latter's 'instant' o) may be set in place of its fluent's contemporaneous increment. Through the quirk of fate—and Leibniz' subsequent relentless insistence upon the superiority of his own brand of *calculus differentialis*—scholarly attention was soon to focus not so much upon the relative adequacy and rigour of Newton's definition of fluxional speed as on the efficiency and conciseness of the notation by means of

(47) See page 26 below.

(48) See §2: note (55).

(49) Fuller details are given in §2: note (68).

(50) In the first state whose text is reproduced in §1 below (with minor variants in Appendix 1.1–5 following).

(51) The unfinished tract in thirteen propositions whose text is, with some slight editorial licence, restored in §2 from a mass of not everywhere consistent or fully completed drafts; significant portions of the latter which offer non-trivial preparatory insights or otherwise complement the main version are given in Appendix 1.6–12 and in Appendix 2.

which he expressed it. Here, quickly abandoning the literal symbols p, q, r, ... for the fluxions of x, y, z, ... which he had employed in his private papers for a quarter of a century, after a momentary return to the equivalent superscription of a double point on the fluent quantities which he had occasionally used in 1665 he abruptly in their place for the first time introduced the 'pricked' letters $\dot{x}, \dot{y}, \dot{z}, \ldots$.[(52)] These singly dotted forms (and the analogously replicated denotations for higher-order fluxions) appear ubiquitously in the remainder of the 1691 tract and correspondingly in the extracts from it which Newton communicated to Wallis the next summer and duly went into print in the latter's *Algebra* in 1693:[(53)] once thus published, the notation came rapidly to be accepted by the world at large as his standard fluxional nomenclature and preferred counterpart to the Leibnizian d-symbol. In later years when, at the time of the squabble over priority of invention of the calculus, it became important (or so he judged) to insist that the fluxional method expounded by him in the 'De quadratura Curvarum' had been contrived well before Leibniz attained its differential equivalent in the middle 1670's, Newton came to tread a tightrope between allowing that in his 'old papers...there are no prickt letters'[(54)] and maintaining, not unfairly, that the series quadrature of an algebraic binomial which he had passed on to Leibniz in 1676 was drawn (with but minimal refinement and polishing of its detail) from a lavish treatise on infinite series and fluxions written five years before.[(55)] Step by step he came to depart from the historical truth that by October 1676 he had attained a general technique for constructing 'Series...which brake off and gave the Quadrature of Curves in finite Equations when it might be'[(56)]—first to infer with only slight exaggeration that 'the Propositions in his Book of Quadratures' were

(52) See §2: notes (35) and (36), and Appendix 1.1: notes (5) and (17); compare also I: 363, note (2).

(53) See §2, Appendix 3: note (12).

(54) As in an unpublished gloss (ULC. Add. 3968.41: 85ᵛ) on his *Commercium Epistolicum D. Johannis Collins et Aliorum de Analysi Promota* (London, 1712) he afterwards admitted to be true of the 'old papers', notably his 1669 'De Analysi' (*ibid.*: 3–20; see II: 206–46), which he there cited. He there went on to lay the smokescreen that 'indeed I seldom used prickt letters when I considered only first fluxions...but when I considered also second third & fourth fluxions as in the body of the book [of Quadratures] I distinguished them by the number of pricks.'

(55) See III: 260, notes (540) and (574), and compare §1: note (21) below. In his gloss on his 1712 *Commercium Epistolicum* (see the previous note) Newton went on: 'In my Letter of 24 Octob. 1676 I...said that...the method there concealed...gave me general Theorems for squaring of figures by series wch sometimes brake of & became finite$_{[,]}$ & how it gave me such Series is explained in...this Book [of Quadratures] & I know no other way of finding them.'

(56) As he succinctly described them on page 193 of his anonymous review of his own *Commercium* ('An Account of the Book entituled *Commercium Epistolicum Collinii & aliorum*...', *Philosophical Transactions*, **29**, No. 342 [January/February 1714/15]: 173–224).

'then known to him';(57) thereafter much more distortedly to attest that he had 'extracted' the latter 'from this [1671] tract...in the year 1676';(58) and finally to make the detailed but equally unsupportable claim that 'In the year 1676 in summer I composed the Book of Quadratures, & then in my Letters of Octob. 24 & Novem 8(59) 1676...I cited severall things out of it, & other Papers from whence I had extracted it; particularly the very words of the first Pro[blem]; the Theorems for squaring figures by converging series...; the ordinates of Curves wch I could compare with the Conic Sections; & the squaring of all Curves exprest by equations of three terms, or comparing them wth the simplest Curves [w^th] wch they can be compared'.(60) So was born the now deeply entrenched myth, parroted in every historical account of Newton's mathematical development down to the present time, that the text of the printed *Tractatus de Quadratura Curvarum* given to the public in 1704(61) was already in finished form a full decade and a half before Newton in fact began tentatively to gather his findings on the quadrature of algebraic curves in the successive drafts of the unfinished treatise which we reproduce in sequel.

If we exclude the excerpts which Newton sent to Wallis in 1692, the ten reworked opening propositions presented in the 1704 *Tractatus*(62) and a few minor, disconnected sheets of the earlier portion of its autograph which his

(57) See *ibid.*; in the *Commercium* itself (page 76, note *) he had similarly 'deduced' that 'Ex his patet Propositiones *Newtoni* de Quadratura Curvarum diu ante annum 1676 inventas esse.'

(58) In a draft (ULC. Add. 3968.41: 16^r) of a preface to an intended re-edition of the printed *Tractatus De Quadratura Curvarum* in about 1719. In a revised version (Add. 3968.3: 26^r) he asserted more fully that 'in the year 1676 [I] extracted the Book of Quadratures from the Tract wrote in ye year 1671 & from other older papers', having recalled in his previous sentence that 'Between the years 1671 & 1676 [I] medled not w^th [mathematical] studies being tyred wth them before'.

(59) See *The Correspondence of Isaac Newton*, **2**, 1960: 179–80. As we have earlier stressed (III: 19, note (60)), his claim therein that 'there is no curve line exprest by any æquation of three terms...but I can in less than half a quarter of an hower tell whether it may be squared, or what are ye simplest figures it may be compared w^th, be those figures Conic sections or others. And then by a direct & short way...compare them' is founded on his contemporary paper dealing with 'The Quadrature of all curves whose æquations consist of but three terms' (see note (35) above).

(60) As he wrote in yet one more draft preface (ULC. Add. 3968.41: 86^r) to his intended 1719 re-edition of the *De Quadratura.*

(61) Namely, as the second of the 'Two Treatises of the Species and Magnitude of Curvilinear Figures' appended by Newton to the *editio princeps* of his *Opticks* (London, 1704: $_2$163–211). Its 'Introductio' and terminal scholium (there newly composed) apart, this is a but lightly reshaped version of the considerably truncated 'Geometriæ Liber secundus' which is reproduced in **2**, 3, §2 below.

(62) Of which David Gregory had already taken a partial copy during his visit to Cambridge in May 1694; see **2**, 2, §2: note (1) following. We will return to consider the novelties of this printed *De Quadratura Curvarum* in our final volume.

mathematical editor, William Jones, acquired about 1710 but never published,[63] none of these drafts—in original or in transcript—of the initial version of his 'De quadratura Curvarum' ever passed out of Newton's hands[64] and, since his death, its very existence has gone all but unsuspected down to our own day. We will not pretend that the onward advance of mathematical discovery was more than temporarily slowed by this failure on Newton's part to communicate the totality of the notational and technical novelties of his treatise to his contemporaries. Within a very few years (and sometimes in but months) all came to be independently rediscovered and quickly absorbed into the scientific mainstream. The 'Q' (for 'Quadratum') which, as a verbal variant on his more usual pictograph '□,' Newton at one point[65] employed to symbolize the operation 'square of' is but the geometrical analogue of Leibniz' standard sign '∫' (= 'S[umma]') for integration which had made its bow in print five years before.[66] His elegant contracted notation for higher-order fluxions whereby, for instance, $12\overset{\vdots\vdots}{z}\ddot{z}$ is written $12\overset{6}{\dot{z}}\overset{2}{\dot{z}}$[67] soon came to have a Leibnizian counterpart in which the corresponding differential $12ddddddz\,ddz$ was shortened to be $12d^6z\,d^2z$,[68] and was itself afterwards exactly re-invented by Brook

(63) See §1: note (1) and §2: note (1) below.

(64) We can place no great faith in a later assertion by Newton that 'My tract of Quadratures was handed about in London in 1691 [January 1691/2?]' (ULC. Add. 3968.41: 20^r, a hasty note written in counterblast to Leibniz' Bernoullian *Charta Volans* in late 1713, here invoked to buttress his wholly untenable following claim that 'I made much use of it in writing my book of Principles & composed it many years before'). While, as we have seen, Fatio was in January 1692 allowed to glance through those sheets of the 'De quadratura', there is nothing to show that others such as Halley or David Gregory were on that occasion accorded an equal privilege. It may well be that Newton is here confusing his 1691 tract with his earlier fluxional treatise which, in Joseph Raphson's words, 'in the Year 1671. he had prepared for the Press...; and which in 1691. Mr. Professor *Halley*, and I, had in our Hands at *Cambridge*, and which was then very much worn by having been lent out' (*The History of Fluxions*, London, 1715: 2; compare III: 11–12).

(65) See §2, Appendix 1.10: note (105).

(66) In Leibniz' widely read article 'De Geometria Recondita et Analysi Indivisibilium et Infinitorum' (*Acta Eruditorum* (June 1686): 292–300): 297, 299. The symbol made its first appearance among his (then) unpublished private papers on 29 October 1675 (N.S.) in his 'Analyseos Tetragonisticæ pars 2da'; see C. I. Gerhardt, *Der Briefwechsel von G. W. Leibniz mit Mathematikern*, **1** (Berlin, 1899): 154. (The original manuscript folio, now Niedersächsische Landesbibliothek, Hanover. L.Hs. **35**, VIII, 18, is reproduced in photofacsimile both facing *Briefwechsel*: 152 and also on the terminal end-paper of M. E. Baron, *The Origins of the Infinitesimal Calculus*, Oxford, 1969.)

(67) See §2, Appendix 1.10: note (132). The convention was at once further contracted to be $\overset{6\,2}{12}$.

(68) F. Cajori in his *History of Mathematical Notations*, **2** (Chicago, 1929): 204–5 credits this form of contraction to Leibniz himself in his letter to Johann Bernoulli of 30 October 1695 (*Leibnizens Mathematische Schriften*, **3** (Halle, 1855): 221). That Newton had employed an equivalent shortening almost four years earlier should silence the conventional criticism that superscript fluxional dots are intrinsically inferior to replicated differential *d*'s in being

Taylor in his *Methodus Incrementorum* in 1715.[69] The *ad hoc* techniques for the exact solution of fluxional equations which Newton presented in his Proposition XI were quickly superseded by more powerful equivalent methods for resolving Leibnizian differential equations which were independently developed by the Bernoulli brothers, Jakob and Johann,[70] and their adherents in the early eighteenth century. The two procedures for extracting the roots of such equations in 'converging' series which he sketched in his following Proposition XII—namely, the straightforward extension of his term-by-term series resolution of ordinary numerical equations by identifying the dominant 'fictitious equation' at each stage,[71] and the alternative method of presupposing the form of the resulting series and thence, by substitution in the given equation and identifying coefficients of like powers to zero, determining its indices and constants[72]—were not long to wait in limbo before being taken up again by George Cheyne,[73] Leibniz[74] and others. Even Newton's most important

incapable of efficiently denoting their function—'as if...12 points would not be as easily told as 12 *d*'s' (to quote James Wilson's pertinent remark in an equivalent context in his *Mathematical Tracts of the late Benjamin Robins*..., **2** (London, 1761): 363).

(69) 'si sit *n* distantia termini alicujus $\dot{s}$, $\ddot{s}$, &c. a termino primo *s*, exprimetur $\overset{n}{s}$, (hoc est, $\dot{s}$ si *n* sit 1, $\ddot{s}$ si *n* sit 2, &c.)...' (*Methodus Incrementorum Directa & Inversa* (London, 1715): 111). Earlier in his book (*ibid.*: 46, 48) Taylor had used an identical superscript notation to represent the *n*-th integral of a variable *Q*.

(70) See J. E. Hofmann, 'Über Auftauchen und Behandlung von Differentialgleichungen im 17. Jahrhundert' (*Humanismus und Technik*, **15**, 1972: 1–40): 18–31.

(71) Of which he had communicated a fully worked first-order example to Wallis in 1692 for publication in the *Volumen alterum* of his *Opera Mathematica*; see §2, Appendix 3: note (20).

(72) The summary enunciation of this which, all but irretrievably anagrammatized into its constituent letters, he had passed to Leibniz in October 1676 Newton likewise permitted Wallis to publish in 1693; see §2, Appendix 3: note (18).

(73) Taking the worked first-order example published by Wallis in 1693 (see note (71)) as his model, Cheyne in his *Fluxionum Methodus Inversa; sive Quantitatum Fluentium Leges Generaliores* (London, 1703) not only elaborated in its full generality (pages 42–6) Newton's first method for resolving the 'Problema. Ex æquatione fluxionem radicis involvente radicem extrahere' but, taking Leibniz' 1693 article (see next note) as his basis, went on (pages 46–50) similarly to expound his alternative approach consisting 'in Assumptione seriei pro qualibet Quantitate incognitâ ex qua cætera commode derivari possint, & in collatione terminorum homologorum Æquationis resultantis ad eruendos terminos assumptæ Seriei'. Earlier in his work (pages 2–13) Cheyne had demonstrated Newton's 1676 series quadrature of an algebraic binomial, giving an inadequate discussion of its generalization to the multinomial case. According to David Gregory (Christ Church, Oxford. MS 364: 94; see W. G. Hiscock, *David Gregory, Isaac Newton and their Circle* (Oxford, 1937): 15) it was the appearance of Cheyne's book which, more than all the persuasion of his mathematical friends, led Newton to publish his *Tractatus De Quadratura Curvarum* in 1704. We will return to discuss the point in our final volume.

(74) 'Supplementum Geometriæ practicæ sese ad Problemata transcendentia extendens, ope novæ Methodi generalissimæ per series infinitas', *Acta Eruditorum* (April 1693): 178–80; on this development of Newton's alternative method (which appeared at about the same time that the second volume of Wallis' *Opera* issued from the press in Oxford) see §2, Appendix 3: note (26). Brook Taylor's supposition (*Methodus Incrementorum* (note (69)): Propositio IX: 28)

achievement in his 'De quadratura', the first explicit enunciation of the Taylor expansion of a general function,(75) was straightaway to be published in an equivalent (if not immediately cognate) form by Johann Bernoulli(76) and to await only twenty years more to be made the basis of a developed treatise on infinitesimal analysis by the man after whom it is now, not without justice, named.(77)

---

that it was possible to assume a single *forma generalis* for the required series was neatly countermanded by James Stirling who, after setting out Taylor's method in his own *Lineæ Tertii Ordinis Neutonianæ* (Oxford, 1717): 24–7, gave a simple instance to underline that 'pendet... forma seriei tam ex coefficientibus, quàm ex exponentibus indeterminatarum in æquatione' (*ibid.*: 28). Taylor at once accepted the criticism; see his letter to John Keill of 17 July 1717 (now in ULC. Res. 1893.3).

(75) Newton deduced the particular 'Maclaurin' form in his Corollary 3 (by successive differentiation, it would seem) and then passed to the general theorem in his Corollary 4; see §2: notes (107) and (109).

(76) 'Additamentum effectionis omnium quadraturarum & rectificationum curvarum per seriem quandam generalissimam,' *Acta Eruditorum* (November 1694): 437–41. By in effect successively integrating by parts Bernoulli there (page 438) deduced that

$$\text{'integr. } n\,dz\text{'}\left[\int_0^z n.dz\right] = nz - \frac{1}{1.2}z^2(dn/dz) + \frac{1}{1.2.3}z^3(d^2n/dz^2) - \ldots.$$

In equivalent modern terms, on setting $n \equiv f'_z$ this is

$$f_z - f_0 = zf'_z - \frac{1}{2!}z^2f''_z + \frac{1}{3!}z^3f'''_z - \ldots;$$

which, since similarly $f'_z - f'_0 = zf''_z - \frac{1}{2!}z^2f'''_z + \ldots$, $f''_z - f''_0 = zf'''_z - \ldots$, $f'''_z - f'''_0 = \ldots$, on eliminating $f'_z, f''_z, f'''_z, \ldots$ reduces to be the 'Maclaurin' expansion

$$f_z - f_0 = zf'_0 + \frac{1}{2!}z^2f''_0 + \frac{1}{3!}z^3f'''_0 + \ldots.$$

(Alternatively, if less straightforwardly, one may transpose $z \to z+x$ and then put $z = -x$.)

(77) Fresh from his discovery, Brook Taylor announced both the general expansion and its particular 'Maclaurin' case in a letter to John Machin on 26 July 1712 (reproduced by H. Bateman in *Bibliotheca Mathematica* (3) **7**, 1906–7: 368–71) as 'a general method of applying Dr Halley's Extraction of roots [see v: 12, note (50)] to all Problems' (page 371), but without giving any proof of its derivation. When three years later he published the expansion as the second Corollary to Proposition VII of his *Methodus Incrementorum* (note (69)), he derived it 'incrementis evanescentibus' (page 23) as the limit of the advancing differences interpolation formula for equal intervals of argument (compare IV: 4) when the latter become vanishingly small. We know that Abraham de Moivre had attained the expansion in exactly this way some four years before Taylor; see his letter to Johann Bernoulli on 6 July 1708 (K. Wollenschläger, 'Der mathematische Briefwechsel zwischen Johann I Bernoulli und de Moivre' (*Verhandlungen der Naturforschenden Gesellschaft in Basel*, **43**, 1933: 151–317): 241–59, especially 242–4, 248–50) where he proceeds by analogously supposing that '[l'abscisse] x contienne un nombre infini d'unités' and pointing out to Bernoulli that the ensuing series 'revient à votre théoreme pour trouver l'intégrale de $y\dot{x}$'. On the question whether, as Stirling later averred in his *Methodus Differentialis* (London, 1730): 102, Jakob Hermann is yet one more valid claimant to have, in

There is no need to go on. Newton's historical importance as author of the 'De quadratura Curvarum' is the minimal one of a lone genius who was able, somewhat uselessly in the long view, to duplicate the combined expertise and output of his contemporaries in the field of calculus. What is not communicated at its due time to one's fellow men is effectively stillborn. But in an edition which seeks to print the extant record of Newton's mathematical achievement rather than assess its temporal impact we may be lavish in our praise of the intrinsic qualities of originality, insight and penetration manifested in his present treatise. Its detailed delights and elegances the reader may now pass on to savour and assess in Newton's own words, aided (we may hope) by our running commentary upon them.

---

the appendix to his *Phoronomia* (Amsterdam, 1716), independently found the expansion, see G. A. Gibson, 'Taylor's Theorem and Bernoulli's Theorem: A Historical Note', *Proceedings of the Edinburgh Mathematical Society* (1) **39**, 1921–2: 25–33, especially 32–3. The traditional attribution of the particular case where the base is zero to Colin Maclaurin reflects, of course, only the success in the middle eighteenth century of his highly eclectic *Treatise of Fluxions* (Edinburgh, 1742), where it is (§751: 610–11) derived by repeated differentiation and openly accredited as 'given by Dr. Taylor, *Method. increm.*'.

It is now, we should add, universally agreed that, twenty years before even Newton attained his formulation of the general expansion, James Gregory had in the early winter of 1670/1—without explicitly enunciating the development in any surviving document or communicating any hint of his technique to the outside world—privately adduced the 'Maclaurin' case to derive the trigonometrical series expansions which he sent to John Collins on the following 15 February (*James Gregory Tercentenary Memorial Volume*, London, 1939: 170–1); see H. W. Turnbull's revealing analysis of Gregory's rough jotted notes pertaining to these (*ibid.*: 353–62) where the necessary computations of the successive derivatives are clearly performed. Given some charity in interpretation, it is indeed possible to regard any expansion of a function of a variable in an ascending series of its powers as a Taylor expansion. In some historically significant cases—notably that of the series developments derived by Newton in the examples illustrating Proposition X of Book 2 of his *Principia* ($_1$1687: 263–9) which we shall take up in our final volume—we would not care to draw any rigid line of demarcation, but a recent identification by R. C. Gupta of some rules for computing incremented sines and cosines attributed to the fourteenth-century Hindu mathematician Mādhava as 'equivalent to modern Taylor series approximations up to the second order of small quantities' ('Second-order Interpolation in Indian Mathematics up to the Fifteenth Century', *Indian Journal of the History of Science*, **4**, 1969: 86–98, especially 92–4) seems to us more than a little anachronistic.

APPENDIX. NEWTON'S DRAFT REPLY TO DAVID GREGORY'S COMMUNICATION ON 7 NOVEMBER 1691 OF HIS METHOD OF QUADRATURE BY SERIES.[1]

From the original [2] in the University Library, Cambridge

[3]Viro Clarissimo Davido Gregorio in Academia Edinburgensi Matheseos Prof. salutem.

Vir Clarissime

Epistolam tuam perlegi et seriem illic positam laudo,[4] juxta et consilium tuum eandem exemplis illustratam in lucem denuò mittendi. Methodus autem qua in eam incidisti perelegans est et concinna[5] et Lectoribus proculdubio placebit. Sed cùm seriem meam pro humanitate tua a me mutuo postulas,[6]

---

(1) We have set out at length in the previous introduction the circumstances in which Gregory came to send his quadrature series to Cambridge in autumn 1691, and also (see our preceding note (25)) have in large part rendered into English the present Latin draft where Newton early in November began to compose his formal response—never, it would appear, completed (compare our previous note (31))—to Gregory's introductory request on 10 October that 'ye will give me your opinion of [my series for quadratures] and allow that I transmitt it to you to be...published in form of a letter to you' and especially to his further entreaty that, 'since I know that ye have such a series long agoe[,]...ye'l tell me so much of the historie of it as ye think fitt I should know and publish in this paper' (*Correspondence of Isaac Newton*, **3**: 170; compare our previous note (22)). Here we are concerned only to make accurate reproduction of Newton's text, citing its principal cancelled variants in footnote according to our usual practice, so as thereby visually to emphasize how lightly he afterwards reshaped it in adapting it to be the opening of the 'De quadratura Curvarum' which follows.

(2) Add. 3980.11, a roughly penned sheet first printed—along with a photo-facsimile of its first page—in *Correspondence*, **3**: 181–2 (but with a mistaken transposition there of its final sentence). Earlier, in sequel to the letters of 10 October and 7 November 1691 (now ULC. Add. 3980.4–6) 'from Dr David Gregory to Sr I. N. concerning the Series for binomial Curves which Gregory published as an invention of his own after it had been describd to him by Mr Craig', it had been accurately listed by Samuel Horsley on 'Octr 23^d^ 1777' (on an obsolete cover now loose in ULC. Add. 4005) as 'a fragment only of Sr I. N. answer to the letter containing this series' with the added opinion that 'I think these Letters very fit to be published as a supplement of the *Commercium Epistolicum* [see II: 20, note (1)] together with Mr Craigs Preface to his work *De calculo fluentium* in which [see notes (8), (11), (12), (14) and (16) of the preceding introduction] he gives the History of that scandalous theft of Dr Gregory's'.

(3) When he examined the manuscript in the early nineteenth century (compare I: xxx), David Brewster here inserted the date '1691' in his characteristically spidery hand.

(4) Newton first went on: 'et methodum illã perelegantem qua in eam incidisti, et rei publicæ mathematicæ rem gratam procul dubio facis quod inventum tam bene exemplis illustratum [afterwards changed to be 'Theorema tam utile'] in lucem denuò mittis'.

(5) This was initially qualified by the proviso 'licet mea generalior sit et habito quodam Theoremate non minus succincta'.

(6) In immediate sequel Newton has cancelled 'eo me adigis ut historiam ejus prius expo-

necesse est ut aliqua prius exponā quæ ad eam spectant. Nam cùm vir celeberrimus[7] D. G. G. Leibnitius epistolare commercium annis abhinc quindecim procurante D. Olden[b]ergo mecum haberet et ea occasione methodum meam serierum infinitarum[8] exponerem: in epistolarum mearum secunda data Octob. 24 1676, hanc seriem descripsi. Hæ sunt epistolæ illæ duæ ex quibus Geometra Celeberrimus Dr Joannes Wallisius in Algebra sua methodum meā serierum infinitarum exposuit[,][9] testatus[10] se multa prætermisisse quæ in illis tradideram.[11] Sed et Craigius vester annis abhinc sex diutius apud nos[12] commoratus MSS mea inspexit[13] (ut ipse in libro suo tunc edito fatetur) qua tempestate quadraturam meam curvæ illius ad vos transmisit de qua cum disputatio epistolaris suborta esset quadraturam curvarum denuò aggressus es et in seriem tuam incidisti. Nosti autem quod Craigius subinde in Scotiam redux se seriem meam vidisse confirmaret[14] cum ipse nec litem ea de re ab illo motam esse nec te seriem invenisse nondum audiveram, neqꝫ tu, ut opinor, me in seriem consimilem incidisse prius didiceras. Sed ut postulato tuo jam satisfaciam describenda sunt verba epistolæ.[15] Et quoniam ea quæ de quadratura cu[r]varum aliquando forte an communicabo[16] pendent a methodo quadam analytica quam subo[b]scure illic attingebam[,][17] describam quoqꝫ quæ ad hanc

nam' and then the tentative interjection 'ut id tuto possim facere' (afterwards minimally altered to be 'ut id jam tuto fieri possit').

(7) Originally 'clarissimus'.

(8) 'convergentium' is deleted.

(9) A cancelled clause reads in continuation 'et series meas desumpsit'.

(10) A needless parenthesis 'si recte memini' is here deleted; the precise words used by Wallis in 1685, 'There is a great deal more (in these Papers) of like nature' (see note (27) of our preceding introduction), are more exactly rendered in Newton's initial version, 'alia multa in ijsdem contineri quæ prætermisit', of Wallis' following testimony.

(11) Newton first continued: 'Curva autem cujus quadraturam in exemplorū secundo ad illustrandam hanc seriem posueram illa ipsa est (si recte memini) cujus quadraturam a me acceptam D. Joannes Craigius annis abhinc sex ad vos per epistolam hinc transmisit: qua occasione excitatus quadraturam curvarum iterum [aggressus es].' The second of Newton's illustrations of his quadrature series in his 1676 *epistola posterior*—namely '$a^4z \times \overline{cc-zz}|^{-2}$' (see *Correspondence*, 2: 116)—was not, according to John Craige's own testimony (see note (8) of our preceding introduction), either of the two instances of 'non-Tschirnhausian' quadrable *curvæ* which he received at Cambridge in 1685. So much for the accuracy of Newton's memory of an event only six years old.

(12) 'Cantabrigiæ' was added in an otherwise trivially variant preliminary draft at this point.

(13) Initially 'chartas meas vidit'.

(14) Thus changed to a more hesitant subjunctive from an over-blunt 'confirmabit'. Is Newton dubious that Craige would so corroborate his present 'historia' of the episode?

(15) A superfluous (and verbally unattached) 'ad Leibnitium' is deleted in sequel.

(16) Initially Newton affirmed more definitely '...aliquando dicero'.

(17) Probably realizing that Leibniz could scarcely culture what had been but shadowily communicated to him sealed in an anagram (see *Correspondence*, 2: 115, note (25)), Newton has here cancelled a following phrase 'quamqꝫ Leibnitius excolit'.

methodum spectant. Methodi hujus initia[18] excogitavit Fermatius et Slusius promovit.[19] Eandem quocꝫ a me auctam fuisse sic commemorabam.

— *Quinetiam hic [non hæretur...]*[20]

De quodam igitur Tractatu quem annis abhinc viginti de seriebus et methodo fluentium quantitatum conscripseram, sed ob ingruentes quasdem de luce et coloribus disputationes imperfectum reliqueram verba faciens et methodum illam cum Slusiana quoad curvarum tangentes ducendas coincidere præterquam quod mea[21] ad quantitates surdas non hæreret [asseverans][22]

---

(18) Originally 'vestigia'.

(19) This replaces a more precise initial choice of phrase 'deinde Slusius quoad tangentes curvarum auxit'.

(20) Newton begins to cite from his *epistola posterior* to Leibniz in October 1676 (see *Correspondence*, **2**: 151, ll. 6 ff.) the 'fundamentum' on which 'conatus sum reddere speculationes de Quadratura curvarum simpliciores'. Presumably, much as in his later treatise 'De quadratura' (see §2 below), he would here wish to illustrate his 'primum Theorema generale' not only with its four accompanying explicit *exempla* (*Correspondence*, **2**: 115–17) but also with the ensuing instance of the cissoid $y^2z = (a-z)^3$, the accurate construction of whose general arc-length (*ibid.*: 117) depends on the prior evaluation of the integral $\int_z^a \frac{1}{2}az^{-\frac{3}{2}}(a+3z)^{\frac{1}{2}}.dz$ (compare III: 321–2, note (733)). In a cancelled first version of the present passage he initially began his quotation from the *epistola* at the opening (*Correspondence*, **2**: 114, *lin. ult.*) of the pertinent paragraph, there writing: 'Ipse quocꝫ eandem in Tractatu quodam incompleto ['quem per ea tempora annis abhinc viginti...de seriebus infinitis composueram' is deleted] a me quocꝫ excultam fuisse sic commemorabam. *Cæterum in tractatu isto series infinitæ non magnam partem obtinebant. Alia [haud pauca] congessi inter quæ erat methodus ducendi tangentes...*'.

(21) An unsubstantiated 'generalior' is here cancelled.

(22) Newton breaks off without attaining his main verb, passing immediately on—or so we reckon—to remodel his unfinished draft letter into the opening of the 'De quadratura Curvarum' which here immediately follows (in §1).

THE FIRST TRACT 'DE QUADRATURA CURVARUM'[1]

Restored[2] from original drafts in the University Library, Cambridge, and in private possession

§1. THE UNFINISHED PRELIMINARY TEXT.[3]

[November 1691]

De quadratura Curvarum.

Annis abhinc quindecim[4] cum D. Leibnitius epistolare commercium procurante D. Oldenburgo, mecum haberet, et ea occasione methodum meam serierum convergentium exponerem: in epistolarum mearum secunda[5] data Octob 24 1676 postquam de methodo quadam ducendi tangentes et enodandi difficiliora problemata locutus essem quam literis transpositis hanc sententiam involventibus (*Data æquatione quotcunꝗ fluentes quantitates involvente fluxiones invenire et vice versa*) celabam;[6] statim subjunxi.[7]

Hoc fundamento conatus sum etiam methodum[8] quadrandi curvas simpliciorem reddere, perveniꝗ ad theoremata quædam generalia, et ut candide agam, ecce primum theorema.[9]

---

(1) We have set out at length in the preceding introduction (see pages 3–8 above) the six-year-long chain of events which early in November 1691 led David Gregory—having discovered it for himself, independently as he claimed, and seen it published at Edinburgh in 1688 as his own invention—to submit to Newton an ameliorated account of the series for squaring a general algebraic binomial which its prime author had, as Gregory well knew, communicated long before to Leibniz in his *epistola posterior* of 1676. To Gregory's accompanying request that he should now sanction its appearance before an English public by telling 'so much of the historie of it as ye think fitt' Newton privately acceded so far as to begin to draft a formal reply (see pages 21–3 above) recounting how John Craige during an extended stay at Cambridge in 1685 had been shown the series by him and thereafter, on his return to Scotland, had transmitted it to Gregory. As soon, however, as he there began to cite the words in which he had passed the series on to Leibniz in October 1676, Newton broke off his response to Gregory and, with his interest in the quadrature of curves thus freshly stimulated, he straightaway—or so we presume—started to reshape it into the opening of a much more broadly comprehensive treatise of his own on the subject. The texts of the two prinicpal states of this unpublished first tract 'De quadratura Curvarum' which we now edit were dispersed by Newton himself after he subsequently absorbed much of their content into the revised 'tractatus' of the same title which he printed in 1704 as the second of the mathematical appendixes to his *Opticks* (see **2**, 3, §2: note (1) below); a number of their manuscript sheets came, in particular, into the possession of William Jones in about 1710 when (compare I: xvii) he was preparing a comprehensive edition of Newton's mathematical papers, but he made no use of them in his brief commentary upon the printed *De Quadratura* when he republished it (with minor corrections) in his *Analysis per Quantitatum Series, Fluxiones ac Differentias* in 1711. They are here re-united in their original sequence for the first time in more than two and a half centuries.

Translation

On the quadrature of curves

Fifteen years ago,[4] when Mr Leibniz held an exchange of letters with me through the agency of Mr Oldenburg and I took the occasion to expose my method of converging series, I spoke in the second of my letters[5] (dated 24 October 1676) of a certain method of drawing tangents and unknotting more difficult problems which I concealed in an anagram enfolding this transposed sentence: *Given an equation involving any number of fluent quantities, to find the fluxions; and vice versa,*[6] and added immediately afterwards:[7]

On this basis I have tried also to render the method[8] of squaring curves simpler, and have attained certain general theorems. To be open about it, here is my first theorem.[9]

---

(2) Out of the many dozens of extant sheets 'De quadratura Curvarum' (now mainly in ULC. Add. 3960.7–10 and 3962.2/3 and in private possession) which Newton penned at this time we have selected those which form his first extended preliminary scheme for the tract, and also those which compose its (unfinished) final state, appending to these our choice of significant portions of further preparatory worksheets and initial verbal drafts, and broadly epitomizing the content of yet other related autograph fragments in pertinent footnotes.

(3) That found on ULC. Add. 3962.2: $31^r/40^r/56^r/56^v/38a^r$–$38b^v$ and on a sheet, originally intervening in this sequence, which is now (compare note (1)) in private possession. This last (11–23(1): 13/14) has already been published by H. W. Turnbull in *The Correspondence of Isaac Newton*, **2**, 1960: 171–2, l. 23 [terminating at '...quadrandæ'] as the first half of a composite 'Manuscript by Newton on Quadratures' whose date he tentatively but erroneously conjectured to be '?1676'.

(4) That is, in 1676.

(5) The text of this celebrated 'epistola posterior' of Newton's to Leibniz in October 1676—to use the now familiar epithet which Newton himself afterwards bestowed on it in his *Commercium Epistolicum D. Johannis Collins et Aliorum de Analysi Promota* (London, 1712): 67—is most recently and accurately reproduced in *Correspondence*, **2**: 110–29 from the autograph (now British Museum. Add. 4294) sent to Oldenburg for onward transmission. In following footnotes we shall for convenience refer our citations of its points of detail to this standard edition rather than to the corrected amanuensis copy (Add. 3977.4, reproduced with amendments on IV: 618–33) which Newton himself retained of the letter and from which he drew the following quoted passage.

(6) Namely, '6a cc d æ 13e ff 7i 3l 9n 4o 4q rr 4s 8t 12v x'; see *Correspondence*, **2**: 115 and the related 'Memorandum', dated 'Octob. 1676', which Newton earlier set down in his Waste Book (ULC. Add. 4004: 81^v, quoted in II: 191, note (25)). His unscrambled sentence effectively subsumes, of course, the titles which he had five years before set down on 'Prob: 1' and 'Prob: 2' of his 1671 tract (see III: 74, 82), and the 'method of drawing tangents and unknotting more difficult problems' is the fluxional *methodus* there expounded and applied in sequel 'Curvarum tangentes ducere' (*ibid.*: 120–48) and to resolving such *problemata difficiliora* as 'Curvæ alicujus ad datum punctum curvaturam invenire' (*ibid.*: 150–76).

(7) See *Correspondence*, **2**: 115.

(8) This replaces 'theoriam' (theory). At this point in his 1676 *epistola* (*Correspondence*, **2**: 115, l. 14) Newton had earlier made vaguer reference to 'speculationes de Quadratura curvarum' (speculations on the squaring of curves).

(9) Compare III: 237, note (540) and IV: 621, note (10).

Ad curvam quamvis propositam sit $dz^{\theta}\times\overline{e+fz^{\eta}}|^{\lambda}$ ordinatim applicata termino abscissæ[(10)] seu basis z normaliter insistens: ubi literæ d, e, f denotant quaslibet quantitates datas et θ, η, λ indices potestatum sive dignitatum quantitatum quibus affixæ sunt. Fac $\frac{\theta+1}{\eta}=r$. $\lambda+r=s$, $\frac{d}{\eta f}\times\overline{e+fz^{\eta}}|^{\lambda+1}=Q$, et $r\eta-\eta=\pi$: et area Curvæ erit

$$Q \text{ in } \frac{z^{\pi}}{s}-\frac{r-1}{s-1}\times\frac{eA}{fz^{\eta}}+\frac{r-2}{s-2}\times\frac{eB}{fz^{\eta}}-\frac{r-3}{s-3}\times\frac{eC}{fz^{\eta}}+\frac{r-4}{s-4}\times\frac{eD}{fz^{\eta}}-\text{ \&c:}$$

literis A, B, C, D &c denotantibus terminos proxime antecedentes,[(11)] nempe A terminum $\frac{z^{\pi}}{s}$, B terminum $\frac{r-1}{s-1}\times\frac{eA}{fz^{\eta}}$ &c. Hæc series ubi r fractio est vel numerus negativus continuatur in infinitum[,] ubi verò r integer est et affirmativus continuatur ad tot terminos tantum quot sunt unitates in eodem r, & sic exhibet Geometricam quadraturam Curvæ. Rem exemplis illustro.

Exempl. 1. Proponatur Parabola [...][(12)]

NB. Annis abhinc quinqꝫ[(13)] cum Joannes Craige Cantabrigiæ diutius commoratus seriem[(14)] hic positam vidisset et quadraturam curvæ quæ in exemplo hoc secundo habetur[(15)] cum Scotis suis communicasset, D. David Gregorius Matheseos Professor Edinburgensis acuto[(16)] vir ingenio, eandem seriem sed minus concinnam sua[(17)] methodo invenit.

[(18)]Et hactenus Epistola.

Diximus igitur quod beneficio hujus seriei figuræ curvilineæ non solum quadrari possunt sed etiam cum Conicis Sectionibus et alijs figuris comparari.[(19)] Nam quemadmodum termini omnes in serie tota in infinitum continuata sunt

---

(10) Thus altered by Newton in the manuscript from an initial 'diametri' (diameter) accurately transcribed from his 1676 *epistola.*

(11) In minimal clarification Newton here began to add 'cum signis suis [? + et − probè observatis]' (with their signs [+ and − appropriately observed]), but straightaway broke off to delete this interpolation in the text of his 1676 letter.

(12) Much as in his revised version (§2 below), Newton's present quotation from his 1676 *epistola* would evidently here extend at least to include all four of its following examples (*Correspondence*, 2: 115–17) illustrating how the preceding quadrature series may be applied, but not perhaps to embrace the construction for rectifying the cissoid's arc—less obviously dependent upon it—which ensues (*ibid.*: 117).

(13) Read 'sex' (six) more accurately, since Craige visited Cambridge in 1685 (see page 3 above). In revise (see §2: note (6) below) Newton will write more vaguely 'quinqꝫ vel sex' (five or six).

(14) In his manuscript Newton was quick here to delete the over-hasty qualification 'convergentem' (convergent), one far from necessarily true of individual instances of his quadrature expansion.

(15) Newton converts into a bald—and erroneous—factual statement what in his tentative response to Gregory was previously only an unsupported impression (see note (11) to our reproduction of his unfinished Latin draft in the appendix to the preceding introduction). At the equivalent point in his revise (see §2: note (7) below) he will reinsert here a very necessary 'si rectè memini' (if I rightly remember) to cover his possible lapse of memory.

In any curve proposed let $dz^{\theta}(e+fz^{\eta})^{\lambda}$ be the ordinate standing perpendicularly at the end of the abscissa[10] or base z, where the letters d, e, f denote any given quantities you please and θ, η, λ the indices of the powers or dignities of the quantities to which they are attached. Make $(\theta+1)/\eta = r$, $\lambda+r = s$, $(d/\eta f)(e+fz^{\eta})^{\lambda+1} = Q$ and $(r-1)\eta = \pi$, and the area of the curve will then be

$$Q\times\left(\frac{z^{\pi}}{s}-\frac{(r-1)\,eA}{(s-1)\,fz^{\eta}}+\frac{(r-2)\,eB}{(s-2)\,fz^{\eta}}-\frac{(r-3)\,eC}{(s-3)\,fz^{\eta}}+\frac{(r-4)\,eD}{(s-4)\,fz^{\eta}}-\ldots\right),$$

with the letters A, B, C, D, ... denoting the terms immediately next preceding,[11] namely A the term z^{π}/s, B the term $(r-1)\,eA/(s-1)\,fz^{\eta}$, and so on. This series, when r is a fraction or a negative number, is to be continued indefinitely, but when r is an integer and positive, merely to as many terms as there are units in this same r, and in this way it exhibits the geometrical quadrature of the curve. I illustrate the point with examples.

Example 1. Let a parabola be proposed...[12]

N.B. Five[13] years ago, after John Craige had, during an extended stay in Cambridge, seen the[14] series here set out and communicated the quadrature of the curve presented in this second example[15] to his fellow Scots, Mr David Gregory, Professor of Mathematics at Edinburgh and a man of keen[16] intelligence, came upon the same series, but in a less neat form, by a[17] method of his own.

[18]And thus far the letter.

I stated, therefore, that with the aid of this series curvilinear figures can be not only squared but also compared with conics and other figures.[19] For, just as all the terms in the whole series continued to infinity are the area of this curve,

(16) Originally—and with an intended nuance?—'subtili' (subtle).

(17) An unintended 'diversa' (different) is here deleted. The question of the degree of Gregory's independence from Newton in discovering his variant of the quadrature series in the late 1680s is discussed at length in the preceding introduction.

(18) A first verbal draft of the following novel extension of Newton's previous 'prime' quadrature expansion and of the text of the two ensuing *Regulæ* is reproduced in §2, Appendix 1.2 below; certain of the many preliminary computations which exist for the latter are given in the immediately preceding Appendix 1.1.

(19) Newton initially went on to repeat from his preliminary draft (ULC. Add. 3962.2: 33r, reproduced in Appendix 1.2 below) the additional assertion 'et figuræ simplicissimæ inveniri quibuscum comparari possunt' (and the simplest figures found with which they can be compared). This affirmation of the power of his 'prime' theorem to effect (when this is possible) the quadrature of algebraic binomials in simplest form accurately reflects his earlier confident assertion in his October 1676 *epistola* that 'quando hujusmodi Curva aliqua non potest geometricè quadrari, sunt ad manus alia Theoremata pro comparatione ejus cum Conicis Sectionibus, vel saltem cum alijs figuris simplicissimis quibuscum potest comparari; ad quod sufficit etiam hoc ipsum unicum jam descriptum Theorema, si debitè concinnetur' (*Correspondence*, 2: 117); it may be that his deletion of it from his present revised text indicates some growing doubt in his mind about the effectiveness of his expansion universally to attain such simplest quadrature.

area hujus curvæ, sic termini omnes dempto primo sunt area alterius cujusdam curvæ, et omnes demptis duobus primis sunt area curvæ tertiæ et omnes demptis tribus primis sunt area quartæ, et sic deinceps in infinitum. Et hæ omnes curvæ inter se comparari possunt siquidem primus seriei terminus differentia est inter areas curvarum duarum primarum, et aggregatum termini primi et secundi differentia est inter areas curvæ primæ ac tertiæ, & sic deinceps. Quinetiam series utrinq; pergit juxta formulam sequentem.

$$Q \text{ in } [\&c] - \frac{s+2}{r+2}\times\frac{fz^{\eta}}{e\text{Ɔ}} + \frac{s+1}{r+1}\times\frac{fz^{\eta}}{e\text{ꓭ}} - \frac{s}{r}\times\frac{fz^{\eta}}{eA} + \frac{z^{\pi}}{s} - \frac{r-1}{s-1}\times\frac{eA}{fz^{\eta}} + \frac{r-2}{s-2}\times\frac{eB}{fz^{\eta}} - \&c:$$

Ubi A denotat terminum $\frac{z^{\pi}}{s}$ ut supra et litteræ inversæ ꓭ, Ɔ, ᗡ &c(20) terminos inde numeratos sinistrorsum, nempe ꓭ terminū primum $-\frac{sfz^{\eta}}{reA}$, Ɔ secundum, et sic deinceps.(21) Exhibet igitur series uno intuitu quadraturas curvarum numero infinitarum utrinq;, quæ omnes ex area unius data(22) (præterquam ubi arearum differentia terminum aliquem infinitum involvit) quadrari possunt: ideoq; ut omnes quadrentur, quærenda est illarum omnium simplicissima, et hujus area primùm investiganda. Designet Π quemlibet terminorum (23)B, C, D &c in serie, et sit m numerus terminorum ab A ad Π, id est 1 si Π sit B, vel 2 si Π sit C[,] vel -1 si Π sit ꓭ, & sic in cæteris. Et si ponatur $\overline{s-m}\times d\Pi z^{m\eta-\pi} = h$, erit $hz^{\theta-m\eta}\times\overline{e+fz^{\eta}}|^{\lambda}$ ordinatim applicata Curvæ cujus area est seriei pars illa quæ pergit a termino Π dextrorsum(24) in infinitum et terminum illum includit.(25)

---

(20) It is unnecessary to introduce an inverted 'V' into the left-hand expansion since, of course, it is identical with $(z^{\pi}/s =)A$.

(21) Newton errs considerably in this attempt to comprehend both the 'leftwards' and the 'rightwards' expansions of his binomial integral in a single formula departing 'each way' from the central term 'Q in $\frac{z^{\pi}}{s}$'. While the latter accurately repeats the prime theorem of his 1676 *epistola* as he cites it above, the former should—on reversing its progression into an analogous 'rightwards' sequence—read

$$`Q \text{ in} \frac{z^{\pi}}{s} - \frac{s+1}{r+1}\times\frac{fz^{\eta}}{eA} + \frac{s+2}{r+2}\times\frac{fz^{\eta}}{e\text{ꓭ}} - \frac{s+3}{r+3}\times\frac{fz^{\eta}}{e\text{Ɔ}} + \&c'$$

where now '$\frac{d}{\eta e}\times\overline{e+fz^{\eta}}|^{\lambda+1} = Q$' and 'ꓭ' denotes the corrected 'first' term$-\frac{s+1}{r+1}\times\frac{fz^{\eta}}{eA}$. In modern Leibnizian equivalent, on setting $I_i = d\int z^{\theta-i\eta}(e+fz^{\eta})^{\lambda}.d\theta$ and taking (with Newton) $r = (\theta+1)/\eta$ and $s = \lambda+r$, both expansions of the integral $I_0 = d\int z^{\theta}(e+fz^{\eta})^{\lambda}.d\theta$ ensue from the basic identity $(d/\eta)z^{(r-i)\eta}(e+fz^{\eta})^{\lambda+1} \equiv (r-i)\,eI_i + (s-i+1)\,fI_{i-1}$ by way of the respective 'rightwards' and 'leftwards' recursions

$$I_{i-1} = Qz^{(r-i)\eta}/(s-i+1) - ((r-i)\,e/(s-i+1)f)\,I_i, \quad Q = (d/\eta f)\,(e+fz^{\eta})^{\lambda+1}$$

and

$$I_i = Q'z^{(r-i)\eta}/(r-i) - ((s-i+1)\,f/(r-i)\,e)\,I_{i-1}, \quad Q' = (d/\eta e)\,(e+fz^{\eta})^{\lambda+1}.$$

so all the terms less the first are the area of another particular curve, all less the first two are the area of a third curve, all less the first three the area of a fourth, and so on in turn indefinitely. And all these curves can be compared with one another, inasmuch as the first term of the series is the difference between the areas of the first two curves, and the total of the first and second terms is the difference between the areas of the first and third curves, and so forth. To be sure, the series proceeds both ways according to the following pattern:

$$Q\times\left([\ldots]-\frac{(s+2)fz^{\eta}}{(r+2)\,e\text{Ɔ}}+\frac{(s+1)fz^{\eta}}{(r+1)\,e\text{ꓭ}}-\frac{sfz^{\eta}}{reA}+\frac{z^{\pi}}{s}-\frac{(r-1)\,eA}{(s-1)fz^{\eta}}+\frac{(r-2)\,eB}{(s-2)fz^{\eta}}-\ldots\right),$$

where A denotes the term z^{π}/s, as above, and the inverted letters ꓭ, Ɔ, ᗡ, ...[20] the terms numbered therefrom to the left, namely ꓭ the first term $-sfz^{\eta}/reA$, Ɔ the second, and so on.[21] The series therefore exhibits at a single glance the quadrature of an infinity of curves each way, all of which can (except when the difference of areas involves some infinite term) be squared in terms of the area of one given;[22] consequently, to square all, you need to seek out the simplest of all those and first discover the area of this. Let Π denote any of the terms [23]B, C, D, ... in the series, and let m be the number of terms from A to Π; that is, 1 if Π be B, or 2 if Π be C, or -1 if Π be ꓭ, and so in other cases. And, if there be put $(s-m)\,d\Pi z^{m\eta-\pi}=h$, then $hz^{\theta-m\eta}(e+fz^{\eta})^{\lambda}$ will be the ordinate of the curve whose area is that part of the series proceeding rightwards[24] from the term Π to infinity and including that term.[25]

By repeated appeal to these formulas Newton's October 1676 expansion *dextrorsum* results in the form (with remainder)

$$I_0=Q\,\frac{z^{(r-1)\eta}}{s}\left(1-\frac{(r-1)\,e}{(s-1)\,fz^{\eta}}+\ldots+\prod_{1\leqslant i\leqslant k-1}\left(\frac{-(r-i)\,e}{(s-i)\,fz^{\eta}}\right)\right)+\prod_{1\leqslant i\leqslant k}\left(\frac{-(r-i)\,e}{(s-i+1)\,f}\right).I_k$$

(compare III: 260, note (574)), while his intended present complementary expansion *sinistrorsum* comes similarly to be

$$I_0=Q'\,\frac{z^{r\eta}}{r}\left(1-\frac{(s+1)\,fz^{\eta}}{(r+1)\,e}+\ldots+\prod_{1\leqslant i\leqslant k-1}\left(\frac{-(s+i)\,fz^{\eta}}{(r+i)\,e}\right)\right)+\prod_{1\leqslant i\leqslant k}\left(\frac{-(s+i)\,f}{(r+i-1)e}\right).I_{-k}.$$

Newton's mistaken enunciation of the successive terms of the latter series development produces no further confusion, since he omits in the sequel to give any instance of its application in a particular case.

(22) 'per differentiam prædict[am]' (by means of the aforesaid difference) is cancelled in continuation.

(23) The equivalent 'leftwards' terms 'ꓭ, A' were also cited here initially. Newton evidently now restricts his attention to those to the right of z^{π}/s so as not to confuse his reader.

(24) Or, of course, 'sinistrorsum' (leftwards); but see the previous note.

(25) This is immediate since, on changing $\theta\rightarrow\theta-m\eta$ (and hence $s\rightarrow s-m$) in the preceding 'rightwards' series expansion, and then multiplying through by d/h, Newton's present m-th term Π corresponds to the initial term $(h/d)\,z^{\pi-m\eta}/(s-m)$.

Ut si quadranda sit Curva cujus ordinatim applicata est $dz^4 \times \overline{e+fz^2}|^{\frac{1}{2}}$, quæro seriem ut supra. Dein quoniam θ hic est 4 et η 2, adeoq; $\theta - m\eta$ seu $4-2m$ scribendo 2 pro m evanescit, quo in casu Curva illa altera cujus ordinata est $hz^{\theta-m\eta} \times \overline{e+fz^\eta}|^\lambda$ fit Conica sectio: usurpo 2 pro m et huic numero respondentem seriei terminum C pro Π, et inde colligo h, et dico quod si area sectionis hujus conicæ cujus ordinatim applicata est $h \times \overline{e+fz^2}|^{\frac{1}{2}}$, addatur ad seriei terminos duos primos $A-B$ ductos in Q: summa erit series tota, id est area curvæ quadrandæ.(26)

Usus autem seriei latius patebit per Regulas sequentes.

Reg. I. Theor II.(27)

Si curvæ alicujus cujus ordinata sit $dz^\theta \times \overline{e+fz^\eta}|^\lambda$ area dicatur A et curvæ alterius cujus ordinata sit $dz^\theta \times \overline{e+fz^\eta}|^{\lambda-1}$ area dicatur B: erit

$$\overline{\lambda\eta+\theta+1} \times A - \lambda\eta e B = dz^{\theta+1} \times \overline{e+fz^\eta}|^\lambda.$$

Et per hanc regulam colligendo aream alterutram ex altera potest index(28) dignitatis ordinatim applicatæ semel vel sæpius unitate augeri vel minui.(29)

Reg II. Theor III.(30)

Si curvām(31) duarum ordinatim applicatæ y et v normaliter insistāt(31) basibus z et x et relatio inter basem(32) et ordinatam unius definiatur per æquationem quamvis

$$ez^\alpha y^\beta + fz^\gamma y^\delta + gz^\epsilon y^\zeta + hz^\eta y^\theta + \&c = 0;$$

ubi e, f, g, h &c designant quantitates datas cum signis suis & α, β, γ [&c] indices dignitatum y et z: et si ponatur $z^\nu = x$, et relatio inter basem et ordinatam curvæ alterius sit

$$ev^\beta x^{\frac{\alpha+\beta\nu-\beta}{\nu}} + fv^\delta x^{\frac{\gamma+\nu\delta-\gamma}{\nu}} + gv^\zeta x^{\frac{\epsilon+\nu\zeta-\zeta}{\nu}} + hv^\theta x^{\frac{\eta+\nu\theta-\theta}{\nu}} + \&c = 0:$$

æquales erunt harum curvarum areæ inter se.(33)

(26) Newton initially omitted the two following *Regulæ* I/II, proceeding immediately to the generalization of his present prime quadrature theorem which is reproduced in Appendix 1.5—and which is here delayed to be 'Theor IV'—with the words 'Hæc de curvis quarum ordinatæ ex binomijs conflantur. Ad trinomia spectat hoc Theorema' (So much for curves whose ordinates are made up of binomials. As for trinomials this [following] theorem regards them).

(27) 'Theor. I' is, we need scarcely say, Newton's preceding 'prime' theorem for the quadrature of binomials.

(28) Namely λ.

(29) In contrast to the analogous rule underlying the preceding 'Theorema I' where (see note (21) above) the index θ of the single quantity z is varied by the addition or subtraction of multiples of η. The present *regula* is proved straightforwardly by differentiating $dz^{\theta+1}(e+fz^\eta)^\lambda$, setting $fz^\eta = (e+fz^\eta)-e$ and collecting the ensuing terms in powers of z and of the binomial $(e+fz^\eta)$.

If I need, for instance, to square a curve whose ordinate is $dz^4(e+fz^2)^{\frac{1}{2}}$, I seek its series as above. Then, because θ is here 4 and η 2, and hence $\theta-m\eta$ or $4-2m$ vanishes on writing 2 in m's place, and in this case the other curve having ordinate $hz^{\theta-m\eta}(e+fz^\eta)^\lambda$ comes to be a conic, I employ 2 in m's place and the term C in the series corresponding to this number in place of Π, and thence gather h, and then assert that, if the area of this conic having ordinate $h(e+fz^2)^{\frac{1}{2}}$ be added to the first two terms, $A-B$, multiplied by Q, the sum will be the entire series, that is, the area of the curve to be squared.(26)

The use of the series will, however, be more broadly evident through the following rules.

Rule I, Theorem II.(27)

If of some curve whose ordinate is $dz^\theta(e+fz^\eta)^\lambda$ the area be called A and of another curve whose ordinate is $dz^\theta(e+fz^\eta)^{\lambda-1}$ the area be called B, then $(\lambda\eta+\theta+1)A-\lambda\eta eB = dz^{\theta+1}(e+fz^\eta)^\lambda$. And by this rule, on gathering one or other area in terms of the second, the index(28) of the ordinate's power can be increased or diminished once or more times by unity.(29)

Rule II, Theorem III.(30)

If the ordinates y and v of two curves stand perpendicularly upon their bases z and x, and the relationship between the base(32) and ordinate of one be defined by any equation $ez^\alpha y^\beta+fz^\gamma y^\delta+gz^\epsilon y^\zeta+hz^\eta y^\theta+\ldots=0$, where $e, f, g, h, \ldots$ denote given quantities with their signs and $\alpha, \beta, \gamma, \ldots$ the indices of powers of y and z, and if there be set $z^\nu = x$ and the relationship between the base and ordinate of the other curve be

$$ev^\beta x^{(\alpha+\nu\beta-\beta)/\nu}+fv^\delta x^{(\gamma+\nu\delta-\delta)/\nu}+gv^\zeta x^{(\epsilon+\nu\zeta-\zeta)/\nu}+hv^\theta x^{(\eta+\nu\theta-\theta)/\nu}+\ldots=0,$$

then the areas of these curves will be equal to one another.(33)

(30) This revises and considerably extends 'Theor. II' in Appendix 1.2 below.

(31) Read 'curvarum' and 'insistant' respectively.

(32) 'seu abscissam' (that is, abscissa) is cancelled.

(33) The preliminary calculations in which—with a, b, c, λ there replacing e, f, g, ϵ and y, z trivially interchanged with v, x—Newton computed this result of transforming the given equation in y and z by the substitution $z = x^{1/\nu}$ under the area-preserving condition

$$\nu\int y.dz = \int v.dx \quad \text{and so} \quad y = (v/\nu(dz/dx) \quad \text{or})\ vx^{1-1/\nu}$$

are reproduced in Appendix 1 below, together with those in which he first effected the following triple reduction in the simplified trinomial instance which he proceeds to specify. There is, of course, nothing in these present paragraphs which he had not already explored at length fifteen years before in his researches into 'The Quadrature of all curves whose æquations consist of but three terms' (III: 373–82) and the culminating 'Quadrature of many Curves whose æquations consist of more then three terms' (III, 383–5): our earlier comment (*ibid.*: 383, note (23)) on the unjustified optimism evinced in the latter's title here continues to hold good.

[34]Ubi notandum est quod indices duarum dignitatum ipsius z poni possunt æquales[35] et quantitas v qu[a] hoc eveniet inde colligi.

Et hoc pacto æquatio omnis trium terminorum $ez^{\alpha}+fz^{\gamma}y^{\delta}+gz^{\zeta}=0$ reduci potest ad aliquam ex his tribus formis. Ponatur $\gamma\zeta+\alpha\delta-\alpha\zeta=\pi$. Dein

$$1.\ \alpha+\zeta=\rho.\quad z^{\frac{\rho}{\zeta}}=x.\quad ev^{-\delta}+fx^{\frac{\pi}{\rho}}+gv^{\zeta-\delta}=0.$$

$$2.\ \gamma-\alpha-\delta=\sigma.^{(36)}\quad z^{\frac{\sigma}{\delta}}=x.\quad ev^{-\zeta}+fv^{\delta-\zeta}+gx^{\frac{\pi}{\sigma}}=0.$$

$$3.\ \delta-\gamma-\zeta=\tau.\quad z^{\frac{\tau}{\zeta-\delta}}=x.\quad ex^{\frac{\pi}{\tau}}+fv^{\delta}+gv^{\zeta}=0.^{(37)}$$

Deinde si curva quadrari potest[,] quadratura ejus prodibit ex his æquationibus vel per seriem prædictam, et sufficit duas quasvis ex æquationibus tentare. Quinetiam si curva quadrari potest prodit ejus quadratura ex his æquationibus absq; serie. Nam aliquis ex indicibus dignitatum ipsius x hoc est ex numeris $\frac{\pi}{\rho}, \frac{\pi}{\sigma}, \frac{\pi}{\tau}$ erit numerus fractus affirmativus unitatem habens pro numeratore.

Ideoq; si nullius ex numeris $\frac{\pi}{\rho}, \frac{\pi}{\sigma}, \frac{\pi}{\tau}$ numerator sit unitas, id est si nullus ex numeris $\frac{\rho}{\pi}, \frac{\sigma}{\pi}, \frac{\tau}{\pi}$ integer sit et affirmativus: curva[38] quadrari nequit.

Si e tribus numeris duo sint integri et eorum alter negativus sit et alter vel affirmativus vel negativus: curva comparabitur cum Hyperbola.

Si tres illi numeri sint fracti irreducibiles et eorum duo habent numerum binarium pro denominatore, Curva comparabitur cum Hyperbola vel Ellipsi. Alijs in casibus figura simplicissima quacum comparabitur erit adhuc magis composita. Et quæcunq; sit sic invenietur per Theorema [II] et [III]. Nam hæc sunt Theoremata per quæ dixi hoc idem præstari posse.[39]

E tribus numeris $\frac{\pi}{\rho}, \frac{\pi}{\sigma}, \frac{\pi}{\tau}$ seligātur isti [d]uo quorum numeratores sunt minimi, puta $\frac{\pi}{\rho}$ et $\frac{\pi}{\tau}$. Deinde e tribus æquationibus seligatur ista quæ habet alterutrum ex his numeris pro indice dignitatis x, puta $ex^{\frac{\pi}{\tau}}+fv^{\delta}+gv^{\zeta}=0$. et sit f coefficiens termini qui habet alterum numerorum pro indice in altera æquationum sitq; δ

(34) Newton first continued: 'Et hæc sunt Theoremata per quæ dixi figuras simplicissimas prodire quibuscum curvæ propositæ comparari possunt' (And these are the theorems through which I said there result the simplest figures with which propounded curves can be compared). The passage in his 1676 *epistola posterior* to which he here refers asserts that 'quando hujusmodi [binomialis] Curva aliqua non potest geometricè quadrari, sunt ad manus alia Theoremata pro comparatione ejus cum...figuris simplicissimis quibuscum potest comparari....Pro trinomijs etiam et alijs quibusdam Regulas quasdem concinnavi' (*Correspondence*, **2**, 117); see also note (19) above) and compare III: 18, where we make it clear that

[34]Here note that the indices of two powers of z can be set equal,[35] and the quantity v whereby this shall happen thence derived.

And in this manner every equation of three terms $ez^{\alpha}+fz^{\gamma}y^{\delta}+gz^{\zeta}=0$ can be reduced to some one of these three following forms. Set $\gamma\zeta+\alpha\delta-\alpha\zeta=\pi$, and then

1. $\alpha+\zeta=\rho$, $z^{\rho/\zeta}=x$: $ev^{-\delta}+fx^{\pi/\rho}+gv^{\zeta-\delta}=0$;
2. $\gamma-\alpha-\delta=\sigma$,[36] $z^{\sigma/\delta}=x$: $ev^{-\zeta}+fv^{\delta-\zeta}+gx^{\pi/\sigma}=0$;
3. $\delta-\gamma-\zeta=\tau$, $z^{\tau/(\zeta-\delta)}=x$: $ex^{\pi/\tau}+fv^{\delta}+gv^{\zeta}=0$.[37]

And, if the curve can be squared, its quadrature will ensue from these equations or by the above-stated series, and it is enough to try any two of the equations. Indeed, if the curve can be squared, its quadrature will result from these equations without the series: for some one of the indices of the powers of x, that is, of the numbers π/ρ, π/σ, π/τ, will then be a positive fraction having unity for its numerator.

Consequently, should the numerator of none of the numbers π/ρ, π/σ, π/τ be unity, that is, if none of the numbers ρ/π, σ/π, τ/π be integral and positive, the curve cannot[38] be squared.

If two of these three numbers be integral, and of these one be negative and the other either positive or negative, the curve is comparable with a hyperbola.

If those three numbers be irreducible fractions and two of them have the number 2 as denominator, the curve is comparable with a hyperbola or ellipse. In other cases the simplest figure with which it shall be compared will be still more compound. And, whatever it may be, it will be ascertained in the following way by means of Theorems II and III; for these are the theorems by which I said this end can be accomplished.[39]

Of the three numbers π/ρ, π/σ, π/τ select those two, say π/ρ and π/τ, whose numerators are least. Then of the three equations select one, say $ex^{\pi/\tau}+fv^{\delta}+gv^{\zeta}=0$, having one or other of these numbers as index of the power

---

Newton's reference in 1676 was to his contemporary 'Quadrature of curves whose æquations consist of but three terms' (on which see the previous note).

(35) An additional phrase 'vel in data proportione ad invicem' (or in a given ratio to each other) is deleted. Why Newton should even momentarily have thought that this might yield a viable reduction is wholly unclear to us.

(36) Read '$-\sigma$' strictly.

(37) The immediately preliminary calculations by which Newton attained this triple reduction are given in Appendix 1.1 below. In addition there are a number of equivalent prior worksheets—ULC. Add. 3962.2: 38v/49v–50v and yet others in private possession—whose content differs so minimally in substance that their detailed reproduction would seem to fulfil no useful purpose here.

(38) In an otherwise unaltered preceding draft of this phrase Newton here accurately specified 'geometricè' (geometrically).

(39) See note (34) above.

index termini alterius. Pone $\zeta+\frac{\pi}{\tau}=\nu$ & $v^{\nu}=y^{\left[\frac{\pi}{\tau}\right]}$ et per Regulam secundam prodibit hæc æquatio $ez^{\frac{\pi}{\tau}}+fy^{\frac{\pi}{\rho}}+g=0$. Ubi ordinata z super basi y aream eandem describit atq; ordinata x super basi v in æquatione priore.(40) Si fractionum $\frac{\pi}{\tau}$, $\frac{\pi}{\rho}$ denominatores sunt numeratoribus majores, minuantur subducendo numeratores semel vel sæpius per Regulam primam donec sint ipsis minores, et si fieri possit, affirmativi: et habebitur figura simplicissima quacum curva sub initio proposita comparari potest,(41) sive figura illa sit Sectio conica sive generis magis compositi.(42) Simplicissimas autem hic voco quarū æquationes sunt minimè affectæ et eo nomine maximè tractabiles et maximè aptæ quæ per methodum meam serierū(43) infinitarum quadrentur, licet una ignotarum quantitatum ad plures dimensiones nonnunquam ascendat. Nam quo altior est dignitas baseos et depressior est ea ordinatæ, eo celerius series converget.(44)

Et sic Curvæ omnes quarum æquationes sunt trium terminorū vel geometricè quadrantur vel comparantur cum sectionibus Conicis vel transmutantur in alias curvas formarum quatuor simplicissimarum quas Conicæ sectiones et Parabola cubica induunt. Nam si indices dignitatum sunt affirmativæ et unitate majores (qui casus in aliqua trium æquationum novissimarum semper incidit,)(45) et si præterea numeratores indicum sunt numeri pares & coefficientes ipsarum sunt ejusdem signi, Curva erit ellipsis:(46) Sin coefficiens altera affirmativa sit et altera negativa, Curva erit Hyperbola in angulis oppositis duarum asymptotωn, ad modum Hyperbolæ Conicæ. Et centrum harum figurarum incidet semper in principium baseos.(47) Si vero numeratorum alter par sit et alter impar, Curva erit Parabola cum cruribus duobus infinitis in eandem plagam pergentibus ad modum Parabolæ Conicæ. Ac deniq; si numerator uterq; sit impar Curva erit Parabola cum cruribus duobus infinitis flexu contrario in plagas oppositas abeuntibus ad modum Parabolæ cubicæ. Ad has

(40) On converting the previous trinomial to be $ex^{\pi/\tau}v^{-\zeta}+fv^{\delta-\zeta}+g=0$, this is transmuted by the area-preserving condition $k\int x.dv=\int z.dy$, $y=v^k$ and so $x=z/k(dv/dy)=zy^{1-1/k}$, to become $ez^{\pi/\tau}+fy^{(\delta-\zeta)/k}+g=0$ if $(\pi/\tau)(1-1/k)-\zeta/k=0$, that is, $k=(\zeta+\pi/\tau)/(\pi/\tau)$; whence, since $\pi=(\delta-\zeta)\rho-\zeta\tau$, there is $(\delta-\zeta)/k=\pi/\rho$.

(41) A cancelled first conclusion reads in sequel 'hoc est Conica sectio si figura proposita cum conica sectione comparari potest, sin minus figura simplicissima' (that is, a conic if the propounded figure can be compared with a conic, but if not, the simplest figure).

(42) Newton first went on: 'Sin comparatio illa impossibilis sit, incidetur tamen in figuras simplicissimas quibuscum curva illa comparari potest' (But if that comparison should be impossible, you will yet fall upon the simplest figures with which that curve can be compared); this observation—and the similar one cited in the previous note—is taken up by him again in the next paragraph after he has made clear what he means by a 'figura simplicissima'.

(43) 'convergentium' (convergent) is deleted. Newton evidently assumes that the qualification is superfluous in present context; he had, as we have seen (IV: 554, note (88)), made

of x, and let f be the coefficient of the term having the second of the numbers for its index in the other of the equations, and δ the index of this other term. Set $\zeta+\pi/\tau = \nu$ and $v^{\nu} = y^{\pi/\tau}$, and by the second Rule there will result this equation, $ez^{\pi/\tau}+fy^{\pi/\rho}+g = 0$, where the ordinate z on base y describes the same area as ordinate x on base v in the previous equation.[40] If the denominators of the fractions π/τ, π/ρ are greater than the numerators, diminish them by subtracting the numerators once or more times by the first Rule till they are less than them and, if it is possible, positive: there will then be had the simplest figure with which the curve initially proposed can be compared,[41] whether it be a conic or a figure of a more compound kind. [42]The 'simplest', however, I here call those whose equations are least complicated and on that head most tractable and fittest to be squared by my method of infinite[43] series, even though one of the unknown quantities should sometimes rise to several dimensions; for the higher the power of the base and the lower that of the ordinate, the more rapidly will the series converge.[44]

And in this way all curves whose equations are of three terms are either geometrically squared, or compared with conics, or transmuted into other curves of the four simplest forms which conics and the cubic parabola assume. For, if the indices of the powers are positive and greater than unity (a case always occurring in some one of the three most recent equations)[45] and, further, the numerators of the indices are even numbers and their coefficients are of the same sign, the curve will be an ellipse;[46] but if one coefficient be positive and the other negative, the curve will be a hyperbola in the opposing angles of two asymptotes, in the style of a conic hyperbola, and the centre of these figures will fall always at the base's origin;[47] if, however, one of the numerators be even and the other odd, the curve will be a parabola with two infinite branches proceeding in the same direction, in the manner of a conic parabola; while, finally, if both numerators be odd, the curve will be a parabola with two infinite branches departing in opposite directions with a contrary flexure, in the style

a similar emendation seven years earlier in the title of Chapter 3 of his unfinished 'Matheseos Univeralis Specimina'.

(44) Before we criticize the inadequacy of this assertion in particular contrary cases, we should remember that Newton has in mind only simple algebraic integrands.

(45) This is only true in general if each of the indices α, γ, δ and ζ are positive 'numeri' (integers): in which case the derivable equalities $\pi = \delta\rho+\zeta\sigma$, $\tau = \sigma-\rho$ readily determine at least one of the indices π/ρ, π/σ and π/τ to be greater than $+1$. As a counter-instance to Newton's claim if this implicit restriction be not obeyed, set $\alpha = 7$, $\gamma = 9$, $\delta = 7$ and $\zeta = -4$, when there is in consequence $\pi = 1$, $\rho = 3$, $\sigma = 5$ and $\tau = 2$, with each of π/ρ, π/σ and π/τ less than unity.

(46) Understand a general higher-order oval and not necessarily an Apollonian ellipse.

(47) In sequel Newton has cancelled a superfluous 'atq; omnes rectæ intra figuram per centrũ ductæ bisecantur in centro' (while all straight lines drawn within the figure through the centre are bisected at the centre).

curvas igitur ut conceptu simplicissimas et quadratu facillimas volui aliarū quadraturas referre.

Si Curva aliqua definitur per æquationem(48) quatuor vel plurium terminorum, et una ignotarum quantitatum in unico tantum æqu[ationis] termino reperiatur: hæc Curva si quadrari potest quadrabitur per series(49) sequentes.

Theor IV.

(50)[Si Curvæ alicujus ordinatim applicata sit $dz^\theta \times \overline{e+fz^\eta+gz^{2\eta}}|^\lambda$ et ponatur $\frac{\theta+1}{\eta}=r$, $r+\lambda=s$, $r+2\lambda=t$, $\frac{d}{\eta g}\times\overline{e+fz^\eta+gz^{2\eta}}|^{\lambda+1}=Q$ et $r\eta-2\eta=\pi$: area Curvæ hujus erit

$$Q \text{ in } \frac{z^\pi}{t}-\frac{\overline{s-1}\times fA}{\overline{t-1}\times gz^\eta}-\frac{\overline{r-2}\times eA+\overline{s-2}\times fB}{\overline{t-2}\times gz^\eta}-\frac{\overline{r-3}\times eB+\overline{s-3}\times fC}{\overline{t-3}\times gz^\eta}-\&c.]$$

Ubi A denotat primū terminum $\frac{z^\pi}{t}$, & B sec[undum] $-\frac{\overline{s-1}\times fA}{\overline{t-1}\times gz^\eta}$ cum signo suo & sic deinceps.(51) Et hæc est series pro trinomijs cujus memini in Epistola.(52)

Theor V.

(53)[Si curvæ alicujus ordinatim applicata sit $dz^\theta\times\overline{e+fz^\eta+gz^{2\eta}+hz^{3\eta}}|^\lambda$, et ponatur $\frac{\theta+1}{\eta}=r, r+\lambda=s, r+2\lambda=t, r+3\lambda=v$, $\frac{d}{\eta h}\times\overline{e+fz^\eta+gz^{2\eta}+hz^{3\eta}}|^{\lambda+1}=Q$,

(48) 'irreducibilem' (irreducible) is deleted.

(49) Newton first began here specifically to cite 'Th[eoremata IV et V]', but quickly changed to this equivalent description of their content. Preliminary versions of the following series expansions of trinomial and quadrinomial algebraic integrals are reproduced in Appendix 1.3–5 below.

(50) A 2-inch strip of the manuscript page (f. 38a^v) bearing the following enunciation has been carefully torn away, presumably by Newton (perhaps because he introduced some systematic error in there transcribing it?). We restore its text merely by repeating that of the equivalent preliminary draft 'Theorema II' reproduced in Appendix 1.5 below.

(51) Read correctly '& *B* secundi coefficientem cum signo suo & sic deinceps' (*B* the coefficient of the second with its sign, and so forth), that is, successively

$$-\frac{(s-1)\,fA}{(t-1)\,g}=B,\quad -\frac{(r-2)\,eA+(s-2)\,fB}{(t-2)\,g}=C,\quad -\frac{(r-3)\,eB+(s-3)\,Cf}{(t-3)\,g}=D,$$

and so on. For, much as in note (21) preceding, on setting there to be $I_i=d\int z^{\theta-i\eta}R^\lambda.d\theta$, $R=e+fz^\eta+gz^{2\eta}$, we may readily deduce the identity

$$(d/\eta)\, z^{\pi-i\eta}R^{\lambda+1}\equiv(r-i-2)\,eI_{i+2}+(s-i+1)\,fI_{i+1}+(t-i)\,gI_i$$

and therefrom the basic recursion

$$I_i=\frac{z^{\pi-i\eta}}{t-i}Q-\frac{(s-i-1)\,f}{(t-i)\,g}I_{i+1}-\frac{(r-i-2)\,e}{(t-i)\,g}I_{i+2}$$

of the cubic parabola. To these curves, therefore, as being the simplest to conceive and easiest to square, I wanted to refer the quadratures of others.

If some curve is defined by an(48) equation of four or more terms, and one of the unknown quantities should be found in but a single term of the equation, this curve will, if it can, be squared by means of the following series.(49)

Theorem IV.

(50)If of some curve the ordinate be $dz^{\theta}(e+fz^{\eta}+gz^{2\eta})^{\lambda}$ and there be set $(\theta+1)/\eta = r$, $r+\lambda = s$, $r+2\lambda = t$, $(d/\eta g)\,(e+fz^{\eta}+gz^{2\eta})^{\lambda+1} = Q$ and $(r-2)\eta = \pi$, then the area of this curve will be

$$Q\times\left(\frac{z^{\pi}}{t}-\frac{(s-1)\,fA}{(t-1)\,gz^{\eta}}-\frac{(r-2)\,eA+(s-2)\,fB}{(t-2)\,gz^{\eta}}-\frac{(r-3)\,eB+(s-3)\,fC}{(t-3)\,gz^{\eta}}-\ldots\right),$$

where A denotes the first term z^{π}/t, B the second one $-(s-1)fA/(t-1)\,gz^{\eta}$ with its sign, and so forth.(51) And this is the series for trinomials which I made mention of in the letter.(52)

Theorem V.

(53)If of some curve the ordinate be $dz^{\theta}(e+fz^{\eta}+gz^{2\eta}+hz^{3\eta})^{\lambda}$ and there be set $(\theta+1)/\eta = r$, $r+\lambda = s$, $r+2\lambda = t$, $r+3\lambda = v$, $(d/\eta h)\,(e+fz^{\eta}+gz^{2\eta}+hz^{3\eta})^{\lambda+1} = Q$

---

where $Q = (d/\eta g)\,R^{\lambda+1}$; whence in succession

$$I_0 = AQ-\frac{(s-1)\,fA}{gz^{\pi}}I_1-\frac{(r-2)\,eA}{gz^{\pi}}I_2$$

$$= (A+Bz^{-\eta})\,Q-\frac{(r-2)\,eA+(s-2)\,fB}{gz^{\pi}}I_2-\frac{(r-3)\,eB}{gz^{\pi}}I_3$$

$$= (A+Bz^{-\eta}+Cz^{-2\eta})\,Q-\frac{(r-3)\,eB+(s-3)\,fC}{gz^{\pi}}I_3-\frac{(r-4)\,eC}{gz^{\pi}}I_4$$

and so on, yielding—with some minor adjustment of the powers of z in his enunciation of it—both Newton's expansion $I_0 = Q(A+Bz^{-\eta}+Cz^{-2\eta}+\ldots)$ and the remainder by which it errs from true at any given stage.

(52) Despite this firm assertion we know no evidence that his remark to Leibniz in October 1676 stating that 'Pro trinomijs etiam et alijs quibusdam Regulas quasdem concinnavi' (*Correspondence*, 2: 117) referred other than to his contemporary researches into the reduction of trinomials and higher multinomials by area-preserving transformations; compare note (33) above, and III: 373–85. Certainly, the preliminary worksheets and drafts reproduced in Appendix 1.3–5 heavily underscore just how novel his present extension of the 'prime' quadrature series in his earlier *epistola posterior* was. Let us be charitable and allow that Newton has, fifteen years later, forgotten the precise context of his previous statement.

(53) Newton has left a gap at this point in the manuscript for later entry of the enunciation of his *Theorema*; we fill it, as doubtless he himself intended, with the text of the preliminary 'Theorema III' which is reproduced in Appendix 1.5 below.

et $r\eta - 3\eta = \pi$: area curvæ hujus erit

$$Q \text{ in } \frac{z^\pi}{v} - \frac{\overline{t-1}\times gA}{\overline{v-1}\times hz^\eta} - \frac{\overline{s-2}\times fA + \overline{t-2}\times gB}{\overline{v-2}\times hz^\eta}$$

$$- \frac{\overline{r-3}\times eA + \overline{s-3}\times fB + \overline{t-3}\times gC}{\overline{v-3}\times hz^\eta}$$

$$- \frac{\overline{r-4}\times eB + \overline{s-4}\times fC + \overline{t-4}\times gD}{\overline{v-4}\times hz^\eta} - \&\text{c.}\Big]^{(54)}$$

(55)Nota autem quod in hac Theorematum progressione quæ ex jam positis facile continuatur in infinitum,(56) si termini aliqui intermedij in ordinatim applicata desiderantur, eorum loca vacua sunt circulis implenda. Ut si ordinatim applicata esset $\overline{az+bz^3[+]cz^7}|^{\frac{1}{2}}$ id est $z^{\frac{1}{2}}\times\overline{a+bz^2+cz^6}|^{\frac{1}{2}}$, scriberem

$$z^{\frac{1}{2}}\times\overline{a+bz^2+0z^4+cz^6}|^{\frac{1}{2}}$$

dein quadrarem figuram non per Theorema IV quod pro Trinomijs est sed per Theorema quintum perinde ac si ordinata ex quatuor terminis constaret.

Nota præterea quod ordinatæ cujusqȝ duæ sunt expressiones quæ dant series totidem.(57) Sic in Theoremate quarto ordinata vel sic scribitur

$$dz^\theta\times\overline{e+fz^\eta+gz^{2\eta}}|^\lambda, \quad \text{vel sic} \quad dz^{\theta+2\eta\lambda}\times\overline{g+fz^{-\eta}+ez^{-2\eta}}|^\lambda,$$

et prior expressio dat seriem in Theoremate illo positam, secunda vero scribendo in hac ser[i]e $\theta+2\eta\lambda$ pro θ, hoc est ponendo $\dfrac{\theta+2\eta\lambda+1}{\eta} = r$, $r+\lambda = s$, & $[r]+2\lambda = t$, ac deinde mutando ubiqȝ signum ipsius η dat seriem novam consimilem. Series utraqȝ tentanda est. Et in earum alterutra duo termini simul evanescunt & seriem abrumpunt ac terminant in Theoremate IV^to vel tres in quinto &c(58) si modo curva geometrice quadrari potest.

(54) As in the previous theorem (compare note (51)), understand that the quantities $A, B, C, D, \ldots$ are the coefficients of the successive powers of z (whose indices should, in correction of the present enunciation, be $i\eta$, $i = 0, 1, 2, 3, \ldots$); namely,

$$\frac{z^\pi}{v} = A,\ -\frac{(t-1)\,gA}{(v-1)\,h} = B,\ -\frac{(s-2)\,fA+(t-2)\,gB}{(v-2)\,h} = C,\ -\frac{(r-3)\,eA+(s-3)\,fB+(t-3)\,gC}{(v-3)\,h} = D,$$

and so forth. In similar proof, on again setting $I_i = d\int z^{\theta-i\eta}R^\lambda.d\theta$ with now

$$R = e+fz^\eta+gz^{2\eta}+hz^{3\eta},$$

from the basic identity

$$(d/\eta)\ z^{\pi-i\eta}R^{\lambda+1} \equiv (r-i-3)\,eI_{i+3}+(s-1-2i)\,fI_{i+2}+(t-i-1)\,gI_{i+1}+(v-i)\,hI_i$$

there ensues the recursion

$$I_i = \frac{z^{\pi-i\eta}}{v-i}Q - \frac{(t-i-1)\,g}{(v-i)\,h}I_{i+1} - \frac{(s-i-2)\,f}{(v-i)\,h}I_{i+2} - \frac{(r-i-3)\,e}{(v-i)\,h}I_{i+3},$$

and $(r-3)\eta = \pi$, the area of this curve will then be

$$Q\times\left(\frac{z^\pi}{v}-\frac{(t-1)\,gA}{(v-1)\,hz^\eta}-\frac{(s-2)\,fA+(t-2)\,gB}{(v-2)\,hz^\eta}-\frac{(r-3)\,eA+(s-3)\,fB+(t-3)\,gC}{(v-3)\,hz^\eta}\right.$$
$$\left.-\frac{(r-4)\,eB+(s-4)\,fC+(t-4)\,gD}{(v-4)\,hz^\eta}-\ldots\right).^{(54)}$$

(55)Note, however, in this sequence of theorems—now, from what has been set out, easily continued to infinity(56)—that, if some intermediate terms be lacking in the ordinate, their vacant places are to be filled with noughts. For instance, if the ordinate were $(az+bz^3+cz^7)^{\frac{1}{2}}$, that is, $z^{\frac{1}{2}}(a+bz^2+cz^6)^{\frac{1}{2}}$, I would write $z^{\frac{1}{2}}(a+bz^2+0z^4+cz^6)^{\frac{1}{2}}$ and then square the figure not by Theorem IV, which is for trinomials, but by Theorem V, exactly as though the ordinate consisted of four terms.

Note, moreover, that for each ordinate there are two expressions, yielding as many series.(57) Thus in Theorem IV the ordinate is either written in this way, $dz^\theta(e+fz^\eta+gz^{2\eta})^\lambda$, or in this, $dz^{\theta+2\eta\lambda}(g+fz^{-\eta}+ez^{-2\eta})^\lambda$; the former expression yields the series set out in that Theorem, but the second, on writing $\theta+2\eta\lambda$ in θ's place in this series—that is, on setting $(\theta+2\eta\lambda+1)/\eta = r$, $r+\lambda = s$, and $r+2\lambda = t$—and then changing the sign of η throughout, gives a new, closely similar series. Both series are to be tried. And in one or other of them two terms simultaneously vanish in Theorem IV, breaking off and terminating the series, or three in Theorem V and so on, (58) if only the curve can be geometrically squared.

---

where $Q = (d/\eta h)\, R^{\lambda+1}$; in succession, therefore,

$$I_0 = AQ-\frac{(t-1)\,gA}{hz^\pi}I_1-\frac{(s-2)\,fA}{hz^\pi}I_2-\frac{(r-3)\,eA}{hz^\pi}I_3$$
$$= (A+Bz^{-\eta})\,Q-\frac{(s-2)\,fA+(t-2)\,gB}{hz^\pi}I_2-\frac{(r-3)\,eA+(s-3)\,fB}{hz^\pi}I_3-\frac{(r-4)\,eB}{hz^\pi}I_4$$
$$= (A+Bz^{-\eta}+Cz^{-2\eta})\,Q-\frac{(r-3)eA+(s-3)\,fB+(t-3)\,gC}{hz^\pi}I_3$$
$$-\frac{(r-4)\,eB+(s-4)\,fC}{hz^\pi}I_4-\frac{(r-5)eC}{hz^\pi}I_5$$
$$= (A+Bz^{-\eta}+Cz^{-2\eta}+Dz^{-3\eta})\,Q-\frac{(r-4)\,eB+(s-4)\,fC+(t-4)\,gD}{hz^\pi}I_4$$
$$-\frac{(r-5)\,eC+(s-5)\,fD}{hz^\pi}I_5-\frac{(r-6)\,eD}{hz^\pi}I_6$$

and so on indefinitely—unless, as he initially went on to remark (see the next note), some adjacent trio of coefficients, $B, C, D, \ldots$ prove to be zero, thereby yielding the exact quadrature of I_0.

Nota etiam quod si terminus r(59) in neutro casu sit numerus integer et affirm[ativus], Curva qu[adrari] non potest, ideoq non opus fuerit calculum ultra tentare.

Si Curva(60) per has series quadrari nequit[,] comparabitur cum alijs figuris per Theorema sequens.(61)

Theor VI.

(62)[Si curvæ alicujus abscissa sit z et pro $e+fz^{\eta}+gz^{2\eta}+hz^{3\eta}$ &c scribatur R, sit autem area curvæ $z^{\theta}R^{\lambda}$: erit ordinatim applicata

$$\theta e \begin{matrix}+\theta\\+\lambda\eta\end{matrix} fz^{\eta} \begin{matrix}+\theta\\+2\lambda\eta\end{matrix} gz^{2\eta} \begin{matrix}+\theta\\+3\lambda\eta\end{matrix} hz^{3\eta} \text{ \&c in } z^{\theta-1}R^{\lambda-1}.]^{(63)}$$

Theor VII.

Si in Curvæ alicujus ordinata $z^{\theta\pm\eta r}\times\overline{e+fz^{\eta}+gz^{2\eta}+\&c}|^{\lambda\pm s}$ maneant quantitates datæ θ, η, λ, e, f, g, &c et pro r et s successive scribantur numeri quicunq integri ut ordinatæ novæ prodeant, ac dentur areæ duarum ex Curvis quæ per has ordinatas innumeras(64) designantur si ordinatæ sunt trium nominum in vinculo radicis, vel areæ trium ex curvis si ordinatæ sunt nominum quatuor, vel areæ quatuor ex curvis si ordinatæ sunt nominum quinq et sic deinceps in infinitum: dico quod dabuntur areæ curvarum omnium. Pro nominibus verò hic habeo terminos omnes tam deficientes in locis vacuis (ut ad Theor V annotatum est) quam extantes in plenis. Indices baseos in nominibus debent esse in progressione arithmetica, et termini omnes in progressione seu plani seu vacui pro nominibus haberi. Propositio verò sic probatur.

(55) Newton initially began to observe 'Et nota quod tres termini simul evanescent & seriem terminabunt si modo Curva quadrari potest' (And note that three terms will vanish together and terminate the series provided the curve can be squared); see the preceding note.

(56) Compare the three theorems of the 'Theorematum Progressio secunda' in Appendix 1.5 below.

(57) Newton had made this point in October 1676 in Example 2 illustrating the prime binomial case as he communicated it to Leibniz in his *epistola posterior* (*Correspondence*, **2**: 116), there expressing $a^4z/(c^2-z^2)^2$ both as $a^4z(c^2-z^2)^{-2}$ and as $a^4z^{-3}(-1+c^2z^{-2})^{-2}$.

(58) An equivalent '& sic deinceps' (and so forth) is replaced.

(59) Originally 'terminorum r, s, t, v &c penultimus[,] hoc est, terminus r in Theoremate priore vel te[rminus] s in Th: IV, vel terminus t in Th: V, &c' (the last but one of the terms $r, s, t, v, \ldots$, that is, the term r in the first Theorem, or the term s in Theorem IV, or the term t in Theorem V).

(60) Newton first went on to write 'definitur per æquationem terminorum quatuor vel plurium et alterutra ignotarum quantitatum' (is defined by an equation of four or more terms, and one or other of the unknowns) but broke off, leaving his intended sentence—and its sense —incomplete. Perhaps he initially had it in mind to append a 'Theor. VI' repeating the gist of his 'Quadrature of many Curves whose æquations consist of more then three terms' as he had set it down fifteen years before? (See III: 383–5, and compare note (33) above.)

Note also that, if the term r[59] be in neither case integral and positive, the curve cannot be squared and there will in consequence be no need to attempt further calculation.

If the curve[60] cannot be squared by these series, it will be compared with other figures by means of the following theorem.[61]

Theorem VI.

[62]If of some curve the abscissa be z and for $e+fz^{\eta}+gz^{2\eta}+hz^{3\eta}\ldots$ there be written R, let the area of the curve be $z^{\theta}R^{\lambda}$ and then the ordinate will be

$$(\theta e+(\theta+\lambda\eta)fz^{\eta}+(\theta+2\lambda\eta)\,gz^{2\eta}+(\theta+3\lambda\eta)\,hz^{3\eta}\ldots)\,z^{\theta-1}R^{\lambda-1}.$$ [63]

Theorem VII.

If in the ordinate $z^{\theta\pm\eta r}(e+fz^{\eta}+gz^{2\eta}+\ldots)^{\lambda\pm s}$ of some curve the quantities $\theta, \eta, \lambda, e, f, g, \ldots$ remain given and in place of r and s there be written in succession any integers so as to produce new ordinates, and if there be given the areas of two of the curves designated by these innumerable ordinates[64] if the ordinates are of three terms within the root-bracket, or the areas of three of the curves if the ordinates are of four terms, or the areas of four curves if the ordinates are of five terms, and so on indefinitely, then I assert that the areas of all the curves will be given. As terms, however, I here consider the entirety both of those lacking in vacant places (as noted at Theorem V) as well of those present in the filled ones: the indices of the base in the terms must be in arithmetical progression, and all in the progression, whether filled or vacant, considered as terms. The proposition is, however, proved as follows.

---

(61) Read 'Theoremata sequentia' (...theorems), since Newton at once splits the immediate sequel into two complementary 'Theor. VI' and 'Theor. VII'.

(62) We restore the following enunciation—left blank in the manuscript (f. 38b^r)—from the much fuller prior expansion reproduced in Appendix 1.3 below. An identical *Theorema* was subsequently introduced into the revised 'De quadratura' (§2 following) as its Proposition IX.

(63) In Leibnizian terms this 'ordinata' is merely the derivative (with respect to the 'basis' z) of the preceding 'area' $z^{\theta}R^{\lambda}$. In sequel Newton has deleted a paragraph beginning 'Usus Theorematis talis est' (The use of the theorem is as follows) and continuing thereafter: 'Sit $z^{\theta-1}\times\overline{e+fz^{\eta}+gz^{2\eta}}|^{\lambda-1} = z^{\theta-1}R^{\lambda-1}$ ordinatim applicata Curvæ propositæ. Pro hujus area scribatur A et pro area curvae cujus ordinata est $z^{\theta+\eta[-1]}\times\overline{e+fz^{\eta}+gz^{2\eta}}|^{[\lambda}$ scribatur B. Et cum per Theorema prædictum sit $z^{\theta-1}R^{\lambda+1}$ area curvæ cujus ordinata est

$$\begin{matrix}\theta\\-1\end{matrix}ez^{\theta-1}\begin{matrix}+\theta-1\\+\lambda\eta\end{matrix}fz^{\theta-1[+]\eta}\begin{matrix}+\theta-1\\+2\lambda\eta\end{matrix}gz^{\theta-1[+2]\eta}\times R^{\lambda}.\text{'}$$

At that point he broke off and began straightaway to draft the following separate 'Theor. VII' (compare note (61)) where this approach is greatly extended.

(64) 'sic prodeuntes' (thus ensuing) is cancelled.

Cas. 1.[65] Curvarum duarum ordinatæ sint $pz^{\theta-1}R^{\lambda-1}$ & $qz^{\theta+\eta-1}R^{\lambda-1}$ & areæ pA ac qB.[66] Et cum per theorema predictum[67] sit $z^{\theta}R^{\lambda}$ area curvæ cujus ordinata est

$$\theta e \; \begin{matrix}+\theta\\+\lambda\eta\end{matrix} fz^{\eta} \; \begin{matrix}+\theta\\+2\lambda\eta\end{matrix} gz^{2\eta} \;\text{ in }\; z^{\theta-1}R^{\lambda-1}{}_{[,]}$$

subduc ordinatas et areas priores de posterioribus & manebit

$$\begin{matrix}\theta e\\-p\\ \end{matrix} \; \begin{matrix}+\theta\\+\lambda\eta\\-q\end{matrix} fz^{\eta} \; \begin{matrix}+\theta\\+2\lambda\eta\end{matrix} gz^{2\eta} \;\text{ in }\; z^{\theta-1}R^{\lambda-1}$$

ordinata novæ curvae et $z^{\theta}R^{\lambda}-pA-qB$ ejusdem area. Pone $\theta e=p$ et $\theta f+\lambda\eta f=q$ et ordinata evadet $\begin{matrix}+\theta\\+2\lambda\eta\end{matrix} gz^{2\eta}$ in $z^{\theta-1}R^{\lambda-1}$ et area $z^{\theta}R^{\lambda}-\theta eA-\theta fB-\lambda\eta fB$. Divide utramq; per $\theta g+2\lambda\eta g$ et aream prodeun[t]em dic C et erit rC area curvæ cujus ordinata est $rz^{2\eta+\theta-1}R^{\lambda-1}$. Et qua ratione ex areis pA et qB aream rC ordinatæ $rz^{2\eta+\theta-1}R^{\lambda-1}$ congruentem invenimus, licebit ex areis qB et rC aream sD ordinatæ $sz^{3\eta+\theta-1}R^{\lambda-1}$ congruentem invenire & sic deinceps in infinitum. Et par est ratio progressionis in partem contrariam præterquā ubi terminorum θ & $\theta+2\lambda\eta$ alteruter deficit[68] et seriem abrumpit: in quo casu dabitur relatio arearum A et B ideoq; si assumatur area pA in principio progressionis unius et area qB in principio progressionis alterius, ex his duabus dabuntur areæ omnes in progressione utraq;. Et quemadmodum progreditur ab areis A et B ad alias omnes sic vicissim ex alijs quibuscunq; duabus assumptis regreditur per Analysin ad areas A et B, adeo ut ex duabus datis cæteræ omnes determinentur Geometricè.[69] Q.E.O.

Hic est casus curvarum ubi ipsius z index θ augetur vel diminuitur perpetua additione vel subductione quantitatis η. Casus alter est Curvarum ubi index λ augetur vel diminuitur unitatibus.

Cas. 2. Ordinatæ $pz^{\theta-1}R^{\lambda}$ & $qz^{\theta+\eta-1}R^{\lambda}$, quibus areæ pA et qB respondeant, si in R seu $e+fz^{\eta}+gz^{2\eta}$ ducantur ac deinde ad R vicissim applicentur evadunt $pe+pfz^{\eta}+pgz^{2\eta}\times z^{\theta-1}R^{\lambda-1}$ et $qez^{\eta}+qfz^{2\eta}+qgz^{3\eta}\times z^{\theta-1}R^{\lambda-1}$. Et per Theor [VI] est $az^{\theta}R^{\lambda}$ area Curvæ cujus ordinata est $\theta ae \begin{matrix}+\theta\\+\lambda\eta\end{matrix} afz^{\eta} \begin{matrix}+\theta\\+2\lambda\eta\end{matrix} agz^{2\eta}\times z^{\theta-1}R^{\lambda-1}$ et $bz^{\theta+\eta}R^{\lambda}$ area curvæ cujus ordinata est

$$\begin{matrix}\theta\\ [+\eta]\end{matrix} bez^{\eta} \; \begin{matrix}+\theta[+\eta]\\+\lambda\eta\end{matrix} bfz^{2\eta} \; \begin{matrix}+\theta[+\eta]\\+2\lambda\eta\end{matrix} bgz^{3\eta}\times z^{\theta-1}R^{\lambda-1}.$$

(65) Newton's straightforward preliminary computations for this and the following case—which we see no point in here other than mentioning—are extant on ULC. Add. 3962.3: 40v.

(66) Here and in sequel understand that $R = e+fz^{\eta}+gz^{2\eta}$, as indeed Newton himself spelled out in an equivalent initial opening to this case.

(67) Namely Theor 'VI'.

Case 1.(65) Let the ordinates of two curves be $pz^{\theta-1}R^{\lambda-1}$ and $qz^{\theta+\eta-1}R^{\lambda-1}$, and their areas pA and qB.(66). And since by the aforesaid theorem(67) $z^\theta R^\lambda$ is the area of the curve whose ordinate is $(\theta e+(\theta+\lambda\eta)fz^\eta+(\theta+2\lambda\eta)gz^{2\eta})z^{\theta-1}R^{\lambda-1}$, subtract the former ordinates and areas from the latter ones and there will remain $(\theta e-p+((\theta+\lambda\eta)f-q)z^\eta+(\theta+2\lambda\eta)gz^{2\eta})z^{\theta-1}R^{\lambda-1}$ as the ordinate of a new curve and $z^\theta R^\lambda-pA-qB$ as its area. Put $\theta e=p$ and $(\theta+\lambda\eta)f=q$, and the ordinate will come to be $(\theta+2\lambda\eta)gz^{2\eta}\times z^{\theta-1}R^{\lambda-1}$ and the area

$$z^\theta R^\lambda-\theta eA-(\theta+\lambda\eta)fB.$$

Divide each by $(\theta+2\lambda\eta)g$ and call the resulting area C, and then rC will be the area of a curve whose ordinate is $rz^{2\eta+\theta-1}R^{\lambda-1}$. And by the method whereby from the areas pA and qB we derived the area rC according with the ordinate $rz^{2\eta+\theta-1}R^{\lambda-1}$, we are free from the areas qB and rC to derive the area sD according with the ordinate $sz^{3\eta+\theta-1}R^{\lambda-1}$, and so on in turn indefinitely. And the method for the progression in the opposite sense is the same, except when one or other of the terms θ and $\theta+2\lambda\eta$ is zero(68) and breaks the series off: in this case the relationship between the areas A and B will be given, and consequently, if the area pA be assumed at the start of one progression and the area qB at the start of the other, from these two will be given all the areas in either sequence. And just as progression is made from the areas A and B to all others, so conversely, should any two others be assumed, regression is made by analysis from them to the areas A and B, so that from any two given all the rest may be determined geometrically.(69) As was to be shown.

This is the case of curves where the index θ of z is increased or diminished by the perpetual addition or subtraction of the quantity η. The second case is of curves where the index λ is increased or diminished by units.

Case 2. To the ordinates $pz^{\theta-1}R^\lambda$ and $qz^{\theta+\eta-1}R^\lambda$ let the areas pA and qB correspond: then, if multiplied by R or $e+fz^\eta+gz^{2\eta}$ and then, conversely, divided by R, they become $(pe+pfz^\eta+pgz^{2\eta})z^{\theta-1}R^{\lambda-1}$ and $(qez^\eta+qfz^{2\eta}+qgz^{3\eta})z^{\theta-1}R^{\lambda-1}$. Now, by Theorem VI, $az^\theta R^\lambda$ is the area of a curve whose ordinate is

$$(\theta ae+(\theta+\lambda\eta)afz^\eta+(\theta+2\lambda\eta)agz^{2\eta})z^{\theta-1}R^{\lambda-1},$$

and $bz^{\theta+\eta}R^\lambda$ the area of a curve whose ordinate is

$$((\theta+\eta)bez^\eta+(\theta+\eta+\lambda\eta)bfz^{2\eta}+(\theta+\eta+2\lambda\eta)bgz^{3\eta})z^{\theta-1}R^{\lambda-1}.$$

And the sum of these four areas is $pA+qB+az^\theta R^\lambda+bz^{\theta+\eta}R^\lambda$, with the sum of the corresponding ordinates

$$\begin{pmatrix}(\theta a+p)e+((\theta+\lambda\eta)af+(\theta+\eta)be+pf+qe)z^\eta \\ +((\theta+2\lambda\eta)ag+(\theta+\eta+\lambda\eta)bf+pg+qf)z^{2\eta} \\ +((\theta+\eta+2\lambda\eta)bg+qg)z^{3\eta}\end{pmatrix}\times z^{\theta-1}R^{\lambda-1}.$$

(68) Originally 'nullus sit' (be nil).

(69) A deleted final clause initially went on trivially to add 'ac deinde ab his ad cæteras pergitur' (and then progression is made from these to the rest).

Et harum quatuor arearum summa est $pA+qB+az^{\theta}R^{\lambda}+bz^{\theta+\eta}R^{\lambda}$ et summa respondentium ordinatarum

$$\begin{array}{llll} \theta ae & \begin{array}{c}+\theta\\+\lambda\eta\end{array} afz^{\eta} & \begin{array}{c}+\theta\\+2\lambda\eta\end{array} agz^{2\eta} & \begin{array}{c}+\theta[+\eta]\\+2\lambda\eta\end{array} bgz^{3\eta} \quad \text{in} \quad z^{\theta-1}R^{\lambda-1}. \\ +pe & +\theta[+\eta]be & \begin{array}{c}+\theta[+\eta]\\+\lambda\eta\end{array} bf & +qg \\ & +pf & +pg & \\ & +qe & +qf & \end{array}$$

Si termini omnes in coefficiente ipsius $z^{\theta-1}R^{\lambda-1}$ præter secundum ponantur seorsim æquales nihilo, per primum fiet $\theta ae+pe=0$ seu $\theta a=-p$, per quartum $\theta b[+\eta b]+2\lambda\eta b=-q$, et per tertium (scribendo $-\theta a$ et $-\theta b[-\eta b]-2\lambda\eta b$ pro p et q) [evadet] $\frac{2ag}{f}=b$. Unde secundus fit $\frac{\lambda\eta aff-4\lambda\eta age}{f}$ adeoq; summa quatuor ordinatarum est $\frac{\lambda\eta aff-4\lambda\eta age}{f}z^{\eta+\theta-1}R^{\lambda-1}$ et summa totidem respondentium arearum $az^{\theta}R^{\lambda}+\frac{2ag}{f}z^{\theta+\eta}R^{\lambda}-\overline{\theta[+\eta]}\,aA-\frac{\theta+\eta+2\lambda\eta}{\frac{1}{2}f}agB$. Dividantur hæ summæ per $\frac{\lambda\eta aff-4\lambda\eta age}{f}$ et si Quotum posterius dicatur D, erit D area curvæ cujus ordinata est Quotum prius $z^{\eta+\theta-1}R^{\lambda-1}$. Et eadem ratione ponendo omnes coefficientis terminos præter primum æquales nihilo potest area curvæ inveniri cujus ordinata est $z^{\theta-1}R^{\lambda-1}$. Dicatur area ista C et qua ratione ex areis A et B inventæ sunt areæ C ac D, ex his areis C et D inveniri possunt aliæ duæ E et F ordinatis $z^{\theta-1}R^{\lambda-2}$ et $z^{\eta+\theta-1}R^{\lambda-2}$ congruentes, et sic deinceps in infinitum. Et per ana[l]ysin contrariam regredi potest ab areis E et F ad areas C ac D et inde ad areas A et B et sic deinceps in infinitum. Igitur si index λ perpetua unitatum additione vel subductione augeatur vel minuatur, et ex areis quæ ordinatis sic prodeuntibus respondent duæ quævis habentur, dantur aliæ omnes in infinitum. Q.E.O.

Cas. 3. Et per casus hosce duos conjunctos, si tam index θ perpetua additione vel subductione ipsius η quam index λ perpetua additione vel subductione unitatis utcunq; augeatur vel minuatur[,] dabuntur areæ omnibus prodeuntibus ordinatis respondentes. Q.E.O.

Cas. 4. Et simili argumento si ordinata constat ex quatuor nominibus et dantur tres arearum, vel si constat ex quinq; nominibus et dantur quatuor, & sic deinceps; dabuntur areæ omnes quæ addendo vel subducendo numerum η indici θ vel unitatem indici λ, generari possunt. Q.E.O.

Prop. VIII. Prob.[70]

Invenire figuras simplicissimas cum quibus curvæ jam expositæ comparari possunt.

If all terms in the coefficient of $z^{\theta-1}R^{\lambda-1}$ except the second be set separately equal to nothing, by the first there will come to be $\theta ae+pe=0$ or $\theta a=-p$, by the fourth $(\theta+\eta+2\lambda\eta)\,b=-q$, and by the third (on writing $-\theta a$ and $-(\theta+\eta+2\lambda\eta)\,b$ in place of p and q) will come $2ag/f=b$. Whence the second one becomes $\lambda\eta a\,(f^2-4ge)/f$ and hence the sum of the four ordinates is $(\lambda\eta a(f^2-4ge)/f)\,z^{\eta+\theta-1}R^{\lambda-1}$ and the sum of the equal number of corresponding areas $az^{\theta}R^{\lambda}+(2ag/f)\,z^{\theta+\eta}R^{\lambda}-(\theta+\eta)\,aA-(\theta+\eta+2\lambda\eta)\,(2ag/f)\,B$. Divide these sums by $\lambda\eta a(f^2-4ge)/f$ and, if the latter quotient be called D, then D will be the area of the curve whose ordinate is the former quotient $z^{\eta+\theta-1}R^{\lambda-1}$. And by the same method, on setting all terms in the coefficient except the first equal to nothing, there can be found the area of the curve whose ordinate is $z^{\theta-1}R^{\lambda-1}$. Call that area C, and by the method whereby from the areas A and B were found the areas C and D, from these areas C and D there can be found two others E and F according with the ordinates $z^{\theta-1}R^{\lambda-2}$ and $z^{\theta+\eta-1}R^{\lambda-2}$, and so on in turn indefinitely. And by a contrary analysis regression can be made from the areas E and F to the areas C and D, and thence to the areas A and B, and so on in turn indefinitely. Therefore, if the index λ be increased or diminished by the perpetual addition or subtraction of units, and of the areas which correspond to the ordinates so resulting any two are had, all the others are given without end. As was to be shown.

Case 3. And by these two cases jointly, if both the index θ and the index λ be increased or diminished in any manner whatsoever by the perpetual addition or subtraction of η and unity respectively, there will be given the areas corresponding to all the resulting ordinates. As was to be shown.

Case 4. And by a similar argument, if the ordinate consists of four terms and three of the areas are given, or if it consists of five terms and four are given, and so on, there will be given all areas which can be generated by adding or subtracting the number η to or from the index θ, or unity to or from the index λ. As was to be shown.

Proposition VIII, Problem.(70)

To find the simplest figures with which the curves just now exhibited can be compared.

---

(70) On f. 36v (compare note (75) below) there is a preliminary remodelling of this Problem [1] as 'Prop XV' in a scheme which, by their altered citations, evidently embraced the preceding Theorems I, IV, V and VII as Propositions IX, X, XI (and so, presumably, the intervening Theorem VI as Proposition XII). The first paragraph of the main text there terminates halfway through at '...figuræ magis compositæ' (...and figures more compound), and Newton then continues with the additions quoted in the next note.

Si geometrice quadrari possunt, prodibit quadratura per series in progressione Theorematum I, IV, V & similium. Sin minus transmutanda est index η in unitatem per Theorema III, deinde indices θ et λ unitatibus augendi vel minuendi sunt per [Th.] VII donec evanescant vel fiant simplicissimi[,] et incidetur in figuras simplicissimas quæsitas sive eæ sint Conicæ sectiones sive aliæ quævis figuræ magis compositæ. Deinde ab harum areis assumptis computanda est area curvæ propositæ. Index vero η sic transmutatur in unitatem. Sit $x^{\delta}\times\overline{e+fx^{\eta}+gx^{2\eta}+hx^{3\eta}\ \&c}|^{\lambda}=v$ ordinata curvæ propositæ. Pone $x^{\eta}=z$ & $\dfrac{\delta-\eta+1}{\eta}=\theta$ et erit $z^{\theta}\times\overline{e+fz+gz^{2}+hz^{3}\ \&c}|^{\lambda}$ ordinata curvæ cujus area æqualis est areæ curvæ prioris.

(71)Porro obtinet eadem quadrandi methodus in Curvis universis quarum ordinatæ non sunt æquationum radices affectæ.(72) Id ex sequentibus Propositionibus patebit.

Prop IX.(73)

Prop. X.(73)

Prop XI(73)

Prop. XII.

Si curvæ alicujus ordinata basi z normaliter insistens sit $[\theta ek+\ldots z^{\eta}+\ldots z^{2\eta}$ &c in $z^{\theta-1}R^{\lambda-1}S^{\mu-1}$, ubi pro $e+fz^{\eta}$ &c et $k+lz^{\eta}$ &c scribuntur R et S respective:](74) erit ejus area $z^{\theta}R^{\lambda}S^{\mu}$.

Prop. XIII.

Si in curvæ alicujus ordinata

$$z^{\theta\pm\eta r}\times\overline{e+fz^{\eta}+gz^{2\eta}\ \&c}|^{\lambda\pm s}\times\overline{k+lz^{\eta}+mz^{2\eta}\ \&c}|^{\mu\pm t}$$

maneant quantitates datæ θ, η, λ, μ, e, f, g, k, l, m &c, et pro r, s ac t vel pro eorum aliquo successivè scribantur numeri quicunqꝫ integri ut ordinatæ novæ prodeant: et si dentur areæ duarum ex curvis quæ per has ordinatas designantur si in vinculis radicalium quatuor sunt nomina (videlicet duo in una radicali $e+fz^{\eta}$ et alia duo in altera $k+lz^{\eta}$) vel si dentur areæ trium ex Curvis ubi in

(71) In his preliminary remodelling on f. 36v (see the previous note) Newton went on differently: 'Eadem methodo qua Curvæ quadrantur in quarum ordinatis duæ sunt radicales in se ductæ, pergitur ad radicales plures: et sic curvæ omnes quadrantur quarum ordinatæ non sunt radices affectæ. Quinetiam si ordinatæ quadraticè afficiuntur, tamen curvæ quadrabuntur per has Regulas' (By the same method as curves are squared in whose ordinates there are two radicals multiplied together, progression is made to several radicals: and in this way are squared all curves whose ordinates are not affected radicals. To be sure, even if the ordinates are affected with square roots, the curves will nevertheless be squared by means of these rules).

If they can be geometrically squared, their quadrature will ensue by the series in the progression of Theorems I, IV, V and their like. If not, the index η must be transmuted into unity by Theorem III, and then by Theorem VII the indices θ and λ are to be increased or diminished by units till they vanish or come to be simplest, and you will arrive at the required simplest figures, whether they be conics or any other figures more compound. Then, assuming the areas of these, you must compute from them the area of the curve proposed. The index η, however, is transmuted into unity this way. Let $x^{\delta}(e+fx^{\eta}+gx^{2\eta}+hx^{3\eta}\ldots)^{\lambda}=v$ be the ordinate of the curve proposed. Set $x^{\eta}=z$ and $(\delta-\eta+1)/\eta=\theta$, and then $z^{\theta}(e+fz+gz^{2}+hz^{3}\ldots)^{\lambda}$ will be the ordinate of a curve whose area is equal to that of the previous curve.

(71)The same method of quadrature holds good, moreover, in all curves universally whose ordinates are not affected roots of equations.(72) This will be evident from the following propositions.

Proposition IX.(73)

Proposition X.(73)

Proposition XI.(73)

Proposition XII.

If the ordinate of some curve standing perpendicularly on the base z be $[(\theta ek+(\ldots)\,z^{\eta}+(\ldots)\,z^{2\eta}\ldots)\,z^{\theta-1}R^{\lambda-1}S^{\mu-1}$, where for $e+fz^{\eta}\ldots$ and $k+lz^{\eta}\ldots$ are written R and S respectively],(74) then its area will be $z^{\theta}R^{\lambda}S^{\mu}$.

Proposition XIII.

If in the ordinate $z^{\theta\pm\eta r}(e+fz^{\eta}+gz^{2\eta}\ldots)^{\lambda\pm s}\,(k+lz^{\eta}+mz^{2\eta}\ldots)^{\mu\pm t}$ of some curve the quantities θ, η, λ, μ, e, f, g, k, l, m, ... remain given, and in place of r, s and t or some one of them there be successively written any integers whatever so as to produce new ordinates, then, if there be given the areas of two of the curves designated by these ordinates if there are four terms in the root-brackets (namely, two in the one radical $e+fz^{\eta}$, and the other two in the other $k+lz^{\eta}$), or if there be given the areas of three of the curves when there are five terms in

(72) Understand algebraic surds of any order.

(73) We may only loosely conjecture what Newton planned to enunciate in these three propositions. Perhaps he intended to set out separately what two years later he comprehended in his revised 'Geometriæ Liber secundus' (see **2**, 3, §2 below) as Corollaries 1–3 to its equivalent Proposition X.

(74) We sketchily fill a gap in the manuscript which Newton left for just such an intended insertion. The full expansion here outlined is given by him in Proposition VIII of his revised 'De quadratura Curvarum' (§2 following).

radicalibus quinqꝫ sunt nomina, vel areæ quatuor ubi sex sunt nomina et sic deinceps: dico quod areæ curvarum omnium dabuntur.

Probatur ad modum [Theorematis] VII.(75)

§2. THE REVISED AND AUGMENTED TREATISE ON QUADRATURE.(1)

[early winter 1691/2]

De quadratura Curvarum(2)

Annis abhinc quindecim cum vir celeberrimus D. Leibnitius epistolare commercium, procurante D. Oldenburgo, mecum haberet, et ea occasione methodum meam serierum convergentium exponerem: in epistolarum mearum secunda data Octob. 24 1676 de methodo quadam ducendi tangentes et enodandi difficiliora problemata, hæc habui.(3)

—Quinetiam hic non hæretur ad æquationes radicalibus unam vel utramqꝫ indefinitam quantitatem involventibus utcunqꝫ affectas, sed absqꝫ aliqua talium æquationum reductione (quæ opus plerumqꝫ redderet immensum) tangens confestim ducitur. Et eodem modo se res habet in quæstionibus de maximis et minimis, alijsqꝫ quibusdam de quibus jam non loquor. Fundamentum harum operationum satis obvium quidem quoniam jam non possum explicationem ejus prosequi,

NB. Literæ transpositæ hoc involvunt Problema: *Data æquatione quotcunqꝫ fluentes quantitates involvente, fluxiones invenire; et vice versa.*(4)

(75) The manuscript terminates on f. 36v with the lightly remodelled version of Proposition VIII preceding whose significant variations have been cited in note (71).

(1) Having drafted the preliminary scheme for a short tract 'De quadratura Curvarum' which is reproduced in §1 preceding, Newton began straightaway—about the beginning of December 1691, we assume—lightly to reshape its text into the ten opening propositions which here follow. At some point, however, before the last week in December when he took his already amplified manuscript with him to London, showing it there over the next fortnight to David Gregory, Fatio de Duillier and no doubt others such as Edmond Halley (see the preceding introduction), Newton decided to broaden its theme beyond the quadrature of 'curves' whose 'ordinates' are already given in integrable form. To that end, over the next few weeks—and much in the style of his earlier unfinished tracts 'Matheseos Universalis Specimina' and 'De computo serierum' (see IV: 526–616)—he built it up to be a major exposition of his general calculus of fluxions, both direct and inverse, further adjoining three new and lengthy propositions wherein he expounds a variety of methods for deriving the exact fluent (when this may be) of a given fluxional equation or, alternatively, for effecting its expansion term by term in an 'interminate converging series', and thereafter applying these techniques—notably the 'Taylor' expansion which is here (Proposition XII, Corollaries 3/4) enunciated in its generality for the first time—to resolve a number of basic problems in elementary differential geometry and the dynamics of enforced motion in a plane. Though Newton went on further to refine—and even minimally to augment (see Appendix 2)—this revised treatise, by early March of 1692 if we may believe Fatio's contemporary testimony

the radicals, or the areas of four when there are six terms, and so on, I assert that the areas of all the curves will be given.

This is proved in the manner of Theorem VII.[75]

Translation

ON THE QUADRATURE OF CURVES[2]

Fifteen years ago, when the very celebrated Mr Leibniz held an exchange of letters with me through the agency of Mr Oldenburg and I took the occasion to expose my method of converging series, in the second of my letters (dated 24 October 1676) I had this to say of a certain method of drawing tangents and unknotting more difficult problems:[3]

...To be sure, there is here no sticking at equations affected in any manner whatever with radicals involving one or both indefinite quantities, but without any reduction of such equations (which would generally render the working immense) a tangent is straightaway drawn. And the same circumstance holds in questions of maxima and minima, and certain others of which I do not now speak. The basis of these operations is obvious enough, indeed, but because I cannot

N.B. The letters when transposed embrace this problem: *Given an equation involving any number of fluent quantities, to find the fluxions; and vice versa.*[4]

---

to Huygens (again see the preceding introduction) his interest not only in publishing its content, as his initial enthusiasm briefly fired him to propose, but even in completing its text was at an end. Soon afterwards (see §1: note (1)) he gave portions of its opening propositions to William Jones, and its remaining portion, having been cannibalized to produce the curtailed 'Tractatus de Quadratura Curvarum' which Newton published in appendix to his *Opticks* in 1704, thereafter disappeared into limbo along with his other private papers. Its text is here for the first time reproduced in full and in its proper sequence—that now found in ULC. Add. 3962.2: 29ʳ–30ᵛ, followed by a sheet in private possession (11–23(1): 17–20, printed as the latter half of a composite erroneously dated as '?1676' in *Correspondence*, 2: 171, ll. 19–174, l. 19; compare §1: note (3)), and then taking up (in order) ULC. Add. 3962.3: 55ʳ, 56ᵛ–57ᵛ /3960.10: 167–71 /3960.7: 121–35 /3960.10: 173–9 /3960.11: 181–95 /3965.6: 38ʳ, 39ʳ. Related preliminary calculations and verbal drafts are cited at pertinent places in following footnotes, while more significant individual items from among them are reproduced in Appendix 1.6–12 below.

(2) We repeat this title from §1 preceding; the present manuscript sequence is without any similar epitomizing head.

(3) The lengthy extract from his *epistola posterior* which follows is, of course, quoted by Newton from the secretary copy of it which he retained in October 1676 (now U.L.C. Add. 3977–4; see IV: 671, note (52) and compare *ibid.*: 618, note (2)), but this differs only trivially from the corrected text which he then passed to Oldenburg for onward transmission to Leibniz (reproduced in *Correspondence*, 2: 110–29; the passage here cited is from 115, l. 6–117, l. 31).

(4) See §1: note (6).

sic potius celavi. 6a cc d æ 13e ff 7i 3l 9n 4o 4q rr 4s 9t 12vx. Hoc fundamento conatus sum etiam reddere methodum quadrandi curvas simpliciorem, perveniꝗ ad Theoremata quædam generalia, et ut candide agam, ecce primum Theorema.

[*Prop. I. Theor. I.*](5)

Ad Curvam aliquam sit $dz^\theta \times \overline{e+fz^\eta}|^\lambda$ ordinatim applicata termino diametri seu basis z normaliter insistens: ubi literæ d, e, f denotant quaslibet quantitates datas, & θ, η, λ indices potestatum sive dignitatum quantitatum quibus affixæ sunt. Fac $\dfrac{\theta+1}{\eta} = r$. $\lambda + r = s$. $\dfrac{d}{\eta f} \times \overline{e+fz^\eta}|^{\lambda+1} = Q$, & $r\eta - \eta = \pi$. Et area Curvæ erit

$$Q \text{ in } \frac{z^\pi}{s} - \frac{r-1}{s-1} \times \frac{eA}{fz^\eta} + \frac{r-2}{s-2} \times \frac{eB}{fz^\eta} - \frac{r-3}{s-3} \times \frac{eC}{fz^\eta} + \frac{r-4}{s-4} \times \frac{eD}{fz^\eta} - \text{ \&c:}$$

liter[i]s A, B, C, D &c denotantibus terminos proxime antecedentes, nempe A terminum $\dfrac{z^\pi}{s}$, B terminum $-\dfrac{r-1}{s-1} \times \dfrac{eA}{fz^\eta}$ &c.

NB. Si literæ A, B, C, D &c denotent terminos cum signis suis + et −, debent termini omnes in serie post primum negativè sumi.

Hæc series ubi r fractio est vel numerus negativus, continuatur in infinitum: ubi verò r integer est et affirmativus continuatur ad tot terminos tantum quot sunt unitates in eodem r, et sic exhibet geometricam quadraturam Curvæ. Rem exemplis illustro.

Exempl 1. Proponatur Parabola cujus ordinatim applicata sit $\sqrt{az}$. Hæc in formam Regulæ reducta fit $z^0 \times \overline{0+az^1}|^{\frac{1}{2}}$. Quare est $d = 1$, $\theta = 0$, $f = a$, $\eta = 1$, [$\lambda = \frac{1}{2}$] ideoꝗ $r = 1$, $s = 1\frac{1}{2}$, $Q = \dfrac{1}{a} \times \overline{az}|^{\frac{3}{2}}$, $\pi = 0$, et area quæsita $\dfrac{1}{a} \times \overline{az}|^{\frac{3}{2}}$ in $\dfrac{1}{1\frac{1}{2}}$, hoc est $\frac{2}{3}z\sqrt{az}$. Et sic in genere si cz^η ponatur ordinatim applicata, prodibit area $\dfrac{c}{\eta+1} z^{\eta+1}$.

Exempl. 2. Sit ordinatim applicata $\dfrac{a^4z}{c^4 - 2cczz + z^4}$. Hæc per reductionem fit $a^4z \times \overline{cc-zz}|^{-2}$, vel etiam

$$a^4z^{-3} \times \overline{-1+ccz^{-2}}|^{-2}.$$

In priori casu est $d = a^4$, $\theta = 1$, $e = cc$, $f = -1$, $\eta = 2$, $\lambda = -2$, ideoꝗ $r = 1$, $s = -1$,

$$Q = \frac{a^4}{-2} \times \overline{cc-zz}|^{-1}$$

hoc est $= \dfrac{-a^4}{2cc-2zz}$, $\pi = 0$, et area Curvæ $= Q$ in $\dfrac{z^0}{-1}$ id est $= \dfrac{a^4}{2cc-2zz}$. In secundo autem casu est $d = a^4$, $\theta = -3$, $e = -1$, $f = cc$, $\eta = -2$, $r = 1$, $s = -1$,

$$Q = \frac{a^4}{-2cc} \times \overline{-1+ccz^{-2}}|^{-1},$$

id est $= \dfrac{-a^4zz}{2c^4 - 2cczz}$, $\pi = 0$, et area $= Q$ in $\dfrac{z^0}{-1}$ hoc est $= \dfrac{a^4zz}{2c^4-2cczz}$.(9) Area his casibus

NB. Annis abhinc quinꝗ vel sex(6) cum D. Joannes Craige Cantabrigiæ diutius commoratus seriem hic positam vidisset, et quadraturam Curvæ, quæ (si rectè memini)(7) in exemplo hoc secundo habetur, cum Scotis suis(8) per literas hinc datas communicasset; D. David Gregorius Matheseos Professor Edinburgensis, eandem seriem sed minus concinnam alia methodo sane non inelegāti invenit.

now pursue its exposition, I have preferred to conceal it in this anagram: ... . On this basis I have tried also to render the method of squaring curves simpler, and have attained certain general theorems. To be open about it, here is my first theorem.

[*Proposition I, Theorem I.*][5]

In any curve let $dz^\theta(e+fz^\eta)^\lambda$ be the ordinate standing perpendicularly on the end of the diameter or base z, where the letters d, e, f denote any given quantities you please and θ, η, λ the indices of the powers or dignities of the quantities to which they are attached. Make $(\theta+1)/\eta = r$, $\lambda+r = s$, $(d/\eta f)\ (e+fz^\eta)^{\lambda+1} = Q$ and $(r-1)\ \eta = \pi$, and the area of the curve will then be

$$Q\times\left(\frac{z^\pi}{s}-\frac{(r-1)\,eA}{(s-1)\,fz^\eta}+\frac{(r-2)\,eB}{(s-2)\,fz^\eta}-\frac{(r-3)\,eC}{(s-3)\,fz^\eta}+\frac{(r-4)\,eD}{(s-4)\,fz^\eta}-\ldots\right),$$

with the letters A, B, C, D, ... denoting the terms immediately next preceding, namely A the term z^π/s, B the term $-(r-1)\ eA/(s-1)\ fz^\eta$, and so on. This series, when r is a fraction or a negative number, is continued indefinitely, but when r is an integer and positive, merely to as many terms as there are units in this same r, and in this way it exhibits the geometrical quadrature of the curve. I illustrate the point with examples.

N.B. If the letters A, B, C, D, ... denote the terms with their signs $+$ and $-$, all terms in the series after the first must be set negative.

Example 1. Let a parabola be proposed whose ordinate be $\sqrt{(az)}$. This, when reduced to the form of the rule, becomes $z^0(0+az^1)^{\frac{1}{2}}$. Therefore $d = 1$, $\theta = 0$, $f = a$, $\eta = 1$, $[\lambda = \frac{1}{2}]$ and consequently $r = 1$, $s = 1\frac{1}{2}$, $Q = (1/a)\ (az)^{\frac{3}{2}}$, $\pi = 0$, and so the area sought is $(1/a)\ (az)^{\frac{3}{2}}\times 1/1\frac{1}{2}$, that is, $\frac{2}{3}z\sqrt{(az)}$. And thus in general, if cz^η be set as the ordinate, the area $(c/(\eta+1))\ z^{\eta+1}$ will result.

Example 2. Let the ordinate be $a^4z/(c^4-2c^2z^2+z^4)$. This comes by reduction to be $a^4z(c^2-z^2)^{-2}$, or again $a^4z^{-3}(-1+c^2z^{-2})^{-2}$. In the former case there is $d = a^4$, $\theta = 1$, $e = c^2$, $f = -1$, $\eta = 2$, $\lambda = -2$ and consequently $r = 1$, $s = -1$,

$$Q = (a^4/-2)\ (c^2-z^2)^{-1},$$

that is, $-a^4/2(c^2-z^2)$, $\pi = 0$, and so the area of the curve is equal to $Q\times z^0/-1$, that is, $a^4/2(c^2-z^2)$. In the second case, however, there is $d = a^4$, $\theta = -3$, $e = -1$, $f = c^2$, $\eta = -2$, $r = 1$, $s = -1$, $Q = (a^4/-2c^2)\ (-1+c^2z^{-2})^{-1}$, that is, $-a^4z^2/2(c^4-c^2z^2)$, $\pi = 0$ and the area is equal to $Q\times z^0/-1$, that is, $a^4z^2/2(c^4-c^2z^2)$.[9] The area is exhibited in different fashion in these

N.B. Five or six[6] years ago, after John Craige had during an extended stay in Cambridge seen the series here set out and communicated the quadrature of the curve presented (if I rightly recall)[7] in this second example to his fellow Scots,[8] Mr David Gregory, Professor of Mathematics at Edinburgh, came upon the same series, but in less neat form, by another though indeed not inelegant method.

(5) We interpolate this heading to Newton's 'prime' quadrature theorem in his 1676 *epistola* so as to clear the way for the 'Prop. II. Theor. II' with which he titles his next proposition.

(6) More precisely, 'sex' (Six); see §1: note (13).

(7) He does not, it would appear; see note (11) of the preceding introduction, and compare §1: note (15).

(8) Originally just 'cum amicis' (to his friends).

(9) See IV: 622, note (11) and compare §1: note (57).

diversimodè exhibetur quatenus computatur a diversis finibus quorum assignatio per hos inventos valores arearum facilis est.

Exempl. 3. Sit ordinatim applicata $\frac{a^5}{z^5}\sqrt{bz+zz}$, hoc est per reductionem ad debitam formam, vel $a^5z^{-\frac{9}{2}}\times\overline{b+z}|^{\frac{1}{2}}$ vel $a^5z^{-4}\times\overline{1+bz^{-1}}|^{\frac{1}{2}}$. Et erit in priori casu $d=a^5$, $\theta=-\frac{9}{2}$, $e=b.f=1$, $\eta=1$, $\lambda=\frac{1}{2}$, ideoq; $r=-\frac{7}{2}$ &c. Quare cùm r non sit numerus affirmativus, transeo(10) ad alterum casum. Hic est $d=a^5$, $\theta=-4$, $e=1$, $f=b$, $\eta=-1$, $\lambda=\frac{1}{2}$, ideoq; $r=3$, $s=3\frac{1}{2}$, $Q=\frac{a^5}{-b}\times\overline{1+bz^{-1}}|^{\frac{3}{2}}$, seu $=-\frac{a^5z+a^5b}{bzz}\sqrt{zz+bz}$. $\pi=-2$ et area $=Q$ in $\frac{z^{-2}}{3\frac{1}{2}}-\frac{2}{2\frac{1}{2}}\times\frac{z^{-1}}{3\frac{1}{2}b}+\frac{1}{1\frac{1}{2}}\times\frac{2}{2\frac{1}{2}}\times\frac{z^0}{3\frac{1}{2}bb}$, hoc est =

$$\frac{-30bb+24bz-16zz}{105bbzz}\times\frac{a^5z+a^5b}{bzz}\sqrt{zz+bz}.$$

Exempl. 4. Sit deniq; ordinatim applicata $\frac{bz^{\frac{1}{3}}}{\sqrt{⑤}\,\overline{c^3-3accz^{\frac{2}{3}}+3aacz^{\frac{4}{3}}-a^3z^2}}$. Hæc ad formam Regulæ reducta fit $bz^{\frac{1}{3}}\times\overline{c-az^{\frac{2}{3}}}|^{-\frac{3}{5}}$. Indeq; est $d=b$, $\theta=\frac{1}{3}$, $e=c$, $f=-a$, $\eta=\frac{2}{3}$, $\lambda=-\frac{3}{5}$, $r=2$, $s=\frac{7}{5}$, $Q=\frac{3b}{-2a}\times\overline{c-az^{\frac{2}{3}}}|^{\frac{2}{5}}$, $\pi=\frac{2}{3}$ et area =

$$Q \text{ in } \frac{5z^{\frac{2}{3}}}{7}-\frac{5}{2}\times\frac{5c}{-7a},$$

id est $\frac{-30abz^{\frac{2}{3}}+75bc}{28aa}\times\overline{c-az^{\frac{2}{3}}}|^{\frac{2}{5}}$. Quod si res non successisset in hoc casu, existente r vel fractione vel numero negativo, tentassem alterū casum purgando terminum $-az^{\frac{2}{3}}$ in ordinatim applicata a coefficiente $z^{\frac{2}{3}}$, hoc est reducendo ordinatim applicatam ad hanc formam $bz^{-\frac{1}{15}}\times\overline{-a+cz^{-\frac{2}{3}}}|^{-\frac{3}{5}}$. Et si r in neutro casu fuisset numerus integer et affirmativus, conclusissem Curvam ex earum numero esse quæ non possunt Geometricè quadrari. Nam quantum animadverto, hæc Regula exhibet æquationibus finitis areas omnium Geometricam quadraturam admittentium Curvarum quarum ordinatim applicatæ constant ex potestatibus, radicibus vel quibuslibet dignitatibus binomij cujuscunq;.

At quando hujusmodi curva aliqua non potest geometricè quadrari, sunt ad manus alia Theoremata pro comparatione ejus cum Conicis Sectionibus vel saltem cum alijs figuris simplicissimis quibuscum potest comparari; ad quod sufficit etiam hoc ipsum jam descriptum Theorema si debite concinnetur. Pro trinomijs etiam et alijs quibusdam Regulas quasdam concinnavi. Sed in simplicioribus vulgoq; celebratis figuris vix aliquid relatu dignum reperi quod evasit aliorum conatus, nisi forte longitudo Cissoidis ejusmodi censeatur. Ea sic construitur.

Sit *VD* Cissois, *AV* diameter circuli ad quem aptatur, *V* vertex et *AF* Asymptotos ejus, *D* punctum quodvis in ea datum(11) ac *DB* perpendiculare quodvis ad *AV* demissum. Cum semiaxe $AF=AV$, et semiparametro $AG=\frac{1}{3}AV$, describatur Hyperbola *FkK*, et inter *AB* et *AV* sumpta *AC* media proportionali erigantur ad *C* et *V* perpendicula

cases in as far as it is computed from different bounds, the assigning of which by these values now found for the areas is easy.

Example 3. Let the ordinate be $(a^5/z^5)\sqrt{[bz+z^2]}$, that is, by reduction to due form, either $a^5z^{-\frac{9}{2}}(b+z)^{\frac{1}{2}}$ or $a^5z^{-4}(1+bz^{-1})^{\frac{1}{2}}$. Then in the former case there will be $d = a^5$, $\theta = -\frac{9}{2}$, $e = b$, $f = 1$, $\eta = 1$, $\lambda = \frac{1}{2}$ and consequently $r = -\frac{7}{2}$, and so on. Therefore, since r is not a positive number, I pass on[10] to the other case. Here there is $d = a^5$, $\theta = -4$, $e = 1$, $f = b$, $\eta = -1$, $\lambda = \frac{1}{2}$ and consequently $r = 3$, $s = 3\frac{1}{2}$,

$$Q = (a^5/-b)\,(1+bz^{-1})^{\frac{3}{2}},$$

that is, $-(a^5/bz^2)\,(z+b)\,\sqrt{[z^2+bz]}$, $\pi = -2$ and so the area is equal to

$$Q \times \left(\frac{z^{-2}}{3\frac{1}{2}} - \frac{2}{2\frac{1}{2}} \times \frac{z^{-1}}{3\frac{1}{2}b} + \frac{1}{1\frac{1}{2}} \times \frac{2}{2\frac{1}{2}} \times \frac{z^0}{3\frac{1}{2}b^2}\right),$$

that is,

$$-\frac{30b^2-24bz+16z^2}{105b^2z^2} \times \frac{a^5(z+b)}{bz^2}\sqrt{[z^2+bz]}.$$

Example 4. Let the ordinate be, finally, $bz^{\frac{1}{3}}/\sqrt[5]{[c^3-3ac^2z^{\frac{2}{3}}+3a^2cz^{\frac{4}{3}}-a^3z^2]}$. This, when reduced to the form of the rule, becomes $bz^{\frac{1}{3}}(c-az^{\frac{2}{3}})^{-\frac{3}{5}}$. And thence there is $d = b$, $\theta = \frac{1}{3}$, $e = c$, $f = -a$, $\eta = \frac{2}{3}$, $\lambda = -\frac{3}{5}$, $r = 2$, $s = \frac{7}{5}$, $Q = (3b/-2a)\,(c-az^{\frac{2}{3}})^{\frac{2}{5}}$, $\pi = \frac{2}{3}$ and the area is equal to $Q \times \left(\frac{5z^{\frac{2}{3}}}{7} - \frac{5}{2} \times \frac{5c}{-7a}\right)$, that is, $-\frac{30abz^{\frac{2}{3}}-75bc}{28a^2}(c-az^{\frac{2}{3}})^{\frac{2}{5}}$. But if the approach had not succeeded in this case, with r proving to be either a fraction or a negative number, I should have tried the other case by clearing the term $-az^{\frac{2}{3}}$ in the ordinate of the coefficient $z^{\frac{2}{3}}$, that is, by reducing the ordinate to this form $bz^{-\frac{1}{15}}(-a+cz^{-\frac{2}{3}})^{-\frac{3}{5}}$. And if r had in neither case been an integer and positive, I should have concluded that the curve is numbered among those which cannot geometrically be squared. For, as far as I observe, this rule exhibits in finite equations the areas of all curves admitting geometrical quadrature, the ordinates of which consist of powers, roots or any elevations of any binomial whatever.

But when any curve of this type cannot be geometrically squared, there are to hand other theorems for its comparison with conics or at least other simplest figures with which it is comparable; sufficient for this also is the very theorem just now described if it be appropriately restyled. For trinomials also and certain others I have contrived certain rules. In the simpler and commonly familiar figures, however, I have found scarcely anything worthy to relate which has eluded the efforts of others, unless maybe the length of the cissoid should be reckoned of this sort. That is constructed in this way.

Let VD be a cissoid, AV the diameter of the circle to which it is fitted, V its vertex and AF its asymptote, D any point given in it,[11] and DB any perpendicular let fall to AV. With semi-axis $AF = AV$ and semi-parameter $AG = \frac{1}{3}AV$ describe the hyperbola FkK and, taking AC a mean proportional between AB and AV, erect at C and V perpendicu-

(10) Newton first began accurately to copy 'proced[o]' (proceed) from his 1676 *epistola* before thus minimally altering it.

(11) This phrase—a later interlineation by Newton in the present manuscript—is not in the 1676 *epistola*.

Ck et *VK* Hyperbolæ occurrentia in *k* et *K*, et agantur rectæ *KT* et *kt* tangentes Hyperbolam in ijsdem *K* et *k* et occurrentes *AV* in *T* et *t*, et ad *AV* constituatur rectangulum *AVNM* æquale spatio *TKkt*: et Cissoidis longitudo *VD* erit sextupla altitudinis *VN*. Demonstratio perbrevis est,(12) sed ad infinitas series redeo.

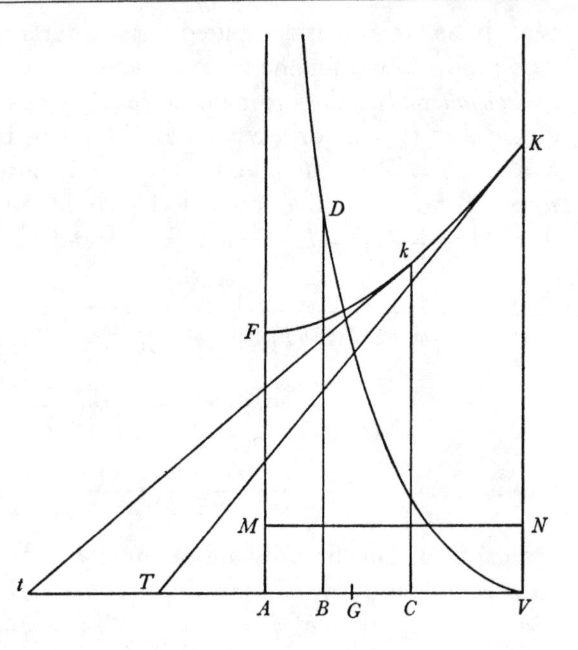

Et hactenus Epistola de methodo quam literis transpositis celabam.

Dixi igitur quod beneficio seriei hic positæ figuræ curvilineæ non solum quadrari possunt sed etiam cum Conicis sectionibus et alijs figuris comparari. Nam quemadmodum termini omnes in serie tota in infinitum continuata sunt area hujus Curvæ, sic termini omnes dempto primo sunt area alterius cujusdam curvæ, et omnes demptis duobus primis sunt area Curvæ tertiæ & omnes demptis tribus primis sunt area quartæ et sic deinceps in infinitum. Et hæ omnes curvæ inter se comparari possunt siquidem primus seriei terminus differentia est inter areas curvarum duarum primarum, et aggregatum termini primi et secundi differentia est inter areas curvæ primæ ac tertiæ et sic deinceps. Quinetiam series pergit utrinqꝫ in infinitum juxta formulam sequentem.

$$[\&c]+\frac{s+3}{r+3}\times\frac{fz^{\eta}}{eV}-\frac{s+2}{r+2}\times\frac{fz^{\eta}}{eX}+\frac{s+1}{r+1}\times\frac{fz^{\eta}}{eY}-\frac{s}{r}\times\frac{fz^{\eta}}{eA}$$

$$+\frac{Qz^{\pi}}{s}-\frac{r-1}{s-1}\times\frac{eA}{fz^{\eta}}+\frac{r-2}{s-2}\times\frac{eB}{fz^{\eta}}-\&c.$$

Ubi A denotat terminum $\frac{Qz^{\pi}}{s}$, et Y, X, V &c terminos ordine proximos(13) ad sinistram. Exhibet igitur series uno intuitu quadraturas curvarum numero infinitarum utrinqꝫ quæ omnes ex area unius data (præterquam ubi arearum differentia terminum aliquem [infinitum](14) involvit) quadrari possunt: ideoqꝫ ut omnes quadrentur, quærenda est illarum omnium simplicissima & hujus area primum investiganda. Designet Φ quemlibet terminorum B, C, D, Y, X, V &c in serie, et sit m numerus terminorum ab A ad Φ[,] id est 1 si Φ sit B, vel 2 si Φ sit C, vel -1 si Φ sit Y & sic in cæteris. Et si ponatur $\overline{s-m}\times d\Phi z^{m\eta-\pi}=h$, erit $hz^{\theta-m\eta}\times\overline{e+fz^{\eta}}|^{\lambda}$ ordinatim applicata Curvæ cujus area est seriei pars illa quæ pergit a termino Φ dextrorsum in infinitum et terminum illum includit.(15)

Ut si quadranda sit Curva cujus ordinatim applicata est $dz^4\times\overline{e+fz^2}|^{\frac{1}{2}}$ quæro

lars Ck and VK meeting the hyperbola in k and K, and draw straight lines KT and kt tangent to the hyperbola at these same K and k, meeting AV in T and t, then on AV form the rectangle $AVNM$ equal to the area $(TKkt)$: the length $\widehat{VD}$ of the cissoid will then be six times the height VN. The proof is very short,(12) but I return to infinite series.

And thus far the letter on the method which I was concealing in a transposed anagram.

I stated, therefore, that with the aid of the series here set out curvilinear figures can not only be squared but also compared with conics and other figures. For, just as all terms in the whole series continued to infinity are the area of this curve, so all the terms less the first are the area of another particular curve, all less the first two are the area of a third curve, all less the first three are the area of a fourth, and so on in turn indefinitely. And all these curves can be compared with one another, inasmuch as the first term of the series is the difference between the areas of the first two curves, and the total of the first and second terms is the difference between the areas of the first and third curves, and so forth. To be sure the series proceeds both ways to infinity according to the following pattern:

$$[\ldots]+\frac{(s+3)fz^{\eta}}{(r+3)\,eV}-\frac{(s+2)fz^{\eta}}{(r+2)\,eX}+\frac{(s+1)fz^{\eta}}{(r+1)\,eY}-\frac{sfz^{\eta}}{reA}$$

$$+\frac{Qz^{\pi}}{s}-\frac{(r-1)\,eA}{(s-1)fz^{\eta}}+\frac{(r-2)\,eB}{(s-2)fz^{\eta}}-\ldots,$$

where A denotes the term Qz^{π}/s, and $Y, X, V, \ldots$ the terms in order immediately next(13) on the left. The series therefore exhibits at a single glance the quadrature of an infinity of curves each way, all of which can (except when the difference of areas involves some infinite term) be squared in terms of the area of one given; consequently, to square all, you need to seek out the simplest of all those and first discover the area of this. Let Φ denote any of the terms $B, C, D, Y, X, V, \ldots$ in the series, and let m be the number of terms from A to Φ; that is, 1 if Φ be B, or 2 if Φ be C, or -1 if Φ be Y, and so in other cases. And if there be put $(s-m)\,d\Phi z^{m\eta-\pi}=h$, then $hz^{\theta-m\eta}(e+fz^{\eta})^{\lambda}$ will be the ordinate of the curve whose area is that part of the series proceeding rightwards from the term Φ to infinity and including that term.(15)

If I need, for instance, to square a curve whose ordinate is $dz^{4}(e+fz^{2})^{\frac{1}{2}}$,

(12) See III: 321, note (733) and compare IV: 623, note (12).

(13) 'sibi' (to one another) is deleted.

(14) This necessary adjective is omitted in the manuscript.

(15) On this extension of the 'prime' theorem as communicated to Leibniz in 1676 see §1: notes (21) and (25).

seriem ut supra. Dein quoniam θ hic est 4 et η 2, adeoq; $\theta-m\eta$ seu $4-2m$, scribendo 2 pro m evanescit; quo in casu curva illa altera cujus ordinata est $hz^{\theta-m\eta}\times\overline{e+fz^{\eta}}|^{\lambda}$, fit Conica Sectio: usurpo 2 pro m, et huic numero respondentem seriei terminum C pro Φ, et inde colligo h, et dico quod si area Sectionis hujus Conicæ cujus ordinatim applicata est $hz^{\theta-m\eta}\times\overline{e+fz^{\eta}}|^{\lambda}$ seu $h\times\overline{e+fz^{\eta}}|^{\lambda}$ addatur ad seriei terminos duos primos $A-B$ ductos in Q; summa erit series tota ideoq; æqualis est areæ Curvæ quadrandæ.

Usus autem seriei latius patebit per Theoremata duo sequentia.

Prop. II. Theor. II.(16)

Si curvarum duarum ordinatæ basi z normaliter insistentes sint $dz^{\theta}\times\overline{e+fz^{\eta}}|^{\lambda}$ et $dz^{\theta}\times\overline{e+fz^{\eta}}|^{\lambda-1}$ et earundem areæ dicantur R et S: erit

$$\overline{\lambda\eta+\theta+1}\times R-\lambda\eta eS = dz^{\theta+1}\times\overline{e+fz^{\eta}}|^{\lambda}.$$

Et per hanc æquationem colligendo aream alterutram ex altera, potest index dignitatis ordinatim applicatæ semel vel sæpius unitate augeri vel minui.

Prop. III. Theor. III.(17)

Si Curvarum duarum ordinatæ y et v normaliter insistant basibus z et x, et relatio inter basem et ordinatam unius definiatur per æquationem quamvis $ez^{\alpha}y^{\beta}+fz^{\gamma}y^{\delta}+gz^{\epsilon}y^{\zeta}+hz^{\eta}y^{\theta}+\&c=0$; ubi e, f, g &c designant quantitates datas cum signis suis et α, β, γ &c indices dignitatum y et z: et si ponatur baseos z dignitas quælibet z^{m} æqualis basi alteri x et relatio inter basem et ordinatam curvæ alterius sit

$$em^{\beta}v^{\beta}x^{\frac{\alpha+m\beta-\beta}{m}}+fm^{\delta}v^{\delta}x^{\frac{\gamma+m\delta-\delta}{m}}+gm^{\zeta}v^{\zeta}x^{\frac{\epsilon+m\zeta-\zeta}{m}}+hm^{\theta}v^{\theta}x^{\frac{\eta+m\theta-\theta}{m}}+\&c=0:$$

æquales erunt harum curvarum areæ inter se.(18)

Ubi notandum est quod indices duarum dignitatum ipsius x poni possunt vel æquales vel in alia quavis relatione ad invicem et quantitas m qua hoc eveniet, inde colligi.

Et hoc pacto æquatio omnis trium terminorum $ez^{\alpha}+fz^{\gamma}y^{\delta}+gy^{\zeta}=0$ reduci potest ad aliquam ex his tribus formis. Ponatur $\gamma\zeta+\alpha\delta-\alpha\zeta=\pi$, deinde

1. $\alpha+\zeta=\rho$. $\frac{\rho}{\zeta}=m$. $z^{m}=x$. $ev^{-\delta}+fm^{\delta}x^{\frac{\pi}{\rho}}+gm^{\zeta}v^{\zeta-\delta}=0$.

2. $\gamma-\alpha-\delta=\sigma$.(19) $\frac{\sigma}{\delta}=m$. $z^{m}=x$. $ev^{-\zeta}+fm^{\delta}v^{\delta-\zeta}+gm^{\zeta}x^{\frac{\pi}{\sigma}}=0$.

3. $\delta-\gamma-\zeta=\tau$.(19) $\frac{\tau}{\zeta-\delta}=m$. $z^{m}=x$. $ex^{\frac{\pi}{\tau}}+fm^{\delta}v^{\delta}+gm^{\zeta}v^{\zeta}=0$.

(16) A minimally reworded repeat of the equivalent 'Reg. I' in §1 preceding.

(17) The previous 'Reg. II' (in §1 above) lightly reshaped and considerably extended.

I seek its series as above. Then, because θ is here 4 and η 2, and hence $\theta-m\eta$ or $4-2m$ vanishes on writing 2 in m's place, and in this case the other curve having ordinate $hz^{\theta-m\eta}(e+fz^{\eta})^{\lambda}$ comes to be a conic, I employ 2 in m's place and the term C in the series corresponding to this number in place of Φ, and thence I gather h, and then assert that, if the area of this conic having ordinate $hz^{\theta-m\eta}(e+fz^{\eta})^{\lambda}$, that is, $h(e+fz^2)^{\frac{1}{2}}$, be added to the first two terms in the series, $A-B$, multiplied by Q, the sum will be the entire series and consequently equal to the area of the curve to be squared.

The use of the series will, however, be more broadly evident through the two following theorems.

Proposition II, Theorem II.(16)

If of two curves the ordinates standing perpendicularly on the base z be $dz^{\theta}(e+fz^{\eta})^{\lambda}$ and $dz^{\theta}(e+fz^{\eta})^{\lambda-1}$, and their areas be called R and S, then $(\lambda\eta+\theta+1)R-\lambda\eta eS = dz^{\theta+1}(e+fz^{\eta})^{\lambda}$. And by this equation, on gathering one or other area in terms of the second, the index of the ordinate's power can be increased or diminished once or more times by unity.

Proposition III, Theorem III.(17)

If the ordinates y and v of two curves stand perpendicularly upon their bases z and x, and the relationship between the base and ordinate of one be defined by any equation $ez^{\alpha}y^{\beta}+fz^{\gamma}y^{\delta}+gz^{\epsilon}y^{\zeta}+hz^{\eta}y^{\theta}+\ldots = 0$, where $e, f, g, \ldots$ denote given quantities with their signs and $\alpha, \beta, \gamma, \ldots$ the indices of powers of y and z, then, if any power whatever z^m of the base z be set equal to the other base x and the relationship between the base and ordinate of the other curve be

$$em^{\beta}v^{\beta}x^{(\alpha+m\beta-\beta)/m}+fm^{\delta}v^{\delta}x^{(\gamma+m\delta-\delta)/m}+gm^{\zeta}v^{\zeta}x^{(\epsilon+m\zeta-\zeta)/m}+hm^{\theta}v^{\theta}x^{(\eta+m\theta-\theta)/m}+\ldots = 0,$$

the areas of these curves will be equal to one another.(18)

Here note that the indices of two powers of z can be set either equal or in any other relationship to one another, and the quantity m whereby this shall happen thence gathered.

And in this manner every equation of three terms $ez^{\alpha}+fz^{\gamma}y^{\delta}+gy^{\zeta} = 0$ can be reduced to some one of these three following forms. Set $\gamma\zeta+\alpha\delta-\alpha\zeta = \pi$, and then

1. $\alpha+\zeta = \rho$, $\rho/\zeta = m$, $z^m = x$: $ev^{-\delta}+fm^{\delta}x^{\pi/\rho}+gm^{\zeta}v^{\zeta-\delta} = 0$;
2. $\gamma-\alpha-\delta = \sigma$,(19) $\sigma/\delta = m$, $z^m = x$: $ev^{-\zeta}+fm^{\delta}v^{\delta-\zeta}+gm^{\zeta}x^{\pi/\sigma} = 0$;
3. $\delta-\gamma-\zeta = \tau$,(19) $\tau/(\zeta-\delta) = m$, $z^m = x$: $ex^{\pi/\tau}+fm^{\delta}v^{\delta}+gm^{\zeta}v^{\zeta} = 0$.

---

(18) Slightly emending his earlier approach (see §1: note (33)) Newton here presupposes equality—rather than proportion—between the curve areas $\int y.dz$ and $\int v.dx$ under the transformation $z = x^{1/m}$, so that now $y = (v/(dz/dx)$ or) $mvx^{1-1/m}$.

(19) In strict truth these should (compare §1: note (36)) read '$-\sigma$' and '$-\tau$' respectively; in Newton's form it follows that $\pi = \delta\rho+\zeta\sigma$ with $\rho+\sigma+\tau = 0$.

Et si curva quadrari potest, quadratura ejus prodibit ex his æquationibus per seriem prædictam, et sufficit unam aliquam ex æquationibus tentare.[20]

[21]Quinetiam si curva quadrari potest prodit quadratura ejus ex his æquationibus absq; serie. Namq; aliquis ex indicibus dignitatum ipsius x hoc est ex numeris $\frac{\pi}{\rho}, \frac{\pi}{\sigma}, \frac{\pi}{\tau}$ erit numerus fractus affirmativus unitatem habens pro numeratore. Ideoq; Curvæ Ordinata erit binomij potestas cujus index est numerus integer et affirmativus $\frac{\rho}{\pi}$, vel $\frac{\sigma}{\pi}$ vel $\frac{\tau}{\pi}$. Unde Quadratura in promptu est per methodos vulgares. At si numerorum $\frac{\rho}{\pi}, \frac{\sigma}{\pi}, \frac{\tau}{\pi}$ nullus est integer et affirmativus curva quadrari nequit.[22] Comparabitur tamen cum alijs figuris.

Nam si e tribus numeris $\frac{\rho}{\pi}, \frac{\sigma}{\pi}, \frac{\tau}{\pi}$ duo sint integri et eorum alter negativus sit et alter vel affirmativus vel negativus: curva comparabitur cum Hyperbola.

Si tres illi numeri sint fracti irreducibiles & eorum duo habent numerum binarium pro denominatore, curva comparabitur cum Hyperbola vel Ellipsi. Alijs in casibus figura simplicissima quacum comparabitur erit adhuc magis composita. Et quæcunq; sit sic invenietur per Theorema secundum ac tertium. Nam hæc sunt Theoremata per quæ dixi in Epistola hoc idem præstari posse.[23]

---

(20) In a preliminary draft of this paragraph on f. 40^r Newton, there making a trivial interchange of y, z with v, x and assuming the unmodified area-preserving proportion $\nu\int v.dx = \int y.dz$, further successively sets $\rho/\alpha = \nu$, $\tau/\zeta = \nu$ and $\sigma/(\alpha-\gamma) = \nu$ to transform the given trinomial $ex^\alpha + fx^\nu y^\delta + gv^\zeta = 0$ under the substitution $x^\nu = z$ into three more reduced forms $ez^{\alpha-\gamma} + fy^{\pi/\rho} + gz^{-\gamma} = 0$, $ey^{-\pi/\tau} + fz^{\gamma-\alpha} + gz^{-\alpha} = 0$ and $ez^\alpha + fz^\gamma + gy^{\pi/\sigma} = 0$ respectively. He thereafter went on: 'Quinetiam hæ sex æquationes per eandem secundam regulam [*sc.* 'Reg. II' in §1 preceding] operatione repetita reduci possunt ad alias duodecim, singulæ ad binas formæ hujus simpliciores $pz^\sigma + q + ry^\tau = 0$. Sit earum aliqua ' (Indeed, these six equations can by the same *Reg II* with the operation repeated be reduced to twelve further ones, each to a pair of this simpler form $pz^\sigma + q + ry^\tau = 0$. Let some one of them...). This amplification is neatly absorbed into the present extension on pages 60–2 below.

On f. 47^r Newton has sketched an '*Exempl.* Quadranda sit Cissois cujus hæc est æquatio $z^3 = [e]yy - zyy$ seu $-[e]z^{-3} + z^{-2} + y^{-2} = 0$' (*Example.* Let there need to be squared the cissoid whose equation is this: $z^3 = ey^2 - zy^2$, that is, $-ez^{-3} + z^{-2} + y^{-2} = 0$). When he proceeded to apply the present reduction 'ponendo [$-3 = \alpha$. $-2 = \gamma$. $0 = \delta$. $-2 = \zeta$ adeoq;] $-2 = \pi$' he found '*1.* $-5 = \rho . \frac{5}{2} = m$', '*2.* $1 = \sigma . \frac{1}{0} = m =$ infin.' and '*3.* $4 = \tau . -2 = m$'; whence there ensues $-e + x^{2/5} + \frac{4}{25}v^{-2} = 0$ and $-ex^{-1/2} + 1 + \frac{1}{4}v^{-2} = 0$. The unfortunate passage to $m = \infty$ by the second mode of reduction evidently led Newton to decide against including this illustrative example in his final text.

(21) A considerably variant preliminary version of the sequel exists on f. 53^r. Its major differences are cited in notes (22) and (23) immediately following.

(22) Newton's draft on f. 53^r (see the previous note) considerably telescopes the next few lines to read: 'Quod si quadrari nequit comparabitur tamen cum conicis sectionibus si

And, if the curve can be squared, its quadrature will ensue from these equations by the above-stated series, and it is enough to try any one of the equations.(20)

(21)Indeed, if the curve can be squared, its quadrature will result from these equations without the series: for some one of the indices of the powers of x, that is, of the numbers π/ρ, π/σ, π/τ, will then be a positive fraction having unity for its numerator. Consequently, the curve's ordinate will be a binomial power whose index is an integral and positive number ρ/π, σ/π or τ/π. Whence the quadrature is immediate by common methods. But if none of the numbers ρ/π, σ/π, τ/π is integral and positive, the curve cannot be squared.(22) It is, however, comparable with other figures.

For if two of the three numbers ρ/π, σ/π, τ/π be integral, and of these one be negative and the other either positive or negative, the curve is comparable with a hyperbola.

If those three numbers be irreducible fractions, and two of them have the number 2 as denominator, the curve is comparable with a hyperbola or ellipse. In other cases the simplest figure with which it is comparable will be still more compound. And, whatever it may be, it will be ascertained in the following way by means of Theorems II and III; for these are the theorems by which I said in my letter this end can be accomplished.(23)

---

modo numerorum $\frac{\pi}{\rho}, \frac{\pi}{\sigma}, \frac{\pi}{\tau}$ numeratores non sunt majores binarijs. Sin numeratorum aliquis binario major sit figura simpl[icissima] quacũ comparabitur erit adhuc magis composita et quæcunqȝ sit invenietur per Theorema secundum ac tertium. Namqȝ hæc sunt Theoremata per quæ dixi hoc idem præstari posse. Reduci enim possunt æquationes tres præcedentes per Theor III ad hasce tres formas $az^{\frac{\pi}{\rho}}+by^{\frac{\pi}{\sigma}}+c=0$, $az^{\frac{\pi}{\rho}}+by^{\frac{\pi}{\tau}}+c=0$. $az^{\frac{\pi}{\sigma}}+by^{\frac{\pi}{\tau}}+c=0$, deinde per Theorema II subducendo numeratores indicum $\frac{\pi}{\rho}, \frac{\pi}{\sigma}$ et $\frac{\pi}{\tau}$ e denominatoribus donec indices illi fiant simplicissimi et, si commode fieri possit, affirmativi: habebuntur Curvæ tres simplicissimæ earum quarum bases et ordinatæ non sunt æquationum radices affectæ' (But if it cannot be squared, it will yet be compared with conics provided the numerators of the numbers π/ρ, π/σ and π/τ are not greater than 2. If, however, some one of the numerators be greater than 2, the simplest figure with which it is comparable will be yet more compound, and whatever it be it will be ascertained by the second and third theorems. For these are the theorems by means of which I stated this end could be achieved. To be explicit, the three preceding equations can be reduced by means of Theorem III to these three forms:

$$az^{\pi/\rho}+by^{\pi/\sigma}+c=0, \quad az^{\pi/\rho}+by^{\pi/\tau}+c=0 \quad \text{and} \quad az^{\pi/\sigma}+by^{\pi/\tau}+c=0;$$

and thereafter by Theorem II, on taking the numerators of the indices π/ρ, π/σ and π/τ from the denominators till those indices come to be simplest and also (if this can conveniently be done) positive, there will be had the three simplest curves of those whose bases and ordinates are not the affected roots of equations).

(23) See §1: note (34). On his draft f. 53r (see the two previous notes) Newton initially here continued: 'Reduci etiam possunt ad æquationes hujus formæ $dz^{\theta}\times\overline{e+fz^{[\eta]}}|^{\lambda}=y$. Sit enim æquatio reducenda $pv^{\alpha}+qv^{\beta}+rx^{\gamma}=0$. sitqȝ jam v basis et x ordinata & ponendo $v^{\nu}=y$

[24]In æquationibus tribus prædictis concipe jam v basem esse trium Curvarum et x ordinatim applicatam et æquationes illæ per Theorema tertium transmutabuntur in alias sex cum areis æqualibus ponendo[25]

In prima.

$$1.\quad \frac{\sigma\zeta}{\pi} = n.\ v^n = y.\ e + fm^\delta n^{\frac{\pi}{\rho}} z^{\frac{\pi}{\rho}} + gm^\zeta y^{\frac{\pi}{\sigma}} = 0.$$

$$2.\quad -\frac{\tau\zeta}{\pi} = n.\ v^n = y.\ ey^{\frac{\pi}{\tau}} + fm^\delta n^{\frac{\pi}{\rho}} z^{\frac{\pi}{\rho}} + gm^\zeta = 0.$$

In secunda.

$$1.\quad -\frac{\delta\tau}{\pi} = n.^{(26)}\ v^n = y.\ ey^{\frac{\pi}{\tau}} + fm^\delta + gm^\zeta n^{\frac{\pi}{\sigma}} z^{\frac{\pi}{\sigma}} = 0.$$

$$2.\quad \frac{\delta\rho}{\pi} = n.^{(26)}\ v^n = y.\ e + fm^\delta y^{\frac{\pi}{\rho}} + gm^\zeta n^{\frac{\pi}{\sigma}} z^{\frac{\pi}{\sigma}} = 0.$$

In tertia.

$$1.\quad \frac{\rho\delta - \rho\zeta}{\pi} = n.^{(26)}\ v^n = y.\ en^{\frac{\pi}{\tau}} z^{\frac{\pi}{\tau}} + fm^\delta y^{\frac{\pi}{[\rho]}} + gm^\zeta = 0.$$

$$2.\quad \frac{\sigma\zeta - \sigma\delta}{\pi} = n.^{(26)}\ v^n = y.\ en^{\frac{\pi}{\tau}} z^{\frac{\pi}{\tau}} + fm^\delta + gm^\zeta y^{\frac{\pi}{\sigma}} = 0.$$

Ubi y est basis et z ordinata curvarum sex novarum prioribus tribus æqualium.[27] Harum sex æquationum tres posteriores sunt ejusdem signi cum tribus prioribus ideoq; negligi possunt. In tribus prioribus vel in earum aliqua

---

prodibit per Theor III [on supposing the area-proportion $\nu \int x.dv = \int y.dz$] æquatio nova $py^{\frac{\alpha}{\nu}} + qy^{\frac{\beta}{\nu}} + rz^\gamma y^{\frac{\nu\gamma-\gamma}{\nu}} = 0$ ubi assumendo $\frac{\alpha}{\nu} = \frac{\nu\gamma - \gamma,}{\nu}$ vel $\frac{\alpha}{\nu} - \frac{\beta}{\nu} = 1$ obtinebitur $\nu = \frac{\alpha+\gamma}{\gamma}$ vel $= \alpha - \beta$, et substituendo hunc valorem ipsius ν prodibunt æquationes desideratæ. Ponendo $\nu = \frac{\alpha-\beta}{\eta}$ prodit æquatio hujus formæ $dz^{\theta\eta+\eta-1} \times \overline{e+fz^\eta}|^\lambda = y$, ubi numerus quilibet pro η in valore ipsius ν assumi potest quo æquatio evadat simplicissima. Deinde index λ per Theor II addendo vel subducendo unitates, et index $\theta\eta + \eta - 1$ per seriem in Theor I addendo vel subducendo aliquoties quantitatem η reduci debent ad formas simplicissimas. Atq; hoc modo incidetur semper in sectionem Conicam si modo Curva cum Sectione conica comparari potest[,] sin minus incidetur semper in figuras simplicissimas cum quibus comparari potest' (They can also be reduced to equations of this form: $dz^\theta(e+fz^\eta)^\lambda = y$. For let the equation to be reduced be $pv^\alpha + qv^\beta + rx^\gamma = 0$ and now let v be the base and x the ordinate; there will then, on setting $v^\nu = y$, ensue by Th. III a new equation $py^{\alpha/\nu} + qy^{\beta/\nu} + rz^\gamma y^{(\nu-1)\gamma/\nu} = 0$ in which by assuming either $\alpha/\nu = (\nu-1)\gamma/\nu$ or $\alpha/\nu - \beta/\nu = 1$ there will be obtained either $\nu = (\alpha+\gamma)/\gamma$ or $\nu = \alpha - \beta$, and on substituting these values for ν the equations desired will result. On setting $\nu = (\alpha-\beta)/\eta$ there ensues an equation of this form: $dz^{\theta\eta+\eta-1}(e+fz^\eta)^\lambda = y$, where any number you please can be assumed in η's place in the value of ν so as to make

(24)Conceive now in the three aforesaid equations that v is the base of the three curves and x the ordinate, and those equations shall by Theorem III be transmuted into six others with equal areas by setting(25)

In the first:

$$1.\ \sigma\zeta/\pi = n,\ v^n = y:\ e+fm^\delta n^{\pi/\rho} z^{\pi/\rho}+gm^\zeta y^{\pi/\sigma} = 0;$$

$$2.\ -\tau\zeta/\pi = n,\ v^n = y:\ ey^{\pi/\tau}+fm^\delta n^{\pi/\rho} z^{\pi/\rho}+gm^\zeta = 0.$$

In the second:

$$1.\ -\delta\tau/\pi = n,\ v^n = y:\ ey^{\pi/\tau}+fm^\delta+gm^\zeta n^{\pi/\sigma} z^{\pi/\sigma} = 0;$$

$$2.\ \delta\rho/\pi = n,\ v^n = y:\ e+fm^\delta y^{\pi/\rho}+gm^\zeta n^{\pi/\sigma} z^{\pi/\sigma} = 0.$$

In the third:

$$1.\ \rho(\delta-\zeta)/\pi = n,\ v^n = y:\ en^{\pi/\tau} z^{\pi/\tau}+fm^\delta y^{\pi/\rho}+gm^\zeta = 0;$$

$$2.\ \sigma(\zeta-\delta)/\pi = n,\ v^n = y:\ en^{\pi/\tau} z^{\pi/\tau}+fm^\delta+gm^\zeta y^{\pi/\sigma} = 0.$$

Here y is the base and z the ordinate of the six new curves equal to the three former ones.(27) Of these six equations the three latter ones are of the same sign as the three former and may consequently be neglected. In the three former, or

---

the equation come out simplest. Thereafter, the index λ, on adding or taking away units by means of Theorem II, and the index $\theta\eta+\eta-1$, on adding or taking away the quantity η some (appropriate) number of times by the series in Theorem I, ought to be reduced to their simplest forms. And in this manner you will arrive always at a conic, provided the curve can be compared with a conic, or else always at the simplest figures with which it is comparable).

(24) Minor preliminary drafts of the extension of his previous 'Reg. II' (in §1 preceding) which Newton now in sequel adjoins are preserved on ff. $54^r/55^r$. Since to cite their verbal variants would be greatly to overstress their significance, we here choose to pass them by.

(25) We omit a first version of the ensuing tabulation in which Newton began initially to posit that 'In prima 1. $\dfrac{\pi-\delta\rho}{\pi} = n.\ v^n = y.\ \ldots$'; the following revise improves only minimally upon this by making appropriate use of the simplifying substitutions $\pi = \delta\rho+\zeta\sigma$ and $\sigma = -(\rho+\tau)$ (on which see note (19) above).

(26) Correctly to relate these to the reduced trinomials alongside, we have transposed these pairs of substitutions in each of the second and third cases back from their order in the manuscript, where Newton has accidentally inverted their sequence.

(27) This extension of the single equivalent reduction in the preliminary 'De quadratura' (see §1: note (40)) proceeds straightforwardly enough from the given general substitution $v = y^{1/n}$ under the area-preserving condition $x = z.dy/dv = nzy^{1-1/n}$, provided we keep a weather-eye on the basic equalities $\pi = \delta\rho+\zeta\sigma$ and $\rho+\sigma+\tau = 0$. In the first case, in consequence, the implied substitution $v^\zeta = y^{\pi/\sigma}$ determines also that $v^{-\delta} = y^{(1/n-1)\pi/\rho}$, while $v^{-\zeta} = y^{\pi/\tau}$ decrees that $v^{\delta-\zeta} = y^{(1/n-1)\pi/\sigma}$; in the second, $v^\delta = y^{-\pi/\tau}$ yields also $v^{\zeta-\delta} = y^{(1/n-1)\pi/\sigma}$, while $v^\delta = y^{\pi/\rho}$ determines $v^\zeta = y^{(1/n-1)\pi/\sigma}$; and in the third case $v^{\delta-\zeta} = y^{\pi/\rho}$ ordains also that $v^{-\zeta} = y^{(1/n-1)\pi/\tau}$, while $v^{\delta-\zeta} = y^{-\pi/\sigma}$ requires that $v^{-\delta} = {}^{(1/n-1)\pi/\tau}$. None of these elegant variant reductions are of much practical importance of course.

quam velis ad formam simplicissimam[28] reducere si denominator indicis alicujus dignitatum y et z major sit quam numerator[,] subduc numeratorem affirmativum vel adde negativum semel vel sæpius per Theorema secundum donec denominator relinquatur minor numeratore & ejusdem signi. Et hoc pacto incidetur semper in Conicam sectionem si modo Curva sub initio proposita cum sectione Conica geometricè comparari potest. Sin minus incidetur in figuras simplicissimas quibuscum Curva illa comparari potest nisi forte figuræ jam inventæ transmutantur (per Theorema secundum et tertium) in alias quarum æquationes sunt magis affectæ, et eo nomine minus tractabiles.

Sit $pz^{\frac{\alpha}{\beta}}+qy^{\frac{\gamma}{\delta}}+r=0$ aliqua æquationum [sex][29] novissimarum et si indicum $\frac{\alpha}{\beta}$, $\frac{\gamma}{\delta}$ numeratores communem habent divisorem sit iste m. Sit etiam $\begin{Bmatrix}+n\\-n\end{Bmatrix}$ divisor termini alicujus $\begin{Bmatrix}\text{affirmativi}\\\text{negativi}\end{Bmatrix}$ in hac serie $\beta-3\alpha$, $\beta-2\alpha$, $\beta-\alpha$, β, $\beta+\alpha$, $\beta+2\alpha$ &c sitq; terminus iste cum signo suo $\beta\pm p\alpha=\phi$. Et si $m-n$ dividit terminum aliquem $\begin{Bmatrix}\text{affirmativum}\\\text{negativum}\end{Bmatrix}$ in hac serie $\delta-3\gamma$, $\delta-2\gamma$, $\delta-\gamma$, δ, $\delta+\gamma$, $\delta+2\gamma$, $\delta+3\gamma$ &c vel si $m+n$ dividit terminum aliquem $\begin{Bmatrix}\text{negativum}\\\text{affirmativum}\end{Bmatrix}$ in eadem[,] sit terminus iste cum signo suo $\delta\pm q\gamma=\omega$. Quærantur autem tales m et n ut $\frac{n\alpha}{m\phi}$ et $\frac{\overline{n\mp m}\times\gamma}{m\omega}$ sint numeri omnium simplicissimi qui hac ratione inveniri possunt. Deinde per Theorema secundum augeantur vel minuantur indicum denominatores β, δ additis vel subductis numeratoribus α, γ donec æquatio data evadat $pz^{\frac{\alpha}{\phi}}+qy^{\frac{\gamma}{\omega}}+r=0$[,] et hæc æquatio per Theorema tertium ponendo $z^{\frac{m}{n}}=x$ vel $y^{\frac{m}{n\mp m}}=v$[30] [transmutabitur in formam simplicissimam].

Sed nondum aperui methodum qua dixi me ad hujusmodi Theoremata pervenisse. Eam celabam literis transpositis quæ hoc Problem[a] involvunt.

Prop [IV]. Probl. [I][31]

Data æquatione quotcunq; fluentes quantitates involvente, invenire fluxiones: et vice versa.[32]

Prior pars problematis sic solvitur. Multiplicetur omnis æquationis terminus per indicem dignitatis quantitatis cujusq; fluentis quam involvit et singulis

(28) The revised text on 11–23(1): 19 here (compare *Correspondence*, 2: 174) abruptly terminates in mid-sentence. We complete the remainder of the present 'Prop. III. Theor. III' from Newton's preliminary draft on ULC. Add. 3962.3: 55r (itself not quite complete; see note (30) following).

(29) By a pardonable confusion the manuscript reads 'trium' (three).

in any one of them which you wish to reduce to simplest form,[28] if the denominator of some index of the powers of y and z be greater than its numerator, subtract a positive numerator or add a negative one once or more times by Theorem II till a denominator be left which is less than the numerator and of the same sign as it. And in this manner you will always arrive at a conic provided the curve initially proposed may geometrically be compared with a conic. If not, you will always arrive at the simplest figures with which that curve is comparable, unless perchance the figures just now ascertained are (by Theorems II and III) transmuted into others whose equations are more affected, and on that head less tractable.

Let $pz^{\alpha/\beta}+qy^{\gamma/\delta}+r=0$ be some one of the six[29] most recent equations and, if the numerators of the indices α/β, γ/δ have a common divisor, let that be m. Let also $\pm n$ be a divisor of some positive/negative term in this series [...] $\beta-3\alpha$, $\beta-2\alpha, \beta-\alpha, \beta, \beta+\alpha, \beta+2\alpha, \ldots$, and let that term with its sign be $\beta\pm p\alpha=\phi$. And if $m-n$ divides some positive/negative term in this series [...] $\delta-3\gamma, \delta-2\gamma, \delta-\gamma$, δ, $\delta+\gamma$, $\delta+2\gamma$, $\delta+3\gamma$, ..., or if $m+n$ divides some negative/positive term in it, let that term with its sign be $\delta\pm q\gamma=\omega$. Look, however, for such m and n as make $n\alpha/m\phi$ and $(n\mp m)\,\gamma/m\omega$ the simplest of all numbers which can be found by this method. Then, by Theorem II, increase or diminish the indices' denominators β, δ by adding or subtracting their numerators α, γ till the given equation comes to be $pz^{\alpha/\phi}+qy^{\gamma/\omega}+r=0$, and this equation will, on setting $z^{m/n}=x$ and $y^{m/(n\mp m)}=v$, by Theorem III[30] [be transmuted into simplest form].

But I have not yet disclosed the method by which I stated that I arrived at theorems of this sort. That I concealed in transposed letters which embrace this problem.

Proposition IV, Problem I.[31]

Given an equation involving any number of fluent quantities, to find the fluxions; and vice versa.[32]

The first part of the problem is solved in this way. Multiply every term of the equation by the index of the power of each fluent quantity which it involves

(30) We suitably—if maybe a little too concisely—restore the ensuing conclusion of Newton's paragraph, evidently penned by him on a (now lost) following sheet or included slip in immediate continuation of the page (f. 55r) at whose end the preceding text terminates.

(31) Originally 'Theor 4'.

(32) See the marginal note on page 50 above, and compare §1: note (6). The ensuing construction of the basic fluxional algorithm which resolves its first, direct part, is, as will be clear, but lightly adapted by Newton from the equivalent one (see IV: 564–70) earlier given by him in the fourth chapter of his 'Matheseos Universalis Specimina', but with 'dotted' fluxions now for the first time (see note (35) below) taking the place of his previous 'literal' ones.

multiplicationibus (33)mutetur quantitas illa unius dimensionis in suam fluxionem, et aggre[g]atū factorum omnium sub proprijs signis erit æquatio nova.

Cas. 1. Sunto a, b, c, d, e quantitates determinatæ et immutabiles,(34) v, x, y, z quantitates fluentes id est indeterminatæ ac perpetuo motu crescentes vel decrescentes: $\dot{v}$, $\dot{x}$, $\dot{y}$, $\dot{z}$ (35) earū fluxiones seu velocitates crescendi vel decrescendi: o quantitas infinite parva & $o\ddot{v}$, $o\ddot{x}$, $o\ddot{y}$, $o\ddot{z}$(36) quantitatum momenta id est incrementa vel decrementa momentanea. Et si proponatur æquatio quævis rationalis(37) fluentes involvens $x^3-xyy+aav-b^3=0$. Multiplicentur termini primo per indices dignitatum x & in singulis pro x unius dimensionis scribatur $\dot{x}$ & aggregatum factorum erit $3\dot{x}x^2-\dot{x}yy$. Idem fiat in y et prodibit $-2x\dot{y}y$. Idem fiat in v et prodibit $+aa\dot{v}$. Ponatur summa factorum $=0$ et habebitur æquatio $3\dot{x}x^2-\dot{x}yy-2x\dot{y}y+aa\dot{v}=0$. Qua ratione definitur relatio fluxionū. Q.E.I.

Nam si quantitates fluentes jam sunt x, y, et v, hæ post momentum temporis incrementis infinite parvis $o\dot{x}$, $o\dot{y}$, $o\dot{v}$ auctæ evadent $x+o\dot{x}$, $y+o\dot{y}$, $v+o\dot{v}$[,] quæ in æquatione prima pro x y et v scriptæ dant æquationem

$$x^3+3o\dot{x}x^2+3oo\dot{x}\dot{x}x+\dot{x}^3o^3-xyy-o\dot{x}yy-2xo\dot{y}y-2\dot{x}oo\dot{y}y+xoo\dot{y}\dot{y}$$
$$+\dot{x}o^3\dot{y}\dot{y}+aav+aao\dot{v}-b^3=0.$$

Subducatur æquatio prior et residuum divisum per o, erit

$$3\dot{x}x^2+3\dot{x}\dot{x}ox+\dot{x}^3oo-\dot{x}yy-2x\dot{y}y-2\dot{x}o\dot{y}y+xo\dot{y}\dot{y}+\dot{x}oo\dot{y}\dot{y}+aa\dot{v}=0.$$

Minuatur quantitas o in infinitum ut momenta fiant infinitissimè parva et neglectis terminis evanescentibus(38) restabit $3\dot{x}x^2-\dot{x}yy-2x\dot{y}y+aa\dot{v}=0$. Q.E.D.(39)

Cas. 2. Si proponatur æquatio irrationalis $x^3-xyy+aa\sqrt{ax-yy}-b^3=0$: pone $\sqrt{ax-yy}=v$, et habebuntur duæ æqu[ationes] $x^3-xyy+aav-b^3=0$ & $ax-yy=vv$. Et inde ut in casu priore prodeunt fluxionum relationes

$$3\dot{x}xx-\dot{x}yy-2x\dot{y}y+aa\dot{v}=0 \quad \text{et} \quad a\dot{x}-2y\dot{y}=2v\dot{v}.$$

(33) An unnecessary 'simul' (simultaneously) is here deleted.

(34) Newton first wrote 'stabiles' (stable).

(35) This first known introduction by Newton of 'dotted' fluxions into a text intended for public circulation—he had in the privacy of his early papers briefly employed a variant 'double-dotted' form in autumn 1665 (see I: 363, note (2)) and much more recently had made use of the present novel 'single-dotted' letters in preliminary computations for Theorem III preceding which are reproduced in Appendix 1.1 below (see especially note (5))—is a late replacement in the manuscript of the equivalent 'literal' nomenclature 'p, q, r, s'. Except (see the next note) for a momentary hesitation between the double- and single-dotted forms, he was henceforth without further essential variation to cleave to the latter notation (and the comparable extensions of it to denote higher-order fluxions which he was speedily to contrive in Propositions XI–XIII below) and this, after it was fixed indelibly in print in 1693 in Wallis' Latin *Algebra* (see Appendix 3 following), soon came to be universally accepted as the one in which Newton had initially framed the bases of his calculus of fluxions—an unjustified

and at each separate multiplication [33]change that quantity in one dimension into its fluxion, and the total of all the products under their proper signs will be the new equation.

Case 1. Let a, b, c, d, e be determinate, unalterable[34] quantities; v, x, y, z fluent quantities, that is, indeterminate ones increasing or decreasing by a perpetual motion; $\dot{v}$, $\dot{x}$, $\dot{y}$, $\dot{z}$[35] their fluxions, namely, their speeds of increasing or decreasing; o an indefinitely small quantity, and $o\ddot{v}$, $o\ddot{x}$, $o\ddot{y}$, $o\ddot{z}$[36] the moments of those quantities, that is, their momentaneous increases or decreases. And so, if any rational[37] equation $x^3-xy^2+a^2v-b^3=0$ involving fluents be proposed, multiply its terms first by the indices of the powers of x and in each in place of x in one dimension write $\dot{x}$, and the total of such products will be $3\dot{x}x^2-\dot{x}y^2$; do the same in y's case and there will result $-2x\dot{y}y$; do the same in v's case and there will result $+a^2\dot{v}$: put the sum of the products equal to 0 and there will be obtained the equation $3\dot{x}x^2-\dot{x}y^2-2x\dot{y}y+a^2\dot{v}=0$. And by this procedure the relationship between the fluxions is defined. As was to be found.

For, if the fluent quantities are now x, y and v, these will after a moment of time, when enlarged by the indefinitely small increments $o\dot{x}$, $o\dot{y}$, $o\dot{v}$, come to be $x+o\dot{x}$, $y+o\dot{y}$, $v+o\dot{v}$; and these latter, when written in the equation in place of x, y and v, give the equation

$$x^3+3o\dot{x}x^2+3o^2\dot{x}^2x+o^3\dot{x}^3-xy^2-o\dot{x}y^2-2xo\dot{y}y-2o^2\dot{x}\dot{y}y+xo^2\dot{y}^2$$
$$+o^3\dot{x}\dot{y}^2+a^2v+a^2o\dot{v}-b^3=0.$$

Subtract the first equation and the residue will, when divided by o, be $3\dot{x}x^2+3o\dot{x}^2x+o^2\dot{x}^3-\dot{x}y^2-2x\dot{y}y-2o\dot{x}\dot{y}y+xo\dot{y}^2+o^2\dot{x}\dot{y}^2+a^2\dot{v}=0$. Let the quantity o diminish infinitely so that the moments shall become infinitesimally small, and when vanishing terms[38] are neglected there will remain

$$3\dot{x}x^2-\dot{x}y^2-2x\dot{y}y+a^2\dot{v}=0.$$

As was to be proved. [39]

Case 2. If the irrational equation $x^3-xy^2+a^2\sqrt{[ax-y^2]}-b^3=0$ be proposed, set $\sqrt{[ax-y^2]}=v$ and there will be had the two equations $x^3-xy^2+a^2v-b^3=0$ and $ax-y^2=v^2$. And therefrom, as in the first case, there result the fluxional relationships $3\dot{x}x^2-\dot{x}y^2-2x\dot{y}y+a^2\dot{v}=0$ and $a\dot{x}-2y\dot{y}=2v\dot{v}$. Then, if by the

extrapolation which was further reinforced when his 1671 tract appeared posthumously in print in the eighteenth century with its original literal fluxions recast into standard dotted equivalents (see III: 72–3, note (86)).

(36) Read '$o\dot{v}$, $o\dot{x}$, $o\dot{y}$, $o\dot{z}$'. Newton fleetingly—and probably unconsciously—returns to his earlier double-dot nomenclature for first-order fluxions (see the previous note, and compare Appendix 1.1: note (17) below).

(37) Free from algebraic 'surd' irrationality, that is.

(38) Originally 'quantitatibus infinite parvis' (infinitely small quantities).

(39) In sequel Newton has cancelled 'Et par est ratio æquationum omniũ rationalium' (And the method for all rational equations is equivalent).

Deind[e] si per æquationum posteriorem quæratur valor fluxionis $\dot{v}_{[,]}$ nempe $\frac{a\dot{x}-2y\dot{y}}{2v}$ seu $\frac{a\dot{x}-2y\dot{y}}{2\sqrt{ax-yy}}$ et scribatur idem pro $\dot{v}$ in priore prodibit æquatio quæsita

$$3\dot{x}xx-\dot{x}yy-2xy\dot{y}+\frac{a^3\dot{x}-2aay\dot{y}}{2\sqrt{ax-yy}}=0.$$

Prop [*V*]. *Prob* [*II*].(40)

Invenire curvas quæ quadrari possunt.

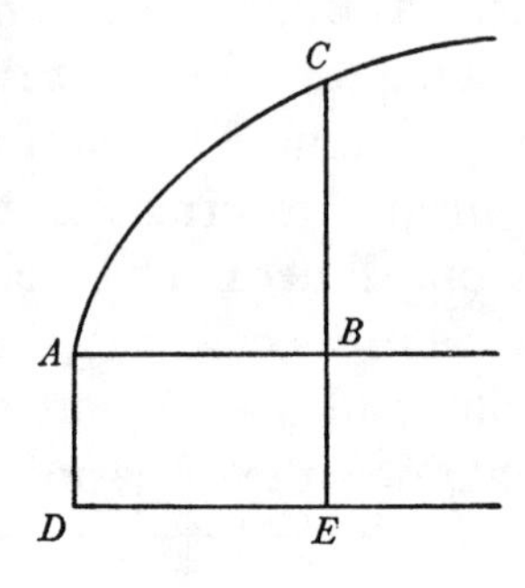

Sit *ABC* figura invenienda, *BC* ordinata, & *AB* abscissa. Producatur *CB* ad *E* ut sit *BE* = 1 et compleatur pgrū(41) *ABED* et arearum *ABC*, *ABED* fluxiones erunt ut *BC* et *BE*. Assumatur igitur æquatio quævis definiens relationem arearum et inde dabitur relatio ordinatarum *BC*, *BE* per Prop. 4.

Prop. [*VI*]. *Prob*. [*III*].

Invenire Curvam GHI datæ curvæ ABC æqualem.(42)

Assumatur æquatio quævis definiens relationē inter Curvarum abscissas *AB* ac *GH* et inde dabitur relatio fluxionum per Prop. 4. Ponantur ordinatæ *BC*, *HI* fluxionibus abscissarū reciprocè proportionales et areæ erunt æquales.(43)

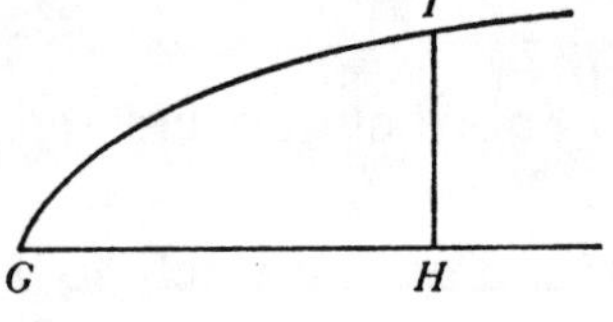

Ex hac Propositione colligitur Theorema III ponendo dignitatem aliquam abscissæ *AB* proportionalem abscissæ *GH*. Ex Propositione verò quinta(44) sub initio studiorum meorum mathematicorum primùm colligebam quadraturas particulares, deinde per inductionem perveni ad hujusmodi generalia.

(40) This repeats 'Prob: 7' of Newton's 1671 tract (III: 194–6) without significant variation.

(41) Read 'parallelogrammum', by which we take Newton here, as in his parent text (III, 196, l. 4), to mean 'rectangulum' (compare VI: 109, note (42)) and so we render it in our English version—though, of course, the present proposition continues to hold true when the angle $\widehat{ABC}$ of ordination is no longer right.

(42) Referring to a first version of the accompanying figure in which the points *G*, *H*, *I* are marked *D*, *E*, *F* respectively and which is preceded by a repeat of the upper portion only of the previous one, Newton's first enunciation of this problem was '*Invenire figuras curvilineas DEF figuræ datæ curvilineæ ABC æqualem*' (*To find curvilinear figures DEF equal to the given curvilinear figure ABC*). He first went on correspondingly: 'Dicatur $AB = z$, $BC = y$, $DE = x$, $EF = v$. Assumatur æquatio quævis definiens relationē inter bases Curvarum *AB* ac *DE*...' (Call $AB = z$, $BC = y$, $DE = x$ and $EF = v$. Assume any equation defining the relationship between the bases *AB* and *DE* of the curves).

(43) For, in Leibnizian equivalent, the inverse proportion $BC:HI = d(GH):d(AB)$ at once determines the equality of the areas $(ABC) = \int BC.d(AB)$ and $(GHI) = \int HI.d(GH)$. As

latter of the equations the value of the fluxion $\dot{v}$, namely, $(a\dot{x}-2y\dot{y})/2v$ or $(a\dot{x}-2y\dot{y})/2\sqrt{[ax-y^2]}$, be ascertained and written in $\dot{v}$'s place in the former, there will result the required equation

$$3\dot{x}x^2-\dot{x}y^2-2xy\dot{y}+a^2(a\dot{x}-2y\dot{y})/2\sqrt{[ax-y^2]}=0.$$

Proposition V, Problem II.(40)

To find curves which can be squared.

Let *ABC* be the figure to be found, *BC* its ordinate and *AB* its abscissa. Produce *CB* to *E* so that $BE=1$ and complete the rectangle *ABED*, and then the fluxions of the areas will be as *BC* and *BE*. Assume, therefore, any equation defining the relationship of the areas, and therefrom will be given the relationship of the ordinates *BC*, *BE* by Proposition IV.

Proposition VI, Problem III.

To find a curve GHI equal to the given curve ABC.(42)

Assume any equation defining the relationship between the abscissas *AB* and *GH* of the curves, and therefrom will be given the relationship of their fluxions, by Proposition IV. Set the ordinates *BC*, *HI* reciprocally proportional to the fluxions of the abscissas, and the areas will be equal.(43)

From this proposition Theorem III is gathered by setting any power of the abscissa *AB* proportional to the abscissa *GH*. From Proposition V,(44) indeed, at the start of my mathematical studies I first derived particular quadratures and then by induction arrived at general ones of this following sort.

Newton hastens to observe in sequel, the particular case in which the posited relationship between the ordinates is $BC = GH^n$ is elaborated at length in Theorem III above. He had earlier enunciated in 'Prob: 8' of his 1671 tract (see III: 198) a yet more general proposition in which the given relationship between the areas (*ABC*) and (*GHI*) is not necessarily, as here, one of equality but also an arbitrary algebraic function.

(44) Newton here unnecessarily abridges an initial citation 'Ex Propositione...tam quinta quam hacce sexta' (From both Proposition V and the present Proposition VI). While Proposition V does indeed derive, by way of Problem 7 of the 1671 tract (see note (40) above) and an equivalent passage in his 'De Analysi' of two years before (see II: 232), from 'Probl: 7' of his October 1666 tract (see I: 430, and compare *ibid.*: 412), it is no less true that the present Proposition VI similarly finds its source, by way of Problem 8 of his 1671 tract (see the previous note), in his 'Prob 6' in October 1666 (see I: 428). The particular case, developed in Theorem III preceding, of the area-preserving transformation of a curve under the substitution $z = x^{1/m}$ is likewise narrowly foreshadowed in exactly similar simplifying substitutions—and, more generally, ones of form $x^{1/m}(b+cx^n)^p$—which Newton had invoked at the beginning of his mathematical studies equivalently to effect long sequences of exact algebraic quadratures both in autumn 1665 (see I: 348–63) and again in October 1666 (I: 405–9).

Prop [VII]. Theor. [IV].

Si curvæ alicujus abscissa sit [z] et pro $e+fz^{\eta}+gz^{2\eta}+hz^{3\eta}$ &c scribatur R; sit autem area curvæ $z^{\theta}R^{\lambda}$[:] erit ordinatim applicata

$$\theta e \quad {+\theta \atop +\lambda\eta} fz^{\eta} \quad {+\theta \atop +2\lambda\eta} gz^{2\eta} \quad {+\theta \atop +3\lambda\eta} hz^{3\eta} \text{ [\&c] in } z^{\theta-1}R^{\lambda-1}.$$ (45)

Prop. [VIII]. Theor [V].

Si pro $e+fz^{\eta}[+gz^{2\eta}]$ &c scribatur R et pro $k+lz^{\eta}+mz^{2\eta}$ &c scribatur S: sit autem area curvæ $z^{\theta}R^{\lambda}S^{\mu}$: erit ordinatim applicata

$$\theta ek \quad {+\theta \atop +\lambda\eta} fkz^{\eta} \quad {+\theta \atop +2\lambda\eta} gkz^{2\eta} \text{ \&c in } z^{\theta-1}R^{\lambda-1}S^{\mu-1}.$$ (46)

$$ {+\theta \atop +\mu\eta} el \quad {+\theta \atop {+\lambda\eta \atop +\mu\eta}} fl$$

$$\phantom{\theta ek \quad {+\theta \atop +\mu\eta} el \quad} {+\theta \atop +2\mu\eta} em$$

Tandem(47) ex his Theorematis collegi series infinitas [quadraturarum] ut sequitur.

[Prop. IX. Theor. VI.](48)

[Si Curvæ abscissa sit z et pro $e+fz^{\eta}+gz^{2\eta}+hz^{3\eta}$ &c scribatur R: sit autem ordinatim applicata $z^{\theta-1}R^{\lambda-1}$ in $a+bz^{\eta}+cz^{2\eta}+dz^{3\eta}$ &c et ponatur $\frac{\theta}{\eta}=r$. $r+\lambda=s$. $s+\lambda=t$. $t+\lambda=v$ &c: erit area

$$z^{\theta}R^{\lambda} \text{ in } \frac{\frac{1}{\eta}a}{re}+\frac{\frac{1}{\eta}b-sfA}{\overline{r+1},e}z^{\eta}+\frac{\frac{1}{\eta}c-\overline{s+1},fB-tgA}{\overline{r+2},e}z^{2\eta}$$

$$+\frac{\frac{1}{\eta}d-\overline{s+2},fC-\overline{t+1},gB-vhA}{\overline{r+3},e}z^{3\eta} \text{ \&c.}$$

Ubi A, B, C, D &c denotant coefficientes terminorum singulorum in serie cum

(45) Theorem VI in the prior version, as we there restored it at least (see §1: note (62)) in line with 'Theor. III' of the preliminary draft reproduced in Appendix 1.2 below.

(46) This is (compare the previous note) essentially the ensuing 'Theor IV' in Appendix 1.2.

(47) Understand 'Nunc' (Now)! It will be clear that the two following propositions—which respectively comprehend the two sequences of quadrature theorems, 'Theorematum

Proposition VII, Theorem IV.

If of some curve the abscissa be z and for $e+fz^\eta+gz^{2\eta}+hz^{3\eta}\ldots$ there be written R, let however the area of the curve be $z^\theta R^\lambda$ and its ordinate will then be

$$(\theta e+(\theta+\lambda\eta)fz^\eta+(\theta+2\lambda\eta)gz^{2\eta}+(\theta+3\lambda\eta)hz^{3\eta}\ldots)z^{\theta-1}R^{\lambda-1}.$$ (45)

Proposition VIII, Theorem V.

If for $e+fz^\eta+gz^{2\eta}\ldots$ there be written R and for $k+lz^\eta+mz^{2\eta}\ldots$ there be written S, let however the area of the curve be $z^\theta R^\lambda S^\mu$ and the ordinate will then be

$$\begin{pmatrix}\theta ek+((\theta+\lambda\eta)fk+(\theta+\mu\eta)el)z^\eta \\ \quad +((\theta+2\lambda\eta)gk+(\theta+\lambda\eta+\mu\eta)fl+(\theta+2\mu\eta)em)z^{2\eta}\ldots\end{pmatrix}\times z^{\theta-1}R^{\lambda-1}S^{\mu-1}.$$ (46)

At length(47) I have from these theorems gathered infinite sequences (of quadratures) as follows.

Proposition IX, Theorem VI.(48)

[If a curve's abscissa be z, and for $e+fz^\eta+gz^{2\eta}+hz^{3\eta}\ldots$ there be written R, let however the ordinate be $z^{\theta-1}R^{\lambda-1}(a+bz^\eta+cz^{2\eta}+dz^{3\eta}\ldots)$ and put $\theta/\eta=r$, $r+\lambda=s$, $s+\lambda=t$, $t+\lambda=v,\ldots$, and the area will then be

$$z^\theta R^\lambda\times\left(\begin{array}{l}\frac{a/\eta}{re}+\frac{b/\eta-sfA}{(r+1)e}z^\eta+\frac{c/\eta-(s+1)fB-tgA}{(r+2)e}z^{2\eta} \\ \qquad +\frac{d/\eta-(s+2)fC-(t+1)gB-vhA}{(r+3)e}z^{3\eta}\ldots\end{array}\right)$$

where $A, B, C, D,\ldots$ denote the coefficients of individual terms in the series

Progressio prima/secunda', reproduced in Appendix 1.5 below—are here enunciated for the first time in their present generality.

(48) In his manuscript (f. 57v) Newton has merely left a half-page blank for the intended later insertion of these two related theorems. The text introduced to fill the place of the missing Proposition IX is that of the equivalent Proposition V of the revised 'Geometriæ Liber secundus' (ULC. Add. 3962.1: 5r–19r, particularly 9r/10r) reproduced in 2, 3, §2 following, but omitting its lengthy appended demonstration (*ibid.*: 10r–12r): only minimal violence is thereby done to the true historical sequence since, only a few months after we suppose that Newton here had it in mind to insert such a theorem, he communicated it to Wallis in the summer of 1692 (see the next note) with but insignificant verbal differences. With a deal more trepidation we similarly restore the companion Proposition X from the draft on ULC. Add. 3962.3: 47v, in line with Proposition VI of the ensuing 'Geometriæ Liber secundus' (Add. 3962.1: 12r/13r).

signis suis + et −, nempe A primi termini coefficientem $\frac{\frac{1}{\eta}a}{re}$, B secundi coefficientē $\frac{\frac{1}{\eta}b-sfA}{\overline{r+1},e}$, C tertij coefficientem $\frac{\frac{1}{\eta}c-\overline{s+1},fB-tgA}{\overline{r+2},e}$, et sic deinceps.][49]

[*Prop. X. Theor. VII.*][48]

[Si pro $e+fz^{\eta}+gz^{2\eta}$ &c scribatur R et pro $k+lz^{\eta}+mz^{2\eta}$ &c scribatur S; sit autem ordinatim applicata $z^{\theta-1}R^{\lambda-1}S^{\mu-1}$ in $a+bz^{\eta}+cz^{2\eta}$ &c et ponatur $\frac{\theta}{\eta}=r$. $r+\lambda=s$. $s+\lambda=t$ &c. $r+\mu=s'$. $s+\mu=t'$ &c. $s'+\mu=t''$ &c: erit area

$$z^{\theta}R^{\lambda}S^{\mu} \text{ in } \frac{\frac{1}{\eta}a}{rek}+\frac{\frac{1}{\eta}b \begin{matrix}-sfk\\-s'el\end{matrix}A}{\overline{r+1},ek}z^{\eta}+\frac{\frac{1}{\eta}c \begin{matrix}-\overline{s+1},fk\\-\overline{s'+1},el\end{matrix}B \begin{matrix}-tgk\\-t'fl\\-t''em\end{matrix}A}{\overline{r+2},ek}z^{2\eta} \text{ \&c.}$$

Ubi A denotat termini primi coefficientem $\frac{\frac{1}{\eta}a}{rek}$, B coefficientem datam secundi, C coefficientem datam tertij, et sic deinceps.][50]

Prop. XI.[51]

Data æquatione fluxiones duarum simplicium quantitatum involvente[,] *invenire relationem quantitatum.*

---

(49) Much as in §1: note (54), but now setting $I_i=\int z^{\theta+i\eta-1}R^{\lambda-1}.dz$, by differentiating $z^{\theta+i\eta}R^{\lambda}$, suitably ordering the ensuing powers of z, and then once more integrating we readily derive the basic identity

$$z^{\theta+i\eta}R^{\lambda} \equiv (\theta+i\eta)\,eI_i+(\theta+(\lambda+i)\,\eta)\,fI_{i+1}+(\theta+(2\lambda+i)\,\eta)\,gI_{i+2}\ldots,$$

whence in Newton's terms there ensues the recursion

$$(r+i)\,eI_i=(1/\eta)\,z^{\theta+i\eta}R^{\lambda}-(s+i)\,fI_{i+1}-(t+i)\,gI_{i+2}-(v+i)\,hI_{i+3}\ldots.$$

Newton's given integral may then be successively evaluated to be

$$\begin{aligned} &aI_0+bI_1+cI_2+dI_3\ldots\\ &=(a/\eta re)\,z^{\theta}R^{\lambda}+(b-asf/re)\,I_1+(c-atg/re)\,I_2+(d-avh/re)\,I_3\ldots\\ &=Az^{\theta}R^{\lambda}+\eta(b/\eta-sfA)\,I_1+\eta(c/\eta-tgA)\,I_2+\eta(d/\eta-vhA)\,I_3\ldots\\ &=(A+Bz^{\eta})\,z^{\theta}R^{\lambda}+\eta(c/\eta-(s+1)\,fB-tgA)\,I_2+\eta(d/\eta-(t+1)\,gB-vhA)\,I_3\ldots\\ &=(A+Bz^{\eta}+Cz^{2\eta})\,z^{\theta}R^{\lambda}+\eta(d/\eta-(s+2)\,fC-(t+1)\,gB-vhA)\,I_3\ldots \end{aligned}$$

and so on. The remainder at each stage is evidently determinable without difficulty.

A few months afterwards Newton passed a minimally reworded enunciation of this quadrature theorem on to John Wallis in one or other of two (now lost) letters of 27 August and 17 September 1692, and the latter subsequently incorporated it the next year in a brief survey

together with their signs + and −, namely A the first term's coefficient $(a/\eta)/re$, B the second's coefficient $(b/\eta - sfA)/(r+1)\,e$, C the third's coefficient

$$(c/\eta - (s+1)fB - tgA)/(r+2)\,e,$$

and so on.][49]

Proposition X, Theorem VII.[48]

[If for $e+fz^{\eta}+gz^{2\eta}\ldots$ there be written R and for $k+lz^{\eta}+mz^{2\eta}\ldots$ there be written S, let however the ordinate be $z^{\theta-1}R^{\lambda-1}S^{\mu-1}(a+bz^{\eta}+cz^{2\eta}\ldots)$ and put $\theta/\eta = r$, $r+\lambda = s$, $s+\lambda = t, \ldots$; $r+\mu = s'$, $s+\mu = t', \ldots$; $s'+\mu = t'', \ldots$; and the area will then be

$$z^{\theta}R^{\lambda}S^{\mu}\times\left(\begin{array}{l}\dfrac{a/\eta}{rek}+\dfrac{b/\eta-(sfk+s'el)\,A}{(r+1)\,ek}z^{\eta}\\ +\dfrac{c/\eta-((s+1)fk+(s'+1)\,el)\,B-(tgk+t'fl+t''em)\,A}{(r+2)\,ek}z^{2\eta}\ldots\end{array}\right),$$

where A denotes the first term's coefficient $(1/\eta)\,a/rek$, B the given coefficient of the second, C the given coefficient of the third, and so on.][50]

Proposition XI.[51]

Given an equation involving the fluxions of two simple quantities, to find the relationship of the quantities.

---

of Newton's techniques for attaining the fluents of given fluxions—and fluxional equations, more generally—which he appended to the Latin edition of his *Algebra* (*Opera Mathematica*, **2** (Oxford, 1693): 390–6, especially 391; the whole passage is reproduced in Appendix 3 below).

(50) Here, little differently, on putting $I_i = \int z^{\theta+i\eta-1}R^{\lambda-1}\,S^{\mu-1}.dz$, we deduce in Newton's terms that

$$(r+i)\,ekI_i = (1/\eta)\,z^{\theta+i\eta}R^{\lambda}S^{\mu} - ((s+i)\,fk+(s'+i)\,el)\,I_{i+1} - ((t+i)\,gk+(t'+i)\,fl+(t''+i)\,em)\,I_{i+2} - \ldots,$$

and thence successively

$$\begin{aligned}aI_0+bI_1+cI_2\ldots &= (a/\eta rek)\,z^{\theta}R^{\lambda}S^{\mu}+(b-a(sfk+s'el)/rek)\,I_1+(c-a(tgk+t'fl+t''em)/rek)\,I_2\ldots\\ &= Az^{\theta}R^{\lambda}S^{\mu}+\eta(b/\eta-(sfk+s'el)\,A)\,I_1+\eta(c/\eta-(tgk+t'fl+t''em)\,A)\,I_2\ldots\\ &= (A+Bz^{\eta})\,z^{\theta}R^{\lambda}S^{\mu}+\eta(c/\eta-((s+1)fk+(s'+1)\,el)B-(tgk+t'fl+t''em)A)I_2\ldots\\ &= (A+Bz^{\eta}+Cz^{2\eta})\,z^{\theta}R^{\lambda}S^{\mu}+\ldots\end{aligned}$$

and so on, again with readily determinable remainders at each stage.

(51) The manuscript at this point (ULC. Add. 3960.10: 167) has only the unspecified head 'Prop. ', but we borrow the number from Newton's preliminary draft of the first three cases of the following problem on Add. 3960.7: 105–12 (reproduced in Appendix 1.6 below). Some time afterwards—perhaps a year or so later when he recast the present treatise 'De quadratura Curvarum' into the 'Geometriæ Liber secundus' whose surviving folios are reproduced in **2**, 3, §2 following—Newton began tentatively to separate this single proposition into a complementary pair which teach how 'per Prob. 1' to find the fluxions of 'complex quantities' (complicated algebraical expressions), and conversely, given these fluxions, how to attain the 'complex' fluent quantities themselves. The texts of this uncompleted bifurcation (Add. 3960.10: 165–7, reworked on 3960.6: 117–119 [+99]) are given in full in Appendix 2 below.

Problema utilissimum est eorum quæ in Mathesi proponi solent. Nam cùm Mathesis omnis quæ circa figuras versatur, in his duobus consi[s]tat, ut figurarum cognitarum inveniantur conditiones, [52] et contra ut inveniantur figuræ quæ assignatas conditiones & proprietates habebunt[,] et curvilinearum inventio difficillima sit; spectat hæc Propositio ad harum inventionem. Solvitur autem in hunc modum.

Distinguenda est æquatio in partes duas æquales quarum una multip[l]icetur per fluxionem unius quantitatis puta z et altera per fluxionem alterius puta y.

Cas. 1. Deinde si æquatio careat alterutra fluentium quantitatum puta y: quadranda est curva cujus ordinata est valor ejus fluxionis $\dot{y}$ divisus per fluxionem alteram $\dot{z}$, et cujus abscissa est quantitas altera fluens z.[53] Nam ejus area vel sola vel una cum data quavis quantitate erit quantitas fluens y. Q.E.I.

Ut si æquatio sit $2z\dot{z}-3z\dot{y}+a\dot{y}=0$, scribendum erit $\frac{2z\dot{z}}{3z-a}=\dot{y}$, et area curvae cujus abscissa est z et ordinata $\frac{2z}{3z-a}$, ponenda erit æqualis y.[54]

Cas. 2. Si æquatio involvat utramq fluentem quantitatem et fieri possit ut una pars æquationis involvat unicam tantum cum fluxione ejus et altera involvat alteram solam cum fluxione ejus: quadrandæ sunt curvæ duæ quarum ordinatæ sunt partes illæ æquationis fluxionibus nudatæ, & abscissæ sunt fluentes quantitates quæ in ordinatis proprijs involvuntur: et ponendo earum areas æquales vel summam unius et quantitatis cujusvis datæ æqualem alteri, habebitur relatio fluentium quantitatum. Q.E.I.[55]

(52) '& proprietates' (and properties) is deleted.

(53) Understand by the unimplemented 'vice versa' of Proposition IV. As part of an intended later scheme in which this converse was made a separate Proposition II in a 'Liber tertius'—following on the 'Geometriæ Liber secundus' in which (see 2, 3, §2 below) he afterwards subsumed the preceding Propositions I–X of the present 'De quadratura Curvarum'?—Newton subsequently revised (and slightly generalized) this 'Cas. 1' to read: 'Si æquatio careat alterutra fluente quantitate, puta y, involvat autem ejus fluxionem aliquam, seu primam $\dot{y}$ seu secundam $\ddot{y}$ seu tertiam $\dddot{y}$ &c: ex valore fluxionis obtinebitur valor fluentis quantitatis per Cas. 1. Prop. 2. Lib. 3' (If the equation should lack one or other, y say, of the fluent quantities, yet involve some fluxion of it, either the first one $\dot{y}$, or the second $\ddot{y}$, or the third $\dddot{y}$, and so on: from the value of the fluxion there will be obtained the value of the fluent quantity by Case 1 of Proposition 2 of Book 3).

(54) In the revision on the facing p. 166 of the manuscript (see the previous note) Newton has 'doubled' this example to read: 'Ut si fuerit $2z\dot{z}-3z\dot{y}+a\dot{y}=0$, seu $\frac{2z\dot{z}}{3z-a}=\dot{y}$, erit $\boxed{\frac{2z}{3z-a}}+g=y$, et $z\boxed{\frac{2z}{3z-a}}-\boxed{\frac{2zz}{3z-a}}+gz+h=\acute{y}$. Et per reductionem

$$\frac{3z-a}{3}\boxed{\frac{2z}{3z-a}}-\tfrac{1}{3}zz+gz+h=\acute{y}.$$

Ideoq si fuerit $2z\dot{z}-3z\ddot{y}+a\ddot{y}=0$ seu $\frac{2z\dot{z}}{3z-a}=\ddot{y}$, erit

The problem is the most useful of those commonly propounded in mathematics. For since all mathematics which is concerned with figures consists in these two components: to ascertain the circumstances(52) of known figures, and, conversely, to ascertain figures which shall possess assigned circumstances and properties, and because the finding of curvilinear (figures) is very difficult, this proposition has regard to the latter's discovery. It is resolved, however, in this manner:

The equation is to be separated into two equal sides, one of which shall be multiplied by the fluxion of one quantity, say z, and the second by the fluxion of the other, say y.

Case 1. Then, if the equation should lack one or other, y say, of the fluent quantities, there needs to be squared the curve whose ordinate is the value of the fluxion $\dot{y}$ divided by the other fluxion $\dot{z}$ and whose abscissa is the other fluent quantity z.(53) For its area, either alone or together with any given quantity, will be the fluent quantity y. As was to be found.

If, for instance, the equation be $2z\dot{z}-3z\dot{y}+a\dot{y}=0$, you will need to write $2z\dot{z}/(3z-a)=\dot{y}$ and set the area of the curve whose abscissa is z and ordinate $2z/(3z-a)$ equal to y.(54)

Case 2. Should the equation involve both fluent quantities, but can be arranged so that one side of the equation involves but a single one together with its fluxion and the other the second alone with its fluxion, there are to be squared the two curves whose ordinates are the sides of the equation stripped of their fluxions and abscissas the fluent quantities involved in the individual ordinates, and then by setting their areas equal, or the sum of one and any given quantity equal to the other, the relationship between the fluent quantities will be obtained. As was to be found.(55)

$$\frac{3z-a}{3}\boxed{\frac{2z}{3z-a}}-\tfrac{1}{3}zz+gz+h=y\text{'},$$

whence 'Datur igitur y in exemplo utroq; per quadraturam Hyperbolæ' (y is therefore given in each example by the quadrature of a hyperbola)—namely, $x(3z-a)=2z$, in which $\int x.dz=\frac{2}{3}z+\frac{2}{9}a\log(z-\frac{1}{3}a)$. On the notation $\acute{y}$ for the fluent ($\int y.dz$) of y and the enveloping 'square' integration operator, and also for the algorithm for evaluating multiple fluents (by reducing them to simple integrals) which is here silently invoked, see Appendix 2 below, especially notes (9) and (13).

(55) On the manuscript's facing page (see note (53)) this was—by way of a preliminary tentative drafted, as we will in no way find surprising, on the still virgin areas of a letter of Humphrey Newton to him on 19 December 1691 (now Add. 3960.12: 219r, reproduced in *Correspondence*, **3**, 1961: 190) whose entirely commonplace original message it has thus preserved for a no doubt suitably grateful posterity—soon afterwards restyled to read: 'Si æquatio involv[a]t utramq; fluentem quantitatem & utriusq; fluxio sit prima et unius tantum dimensionis & termini omnes æqu[e]ntur nihilo: distinguenda est æquatio in tres partes quarum una sit terminorum omnium solitariorum qui multiplicantur per unam fluxionem $\dot{z}$, altera terminorum solitariorum qui multiplicantur per alteram fluxionem $\dot{y}$, ac tertia terminorum

Ut si æquatio sit $aa\dot{x}yy - aa\dot{x}yy - ax\dot{x}yy = a^4\dot{y} + a^3x\dot{y}$:(56) ea per reductionem fiet $\frac{aa\dot{x}}{a+x} - a\dot{x} = \frac{a^3\dot{y}}{yy}$. Quæro igitur aream Curvæ cujus ordinata est $\frac{a^3}{yy}$ et abscissa y, ut et aream curvæ cujus ordinata est $\frac{aa}{a+x} - a$ et abscissa $x_{[,]}$ et aream priorem quæ est $-\frac{a^3}{y}$ pono æqualem posteriori quæ est $\boxed{\frac{aa}{a+x}} - ax$. Et sic habeo æquationem hujusmodi $ax - \frac{a^3}{y} = \boxed{\frac{aa}{a+x}}$ id est = areæ hyperbolæ cujus ordinata est $\frac{aa}{a+x}$ et abscissa x et quæ in serie infinita fit $ax - \frac{1}{2}xx + \frac{x^3}{3a} - \frac{x^4}{4aa} + \&c$. Est igitur $\frac{a^3}{y} = \frac{1}{2}xx - \frac{x^3}{3a} + \frac{x^4}{4aa} - \&c$ æquatio quam invenire opportuit. Vel etiam si cc sit data quævis quantitas$_{[,]}$ est $\frac{a^3}{y} = cc + \frac{1}{2}xx - \frac{x^3}{3a} + \frac{x^4}{4aa} - \&c$ æquatio invenienda.

Cas. 3. Si partes æquationis ab alternis fluentibus quantitatibus liberari nequeunt: distinguendum est inter terminos solitarios et relativos.(58) Relativum voco cui similis existit quoad literas & literarum dignitates in altera parte æquationis, uti terminum $3azz\dot{y}$ si in altera æquationis parte adsit terminus $az\dot{z}y$ per numerum quemvis multiplicatus. Solitarium voco cui nullus alius est similis.(59) Si æquatio quantitatem fractam cum denominatore ex duobus vel pluribus quantitatibus constantem vel quantitatem surdam similiter com-

omnium sociorum: dein pars prima et secunda a fluentibus quantitatibus quarum fluxiones non involvunt, per multiplicationem vel divisionem si fieri possit liberandæ sunt et tum demum per Prop 2 lib 3 investigandæ tres fluentes quantitates quarum hæ tres partes sunt fluxiones et summa quantitatum inventarum ponenda est æqualis nihilo' (If the equation involves both fluent quantities, and the fluxion of each be the first and all the terms are [put] equal to zero: the equation needs to be divided into three, one to contain all the solitary terms which are multiplied by one of the fluxions $\dot{z}$, the second the solitary terms which are multiplied by the other fluxion $\dot{y}$, and the third all the allied ones; and then the first and second parts must, if possible, be freed by multiplication or division from the fluent quantities whose fluxions they do not involve; and, lastly, by means of Proposition 2 of Book 3 there need to be discovered the three fluent quantities of which these three parts are the fluxions, and the sum of the quantities so found set equal to nothing). The *terminus technicus* 'socius' (allied) is a late replacement in the manuscript for his earlier 'relativus' (relative) which, in the above, he goes on to introduce in equivalent circumstances in 'Cas. 3', and he so similarly defined it in his subsequent revise (see note (59) following),

(56) With his attention fixed on the reduced form which follows, Newton omits to notice that the first two terms cancel, leaving only $-ax\dot{x}yy = a^4\dot{y} + a^3x\dot{y}$; and so we 'translate' it in our English version. The revised '*Exempl. 1*' which he afterwards set on the facing page of the manuscript subtly changes this illustration to read:

'Sit æquatio $aa\dot{z}yy - az\dot{z}yy - z^2\dot{z}yy - a^4\dot{y} - a^3z\dot{y} = 0$',

continuing thereafter: 'et quoniam termini omnes solitarij sunt, distinguo eandem in partes duas$_{[,]}$ primam $aa\dot{z}yy - az\dot{z}yy - z^2\dot{z}yy$ per $\dot{z}$ multiplicatam et secundam $-a^4\dot{y} - a^3z\dot{y}$ per $\dot{y}$

If, for instance, the equation be $-ax\dot{x}y^2 = a^4\dot{y}+a^3x\dot{y}$,(56) this by reduction will become $(a^2/(a+x)-a)\,\dot{x} = (a^3/y^2)\,\dot{y}$. I therefore seek the area of the curve whose ordinate is a^3/y^2 and abscissa y, and also the area of the curve whose ordinate is $a^2/(a+x)-a$ and abscissa x, and set the former area, that is, $-a^3/y$, equal to the latter one, which is $[\int a^2/(a+x)\,.\,dx]$(57)$-ax$. And thus I have an equation of this kind, $ax-a^3/y = [\int a^2/(a+x)\,.\,dx]$, that is, the area of a hyperbola whose ordinate is $a^2/(a+x)$ and abscissa x, and which in infinite-series form becomes $ax-\frac{1}{2}x^2+\frac{1}{3}x^3/a-\frac{1}{4}x^4/a^2+\ldots$. Therefore $a^3/y = \frac{1}{2}x^2-\frac{1}{3}x^3/a+\frac{1}{4}x^4/a^2-\ldots$ is the equation which it was required to find. Or again, if c^2 be any given quantity, then is $a^3/y = c^2+\frac{1}{2}x^2-\frac{1}{3}x^3/a+\frac{1}{4}x^4/a^2-\ldots$ the equation to be found.

Case 3. If each side of the equation cannot be freed from the other's fluent quantity, distinction is to be made between solitary and relative(58) terms. I call 'relative' one to which a similar exists in regard to its letters and powers of those letters on the other side of the equation: as, for example, the term $3az^2\dot{y}$ if on the other side of the equation there is present a term $az\dot{z}y$ multiplied by any number; 'solitary' I call one to which there is no similar.(59) Should the equation involve a fractional quantity with a denominator consisting of two or more quantities, or a surd quantity which is similarly complicated, terms in the numerator of the fraction and the coefficient of the radical are to be considered

---

multiplicatam[,] et libero has partes ab alienis fluentibus quantitatibus dividendo æquationem per yy et $a+[z]$. Sic enim fit $\frac{aa\dot{z}}{a+z}-z\dot{z}-\frac{a^3\dot{y}}{yy}[=0]$. Pars prima $\frac{aa\dot{z}}{a+z}-z\dot{z}$ fluxio est quantitatis $\boxed{\frac{aa}{a+z}}-\frac{1}{2}z^2+g$, & pars secunda $-\frac{a^3\dot{y}}{yy}$ fluxio est quantitatis $\frac{a^3}{y}+g$: ideoq $\boxed{\frac{aa}{a+z}}-\frac{1}{2}z^2+\frac{a^3}{y}+2g=0$ æquatio quæsita est per quam relatio inter z et y definitur. Hic autem g est quantitas quælibet assumpta et $\boxed{\frac{aa}{a+z}}$ area Hyperbolæ cujus abscissa est z et ordinata rectangula $\frac{aa}{a+z}$.'

(57) As on II: 227 and elsewhere, we borrow this Leibnizian notation to render Newton's 'square' and its boxed-in integrand.

(58) Here—and several times *mutatis mutandis* in sequel—Newton first wrote 'confœderatos' (allied); in further revise (see the next note) he was to employ the synonym 'socios'. From the revised text quoted in note (55) it will be evident that he there subsequently chose to combine this present 'Cas. 3' with the preceding Case 2, which is indeed but its particular instance where no 'allied' terms are present in the given equation. To fill the gap in sequence thus left, in the revised text 'Cas. 4' below was renumbered to be 'Cas. 3', and correspondingly so for the rest.

(59) These definitions were amended slightly on the manuscript's facing verso—in immediate sequel to the text quoted in note (55)—to read: 'Terminos autem voco socios qui similes sunt quoad literas & literarum dignitates ubi fluxiones in fluentes quantitates mutantur: uti terminos $3az\dot{z}y$ & $az\dot{z}\dot{y}$ vel $2cz^3\dot{z}y^3$ & $7cz^4y^2\dot{y}$; & terminum solitarium voco cui nullus alius est similis' (I call 'allied' terms, however, ones which are similar in regard to their letters and powers of letters when the fluxions are changed into fluent quantities: as, for instance, the terms $3az\dot{z}y$ and $az^2\dot{y}$, or $2cz^3\dot{z}y^3$ and $7cz^4y^2\dot{y}$; and I call 'solitary' a term to which no other is similar). The next sentence is then repeated in continuation with the trivial replacement of 'pro relativis' by 'pro sociatis', but the remainder of the present paragraph is discarded.

plexam involvat, termini in numeratore fractionis et coefficiente radicalis pro relativis habendi sunt qui ex ijsdem literis constant cum terminis in denominatore vel radice per quantitatem alterutram fluentem vel per utramq; semel vel sæpius multiplicatis, uti terminos in numeratore fractionis hujus surdæ $\frac{3x\dot{x}y+xx\dot{y}-2aa\dot{x}}{\sqrt{xy-aa}}$.(60) Igitur termini omnes solitarij(61) in utraq; æquationis parte liberandi sunt a quantitate fluente cujus fluxionem non involvunt. Deinde termini omnes relativi in una æquationis parte quæ involvit alterutram fluxionem puta $\dot{y}$, subducendi sunt de terminis omnibus relativis in altera æquationis parte quæ involvit fluxionem alteram $\dot{z}$, Et per Propositionem præcedentem(62) quantitas fluens quærenda est cujus hæc differentia sit fluxio. Dicatur ista quantitas Q. Quadranda insuper est curva cujus abscissa est quantitas z et ordinata aggregatum terminorum solitariorum qui per abscissæ fluxionem $\dot{z}$ in una æquationis parte multiplicantur, ut et curva cujus abscissa est quantitas altera y et ordinata aggregatum terminorum solitariorum qui per abscissæ hujus fluxionem $\dot{y}$ in altera æquationis parte multiplicantur. Sit area prior P et posterior R et sit c data quævis quantitas et æquatio $P+Q = R+c$ exhibebit relationem inter fluentes quantitates y et z. Q.E.I.

Ut si æquatio sit $a^4y^3\dot{z}+aay^3z^2\dot{z}-2y^4z^3\dot{z}=z^4y^3\dot{y}+a^5zz\dot{y}$, quoniam termini $a^4y^3\dot{z}+aay^3z^2\dot{z}$ ex una æquationis parte et terminus $a^5zz\dot{y}$ ex altera solitarij sunt, libero primos a y, et ultimum ab z dividendo æquationem per y^3zz et fit $\frac{a^4\dot{z}}{zz}+aa\dot{z}-2yz\dot{z}=zz\dot{y}+\frac{a^5\dot{y}}{y^3}$. Deinde per Propositionem præcedentem(62) quæro quantitatem cujus fluxio est terminorum relativorum $-2yz\dot{z}$ et $zz\dot{y}$ differentia $-2yz\dot{z}-zz\dot{y}$, et prodibit $-zzy$. Hæc est quantitas Q. Postea quæro areas Curvarum quarū ordinatæ sunt $\frac{a^4}{zz}+aa$ et $\frac{a^5}{y^3}$ et abscissæ z et y. Hæ prodeunt $-\frac{a^4}{z}+aaz=P$ et $-\frac{2a^5}{yy}$(63) $=R$. Ideoq; æquatio $P+Q=R+c$ est

$$-yzz-\frac{a^4}{z}+aaz=-\frac{2a^5}{yy}{}^{(63)}+c,$$

seu $yzz+\frac{a^4}{z}-aaz=\frac{2a^5}{yy}$(63) $-c$. Ubi quantitas data c vel adesse vel deesse potest vel pro lubitu augeri vel minui.(64)

(60) The fluent of which this is the fluxion is determined to be $2x\sqrt{[xy-a^2]}$ in the next paragraph but one.

(61) 'qui involvunt fluxionem' (which involve the fluxion) is deleted in sequel.

(62) Understand Proposition IV above (or, it may be, some revised version of it which specifically embraces the inverse method of fluxions).

(63) Read '$-\frac{a^5}{2yy}$' and '$\frac{a^5}{2yy}$' respectively, and we so mend the English version at this point. Though far from being without precedent in his earlier papers (see, for example, the instances

as relative when they consist of the same letters as terms in the denominator or root, or are multiplied once or more times by one or other fluent quantity or by both: as, for example, the terms in the numerator of this surd fraction $(3x\dot{x}y+x^2\dot{y}-2a^2\dot{x})/\sqrt{[xy-a^2]}$.(60) In consequence, all solitary terms(61) on each side of the equation are to be freed of the fluent quantity whose fluxion they do not involve. Then all relative terms on one side of the equation which involve one or other of the fluxions, say $\dot{y}$, are to be taken away from all the relative terms on the other side of the equation which involve the other fluxion $\dot{z}$, and by the preceding Proposition(62) the fluent quantity of which this difference shall be the fluxion is to be sought. Call that quantity Q. There must, in addition, be squared the curve whose abscissa is the quantity z and ordinate the total of the solitary terms which are multiplied by the fluxion $\dot{z}$ on one side of the equation, and also the curve whose abscissa is the other quantity y and ordinate the total of solitary terms which are multiplied by the fluxion $\dot{y}$ of this abscissa on the other side of the equation. Let the former area be P and the latter one R and c be any given quantity, and the equation $P+Q=R+c$ will then exhibit the relationship between the fluent quantities y and z. As was to be found.

If, for instance, the equation be $a^4y^3\dot{z}+a^2y^3z^2\dot{z}-2y^4z^3\dot{z}=z^4y^3\dot{y}+a^5z^2\dot{y}$, because the terms $a^4y^3\dot{z}+a^2y^3z^2\dot{z}$ on the one side of the equation and the term $a^5z^2\dot{y}$ on the other are solitary, I free the former from y and the last from z by dividing the equation by y^3z^2, and there comes to be $(a^4/z^2)\,\dot{z}+a^2\dot{z}-2yz\dot{z}=z^2\dot{y}+(a^5/y^3)\,\dot{y}$. Then by the preceding Proposition(62) I seek the quantity whose fluxion is the difference $-2yz\dot{z}-z^2\dot{y}$ of the relative terms $-2yz\dot{z}$ and $z^2\dot{y}$, and it will prove to be $-z^2y$. This is the quantity Q. Thereafter I seek the areas of the curves whose ordinates are a^4/z^2+a^2 and a^5/y^3 and abscissas z and y. These prove to be $-a^4/z+a^2z=P$ and $-[\frac{1}{2}]a^5/y^2=R$. Consequently the equation $P+Q=R+c$ is $-yz^2-a^4/z+a^2z=-[\frac{1}{2}]a^5/y^2+c$, or $yz^2+a^4/z-a^2z=[\frac{1}{2}]a^5/y^2-c$, where the given quantity c may or may not be present or may be increased or diminished at will.(64)

we have noted on I: 130, 302 and VI: 344), Newton's appeal to the false (if seductive) integration rule $\int x^n.dx=(n+1)\,x^{n+1}$ (with here $n=-3$) is merely a careless slip.

(64) In the revised equivalent '*Exempl. 2*' which Newton inserted on the facing versos (Add. 3960.10: 168/170)—compare notes (53) and (56) above—this illustration was minimally reworded to be: 'Sit æquatio $a^4y^3\dot{z}+aay^3z^2\dot{z}-2y^4z^3\dot{z}-z^4y^3\dot{y}-a^5z^2\dot{y}=0$. Et quoniam termini $a^4y^3\dot{z}+aay^3z^2\dot{z}-a^5z^2\dot{y}$ solitarij sunt libero eosdem ab alienis fluentibus quantitatibus dividendo æquationem per y^3z^2, & fit $\frac{a^4\dot{z}}{z^2}+aa\dot{z}-2yz\dot{z}-zz\dot{y}-\frac{a^5\dot{y}}{y^3}=0$. Pars prima $\frac{a^4\dot{z}}{z^2}+aa\dot{z}$ fluxio est quantitatis $-\frac{a^4}{z}+aaz+g$. Pars secunda $-\frac{a^5\dot{y}}{y^3}$ fluxio est quantitatis $\frac{2a^5}{yy}$[!]$+g$. Pars tertia $-2yz\dot{z}-zz\dot{y}$ fluxio est quantitatis $-yzz+g$. Et summa omnium æquationem quæsitam componit $-\frac{a^4}{z}+aaz-\frac{2a^5}{yy}-yzz+3g=0$.'

Si termini solitarij ab alternis(65) quantitatibus fluentibus, dividendo ips[o]s per quantitates simplices unius nominis liberari non possunt, tentandæ sunt divisiones per coefficientes ex duobus vel pluribus nominibus compositas sive rationales sive surdas. Sic æquatio $3x\dot{x}y - 2aa\dot{x} = 4y\dot{y}\sqrt{xy-aa} - xx\dot{y}$ si dividatur per $\sqrt{xy-aa}$, fit $\dfrac{3x\dot{x}y-2aa\dot{x}}{\sqrt{xy-aa}} = 4y\dot{y} - \dfrac{xx\dot{y}}{\sqrt{xy-aa}}$. Et hic(66) quantitas Q cujus fluxio est terminorum relativorum differentia $\dfrac{3x\dot{x}y+xx\dot{y}-2aa\dot{x}}{\sqrt{xy-aa}}$ prodit $2x\sqrt{xy-aa}$. Et curvæ area R cujus ordinata est $4y$ prodit $2yy$. Ideoqȝ æquatio quæsita $P+Q=R+c$ fit $2x\sqrt{xy-aa} = 2yy+c$. Nam area P in hoc exemplo nulla est.

Si termini omnes ex alterutra vel utraqȝ æquationis parte relativi sint, reducantur ad eandem æquationis partem et multiplicetur æquatio tota per rectangulum sub indefinitis dignitatibus quantitatū fluentium.(67) Dein quæratur quantitas fluens complexa cujus fluxio congruat cum æquatione prodeunte, et ponatur eadem æqualis dato. Hanc Regulam vel huic similem mecum aliquando communicavit D. Fatio.(68)

Ut si fuerit $9x\dot{x}y - 18\dot{x}yy - 18xy\dot{y} + 5xx\dot{y} = 0$. multiplico æquationem per $x^{\mu}y^{\nu}$, et fit $9\dot{x}x^{\mu+1}y^{\nu+1} - 18\dot{x}x^{\mu}y^{\nu+2} - 18x^{\mu+1}y^{\nu+1}\dot{y} + 5x^{\mu+2}y^{\nu}\dot{y} = 0$. Hujus pars quæ per $\dot{x}$ multiplicatur fluxio est quantitatis

$$\frac{9}{\mu+2}x^{\mu+2}y^{\nu+1} - \frac{18}{\mu+1}x^{\mu+1}y^{\nu+2} + g = 0.$$

(65) In an otherwise insignificantly altered revise of the present paragraph which Newton subsequently entered on the facing p. 170 of his manuscript he altered this word to be 'alienis' (foreign).

(66) Understand, as Newton adds in his final sentence of the paragraph, that 'quantitas P deest' (the quantity P is lacking).

(67) 'per cujus fluxionem pars altera æquationis multiplicatur' (by whose fluxion the other part of the equation is multiplied) is deleted in sequel.

(68) In the slightly abridged revision of this rule which he afterwards penned on Add. 3960.10: 170 Newton omitted this rare expression of his debt to another for a technique of which he is here glad to make use, and which indeed he is quick to generalize to higher-order fluxional equations in 'Cas. 5' following. It remains still too little known that Nicolas Fatio de Duillier had, as he later correctly asserted in appendix to his *Lineæ Brevissimi Descensus Investigatio Geometrica Duplex*...(London, 1699: 18: 'equidem [Calculi] Fundamenta universa, ac plerasque Regulas, proprio Marte, Anno 1687, circa Mensem Aprilem & sequentes...inveni; quo tempore neminem eo Calculi genere, præter me ipsum, uti putabam'), acquired by the summer of 1687 entirely by his own unaided efforts a considerable expertise in the elementary procedures of differential and integral calculus, the high quality of which is readily gauged from the long letter discussing the ramifications of the inverse method of tangents which he wrote to Huygens in October 1687 (*Œuvres complètes de Christiaan Huygens*, **22** (The Hague, 1950): 120–51) and from their ensuing exchanges on the topic over the next three years (see especially *Œuvres complètes*, **9**, 1901: 143–4, 154–8, 169–70, 190, 198–302, 471–2, 517–19 and 573–6). The present rule for reducing first-order differential (fluxional) equations into directly

If each set of solitary terms cannot be freed from the other's[65] fluent quantity by dividing them by simple quantities of one term, division must be tried by coefficients compounded of two or more terms, either rational or surd. Thus, if the equation $3x\dot{x}y-2a^2\dot{x}=4y\dot{x}\sqrt{[xy-a^2]}-x^2\dot{y}$ be divided by $\sqrt{[xy-a^2]}$, it becomes $(3x\dot{x}y-2a^2\dot{x})/\sqrt{[xy-a^2]}=4y\dot{y}-x^2\dot{y}/\sqrt{[xy-a^2]}$. Here[66] the quantity Q whose fluxion is the difference $(3x\dot{x}y+x^2\dot{y}-2a^2\dot{x})/\sqrt{[xy-a^2]}$ of the relative terms proves to be $2x\sqrt{[xy-a^2]}$; and the area R of the curve whose ordinate is $4y$ proves to be $2y^2$. Consequently the required equation $P+Q=R+c$ becomes $2x\sqrt{[xy-a^2]}=2y^2+c$. The area P in this example is, of course, nil.

If all terms on one or other side of the equation, or on both sides, be relative, reduce them to be on the same side of the equation and multiply the whole equation by a product of indefinite powers of the fluent quantities;[67] then determine the compound fluent quantity whose fluxion shall agree with the resulting equation, and set it equal to a given magnitude. This rule, or one like it, was communicated to me some while ago by Mr Fatio.[68]

If, for instance, there were $9x\dot{x}y-18\dot{x}y^2-18xy\dot{y}+5x^2\dot{y}=0$, multiply the equation by $x^\mu y^\nu$, and there comes to be

$$9\dot{x}x^{\mu+1}y^{\nu+1}-18\dot{x}x^{\mu}y^{\nu+2}-18x^{\mu+1}y^{\nu+1}\dot{y}+5x^{\mu+2}y^{\nu}\dot{y}=0.$$

The part of this which is multiplied by $\dot{x}$ is the fluxion of the quantity

$$(9/(\mu+2))\,x^{\mu+2}y^{\nu+1}-(18/(\mu+1))\,x^{\mu+1}y^{\nu+2}+g=0.$$

---

quadrable form by multiplying through by (as he styled it) the 'transformateur' $x^\mu y^\nu$ Fatio first communicated—shortly after making its discovery, it would appear —to Huygens in a lost letter of mid-February 1691. (See Huygens' ensuing report to Leibniz on 23 February (N.S.) in C. I. Gerhardt, *Der Briefwechsel von Leibniz mit Mathematikern*, **1** (Berlin, 1899): 635–9 [= *Œuvres complètes*, **10**, 1905: 18–22]. He there remarked that Fatio had applied his rule to find 'les deux mesmes courbes dont je vous avois proposé [in his preceding letter of 24 August 1690; *Briefwechsel*, **1**: 597–8 = *Œuvres*, **9**: 472] les soutangentes'—viz. $-2xy\dot{x}+4x^2\dot{y}-y^2\dot{y}=0$ and $-3a^2y\dot{x}+2xy^2\dot{x}-2x^2y\dot{y}+a^2x\dot{y}=0$ on recasting these into the equivalent fluxional forms to which Fatio himself afterwards returned (in ?January 1692; see *Œuvres*, **10**: 74)—'desquelles la 2. a plus de difficulté'; after multiplication through by the respective transformers y^{-5} and x^4, these prove to be the fluxions of $-x^2y^{-4}+\frac{1}{2}y^{-2}=c$ and $a^2x^{-3}y-x^{-2}y^2=c$.) Precisely when Fatio imparted his rule to Newton is not known but, since he was in Holland during May 1690–September 1691 and is not known to have corresponded with Newton during that period, we may reasonably assume that Newton was still unaware of it when he departed from Cambridge to begin his three-week visit to London in late December 1691.

This rule, needless to say, resolves only the simplest class of fluxional equations which may—ignoring any 'solitary' terms—be reduced to the form $(\partial C/\partial x)\,\dot{x}+(\partial C/\partial y)\,\dot{y}=0$, and cannot touch the more usual one where the fluent $C=0$ also enters the equation. In his 'Cas. 4' following Newton will in fact (see note (71)) construct a particular solution in one instance where the 'soutangente deguisée' (as Huygens dubbed the general first-order 'equation differentielle' in elaborating for L'Hospital in July 1693 the details of Fatio's method, there laying heavy stress on its limitations; see his *Œuvres*, **10**: 464–8) has the tractable form $(\partial C/\partial x+\lambda C)\,\dot{x}+(\partial C/\partial y+\mu C)\,\dot{y}=0$.

Et ejusdem quantitatis fluxio reliqua est

$$\frac{9\times\overline{\nu+1}}{\mu+2}x^{\mu+2}y^{\nu}\dot{y}-\frac{18\times\overline{\nu+2}}{\mu+1}x^{\mu+1}y^{\nu+1}\dot{y},$$

quae igitur congruere debet cum altera æquationis parte

$$5x^{\mu+2}y^{\nu}\dot{y}-18x^{\mu+1}y^{\nu+1}\dot{y}.$$

Ergo $\frac{9\times\overline{\nu+1}}{\mu+2}=5$ & $\frac{18\times\overline{\nu+2}}{\mu+1}=18$, et inde $\mu=\frac{5}{2}$ et $\nu=\frac{3}{2}$ & his in æquatione novissima substitutis fit $2x^{\frac{9}{2}}y^{\frac{5}{2}}-\frac{36}{7}x^{\frac{7}{2}}y^{\frac{7}{2}}+g=0$, æquatio quæsita.(69)

Cas. 4.(70) Si res ne sic quidem succedit, terminos qui multiplicantur per fluxionem alterutram, puta $\dot{y}$, redige in ordinem secundum dimensiones ipsius y ac divide respectivè per terminos arithmeticæ progressionis -2, -1, 0, 1, 2, 3 &c aptando hujus terminum 0 ad æquationis terminum quemvis deficientem. Ad terminos prodeuntes adjice æquationis terminum illum deficientem data quavis quantitate multiplicatum et aggregatum pone æquale nihilo. Dein tenta si aggregatum hujus æquationis multiplicatæ per datam aliquam quantitatem et ejusdem æquationis multiplicatæ per indices dignitatum z ductos in $\frac{y}{z}$ producere possit terminos illos æquationis propositæ qui multiplicantur per fluxionem alteram $\dot{z}$.(71)

Ut si proponatur æquatio $ay\dot{z}+2az\dot{z}+2yz\dot{z}-yy\dot{z}=2az\dot{y}+z\dot{y}y$, terminos $2az+zy$ qui multiplicantur per $\dot{y}$ dispono secundum dimensiones ipsius y ac divido per progressionem arithmeticam hoc ordine $\overset{2}{2az}.\ \overset{1}{+zy}.\ \overset{0}{*}.$ et terminos prodeuntes $az+zy+eyy$ pono $=0$. Dein hos terminos multiplico tum per indices dignitatum z ductos in $\frac{y}{z}$ tum per datam quamvis quantitatem f et tento si omnium aggregatum $ay+yy+faz+fzy+feyy$ æquari possit terminis $ay+2az+2yz-yy$ qui in æquatione multiplicantur per $\dot{z}$, et res succedit. Nam facta terminorum correspondentium collatione prodeunt $2az=faz$, $2zy=fzy$, & $-yy=feyy+yy$, et inde $f=2$ & $e=-1$. Unde æquatio $az+zy+eyy=0$ fit $az+zy-yy=0$. Et hæc est relatio quæsita inter z et y.(72)

---

(69) Newton copied this illustration without change in the revision which (compare note (65) above) he penned on Add. 3965.10: 170.

(70) Afterwards renumbered to be 'Cas. 3' on Add. 3965.10: 170 (compare note (58) above), where, in a slight attenuation of the following open phrase, Newton began 'Si res non sic succedit, terminos qui multiplicantur per' (If the attempt does not succeed this way, [rank] the terms which are multiplied by...), understanding the sequel to follow without alteration. Earlier, he had intended in 'Cas. 4' to give—in foretaste of his generalized exposition of the technique in Proposition XII following—an outline of his method for extracting the fluent 'root' of a (first-order) fluxional equation as an infinite 'converging' series; the pertinent manuscript drafts are reproduced *in extenso* in Appendix 1.7 below.

And the remainder of the fluxion of this same quantity is

$$(9(\nu+1)/(\mu+2))\, x^{\mu+2}y^{\nu}\dot{y}-(18(\nu+2)/(\mu+1))\, x^{\mu+1}y^{\nu+1}\dot{y},$$

which ought therefore to agree with the equation's other part

$$5x^{\mu+2}y^{\nu}\dot{y}-18x^{\mu+1}y^{\nu+1}\dot{y}.$$

Therefore $9(\nu+1)/(\mu+2)=5$ and $18(\nu+2)/(\mu+1)=18$, and thence $\mu=\frac{5}{2}$ and $\nu=\frac{3}{2}$, and when these are substituted in the most recent equation there comes $2x^{\frac{9}{2}}y^{\frac{5}{2}}-\frac{36}{7}x^{\frac{7}{2}}y^{\frac{7}{2}}+g=0$, the required equation.(69)

Case 4.(70) If the attempt does not succeed even this way, rank the terms which are multiplied by one or other fluxion, $\dot{y}$ say, in order according to the dimensions of y and divide respectively by the terms of the arithmetical progression -2, -1, 0, 1, 2, 3, fitting its term 0 to any lacking in the equation. To the resulting terms append that term lacking in the equation multiplied by any given quantity, and set the total equal to zero; then test if the total of this equation multiplied by a given quantity and of the same equation multiplied by the indices of the power of z times y/z might produce the terms in the proposed equation which are multiplied by the other fluxion $\dot{z}$.(71)

If, for instance, the equation $ay\dot{z}+2az\dot{z}+2yz\dot{z}-y^2\dot{z}=2az\dot{y}+zy\dot{y}$ be proposed, I arrange the terms $2az+zy$ which are multiplied by $\dot{y}$ according to the dimensions of y and divide them by the arithmetical progression in this order

$$\begin{matrix} 2 & 1 & 0 \\ 2az & +zy & +[0], \end{matrix}$$

and set the resulting terms $az+zy+ey^2=0$. Then I multiply these terms both by the indices of the powers of z times y/z and by any given quantity f, and test if the total aggregate $ay+y^2+faz+fzy+fey^2$ might be equal to the terms $ay+2az+2yz-y^2$ which are multiplied by $\dot{z}$ in the equation, and the trial succeeds. For when comparison of corresponding terms is made there ensues $2az=fz$, $2zy=fzy$ and $-y^2=fey^2+y^2$, and thence $f=2$ and $e=-1$. Whence the equation $az+zy+ey^2=0$ becomes $az+zy-y^2=0$. And this is the required relationship between z and y.(72)

(71) Here in effect, given the fluxional equation $A\dot{z}+B\dot{y}=0$ which is not in immediately quadrable form, Newton tests whether for some $C=0$ there is (preserving homogeneity in dimension between this fluent and its fluxional 'derivative') $y\dot{C}\equiv(A+\lambda C)\,\dot{z}+(B+\mu C)\,\dot{y}=0$, and hence simultaneously $y.\partial C/\partial z=A+\lambda C$ and $y.\partial C/\partial y=B+\mu C$. By the second there is, for some μ, $C=y^{\mu}\int By^{-(\mu+1)}.dy$ and his test is that there shall also be $(y/z)(z.\partial C/\partial z)=A+\lambda C$ for some λ.

(72) In the terms of the previous note there is here

$$A=ay+2az+2yz-y^2 \quad\text{and}\quad B=-2az-yz,$$

and the test is that $(2/\mu)\,ay+(1/(\mu-1))\,y^2=ay+2(1+\lambda/\mu)\,az+(2+\lambda/(\mu-1))\,yz-y^2+\lambda ey^{\mu}$. It is readily computed that $\mu=2=-\lambda$ and so $e=-2$, whence $C=az+yz-y^2=0$, as Newton finds. This is, of course, only a particular solution of the given equation $A\dot{z}+B\dot{y}=0$, for where $C=az+yz-y^2\neq 0$ it follows that $y\dot{C}=(A-2C)\,\dot{z}+(B+2C)\,\dot{y}=2C(\dot{y}-\dot{z})$ and so $C=ky^2e^{-2\int y^{-1}.dz}$, $k\neq 0$ in general.

Si res non sic successisset, permutassem nomina quantitatū y et z et regulam iterum tentassem.

Dignus est hic casus qui pluribus tractetur sed brevitati consulo.

Cas. 5.[73] Si æquatio quantitatem uniformiter fluentem z cujus fluxio sit unitas & quantitatem non uniformiter fluentem y cum pluribus ejus fluxionibus (prima $\dot{y}$, secunda $\ddot{y}$, tertia $\dddot{y}$ &c) involvit: excerpe terminum aliquem in qua fluxio ultima reperitur, muta fluxionem illam ultimam in primam, multiplica terminum per fictitiā dignitatem $z^{\alpha}y^{\beta}$ &[74] sit $cz^{\lambda}y^{\nu}\dot{y}$ terminus qui producitur, et d quantitas fictitia. Pone $\nu cz^{\lambda}y^{\nu-1}\dot{y}^2 = P$, $\overline{\nu d+d+\lambda c}\times z^{\lambda-1}y^{\nu}\dot{y} = Q$ & $\overline{\lambda d-d}\times z^{\lambda-2}y^{\nu+1} = R$ si ipsius y fluxio secunda $\ddot{y}$ est ultima fluxionum quæ in æquatione habentur. Vel $3\nu cz^{\lambda}y^{\nu-1}\dot{y}\ddot{y} = P$, $\overline{\nu d+d+2\lambda c}\times z^{\lambda-1}y^{\nu}\ddot{y} = Q$ et $\overline{\lambda-1}\times\overline{2\nu d+2d+\lambda c}\times z^{\lambda-2}y^{\nu}\dot{y} = R$ si ipsius y fluxio tertia $\dddot{y}$ est fluxionum ultima. Vel $4\nu cz^{\lambda}y^{\nu-1}\dot{y}\dddot{y} = P$, $\overline{\nu d+d+3\lambda c}\times z^{\lambda-1}y^{\nu}\dddot{y} = Q$ et

$$\overline{\lambda-1}\times\overline{3\nu d+3d+3\lambda c}\times z^{\lambda-2}y^{\nu}\ddot{y} = R$$

si ipsius y fluxio quarta $\ddddot{y}$ est fluxionum ultima. Vel $5\nu cz^{\lambda}y^{\nu-1}\dot{y}\ddddot{y} = P$,

$$\overline{\nu d+d+4\lambda c}\times z^{\lambda-1}y^{\nu}\ddddot{y} = Q \quad \text{et} \quad \overline{\lambda-1}\times\overline{4\nu d+4d+6\lambda c}\times z^{\lambda-2}y^{\nu}\dddot{y} = R$$

si $\overset{.....}{y}$ est fluxionum ultima[,] et sic deinceps in infinitum. Ubi nota quod in valore R coefficientes ipsius λc, nempe 1, 3, 6, 10 &c, inveniuntur perpetua additione numerorum 1, 2, 3, 4 &c.[75] Terminos P, Q, R sic inventos divide per $z^{\alpha}y^{\beta}$ ac terminos prodeuntes compara cum æquationis terminis homogeneis factaq; coefficientium æquatione determina dignitatum exponentes α et β.

His inventis, multiplica æquationem per $z^{\alpha}y^{\beta}$ et terminos omnes excerpe in quibus ipsius y fluxio ultima reperitur. Muta fluxionem illam in primam et terminos prodeuntes adserva. Eorundem quære fluxionem primam, secundam, tertiam &c pergendo donec ad ipsius y fluxionem ultimam perveniatur et fluxionum inventarum novissimam aufer ex æquatione.

---

(73) Newton extends the simple Fatian rule expounded in 'Cas. 3' preceding to embrace (where this proves possible) higher-order fluxional equations.

(74) Newton has deleted—evidently as superfluous—an intervening passage where he initially continued 'quære ejusdem fluxionem primam, secundam, tertiam &c donec ad ipsius y fluxionem ultimam perveniatur et ex fluxionum inventarum ultima excerpe terminum qui per ipsius y fluxionem primam et penultimam multiplicatur, et compara hunc terminum cum homogeneo termino æquations per y^{β} multiplicatæ, adæquando eorum coefficientes et inde determinando exponentem β dignitatis fictitiæ y^{β}. Postea' (and seek its first, second, third,... fluxions till you attain y's final fluxion, and then from the last of the fluxions so found pick out a term which is multiplied by the first and next-to-last fluxions of y with the term of the same form in the equation multiplied by y^{β}, equating their coefficients to onc another and thereby determining the exponent β of the fictitious power y^{β}. Afterwards).

(75) On assuming (with Newton) the comparison equation to be of form

$$cz^{\lambda}y^{\nu}\dot{y}+dz^{\lambda-1}y^{\nu+1}+\ldots = 0,$$

If the trial had not succeeded this way, I should have interchanged the denominations of the quantities y and z and again tested the rule.

This case is worthy to be treated in more detail, but it is my intention to be brief.

Case 5.(73) If the equation involves a uniformly fluent quantity z, whose fluxion shall be unity, and a non-uniformly fluent quantity y together with several of its fluxions (the first $\dot{y}$, the second $\ddot{y}$, the third $\dddot{y}$, and so on), pick out some term in which the final fluxion is found, change that final fluxion into the first, multiply the term by a fictitious power $z^{\alpha}y^{\beta}$, and(74) let $cz^{\lambda}y^{\nu}\dot{y}$ be the term which is produced and d a fictitious quantity. Set

$$\nu cz^{\lambda}y^{\nu-1}\dot{y}^2 = P, \quad (\nu d+d+\lambda c)\, z^{\lambda-1}y^{\nu}\dot{y} = Q \quad \text{and} \quad (\lambda-1)\, dz^{\lambda-2}y^{\nu+1} = R$$

if the second fluxion $\ddot{y}$ of y is the final one of the fluxions which are had in the equation; or $3\nu cz^{\lambda}y^{\nu-1}\dot{y}\ddot{y} = P$, $(\nu d+d+2\lambda c)\, z^{\lambda-1}y^{\nu}\ddot{y} = Q$ and

$$(\lambda-1)\,(2\nu d+2d+\lambda c)\, z^{\lambda-2}y^{\nu}\dot{y} = R$$

if the third fluxion $\dddot{y}$ of y is the final one of the fluxions; or $4\nu cz^{\lambda}y^{\nu-1}\dot{y}\dddot{y} = P$, $(\nu d+d+3\lambda c)\, z^{\lambda-1}y^{\nu}\dddot{y} = Q$ and $(\lambda-1)\,(3\nu d+3d+3\lambda c)\, z^{\lambda-2}y^{\nu}\ddot{y} = R$ if the fourth fluxion $\ddddot{y}$ of y is the final one of the fluxions; or $5\nu cz^{\lambda}y^{\nu-1}\dot{y}\ddddot{y} = P$,

$$(\nu d+d+4\lambda c)\, z^{\lambda-1}y^{\nu}\ddddot{y} = Q \quad \text{and} \quad (\lambda-1)\,(4\nu d+4d+6\lambda c)\, z^{\lambda-2}y^{\nu}\dddot{y} = R$$

if $\overset{\cdot\cdot\cdot\cdot\cdot}{y}$ is the last of the fluxions; and so on in turn indefinitely. Here note that in the value of R the coefficients of λc, namely 1, 3, 6, 10, ... are found by perpetual addition of the numbers 1, 2, 3, 4, ...(75) The terms P, Q, R, found in this way divide by $z^{\alpha}y^{\beta}$ and compare the ensuing terms with ones of the same form in the equation, and then by setting their coefficients equal determine the exponents α and β of the powers.

Once these are ascertained, multiply the equation by $z^{\alpha}y^{\beta}$ and pick out all terms in which the final fluxion of y is found. Change that fluxion to a first one and retain the resulting terms. Of these seek out the first, second, third, ... fluxion, proceeding until you arrive at the final fluxion of y, and take the most recent of the fluxions so found from the equation.

---

we may readily check that its first fluxion is

$$cz^{\lambda}y^{\nu}\ddot{y} + (p_1 z^{\lambda}y^{\nu-1}\dot{y} + q_1 z^{\lambda-1}y^{\nu})\, \dot{y} + r_1 z^{\lambda-2}y^{\nu+1} + \ldots \; = 0,$$

where $p_1 = c\nu$, $q_1 = c\lambda+d(\nu+1)$ and $r_1 = d(\lambda-1)$; whence its second fluxion proves to be

$$cz^{\lambda}y^{\nu}\dddot{y} + (p_2 z^{\lambda}y^{\nu-1}\dot{y} + q_2 z^{\lambda-1}y^{\nu})\, \ddot{y} + r_2 z^{\lambda-2}y^{\nu+1}\dot{y} + \ldots = 0,$$

on setting $p_2 = 3c\nu$, $q_2 = 2c\lambda+d(\nu+1)$ and $r_2 = (\lambda-1)\,(c\lambda+2d(\nu+1))$; and in general that its $(i-1)$-th fluxion, $i \geqslant 2$, is $cz^{\lambda}y^{\nu}\overset{[i+1]}{y} + (p_i z^{\lambda}y^{\nu-1}\dot{y} + q_i z^{\lambda-1}y^{\nu})\overset{[i]}{y} + r_i z^{\lambda-2}y^{\nu+1}\overset{[i-1]}{y} + \ldots = 0$, where $\overset{[i]}{y}$ is the i-th fluxion (with respect to z) of y, and

$$p_i = p_{i-1}+c\nu = (i+1)\, c\nu, \quad q_i = q_{i-1}+c\lambda = ic\lambda+d(\nu+1)$$

and

$$r_i = r_{i-1}+(\lambda-1)\, q_i = (\lambda-1)\, (\tfrac{1}{2}i(i-1)\, c\lambda+id(\nu+1)).$$

Dein excerpe ex æquationis parte relicta terminos omnes in quibus ipsius y fluxio penultima reperitur. Muta hanc fluxionem in primam et quære fluentes quantitates quarum termini prodeuntes sunt fluxiones, inventasꝗ adjice terminis asservatis, adjectarum quære fluxionem primam, secundam, tertiam &c pergendo donec ad ipsius y fluxionem penultimam perveniatur & fluxionum inventarum novissimam aufer ex æquationis parte reliqua. [&c]

Si jam termini omnes in quibus y aut aliqua ejus fluxio reperitur ex æquatione evanuerint:(76) quære aream curvæ cujus abscissa est z et ordinata aggregatum terminorum æquationis qui post novissimam subductionem relicti sunt et hanc aream adjice terminis asservatis si modo ipsius y fluxio secunda $\ddot{y}$ est ultima fluxionum quæ in æquatione proposita habebantur. Sin fluxio tertia $\dddot{y}$ est ultima quære adhuc aream curvæ alterius cujus abscissa est z et ordinata est area prior ad unitatem ap[p]licata et aucta quantitate quavis data & hanc aream adjice terminis asservatis. At si fluxio quarta $\ddddot{y}$ est omnium ultima quære insuper aream curvæ tertiæ cujus abscissa est z et ordinata est area secunda(77) una cum data quavis quantitate et hanc aream adjice terminis asservatis. [&c]

Deniꝗ terminos asservatos una cum adjectis pone æquales quantitati cuicunꝗ datæ et hæc æquatio exhibebit relationem inter quantitates y et z et earum fluxiones $\dot{y}$ et 1. Inde vero per casus quatuor primos investiganda est relatio inter quantitates y et z.

Exempl. 1. Proponatur æquatio

$$12zyy+24zzy\dot{y}+6z^3\dot{y}^2+3z^3y\ddot{y}-20zyy\dot{y}-4y^{(78)}-12z^2y\dot{y}^2-4z^2y^2\ddot{y}=0,$$

ubi ipsius y fluxio ultima est $\ddot{y}$. Terminorum igitur $3z^3y\ddot{y}$ vel $-4z^2y^2\ddot{y}$ alterutrum excerpo[,] puta $3z^3y\ddot{y}$, mutoꝗ $\ddot{y}$ in $\dot{y}$ terminumꝗ multiplico per $z^\alpha y^\beta$ et fit $3z^{\alpha+3}y^{\beta+1}\dot{y}$. Ergo $c=3$, $\lambda=\alpha+3$ & $\nu=\beta+1$ & $\frac{P}{z^\alpha y^\beta}=\overline{3+3\beta}\times z^3\dot{y}^2$. Hunc terminum confero cum æquationis termino homogeneo $6z^3\dot{y}^2$ et prodit $3+3\beta=6$ et inde $\beta=1$, et $\nu=2$. Porro est $\frac{Q}{z^\alpha y^\beta}=\overline{3d+3\lambda}\times z^2y\dot{y}$, et

(76) For his subsequent 'Geometriæ Liber tertius' (compare note (58) above) Newton on the facing p. 126 in his manuscript—appealing to a restyled sequence in which (see note (58)) the preceding Cases 2 and 3 were afterwards combined into a single 'Cas. 2'—abridged the remainder of his present overlong sequel to be 'quære quantitatem fluentem cujus fluxio prima est aggregatum terminorum æquationis qui post novissimam subductionem relinquebantur si modo fluxio secunda est fluxionum ultima, vel cujus fluxio secunda est aggregatum illud si fluxio tertia est fluxionum ultima, vel tertia si quarta &c, & hanc quantitatem adjice quantitatibus asservatis & summam pone æqualem nihilo, et hæc æquatio exhibebit relationem inter quantitates y et z earumꝗ fluxiones $\dot{y}$ et 1. Inde vero per casus tres primos investiganda est relatio inter quantitates y et z' (seek out the fluent quantity whose first fluxion is the aggregate of the terms remaining in the equation after the most recent subtraction provided

Then pick out from the part of the equation left all terms in which the last fluxion but one of y is found. Change this fluxion to a first one and seek out the fluent quantities of which the resulting terms are the fluxions; then, once they are found, add them to the terms retained, seek the first, second, third, ... fluxions of the added quantities, proceeding until you arrive at the last but one fluxion of y, and take the most recent of the fluxions so ascertained from the part of the equation left. [And so on.]

If now all terms in which y or some fluxion of it is found shall have vanished from the equation,(76) determine the area of a curve whose abscissa is z and ordinate the total of the terms left in the equation after the most recent subtraction, and add this area to the terms retained provided the second fluxion $\ddot{y}$ of y is the final one of the fluxions which were had in the equation proposed. But if the third fluxion $\dddot{y}$ is the last one, determine further the area of a second curve whose abscissa is z and ordinate the previous area divided by unity and augmented by any given quantity, and add this area to the terms retained. While if the fourth fluxion $\ddddot{y}$ is the final one of all, determine in addition the area of a third curve whose abscissa is z and ordinate the second area(77) together with any given quantity, and add this area to the terms retained. [And so on.]

Finally, set the terms retained together with those added equal to any given quantity whatever, and this equation will exhibit the relationship between the quantities y and z and their fluxions $\dot{y}$ and 1. Therefrom, of course, the relationship between the quantities y and z is to be investigated by the first four Cases.

Example 1. Let there be proposed the equation

$$12zy^2+24z^2y\dot{y}+6z^3\dot{y}^2+3z^3y\ddot{y}-2[4]zy^2\dot{y}-[6]y-12z^2y\dot{y}^2-4z^2y^2\ddot{y}=0,$$

in which the final fluxion of y is $\ddot{y}$. I therefore pick out one or other of the terms $3z^3y\ddot{y}$ or $-4z^2y^2\ddot{y}$, say $3z^3y\ddot{y}$, change $\ddot{y}$ into $\dot{y}$ and multiply the equation by $z^\alpha y^\beta$, and there comes to be $3z^{\alpha+3}y^{\beta+1}\dot{y}$. In consequence $c=3$, $\lambda=\alpha+3$, $\nu=\beta+1$ and so $P/z^\alpha y^\beta=(3+3\beta)\,z^3\dot{y}^2$. This term I compare with the term $6z^3\dot{y}^2$ in the equation which is of the same form, and there results $3+3\beta=6$ and thence

that the second fluxion is the final one of the fluxions, or whose second fluxion is that total if the third is the final one of the fluxions, or the third if the fourth is, and so on, and then adjoin this quantity to those stored up and set the sum equal to zero: this equation will exhibit the relationship between the quantities y and z and their fluxions $\dot{y}$ and 1. Thereby, to be sure, through the first three cases you must determine the relationship between the quantities y and z).

(77) Again understand 'ad unitatem applicata' (divided by unity) to preserve homogeneity in dimension.

(78) For the present equation to be (on multiplying through by zy) the second fluxion of $z^4y^3-z^3y^4=dz+e$, as Newton wishes it to be, these two terms should be '$-24zy y\dot{y}-6y$' and so we have rendered them in our English version, there also introducing the minor adjustments to Newton's argument necessitated in sequel.

$\frac{R}{z^{\alpha}y^{\beta}} = \overline{\lambda d - d} \times zy^2$. Et hos terminos conferendo cum æquationis terminis homogeneis $24z^2y\dot{y}$ et $12zyy$ prodeunt $3d+3\lambda = 24$ et $\lambda d - d = 12$. Unde

$$8-\lambda = d = \frac{12}{\lambda-1} \quad \text{et} \quad \lambda\lambda = 9\lambda - 20.$$

et $\lambda = 4$ vel 5, & $\alpha = \lambda - 3 = 1$ vel 2. Tentandus est numerus uterq*ꝫ*.

(79)Primo usurpo 1 pro α et per $z^{\alpha}y^{\beta}$ seu zy multiplico æquationem propositam et fit

$$12z^2y^3 + 24z^3yy\dot{y} + 6z^4y\dot{y}^2 + 3z^4y^2\ddot{y} - 20z^2y^3\dot{y} - 4zy^4 - 12z^3y^2\dot{y}^2 - 4z^3y^3\ddot{y} = 0.$$

Termini excerpendi sunt $3z^4y^2\ddot{y} - 4z^3y^3\ddot{y}$(80) qui mutando $\ddot{y}$ in $\dot{y}$ fiunt $3z^4y^2\dot{y} - 4z^3y^3\dot{y}$. Hos asservo et eorundem quæro fluxionem primam

$$12z^3y^2\dot{y} + 6z^4y\dot{y}^2 + 3z^4y^2\ddot{y} - 12z^2y^3\dot{y} - 12z^3y^2\dot{y}^2 - 4z^3y^3\ddot{y}.$$

Et quoniam ipsius y fluxio ultima $\ddot{y}$ hic reperitur non pergo ultra, sed aufero hanc inventam fluxionem primam ex æquatione et restat

$$12zzy^3 + 12z^3y^2\dot{y} - 8z^2y^3\dot{y} - 4zy^4.$$

Quoniam ipsius y fluxio penultima est $\dot{y}$, termini hinc excerpendi sunt $12z^3y^2\dot{y} - 8z^2y^3\dot{y}$ et hi sunt fluxiones quantitatum $4z^3y^3 - 2z^2y^4$ ideoqꝫ quantitates hasce adservatis adjicio et harum fluxionem $12z^2y^3 + 12z^3y^2\dot{y} - 4zy^4 - 8z^2y^3\dot{y}$ aufero ex æquationis parte relicta, et quoniam nihil amplius relinquitur pono terminos asservatos una cum adjectis æquales dato cuilibet d et sic habeo æquationem $3z^4y^2\dot{y} - 4z^3y^3\dot{y} + 4z^3y^3 - 2z^2y^4 = d$: quae scribendo subintellectam fluxionem $\dot{z}$ fit $3z^4y^2\dot{y} - 4z^3y^3\dot{y} + 4z^3\dot{z}y^3 - 2z^2\dot{z}y^4 = d\dot{z}$. Et hinc per casum tertium prodit æquatio quæsita $z^4y^3 - z^3y^4 = dz + e$, ubi d et e sunt quantitates pro lubitu assumptæ.

Ubi obiter notandum venit quod fluxiones incognitarum y et z per omnes æquationis cujusqꝫ terminos debent esse ejusdem ordinis, id est vel primi ordinis $\dot{y}$ et $\dot{z}$ vel secundi $\ddot{y}$, $\dot{y}^2$, $\dot{y}\dot{z}$, $\dot{z}^2$ vel tertij $\dddot{y}$, $\ddot{y}\dot{y}$, $\ddot{y}\dot{z}$, $\dot{y}^3$, $\dot{y}^2\dot{z}$, $\dot{y}\dot{z}^2$, $\dot{z}^3$ &c. Et ubi res aliter se habet[,] complendus est ordo altissimus per subintellectas fluxiones quantitatis uniformiter fluentis z.

Exempl. 2. Proponatur æquatio $6z + 3\ddot{y} + z\dddot{y} - 9\dot{y}\ddot{y} - 3y\dddot{y} = 0$ ubi ipsius y fluxio ultima est $\dddot{y}$. Hic excerpendus est alteruter terminorum $z\dddot{y}$ et $-3y\dddot{y}$. Excerpatur $-3y\dddot{y}$, mutetur $\dddot{y}$ in $\dot{y}$ et multiplicetur terminus per $z^{\alpha}y^{\beta}$ et fiet

$$-3z^{\alpha}y^{\beta+1}\dot{y} = cz^{\lambda}y^{\nu}\dot{y}.$$

(79) Newton's preliminary computations for the following—too straightforward to make their detailed reproduction here other than superfluous—are preserved on Add. 3960.7: 102 together with a minimally variant predraft of the remainder of this prime *exemplum*.

(80) The terms containing $\ddot{y}$ as a factor, namely.

$\beta = 1$ and so $\nu = 2$. Moreover, there is $Q/z^{\alpha}y^{\beta} = (3d+3\lambda)\, z^2y\dot{y}$ and

$$R/z^{\alpha}y^{\beta} = (\lambda-1)\, dzy^2.$$

And on comparing these terms with the ones in the equation, $24z^2y\dot{y}$ and $12zy^2$, of like form there ensues $3d+3\lambda = 24$ and $(\lambda-1)\, d = 12$; whence

$$8-\lambda = d = 12/(\lambda-1),$$

and so $\lambda^2 = 9\lambda-20$ and $\lambda = 4$ or 5 and so $\alpha = \lambda-3 = 1$ or 2. Both numbers are to be tried.

(79)First, I employ 1 in α's place and multiply the equation proposed by $z^{\alpha}y^{\beta}$, that is, zy, and there comes

$$12z^2y^3+24z^3y^2\dot{y}+6z^4y\dot{y}^2+3z^4y^2\ddot{y}-2[4]z^2y^3\dot{y}-[6]zy^4-12z^3y^2\dot{y}^2-4z^3y^3\ddot{y} = 0.$$

The terms to be picked out are $3z^4y^2\ddot{y}-4z^3y^3\ddot{y}$,(80) which on changing $\ddot{y}$ into $\dot{y}$ become $3z^4y^2\dot{y}-4z^3y^3\dot{y}$. These I retain and seek their first fluxion

$$12z^3y^2\dot{y}+6z^4y\dot{y}^2+3z^4y^2\ddot{y}-12z^2y^3\dot{y}-12z^3y^2\dot{y}^2-4z^3y^3\ddot{y}.$$

And, because the final fluxion $\ddot{y}$ of y is here found, I proceed no further but take this first fluxion so found away from the equation and there remains

$$12z^2y^3+12z^3y^2\dot{y}-[12]z^2y^3\dot{y}-[6]zy^4.$$

Because the last but one fluxion of y is $\dot{y}$ the terms to be picked out from this are $12z^3y^2\dot{y}-[12]z^2y^3\dot{y}$, and these are fluxions of the quantities $4z^3y^3-[3]z^2y^4$; consequently I add these quantities to those retained and take away their fluxion $12z^2y^3+12z^3y^2\dot{y}-[6]zy^4-[12]z^2y^3\dot{y}$ from the part of the equation left, and because nothing more remains I set the terms retained together with those added equal to any given d, and thus I have the equation

$$3z^4y^2\dot{y}-4z^3y^3\dot{y}+4z^3y^3-[3]z^2y^4 = d:$$

which, on writing in the fluxion $\dot{z}$ understood, becomes

$$3z^4y^2\dot{y}-4z^3y^3\dot{y}+4z^3\dot{z}y^3-[3]z^2\dot{z}y^4 = d\dot{z}.$$

And from this by Case 3 there results the required equation $z^4y^3-z^3y^4 = dz+e$, where d and e are quantities assumed at will.

Here, by the way, it occurs to me to note that the fluxions of the unknowns must be of the same order throughout the terms of each equation; that is, either of the first order $\dot{y}$ and $\dot{z}$, or of the second $\ddot{y}$, $\dot{y}^2$, $\dot{y}\dot{z}$, $\dot{z}^2$, or of the third $\dddot{y}$, $\ddot{y}\dot{y}$, $\ddot{y}\dot{z}$, $\dot{y}^3$, $\dot{y}^2\dot{z}$, $\dot{y}\dot{z}^2$, $\dot{z}^3$, and so on. And when the situation is otherwise, the highest order must be filled up by understood fluxions of the uniformly fluent quantity z.

Example 2. Let there be proposed the equation $6z+3\ddot{y}+z\dddot{y}-9\dot{y}\ddot{y}-3y\dddot{y} = 0$, in which the final fluxion of y is $\dddot{y}$. Here you must pick out one or other of the terms $z\dddot{y}$ and $-3y\dddot{y}$. Pick out $-3y\dddot{y}$, change $\dddot{y}$ into $\dot{y}$ and multiply the term by $z^{\alpha}y^{\beta}$, and there will come to be $-3z^{\alpha}y^{\beta+1}\dot{y} = cz^{\lambda}y^{\nu}\dot{y}$. In consequence $c = -3$, $\lambda = \alpha$,

Ergo $c = -3$, $\lambda = \alpha$, $\nu = \beta+1$ et $\dfrac{P}{z^\alpha y^\beta} = -9\nu\dot{y}\ddot{y}$, quæ quantitas cum æquationis termino homogeneo $-9\dot{y}\ddot{y}$ collata dat $\nu = 1$ et inde $\beta = 0$. Porro

$$\frac{Q}{z^\alpha y^\beta} = \overline{2d-6\alpha} \times z^{-1}y\ddot{y} \quad \text{et} \quad \frac{R}{z^\alpha y^\beta} = \overline{\alpha-1} \times \overline{4d-3\alpha} \times z^{-2}y\dot{y}:$$

quibus nulli respondent termini in æquatione. Est igitur $2d-6\alpha = 0$ et $\overline{\alpha-1} \times \overline{4d-3\alpha} = 0$. Per æquationem priorem fit $d = 3\alpha$ et per posteriorem $\overline{\alpha-1} \times 9\alpha = 0$, seu $\alpha\alpha-\alpha = 0$ et inde $\alpha = 0$ vel 1. Primo sit $\alpha = 0$ et erit $z^\alpha y^\beta = 1$ et æquatio per $z^\alpha y^\beta$ multiplicata manebit eadem quæ prius.

Termini excerpendi sunt $z\dddot{y} - 3y\ddot{y}$ qui mutando $\ddot{y}$ in $\dot{y}$ fiunt $z\ddot{y}-3y\dot{y}$. Hos asservo et eorundem quæro fluxionem primam $\dot{y}+z\ddot{y}-3\dot{y}\dot{y}-3y\ddot{y}$ et secundam $2\ddot{y}+z\dddot{y}-9\dot{y}\ddot{y}-3y\dddot{y}$[,] & quoniam ipsius y fluxio ultima $\dddot{y}$ in hac secunda reperitur non pergo ultra sed aufero hanc secundam ex æquatione et restat $6z+\ddot{y}$. Unicus est hic terminus excerpendus $\ddot{y}$ qui in $\dot{y}$ mutatus fit fluxio quantitatis y. Igitur terminis asservatis adjicio quantitatem y et ejus fluxionem penultimam $\ddot{y}$ de æquationis parte relicta aufero et restat $6z$.(81) Hæc est fluxio secunda quantitatis z^3+gz+h ideoq; hanc etiam quantitatem adjicio terminis asservatis et fit summa $z\dot{y}-3y\dot{y}+y+z^3+gz+h = 0$, id est(82)

$$z\dot{y}-3y\dot{y}+\dot{z}y+\dot{z}z^3+gz\dot{z}+h\dot{z}[=0].$$

Et hæc æquatio per casum tertium dat $zy-\frac{3}{2}yy+\frac{1}{4}z^4+\frac{1}{2}gz^2+hz+i\ [=0]$ quam invenire oportuit.

Exempl. 3. Sit $\overline{2-a\ddot{y}}\sqrt{ay-zz}+a\dot{y}\ddot{y}-2z\ddot{y}+2ay\ddot{y}-2zz\ddot{y} = 0$ vel quod perinde est $2-a\ddot{y}+\dfrac{a\dot{y}\ddot{y}-2z\ddot{y}+2ay\ddot{y}-2zz\ddot{y}}{\sqrt{ay-zz}} = 0$. Nam hujusmodi casus omnes tentare convenit.(83)

Cas. 6.(84) Si quantitas uniformiter fluens z in æquatione non reperiatur, pro quantitate altera y et ejus fluxionibus $\dot{y}$, $\ddot{y}$, $\dddot{y}$, $\ddddot{y}$, si y adsit, vel si etiam y desit, pro $\dot{y}$, $\ddot{y}$, $\dddot{y}$, $\ddddot{y}$, $\overset{.....}{y}$, vel si y et $\dot{y}$ desint, pro $\ddot{y}$, $\dddot{y}$, $\ddddot{y}$, $\overset{.....}{y}$, $\overset{......}{y}$ &c scribe respectivè

(81) Newton first continued: 'Assumantur quantitates quævis determinatæ d, e, f et area curvæ cujus abscissa est z & ordinata $6z$ est $3z^2+d$, et area curvæ alterius cujus abscissa est z et ordinata $3z^2+d$ est z^3+dz+e quæ terminis asservatis adjecta facit summam

$$z\dot{y}-3y\dot{y}+y+z^3+dz+e = 0\text{'}$$

(Assume any determinate quantities d, e, f and the area of the curve whose abscissa is z and ordinate $6z$ is then $3z^2+d$, and the area of a second curve whose abscissa is z and ordinate $3z^2+d$ is z^3+dz+e: add this to the terms retained and it makes the sum

$$z\dot{y}-3y\dot{y}+y+z^3+dz+e = 0).$$

(82) A superfluous repeated specification 'si ipsius z subintellecta fluxio $\dot{z}$ scri[batur]' (if the understood fluxion $\dot{z}$ of z be written) was deleted in sequel before it was completely penned.

$\nu = \beta+1$ and so $P/z^{\alpha}y^{\beta} = -9\nu\dot{y}\ddot{y}$; and this quantity, when compared with the term in the equation, $-9\dot{y}\ddot{y}$, of like form, yields $\nu = 1$ and thence $\beta = 0$. Furthermore, $Q/z^{\alpha}y^{\beta} = (2d-6\alpha)\, z^{-1}y\ddot{y}$ and $R/z^{\alpha}y^{\beta} = (\alpha-1)\,(4d-3\alpha)\, z^{-2}y\dot{y}$: to these no terms in the equation correspond and there is therefore $2d-6\alpha = 0$ and $(\alpha-1)\,(4d-3\alpha) = 0$. By the first equation there comes $d = 3\alpha$ and so by the latter one $(\alpha-1).9\alpha = 0$, that is $\alpha^2-\alpha = 0$ and thence $\alpha = 0$ or 1. First, let $\alpha = 0$ and then $z^{\alpha}y^{\beta} = 1$, and so the equation will, when multiplied by $z^{\alpha}y^{\beta}$, remain the same as before.

The terms to be picked out are $z\dddot{y}-3y\dddot{y}$, which on changing $\dddot{y}$ into $\dot{y}$ become $z\dot{y}-3y\dot{y}$. These I retain and seek out their first fluxion $\dot{y}+z\ddot{y}-3\dot{y}^2-3y\ddot{y}$ and second one $2\ddot{y}+z\dddot{y}-9\dot{y}\ddot{y}-3y\dddot{y}$, and because the final fluxion $\dddot{y}$ of y is found in this second one I go no further, but take this second one away from the equation and there remains $6z+\ddot{y}$. Here there is a single term $\ddot{y}$ to be picked out, and this when changed into $\dot{y}$ becomes the fluxion of the quantity y. Therefore to the terms retained I add the quantity y, and take away its last but one fluxion $\ddot{y}$ from the part of the equation left and there remains $6z$.(81) This is the second fluxion of the quantity z^3+gz+h, and consequently I add this quantity also to the terms retained and their sum becomes $z\dot{y}-3y\dot{y}+y+z^3+gz+h = 0$, that is,(82)

$$z\dot{y}-3y\dot{y}+\dot{z}y+z^3\dot{z}+gz\dot{z}+h\dot{z} = 0.$$

And by Case 3 this yields the equation $zy-\frac{3}{2}y^2+\frac{1}{4}z^4+\frac{1}{2}gz^2+hz+i = 0$ which it was required to find.

Example 3. Let $(2-a\ddot{y})\sqrt{[ay-z^2]}+a\dot{y}^2-2z\dot{y}+2ay\ddot{y}-2z^2\ddot{y} = 0$, or, what is exactly the same, $2-a\ddot{y}+(a\dot{y}^2-2z\dot{y}+2ay\ddot{y}-2z^2\ddot{y})/\sqrt{[ay-z^2]} = 0$. For it is appropriate to test all cases of this sort.(83)

Case 6.(84) If the uniformly fluent quantity z should not be found in the equation, in place of the other quantity y and its fluxions $\dot{y}$, $\ddot{y}$, $\dddot{y}$ [...] if y be present, or, if y also be absent, in place of $\dot{y}$, $\ddot{y}$, $\dddot{y}$, $\ddddot{y}$, $\overset{\cdot\cdot\cdot\cdot\cdot}{y}$ [...], or, if y and $\dot{y}$ be absent, in place of $\ddot{y}$, $\dddot{y}$, $\ddddot{y}$, $\overset{\cdot\cdot\cdot\cdot\cdot}{y}$, $\overset{\cdot\cdot\cdot\cdot\cdot\cdot}{y}$ ..., write respectively x, $v^{\frac{1}{2}}$, $\frac{1}{2}\dot{v}$, $\frac{1}{2}\ddot{v}v^{\frac{1}{2}}$, $\frac{1}{2}\dddot{v}v+\frac{1}{4}\ddot{v}\dot{v}$, ... Then,

(83) Newton abandons this surd example, leaving the rest of his manuscript page blank—doubtless for a future application of the preceding algorithm to some suitably modified equation. The source of his dissatisfaction will be clear. While the substitution $v = \sqrt{[ay-z^2]}$ reduces the given equation to the surd-free form $2-a\ddot{y}+2\dot{v}\dot{y}+2v\ddot{y} = 0$ whose fluent proves, by direct 'integration', to be $2z-a\dot{y}+2v\dot{y}+g = 0$, that is, $2v(-\dot{v}+\dot{y})+g = 0$, this can be integrated a second time by Newton's present techniques only if $g = 0$, whence (since by implication $v \neq 0$) there is $-v+y+h = 0$. Newton, we may all but be sure, contrived his *exemplum* by twice 'differentiating' $y = \sqrt{[ay-z^2]}$, and so would have remained unaware till he here began to consider the reverse double integration that his initial fluent is merely one particular solution of the ensuing fluxional equation.

(84) This subsumes two separate 'Cas. 6' and 'Cas. 7' in the preliminary version (Add. 3960.9: 157) reproduced in Appendix 1.9 below.

x, $v^{\frac{1}{2}}$, $\frac{1}{2}\dot{v}$, $\frac{1}{2}\ddot{v}v^{\frac{1}{2}}$, $\frac{1}{2}\dddot{v}v+\frac{1}{4}\ddot{v}\dot{v}$. Dein [fingendo quod x uniformiter fluit](85) si in æquatione prodeunte desint fluxiones $\dot{v}$, $\ddot{v}$ et $\dddot{v}$, quære areas curvarum duarum quæ habent ordinatas $\frac{1}{v^{\frac{1}{2}}}$ et $\frac{x}{v^{\frac{1}{2}}}$ et abscissam communem x, et pone aream prioris æqualem quantitati uniformiter fluenti z, et aream posterioris æqualem quantitati fluenti cujus fluxio est x, id est quantitati y si x sit $\dot{y}$ vel quantitati $\dot{y}$ si x sit $\ddot{y}$ &c.(86) At si $x = y$, aream posteriorem neglige.

Quod si in æquatione prodeunte fluxiones omnes $\dot{v}$, $\ddot{v}$, $\dddot{v}$ non desint quære per casus quinqꝫ priores æquationem novam in qua solæ habentur x et v absqꝫ $\dot{v}$, $\ddot{v}$ *et* $\dddot{v}$, dein perge [ut] supra.

Exempl. 1. Sit $4\ddot{y}^3 = \dddot{y}^2$ et hæc æquatio scribendo x pro $\ddot{y}$ et $v^{\frac{1}{2}}$ pro $\dddot{y}$ fit $4x^3 = v$. Unde $\frac{1}{v^{\frac{1}{2}}} = \frac{1}{2}x^{-\frac{3}{2}}$ et $\frac{x}{v^{\frac{1}{2}}} = \frac{1}{2}x^{-\frac{1}{2}}$, et areæ curvarum quarum hæ sunt ordinatæ sunt $-x^{-\frac{1}{2}}$ et $x^{\frac{1}{2}}$. Ergo $z = -x^{-\frac{1}{2}}$ et $\dot{y} = x^{\frac{1}{2}}$ et inde exterminando x fit $\dot{y} = -\frac{1}{z}$ seu $= -\frac{\dot{z}}{z}$ et inde per Cas. 1. $y = \boxed{-\frac{1}{z}}$ id est $y =$ areæ Hyperbolæ cujus abscissa e[s]t z et ordinata $-\frac{1}{z}$. Area hæc tota infinita est, sed ejus partes ordinatis parallelis interjectæ(88) sunt finitæ, et partibus analogis ipsius y æquantur.

(89)C[onvenit alipuando sic operari. Inventa æquatione $\dot{y} = x^{\frac{1}{2}}$ pro x restitue $\ddot{y}$ et fiet $\dot{y}^2 = \ddot{y}$. Hic jam pro $\dot{y}$ et $\ddot{y}$ scribe x et $v^{\frac{1}{2}}$ juxta regulam et fiet $x^2 = v^{\frac{1}{2}}$, et area curvæ cujus ordinata est $\frac{x}{v^{\frac{1}{2}}}$ seu $\frac{1}{x}$ erit quantitas cujus fluxio est, hoc est quantitas y. Erat $z = -x^{-\frac{1}{2}}$ et inde est $x = \frac{1}{z^2}$.](90)

Exempl. 2. Sit $4\dot{y}^2\ddot{y} = 2y+a$ et hæc æquatio scribendo x, $v^{\frac{1}{2}}$ et $\frac{1}{2}\dot{v}$ pro y, $\dot{y}$ et $\ddot{y}$ fit $2v\dot{v} = 2x+a$ id est $= 2x\dot{x}+a\dot{x}$ et inde per Cas. 2 prodit $vv = xx+ax+$data

(85) We interpolate this necessary phrase to make clear the subtlety that in effecting the previous substitution Newton silently alters the base variable, so that, while (as usual) there was $\dot{y} = dy/dz$, $\ddot{y} = d(\dot{y})/dz$, $\dddot{y} = d(\ddot{y})/dz$, ... in explicit Leibnizian equivalent, there is now $\dot{v} = dv/dx$, $\ddot{v} = d(\dot{v})/dx$, $\dddot{v}=d(\ddot{v})/dx$, ... It follows that, when y is not absent, the substitutions $y \rightarrow x$ and $\dot{y} \rightarrow v^{\frac{1}{2}}$ determine in sequel $\ddot{y} \rightarrow d(v^{\frac{1}{2}})/dz = v^{\frac{1}{2}}.d(v^{\frac{1}{2}})/dx = \frac{1}{2}\dot{v}$, $\dddot{y} \rightarrow v^{\frac{1}{2}}.d(\frac{1}{2}\dot{v})/dx = \frac{1}{2}v^{\frac{1}{2}}\ddot{v}$, $\ddddot{y} \rightarrow v^{\frac{1}{2}}.d(\frac{1}{2}v^{\frac{1}{2}}\ddot{v})/dx = \frac{1}{2}(\frac{1}{2}\dot{v}\ddot{v}+v\dddot{v})$, ...; and the like holds *mutatis mutandis* when y is absent from the equation and successively $\dot{y}, \ddot{y}, \dddot{y}$, ... is the lowest fluxion of it (with respect to z) present in it. Preliminary computations on Add. 3960.8: 149/153/159 where Newton first introduced this scheme of fluxional transformation are reproduced in Appendix 1.10 below. It there appears that he first set $\dot{y} = \dot{v}/x$ (or $v = \int x.dy$) in an area-preserving transmutation in which, much as in his preceding 'Theor. III' (see note (18) above), he posited that there is also $\dot{y} = v^n$, and then chose $n = \frac{1}{2}$ to yield $\ddot{y} = d(\dot{y})/dz = nv^{n-1}.dv/dz = nv^{2n-1}x$ in the simpler form $\ddot{y} = \frac{1}{2}x = \frac{1}{2}v^{-\frac{1}{2}}\dot{v}$; only subsequently did he come to appreciate the further 'simplification' which ensues on retaining $\dot{y} = v^{\frac{1}{2}}$ but switching the base variable from z to $y = x$.

(86) Whence the required relationship between $z \equiv z_x$ and $y \equiv y_x$ is obtained by eliminat-

supposing that x flows uniformly,(85) if in the ensuing equation the fluxions $\dot{v}$, $\ddot{v}$, $\dot{\ddot{v}}$ [...] are lacking, seek out the areas of two curves which have ordinates $1/v^{\frac{1}{2}}$ and $x/v^{\frac{1}{2}}$ and the common abscissa x, and set the area of the former equal to the uniformly fluent z and the area of the latter equal to the fluent quantity whose fluxion is x, that is, the quantity y if x be $\dot{y}$, or the quantity $\dot{y}$ if x be $\ddot{y}$, and so on.(86) While if $x = y$, neglect the latter area.

But if in the ensuing equation all the fluxions $\dot{v}$, $\ddot{v}$, $\dot{\ddot{v}}$ [...] are not lacking, determine by the previous five Cases a fresh equation in which x and v alone are had without $\dot{v}$, $\ddot{v}$, $\dot{\ddot{v}}$ [...], then proceed as above.

Example 1. Let $4\ddot{y}^3 = \dot{\ddot{y}}^2$ and then this equation comes, on writing x in place of $\ddot{y}$ and $v^{\frac{1}{2}}$ in place of $\dot{\ddot{y}}$, to be $4x^3 = v$; whence $1/v^{\frac{1}{2}} = \frac{1}{2}x^{-\frac{3}{2}}$ and $x/v^{\frac{1}{2}} = \frac{1}{2}x^{-\frac{1}{2}}$, and the areas of the curves of which these are ordinates are $-x^{-\frac{1}{2}}$ and $x^{\frac{1}{2}}$. In consequence, $z = -x^{-\frac{1}{2}}$ and $\dot{y} = x^{\frac{1}{2}}$, and by eliminating x therefrom there comes $\dot{y} = -1/z$, that is, $\dot{y} = -\dot{z}/z$ and thence, by Case 1, $y = [\int -1/z.dz]$;(87) that is, y is equal to the area of a hyperbola whose abscissa is z and ordinate $-1/z$. This total area is infinite, but its portions intercepted by parallel ordinates(88) are finite and equal to the analogous portions of y.

(89)[It is sometimes convenient to work in this way. Once the equation $\dot{y} = x^{\frac{1}{2}}$ has been found, restore $\ddot{y}$ in place of x and there will come to be $\dot{y}^2 = \ddot{y}$. Here now in place of $\dot{y}$ and $\ddot{y}$ write x and $v^{\frac{1}{2}}$ according to the rule and there will come $x^2 = v^{\frac{1}{2}}$, and the area of the curve whose ordinate is $x/v^{\frac{1}{2}}$, that is, $1/x$, will be the quantity whose fluxion it is, namely, the quantity y. There was $z = -x^{-\frac{1}{2}}$ and from this $x = 1/z^2$.](90)

Example 2. Let $4\dot{y}^2\ddot{y} = 2y + a$ and this equation comes, on writing x, $v^{\frac{1}{2}}$ and $\frac{1}{2}\dot{v}$ in place of y, $\dot{y}$ and $\ddot{y}$, to be $2v\dot{v} = 2x + a$, that is, $2v\dot{v} = 2x\dot{x} + a\dot{x}$, and therefrom by Case 2 there ensues $v^2 = x^2 + ax + c$, c a given quantity. Let that given $c = \frac{1}{4}a^2$,

---

ing the parameter. Evidently, as Newton at once proceeds to observe, if the substitution $y = x$ was earlier made, then immediately $z_y = 0$ is the required fluent equation.

(87) We again (compare note (57) preceding) employ modern Leibnizian notation to render Newton's enveloping 'square' symbol in our English version.

(88) Namely, $\int_b^a -1/z.dz = \log(b/a)$ where the parallel ordinates are of length $1/a$ and $1/b$.

(89) In the manuscript Newton has entered only the opening letter of the following paragraph, and we take the rest from his preliminary draft on Add. 3960.9: 158. Whether he himself suddenly decided against doing so, or whether he was merely momentarily disturbed and omitted to go on to include it through sheer oversight, is not clear. The reader will make up his own mind. The convenience gained by this slightly variant approach would seem always to be minimal, if it is in any way real.

(90) Whence, on substituting back (and, with Newton, ignoring possible constants of integration), there is $\ddot{y} = 1/z^2$ and so $\dot{y} = -1/z$ as before. From the strictly practical viewpoint, we may add, it seems simplest of all to reduce the given fluxional equation to be $\frac{1}{2}\ddot{y}^{-\frac{3}{2}}\dot{\ddot{y}} = 1$, whence a first integration yields $-\ddot{y}^{-\frac{1}{2}} = z + k$, that is, $\ddot{y} = (z+k)^{-2}$ and thence the general solution is $y = -\log(z+k) + lz + m$.

quavis c. Sit data illa $c = \frac{1}{4}aa$, et radice extracta fiet $v = x+\frac{1}{2}a$. Hæc est relatio inter x et v. Quæratur area curvæ cujus abscissa est x et ordinata $\frac{1}{v^{\frac{1}{2}}}$ id est $\frac{1}{\sqrt{x+\frac{1}{2}a}}$, & hæc area ad unitatem applicata, nempe $2\sqrt{x+\frac{1}{2}a}$, æqualis erit quantitati uniformiter fluenti z. Pro x scribatur ejus valor y, et fiet

$$z = [2]\sqrt{y+\tfrac{1}{2}a},$$

relatio inter z et y quam invenire oportuit.(91)

Exempl. 3. Sit $3\dot{y}^2\ddot{y}+2c\dddot{y} = 0$ et scribendo x, $v^{\frac{1}{2}}$ et $\frac{1}{2}\dot{v}$ pro $\dot{y}$, $\ddot{y}$ et $\dddot{y}$ fiet $3x^2v^{\frac{1}{2}}+c\dot{v} = 0$. Unde $3x^2\dot{x} = -c\dot{v}v^{-\frac{1}{2}}$ et per Cas. 2, $x^3 = -2cv^{\frac{1}{2}}$. Ergo $\frac{1}{v^{\frac{1}{2}}} = -2cx^{-3}$, et $\frac{x}{v^{\frac{1}{2}}} = -2cx^{-2}$, et areæ curvarum his ordinatis et abscissa x descriptarum sunt $cx^{-2} = z$ et $2cx^{-1} = y$, & hinc exterminando x fit $yy = 4cz$.(92)

Cas. 7. Si relatio inter y et z per regulas horum casuum inveniri nequit, eruenda est quantitas y ex æquatione in serie interminata convergente per Propositionem sequentem et(93) tentanda est hujus reductio in æquationem finitam.

Prop. [*XII*]

Ex æquatione quantitates duas fluentes vel solas vel una cum earum fluxionibus involvente quantitatem alterutram in serie interminata convergente extrahere.(94)

Sit y quantitas fluens extrahenda. Fluat altera z(95) uniformiter et exponatur ejus fluxio per unitatem. Sic enim æquatio involvet solas ipsius y fluxiones $\dot{y}$, $\ddot{y}$, $\dddot{y}$, $\ddddot{y}$, &c: per quas tamen convenit subintelligere fluxionum rationes $\frac{\dot{y}}{\dot{z}}$, $\frac{\ddot{y}}{\dot{z}^2}$, $\frac{\dddot{y}}{\dot{z}^3}$, $\frac{\ddddot{y}}{\dot{z}^4}$ [&c] denominatoribus $\dot{z}$, $\dot{z}^2$, $\dot{z}^3$, $\dot{z}^4$ [&c] qui jam sunt unitates non comparentibus.

Cas. 1. Fac ut æquationis propositæ termini omnes conjunctim æquentur nihilo, utq; indices dignitatum ipsius y et ejus fluxionum passim affirmativi sint, et sit kz^λ terminus infimus (sive terminus infimæ dignitatis) eorum qui neq; per y neq; per aliquam ejus fluxionem multiplicantur. Sit $l[z]^\mu y^\alpha\dot{y}^\beta\ddot{y}^\gamma\dddot{y}^\delta\ddddot{y}^\epsilon$

(91) More generally, an argument *ab initio* would integrate the given equation, converted to the equivalent form $4\dot{y}^3\ddot{y} = 2y\dot{y}+a\dot{y}$, once to yield the first fluent $\dot{y}^4 = y^2+ay+c$ or $(y^2+ay+c)^{-\frac{1}{4}}\dot{y} = 1$, and again to find $z+d = \int(y^2+ay+c)^{-\frac{1}{4}}.dy$. Newton's particular solution not only posits that $c = \frac{1}{4}a^2$ (and so $(y^2+ay+c)^{-\frac{1}{4}} = (y+\frac{1}{2}a)^{-\frac{1}{2}}$) but also silently sets $d = 0$.

(92) Again possible constants of integration are ignored. The general first fluent of the given fluxional equation is $\dot{y}^3+2c\ddot{y} = k$, which has no simple solution when $k \neq 0$; while integration of $\dot{y}^3+2c\ddot{y} = 0$, that is, $1/c = -2\dot{y}^{-3}\ddot{y}$, yields most generally $(1/c)\,z+l = \dot{y}^{-2}$, and thence $y+m = \int(z/c+l)^{-\frac{1}{2}}.dz$.

(93) Newton first concluded with the more detailed instructions that 'progressio seriei investiganda, et simul habebuntur ipsius fluxiones $\dot{y}$, $\ddot{y}$, $\dddot{y}$ &c in seriebus consimilibus. Investi-

and when the root is extracted there comes $v = x+\frac{1}{2}a$. This is the relationship between x and v. Determine the area of the curve whose abscissa is x and ordinate $1/v^{\frac{1}{2}}$, that is, $1/\sqrt{[x+\frac{1}{2}a]}$, and this area, namely, $2\sqrt{[x+\frac{1}{2}a]}$, divided by unity will be equal to the uniformly flowing quantity z. In place of x write its value y and there will come $z = 2\sqrt{[y+\frac{1}{2}a]}$, the relationship between z and y which it was required to find.(91)

Example 3. Let $3\dot{y}^2\ddot{y}+2c\dddot{y} = 0$ and on writing x, $v^{\frac{1}{2}}$ and $\frac{1}{2}\dot{v}$ in place of $\dot{y}$, $\ddot{y}$ and $\dddot{y}$ there will come to be $3x^2v^{\frac{1}{2}}+c\dot{v} = 0$; whence $3x^2\dot{x} = -c\dot{v}v^{-\frac{1}{2}}$ and so, by Case 2, $x^3 = -2cv^{\frac{1}{2}}$. In consequence, $1/v^{\frac{1}{2}} = -2cx^{-3}$ and $x/v^{\frac{1}{2}} = -2cx^{-2}$, and the areas of the curves described with these ordinates and abscissa x are $cx^{-2} = z$ and $2cx^{-1} = y$, and by eliminating x from these there comes $y^2 = 4cz$.(92)

Case 7. If the relationship between y and z cannot be ascertained by these rules, you will need to derive the quantity y from the equation in an unterminated converging series by means of the following proposition, and then(93) to try and reduce it to a finite equation.

Proposition XII.

Out of an equation involving two fluent quantities, either alone or together with their fluxions, to extract one or other quantity in an unterminated converging series.(94)

Let y be the fluent quantity to be extracted. Let the other one, z,(95) flow uniformly and its fluxion be expressed by unity; for in this way the equation shall involve the fluxions $\dot{y}$, $\ddot{y}$, $\dddot{y}$, $\ddddot{y}$, ... of y alone—by which, however, it is convenient to understand the fluxional ratios $\dot{y}/\dot{z}$, $\ddot{y}/\dot{z}^2$, $\dddot{y}/\dot{z}^3$, $\ddddot{y}/\dot{z}^4$, ..., the denominators $\dot{z}$, $\dot{z}^2$, $\dot{z}^3$, $\dot{z}^4$, ... (which are now units) having no effect.

Case 1. Arrange that all terms in the equation proposed be jointly equal to nothing and that the indices of the powers of y and its fluxions be everywhere positive, and let kz^λ be the lowest term (that is, the term of lowest power) of those which are multiplied neither by y nor by any fluxion of it. Let $lz^\mu y^\alpha \dot{y}^\beta \ddot{y}^\gamma \dddot{y}^\delta \ddddot{y}^\epsilon$ be any other term in the remaining ones which are multiplied by y and its

ganda sunt etiam serierum progressiones. Ex una quavis cognita innotescunt cæteræ. Deinde beneficio progressionis tentanda est reductio seriei alicujus in æquationem finitam' (to discover the progression of the series, and there will then simultaneously be had its fluxions $\dot{y}$, $\ddot{y}$, $\dddot{y}$, ... in closely similar series. The progressions of the series also are to be discovered: once one of them is known, the rest are ascertained from it. Thereafter, with the aid of the progression you are to attempt the reduction of some series to a finite equation).

(94) Newton's rough preliminary draft of this—under the shorter head 'Ex æquationibus quantitates ignotas in seriebus interminatis extrahere' (Out of equations to extract the unknown quantities in unterminated series), but otherwise only insignificantly different from its re-written version—is preserved on Add. 3960.8: 142–4. Only a couple of its minor further variants are pin-pointed in following footnotes.

(95) The rough draft (see the previous note) here adds the minimal clarification 'quā æquatio involvit' (which the equation involves).

terminus alius quilibet reliquorum qui per y & ejus fluxiones(96) multiplicantur. Et percurrendo omnes æquationis terminos collige ex unoquoqꝫ numerum $\frac{\lambda-\mu+\beta+2\gamma+3\delta+4\epsilon}{\alpha+\beta+\gamma+\delta+\epsilon}$, sitqꝫ ν maximus(97) numerorum qui sic colliguntur et erit z^ν dignitas primi termini seriei. Sit ejus coefficiens a, et in æquatione resolvenda scribendo az^ν pro y, $\nu az^{\nu-1}$ pro $\dot{y}$, $\overline{\nu-1}\times\nu az^{\nu-2}$ pro $\ddot{y}$, $\overline{\nu-2}\times\overline{\nu-1}\times\nu az^{\nu-3}$ pro $\dddot{y}$, $\overline{\nu-3}\times\overline{\nu-2}\times\overline{\nu-1}\times\nu az^{\nu-4}$ pro $\ddddot{y}$ &c pone terminos omnes æquales nihilo qui prodeunt ejusdem dignitatis ipsius z cum termino kz^λ, et inde collige coefficientem a. Sic habebis primum seriei terminum az^ν. Pro reliquis omnibus nondum inventis seriei terminis scribe p et erit $az^\nu+p=y$, $\nu az^{\nu-1}+\dot{p}=\dot{y}$, $\overline{\nu-1}\times\nu az^{\nu-2}+\ddot{p}=\ddot{y}$, $\overline{\nu-2}\times\overline{\nu-1}\times\nu az^{\nu-3}+\dddot{p}=\dddot{y}$,

$$\overline{\nu-3}\times\overline{\nu-2}\times\overline{\nu-1}\times\nu az^{\nu-4}+\ddddot{p}=\ddddot{y}, \text{ \&c.}$$

In æquatione resolvenda pro y, $\dot{y}$, $\ddot{y}$, $\dddot{y}$, $\ddddot{y}$, &c scribe hos eorum valores et prodibit æquatio nova resolvenda ex qua primum terminum seriei p eadem ratione extrahes qua primum seriei totius y jam modo ex æquatione prima resolvenda extraxisti. Deinde ex æquatione tertia resolvenda quæ similiter prodibit similiter extrahes tertium seriei totius terminum, ex quarta quartum & si[c] in infinitum. Series autem sic inventa vel sola vel una cum data quavis quantitate erit quantitas y. Q.E.I.(98)

Cas. 2. Operatio jam descripta adhiberi debet ubi quantitas z perparva est. Si permagna est debebit altissimus terminus eorum qui non multiplicantur per y, $\dot{y}$, $\ddot{y}$, $\dddot{y}$, $\ddddot{y}$, &c adhiberi pro kz^λ et minimus(99) numerorum

$$\frac{\lambda-\mu+\beta+2\gamma+3\delta+4\epsilon}{\alpha+\beta+\gamma+\delta+\epsilon}$$

adhiberi pro indice ν, cæteraqꝫ peragi ut prius.

---

(96) This replaces the too restrictive initial phrase 'vel per aliquam ejus fluxionem' (or by some fluxion of it).

(97) Newton originally here repeated the mistaken opposite 'minimus' (least) from his preliminary draft on Add. 3960.8: 142. Evidently, if $y = az^\nu+O(z^{\nu+g})$, $g>0$, then the general term $lz^\mu y^\alpha \dot{y}^\beta \ddot{y}^\gamma \dddot{y}^\delta \ldots$ is of dimension $z^p+O(z^{p+h})$, $h>0$, where

$$p = \mu+\alpha\nu+\beta(\nu-1)+\gamma(\nu-2)+\delta(\nu-3)\ \ldots;$$

at once, on setting $A=\lambda-\mu+\beta+2\gamma+3\delta\ \ldots$ and $B=\alpha+\beta+\gamma+\delta\ \ldots$ where λ is the lowest dimension of the y-free terms, there is $p=\lambda-A+B\nu\geqslant\lambda$ and so $\nu\geqslant A/B$, whence ν is the greatest of the fractions A/B so constructible for each term in the equation involving y and its various fluxions.

(98) Newton codifies the 'methodus generalior reducendi Problemata ad æquationes infinitas' which he in 1684 had exemplified in two first-order instances in Caput 5 of his 'Matheseos Universalis Specimina' (see IV: 576–88), here extending it to embrace algebraic fluxional equations of arbitrary order. In the late summer of 1692, some eight months after

fluxions.(96) Then, on running through all the terms in the equation, gather from each one the number $\frac{\lambda-\mu+\beta+2\gamma+3\delta+4\epsilon}{\alpha+\beta+\gamma+\delta+\epsilon}$ and let ν be the greatest(97) of the numbers collected in this way: z^ν will then be the power of the first term in the series. Let its coefficient be a, and on writing in the equation to be resolved az^ν in place of y, $\nu az^{\nu-1}$ in place of $\dot{y}$, $(\nu-1)\,\nu az^{\nu-2}$ in place of $\ddot{y}$, $(\nu-2)(\nu-1)\nu az^{\nu-3}$ in place of $\dddot{y}$, $(\nu-3)(\nu-2)(\nu-1)\,\nu az^{\nu-4}$ in place of $\ddddot{y}$, and so on, set all ensuing terms of the same power of z as the term kz^λ equal to nothing and derive therefrom the coefficient a. In this way you will have the first term az^ν of the series. For all the remaining terms not yet found in the series write p, and there will then be $az^\nu+p = y$, $\nu az^{\nu-1}+\dot{p} = \dot{y}$, $(\nu-1)\,\nu az^{\nu-2}+\ddot{p} = \ddot{y}$,

$$(\nu-2)(\nu-1)\,\nu az^{\nu-3}+\dddot{p} = \dddot{y}, \quad (\nu-3)(\nu-2)(\nu-1)\,\nu az^{\nu-4}+\ddddot{p} = \ddddot{y},$$

and so on. In the equation to be resolved in place of y, $\dot{y}$, $\ddot{y}$, $\dddot{y}$, $\ddddot{y}$, ... write these their values and there will ensue a new equation to be resolved, out of which you shall extract the first term of the series p by the same method as you just now from the first equation to be resolved extracted the first one of the whole series y. Then out of the third equation to be resolved which shall similarly ensue you will similarly extract the third term in the whole series, out of the fourth the fourth one, and so on indefinitely. The series, however, found in this way, either by itself or together with any given quantity, will be the quantity y. As was to be found.(98)

Case 2. The procedure just now described should be applied when z is very small. If it is very large, the highest term of those which are not multiplied by y, $\dot{y}$, $\ddot{y}$, $\dddot{y}$, $\ddddot{y}$, ... must be employed for kz^λ and the least(99) of the numbers $\frac{\lambda-\mu+\beta+2\gamma+3\delta+4\epsilon}{\alpha+\beta+\gamma+\delta+\epsilon}$ used for the index ν, and then the rest accomplished as before.

---

contriving the present 'resolutio generalis', he passed on to John Wallis a light restyling of its particular case $\gamma = \delta = \epsilon = \ldots = 0$ which yields the series-expansion of a first-order fluxional equation, elaborately applying this to extract, by four iterations of the basic *operatio*, the simplest 'root' $y = d-\frac{1}{2}z-\frac{3}{8}z^2/d-\frac{9}{16}z^3/d^2 \ldots$ of the given *æquatio resolvenda*

$$y^2-z^2\dot{y}-d^2+dz = 0;$$

see Appendix 3 below, where we reproduce the text of Newton's solution of this instance of the 'Prob[lema]. Ex æquatione fluxionem radicis involvente radicem extrahere' as Wallis published it the next year in his Latin *Algebra* (*Opera Mathematica*, **2**, Oxford, 1693: 393–5).

(99) Much as in note (97), if $y = az^\nu+O(z^{\nu-g})$, then the general term $lz^\mu y^\alpha \dot{y}^\beta \ddot{y}^\gamma \dddot{y}^\delta \ldots$ is of dimension $lz^p+O(z^{p-h})$ where $p = \lambda-A+B\nu \leqslant \lambda$; at once

$$\nu \leqslant A/B = (\lambda-\mu+\beta+2\gamma+3\delta\ldots)/(\alpha+\beta+\gamma+\delta\ldots),$$

so that ν is the least of the fractions A/B thus constructed.

Cas. 3. At si z debet esse magnitudin[i]s alicujus mediocris, assume datam quamvis quantitatem w magnitudini illi quamproxime æqualem et ponendo $z = w+x$ scribe passim in æquatione $w+x$ pro z et ex æquatione prodeunte (usurpando x pro z) extrahe quantitatem y ut in casu primo.

Et nota quod si quantitas y sit æquationis resolvendæ latus impossibile series prodibit infinite magna: et contra series omnis infinite magna denotat impossibilitatem lateris finiti.(100)

Nota etiam quod hæc operatio includit operationes omnes Analyticas qua dividimus, radices extrahimus et æquationes affectas resolvimus adeoq; ad reductionem quantitatum omnium in series infinitas sufficit. Nam si fluxiones $\dot{y}$, $\ddot{y}$, $\dddot{y}$, $\ddddot{y}$ &c desunt, redit operatio ad extractionem radicum vel affectarum si æquationes sunt affectæ vel simplicium si non sunt affectæ vel deniq; ad divisionem si latus æquationis ad unicam tantum dimensionem ascendit.(101)

Corol. 1. Hinc curvæ omnes per series interminatas convergentes quadrari possunt. Nam si Curvæ abscissa sit z et ejus ordinata reducatur in ejusmodi seriem, dein series tota multiplicetur per z et ejus termini subinde dividantur per indices dignitatum z:(102) series quæ prodit erit area Curvæ, si modo area illa finitæ sit magnitudinis.

Corol. 2. Ubi quantitas(103) y ex æquatione resolvenda extrahitur, ejus fluxiones $\dot{y}$, $\ddot{y}$, $\dddot{y}$, &c simul prodeunt. Nam si termini seriei multiplicentur per indices dignitatum z ac deinde dividantur per z,(104) series quæ prodit æqualis erit $\frac{\dot{y}}{\dot{z}}$ et si termini hujus seriei similiter multiplicentur per indices dignitatum z & subinde dividantur per z(104) series quæ jam prodit æqualis erit $\frac{\ddot{y}}{\dot{z}^2}$ & operatione repetita prodibit series nova æqualis $\frac{\dddot{y}}{\dot{z}^3}$ & sic deinceps in infinitum. Sed hic suppono quod z fluit uniformiter quodq; ejus fluxio $\dot{z}$ & fluxionis dignitates $\dot{z}^2$, $\dot{z}^3$ &c [pro unitatibus] habentur.(105)

Corol. 3. Hinc verò si series(106) prodit hujus formæ

$$y = az+bz^2+cz^3+dz^4+ez^5+\&c$$

(ubi terminorum a, b, c, d &c aliqui vel deesse vel negativi esse possunt,)

(100) One more utterance of a curious notion, as thoroughly mistaken as it appeared to him valid, which Newton had most recently paraded in his 'Matheseos Universalis Specimina'; see IV: 541, note (49) and compare III: 69, note (73).

(101) Newton refers to the general technique for reducing the roots of a given 'affected' algebraic equation to an equivalent infinite series which he had expounded at fulsome length in his 1671 tract (see III: 42–68). To be sure, when $\alpha = \beta = \gamma = \delta = \ldots = 0$ his present fluxional algorithm reduces to choosing the dimension ν of the opening term az^ν in the posited ascending (descending) power-series expansion of y to be the greatest (least) of the fractions $(\lambda-\mu)/\alpha$

Case 3. But if z must be of some middling size, assume any given quantity w approximately equal to that magnitude and, on setting $z = w+x$, write $w+x$ everywhere in z's place in the equation and out of the ensuing equation (employing x in place of z) extract the quantity y as in Case 1.

And note that if y be an impossible 'side', the series will prove to be infinitely great; conversely, every infinitely great series represents an impossibility in a finite root.[(100)]

Note, too, that this procedure includes all analytical operations by which we divide, extract roots and resolve affected equations, and hence suffices to reduce all quantities to infinite series. For, if the fluxions $\dot{y}$, $\ddot{y}$, $\dddot{y}$, $\ddddot{y}$, ... are absent, the procedure reverts to the extraction of roots, either affected if the equations are affected or simple if they are not, or finally to division if the equation's 'side' (variable) rises to but a single dimension.[(101)]

Corollary 1. Hence all curves can be squared by means of unterminated converging series. For, if the curve's abscissa be z and its ordinate be reduced to a series of this sort, and then the whole series be multiplied by z and its terms thereafter divided by the indices of their powers,[(102)] the series which results will be the area of the curve, provided that that area be of finite size.

Corollary 2. When the[(103)] quantity y is extracted out of the equation to be resolved, its fluxions $\dot{y}$, $\ddot{y}$, $\dddot{y}$, ... result at the same time. For, if the terms of the series be multiplied by the indices of their powers of z and then divided by z,[(104)] the series which results will be equal to $\dot{y}/\dot{z}$, and if the terms of this series be similarly multiplied by the indices of their powers of z and thereafter divided by z,[(104)] the series which now results will be equal to $\ddot{y}/\dot{z}^2$, and by repeating the procedure there will result a new series equal to $\dddot{y}/\dot{z}^3$, and so on successively to infinity. But I here suppose that z flows uniformly, and that its fluxion $\dot{z}$ and the powers $\dot{z}^2$, $\dot{z}^3$, ... of that fluxion are had as units.[(105)]

Corollary 3. Hence, indeed, if the series[(106)] proves to be of this form

$$y = az+bz^2+cz^3+dz^4+ez^5+\dots$$

(where any of the terms $a[z]$, $b[z^2]$, $c[z^3]$, $d[z^4]$, ... can either be lacking or be

constructed for each term $lz^\mu y^\alpha$ in the given equation whose y-free terms are of lowest (highest) dimension λ in z; this exactly mirrors Newton's use twenty years before (III: 50–2) of a squared-off rectangular array in which the general term $lz^\mu y^\alpha$ is set in a cell (μ, α) μ up and α to the right from the left bottom-most, and a straight-edge is applied at a corner of the cell $(\lambda, 0)$ containing the y-free term of lowest (highest) dimension λ in z, so as to touch the corresponding corner of the occupied cell (μ, α), $\alpha \neq 0$, for which its slope $(\lambda-\mu)\alpha$ is greatest (least).

(102) If, that is, the series be integrated with respect to z.

(103) An unwanted 'alia' (other) is here deleted.

(104) In other words, if they be differentiated with respect to z in each case.

(105) That is, are constant.

(106) 'inventa' (found) is cancelled.

Fluxiones ipsius y, ubi z evanescit, habentur ponendo $\frac{\dot{y}}{\dot{z}}=a$, $\frac{\ddot{y}}{\dot{z}^2}=2b$, $\frac{\dddot{y}}{\dot{z}^3}=6c$, $\frac{\ddot{\ddot{y}}}{\dot{z}^4}=24d$. $\frac{\dot{\ddot{\ddot{y}}}}{\dot{z}^5}=120e$ [&c].(107)

Corol. 4. Et hinc si in æquatione resolvenda scribatur $w+x$ pro z ut in casu tertio et resolvendo æquationem prodeat series

$$[y=]\,ex+fxx+gx^3+hx^4+\&c,$$

fluxiones ipsius y(108) ex assumpta utcunq; magnitudine ipsius z habebuntur in æquationibus finitis ponendo $x=0$ et $w=z$. Nam tales erunt æquationes $\frac{\dot{y}}{\dot{z}}=e$. $\frac{\ddot{y}}{\dot{z}^2}=2f$. $\frac{\dddot{y}}{\dot{z}^3}=6g$. $\frac{\ddot{\ddot{y}}}{\dot{z}^4}=24h$ &c(109) per Corollarium superius collectæ. Igitur æquatio resolvenda quæ fluxionem secundam aliasq; $\ddot{y}$, $\dddot{y}$, $\ddot{\ddot{y}}$, &c involvebat reducitur ad æquationem finitam $\frac{\dot{y}}{\dot{z}}=e$ quæ fluxionem primam solummodo involvit, et superest ut per Propositionem præcedentem investigetur hujus reductio in æquationem finitam in qua solæ fluentes quantitates y et z inveniantur.(110)

Sed resolutionem æquationum in series convergentes interminatas nondum exemplis illustravimus.

Exempl.(111)

(107) Whence, if we correct a minor omission on Newton's part by positing that $y=Y\equiv y_0$ when $z=0$. we may substitute these fluxional values in the preceding series to make it

$$\text{`}y=[Y+]\frac{\dot{y}}{\dot{z}}z+\frac{1}{2}\frac{\ddot{y}}{\dot{z}^2}z^2+\frac{1}{6}\frac{\dddot{y}}{\dot{z}^3}z^3+\frac{1}{24}\frac{\ddot{\ddot{y}}}{\dot{z}^4}z^4+\frac{1}{120}\frac{\dot{\ddot{\ddot{y}}}}{\dot{z}^5}z^5\ \&\text{c'},$$

which is but a Newtonian form of the 'Maclaurin' series

$$y\equiv y_z=y_0+y_0'z+\frac{1}{2!}y_0''z^2+\frac{1}{3!}y_0'''z^3+\frac{1}{4!}y_0^{iv}z^4+\frac{1}{5!}y_0^{v}z^5\ldots,$$

where $y_0', y_0'', y_0''', \ldots$ are the values at $z=0$ of the first, second, third, ... derivatives of y with respect to z. While, as we have already remarked in the preceding introduction, it is undeniable that James Gregory made some hidden use of this formula in deriving a number of trigonometrical series in the early winter of 1670/71 (see H. W. Turnbull in the *James Gregory Tercentenary Memorial Volume* (London, 1939): 12–13, 350–64), the theorem is here stated for the first time in explicit form and in its full generality.

(108) 'quæcunq;' (whatever [they be]) is cancelled. In first draft on Add. 3960.8: 144 Newton originally continued 'ad omnem magnitudine[m ipsius] z' (corresponding to every magnitude of z).

(109) In the terms of note (107) these are the successive values $y_w', y_w'', y_w''', y_w^{iv}, \ldots$ of the derivatives of $y\equiv y_z$ with respect to z (and so, since $\dot{z}=\dot{x}$, to x) when $x=0$. Much as before, on adding to it the constant term $Y'\equiv y_w$ (the value of y when $x=0$), we may substitute these derivative values in Newton's polynomial to have it read

$$\text{`}y=[Y'+]\frac{\dot{y}}{\dot{x}}x+\frac{1}{2}\frac{\ddot{y}}{\dot{x}^2}x^2+\frac{1}{6}\frac{\dddot{y}}{\dot{x}^3}x^3+\frac{1}{24}\frac{\ddot{\ddot{y}}}{\dot{x}^4}x^4\ \&\text{c'},$$

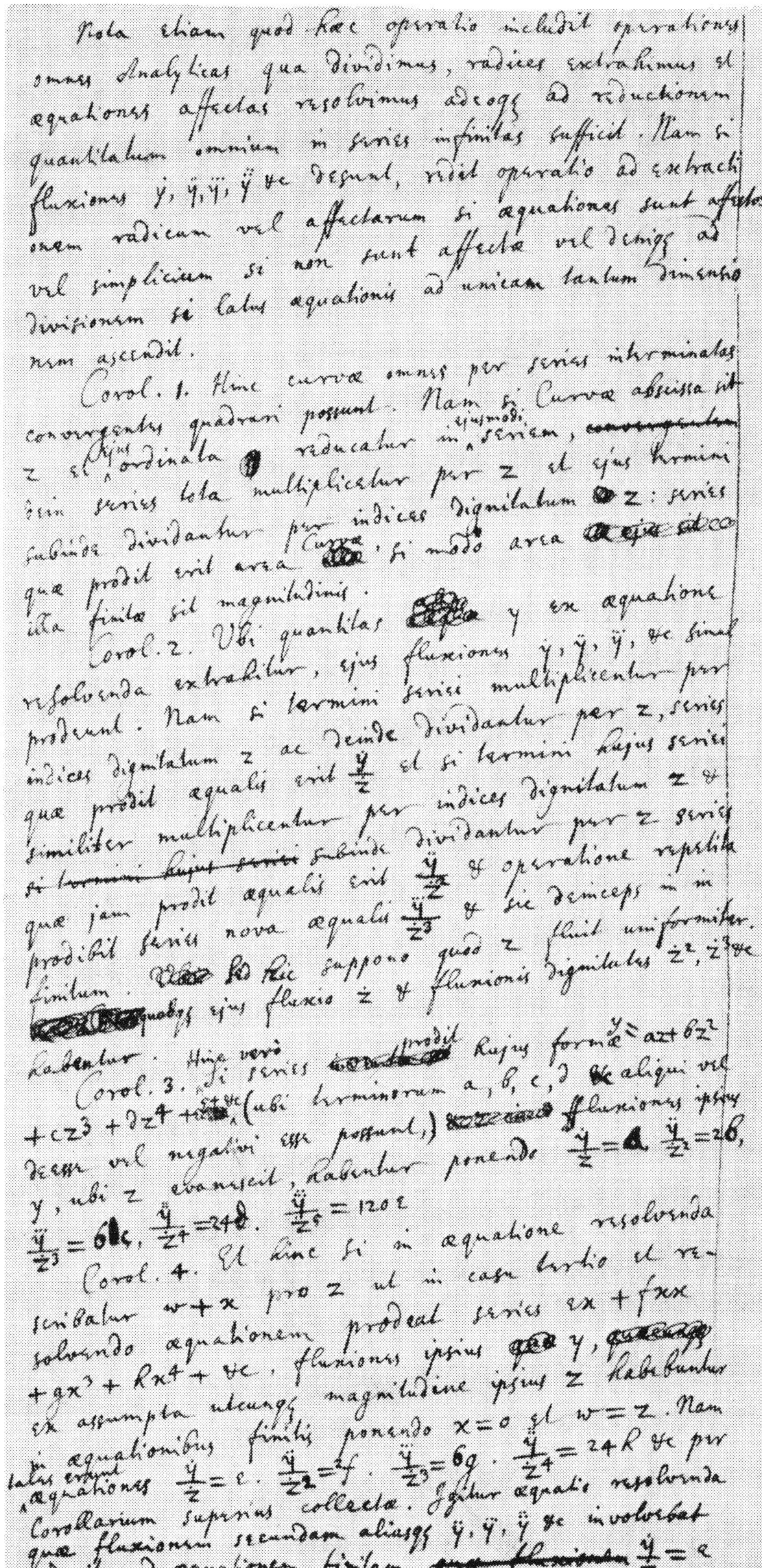

Nota etiam quod hæc operatio includit operationes omnes Analyticas qua dividimus, radices extrahimus et æquationes affectas resolvimus adeoqꝫ ad reductionem quantitatum omnium in series infinitas sufficit. Nam si fluxiones $\dot{y}, \ddot{y}, \dddot{y}, \ddddot{y}$ &c designat, redit operatio ad extractionem radicum vel affectarum si æquationes sunt affectæ vel simplicium si non sunt affectæ vel deniqꝫ ad divisionem si latus æquationis ad unicam tantum dimensionem ascendit.

Corol. 1. Hinc curvæ omnes per series interminatas convergentes quadrari possunt. Nam si Curvæ abscissa sit z et ejus ordinata reducatur in ejusmodi seriem, deinde series tota multiplicetur per z et ejus termini subinde dividantur per indices dignitatum z: series quæ prodit erit area Curvæ si modo area illa finitæ sit magnitudinis.

Corol. 2. Ubi quantitas y ex æquatione resolvenda extrahitur, ejus fluxiones $\dot{y}, \ddot{y}, \dddot{y}$, &c simul prodeunt. Nam si termini seriei multiplicentur per indices dignitatum z ac deinde dividantur per z, series quæ prodit æqualis erit $\frac{\dot{y}}{\dot{z}}$ et si termini hujus seriei similiter multiplicentur per indices dignitatum z & subinde dividantur per z series quæ jam prodit æqualis erit $\frac{\ddot{y}}{\dot{z}^2}$ & operatione repetita prodibit series nova æqualis $\frac{\dddot{y}}{\dot{z}^3}$ & sic deinceps in infinitum. Hic suppono quod z fluit uniformiter quibus ejus fluxio $\dot{z}$ & fluxionis dignitates $\dot{z}^2, \dot{z}^3$ &c habentur. Hinc vero

Corol. 3. Si series prodit hujus formæ $y = az + bz^2 + cz^3 + dz^4 +$ &c (ubi terminorum a, b, c, d &c aliqui vel deesse vel negativi esse possunt,) fluxiones ipsius y, ubi z evanescit, habentur ponendo $\frac{\dot{y}}{\dot{z}} = a$. $\frac{\ddot{y}}{\dot{z}^2} = 2b$, $\frac{\dddot{y}}{\dot{z}^3} = 6c$, $\frac{\ddddot{y}}{\dot{z}^4} = 24d$. $\frac{y}{\dot{z}^5} = 120e$

Corol. 4. Et hinc si in æquatione resolvenda scribatur $w + x$ pro z ut in casu tertio et resolvendo æquationem prodeat series $ex + fxx + gx^3 + hx^4 +$ &c. fluxiones ipsius y, ex assumpta utcunqꝫ magnitudine ipsius z habebuntur in æquationibus finitis ponendo $x = 0$ et $w = z$. Nam æquationes tales erunt $\frac{\dot{y}}{\dot{z}} = e$. $\frac{\ddot{y}}{\dot{z}^2} = 2f$. $\frac{\dddot{y}}{\dot{z}^3} = 6g$. $\frac{\ddddot{y}}{\dot{z}^4} = 24h$ &c per Corollarium superius collectæ. Igitur æquatio resolvenda quæ fluxiones secundas aliasqꝫ $\ddot{y}, \dddot{y}, \ddddot{y}$ &c involvebat reducitur ad æquationem finitam $\frac{\dot{y}}{\dot{z}} = e$ quæ

Plate I. The 'Taylor' power-series expansion of an algebraic function generally enunciated (**1**, §2).

negative), the fluxions of y, when z vanishes, are had by setting $\dot{y}/\dot{z} = a$, $\ddot{y}/\dot{z}^2 = 2b$, $\dddot{y}/\dot{z}^3 = 6c$, $\ddddot{y}/\dot{z}^4 = 24d$, $\dot{\ddddot{y}}/\dot{z}^5 = 120e$, ...(107)

Corollary 4. And hence if in the equation to be resolved there be written $w+x$ for z, as in Case 3, and by resolving the equation there should result the series $(y =)\ ex+fx^2+gx^3+hx^4+...$, the fluxions of y(108) for any assumed magnitude of z whatever will be obtained in finite equations by setting $x = 0$ and so $z = w$. For the equations of this sort gathered by the previous Corollary will be $\dot{y}/\dot{z} = e$, $\ddot{y}/\dot{z}^2 = 2f$, $\dddot{y}/\dot{z}^3 = 6g$, $\ddddot{y}/\dot{z}^4 = 24h$, ...(109) Therefore the equation to be resolved, which before involved the second and other fluxions $\ddot{y}, \dddot{y}, \ddddot{y}, ...$, is reduced to the finite equation $\dot{y}/\dot{z} = e$ involving only the first fluxion, and it remains to investigate by the preceding Proposition how to reduce this to a finite equation in which the fluent quantities y and z alone are to be found.(110)

But we have not yet adduced examples to illustrate the resolution of equations into unterminated converging series.

Example.(111)

---

that is, in modern equivalent

$$y_z = y_{w+x} = y_w + y'_w x + \frac{1}{2!}y''_w x^2 + \frac{1}{3!}y'''_w x^3 + \frac{1}{4!}y^{iv}_w x^4 + \ldots.$$

This 'Taylor' expansion of the function $y \equiv y_{w+x}$ is here enunciated in its generality (the minor omission of y_w apart) some twenty years before Brook Taylor himself discovered it and straightaway communicated it in a virtually identical form to John Machin on 26 July 1712; see the preceding introduction.

(110) In the preliminary draft on Add. 3960.8: 144 this last sentence was initially set to be a separate 'Cor. 5' (compare Appendix 1.7: note (87) below) there announcing equivalently that 'Deniq; per Propositionem superiorem ex æquatione fluxiones involvente $\frac{\dot{y}}{\dot{z}} = e$ investiganda est æquatio finita solas fluentes quantitates y et z involvens' (Lastly, by the preceding Proposition [XI] from the equation $\dot{y}/\dot{z} = e$ involving their fluxions there needs to be discovered a finite equation involving the fluent quantities y and z alone). But Newton for once, here misled by a notation which cannot distinguish between the general derivative $y'_z \equiv \dot{y}/\dot{z}$, z free, and its particular value $y'_w = e$, sees only a mirage: there is no way of passing 'per Prop. XI' from the latter 'equation' to determine the 'æquatio finita' $y = y_z$ 'in qua solæ fluentes y et z inveniuntur'.

(111) For the intended future insertion of one or more worked applications of his preceding 'methodus generalis' for expanding the root y of a given fluxional equation as a power-series in the independent variable z Newton has left empty the remainder of his present manuscript sheet (pp. 179/80, blank except for half a dozen lines at the top of p. 179 where the text which here terminates is penned). We have already observed (note (98)) that a few months afterwards he communicated to John Wallis his detailed resolution of the first-order fluxional *æquatio* $y^2 - z^2\dot{y} - d^2 + dz = 0$ into the equivalent series $y = d - \frac{1}{2}z - \frac{3}{8}z^2/d - \frac{9}{16}z^3/d^2 \ldots$ (compare pages 178–80 below). His worksheets of this period, notably Add. 3960.8: 138–42 and 3960.9: 152–6/164, contain a number of other attempts by him to contrive suitable examples by which to illustrate his method: for instance, he tried several times (pp. 141, 156, 164) to effect the solution of the deceptively 'simple' third-order equation '$\sqrt{1+\dot{y}^2} \times \ddot{y} = n\dddot{y}$', seeking successively to determine the $a, b, c, d, \ldots$ in the posited equivalent first-order equation

$$\dot{y} = a + by + cy^2 + dy^3 \ldots$$

Prop. [*XIII.*]

Per Propositiones præcedentes Problemata solvere.(112)

[*1.*](113) Maximæ non fluunt ne majores fiant neq; defluunt ne majores fuerint. Et eadem est ratio(114) minimarum. Inveniuntur igitur ponendo earum fluxionem nullam esse. Ut si fuerit $xx-ax+by=yy$ et inde $2x\dot{x}-a\dot{x}+b\dot{y}=2y\dot{y}$, pone fluxionem $\dot{y}$ nullam esse et inde fiet $2x\dot{x}-a\dot{x}=0$ seu $x=\frac{1}{2}a$.(115)

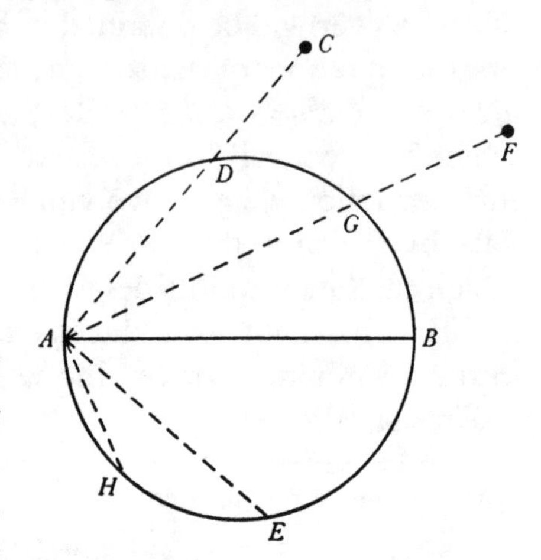

[*2.*] Motus punctorum in machinis(116) quibuscunq; et inde tange[n]tes Curvarum quas puncta describunt determinantur per sequentes tres Regulas.

Reg. 1.(117) *Sciendum est quomodo ex puncti alicujusvis motu recto motus obliqui colligi possint et contra.* Moveatur punctum A directe in linea AB et exponatur ejus motus per longitudinem AB et super diametro AB describatur circulus $ADGBE$[,] et motus obliqui puncti A versus data quælibet puncta C, F exponen-

---

by substituting this and its derivatives

$$\dot{y} = ab+(2ac+b^2)\,y+3(ad+bc)\,y^2\ldots \quad \text{and} \quad \ddot{y} = (2ac+b^2+6(ad+bc)\,y\ldots)\,\dot{y}$$

in the primary equation and identifying coefficients of like powers of y, whence immediately (were the prior operation conveniently to be effected) there would be

$$z = \int (a+by+cy^2\ldots)^{-1}.dy;$$

elsewhere (p. 152), by reverse differentiation and subsequent doctoring, he constructed first- and second-order fluxional equations of which $azy+y^2=c^2$ and $az+zy=y^2$ are particular solutions, and made a half-hearted stab at determining in each case the equivalent series, but abandoned the effort almost immediately. In Appendix 1.11 below we reproduce, with suitable editorial fleshing-out, Newton's more finished computations (pp. 138–9/153) which yield the opening terms in the series resolutions $y=y_z$ of the equations

$$2z^3-3z^2y-y\dot{y}+y^2\ddot{y}=0 \quad \text{and} \quad y^{-2}=\ddot{y}$$

respectively (with x trivially replacing z in each case). At one point (pp. 141–2) Newton also—compare note (101) above—works through the analogous series resolution of the fluxion-free equation $-\log(1-y)+y=\log(1+x)+x$ to determine with some difficulty that '$y = x-\frac{1}{2}xx+\frac{1}{4}x^3-\frac{3}{16}x^4+\frac{5}{32}x^5+\frac{25}{192}x^6$ [&c]'.

(112) Of the four groups of problems whose fluxional resolution Newton here discusses, the two first—dealing respectively with the determination of extremal values of a variable as those at which it has no instantaneous fluxion, and in much greater detail with the limit-motion of points 'in mechanical structures of any type' and thence with the instantaneous tangential direction of the curves which those points are constrained to describe—restate only superficially (as we specify in the following footnotes) the content of his earlier findings in these areas, particularly as he had written it up in his 'Geometria Curvilinea' a decade earlier; the long ensuing account of the curvature of curves and its fluxional variations (and of the

Proposition XIII.

By means of the preceding propositions to solve problems.

[*1.*][113] Maxima flow not onwards, so as not to come to be greater, nor backwards, so as not to have been greater. And the like consideration[114] holds for minima. They are therefore found by setting their fluxion to be nil. Were there, for instance, $x^2-ax+by=y^2$ and thence $2x\dot{x}-a\dot{x}+b\dot{y}=2y\dot{y}$, put the fluxion $\dot{y}$ to be nothing and thereby will there be $2x\dot{x}-a\dot{x}=0$ or $x=\frac{1}{2}a$.[115]

[*2.*] The motions of points in machines[116] of any type, and therefrom tangents of the curves which the points describe, are determined by means of the three rules following.

Rule 1.[117] *To know how from the direct motion of any point its oblique motions may be gathered; and the converse.*

Let the point A move directly in the line AB, and express its motion by the length AB, then on diameter AB describe the circle $ADGBE$: the oblique motions of the point A towards any given points C, F whatever will be expressed

related classes of angles of contact between a curve and its rectilinear tangent) is, with an insight here as novel as it is profound, made narrowly dependent on the Taylor expansion of the osculating curve's incremented ordinate; the two concluding dynamical problems depart from his contemporary refinements of the notions (as set by him in his *Principia* in 1687) that an orbit may, in measuring the diverting central force upon it at any point, be replaced by its circle of curvature there, and that a force is measured by the change in (radial) fluxional speed which its action induces.

(113) Much as in Problem 3 of his 1671 tract (III: 116–18) and in line with Problem 1 of his 'Geometria Curvilinea' (IV: 474), Newton adumbrates the simple test for $y \equiv y_x$ to be a maximum or minimum that $(dy/dx =)\ \dot{y}$ shall instantaneously be zero. As before he omits any consideration of the anomalous case when simultaneously $\ddot{y}=0$ (and the corresponding Cartesian curve $y=y_x$ has an inflexion at the point).

(114) Understand 'mutatis mutandis'. More precisely 'minimæ non fluunt ne minores fiant neq defluunt ne minores fuerint' (Minima flow neither onwards, not to become less, nor backwards, not to have been less).

(115) From which abrupt termination the reader is to conclude that the rest is straightforward? In fact, if we convert the given equation to the equivalent explicit form

$$y=\tfrac{1}{2}b\pm\sqrt{[(x-\tfrac{1}{2}a)^2-\tfrac{1}{4}(a^2-b^2)]},$$

it will be clear that we must be subtler than Newton allows in such distinctions: for, when $x=\frac{1}{2}a$ and $a \neq b$, y simultaneously attains both the local maximum $\frac{1}{2}(b-\sqrt{[b^2-a^2]})$ and the local minimum $\frac{1}{2}(b+\sqrt{[b^2-a^2]})$, though neither is real for $a>b$; but, when $a=b$, the given equation breaks into the pair of $y=x$ and $y=a-x$ which merely share the common values $x=y=\frac{1}{2}a$ without singly possessing either maximum or minimum. Had he here paused to think about his solution, Newton would surely have clarified its vagueness before passing on.

(116) That is, 'Engines'—to use the name which Newton had given to one such species in 1665 (see I: 264, note (52))—made up of interconnected and mutually pivoted/sliding lines which have a degree of freedom to move about in a given plane.

(117) This was initially enunciated as '*Distinguendum est inter puncti motum directum et motus relativos obliquos*' (*To distinguish between the direct motion of a point and its oblique relative ones*).

tur per rectarū *AC AF* partes, *AD*, *AG* quæ intra circulum jacent. Nam defluxiones linearum *AB*, *AC*, *AF* sunt ut *AB*, *AD*, *AG*. Ad *AC*, *AF* erigantur perpendicula *AE*, *AH* & motus angulares puncti *A* circa centra *C*, *F*, hoc est ejus motus obliqui versus *E* et *H* exponentur per *AE* et *AH*. Et contra si habentur motus quilibet obliqui *AD*, *AG*, *AE* &c invenietur motus directus *AB* erigendo(118) *DB*, *GB*, *EB* &c et ad eorum concursum *B* agendo rectam *AB*.(119)

Reg. 2. Intelligenda est compositio motus totius ex partibus. Super recta *AB* moveatur recta *BC* motu quocunq̧. Et(120) motus puncti *C* orietur partim a motu puncti *B* seu fluxione longitudinis *AB*, partim a rotatione rectæ *BC* circa punctum *B* et partim a fluxione longitudinis *BC*. Exponatur motus primus per longitudinem *CE* ductam parallelam *AB*, secundus per longitudinem *CD* ductam perpendicularem ipsi *BC* ac tertius per *CF* sumptam in *BC* producta; et ab exponentis alicujus *CE* termino *E* reliquis exponentibus *CD*, *CF* parallelæ et æquales agantur *EG*, *GH*, et juncta *CH* exponens erit motus totius. Movebitur igitur punctum *C* in recta *CH* ea cum velocitate quæ per longitudinem *CH* exponitur, et [propterea *CH* tanget] curvā quam punctum *C* describit.(121)

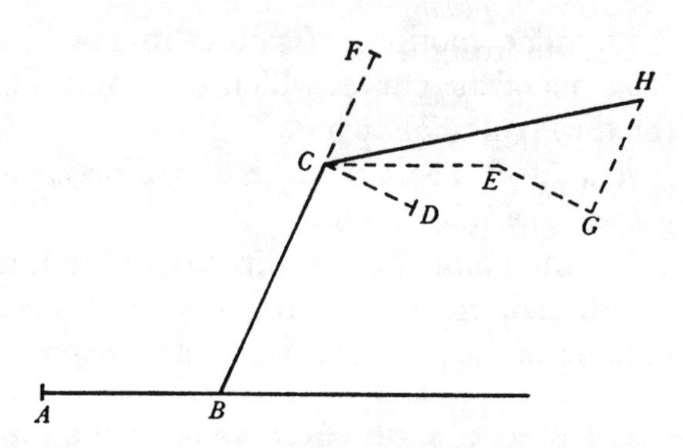

Reg. 3. Intelligenda est comparatio motuum diversorū punctorum. Ut si recta *ABC* circa polum *A* revolvens secuerit rectas duas positione datas *DB*, *DC* in *B* et *C* et comparandi sint motus intersectionum, per intersectionum alterutram *B* duco rectam positione datam(122) *BE*, rectæ *DC* per alteram intersectionem tran[s]eunti parallelam, ut et rectam *EF* ipsi *AB* parallelam ad constituendum triangulum quam minimum *BEF*[,] et motus intersectionis *C* erit ad motum intersectionis linearum *AB* et *EB*(123) ut *AC* ad *AB* et motus hujus intersectionis

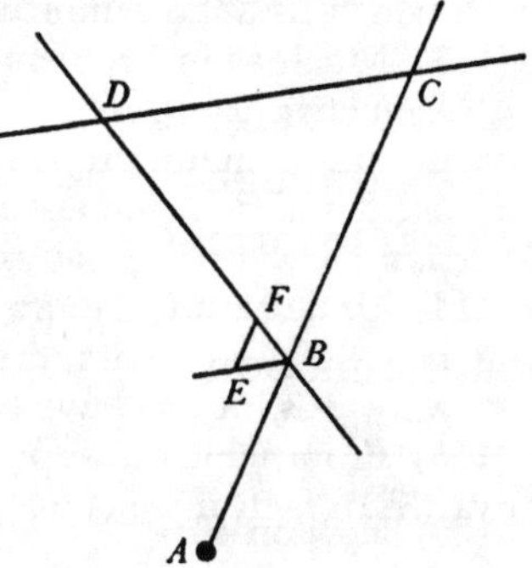

(118) Understand 'perpendiculares' (the perpendiculars). Newton initially wrote a still less explicit 'ducendo' (by drawing).

(119) Suddenly deciding to make no explicit mention of tangents in his rules for compounding limit-motions, Newton here deleted an intended termination 'hoc est invenietur tangens Curvæ quam punc[tum B describit]' (that is, there will be found the tangent of the curve which the point [*B* describes]) before completing it. This primary rule is essentially the cancelled Proposition 30 (IV: 472) of his earlier 'Geometria Curvilinea', which in turn recasts Proposition 1 (I: 400) of his October 1666 tract.

by the portions *AD*, *AG* of the straight lines *AC*, *AF* which lie within the circle. For the fluxions of the lines *AB*, *AC*, *AF* are as *AB*, *AD*, *AG*. To *AC*, *AF* erect perpendiculars *AE*, *AH*, and the angular motions of point *A* round the centres *C* and *F*, that is, its oblique motions towards *E* and *H*, will be expressed by *AE* and *AH*. And conversely, if any oblique motions *AD*, *AG*, *AE*, ... whatever are had, the direct motion *AB* will be ascertained by erecting(118) *DB*, *GB*, *EB*, ... and to their common meet *B* drawing the straight line *AB*.(119)

Rule 2. To comprehend the composition of the total motion from its parts.

Let the straight line *BC* move on the straight line *AB* with any motion whatever. Then(120) the motion of its point *C* will arise partly from the motion of the point *B*, that is, the fluxion of the length *AB*; partly from the rotation of the line *BC* round the point *B*; and partly from the fluxion of the length *BC*. Express the first motion by the length *CE* drawn parallel to *AB*, the second by the length *CD* drawn perpendicular to *BC*, and the third by *CF* taken in *BC* produced; then from the end-point *E* of any of these exponents *CE* parallel and equal to the remaining exponents *CD*, *CF* draw *EG*, *GH*, and *CH* when joined will be the exponent of the total motion. The point *C* will therefore move in the straight line *CH* with the speed expressed by the length *CH*, and accordingly *CH* shall touch the curve which the point *C* describes.(121)

Rule 3. To comprehend the comparison of the differing motions of points.

If, for instance, the straight line *ABC* shall, as it revolves round the pole *A*, cut two straight lines *DB*, *DC* given in position in *B* and *C*, and the motions of the intersections need to be compared, through either one of the intersections *B* I draw a straight line *BE*, given in position,(122) parallel to the straight line *CD* passing through the other intersection, and also the straight line *EF* parallel to *AB* so as to form the minimal triangle *BEF*: the motion of the intersection *C* will then be to the motion of the intersection of the lines *AB* and *EB*(123) as *AC* to *AB*, and the motion of the latter intersection to the motion of the intersection of

(120) Newton first continued by asserting 'componetur motus puncti *C* partim ex motu rectæ *BC* super recta *AB*, hoc est ex fluxione rectæ *AB*, partim ex rotatione rectæ *BC* circa punctum *B*, et partim ex fluxione longitudinis *BC*' (the motion of its point *C* will be compounded partly from the motion of the straight line *BC* upon the straight line *AB*, that is, the fluxion of the line *AB*; partly from the rotation of the line *BC* round the point *B*; and partly from the fluxion of the length *BC*).

(121) This is essentially but a reshaping of the equivalent Proposition 27 (IV: 468–70) of the earlier 'Geometria Curvilinea'.

(122) A curious phrase: Newton does not mean that the line *BE* is given absolutely in position, but merely intends it to be drawn at each point *B* of *DB* 'in plagam datam' (in given direction), namely—as he goes on to specify—parallel to *DC*.

(123) This replaces the equivalent 'ad motum puncti in quo recta *AC* secat rectam *EB*' (to the motion of the point in which the line *AC* cuts the line *EB*).

ad motum intersectionis linearum AB et DB ut BE ad BF id est ut DC ad DB. Ideoq; motus intersectionis C erit ad motum intersectionis linearum AB et DB seu fluxio lineæ DC ad fluxionem lineæ DB ut AC ad AB et DC ad DB conjunctim, id est ut $AC \times DC$ ad $AB \times DB$. Et simili argumento motus punctorum & fluxiones linearum in alijs casibus pro lubitu comparari possunt.(124)

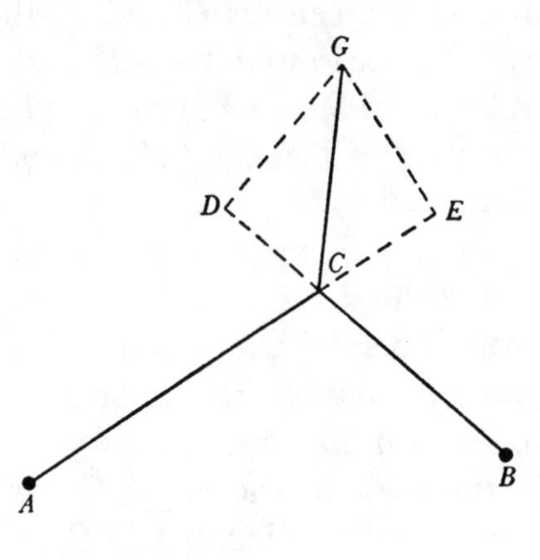

Usus harum regularum exemplo uno et altero satis patebit.

Exempl. 1.(125) Si a datis punctis A, B ad curvam quamvis concurrant rectæ AC, BC, exponantur harum fluxiones per CE et CD, erigantur perpendicula DG, EG concurrentia in G et motus puncti C in Curva quoad situm et velocitatem exponetur per rectam CG per Reg. 1, et hæc recta propterea curvam tanget.

Exempl. 2.(126) Polo P, regula AC, intervallo AB describitur Conchois,(127) ad punctum quodvis D ducenda est tangens. Jungatur DP regulæ occurrens in C et ad Conchoidis axem PB demittatur normalis $D[F]$(128) et ob datam(129) CD,

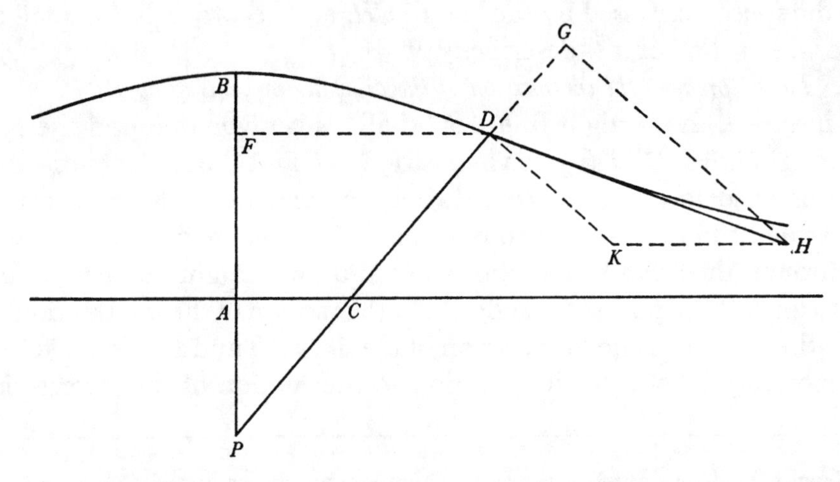

fluxiones ipsarū $[P]D$ et CP æquales erunt estq; fluxio ipsius CP ad motum angularem puncti C circa polum P ut AC ad AP et hic motus ad motum angularem puncti D(130) ut PC ad PD seu AP ad $[P]F$. Exponatur igitur fluxio ipsius PD per $DG = AC$, et motus angularis puncti D circa polum per $GH = P[F]$. Hoc est producatur PD ad G ut sit $DG = AC$ et erigatur perpendiculum $GH = P[F]$ et juncta DH exponens erit motus puncti D in Chonchoide per Reg. 1 ideoq; Chonchoidem tanget.(131)

Idem aliter. Fluxio rectæ AC est ad motum angularem puncti C circa polum P ut PC ad PA et hic motus angularis ad motum angularem puncti D circa punc-

the lines AB and DB as BE to BF, that is, as DC to DB; consequently, the motion of the intersection C will be to the motion of the intersection of the lines AB and DB, that is, the fluxion of the line DC to the fluxion of the line DB, as AC to AB and DC to DB jointly, and so as $AC \times DC$ to $AB \times DB$. And by a similar argument the motions of points and the fluxions of lines may in other cases be compared at will.(124)

The use of these rules will be evident enough from an example or two.

Example 1.(125) If straight lines, AC, BC issuing from the given points A, B meet at any curve, express their fluxions by CE and CD and erect the perpendiculars DG, EG meeting in G: the motion of the point C in the curve will then in regard to its position and speed, by Rule 1, be expressed by the straight line CG, and this line shall accordingly touch the curve.

Example 2.(126) A conchoid is described with pole P, ruler AC and radius AB, and at any point D on it a tangent has to be drawn. Join DP meeting the ruler in C and to the conchoid's axis PB let fall the normal DF:(128) then, because CD is given,(129) the fluxions of PD and PC will be equal, and the fluxion of PC to the angular motion of point C round the pole P is as AC to AP, while this motion to the angular motion of point D(130) is as PC to PD or AP to PF. Express, therefore, the fluxion of PD by $DG = AC$, and the angular motion of point D round the pole by $GH = PF$, that is, extend PD to G so that $DG = AC$ and erect the perpendicular $GH = PF$, and DH when joined will be the exponent of the motion of point D in the conchoid, by Rule 1, and shall consequently touch the conchoid.(131)

The same another way. The fluxion of the straight line AC is to the angular motion of point C round the pole P as PC to PA, while this angular motion is to the angular motion of point D round the point C as PC to CD, that is, as PA to

(124) Compare the cancelled Corollary to Proposition 26 (IV: 468) and also Proposition 30 (*ibid.*: 474) of the 'Geometria Curvilinea'.

(125) The construction of tangents to curves defined in bipolar coordinates; compare Mode 3 of Problem 4 of Newton's 1671 tract (III: 136–8) and Case 1 of Proposition 29 of the subsequent 'Geometria Curvilinea' (IV: 472).

(126) While the detail of these two following variant constructions of the conchoid's general tangent is new, their basis is adumbrated in Problem 7 (IV: 480) of the earlier 'Geometria'.

(127) Originally 'Chonchois', an idiosyncratic spelling which Newton also employs in the sequel without there bothering to delete the intrusive 'h'.

(128) The manuscript reads 'DB' in error—a confusion between 'F' and 'B' explained, here and in analogous occurrence below, by the fact that in Newton's hurriedly drawn figure the latter letter seems to mark the former's point. In the text reproduced we have corrected all such visual slips on Newton's part, and also his similarly explainable erroneous interchange of 'P' with its neighbouring 'A' and 'C'.

(129) Only 'longitudine' (in length) of course.

(130) Again understand 'circa polum P' (round the pole P).

(131) The equivalent proof by infinitesimal increments is perhaps more revealing to the modern eye. If PCD is conceived 'instantaneously' to pass into the indefinitely near position

tum C ut PC ad CD hoc est ut PA ad $A[F]$. Ad PD erigatur igitur perpendiculum DK ipsi $A[F]$ æquale quo motus angularis puncti D circa punctum C exponatur. Deinde a puncto K ipsi AC parallela et ipsi PC æqual[is] agatur KH qua fluxio ipsius AC exponatur et juncta DH exponens erit motus puncti D per Reg. 2 et Conchoidem tanget.(132)

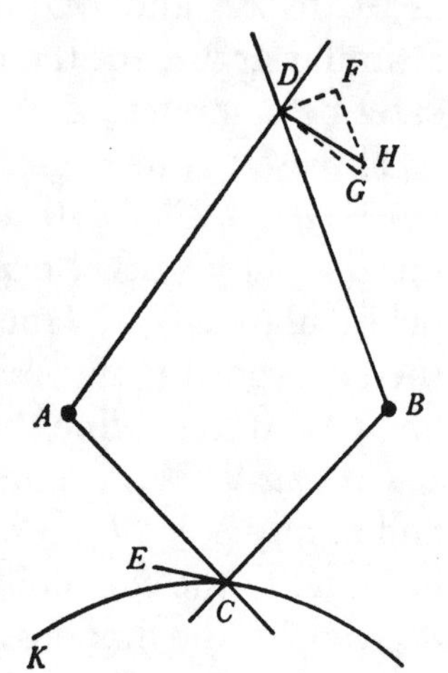

Exempl. 3.(133) Si anguli duo dati DAC, DBC circa vertices positione datos A, B rotentur et concursu crurum AC, BC curvam quamvis datam CK contingunt in C, et requiritur tangens curvæ ad quam altera duo crura AD, BD concurrunt in D: exponatur motus puncti C per rectam EC quæ curvam CK tangit in K et motus angularis puncti C circa polum A exponetur per distantiam puncti E a recta AC. Ad AD erige perpendiculum DG quod sit ad hanc distantiam ut AD ad AC et exponetur motus angularis puncti D circa polum A per perpendiculum illud DG. Simili ratione exponatur motus angularis puncti D circa polum B per erectum ad BD perpendiculum DF. Deinde erecta ad DG et DF perpendicula GH FH concurrant ad H et acta DH exponens erit motus puncti D in Curva et Curvam tanget.(134)

Et simili ratione ducuntur tangentes curvarum omnium absq; usu æquationum ubi Curvæ absq; æquationibus definiuntur. Ubi vero æquationes defini-

---

Pcd, and $c\gamma$, $d\delta$ are let fall perpendicular to PCD, then the conchoid's defining property $CD = cd$ determines that $C\gamma = D\delta$ and so $\text{fl}(PC) = \text{fl}(PD)$; moreover,

$$\text{fl}(PC):PC\times\text{fl}(\widehat{APC}) = C\gamma:(PC\times\widehat{CPc}\text{ or) }\gamma c = AC:PA$$

and
$$PC\times\text{fl}(\widehat{APC}):PD\times\text{fl}(\widehat{APC}) = \gamma c:\delta d = PC:PD = PA:PF,$$

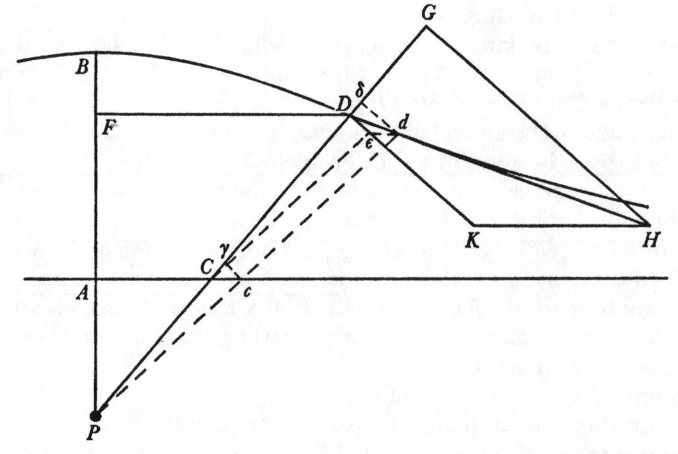

so that $\text{fl}(PD):PD\times\text{fl}(\widehat{APC}) = AC:PF$ is equal to $D\delta:\delta d = DG:GH$. As we have seen (1: 394)

AF. To *PD* erect, therefore, the perpendicular *DK* equal to *AF*, whereby the angular motion of point *D* round the point *C* shall be expressed. Then from the point *K* parallel to *AC* and equal to *PC* draw *KH* whereby the fluxion of *AC* shall be expressed, and *DH* will, when joined, be the exponent of the motion of point *D*, by Rule 2, and so shall touch the conchoid.(132)

Example 3.(133) If the two given angles $D\widehat{A}C$, $D\widehat{B}C$ rotate round the vertices *A*, *B* given in position and at the meet of their legs *AC*, *BC* in *C* trace any given curve, and there is required the tangent at *D* to the curve at which the other two legs meet, express the motion of point *C* by the straight line *EC* which touches the curve *CK* at *K* and then the angular motion of point *C* round the pole *A* will be expressed by the distance of the point *E* from the straight line *AC*. To *AD* erect a perpendicular *DG* which shall be to this distance as *AD* to *AC*, and the angular motion of point *D* round the pole *A* will be expressed by that perpendicular *DG*. By a similar reasoning express the angular motion of point *D* round the pole *B* by the perpendicular *DF* erected to *BD*. Then, when perpendiculars *GH*, *FH* are erected to *DG* and *DF*, let these meet at *H*, and *DH* will, when drawn, be the exponent of the motion of point *D* in the curve and so shall touch the curve.(134)

And by a similar method tangents are drawn to all curves without employing equations when the curves are defined without equations. When, indeed,

---

Newton had more than a quarter of a century earlier discovered the easy construction of the point H' in which the tangent *DH* meets the *regula* *AC* by setting $DG' = PC$ along *DG* and raising the perpendicular $G'H'$: this is an immediate corollary to the present variant, for at once $CG':G'H' = AC:PA$ and

$$DG':G'H' = (DG \text{ or}) \ AC:(GH \text{ or}) \ PF,$$

so that
$$CG':DG' = PF:PA = PD:PC$$

and *dividendo*
$$CD:DG' = CD:PC.$$

(132) In the equivalent argument which relates vanishing increments (see the previous note) if follows much as before, on drawing $C\epsilon$ equal and parallel to *cd* (so that $\widehat{DC\epsilon} = \widehat{CPc}$ and $D\epsilon$ is, in the limit, parallel to γc), that $\text{fl}(AC):PC\times\text{fl}(\widehat{APC}) = Cc:\gamma c = PC:PA$ and also

$$PC\times\text{fl}(\widehat{APC}):CD\times\text{fl}(\widehat{APC}) = \gamma c:D\epsilon = PC:CD = PA:AF,$$

whence $\text{fl}(PD):CD\times\text{fl}(\widehat{APC}) = AC:AF$ is equal to $D\delta:D\epsilon = DG:DK$ and therefore $DK:AF = DG:AC = KH:PC$ (since *KH* is perpendicular to *PA* and *GH* perpendicular to *PC*, so that $\widehat{APC} = \widehat{KHG}$ and so *KH*, *PC* are equally inclined to *DG*, *AC*).

(133) This adapts the construction of the tangent to the *curva describenda* in a general organic description as Newton had set it out more than twenty years before in Lemma 11 (II: 140–2; see also 142, note (25)) of his 'De Modo describendi Conicas sectiones et curvas trium Dimensionum...'.

(134) Specifically, where *Ee* and $E\epsilon$ are the perpendiculars from *E* onto *AC* and *BC*, in the limit as *E* passes into coincidence with *C* the ratio $Ee/E\epsilon$ is given, and so also therefore is $DG/DF = (AD/BD)\times(Ee/E\epsilon)/(AC/BC)$.

tionem ingrediuntur, his fluxiones indeterminatarum quantitatum determinantur[,] dein æquationibus rejectis cætera peraguntur absq; calculis.(135)

Ex tangentibus determina[n]tur Curvarum omnium Asymptoti faciendo ut puncta contactus abea[n]t in infinitum. Nam tangens tunc in Assymptoton vertitur.

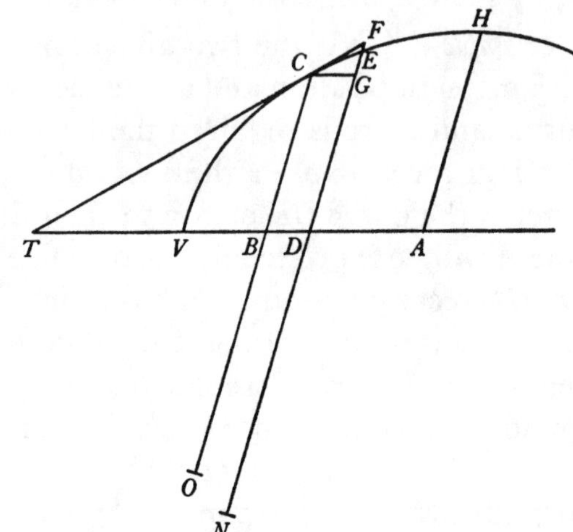

[*3.*] Curvatura curvarum proportionalis est angulo contactûs quem Curvæ continent cum tangentibus rectilineis estq; in circulo reciprocè ut circuli diameter.(136) Si circulus et altera quævis curva tangant lineam rectam in æqualibus angulis contactus adeoq; in puncto illo sint æqualiter curvæ; variatio curvaturæ curvæ alterius erit proportionalis angulo contactus(137) quem circulus continet cum curva ad idem punctum. Est et variationis variatio et hujus vari[ati]onis variatio &c.(138) Quæ omnia sic determinantur. Sit curvæ alicujus [V]C abscissa $AB = z$ & ordinata $BC = y$,(139) sintq; harum fluxiones primæ $\dot{z}, \dot{y}$, secundæ $\ddot{z}, \ddot{y}$, tertiæ $\dddot{z}, \dddot{y}$, ut supra.(140) Capiendo BT ad BC ut [$\dot{z}$] ad $\dot{y}$ ducetur tangens CT. Dein si producatur ordinata CB versus concavas Curvæ partes ad usq; O ut sit $CO = \frac{\dot{z}\dot{z}.}{\frac{1}{2}\ddot{y}}$ in $\frac{CT^q}{BT^q}$, et circulus ducatur per punctum O qui curvam tangat in C, hic circulus eandem habebit curvaturam cum Curva in puncto C.

(135) Adjusting a nuance—perhaps his first choice of word had too numerical a tone for him here?—Newton here replaces an initial 'computis' (computations).

(136) Compare his earlier equivalent definitions in Problem 2 of his October 1666 tract (I: 419) and in Problem 5 of his 1671 tract (III: 150).

(137) Newton first began here to specify 'infinite m[inori]' (infinitely less).

(138) This further extension to embrace higher-order variations in curvature was made when Newton, as soon as he had penned it, replaced a first version of the following construction of the osculating circle at an arbitrary point of a given Cartesian curve where the ordinate angle $\widehat{ABC}$ is restricted to be right; there, much as he had done a quarter of a century before (see I: 245–71, 387–9), he initially argued: 'Curvatura autem sic invenitur. Sit abscissa $AB = w$, ordinata $B[C] = v$. Augeatur abscissa parte quàm minima $Bb = z$, et sit ordinata bc in serie convergente $v + az + bz^2 + cz^3 +$ &c. et terminus az determinabit tangentem Cd complendo parallelogrammū $BbgC$ et producendo bg ad d ut sit gd ad Bb [hoc est] az ad z [ut a ad 1]. Deinde terminus bz^2 determinabit curvaturā curvæ ad punctum C producendo db

equations enter into their definition, the fluxions of the indeterminate quantities may be determined by them and then, with the equations rejected, the rest is accomplished without calculations.(135)

From their tangents the asymptotes are determined in all curves by making the points of contact go off to infinity. For the tangent then turns into an asymptote.

[*3.*] Curvature in curves is proportional to the angle of contact which curves contain with their rectilinear tangents, and in a circle is reciprocally as the circle's diameter.(136) Should a circle and any other curve touch a straight line in equal angles of contact, and hence at that point be equally curved, the variation of curvature in the other curve will be proportional to the (137)angle of contact which the circle contains with the curve at this same point. There is also the variation of the variation, and the variation of this variation, and so on.(138) All of which are determined in this manner. In some curve VC let the abscissa be $AB = z$ and the ordinate $BC = y$,(139) and let the first fluxions of these be $\dot{z}$, $\dot{y}$, their second ones $\ddot{z}$, $\ddot{y}$, their third $\dddot{z}$, $\dddot{y}$, (and so on) as above.(140) By taking BT to BC as $\dot{z}$ to $\dot{y}$ there will be drawn the tangent CT. Then if the ordinate CB be extended on the concave side of the curve as far as O so that

$$CO = (\dot{z}^2/\tfrac{1}{2}\ddot{y}).(CT^2/BT^2),$$

and a circle be drawn through the point O to touch the curve at C, this circle will have the same curvature as the curve at the point C. For if any other

---

ad O ut sit dO ad Cd hoc est ad $\sqrt{z^2+aaz^2}$ ut Cd ad dc seu bz^2 [id est] $= \dfrac{1+aa}{b}$ & circulum ducendo per O qui tangat curvam in puncto C. Nam circulus ille ob continue proportionales dO, dC, dc, ejusdem erit curvaturæ cum curva in puncto C' (The curvature, however, is ascertained in this way. Let abscissa $AB = w$, ordinate $BC = v$. Augment the abscissa by the indefinitely small part $Bb = z$ and let the ordinate bc be in [the form of] the converging series $v+az+bz^2+cz^3+\ldots$. And then the term az will determine the tangent Cd by completing the rectangle $BbgC$ and extending bg to d so that gd to Bb, that is, az to z, be as a to 1. Thereafter the term bz^2 will determine the curvature of the curve at the point C by extending db to O so that dO be to Cd $(= \sqrt{[z^2+a^2z^2]})$ as Cd to dc (or bz^2), that is, equal to $(1+a^2)/b$, and drawing a circle through O to touch the curve at point C. For because dO, Cd and dc are in continued proportion that circle will be of the same curvature as the curve at the point C).

(139) The figure reproduced is that which Newton afterwards set in juxtaposition on the facing verso (p. 186) of his manuscript, thereby superseding a prior (but uncancelled) equivalent diagram accompanying the text itself on p. 187; as it was originally his mind to do (see the previous note) but in the present form of argument is no longer necessary, the latter restricts the ordinate angle $\widehat{ABC}$ to be right, and also has a line CP drawn normal to the curve at C (and so to its tangent CT) and meeting the abscissa in P, with the origin A taken to be the curve's intersection (here marked V) with the latter base line and not its present position at the foot of the prime ordinate.

(140) In sequel, however, z is taken to be the base variable, so that $\ddot{z} = \dddot{z} = \ldots = 0$.

Nam si ducatur et utrinq; producatur alia quævis ordinata DE Curvæ occurrens in E, & sumatur $FN = \frac{FC^q}{FE} = \frac{BD^q}{FE} \times \frac{CT^q}{BT^q}$ [:] incidet punctum [N] in circulum qui tangit curvam in C et transit per punctum E. Compleatur parallelogrammum $CBDG$ et dicatur BD vel [C]$G = x$ sitq; series infinita

$$ax + bxx + cx^3 + dx^4 + \&c$$

ordinatarum differentia GE et erit $GF = ax$ & $FE = -bxx - cx^3 - dx^4$ &c[(141)] et $FN = \frac{FC^q}{FE} = \frac{CT^q}{BT^q} \times \frac{CG^q}{FE} = \frac{CT^q}{BT^q} \times \frac{-1}{b + cx + dxx + \&c}$. Coeant puncta E et C ut x evanescat et anguli contactuum circuli et curvæ æquentur et fiet

$$FN = \frac{CT^q}{BT^q} \times \frac{-1}{b}.$$

Sed per Corol [4] Prop. [XII] est $\frac{-1}{b} = \frac{-\dot{z}\dot{z}}{\frac{1}{2}\ddot{y}}$,[(142)] ideoq;

$$FN = CO\left[= \frac{CT^q}{BT^q} \times \frac{-\dot{z}\dot{z}}{\frac{1}{2}\ddot{y}}\right].$$

Hoc est circulus ejusdem curvaturæ cum Curva ad C transit per punctum O.

Ut si fuerit ad Ellipsin VCH æquatio $aa - bzz = yy$, et inde $-2bz\dot{z} = 2y\dot{y}$ et $-b\dot{z}\dot{z} = \dot{y}\dot{y} + y\ddot{y}$ hoc est $= \frac{bbz^2\dot{z}^2}{yy} + y\ddot{y}$, erit $\frac{\dot{z}^2}{\frac{1}{2}\ddot{y}} = \frac{-2y^3}{byy + bbzz}$ id est $= \frac{-2y^3}{baa}$. Et inde $CO = \frac{TC^q}{TB^q} \times \frac{-2y^3}{baa}$. Signum negativum denotat rectam CO ducendam esse ad partes puncti C versus B. Evanescat z et TC ac TB evadent æquales et y æqualis erit a, adeoq; CO æqualis $-\frac{2a}{b}$. Incidit jam punctum B in ellipseos centrum A et ordinata BC coincidit cum semidiametro $AH = a$ existente $\frac{2a}{b}$ latere recto ad hanc semidiametrum pertinente.[(143)] Unde cum angulus BAH indeterminatus

---

(141) In the terminology of note (109) above, but where now $BC = y \equiv y_z$ is the *ordinata* corresponding to the *abscissa* $AB = z$, Newton assumes that, where the base receives the increment $BD = x$, the corresponding augmented ordinate $DE = y_{z+x}$ can be expanded as a power-series $y_z + ax + bx^2 + cx^3 + dx^4 \ldots$ in which (as he has stated above in Corollary 4 to the preceding Proposition XII) $a = \dot{y}/\dot{z}$, $b = \frac{1}{2}\ddot{y}/\dot{z}^2$, $c = \frac{1}{6}\dddot{y}/\dot{z}^3$, $d = \frac{1}{24}\ddddot{y}/\dot{z}^4 \ldots$,; at once, since $\dot{y}/\dot{z}(= dy/dz) = GF/(CG$ or$)$ BD, there ensues $GF = ax$ and so

$$EF = (DF - DE \text{ or})\ y_z + ax - y_{z+x}.$$

(142) See the previous note. There is, we may add, no need in thus constructing the chord of the osculating circle at a general point on a curve defined in oblique Cartesian coordinates separately to compute $b = \lim_{BD\to 0} (EF/BD^2)$ and thereafter set $CO = (CT^2/BT^2)/b$; it is enough directly to evaluate $CO = \lim_{BD\to 0} (FN) = \lim_{BD\to 0} (CF^2/EF)$, and indeed we have reproduced in the previous volume (VI: 582) Newton's contemporary determination in this way of the conjugate

ordinate DE be drawn and produced either way, meeting the curve in E, and there be taken $FN = FC^2/FE = (BD^2/FE).(CT^2/BT^2)$, the point N will fall on the circle which touches the curve at C and passes through the point E. Complete the parallelogram $CBDG$, then call BD or $CG = x$ and let the infinite series $ax+bx^2+cx^3+dx^4+\ldots$ be the difference GE of the ordinates: there will be $GF = ax$ and $EF = -(bx^2+cx^3+dx^4\ldots)$,(141) and so

$$FN = FC^2/EF = (CT^2/BT^2).(CG^2/FE) = (CT^2/BT^2).-1/(b+cx+dx^2\ldots).$$

Let the points E and C coincide, so that x vanishes and the angles of contact of the circle and the curve are equal, and there will come $FN = (CT^2/BT^2).-1/b$. But by Corollary 4 of Proposition XII there is $-1/b = -\dot{z}^2/\frac{1}{2}\ddot{y}$,(142) and hence $FN = CO = (CT^2/BT^2).[-]\dot{z}^2/\frac{1}{2}\ddot{y}$; that is, the circle of the same curvature as the curve at C passes through the point O.

If, for instance, there were to an ellipse VCH the equation $a^2-bz^2 = y^2$, and thence $-2bz\dot{z} = 2y\dot{y}$ and so $-b\dot{z}^2 = \dot{y}^2+y\ddot{y}$, that is, $b^2z^2\dot{z}^2/y^2+y\ddot{y}$, there will be $\dot{z}^2/\frac{1}{2}\ddot{y} = -2y^3/(by^2+b^2z^2)$, that is, $-2y^3/ba^2$, and from this

$$CO = (CT^2/BT^2).-2y^3/ba^2.$$

The negative sign denotes that the straight line CO is to be drawn on the side of the point C towards B. Let z vanish, and CT and BT will prove equal and y will be equal to a, and hence CO equal to $-2a/b$: the point B now falls at the ellipse's centre A and the ordinate BC coincides with the semi-diameter $AH = a$, with $2a/b$ the *latus rectum* pertaining to this semi-diameter.(143) Whence, since

chord of the osculating circle at an arbitrary point on an ellipse which he here proceeds to adopt as his following example.

(143) That is, $2VA^2/AH$ where (since, at V, $a^2-bz^2 = 0$)

$$VA = \sqrt{(a^2/b)}.$$

It follows that, if (as Newton below supposes) HA be produced to meet the circle of curvature at H in K, there will be $HK = 2VA^2/AH$: which is Newton's result—there derived from first principles (compare the preceding note)—on VI: 582.

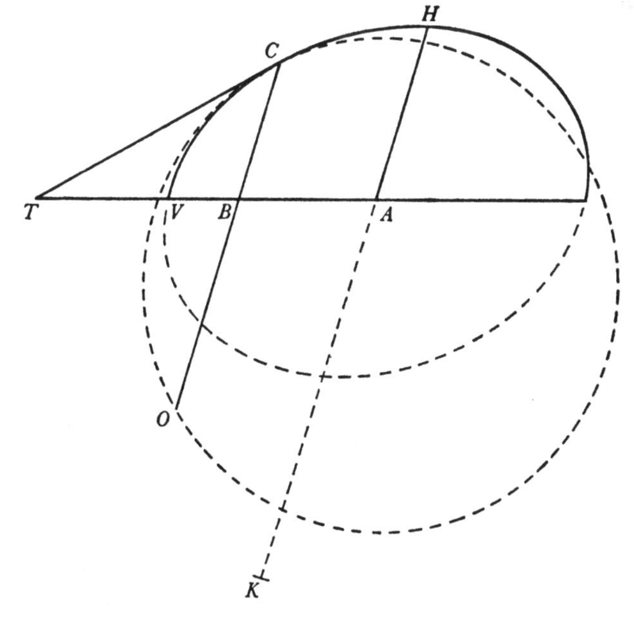

sit et punctum H pro quolibet Ellipseos cujuscunq; puncto sumi possit, sed et Ellipsis mutando signum literæ b in Hyperbolam vertitur,(144) et Hyperbola ubi conjugata sua in infinitum abit migrat in Parabolā: consequens est si a Sectionis cujuscunq; conicæ puncto quocunq; H ducatur ejusdem [semi]diameter HA et in eadem capiatur longitudo HK lateri recto ad diametrum illam pertinenti æqualis, et per puncta H & K describatur circulus qui sectionem conicam tangat in H: hic circulus erit ejusdem curvaturæ cum sectione conica in puncto H.

Si ordinata BC ad abscissam AB uniformiter fluentem perpendicularis est, et curvæ perpendiculariter occurrit in puncto quovis H (quod fit ubi ordinatæ fluxio $\dot{y}$ nulla est;) diameter circuli cujus curvatura eadē est cum curvatura Curvæ in puncto H est $\frac{\dot{z}\dot{z}}{\frac{1}{2}\ddot{y}}$,(145) et exponens curvaturæ est $\frac{\ddot{y}}{2\dot{z}\dot{z}}$, exponens variationis curvaturæ $\frac{\dddot{y}}{6\dot{z}^3}$, exponens variationis variationis sive variationis secundæ $\frac{\ddddot{y}}{24\dot{z}^4}$, exponens variationis variationis variationis sive variationis tertiæ $\frac{\dot{\ddddot{y}}}{120\dot{z}^5}$ & sic deinceps.(146) Nam hi exponentes (per Corol. [4] Prop. [XII]) sunt coefficientes

---

(144) Newton's parallel calculations for this hyperbolic instance, departing from the defining equation '$aa+bz^2 = v^2$' (in which v trivially replaces y) and referred to a partially relettered figure, are preserved on ULC. Add. 3963.15: 180v. With minimal adjustment to the present context these would read: '[Sit] $aa+bz^2 = y^2$. [Erit] $2bz\dot{z} = 2y\dot{y}$. [adeoq;] $b\dot{z}\dot{z} = y\ddot{y}+\dot{y}\dot{y} = y\ddot{y}+\frac{bbzz\dot{z}\dot{z}}{yy}$. [unde] $\frac{byy\dot{z}\dot{z}-bbzz\dot{z}\dot{z}}{y^3} = \ddot{y}$. [hoc est] $\frac{\dot{z}\dot{z}}{\ddot{y}} = \frac{y^3}{byy-bbzz} = \frac{y^3}{baa}$. [Quare] $CO = \frac{2CT^q}{BT^q}\times\frac{y^3}{baa}$. Si B incidit in A est $CO = \frac{2AH^3}{baa} = \frac{2a^3}{baa}$. [ubi] $b = -\frac{AH^2}{VA^2}$. [hoc est] $CO = \frac{2VA^2}{AH}$.'

(145) In the preceding construction of the chord $CO = (CT^2/BT^2)/(\frac{1}{2}\ddot{y}/\dot{z}^2)$ of the osculating circle at C, when C comes to coincide with H (and so, with T passing to infinity, when $\dot{y}/\dot{z} = CT/BT$ is unity) this becomes $HK = 1/(\frac{1}{2}\ddot{y}/\dot{z}^2)$; and when, furthermore, the ordinate angle $\widehat{VBC} = \widehat{VAH}$ is right, it will be clear that HK is the diameter of the circle.

(146) While it is natural to define the curvature of a curve as the reciprocal of the radius ρ—or, as here (where $z = \dot{y} = 0$), of the diameter $2\rho = 2/(\ddot{y}/\dot{z}^2)$—of the osculating circle, we should notice the essential arbitrariness of Newton's present proposal to extend this intrinsic measure $b = \frac{1}{2}\ddot{y}/\dot{z}^2$ at the point H by defining the successive 'variations' in that curvature to be 'represented' by its adjusted fluxional derivatives (with respect to the base z only) $c = \frac{1}{3}\dot{b}/\dot{z}$, $d = \frac{1}{4}\dot{c}/\dot{z}, \ldots$ simply because, as he hastens to add, these are the analogous coefficients of higher-order terms in the 'Taylor' expansion $bz^2+cz^3+dz^4\ldots$ of the difference in magnitude between the general ordinate $BC = y_z$ and the *ordinata prima* $AH = y_0$ when (see the next note) their distance $AB = z$ apart becomes vanishingly small. In Problem 6 of his 1671 tract (see III: 186–92) Newton had proposed the variant measure ds/dp of the first variation in 'quality' of curvature at the point C, where s is the related arc-length measured from some given point in the curve; though this intrinsic measure is not (compare III: 187, note (374)) an absolute invariant of the coordinate system in which the curve is defined, it has the considerable advantage of being essentially independent of any such extrinsic system.

We should add that Newton had already conceived his present generalized notion of

the angle $B\widehat{A}H$ is indeterminate and the point H may be taken as any point whatever of any ellipse at all, while also by changing the sign of the letter b the ellipse turns into a hyperbola,(144) and the hyperbola when its conjugate goes off to infinity passes into a parabola, the consequence is that, if from any point whatever H of any conic there be drawn its semi-diameter HA and in this be taken the length HK equal to the *latus rectum* pertaining to that diameter, and then through the points H and K be described a circle to touch the conic at H, this circle will be of the same curvature as the conic at the point H.

If the ordinate BC is perpendicular to the uniformly fluent abscissa AB and meets the curve perpendicularly at any point H (which happens when the ordinate's fluxion $\dot{y}$ is nil), the diameter of the circle whose curvature is the same as the curvature of the curve at the point H is $\dot{z}^2/\frac{1}{2}\ddot{y}$,(145) and the exponent of curvature is $\ddot{y}/2\dot{z}^2$, the exponent of the variation of curvature is $\dddot{y}/6\dot{z}^3$, the exponent of the variation of the variation or second variation is $\ddddot{y}/24\dot{z}^4$, the exponent of the variation of the variation of the variation or third variation $\overset{\cdot\cdot\cdot\cdot\cdot}{y}/120\dot{z}^5$, and so forth.(146) For these exponents are (by Corollary 4 of Proposition XII)

---

curvature variation when he came in the middle 1680's to compose *Exempl. 1* of Proposition X of the second book of his *Principia* (London, $_1$1687: 263–6). Having there, with respect to abscissa $OB = a$ and perpendicular ordinate $BC = e$, derived the Cartesian equation $e \equiv e_a = \sqrt{[n^2-a^2]}$ of the semicircle of radius n, and thence determined the series expansion $e-(a/e)\,o-\frac{1}{2}(n^2/e^3)\,o^2-\frac{1}{2}(n^2a/e^5)\,o^3\ldots$ of the augmented *ordinata* $DG = e_{a+o}$ corresponding to the increment $BD = o$ of the base, he there went on (*ibid.*: 263–4): 'Hujusmodi Series distinguo in terminos successivos in hunc modum. Terminum primum appello in quo quantitas infinite parva o non extat; secundum in quo quantitas illa extat unius dimensionis; tertium in quo extat duarum, quartum in quo trium est, & sic in infinitum. Et primus terminus, qui hic est e, denotabit semper longitudinem ordinatæ BC insistentis ad indefinitæ quantitatis initium B; secundus terminus $[e'_a o]$ qui hic est $\frac{ao}{e}$, denotabit differentiam inter BC & DF,... atq; adeo positionem Tangentis CF semper determinat.... Terminus tertius $[\frac{1}{2}e''_a o^2]$, qui hic est $\frac{nnoo}{2e^3}$ designabit lineolam FG, quæ jacet inter Tangentem & Curvam, adeoq; determinat angulum contactus FCG, seu curvaturam quam curva linea habet in C.... Terminus quartus $[\frac{1}{6}e'''_a o^3]$, qui hic est $\frac{anno^3}{2e^5}$, exhibet variationem Curvaturæ; quintus variationem variationis, & sic deinceps. Unde obiter patet usus non contemnendus harum Serierum in solutione Problematum, quæ pendent a Tangentibus & curvatura Curvarum'. When thirty years afterwards he came under heavy attack from Johann Bernoulli for not being able correctly to form second differences—an onslaught founded on his nephew Niklaus' ingenious but fundamentally mistaken interpretation of the seat of the error in the main Proposition X which, in this same *Exempl. 1*, led to a provably impossible assignation of the proportion of resistance to gravity necessary to allow motion in a semicircle under constant downwards gravity (see D. T. Whiteside, 'The Mathematical Principles underlying Newton's *Principia Mathematica*' (*Journal for the History of Astronomy*, **1**, 1970: 116–38): 129–8)—Newton was happy in anonymous riposte to seize on this paragraph as evidence that, since he had there put the line FG to be 'but half the second Difference $[e''_a o^2]$ of the Ordinate $[BC = e_a]$... therefore Mr. *Newton* when he wrote his *Principia*, put the third Term of the Series equal to

terminorum seriei convergentis per quam longitudo ordinatæ definitur ubi abscissa z nulla(147) est, et harum coefficientium prima per z multiplicata determinat tangentem, et ubi ordinata tam ad abscissam quam ad curvam perpendicularis est secunda per z^2 multiplicata determinat curvaturam Curvæ, tertia(148) determinat variationem curvaturæ et sic deinceps.(149)

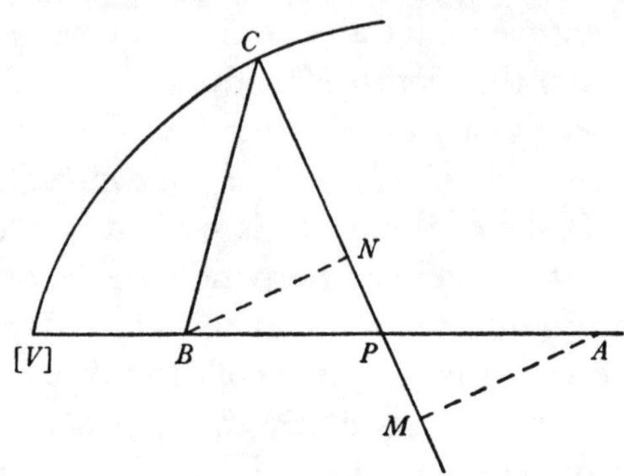

Unde si recta $CB = y$ ad abscissam $AB = z$ in angulo quocunqꝫ dato ABC ordinatim applicetur, et ad Curvæ cujusvis punctum quodcunqꝫ C erigatur ad concavas curvæ partes perpendiculum CP et a punctis A et B ad CP demittantur perpendicula AM, BN, sint autem c ad d ut BC ad CN, c ad e ut BC ad BN, c ad f ut AB ad MN et c ad g ut AB ad $AM+BN$,(150) adeoqꝫ $\frac{g}{c}z-\frac{e}{c}y = AM = x$, et $\frac{f}{c}z+\frac{d}{c}y = CM = v$. et (per Pro[b] 1) $\frac{g}{c}\dot{z}-\frac{e}{c}\dot{y} = \dot{x}$, $\frac{g}{c}\ddot{z}-\frac{e}{c}\ddot{y} = \ddot{x}$, $\frac{g}{c}\dddot{z}-\frac{e}{c}\dddot{y} = \dddot{x}$, $\frac{g}{c}\ddddot{z}-\frac{e}{c}\ddddot{y} = \ddddot{x}$ &c et $\frac{f}{c}\dot{z}+\frac{d}{c}\dot{y} = \dot{v}$, $\frac{f}{c}\ddot{z}+\frac{d}{c}\ddot{y} = \ddot{v}$, $\frac{f}{c}\dddot{z}+\frac{d}{c}\dddot{y} = \dddot{v}$, $\frac{f}{c}\ddddot{z}+\frac{d}{c}\ddddot{y} = \ddddot{v}$ &c. Et si x

---

half of the second Difference of the first Term, and by consequence had not then forgotten the Method of second Differences' (see his 'An Account of the Book entituled *Commercium Epistolicum*...' (*Philosophical Transactions*, **29**, No. 342 [for January/February 1714/15]: 173–224): 207). We shall return to this point in our final volume when we reproduce the several differing revised forms of Proposition X which Newton was impelled to draft in correction of its subtly erroneous first version in the autumn of 1712.

(147) Understand rather 'infinite parva' (infinitely small) since when $z = 0$ the ordinate BC comes absolutely to coincide with the prime ordinate AH, so that these can have no difference of any infinitesimal order.

(148) It is presumably by an oversight that Newton has here omitted analogously to specify 'per z^3 multiplicata' (multiplied by z^3)—as indeed he had added at the equivalent point in an immediately preceding, cancelled first draft of the present phrase.

(149) A first draft by Newton on p. 189 of the manuscript originally combined the content of this and the following paragraphs to be: 'Porro curvarum curvatura juxta et omnes variationes curvaturæ sic inveniuntur. Ad curvæ cujusvis punctum quodcunqꝫ C erigatur perpendiculum CP et ad hanc demittantur perpendicula AM, BN et nominetur $AM = x$ et $CM = v$. Sitqꝫ c ad e ut BC seu y ad BP et inde $BP = \frac{e}{c}y$ & $AP = z-\frac{e}{c}y$. Sit insuper f ad c ut AP ad AM et g ad c ut AB ad MN et g ad d ut BC ad CN. et inde $AM = \frac{c}{f}z-\frac{e}{f}y = x$. $MN = \frac{c}{g}z$. $CN = \frac{d}{g}y$ et $C[M] = \frac{d}{g}y+\frac{c}{g}z = v$. Deinde fluxionibus $\dot{z}$, $\ddot{z}$, $\dddot{z}$, $\ddddot{z}$, $\dot{y}$, $\ddot{y}$, $\dddot{y}$, $\ddddot{y}$ &c per æquationem qua curva determinatur inventis ponantur juxta Pro[b] 1,

the coefficients of the terms in the converging series by which the length of the ordinate is defined when the abscissa z is nil,[147] and of these coefficients the first multiplied by z determines the tangent, and—when the ordinate is perpendicular both to the abscissa and the curve—the second multiplied by z^2 determines the curvature of the curve, the third[148] determines the variation of the curvature, and so on.[149]

Whence, if the straight line $CB = y$ be ordinate to the abscissa $AB = z$ at any given angle $A\widehat{B}C$, and at any point whatever of any curve a perpendicular CP be erected on the curve's concave side and from the points A and B perpendiculars AM, BN be let fall to CP, let however c be to d as BC to CN, c to e as BC to BN, c to f as AB to MN, and c to g as AB to $AM+BN$,[150] and therefore $(g/c)\,z-(e/c)\,y = AM = x$ and $(f/c)\,z+(d/c)\,y = MC = v$: then (by Problem 1) $(g/c)\,\dot{z}-(e/c)\,\dot{y} = \dot{x}$, $(g/c)\ddot{z}-(e/c)\,\ddot{y} = \ddot{x}$, $(g/c)\,\dddot{z}-(e/c)\,\dddot{y} = \dddot{x}$, $(g/c)\,\ddddot{z}-(e/c)\,\ddddot{y} = \ddddot{x}$,... and likewise $(f/c)\,\dot{z}+(d/c)\,\dot{y} = \dot{v}$, $(f/c)\ddot{z}+(d/c)\,\ddot{y} = \ddot{v}$, $(f/c)\,\dddot{z}+(d/c)\,\dddot{y} = \dddot{v}$,

---

$$\frac{c}{f}\dot{z}-\frac{e}{f}\dot{y}=\dot{x}.\quad \frac{c}{f}\ddot{z}-\frac{e}{f}\ddot{y}=\ddot{x}.\quad \frac{c}{f}\dddot{z}-\frac{e}{f}\dddot{y}=\dddot{x}.\quad \frac{c}{f}\ddddot{z}-\frac{e}{f}\ddddot{y}=\ddddot{x}.\ \&c$$

et rur[s]us $\quad \frac{d}{g}\dot{y}+\frac{c}{g}\dot{z}=\dot{v}.\quad \frac{d}{g}\ddot{y}+\frac{c}{g}\ddot{z}=\ddot{v}.\quad \frac{d}{g}\dddot{y}+\frac{c}{g}\dddot{z}=\dddot{v}.\quad \frac{d}{g}\ddddot{y}+\frac{c}{g}\ddddot{z}=\ddddot{v}.$ [&c]

et $\frac{2\dot{x}^2}{\ddot{v}}$ diameter erit circuli eandem habentis curvaturam cum curva in puncto C adeoq; $\frac{\ddot{v}}{2\dot{x}^2}$ proportionalis erit angulo contactus et curvaturæ curvæ. Et sim[iliter] $\frac{\dddot{v}}{6\dot{x}^3}$ proportionalis erit variationi curvaturæ et $\frac{\ddddot{v}}{24\dot{x}^4}$ proportionalis erit variationis variationi seu variationi secundæ, et $\frac{\overset{....}{\ddot{v}}}{120\dot{x}^5}$ proportionalis erit hujus variationis variationi seu variationi tertiæ: et sic deinceps in infinitum' (Furthermore, the curvature of curves in line with all variations of the curvature may be found in this way. At any arbitrary point C of any curve erect the normal CP and to this drop the perpendiculars AM, BN and name $AM = x$ and $MC = v$. Let c be to e as BC (or y) to BP, and thence $BP = (e/c)\,y$ and so $AP = z-(e/c)\,y$. Let there in addition be f to c as AP to AM, g to c as AB to MN and g to d as BC to CN, and so $AM = (c/f)\,z-(e/f)\,y = x$, $MN = (c/g)\,z$, $CN = (d/g)\,y$ and $CM = (d/g)\,y+(c/g)\,z = v$. Next, once the fluxions $\dot{z}, \ddot{z}, \dddot{z}, \ddddot{z},\ldots, \dot{y}, \ddot{y}, \dddot{y}, \ddddot{y},\ldots$ have been ascertained by means of the equation whereby the curve is determined, let there in line with Problem 1 be put $(c/f)\,\dot{z}-(e/f)\,\dot{y} = \dot{x}$, $(c/f)\,\ddot{z}-(e/f)\,\ddot{y} = \ddot{x}$, $(c/f)\,\dddot{z}-(e/f)\,\dddot{y} = \dddot{x}$, $(c/f)\,\ddddot{z}-(e/f)\,\ddddot{y} = \ddddot{x},\ldots$ and again $(d/g)\,\dot{y}+(c/g)\,\dot{z} = \dot{v}$, $(d/g)\ddot{y}+(c/g)\,\ddot{z} = \ddot{v}$,

$$(d/g)\,\dddot{y}+(c/g)\,\dddot{z} = \dddot{v},\quad (d/g)\,\ddddot{y}+(c/g)\,\ddddot{z} = \ddddot{v},\ldots.$$

And then $2\dot{x}^2/\ddot{v}$ will be the diameter of a circle having the same curvature as the curve at the point C, and hence $\frac{1}{2}\ddot{v}/\dot{x}^2$ will be proportional to the angle of contact and so to the curvature of the curve. And similarly $\frac{1}{6}\dddot{v}/\dot{x}^3$ will be proportional to the variation of the curvature, $\frac{1}{24}\ddddot{v}/\dot{x}^4$ proportional to the variation of the variation or second variation, $\frac{1}{120}\overset{....}{\ddot{v}}/\dot{x}^5$ to the variation of this variation or third variation, and so on in turn indefinitely).

(150) These ratios will not, of course, be constant for different points C on the curve, but, since Newton in the sequel is concerned only with fluxional variations in the curve's arc within a vanishingly small distance from some fixed C, he permissibly assumes them to be so for his purpose.

fluat uniformiter ut $\ddot{x}$, $\dddot{x}$, $\ddddot{x}$ &c evanescant, & ipsius v fluxio $\dot{v}$ nulla sit,(151) deinde ex his æquationibus et æquatione qua relatio inter abscissam z et ordinatam y [definitur] & æquationibus inde per Pro[b] 1 collectis inveniantur fluxionū rationes(152) $\frac{\ddot{v}}{\dot{x}^2}, \frac{\dddot{v}}{\dot{x}^3}, \frac{\ddddot{v}}{\dot{x}^4}$ &c: Ratio prima divisa per 2 nempe $\frac{\ddot{v}}{2\dot{x}^2}$ exponens erit curvaturæ curvæ ad punctum C, et ejus reciprocum $\frac{2\dot{x}^2}{\ddot{v}}$ diameter erit circuli eandem habentis curvaturam cum curva ad idem punctum C. Ratio secunda divisa per 2×3 nempe $\frac{\dddot{v}}{6\dot{x}^3}$ exponens erit variationis curvaturæ ad idem punctum. Ratio tertia divisa per $2\times 3\times 4$ nempe $\frac{\ddddot{v}}{24\dot{x}^4}$ exponens erit variationis illius variationis sive variationis secundæ et sic deinceps in infinitum. Nam fluxiones $\dot{x}\ \ddot{x}\ \dddot{x}$ &c(153) $\dot{v}\ \ddot{v}\ \dddot{v}$ &c eædem sunt ac si recta CM ordinatim applicaretur ad abscissam uniformiter fluentem AM. Exponentes autem curvaturæ et variationū curvaturæ hic voco fluxionum rationes(154) curvaturæ et variationibus ejus proportionales, quæ quidem rationes hic nihil aliud sint quam numeri. Rationes verò divido per numeros 2, 2×3, $2\times 3\times 4$ &c ut sint etiam exponentes angulorum contactus ex quibus curvatura et ejus variationes oriuntur. Nam si ad distantiam quamvis(155) r subtendatur angulus contactus, et subtensa distantiæ perpendicularis existens nominetur s[,] dein distantia et subtensa in infinitum diminuantur, ratio ultima $\frac{s}{rr}$ erit $\frac{\ddot{v}}{2\dot{x}^2}$ ubi angulus contactus is est quem recta et curva continent et a quo curvatura oritur, et $\frac{s}{r^3}$ erit $\frac{\dddot{v}}{6\dot{x}^3}$ ubi angulus contactus is est quem circulus et curva æqualiter curvæ continent et a quo variatio curvaturæ oritur, et $\frac{s}{r^4}$ erit $\frac{\ddddot{v}}{24\dot{x}^4}$ ubi angulus contactus is est a quo variatio illius variationis oritur & sic in reliquis.

In exemplum hujus Regulæ proponatur iterum ad Ellipsin æquatio $aa-bzz=yy$.(156) et erit $-bz\dot{z}=y\dot{y}$, $-b\dot{z}\dot{z}-bz\ddot{z}=\dot{y}\dot{y}+y\ddot{y}$,

$$-3b\dot{z}\ddot{z}-bz\dddot{z}=3\dot{y}\ddot{y}+y\dddot{y}\ \&c$$

et per regulam $\frac{g}{c}z-\frac{e}{c}y=x$. $\frac{g}{c}\dot{z}-\frac{e}{c}\dot{y}=\dot{x}$. $\frac{g}{c}\ddot{z}-\frac{e}{c}\ddot{y}=0$. $\frac{g}{c}\dddot{z}-\frac{e}{c}\dddot{y}=0$ &c.(157) Item

(151) Necessarily so, since in the immediate vicinity of C the tangential direction of the curve is parallel to the abscissa AM, which therefore instantaneously 'flows' in length but not in position, carrying with it the perpendicular ordinate MC, which therefore *vice versa* 'flows' instantaneously parallel to itself without (to a first infinitesimal order at least) varying its length v.

(152) Originally 'proportiones' (proportions). Newton has made three corresponding replacements of 'proportio' by 'ratio' in the next few lines of his manuscript text.

(153) Here, since he has earlier taken x to be the base variable (so that $\ddot{x}=\dddot{x}=\ldots=0$),

$(f/c)\ddddot{z}+(d/c)\,\ddddot{y}=\ddddot{v}$, ... And if x should flow uniformly, so that $\ddot{x}$, $\dddot{x}$, $\ddddot{x}$, ... vanish, and the fluxion $\dot{v}$ of v be nil,(151) and then from these equations and the equation whereby the relationship between the abscissa z and ordinate y is defined and the equations collected therefrom by Problem 1 there be found the fluxional ratios(152) $\ddot{v}/\dot{x}^2$, $\dddot{v}/\dot{x}^3$, $\ddddot{v}/\dot{x}^4$, ..., the first ratio divided by 2, namely $\ddot{v}/2\dot{x}^2$, will be the exponent of the curvature of the curve at the point C, and its reciprocal $2\dot{x}^2/\ddot{v}$ will be the diameter of the circle having the same curvature as the curve at the same point C; the second ratio divided by 2×3, namely $\dddot{v}/6\dot{x}^3$, will be the exponent of the variation of curvature at the same point; the third ratio divided by $2\times3\times4$, namely $\ddddot{v}/24\dot{x}^4$, will be the exponent of the variation of that variation or second variation; and so on indefinitely. For the fluxions $\dot{x}$, $\dot{v}$, $\ddot{v}$, $\dddot{v}$, ... are the same as though the straight line CM were ordinate to the uniformly fluent abscissa AM. The exponents, however, of curvature and the variations of curvature I here call fluxional ratios(154) proportional to the curvature and its variations, since indeed these ratios are here nothing other than numbers. These ratios, may I add, I divide by the numbers 2, 2×3, $2\times3\times4$, ... so that they may also be exponents of the angles of contact from which the curvature and its variations arise. For, if an angle of contact be subtended at any distance(155) r and the subtense—perpendicular to the distance—be named s, and then the distance and subtense be indefinitely diminished, the last ratio of s/r^2 will be $\ddot{v}/2\dot{x}^2$ when the angle of contact is that contained by a straight line and a curve, that of s/r^3 will be $\dddot{v}/6\dot{x}^3$ when the angle of contact is that contained by a circle and a curve equally curved and the variation in curvature arises from it, that of s/r^4 will be $\ddddot{v}/24\dot{x}^4$ when the angle of contact is that from which the variation of that variation arises, and so on in remaining cases.

In exemplification of this rule let there again be proposed the equation $a^2-bz^2=y^2$ to an ellipse.(156) Then $-bz\dot{z}=y\dot{y}$, $-b\dot{z}^2-bz\ddot{z}=\dot{y}^2+y\ddot{y}$,

$$-3b\dot{z}\ddot{z}-bz\dddot{z}=3\dot{y}\ddot{y}+y\dddot{y},\ \ldots,$$

and by the rule $(g/c)\,z-(e/c)\,y=x$, $(g/c)\,\dot{z}-(e/c)\,\dot{y}=\dot{x}$, $(g/c)\ddot{z}-(e/c)\ddot{y}=0$,

$$(g/c)\,\dddot{z}-(e/c)\,\dddot{y}=0,\ \ldots;^{(157)}$$

Newton clearly meant to write '$\dot{x}$' only; we have forborne to render the superfluous '$\ddot{x}$, $\dddot{x}$ &c' in our English version.

(154) Newton initially began to add 'sive numeros expo[nentes]' (or exponent numbers) before changing his mind and deleting this unfinished adjunction.

(155) Originally 'ad radium quemvis' (at any radius).

(156) Newton's extensive preliminary calculations for this example, here drastically curtailed, exist on ULC. Add. 3960.9: 161/163. Because of their considerable intrinsic interest—and not least for the abridged notation for fluxions of high order which Newton there at one point uniquely employs—we reproduce them in full in Appendix 1.12 below.

(157) Since x is the base variable there is $\ddot{x}=\dddot{x}=\ldots=0$.

$\frac{f}{c}z+\frac{d}{c}y=v.\ \frac{f}{c}\dot{z}+\frac{d}{c}\dot{y}=\dot{v}=0.\ \frac{f}{c}\ddot{z}+\frac{d}{c}\ddot{y}=\ddot{v}.\ \frac{f}{c}\dddot{z}+\frac{d}{c}\dddot{y}=\dddot{v}$ &c. Ex his æquationibus exterminando $\dot{z}$, $\ddot{z}$, $\dddot{z}$, $\dot{y}$, $\ddot{y}$, $\dddot{y}$, &c invenientur $\frac{\ddot{v}}{2\dot{x}^2}$, $\frac{\dddot{v}}{6\dot{x}^3}$ &c.[158]. Sed calculus abbreviabitur faciendo ut punctum B coincidat cum Ellipseos centro A, et ordinata BC cum [semi]diametro $A[H]$.[159] Sic enim quantitates z et f et fluxio $\dot{y}$ evanescent et y[160] et v adeoq etiam c & g æquales evadent, et æquationes superiores in has vertentur $a=y$, $-b\dot{z}^2=\ddot{y}y$. $-3b\dot{z}\ddot{z}=y\dddot{y}$. &c. $z-\frac{e}{c}y=x$. $\dot{z}=\dot{x}$. $\ddot{z}-\frac{e}{c}\ddot{y}=0$. $\dddot{z}-\frac{e}{c}\dddot{y}=0$. [&c] $\frac{d}{c}y=v$. $\frac{d}{c}\ddot{y}=\ddot{v}$. $\frac{d}{c}\dddot{y}=\dddot{v}$. [&c]. Et ex his æquationibus facile prodeunt[161] $-\frac{bd}{2ac}=\frac{\ddot{v}}{2\dot{x}^2}$ exponens curvaturæ, et $\frac{bbde}{2a^2cc}=\frac{\dddot{v}}{6\dot{x}^3}$ exponens variationis curvaturæ. [162]Unde cùm $\frac{2a}{b}$ sit latus rectum Ellipseos ad verticem $[H]$ pertinens, si pro hoc latere scribatur R, exponentium reciproca erunt $\frac{c}{d}R$ et $\frac{cc}{2de}RR$. Ideoq $\frac{c}{d}R$ diameter est circuli eandem habens curvaturam cum Ellipsi in puncto $[H]$, estq hæc diameter ad latus rectum R ut c ad d hoc est ut CB ad CN:[163] et propterea hic circulus aufert ab Ellipseos semidiametro CB vel $[H]A$ producta longitudinem lateri recto R æqualem, quemadmodum etiam supra[164] ostensum est. Variatio autem curvaturæ est reciprocè ut quadratum hujus [semi]diametri[165] ductum in $\frac{CN}{2BN}$.[163] Et cum punctum $[H]$ ubivis in Ellipsi sumi possit et Ellipsis mutando signum litteræ b in Hyperbolam verti, Hyperbola autem ubi conjugata sua in infinitū abit et evanescit in Parabolam mutatur, hæ Ellipseos proprietates ad omnia omnium Sectionum Conicarum puncta obtinent.

His methodis determinantur proprietates curvarum. Si vicissim a proprietatibus ad Curvas pergendum est; debent proprietates ad fluxiones rectarum a quibus curvæ determinantur reduci ac deinde rectæ illæ a fluxionibus suis

(158) In preliminary computation (see Appendix 1.12 below) Newton in fact determined the 'exponent' of curvature $\frac{1}{2}\ddot{v}/\dot{x}^2$ to be $-\frac{1}{2}a^2bc(ef+dg)/(gy+ebz)^3$ before proceeding to the following abbreviated method of calculation.

(159) Compare the figure in note (143) above. Here likewise it is assumed that the chord CO of the osculating circle at C passes into the parallel one—which Newton earlier denominated as HK (see page 112 above)—of the circle of curvature at H.

(160) Read '$\frac{d}{c}y$'. Newton here carelessly repeats a momentary slip in his preliminary computations (see Appendix 1.12: note (126)) but does not, fortunately, carry it into the present sequel.

likewise

$$(f/c)\,z+(d/c)\,y=v,\quad (f/c)\,\dot z+(d/c)\,\dot y=\dot v=0,\quad (f/c)\,\ddot z+(d/c)\,\ddot y=\ddot v,$$
$$(f/c)\,\dddot z+(d/c)\,\dddot y=\dddot v,\ \ldots.$$

And on eliminating $\dot z$, $\ddot z$, $\dddot z$, ..., $\dot y$, $\ddot y$, $\dddot y$, ... from these equations there will be found $\ddot v/2\dot x^2$, $\dddot v/6\dot x^3$, ...(158) But the computation will be shortened by making the point *B* coincide with the ellipse's centre *A* and the ordinate *BC* with the semi-diameter *AH*.(159) For in this way the quantities z and f and the fluxion $\dot y$ shall vanish, and so y(160) and v and hence also c and g come to be equal, and the previous equations turn into these: $a=y$, $-b\dot z^2=y\ddot y$, $-3b\dot z\ddot z=y\dddot y$, ...; $z-(e/c)\,y=x$, $\dot z=\dot x$, $\ddot z-(e/c)\,\ddot y=0$, $\dddot z-(e/c)\,\dddot y=0$, ...; $(d/c)\,y=v$, $(d/c)\,\ddot y=\ddot v$, $(d/c)\,\dddot y=\dddot v$, ... And from these equations there easily ensue(161) $-bd/2ac=\ddot v/2\dot x^2$ for the exponent of curvature, and $b^2de/2a^2c^2=\dddot v/6\dot x^3$ for the exponent of the variation of curvature.(162) Whence, since $2a/b$ is the *latus rectum* of the ellipse pertaining to the vertex *H*, if *R* be written for this *latus* the reciprocals of the exponents will be $(c/d)\,R$ and $(c^2/2de)\,R^2$. Consequently, $(c/d)\,R$ is the diameter of a circle having the same curvature as the ellipse at the point *H*, and this diameter is to the *latus rectum R* as c to d, that is, as *CB* to *CN*;(163) accordingly, this circle abstracts from the ellipse's semi-diameter *CB* or *HA* produced a length equal to the *latus rectum R*, as has also been shown above.(164) The variation of the curvature, however, is reciprocally as the square of this semi-diameter(165) multiplied by $CN/2BN$.(163) And since the point *H* may be taken anywhere on the ellipse, and the ellipse by changing the sign of the letter b turned into a hyperbola, while the hyperbola when its conjugate goes off and vanishes to infinity is changed into a parabola, these properties of the ellipse hold good for all points on all conics.

By these methods the properties of curves are determined. If conversely you need to proceed from the properties to the curves, the properties must be reduced to fluxions of the straight lines by which the curves are determined, and then those lines derived from their fluxions. If, for instance, the curve *CE* be sought

(161) Newton's calculations of the two following 'exponents' (of the ellipse's curvature at *H* and its first variation) are reproduced in Appendix 1.12. With some difficulty—and after two false starts—he there went on to compute that the exponent $\frac{1}{24}\ddddot v/\dot x^4$ of second variation of curvature at *H* is $-(5b^3e^2+b^2c^2)\,d/8a^3c^3$ (see *ibid.*: note (138)).

(162) A rough but identical draft of the two following sentences exists on Add. 3960.9: 163.

(163) Understand when *BC* comes to be *AH*, in which case the subtense *BN* (ever parallel to the ellipse's tangential direction at *C*) coincides with the (extended) abscissa *VA*, and so is the projection of *AH* upon it; in consequence $CN/CB=c/d$ and $CN/BN=d/e$ are then the sine and tangent respectively of the ordinate angle $A\widehat{B}C$.

(164) See pages 110–12 preceding.

(165) Namely, $(c/d)^2R^2=4a^2c^2/b^2d^2$.

derivari. Ut si quæratur curva CE quam recta CT tangit in C(166) ac detur æquatio qua relatio inter curvæ ordinatam $BC = y$, et rectas AB, AT [definitur] quarum altera $AB = z$ ad ordinatam altera $AT = v$ ad tangentem terminatur: quoniam ipsarum z et y fluxiones $\dot{z}$ et $\dot{y}$ sunt ad invicem ut BT ad BC hoc est ut $z+v$ ad y[,] adeoq; est $y\dot{z} = \dot{y}z+\dot{y}v$ & $v = \dfrac{y\dot{z}-\dot{y}z}{\dot{y}}$;

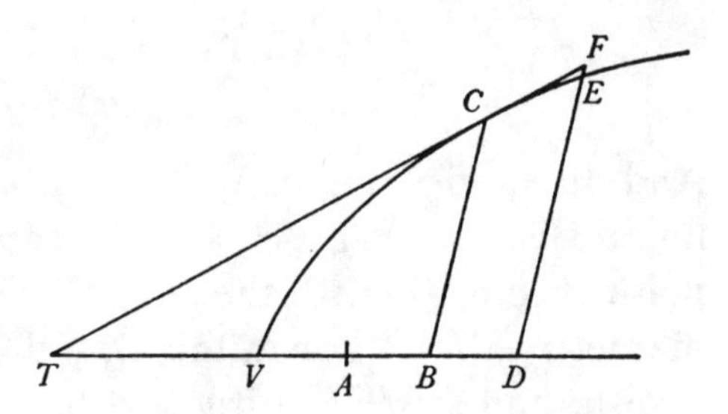

in æquatione data scribendo hunc $\dfrac{y\dot{z}-\dot{y}z}{\dot{y}}$ pro v producetur æquatio nova quæ solas quantitates fluentes y et z una cum earum fluxionibus $\dot{y}$ et $\dot{z}$ involvet. Et ex hac æquatione per Prop [XI] eruendo quantitates y et z solvetur Problema.

Si a tangentis CF puncto quovis F ad Curvam quamvis ducatur FE quæ vel datæ positione rectæ [CB] parallela sit(167) vel ad datum punctum [P](168) convergat, et longitudo $\dfrac{CF^q}{FE}$ quæ ultimò fit ubi punctum F coincidit cum puncto contactûs C æquationem quamvis ingrediatur: substituendæ sunt rectarum fluxiones analogæ, et ex fluxionibus rectæ colligendæ.(169)

Si area Curvæ investigandæ æquationem ingrediatur[,] dicendum est quod fluxio areæ sit ad fluxionem rectanguli sub abscissa et recta data ut ordinatim applicata ad rectam illam datam, et ex æquatione quæ ab hac proportione oritur et æquatione priore determinandæ erunt quantitates fluentes.

Si longitudo curvæ æquationem ingrediatur, relatio fluxionis longitudinis illius ad fluxiones abscissæ et ordinatæ dabunt æquationē novam, et ex æquatione utraq; determinabuntur quant[it]ates fluentes.

Et universaliter si detur Curvæ proprietas quæcunq;, fluxiones rectarum ex hac proprietate derivari debent et ex fluxionibus ipsæ rectæ.

Ut si invenienda sit curva cujus longitudo v sit media proportionalis inter abscissam z et rectam datam a, erit $vv = az$ et $2v\dot{v} = a\dot{z}$, hoc est $2\dot{v}\sqrt{az} = a\dot{z}$. Sit

(166) The manuscript lacks any corresponding figure; that accompanying the Latin text is our editorial interpolation, but it cannot be too far removed from Newton's intention.

(167) See the preceding figure.

(168) So Newton specified in an immediately preceding, cancelled equivalent phrase which mentioned only this latter, general case in which the line EF rotates round a given pole. Again (compare note (166)) to repair the manuscript's deficiency in this regard, we have introduced an appropriate illustrative figure into our English version at this point.

(169) This replaces the clumsier equivalent direction that 'quærenda est relatio ipsarum CF et EF [et] exponenda est hæc ratio $\dfrac{CF^q}{FE}$ per rationem fluxionis' (the relationship between

which the straight line CT touches at C,(166) and there be given the equation whereby the relationship between the curve's ordinate $BC = y$ and the straight lines AB, AT is defined, one of which $AB = z$ terminates at the ordinate and the other $AT = v$ at the tangent, because the fluxions $\dot{z}$ and $\dot{y}$ of z and y are to each other as BT to BC, that is, as $z+v$ to y, and hence $y\dot{z} = \dot{y}(z+v)$ and so $v = (y\dot{z}-\dot{y}z)/\dot{y}$, by writing this $(y\dot{z}-\dot{y}z)/\dot{y}$ in v's place in the given equation there will be produced a new equation involving the fluent quantities y and z alone together with their fluxions $\dot{y}$ and $\dot{z}$. And on eliciting the quantities y and z from this equation by means of Proposition XI the problem will be solved.

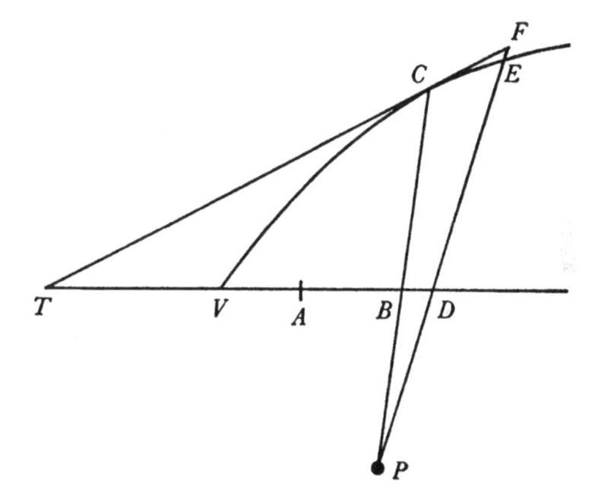

If from any point F of the tangent CF to any curve there be drawn FE either parallel to the straight line CB given in position(167) or converging on a given point P,(168) and the ultimate value of the length CF^2/FE when the point F comes to coincide with the point C of contact should enter into any equation, analogous fluxions of the straight lines are to be substituted for it, and then from their fluxions the lines themselves obtained.(169)

If the area of the curve to be discovered should enter into the equation, you will need to state that the fluxion of the area is to the fluxion of the rectangle contained by the abscissa and a given straight line as the ordinate to that given line, and then determine the fluent quantities from the equation arising from this proportion and from the previous equation.

If the length of the curve enters into the equation, the relationship of the fluxion of that length to the fluxions of abscissa and ordinate will yield a fresh equation, and the fluent quantities will be determined from both equations.

And universally, if there be given any property whatever of a curve, the fluxions of the straight lines must be derived from this property and then the lines themselves from their fluxions.

If, for instance, there has to be found a curve whose length v shall be a mean proportional between the abscissa z and a given straight line a, there will be $v^2 = az$ and so $2v\dot{v} = a\dot{z}$, that is, $2\dot{v}\sqrt{(az)} = a\dot{z}$. Let the ordinate be y and its angle

---

CF and EF needs to be ascertained, and then this ratio CF^2/FE expressed by means of the ratio of their fluxion). Much as before, of course, the 'length' $\lim_{E, F\to C} (CF^2/FE)$ is that of the chord of the osculating circle at C which passes through B.

ordinatim applicata y & angulus applicationis rectus et erit

$$\dot{y}^2+\dot{z}^2=\dot{v}^2=\frac{aa\dot{z}^2}{[4]az} \quad \text{et} \quad \dot{y}=\dot{z}\sqrt{\frac{aa-[4]az}{[4]az}}.^{(170)}$$

Haud secus si alia aliqua curvaturæ vel variationū ejus proprietas habeatur vel si curvæ longitudo, vel area vel centrum gravitatis aut horum relatio vel proprietas aliqua vel quæstionem ingrediatur vel ex conditionibus quæstionis derivari possit; ex his colligendæ sunt rectarum fluxiones primæ, secundæ, tertiæ &c et ex fluxionibus ipsæ rectæ. Nam per rectas sic inventas curvæ quæsitæ determinari solent.

Tandem ut horum omnium usus clarior sit, lubet Problema unum et alterum subjungere quod cum rerum natura conjunctum sit.(171)

Prob. 1.

Corpora gravia sunt in centrum positione datum et per cælos trajecta incurvantur vi gravitatis in idem centrum. Datur linea curva quam corpus tali motu in medio non resistente describit & requiritur quantitas gravitatis in locis singulis.(172)

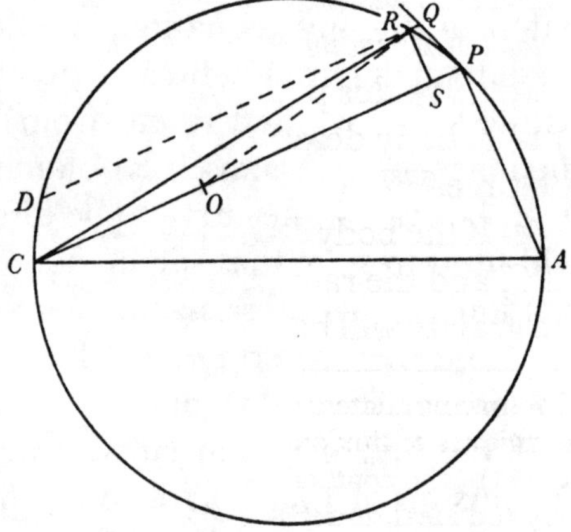

Ad solutionem Problematis hujus utimur his Propositionibus Lemmaticis,

1. quod corpus radio ad centrum ducto aream tempori proportionalem describit.(173)

2. Quod si corpus P moveatur circa centrum $[O]$ in perimetro circuli APC, et radius OP circulo occurrat in C sitq; CA diameter circuli[,] gravitas erit reciproce ut

$$\frac{OP^{\text{quad}}\times CP^{\text{cub}}}{CA^{\text{quad}}}.^{(174)}$$

(170) Though he does not go on to specify the fluent equation

$$y-k = \sqrt{[z(\tfrac{1}{4}a-z)]}+\tfrac{1}{8}a\cos^{-1}(1-8z/a)$$

which ensues straightforwardly from it by simple integration, Newton was well aware (see III: 162 and compare *ibid.*: 163, note (307)) that this states the 'Fermatian' defining property of the cycloid generated by rolling the circle $x^2 = \frac{1}{4}az-z^2$ along the line $z = \frac{1}{4}a$: namely, that the slope $\dot{y}/\dot{z}$ of the cycloid's tangent at its general point (z, y) is equal to that, x/z, of the corresponding chord of the generating circle. We shall see in the final volume that when, on returning home from a hard day's work at the Mint in late January 1697, Newton was confronted with Bernoulli's challenge to identify the brachistochrone under simple gravity and then succeeded in reducing its defining dynamical condition to a similar fluxional equation (see VI: 460–1, note (14)), he found no difficulty in at once naming the corresponding 'fluent' curve. Unaware five years earlier of such an application to the 'real' world, and grown suddenly impatient of such pure mathematical subtleties, he here breaks off to cancel the preceding paragraphs, briefly sketching their general theme and fluxional solution in a couple

of application right, and then $\dot{y}^2+\dot{z}^2 = \dot{v}^2 = a^2\dot{z}^2/4az$ and so

$$\dot{y} = \dot{z}\surd[(a^2-4az)/4az].^{(170)}$$

No differently, if any other property of curvature or its variations be had, or if the curve's length or area or centre of gravity or their relationship or some property of them should either enter into the question or be derivable from the question's conditions, from these must be gathered the first, second, third, ... fluxions of straight lines, and then from these fluxions the lines themselves. For the curves required are usually determined by straight lines ascertained in this way.

Lastly, to make the use of all these precepts clearer, it is agreeable to append a problem or two which shall have some connection with the physical world.[171]

Problem 1.

Bodies gravitate to a centre given in position and, as they shoot across the heavens, are curved by the force of their gravity towards the same centre. There is given the curve which a body describes with such a motion in a non-resisting medium, and the 'quantity' of gravity at each of its individual places is required.[172]

To solve this problem we make use of the following lemmatical propositions:

1. A body describes by a radius drawn to the centre an area proportional to the time.[173]

2. If the body P should move round the centre O in the perimeter of the circle APC, and the radius OP meet the circle in C, let CA be the circle's diameter and the gravity will be reciprocally as $OP^2 \times CP^3/CA^2$.[174] For draw the straight line

of following sentences before quickly passing on to instance the usefulness of his exhibited techniques of fluxional analysis in resolving problems relating to central forces.

(171) The contact with physical reality is slim, however, since Newton proceeds to discuss only the ideal, unresisted motions of mass-less 'heavenly bodies' as they orbit freely in otherwise empty space under their instantaneously uniform *vis inertiæ* and the diverting tug of their 'gravity' to a central point-'sun'. In immediate sequel, without specifying its number or giving it a preliminary general enunciation, he initially went straight on to consider the particular problem in which 'Datur cent[r]um C ad quod corpora omnia vi gravitatis tendunt' (There is given the centre C to which all bodies by the force of gravity tend...), but his instinct to present it more formally rapidly prevailed.

(172) The direct problem of measuring the law of gravity under which free motion in a given trajectory round the force-centre may be induced. In his ensuing 'lemmatical' propositions Newton summarily enunciates his generalization of the first Keplerian law of planetary motion, and then employs it to derive the measure of the force to an arbitrary point in its plane under which a given circle—and so the curve which it osculates—may be traversed.

(173) The fundamental generalized Keplerian theorem which Newton had rightly set in grand place as Proposition I of the first book of his 1687 *Principia* (compare VI: 34 and 122–4).

(174) As we have seen (VI: 550) Newton had only recently attained this generalization of Proposition VII of his *Principia*'s first book ($_1$1687: 45–6; compare VI: 42–4 and 134). In applying this formula Newton below replaces CA by the equal diameter PB through P.

Nam ducatur recta PQ circulum tangens in P ipsiqȝ PC parallela agatur QD circulo occurrens in R ac D, jungatur RO et ad PO demittatur perpendiculum RS, et si tempus omne dividatur in æqualia momenta,(175) et corpus absqȝ gravitate unico temporis momento ea cum velocitate quam habet in P describere posset tangentem PQ et vi gravitatis describat arcum PR: gravitas qua deducitur a tangente in arcum erit ut intervallum $QR_{[,]}$ hoc est, ut $\frac{PQ^q}{DQ}$; hoc est (ob proportionales CP ad CA ut RS ad PQ) ut $\frac{CA^q \times RS^q}{CP^q \times DQ}$, hoc est (cùm sit area trianguli OPR per Lemma 1, dato tempori proportionalis, adeoqȝ OP ipsi RS reciproce proportionalis) ut $\frac{CA^q}{OP^q \times CP^q \times DQ}$, id est (si ob infinite parvū temporis momentum coeant DQ, CP) reciproce ut $\frac{OP^{\text{quad.}} \times CP^{\text{cub.}}}{CA^{\text{quad.}}}$. Q.E.O.

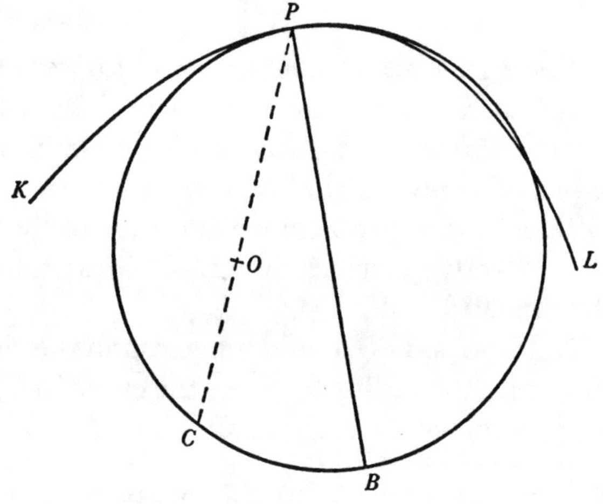

His præmissis Problema sic solvitur.(176)

Designet KPL lineam curvam in qua corpus movetur, P corpus ipsum, O centrum in quod grave est. Describatur circulus PCB qui Curvam tangat in P et sit ejusdem curvaturæ cum Curva in puncto contactus. Jungatur et producatur PO donec circulo occurrat in C et sit PB circuli diameter et ob æqualem curvaturam curvæ et circuli gravitas corporis in loco P (per Lem. 2) erit reciprocè ut $\frac{OP^{\text{qu.}} \times CP^{\text{cub.}}}{[PB]^{\text{quad.}}}$.

(177)*Corol. 1.* Igitur si curva KPL sit Ellipsis centrum habens D et si a corpore P per centrum illud ducatur Ellipseos diameter PDQ, sitqȝ diameter conjugata KDL rectæ PO occurrens in F: vis gravitatis qua corpus P revolvetur in hac figura circa centrum quodvis datum O, erit in locis singulis ut $\frac{FP^{\text{cub.}}}{OP^{\text{quad.}}}$.(178) Nam circulus PCB qui ejusdem est curvaturæ cum Ellipsi in P, occurrat Ellipseos diametro PQ in R et circuli hujus diameter PB occurrat Ellipseos diametro conjugatæ KL in G, et PR erit latus rectum Ellipseos ad verticem P pertinens ut supra(179) invenimus ideoqȝ ex natura Ellipseos æquabitur $\frac{KL^q}{PQ}$ et rectangulum

(175) The necessary requirement that these be 'quam minima' (minimal) was specified in a preceding cancellation which Newton's present phrase otherwise repeats.

PQ touching the circle at P and parallel to PC draw QD meeting the circle in R and D, join RO and to PO let fall the perpendicular RS, and then, if every (interval of) time be divided into equal moments,(175) and the body were able in the absence of gravity, in a single moment of time and with the speed which it has at P, to describe the tangent PQ whereas by the force of gravity it shall describe the arc $\widehat{PR}$, the gravity whereby it is drawn aside from the tangent into the arc will be as the interval QR, or as PQ^2/DQ; that is (because CP is in proportion to CA as RS to PQ) as $CA^2 \times RS^2/CP^2 \times DQ$, or (since the area of the triangle OPR is, by Lemma 1, proportional to the given (moment of) time, and hence OP is reciprocally proportional to RS) as $CA^2/OP^2 \times CP^2 \times DQ$, that is (if, through the moment of time coming to be infinitely small, DQ and CP coincide) reciprocally as $OP^2 \times CP^3/CA^2$. As was to be shown.

With these premisses the problem is thus solved.(176)

Let KPL denote the curve in which the body moves, P the body itself, O the centre to which it gravitates. Describe the circle PCB to touch the curve at P and be of the same curvature as the curve at the point of contact. Join and extend PO till it meets the circle in C and let PB be the circle's diameter, and then, because of the equal curvature of the curve and circle, the gravity of the body at the place P will (by Lemma 2) be reciprocally as $OP^2 \times CP^3/PB^2$.

(177)*Corollary 1.* In consequence, if the curve be an ellipse having centre D and if from the body P through that centre the ellipse's diameter PDQ be drawn, let KDL be the conjugate diameter meeting the straight line PO in F, and the force of gravity whereby the body P shall orbit in this figure round any given centre O will at each individual place be as FP^3/OP^2.(178) For let the circle PCB which is of the same curvature as the ellipse at P meet the ellipse's diameter PQ in R, and the diameter PB of this circle meet the ellipse's conjugate diameter KL in G, and then PR will be the *latus rectum* of the ellipse pertaining to the vertex P, as we have shown above,(179) and consequently will from the nature of an ellipse be equal to KL^2/PQ, and so the rectangle $PQ \times PR$ will be equal to KL^2. Join

(176) On making the crucial assumption that the preceding circle may to sufficient accuracy 'ob æqualem curvaturam' replace a given trajectory which it osculates in the immediate neighbourhood of the point C of contact; the same premiss is, we may observe, made in the equivalent Corollary 3 which he appended at about this time to the extended version of Proposition VII of the *Principia*'s first book (see VI: 565).

(177) A trivially variant preliminary draft of these three following corollaries exists on Add. 3960.9: 163.

(178) This result is already obtained in Prop. X on VI: 584–6, but the following proof is that 'per Cor. 3 Prop. VII' (compare note (176) above) which he afterwards (see VI: 588) outlined in the scholium subsequently added after Proposition XVII in the *Principia*'s second edition ($_2$1713: 58).

(179) See pages 110–12 preceding.

QPR æquabitur KL^q. Jungantur BC, BR et ob similitudinem triangulorum rectangulorum PRB, PGD & PCB, PGF rectangula DPR, GPB FPC sibi mutuo æqualia erunt; ideocꝫ si pro DPR seu $\frac{1}{2}QPR$ scribatur $\frac{1}{2}KL^q$, erit

$$\frac{\frac{1}{2}KL^q}{GP} = PB \quad \text{et} \quad \frac{\frac{1}{2}KL^q}{FP} = PC.$$

Et inde $\frac{OP^q \times PC^{cub}}{PB^q}$ cui gravitatem corporis in P reciprocè proportionalem esse monstravimus, fiet $\frac{OP^q \times \frac{1}{2}KL^q \times GP^q}{FP^{cub}}$. Sed rectangulum $KL \times GP$ semissis est parallelogrammi quod circa Ellipsin describitur adeocꝫ datur et propterea gravitas corporis in P est reciprocè ut $\frac{OP^q}{FP^{cub}}$ et directe ut $\frac{FP^{cub}}{OP^{quad}}$.

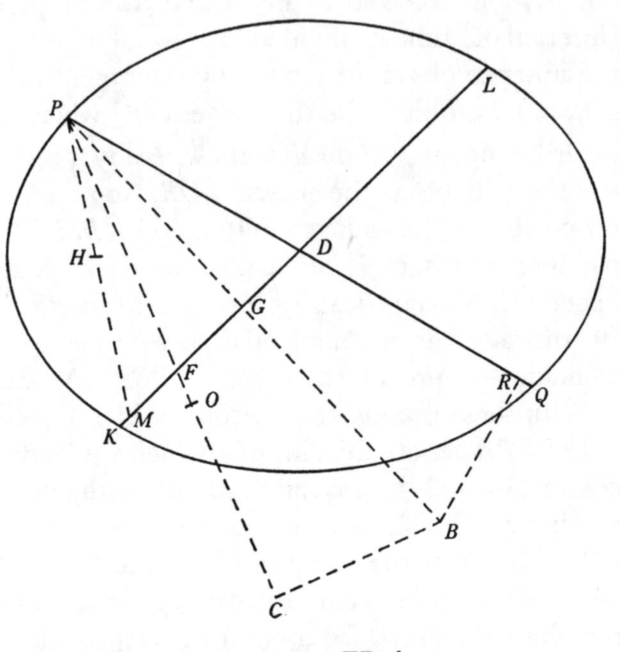

Corol. 2. Si gravitatis centrum O coincidat cum Ellipseos centro D, æquales erunt FP et OP et propterea gravitas corporis in P erit ut FP distantia corporis a centro.(180)

Corol. 3. Si gravitatis centrum O incidat in focum(181) Ellipseos erit FP æqualis semissi axis majoris Ellipseos, et propterea gravitas quæ fuit ut $\frac{FP^{cub}}{OP^{quad}}$ jam erit ut $\frac{1}{OP^q}$ hoc est reciprocè ut quadratum distantiæ corporis a centro.(182) Et hæc est lex gravitatis qua Planetæ majores circa Solem revolvuntur et minores(183) circa Terram Jovem et Saturnum ut alibi(184) demonstravimus.

Prob. 2.

Si corpora gravitent(185) in centrum positione datum & lex gravitatis(186) habetur, invenire

(180) Equally 'gravitatis' (of gravity) and 'Ellipseos' (of the ellipse), namely. Newton here repeats *mutatis mutandis* his earlier Corol. 1 to Proposition X on VI: 586.

(181) The point H in Newton's figure; in which case FP passes into MP, equal in length (see VI: 47, note (45)) to the ellipse's major semi-axis, as he hastens to mention. Observe that he here returns to the standard Keplerian *terminus technicus* which he had universally made use of in his youth, but which in the middle 1680's he came to reject for the anatomically curious equivalent 'umbilicus' (see VI: 46, note (42)). We have elsewhere (VI: 601, note (7) and 603, note (9)) remarked that, at about the present time or perhaps a little later, Newton also—in

BC, BR and then, because the right-angled triangles PRB, PGD and PCB, PGF are similar, the rectangles $PD \times PR$, $PG \times PB$ and $PF \times PC$ will be mutually equal one to another; consequently, if in place of $PD \times PR$, that is, $\frac{1}{2}PQ \times PR$, there be written $\frac{1}{2}KL^2$, there will be $\frac{1}{2}KL^2/PG = PB$ and $\frac{1}{2}KL^2/PF = PC$, and thence $OP^2 \times PC^3/PB^2$, to which we have demonstrated the gravity at P to be reciprocally proportional, will become $OP^2 \times \frac{1}{2}KL^2 \times GP^2/FP^3$. But the rectangle $KL \times GP$ is half the parallelogram which may be described round the ellipse, and hence is given; accordingly, the gravity of the body at P is reciprocally as OP^2/FP^3 and so directly as FP^3/OP^2.

Corollary 2. Should the centre O of gravity coincide with the ellipse's centre D, then FP and OP will be equal, and accordingly the gravity of the body at P will be as the distance FP of the body from the centre.(180)

Corollary 3. Should the centre O of gravity fall at the ellipse's focus,(181) then FP will be equal to half the major axis of the ellipse, and accordingly the gravity, which was as FP^3/OP^2, will now be as $1/OP^2$, that is, reciprocally as the square of the distance of the body from the centre.(182) And this is the law of gravity whereby the major planets orbit round the Sun and the minor ones(183) round the Earth, Jupiter and Saturn, as we have elsewhere(184) demonstrated.

Problem 2.

If bodies gravitate to a centre given in position and the law of gravity is had,(186) *to find*

---

a further twist to his navel cord?—briefly toyed with employing the word 'nodus' (knot) to denote this primary centre of 'gravity' in the planetary ellipse.

(182) Understand 'gravitatis' (of gravity): the geometrically equivalent 'a foco Ellipseos' (from the focus of the ellipse) was first written. With but minimal emendation of its verbal surface Newton here repeats Corol. 2 to Proposition X on VI: 586–8.

(183) Namely, the Moon (round the Earth) and the satellites of Jupiter and Saturn respectively.

(184) By which, we may presume, Newton intends 'in Libro tertio *Principiorum*' (in the *Principia*'s third book). There, however, while in his *editio princeps* he both gave explicit numerical corroboration ($_1$1687: 405–7) of his Proposition III 'Vim qua Luna retinetur in Orbe suo...esse reciprocè ut quadratum distantiæ locorum ab ipsius centro' and in his Hypothesis V asserted, with an airy 'Constat ex observationibus Astronomicis' (*ibid.*: 402), that the *planetæ circumjoviales* equivalently obey Kepler's third planetary law ('tempora periodica esse in ratione sesquialtera distantiarum a [Jovis] centro') in their periodic orbits round Jupiter, he intentionally (see VI: 40, note (26)) made no mention of Saturn's satellites, and these *planetæ circumsaturnii* were affirmed analogously to satisfy the third Keplerian law only in Phænomenon II of the *Principia*'s second edition ($_2$1713: 359–60) after Cassini had made accurate observation of their periods of revolution 'circa centrum Saturni'.

(185) In the manuscript this is written in over the equivalent phrase 'gravia sunt' (employed in the enunciation of Problem 1 without amendment), thereby replacing it.

(186) In sequel Newton has set cancellation brackets round the restriction 'pro ratione distantiarum a centro' (in relation to their distances from the centre), though his reluctance physically to strike out this very necessary phrase may reflect some hesitation on his part in doing so.

motum corporis in spatijs non resistentibus de loco dato in plagam datam data cum velocitate egressi.(187)

Exponatur tempus uniformiter fluens per longitudinem quamvis z,(188) & sit y altitudo seu distantia corporis a centro.

Cas. 1. Et si corpus recta ascendit vel recta descendit erit velocitas ejus $\dot{y}$ et gravitas $\ddot{y}$. Nam altitudinis fluxio est corporis velocitas et velocitatis fluxio est ut corporis gravitas. Ideoqȝ ex data gravitatis lege dabitur $\ddot{y}$ et inde eruendæ erunt $\dot{y}$ et y.(189)

Cas. 2 Sin corpus oblique moveatur e(190)

---

(187) The general inverse problem of central forces. Much as in Propositions XXXIX and XL/XLI of the first book of his published *Principia* ($_1$1687: 122–5/125–9; compare VI: 336–8/340–8) but with the novelty that their arguments from the ratios of related vanishing infinitesimals of variable geometrical line-lengths and areas are here transposed into the equivalent 'dotted' language of the limit-fluxions of the algebraic variables which express their magnitude, Newton proceeds to distinguish the two component cases of rectilinear motion through the centre of motion, and of general 'oblique' curvilinear motion about it.

(188) Whence, of course, $\dot{z}$ is constant—in sequel Newton sets it to be unity—and $\ddot{z} = 0$.

(189) More precisely, if the 'gravity' of the descending body at height y above the centre of force is $f(y)$, then $\ddot{y} = -f(y)$ and so $2\dot{y}\ddot{y} = -2f(y).\dot{y}$, whence there is $\dot{y}^2 = V^2 - 2\int_Y^y f(y).dy$ on taking V to be the speed of descent at some given height Y; and in immediate consequence the time of descent z from height Y to height y is $\int_Y^y 1/\dot{y}.dy$ where $\dot{y} = \sqrt{\left[V^2 - 2\int_Y^y f(y).dy\right]}$ is the instantaneous speed of descent at height y. (Compare VI: 339, note (192).)

(190) Understand 'e loco dato in plagam datam cum data velocitate,...' (from a given place in given direction with a given speed,...) in continuation. What impelled Newton to break thus abruptly off in mid-sentence is not clear to us, since the equivalent Propositions XL and XLI of his *Principia*'s first book (see note (187) above) are readily recast into his present fluxional terms on now setting z to be the time of orbit over the arc s through a central angle x from an initial position at a distance Y from the force-centre to a point at a distance y from it, whence the orbital speed at the latter place is $\dot{s} = \sqrt{[\dot{y}^2 + y^2\dot{x}^2]}$. The only subtlety is that the

the motion of a body setting off in non-resisting space from a given place in a given direction and with a given speed.(187)

Express the uniformly fluent time by any length z,(188) and let y be the height or distance of the body from the centre.

Case 1. Then if the body ascends or descends straight up or down its speed will be $\dot{y}$ and gravity $\ddot{y}$. For the fluxion of the height is the body's speed and the fluxion of the speed is as the body's gravity. Consequently, given the law of gravity, from it there will be given $\ddot{y}$ and you will need therefrom to elicit $\dot{y}$ and y.(189)

Case 2. But should the body move obliquely off(190)

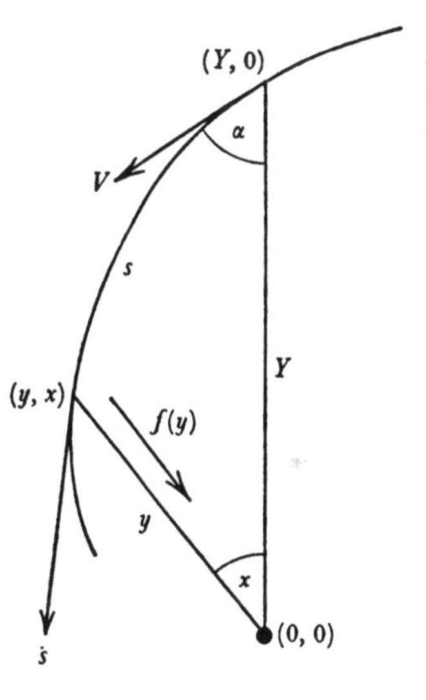

central 'gravity' $f(y)$ at (y, x) is measured by the *difference* between the radial acceleration $\ddot{y}$ at the point (y, x) in the orbit and the radial acceleration Q^2/y^3 (where $Q = y^2\dot{x}$ is the Keplerian constant) in the instantaneous tangential motion there; accordingly, much as Leibniz found in 1689 (see VI: 11, note (32)), there is $\ddot{y}-Q^2y^{-3} = -f(y)$ and so

$$2\dot{y}\ddot{y}-2Q^2y^{-3}\dot{y} = -2f(y).\dot{y},$$

whence $\dot{s}^2 = \dot{y}^2+Q^2y^{-2} = V^2-2\int_Y^y f(y).dy$ on taking V to be the orbital speed at distance Y from the force-centre. At once the time of orbit is

$$z = \int_Y^y 1/\dot{y}.dy = \int_Y^y 1/\sqrt{[\dot{s}^2-Q^2y^{-2}]}.dy$$

and the central orbital angle is

$$x = \int_{y=Y}^{y=y} Q/y^2.dz = \int_Y^y Q/y^2\sqrt{[\dot{s}^2-Q^2y^{-2}]}.dy$$

where $\dot{s} = \sqrt{\left[V^2-2\int_Y^y f(y).dy\right]}$ is the orbital speed at (y, x) and, on taking α to be the angle of projection at $(Y, 0)$, $Q = VY\sin\alpha$ is the Keplerian constant. (Compare VI: 348, note (209), in which $\dot{s} = ds/dz$ is put to be v.) The particular instance in which $\alpha = 0$ (and so $Q = \dot{x} = 0$ and $s = Y-y$, whence $\dot{s} = -\dot{y}$) is, of course, that of 'Cas. 1' preceding; see the previous note.

APPENDIX 1. PRELIMINARY NOTES AND DRAFTS FOR THE TWO STATES OF THE 1691 'DE QUADRATURA'.[1]

From autographs in the University Library, Cambridge and in private possession

[1][2]

$$[\text{Sit}]\ ax^m + bx^ny^r - cy^s = 0. \quad [\text{Pone}]\ y^s = \left[\frac{a}{c} \times \overline{x+p}|^m\right].^{(3)}$$

ax^m	ax^m
bx^ny^r	$b \times \left.\frac{a}{c}\right\}^{\frac{r}{s}} x^{\frac{mr+ns}{s}} + \left[\frac{m}{s}\right] rb \times \left.\frac{a}{c}\right\}^{\frac{r}{s}} x^{\frac{mr+ns-s}{s}} p + \&c$
$-cy^s$	$-ax^m - [m]\, ax^{m-1}p - \frac{[mm]-[m]}{2} ax^{m-2}pp[-\&c]$

(1) We have already remarked (see §1: note (2) above) that there exists both in Cambridge University Library and in private possession a considerable mass of fragmentary prior calculations and incomplete preparatory drafts for the two principal states of Newton's first treatise 'De quadratura Curvarum'. While it would be largely superfluous to print these in their totality—and to do so would certainly spoil the balance and proportion of the present volume—a minority of these preliminary notes and computations have some degree of interest and significance in their own right and we make no apology for here reproducing their essential content (filled out, as occasion requires, with appropriate editorial interpolations in the pattern of previous volumes).

(2) ULC. Add. 3962.2: 38ᵛ/37ʳ. These scattered calculations represent Newton's initial attempts in mid-November 1691 to reshape and extend his researches of fifteen years before into 'The Quadrature of all curves whose æquations consist of but three termes' and—a deal less optimistically—also into those 'whose æquations consist of more then three termes' (see III: 373–85; compare also VI: 45–0). These area-preserving transformations of trinomial Cartesian defining equations were, with minimal refinements and trivial changes in the denotation of variables, quickly absorbed into the mature 'Theor. II' whose bare enunciation Newton gives in [2] following. These computations are more generally significant, as we have previously observed (§2: note (35)), because in penning them Newton for, as far as we know, the first time there (see note (5))—without any insistence on his novelty in so doing—denotes the fluxion of a variable quantity by entering a superscript dot over it in a style which he was at once universally to adopt in his extended treatise 'De quadratura Curvarum' (§2 preceding), and likewise only a few months afterwards to employ in the excerpts from his new tract which in late August 1692 he communicated to John Wallis, allowing him to publish it in his Latin *Algebra* the next year (see Appendix 3 below) as his standard fluxional notation without remark upon its recent invention.

(3) Having thus optimistically assumed that the preceding trinomial equation can be reduced to this simpler binomial form, Newton in sequel introduces into it in place of y^r and y^s the equivalent series expansions of $(a/c)^{1/s}\,(x+p)^{m/s}$ and $(a/c)\,(x+p)^m$ respectively, but at once breaks off when he realizes the ineffectiveness of the procedure.

$ax^m+bx^ny^r+cy^s=0$. [Pone $y\dot{x}=v\dot{z}$ et sit] $ex^t=z$.(4) [Fit] $et\dot{x}x^{t-1}=\dot{z}$.(5) [adeoq] $\frac{y}{etx^{t-1}}=v$. [Unde] $x=e^{\frac{1}{t}}z^{\frac{1}{t}}$.(6) [ut et]

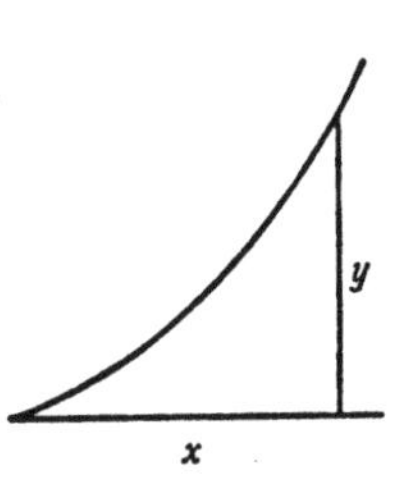

$$y=evtx^{t-1}=vtz^{\frac{t-1}{t}}\times\frac{1}{e^{\frac{1}{t}}}.$$ [Itaq facta substitutione]

$$ae^{\frac{m}{t}}z^{\frac{m}{t}}+be^{\frac{n}{t}}z^{\frac{n}{t}}\times e^rv^rt^rx^{tr-r}+ce^sv^st^sx^{st-s}[=0].^{(7)}$$

Curva $ax^my^q+bx^ny^r+cx^py^s$ &c $=0$. $x^t=z$. $\frac{y}{x^{t-1}}=v$.(8)

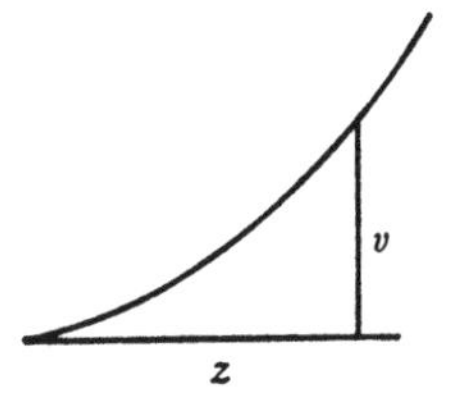

‖

Curva $az^{\frac{m+tq-q}{t}}v^q+bz^{\frac{n+tr-r}{t}}v^t+cz^{\frac{p+ts-s}{t}}v^s$ &c $=0$.

Sit ergo $q=0=p$.(9) et $m=n+tr-r$ vel $=ts-s$.(10)

$$[ax^\alpha y^\beta+bx^\gamma y^\delta+cx^\epsilon y^\zeta=0.\quad x^\lambda=z.\quad \frac{y}{x^{\lambda-1}}=v.^{(11)}$$

Prodit $ay^\beta z^{\frac{\alpha+\lambda\beta-\beta}{\lambda}}+by^\delta z^{\frac{\gamma+\lambda\delta-\delta}{\lambda}}+cy^\zeta z^{\frac{\epsilon+\lambda\zeta-\zeta}{\lambda}}=0.$]

[Ubi] $\beta=\delta=\epsilon=0$. [fit $az^{\frac{\alpha}{\lambda}}+bz^{\frac{\gamma}{\lambda}}+cy^\zeta z^{\frac{\lambda\zeta-\zeta}{\lambda}}=0$. Cape]

$$\frac{\lambda\zeta-\zeta}{\lambda}=\frac{\alpha}{\lambda}.\qquad \alpha+\zeta-\lambda\zeta=0.$$

[prodit]

[vel $=]\frac{\gamma}{\lambda}$. $\qquad \gamma+\zeta-\lambda\zeta=0$.

[Hoc est] $\frac{\alpha+\zeta}{\zeta}=\lambda$. $x^\lambda=z$. $a+bz^{\frac{\gamma\zeta-\alpha\zeta}{\alpha+\zeta}}+cy^\zeta=0$.

---

(4) Where e and t are free, that is. Newton now has recourse, much as in his earlier researches into the quadrature of curves defined by trinomial Cartesian equations (see note (2)), to the transformation $(x, y)\rightarrow(v, z)$ which preserves (in equivalent Leibnizian terms) the areas $\int y.dx=\int v.dz$ unchanged.

(5) The first known occurrence of what was ever afterwards to be Newton's standard notation for (first-order) fluxions. In the present text, however, he will momentarily revert to the 'double-dot' superscript which he had employed in an equivalent sense more than twenty-five years earlier (see note (17) below).

(6) Strictly, '$x=e^{-\frac{1}{t}}z^{\frac{1}{t}}$'. The trivial slip is carried into the sequel.

(7) Newton breaks off to simplify his approach.

(8) Understand that the area-transformation is now (in Leibnizian terms) $t\int y.dx=\int v.dz$, whence '$ty\dot{x}=v\dot{z}$'.

(9) Whence the defining equation of the 'Curva' reduces to be $ax^m+bx^ny^r+cy^s=0$.

(10) Thus restricting t to be respectively $(m-n+r)/r$ or $(m+s)/s$: in the former case the transformed equation after division by $z^{m/t}$ becomes $a+bv^t+cz^{-m/t}v^s$ '&c' $=0$.

(11) Much as before (see note (8)), the implicit area-transformation is now '$\lambda y\dot{x}=v\dot{z}$'.

$$[\text{vel}]\quad \frac{\gamma+\zeta}{\zeta}=\lambda.\quad x^{\lambda}=z.\quad az^{\frac{\alpha\zeta-\gamma\zeta}{\gamma+\zeta}}+b+cy^{\zeta}=0.$$

$$\left[\text{Si}\quad \beta=\epsilon=0.\quad az^{\frac{\alpha}{\lambda}}+by^{\delta}z^{\frac{\gamma+\lambda\delta-\delta}{\lambda}}+cy^{\zeta}z^{\frac{\lambda\zeta-\zeta}{\lambda}}=0.\quad \text{pone}\ \frac{\alpha+\zeta}{\zeta}=\lambda.\quad x^{\lambda}=z.\right]$$

$$az^{\frac{\alpha\zeta}{\alpha+\zeta}}+by^{\delta}z^{\frac{\gamma\zeta+\delta\alpha}{\alpha+\zeta}}+cy^{\zeta}z^{\frac{\alpha\zeta}{\alpha+\zeta}}=0.$$

$$[\text{sive}]\ a+by^{\delta}z^{\frac{\gamma\zeta-\alpha\delta-\alpha\zeta}{\alpha+\zeta}}+cy^{\zeta}=0.$$

$$\left[\frac{\gamma+\lambda\delta-\delta}{\lambda}=\frac{\alpha}{\lambda}.\right]\quad \frac{\alpha+\delta-\gamma}{\delta}=\lambda.\quad \left[\frac{\lambda\zeta-\zeta}{\lambda}=\right]\frac{\alpha\zeta-\gamma\zeta}{\alpha+\delta-\gamma}.$$

$$\frac{\gamma+\lambda\delta-\delta}{\lambda}=\frac{\lambda\zeta-\zeta}{\lambda}.\quad \frac{\gamma+\zeta-\delta}{\zeta-\delta}=\lambda.\quad \left[\frac{\gamma+\lambda\delta-\delta}{\lambda}=\right]\frac{\gamma-\delta+\zeta}{\gamma\zeta}.$$

$$[\text{unde}]\ az^{\frac{\alpha\gamma^{(12)}+\alpha\zeta-\alpha\delta-\gamma\zeta}{\gamma[-\delta+\zeta]}}\ [+by^{\delta}+cy^{\zeta}=0.]^{(13)}$$

$$[\text{Si}]\ ax^{m}+b+cv^{n}=0.\quad [\text{pone}]\ e+fx^{p}=z^{q}.\quad [\text{eritq}]\ pfx^{p-1}=q\dot{z}z^{q-1}.^{(14)}$$

$$[\text{sive}]\ \dot{z}=\frac{pf}{q}x^{p-1}z^{1-q}.\quad [\text{Est etiam}]$$

$$1.\dot{z}::y.v.^{(15)}\quad [\text{adeoq}]\ y=\frac{qvz^{q-1}}{pfx^{p-1}}.\quad [\text{hoc est}]\ v=\frac{pfy}{q}x^{p-1}z^{1-q}.$$

$$[\text{Cape}]\ p=m.\quad f=a.\quad [\text{Erit}\ v^{n}=]\left.\frac{ma}{q}\right\}^{n}y^{n}z^{n-qn}\ ^{(16)}$$

$$x^{p}=z.\quad x^{q}=y.\quad [\text{adeoq}]\ px^{p-1}=\dot{z}.\quad qx^{q-1}=\dot{y}.^{(17)}$$

$$[\text{Quia}]\ v=\dot{z}y.^{(18)}\quad [\text{evadit}]\ \dot{y}z=v=qzx^{q-1}.^{(19)}$$

---

(12) This extraneous term needs to be deleted.

(13) Newton afterwards copied these results unaltered into his draft 'Theor. II' (reproduced in [2] following). Here on his worksheet (f. 38v) he passes straightaway on to consider further possibilities of transforming the reduced trinomial defining equation $ax^m+b+cv^n = 0$.

(14) Understand that x is the base variable, so that its fluxion $\dot{x}$ is unity.

(15) That is, ($v\dot{x}$ or) $v = y\dot{z}$, the assumed area-preserving transmutation of the coordinates $(v, x) \rightarrow (y, z)$.

(16) Newton abandons this unpromising variant approach without entering the final factor x^{nm-n} in this value of v^n; it was doubtless his initial intention similarly to compute

$$x^m = f^{-m/p}(z^q-e)^{m/p},$$

and thence, on substituting these values in the given trinomial equation, to derive that defining the 'curva' of equal area $\int y.dz = \int v.dx$.

$$ax^m+b+cv^n=0. \quad \text{[Capiendo fluxiones erit]}^{(20)}$$

$$max^{m-1}+ncv^{n-1}\dot{v}=0. \quad \text{[In figura] } \dot{v}x=\dot{DC}.$$

$$v=\dot{BC}. \quad \text{[adeoq}] -\frac{max^m}{ncv^{n-1}}=\dot{DC}.$$

$$\text{[Pone] } x^p=z. \quad px^{p-1}=\dot{z}. \quad \text{[Erit]}^{(21)}$$

$$pyx^{p-1}=\dot{z}y=-\frac{max^m}{ncv^{n-1}}. \quad \text{[sive]}$$

$$v^{n-1}=-\frac{maz^{\frac{m-p+1}{p}}}{pncy}. \quad \text{[Prodit igitur]}$$

$$az^{\frac{m}{p}}+b+c\times\overline{-\frac{ma}{pnc}}\Big|^{\frac{n}{n-1}}\times y^{\frac{n}{n-1}}\times z^{\frac{mn-pn+n}{pn-p}}[=0].$$

$$\text{[At si] } v^q=z. \quad q\dot{v}v^{q-1}=\dot{z}=-q\frac{ma}{nc}x^{m-1}v^{q-n}.$$

$$\text{[erit]} -\frac{qyma}{nc}x^{m-1}v^{q-n}=\dot{z}y=\frac{max^m}{ncv^{n-1}}. \quad \text{[hoc est]}$$

$$-qyx^{-1}=v^{1-[q]}. \quad \text{[sive] } -qyv^{[q]-1}=x=-qyz^{\frac{q-1}{q}}.$$

$$\text{[Unde] } -^{(22)}aq^my^mz^{\frac{mq-m}{q}}+b+cz^{\frac{n}{q}}=0. \quad \text{[hoc est]}$$

$$-aq^my^m+bz^{\frac{m-mq}{q}}+cz^{\frac{n+m-mq}{q}}[=0].^{(23)}$$

(17) A momentary lapse into the double-dot superscript notation for fluxions which Newton had briefly employed long before in the autumn of 1665, in modification of an earlier partial-derivative notation; see I: 363, note (2).

(18) This is, in standard Newtonian dot-notation, the basic area-preserving transformation $v\dot{x}=\dot{z}y$, where x is taken to be the fundamental variable and therefore $\dot{x}=1$.

(19) Whence $v=qx^{p+q-1}$ must be the defining equation of the given Cartesian *curva*, in conflict with the general trinomial equation $ax^m+b+cv^n=0$ which Newton manifestly continues here to assume. Thus forewarned, he in sequel assumes only one restriction of the type $x^p=z$ on his basic area-preserving transformation of the given equation.

(20) Again understanding that $\dot{x}=1$.

(21) Now assuming the variant fluxional equation $x\dot{v}=y\dot{z}$ which in the transformation $(v,x)\to(y,z)$ preserves the complementary area $(ACD)=\int x.dv=\int y.dz$ unchanged in magnitude. Notice that Newton here uses $\dot{DC}$ to denote fl (ACD).

(22) Strictly $(-1)^m$.

(23) Which is again more complicated than the given trinomial equation! Evidently disappointed in this approach also, Newton breaks off his computation. We omit in sequel a few lines of calculation which relate directly to the variant reductions of the trinomial $dz^\theta+ey^\mu+fy^{\mu+\nu}=0$ as he at once proceeded to list them in 'Theor. II' of the verbal draft which now follows.

[2][(24)]

Et hactenus Epistola.

Cæterùm[(25)] beneficio hujus seriei curvæ lineæ non solum quadrari possunt sed etiam cum conicis sectionibus & alijs figuris comparari: &[(26)] figuræ simplicissimæ inveniri quibuscum comparari possunt. Nam quemadmodum termini omnes in serie tota infinitè continuata sunt area hujus curvæ, sic termini omnes dempto primo sunt area alterius cujusdam curvæ, et omnes demptis duobus primis sunt area quartæ et sic deinceps in infinitum: et hæ omnes curvæ inter se comparari possunt siquidem primus seriei terminus QA differentia est inter areas curvarū duarū primarū[,] et aggregatū termini primi et secundi $QA+QB$ differentia est int^r areas curvæ primæ ac tertiæ: et sic in cæteris.

[(27)]Exhibet igitur series unaquæqȝ uno intuitu quadraturas curvarum numero infinitarum utrinqȝ quæ omnes ex area unius data quadrari possunt: ideoqȝ ut omnes quadrentur quærenda est (præterquam ubi earum differentia terminum aliquem inf[initum] involvit) illarum omnium simplicissima et hujus area primum[(28)] investiganda. [(29)]Tota autem curvarum series coram exhiberi potest et curva simplicissima eligi si in ordinatim applicata $dz^\theta \times \overline{e+fz^\eta}|^\lambda$ termini hujus progressionis arithmeticæ $\theta-2\eta$, $\theta-\eta$, θ, $\theta+\eta$, $\theta+2\eta$, $\theta+3\eta$ &c utrinqȝ in infinitum productæ scribantur pro θ.

(24) ULC. Add. 3962.2: 33^r/34^r, completed by a sheet (11–23(1): 20/21) now in private possession. A transitional draft (ULC. Add. 3962.2: 37^v/38^r; extracts are cited in sequel) converts the following first extension of Newton's 'prime' quadrature theorem in his 1676 *epistola posterior* into the corresponding revised text in the primary state of Newton's 1691 'De quadratura' reproduced above (compare §1: note (18)). To avoid unnecessary repetition here of technical footnotes we largely restrict our commentary upon this manuscript draft to noticing significant cancelled verbal variants and preliminary revisions.

(25) In sequel Newton first began to write 'de seriebus hisce observandum [? venit quod...]'; compare his subsequent opening 'Diximus igitur quod...' in §1 (page 26 above).

(26) 'hoc pacto' was added at this point in an otherwise identical preceding draft of the following sentence.

(27) In the manuscript Newton here subsequently inserted a double dagger referring to f. 37^v where (compare note (24)) he here interpolates the sentence 'Quinetiam series utrinqȝ in infinitum continuari potest ut fit in hac formula...ubi A denotat terminum initialem $\frac{z^\pi}{s}$ ut supra et litteræ inversæ Ɐ, Ɔ, Ꟶ &c denotant secundum, tertium, quartum cæterosqȝ sinistrorsum: nempe Ɐ terminum $-\frac{s+1}{r+1}\times\frac{fz^\eta}{eA}$, Ɔ terminum proximum sinistrorsum et sic deinceps', much as in the revised text in §1 above.

(28) 'per methodum' is deleted.

(29) Newton first concluded: 'Invenitur autem curvarum simplicissima ponendo $\eta r-1-\eta\phi=\theta$ et quærendo talem numerum integrum pro ϕ, quo θ vel nihil vel æqualis $-\eta\lambda$, vel numerus simplicissimus sive integer sive fractus evadat. Nam $dz\times\overline{e+fz^\eta}|^\lambda$ erit ordinatim applicata hujus curvæ simplicissimæ et hæc curva in serie curvarum erit secunda si $\phi+1$ sit 2, tertia si 3, quarta si 4 & sic deinceps. Debet autem ϕ non major esse quàm s ubi s est numerus integer & affirmativus'. In his later preliminary revise on f. 37^v (compare note (24)) he after-

Exempl 1. Proponatur curva cujus ordinatim applicata sit

$$\frac{d}{z^4}\sqrt{e+fz^3} \text{ seu } dz^{-4}\times\overline{e+fz^3}|^{\frac{1}{2}} \text{ et erit}^{(30)}\ r\left(=\frac{\theta+1}{\eta}\right)=-1.$$

$$s(=\lambda+r)=-\tfrac{1}{2}.\quad Q\left(=\frac{d}{\eta f}\times\overline{e+fz^\eta}|^{\lambda+1}\right)=\frac{d}{3f}\times\overline{e+fz^3}|^{\frac{3}{2}}.\quad \&$$

$\pi(=r\eta-\eta)=-6$. adeoq; Area =

$$Q \text{ in } 2z^{-6}-\frac{-1-1}{-\frac{1}{2}-1}\times\frac{eA}{fz^3}+\frac{-1-2}{-\frac{1}{2}-2}\times\frac{eB}{fz^3}\quad \&\text{c.}^{(31)}$$

Exempl. Quadranda sit curva cujus ordinatim applicata $dz^4\times\overline{e+fz^2}|^{\frac{1}{2}}$. [32]Quoniam η hic est 2 & θ est 4 adeoq; progressio arithmetica 6, 4, 2, 0, -2, -4, &c[,] series ordinatarum erit $dz^4\times\overline{e+fz^2}|^{\frac{1}{2}}$, $dz^2\times\overline{e+fz^2}|^{\frac{1}{2}}$, $d\times\overline{e+fz^2}|^{\frac{1}{2}}$ &c quarum tertia est ordinata Sectionis conicæ. Quare ut quadretur curva proposita cujus ordinata est $dz^4\times\overline{e+fz^2}|^{\frac{1}{2}}$, quæro hujus aream in serie indefinita ut supra, nempe[33] Q in $\frac{z^3}{3}-\frac{1\frac{1}{2}}{2}\times\frac{eA}{fz^2}+\frac{\frac{1}{2}}{1}\times\frac{eB}{fz^2}-\frac{-\frac{1}{2}}{0}\times\frac{eC}{fz^2}$ &c. Ubi Q valet $\frac{d}{2f}\times\overline{e+fz^2}|^{\frac{3}{2}}$. Deinde quæro etiam aream simplicissimæ curvarum in præfata serie[,] puta conicæ sectionis cujus ordinata est $d\times\overline{e+fz^2}|^{\frac{1}{2}}$, eamq; invenio

$$Q \text{ in } \frac{z^{-1}}{1}-\frac{-\frac{1}{2}}{0}\times\frac{eA}{fz^2}\&\text{c.}$$

---

wards further altered the sequel to read, much as in §1 preceding: 'Sit Π terminus quilibet in serie, et m numerus terminorum ab A ad Π. id est 1 si Π sit B, vel 2 si Π sit C, vel -1 si Π sit $\mathcal{A}$ et sic in cæteris. Et si ponatur $\overline{s-m}\times d\Pi z^{m\eta-\pi}=h$ erit $hz^{\theta-m\eta}\times\overline{e+fz^\eta}|^\lambda$ ordinatim applicata Curvæ cujus area est seriei pars illa quæ pergit a termino Π dextrorsum in infinitum. Et hæc curva substituendo seriem numerorū integrorū -2, -1, 0, 1, 2, 3 &c successive pro m, convertitur in seriem curvarum illarū omniū quæ inter se comparari possint; et quarum simplicissima invenietur substituendo numerum integrum quo $\theta-m\eta$ vel evanescat vel numerus omnium simplicissimus evadat'.

(30) Since here $\theta=-4$, $\eta=3$ and $\lambda=\frac{1}{2}$.

(31) Newton proceeds no further with this first chosen *exemplum*, evidently because its series, no matter how far rightwards or leftwards it be continued, will manifestly not terminate.

(32) In sequel to the passage quoted in note (29) Newton afterwards continued from this point on his revised ff. 37ᵛ/38ʳ (when, at least, we make good certain gaps in the now eroded manuscript from the minimally redrafted text given in §1 above): '[quæro curvam] ut supra[,] dein [quoniam θ] hic est 4 et η 2 adeoq; $\theta-m\eta$ seu $4-2m$ scribendo 2 pro m [evanescit, quo] in casu hæc curva fit transmutatim conica sectio, usurpo 2 pro m et huic numero respondentem seriei terminum C pro Π et inde colligo h[,] et quoniam m est 2 dico quod si area sectionis hujus conicæ cujus ordinata est $h\times\overline{e+fz^2}|^{\frac{1}{2}}$, addatur ad ser[iei termin]os duos $A[-]B$ duc[tos in Q, summa] erit area curvæ quadrandæ'.

(33) Newton has deleted, evidently as *de trop*, a following intervening passage 'ponendo $r\left(=\frac{\theta+1}{\eta}\right)=2\frac{1}{2}.s(=\lambda+r)=3.\quad Q\left(=\frac{d}{\eta f}\times\overline{e+fz^\eta}|^{\lambda+1}\right)=\frac{d}{2f}\times\overline{e+fz^2}|^{\frac{3}{2}}.\quad \pi(=r\eta-\eta)=3.$ & aream = '.

Et quoniam hujus areæ primus terminus congruit cum tertio superioris & secundus cum quarto, quæro aream quæ sit ad hanc ut illius tertius terminus C ad hujus prim[u]m A et huic inventæ addo duos primos terminos illius. Nam summa erit area illa tota quam invenire oportuit.

Methodus quadrandi curvas adhuc generalior redditur per Theorema sequens.

Theor. II.(34)

Propositis quotcunq; datis quantitatibus a, b, c, d, e cum signis suis & indeterminatis [sex](35) z, y, x, v, t, s quarum z, x et t sint bases & y, v et s ordinatim applicatæ; & quod relatio inter ordinatam et basem curvæ alicujus sit

$$ax^{\alpha}y^{\beta}+bx^{\gamma}y^{\delta}+cx^{\epsilon}y^{\zeta}+dx^{\eta}y^{\theta} \text{ \&c} = 0,$$

ubi α, β, γ, &c sunt indices dignitatum. Fiat $x^{\lambda} = z$ et sit relatio inter basem et ordinatam alterius curvæ

$$ay^{\beta}z^{\frac{\alpha+\lambda\beta-\beta}{\lambda}}+by^{\delta}z^{\frac{\gamma+\lambda\delta-\delta}{\lambda}}+cy^{\zeta}z^{\frac{\epsilon+\lambda\zeta-\zeta}{\lambda}}+dy^{\theta}z^{\frac{\eta+\lambda\theta-\theta}{\lambda}} \text{ \&c} = 0.$$

Et areæ harum duarum curvarum æquales erunt inter se.

Unde si relatio inter basem et ordinatam curvæ alicujus definiatur per æquationē quāvis trium terminorum $ax^{\alpha}+bx^{\gamma}v^{\delta}+cv^{\zeta} = 0_{[,]}$ hæc curva transformari potest in aliam(36) sibi æqualem, idq; tripliciter: nempe

$$\text{vel ponendo } z = x^{\frac{\alpha+\zeta}{\zeta}} \text{ et } a+by^{\delta}z^{\frac{\gamma\zeta-\alpha\delta-\alpha\zeta}{\alpha+\zeta}}+cy^{\zeta} = 0;$$

$$\text{vel ponendo } z = x^{\frac{\alpha+\delta-\gamma}{\delta}} \text{ et } a+by^{\delta}+cy^{\zeta}z^{\frac{\alpha\zeta-\gamma\zeta-\alpha\delta}{\alpha+\delta-}} = 0,$$

vel deniq; ponendo $z = x^{\frac{\gamma+\zeta-\delta}{\zeta-\delta}}$ et $az^{\frac{\alpha\gamma+\alpha\zeta-\alpha\delta-\gamma\zeta}{\gamma+\zeta-\delta}}+by^{\delta}+cy^{[\zeta]} = 0.$(37) Et hæ tres æquationes(38) si dividatur prima per y^{δ} et secunda per y^{ζ}, redeunt ad hanc formam, $dz^{\theta}+ey^{\mu}+fy^{\mu+\nu} = 0$. Hæc autem si quadrari potest quadrabitur per seriem indefinitam in Theoremate primo$_{[,]}$ sin minus transmutabitur per eandem seriem in curvas innumeras hac æquatione designatas

$$dz^{\theta}+ey^{\mu+n\nu}+fy^{\mu+\nu+n\nu} = 0$$

(34) Much revised, this reappears as the opening paragraphs of 'Reg. II. Theor. III' in §1 preceding.

(35) We trivially amend an unaltered 'quatuor' in the manuscript relating to a first choice by Newton in sequel of four indeterminates 'z, y, x, v quarum z et x sint bases & y et v ordinatim applicatæ'.

(36) A far from necessarily true 'simpliciorem' is here justly replaced.

(37) The prior calculations on which Newton bases this triplicate reduction are reproduced in [1] preceding. Notice that the extraneous '$\alpha\gamma$' in the numerator of the index of z in the last equation's initial term passes, as before (see note (12) above), still undetected.

(38) Originally 'formæ'.

ubi n denotat numerum quemvis integrum tam negativum quam affirmativum.(39) Deinde hæ omnes per Theorema secundum abeunt in formas duas sequentes.

1. Ponatur $\frac{\theta+\mu+n\nu}{\theta}=\lambda$ & $y^\lambda=v$, et erit $dx^\theta+e+fv^{\frac{\nu}{\lambda}}=0$.

2. Ponatur $\frac{\theta+\mu+\nu+n\nu}{\theta}=\lambda$ & $y^\lambda=v$, et erit $dx^\theta+ev^{-\frac{\nu}{\lambda}}+f=0$.

Et hæ per Theorema primum transmutantur in curvas innumeras hisce duabus æquationibus designatas

$$dx^{\theta+m\theta}+ex^{m\theta}+fv^{\frac{\nu}{\lambda}}=0,\quad \&\ dx^{\theta+m\theta}+ev^{-\frac{\nu}{\lambda}}+fx^{m\theta}=0:$$

ubi m denotat numerū quēlibet integrū. Hæ vero per theorema secundum abeunt in quatuor sequentes.(40)

1. Ponatur $\frac{m\theta\lambda+\nu}{\nu}=\pi$ & $x^\pi=z$: et erit $dz^{\frac{\theta}{\pi}}+e+fy^{\frac{\nu}{\lambda}}=0$.

2. Ponatur $\frac{\theta\lambda+m\theta\lambda+\nu}{\nu}=\pi$ & $x^\pi=z$: et erit $d+ez^{\frac{-\theta}{\pi}}+fy^{\frac{\nu}{\lambda}}=0$.

3. Ponatur $\frac{m\theta\lambda-\nu}{\nu}=\pi$, & $x^\pi=z$: et erit $dz^{\frac{\theta}{\pi}}+ey^{\frac{-\nu}{\lambda}}+f=0$.

4. Ponatur $\frac{\theta\lambda+m\theta\lambda-\nu}{\nu}=\pi$, & $x^\pi=z$: et erit $d+ey^{\frac{-\nu}{\lambda}}+fz^{\frac{-\theta}{\pi}}=0$.

Hæ autem æquationes substituendo alios atq; alios numeros integros pro n et m, exhibent innumeras curvas quæ cum curva quadranda quam æquatione generali $ax^\alpha+bx^\gamma v^\delta+cv^\zeta=0$ sub initio designavimus, geometrice comparari possunt, et quarum simplicissima est omnium curvarum simplicissima quibuscū curva illa

(39) Newton first wrote that 'Area autem curvæ quam hæc æquatio definit inveniri potest per seriem indefinitam quam posuimus sub initio. Et inde datur area Curvæ quam æquatio quævis trium terminorum $ax^\alpha+bx^\nu v^\delta+cv^\zeta=0$ definit'. In sequel he continued thereafter: 'Beneficio ejusdem Theorematis fieri potest ut indices ipsius y in æquatione $dz^\theta+ey^\mu+fy^{\mu+\nu}=0$ datam habeant summam vel differentiam vel proportionem vel ut earum alterutra evanescat.

1. Ponatur $\frac{\theta+\mu+\nu}{\theta}=\lambda$, & $y^\lambda=v$: Et erit $dx^\theta+ev^{\frac{-\theta\nu}{\theta+\mu+\nu}}+f=0$.

2. Ponatur $\frac{\theta+\mu}{\theta}=\lambda$ & $y^\lambda=v$: Et erit $dx^\theta+e+fv^{\frac{\theta\nu}{\theta+\mu}}=0$.

Et hoc modo æquationes omnes trium terminorum reducuntur ad hanc formam $dx^\theta+g+hv^\pi=0$. Hæc autem ulteriorem admittit reductionem per theorema sequens [*Theor. III*].'

(40) A preliminary listing on ULC. Add. 3962.2: 37r orders these in the sequence 1, 2, 4, 3 but is otherwise only trivially variant in its designation of quantities.

quadranda comparationem geometricam(41) admittit.(42) Loquor hic de comparatione non particulari totius cum toto vel partis alicujus cum parti, sed generali totius cum toto et partium omnium cum omnibus. Nam comparatio particularis nonnunquam incidit ubi hæ regulæ generales locum non habent. Eadem methodo qua incidi in seriem interminatam quam exposui,(43) perveni etiam ad series magis compositas et methodus fundatur in hoc Theoremate quod prius(44) per inductionem inveneram.

Theor. III.(45)

Si Series terminorum quotcunꝗ $e+fz^{\eta}+gz^{2\eta}+hz^{3\eta}$ &c dicatur R, erit $z^{\theta}R^{\lambda}$ area curvæ cujus ordinatim applicata est

$$\theta e \begin{matrix}+\theta\\+\lambda\eta\end{matrix} fz^{\eta} \begin{matrix}+\theta\\+2\lambda\eta\end{matrix} gz^{2\eta} \begin{matrix}+\theta\\+3\lambda\eta\end{matrix} hz^{3\eta}\,[\&c] \text{ in } z^{\theta-1}R^{\lambda-1}.$$

Et ejusdem generis est Theorema sequens.

Theor IV.(46)

Si series terminorum quotcunꝗ $e+fz^{\eta}$ &c dicatur R et alia series terminorum quotcunꝗ $k+lz^{\eta}$ &c dicatur S: erit $z^{\theta}R^{\lambda}S^{\mu}$ area curvæ cujus ordinatim applicata est

$$\begin{matrix} \theta ek & \begin{matrix}+\theta\\+\lambda\eta\end{matrix} fkz^{\eta} & \begin{matrix}+\theta\\+\lambda\eta\end{matrix} flz^{2\eta}\,[\&c] \text{ in } z^{\theta-1}R^{\lambda-1}S^{\mu-1}. \\ & \begin{matrix}+\theta\\+\mu\eta\end{matrix} el & +\mu\eta \\ & & [\&c] \end{matrix}$$

Invenitur et probatur Theorema regrediendo ab area ad ordinatam.

(41) 'generaliter' is deleted in sequel.

(42) A cancelled first continuation reads: 'Si curva illa quadrari potest, prodibit ejus quadratura per Theorema primum absꝗ reductione ad hasce quatuo[r] formas ut supra dictum est. Sin quadrari non potest sed cum conica tamen sectione comparari, simplicissima curvarum inventarum erit Conica illa sectio. Si neꝗ quadrari potest neꝗ cum Conica sect. comparari, curva simplicissima altioris generis quacum comparari [potest sic invenietur].'

(43) The unnecessary precise citation 'in Theoremate primo' was added and then afterwards deleted.

(44) Namely, in his 1671 tract; Newton's following general 'Theor. III' embraces the fifth, sixth and seventh orders of 'magis generalia Theoremata quibus via ad altiora sternitur' which are there listed in its 'Catalogus Curvarum aliquot ad rectilineas figuras relatarum, ope Prob 7 constructus' (III: 236–40, especially 238–40; see also *ibid.*: 260–2).

(45) Afterwards verbally recast to be 'Theor VI' in §1 above.

(46) This straightforward extension of the previous theorem finds no place in the ensuing prior version of the 'De quadratura' reproduced in §1, but Newton subsequently set it (in minimally altered form) to be 'Theor. V' of his augmented treatise; see §2: note (45) above.

[3][47] [Si curvæ alicujus abscissa sit z et pro $e+fz^{\eta}+gz^{2\eta}+hz^{3\eta}$ &c scribatur R, sit autem area curvæ $z^{\theta}+az^{\theta-\eta}+bz^{\theta-2\eta}+cz^{\theta-3\eta}+dz^{\theta-4\eta}$ &c in R^{λ}; erit ordinata]

$$\begin{matrix}\theta \\ +3\lambda\eta\end{matrix} hz^{\theta+3\eta-1} \begin{matrix}+\theta \\ +2\lambda\eta\end{matrix} gz^{\theta+2\eta-1} \begin{matrix}+\theta \\ +\lambda\eta\end{matrix} fz^{\theta+\eta-1} + \theta ez^{\theta-1} \begin{matrix}+\theta \\ -\eta\end{matrix} ae[z^{\theta-\eta-1} \&c] \text{ in } R^{\lambda-1}:$$

$$\begin{matrix}+\theta-\eta \\ +3\lambda\eta\end{matrix} ah \quad \begin{matrix}+\theta-\eta \\ +2\lambda\eta\end{matrix} ag \quad \begin{matrix}+\theta-\eta \\ +\lambda\eta\end{matrix} af \quad \begin{matrix}+\theta-2\eta \\ +\lambda\eta\end{matrix} bf$$

$$\begin{matrix}+\theta-2\eta \\ +3\lambda\eta\end{matrix} bh \quad \begin{matrix}+\theta-2\eta \\ +2\lambda\eta\end{matrix} bg \quad \begin{matrix}+\theta-3\eta \\ +2\lambda\eta\end{matrix} cg$$

$$\begin{matrix}+\theta-3\eta \\ +3\lambda\eta\end{matrix} ch \quad \begin{matrix}+\theta-4\eta \\ +3\lambda\eta\end{matrix} dh$$

[Unde si ordinata est $dz^{\theta+3\eta-1}R^{\lambda-1}$[48], hoc est si]

$$-a=\frac{\overline{\theta+2\lambda\eta}\times g}{\overline{\theta-\eta+3\lambda\eta}\times h}. \quad -b=\frac{\overline{\theta+\lambda\eta}\times f+\overline{\theta-\eta+[2]\lambda\eta}\times ag}{\overline{\theta-2\eta+3\lambda\eta}\times h}.$$

$$-c=\frac{\theta e+\overline{\theta-\eta+\lambda\eta}\times af+\overline{\theta-2\eta+2\lambda\eta}\times bg}{\overline{\theta-3\eta+3\lambda\eta}\times h}.$$

$$-d=\frac{\overline{\theta-\eta}\times ae+\overline{\theta-2\eta+\lambda\eta}\times bf+\overline{\theta-3\eta+[2]\lambda\eta}\times cg}{\overline{\theta-4\eta+3\lambda\eta}\times h}. \quad \&c.$$

$$[\text{erit}]\ \text{Area}=\frac{d}{\overline{\theta+3\lambda\eta}\times h}\times R^{\lambda} \text{ in } z^{\theta}+az^{\theta-\eta}+bz^{\theta-2\eta}+cz^{\theta-3\eta} \ \&c.$$

[Pro θ et λ scribantur $\theta-3\eta+1$ et $\lambda+1$ respective[49] et evadet]

$$\text{Area}=\frac{d}{\overline{\theta+1+3\lambda\eta}\times h}R^{\lambda+1} \text{ in } z^{\theta-3\eta+1}+az^{\theta-4\eta+1}+bz^{\theta-5\eta+1} \ \&c$$

$$[\text{ubi}]-a=\frac{\overline{\theta-3\eta+1+2\lambda\eta+2\eta}\times g}{\overline{\theta-\eta+1+3\lambda\eta}\times h}. \ [\&c.\ \text{Pone}]\ \frac{\theta+1}{\eta}=r.$$

---

(47) ULC. Add. 3962.3: 39r. In a first broad extension of the prime theorem in his October 1676 *epistola posterior* for deriving the quadrature of a binomial integrand $dz^{\theta}(e+fz^{\eta})^{\lambda}$ as a *series convergens* (see *Correspondence*, **2**: 115; and compare §1: notes (9) and (21) above) Newton plunges straightaway in to attack the general problem of integrating the polynomial

$$dz^{\theta}(e+fz^{\eta}+gz^{2\eta}+hz^{3\eta}+\ldots)^{\lambda}$$

by a similar series expansion. Much as before, we fill out Newton's jejune equations in his worksheet with verbal interpolations which will better convey the sequence of his argument.

(48) The second d here introduced (in parallel with that used to express the coefficient of the preceding binomial *ordinata* for which $g = h = \ldots = 0$) will cause no great confusion. Having doubtless in the meantime come to notice this duplicate designation, Newton will in his final sentence silently replace the preceding coefficients $a, b, c, d, \ldots$ by their upper-case equivalents.

(49) Thus converting the *ordinata* to the form $dz^{\theta}(e+fz^{\eta}+\ldots)^{\lambda}$ which exactly generalizes the binomial instance in Newton's 1676 'prime' theorem.

$$\frac{\theta+1+\lambda\eta}{\eta}=s.\quad \frac{\theta+1+2\lambda\eta}{\eta}=t.\quad \frac{\theta+1+3\lambda\eta}{\eta}=v.\quad \frac{d}{\eta h}R^{\lambda+1}=Q.\quad \theta+1-3\eta=\pi.$$

[fit] Area $[=]\frac{d}{\eta vh}R^{\lambda+1}$ in $z^{\pi}+\frac{az^{\pi}}{z^{\eta}}+\frac{bz^{\pi}}{z^{2\eta}}$ &c

$$[\text{hoc est}]=Q\times\frac{z^{\pi}}{v}-\frac{\overline{t-1}\times gA}{\overline{v-1}\times hz^{\eta}}-\frac{\overline{s-[2]}\times fA+\overline{t-2}\times gB}{\overline{v-2}\times hz^{\eta}}$$
$$-\frac{\overline{r-[3]}\times eA+\overline{s-[3]}\times fB+\overline{t-3}\times gC}{\overline{v-3}\times hz^{\eta}}\,[\&c].^{(50)}$$

---

[4](51) [Si curvæ alicujus abscissa sit z et pro $e+fz^{\eta}+gz^{2\eta}+hz^{3\eta}$ &c scribatur R, sit autem area curvæ $z^{\theta}R^{\lambda}$; erit ordinata

$$\theta ez^{\theta-1}\begin{matrix}+\theta\\+\lambda\eta\end{matrix}fz^{\theta+\eta+1}\begin{matrix}+\theta\\+2\lambda\eta\end{matrix}gz^{\theta+2\eta-1}\begin{matrix}+\theta\\+3\lambda\eta\end{matrix}hz^{\theta+3\eta-1}\ \&\text{c in } R^{\lambda-1}.]^{(52)}$$

Hæc ordinata ponendo $g=0=h$ [&c] evadit $\theta ez^{\theta-1}\begin{matrix}+\theta\\+\lambda\eta\end{matrix}z^{\theta+\eta-1}$ in $R^{\lambda-1}$, existente area $z^{\theta}R^{\lambda}$:(53) et si pro indefinito θ scribantur successive $\theta-\eta$, $\theta-2\eta$, $\theta-3\eta$ &c & ordinatæ omnes inde prodeuntes ducantur respective in quasvis p, q, r &c et addantur ordinatæ primæ, summa ordinatarum erit

$$R^{\lambda-1}\text{ in }\begin{matrix}\theta\\-3\eta\end{matrix}erz^{\theta-3\eta-1}\left.\begin{matrix}+\theta\\-3\eta\\+\lambda\eta\end{matrix}\right\}frz^{\theta-2\eta-1}\left.\begin{matrix}+\theta\\-2\eta\\+\lambda\eta\end{matrix}\right\}fqz^{\theta-\eta-1}\left.\begin{matrix}+\theta\\-\eta\\+\lambda\eta\end{matrix}\right\}fpz^{\theta-1}\begin{matrix}+\theta\\+\lambda\eta\end{matrix}fz^{\theta+\eta-1}$$
$$\left.\begin{matrix}\theta\\-2\eta\end{matrix}\right\}eq\qquad\left.\begin{matrix}\theta\\-\eta\end{matrix}\right\}ep\qquad+\theta e$$

et summa arearum his respondentium $rz^{\theta-3\eta}+qz^{\theta-2\eta}+pz^{\theta-\eta}+z^{\theta}$ in R^{λ}. Jam ut termini intermedij in Ordinata evanescant pono eos $=0$ et sic prodeunt

---

(50) Where now the capitals A, B, C, ... replace the previous lower-case letters a, b, c, ... (compare note (48)). In square brackets we have mended a trivial slip on Newton's part in assigning the coefficients of fA, eA and fB in the two last numerators to be $s-1$, $r-1$ and $s-2$ respectively: this was at once corrected by Newton himself in the immediate revise which we reproduce in [4] following.

(51) ULC. Add. 3962.3: 55v/54v, a first verbal amplification of the preceding worksheet computations.

(52) We borrow this enunciation from Proposition VII of the revised 'De quadratura Curvarum' (§2 above).

(53) Where, that is, $R=e+fz^{\eta}$. This binomial instance is, with the trivial replacement of λ by $\lambda+1$ and θ by $\theta+\eta-1$, the prime example of such quadrature which Newton had communicated to Leibniz in 1676 (compare §1: note (9)).

(54) Newton first wrote out equivalently '& sic deinceps in infinitum'.

$$\frac{\theta\times -e}{\theta+\lambda\eta-\eta\times f}=p.\quad \frac{\theta-\eta}{\theta+\lambda\eta-2\eta}\times\frac{-ep}{f}=q.\quad \frac{\theta-2\eta}{\theta+\lambda\eta-3\eta}\times\frac{-e[q]}{f}=r.\quad \&\text{c.}^{(54)}$$

Unde area erit

$$R^{\lambda}\text{ in }z^{\theta}+\frac{\theta}{\theta+\lambda\eta-\eta}\times-\frac{eA}{fz^{\eta}}+\frac{\theta-\eta}{\theta+\lambda\eta-2\eta}\times-\frac{eB}{fz^{\eta}}+\frac{\theta-2\eta}{\theta+\lambda\eta-3\eta}\times-\frac{eC}{fz^{\eta}}+\&\text{c.}$$

Quæ series scribendo r pro $\frac{\theta+\eta}{\eta}$ et s pro $\frac{\theta+\lambda\eta}{\eta}$ et ubiq̧ $\lambda+1$ pro λ, et $\theta+\eta-1$ pro θ reducitur ad formam Theorematis sub initio positi.(55) Et notandum est quod ubi intermedij termini ex ordinata tolluntur, terminus primus et ultimus qui restant exhibent ordinatas duarum illarum curvarum quæ inter se geometricè comparari possint.

[Et eadem m]ethodo [areæ omnes curvarum mag]is compositarum ad series deduci possunt. [Sic in n]ovissimo Theoremate ponendo $h=0$.(56) ordinata fit $\theta ez^{\theta-1}\begin{matrix}+\theta\\+\lambda\eta\end{matrix}fz^{\theta+\eta-1}\begin{matrix}+\theta\\+2\lambda\eta\end{matrix}gz^{\theta+2\eta-1}$ in $R^{\lambda-1}$, & area $z^{\theta}R^{\lambda}$. Hic pro θ scribe successive $\theta-\eta$, $\theta-2\eta$, $\theta-3\eta$ &c & ordinatas prodeuntes in p, q, r &c ductas, adde præcedenti ordinatæ et summa erit

$$[R^{\lambda-1}\text{ in}]\begin{array}{llllll}
\begin{matrix}\theta\\-3\eta\end{matrix}erz^{\substack{\theta\\-3\eta\\-1}} & \begin{matrix}+\theta\\+\lambda\eta\\-3\eta\end{matrix}frz^{\substack{\theta\\-2\eta\\-1}} & \begin{matrix}+\theta\\+2\lambda\eta\\-3\eta\end{matrix}grz^{\substack{\theta\\-\eta\\-1}} & \begin{matrix}+\theta\\+2\lambda\eta\\-2\eta\end{matrix}gqz^{\substack{\theta\\-1}} & \begin{matrix}+\theta\\+2\lambda\eta\\-\eta\end{matrix}gpz^{\substack{\theta\\-1\\+\eta}} & \begin{matrix}+\theta\\+2\lambda\eta\end{matrix}gz^{\substack{\theta\\-1\\+2}} \\
 & \begin{matrix}+\theta\\-2\eta\end{matrix}eq & \begin{matrix}+\theta\\+\lambda\eta\\-2\eta\end{matrix}fq & \begin{matrix}+\theta\\+\lambda\eta\\-\eta\end{matrix}fp & \begin{matrix}+\theta\\+\lambda\eta\end{matrix}f & \\
 & & \begin{matrix}+\theta\\-\eta\end{matrix}ep & +\theta e & &
\end{array}$$

[et summa arearum respondentium $z^{\theta}+pz^{\theta-\eta}+qz^{\theta-2\eta}+rz^{\theta-3\eta}$ in R^{λ}. Jam pone terminos intermedios $=0$ et sic prodeunt]

$$\frac{\theta+\lambda\eta}{\theta+2\lambda\eta-\eta}\times-\frac{f}{g}=p.\quad \frac{\theta e+\theta fp+\lambda\eta fp-\eta fp}{\theta+2\lambda\eta-2\eta\times -g}=q.$$

$$\frac{\overline{\theta-\eta}\times ep+\overline{\theta+\lambda\eta-2\eta}\times fq}{\theta+2\lambda\eta-3\eta\times -g}=r.\quad \frac{\overline{\theta-2\eta}\times eq+\overline{\theta+\lambda\eta-3\eta}\times fr}{\theta+2\lambda\eta-4\eta\times -g}=s.\quad \&\text{c.}$$

(55) See note (53), and compare §1: note (21) above.

(56) Understand more precisely that all such coefficients are zero, when $R=e+fz^{\eta}+gz^{2\eta}$.

Unde summa arearum respondentium $\overset{\theta}{z^{\theta}}+\overset{\theta}{pz^{-\eta}}+\overset{\theta}{qz^{-2\eta}}+\overset{\theta}{rz^{-3\eta}}+sz^{-4\eta}$ &c [in R^{λ}].

$$[\text{Pone}]\ \frac{\theta}{\eta}=\tau.\ \tau+\lambda=\sigma.\ \tau+2\lambda-1=\phi.\ [\text{Erit}]\ \frac{\sigma f}{-\phi g}=p.$$

$$\frac{\tau e+\overline{\sigma-1}fp}{\overline{\phi-1}\times -g}=q.\quad \frac{{}^{\tau}_{-1}ep+{}^{\sigma}_{-2}fq}{{}^{\phi}_{-2}\times -g}=r.\quad \frac{{}^{\tau}_{-2}eq+{}^{\sigma}_{-3}fr}{\left.{}^{\phi}_{-3}\right\}\times -g}=s.\quad \&\text{c.}^{(57)}$$

Sit $dz^{\theta}\times\overline{e+fz^{\eta}+gz^{2\eta}}|^{\lambda}$ ordinatim applicata.[58] Pone $\dfrac{\theta+1}{\eta}=r.\ r+\lambda=t.$

$$r+2\lambda=s.\ [\text{ut et}]\ \frac{tf}{sg}=k.\ \frac{{}^{r}_{-1}e+{}^{+t}_{-1}fk}{{}^{s}_{-1}g}=l.\ \frac{{}^{r}_{-2}ek+{}^{+t}_{-2}fl}{{}^{s}_{-2}g}=m.\ \frac{{}^{+r}_{-3}el+{}^{+t}_{-3}fm}{{}^{s}_{-3}g}=n.\ \&\text{c.}$$

$$\frac{d}{\overline{\theta+2\lambda\eta+1}\times g}R^{\lambda+1}=Q.$$

[Evadet] $Qz^{\theta-2\eta+1}$ in $1-kz^{-\eta}+lz^{-2\eta}-mz^{-3\eta}+nz^{-4\eta}$ &c = areæ.

Theor IV.[59]

Si sit $dz^{\theta}\times\overline{e+fz^{\eta}+gz^{2\eta}}|^{\lambda}$ ordinatī applicata, pone $\dfrac{\theta+1}{\eta}=r,\ r+\lambda=t,$ $r+2\lambda=s,\ \dfrac{dR^{\lambda+1}}{\eta g}=Q$ et area erit

$$Q\times\frac{z^{r\eta-2\eta}}{s}-\frac{[\overline{t-1}]\times fA}{[\overline{s-1}\times g]\,z^{\eta}}-\frac{\overline{r-2}\times eA+\overline{t-2}\times fB}{\overline{s-2}\times gz^{\eta}}$$
$$-\frac{\overline{r-3},eB+\overline{t-3},fC}{\overline{s-3},gz^{\eta}}-\frac{\overline{r-4},eC+\overline{t-4},fD}{\overline{s-4},gz^{\eta}}\ [\&\text{c}].$$

[Ubi] A denotat primum terminum $\dfrac{z^{r\eta-[2]\eta}}{s}$, [$B$ terminum secundum $-\dfrac{\overline{t-1}\times fA}{\overline{s-1}\times gz^{\eta}}$]

(57) With the minimal adjustments introduced by Newton in sequel, this trinomial instance was to be inserted into the preliminary version of the 'De quadratura' (§1 above) as its Theorem IV.

(58) On replacing λ and θ in the preceding by $\lambda+1$ and $\theta+1$ respectively, that is.

(59) Newton proceeds to draft the finished theorem as he will insert it into his 'De quadratura'; see §1: note (50).

et sic deinceps.(60) Et nota quod in hac area [termini duo simul] evanescent et seriem terminabunt ubi curva geometricam [quadraturam admittit.]

---

[5](61) ——area curvæ quadrandæ. Et his similia observanda veniunt in Theorematis sequentibus.(62)

Theorema II.(63)

Si Curvæ alicujus ordinatim applicata sit $dz^{\theta}\times\overline{e+fz^{\eta}+gz^{2\eta}}|^{\lambda}$ et ponatur $\frac{\theta+1}{\eta}=r,\ r+\lambda=s,\ r+2\lambda=t,\ \frac{d}{\eta g}\times\overline{e+fz^{\eta}+gz^{2\eta}}|^{\lambda+1}=Q$ et $r\eta-2\eta=\pi$: area Curvæ hujus erit

$$Q \text{ in } \frac{z^{\pi}}{t}-\frac{\overline{s-1}\times fA}{\overline{t-1}\times gz^{\eta}}-\frac{\overline{r-2}\times eA+\overline{s-2}\times fB}{\overline{t-2}\times gz^{\eta}}$$
$$-\frac{\overline{r-3}\times eB+\overline{s-3}\times fC}{\overline{t-3}\times gz^{\eta}}-\frac{\overline{r-4}\times eC+\overline{s-4}\times fD}{\overline{t-4}\times gz^{\eta}}-\&c.$$

Ubi A denotat primum terminum $\frac{z^{\pi}}{t}$, & B secundum $-\frac{\overline{s-1}\times fA}{\overline{t-1}\times gz^{\eta}}$ cum signo suo & sic deinceps.(64) Et nota quod termini duo nonnunquam simul evanescunt et seriem abrumpunt: et tunc Curva geometricam quadraturam admittit. Quæ omnia mutatis mutandis in sequenti Theoremate observanda sunt.

Theorema III.(65)

Si curvæ alicujus ordinatim applicata sit $dz^{\theta}\times\overline{e+fz^{\eta}+gz^{2\eta}+hz^{3\eta}}|^{\lambda}$, et ponatur $\frac{\theta+1}{\eta}=r,\ r+\lambda=s,\ r+2\lambda=t,\ r+3\lambda=v,\ \frac{d}{\eta h}\times\overline{e+fz^{\eta}+gz^{2\eta}+hz^{3\eta}}|^{\lambda+1}=Q$, et $r\eta-3\eta=\pi$: area curvæ hujus erit

---

(60) As we have similarly observed of the theorem as introduced into the 'De quadratura' (see §1: note (51)), this should read 'B termini secundi coefficientem $\frac{-\overline{t-1}\times fA}{\overline{s-1}\times g}$, et sic deinceps'; correspondingly, the denominators in the preceding expansion (after the first) should be $\overline{s-1}\times gz^{\eta}$, $\overline{s-2}\times gz^{2\eta}$, $\overline{s-3}\times gz^{3\eta}$, $\overline{s-4}\times gz^{4\eta}$, and so on.

(61) From the original (11–23(1): 14–16) in private possession. This revision and extension of the preceding was initially designed by Newton to follow straight after Theorem I of his 'De quadratura Curvarum'; see §1: note (26).

(62) A first cancelled version of this transitional sentence has already been cited in §1: note (26).

(63) With slight verbal adjustment this became 'Theor IV' in §1; compare note (57) above.

(64) See note (60) for the necessary corrections which have to be made to this enunciation.

(65) We have already reproduced the text of this quadrinomial case in filling out the equivalent 'Theor. V'—lacunary in the manuscript (see §1: note (53))—which it soon afterwards became in the preliminary 'De quadratura Curvarum'.

$$Q \text{ in } \frac{z^\pi}{v} - \frac{\overline{t-1}\times gA}{\overline{v-1}\times hz^\eta} - \frac{\overline{s-2}\times fA+\overline{t-2}\times gB}{\overline{v-2}\times hz^\eta}$$

$$-\frac{\overline{r-3}\times eA+\overline{s-3}\times fB+\overline{t-3}\times gC}{\overline{v-3}\times hz^\eta}$$

$$-\frac{\overline{r-4}\times eB+\overline{s-4}\times fC+\overline{t-4}\times gD}{\overline{v-4}\times hz^\eta} - \&c.$$

Atq; hæc est Theorematum Progressio prima. Nam Progressio[(66)] absq; ulteriori calculo in infinitum jam facile continuatur.[(67)] Sunt autem Theorematum Progressiones numero infinitæ quarum secunda est hæc.

THEOREMATUM PROGRESSIO SECUNDA.[(68)]

Theorema I.

Si Curvæ alicujus ordinata sit $dz^\theta \times \overline{e+fz^\eta}|^\lambda \times \overline{h+iz^\eta}|^\mu$ et ponatur $\frac{\theta+1}{\eta} = r$, $r+\lambda = s$, $r+\mu = \sigma$, $r+\lambda+\mu = t$, $\frac{d}{\eta fi} \times \overline{e+fz^\eta}|^{\lambda+1} \times \overline{h+iz^\eta}|^{\mu+1} = Q$, $r\eta - 2\eta = \pi$:

---

(66) Newton inserted 'observato tenore' at this point in a preceding first draft of the present sentence.

(67) This primary sequence of theorems, yielding the series quadrature of curves whose ordinate is of the general form $dz^\theta R^\lambda$, $R = e+fz^\eta+gz^{2\eta}+hz^{3\eta}+\ldots$, was soon to be subsumed into the revised 'De quadratura' (§2 above) as its Proposition IX. In sequel Newton first continued: 'Nota autem quod in Theoremate primo æque ac in his duobus debent termini omnes post primum negative poni, si modò literæ *A*, *B*, *C* &c terminos cum signis suis + et − tam in illo Theoremate quam in hisce significent.'

(68) Deriving, namely, the series quadrature of curves whose ordinate is of form $dz^\theta R^\lambda S^\mu$, $R = e+fz^\eta+gz^{2\eta}\ldots$ and $S = h+iz^\eta+kz^{2\eta}\ldots$. This second sequence was little afterwards subsumed into the revised 'De quadratura'—in sequel to its immediately preceding generalization of the present 'Theorematum Progressio prima' (see note (67))—as its Proposition X. We scarcely need to remark that the following theorems were obtained in a manner entirely analogous to that employed by him in deriving the previous sequence in [3] above; indeed on the same worksheet (f. 39r) he went on to compute in terms of powers

$$z^{\theta+k\eta-1}, \quad k = 3, 2, 1, 0, -1, -2, \ldots,$$

the 'ordinata' of a curve whose 'area' is '$z^\theta+az^{\theta-\eta}+bz^{\theta-2\eta}+cz^{\theta-3\eta}+dz^{\theta-4\eta}$ [&c in] $R^\lambda S^\mu$', where $R = e+fz^\eta+gz^{2\eta}$ and $S = h+iz^\eta$; and thereafter straightforwardly to derive the required series expansion of the integral of $z^\theta R^\lambda S^\mu$ by successively calculating the conditions which ensue from setting $a = b = c = \ldots = 0$. A similar preliminary computation in which is deduced the most general 'Theorema III' following is preserved in private possession.

area curvæ hujus erit

$$Q \text{ in } \frac{z^\pi}{t} - \frac{\overline{s-1}\times fhA+\overline{\sigma-1}\times eiA}{\overline{t-1}\times fiz^\eta} - \frac{\overline{r-2}\times ehA+\overline{s-2}\times fhB+\overline{\sigma-2}\times eiB}{\overline{t-2}\times fiz^\eta}$$

$$-\frac{\overline{r-3}\times ehB+\overline{s-3}\times fhC+\overline{\sigma-3}\times eiC}{\overline{t-3}\times fiz^\eta} - \&c.$$

Theorema II.

Si Curvæ alicujus ordinata sit $dz^\theta\times\overline{e+fz^\eta+gz^{2\eta}}|^\lambda\times\overline{h+iz^\eta}|^\mu$ et ponatur

$$\frac{\theta+1}{\eta}=r.\ r+\lambda=s,\ r+2\lambda=t,\ r+\mu=\sigma,\ s+\mu=\tau,\ t+\mu=v.$$

$$\frac{d}{\eta gi}\times\overline{e+fz^\eta+gz^{2\eta}}|^{\lambda+1}\times\overline{h+iz^\eta}|^{\mu+1}=Q\ \&\ r\eta-3\eta=\pi:$$

area curvæ hujus erit

$$Q \text{ in } \frac{z^\pi}{v} - \frac{\overline{\tau-1}\times fiA+\overline{t-1}\times ghA}{\overline{v-1}\times giz^\eta}$$

$$-\frac{\overline{\sigma-2}\times eiA+\overline{s-2}\times fhA+\overline{\tau-2}\times fiB+\overline{t-2}\times ghB}{\overline{v-2}\times giz^\eta}$$

$$-\frac{\overline{r-3}\times ehA+\overline{s-3}\times fhB+\overline{\sigma-3}\times eiB+\overline{\tau-3}\times fiC+\overline{t-3}\times ghC}{\overline{v-3}\times giz^\eta}$$

$$-\frac{r-4,\ ehB+s-4,\ fhC+\sigma-4,\ eiC+\&c}{v-4,\ giz^\eta}[-\&c].$$

Theorema III.

Si Curvæ alicujus ordinata sit

$dz^\theta\times\overline{e+fz^\eta+gz^{2\eta}}|^\lambda\times\overline{h+iz^\eta+kz^{2\eta}}|^\mu$, et ponatur

$$\frac{dz^{\theta+1-4\eta}}{\eta gk}\times\overline{e+fz^\eta+gz^{2\eta}}|^{\lambda+1}\times\overline{h+iz^\eta+kz^{2\eta}}|^{\mu+1}=Q:$$

et terminorum $z^{\frac{\theta+1}{\eta}}$ & $e+fz^\eta+gz^{2\eta}$ & $h+iz^\eta+kz^{2\eta}$, in se ductorum

coefficientes sint	& indices dignitatum		
eh, fh, gh	$\frac{\theta+1}{\eta}=r.$	$r+\lambda=s.$	$r+2\lambda=t.$
ei, fi, gi	$r+\mu=s'.$	$s+\mu=t'.$	$t+\mu=v'.$
ek, fk, gk	$r+2\mu=t''.$	$s+2\mu=v''.$	$t+2\mu=w''.$

Area Curvæ erit

$$Q \text{ in } \frac{1}{w''} - \frac{v''-1, fkA + v'-1, giA}{w''-1, gkz^{\eta}}$$

$$-\frac{t''-2, ekA+t'-2, fiA+t-2, ghA+v''-2, fkB+v'-2, giB}{w''-2, gkz^{\eta}}$$

$$-\frac{s'-3, eiA+s-3, fhA+t''-3, ekB+t'-3, fiB+t-3, ghB+v''-3, fkC+v'-3, giC}{w''-[3], gkz^{\eta}}$$

$$-\frac{r-4, ehA+s'-4, eiB+s-4, fhB+t''-4, ekC+t'-4, fiC+t-4, ghC+v''-4, fkC+v'-4, giD}{w''-[4], gkz^{\eta}}$$

[−&c.](69)

---

[6](70) *Prop. XI.*

Data æquatione fluxiones duarum quantitatum involvente invenire relationem fluentium quantitatum.

Problematum mathematicorum utilissimum est. Sunto fluentes quantitates y & z, & earum fluxiones $\dot{y}$ & $\dot{z}$. Fluat autem alterutra z uniformiter & exponatur eadem per rectangulū sub abscissa Curvæ et unitate et ejus fluxio $\dot{z}$ per unitatem; et altera y per aream ejusdem Curvæ et ejus fluxio $\dot{y}$ per ordinatam: ac detur æquatio definiens fluxionem $\dot{y}$, sitq; $\dot{y}$ in æquatione illa unius tantū dimensionis.

Cas. 1. Si æquatio illa involvit z et $\dot{y}$ sine y, quadranda est curva cujus ordinata est valor $\frac{\dot{y}}{\dot{z}}$ et abscissa z, et ejus area erit y. Q.E.I.

Cas. 2. Sin æquatio involvit etiam y, sed numerator valoris $\dot{y}$ careat y et denominator ejusdem careat z: quadranda est curva cujus abscissa est z et ordinata numerator ille, ut et curva cujus abscissa est y et ordinata denominator

---

(69) Newton breaks off at the top of an otherwise clean page; whether he had it initially in mind further to continue this present sequence of theorems, or maybe to adjoin a yet more general 'Theorematum Progressio tertia', must remain anyone's guess. We have trivially amended the two last denominators, which in the manuscript are both set to be $w''-2, gkz^{\eta}$. Throughout, of course, Newton continues his earlier slip (see note (60) above) of omitting to augment the indices of z in all denominators after the second to be successively 2η, 3η, 4η, ...

(70) ULC. Add. 3960.7: 105–9, a first version of Cases 1–5 of the revised Proposition XI above (see §2: note (51)) where its several loosenesses and occasional slips (compare note (77) following in particular) are respectively tightened and adjusted.

ille; et harum areæ ponendæ sunt æquales. Nam hæc æquatio dabit relationem inter y et z. Q.E.I.(71)

Cas. 3. Si numerator ab y et denominator ab z liberari nequeunt, multiplicandus est numerator per z et denominator per y, (72) deinde si opus est multiplicandus vel dividendus est uterꝗ per y vel per aliam ejus dignitatem et uterꝗ per z vel per aliquā ejus dignitatem ut termini omnes in numeratore quibus nulli sunt similes in denominatore liberentur a coeffi[ci]ente y et omnes in denominatore quibus nulli sunt similes in numeratore liberentur a coefficiente z. Si hoc fieri possit dividatur vicissim numerator per z ac denominator per y, et quæratur area curvæ cujus abscissa est z & ordinata est pars illa numeratoris in qua y jam non reperitur; sitꝗ area ista A. Quæratur etiam area curvæ cujus abscissa est y et ordinata est pars illa denominatoris in qua z jam non reperitur; sitꝗ area ista C. Deniꝗ de parte reliqua numeratoris in cujus terminis omnibus y reperitur ducta in $\dot{z}$ subduc partem reliquam denominatoris in cujus terminis omnibus z reperitur ductam in $\dot{y}$: et per Prop. [IV] regrediendo quære quantitatem cujus fluxio est illa differentia, sitꝗ ista B. Et erit $A+B=C$ æquatio qua relatio inter y et z determinatur. Q.E.I.

Cas. 4. Si quantitas B inveniri non potest[,] transmutandæ sunt fluentes quantitates in alias hac methodo. Sit $z=x^{\nu}$(73) et $y=v^{\lambda}$ et erit $\dot{z}=\nu\dot{x}x^{\nu-1}$ et $\dot{y}=\lambda\dot{v}v^{\lambda-1}$. In

(71) Newton has cancelled a following illustration: 'Ut si sit

$$aazyy-bbyy\sqrt{aa-zz}+a^3z\dot{y}-y^3z\dot{y}=0. \quad \text{hoc est} \quad \dot{y}=\frac{a^2-\dfrac{bb}{z}\sqrt{aa-zz}}{y-a^3y^{-2}}.$$

quadranda est tum curva cujus ordinata est $a^2-\dfrac{bb}{z}\sqrt{aa-zz}$ tum etiam curva cujus ordinata est $y\dot{y}-a^3y^{-2}\dot{y}$ et aream illam huic quantitati æqualem ponendo habebitur relatio inter z et y. Q.E.I.' Having thus divided through his given equation by zy^2 to separate the variables y and z, Newton's only remaining difficulty is to effect the quadrature of $b^2z^{-1}\sqrt{[a^2-z^2]}$ and this is readily achieved (by means of the substitution $z=1/x$) in terms of the area under the ellipse $v=\sqrt{[1-a^2x^2]}$; compare Ordo 5.1 of the latter 'Catalogus' of integrals in his 1671 tract (III: 248).

(72) An uncompleted first continuation here reads: 'et si terminus aliquis reperiatur in denominatore per z vel per aliquam ejus dignitatem mult[iplicatus] vel div[isus] cui nullus est similis et homogeneus in numeratore, efficiendum est per multiplicationem vel divisi[onem numeratoris] ac denominatoris per dignitatem aliquam ipsius z ut [in denominatore] illo non reperiatur z. Et similiter si in numeratore [terminus] est aliquis per y vel per ejus dignitatem qu[em]vis mul[ti]plicatus vel divisus cui nullus est similis in denominatore, delenda [est y]'.

(73) Newton initially went on: 'et erit $\dot{z}=\nu\dot{x}x^{\nu-1}$. Tempus autem quo pars minima ipsius z fluxu uniformi generatur est ad tempus quo pars correspondens ipsius x æquali fluxu generatur reciproce ut pars ipsius x ad partem ipsius z hoc est ut $\dot{x}$ ad $\dot{z}$ sive ut 1 ad $\nu x^{\nu-1}$. Si fluxio ipsius y minuatur in ratione temporis aucti vel augeatur in ratione temporis diminuti, (hoc est si pro $\dot{y}$ scribatur $\dfrac{\dot{y}}{\dot{z}}$ seu $\dfrac{\dot{y}}{\nu[\dot{x}]\,x^{\nu-1}}$) hæc fluxio toto tempore generabit eandem quantitatem y ac prius. Igitur in æquatione qua relatio inter y, z, $\dot{y}$ et $\dot{z}$ definitur scribatur x^{ν} pro z et $\nu\dot{x}x^{\nu-1}$ pro $\dot{z}$. Rursus sit $y=v^{\lambda}$ et erit $\dot{y}=\lambda\dot{v}v^{\lambda-1}$.'

æquatione qua relatio inter y, z, $\dot{y}$ et $\dot{z}$ definitur scribantur hi earum valores et prodibit æquatio nova qua relatio inter v x $\dot{v}$ et $\dot{x}$ definitur. Tentandum est per casum tertium si indices ν et λ tales esse possint ut inveniatur quantitas B. Si res succedit habebitur per æquationem $A+B=C$ relatio inter v et x et inde per æquationes $z=x^{\nu}$ et $y=v^{\lambda}$ habebitur relatio inter y et z. Q.E.I.

[*Corol*] *1.* Hinc fluxiones quantitatum compositarum facile inveniun[tur]. Proponatur quantitas quævis z^3. Dic istam v et erit $3z^2\dot{z}=\dot{v}$ [adeoqꝫ $3z^2\dot{z}$] est fluxio ipsius z^3. Similiter $\dot{z}xx+2zx\dot{x}$ est fluxio ipsius zxx et $\dfrac{2y\dot{y}-a\dot{z}}{2\sqrt{yy-az}}$ est fluxio ipsius $\sqrt{yy-az}$.

Corol. 2. Et per contrarias operationes regreditur ad fluentes quantitates: ut a fluxione $3\dot{z}z^2$ ad quantitatem z^3, a fluxione $\dot{z}xx+2zx\dot{x}$ ad quantitatem zxx, a fluxione $3z^2\dot{z}xxy+2z^3\dot{x}xy+z^3xx\dot{y}$ ad quantitatem z^3xxy. Fluxio a qua fit regressus tot involvit terminos quot sunt simplices fluentes quantitates. A fluxionib^s fractis et surdis regressus fit per sequentia theoremata. Sit R quælibet terminorum compositio et erit $\lambda\dot{R}R^{\lambda-1}$ fluxio quantitatis surdæ R^{λ}. Sit S alia quævis terminorum compositio et erit $\dfrac{\dot{R}S-R\dot{S}}{SS}$ fluxio quantitatis fractæ $\dfrac{R}{S}$. Per hæc Theoremata regreditur a fluxionibus quoties casus possibiles incidunt,(74) præsertim si duæ vel plures sunt quantitates simplices fluentes in vinculo radicis fluxionis compositæ. Sed brevitati consulo.

Distinguantur complexæ quantitates et earum fluxiones in var[ia genera] ut eæ sint *primi generis* quæ ab una quantitate simplici fluente [dependent] et eæ *secundi* quæ dependent a duabus ac eæ *tertij* quæ dependent a tribus & sic deinceps: et cum fluxio quantitatis secundi generis semper constet ex partibus duabus quarum una multiplicatur per unam fluxionem simplicem et altera per alteram, dicantur partes istæ *conjugatæ* et similiter *fluxiones conjugatæ* dicantur partes tres ex quibus fluxio omnis quantitatis tertij generis constat, et partes quatuor ex quibus fluxio omnis quantitatis quarti generis constat & sic porro. Fluxiones autem quæ conjugatis proportionales sunt sed non sunt conjugatæ dicantur tantum *cognatæ*. Sic quantitatis secundi generis zyy fluxio $\dot{z}yy+2zy\dot{y}$ ex fluxionibus conjugatis $\dot{z}yy$ et $2z\dot{y}$ componitur et his proportionales omnes fluxiones $a\dot{z}yy$ et $bzy\dot{y}$ sunt sibi invicem cognatæ. Si quantitatum cognatarum coefficientes sunt ut indices dignitatum quantitatum simplicium fluxionibus suis affectarum unitate aucti ([velut hic] a ad b ut $0+1$ ad $1+1$) quantitates illæ cognatæ sunt etiam conjugatæ.

Et his præmissis regressus a fluxionibus ad quantitates sic instituetur.

(74) Originally 'Et per hæc Theoremata regressus a fluxionibus surdis vel fractis non sunt admodum difficiles'.

Cas. 1. Fluxiones complicatæ rationales et integræ, si adsint omnes earum conjugatæ, in quantitates redeunt, mutando fluxionem simplicem in quantitatem suam ac dividendo fluxionem complicatam per indicem dignitatis quantitatis illius simplicis, ut fluxio $3y^2\dot{y}$ in quantitatem y^3 scribendo y pro $\dot{y}$ et dividendo $3y^3$ per indicem $3_{[,]}$ et fluxio $\dot{z}yy+2zy\dot{y}$ in quantitatem zyy mutando fluxionem simplicē $\dot{z}$ vel $\dot{y}$ in conjugatarum alterutra, in quantitatem suam z vel y, ac dividendo quantitatem prodeuntem zyy vel $2zyy$ per indicem dignitatis quantitatis illius simplicis z vel y. Et eadem ratione fluxio $3y^2\dot{y}+\dot{z}yy-4ay\dot{y}+2zy\dot{y}$ redit in quantitatem $y^3+zyy-2ayy$. Nam fluxio $3y^2\dot{y}$ cùm sit primi generis redit sola in quantitatem y^3 et fluxio $\dot{z}yy$ redit in quantitatem zyy sed non sola cum sit secundi generis. Quantitatis inventæ zyy fluxio est $\dot{z}yy+2zy\dot{y}$ ideoq̧ fluxionis $\dot{z}yy$ conjugata est $2zy\dot{y}$ et hæ duæ fluxiones simul redeunt in quantitatem zyy. Restat fluxio $-4ay\dot{y}$ quæ cum sit primi generis redit sola in quantitatem $-2ay\dot{y}$.(75)

Cas. 2. Si fluxio est fractio irreducibilis cum denominatore duorum vel plurium nominum, quærantur denominatoris divisores omnes primi fluentes, et si divisor est aliquis cui nullus alius est æqualis, regressus ab hac fluxione ad quantitatem fluentem fieri non potest geometricè. Sin talis divisor non sit, rejiciendus est divisor unus magnitudinis cujusq̧ sic ut ex duobus æqualibus divisoribus relinquatur unus$_{[,]}$ ex tribus relinquantur duo$_{[,]}$ ex quatuor relinquantur tres &c et contentum sub divisoribus omnibus qui relinquuntur ductum in maximum denominatoris divisorem non fluentem erit denominator quantitatis quæsitæ. Dein per Cas. 1 quæratur quantitas cujus fluxio est numerator fluxionis propositæ per divisores rejectos et max[imum] div[isorem] non fl[uentem] multiplicatus et hæc quantitas divisa per denominatorem inventam erit numerator quantitatis quæsitæ. Ut si fluxio sit fractio

$$\frac{2a\dot{x}y^5-2\dot{x}y^6+5axy^4\dot{y}-6ay^5\dot{y}\ ^{(76)}}{2a^3x^3-[4]aax^3y+[2]ax^3yy},$$

(75) In a slight generalization of this mutation technique on Add. 3960.13: 221 *bis*r Newton afterwards added that 'Methodus mutando generalior est ut quærantur propositarum fluxionum genitores sub indefinitis coefficientibus, dein ex genitoribus vicissim eruantur fluxiones et conferantur cum homogeneis fluxionibus propositis ad determinandas coefficientes, et augendo genitores quantitatibus quæ in genesi evanescere possunt, id est quantitate quavis data g si unica est fluxio generationis primæ, vel quantitatibus $gz+h$ si fluxio est secundæ generationis, vel gz^2+hz+i si tertiæ'. In sequel he there adjoined two instances illustrating its application: '*Exempl. 1.* Sit fluxio $3\dot{x}x^2-\dot{x}yy-2xy\dot{y}+2aa\dot{v}$, et genitores erunt $px^3-fxyy+rv$, et ex his prodeunt fluxiones $3p\dot{x}x^2-f\dot{x}yy-2fxy\dot{y}+r\dot{v}$. Et conferendo terminos homogeneos fiunt ...$p=1$, ..$f=1$ et $r=2aa$. Ergo [fluens quæsita est] $x^3-xyy+2aav+g$'; and, at a second level, '*Exempl. 2.* Sit fluxio $9az+4by\dot{y}+2by\ddot{y}+12y\dot{y}^2+6y^2\ddot{y}$, ...et genitores erunt $pz^3+qzyy+ry^3$. Et ex his prodit fluxio secundæ generationis $6pz+4qy\dot{y}+2qz\dot{y}^2+2qzy\ddot{y}+6ry\dot{y}^2+3ry^2\ddot{y}$. Unde conferendo terminos homogeneos fit ... $p=\frac{3}{2}a$, ... $b=q$ et ... $r=2$. Ergo... fluens quantitas quæsita est $\frac{3}{2}az^3+bzyy+2y^3+gz+h$'.

(76) By a trivial slip Newton in his manuscript set the denominator to be

$$`2a^3x^3-2aax^3y+ax^3yy'.$$

quære denominatoris divisores primos fluentes x, x, x, $a-y$, $a-y$ et maximum non fluentem $2a$. De fluentibus rejice divisorem unum magnitudinis utriusque x et $a-y$ et contentum sub reliquis x, x et $a-y$ ac divisore dato $2a$, nempe $2aaxx-2axxy$ erit denominator quantitatis quæsitæ. Reduc fractionem ad hunc denominatorem quadratum et ejus numerator erit

$$4aax\dot{x}y^5-4ax\dot{x}y^6+10a^2x^2y^4\dot{y}-12ax^2y^5\dot{y}.$$

Per casum primum quære quantitatem cujus hæc est fluxio et prodibit $2aaxxy^5-2aaxxy^6$. Divide hanc quantitatem per denominatorem inventum et prodibit quantitatis quæsitæ numerator y^5. Ergo $\dfrac{y^5}{2aaxx-2axxy}$ est quantitas fluens quam invenire oportuit.(77)

Cas. 3. Si fluxio involvit factorem(78) surdum, extrahatur ex radice quicquid rationale est et sit radix reliqua R^α. Dividatur fluxio per ipsius R semel vel sæpius si fieri possit et sit ν numerus divisionum & Q quotum ultimum, et erit fluxio $QR^{\alpha+\nu}$.(79)

---

[7](80) *Cas. 4.* Si res ne sic quidem succedit: solvendum est Problema per methodum serierum infinitarum(81) hoc modo. Sit y quantitas quam ex assumpta altera quantitate z invenire oportet. Fluat illa altera quantitas z uniformiter et sit ejus fluxio 1. Sic æquatio tres tantum quantitates indeterminatas involvet nempe z, y et $\dot{y}$. Ut primus seriei terminus inveniatur scribe passim in æquatione $\dfrac{y}{z}$ (82)

---

(77) If, as Newton, we carelessly appeal to the 'rule' that the fluxion of A/B is $(A\dot{B}+B\dot{A})/B^2$ (where here $A = y^5$ and $B = 2ax^2(a-y)$).

(78) 'vel divisorem' is deleted.

(79) The manuscript draft here terminates in mid-page, to be immediately reshaped (Add. 3960.7: 109–11) into a single set of Cases 1–3 which differ little from the corresponding portions of their similarly numbered revises in Proposition XI in §2 above; after which ensue in quick succession two minimally variant drafts of the preliminary 'Cas. 4'—soon to be delayed to become the separate Proposition XII in §2—whose text we reproduce in [7] next following.

(80) ULC. Add. 3960.7: 111–12, with a concluding 'Exempl. 1' adjoined from the minimal revise on *ibid.*: 137. Almost at once (compare §2: note (70) above) this exposition of Newton's general method for extracting the fluent root of a given fluxional equation as an infinite 'converging' sum-series was postponed—and greatly elaborated—to be a separate following Proposition XII. On Add. 3960.10: 173 this 'Cas. 4' is drastically curtailed to be but a single sentence referring to the future discussion of its theme (see the next note), and wholly omitted in the ensuing final text of Proposition XI of the 'De qudratura' (reproduced in §2 preceding).

(81) In the rump of this 'Cas. 4' on Add. 3960.10: 173 (see the previous note) the whole of the remainder of the text which follows is subsumed, in anticipation of the later augmented discussion of its content, into the single phrase 'in sequenti Propositione [*sc.* XII] expositam'.

(82) In which z is of dimension -1, that by which Newton—neglecting consideration of

pro fluxione $\dot{y}$(83) ac terminos æquationis prodeuntis colloca in cellulis tabulæ tessaratæ(84) ut in extractione radicum affectarum & ex omnibus selige terminos illos quos applicata regula tangit. Assumatur jam dignitas quælibet fictitia az^{ν} pro primo seriei termino et pro indice ν substitue numerum quo termini omnes selecti ad easdem(85) dimensiones ascendant si modo z^{ν} in ipsis pro y et $z^{2\nu}$, $z^{3\nu}$ &c pro y^2, y^3 &c scribantur. Dein ex æquatione tres indeterminatas quantitates z, y et $\dot{y}$ involvente selige terminos ex quibus termini selecti priores prodiere et pro a scribe quantitatem qua termini novissime selecti evanescant, si modo in ipsis passim scribatur az^{ν} pro y et $\nu az^{\nu-1}$ pro $\dot{y}$. Et quantitas az^{ν} erit primus terminus seriei. Pro reliquis suis terminis scribe p et erit $y = az^{\nu}+p$. In æquatione tres ignotas involvente scribe $az^{\nu}+p$ pro y et habebitur æquatio nova ex qua primum terminum seriei p iterando operationem jamjam descriptam elicias, et sic deinceps in infinitum.

(86)*Exempl. 1.* Invenienda sit linea curva AC cujus longitudo x sit media proportionalis inter ejus abscissam AB seu z et longitudinem datam l. Et cum sit $xx = lz$, erit $2x\dot{x} = l\dot{z}$, ideoqꝫ $l.2x::\dot{x}.\dot{z}$. hoc est si ducatur tangens CT abscissæ(87)

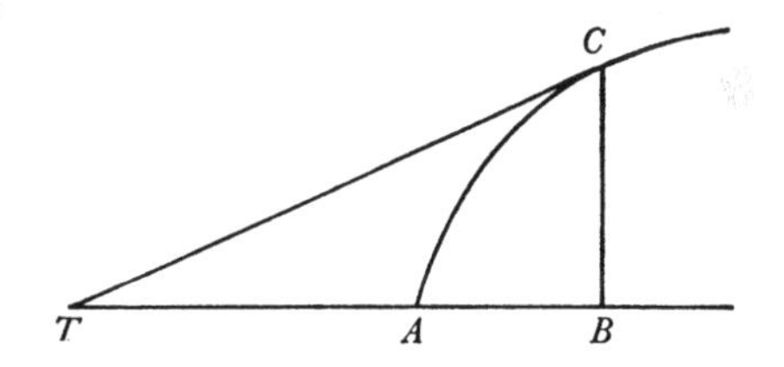

---

non-algebraic quantities y!—assumes the operation of fluxional 'differentiation' to reduce the power of any given fluent.

(83) Newton first went on: 'et æquatio jam duas tantum quantitates involvet nempe z et y. Qua ratione radices ex æquationibus per methodū serierum convergentium extrahere docui extrahe ex hac æquatione primum terminum radicis y. Sit iste az^n & hic erit primus seriei terminus. Pro terminis omnibus reliquis scribe v et erit $y = az^n+v$ et $\dot{y} = naz^{n-1}+\dot{v}$. In æquatione tres indeterminatas quantitates z y et $\dot{y}$ involvente pro y et $\dot{y}$ scribe az^n+v et $naz^{n-1}+\dot{v}$ et prodibit æquatio nova ubi si pro $\dot{v}$ scribatur $\frac{v}{z}$ et per terminum infimum involventem v dividatur terminus infimus eorū in quibus v non reperitur, habebitur secundus seriei terminus. Sit iste bz^m et si pro reliquis seriei terminis scribatur t erit

$$y = az^n+bz^m+t \quad \& \quad \dot{y} = naz^{n-1}+mbz^{m-1}+\dot{t}.\text{'}$$

This he began to amend to read 'ac terminos prodeuntes duc in excessum indicis infimæ dignitatis z non multiplicatæ per y supra indices dig[nitatum multiplicatarum]' and then 'Subduc indices dignitatū z in his terminis ab indicibus dignitatis infimæ ipsius z in terminis non multiplicatis per y et per excessus multiplica terminos novissime inventos et ex æquatione prodeunte quæ jam duas tantum involvit quantitates incognitas, y et z, eadem ratione qua quotos et radices ex æquationibus per methodum serierum infinitarum eruere docui erue primum terminum valoris y' before opting to continue as in his final sequel.

(84) That is, composed of *tessaræ*, 'chequered'; in his revise (on p. 137 of the manuscript) Newton afterwards converted this to the classical form 'tessellatæ'.

(85) In his revise Newton here inserted 'ubiqꝫ'.

(86) As we have said (note (80) above) the following *exemplum* is adjoined from the revised text on Add. 3960.7: 137.

(87) The manuscript text ends at the bottom of p. 137 and its continuation is now seemingly

[8][88]

Cas. 5. Si æquatio quantitatem uniformiter fluentem z cujus fluxio sit unitas, et quantitatem non uniformiter fluentem y cum pluribus ejus fluxionibus (prima $\dot{y}$ secunda $\ddot{y}$ tertia $\dddot{y}$ &c) involvit: multiplica æquationem per $y^{\mu}z^{\nu}$. Tum excerpe terminos omnes in quibus flux[i]onum ultima reperitur, & in ijs muta fluxionem illam ultimam in primam & terminos adserva.[89] Dein horum fluxiones collige usq; ad ultimam & aufer ex æquatione.[90] Fac autem ut dignitatum exponentes μ et ν tales sint ut quantitates omnes in quibus y et ejus fluxiones reperiuntur ex æquatione evanescant, & tam in terminis adservatis quam in æquatione extermina μ et ν. Postea quære aream curvæ cujus abscissa est z et ordinata aggregatum terminorum æquationis qui post novissimam sub-

lost, but understand in sequel something like 'incidens in T et capiatur ordinata $BC = y$, erit $CT^2.BT^2::l.4z$ adeoq; $\dot{y}^2.\dot{z}^2::CB^2.BT^2::l-4z.4z$, unde $z\dot{y}^2 = \frac{1}{4}l-z$'. While Newton would then have gone on to extract the fluent y as a series in z by substituting $y = az^{\nu}+p$ to find $a^2\nu^2z^{2\nu-1} = \frac{1}{4}l$ and so $y = l^{\frac{1}{2}}z^{\frac{1}{2}}+O(z^{\frac{3}{2}})$, the result is more directly and straightforwardly derived by extracting $(\dot{y}/z^{-\frac{1}{2}} =)\ \sqrt{[\frac{1}{4}l-z]}$ as a binomial series and integrating its product with $z^{-\frac{1}{2}}$ term by term to obtain y.

At the top of p. 138 immediately following—the manuscript is now partially eroded away—Newton has tentatively penned a concluding '*Cas. 5.* Si [æquatio] fluxiones non solum primas sed etiam secundas ac [tertias] aliasq; involvit: reducenda est æquatio per Corol. 5 Propositionis sequentis ad aliam æquationem finitam quæ solam fluxionem primam involvit: ac deinde tentanda est ejus reductio (ut supra in casu primo, secundo vel tertio) ad aliam æquationem finitam quæ solas quantitates fluentes y et z involvit'. This mysterious citation of 'Corol. 5 Propositionis sequentis [*sc.* XII]' is clarified in the otherwise trivially variant rewrite of this on Add. 3960.10: 173, where 'Corol. 5' is amended in afterthought to be (as one would have expected) 'Corol. 4'; compare §2: note (110) above.

(88) ULC. Add. 3960.7: 123/101/128; minor redrafts on *ibid.*: 122 are quoted in following footnotes. This first version of the equivalent 'Cas. 5' in §2 above was evidently abandoned when Newton could not persuade its appended *exemplum* to yield his desired result; compare note (91) below.

(89) In a later restyling of the sequel on the facing p. 122 in the manuscript—evidently as part of the revised 'Geometriæ Liber [?3]' for which he had similarly revamped the earlier cases (see §2: note (53))—Newton continued: 'Postea per Cas 1 Prop. 2 hujus lib[ri] quære quantitatem fluentem cujus fluxio prima est aggregatum terminorum æquationis qui post novissimam subductionem relinquebantur si modo fluxio secunda est fluxionum ultima, vel cujus fluxio secunda est aggregatum illud si fluxio tertia est fluxionum ultima, vel tertia si quarta &c[,] hanc quantitatem adjice quantitatibus asservatis et summam pone $= 0$. Nam hæc æquatio [...]'. (The two final sentences of the paragraph are understood to follow unchanged.)

(90) Afterwards expanded on the manuscript's facing page to read: 'Excerpe etiam terminos in quibus fluxio penultima reperitur non multiplicata per primam, et muta hanc fluxionem in primam divideq; hos terminos per 2 si fluxionum ultima est fluxio secunda, per 3 si eadem est fluxio tertia, per 4 si quarta est, et sic deinceps. Postea fluentes quantitates quarum hi termini sunt fluxiones investiga et adjice excerptis primis. Dein horum omnium fluxiones collige donec veniatur ad ultimā fluxionum quæ sunt in æquatione & ultimo collectas aufer ex æquatione.' The sequel was correspondingly converted to be: 'Fac autem ut dignitatum exponentes μ et ν sint ut termini omnes in quibus y aut ejus fluxio aliqua reperitur ex æquatione evanescant....'

ductionem relinquebantur: et rursus si opus est quære aream alterius curvæ cujus abscissa est z et ordinata area modo inventa ad unitatem applicata et data longitudine aucta[,] et hoc fac tertio & sæpius si opus fuerit. Deniq; si fluxio secunda est fluxionum ultima adjice aream primam quantitatibus asservatis, si tertia adjice quartam[,] si quarta quintam &c et summam pone $= 0$. Nam hæc æquatio exhibebit relationem primarum fluxionum quantitatum y et z. Et ex hac relatione per casus quatuor primos eruenda est relatio quantitatum fluentium.

Ut si proponatur æquatio

$$24z - 6a\dot{y}^2 - 6ay\ddot{y} - 2az\dot{y}\ddot{y} - 2azy\dddot{y} + 4\dddot{y}y^3 + 36\ddot{y}\dot{y}y^2 + 24y\dot{y}^3 = 0,$$

multiplico eandem per $y^\mu z^\nu$ et excerpo terminos omnes $-2azy\dddot{y} + 4y^3\dddot{y}$ in $y^\mu z^\nu$ in quibus fluxionum ultima $\dddot{y}$ reperitur, mutoq; fluxionem illam in primam $\dot{y}$ et terminos $-2azy\dot{y} + 4y^3\dot{y}$ in $y^\mu z^\nu$ adservo. Dein colligo horum fluxiones usq; ad ultimā $\dddot{y}$ et aufero. Termini $4y^{\mu+3}z^\nu\dot{y}$ fluxio est primo

$$\overline{4\mu+12}\,y^{\mu+2}z^\nu\dot{y}^2 + 4\nu y^{\mu+3}z^{\nu-1}\dot{y} + 4y^{\mu+3}z^\nu\ddot{y}_{[,]}$$

deinde

$$\overline{4\mu+12} \times \overline{\mu+2}\,y^{\mu+1}z^\nu\dot{y}^3 + \overline{\mu+3} \times 8\nu y^{\mu+2}z^{\nu-2}\dot{y}^2 + \overline{\mu+3} \times 12y^{\mu+2}z^\nu\dot{y}\ddot{y}$$
$$+ \overline{\nu-1} \times 4\nu y^{\mu+3}z^{\nu-2}\dot{y} + 8\nu y^{\mu+3}z^{\nu-1}\ddot{y} + 4y^{\mu+3}z^\nu\dddot{y}.$$

Et hæc posterior fluxio de æquatione sublata delere debet terminos relativos $4\dddot{y}y^3 + 36\ddot{y}\dot{y}y^2 + 24y\dot{y}^3$ in $y^\mu z^\nu$. Collatis igitur utrinq; terminis homogeneis animadverto quod terminus per fluxionem solam $\ddot{y}$ multiplicatus evanescere debet ideoq; ejus coefficiens 8ν nihil est & $\nu = 0$. Animadverto etiam quod ex æqualitate terminorum qui per $\dot{y}\ddot{y}$ multiplicantur prodit coefficientium hæc æquatio $\overline{\mu+3} \times 12 = 36$ et inde $\mu = 0$. Deleo igitur ubiq; dignitatum exponentes μ et ν et sic termini adservati evadunt $-2azy\dot{y} + 4y^3\dot{y}$. Horum fluxiones primæ sunt $-2az\dot{y}^2 - 2azy\ddot{y} - 2ay\dot{y} + 12y^2\dot{y}^2 + 4y^3\ddot{y}$ & secundæ seu ultimæ

$$-6az\dot{y}\ddot{y} - 4a\dot{y}^2 - 2azy\dddot{y} - 4ay\ddot{y} + 24y\dot{y}^3 + 36\dot{y}\ddot{y}y^2 + 4y^3\dddot{y}.$$

Has aufero ex æquatione proposita et fluxio ablatitia delet terminos illos relativos et manet æquationis pars $24z -$ [2] $a\dot{y}^2 -$ [2] $ay\ddot{y}$[+4] $az\dot{y}\ddot{y}$, ac termini adservati fiunt $-2azy\dot{y} + 4y^3\dot{y}$. Excerpo etiam terminum $-$[2] $ay\ddot{y}$ in quo fluxio penultima $\ddot{y}$ reperitur non multiplicata per primam $\dot{y}$ et muto $\ddot{y}$ in y[,] et quoniam fluxionum ultima [hic] est fluxio [secunda $\ddot{y}$] divido hunc terminum $-$[2]ayy per [2] & quantitatem $-ayy$ excerptis primis [addo.](91)

---

(91) We borrow this last, unfinished sentence from the preliminary revise on p. 128, here making some necessary amendments of its numerical coefficients in line with those introduced —in correction of a trivial slip by Newton—in the preceding 'æquationis pars'. It will be clear that, if the fourth term in the given equation had been $-6az\dot{y}\ddot{y}$, then the whole would have been the third fluxion of $y^4 + z^4 - azy^2 = gz^2 + hz + i$, and we may presume that Newton

[9][92]

Cas. 6. Si quantitas y quæ non fluit uniformiter in æquatione non reperiatur:[93] pro prima fluxionum quæ in æquatione habentur (sive ea sit $\dot{y}$ sive $\ddot{y}$ sive $\dddot{y}$ &c) scribe y, pro earum secunda scribe $\dot{y}$, pro tertia scribe $\ddot{y}$, pro quarta scribe $\dddot{y}$ & sic deinceps. Tum per casus præcedentes quære relationem inter y et z. Dein pro y scribe $\dot{y}$ et inde collige y et habebis quæsitam relationem inter y et z si modo prima fluxionum erat $\dot{y}$. Sin earum prima erat $\ddot{y}$, scribe iterum $\dot{y}$ pro y et inde collige y. Et hoc fac tertio si prima fluxionum erat $\dddot{y}$ &c. Sic enim habebis relationem quæsitam inter y et z.

Ut si fuerit $4\ddot{y}^3 = \dddot{y}^2$, hæc æquatio scribendo y pro $\ddot{y}$ et $\dot{y}$ pro $\dddot{y}$ fit $4y^3 = \dot{y}^2$, seu $4\dot{z}^2y^3 = \dot{y}^2$. Unde $\dot{z} = \frac{1}{2}\dot{y}y^{-\frac{3}{2}}$, & per casum primum $z = y^{-\frac{1}{2}}$ + vel − data quavis c[,] seu $z-c = y^{-\frac{1}{2}}$ et $\frac{1}{z^2-2cz+cc} = y$. Hic pro y scribatur $\dot{y}$ et fiet $\frac{\dot{z}}{z^2-2cz+cc} = \dot{y}$ et inde $d-\frac{1}{z-c} = y$. Rursus pro y scribatur $\dot{y}$ et fiet $d\dot{z}-\frac{\dot{z}}{z-c} = \dot{y}$ seu

$$d\dot{z}+\frac{\dot{z}}{c-z} = \dot{y}$$

et inde $e+dz+\boxed{\frac{1}{c-z}} = y$. Ubi c d et e pro lubitu assumuntur et $\boxed{\frac{1}{c-z}}$ est area Hyperbolæ cujus abscissa est z et ordinata $\frac{1}{c-z}$, et y est quantitas fluens quam invenire oportuit.[94]

miscomputed at his third 'differentiation' thereof in contriving his present example. Rather than retrieve his none too obvious error, he henceforth preferred to abandon the whole draft so as to make way for the revised version reproduced on pages 82–8 above (whose revamped 'Exempl. 1' is, however, faulty in not one but two of its numerical coefficients; see §2: note (78)).

(92) ULC. Add. 3960.9: 157. These two initially separate, complementary Cases 6/7 were afterwards combined, as an amplified version of the latter 'Cas. 7' and by way of rough intermediary revises on Add. 3960.8: 149–51/155, in the finished 'Cas. 6' reproduced on pages 88–92 above; in this restyling, it will be clear, the illustrative example in the present 'Cas. 6' was suitably remodelled to be 'Exempl. 1' of the new case, while the two worked instances of 'Cas. 7' are incorporated there with minimal change as the ensuing 'Exempl. 2' and 'Exempl. 3'.

(93) Newton here first wrote 'desideratur' (is wanting).

(94) In a first draft of this initial sixth case on Add. 3960.8: 149 (where it makes its *début* as 'Reg. 1'—the only one developed—of a miscellany of rules intended to expound a variety of techniques of solution of fluxional 'æquationes quæ quantitatem unam fl[uentem] cum ejus fluxionibus involvant') two simpler illustrations are given of the method, namely 'si æquatio fuerit $\dot{y}^3+2bb\ddot{y} = 0$. area curvæ cujus abscissa est $\dot{y}$ et ordinata $\frac{1}{\ddot{y}}$ seu $-\frac{2bb}{\dot{y}^3}$ æqualis est quantitati non fluenti z. Deinde æquatio sic inventa $z = \frac{bb}{\dot{y}^2}$ seu $\dot{y} = bz^{-\frac{1}{2}}$ per casum primum dat æquationem quæsitam $y = 2bz^{\frac{1}{2}}$. Rursus si æquatio fuerit $\ddot{y}^4+3c^3\dddot{y} = 0$ area curvæ cujus basis est $\ddot{y}$ et ordinata $\frac{1}{\dddot{y}} = -\frac{3c^3}{\ddot{y}^4}$ erit $\frac{c^3}{\ddot{y}^3} = z$. Et inde est $\ddot{y} = cz^{-\frac{1}{3}}$ et per Cas. 1 $\dot{y} = \frac{3}{2}cz^{\frac{2}{3}}$ et $y = \frac{9}{10}cz^{\frac{5}{3}}$'.

Cas. 7. Si quantitas uniformiter fluens z in æquatione non reperiatur, pro quantitate altera y scribe x et pro ejus fluxionibus $\dot{y}$, $\ddot{y}$, $\dddot{y}$ & $\ddddot{y}$ scribe $v^{\frac{1}{2}}$, $\frac{1}{2}\dot{v}$, $\frac{1}{2}\ddot{v}v^{\frac{1}{2}}$ & $\frac{1}{2}\dddot{v}v+\frac{1}{4}\ddot{v}\dot{v}$ respective. Vel si quantitas fluens y in æquatione etiam desit, pro ejus fluxione prima $\dot{y}$ scribe x et pro reliquis ejus fluxionibus $\ddot{y}$, $\dddot{y}$, $\ddddot{y}$, $\overset{\cdot\cdot\cdot\cdot\cdot}{y}$ scribe $v^{\frac{1}{2}}$, $\frac{1}{2}\dot{v}$, $\frac{1}{2}\ddot{v}v^{\frac{1}{2}}$ & $\frac{1}{2}\dddot{v}v+\frac{1}{4}\ddot{v}\dot{v}$ respectivè.(95). Dein per casus quatuor primos si $\ddot{v}$ et $\dddot{v}$ desint, vel per casum quintum si adsint[,] erue ex æquatione relationem inter x et v perinde ac si x uniformiter flueret et ejus fluxio esset unitas. Postea per casum sextum quære aream Curvæ cujus abscissa est x et ordinata $\frac{1}{v^{\frac{1}{2}}}$ et pone hanc æqualem quantitati uniformiter fluenti z + quavis data. Et si x æquetur y, habebitur relatio quæsita inter z et y. Sin x æquetur $\dot{y}$, quære aream Curvæ cujus abscissa est x et ordinata $\frac{v^{\frac{1}{2}}}{x}$ & pone hanc æqualem y + quavis data.

Ut si fuerit $4\dot{y}^2\ddot{y} = 2y+a$, hæc æquatio scribendo x pro y, ... quam invenire oportuit.(96)

Rursus si fuerit $3\dot{y}^2\ddot{y}+2c\dddot{y} = 0$ hæc æquatio scribendo x pro $\dot{y}$ et $v^{\frac{1}{2}}$ pro $\ddot{y}$ et $\frac{1}{2}\dot{v}$ pro $\dddot{y}$, evadit $3xxv^{\frac{1}{2}}+c\dot{v} = 0$, seu $3xx\dot{x}v^{\frac{1}{2}}+c\dot{v} = 0$. Unde $3xx\dot{x} = -c\dot{v}v^{-\frac{1}{2}}$ & per casum secundum $x^3 = [-]\,2cv^{\frac{1}{2}}$.(97) Hoc est $y^3 = [-]\,2c\dot{y}$, seu $y^3\dot{z} = [-]\,2c\dot{y}$. Unde $\dot{z} = [-]\,2c\dot{y}y^{-3}$ et per casum primum $z = cy^{-2}+d$. Pro y scribatur $\dot{y}$ et fiet $z = c\dot{y}^{-2}+d$ et

$$\sqrt{\frac{c}{z-d}} = \dot{y} \quad \text{seu} \quad \dot{z}\sqrt{\frac{c}{z-d}} = \dot{y}.$$

Et inde per Cas. 1 $2\sqrt{cz-cd} = y+e$. Ubi c d et e sunt quantitates quælibet pro lubitu assumendæ, et y est quantitas fluens quam invenire oportuit.(98)

---

[10](99) [Sit] $AB = y$. $BC = x$. $AD[C]B = v$. $BE = w$.(100) [incrementum] $\dot{y} = Bb$. [erit] $BbcC = \dot{y}x = \dot{v}$.(101)

---

(95) See §2: note (85) and Newton's preliminary computations in [10] following. It is again understood that x is the base variable, so that $\dot{x} = 1$.

(96) The omitted text is identical with that of the equivalent 'Exempl. 2' on pages 90–2 above. An insignificantly variant initial draft exists on Add. 3960.8: 150.

(97) Understand '+ data quavis d' more generally. We have introduced some necessary minus signs into Newton's text here and in the next sentence.

(98) For this worked illustration also (compare note (96) above) a minimally differing draft is found—there listed as 'Exempl. 2'—on Add. 3960.8: 150.

(99) ULC. Add. 3960.8: 153/[149+] 152, 155. Newton in these rough calculations—here filled out with appropriate editorial interpolations to smooth the sequence of deduction—develops two variant schemes of substitution for simplifying fluxional equations in which the 'ordinate' variable y does not itself appear. The latter, as we have seen (§2: note (85)), was afterwards expanded into 'Cas. 6' of Proposition XI of his 1691 'De quadratura Curvarum'.

(100) Whence in equivalent Leibnizian terms $v = \int x.dy$ and $w = x.dy/dx$. Newton makes no use of the latter *subtangens* in his ensuing calculations.

(101) Understand throughout this first section that the superscript dots denote fluxions with respect to the (here unexpressed) independent base variable z.

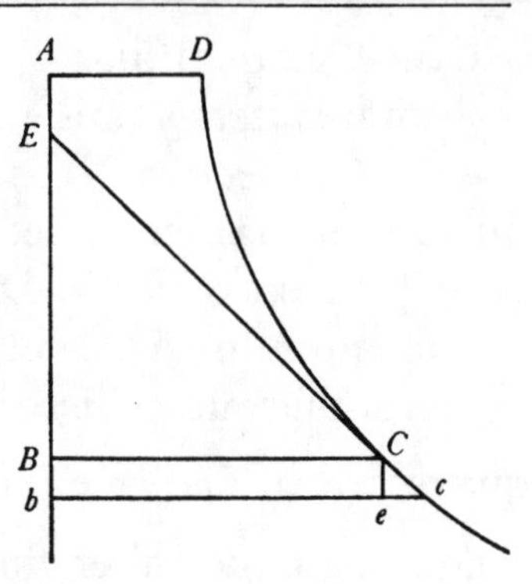

[Pone] $v^n = \dot{y}$. [prodit]

$$nv^{n-1}\dot{v} = \ddot{y} = nv^{n-1}\dot{y}x = nv^{2n-1}x. \text{ [ut et]}$$

$$\overline{n-1},\, nv^{n-2}\dot{v}^2 + nv^{n-1}\ddot{v} = \dddot{y}. \text{ [hoc est]}$$

$$\overline{n-1},\, nv^{n-2}v^{2n}x^2 + nv^{n-1},\, \overline{\ddot{y}x + \dot{y}\dot{x}} = \dddot{y}. \text{ [sive]}$$

$$\overline{nn-n},\, v^{3n-2}x^2 + nnv^{3n-2}x^2 + nv^{2n-1}\dot{x}$$
$$= \overline{2nn-n},\, v^{3n-2}x^2 + nv^{2n-1}\dot{x} = \dddot{y}$$

[Ergo si] $n = \frac{1}{2}$.(102) [adeoq͡ʒ] $\ddot{y} = nx$. [erit] $v^{\frac{1}{2}} = \dot{y}$. [id est] $v = \dot{y}^2$. $\dot{v} = 2\dot{y}\ddot{y}$. [vel] $\frac{1}{2}\dot{v}v^{-\frac{1}{2}} = \ddot{y}$. $+\frac{1}{2}\ddot{v}v^{-\frac{1}{2}} - \frac{1}{4}\dot{v}^2v^{-\frac{3}{2}} = \dddot{y}$. [sive] $\frac{1}{2}\frac{\ddot{v}}{\dot{y}} - \frac{1}{4}\frac{xx}{\dot{y}} = \dddot{y}$. [Etiam] $\frac{1}{2}x = \ddot{y}$. $\frac{1}{2}\dot{x} = \dddot{y}$. $\frac{1}{2}\dot{y}\dot{x} = [\triangle]\ Cec = [\frac{1}{2}]\ \ddot{v}$(103)

[Ubi] z fl[uit] uniform[iter cape] $y, \dot{y}, \ddot{y}, \dddot{y}, \ddddot{y}$ [&c]. [et ubi] y fl. uniform. [cape] $v, \dot{v}, \ddot{v}, \dddot{v}$ [&c]. [Tum] Si $y, \dot{y}, \ddot{y}, \dddot{y}$ [&c] æquat[ionem] ingr[ediantur,]

‖ pro $y, \dot{y}^2, \ddot{y}, \dddot{y}, \ddddot{y}$ &c scribe $t,\ v,\ \frac{1}{2}\dot{v},\ \frac{1}{2}\ddot{v}v^{\frac{1}{2}},\ \frac{1}{2}\dddot{v}v,\ \frac{1}{2}\ddddot{v}v^{\frac{3}{2}}$ &c.

$y\,.\,\dot{y}\,.\,\ddot{y}\,.\,\dddot{y}\,.\,\ddddot{y}$ sunto $x,\ v^{\frac{1}{2}},\ \frac{1}{2}\dot{v},\ \frac{1}{2}\ddot{v}v^{\frac{1}{2}}\ \frac{1}{2}\dddot{v}v + \frac{1}{4}\ddot{v}\dot{v},\ \frac{3}{4}\ddot{v}v^{\frac{3}{2}} + \frac{3}{4}\dddot{v}\dot{v}v^{\frac{1}{2}}$.(104)

[Ut si fuerit aequatio] $\sqrt{\dot{y}\dot{y}+\dot{z}\dot{z}} \times \ddot{y} = \dddot{y}$. [sive posito $\dot{z} = 1$] $\sqrt{1+\dot{y}\dot{y}} \times \ddot{y} = \dddot{y}$.
‖ [cape $\dot{y} = v^{\frac{1}{2}}, \ddot{y} = \frac{1}{2}\dot{v}, \dddot{y} = \frac{1}{2}\ddot{v}v^{\frac{1}{2}}$. erit] $\sqrt{1+v} \times \dot{v} = \ddot{v}v^{\frac{1}{2}}$. [Pone insuper $v = t, \dot{v} = s^{\frac{1}{2}}$, $\ddot{v} = \frac{1}{2}\dot{s}$. fit] $\sqrt{\frac{1+t}{t}} \times s^{\frac{1}{2}} = [\frac{1}{2}]\,\dot{s}$. [sive] $\sqrt{\frac{1+t}{t}} = [\frac{1}{2}]\,s \times s^{-\frac{1}{2}}$. [adeoq͡ʒ] $Q\sqrt{\frac{1+t}{t}} = s^{\frac{1}{2}}$. [id est $Q\sqrt{\frac{1+v}{v}} = \dot{v}$.](105)

[pone $\dot{y} = x$, $\ddot{y} = v^{\frac{1}{2}}$, $\dddot{y} = \frac{1}{2}\dot{v}$. Erit] $\sqrt{1+x^2} \times v^{\frac{1}{2}} = \frac{1}{2}\dot{v}$. [seu] $2\sqrt{1+x^2} = \dot{v}v^{-\frac{1}{2}}$.

---

(102) This to yield $\ddot{y}$ in reduced form, as Newton at once makes clear.

(103) The text breaks off with a final term '$-\frac{1}{4}xx$' unentered. In sequel Newton will direct his attention to the variant scheme of substitution which (see §2: note (85)) retains $\dot{y} = v^{\frac{1}{2}}$ but replaces the condition $\dot{v} = \dot{y}x$ by a change of base variable from z to y.

(104) Read '$\frac{1}{2}\dddot{v}v^{\frac{3}{2}} + \frac{3}{4}\dddot{v}\dot{v}v^{\frac{1}{2}} + \frac{1}{4}\ddot{v}^2v^{\frac{1}{2}}$'. Notice that the superscript dots now denote fluxions with respect to x. Here in succession (compare §2: note (85)) there is

$$\tfrac{1}{2}\dot{v} = v^{\frac{1}{2}}.d(v^{\frac{1}{2}})/dx, \quad \tfrac{1}{2}\ddot{v}v^{\frac{1}{2}} = v^{\frac{1}{2}}.d(\tfrac{1}{2}\dot{v})/dx, \quad \tfrac{1}{2}\dddot{v}v + \tfrac{1}{4}\ddot{v}\dot{v} = v^{\frac{1}{2}}.d(\tfrac{1}{2}\ddot{v}v^{\frac{1}{2}})/dx, \ldots.$$

In the sequel Newton applies this scheme of substitution to reduce two simple third-order fluxional equations.

(105) Newton makes unique use here of the capital Q (for 'Q[uadratum]'?) as an integration symbol. He will at once replace it in the following revised *exemplum* by his more familiar equivalent pictograph □.

[adeoq;] $2\boxed{\sqrt{1+x^2}} = 2v^{\frac{1}{2}}$. [hoc est] $\boxed{\sqrt{1+\dot{y}^2}} = \ddot{y}$. [unde] $\Box\dfrac{1}{\boxed{\sqrt{1+\dot{y}^2}}} = z$. [et]

$\Box\dfrac{\dot{y}}{\boxed{\sqrt{1+\dot{y}\dot{y}}}} = y$.

[Si] $a\dot{y}^3 = b\dddot{y}$. [pro $\dot{y}, \ddot{y}, \dddot{y}, \ddddot{y}$ &c scribe $v^{\frac{1}{2}}, \frac{1}{2}\dot{v}, \frac{1}{2}\ddot{v}v^{\frac{1}{2}}, \frac{1}{2}\dddot{v}v+\frac{1}{4}\dot{v}\ddot{v}$ &c. erit] $av^{\frac{3}{2}} = \frac{1}{2}b\ddot{v}v^{\frac{1}{2}}$. [sive] $av = \frac{1}{2}b\ddot{v}$. [Similiter capiendo fluxiones fit $3a\dot{y}^2\ddot{y} = b\ddddot{y}$.] [sive]

$$\tfrac{3}{2}a\dot{v}v = \tfrac{1}{2}b\dddot{v}v+\tfrac{1}{4}b\ddot{v}\dot{v}. \quad \text{[id est] } a\dot{v} = \tfrac{1}{2}b\dddot{v}.$$

[ut et] $6a\dot{y}\ddot{y}^2+3a\dot{y}^2\dddot{y} = b\ddddot{y}$. [sive] $\frac{3}{2}a\dot{v}^2v^{\frac{1}{2}}[+]\,\frac{3}{2}a\ddot{v}v^{\frac{3}{2}} = \frac{3}{[4]}\,b\ddddot{v}v^{\frac{3}{2}}+\frac{3}{[4]}\,b\dddot{v}\dot{v}v^{\frac{1}{2}}$. [id est] $a\ddot{v} = \frac{1}{2}b\ddddot{v}$.(106) [necnon] $6a\ddot{y}^3+12a\dot{y}\ddot{y}\dddot{y}+3a\dot{y}^2\ddddot{y} = [b]\,\overset{\cdots}{\ddot{y}}$.(107)
6

---

[11](108) [Sit] x^m(109) [terminus infimus(110) non multiplicatus per y et] $x^ny^p\dot{y}^q\ddot{y}^r\dddot{y}^s$ [terminus quilibet. Substituendo] $x^t = y$. [adeoq;] $tx^{t-1} = \dot{y}$. $\overline{t-1},tx^{t-2} = \ddot{y}$. [&c terminus fit] $x^nx^{pt}x^{qt-q}x^{rt-2r}x^{st-3s}$. [Hunc pone $= x^m$. id est]

$$x^{n+pt+qt-q+rt-2r+st-3s} = x^m.$$

[erit] $t = \dfrac{m-n+q+2r+3s}{p+q+r+s}$. [Maximus(110) seligendus est.]

[Ut si fuerit æquatio] $2x^3-3xxy-y\dot{y}+y^2\ddot{y}[=0]$. [Infimus terminus
[$t=$] 1. 2. 2.

---

(106) Since $a\dot{v} = \frac{1}{2}b\dddot{v}$ and so $\frac{3}{2}a\dot{v}^2v^{\frac{1}{2}} = \frac{3}{4}b\dddot{v}\dot{v}v^{\frac{1}{2}}$. The two coefficients here adjusted in the previous equation were both mistakenly set by Newton in the manuscript to be '$\frac{3}{2}$', but the slip would have been immediately detected upon his checking back.

(107) 'quæ reducta fit $a\dddot{v} = \frac{1}{2}b\overset{\cdots}{\ddot{v}}$'. Evidently satisfied that his revised scheme of substitution yields consistent results for the successive fluxional derivatives of the basic equation $a\dot{y}^3 = b\dddot{y}$, Newton forbears to specify the ensuing reduction.

(108) ULC. Add. 3960.8: 138–9/3960.9: 153. In these jotted notes and rough calculations —here fleshed out with a number of our verbal interpolations to point their sense—Newton briefly expounds his algorithm for selecting the dimension of the base variable, here x, in the opening term of the (ascending) series expansion of the 'root' y of a given fluxional equation, and then applies it in two particular instances, the first of which is well fit after some necessary minor correction of its detail (see notes (112) and (113) below) to have been one of the worked *exempla* for which in his manuscript (see §2: note (111) preceding) he made tentative provision at the end of Proposition XII of his 'De quadratura Curvarum'.

(109) For simplicity of exposition Newton omits the (not necessarily unit) coefficient of this 'terminus'—permissibly so since, for the moment, only its dimension is significant. To be pedantically strict, read 'kx^m'—or some equivalent—and in the sequel, correspondingly, set there to be '$lx^ny^p\dot{y}^q$ ...' and '$ax^t = y.atx^{t-1} = \dot{y}$...'.

(110) Understanding, as in the two following instances, that y is to be expanded as a series in ascending powers of x. When, conversely, it needs to be expressed as a descending power-series, take x^m to be the 'terminus altissimus' and select 'minimus' t respectively. (Compare §2: notes (97) and (99).)

non multiplicatus per y est $2x^3$, unde $m = 3$, ut et maxima dignitas $t = 2$. Pone $ax^2 = y$. $2ax = \dot{y}$. $0 = \ddot{y}$. et evadit $2x^3 - 2ax^3 = 0$ sive $a = 1$. Operatio sic procedit.][(111)]

		$y =$	$xx+2x^3-\frac{15}{11}x^4+$ &c.
$xx+p=y$.		$2x^3$	$2x^3$
$2x+\dot{p}=\dot{y}$.	1.	$-3xxy$	$-3x^4-3xxp$
$2+\ddot{p}=\ddot{y}$.	2.	$-y\dot{y}$	$-2x^3-2xp-xx\dot{p}-p\dot{p}$
$\dddot{p}=\dddot{y}$	2.	$+y^2\dddot{y}$	$+x^4\dddot{p}+2x^2p\dddot{p}[+]pp\dddot{p}$
$2x^3+q=p$.		$-2x^4$	$-2x^4$[(112)] [Pone] $+bx^3=p$. $+3bx^2=\dot{p}$.
$6xx+\dot{q}=\dot{p}$.	2.	$-3xxp$	$-6x^5-3xxq$ $6bx=\ddot{p}$. $6b=\dddot{p}$ [Fit]
$12x+\ddot{q}=\ddot{p}$.	3.	$-2xp$	$-4x^4-2x$ $-2x^4-2bx^4-3bx^4+6bx^4=0$.
$12+\dddot{q}=\dddot{p}$.	3.	$-xx\dot{p}$	$-6x^4-xx\dot{q}$ [id est] $2=b$.
	$\frac{5}{2}$.	$-p\dot{p}$	$-12x^5-6xxq-2x^2\dot{q}-q\dot{q}$
	3.	$+x^4\dddot{p}$	$+12x^4+x^4\dddot{q}$
	$2\frac{1}{2}$.	$+2x^2p\dddot{p}$	$+48x^5+24x^2q+4x^5\dddot{q}+2x^2q\dddot{q}$
	$\frac{7}{3}$.	$+pp\dddot{p}$	$+48x^6+48x^3q+12qq+4x^6\dddot{q}+4x^3q\dddot{q}+qq\ddot{q}$
[$-\frac{15}{11}x^4+r=q$.]		$48x^6$	
		$+30x^5$	[Pone] $cx^4=q$. $4cx^3=\dot{q}$.
	3.	$+15xxq$	$12cx^2=\ddot{q}$. $24cx=\dddot{q}$. [Fit]
	4.	$-2xq$	$-2cx^5+24cx^5+30=0$.
	2.	$-xx\dot{q}$	(113) [id est] $c=-\frac{15}{11}$.[(114)]
	3.	$+46x^3\dot{q}$	
	$2\frac{1}{2}$.	$+11q\dot{q}$	(113)
	4.	$+x^4\dddot{q}$	
	3.	$+4x^5\dddot{q}$	
	$2\frac{1}{2}$.	$+2x^2q\dddot{q}$	(113)
	2.	$+4x^6\dddot{q}$	
		&c	

(111) Several small errors (see the next two notes) in the scheme which follows vitiate its accuracy, but we reproduce it nonetheless to illustrate the way in which, on the pattern of his earlier resolution of non-fluxional 'affected' equations (compare II: 224, 228, 234 and III: 54–6, 58, 62), Newton typically laid out a computation of this kind. The left-hand column, we may observe, lists the values of $t = (m-n+\beta+3\gamma)/(\alpha+\beta+\gamma)$, $m = 3, 4, 5, \ldots$, corresponding to the fluxional term ($m = 3$) $lx^n y^\alpha \dot{y}^\beta \ddot{y}^\gamma$, ($m = 4$) $l'x^n p^\alpha \dot{p}^\beta \ddot{p}^\gamma$, ($m = 5$) $l''x^n q^\alpha \dot{q}^\beta \ddot{q}^\gamma$, ... on its immediate right.

[Sit jam] $\frac{1}{y^2} = \ddot{y}$.(115) [Pone $y = w+v$. $\ddot{y} = \ddot{v}$. erit] $1 = y^2\ddot{y} = w^2+2wv+v^2$ in $\ddot{v}$. [ubi est x^0 infimus terminus non multiplicatus per v. Itaq; scribe] $v = x^r$. [et pone terminum quemlibet $v^q\ddot{v} = x^0$. unde $qr+r-2 = 0$ Tum percurrendo omnes terminos æquationis] $1 = w^2\ddot{v}+2wv\ddot{v}+v^2\ddot{v}$. [quære maximam dignitatem

$$\frac{2}{1}.\quad \frac{2}{2}.\quad \frac{2}{3}.$$

$r = \frac{2}{q+1}$, quæ hic est 2. Pone] $ax^2 = v$. $2ax = \dot{v}$. $2a = \ddot{v}$. [fit]

$$1 = 2w^2a+4a^2x^2w+2a^3x^4.$$

[adeoq;] $a = \frac{1}{2w^2}$. [Sin quantitas x permagna sit, adhibenda est minima dignitas $r = \frac{2}{3}$, ponendo] $ax^{\frac{2}{3}} = v$. [Pro reliquis omnibus seriei terminis nondum inventis scribe p, et erit] $\frac{1}{2w^2}x^2+p = v$. $\frac{1}{w^2}x+\dot{p} = \dot{v}$. $\frac{1}{w^2}+\ddot{p} = \ddot{v}$. [Quibus substitutis in æquatione novissima resolvenda fit]

$$[0] = w^2\ddot{p}+\frac{1}{w^3}x^2+\frac{2p}{w}+\frac{1}{2w^2}x^2\ddot{p}+\frac{2p\ddot{p}}{w}^{(116)}+\frac{1}{4w^6}x^4+\frac{1}{4w^4}x^4\ddot{p}+\frac{1}{w^4}xxp$$
$$+\frac{1}{w^2}xxp\ddot{p}+\frac{1}{w^2}pp+\frac{1}{w^2}pp\ddot{p}.^{(117)}$$

[Hic est $\frac{1}{w^3}x^2$ infimus terminorum qui neq; per p neq; per aliam ejus fluxionem

(112) The two terms in this line should both be '$-3x^4$', so determining the equation on the right to be '$-3x^4-2bx^4-3bx^4+6bx^4 = 0$. [id est] $3 = b$'; in consequence the substitution at the left needs to be amended to read '$3x^3+q = p$. $9xx+\dot{q} = \dot{p}$. $18x+\ddot{q} = \ddot{p}$. $18+\dddot{q} = \dddot{p}$' with necessary widespread adjustment of coefficients in the array of following terms.

(113) The values of t in the left-hand column corresponding to these terms should (compare note (111)) be '4', '3' and '3' respectively. The two latter emendations are insignificant, but the first requires an additional term '$-4cx^5$' to be inserted in the equation on the right in determining the substitution $cx^4+r = q$ to be effected at this stage.

(114) Newton here abandons his computation unfinished and without indicating in his manuscript any correction of the numerical slips which (see the two previous notes) he had unwittingly introduced *en route*.

(115) On multiplying through by $2\dot{y}$ we will readily compute that this second fluxional equation has the exact fluent $2w^{-1}-2y^{-1} = \dot{y}^2$ (supposing, with Newton, that $y = w$ when $x = 0 = \dot{y}$), so that $x = \int(2(y-w)/wy)^{-\frac{1}{2}}.dy$. Even had Newton noticed as much, he would not necessarily have made haste to abandon this neatly simple second-order equation as a test-bed for his series method of solution; neither the equivalent first-order fluxional equation nor its fluent integral so easily produce the required expansion of y in powers of x.

(116) These two terms should be '$\frac{1}{w}x^2\ddot{p}+2wp\ddot{p}$'. In making substitution of $v = (1/2w^2)\, x^2+p$ Newton fails accurately to evaluate the product $2wv\ddot{p}$.

(117) Read '$+pp\ddot{p}$' simply.

multiplicantur et ponendo $p = x^r$ in termino quolibet $x^s p^t \ddot{p}^u = x^2$ invenitur maxima dignitas $r = \frac{2-s+2u}{t+u}$ esse 4.(118) Ergo ponendum est] $p = ax^4+q$. $\dot{p} = 4ax^3+\dot{q}$. $\ddot{p} = 12axx+\ddot{q}$. [His valoribus in æquatione novissima substitutis determinatur a, et sic deinceps iterando operationem prodit series

$$y = w+\frac{1}{2w^2}x^2-\frac{1}{12w^5}x^4 \text{ \&c.}^{(119)}]$$

---

[12](120) Per æquationes $-bz\dot{z} = y\dot{y}$ et $\frac{g}{c}\dot{z}-\frac{e}{c}\dot{y} = \dot{x}$ fiunt $\dot{z} = \frac{cy\dot{x}}{gy+ebz}$. [et] $\dot{y} = \frac{-bcz\dot{x}}{gy+ebz}$. Deinde per æquatione[s] $-b\dot{z}^2-bz\ddot{z} = \dot{y}^2+y\ddot{y}$ & $\frac{g}{c}\ddot{z}-\frac{e}{c}\ddot{y} = 0$, exterminatis $\dot{z}$, $\ddot{z}$ et $\dot{y}$ fit $\ddot{y} = \frac{-gbccaa\dot{x}^2}{\overline{gy+ebz}|^3}$. Et per æquationem $\frac{f}{c}\ddot{z}+\frac{d}{c}\ddot{y} = \ddot{v}$ exterminata $\ddot{y}$ et $\ddot{z}$ fit $\frac{gc\ddot{v}}{ef+dg} = \ddot{y} = \frac{-gbccaa\dot{x}^2}{\overline{gy+ebz}|^3}$ et inde $\frac{\overline{ef+dg} \text{ in } -bcaa}{2\times\overline{gy+ebz}|^3} = \frac{\ddot{v}}{2\dot{x}^2}$. Hæc est igitur exponens curvaturæ Ellipseos ad punctum datum C.(121) Et hujus reciproca $\frac{2\times\overline{gy+ebz}|^3}{ef+dg \text{ in } -bcaa}$ diameter est circuli eandem habentis curvaturā cum Ellipsi ad

---

(118) Namely, in the case of the term $w^2\ddot{p}$. (In editorial honesty we should add that in a cancelled listing of the separate values of r which he began to insert under each term in the preceding equation Newton made this number to be $\frac{`2'}{1}$.) It follows immediately that to $O(x^6)$ there is $0 = w^2\ddot{p}+w^{-3}x^2$ and so $\ddot{p} = -w^{-5}x^2$, or $p = -\frac{1}{12}w^{-5}x^4$.

(119) Newton does not give this explicit series expansion in the manuscript, but it is in the spirit of his text so to append it.

(120) ULC. Add. 3960.9: 161/163, Newton's preliminary computations—more elaborate and extensive than those he introduced into his final text on page 118 above (see §2: note (156))—of the curvature of the ellipse (C) defined, with respect to its centre A as origin and oblique coordinates $AB = z$ and $BC = y$, by the Cartesian equation $a^2-bz^2 = y^2$, and also equivalently in relation to the semi-intrinsic perpendicular coordinates

$$AM = (g/c)\,z-(e/c)\,y = x \quad\text{and}\quad MC = (f/c)\,z+(d/c)\,y = v,$$

where MC is instantaneously normal to the curve at C, and in which

$$CN/BC = d/c,\quad BN/BC = e/c \quad\text{and}\quad MN/AB = f/c,\quad (AM+BN)/AB = g/c$$

are the respective cosines and sines of the angles at which BC and AB are inclined to that normal.

(121) Where, as Newton goes on implicitly to specify, the 'exponens curvaturæ ad C' is taken to be the reciprocal of the diameter, $2\dot{x}^2/\ddot{v}$, of the osculating circle at $C(z, y) \equiv (x, v)$.

(122) Understand 'puncto B coincidente cum Ellipseos centro A, et ordinata BC cum semidiametro AH', much as Newton afterwards specified in his revised text (page 118 above). Here, however, he at first fails to notice that, because the tangential direction of the ellipse at H (and so its parallel AM) is that of the conjugate diameter VA, in the limit as B comes to coincide with A the angle $B\widehat{A}M$ ($= A\widehat{B}N$) vanishes, and therefore $f/c = 0$ and $g/c = 1$, that is, (since $c \neq 0$) $f = 0$ and $g = c$.

idem punctum. Pone $z=0$ et fiet $y=a$,(122) et circuli diameter $=\dfrac{-2g^3a}{efbc+dgbc}$,(123) quæ longitudo est ad $\dfrac{2a}{b}$ hoc est ad illud Ellipseos latus rectum quod pertinet ad verticem [H] ut $-g^3$ ad $efc+dgc$,(123) hoc est ut BC ad CN.(124)

[Pone $z=0$.] $a=y$. [$f=0$.(125) $\dot{y}=0$. Fit] $\frac{g}{c}z-\frac{e}{c}y=x$. $\frac{g}{c}\dot{z}=\dot{x}$. $\frac{g}{c}\ddot{z}-\frac{e}{c}\ddot{y}=0$. $\frac{g}{c}\dddot{z}-\frac{e}{c}\dddot{y}=0$. [&c] $y=v$.(126) $\dot{y}=\dot{v}=0$. $\ddot{y}=\ddot{v}$. $\dddot{y}=\dddot{v}$. [&c]

Æquationes $-b\dot{z}^2=\ddot{y}[y]$, $\frac{g}{c}\dot{z}=\dot{x}$ et $\ddot{y}=\ddot{v}$ evadunt $-\dfrac{bcc}{gg[a]}\dot{x}^2=\ddot{v}$. Unde facile prodeunt $\dfrac{\ddot{v}}{2\dot{x}^2}=-\dfrac{bcc}{2gga}$ exponens curvaturæ. [Porro]

$$-3b\dot{z}\ddot{z}=y\dddot{y}=a\dddot{y}=a\dddot{v}=-3b\dot{z},\frac{e}{g}\ddot{v}$$

[hoc est] $=-\dfrac{3bec\dot{x}\ddot{v}}{gg}=+\dfrac{3bec,bcc\dot{x}^3}{gg,gga}$. [Unde] $\dfrac{\dddot{v}}{\dot{x}^3}=\dfrac{3bbc^{[3]}e}{ag^4}$ [adeoq*ꝫ*]

$$\frac{\dddot{v}}{6\dot{x}^3}=\frac{bbc^{[3]}e}{2ag^4}$$

[exponens variationis curvaturæ].(127)

[Pone $y=a$. $z=0$. $f=0$. $g=c$. $\dot{y}=0$. Fit] $-b\dot{z}\dot{z}=a\ddot{y}$. $-3b\dot{z}\ddot{z}=a\dddot{y}$ $z-\frac{e}{c}a=x$. $\dot{z}=\dot{x}$. $\ddot{z}=\frac{e}{c}\ddot{y}$. $\dddot{z}=\frac{e}{c}\dddot{y}$. [&c] $\frac{d}{c}a=v$. [$0=\dot{v}$.] $\frac{d}{c}\ddot{y}=\ddot{v}$. $\frac{d}{c}\dddot{y}=\dddot{v}$. [&c].

---

(123) That is, since (see the previous note) $f=0$ and $g=c$, 'circuli diameter $=-\dfrac{2ac}{bd}$, quæ est ad $\dfrac{2a}{b}$... ut $-c$ ad d'. The *latus rectum* at C pertaining to the conjugate semidiameters VA and AH is of course $2VA^2/AH$ where (here) $VA=\sqrt{(a^2/b)}$ and $AH=a$.

(124) Understand as BC comes to be AH, whence this ratio is the secant $-c/d$ of the ordinate angle ($\widehat{VAB}=$) $\widehat{VAH}$. Thus suddenly brought to realize that there must consequently be $c=g$ and $f=0$ (see note (122)) Newton breaks off to introduce this simplification *ab initio* together with the basic conditions $z=0=\dot{y}$ and $y=a$ which determine the present particular case. As before (see §2: note (143)), where HA is extended to meet the osculating circle at H in the point K, the chord HK of this circle of curvature is equal to the *latus rectum* $2a/b$.

(125) For the moment Newton does not see that this necessarily implies that $g(=\sqrt{[c^2-f^2]})$ shall be equal to c.

(126) Read '$\frac{d}{c}y=v$', and correspondingly in sequel '$\frac{d}{c}\dot{y}=\dot{v}=0$. $\frac{d}{c}\ddot{y}=\ddot{v}$. $\frac{d}{c}\dddot{y}=\dddot{v}$. &c' Newton will, without serious effect, carry this slip into his main text above (see §2: note (160)).

(127) Newton breaks off to start again, now setting $g=c$ (compare notes (120) and (125)) and introducing the previously omitted factor d/c into equations connecting like fluxions of y and v (see note (126)).

[Unde] $-b\dot x^2 = a\ddot y = \frac{ac}{d}\ddot v$. [sive] $-\frac{bd}{ac}\dot x^2 = \ddot v$. [adeoq; $\frac{\ddot v}{2\dot x^2} = -\frac{bd}{2ac}$ exponens curvaturæ].(128)

[Porro] $-3\frac{be}{c}\ddot y\dot x = \dddot y a = -3\frac{be}{d}\ddot v\dot x = \frac{3bbe}{ac}\dot x^3 = \frac{c}{d}\dddot v a$. [unde] $\frac{3bbde}{a[a]\,cc} = \frac{\dddot v}{\dot x^3}$ [itaq; $\frac{\dddot v}{6\dot x^3} = \frac{bbde}{2a^2cc}$ exponens variationis curvaturæ].(128)

[Rursus] $-4b\dddot z\dot x - 3b\ddot z\ddot z = a\ddddot y$.(129) $\frac{c}{e}\ddddot z = \ddddot y$. $\frac{d}{c}\ddddot y = \ddddot v$. [unde] $-\frac{4be}{c}\dddot y\dot x - 3b\frac{ee}{cc}\ddot y\ddot y = a\ddddot y$. [hoc est] $\frac{-12b^3ee}{a^2cc}\dot x^4 - \frac{3b^3ee}{aacc}\dot x^4 = a\ddddot y = \frac{ac}{d}\ddddot v$. [sive] $\frac{-15b^3eed}{a^3c^3} = \frac{\ddddot v}{\dot x^4}$ [itaq;]

$$\left[\frac{\ddddot v}{24\dot x^4} = \right] -\frac{5b^3eed}{8a^3c^3}$$

[exponens variationis secundæ].

$$\text{[Iterum]}\ -4b\ddddot z\dot x - 7b\dddot z\ddot z = a\overset{\cdot\cdot\cdot\cdot\cdot}{y}.^{(129)}\ \frac{c}{e}\overset{\cdot\cdot\cdot\cdot\cdot}{z} = \overset{\cdot\cdot\cdot\cdot\cdot}{y}.\ \frac{d}{c}\overset{\cdot\cdot\cdot\cdot\cdot}{y} = \overset{\cdot\cdot\cdot\cdot\cdot}{v}.$$

$$-4\frac{be}{c}\ddddot y\dot x - 7\frac{bee}{cc}\dddot y\ddot y = \frac{60b^4e^3}{a^3c^3}\dot x^5 + \frac{21b^3e^3}{aac^3}\dot x^3 \times \frac{b}{a}\dot x^2 = \frac{81b^4e^3}{a^3c^3}\dot x^5 = \frac{ac}{d}\overset{\cdot\cdot\cdot\cdot\cdot}{v}.$$

$$\text{unde}\ \frac{\overset{\cdot\cdot\cdot\cdot\cdot}{v}}{120\dot x^5} = \frac{27b^4e^3d}{40a^4c^4}\ \text{exponens variationis tertiæ.}^{(130)}$$

[Dato $-bz\dot z = \dot y y$. fit]

$-b\dot z\dot z = \ddot y y$.(131)	1.			
$-2b\ddot z\dot z = \dddot y[y]$.(131)	2.			
$-2b\dddot z\dot z - 2b\ddot z\ddot z = \ddddot y[y]$.	2.	2.		
$-2b\ddddot z\dot z - 6b\dddot z\ddot z = \overset{\cdot\cdot\cdot\cdot\cdot}{y}[y]$.	2.	6.		
$-2b\overset{\cdot\cdot\cdot\cdot\cdot}{z}\dot z - 8b\ddddot z\ddot z - 6b\dddot z\dddot z = \overset{\cdot\cdot\cdot\cdot\cdot\cdot}{y}[y]$.	2.	8.	6.	
$-2b\overset{\cdot\cdot\cdot\cdot\cdot\cdot}{z}\dot z - 10b\overset{\cdot\cdot\cdot\cdot\cdot}{z}\ddot z - 20b\ddddot z\dddot z = \overset{\cdot\cdot\cdot\cdot\cdot\cdot\cdot}{y}[y]$.	2.	10.	20.	
$-2b\overset{7\,1}{zz}$(132)$-\overset{6\,2}{12}-\overset{5\,3}{30}-\overset{4\,4}{20}\,[=\overset{8}{y}y]$.(133)	2.	12.	30.	20.
	2.	14.	50.(134)	70.(135)

---

(128) And so Newton afterwards made them to be in his main text (see §2: note (161)).

(129) As Newton will come almost at once to see, he here omits the terms '$+3\ddot y^2$' and '$+10\ddot y\dddot y$' respectively necessary to complete these fluxional equations.

(130) Newton halts his sequence of computations to check their accuracy, and—after an initial oversight (see note (131) following)—speedily discovers his omissions in the basic fluxional equations which are derived over-hastily 'by sight' in his two most recent paragraphs (see the previous note).

(131) Correctly so since, when $y = a$ (as Newton here supposes; compare note (124) above), there is $\dot y = 0$. He will, however, pay drastically in immediate sequel for failing to observe that the zero terms '$+\dot y^2$' and '$+3\dot y\ddot y$' here respectively omitted produce in the ensuing fluxional derivatives terms not involving $\dot y$ and which are therefore not zero when $y = a$.

$$[-b\dot{z}\dot{z}-b\ddot{z}z = \ddot{y}y+\dot{y}^2.]$$

$$-3b\ddot{z}\dot{z}-bz\dddot{z} = 3\ddot{y}\dot{y}+y\dddot{y}.$$

$$-4b\dddot{z}\dot{z}-3b\ddot{z}\ddot{z}-bz\ddddot{z} = 4\dddot{y}\dot{y}+3\ddot{y}\ddot{y}+\ddddot{y}y.$$

$$-5b\ddddot{z}\dot{z}-10b\dddot{z}\ddot{z}-bz\overset{\cdot\cdot\cdot\cdot\cdot}{z} = 5\ddddot{y}\dot{y}+10\dddot{y}\ddot{y}+\overset{\cdot\cdot\cdot\cdot\cdot}{y}y. \quad [\&c]$$

[Pone $\dot{y} = z = 0$. evadit] $-4b\dddot{z}\dot{z}-3b\ddot{z}^2 = 3\ddot{y}\ddot{y}+\ddddot{y}y$. [hoc est[136]]

$$-4\frac{be}{c}\dddot{y}\dot{x}-3b\frac{e[e]}{c[c]}\ddot{y}^2 = 3\ddot{y}^2+\frac{ca}{d}\ddddot{v}.$$

[sive[137]] $-\frac{12b^3ee}{aacc}\dot{x}^4-\frac{3b^3ee}{aacc}\dot{x}^4 = +3\frac{bb}{aa}\dot{x}^4+\frac{ca}{d}\ddddot{v}$. [id est] $-\frac{15b^3eed}{a^3c^3}\dot{x}^4-\frac{3bbdcc}{a^3c^3}\dot{x}^4 = \ddddot{v}$

[recte]. [Quare est $\frac{\ddddot{v}}{24\dot{x}^4} = -\frac{5b^3eed+bbccd}{8a^3c^3}$ exponens variationis secundæ.][138]

---

(132) Newton initially wrote '$-2b\overset{\cdot\cdot\cdot\cdot\cdot\cdot\cdot}{z}\dot{z}$' in the pattern of his preceding first terms before introducing this elegant contracted equivalent form. In continuation he rather confusingly omits to cite either the constant b or the base variable z; thus '$-\overset{5\,3}{30}$' signifies in the preceding 'standard' dot-notation '$-30b\overset{\cdot\cdot\cdot\cdot\cdot}{z}\dddot{z}$'. But how much better to have gone on likewise to write '$-12b\overset{6}{z}\overset{2}{z}-30\overset{5}{z}\overset{3}{z}-20\overset{4}{z}\overset{4}{z}$'!

(133) We adapt Newton's newly introduced contracted terminology to avoid setting some cumbrous pattern of 8 dots over y to denote its eighth fluxion.

(134) Read '42' since, when $\dot{y} = 0$, the derivative of the fluxional equation in the previous line is (again to adapt Newton's contracted notation there)

$$`-2b\overset{8}{z}\overset{1}{z}-14b\overset{7}{z}\overset{2}{z}-42b\overset{6}{z}\overset{3}{z}-70b\overset{5}{z}\overset{4}{z} = \overset{9}{y}y'.$$

(135) This array groups the numerical coefficients of the terms on the left-hand sides of the equations alongside. At this point Newton suddenly realizes the deficiencies in his allocation of the terms in y on their right-hand sides (compare note (131) preceding), and so breaks off to seek afresh to evaluate them correctly.

(136) Since $y = a$ and also $\ddot{z} = (e/c)\ddot{y}$, $\dddot{z} = (e/c)\dddot{y}$ and $\ddddot{y} = (c/d)\ddddot{v}$.

(137) On substituting $\ddot{y} = -(b/a)\dot{x}^2$ and $\dddot{y} = (-3(be/ac)\,\ddot{y}\dot{x}$ or) $3(b^2e/a^2c)\,\dot{x}^3$.

(138) This corrected evaluation of the exponent of the second variation of curvature at the point $(0, a)$ of the ellipse $a^2-bz^2 = y^2$ did not (compare §2: note (161) above) subsequently find its way into Newton's main text.

APPENDIX 2. TWO REWORKED PROPOSITIONS ON THE FLUXIONS AND FLUENTS OF 'COMPLEX' QUANTITIES (? 1693).(1)

From the original autographs in the University Library, Cambridge

[1](2) *Prop* [*XI*].(3)

Invenire fluxiones quantitatum complexarum.

Proponatur quantitas z^3. Dic istam v et per Prob. 1(4) erit $3z^2\dot{z} = \dot{v}$. Ergo $3z^2\dot{z}$ est fluxio ipsius v id est ipsius z^3. Rursus per Prob. 1(4) est $6z\dot{z}^2 + 3z^2\ddot{z} = \ddot{v}$ ideoq $6z\dot{z}^2 + 3z^2\ddot{z}$ est ipsius z^3 fluxio secunda vel secundæ gene[rationis].(5) Et eodem argumento $\dot{z}x^2 + 2zx\dot{x}$ fluxio est ipsius zx^2 et $\ddot{z}x^2 + 4\dot{z}\dot{x}x + 2z\dot{x}^2 + 2zx\ddot{x}$ est ejus fluxio secunda, et $\dfrac{2y\dot{y} - a\dot{z}}{2\sqrt{yy - az}}$ fluxio est ipsius $\sqrt{yy - az}$. Et sic in cæteris.

Prop. [*XII*].

Ex fluxionibus quantitatum complexarum invenire quantitates.

Cas. 1.(6) Si fluxio dependet ab unica tantum quantitate simplici fluente,

(1) Of this pair of related problems it will be clear that Newton's new Proposition XII is a much modified and augmented replacement for Proposition XI in his preceding tract 'De quadratura Curvarum' (compare §2: note (51) above) and that the intervening new Proposition XI is but a rider to it, merely summarizing the two cases of his earlier 'Prop. IV. Prob. I' to which it makes explicit appeal (see note (4) following) in justification of its fluxional derivations. What solely distinguishes this otherwise minor textual revise—and what has led us to set it in a place of honour in a separate appendix—is the subtle method of reducing multiple algebraic integrals to simple quadratures which Newton for the first time here (in Case 1 of Proposition XII) expounds, and which even after its publication in 1704, in an equivalent geometrical garb but without demonstration (see note (9) below), remained unproved by the world at large till the year of his death. In [2] is adjoined Newton's working of the simplest instance of the algorithm (when it is applied to compute the multiple integrals of z), in itself near-trivial but hinting suggestively (see note (21)) at its subsequent extension.

(2) Add. 3960.7: 117–19/ 3960.10: 165–7. A preliminary version of Proposition XI and the enunciation of Case 1 of Proposition XII exists on Add. 3960.10: 165; the latter was subsequently reshaped and greatly extended into its present form in a prior draft on Add. 3962.3: 59r/59v which differs only minimally from the final version reproduced.

(3) The manuscript proposition is unnumbered, but our interpolation is that (see note (17) below) by which it is cited in the following proposition.

(4) That is, 'Prop. IV. Prob. I' of the preceding 'De quadratura Curvarum' (see pages 62–6 above).

(5) We insert this alternative denomination from an otherwise unchanged revision of the present proposition on Add. 3960.7: 99, which here abruptly terminates.

(6) In his initial draft of this first case on Add. 3960.10: 165 Newton went on merely to announce: 'Si fluxio dependet ab unica tantum quantitate simplici fluente, concipe hanc quantitatem uniformiter fluere et esse abscissam Curvæ alicujus et ejus fluxionem per unitatem exponi et fluxionem propositam esse ordinatam ejusdē Curvæ, et area Curvæ per præcedentes Propositiones inventa erit quantitas quæsita.'

puta z, concipe hanc uniformiter fluere et ejus fluxionem esse unitatem. Sitq fluxio quantitatis complexæ æqualis $\dot{y}$, et quærantur areæ curvarum quæ ordinatim applicatas habent $\dot{y}$, $\dot{y}z$, $\dot{y}z^2$, $\dot{y}z^3$, $\dot{y}z^4$ &c et abscissam communem z.

(7)Sunto areæ illæ A, B, C, D &c id est A area curvæ cujus ordinata e[s]t $\dot{y}$, B ordinata curvæ cujus area est $z\dot{y}$ &c[,] et sit $n+1$ exponens fluxionis id est 1 si $\dot{y}$ sit fluxio prima adeoq y quantitas fluens invenienda[,] vel 2 si $\dot{y}$ sit fluxio secunda & $\acute{y}$ quantitas fluens invenienda, vel 3 si $\dot{y}$ sit fluxio tertia & $\acute{\acute{y}}$ quantitas fluens invenienda &c. Et quantitas illa fluens erit

$$\frac{\begin{array}{l} Az^n - nBz^{n-1} + \dfrac{n\times n-1}{2}Cz^{n-2} - \dfrac{n\times n-1\times n-2}{2\times 3}Dz^{n-3} \\ \qquad + n\times\dfrac{n-1}{2}\times\dfrac{n-2}{3}\times\dfrac{n-3}{4}Ez^{n-4}\&c \end{array}}{1\times 2\times 3\times 4\ \&c} + a + bz + cz^2 + dz^3 + ez^4\ \&c.$$

ubi numerorum 1, 2, 3, 4 &c tam in denominatore seriei quam [in] dignitatum ultimarum exponentibus ultimus idem semper est cum numero n et quantitates a, b, c, d, e [&c] sunt datæ quævis cum signis suis + et −. Igitur si quæritur y ponendus est $n=0$, et series evadet $A+a=y$. Si quæritur $\acute{y}$ ponendus est $n=1$ et series evadet $Az-B+a+bz=\acute{y}$. Si quæritur $\acute{\acute{y}}$ ponendu[s] est $n=2$ et series evadet $\dfrac{Az^2-2Bz+C}{1\times 2}+a+bz+cz^2=\acute{\acute{y}}$. Si quæritur $\acute{\acute{\acute{y}}}$ ponendus est $n=3$ et series evadet $\dfrac{Az^3-3Bz^2+3Cz-D}{1\times 2\times 3}+a+bz+cz^2+dz^3=\acute{\acute{\acute{y}}}$ & sic deinceps in infinitum.

Sunto areæ illæ A, B, C, D, &c et erit $A+g=y$, $Az-B+gz+h=\acute{y}$.

$$\frac{Az^2-2Bz+C}{2}+\tfrac{1}{2}gz^2+hz+i=\acute{\acute{y}}.$$

$$\frac{Az^3-3Bz^2+3Cz-D}{6}+\tfrac{1}{6}gz^3+\tfrac{1}{2}hz^2+iz+k=\acute{\acute{\acute{y}}}.$$

$$\frac{Az^4-4Bz^3+6Cz^2-4Dz+E}{24}+\tfrac{1}{24}gz^4+\tfrac{1}{6}hz^3+\tfrac{1}{2}iz^2+kz+l=\acute{\acute{\acute{\acute{y}}}}.\quad [\&c.]$$

Ubi g, h, i, k, l &c sunt quantitates quælibet assumptæ cum signis suis + et −, et denominatores 2, 6, 24 &c generantur per continuam multiplicationem numerorum 1, 2, 3, 4, 5 &c[,] et arearum A, B, C, D, E &c coefficientes sunt potestates binomij $z-1$.(8) Et y, $\acute{y}$, $\acute{\acute{y}}$, $\acute{\acute{\acute{y}}}$, $\acute{\acute{\acute{\acute{y}}}}$ &c sunt fluentes quantitates quarum quælibet

(7) The following superseded paragraph is that of the preliminary revise on Add. 3962.3: 59r/59v; we reproduce it because of its several interesting (if minor) variations from the uncancelled sequel.

(8) More precisely, in the evaluation of the $(n+1)$-th integral of y the 'coefficients' of $A, B, C, D, E, \ldots$ are respectively z^n, $-\binom{n}{1}z^{n-1}$, $+\binom{n}{2}z^{n-2}$, $-\binom{n}{3}z^{n-3}$, $+\binom{n}{4}z^{n-4}$, ...

prior est fluxio posterioris, adeo ut $\dot{y}$ sit ipsius y fluxio prima et ipsius $\acute{y}$ fluxio secunda et ipsius $\acute{\acute{y}}$ fluxio tertia et sic deinceps.(9)

Exempl. 1. Proponatur fluxio $z^3\dot{z}$ seu z^3 (nam $\dot{z}$ est 1) et erit $z^3 = \dot{y}$, et ordinatim applicatæ $\dot{y}$, $z\dot{y}$, $zz\dot{y}$, $z^3\dot{y}$ &c erunt z^3, z^4, z^5, z^6 &c et areæ curvarum quarum hæ sunt ordinatæ, erunt $\frac{1}{4}z^4 = A$. $\frac{1}{5}z^5 = B$. $\frac{1}{6}z^6 = C$. $\frac{1}{7}z^7 = D$. $\frac{1}{8}z^8 = E$. &c. Adeoq;

(9) Newton subsequently deleted this final sentence, evidently with the intention of advancing its explanation of this (here) novel notation for first- and higher-order fluents to some now unknown prior position. In modern Leibnizian restatement of his suppositions, he here sets there to be

$$dA/dz = \dot{y} = dy/dz, \quad dB/dz = z.dA/dz, \quad dC/dz = z^2.dA/dz (= z.dB/dz),$$
$$dD/dz = z^3.dA/dz (= z^2.dB/dz = z.dC/dz), \ldots,$$

whence straightforwardly

$$\int A.dz = Az - B, \quad \int B.dz = Bz - C, \quad \int C.dz = Cz - D, \ldots; \quad \int Az.dz = \tfrac{1}{2}(Az^2 - C),$$
$$\int Bz.dz = \tfrac{1}{2}(Bz^2 - D), \ldots; \quad \int Az^2.dz = \tfrac{1}{3}(Az^3 - D), \ldots;$$

and so forth. Accordingly, where $y = g$ when $z = Z$, say, there is $A = \int_Z^z \dot{y}.dz = y - g$ or $y = A + g$, and thence successively

$$\acute{y} = \int_Z^z y.dz = [Az - B]_{z=Z}^{z=z} + gz + h,$$

$$\acute{\acute{y}} = \int_Z^z \acute{y}.dz = [\tfrac{1}{2}(Az^2 - C) - (Bz - C)]_{z=Z}^{z=z} + \tfrac{1}{2}gz^2 + hz + i,$$

$$\acute{\acute{\acute{y}}} = \int_Z^z \tfrac{1}{2}\acute{\acute{y}}.dz = \tfrac{1}{2}[\tfrac{1}{3}(Az^3 - D) - \tfrac{2}{2}(Bz^2 - D) + (Cz - D)]_{z=Z}^{z=z} + \tfrac{1}{6}gz^3 + \tfrac{1}{2}hz^2 + iz + k, \ldots$$

and so on, where $h = -gZ$, $i = -(\frac{1}{2}gZ^2 + hZ)$, $k = -(\frac{1}{6}gZ^3 + \frac{1}{2}hZ^2 + iZ)$, ... This direct manner of generating the values of the successive fluents $\acute{y}$, $\acute{\acute{y}}$, $\acute{\acute{\acute{y}}}$, ... is essentially that presented in 1727 by Benjamin Robins in proof of the equivalent theorem afterwards introduced into the printed *Tractatus de Quadratura Curvarum* in 1704; see his 'Demonstration of the 11th Proposition of Sir Isaac Newton's Treatise of Quadratures', *Philosophical Transactions*, **34**, No. 397 [for January–March 1727]: 230–6, especially 232–4, and compare **2**, 3, §2: note (96) below. Once the pattern of the coefficients is recognized, it is easily generalized by induction; for, since

$$\frac{dA}{dz}.z^n - \binom{n}{1}\frac{dB}{dz}.z^{n-1} + \ldots + (-1)^n.\frac{d}{dz}\left[\int_Z^z \dot{y}z^n.dz\right] = \left(1 - \binom{n}{1} + \ldots (-1)^n\right).\dot{y}z^n = 0,$$

at once for all positive integral n there is

$$\int_Z^z \frac{1}{(n+1)!}\left(Az^{n-1} - \binom{n-1}{1}Bz^{n-2} + \binom{n-1}{2}Cz^{n-3} \ldots + (-1)^{n-1}. \int_Z^z \dot{y}z^{n-1}.dz\right).dz.$$
$$= \frac{1}{n!}\left(Az^n - \binom{n}{1}Bz^{n-1} + \binom{n}{2}Cz^{n-2} \ldots + (-1)^n. \int \dot{y}z^n.dz\right).$$

This recursion replaced (pp. 40–1) the preceding step-by-step derivation in the amended version of Robins' 'Demonstration' which appeared six years afterwards in Reid and Gray's *Philosophical Transactions...Abridged, and Disposed under General Heads*, **6** (London, 1733): 39–42; see also Samuel Horsley, *Isaaci Newtoni Opera quæ exstant Omnia*, **1** (London, 1779): 380–3, note (rr), where due recognition is given to Robins for his elegant exegesis of a 'Propositio reconditissima'.

est $\frac{1}{4}z^4+g=y$. $\frac{1}{20}z^5+gz+h=\dot{y}$. $\frac{1}{120}z^6+\frac{1}{2}gz^2+hz+i=\ddot{y}$

$$\tfrac{1}{840}z^7+\tfrac{1}{6}gz^3+\tfrac{1}{2}hz^2+iz+k=\dddot{y}.\quad [\&c.]$$

Exempl. 2. Proponatur fluxio $\frac{a}{1+z}$[10] $=\dot{y}$ et ordinatæ curvarum erunt $\frac{a}{1+z}$, $\frac{az}{1+z}$, $\frac{azz}{1+z}$, $\frac{az^3}{1+z}$ &c id est $\frac{a}{1+z}$, $a-\frac{a}{1+z}$, $az-a+\frac{a}{1+z}$, $azz-az+a-\frac{a}{1+z}$ &c et areæ curvarum earundem erunt $\frac{a}{1+z}\square$[11] $=A$, $az-\frac{a}{[1]+z}\square=B$.

$$\tfrac{1}{2}az^2-az+\frac{a}{[1]+z}\square=C.\ \tfrac{1}{3}az^3-\tfrac{1}{2}azz+az-\frac{a}{1+z}\square=D.\ \&c.$$

Ideoq est $\frac{a}{1+z}\square+g=y$. $\overline{1+z},\frac{a}{1+z}\square-az+gz+h=\dot{y}$.

$$\frac{1+[2]\,z+z^2}{2},\ \frac{a}{1+z}\square-\tfrac{3}{4}az^2-\tfrac{1}{2}az+\tfrac{1}{2}gz^2+hz+i=\ddot{y}.$$

$$\frac{1+[3]\,z+[3]z^2+z^3}{6}\boxed{\frac{a}{1+z}}-$$ [12]

Exempl. 2. Proponatur fluxio $\sqrt{rr-zz}=\dot{y}$, et ordinatæ curvarum erunt $\sqrt{rr-zz}$, $z\sqrt{rr-zz}$, $z^2\sqrt{rr-zz}$, $z^3\sqrt{rr-zz}$ &c et earum areæ $\boxed{\sqrt{rr-zz}}=A$. $-\frac{1}{3}\dot{y}^3=B$. $-\frac{1}{4}z\dot{y}^3+\frac{1}{4}rr\boxed{\sqrt{rr-zz}}=C$. $-\frac{2}{15}rr\dot{y}^3-\frac{1}{5}zz\dot{y}^3=D$. &c.[13] Et hinc fit

(10) Originally '$\frac{1}{a+z}$'; Newton's preparatory calculations for this dimensionally better balanced variant example are preserved on Add. 3962.3: 59^v.

(11) Notice the curious rearwards position of this 'square' integration operator; Newton's earlier, more logical usage (see I: 302 and 404–11, for example) would have here led him to write in natural sequence '$\square\frac{a}{1+z}$' (that is, $\int a/(1+z)\,.dz$ in modern Leibnizian equivalent) and the like *mutatis mutandis* in the sequel.

(12) Newton breaks off to cancel this first 'Exempl. 2'. To complete this evaluation of the fourth-order fluent, read in continuation '$-\frac{11}{36}az^3-\frac{5}{12}az^2-\frac{1}{6}az+\frac{1}{6}gz^3+\frac{1}{2}hz^2+iz+k=\dddot{y}$', as may readily be seen.

(13) While these integrals '$\square z^i\sqrt{rr-zz}$', $i=1, 2, 3, \ldots$, are each readily computable from first principles, they may also be derived straightforwardly by suitably applying Theorem I of the preceding 'De quadratura' or by appealing alternatively to Ordo 3 of the first and Ordo 5 of the second of the two catalogues of curve areas which Newton inserted in his 1671 tract (see III: 238, 248; compare I: 406, 408 when $n=2$). In immediate sequel at this point he has deleted an initial geometrical restriction (uncancelled in the preliminary draft on Add. 3962.3: 59^r/59^v): 'Ubi $\boxed{\sqrt{rr-zz}}$ ponitur pro area circuli cujus radius est r, abscissa z et ordinata $\sqrt{rr-zz}$.' As we have seen (II: 226, note (78)) Newton had originated this clumsy practice of enclosing the integrand within the '$\square$' operator in his 'De Analysi' in the summer of 1669.

$$\boxed{\sqrt{rr-zz}}+g=y.\ z\boxed{\sqrt{rr-zz}}+\tfrac{1}{3}\dot{y}^3+gz+h=\acute{y}.$$

$$\frac{rr+4zz}{8}\boxed{\sqrt{rr-zz}}+\tfrac{5}{24}\dot{y}^3z+\tfrac{1}{2}gz^2+hz+i=\ddot{y}.$$

$$\frac{6rrz+8z^3}{48}\boxed{\sqrt{rr-zz}}+\frac{8rr+27zz}{360}\dot{y}^3+\tfrac{1}{6}gz^3+\tfrac{1}{2}hz^2+iz+k=\dddot{y}.^{(14)}$$

[&c.]

Cas. 2. Si fluxio proposita(15) duas vel plures involvat quantitates simplices fluentes, puta v, x, y, z aut earum aliquas, scribenda est una earum in datas coefficientes ducta pro reliquis, puta rx, pro v, et sx pro y & tx pro z. Dein quærenda est quantitas fluens ut in casu primo, et coefficientes r s ac t exterminandæ,(16) et si fluxio quantitatis inventæ congruit cum fluxione proposita solvitur Problema. Nam congruet si Problema solvi potest.

Exempl. 1. Proponatur fluxio $3\dot{x}x^2-\dot{x}yy-2xy\dot{y}+aa\dot{v}$, et scribendo rx pro v et sx pro y adeoq $r\dot{x}$ pro $\dot{v}$ et $s\dot{x}$ pro $\dot{y}$, fluxio proposita evadet $3\dot{x}x^2-3ss\dot{x}x^2+raa\dot{x}$. Scribatur 1 pro $\dot{x}$ et area Curvæ cujus hæc ordinata est et x abscissa, erit x^3-ssx^3+raax. Exterminantur coefficientes ss et r scribendo yy pro $ssxx$ et v pro rx et quantitas quæsita erit $x^3-xyy+aav$. Nam hujus fluxio per Prop. XI(17) inventa, congruit cum fluxione proposita.

Exempl 2. Proponatur fluxio $\dfrac{2y\dot{y}-a\dot{z}}{2\sqrt{yy-az}}$. Et hæc scribendo ry pro z et $r\dot{y}$ pro $\dot{z}$ evadet $\dfrac{2y\dot{y}-ar\dot{y}}{2\sqrt{yy-ary}}$. Scribatur 1 pro $\dot{y}$ et area Curvæ cujus hæc est fluxio et y abscissa prodit $\sqrt{yy-ary}$ per Prop [IV].(18) Exterminetur r scribendo z pro ry, et habebitur quantitas quæsita $\sqrt{yy-az}$. Nam hujus fluxio per Prop. XI(17) inventa congruit cum fluxione proposita.

Siquando substituendo quantitatem unam fluentem pro alia termini aliqui se mutuo destruant debet dignitas indeterminata unius substitui pro reliquis.

(14) In draft on Add. 3962.3: 59v Newton first made the numerator of the coefficient of $\dot{y}^3$ to be '$16rr+39zz$', but caught his slip in a checking computation which there follows.

(15) In afterthought on Add. 3960.10: 165, doubtless in parallel with his intended following 'Cas. 3' (see note (19) below), Newton here inserted 'sit primæ generationis et'.

(16) This replaces an equivalent but lengthier direction 'et in hujus terminis singulis restituendæ dignitates ipsarum x et y dignitatibus coefficientium r et s congruentes'.

(17) That is, the revised problem 'Invenire fluxiones quantitatum complexarum' which immediately precedes.

(18) More precisely, by its 'Cas. 2'. The manuscript text here—by a simple slip?—cites 'Prop. V', but we amend this number to be that of the 'Prob. I' of the 'De quadratura Curvarum' to which (see pages 64–6 above) Newton manifestly appeals.

(19) Understand probably 'secundæ generationis'. This abandoned opening to a parallel third case of the present Prob. XII is a late addition to the preceding on a facing verso (3960.11: 166) in the manuscript.

Sic fluxio $4x^3\dot{x}-\frac{3a^4x^2\dot{x}}{y^3}+\frac{3a^4x^3\dot{y}}{y^4}$ scribendo $s x$ pro y et $s\dot{x}$ pro $\dot{y}$ evadit $4x^3\dot{x}$ terminis reliquis se mutuo delentibus. At si scribatur x^ν pro y et $\nu\dot{x}x^{\nu-1}$ pro $\dot{y}$: Fluxio illa evadit $4x^3\dot{x}\,{}^{+3\nu}_{-3}\,a^4\dot{x}x^{2-3\nu}$ et area curvæ cujus hæc est ordinata prodit $x^4-a^4x^{3-3\nu}$. Pro $x^{-3\nu}$ seu $\frac{1}{x^{3\nu}}$ scribe $\frac{1}{y^3}$ et quantitas quæsita prodibit $x^4-\frac{a^4x^3}{y^3}$.

Cas. 3. Si fluxio est sec[19]

---

[2][20] [Proponatur $y=z$. Erit] $\dot{y}=1$. $A=z$. $B=\frac{1}{2}z^2$. $C=\frac{1}{3}z^3$. $D=\frac{1}{4}z^4$. [&c. Pone] $P=z$. $Q=\frac{zz}{2}$. $R=\frac{z^3}{3}$. $S=\frac{z^4}{4}$. [&c. Prodit] $z=A=P$. $Az-B=\frac{1}{2}z^2=Q$.

$$\frac{Az^2-2Bz+C}{2}=\frac{z^3-z^3+\frac{1}{3}z^3}{2}=\frac{1}{6}z^3=\frac{R}{2}.$$

$$\left[\frac{Az^3-3Bz^2+3Cz-D}{6}=\right]\frac{z^4-\frac{3}{2}z^4+z^4-\frac{1}{4}z^4}{[6]}=\frac{1}{24}z^4=\frac{S}{6}.\quad [\&c.]^{(21)}$$

---

(20) Add. 3960.8: 158. Preliminary calculations for a first intended 'Exemplum' of the algorithm for evaluating multiple integrals expounded in Cas. 1 of Prop. XII in [1] preceding.

(21) Newton's singling out here of the sequence of numerators

$$P=A,\quad Q=Az-B,\quad R=Az^2-2Bz+C,\quad S=Az^3-3Bz^2+3Cz-D,\ \ldots$$

suggests that he was intending to go on to express them in more concise form. In Proposition XI of his revised 'Geometriæ Liber secundus' (**2**, 3, §2 below) he was soon, indeed, to refine his preceding algorithm for reducing multiple integrals in just such a manner. In effective résumé, if in note (9) above we set $z=t$ and put $Z=0$ $(=g=h=i=k=\ldots)$, then the n-th fluent of $\dot{y}$ proves to be

$$\frac{1}{n!}\left(t^n\int_0^t\dot{y}\,.\,dz-\binom{n}{1}t^{n-1}\int_0^t z\dot{y}\,.\,dz+\binom{n}{2}t^{n-2}\int_0^t z^2\dot{y}\,.\,dz\ldots+(-1)^n\int_0^t z^n\dot{y}\,.\,dz\right)=\frac{1}{n!}\int_0^t(t-z)^n\,\dot{y}\,.\,dz,$$

that is, $\frac{1}{n!}\int_0^t z^n\dot{y}\,.\,dx$ on substituting $t-z=x$; whence in Newton's present terms there is $P=\int_0^t\dot{y}\,.\,dx$, $Q=\int_0^t x\dot{y}\,.\,dx$, $R=\int_0^t x^2\dot{y}\,.\,dx$, $S=\int_0^t x^3\dot{y}\,.\,dx$, and so on. This extension, too, was first given general published proof by Robins on pages 234–6 of his 1727 'Demonstration...', though—almost certainly unbeknown to Newton—its instances $n=1$ and $n=2$ had been verbally derived by Blaise Pascal, in equivalent indivisibles form as the 'triangular' and 'pyramidal' limit-sums of magnitudes, in his excursus of the 'Proprietez des Sommes Simples, Triangulaires, & Pyramidales' in the third tract of the *Lettres de A. Dettonville* which he brought out anonymously at Paris in the winter of 1658/9, 'tout cela sans écrire une seule formule, mais dans un langage si transparent et si précis qu'on peut immédiatement le transcrire en formules' (N. Bourbaki, *Éléments d'histoire des mathématiques* (Paris, 1960): 207).

APPENDIX 3. THE REWORKED EXCERPTS FROM THE 'DE QUADRATURA' COMMUNICATED BY NEWTON TO JOHN WALLIS IN AUGUST/SEPTEMBER 1692 FOR PUBLICATION.(1)

Extracted from Wallis' *De Algebra Tractatus; Historicus & Practicus* (1693)(2)

Clarissimus *Newtonus* in Epistola sua ad D. *Oldenburgium* Oct. 24. 1676. data,(3) scribit se tunc compotem fuisse methodi cujusdam qua *Slusius* Curvarum

(1) As we remarked in note (33) to the preceding Introduction, when in July 1692 David Gregory formally communicated to his fellow Savilian Professor the newly augmented variant on Newton's series expansion for squaring a general algebraic binomial which he had initially attained in 1686 (and allowed Pitcairne to publish as his discovery two years later), it is clear that Wallis lost little time in sounding out the reactions of its prime author to having it appear as an *addendum* to the Latin edition of his *Algebra* which he was then guiding through press in Oxford. The texts neither of Wallis' letter(s) to Cambridge (in mid-August?) nor of Newton's replies on 27 August and 17 September following have survived, but the present extract from the second volume of Wallis' *Opera* published the year afterwards—in which (see the next note) Newton's *ipsissima verba* are in large part repeated—accurately conveys the essence of the latter's response. Wallis did not, it will be evident, restrict himself merely to applying for permission to print the 'Theorema primum' of quadrature by infinite series which Newton had set down for Leibniz sixteen years before in his *epistola posterior*, or even to seeking confirmation of the circumstances in which John Craige had afterwards in the middle 1680's outlined its structure and scope to Gregory, but more daringly went on to request an elucidation and some illustration of the pair of Latin anagrams in which (see IV: 673, note (60)) Newton had in October 1676 encapsulated the verbal essence of his method of fluxions. That Newton obliged him in every respect will be seen from the present summary account which Wallis prepared on the basis of his response. Not only did Newton furnish his version of the manner in which his 'prime' theorem for binomial quadrature had been passed to Gregory, but for good measure enunciated its extension to the trinomial case as he had set it out just a few months before in Proposition IX of his 'De quadratura Curvarum' (§2 above). And, even more importantly, he communicated the unenciphered originals (see II: 191, note (25)) of his two 1676 anagrams, elaborating their content as two fluxional *Problemata*: in the first he explains, in a minimal variation on Case 1 of Proposition IV of his 'De quadratura', how to compute the fluxion of a given algebraic quantity; and in the second, in a refinement of the kernel of his preceding Proposition XII, he specifies—but does not make completely intelligible to the innocent reader—his technique for extracting the fluent 'root' of a given first-order fluxional equation as a 'converging' infinite series in powers of the base variable. The central purpose of this display of mathematical expertise was, of course, to give the student of Wallis' Latin *Algebra* a tantalizing foretaste of the fuller riches of Newton's extended treatise on fluxional quadrature, and in ensuing years Wallis himself never tired in his efforts to persuade Newton to permit a more complete publication of its content: 'You are not', he wrote on 10 April 1695, 'so kind to your Reputation (& that of the Nation) as you might be, when you let things of worth ly by you so long' (*Correspondence*, **4**, 1967: 100). But Newton was not to publish his 'De quadratura' till the year after Wallis' death (in 1703), and then only, as we shall make clear in our final volume, in a slight amplification of the truncated version which in 1693 he had abstracted to be the 'Liber secundus' of a general work on 'Geometria' (see **2**, **3**, §2: note (1) below).

tangentes ducit, sed quæ Slusiana generalior est. In hac enim non hæreri ad æquationes radicalibus unam vel utramq; indefinitam quantitatem (Ordinatam scilicet & Abscissam) involventibus utcunq; affectas; sed absq; aliqua talium æquationum reductione (quæ opus plerumq; redderet immensum) tangentem confestim duci. Nec in *tangentibus* tantum & *maximis minimisque* determinandis, sed etiam in *quadratura curvarum* & quæstionibus quibusdam aliis hanc methodum locum habere. Nam & ejus beneficio se generalia quædam de Quadratura Curvarum Theoremata invenisse; quorum Primum est hujusmodi.(4)

Ad Curvam aliquam quadrandam sit z abscissa curvæ, & $dz^{\theta}\times\overline{e+fz^{\eta}}|^{\lambda}$ ordinatim applicata termino abscissæ normaliter insistens. Ubi literæ d, e, f denotant quaslibet quantitates datas, & θ, η, λ indices seu exponentes dignitatum quibus affixæ sunt, nempe θ & η indices dignitatum ipsius z, & λ indicem

(2) *Opera Mathematica*, **2** (Oxford, 1693): 390–6, the first of three additions made by Wallis to Cap. XCV in his much amplified Latin version (*ibid.*: 1–482) of his *Treatise of Algebra, Both Historical and Practical* (London, 1685). We have, within square brackets, incorporated in this edited reproduction of Wallis' printed text two groups of minimal typographical 'amendments' noticed by Newton himself in the printed sheets (Ccc and Ddd) sent him by Wallis in advance of publication, but passed back by Newton early in 1693—he speaks of Wallis' Latin *Algebra* as being 'neare finishing' in his unpublished draft covering letter (ULC. Add. 3977.7) — apparently too late to be included in the book's concluding list of *Errata*; these corrections were afterwards unobtrusively entered by him into his library copy of the published *Volumen alterum* (now in the Whipple Science Museum, Cambridge). If proof beyond the direct evidence of the text itself be needed to confirm Newton's authorship of its content, in his own annotated copy of the volume (now Bodleian. Savile Gg. 2) Wallis has written alongside the first paragraph that 'Quæ hic sequuntur, sunt ipsius Newtoni verba, ab ipso scripta, atq; ad me missa, eo animo ut hic inserantur: sed quasi meo nomine. usq; ad pag. 396. l. 19'. (This marginal note was first published by S. P. Rigaud in his *Historical Essay on the First Publication of Sir Isaac Newton's 'Principia'* (Oxford, 1838): 22 with the comment that Newton 'does not appear to have hesitated in complying with Wallis's request [to provide examples of his fluxional method]: he freely furnished him with the account of his method and of the notation which he adopted in the use of it: he did this with the express view of its being given to the public...' (*ibid.*: 22—3).) We would, however, register firm disapproval of the manner in which H. W. Turnbull has given English translation of Wallis' pages in his edition of Newton's correspondence, splitting them into two and arbitrarily allocating the halves to be the 'originals' of Newton's lost letters to Wallis of 27 August and 17 September 1693 which conveyed their content (see *Correspondence*, **3**, 1961: 220–1/222–8): there is no slightest supporting evidence for this division, which seems to have been dictated by the wholly contingent circumstance that he could thereby make use of an existing translation of the latter 'letter', in the hand of John Colson it would appear, which was made some twenty years afterwards for William Jones (compare *ibid.*: 228, note (1)).

(3) See *Correspondence*, **2**, 1960: 110–29. The immediate sequel is a narrow paraphrase, slightly abridged, of the first paragraph on p. 115.

(4) Newton proceeds to cite, with only minimal variation, the binomial 'primum Theorema' of his 1676 *epistola posterior* (*Correspondence*, **2**: 115) which is also, of course, Theorem I of his more recent 'De quadratura Curvarum'.

dignitatis binomii $e+fz^{\eta}$. Fac $\frac{\theta+1}{\eta}=r.\ \lambda+r=s.\ \frac{d}{\eta f}\times\overline{e+fz^{\eta}}|^{\lambda+1}=Q.\ r\eta-\eta=\pi.$ Et area Curvæ erit

$$Q \text{ in } \frac{z^{\pi}}{s}-\frac{r-1}{s-1}\times\frac{eA}{fz^{\eta}}+\frac{r-2}{s-2}\times\frac{eB}{fz^{\eta}}-\frac{r-3}{s-3}\times\frac{eC}{fz^{\eta}}+\frac{r-4}{s-4}\times\frac{eD}{fz^{\eta}}-\&c.$$

Ubi Q ducitur in seriem totam subsequentem; & A, B, C, D, &c. denotant terminos seriei proxime antecedentes, nempe A terminum primum $\frac{z^{\pi}}{s}$, B terminum secundum $\frac{r-1}{s-1}\times\frac{eA}{fz^{\eta}}$, & sic deinceps. Hæc series ubi r fractio est, vel numerus negativus, continuatur in infinitum: ubi vero r integer est & affirmativus, continuatur ad tot terminos tantum quot sunt unitates in eodem r, & sic exhibet geometricam quadraturam Curvæ.

Quadrandi Curvas per hanc seriem exempla quatuor subjungit(5) (quæ brevitatis gratia prætermitto,) & inter ea duos esse casus in omni curva tentandos dicit. Nam Ordinatam $dz^{\theta}\times\overline{e+fz^{\eta}}|^{\lambda}$ reduci posse ad hanc formam

$$dz^{\theta+\eta\lambda}\times\overline{f+ez^{-\eta}}|^{\lambda},$$

& in hoc casu f scribendum esse in Serie vel Theoremate pro e, & e pro f, & $-\eta$ pro η, & $\theta+\eta\lambda$ pro θ. Et si series in neutro casu abrumpatur & finita evadat, concludit Curvam ex earum numero esse quæ non possunt geometrice quadrari. Hanc enim regulam, quantum animadvertit, exhibere in æquationibus finitis areas omnium geometricam quadraturam admittentium curvarum, quarum Ordinatim applicatæ constant ex potestatibus, radicibus vel quibuslibet dignitatibus binomii cujuscunque. Et quando hujusmodi Curva non potest geometrice quadrari, esse ad manus alia Theoremata(6) pro comparatione ejus cum Conicis Sectionibus vel saltem cum aliis figuris simplicissimis, quibuscum potest comparari: Ad quod sufficere etiam hoc ipsum jam descriptum si debite concinnetur.

Annis aliquot post scriptam hanc Epistolam elapsis, D. *David Gregorius* M.D. tunc Matheseos Professor *Edinburgensis*, nunc Collega mea dignissimus, invenit per aliam methodum (uti audio)(7) seriem ejudsem generis cum hacce *Newtoniana*,

(5) See *Correspondence*, **2**: 115–17 and pages 50–2 above.

(6) Namely, Theorems II and III of the 'De quadratura Curvarum' (§2 preceding).

(7) A carefully hesitant interpolation by Wallis which neatly and diplomatically evades any obligation on his part here to pass judgement on the justice of Gregory's claim to have independently discovered Newton's quadrature series; he had, of course, already 'heard' both Gregory's and Newton's testimony on the point, if not that of their intermediary John Craige also. The few known facts of this subdued and short-lived controversy—amicably resolved by its two principal disputants after Newton chose to accept Gregory's word that he had not seen the 1676 'primum theorema' when he allowed Pitcairne to publish his own 'more complicated' variant of it—are reviewed in the preceding introduction (see pages 3–6 above).

quamquam terminis complexioribus involutam; eamque D. *Archibaldus Pitcarnus*, diebus tribus circiter vel quatuor antequam *Newtonianam* viderant, in lucem edidit. Nam D. *Johannes Craig*, qui in Academia *Cantabrigiensi* cum D. *Newtono* conversatus fuerat, & seriem ejus ibi viderat, postulabat à *Newtono* perEpistolam, ut is exemplar seriei suæ ad ipsum tunc *Edinburgi* commorantem mitteret, ut cum *Gregoriana* conferri posset.

Hanc seriem D. *Newtonus* Theorema *primum* vocat. Dicit enim se pro Curvis quarum Ordinatæ ex Trinomiis conflantur, & pro aliis perplexioribus, regulas alias concinnasse. Jamque in Literis ad me datis Aug. 27, & Sept. 17. 1692. scribit progressiones esse horum Theorematum in infinitum pergentes; quæ tamen omnia in hujusmodi Theorematis generalioribus comprehendi possunt.(8)

Si Curvæ abscissa sit z & pro terminis hujus seriei $a+bz^{\eta}+cz^{2\eta}+dz^{3\eta}+$&c. hujusque $e+fz^{\eta}+gz^{2\eta}+hz^{3\eta}+$&c. vel pro eorum quotlibet, scribantur Q & R respective; nempe Q pro primis & R pro secundis: sit autem Ordinatim applicata $z^{\theta-1}\times QR^{\lambda-1}$, & ponatur $\frac{\theta}{\eta}=r$, $r+\lambda=s$, $s+\lambda=t$, $t+\lambda=v$ &c. erit area Curvæ

$$z^{\theta}R^{\lambda} \text{ in } \frac{\frac{1}{\eta}a}{re}+\frac{\frac{1}{\eta}b-sfA}{\overline{r+1}\times e}z^{\eta}+\frac{\frac{1}{\eta}c\frac{-s}{-1}fB-tgA}{\overline{r+2}\times e}z^{2\eta}+\frac{\frac{1}{\eta}d\frac{-s}{-2}fC\frac{-t}{-1}gB-vhA}{\overline{r+3}\times e}z^{3\eta}+\&\text{c.}$$

Ubi A, B, C, &c. denotant coefficientes datas terminorum seriei cum signis suis, nempe A primi termini coefficientem $\frac{\frac{1}{\eta}a}{re}$, B secundi coefficientem $\frac{\frac{1}{\eta}b-sfA}{\overline{r+1}\times e}$ & sic deinceps.(9)

Qua methodo in hujusmodi series incidit, minime explicuit. Eandem tamen literis quibusdam celavit, quæ recto ordine dispositæ conficiunt hanc sententiam,(10), *Data æquatione fluentes quotcunque quantitates involvente, fluxiones invenire: & vice versa.* Per *fluentes quantitates* intelligit indeterminatas, id est quæ in generatione Curvarum per motum localem perpetuo augentur vel diminuuntur, & per earum *fluxionem* intelligit celeritatem incrementi vel decrementi. Nam quamvis *fluentes quantitates* & earum *fluxiones* prima fronte conceptu difficiles videantur, (solent enim nova difficilius concipi,) earundem tamen notionem cito faciliorem evasuram putat, quam sit notio *momentorum* aut partium *minimarum* vel *differentiarum infinite parvarum*; propterea quod figurarum & quanti-

(8) See the 'Theorematum progressiones' elaborated in §2, Appendix 1.5 preceding, and Propositions IX/X of the 'De quadratura' wherein they were afterwards subsumed. The following cited instance is the latter's Proposition IX.

(9) Newton was afterwards to give an explicit 'Demonstratio' of this series expansion in the equivalent Proposition V of his 'Geometriæ Liber secundus' (**2**, **3**, §2 below); for our modern proof see §2: note (49) preceding.

(10) See *Correspondence*, **2**: 153, note (25) and also II: 191, note (25).

tatum generatio per motum continuum magis naturalis est & facilius concipitur, & Schemata in hac methodo solent esse simpliciora, quam in illa partium.(11) Attamen non negligit Theoriam talium partium, sed ea etiam utitur quoties ipsa vel opus brevius reddit & magis perspicuum, vel ad rimandas fluxionum proportiones conducit. Abscissam Curvæ aliamve aliquam quantitatem fluentem, uniformiter augeri supponit, & pro ejus fluxione unitatem ponit; pro reliquis autem quantitatibus fluentibus ipsas ponit quantitates punctis notatas in hunc modum.(12) Sint v, x, y, z fluentes quantitates, & earum fluxiones his notis $\dot{v}$, $\dot{x}$, $\dot{y}$, $\dot{z}$ designabuntur respective. Et quoniam hæ fluxiones sunt etiam indeterminatæ quantitates, & perpetua mutatione redduntur majores vel minores, considerat velocitates quibus augentur vel diminuuntur tanquam earum fluxiones, & punctis binis notat in hunc modum $\ddot{v}$, $\ddot{x}$, $\ddot{y}$, $\ddot{z}$, & perpetuum incrementum vel decrementum harum fluxionum considerat ut ipsarum fluxiones, & punctis ternis sic notat $\dddot{v}$, $\dddot{x}$, $\dddot{y}$, $\dddot{z}$, & harum fluxiones punctis quaternis sic $\ddddot{v}$, $\ddddot{x}$, $\ddddot{y}$, $\ddddot{z}$. Qua ratione $\dot{v}$ est fluxio quantitatis v, & $\ddot{v}$ fluxio ipsius $\dot{v}$, & $\dddot{v}$ fluxio ipsius $\ddot{v}$. Et si quantitates fluentes vel fractæ sunt vel surdæ, fluxiones earum sic notat, Quantitatum $\frac{yy}{b-x}$ & $\sqrt{aa-xx}$, fluxiones sunt $\dot{\overline{\frac{yy}{b-x}}}$ & $\dot{\overline{\sqrt{aa-xx}}}$, & harum fluxiones sunt $\ddot{\overline{\frac{yy}{b-x}}}$ & $\ddot{\overline{\sqrt{aa-xx}}}$, & sic porro.(13) His autem expositis methodum suam in Propositione hacce problematica fundat.

PROB. I.(14)

Data æquatione, fluentes quotcunque quantitates involvente, invenire fluxiones.

Solutio.

Multiplicetur quilibet æquationis terminus separatim per indices singulos dignitatum quantitatum omnium fluentium, quæ in termino illo continentur;

(11) In his *epistola posterior* to Leibniz in 1676 Newton had written that 'Hoc fundamento conatus sum etiam reddere speculationes de Quadratura curvarum simpliciores' (*Correspondence*, 2, 115).

(12) Here for the first time Newton's classical dot-notation for fluxions—which, as we have seen (§2: note (35) above), had been introduced by him in his private papers only a few months before—made its public bow and was at once accepted by the world at large as his standard nomenclature and his preferred equivalent to the Leibnizian *d*-symbol, even though, as Newton was ever quick to point out, the latter in strict truth denoted a vanishingly small *differentia*, and he himself 'never placed the method in the use of prickt letters' (ULC. Add. 3968.41: 20r, a stray fragment penned at the time of his squabble with Leibniz over priority of discovery of the calculus).

(13) This extension to denote the fluxions of compound quantities is not found in any draft of the 1691 'De quadratura Curvarum' known to us. Did Newton perhaps specially here concoct it for Wallis' benefit in August 1692?

(14) A light restyling of Proposition IV of the preceding 'De quadratura Curvarum' (§2 above).

& in singulis multiplicationibus mutetur latus unum dignitatis in ejus fluxionem; & aggregatum productorum omnium sub propriis signis componet æquationem novam, quæ relationem fluxionum involvet.

Explicatio.

Sint a, b, c, d &c. quantitates determinatæ & immutabiles; & proponatur æquatio quævis quantitates fluentes z, y, x, &c. involvens, puta

$$x^3 - xyy + aaz - b^3 = 0.^{(15)}$$

Multiplicentur termini primo per indices dignitatum ipsius x in ipsis contentarum respective; & in qualibet multiplicatione pro uno latere dignitatis (hoc est pro x unius dimensionis,) scribatur $\dot{x}$ & summa productorum erit $3\dot{x}x^2 - \dot{x}yy$. Idem fiat in y(16) & producetur $-2xy\dot{y}$. Idem fiat in z, & producetur $+aa\dot{z}$. Summa productorum omnium æquetur nihilo; & habebitur æquatio

$$3\dot{x}x^2 - \dot{x}yy - 2xy\dot{y} + aa\dot{z} = 0.$$

Dico quod hæc erit æquatio relationem fluxionum involvens.

Demonstratio.

Sit enim o quantitas infinite parva, & sint $o\dot{z}$, $o\dot{y}$, $o\dot{x}$ Synchrona momenta seu incrementa momentanea quantitatum fluentium z, y & x: & hæ quantitates proximo temporis momento per accessum incrementorum momentaneorum evadent $z + o\dot{z}$, $y + o\dot{y}$, $x + o\dot{x}$: quæ in æquatione prima pro z, y & x scriptæ, dant hanc æquationem

$$x^3 + 3xxo\dot{x} + 3xoo\dot{x}\dot{x} + \dot{x}^3o^3 - xyy - \dot{x}oyy - 2xo\dot{y}y - 2\dot{x}oo\dot{y}y$$
$$[-]xoo\dot{y}\dot{y}[-]\dot{x}o^3\dot{y}\dot{y} + aaz + aao\dot{z} - b^3 = 0.$$

Subduc æquationem primam, & residuum divide per o: erit

$$3\dot{x}x^2 + 3\dot{x}\dot{x}ox + \dot{x}^3oo - \dot{x}yy - 2xy\dot{y} - 2\dot{x}oy\dot{y}[-]xo\dot{y}\dot{y}[-]\dot{x}o^2\dot{y}\dot{y} + aa\dot{z} = 0.$$

Terminos multiplicatos per o tanquam infinite parvos dele, & manebit æquatio $3\dot{x}x^2 - \dot{x}yy - 2xy\dot{y} + aa\dot{z} = 0$. Q.E.D.

Simili methodo æquatio $x^3 - xyy + aa\sqrt{ax - yy} - b^3 = 0$. evadit

$$3x^2\dot{x} - \dot{x}yy - 2xy\dot{y} + aa\,\dot{\overline{\sqrt{ax - yy}}} = 0.$$

(15) Newton's equivalent worked example in the 'De quadratura' (Proposition IV, Case 1) had been $x^3 - xyy + aav - b^3 = 0$; see page 64 above. The present substitution of '$+aaz$' for '$+aav$' is without significance.

(16) The printed text here reads '$\dot{y}$', a trivial slip—but, for the untutored reader of Wallis' *Algebra*, a potentially confusing one—which Newton has failed to amend in his library copy (on which see note (2)).

Ubi si fluxionem $\dot{\sqrt{ax-yy}}$ eliminare velis, pone $\sqrt{ax-yy}=z$, & per consequens $ax-yy=z^2$, & Propositio dabit $a\dot{x}-2y\dot{y}=2z\dot{z}$, id est $\frac{a\dot{x}-2y\dot{y}}{2z}=\dot{z}$, seu $\frac{a\dot{x}-2y\dot{y}}{2\sqrt{ax-yy}}=\dot{\sqrt{ax-yy}}$. Et inde æquatio primo inventa fit

$$3x^2\dot{x}-\dot{x}yy-2xy\dot{y}+\frac{a^3\dot{x}-2aay\dot{y}}{2\sqrt{ax-yy}}=0.$$

Et eadem methodo pergere licet ad fluxiones secundas ac tertias. Ponatur æquatio $zy^3-z^4+a^4=0$, & Propositio inde dabit $\dot{z}y^3+3z\dot{y}y^2-4\dot{z}z^3=0$, & hæc æquatio per operationem repetitam evadet

$$\ddot{z}y^3+6\dot{z}\dot{y}y^2+3z\ddot{y}y^2+6zy\dot{y}^2-4\ddot{z}z^3-12\dot{z}^2z^2=0,$$

& sic deinceps.

Ut applicetur hæc methodus ad quadraturam Curvarum, D. *Newtonus* pro fluxione abscissæ uniformi posita, unitatem assumit; & ordinatam ponit pro fluxione areæ; deinde relationem inter abscissam & aream per æquationem quamvis assumptam definit; & inde colligendo earum fluxiones obtinet æquationem novam, qua relatio inter abscissam & ordinatam definitur; & tum demum ad quadrandas Curvas efficit ut hæc æquatio formam quamvis desideratam acquirat. Et hac methodo dicit se computo brevissimo & facillimo Theoremata supra posita invenisse. Sed & eadem methodo comparat diversas curvas inter se, ponendo scilicet earum ordinatas fluxionibus abscissarum reciproce proportionales.

Sub finem Epistolæ anni 1676[(17)] scribit etiam Problema determinandi Curvas per conditiones tangentium, (quod Leibnitius inversum de tangentibus Problema appellat) in sua potestate esse, una cum aliis difficilioribus; ad quæ solvenda se usum esse dicit duplici methodo, una concinniore, altera generaliore; & utramque literis transpositis celat: quæ in ordinem redactæ hanc sententiam exhibent.[(18)] *Una methodus consistit in Extractione fluentis quantitatis ex æquatione simul involvente fluxionem ejus. Altera tantum in assumptione seriei pro quantitate qualibet incognita ex qua cætera commode derivari possunt, Et in collatione terminorum homologorum æquationis resultantis ad eruendos terminos assumptæ Seriei.* Harum methodorum Secunda ex verbis jam recitatis absque ulteriore explicatione intelligi potest; priorem ab Authore jam accepi ut sequitur.

Hæc methodus, ait, ejusdem est generis cum ea pro extrahendo radices ex

(17) Where Newton baldly asserted to Leibniz that 'ne nimium dixisse videar, inversa de tangentibus Problemata sunt in potestate, aliaqꝫ illis difficiliora...' (*Correspondence*, **2**: 129).

(18) Again (compare note (10)) see II: 191, note (25).

æquationibus affectis superius(19) descripta. Pone quod Problema resolvendum reducatur ad æquationem fluentes quantitates y & z una cum earum fluxionibus $\dot{y}$ & $\dot{z}$ involventem, & quod fluxio ipsius z uniformis sit. Ut hæc fluxio ex æquatione evanescat, pro ea ponatur unitas, & manebit æquatio solas y, z & $\dot{y}$ involvens, quam *Resolvendam* vocat. Proponitur inventio ipsius y in Serie infinita convergente, quæ solam z involvet. Hoc in aliquibus æquationibus impossibile est, in aliis præparationem æquationum requirit, ubi vero directe confici possit resolutio est hujusmodi.

PROB. II.(20)

Ex æquatione fluxionem radicis involvente radicem extrahere.

Resolutio.

Termini omnes, ex eodem æquationis latere consistentes, æquentur nihilo, & ipsarum y & $\dot{y}$ dignitates (si opus sit) exaltentur vel deprimantur, sic ut earum indices nec alicubi negativi sint, nec tamen altiores quam ad hunc effectum requiritur; & sit kz^{λ} terminus infimæ(21) dignitatis eorum qui neque per y neque per ejus fluxionem $\dot{y}$ neque per earum dignitatem quamvis multiplicantur. Sit $lz^{\mu}y^{\alpha}\dot{y}^{\beta}$ terminus alius quilibet, & omnes ordine terminos percurrendo collige ex singulis seorsim numerum $\frac{\lambda-\mu+\beta}{\alpha+\beta}$ sic, ut tot habeas ejusmodi numeros quot sunt termini. Horum numerorum maximus(22) vocetur ν, & z^{ν} erit dignitas primi termini Seriei. Pro ejus coefficiente ponatur a, & in æquatione quæ *resolvenda*

(19) Wallis refers to the preceding Caput XCIV of his Latin *Algebra*, where (*Opera Mathematica*, **2**: 381–3) he had drawn upon both of Newton's 1676 *epistolæ* to Leibniz (see *Correspondence*, **2**: 23–4/126–7) to expound and exemplify his 'Nova Methodus extrahendi Radices Affectarum Æquationum' much as Newton had set it down in his (yet unpublished) 1671 fluxional tract (see III: 42–4/50–2).

(20) An elaborate reworking of the main portion of Proposition XII of the preceding 'De quadratura Curvarum' (§2 above). As we shall see in our final volume, when in the early summer of 1712 Newton planned to issue an augmented version of the abridged *Tractatus de Quadratura Curvarum* which he had published in appendix to his *Opticks* in 1704, he copied the present mode of resolution of a first-order fluxional equation out virtually *verbatim* as Proposition XIV of a re-titled 'Analysis per quantitates fluentes et earum momenta' (ULC. Add. 3960.6: 79–81) and went on to repeat the bulk of its text—there even forgetfully transcribing in its final paragraph the 'dicit *Newtonus*' which Wallis here interpolates!—in a following Proposition XV (*ibid.*: 113) which, in line with Proposition XII of the 1691 'De quadratura', elaborates the equivalent technique of series-resolution in the case of higher-order fluxional equations.

(21) Where, it is understood, the series expansion of y shall be in ascending powers of z. When (compare §2: note (99) preceding) a descending power-series is required, read 'altissimæ'.

(22) See §2: note (97) above when $\gamma = \delta = \ldots = 0$. But where (see *ibid.*: note (99) and compare the previous note) the 'root' y needs to be expanded as a descending power-series in z, read 'minimus'.

dicitur scribe az^{ν} pro y, & $\nu az^{\nu-1}$ pro $\dot{y}$; ac termini omnes resultantes in quibus z ejusdem est dignitatis ac in termino kz^{λ}, sub propriis signis collecti, ponantur æquales nihilo. Nam hæc æquatio debite reducta dabit coefficientem a. Sic habes az^{ν} terminum primum Seriei.

Operatio secunda.

Pro reliquis omnibus hujus Seriei terminis nondum inventis pone p, & habebis Æquationem $y = az^{\nu}+p$, & inde etiam per Prob. I. Æquationem $\dot{y} = \nu az^{\nu-1}+\dot{p}$. In resolvenda, pro y & $\dot{y}$ scribe hos eorum valores & habebis Resolvendam novam, ubi p officium præstat ipsius y: & ex hac Resolvenda primum extrahes terminum Seriei p eodem modo atque terminum primum Seriei totius $y = az^{\nu}+p$ ex Resolvenda prima extraxisti

Operatio tertia et sequentes.

Dein tertiam Resolvendam eadem ratione invenias atque secundam invenisti, & ex ea terminum tertium Seriei totius extrahas. Et similiter Resolvendam quartam invenies, & ex ea quartum Seriei terminum, & sic in infinitum. Series autem sic inventa erit radix Æquationis quam extrahere oportuit.

EXEMPLUM.(23)

Ex Æquatione $y^2\dot{z}^2-z^2\dot{z}\dot{y}-dd\dot{z}\dot{z}+dz\dot{z}^2 = 0$. *extrahenda sit radix* y.

Pone $\dot{z} = 1$, & Æquatio evadet $y^2-z^2\dot{y}-dd+dz = 0$, quæ est Resolvenda. Jam vero terminus infimus in quo nec y neque $\dot{y}$ reperitur, est dd, qui ipsi kz^{λ} æquatus dat $\lambda = 0$. Terminis reliquis y^2, $-z^2\dot{y}$ pone $lz^{\mu}y^{\alpha}\dot{y}^{\beta}$ æqualem successive, & inde in primo casu habebis $\mu = 0$, $\alpha = 2$, $\beta = 0$, & in secundo $\mu = 2$, $\alpha = 0$ & $\beta = 1$. Et hinc $\dfrac{\lambda-\mu+\beta}{\alpha+\beta}$ sit in primo casu 0, in secundo -1. Unde ν est 0, & az^{ν} & $\nu az^{\nu-1}$ sunt a & 0; quarum ultimæ duæ a & 0 in Resolvenda pro y & $\dot{y}$ scriptæ, producunt $aa+0z^2-dd+dz[=0]$; & termini aa & $-dd$, in quibus index dignitatis z est λ seu 0, positi æquales nihilo dant $a = d$. Unde primus Seriei terminus az^{λ} evadit d.

Operatio secunda.

Pro terminis reliquis pone p, & habebis æquationem $y = d+p$, & inde (per Prob. I.) $\dot{y} = \dot{p}$; qui valores in Resolvenda pro y & $\dot{y}$ substituti dant Resolvendam novam $2dp+pp-zz\dot{p}+dz = 0$. ubi p & $\dot{p}$ vices subeunt ipsarum y & $\dot{y}$. Terminus

(23) This worked example is (see §2: note (98) preceding) here brand new. A few months before, Newton had found some difficulty in contriving an appropriate illustration of his series-resolution of a given fluxional equation; although in his rough notes (see Appendix 1.11 above) he had shakily outlined the opening stages in deriving the solution in this way of two simple higher-order equations, when he came to write up Proposition XII of the 'De quadratura' he had been forced to leave a gap in his manuscript after optimistically penning its concluding word '*Exempl.*' (see §2: note (111)).

unicus in quo nec p neque $\dot{p}$ reperitur est dz, qui cum termino kz^{λ} collatus dat $\lambda = 1$. Terminis reliquis $2d\dot{p}$, pp & $-zz\dot{p}$ pone $lz^{\mu}p^{\alpha}\dot{p}^{\beta}$ æqualem successive; & inde in primo casu habebis $\mu = 0$, $\alpha = 1$ & $\beta = 0$; in secundo $\mu = 0$, $\alpha = 2$, & $\beta = 0$; & in tertio $\mu = 2$, $\alpha = 0$, & $\beta = 1$. Et hinc $\frac{\lambda-\mu+\beta}{\alpha+\beta}$ evadit in primo casu 1, in secundo $\frac{1}{2}$, in tertio 0. Unde ν est 1, & az^{ν} & $\nu az^{\nu-1}$ sunt az & a. Termini duo ultimi az & a in Resolvenda pro p & $\dot{p}$ respective scripti, producunt

$$2daz + a^2z^2 - az^2 + dz[= 0].$$

Et termini $2daz$ & dz in quibus index dignitatis z est λ seu 1, positi æquales nihilo, dant $a = -\frac{1}{2}$. Unde az^{λ} terminus primus Seriei p fit $-\frac{1}{2}z$.

Operatio tertia.

Pro terminis reliquis nondum inventis pone q & habebis æquationem $p = -\frac{1}{2}z + q$, & inde (per Prob. I.) $\dot{p} = -\frac{1}{2} + \dot{q}$: qui valores pro p & $\dot{p}$ in Resolvenda novissima substituti producunt Resolvendam novam

$$2d\dot{q} - zq + qq[+]\,\tfrac{3}{4}zz - zz\dot{q} = 0.$$

Ubi q & $\dot{q}$ vices supplent ipsorum y & $\dot{y}$. Terminus unicus in quo neque q nec $\dot{q}$ reperitur est $\frac{3}{4}zz$, qui cum az^{λ} collatus dat $\lambda = 2$. Terminis reliquis $2d\dot{q}$, $-zq$, $+qq$, $-zz\dot{q}$ pone $lz^{\mu}q^{\alpha}\dot{q}^{\beta}$ æqualem successive; & inde in primo casu habebis $\mu = 0$, $\alpha = 1$, & $\beta = 0$; in secundo, $\mu = 1$, $\alpha = 1$, $\beta = 0$; in tertio, $\mu = 0$, $\alpha = 2$, $\beta = 0$; in quarto $\mu = 2$, $\alpha = 0$, $\beta = 1$: & inde $\frac{\lambda-\mu+\beta}{\alpha+\beta}$ evadit in primo casu 2, in secundo tertio & quarto 1. Et hinc ν est 2, vel az^{ν} & $\nu az^{\nu-1}$ sunt az^2 & $2az$: qui valores in Resolvenda pro q & $\dot{q}$ substituti dant

$$2daz^2 - az^3 + aaz^4 + \tfrac{3}{4}zz - 2az^3[= 0];$$

& termini $2dazz + \frac{3}{4}zz$ in quibus index dignitatis z est λ seu 2, positi æquales nihilo, dant $a = -\frac{3}{8d}$. Unde az^{λ} terminus primus Seriei q evadit $-\frac{3zz}{8d}$.

Operatio quarta.

Pro reliquis Seriei terminis nondum inventis pone r, & habebis æquationes $q = -\frac{3zz}{8d} + r$, & $\dot{q} = -\frac{3z}{4d} + \dot{r}$; & inde Resolvendam novam

$$2d\dot{r} + \frac{9z^3}{8d} + zr + \frac{9z^4}{64dd} - \frac{3zzr}{8d} + rr - zz\dot{r} = 0;$$

& ex ea per Methodum superiorem habebis $-\frac{9z^3}{16dd}$ terminum primum Seriei r. Et sic pergitur in infinitum.

Est igitur radix extrahenda

$$y = d+p = d-\tfrac{1}{2}z+q = d-\tfrac{1}{2}z-\frac{3zz}{8d}+r = d-\tfrac{1}{2}z-\frac{3zz}{8d}-\frac{9z^3}{16dd}-\&\text{c.}$$

Et operationem continuando producere licet radicem ad terminos plures.

Et eadem methodo, dicit *Newtonus*, radices æquationum, fluxiones secundas, tertias, quartas, $(\ddot{y}, \dddot{y}, \ddddot{y})$ aliasque involventium, extrahi posse.(24)

His utitur radicum extractionibus ubi aliæ Methodi nil prosunt. Nam in Epistola prædicta anni 1676 docet,(25) quod in Solutione problematum de tangentibus inversorum, casus aliqui dantur in quibus hæc Methodus generalis non requiritur: & particulariter si in triangulo rectangulo quod ab ordinata, tangente, & interjacente parte abscissæ, constituitur, relatio duorum quorumlibet è lateribus tribus per æquationem quamvis definiatur; Problema absq Methodo hacce generali solvi poterit.

Methodi autem hæ omnes tam particulares quam general[e]s collectim sumptæ, solutionem exhibent Secundæ partis problematis quod *Newtonus* sub initio istius Epistolæ his verbis proposuit. *Data æquatione quotcunque fluentes quantitates involvente fluxiones invenire, & vice versa.* Nam tota fluxionum Methodus in hujus directa & inversa solutione consistit.(26)

---

(24) This extension is comprehended, of course, in Newton's exposition of the general technique of extracting the root of a given fluxional equation of arbitrary order as he set it down in Proposition XII of the preceding 'De quadratura Curvarum'. His basic assumption throughout that one (at least) of the significant terms in the equation at each stage of working is free of y and its fluxions is, we should notice, far from always valid; indeed, the oversight proves crucial in the present example, for instance, when we seek a series expansion of y in *descending* powers of z. On substituting $y = kz^\lambda + O(z^{\lambda-1})$ and hence $\dot{y} = k\lambda z^{\lambda-1}+O(z^{\lambda-2})$ in the given equation $y^2-z^2\dot{y}+dz-d^2 = 0$, the 'fictitious' equation—to employ the name used by Newton in the non-fluxional technique (see III: 54)—here proves to be $y^2-z^2\dot{y} = 0$, whence to $O(z^0)$ there is $y = z$; we may thereafter substitute $y = z+p$ to derive the secondary equation $2pz+p^2-z^2\dot{p}+dz-d^2 = 0$ and thence, on isolating the second 'fictitious' equation $2pz+dz = 0$ therein, determine that to $O(z^{-1})$ there is $p = -\tfrac{1}{2}d$; and then $p = -\tfrac{1}{2}d+q$ to deduce by means of the third 'fictitious' equation $2qz-z^2\dot{q}-\tfrac{3}{4}d^2 = 0$ that to $O(z^{-2})$ there is $q = \tfrac{1}{4}d^2z^{-1}$; and so on, thereby attaining the expansion $y = z-\tfrac{1}{2}d+\tfrac{1}{4}d^2z^{-1}\ldots$. Here, however, direct 'assumption' of the series $y = z+e+fz^{-1}+gz^{-2}\ldots$ (and so $\dot{y} = 1-fz^{-2}-2gz^{-3}\ldots$) according to the 'alternate' method specified by Newton in his 1676 anagram (see note (18) above, and compare IV: 576 ff.) will, by 'collation of homologous terms' in the resulting equation

$$(d+2e)\,z+(-d^2+e^2+3f)+(2ef+4g)\,z^{-1}\ldots = 0,$$

more rapidly evaluate the successive coefficients to be $e = -\tfrac{1}{2}d$, $f = \tfrac{1}{4}d^2$, $g = \tfrac{1}{16}d^3, \ldots$.

(25) As we have already had occasion to remark (IV: 576, note (147)), Newton stated in the penultimate paragraph of his *epistola posterior* to Leibniz that 'Inversum hoc Problema de tangentibus...non indiget his methodis....Sed... quando in triangulo rectangulo quod ab illa axis parte & tangente ac ordinatim applicata constituitur, relatio duorum quorumlibet laterum per æquationem quamlibet definitur, Problema solvi potest absq mea methodo generali' (*Correspondence*, **2**: 129).

(26) Wallis concluded the paragraph by informing his reader that 'Huic Methodo affinis

est tum Methodus differentialis *Leibnitii*, tum utraque antiquior illa quam D[r] *Is. Barrow* in Lectionibus Geometricis exposuit. Quod agnitum est in *Actis Leipsicis* (Anno 1691, mense Jan.) à quodam qui methodum adhibet *Leibnitii* similem. Quodque ab his duobus est superadditum, est formularum Analyseos brevium & commodarum adaptatio illius Theoriis' (*Opera Mathematica*, **2**: 396)—to which he added in his usual heavily paternalistic way 'Et quidem superstruuntur omnes *Arithmeticæ* [*sc.* meæ] *Infinitorum*'. We have already quoted (IV: 491, note (25)) the Latin phrases in which Jakob Bernoulli in *Acta Eruditorum* (January 1691): 14 earnestly commended the 'calculus' which Barrow 'adumbrated' in his *Lectiones Geometricæ* (London, 1670) and upon which, so he firmly believed, Leibniz' own *Calculus differentialis* was 'founded'.

When in late August 1694 Huygens passed on to Leibniz the 'extrait' of the present pages 'du livre de Wallis' which David Gregory had made for him the year before (see note (33) of the preceding introduction), Leibniz replied on 4 September (N.S.) that 'Il me semble que M. Wallis parle assez froidement de M. Newton et comme s'il estoit aisé de tirer ces methodes des leçons de Mr Barrow. Quand les choses sont faites il est aisé de dire: *et nos hoc poteramus*. Les choses composées ne sçauroient estre si bien démelées par l'esprit humain sans aide de caracteres' (*Œuvres complètes de Christiaan Huygens*, **10**, 1905: 675). But of the various Newtonian quadrature and fluxional methods there expounded he was himself not a little critical: 'Je voy que son calcul s'accorde avec le mien, mais ie pense que la consideration des differences et des sommes est plus propre à éclaircir l'esprit; ayant encor lieu dans les series ordinaires des nombres, et repondant en quelque façon aux puissances et aux racines....Je suis bien aise... de voir enfin le dechifrement des enigmes contenus dans la lettre de M. Newton à feu Mons. Oldenbourg. Mais je suis fasché de n'y point trouver les nouvelles Lumieres que je me promettois pour l'inverse des Tangentes. Car ce n'est qu'une methode d'exprimer la valeur de l'ordonnée de la courbe demandée *per seriem infinitam*, dont je sçavois le fond dés ce temps là, comme je témoignay alors à Mons. Oldenbourg' [in his letter of 11/21 June 1677; see *Correspondence*, **2**: 216: 'Quod ait Problemata Methodi Tangentium inversæ esse in potestate, hoc arbitror ab eo intelligi per Series scilicet Infinitas. Sed a me ita desiderantur, ut Curvæ exhibeantur Geometrice, quatenus id fieri potest, suppositis (minimum) quadraturis']. Nevertheless, in his public review (*Acta Eruditorum* (June 1696): 256–9) of Wallis' 'Alterum *Operum* Volumen' Leibniz elected diplomatically to say nothing of its pages 390–6 beyond reiterating their concluding editorial observation that 'Calculo quoque differentiali *Leibnitii* affinem esse methodum fluxionum *Newtoni* (in *Principiis Naturæ Mathematicis* primum editam) tum utraque esse antiquiorem *Barrovii*, & omnes Wallisianæ *Arithmeticæ Infinitorum* superstrui' (*ibid.*: 257). Writing privately to Johann Bernoulli he was less reluctant to evidence his disappointment with these '*Newtoniana*, in quibus sperabam reperire aliquid amplius pro Methodo tangentium inversa' though adding that 'virum esse egregium fatendum est' (*Got. Gul. Leibnitii et Johan. Bernoullii Commercium Philosophicum et Mathematicum*, **1** (Lausanne/Geneva, 1745): 185). In his response on 15 August (*ibid.*: 190–1) Bernoulli showed his usual grudging impatience of admitting any merit or originality in another mathematician's work: 'Wallisius Newtoni Methodum paucis quidem explicat; ex illis paucis tamen video, quod in re neutiquam differat a Calculo differentiali, ut ipse Newtonus fatetur in suis *Princ. Nat.* [$_1$1687] pag. 254 [= Liber 2, Lemma II, Scholium; see IV: 524, note (11)]. Quod in hoc dicitur *Differentiale*, ibi est *Fluxio*, & quod in hoc *Summa*, ibi *Fluens*. Et Nervus hujus Methodi, ut & calculi differentialis ad duo hæc Problemata redit: *Datis quantitatibus fluentibus, invenire earum fluxiones*; & vicissim, *Datis fluxionibus, invenire earum fluentes*. Loco Litteræ *d* ad designandam differentialem primam, vel fluxionem, utitur puncto supra scripto; pro differentiali secunda denotanda utitur duobus punctis, & ita porro. Sic dx est $\dot{x}$, ddx est $\ddot{x}$, $dddx$ est $\dddot{x}$, &c. Cæterum processus ipse operationis est utrobique idem, adeo ut nesciam annon Newtonus, Tuo calculo viso, suam demum Methodum fabricaverit; præsertim cum ex loco citato videam Te ipsi tuum Calculum communicasse, antequam ipse suam edidisset Methodum. De cætero Wallisius Tom. II. pag. 394 [namely in 'Prob. II'] modum explicat, quo utitur Newtonus ad radicem extrahendam ex æquatione fluxionem radicis involvente, & quidem per seriem. Sed universali Seriei [the 'most general' quadrature series—effectively a Taylor expansion—delivered by him in *Acta Eruditorum* (November 1694): 437–40; see page 19 above] palmam non præripiet: Est enim ille modus *Newtonianus* admodum operosus, & fere idem cum Tuo, quem vero longe succinctiorem & ad praxin aptiorem reddidisti in *Act.* 1693. pag. 178....'. In this last remark Bernoulli was less than fair to Newton, since Leibniz in his 'Supplementum Geometriæ practicæ sese ad Problemata transcendentia extendens, ope novæ Methodi generalissimæ per series infinitas' (*Acta Eruditorum* (April 1693): 178–80 [= (ed.) C. I. Gerhardt, *Leibnizens Mathematische Schriften*, **5** (Halle, 1858): 285–8]) had merely—without, of course, any foreknowledge of the solution of the latter anagram in his 1676 *epistola posterior*—reproduced Newton's 'alternative' method for deriving the 'root' z of the given first-order differential equation $f(z, y, dz/dy) = 0$ by arbitrarily positing the series expansion '$z = by + cy^2 + ey^3 + fy^4$ etc.' and then evaluating the quantities $b, c, e, f, \ldots$ by substituting the series in the given equation and equating the coefficients of the ensuing powers of y identically to zero: in this way, for instance, he was able (*ibid.*: 180) to resolve the de Beaune equation $a\,dz/dy = y - z$ as the infinite series

$$z = \tfrac{1}{2}y^2/a - \tfrac{1}{6}y^3/a^2 + \tfrac{1}{24}y^4/a^3 - \tfrac{1}{120}y^5/a^4 \ldots,$$

that is, $y - x$ where (as he proceeded equivalently to state) $y = -a \log(1 - x/a)$. (Compare III: 84, note (109).) Bernoulli's preceding assertion that Newton's superscript fluxional dots are exactly equivalent to Leibniz' prefixed *d*[*ifferentiæ*] and his ancillary suggestion that Newton might well have 'fabricated' his method after having had a sight of Leibniz' 'calculus' were to be repeated by him in a letter to Leibniz on 7 June 1713 (N.S.) whose kernel the latter shortly after had printed as an anonymous *Charta volans* which he circulated among the leading scientists of Western Europe: 'Videtur *N*[*ewtonus*] occasionem nactus serierum opus multum promovisse per Extractiones Radicum, ... & quidem in iis excolendis ut verisimile est ab initio omne suum studium posuit, nec credo tunc temporis vel somniavit adhuc de Calculo suo fluxionum et fluentium, vel de reductione ejus ad generales operationes Analyticas ad instar Algorithmi.... Ejusque meæ coniecturæ validissimum *indicium* est, quod de literis x vel y punctatis, uno, duobus, tribus, &c punctis superpositis, quas pro dx, ddx, d^3x; dy, ddy, &c nunc adhibet, in omnibus istis Epistolis [*Commercii Epistolici Collinsiani*...] nec volam, nec vestigium invenias' (compare *Leibnitii et Bernoullii Commercium Epistolicum*, **2**, 1745: 309). In his subsequent 'Annotatio' upon this passage of the fly-sheet Newton swiftly demolished Bernoulli's presumptions: 'Methodus fluxionum utique non consistit in forma symbolorum. Pro fluxionibus ipsarum x, y, z, *Newtonus* quandoque ponit easdem literas punctis notatas $\dot{x}$, $\dot{y}$, $\dot{z}$; ... quandoque literas alias ut p, q, r; quandoque lineas exponentes.... Pro Fluxionibus D. *Leibnitius* nulla habet symbola. Is Momentorum sive Differentiarum symbola dx, dy, dz primo cœpit adhibere Anno 1677; *Newtonus* momenta denotabat per rectangula sub fluxionibus & momento o cum *Analysin* suam scriberet, anno scilicet 1669 aut antea.... *Newtonus* nunquam mutavit literam o in literam $\dot{x}$ punctatam uno puncto, sed litera illa o usus est... & adhuc utitur in eodem sensu ac sub initio, idque maximo cum fructu. Est enim o symbolum unicum quo *Newtonus* utitur pro quantitate infinite parva: At symbolum $\dot{x}$ quantitatem finitam designat' (*Commercium Epistolicum D. Johannis Collins, et Aliorum, de Analysi Promota* (London, $_2$1722): 247, 249; compare Newton's review of the first (1712) edition of the *Commercium* in *Philosophical Transactions*, **29**, No. 342 [for January/February 1714/15]: 173–224, especially 204–5). Leibniz in his 1712 review of Newton's 'De Analysi' (reproduced in II: 259–62) had, as we have seen, remarked more percipiently that the true identity between Newtonian fluxions and his own *differentiæ* sets $o\dot{x} \equiv dx$ (and likewise $o^2\ddot{x} \equiv d^2x$, $o^3\dddot{x} \equiv d^3x, \ldots$) where o is an infinitesimal 'moment' of the base variable of 'time'—a point further underlined by Newton in his own counter-observations upon the review (see *ibid.*: 266 ff.).

PART 2

RESEARCHES IN PURE GEOMETRY AND THE QUADRATURE OF CURVES

(*c.* 1693)

INTRODUCTION

In this present part we reproduce the bulky autograph record, all but unknown to modern mathematical historians,[1] of Newton's elaborate studies in geometry during a long period in the early 1690's when his attention was given over both to minutely exploring and interpreting the riches of classical geometrical analysis, and also to systematizing and developing his own scattered earlier investigations and instrumental constructions of curves, and he came to present the essence of his insights in both areas in a rapidly evolving succession of drafts, none ever fully finished, of a comprehensive multi-partite treatise of 'Geometria'. With one notable exception—his detailed justification, in the final version of his 'Liber primus', of the possibility of generating each of the numerous species of cubic curve as the perspective 'shadow' of one or other of the five divergent parabolas[2]—these uneven writings are memorable less, perhaps, for their individual flashes of innovatory brilliance and sustained displays of masterly technical expertise than for the finely textured picture which they collectively project of Newton's mature attitude *vis-à-vis* the 'ancient' geometers and the 'new' breed of Cartesian mathematicians of whom, by upbringing and working practice, he himself was one for all that he might defer to the superior logical authority of the former.[3] To him, as to many others of his period, it was incredible that the Greeks could have attained their wealth of geometrical discoveries solely by way of the elegantly composed arguments with which they rigorously demonstrated their truth, and he had long been ready to subscribe to the widely popular contemporary view[4] that these classical synthetic proofs were founded on a prior analysis, traces of which were afterwards erased, for whose efficient pursual there had been developed special exploratory techniques which differed little from modern algebraic

(1) A glimpse of the richness and penetration of Newton's manuscript researches into the 1,1 linear correspondence of points is given in our 'Patterns of Mathematical Thought in the later Seventeenth Century' (*Archive for History of Exact Sciences*, **1**, 1961: 179–388): 278–81. In his more impressionistic account of 'Newton and Greek Geometry' (*Harvard Library Bulletin*, **13**, 1959: 354–61) G. L. Huxley is severely handicapped in his assessments by his lack of knowledge of Newton's autograph papers on the topic, but in compensation makes a number of pertinent quotations from David Gregory's contemporary records of his conversations with Newton on aspects of classical geometry (from the selection published by W. G. Hiscock in *David Gregory, Isaac Newton and Their Circle: Extracts from David Gregory's Memoranda, 1677–1708*, Oxford, 1937).

(2) See Newton's elaborate 'Exemplum in lineis tertij ordinis' of his distinction of curves into species 'per casus projectionum' in 3, §1 below.

(3) 'Et si authoritas novorum Geometrarum contra nos facit, tamen major est authoritas Veterum', he wrote at one point; see 2, §2, Appendix 2.1 below.

(4) See IV: 222, note (20) and 277, note (9).

counterparts 'unless in the trivial manner of their expression'.(5) But the earnestness with which he now seeks out the vestiges of the ancients' geometrical analysis, painstakingly endeavouring therefrom accurately to restore and faithfully to reconstruct their 'long-neglected *genre* of discovery' as he termed it,(6) has no precedent in his earlier attempts to penetrate the secret of the Greek mathematical heuristic.

Apart from Euclid's introductory *Elements* and a variety of ingenious particular modes of solution to more difficult special problems—notably those requiring the finding of two mean proportionals between a given pair of magnitudes, the trisecting of a given angle, and the squaring of a circle's area(7)—the theoretical geometry of the classical world as it had survived into public awareness in Europe in the last decade of the seventeenth century effectively, with the exception of Archimedes' severely technical treatises which in their themes, style of proof and methods of reduction(8) stood sharply apart from the other works in the Greek mathematical corpus, comprise but a short further tract by Euclid on *Data*(9) and seven (of the original eight) books of Apollonius on *Conics*(10) together with Pappus' bulky *Collection*(11) of extracts,

(5) 'Nec differre videtur Algebra nostra ab illorum [*sc.* Veterum] Analysi nisi in modulo expressionis', he stated in a draft preface (2, §1 following; compare *ibid.*: note (8)) to his projected treatise on 'Geometria'.

(6) 'genus inventionis diu neglectum' in his original Latin phrase at the end of the first 'Liber' of the revised text of Newton's 'Geometria' reproduced in 2, §3 below.

(7) The various techniques of solution contrived for these three special problems are gathered and collated by T. L. Heath in his *History of Greek Mathematics*, **1** (Oxford, 1921): 218–70; the related Greek texts are set out with facing English translations by I. Thomas in his *Greek Mathematical Works*, **1** (London, 1939): 256–363.

(8) These are examined at length by E. J. Dijksterhuis in his *Archimedes* (Copenhagen, 1956); see especially his Chapter III: 49–141: 'The Elements of the Work of Archimedes'. Though he was himself primarily interested only in his predecessor's neusis constructions of such problems as that of angle-trisection (compare V: 464, 471) and the main result of his *On the Sphere and Cylinder*—this probably derived through Barrow (see VI: 214–15, note (8))—Newton in his present papers is very ready to continue his earlier estimation of Archimedes as (see V: 428) 'Mathematicorum princeps'; compare 2, §1: note (54) following.

(9) Isaac Barrow's Latin epitome of this work (Cambridge, $_1$1657) was conveniently available for Newton's use; see 1, §1: note (10).

(10) The latter three of these, extant only in Arabic versions, had recently been given a Latin *editio princeps* by Borelli at Florence in 1661; see 2, §1: note (30) below. The eighth book was soon to be restored by Edmond Halley, to Pappus' outline, in his magnificent folio edition *Apollonii Pergæi Conicorum Libri Octo* (Oxford, 1710). The content of the *Conica* is well summarized by T. L. Heath in his English paraphrase *Apollonius of Perga*: *Treatise on Conic Sections edited in modern notation*... (Cambridge, 1896).

(11) Federigo Commandino's Latin translation of the surviving Greek text of this voluminous *Syntaxis* was brought out posthumously as *Pappi Alexandrini Mathematicæ Collectiones a Federico Commandino Vrbinate in Latinum conuersæ, & Commentarijs illustratæ* (Pesaro, $_1$1588); we have already remarked (IV: 218, note (6)) that Newton himself very likely drew his deep knowledge of Pappus' work from Manolessi's recent 'improved' edition of it (Bologna, 1660).

appropriately refurbished and partially augmented by him, from the cream of his predecessors' writings on geometry. Ever since the appearance in 1588 of the Latin *editio princeps* of Pappus' late Hellenistic compilation most serious mathematicians had felt the need to look at the lengthy preamble to its seventh book where he surveyed the ancient *τόπος ἀναλυόμενος*, characterizing the common analytical basis of this obscure field of the Greek 'resolving locus'—*locus resolutus* as Commandino rendered it(12)—and briefly digesting the content not only of Euclid's *Data* and Apollonius' *Conics* but of a number of their other works, since lost,(13) on the topic. Many—Harriot, Descartes, Huygens, Newton himself at an earlier period(14)—were concerned in the main, of course, merely to cannibalize the dead carcase of Greek geometrical analysis there displayed for its several stimulating locus-problems and the skilful techniques of solution which Pappus went on to append in the propositions of his following book. Yet others—Ghetaldi, Anderson, Snell, Schooten, Fermat(15)—had been lured by the more positive ideal of reviving this all but forgotten branch of classical geometry into attempting individual restorations, following the guide-lines set down by

(12) *Mathematicæ Collectiones* ($_2$1660): 240; see IV: 223–4, note (22) and compare 2, §1: note (27) below.

(13) Unbeknown to Newton at this time, however, Edward Bernard had just identified among the Selden manuscripts in the Bodleian an Arabic version of Apollonius' books *On cutting off a ratio*—'Codex...non solum pessima manu exaratus...punctisque diacriticis plerumque destitutus, quibus in scripturâ Arabicâ literæ quamplurimæ solent distingui; sed & gravioribus adhuc vitiis laborabat, quod verba sæpiuscule & integras nonnunquam periodos omiserit, & Diagrammatum lineas literis male signatas & distinctas habuerit'. After Bernard's death Halley determined to complete the task, then yet scarcely begun, of turning it into Latin and at length, after a Herculean effort which required him not least to teach himself enough Arabic to master the mathematical essence of the manuscript, was able to give to the world not only *Apollonii Pergæi De Sectione Rationis Libri Duo Ex Arabico MSto. Latine Versi* (Oxford, 1706) but his parallel restoration of Apollonius' two irretrievably lost books *De Sectione Spatii*, further premissing to this 'Opus Analyseos Geometricæ studiosis apprime Utile' a first printing of the Greek text of Pappus' preface to the seventh book of his *Collection*, along with a new rendering of its text into Latin.

(14) Thomas Harriot's numerous extracts and borrowings 'ex Pappo' will be found in British Museum. Add. 6782–9: *passim*. Descartes, of course, founded his whole method of analytical geometry on his generalized solution, in Books 1 and 2 of his *Geometrie*, of the problem of showing the identity of the Greek 4-line locus with the conic; see IV: 218–20. Christiaan Huygens gave variant, simplified demonstration of a couple of Pappus' neusis constructions in the *Illustrium Quorundam Problematum Constructiones* which he appended to his treatise *De Circuli Magnitudine Inventa* (Leyden, 1654); see his *Œuvres complètes*, **12** (The Hague, 1910): 183–215, especially 191/3, 211. In the paper on 'Veterum Loca solida restituta' which he drafted in the late 1670's or thereabouts Newton, as we have seen (IV: 276), explicitly referred his geometrical construction of an ellipse through five given points to Pappus, viz. *Mathematical Collection* VIII, 13.

(15) For the titles of their individual restorations see 2, §1: notes (31) and (32) below, where we give reason for thinking that Newton himself was familiar only with Frans van Schooten's *Apollonii Pergæi Loca Plana Restituta* (issued by the latter as the third book of his *Exercitationum Mathematicarum Libri Quinque*, Leyden, 1657).

Pappus, of the vanished treatises of the τόπος ἀναλυόμενος: notably Apollonius' books on the 'triple section' (*De sectione determinata/rationis/spatii*) and his works on 'plane [rectilinear and circular] loci',(16) but most difficult of all, not least because the surviving Greek manuscript of Pappus' *Collection* was (and is) at this point so lacunary, the three large books of Euclid on the elusive topic of *Porisms*.(17) Though these restitutions do not lack technical expertise—considerable ingenuity was shown in particular in solving by a variety of ways Apollonius' problems of constructing a circle to be tangent (in any of the eight positions generally possible) to three given circles, and inclining a given line-segment in an angle of a given rhombus so as to pass (extended if necessary) through the opposite corner(18)—they failed, singly or collectively, to shed light on their common methodological rôle as components of a 'discovering locus', and the central treatise by Euclid on porisms remained, despite a lone brave try by Fermat,(19) essentially unexplained. By the last decade of the seventeenth

(16) Two books *De locis planis* in general, that is, along with two each on the particular problems (see note (18) below) *De tactionibus* and *De inclinationibus*. Pappus' summaries of the content of Apollonius' lost works will be found in his *Mathematicæ Collectiones*, $_2$1660: 242–4, 247–9.

(17) The most lavish attempt to date to recover this key treatise of the ancient *locus resolutus* is that set out at length by Michel Chasles in his *Les Trois Livres de Porismes d'Euclide, rétablis pour la première fois, d'après la Notice et les Lemmes de Pappus*... (Paris, 1860). Though he was, of course, ignorant of Newton's preceding endeavours to restore the kernel of Euclid's work, Chasles in his introduction gave what remains otherwise a valuable 'Exposé historique' (*ibid.*: 1–10) of previous 'essais de divination de la doctrine des Porismes', singling out the important contributions made in the mid-eighteenth century by Robert Simson (compare 2, §1: note (55) below) but criticizing even him for failing to explore 'quelle avait pu être la pensée qui a dirigé [Euclide] dans sa conception originale; il n'a pas fait voir non plus comment cette doctrine des Porismes devait être si utile, nécessaire même pour la résolution des problèmes, comme le dit Pappus, et quels rapports elle pouvait avoir avec les propositions et les méthodes modernes' (page 8). Had he known Newton's discussion of the 'Inventio Porismatum' (1, §2 below), he would undoubtedly have warmly praised its stress on the central rôle of the 1,1 point-correspondence in reconstructing the basic design of Euclid's treatise and explaining the wide use of its propositions in geometrical analysis.

(18) On Newton's own resolutions of these classical problems 'Of contacts' and 'On inclinations' see 1, §2: note (32) and 2, §3: notes (62) and (65). His knowledge of preceding restorations of Apollonius' works *Tactionum* and *De inclinationibus* was not of the best: he was certainly ignorant of the particular solution of the former problem with which Alexander Anderson concluded his *Variorum Problematum Practice* (pages 49–54) in 1612 (compare 1, §1: note (32) below), and seemingly unaware of Apollonius' own general construction of the latter as it is repeated by Pappus in his *Mathematical Collection* VII, 70 (on which see 2, §3: note (70) following).

(19) In his *Porismatum Euclidæorum Renovata Doctrina, & sub formâ Isagoges recentioribus Geometris exhibita*, published only posthumously in his *Varia Opera Mathematica* (Toulouse, 1697): 116–19. Fermat succeeded in reconstructing two of the Euclidean porisms reducible to 1,1 point-correspondences in straight lines and circles; see 1, §2: note (38) below. Newton, however, does not seem to have studied Fermat's tract; see 2, §1: note (33).

century the initial enthusiasm for reconstructing the lost works of ancient analysis to Pappus' ground plan had ebbed away, to be replaced by a *fin-de-siècle* pessimism that any such adequate restitution of the edifice of Greek higher geometry would ever be possible on such crumbling foundations.(20) It was to be Newton's signal achievement in his present reappraisal of the 'Analysis Geometrica' of the ancients not only to pose the eminently plausible conjecture that the primary function of the τόπος ἀναλυόμενος was to provide a *penus analytica*,(21) a well-stocked 'analytical larder' of basic locus-constructions and curve properties of practical service to the working geometer in his various resolutions of problems, but to draft a scheme for restoring Euclid's *Porisms* under Pappus' subheads which brought out the point that this was itself a miniature tool-chest of the indeterminate 'plane' loci of points defining or defined by standard metrical or incidence conditions imposed on simple configurations of straight lines and circles.

In his opening discussion(22) of the manner of analytically resolving any problem Newton outlines, much as Pappus before him, how the basic aim is, by postulating as given what is sought, to proceed therefrom to derive things already known; after which one searches, by suitably reversing the sequence, to compose a direct argument passing from only what is known to attain or construct what is required. This is of course, as Newton is well aware, only to suggest a general strategy: the structure of the particular problem will determine how complicated the composition of the analysis must be, and only in the simplest instances will a straight inversion of its progression suffice. In other cases, where one or more of its unknowns are permitted to vary, a problem will come to have an interlocking pattern of possibilities of solution. In a quantitative context this will be embodied in arithmetical ties between the variables reducible

(20) Even in 1706, in preface to his *editio princeps* of Apollonius' *De sectione rationis* (see note (13) above) where he gave first printing of the Greek text—and a fresh Latin rendering—of Pappus' preamble to the seventh book of his *Collection*, a similarly bleak mood could grip Edmond Halley when he came to take account of the heavily fragmented text of his predecessor's digest of Euclid's work: 'Hactenus *Porismatum* descriptio nec mihi intellecta nec lectori profutura...; unde quid sibi velit *Pappus* haud mihi datum est conjicere' he wrote in a despairing footnote (page xxxvii; see 2, §1: note (46), where we give more detailed quotation of Halley's *cri de cœur*). Little more than a decade later, fortunately, Robert Simson was to deal a death blow to such an extreme of pessimism by giving a full and careful restoration of Pappus' main Euclidean porism (see 2, §1: note (55)).

(21) This familiar modern notion of a classical Greek 'Treasury of Analysis' is, we have already remarked (IV: 224, note (22)), yet one more nugget among the many mined by David Gregory during his visit to Cambridge in May 1694; for among a miscellany of other 'Adnotata Math: ex Neutono' he afterwards recorded that 'Locus resolutus veterum est penus Analytica. Exemplum, Euclidis *data* et *porismata* si extarent. Pappus recitat prop: Apollonij *de sectione lineæ determinata*' (ULE. Gregory C43, more fully printed—with some errors of transcription—in *The Correspondence of Isaac Newton*, **3**, 1961: 331).

(22) 1, §1 below.

to a nexus of algebraic equations usually determinate in their solution;[23] in a geometrical configuration the pattern will manifest itself as a comparable number of intersecting point-loci whose cuts determine those points which solve the problem—whence, in Newton's words, 'the determinate sections of the ancients (which are the equations of more modern mathematicians) ought to be ready to hand, as also the determinations of loci; for geometry in its entirety is nothing else than the finding of points by the intersections of loci'.[24] In particular, every problem constructible by the intersection of a straight line and a circle—'Euclidean', as we now say—is, in the ancient scheme of the τόπος ἀναλυόμενος, resolvable by an analysis from 'givens' in the style prescribed by Euclid himself in his *Data*.[25] More generally, where a straight line cuts any other species of conic, because the latter is always representable as a classical 4-line locus, the pairs of points of intersection will be definable as an analogous quadratic *sectio determinata* whose four base points are the meets of the given straight line with the four lines with respect to which the conic locus is defined—and this was no doubt the reason, as Newton here perceptively suggests (and afterwards similarly remarked to David Gregory and Halley), why Apollonius devoted a whole treatise to enumerating the variety of possible singular cases of this species of determinate section.[26]

Of the other works by Apollonius which Pappus summarized, his books *De*

(23) Indeterminate solutions will exhibit themselves geometrically as (one species of) porismatic propositions. Thus, if in an Apollonian *sectio spatii* a straight line is drawn through the fixed Cartesian point (a, b) in a plane to meet the axes in points $(-\lambda, 0)$ and $(0, \mu)$, the product of whose distances $c-\lambda$ and $d-\mu$ from the base points $(c, 0)$ and $(0, d)$ is some given magnitude k, then simultaneously $(c-\lambda)\ (d-\mu) = k$ and $\lambda\mu = b\lambda+a\mu$, so that the line is fixed in one of two possible positions; except when $a = c$, $b = d$ and $k = ab$, in which case $(a-\lambda)\ (b-\mu) \equiv ab$ for all λ and μ, and therefore all straight lines through (a, b) cut out this particular *sectio spatii* (measured from the feet $(a, 0)$ and $(0, b)$ of the projections of (a, b) on the axes). The result is Newton's Porism 1 in 1, §2 below.

(24) 'Determinatæ item sectiones veterum quæ sunt æquationes recentiorum in promptu esse debent. Ut et Locorum determinationes. Nam Geometria tota nihil aliud est quam inventio punctorum per intersectiones Locorum' (see 1, §1.3 following).

(25) *Data*, 58/59 (= *Elements*, VI, 28/29); compare 1, §1: note (14) below.

(26) We will be disappointed, however, that Newton did not at this point explore any of the properties of the 'involution'—to use the now standard *terminus technicus* introduced by Girard Desargues in 1639 in his little-read *Brouillon Proiect* (see R. Taton, *L'Œuvre Mathématique de G. Desargues* (Paris, 1951): 110–11)—formed by the totality of pairs of points cut out by such a quadratic *sectio determinata* in a given straight line. Had he looked at the matter in this way, it would have been immediately evident to him that, because each member of such a pair x, y uniquely determines the other 'by simple geometry' (as he would put it), the correspondence $x \leftrightarrow y$ is 1,1 and therefore $(ABxX) = (ABy\ \infty)$, where the points A, B are invariants of the involution and X corresponds to the point at infinity in the line. The great majority of Pappus' lemmas (*Collection* VII, 22–56) on Apollonius' *De sectione determinata* are straightforward 'plane' applications of this basic cross-ratio equality; compare Michel Chasles, *Aperçu Historique sur l'Origine et le Développement des Méthodes en Géométrie* (Paris, $_2$1875): 39–41.

sectione rationis and *De sectione spatii*(27) afforded Newton a yet profounder insight into the ancient 'analytical locus'. At a superficial glance the complementary problems which they treated were straightforward: through a given point to draw a straight line to intersect two given ones such that the segments cut off in each of the latter, terminated at given points within them, shall be in a given direct or inverse proportion. Regarded as such they are, as Apollonius went on to show, readily reducible to a quadratic *sectio determinata* of one or other of the given lines. But Newton, in pondering upon why such a dual problem should have been admitted into the τόπος ἀναλυόμενος, was led to ignore the restriction that the transversal shall pass through a given point and to pay heed only to the 'simple' indeterminate relationship mutually connecting the end-points of all pairs of segments defined in this way to have a constant ratio or product. As he well knew, a straight line drawn tangent to a given parabola or central conic will meet a pair of its tangents in points whose distances from given points in them(28) have just such a fixed ratio or product, or more generally stand in some mutually determinable 1,1 relationship to each other. The converse of this porismatic theorem which is our modern anharmonic definition of the conic as an envelope—namely, that a straight line drawn through pairs of points in 1,1 correspondence to each other in given lines ever touches a unique conic tangent to these, other than when their junction is a self-corresponding point and the transversal passes always through a fixed point(29)—he now assumes without further ado not merely in restoring the hidden basis of Apollonius' problems of 'cutting off a ratio' and 'cutting off a space' but in beginning a broader exploration of the ancient art of 'inventing' such generalized porisms.

Although in the preliminary paper on this topic(30) where he first began tentatively to investigate the fruitful possibilities of the 1,1 correspondence defining (or, conversely, defined by) 'simple' proportions of lines he went on to

(27) We would remind that in the early 1690's Halley was still more than ten years away from publishing his Latin *editio princeps* of the former from its still surviving Arabic version, together with his complementary restoration of the latter; see note (13) preceding.

(28) In the case of a *sectio rationis* $Ax/By = AX/BY$, constant, or $(A \infty xX) = (B \infty yY)$, where the defining condition is that the line at infinity shall be one of the transversals xy, the given points may be chosen to be any corresponding points A, B; the analogous theorem on the parabola's tangent is Apollonius, *Conics* III, 41. (Compare 1, §2: note (26) below.) For a *sectio spatii* $Ax \times By = AX \times BY$, constant, or $(A \infty xX) = (\infty ByY)$, it is manifestly necessary that each of the given points A, B should correspond to infinity in the other's line; the related theorem on the tangent to a (central) conic was first derived by Newton himself as Corollary 2 to Lemma XXV of the first book of his *Principia* (see VI: 280, and compare 1, §2: note (24) following).

(29) At infinity in the case of a *sectio rationis*, whence the transversals become a set of parallels. In the complementary instance of a porismatic *sectio spatii* any finite point will serve as the fixed pole provided that the given base points are correctly aligned with it.

(30) The 'Inventio Porismatum' reproduced as 1, §2 below.

hazard a fairly disastrous extension[31] to the analogous case of the 2,1 and 2,2 correspondences cut out by tangents to a conic in lines, one or both of which no longer touch it, he there laid a firm basis for treating all 'other relationships of unknown quantities' which mutually determine these *per simplicem geometriam*; 'likewise the species of figures and the... loci of points, and also lengths, angles and points which either regard the determination of loci or otherwise contribute to the resolution of a problem'.[32] That broad specification of the range of the classical porism embraced, of course, not merely the numerous lemmas set by Pappus in his *Mathematical Collection* to clarify and augment Euclid's lost work on the topic, but a wealth of other theorems and problems in the field of the geometry of 'plane' (straight-line/circle) and 'solid' (conic) loci. And of this rich potential Newton took full advantage when in his succeeding full-fledged drafts for his proposed treatise on 'Geometria'[33] he came to attempt a narrower restoration of the main content of Euclid's *Porisms* to Pappus' cryptic, lacunary outline, both accurately reconstructing the latter's generalization of Euclid's component theorems on the 'Desargues locus'[34] and giving due prominence to his proof of the constancy of cross-ratio on a line-pencil,[35] and passing thereafter to adduce (in the particular instance of the circle to which he followed Euclid in confining his attention) appropriate restylings[36] of the analogous homographic definition of a conic point-locus as the intersection of two such equi-cross pencils which he had himself derived some dozen years before as an unforeseen bonus of his researches into the Greek 3/4-line locus.[37] Had he given consideration to the general conic case in these amplified restitutions of Euclid's lost book, and systematically revised these essays into the higher analysis of the ancients, we would here have had a treatise on elementary projective geometry fit to place besides Desargues' *Brouillon Proiect*,[38] La Hire's *Nouvelle Methode en Geometrie*[39] and the other founding treatises of our modern theory.

(31) See 1, § 2: notes (27) and (29) following.

(32) See pages 230–2 below.

(33) Those reproduced in 2, §§1/2.

(34) See 2, §1: note (55) below.

(35) *Collection* VII, 129/142; compare 2, §1: note (59).

(36) See 2, §2: notes (109), (112) and (114).

(37) In the terminal propositions of his 'Solutio Problematis Veterum de Loco solido', that is; see IV: 306–14, and compare 321, note (90).

(38) *Brouillon Proiect d'Vne Atteinte aux Euenemens des Rencontres du Cone auec un Plan* (Paris, 1639). The text of the sole surviving copy (Bibliothèque Nationale, Paris. Vp 1209) of this formative work—which had its historical impact through a manuscript transcript made in 1679 by La Hire (afterwards printed by N.-G. Poudra in his *Œuvres de Desargues réunies et analysées*, 1 (Paris, 1864): 103–230)—is now conveniently reproduced by Taton in his *L'Œuvre Mathématique de G. Desargues* (see note (26) above): 99–180.

(39) *Nouvelle Methode en Geometrie pour les Sections des Superficies coniques et cylindriques*; *Qui ont pour bases des cercles, ou des Paraboles, des Elipses, & des Hyperboles* (Paris, 1673). The work also includes at its end (pages 75–94) an account by La Hire of his revolutionary method of 'Les

Along with these penetrating insights into the structure of the 1,1 point-correspondence, Newton incorporated into his middle drafts(40) of the 'Geometria' many of the more familiar metrical propositions of Euclidean geometry, sketching—as tradition demanded(41)—their (undocumented) practical origin in problems of land-measurement, and probing the 'perfection' of mechanical description which it demands in its postulates and prescribes in its constructions of problems by compounding the motions of lines upon lines; then proceeding(42) in more strictly Euclidean manner to develop the whole of plane geometry from a set of 'rectilinear principles', embracing the generalized Apollonian notion of a *sectio determinata* as a 'given', which he applies by way of a group of prescriptive rules to give elegant resolution of a number of standard problems; and finally gives an extended discussion(43) of the possibilities of instrumental constructions of conics, cissoids, ovals and other higher curves. Much of the illustrative material here, however, was his own, often given novel twists or extensions which we need not here specify. And, as he took up more and more points of interest in his own earlier researches into the description of curves and the construction of problems in geometry, his initial zeal to recover the foundations of ancient analysis and its lost techniques and results waned. One draft(44) is abandoned in the middle of an exegesis of the generalized Greek $(2n-1)/2n$-line locus, never again to be taken up by Newton; another(45) peters thinly out as he begins one more exposition of the ancients' distinction between analysis and synthesis. Then suddenly the 'Geometria' threw off all its classical ties to become a thorough-going contemporary treatise on curves and their properties. As he slips into this accustomed rôle of modern innovator once more one can sense Newton's relief at relinquishing the frock of classical expositor and interpreter, relentlessly stressing—and overstressing(46)—the superiority of the

Planiconiques' which, we have already remarked (VI: 271, note (70)), has close—if probably fortuitous—structural similarities with the equivalent plane projective 'transmutation of figures' outlined by Newton in Lemma XXII of the first book of his *Principia*. See also René Taton, 'La première œuvre géométrique de Philippe de La Hire' (*Revue d'Histoire des Sciences*, **5**, 1952: 93–111), and also his 'La préhistoire de la "géométrie moderne"' (*ibid.*, **6**, 1953: 197–224, especially 204–5).

(40) 2, §§2/3 below.

(41) See 2, §2: note (6).

(42) In Book 1 of the draft 'Geometria' reproduced in 2, §3 below.

(43) In the ensuing Book 2 (pages 382–400 following).

(44) 2, §2 below.

(45) 2, §3 following.

(46) See especially 2, §1: notes (14)–(17), where we point to several gross over-estimates on Newton's part of the relative inferiority of Cartesian algebraic geometry. It was of course his own excuse for so belligerently insisting upon the elegance and simplicity of classical geometrical composition that in his present treatise 'ad eos jam loquor qui Algebram aut non intelligunt aut propter ejus complexas atq; adeo inelegantes operationes minus colunt' (as he puts it in

ancient polished syntheses over the cruder 'algebraic' resolutions of more recent Cartesian geometry.

Over this final version of the 'Geometria' we may speed very lightly. In its first book[47] Newton builds carefully up from fundamentals to compound algebraic equations from their linear factors, and then quickly drafts out a general theory of curves, ordering their 'grade' by the number of their possible individual intersections with a given straight line,[48] as he had long done, and thereafter, on the analogy of the three projectively distinct species of conic, further subdividing each grade into its component, perspectively 'coordinate' classes. Though its details have never hitherto been published, his long ensuing exemplification of the cubic case of the last where he shows in fine—with a few oversights, but here taking account of the six species omitted from his subsequent algebraic enumeration[49]—how all known curves of third degree can be generated as some readily definable optical 'shadow' of some one of the five species of divergent parabola, has come universally to be recognized as perhaps the outstanding product of Newton's geometrical genius. In the later pages of this opening book he enunciates several of the general properties of algebraic curves which he had discovered in his earlier years, and then, after a brief excursion into the elements of geometrical fluxions (of straight lines, arc-lengths

---

2, §3; see page 372 below).

Let us take this opportunity to retract our earlier reluctance (see I: 17–18, note (11)) to believe in the past existence of a copy 'of Descartes' Geometry...marked in many places with his [Newton's] own hand, *Error, Error, non est Geom.*'(as David Brewster recorded in his *Memoirs...of Sir Isaac Newton*, **1** (Edinburgh, 1855): 22, note 1). The book in question, a copy of the second Latin edition of the *Geometria* (two volumes [Amsterdam, 1659/1661] bound as one; see I: 20, note (6)) signed in a front tuck with the autograph 'Isaac Newton', surfaced in August 1971 in the Wren Library at Trinity College, Cambridge (where it is now shelf-marked NQ. 16. 203) during its recent extensive restoration, having been lost to public knowledge there since the early nineteenth century, shortly after it came to Trinity in a bequest of books from its late master Robert Smith. As we formerly conjectured, Brewster himself could never have seen the copy 'among the family papers' (where it never was)—or indeed, most probably, even in the Wren Library, where it had already been taken off the public shelves as a 'worthless' duplicate long before Brewster ever visited Trinity—but must have taken his information solely from the Conduitt memorandum which we quoted in I: 17, note (11). We will give a full account of Newton's *marginalia* to the book (which are closely related to—and contemporary with—his listing of 'Errores Cartesij Geometriæ' on IV: 336–44) in our final volume, but may here briefly notice that each of his eight charges of 'error' and three (separate) comments of 'non Geom' are to be treated individually in their own right, and nowhere are they connected to form any imputation that Descartes errs because he adopts an algebraic method of geometrical analysis. Let that ubiquitous phantom be laid to rest.

(47) 3, §1 below.

(48) That is, by the algebraic degree of their defining Cartesian equation in some standard coordinate system. We have already remarked (see IV: 341, note (23)) that Newton—and not, as it is tempting to assume, Descartes—was the first systematically thus to grade algebraic curves.

(49) See 3, §1: notes (54) and (65) below.

and angles), concludes with a long discussion of defining and computing the 'instantaneous' tangential direction of motion in curves specified in a variety of coordinate systems: generalized Cartesian, bipolar and more special types. The second book(50) of the 'Geometria' in this final draft has little to do with geometry at all, being a minimally altered and extended reworking of the first ten propositions of the tractate 'De quadratura Curvarum' which David Gregory had stimulated him to compile in the early winter of 1691/2,(51) supplemented by two tables of integrals which he took over virtually unchanged from his 1671 treatise on fluxions(52) and a concluding eleventh proposition reducing a simple class of multiple integrals to first-order form which he had similarly adumbrated in an intended addition to his earlier 'De quadratura'.(53) The relentlessly Cartesian nature of these expounded techniques of algebraic quadrature is revealing of how far he had come from his previous aim of renewing the ancient geometrical analysis.

With the exception of the last, alien tract on quadrature—quickly to revert to its original title(54) and, further recast and augmented, to appear in printed form some ten years afterwards(55)—none of these rough, unfinished preliminary studies were ever revised by Newton for the press, nor did he find time or inclination in his later years to pull them together into some more publishable composite. After his death they passed swiftly into the limbo of his private papers,(56) where they have remained (through several changes of geographical location) to the present day, only rarely touched by the quick-sifting hands of modern cataloguers(57) and scholars with other interests to the forefront.

(50) Reproduced in 3, §2 following.

(51) See pages 24–70 above. As we below observe, Gregory was already able to record Newton's intention 'to publish two parts of Geometry, the first the Geometry of the Ancients, the second of Quadratures' among other 'Varia Astronomica et Philosophica' which he set down after conversing with him 'Londini...28 Decr 1691' (Gregory C85 = Royal Society MS 247: 70, printed in *The Correspondence of Isaac Newton*, **3**: 191).

(52) Of these two tables (see III: 236–40/240–54) the prior 'Catalogus Curvarum aliquot ad rectilineas figuras relatarum' was abridged in the transplant to its first four *Ordines* only, the ensuing 'Theoremata magis generalia' (*ibid.*: 238–40) being absorbed into the preceding propositions of the revised 'De quadratura Curvarum', but the latter 'Catalogus Curvarum aliquot ad Conicas Sectiones relatarum' was taken over essentially intact; compare 3, §2: notes (88) and (90).

(53) See pages 164–9 above; and compare 3, §2: note (92) following.

(54) Namely, 'Tractatus De Quadratura Curvarum', in which form it was copied by David Gregory in May 1694; see 3, §2: notes (2) and (3) below.

(55) As the second of the 'Two Treatises of the Species and Magnitude of Curvilinear Figures' which he appended to his *Opticks* in 1704. We shall return to specify the fine detail of these further reshapings and additions in our final volume.

(56) Initially in the possession of the Portsmouth family and latterly, since 1888, on deposit in Cambridge University Library; see I: xviii–xxxiii.

(57) Even in the official printed *Catalogue of the Portsmouth Collection of Books and Papers*

Which is not to say that Newton in his later life was ever greatly reluctant to share with others the fruits of his restoration of Greek higher geometry or communicate his deep insight into the range and power of classical analysis. In particular, when David Gregory paid his now celebrated visit to Newton in Cambridge early in May 1694 and was shown a mass of the latter's scientific papers, he not only made notes of particular problems(58) in the present corpus of geometrical writings but, primed by Newton, took away with him some firm impression of the purpose and content of the ancient 'treasury of analysis' and how it might be revived from Pappus' preface to the seventh book of his *Collection*.(59) On his return to Oxford, in a lengthy retrospective of his Cambridge stay which he took considerable care to polish in its detail, he later recollected (among a multitude of other matters) that, to the revised edition of his *Principia* which Newton was then busy preparing,

'the author will adjoin two treatises: One about the geometry of the ancients, in which the errors of the moderns about the mind [*sc.* intention] of the ancients are disclosed in regard to the problem [of the 3/4-line locus] proposed by Pappus and recited by Descartes, 'plane' and 'solid' loci, analysis and porisms, and to the refinement of the ancients' method of discovering and recording; and where it will be shown that our specious algebra is fit enough to find out, but entirely unfit to consign to writing and commit to posterity. In this treatise the authentic purpose of the ancients is explained; the book of Euclid's *Data*, his book of *Porisms*, the lost books of Apollonius and the further ones of the ancients will be explained from what is said by Pappus and other commentators about them but has thus far been understood by no one....

'The second tract will contain his method of Quadratures, and this will augment and promote that topic wonderfully. It proceeds by way of series which break off, and compares non-quadrable areas with simplest ones. Thereto he appends tables for the different forms and grades of [quadrable] figures up to the tenth order, in whose use

written by or belonging to Sir Isaac Newton...Drawn up by the Syndicate appointed [*in*] *1872* (Cambridge, 1888; see I: xxx–xxxiii on its background), in the section (now ULC. Add. 3963) of 'Papers relating to Geometry' J. C. Adams—no geometer himself—could spare only a few brief lines to record that they were 'Geometri[æ] Liber 1. A fragment', 'Analysis Geometrica', 'Regula Datorum', a 'Fragment relating to Curves', 'Part of a Treatise on Geometry (in Latin)' and —most fulsome of all merely because it parrots an earlier editorial phrase of Samuel Horsley's on 26 October 1777 (see 1, §1: note (3) following)—'Fragments relating to the writings of the Ancients in general, but especially to the Porisms of Euclid, and the Loci of Apollonius' (*Catalogue:* 3–4; equivalent titles in Adams' hand are preserved on various slips and folders with the papers themselves).

(58) See especially 1, §1, Appendix 1: notes (1), (10) and (20).

(59) Among other 'Adnotata Math: ex Neutono. 1694. Maio' he recorded in particular that 'Libris...Apollonij [de *Conicis*] edendis Oxonij prefandum est de Veterum Geometria (in quem finem consulenda est prefatio ad...7^mum^ Lib: Pappi [quæ] extat Græce ac Arabice)' (Gregory C43 = Royal Society MS 247: 68^v^, printed in *The Correspondence of Isaac Newton*, **3**: 327).

anyone could substitute numbers in place of the letters to measure a given space; and likewise other tables up to the eleventh class, where non-quadrable spaces are compared with conic ones....'[60]

The tables of integrals which Gregory saw were in fact, as his sequel makes clear, those which Newton had introduced more than twenty years earlier into his general 1671 tract on fluxions and infinite series, rather than their lightly refashioned equivalent which he had much more recently appended to his revised text 'De quadratura Curvarum'.[61] Let us not insist too harshly on a confusion which Newton himself may well have provoked and failed to dispel. There can be no doubting that Gregory's précis of the first of Newton's 'treatises' exhibits a reasonably accurate awareness of the range and penetration of the latter's understanding of the ancient mode of geometrical analysis, and also a degree of familiarity with Newton's plan to restore the essence of the lost works of Euclid and Apollonius on the 'resolving locus' in accord with Pappus' digest of their content. And when, half a dozen years afterwards, Gregory himself began to think seriously of preparing a standard Greek/Latin *editio princeps* of

(60) In Gregory's original Latin: 'Huic Editioni subjunget Auctor duos tractatus[,] alterum de Veterum Geometria, ubi Neotericorum errores deteguntur circa Veterum mentē de problemate a Pappo proposito, et a Cartesio recitato, de locis planis et solidis, de Analysi et Porismatibus, de concinna Veterum methodo investigandi et scribendi, ubi ostendetur Algebram nostram speciosam esse ad inveniendum aptam satis at literis posterisqꝫ consignandum prorsus ineptam. In hoc tractatu genuinum Veterum institutum explicatur, *Datorum* Euclideorum liber, *Porismatum* liber, Apollonij libri deperditi reliquique Veterum explicabuntur ex illis quæ a Pappo alijsqꝫ de eis dicuntur quæqꝫ a nemine hactenus intellecta....Secundus Tractatus Methodum suam Quadraturarum continebit quæ rem istam mire augebit et promovebit. Hæc per series abrumpentes procedit et spatia non quadrabilia cum simplicissimis confert. Huic subjungit tabulas pro diversis formis et gradibus figurarum usque ad ordinem decimum. Ad usum poterit quispiam numeros loco literarum substituere, ut datum spatium mensuretur. Item alias tabulas ad usque classem undecimum ubi spatia non quadrabilia cum coni sectionibus comparantur' (ULE. Gregory C 42; an alternative English translation is provided by H. W. Turnbull in *The Correspondence of Isaac Newton*, **3**: 385–6). On the latter tables, Gregory accurately added, 'innituntur quædam abstrusiora in *Philosophia* sua hactenus edita ut [Lib.1] Corol: 3. prop. XLI et Corol: 2 prop. XCI'; see VI: 354, note (214) and 226, note (32) respectively for our more detailed justification of the point.

(61) We have already stated (note (52) above) that the prior table of exact integrals which Newton took over in his 'Geometriæ Liber secundus' (3, §2 following) was extended only as far as its fourth order. The sequel to the above passage, indeed, makes it clear that Gregory here thoroughly confused Newton's recent 'Secundus Tractatus' with his 1671 fluxional tract, for he there went on: 'In Auctoris M.S. ubi hæc consignantur videre est Methodus Infinitarum Serierum, de Tangentibus, de Maximis et Minimis, de Curvatura deqꝫ Rectificatione Curvarum. Atque hæc omnia tam in figuris transcendentibus ut Quadratrice Nicomedis cujus meminit Dinostratus, Trochoide &c (quarum nempe infinitæ sunt conjugatæ, et proinde Æquationes ad illas quæ ad omnes conjugatas itidem pertine[n]t trans[c]endentes erunt) quam determinatis. Hæc schedula [!] continet etiam abstrusissima alia problemata particularia de loco focorum imaginum &c'. (See III: 32–328, *passim*, but on the last 'problemata particularia dioptrica' especially 120, note (186) and 150, note (266).)

the totality of Euclid's surviving works, we will not be surprised that it was to Newton that he turned for advice and enlightenment, reminding himself in a private memorandum in May 1701 to go to Cambridge 'To talk about Euclid especially the *data*; & if I should write a preface, & what instances put in it'.(62) In not narrowly following Newton's response in the published preface (and badly stumbling at a crucial point in his own editorial remarks upon Euclid's *Data*)(63) Gregory subsequently had only himself to blame.

In other ways, too, one could document Newton's continuing willingness right into extreme old age to impart respect for the elegance and power and strength of Greek geometry. When de Moivre wrote to Johann Bernoulli in March 1705 regarding Edmond Halley's newly completed Latin translation of the recently unearthed Arabic version of Apollonius' *De sectione rationis* that Newton had 'no slightest doubt' of its authenticity, he also conveyed the latter's further dictum that 'this treatise is yet the more valuable in that it is a geometrical analysis'.(64) When, again, someone asked him to report upon the quality of Hugo de Omerique's *Analysis Geometrica*,(65) it was entirely typical that Newton should write back that 'I...find it a judicious & valuable piece answering to y^{e} Title. For therein is laid a foundation for restoring the Analysis of the Ancients wch is more simple[,] more ingenious & more fit for a Geometer then the Algebra of the Moderns'.(66) That predilection for the neatness,

(62) Gregory A68$_2$ = Royal Society MS 247: 74, first printed by S. P. Rigaud 'From the original [then] in the possession of D[uncan] F[arquharson] Gregory' in his *Historical Essay on the First Publication of Sir Isaac Newton's Principia* (Oxford, 1838): $_2$79.

(63) See 1, §1, Appendix 1: note (10) below.

(64) 'M. Halley...a traduit tout nouvellement en latin un manuscrit arabe qui appartenoit à la Bibliotheque d'Oxford: c'est un traité d'Apollonius *de sectione rationis*...tout à fait conforme à ce qu'en dit Pappus, et M. Newton ne doute nullement que le traité arabe ne soit une traduction fidele de l'original grec.... Ce traité est d'autant plus estimable, que c'est une analyse géométrique' (see K. Wollenschläger, 'Der mathematische Briefwechsel zwischen Johann I Bernoulli und Abraham de Moivre', *Verhandlungen der Naturforschenden Gesellschaft in Basel*, **43**, 1933: 151–317, especially 198).

(65) *Analysis Geometrica, sive Nova, et Vera Methodus Resolvendi tam Problemata Geometrica, quam Arithmeticas Quæstiones* (Cadiz, 1698). For all its 440 pages Hugo's book is hardly a profound venture into geometrical analysis, since he restricts its scope entirely to 'plane' (straight-line/circle) problems. His opening statement (pages 1–5) of the nature and purpose of analysis—defined to be the 'Assumptio quæsiti tanquam concessi, per ea, quæ deinceps consequuntur ad aliquod concessum procedens'—is clear and precise enough, however, and his choice of illustrative problems in the sequel is laudably eclectic, giving evidence of his wide reading not only in Euclid and Pappus but in their modern successors Viète, Ghetaldi, Grégoire de Saint-Vincent and Frans van Schooten.

(66) Bodleian. New College MS 361.2: 19^{r}; compare J. Pelseneer, 'Une opinion inédite de Newton sur l' "Analyse des Anciens" à propos de l' *Analysis geometrica* de Hugo de Omerique', *Isis* **14**, 1930: 155–65. To his unknown correspondent—whom from the following we might readily guess to be Henry Pemberton (if indeed the letter was ever sent off at all)—Newton

elegance and relative simplicity of classical geometrical resolution was as strong as ever at the end of his life when 'I have often heard him', wrote Henry Pemberton a little while after his death in a celebrated passage,

'censure the handling geometrical subjects by algebraic calculations; [and] praise *Slusius*, *Barrow* and *Huygens* for not being influenced by the false taste, which then began to prevail. He used to commend the laudable attempt of *Hugo de Omerique* to restore the ancient analysis, and very much esteemed Apollonius' book *De sectione rationis* for giving us a clearer notion of that analysis than we had before.... Sir *Isaac Newton* has several times particularly recommended to me *Huygens*'s stile and manner. He thought him the most elegant of any mathematical writer of modern times, and the most just imitator of the ancients. Of their taste, and form of demonstration Sir *Isaac* always professed himself a great admirer: I have heard him even censure himself for not following them yet more closely than he did....'(67)

For our part we may be grateful that he did not! If he had immersed himself in the styles and techniques of ancient geometry in his youth, he would surely never have gone on to make the magnificent advances in Cartesian geometry and calculus which were the highlight of his first *annus mirabilis* of mathematical invention. On that cautionary note let us now turn to the texts wherein, in his late maturity, Newton came most closely to understand the spirit and identify the essence of the antique analysis which he so abundantly and unstintingly admired.

---

added in justification of this warm testimonial that 'it leads him more easily & readily to the composition of Problems[,] & the Composition wch it leads him to is usually more simple & elegant then that wch is forct from Algebra'.

(67) *A View of Sir Isaac Newton's Philosophy* (London, 1728): Preface: [B2^v/a1^r]. Pemberton concluded in the same vein: '...and speak with regret of his mistake at the beginning of his mathematical studies, in applying himself to the works of *Des Cartes* and other algebraic writers before he had considered the elements of *Euclide* with that attention, which so excellent a writer deserves'.

1

RESTORING THE GREEKS' GEOMETRICAL METHOD[1]

§1. THE GEOMETRICAL ANALYSIS OF THE 'ANCIENTS'[2]

From the original drafts in the University Library, Cambridge

[1][3] ANALYSIS GEOMETRICA.

[4]Arithmetica rationes abstractas tractat & Problemata in genera distinguit secundum dimensiones æquationum quæ nihil aliud sunt quam rationes compositæ. Geometria circa figuras versatur et Problemata distinguit in genera secundum harum descriptiones magis aut minus faciles et utiles[5] adeo ut Problemata omnia quæ per rectas et circulos construi pos[s]int sint ejusdem generis geometrici quamvis æquationum dimensionibus differant, et quæ per circulos et alias sectiones conicas construantur sint diversi generis quamvis figuræ dimensionibus æquationum suarum congruant.[6]

Figuræ in Geometriam primitus receptæ vel plana sunt rectilinea et circularia, vel solida superficiebus planis, sphæricis, cylindricis & conicis terminata. Et

---

(1) We have earlier (see IV: 222 ff.) noticed how, after reading Commandino's Latin translation of Pappus' *Mathematical Collection*—probably (see IV: 218, note (6)) in the lightly emended edition which Manolessi published at Bologna in 1660 as *Pappi Alexandrini Mathematicæ Collectiones a Federico Commandino...in Latinam conversæ, & Commentarijs illustratæ. In hac nostra editione ab innumeris, quibus scatebant mendis...diligenter vindicatæ*—Newton came increasingly from the late 1670's to broaden his geometrical interest from the Cartesian algebraic analysis which he had mastered in his youth to study and then to imitate the more wordy resolutions and formal compositions of the Greek masters which he there found summarized. After a crowded intervening decade during which problems of pure mathematics claimed little of his attention, in the ensuing quiet of the early 1690's he returned to explore the content and method of classical geometry, first to gather and elaborate the techniques of analysis which (following Pappus) he conceived to underlie the elegant but prolix and stiffly uninviting treatises of Euclid, Apollonius and their coevals, and thereafter to plan—but never, unfortunately, to complete—a general work on 'Geometria' in which these explorations and restorations of the poristic expertise of the 'ancients' were to have pride of place. The many, considerably variant surviving drafts of this project are reproduced in Sections 2 and 3 below. Here we give Newton's preliminary glosses on Euclid's *Data* and, by way of Pappus' summary of them in his seventh book, Apollonius' *De sectione determinata*, *De sectione rationis* and *De sectione spatii* which together with Euclid's *Porismata* were the key works in this classical field of the τόπος ἀναλυόμενος (on which see IV: 223, note (22)).

Translation

[1][3] GEOMETRICAL ANALYSIS

[4]Arithmetic treats of abstract ratios and distinguishes problems into kinds according to the dimensions of their equations, which are nothing other than compounded ratios. Geometry is concerned with figures and distinguishes problems into kinds according to the greater or less ease and usefulness of their descriptions,[5] so that all problems constructible by means of straight lines and circles shall be of the same geometrical kind though they differ in the dimensions of their equations, while those which are to be constructed by circles and other conic sections shall be of a diverse kind though these figures are in accord in the dimensions of their equations.[6]

The figures originally accepted into geometry are either rectilinear and circular plane ones, or solids terminated in plane, spherical, cylindrical and

(2) Three preliminary pieces in which Newton outlines the basic principles of classical Greek geometrical analysis, and then lists general rules for its successful application to resolve problems of geometrical construction (graded according to the degree of difficulty in describing the auxiliary lines and curves employed therein) and of identifying curvilinear point-loci. Though their present juxtaposition is not dictated by the internal evidence of their texts, it will be clear that in much of their content they reinforce and subtly complement each other. As with the other geometrical autographs which make their first appearance in print in following pages, Newton's handwriting in the manuscript agrees broadly with its having been penned in or shortly after 1691; this date is more narrowly confirmed by the fact that the second of these drafts is (see note (43) below) written around a draft calculation for Proposition XIII of the preceding 'De quadratura Curvarum', while a firm *terminus post quem non* for their composition is afforded by the external testimony of David Gregory, who during his visit to Cambridge in early May 1694 saw *inter alia* the related fragment on the 'Analysis veterum' here set in sequel as Appendix 1.

(3) Add. 3963.14: 154^r–155^v, a single much cancelled and overwritten folded sheet. When Samuel Horsley afterwards on 'Octr. 26. 1777' (compare I: xxv–xxvii) came to examine this and the related folios—notably the 'Inventio Porismatum' reproduced in §2 below—which were gathered together in its original paper wrapper, he described them thereon accurately enough as 'Fragments relating to the Writings of the Antients in General but especially to the Porisms of Euclid & the loci of Apollonius. Very curious & fit to be published'. In Appendix 1 following we print a more jejune piece (Add. 3963.4: 27^r–28^r) bearing the same title; in this, after briefly surveying the ways in which the propositions of Euclid's *Data* are 'of the most frequent use in resolving problems of quadratic grade', Newton restricts himself to illustrating their particular application by reworking three geometrical problems from his earlier 'Arithmetica Universalis'.

(4) The two introductory paragraphs which follow are a late insertion on f. 155^v of the manuscript. The better to dovetail them in, Newton minimally adjusted the first phrase of his original opening; see note (10) below.

(5) Initially just 'secundum harum descriptiones faciliores' (according to the comparative ease of their descriptions).

(6) Newton returns to a theme on which he had waxed strong in his 'Arithmetica Universalis' a decade earlier; see V: 422–6, 470–6.

primum [Veteres] solverunt Problemata per figuras[7] in plano ope postulatorum[,] deinde problema deliacum[8] et similia per intersectiones circulorum et superficierum solidarum quæ in Geometriam receperant ac tandem lineas magis compositas quam sunt sectiones conicæ adhibere cœperunt[,] unde orta est distinctio Problematum in plana solida et linearia.[9] At quali Analysi adjuti eousq; pervenerunt satis constat.

Inter libros veterum ad hanc Analysin[10] spectantes Pappus recenset Euclidis *data*. Tradit hic Euclides principia colligendi ex assumptis sive notis sive ignotis tanquam ex datis quantitates alias et ex his alias donec ad quæsitum vel sic progrediendo vel regrediendo perveniatur. Et hujus analyseos extant varia exempla apud Pappum.[11] Continet vero hic Euclidis Liber methodum perfectam resolvendi Problemata omnia plana. Nam si[12] incidatur in assumptarum datum rectangulum[13] hoc est in æquationem quadraticam[,] dantur æquationis latera[13] per *Datorum* Prop 58, 59, 60, 84, 85.[14] Et hujus rei extant exempla quædam apud Pappum Lib Prop .[15] Quinetiam si incidatur in æquationem biquadraticam imparibus terminis carentem dantur latera per *Datorum* Prop 86 & 87[14] si modo Propositio 87 sic restituatur

Si duæ rectæ x, y datum spatium ab comprehendant in angulo dato, quadratum autem unius xx quadrato alterius yy minus sit dato cc quam in ratione d ad e, earum utraq; data erit.[16]

(7) This replaces a too restrictive 'rectas' (straight lines).

(8) That of doubling a cube but 'keeping its [cubic] shape' in Eutocius' words (*On the Sphere and Cylinder*, II, 1). T. L. Heath in his *History of Greek Mathematics*, 1 (Oxford, 1921): 244–70 gives an informative account of the prehistory of the problem and of the many solutions which it received in classical times: according to Eutocius it was Hippocrates of Chios who first reduced the duplication of a cube to be a particular case of the more general problem of finding two mean proportionals between two given lines (see *ibid.*: 245), and so paved the way to Archytas' ingenious geometrical construction of it as the common point of a cone, a cylinder and a torus (*ibid.*: 246–9; see also B. L. van der Waerden, *Science Awakening*, Groningen, 1954: 150–2).

(9) Compare V: 423, note (617) and 469, note (687).

(10) Originally—when (see note (4)) this was the opening sentence of the piece—Newton specified 'ad Analysin Geometrarum' (the analysis of geometers). Among the books listed by him in the prolegomenon to his seventh book as 'pertaining to the resolved locus' Pappus in fact gave pride of place to the *Data*: 'Dictorum librorum...ordo talis est. Euclidis datorum liber unus. Apollonius...de proportionis sectione libri duo, ...de spacij sectione duo, ...tactionum duo. Euclidis porismatum tres. Apollonij...inclinationum duo, ...planorum locorum duo, conicorum octo. Aristæi...locorum solidorum quinque. Euclidis...locorum ad superficiem duo. Eratosthenis de medietatibus duo' (*Mathematicæ Collectiones* (note (1)): 241). Pappus' following summary (*ibid.*: 241–2) of the content of this 'Primus liber, qui est datorum' agrees only broadly in the number and order of its propositions with the extant Greek texts; further variations occur in the many printed editions of its several medieval and modern Latin and English translations, notably that by Robert Simson (*The Book of Euclid's Data...Corrected*, Glasgow, 1762). The edition here used by Newton—one standard in Britain over the preceding century, despite the appearance of David Gregory's would-be *editio princeps* of *Euclidis quæ supersunt omnia* (Oxford, 1703)—is, we need scarcely say, Barrow's *Euclidis Data succinctè Demonstrata; Una cum Emendationibus quibusdam* (Cambridge, $_1$1657).

cone surfaces. At first, the [ancients] solved problems by means of figures(7) in a plane with the aid of postulates, thereafter accomplished the Delian(8) and similar problems through the intersections of circles and solid surfaces which they had received into geometry, and at length began to employ lines more complex than conics are: whence arose the distinction of problems into 'plane', 'solid' and 'linear'.(9) The mode of analysis, however, which assisted them to reach so far is well enough settled.

Among the books of the ancients regarding this analysis(10) Pappus reviews Euclid's *Data*. Here Euclid presents principles for gathering from ones assumed as given—whether known or no—further quantities, and from these others, till by advancing in this way or by regressing there is attained what is sought. And of this analysis various examples are extant in Pappus.(11) This book of Euclid's contains in fact a complete method for resolving all plane problems. For should you(12) meet with a given 'rectangle'(13) of assumed things, that is, with a quadratic equation, the equation's 'sides'(13) are given by *Data*, Propositions 58, 59, 60, 84 and 85.(14) And of this there are instances in Pappus, Book Prop. .(15) To be sure, if you fall in with a quartic equation lacking its odd terms, its 'sides' are given by *Data*, Prop. 86 and 87(14) provided Proposition 87 be thus restored:

If two straight lines x, y should at a given angle comprehend a given space ab, the square x^2 of one being less by a given magnitude c^2 than the square y^2 of the other in the ratio d to e, each of them will be given.(16)

---

(11) For instance, in Propositions 72, 85, 87, 105, 108, 117, 164, 204, 218 and 236–8 of Book VII of the *Collection*.

(12) Newton has deleted 'ex assumptis' (departing from the assumed) here.

(13) That is, 'product' and 'roots' respectively in the arithmetical equivalent.

(14) Understand in Barrow's edition of Euclid's work (see note (10) above); his Propositions 84–87 are, for instance, Propositions LXXXVII–XC in Simson's 1762 ordering. Propositions 58/59 essentially repeat *Elements* VI, 28/29 in geometrically solving the general quadratic equation $ax^2 \mp bx = c^2$ by completing the square $a(\frac{1}{2}b/a \mp x)^2 = \frac{1}{4}b^2/a + c^2$; Proposition 60 is the related particular case where the quadratic is $a(x^2 - b^2) = \pm c^2$. Propositions 84/85 pose in geometrical form the problem of solving the pair of simultaneous equations $xy = ab$ and $x \mp y = c$ (when there ensues the quadratic $x^2 - cx = \pm ab$). Propositions 86/87—on amending the latter's enunciation to be as Newton now 'restores' it—present the analogous problem of geometrically resolving the simultaneous pair $xy = ab$ and $x^2 \mp c^2 = (d/e)\, y^2$ (whence there results $x^4 \mp c^2x^2 = a^2b^2d/e$, quadratic in x^2).

(15) 'Lib 7 Prop. 72, 87' ([*Collection*] VII, 72, 87) will illustrate Newton's point as well as any other.

(16) This improves upon Barrow's otherwise identical enunciation by replacing the latter's defining phrase '...quadratum autem unius quadrato alterius majus sit dato; ...' (the square of one being greater by a given magnitude than the square of the other) which asserts but a particular case ($d = e$ in the terms of note (14) above) of the preceding Proposition 86. This crude separate presentation of a particular instance in sequel (rather than as a preliminary) to the proof of its generalization is, set alongside the omission from the surviving text of the *Data* of the balancing complementary theorem, 'clear' evidence in Newton's eyes that Proposition 87 as it there stands is vitiated and needs to be appropriately amended (as now he

i.e. si fuerit $xy = ab$ et $xx+[c]c = \frac{d}{e}yy$, dantur x et y. Ubi nota quod Veteres loco æquationis unius biquadraticæ ob simplicitatem expressionis ponebant duas quadraticas. Et similiter pro una æquatione quadratica ponere solebant binas, uti $xy = bc$ et $\frac{d}{e}x \pm a = y$.(17) Et pro una æquatione biquadratic[a] cum terminis imparibus ponebant hujusmodi binas æquationes quadraticas vel quod perinde est binas datas rationes inter rectangula hujus formæ $y+\frac{f}{e}x-a$ in $y-\frac{g}{e}x+b$;(18) et hujusmodi rectangula ut commode enunciari possent sic designabant. A dato puncto A in data positione ducatur recta $AC = x$ et in dato angulo erigatur $CD = y$. In alio dato angulo ducatur $A[E]$ occurrens CD productæ in E sic ut fuerit CE ad AC ut f ad e hoc est $CE = \frac{fx}{e}$. A DE aufer $EG = a$ et ipsi AE agatur parallela FG. Et erit $D[G] = y+\frac{f}{e}x-a$. Eadem ratione ex-

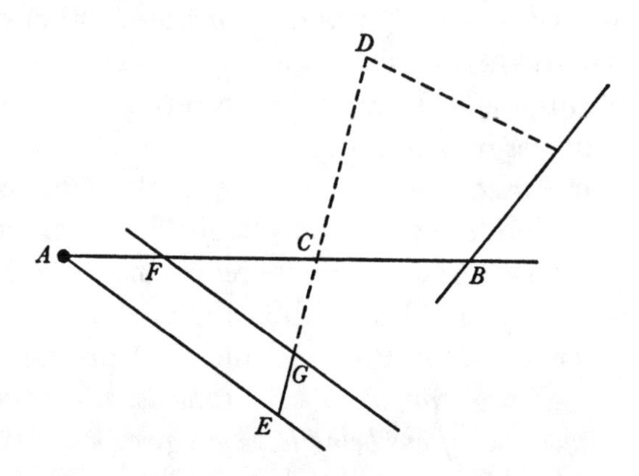

ponendo longitudinem $y-\frac{g}{e}x+b$ ut et alias duas ejusmodi longitudines per lineas totidem a puncto D ad alias totidem positione datas rectas in datis angulis ductas: æquationem quadraticam enunciabant dicendo quod rectangulum sub DG et alia ducta esset ad rectangulum sub reliquis duabus ductis in data ratione.(19) Et hinc Euclides aggressus est celeberrimum illud Problema determinandi

corrects its enunciation) to fill the gap—'Propositio illa...liquido corrupta fuit cum jam nihil differat a præcedente nisi quod sit minus generalis' he argues at the corresponding place in his variant discussion of 'Analysis Geometrica' in Appendix 1 (see page 222 below). Seventy years later, knowing Newton's prior opinion in this matter only in the somewhat garbled form in which (see Appendix 1: note (10)) David Gregory subsequently published it, without mention of its author, in preface to his *Euclidis quæ supersunt omnia* in 1703, Robert Simson was not prepared so blithely to abandon a Euclidean theorem whose 'Enuntiation and Demonstration are quite entire and right' in favour of a replacement which 'tho'...true...has no connexion in the least with the Greek text'; instead, he evolved a happy compromise in which the original Propositions 86 and 87 (determining the solution of $xy = ab$ where $x^2-c^2 = (d/e)\,y^2$ and $x^2-c^2 = y^2$ respectively) are retained but in inverted sequence—these become Propositions 89 and 87 in his revised listing—while 'the deficiency [regarding the complementary cases in which $x^2+c^2 = y^2$ and, more generally, $x^2+c^2 = (d/e)\,y^2$] which the Doctor [Gregory] justly observes to be in this part of Euclid's *Data*, and which no doubt is owing to the carelessness and ignorance of the Greek Editors' is now 'supplied...by adding the two Propositions which are the 88. and 90. in this Edition' (*The Book of Euclid's Data* (note (10)): 461). It will be evident from this that Simson's interpolated Proposition 90 is identical with Newton's present 'restored' Proposition 87.

that is, were there $xy = ab$ and $x^2+c^2 = (d/e)\,y^2$, then x and y are given. Here notice that the ancients in place of a single quartic equation used, because of the simplicity of representation, to put two quadratics. And similarly in preference to a single quadratic equation it was their habit to put two, such as $xy = bc$ and $(d/e)\,x \pm a = y$.[17] And in place of a single quartic equation possessed of odd terms they were wont to put a pair of quadratic equations of this sort, or equivalently a pair of given ratios between rectangles of this form

$$(y+(f/e)\,x-a)\times(y-(g/e)\,x+b);^{(18)}$$

and so that they might conveniently express rectangles of this sort they represented them in this manner: From the given point A in given position draw the straight line $AC = x$ and at a given angle to it erect $CD = y$; at another given angle draw AE meeting CD produced in E such that CE be to AC as f to e, that is, $CE = (f/e)\,x$, and from DE take away $EG = a$, drawing FG parallel to AE; there will then be $GD = y+(f/e)\,x-a$. In the same say, by denoting the length $y-(g/e)\,x+b$ together with two other lengths of the same type by an equal number of lines drawn from the point D at given angles to a corresponding number of other straight lines given in position, they expressed a quadratic equation by stating that the rectangle contained by GD and another drawn line should be to the rectangle beneath two remaining drawn lines in a given ratio.[19] And on this basis Euclid attacked the much discussed problem of determining

(17) In particular, of course, when $d = \mp e$ (and so there is $x \mp y = a$) this simultaneous pair reduces to be the algebraic equivalent of *Data*, 84/85 (compare note (14) above).

(18) Newton draws this out of his fertile imagination! There is no evidence that the ancient Greek mathematicians ever attempted the solution of an irreducible quartic equation in this or any other manner. Even by 1691 there were, as far as we can discover, no more recent precedents on which he might have based the remark. While Newton himself, following in Descartes' footsteps, had in his 'Problems for construing æquations' (II: 450-516; see especially 490–8) shown how to construct a general quartic equation *modis infinitis* by the intersections of a circle and a conic, he had never attempted to define these as 4-line loci defined by any equivalent to the Cartesian equation

$$\text{`}x+\frac{f}{e}y+a \text{ in } x+\frac{g}{e}y-b = [\lambda \text{ in}]\; x+\frac{h}{e}y-c \text{ in } x+\frac{k}{e}y-d\text{'}$$

which he first wrote down in the manuscript at this point.

(19) An equally anachronistic observation on Newton's part. We need scarcely remark that this mode of reduction of the oblique distance of a point from a given line to a linear function of its Cartesian coordinates (with respect to some fixed origin, abscissa and ordination angle) was pioneered by Descartes himself in the first book of his *Geometrie* [= *Discours de la Methode* (Leyden, 1637): 297–413]: 310–12, where he adds (*ibid.*: 312) that 'ainsi vous voyés, qu'en tel nombre de lignes données par position qu'on puisse auoir, toutes les lignes tirées dessus du point...a angles donnés...se peuuent tousiours exprimer chascune par trois termes; dont l'vn est composé de la quantité inconnue y, multipliée, ou diuisée par quelque autre connue; & l'autre de la quantité inconnue x, aussy multipliée ou diuisée par quelque autre connuë, & le troisiesme d'vne quantité toute connuë'.

locum puncti a quo si quatuor rectæ ad alias totidem positione datas in datis angulis ducantur rectangulum sub duabus ductarū sit ad rectangulum sub alijs duabus in data ratione.[20] Nam si punctum illud determinari possit habebuntur longitudines *DC AC* hoc est æquationum latera x et y. Sed cum punctum illud quatenus ab unica tantum æquatione[21] determinatur non unicum sit sed ubivis in linea quadā curva sumi possit & quatenus ab altera æquatione determinatur ubivis sumi possit in alia quadam curva et quatenus ab utraq; determinatur incidit in communem curvarum intersectionem[,] ideo Euclides[22] determinationem harum curvarum aggressus est ut quemadmodum Problemata plana[23] per descriptiones rectarum et circulorū sic solida[23] per descriptiones harum curvarum solvi possent. Et hæc fuit origo tractatuum de locis solidis. Nam Curvæ in quibus punctum illud locatur sunt sectiones Conicæ.

Post Euclidem Apollonius Problema aggressus est[24] & cum solutio ejus in duobus consisteret[,] primo ut invenirentur puncta quædam per quæ Conica sectio[25] transiret, deinde ut quemad[mod]um Geometræ rectam per puncta duo ducere postularant et circulum per tria puncta describere docuerant, sic Conicam sectionem per puncta inventa determinarent: docuit Apollonius inventionem punctorum in libro quem scripsit *de sectione determinata*. Nam Conicam Sectionem per data quinq; puncta describere innotuerat ut ex Pappo colligo qui lib. [VIII] Prop. [13][26] docuit ex eorum mente descriptionem Ellipseos per tot puncta.[27] Sunto lineæ quatuor positione datæ *AB*, *BC*, *CD*, *DA* a recta quavis quinta *IM* in punctis *I*, *K*, *L*, *M*, respective intersectis et in recta quinta inveniendum sit punctum *P* a quo si aliæ quatuor rectæ *PE*, *PF PG PH* ad quatuor priores in datis angulis ducantur sit rectangulum sub duabus ductis $PE \times PF$

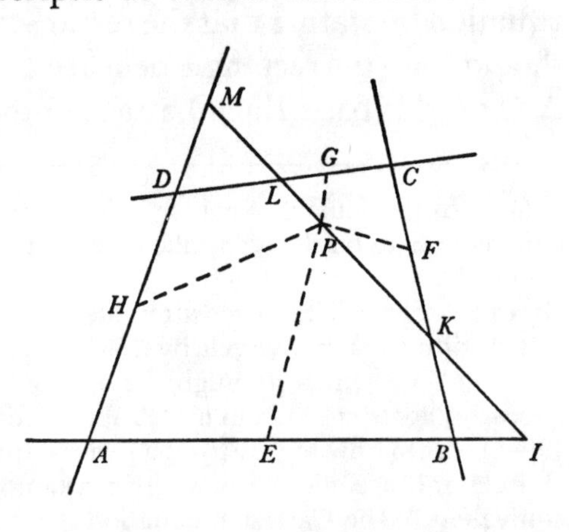

(20) By Apollonius' later testimony (see IV: 220) it was in one of his lost books of *Conics* that Euclid broached this problem of the 4-line locus. Newton's following conjecture that it was Euclid's identification of this locus as a general conic which led him and his successors systematically to explore the properties of the 'locus solidus' is purely his reasonable guess in the absence of any historical evidence which could either confirm or deny it. Otto Neugebauer has recently suggested a far different source for this topic of enduring interest in Greek mathematics ('The Astronomical Origin of the Theory of Conic Sections', *Proceedings of the American Philosophical Society*, **92**, 1948: 136–8).

(21) In the two variables x and y, that is.

(22) Newton has deleted 'et qui eum secuti sunt' (and those who followed him) in sequel.

(23) Originally, the contrast was more loosely between 'simpliciora' (simpler) and 'magis composi[ta]' (more compound). On the classical distinction between 'plane' (straight-line/circle) and 'solid' (conic) problems see V: 423, note (617).

the locus of a point such that, if four straight lines be drawn from it at given angles to an equal number of others given in position, the rectangle contained by two of the lines drawn shall be to the rectangle beneath the other two in a given ratio.[20] For if that point be determinable, there will be had the lengths *DC* and *AC*, that is, the 'sides' x and y of the equations. However, that point, inasmuch as it is determined by merely a single equation,[21] is not unique but may be taken anywhere on a certain curve, and insofar as it is determined by a second equation may be taken anywhere on a certain other curve, and so, inasmuch as it is determined by both, falls at the common intersection of the curves; Euclid[22] consequently attacked the determination of these curves in order that, just as 'plane'[23] problems are resolved by describing straight lines and circles, so 'solid'[23] ones might be solved by their description. And this was the origin of his tracts on solid loci; for the curves in which that point is located are conics.

After Euclid, Apollonius attacked the problem.[24] Since his solution consisted in two parts, first to find certain points through which the conic[25] should pass, and then—just as geometers had postulated the drawing of a straight line through two points and explained the description of a circle through three—likewise to determine the conic through the points so found, Apollonius explained the discovery of these points in the book which he wrote *On determinate section*. For it was known (to both) how to describe a conic through five given points, as I collect from Pappus who in Book VIII, Proposition 13[26] explained in their spirit the description of an ellipse through that number of points. [27]Let there be four lines *AB*, *BC*, *CD*, *DA* given in position and intersected by any fifth straight line *IM* in the points *I*, *K*, *L*, *M* respectively, and let it be required to find a point *P* in the fifth line such that, if four other straight lines *PE*, *PF*, *PG*, *PH* be drawn from it at given angles to the four first, the rectangle $PE \times PF$

(24) Apollonius' solution of the reduced 3-line locus (where two of the base-lines coincide) is in essence contained in his *Conics*, III, 54–6; his discussion of the general 4-line locus has not survived, but its demonstration readily follows as an extension of the particular 3-line case (see IV: 221, note (16) and 276, note (7)).

(25) Newton first wrote 'Curva' (curve). In thus specifying its type he slides over the question of how Euclid and his successors might have reached this conclusion. Perhaps he takes it to have been self-evident from its definition that the 4-line locus is quadratic in form?

(26) Compare IV: 276–8. When Newton acquainted him of this fact on a visit to London during June 1706, David Gregory justly observed in his private diary that 'Pappus's Maner of describing a Conick-Section throug[h] five points Prop. 13. Lib. 8 is very pretty; yet not near so simple as Mr Newtons Prop. XXII. Pag. 79 & 80 *Princip*: [$_1$1687 = VI: 256–8], that proceeding by the proportion of rectangles, this by the proportion of Lines' (Christ Church, Oxford. MS 346: 172, printed in a modernized transcription by W. G. Hiscock in his *David Gregory, Isaac Newton and their Circle* (Oxford, 1937): 35).

(27) A first version of the next two sentences is entered on ff. 155^r/154^v of the manuscript, but the few (mostly verbal) variants occurring in it are, we adjudge, too minor to warrant being spelled out here.

ad rectangulum sub alijs duabus ductis $PG \times PH$ in data ratione. Et ob datas rationes PE ad PI, PF ad PK, PG ad PL & PH ad PM dabitur ratio rectanguli $PI \times PK$ ad rectangulum $PL \times PM$ adeoq; datur punctum P per librum Apollonij *de sectione determinata.* Nam Liber ille accurate enumerat casus omnes puncti P in qua recta IM secanda est eosq; per omnes ejus Propositiones determinat, adeo ut vel inde sciri potest quod Apollonius eundem scripsit ut Problema Euclideum de quatuor lineis absolveret.(28) Veteres igitur Problema illud ad finem perduxerunt quamvis Pappus vir non admodum perspicax mentem eorū minime(29) perciperet.

Spectant etiam libri Apollonij *de sectione rationis* et *sectione spatij*(30) ad constructionem solidorum Problematum. In regula BC circa polum B revolvente et

(28) In his account of this lost work of Apollonius, Pappus had made no such connection, merely specifying that it elaborated the separate cases of the problem 'Datam infinitam rectam lineam vno puncto secare, ita vt interiectarum linearum ad data ipsius puncta, vel vnius quadratum, vel rectangulum duabus contentum, vel rectangulum duabus contentum datam proportionem habeat, vel ad rectangulum contentum vna ipsarum interiecta, & alia extra data, vel duabus interiectis contentum punctis...datis.... Demonstravit autem hanc... Apollonius, in nudis rectis lineis vsitato ac peruulgato modo tentans, vt in secundo libro... elementorum Euclidis, & rursus eandem demonstrauit ad institutionem magis accommodate & ingeniose per semicirculos' (*Mathematicæ Collectiones* (note (1)): 243). Whatever the historical truth of Newton's eminently plausible conjecture—one whose authenticity, even in our improved modern knowledge of the classical past, remains wholly untestable against any documented evidence—he himself in later years was in no way reluctant to share his insight with others. We have earlier remarked (see IV: 280, note (19)) that he brought it to David Gregory's notice during a visit by the latter to Cambridge in May 1694—which did not prevent him reiterating the point during more than one of their subsequent meetings in London. On '22 Oct^{ris} 1704', for instance, Gregory noted in his diary: 'Mr Newton says that Apollonius *de* [*determinata*] *sectione* was in order to the solution of the Problem that Euclid proposed, & which [compare IV: 221] Des Cartes pretended to have solved' (Christ Church. MS 346: 127 = Hiscock, *Gregory*: 20); and on 16 June 1706 he added in the same vein that the 'Problema Veterum having four Points already determined by...the Intersections of the given lines: If a fifth be found, the Conick-Section desired is drawn by Pappus's Problem, To describe a Conick-Section through five given points. All the matter then is to find that fifth point. To this end Apollonius wrote his Book *De Sectione determinata*' (MS 346: 172). The remark found its way into print just a few weeks later in the *princeps* Latin edition of Apollonius' *De Sectione Rationis* (see note (30) following) when Gregory's fellow Savilian Professor, Edmond Halley, stated in his 'Præfatio ad Lectorem' (signature a4^{r}): '...ad problema de *Sectione Determinatâ*, ab *Apollonio* plenissime resolutum, tota red[i]t difficultas inveniendi punctum quintum in Loco [ad quatuor rectas] describendo. Datis autem quinque punctis docet *Pappus* Locum Ellipticum perficere, Lib. VIII. Prop. 13, 14. Eodemque modo, nec difficilius, mutatis mutandis, Locus Hyperbolicus per data quinque puncta describitur'. (Compare IV: 223, note (21).)

(29) In the first instance, specifying 'mentem eorū librorum' (the design of their books), Newton opted for the milder adverb 'parum' (inadequately). Even this more temperate castigation is perhaps too severe. For while, in stressing that Apollonius omits to give a full solution of the problem of the 4-line locus in his *Conics*, 'pleraque imperfecta relinquens' (*Mathematicæ Collectiones* (note (1)): 251; compare IV: 221), Pappus had failed to go on to support Newton's present thesis that his propositions *De determinata sectione* provide for its

contained by two of these drawn lines shall be to the rectangle $PG \times PH$ beneath the other two in a given ratio Then because of the given ratios PE to PI, PF to PK, PG to PL and PH to PM there shall be given the ratio of the rectangle $PI \times PK$ to the rectangle $PL \times PM$, and hence the point P is given by Apollonius' book *On determinate section*. For that book accurately enumerates all the cases of the point P in which the straight line IM is to be cut, and determines them by the totality of its propositions—so much so that it can even be inferred therefrom that Apollonius wrote it to complete the Euclidean problem of the four lines.(28) The ancients, therefore, pursued that problem to its finish notwithstanding that Pappus, a not overly perspicacious man, may not in the least (29) have appreciated their purpose.

Apollonius' books *On cutting off a ratio* and *On cutting off a space*(30) also have regard to the construction of 'solid' problems. In the ruler BC revolving round the

complete analysis—indeed, we may well ask, how could they possibly differentiate between its component conic species?—he had made it clear in his summary of Apollonius' work (see the previous note) that its author restricted its application solely to cases involving straight lines and circles.

(30) While the former of these survives in a mediocre Arabic version made about the beginning of the ninth century, we should remember that when Newton wrote down his following outline of their scope and purpose Edmond Halley's *princeps* Latin edition of its text (*Apollonii Pergæi de Sectione Rationis Libri Duo Ex Arabico MS*^{to}. *Latine Versi....Opus Analyseos Geometricæ studiosis apprime Utile*, Oxford, 1706)—and his appended restoration of the *de Sectione Spatii Libri Duo* to an analogous pattern—was still fifteen years in the future. All he could draw upon in 1691 was Pappus' summary in the seventh book of the *Collection*: 'Librorum de [pro]portionis sectione, qui duo sunt, propositio est vna subdiuisa[:] Per datum punctum rectam lineam ducere secantem à duabus rectis lineis positione datis ad data in ipsis puncta lineas, quæ proportionem habeant eandem datæ proportioni. Contingit autem figuras differentes esse & numero plures ob subdiuisionem factam, & linearum datarum inter se positionem, & differentes casus puncti dati: & ob resolutiones, compositionesq; ipsorum, & determinationum.... Libri de spacij sectione duo sunt; problema autem vnum bis subdiuisum, & vna propositio, quæ alia quidem habet superiori similia, sed eo tantum differt, quòd... oportet duas rectas lineas abscissas... datum spacium continere. Dicetur enim sic[:] Per datum punctum rectam lineam ducere secantem à duabus rectis lineis positione datis ad data puncta, lineas, quæ spacium contineant dato spacio æquale. Et hæc ob eandem causam multitudinem habet figurarum' (*Mathematicæ Collectiones:* 242). In modern terms, where the transversal through the external point meets the given lines in pairs of points distant x, y respectively from the given points in them, it will be clear that the *sectio rationis:* $x/y = x'/y'$ defines the correspondence $(0 \infty xx') = (0 \infty yy')$, while the *sectio spatii*: $ay = x'y'$ defines the correspondence

$$(0 \infty xx') = (\infty 0yy');$$

furthermore, that the transversal in the two cases envelops a parabola and a central conic respectively (compare Halley, *Apollonii de Sectione Rationis...*: 60–5/135–8 and 143/147/157; see also IV: 223, note (21)). However tempting it may now be so to conclude in hindsight, it is impossible to say how far Apollonius' own purpose embraced these more general considerations and we will not here dogmatize. Newton, we may note, chooses in sequel—consciously or no—wholly to ignore Pappus' very clear specification that in each of the Apollonian problems the line-segments cut off by the transversal shall lie in the given straight lines.

rectæ positione datæ AC occurrente in puncto C sumatur CD ipsi AC vel directe vel reciproce proportionalis et incidet punctum D in locum solidum.(31) Agatur recta quævis tertia positione data ED et punctum D in quo hæc recta secat locum illum solidum determinabitur per libros illos Apollonij *de sectione rationis* et *sectione spatij*. Ex punctis autem sic inventis determinantur loca ut supra. Problemata autem per hæc loca solvuntur deducendo æquationes duas quadraticas ad linearum AC, CD rationes vel directas vel reciprocas. Et quemadmodum Apollonius loca solida per punctorum inventionem determinavit: sic etiam loca plana per puncta et tangentes in libro *Tactionum*(32) determinare docuit. Ad locorum determinationem spectabant etiam libri Apollonij et aliorum(33) *de locis planis* et *solidis*. Quæ vero de hac re tractarunt sic brevius et generalius enunciantur.(34)

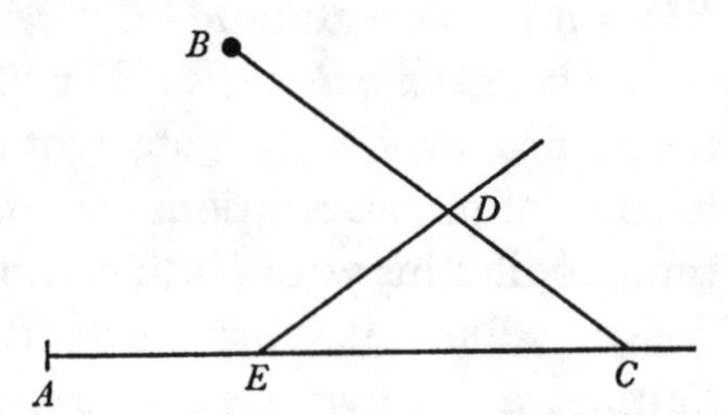

Si lineæ pro numero punctorum in quibus a linea recta secari possint distinguantur in genera sic, ut linea primi sit generis quam recta in unico tantum puncto secare potest[,] ea secundi quam secare potest in duobus[,] ea tertij quam in tribus: et a lineæ alicujus puncto indeterminato ad rectas quotcunqȝ positione datas quæ non sunt omnes parallelæ(35) ducantur aliæ totidem rectæ in datis angulis, et ex rectis ductis quibuscunqȝ confletur æquatio.(36) Et si rectæ ductæ ad unicam tantum dimensionem in æquatione illa

(31) This is true in neither case, unfortunately. If B is the origin of a system of rectangular Cartesian coordinates in which D is the general point (x, y), A is (a, b) and the line AC through it is defined by $x = a$, it is readily shown that BD intersects AC in the point $(a, ay/x)$ and hence $AC = (bx-ay)/x$, while $CD = (a-x)\sqrt{[x^2+y^2]}/x$; accordingly, the loci (D) defined by $CD = k.AC$ and $CD = k/AC$ are respectively the irreducible 'circular' quartic

$$(x-a)^2\,(x^2+y^2) = k^2(bx-ay)^2$$

and sextic $(x-a)^2\,(bx-ay)^2\,(x^2+y^2) = k^2x^4$. As we observed in the previous note, neither of these Newtonian *sectiones*—in which CD is neither fixed in direction nor terminated at one end by a given point—is strictly Apollonian in character. Since in their classical specification (again see the previous note) the transversal cutting off the points by a *sectio rationis* or a *sectio spatii* in each case envelops some species of conic, and because from any given point in its plane it will in general be possible to draw two tangents to that 'solid' locus (though these—and the solutions they determine—need not, of course, be real), there will be two pairs of section-points which answer the requirement and these will in turn be constructible as the roots of a quadratic equation—and Apollonius in his extant *De Sectione Rationis* does indeed so construct them (compare T. L. Heath's summary of the method in his *Greek Mathematics*, **2**: 176–9).

(32) On circle tangencies, to be more precise. As ever, Pappus had embraced the two books of this lost work of Apollonius in a single general problem: 'Punctis, & rectis lineis, & circulis tribus quibuscumque positione datis circulum describere per vnumquodque datorum punctorum, qui vnamquamqȝ linearum datarum contingat' (*Mathematicæ Collectiones*: 243). Newton had already (see IV: 254–8; V: 254–66, 614–18; VI: 164, note (154)) given a variety of solutions to particular cases of this.

(33) Notably Euclid and Aristæus, upon whose (long lost) works on conics Apollonius himself plentifully drew: 'Euclides autem secutus Aristæum scriptorem luculentum in ijs, quæ

pole B and meeting the straight line AC given in position in the point C take CD either directly or reciprocally proportional to AC, and the point D will fall on a solid locus.(31) Draw any third straight line ED given in position, and the point D in which this line cuts that solid locus will be determined by those books of Apollonius *On cutting off a ratio* and *On cutting off a space*. From the points thus found, however, the loci are determined as above. Problems, let me add, are solved through these loci by reducing two quadratics to either direct or reciprocal ratios of the lines AC and CD. And, just as he determined solid loci by the finding of points, so also Apollonius in his book *Of contacts*(32) taught how to determine plane loci by means of points and tangents. To the determining of loci the books of Apollonius and others(33) *On plane/solid loci* also have regard. Their treatment of this topic may, however, be more shortly and generally expressed in this way.(34)

Let lines be distinguished into classes according to the number of points in which they may be cut by a straight line, so that a line of first kind shall be one which a straight line can cut in but a single point, one of second intersectable by it in two, one of third in three; and from some indeterminate point of a line draw to any number of straight lines given in position (and not all parallel(35)) at given angles an equal number of other straight lines, and from out of any of the straight lines so drawn let an equation be constituted.(36) Then if the straight

de conicis tradiderat; ...quantum ostendi potuit de loco [ad tres & quatuor lineas] per eius conica memoriæ prodidit....Adjicere autem loco, quæ deerant, [Apollonius] facile potuit, animo comprehendens ea, quæ ab Euclide de loco scripta fuerant, & dans operam Euclidis discipulis Alexandriæ longo tempore, ex quo adeo excellentem in mathematicis habitum est assecutus' (Pappus, *Collectiones*: 251).

(34) Originally 'his paucis fere comprehenduntur' (may...be all but comprehended in these few points). Newton's following précis of classical notions of a curvilinear point-locus goes in fact far beyond the surviving evidence upon this topic of ancient geometrical analysis in attributing to Apollonius and his Greek colleagues awareness of the fruitfulness of classifying curves according to their (modern Newtonian) degree, or of the existence of a general rectilinear diameter in curves other than conics which he, too, had pioneered. Compare notes (36), (41) and (42) below.

(35) In which case, of course, the locus reduces to be an appropriate number of straight lines parallel to those given. This scarcely necessary proviso was advanced in the manuscript as an afterthought from an equivalent position which it initially occupied in the final sentence of the present paragraph.

(36) There is no classical precedent for this massive generalization of the simple classical notion of a curve's defining 'symptom' which underlies Pappus' exposition of the Euclidean 4-line conic locus and his extension of it (*Mathematicæ Collectiones*: 251–2; compare Ivor Thomas, *Greek Mathematical Works*, **2** (London, 1941): 603) to the $2n$-line case. It had been Newton's own discovery some dozen years earlier (see IV: 340–2)—one which he then faulted Descartes for not attaining in his *Geometrie*—that the latter (for $n > 5$ at least; see *ibid.*: 342–3, note (27)) lacks enough independent constants in its defining property to represent the general algebraic curve of n-th degree. The very idea of classifying curves according to such degrees was, we have remarked more than once (IV: 341, note (23); V: 424, note (620)), yet one more innovation due to Newton.

ascendant, linea in qua punctum indeterminatum locatur erit primi generis, id est recta. Sin rectæ ductæ vel seorsim vel conjunctim ad duas dimensiones ascendant[,] linea in qua punctum indeterminatum locatur erit secundi generis id est circulus vel sectio Conica. Et idem eveniet si præterea rectarum quo[t]cunq; quæ a datis punctis ducuntur quadrata æquationem ingrediuntur.(37) Ubi vero harum quadrata æquationem ingrediuntur sed lineæ in datis angulis ductæ ad unicam tantum dimensionem ascendunt vel æquationem non ingrediuntur[,] punctum indeterminatum locabitur in circulo.(38) Porro si rectæ in datis angulis ductæ vel seorsim vel una cum quadratis ductarum ad data puncta ascendunt ad tres dimensiones[,] locabitur punctū indeterminatum in linea tertij generis. Si ascendunt ad quatuor dimensiones locabitur in linea quarti. Sed hæc intelligenda sunt de æquationibus quæ non possunt per divisionē in binas(39) æquationes resolvi.

Si linea curva generis cujuscunq; a parallelis quotcunq; in pleno punctorum numero secetur[,] id est in duobus si secundi est generis, in tribus si tertij, in quatuor si quarti, et parallelæ duæ a recta tertia transversim secentur sic ut segmenta hinc inde ad curvam terminata sibi invicem æquentur[,] id est ut parallelæ utriusq; segmentum ex una parte æquetur segmento ex altera si linea curva secundi sit generis [,] vel summa duorum segmentorum ex una parte segmento ex altera si curva sit tertij generis linea[,] vel summa trium vel duorum segmentorum ex una parte segmento quarto vel tertio et quarto ex altera[:](40) linea transversim ducta erit diameter figuræ curvilineæ, et secabit parallelas omnes sic ut segmentum vel segmenta ex una parte æquentur segmento vel segmentis ex altera.(41) Et hæc segmenta erunt ordinatim applicatæ ad diametrum. Et punctum in quo diametri omnes concurrunt siquod sit, erit centrum figuræ.(42)

---

[2](43) In Analysi veterum observandæ sunt hæ regulæ.

1. Ex datis colligenda sunt ignota et his datis alia quæcunq;(44) inveniri possunt

---

(37) This is merely the particular case in which the straight lines whose product enters into the defining equation of the locus come two by two to coincide.

(38) Pappus lists this latter instance in his summary of the second book of Apollonius' lost work on *Plane Loci* under the universal enunciation 'Si à quotcumq; datis punctis ad punctum vnum inflectantur rectæ lineæ: & sint species, quæ ab omnibus fiunt dato spacio æquales[,] punctū continget positione datam circumferentiā' (*Mathematicæ Collectiones*: 248; compare IV: 233, note (13)). Perhaps treading in the steps of Pierre Fermat (whose geometrical restoration of the general Apollonian problem appeared posthumously in his *Varia Opera Mathematica* (Toulouse, 1679): 33–41; see IV: 234, note (13)), Newton himself had treated the case where three points are given some dozen years previously in his 'Solutio Problematis Veterum de Loco solido' (see IV: 318–20).

(39) Initially 'duas vel p[lures]' (two or more).

(40) Understand 'si curva sit quarti generis &c' (if the curve be of fourth kind, and so on).

(41) This concept of the rectilinear diameter of an arbitrary (algebraic) curve is yet again Newton's generalization of a property known in classical times only in the simplest instance of

lines drawn rise to merely a single dimension in that equation, the line in which the indeterminate point is located will be one of first kind, that is, a straight line. But if the straight lines drawn rise either individually or jointly to two dimensions, the line in which the indeterminate point is located will be of second kind, that is, a circle or a conic. And the same will happen if, furthermore, the squares of any number of straight lines drawn from given points enter the equation.(37) When indeed the squares of these enter the equation but the lines drawn at given angles rise merely to a single dimension or do not enter the equation at all, the indeterminate point will be located in a circle.(38) Moreover, if the straight lines drawn at given angles either individually or together with the squares of ones drawn to given points rise to three dimensions, the indeterminate point will be placed in a line of third kind. If they rise to four dimensions, it will be located in one of fourth. But these assertions are to be understood for equations which cannot be resolved by division into a pair of(39) equations.

If a line of any kind be cut by any number of parallels in the full number of points, namely in two if it is of second kind, in three if of third, in four if of fourth, and two of the parallels be transversely cut by a third straight line such that the segments on its either side terminating at the curve be equal to one another—that is, so that the segment of each parallel on one side be equal to the segment on the other if the curve be a line of second kind, or the sum of two segments on one side equal to the segment on the other if the curve be a line of third kind, or the sum of three or two segments on one side to the fourth segment or third and fourth ones on the other,(40) the transversal so drawn will be a diameter of the curvilinear figure, and will intersect all parallels such that the segment or segments on one side shall equal the segment or segments on the other.(41) And these segments will be ordinates to the diameter. While the point in which all diameters concur, if any such there be, will be the figure's centre.(42)

---

[2](43) In the ancients' analysis these rules are to be observed:

1. From given entities unknown ones are to be gathered, and with these as given any others whatever(44) that can be found, whether those which are

---

the conic; see II: 94, note (7) and IV: 394.

(42) Compare IV: 354–6, where Newton calls this centre of a curve of n-th kind—in distinction from its simple geometrical centre (if there be one)—the 'centre of n-th section'.

(43) Add. 3963.15: 180v/179r; the manuscript sheet was originally used by Newton to compute, in parallel to his discussion of the 'curvatura Ellipseos' in Proposition XIII of his contemporary treatise 'De quadratura Curvarum', the curvature at the general point (z, v) of the hyperbola $a^2+bz^2 = v^2$ (see **1**, §2: note (144) above). Here he seeks to set out in some reasonably coherent sequence an extension of the rules for the analysis of geometrical problems which he had earlier adumbrated in his 'Arithmetica Universalis' (compare V: 158–72 especially 162–4).

(44) Initially 'alia quotquot' (any number of others).

sive collecta illa sint quantitates(45) sive quantitatum(45) relationes non nimis complexæ. Et si una unius ignotæ relatio data vel duæ duarum ignotarum relationes datæ ob[v]enerint, ignotæ pro datis habendæ sunt. Hac enim methodo Problema vel solvetur inveniendo ignota omnia, vel deducetur ad Problema simplicius. Nam quo plura dantur eo facilius cætera inveniri solent.

2. Si hæretur considerandum est Problema tanquam confectum et ex quantitatibus ignotis ceu datis colligend[æ] sunt datæ ceu ignotæ. Et ex relationibus ignotarum ad datas sic collectis ignotæ dabuntur ut supra.

3. Si adhuc hæretur considerandi sunt Problematū casus simpliciores et cum his solutis difficiliores comparandi sunt.

4. Comparandæ(46) sunt quantitates indeterminatæ quæ simul infinitæ fiunt & simul evanescunt ut et quantitates quarum alterutra(47) evanescit ubi altera evadit infinita. Nam primo casu proportiones directæ, secundo reciprocæ obvenire solent.(48)

5. Si puncta duo vel quantitates duæ obvenerint quorum eædem sunt conditiones et proprietates omnes, considerandæ sunt horum relationes ad invicem et conditiones communes. Et si puncta sunt, determinandū est vel aliud punctum quod eodem modo se habeat ad utrumqꝫ vel linea seu recta seu circularis quæ per utrumqꝫ transeat.(49)

6. Si ab indeterminato quovis pūcto ad rectas quotcunqꝫ positione datas ducantur aliæ rectæ in datis angulis et hæ ductæ et harum part[e]s rectis positione datis interjectæ et partes rectarum positione datarum tam ad data puncta abscissæ quam rectis ductis interjectæ aut his omnibus proportionales vel harum omnium aliquæ per solam additionem et sub[ductionem] connexæ relationem quamvis (sive relatio illa æquatio sit sive ratio data)(50) ingrediuntur, punctum illud ignotum locatur in linea recta. Et si præterea rectarum a puncto illo ad data quotcunqꝫ puncta ductarum quadrata ad lineas quasvis datas applicata relationem eandem per additionem et subductionem ingrediuntur, vel si quadrata illa sola ad datas applicata relationem ingrediuntur, punctum illud locatur in circulo. Idem eveniet si rectæ duæ a punctis datis ad punctum indeterminatum concurrentes datum contineant angulum vel datam habeant rationem ad invicem, nec non ubi duæ rectæ a dato puncto ductæ vel coincidentes vel datum continentes angulum sint reciproce proportionales et earum una terminatur ad rectam positione datam et altera ad punctum indetermi-

(45) Newton has here deleted his superfluous respective qualification of these as 'singulæ' (individual) and 'duarum ignotarum' (two unknown).

(46) This replaces 'Considerandæ' (...considered).

(47) Initially 'utraqꝫ successivè' (each one of which in succession).

(48) When, namely, these correspondences are 1, 1 and hence (see note (30) above) define an Apollonian *sectio rationis* and *sectio spatii* respectively.

(49) Newton's 'regula fratrum' (rule of brothers); see v: 612 ff.

gathered be quantities(45) or not overly complex relationships of quantities.(45) And if one given relationship of one unknown turns up, or two given relationships of two unknowns, the unknowns are to be considered as given. For by this method a problem will either be solved by finding all its unknowns, or it will be reduced to a simpler problem. And of course the more things there are given, the easier it will usually be to discover the rest.

2. If you become stuck, you must consider the problem as accomplished and out of the unknown quantities viewed as given gather given ones as unknowns; from the relationships of the unknowns to the givens recovered in this way the unknowns will then be given as above.

3. If you still stick, you must consider simpler cases of the problems and once they are solved compare the more difficult ones with them.

4. Indeterminate quantities which simultaneously come to be infinite and simultaneously vanish are to be compared(46), as also (pairs of) quantities, one of which(47) vanishes when the other becomes infinite. For the proportionality usually proves to be direct in the first case, and reciprocal in the second.(48)

5. If there happen to be two points or two quantities whose every circumstance and property are the same, you must consider their relationships to each other and their common circumstances. And if they are points, you must determine either another point which shall be related the same way to both, or a line —be it straight or circular—which shall pass through both.(49)

6. If from any indeterminate point there be drawn to any number of straight lines given in position other straight lines at given angles, and these drawn lines and their portions intercepted by straight lines given in position and the portions of straight lines given in position which are both cut off at given points and also intercepted by the straight lines drawn—or any proportionals to all these, or some of all these connected by addition and subtraction alone —enter into any relationship (whether that relationship be an equation or a given ratio),(50) that unknown point is located in a straight line. And if, furthermore, the squares of straight lines drawn from that point to any number of given points enter, when divided by any given lines, into the same relationship through addition and subtraction, or if those squares divided by givens alone enter the relationship, that point is located in a circle. The same will happen if two straight lines drawn from given points and concurrent at the indeterminate point should contain a given angle or have a given ratio to one another, and also when two straight lines drawn from a given point and either coincident or containing a given angle shall be reciprocally proportional, one of them terminating at a straight line given in position and the other at the indeterminate

(50) Compare note (36) above.

natum. Problema autem ubi hi casus incidunt solvetur quærendo data puncta per quæ locus transibit et datas positione rectas quas tanget & locum per puncta ducendo quæ tanget rectas.

6[*bis*]. Conditio aliqua earum quæ punctum ignotum determinant omittenda est, et locus puncti considerandus quem conditiones reliquæ determinant.(51) Imò conditiones singulæ sunt successive omittendæ ut locus simplicissimus inveniatur.

7. Siquando quantitatum(52) relationes magis complexæ ob[v]eniunt quam quæ verbis enunciari possunt, quantitates per symbola designandæ sunt et computationes ineundæ.

8. Postquam constructio Problematis per Analysin inventa(52) est, enuncianda est constructio illa et ejus demonstratio componenda. Nam Problema resolvitur quidem per Analysin at non solvitur absq; Compositione.

---

[3](53) Problematum solutiones juxta sequentes Regulas(54)

Reg. 1. Circumspicere quid ex datis consequatur ut ex pluribus datis facilius assequamur quod propositum est. Item circumspicere quomodo schemata construantur ut ex datis aliquid colligamus. In hunc [finem] cognoscendæ sunt proportionalium leges et transmutationes, eo quod Geometria [per] proportionales ob simplicitatem magis quam per æquationes amat progredi. Cognoscendæ sunt etiam Figurarum proprietates quæ in *Elementis*(55) sunt & determinationes simpliciores: Et quando triangula vel quadrangula dantur specie, quando specie et magnitudine. Determinatæ item sectiones veterum quæ sunt æquationes recentiorū in promptu esse debent. Ut et Locorum determinationes. Nam Geometria tota nihil aliud est quam inventio punctorum per intersectiones Locorum.(56)

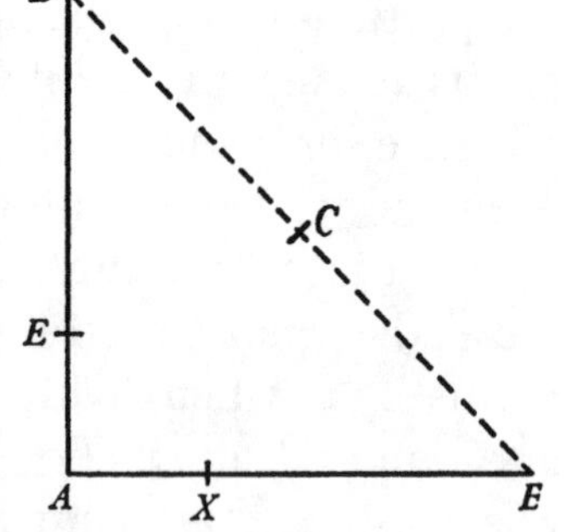

Sectio determinata dici potest simplex duplex triplex &c prout in uno duobus tribus punctis &c fit, vel ut recentes loquuntur prout æquatio unius duarum triū dimensionum est &c.

Si secanda est recta data AB in X ita ut sit

$$AE.AX::BX.DE.$$

(51) This notion is developed more fully in the rough contemporary drafts reproduced in Appendix 2 following. Newton here originally specified 'locus...qui ex reliquis conditionibus resultat' (...which results from the remaining conditions).

(52) Initially 'terminorum' (terms) and 'deductū' (deduced) respectively.

(53) Add. 4004: 108r/108v. In this unfinished piece—entered by him on a nearly blank sheet detached from his Waste Book which was myopically if conscientiously 'restored' to its original position therein by his nineteenth-century editor, J. C. Adams—Newton broaches the generalization of the classical (quadratic) *sectio determinata* to higher orders, but breaks off before adducing any example which is other than an instance of its Apollonian case.

point. A problem, however, in which these cases are met with will be solved by seeking given points through which the locus shall pass and straight lines given in position which it shall touch, and then drawing the locus through the points and to touch the straight lines.

6 [*bis*]. Some circumstance of those which determine the unknown point must be omitted, and consideration paid to the point-locus which the remaining conditions determine.(51) Indeed, single conditions are successively to be omitted in order that the simplest locus shall be discovered.

7. Should ever relationships of quantities(52) occur which are too overly complex to be expressed in words, the quantities must be denoted by symbols and computations instituted (therefrom).

8. After the construction of a problem is found(52) by analysis, that construction is to be enunciated and its demonstration composed. For while a problem is indeed resolved by analysis, it is not solved away from its composition.

---

[3](53) The solutions of problems(54) in accord with the following rules.

Rule 1. Survey what may be the consequence of what is given, so that from several givens we may the more easily attain what is proposed. Likewise survey how schemes are to be constructed so that we may conclude something from the givens. To this end the laws and transformations of proportionals must be learnt, seeing that geometry has a liking to proceed, because of their simplicity, more by way of proportionals than through equations. You must also learn the properties of figures which are in the *Elements*(55) and their simpler determinations; also, when triangles or quadrilaterals are given in species, and when in species and size. The determinate sections of the ancients—the equations, that is, of more recent mathematicians—you likewise ought to have ready to hand. And the determinations of loci, too. For geometry in its entirety is nothing else than the finding of points by the intersections of loci.(56)

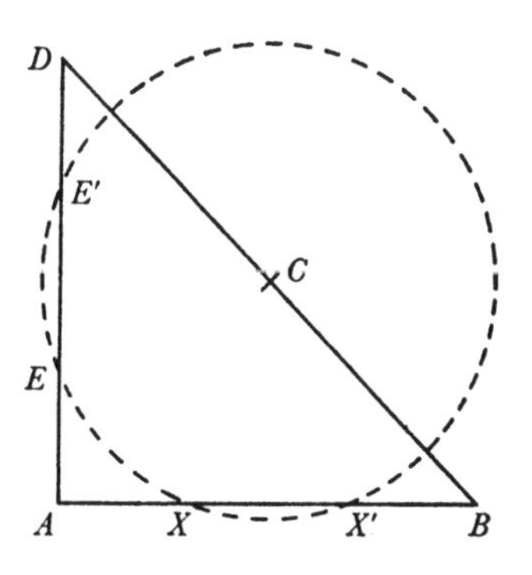

A determinate section is said to be simple, double, triple, ... according as it happens in one, two, three, ... points, or, as moderns say, according as its equation is of one, two, three, ... dimensions.

If the given straight line AB has to be cut in X so that

$$AE:AX = BX:DE,$$

(54) Understand some equivalent to 'quærendæ sunt' (are to be sought).

(55) 'Euclidis' (Euclid's), of course.

(56) A harshly restrictive dictate, to be sure! Would Newton in his present frame of mind reject from geometry all curves defined as the envelopes of lines, or any consideration of such topological concepts as betweenness and inclusion?

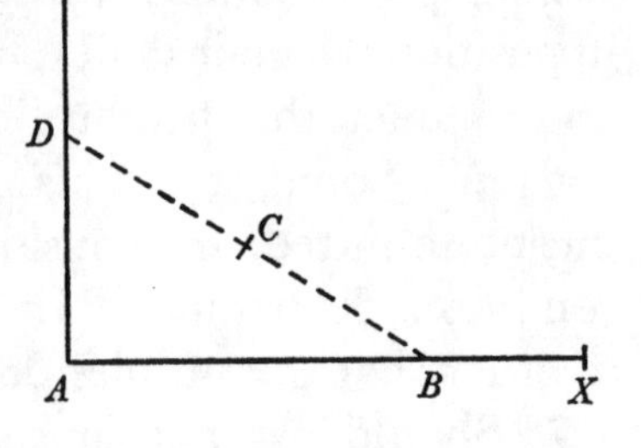

vel rectangulum AXB æquetur dato rectangulo AED: sit angulus BAD rectus. Biseca BD in C. Centro C radio CE describe circulū secantem AB in X. Atq; hoc modo construi potest omnis æquatio quadratica.(57) Sed rem longius prosecutus est Apollonius.(58)

Igitur si in recta aliqua dantur [quatuor] puncta A, B, [E,] F et secanda sit recta in X ita ut sit

$$AE.AX::BX.FX_{[,]}$$

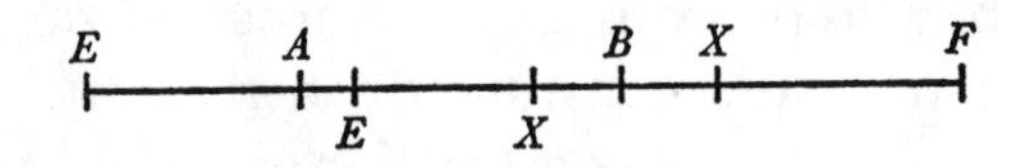

componendo vel dividendo erit

$$AE.EX::BX.BF$$

unde solvetur Problema ut prius.

Si in recta dentur quatuor puncta A, B, F, G et secanda sit recta in X ita ut rectangulū AXB sit ad rectangulū FXG in data ratione AH ad HG, erige perpendiculum HQ quod sit medium proportionale inter AH et HG. Junge AQ, GQ. Super BF constitue triangulum BSF simile triangulo AQG et ad easdem partes rectæ AG si punctum X quæritur vel inter A et B vel inter F et G, aliter ad partes contrarias [et occurrant FS, BS rectis AQ et GQ in P et R].(59) Super diametro PR describe circulum secantem rectam AG in X.(60)

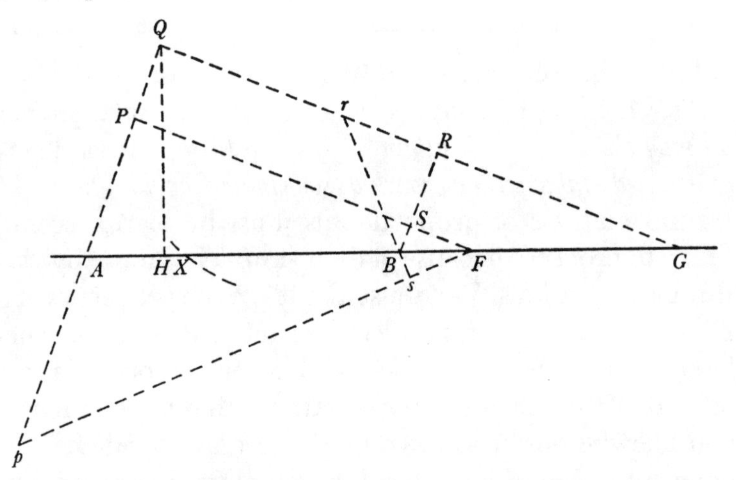

(57) For the section constructs the two solutions AX, AX' of the equation

$$AX(AX-AB) = AE \times DE$$

in all its possible cases. Compare Newton's similar Cartesian construction on 1:492–4.

(58) In his two lost books on the topic, whose content is summarized by Pappus in his *Mathematical Collection*; see note (28) above.

(59) A preceding equivalent construction directing 'In angulo AFP æquali AGQ & GBR æquali GAQ age rectas FP, BR occurrentes AQ et GQ in P et R' (At the angles $\widehat{AFP} = \widehat{AGQ}$ and $\widehat{GBR} = \widehat{GAQ}$ draw the straight lines FP, BR meeting AQ and GQ in P and R) has been cancelled by Newton in the manuscript, but is not replaced. We here make appropriate restoration of its essence.

(60) If H is the origin of a system of rectangular Cartesian coordinates in which $HA = a$, $BH = b$, $HF = f$, $HG = g$ and the general abscissa $HX = x$, then when X falls between A

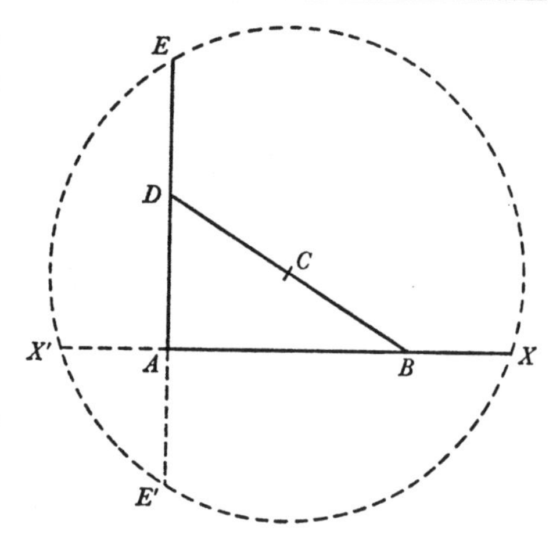

or that the rectangle $AX \times XB$ shall be equal to the given rectangle $AE \times ED$: let the angle $B\widehat{A}D$ be right, bisect BD in C, and with centre C and radius CE describe a circle cutting AB in X. And in this way every quadratic equation can be constructed.(57) But Apollonius has pursued the matter somewhat farther.(58)

If, accordingly, in some straight line there are given the four points A, B, E, F and the straight line shall need to be cut in X so that $AE:AX = BX:FX$, by compounding and dividing there will be $AE:EX = BX:BF$, whence the problem will be solved as before.

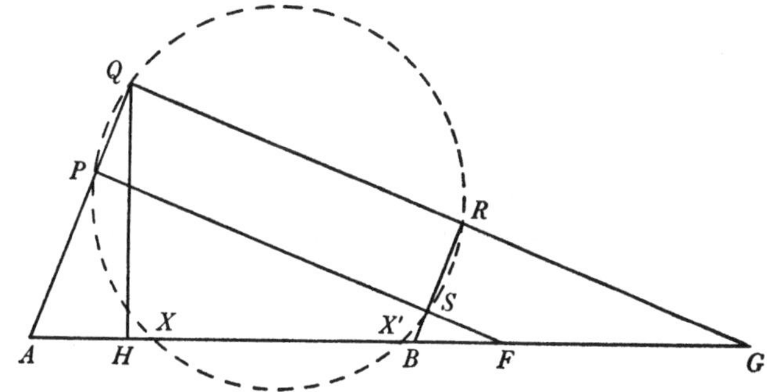

If in a straight line there be given the four points A, B, F, G and the line shall need to be cut in X so that the rectangle $AX \times XB$ be to the rectangle $XF \times XG$ in the given ratio AH to HG, erect the perpendicular HQ which shall be a mean proportional between AH and HG, join AQ and GQ, and on BF set up the triangle BSF similar to the triangle AQG—on the same side of the line AG if the point X is required either between A and B or between F and G, otherwise on the opposite one—and let FS, BS meet the lines AQ and GQ in P and R:(59) on diameter PR describe a circle cutting the line AG in X.(60)

---

and B or between F and G the determinate section reduces to ascertaining x where

$$(x+a)\ (x-b)/(x-f)\ (x-g) = -a/g,$$

that is, $(a+g)\ x^2 - (af+bg)\ x + ag(f-b) = 0$. This quadratic equation Newton constructs as the meet of the abscissa ($y = 0$) with the circle

$$x^2+y^2-((af+bg)/(a+g))\ x+\lambda y+ag(f-b)/(a+g) = 0$$

which is constrained to pass through the point $Q(0, \surd[ag])$, so that

$$\lambda = -\surd[ag]\ (1+(f-b)/(a+g))$$

and the circle's centre is at $(\frac{1}{2}(af+bg)/(a+g))$, $\frac{1}{2}\surd[ag]\ (1+(f-b)/(a+g))$; accordingly, the lines BS and FS drawn from $B(b, 0)$ and $F(f, 0)$ to the further end-point S

$$((af+bg)/(a+g),\ \surd[ag]\ (f-b)/(a+g))$$

of the circle's diameter through Q have perpendicular slopes $\surd[g/a]$ and $-\surd[a/g]$, and so are parallel to AQ and GQ respectively, as Newton's construction implies. Newton proceeds less happily in the contrary case in which X is other than between A and B on F and G. Here the

Addo si recta secanda sit in X ita ut rectangulũ AXB sit ad differentiam inter rectangulum FXG & rectangulum datum $M \times N$ [in ratione data]: seca rectam illam in T et V ita ut FTG et FVG seorsim æquetur dato rectangulo $M \times N$. Dein rursus seca in X ita ut rectangulum AXB sit ad rectangulum TXV in ratione illa data.

Si recta secanda est in X ita ut rectangulum AXB sit ad summam rectanguli FXG et rectanguli dati $M \times N$ [in ratione data], erit dividendo $AXB \div M \times N$[61] ad FXG in ratione data et inverse FXG ad $AXB \div M \times N$[61] in ratione data. Qui casus est superioris propositionis.

Ad hos casus facile est cæteros reducere.[62]

APPENDIX 1. A PRELIMINARY ACCOUNT OF 'GEOMETRICAL ANALYSIS'.[1]

From the original draft[2] in the University Library, Cambridge

Analysis Geometrica.[3]

Antequam resolutionem calamo aggrediamur conditiones[4] Problematis mente evolvendæ sunt colligendo quicquid ex datis consequatur sive data illa sint puncta et quantitates sive proportiones quantitatum aut aliæ quævis earundem relationes. Ac tandem ubi methodus simplicissima et brevissima colligendi quæsitum innotuerit, ineundus erit calculus. In hac methodo constitit Analysis veterum ut ex Euclidis *Datis* et resolutione Problematum apud Pappum manifestum est. Nam librum *Datorum* ad analysin spectare Pappus expressè[5] tradit. Scriptus est utiq; liber iste ut quoties incidimus in quantitates & quantitatum relationes & figuras determinatas sive eæ sint rationes ignotarum quantitatum sive triangula[6] et circuli sive extrema vel media proportionalia

---

determinate section reduces to $(x+a)(x-b)/(x-f)(x-g) = +a/g$ and hence to resolving the quadratic $(a-g)x^2-(2ag+af-bg)x+ag(f+b) = 0$; the circle through Q which constructs the roots HX of this is therefore $x^2+y^2-((2ag+af-bg)/(a-g))x+\mu y+ag(f+b)/(a-g) = 0$ where $\mu = -\sqrt{[ag]}(1+(f+b)/(a-g))$, but the mirror-image s

$$((af+bg)/(a+g), -\sqrt{[ag]}(f-b)/(a+g))$$

of S in BF does not even lie in its circumference and so in consequence neither can the points p and r which Newton constructs therefrom—*à fortiori* the circle on diameter pr will not intersect AH in the required points X.

(61) This should be $M \times N'$ where N'/N is the given ratio.

(62) What other *Regulæ* Newton had it in mind to add in sequel we may only guess, but no doubt they would have dealt at some point with some generalization of the Greek notion of a porism on the lines of his discussion of their 'invention' in §2 following.

(1) In this unfinished piece—much more carefully written out than the equivalent portion of the roughly drafted text reproduced in §1.1 above—Newton further expounds the fertile method of geometrical analysis *ex datis*, again showing how Propositions 58–60 and 84–87 of

If, let me add, a straight line shall need to be cut in X so that the rectangle $AX \times XB$ be to the difference between the rectangle $FX \times XG$ and a given rectangle $M \times N$ in a given ratio, cut that line in T and V so that $FT \times TG$ and $FV \times VG$ shall each separately be equal to the given rectangle $M \times N$, and then again cut it in X so that the rectangle $AX \times XB$ shall be to the rectangle $TX \times XV$ in that given ratio.

If the line is to be cut in X so that the rectangle $AX \times XB$ be to the sum of the rectangle $FX \times XG$ and the given rectangle $M \times N$ in a given ratio, then *dividendo* will there be $AX \times XB - M \times N$(61) to $FX \times XG$ in the given ratio, and *invertendo* $FX \times XG$ to $AX \times XB - M \times N$(61) in a given ratio: which is the case of the preceding proposition.

To these cases it is easy to reduce the ones remaining.(62)

---

sive radices æquationum quadraticarum sive deniq; radices biquadraticarum carentium terminis imparibus, hæ omnes pro datis habeantur. Nam radices quadraticarum omnium dantur cum per Prop 58, 59 tum per Prop 60, tum etiam per Prop 84 et 85 Libri illius: eæ vero biquadraticarum dantur per Prop 86 ejusdem libri.(7)

Euclid's book on the topic yield, when suitably interpreted, standard resolutions of the general quadratic equation and also (in near-trivial extension) of the quartic which lacks terms in odd powers of the variable. As a bonus, to illustrate the power and elegance of the approach he suitably reworks three of the geometrical problems of his earlier treatise on 'Arithmetica Universalis'. Some three years afterwards, we may add, when David Gregory in May 1694 visited Newton, he was shown the present manuscript and made a couple of hurried jottings of its content (see notes (10) and (20) below) on the back of an autograph scrap of Newton's in which he recomputed for Gregory his construction of the cone frustum of least resistance to uniform motion along its axis. (See VI: 416–17. The fragment, later catalogued by Gregory in his Codex C as '48. De Coni truncati resistentia Neutoni manu', is now in private possession in the United States.)

(2) Add. 3963.4: 27r–28r, the two recto sides of a single folded sheet with additions (here incorporated into the text) on the intermediate verso.

(3) Originally 'REGULA DATORUM'.

(4) Newton first wrote 'status'.

(5) By setting it in prime place in his listing of the classical writings in this field of the τόπος ἀναλυόμενος; see §1.1: note (10) above.

(6) 'latera triangulorum' was first written.

(7) See §1.1: note (14). Newton first went on: 'Continet igitur Liber ille methodum generalem resolvendi Problemata plana: adeo ut hodierni mathematici qui tantam peritiam Veteribus denegant, ignorantiam non Veterum sed suam prodant. Quinetiam Problemata omnia solida per Loca data resolvere didicerant. Nam locus omnis solidus eo deduci potest ut quæratur punctum a quo si rectæ tres vel quatuor ad alias totidem positione datas in datis angulis ducantur, rectangulum sub duabus vel æquale sit rectangulo sub tertia et linea data vel datam habeat rationem ad rectangulū sub tertia et quarta. Dein puncti illius [? datur locus per librum Apollonij *de sectione determinata*]'. He then broke off this unfinished citation of the

Ut si incideretur in æquationem aliquam puta $xx+ax = bb$, spectari posset $xx+ax$ ut datum rectangulum cujus latera x et $x+a$ datam habent differentiam a: qua ratione dantur latera illa per Prop 84.(8) Si incideretur in $\frac{d}{e}xx+ax = bb$, spectari posset $\frac{d}{e}xx+ax$ ut parallelogrammum datum ad datam rectam a applicandum, excedens data specie figura $\frac{d}{e}xx$, et inde dantur excessus illius latitudines x & $\frac{d}{e}x$ per Prop 59. Et sic dato $ax-\frac{d}{e}xx$ vel $\frac{d}{e}xx-a[x]$ datur latitudo x per Prop 58. Porro si incideretur in datum $\frac{d}{e}x^4[+]\,aaxx$, daretur rectangulum $x\sqrt{\frac{d}{e}xx+aa}$,(9) & quoniam quadratum lateris unius $\frac{d}{e}xx+aa$ quadrato alterius xx majus sit dato quam in ratione, daretur latus x per Prop 86. Si deniq incideretur in datum $\frac{d}{e}x^4-aaxx$, daretur rectangulum $x\sqrt{\frac{d}{e}xx-aa}$: et quoniam quadratum lateris [unius] $dxx-aa$ [quadrato alterius xx] minus sit dato quam in ratione[,] daretur latus x per Prop 87 si modo Propositio illa incorrupta ad manus nostras pervenisset.(10) Nam liquido corrupta fuit cum jam nihil differat a præcedente nisi

4-line locus problem (compare pages 206–8 above) to replace it with a curt 'Rem paucis illustrabo' and proceeded straight into the second of his worked examples below. Subsequently he cancelled the whole to insert the two further paragraphs of general explanation which follow before passing on to give particular illustration of their content.

(8) Newton initally added in clarification that 'Spectari etiam posset ut datus gnomon [*sc.* $(x+\frac{1}{2}a)^2-(\frac{1}{2}a)^2$] quo datum quadratum $\frac{1}{4}aa$ diminuitur et sic dabitur gnomonis latitudo $[(x+\frac{1}{2}a)-\frac{1}{2}a = x]$ per Prop 60'.

(9) That is, xy where $y^2 = (d/e)\,x^2+a^2$.

(10) Newton's preferred 'restoration' of *Data*, 87 to resolve the slightly more general equation $y^2 = (d/e)\,x^2-a^2$ in which is given the product xy is enunciated, *mutatis mutandis*, on page 202 above. As an interesting historical aside, when David Gregory was shown the present paper in May 1694 (see note (1)), he straightforwardly reduced the geometrical conditions of 'Prop. 86 *Datorum*' to be '$xy = [b^2]$. $yy-aa$ ad xx ut p ad q' and so, on eliminating x, to the problem 'de resolvenda æquatione $[p/q]\,x^4+a^2x^2-b^4 = 0$', and then in immediate sequel noted Newton's present remark that 'Prop. 87 corrupta [est]. Priorem tamen æquationem alijs signis affectam [*sc.* $(p/q)\,x^4-a^2x^2-b^4 = 0$] resolvere probabile est'. Furthermore, in a jotting subsequently crowded in alongside he further noticed that 'Prop: 84, 85 *datorum* æquationes quadraticas affectas resolvunt. Idem prius circa 58'. Some seven years later, however, when he began to plan his collected Greek/Latin edition of Euclid's surviving works, Gregory could not recall the fine detail of Newton's point, and on 21 May 1701 he made a private memorandum 'To consult Mr Newton...To talk about Euclid especially the *data*; & if I should write a preface, & what instances put in it' (Gregory A 68_2 = Royal Society MS 247: 74, first printed by S. P. Rigaud in his *Historical Essay on the First Publication of Sir Isaac Newton's Principia* (Oxford, 1838): $_2$79 and more accurately in *The Correspondence of Isaac Newton*, **4**, 1967: 355). A year later, fresh from conversing with his mentor at London in July

1702, he set down that 'M[r] Newton is for adjusting Euclids *data* by Pappus's account of them [in the proem to the seventh book of his *Mathematical Collection*].... There is a prop: of the *Data* concerning one case of biquadratick Equations, vitiated but which may with some changes be restored from the other (which relates to the other case) which is entire' (Gregory C 36 = Royal Society MS 257.63). But which of the two related Propositions, 86 or 87, had Newton meant? Unwilling to decide for himself without further guidance—and with the proofs of his *ΕΥΚΛΕΙΔΟΥ ΤΑ ΣΩΖΟΜΕΝΑ* now coming thick and fast off the press—Gregory wrote urgently to Newton on 30 September: 'I remember that some time since, you told me that there are some Propositions in Euclids *Data* designed for the Resolution of Biquadratick Equations which want the second and fourth terms, but that one of them is corrupted. And the *Euclid* in Greek & Latin, that is printing by the [*sc.* Oxford] University, being in a good forwardness and they now at work on the *Data*; I give you this trouble to desire of you to acquaint me which Propositions these be, and how (or to what purpose) you think that which is corrupted ought to be restored. Your speedy answer to this, and what else you may think fitt to impart concerning this work, will be a great honour, and of great advantage, to the edition' (*Correspondence*, **4**: 391). Newton seems not to have responded to this plea and, in increasing desperation we may surmise, Gregory made a contrary choice of 'corrupt' proposition in his introduction to the *Euclidis quæ supersunt omnia. Ex Recensione Davidis Gregorii* which ultimately appeared in the summer of 1703: 'Unicum theorema 86[tum] [!] insigniter vitiatum videtur; quod tamen ex MSS. restitui nequit. Illud sic legendum crediderim...optimè: *Si duæ rectæ datum spatium comprehendant in angulo dato*; *earum autem quadrata simul sumpta æqualia sint spatio dato: & utraque ipsarum data erit.* Etenim si sic fuisset enunciatum hoc theorema, tum positis (methodo, sive potius stylo recentiorum) x & y duabus rectis, ...fiet $xy = a^2$, & $b^2-x^2 = y^2$ sive $b^2 = x^2+y^2$; & reductione facta ad unicam æquationem, $x^4-b^2x^2+a^4 = 0$, vel $y^4-b^2y^2+a^4 = 0$. Ideoque cum Euclides ostendat in hoc casu dari x & y, simul ostendit resolutionem æquationis hujusmodi quatuor dimensionum, nempe $z^4 * -dz^2 * +c = 0$, vel hujusmodi $z^4 * -dz^2 * -c = 0$. Unde factum, ut duobus his theorematibus contineretur resolutio omnium æquationum (quas vocant) biquadraticarum secundo & quarto terminis carentium. At si theorema hoc 86[tum] accipiatur modo quo vulgo legitur, tum erit $xy = a^2$ & $x^2-b^2.y^2::e.f$, sive $fx^2-fb^2 = ey^2$; & rejecta x, fiet $ey^4+fb^2y^2-fa^4 = 0$; vel rejecta y, fiet $fx^4-fb^2x^2-ea^4 = 0$. Atque sic resoluta haberetur æquatio hujus formulaæ $z^4 * +dz^2 * -c = 0$, vel hujus $z^4 * -dz^2 * -c = 0$; nempe ejusdem cum illis quæ in sequente 87mo resolutæ sunt; & formula hæc $z^4 * -dz^2 * +c = 0$ omnino omissa esset ab Euclide: quod minime est verisimile, præsertim si consideretur theor. 84to *Datorum* resolutas esse æquationes pure quadraticas harum formularum $z^2+d[z]-c = 0$, & $z^2-dz-c = 0$; & 85to hanc

$$z^2[\pm]\, dz+c = 0;$$

cujusmodi æquatio in biquadraticis est omissa, nisi lectio supraposita (vel æquipollens alia) admittatur. Immo [compare §1.1: note (14) above] constructionis harum æquationum (simpliciter sive una æquatione, non autem duabus, ut in 84to & 85to, enunciatarum) habita est ratio in theorematibus 58vo & 59no *Datorum*; immo & in ipsis *Elementis*, prop. 28va & 29na libri 6...' (Præfatio: [b1[v]/b2[r]]). Half a century afterwards, in his English version of *The Book of Euclid's Data...Corrected* which first appeared as an appendix to the second edition (London, 1762) of his *The Elements of Euclid, viz. The First Six Books together with the Eleventh and Twelfth*, Robert Simson was not—in ignorance of its Newtonian antecedents—impressed by Gregory's argument for replacing a seemingly vitiated theorem by one which 'tho' it contains a true Proposition...has no connexion in the least with the Greek text. [I]t is strange that Dr. Gregory did not observe, that if Prop. 86. was changed with this, the Demonstration of the 86. must be cancelled, and another put in its place, but the truth is, both the Enuntiation and the Demonstration of Prop. 86. are quite entire and right...' (*ibid.*: 461). We have already stated (in §1.1: note (16)) how Simson himself intercalated two extra propositions into the Greek text of the *Data* in order 'best' to supply its putative deficiencies.

quod sit minus generalis.(11) Nam Geometræ non solent particularia proponere quæ in generalibus prius demonstratis continentur. Continent igitur Euclidis *Data* methodum generalem resolvendi Problemata plana.(12) Sed ut methodus uberior sit addiderim hujusmodi Propositionem.(13)

Si duæ ignotæ quantitates duas ex his octo relationibus datas habeant[,] nempe summa, differentia, proportione, rectangulo, summa ac differentia quadratorum, et excessu quo una major aut minor est altera quam in ratione, vel unius quadratū majus aut minus est alterius quadrato quam in ratione: dantur quantitates. Nam hæc ex *Datis* Euclideis facilè consequuntur & usus sunt frequentissimi.

Sed et hoc Porisma(14) usui esse potest. Si tres vel quatuor sunt ignotæ quantitates quarum omnium differentiæ dantur & rectangulum sub duabus æquale est rectangulo sub tertia et recta data vel datam habet rationem ad rectangulū sub alijs duabus, dantur quantitates per *Sectionem determinatam* Apollonij.(15)

Quinetiam loca plana et solida et plusquam solida pro datis habenda sunt quoties dantur conditiones ex quibus absq; ulteriori Analysi describi possunt. Quem in finem Apollonius circulum ex datis punctis [et] tangentibus describere in libro *Tactionum*(16) docuit.

Sed quomodo ex datis ad data progrediendum sit lubet exemplo uno et altero exponere.(17)

Exempl. 1. Invenire triangulum ABC cujus datur area ambitus et angulus A ad verticem.(18)

(11) Newton initially specified the proposition to be 'liquido corruptiva [!]...quod jam nihil aliud sit quàm casus particularis Propositionis præcedentis'; again see §1.1: note (16).

(12) A cancelled continuation at this point reads 'quo spectabant etiam Euclidis *Porismata* una cum Apollonij *locis planis* et libro *tactionum*, alijq; veterum scriptis quæ ad locum resolutum pertinebant'. Newton's subsequent detailed restorations of the main propositions of Euclid's work on *Porisms* are reproduced in the following sections (see especially pages 310–26 below); he has already made brief reference to Apollonius' books on *Plane Loci* and *Tactions* in §1.1 preceding (see page 210).

(13) Newton has in the manuscript altered this to be 'Porisma' before once more specifying it to be a 'Propositio'. He might best have described the following general 'rule of givens' by the title which (see note (3)) he originally set upon the present piece.

(14) Compare the previous note. Newton has in immediate sequel deleted 'non raro usui est'.

(15) Pappus' enunciation of the component cases of this quadratic section of a line, as treated in Apollonius' lost work, is cited in §1.1: note (28).

(16) On this book (which has likewise failed to survive) see §1.1: note (32).

(17) As we specify in following footnotes, Newton in fact adopts all three from the group of geometrical problems (see v: 184 ff.) which he had earlier set to illustrate the (mostly algebraic) reductions expounded in his 'Arithmetica Universalis', though here of course their demonstrations are recast to argue *ex datis*.

(18) Problem 4 of the earlier 'Arithmetica'; see v: 188 (and compare *ibid.*: 574–6 and 600). In the manuscript (ff. $27^r/27^v$) Newton has cancelled an intended initial '*Exempl. 1.* Trianguli datur area et angulorum aliquis et vel summa vel differentia laterū quæ circa angulū

Ex area et ang[u]lo datur rectangulum sub lateribus AB, AC per *Dat* 66. Demittatur perpendiculum CD et ob datam rationem AC ad AD dabitur rectangulum BAD per *Dat* 2 & 70. Hujus duplo æquale est $AC^q + AB^q - DC^q$ (per Pr. 13 lib 2 *Elem.*) et propterea datur. Huic adde datum $2AB \times AC$ et dabitur differentia quadratorum summæ laterum et basis(19) id est rectangulum sub summa et differentia basis et summæ laterum. Et inde datur illa differentia per *Dat* 57. Hanc aufer a perimetro et relinquetur $2BC$. Datur ergo trianguli latus BC, et propterea datur etiam summa reliquorum laterum: a quorum quadrato si auferas quadruplum eorum rectangulum manebit quadratum differentiæ laterum eorū quæ proinde datur. Ex datis autem summa ac differentia dantur latera.

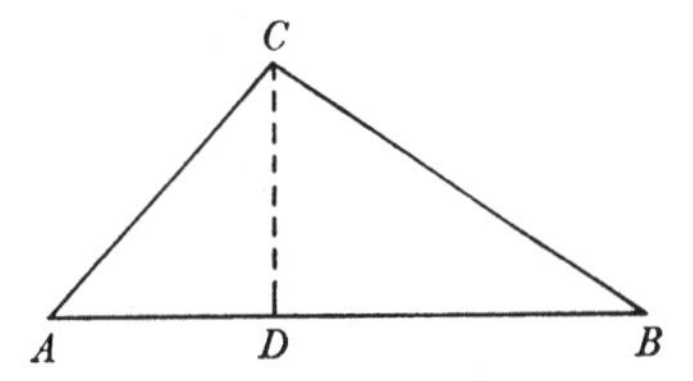

Exempl. 2. Datas quatuor positione rectas AD AE BF CG quinta secare cujus partes con[tin]uo ordine sumptæ datas habeant rationes ad invicem.(20)

Ad datarum aliquam AD age EH FI alteri datæ parallelas et ob datas illas rationes dabuntur etiam rationes CH ad CI et EH ad FI. Dantur autem specie triangula BFI, AEH et inde dātur rationes BI ad IF et EH ad AH[,] et propterea ob datam rationem intermediam IF ad EH datur rationum trium summa BI ad AH. Capiatur BK ad AC in eadem ratione et residua KI et CH erunt etiam in eadem ratione. Dantur ergo rationes KI ad CH et CH ad CI et propterea datur

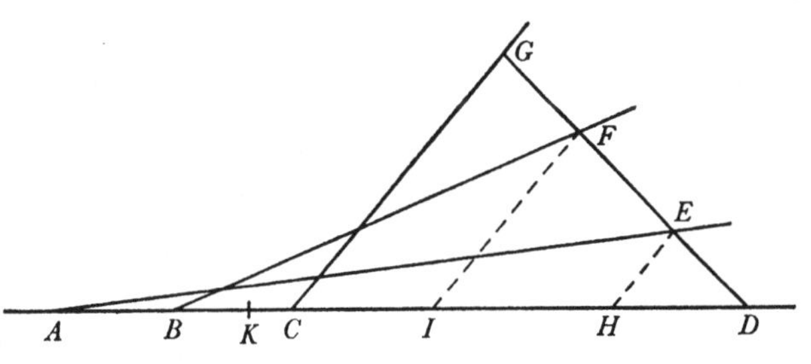

datū [jacent]: requiritur triangulum'. His ensuing demonstration straightforwardly argues: 'Ex datis angulo et area datur rectangulum laterum [*i.e.* $2 \times \text{area}/\sin(\text{angle})$]. Auferatur istud quater a quadrato summæ laterum et habebitur quadratum differentiæ earundem[,] vel addatur idem rectangulū quater ad quadratum differentiæ laterū et habebitur quadratū summæ. Datis autem summa ac differentia laterum dantur latera'. We need not reproduce the scalene triangle ABC which Newton set in isolation alongside, doubtless intending to specify its vertical angle C as the one given along with the adjacent *latera* AC and BC.

(19) That is, $(AB+AC)^2 - BC^2 = (AB+AC+BC)(AB+AC-BC)$ where the difference of the two factors is manifestly twice the 'base' BC.

(20) Problem 52 of the 'Arithmetica' (see v: 298–302), here abstracted from its Wrennian cometary context. In May 1694 (see note (1) above) David Gregory pithily cited it as 'probl: Wallisii [compare v: 299, note (400)] ex veterum porismatis [!] resolutum', having perhaps also then been shown Newton's locus construction *à l'ancienne* (see vi: 295, note (110)) of the equivalent problem as it had appeared in his *Principia* (Liber i, Lemma XXVII, Corol.) in 1687. In strict truth we may only qualify the indeterminate case of this problem—known to Huygens if not to Newton himself at that time (see v: 302–3, note (405))—as a porism in Euclid's sense.

ratio KI ad CI et divisim ratio KC ad CI. Et ob datam KC datur CI et inde datur etiam tum longitudo CH tum puncta E et F per quæ recta DG ducenda est.

Exempl 3. A dato puncto C rectam CF ducere quæ cum alijs duabus positione datis rectis AE et AF triangulum datæ magnitudinis AEF comprehendet.(21)

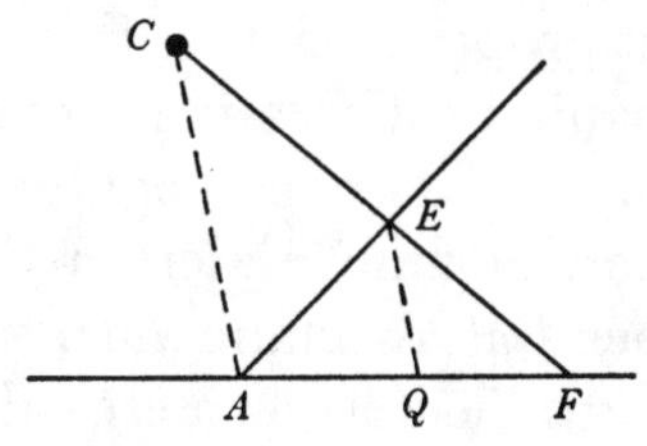

Junge AC eiq; parallela sit EQ occurrens AF in Q et ob datam aream trianguli dabitur rectangulum $AF \times EQ$ per *Dat* 66 ideoq; cum sit CA ad AF ut EQ ad QF dabitur QF per *Dat* 57. Ac deniq; ob datam rationem AQ ad EQ dabitur rectangulum QAF per *Dat* 2 et 70 et inde dantur AQ, AF per *Dat*. 84.

Hujus methodi exempla [quædam extant apud Pappum].(22)

APPENDIX 2. THE DETERMINATION OF POINT-LOCI.(1)

From a stray draft sheet(2) in the University Library, Cambridge

[1] Lineas in quarum intersectionibus erant puncta quæsita Veteres nominabant loca illorū punctorū. Inveniebant autem ejusmodi loca omittendo unam conditionem Problematis et quærendo lineam cujus singula puncta implerent condi[ti]ones reliquas. Nam si singula(3) puncta unius lineæ implebant omnes conditiones præter unam & singula alterius implebant omnes præter aliam, necesse erat omnia utriusq; puncta implere omnes adeoq; quæstioni satisfacere. Ut si a duobus punctis A, B ducendæ essent rectæ duæ AC BC datum continentes angulum ACB(4) datāq; habentes rationem ad invicem: neglige conditionem datæ illius rationis(5) et locus puncti C erit circumferentia circuli transeuntis per

---

(21) Problem 10 of the earlier 'Arithmetica'; see v: 196. The following proof is a deal shorter and more elegant.

(22) We complete Newton's unfinished sentence in tune with his parting remark in an immediately preceding cancelled conclusion that 'Qui plura desiderat exempla Pappum adeat'. The reader may decide whether or not he originally intended to go on similarly to analyse and exemplify Euclid's *Porisms* and Apollonius' several works in the field of the 'resolved locus' as at one point (see note (12)) he seems to have had it in mind here to do.

(1) Two consecutive endeavours by Newton to encapsulate what the 'ancients', notably Euclid and Apollonius, understood by an indeterminate 'locus of points', and the manner in which (compare §1.1: note (51) above) they applied the concept to resolve determinate geometrical problems by permitting one (or more) of the conditions defining them to vary. This was subsequently a topic in a long conversation which he held with David Gregory in London in July 1702 on the subject of the Greeks' geometrical analysis, for *inter alia* (compare Appendix 1: note (10) preceding and 2, §1, Appendix 1: note (4) below) the latter took away the thought that 'The ancients solved problems by the intersection of two Loca satisfying each his own condition' (Gregory C 36 = Royal Society MS 247.63).

(2) Add. 3963.3: 17r.

(3) This—not entirely trivially—replaces 'omnia'.

puncta illa data *A*, *B* ac datum angulum *ACB* capientis per [21] III *Elem* ut et per Apollonij *loca plana* a Pappo breviter com[m]emorata. Nēpe[6] punctum *C* ad quod rectæ *AC BC* ductæ datum illum angulum contineant ubivis in circumferentia illa sumi potest, extra eam nullibi. Neglige verò cōditionem dati anguli et[7] locus pūcti *C* erit alterius circuli circumferentia (per Apoll[onij] *loc. plan.*[8]) cujus diameter est *EF* si modo linea *AB* infinite producta secetur punctis *E* et *F* ea lege ut sit *AE* ad *EB* et *AF* ad *FB* in data illa ipsius *AC* ad *BC* ratione. Nam in hujus circumferentia punctū *C* ad quod actæ rectæ *AC BC* sint in data ratione ubiq; est, extra eam nullibi. Est igitur punctū *C* in utraq; circumferentia atq; adeo in utriusq; intersectione. Unde Problema descriptione circumferentiarum illarū solvetur.

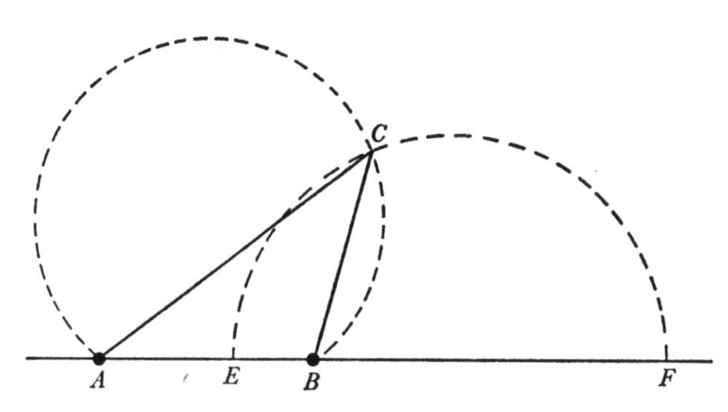

Quod si conditionem dati alteru[t]rius puncti *A* vel *B* omittas prodibunt adhuc aliæ multæ solutiones. Sciendum est autem punctum omne duabus conditionibus dari[,] una[9] qua restringitur a superficie[10] ad lineam aliquam[,] altera qua restringitur a tota linea ad aliquod ejus punctum. Neglige posteriorem conditionem[,] id est acta per alterutrum datum punctum *A* linea[11] quavis *AD* concipe punctum illud *A* non amplius dari sed ubivis in linea illa *AD* sumi posse[,] et prout linea alia atq; alia *AD* agitur habebis alium atq; alium locum puncti *C*.

(4) The manuscript here lacks any figure: that reproduced alongside is our restoration in line with the text.

(5) Originally '...conditionem illam quod lineæ *AC* et *BC* datam habeant illam rationem ad invicem'.

(6) Read 'Nempe'.

(7) In an otherwise identical preceding cancelled phrase Newton initially at this point specified 'ex reliquis condi[tionibus]'.

(8) Again here (and correspondingly below) understand 'a Pappo commemorata'; see his *Mathematicæ Collectiones* (§1.1: note (1)): 248: 'Si à duobus punctis datis rectæ lineæ inflectantur, & sit quod ab vna efficitur eo, quod ab altera dato maior quàm in proportione, punctum positione datam circumferentia[m] continget'. This celebrated Apollonian 'plane' locus was, as we have seen (IV: 231, note (6)), known more than a century earlier to Aristotle, who in his *Meteorologica* (III. v, 376a) elegantly constructed its centre. Though at this point in his work Aristotle gave an explanation of the primary rainbow which was to survive without essential change for more than 1500 years, Newton himself appears never to have looked at it.

(9) A cancelled initial sequel reads 'ut sit in linea aliqua[,] altera ut sit in puncto aliquo illius Lineæ'.

(10) Understand 'plana', of course.

(11) Initially qualified as 'recta vel circulari vel alia'.

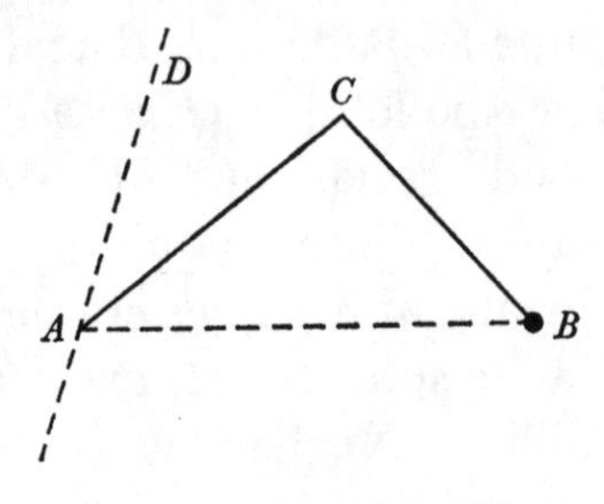

Qui locus (per Apollon. *loc. plan.*) rectilineus erit si acta linea *AD* recta est, circularis si circulus siquidem triangulum *ABC* ob datum angulum *C* datamq rationem *AC* ad *BC* detur specie adeoq angulus *B* ratioq *AB* ad *BC* detur.(12) Et hoc modo habentur loca infinita puncti *C* tam rectilinea quam circularia quorum duo quævis descripta communi intersectione punctum illud dabunt.

Cæterum ex collatione conditionum quæstionis sæpe colliguntur novæ conditiones quarum ope loca facilius determinantur[,] ut in hoc casu ex angulo(13)

---

[2] [Lineas in quarum intersectionibus erant puncta quæsita Veteres nominabant loca illorum punctorum.] Inveniebant autem ejusmodi loca fingendo certam aliquam Problematis conditionem esse incertam omittendo(14) aliquam a conditionibus(15) et quærendo lineam per quam punctum illud jam indeterminatum factū vagetur. Ut si triangulum *ABC* super data basi *AB* datis lateribus *AC BC* constituendum sit, finge latus *BC* incertum esse et punctum *C* vagabitur per circumferentiam circuli centro *A* radio *AC* descripti. Finge latus *AC* incertum esse et punctum *C* vagabitur per circumferentiam circuli centro *B* radio *BC* descripti. Locatur ergo *C* in utraq circumferentia atq adeo describendo circulos illos reperietur in utriusq communi intersectione.

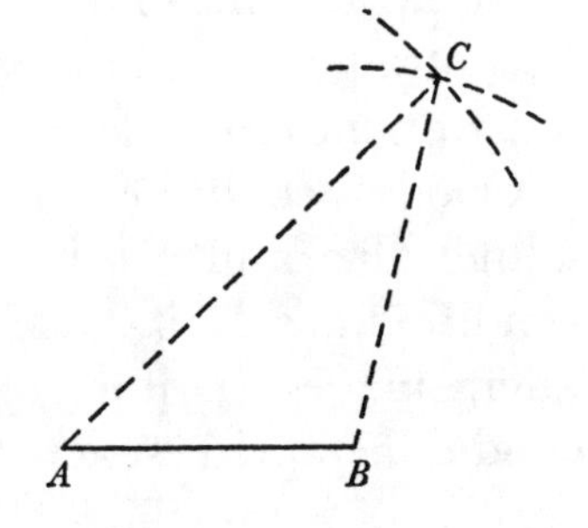

Cæterum ex statu quæstionis non rarò(16) datur unus puncti quæsiti locus et tunc sufficit alium invenire. Ut si a dato puncto *A* ducenda est recta quæ datum circulum *BD* tangat, circumferentia circuli illius est unus locus puncti contactus. Puta factum esse

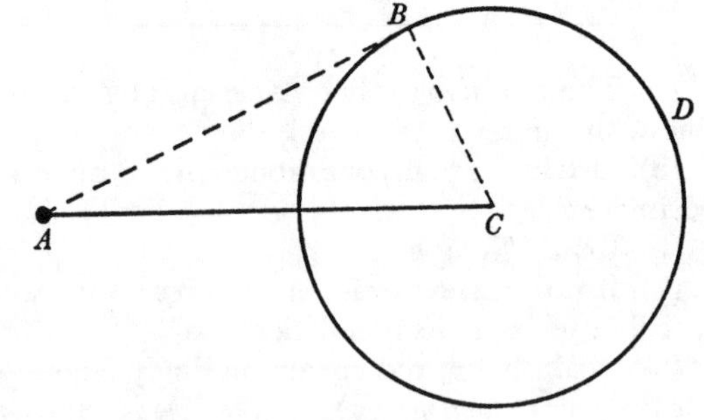

---

(12) Whence, on rotating the locus (*C*) anticlockwise round *B* through the given angle $\widehat{ABC}$, the correspondence $A \leftrightarrow C$ is a homothety of centre *B*.

(13) Newton breaks off in mid-sentence to begin again. Understand '...ex angulo *C* et ratione *AC* ad *BC* datis datur triangulum *ABC* specie'.

(14) 'vel pro incerta hendo [= habendo]' is deleted in sequel.

(15) Originally defined as ones 'quibus punctum quæsitū determinatur', then merely as 'quæstionis', and lastly as 'quæ ad puncti quæsiti determinationem sufficiebant' in a gabble of following attempts by Newton appropriately to specify them.

quod quæritur sitq; B punctum contactus et BC circuli dati radius[,] et erit angulus ABC rectus adeoq; pūctum B si longitu[do] BC fingatur incerta vagabitur per circumferentiam circuli super diametro AC descripti.(17)

---

(16) 'Aliquando' was written at the equivalent point in a cancelled first, near-identical opening to the paragraph in the manuscript's preceding line.

(17) This elegant construction of the tangent from an external point to a circle—one not to be found in Euclid (who in *Elements* III, 17 elaborates a more circuitous method of so doing)—was seemingly first made public by Johann Voegelin in Proposition 7 of the second book of his *Elementale Geometricum ex Euclidis Geometria* (Witeberg, 1528), but it was not widely known till Clavius came in 1574 to introduce it into his edition of the *Elements* (in a scholium to III, 31); see J. Tropfke, *Geschichte der Elementar-Mathematik*, 4 (Berlin/Leipzig, $_2$1923): 116. Even in the seventeenth century it was very far from being the simple textbook exercise at which we no longer cast a second glance. (It does not appear, for example, in Barrow's 1655 *Euclid*, which was Newton's bible in such things.) Whether Newton is here right to insinuate that the construction must have been known to the 'ancients'—finding a place, maybe, in Apollonius' lost treatise on *Plane Loci*?—we may leave the reader to judge.

§2. THE NATURE OF PORISMS AND THEIR INVENTION.[1]

From the original[2] in the University Library, Cambridge

Inventio Porismatum.

Inventio Porismatū ad veterum Analysin apprimè conducebat. Cum hæc ars lateat Geometr[i]s nostri temporis,[3] lubet aliqua huc spectantia proponere.

Porisma est Propositio qua ex conditionibus Problematis datum aliquod ad ejus resolutionem utile colligimus.[4] Formam induit vel Theorematis vel Problematis pro lubitu, ideoq medium quiddam inter utrumq cen[s]etur. Data verò sic colligenda, sunt proportiones directæ et inversæ, aliæq quantitatum ignotarum relationes, species item figurarum, et lineæ in quibus puncta ignota locantur quæq ideo punctorum loca dici solent,[5] nec non longitudines anguliq

(1) Newton here broaches that elusive will-o'-the-wisp which is the classical notion of a 'porism'. (See the preceding introduction to this part.) Basing himself narrowly on Pappus' attempt in his seventh book (*Mathematicæ Collectiones* (§1: note (1)): 245; compare note (4) below) to clarify the nature of a species of mathematical proposition which already in Pappus' day had no uniquely received meaning—and, to his considerable profit, seemingly ignorant of Proclus' essay at its definition which serves little more than to accentuate the confusion over its precise significance—Newton here concentrates on its employment by the 'younger' Greek technical geometers, Euclid paramountly, in the field of the τόπος ἀναλυόμενος to 'discover' indeterminate relationships of general kind: constancies of proportion between lines, similarities (homothetic and anti-homothetic) in correspondences between figures, invariances in point-transformations of any type. He stresses in particular the constant proportions, direct and inverse, which arise in the case of the 1, 1 and 2, 2 (including 1, 2) point-correspondences cut out in given straight lines by rectilinear transversals which are constrained either to pass through given points or to touch given conics, and listing necessary conditions—false in the 1, 2 and 2, 2 cases in fact—for these to be porismatic instances of an Apollonian *sectio rationis* or *sectio spatii*.

(2) Add. 3963.14: 159ʳ–160ᵛ.

(3) Descartes, as we have earlier remarked (IV: 277, note (9)) firmly held the view—in his youthful *Regulæ ad Directionem Ingenii* at least—that the ancient mathematicians deliberately sought to conceal their 'art of discovery' as a professional secret, and a similar belief underlay the efforts of many others of those who in the seventeenth and eighteenth centuries endeavoured to retrieve (and, where necessary, 'restore') the classical techniques of analysis. John Wallis, for instance, in a letter to Kenelm Digby on 21 November 1657—published the following year in a compendium of his correspondence (*Commercium Epistolicum De Quæstionibus quibusdam Mathematicis, nuper habitum*, Oxford, 1658: 33–51) which Newton himself had studied as a newly graduated student (see I: 116)—observed in an aside that 'non ignotum credo fore...id quidem in Archimede à gravissimis viris doctissimisque maximè desiderari, & ta[men] non vitio verti, quod ipse quasi datâ operâ ita occultaverit sua inquisitionis vestigia; quasi invidisset posteris investigandi artem, à quibus tamen assensum inventis extorquere vellet. Sed nec Archimedes solus, verum & veterum plærique omnes, Analyticen suam, (quam habuisse, extra dubium est,) eousque celarunt posteros, ut Recentioribus facilius jam fuerit novam suo marte comminisci...quam indagâsse veterem' (*ibid.*: 43). In his middle years Newton certainly came to be convinced that the ancients possessed an arcane store of physical, chemical and

Translation

THE FINDING OF PORISMS

The finding of porisms was a key element in the ancients' technique of analysis. Since this art lies hidden to geometers of our time,(3) it is agreeable to set out some points regarding it.

A porism is a proposition whereby out of the circumstances of a problem we gather some given thing of use to its resolution.(4) It takes on the form either of a theorem or of a problem at pleasure, and is in consequence reckoned to be a sort of mean between each. The given things, however, which are to be thus gathered are direct and inverse proportions and other relationships of unknown quantities; likewise the species of figures and the lines in which unknown points are located, and which in consequence are usually said to be the loci of the points;(5) and also lengths, angles and points which either regard the deter-

---

astronomical *prisca sapientia* which had only fitfully survived into modern times, notably in the medieval *corpus hermeticum* of Greek alchemical writings (now accepted as spurious). While in a mathematical context it was not in his character entirely to dismiss so seductive an explanation of the rarity of known Greek texts expounding techniques and modes of analysis, he here presents a neutral face on the matter.

(4) Newton first began by asserting 'Porisma est Corollarium ex datis Problematis conditionibus datam novam conditionem [exhibens]' (A porism is a corollary (exhibiting) from the given conditions of a problem some new given condition). He has also cancelled at this point an initial version of the sequel stating 'estqꝫ nec purum Problema quia formam habet Theorematis, nec purum Theorema quia problematice colligitur & quæsitũ aliquod determinat' (and it is neither a pure problem since it has the form of a theorem, nor a pure theorem since it is gathered in the manner of a problem and determines some requirement). In preliminary to summarizing the content of Euclid's lost tripartite work *De Porismatibus* Pappus had long before remarked (*Mathematicæ Collectiones:* 245) that '[Porismatum] species omnes neqꝫ theorematum sunt, neqꝫ problematum, sed mediam quodammodo inter hæc formam, ac naturam habent, ita vt eorum propositiones formari possint, vt theorematum, vel vt problematum', further observing that 'Horum...trium differentiam veteres multo melius cognouisse ex definitionibus perspicuum est. Dixerunt enim theorema esse, quod proponitur in ipsius propositi demonstrationem. Problema, quod affertur in constructionem propositi. Porisma vero, quod proponitur in porismum, hoc est in inuentionem, & inuestigationem propositi. Immutata autem est hæc porismatis definitio à iunioribus, qui nequeunt omnia inuestigare, sed his [vt] elementis vtuntur, & ostendunt solummodo quod hoc est quod quæritur, non autem illud ipsum inuestigant....[Hi] sic porisma definierunt[:] Porisma est quod hypothesi deficit à locali theoremate'. (For no very good reason Hultsch has supposed that the two last sentences are an interpolation in Pappus' original text; compare T. L. Heath, *A History of Greek Mathematics*, **1**, Oxford, 1921: 431–2, where the concluding definition of the 'juniors' is rendered: 'A porism is that which falls short of a locus-theorem in respect of its hypothesis'.)

(5) Pappus had written in immediate sequel to the passage cited in the previous note that 'Huius...generis porismatum loci ipsi sunt vna species, atque de hac ipsa abunde tractatur in resoluto loco; seorsum autem à porismatibus collecta, inscriptaqꝫ ac tradita sunt, quod magis diffusa, & copiosa sit cæteris speciebus' (*Mathematicæ Collectiones*: 245).

et puncta quæ vel ad determinationem locorum(6) spectant vel aliàs ad resolutionem Problematis conducunt. Proportiones autem primum locum obtinent et investigantur per sequentia Theoremata.(7)

1. Si quantitates duæ indeterminatæ, earumve potestates homogeneæ se mutuo simpliciter(8) determinant, eæqꝫ vel simul evanescunt et simul evadunt infinitæ, vel simul evanescunt & semel redeunt ad eandem rationem(9) vel deniqꝫ bis redeunt ad eandem rationem:(9) eædem erunt in data ratione.(10)

2. Sin fieri possit ut utraqꝫ successive evanescit ubi altera fit infinita, vel si earum una evanescit ubi altera fit infinita & ambæ semel redeunt ad idem medium proportionale(11) vel deniqꝫ si bis redeunt ad idem medium proportionale:(11) eædem erunt reciproce proportionales.(12)

3. Si quantitates duæ indeterminatæ vel earum potestates intermediæ homogeneæ se mutuò dupliciter(13) determinant, et plana aliqua ex ipsis seorsim confecta simul evanescunt & simul evadunt infinita bis, vel si harum quatuor conditionum una vel plures desit & defectus suppleatur per tot reditus ad eandem rationem:(14) erunt hæc plana in data ratione.(15)

4. Quod si utrūqꝫ successivè evanescit ubi alterū evadit infinitū idqꝫ bis, vel si harum quatuor conditionum una vel plures desint et defectus suppleatur per tot

(6) 'illorum' (those) is deleted.

(7) Originally 'sequentes regulas' (...rules).

(8) By a 1, 1 correspondence, that is. In Proposition 11 of his earlier 'Solutio Problematis Veterum de Loco solido', as we have seen (IV: 306), Newton had accurately assigned the 'plenissima relatio quantitatum [x & y] quæ ab invicem per simplicem Geometriam determinabiles sunt' to be of the general form $axy+bx+cy+d = 0$, where a, b, c, d are constants.

(9) Newton first wrote 'proportionales sint' (be proportional).

(10) In the algebraic terms of note (8) above, if $x \leftrightarrow y$ by the 1, 1 correspondence

$$axy+bx+cy+d = 0$$

such that $0 \leftrightarrow 0$ and $\infty \leftrightarrow \infty$, then by the first $d = 0$ and by the latter $a = 0$, so that $bx+cy = 0$ and consequently $x/y = -c/b$, constant. If, alternatively, for some $X_1 \neq X_2$ (and so $Y_1 \neq Y_2$) there proves to be $X_1/Y_1 = X_2/Y_2$ ($= k$, say), then at once $akY_1^2+(bk+c)\,Y_1+d=0$ and $akY_2^2+(bk+c)\,Y_2+d=0$; and hence, if either $0 \leftrightarrow 0$ or $\infty \leftrightarrow \infty$ (so that correspondingly $d = 0$ or $a = 0$), there ensues also $a = 0$ or $d = 0$ and $bk+c = 0$, whence again $x/y = -c/b$, constant. If, finally, for some $X_1 \neq X_2 \neq X_3$ (and so $Y_1 \neq Y_2 \neq Y_3$) there proves to be

$$X_1/Y_1 = X_2/Y_2 = X_3/Y_3 = k,$$

then $akY_i^2+(bk+c)\,Y_i+d = 0$, $i = 1, 2, 3$, and therefore yet again $a = bk+c = d = 0$. (Newton had earlier obtained the reduction $a = 0$ which ensues from the condition $\infty \leftrightarrow \infty$ by an elegant equivalent appeal to the geometrical model of a straight line defined in oblique Cartesian coordinates by the equation $bx+cy+d = 0$, where the crucial ratios $b:c:d$ are fixed by two points (X_i, Y_i), $Y_i = kX_i+l$, given upon it; see III: 342–4. He evidently here regards the porismatic instance $d = 0$ of this, in which the coordinates x, y are 'cut off' in the constant ratio $-c/b$, too simple to be set as a following example.)

(11) Newton has here (compare note (9) above) replaced 'sunt reciproce proportionales' (are reciprocally proportional), also deleting a related following clause 'habeantqꝫ idem

mination of[6] loci or otherwise contribute to the resolution of a problem. But proportions hold first place and are tracked down by means of the following theorems:[7]

1. If two indeterminate quantities, or their uniform powers, mutually determine each other simply,[8] and these either simultaneously vanish and simultaneously become infinite, or simultaneously vanish and once return to the same ratio,[9] or finally if they twice return to the same ratio,[9] then they will be in a given ratio.[10]

2. But should it happen that each successively vanishes when the other becomes infinite, or if one of them vanishes when the other becomes infinite and both return once to the same mean proportional,[11] or finally if they twice return to the same mean proportional,[11] then they will be reciprocally proportional.[12]

3. If two indeterminate quantities, or their uniform intermediate powers mutually determine each other doubly,[13] and certain 'planes' (quadratic products) separately made up out of them twice simultaneously vanish and simultaneously prove to be infinite, or should one or more of these four circumstances be lacking and the deficiency be filled by an equal number of returns to the same ratio,[14] then these 'planes' will be in a given ratio.[15]

4. But if each successively vanishes when the other comes to be infinite, and that twice, or if one or more of these four circumstances be lacking and the deficiency be filled by an equal number of returns to the same mean propor-

---

medium proportionale in utraq casu' (and so shall have the same mean proportional in each case) hereby rendered superfluous.

(12) Much as in note (10), if in the general 1, 1 correspondence $x \leftrightarrow y$ determined by $axy+bx+cy+d = 0$, there is now $0 \leftrightarrow \infty$ and $\infty \leftrightarrow 0$, at once $c = b = 0$ respectively, and hence $xy = -d/a$, constant. If, alternatively, one or both of the latter conditions are replaced by one ($i = 1, 2$) or two ($i = 1, 2, 3$) 'returns' ($X_1 \neq X_2 \neq X_3$) to the condition $\sqrt{(X_i Y_i)} = k$, whence $bX_i+cY_i+(d+ak^2) = 0$, then $b = c = d+ak^2 = 0$ and yet again $xy = -d/a$, universally.

(13) That is, (in the algebraic terms of note (8) preceding) if the general quantities x and y are now related by some 2, 2 correspondence

$$ax^2y^2+(bx+cy+d)\,xy+ex^2+fy^2+gx+hy+i = 0.$$

(14) Originally (compare note (9)) 'per tot directas proportionalitates earundem quantitatum' (by an equivalent number of direct proportionalities of the same quantities).

(15) If in the general 2, 2 correspondence $x \leftrightarrow y$ of note (13) there is ∞ (twice) $\leftrightarrow \infty$ (twice), then at once $a = b = c = d = 0$ and so the correspondence reduces to be, for any k,

$$(x-\alpha)\,(x-\beta)/(y-\gamma)\,(y-\delta) = -f/e$$

where α, β and γ, δ are the roots respectively of $ex^2+gx+k = 0$ and $fy^2+hy+(i-k) = 0$. In the present particular instance in which also 0 (twice) $\leftrightarrow$ 0 (twice), and hence $\alpha = \beta = \gamma = \delta = 0$ and so $g = h = i = k = 0$, the defining conditions of the ensuing correspondence $ex^2+fy^2 = 0$ may evidently be replaced by an appropriate number of 'returns' of the ratio x^2/y^2 to the constant value $-f/e$.

reditus ad idem medium proportionale:[16] dabitur planorum medium proportionale.[17] Et sic pergitur in infinitum.

Quantitates autem se mutuo simpliciter, dupliciter, tripliciter determinare dico ubi ex alterutra assumpta prodit altera simpliciter absq; extractione radicum vel dupliciter per extractionem radicis quadraticæ vel tripliciter per extractionem radicis cubicæ et sic deinceps.[18]

5. Addere licet et hujusmodi Theoremata. Si quantitas aliqua x determinat alteram quantitatem y simpliciter, quantitas autem illa altera y determinat quantitatem primam x dupliciter, et ubi illa altera y vel evanescit vel infinita evadit, planum ex hac prima x vel bis evanescit vel bis evadit infinita, planum illud & quantitas illa altera y in aliam quamvis quantitatem datam ducta et xx vel datam habebunt rationem si simul evanescent et simul infinitæ fiunt, vel datum medium proportionale si contra. Et idem eveniet ubi una vel plures ex his conditionibus desunt & defectus suppletur per tot reditus vel ad eandem rationem vel ad idem medium proportionale.[19] Res[20] patebit exemplis.

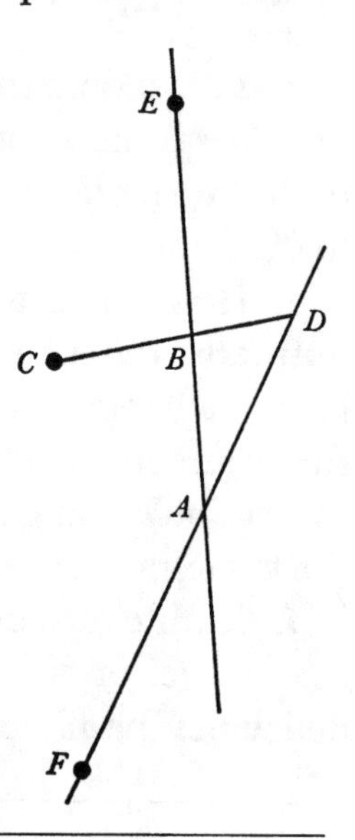

Porisma 1. A dato puncto C ad datas positione rectas AB, AD, ducitur recta mobilis CBD: ad resolutionem Problematis alicujus quæritur relatio simplicissima rectarum AB, AD.

Quoniam hæ se mutuo simpliciter determinant, abeat

(16) Originally (compare notes (11) and (14)) Newton wrote 'per tot reciprocas proportionalitates quantitatum' (by an equal number of reciprocal proportionalities).

(17) In this case, if the 2, 2 correspondence $x \leftrightarrow y$ set out in note (13) simultaneously satisfies the conditions $\alpha, \beta \leftrightarrow \infty$ (twice) and ∞ (twice) $\leftrightarrow \gamma, \delta$, then α, β and γ, δ are respectively the roots of $ax^2+cx+f \equiv k(bx^2+dx+h) = 0$ and $ay^2+by+e \equiv l(cy^2+dy+g) = 0$, so that $ad = bc$, $ag = ce$ and $ah = bf$, and the correspondence reduces to be

$$ax^2y^2+(bx+cy+bc/a)\,xy+ex^2+fy^2+(ce/a)\,x+(bf/a)\,y+i = 0,$$

that is, $(ax^2+cx+f)\,(ay^2+by+e) = ef-ai$ or

$$(x-\alpha)\,(x-\beta)\,(y-\gamma)\,(y-\delta) = (ef-ai)/a^2,$$

constant. In Newton's present instance, there is $\alpha = \beta = \gamma = \delta = 0 = b = c = e = f$ and the correspondence is therefore $ax^2y^2+i = 0$, one or both of whose defining conditions

$$0 \text{ (twice)} \leftrightarrow \infty \text{ (twice)} \quad\text{and}\quad \infty \text{ (twice)} \leftrightarrow 0 \text{ (twice)}$$

may be replaced by an equal number of 'returns' of the mean proportional, xy, between x^2 and y^2 to the constant value $\sqrt{(-i/a)}$.

(18) In the algebraic equivalent (compare notes (8) and (13) above in the cases where $n = 1, 2$) the quantities x and y 'mutually determine each other n-ply' when they are related in the n, n correspondence $x \leftrightarrow y$ defined by

$$ax^ny^n+(bx+cy+d)\,x^{n-1}y^{n-1}+(\ldots)\,x^{n-2}y^{n-2}+\ldots ex^n+fy^n+gx^{n-1}+hx^{n-1}+\ldots+l = 0.$$

(19) In this particular instance of Theorems 3 and 4 preceding (compare notes (15) and (17) when $a = c = f = 0$) the general 2, 1 correspondence $x \leftrightarrow y$ is defined by the equation $(bx^2+dx+h)\,y+ex^2+gx+i = 0$. Accordingly, given that ∞ (twice) $\leftrightarrow \infty$ and also $\alpha, \beta \leftrightarrow \gamma$,

tional,(16) then the mean proportional of the 'planes' will be given.(17) And so you may go on indefinitely.

I say, however, that quantities mutually determine each other simply, doubly, triply when, once either is assumed, the other results from it simply without extraction of roots, doubly by extraction of a square root, or triply by the extraction of a cube root. And so forth.(18)

5. It is permissible to add also theorems of this sort. If some quantity x determines a second quantity y simply, while that other quantity y determines the first quantity x doubly, and when that other one y either vanishes or comes to be infinite a 'plane' from this first one x, either twice vanishes or twice comes to be infinite, then that 'plane' and the other quantity y multiplied by any other given quantity and x^2 shall have either a given ratio (if they simultaneously vanish and simultaneously become infinite) or a given mean proportional (if the inverse). And the same will happen when one or more of these circumstances are lacking and the deficiency is filled by an equal number of returns either to the same ratio or to the same mean proportional.(19) The matter(20) will be evident from examples.

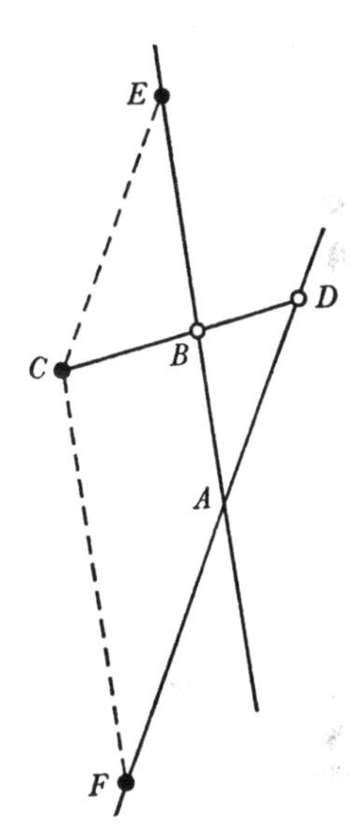

Porism 1. From the given point C to the straight lines AB, AD given in position a mobile straight line CBD is drawn: for the resolution of some problem there is required the simplest relationship of the lines AB, AD.

Seeing that these mutually determine one another simply, let the point D pass off to infinity and in that case let the point

it follows that $b = d = 0$ and that α, β are the roots of $ex^2+gx+i+hy = 0$, whence the correspondence is $(x-\alpha)(x-\beta)/(y-\gamma) = -h/e$, constant. Where, however, it is given that $\alpha, \beta \leftrightarrow \infty$ and also ∞ (twice) $\leftrightarrow \gamma$, then α, β are the roots of $bx^2+dx+h = 0$ and γ is the root of $by+e \equiv k(dy+g) = 0$, so that $de = bg$ and the correspondence reduces to be

$$(bx^2+dx+h)(by+e) = eh-bi \quad \text{or} \quad (x-\alpha)(x-\beta)(y-\gamma) = (eh-bi)/b^2,$$

constant. It will be evident that in Newton's present instances, where $\alpha = \beta = \gamma = 0$, one or both of the given conditions 0 (twice) $\leftrightarrow$ 0, ∞ (twice) $\leftrightarrow \infty$ and 0 (twice) $\leftrightarrow \infty$, ∞ (twice) $\leftrightarrow$ 0 may without loss be replaced by an equal number of returns to the constant values of the ratio x^2/y and product x^2y which respectively ensue.

(20) Newton first misleadingly wrote 'Methodus' (The method). In the seven particular porisms which he propounds in sequel to exemplify the preceding general rules he nowhere in fact attempts to do more than straightforwardly apply them—and even then he fails to observe that the defining conditions laid down on the 2, 2 and 2, 1 correspondences postulated in his Theorems 3–5 are not (see notes (27), (29) and (40) below) duplicated in the circumstances of Porisms 3–6 in which he mistakenly endeavours to exemplify them. Since he never again sought to make a similar application to the case of 'doubly mutual' correspondences generated by tangents to a fixed conic, we may perhaps presume that Newton came subsequently at least to notice his error, if not to correct it.

punctum D in infinitum et eo in casu incidat punctum [B] in punctum E. Abeat jam punctum B in infinitum et eo in casu incidat punctum D in punctum F[,] et rectangulum $EB \times FD$ determinatum erit per Theor. 2. Incidant puncta B ac D in A et rectangulum illud evadet $EA \times FA$, ideoqꝫ datur. Est igitur $EB \times FD = EA \times FA$ et inde $EB.EA::FA.FD$.(21)

Porisma 2. Circulum vel sectionem quamvis Conicam(22) GCH tangant tres rectæ AG, AH, BCD in G, H, C, et earum duæ AG, AH dentur positione [ac] tertia BD mobilis existat: ad resolutionem Problematis alicujus quæritur relatio simplicissima rectarum AB, AD.

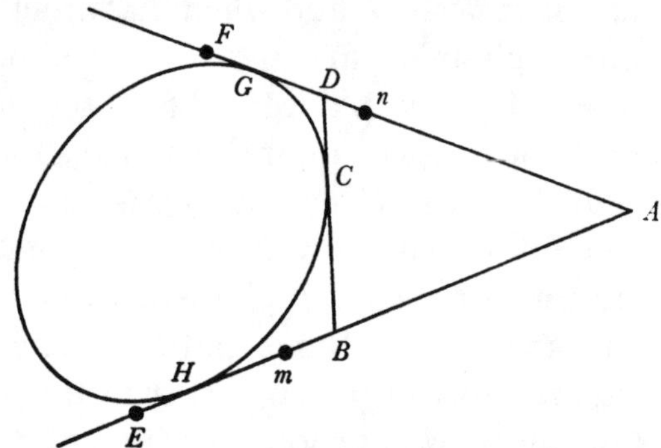

Quoniam puncta B, D se mutuò simpliciter determinant, abeat punctum D versus A in infinitum et eo in casu incidat punctum B in E. Abeat jam B in infinitum versus A et eo in casu incidat punctum D in F[,] et rectangulum $FD \times EB$ determinatum erit per Theor. 2. Incidat punctum D in punctum A et punctum B incidet in contactum(23) H, ideoqꝫ rectangulum illud nunc erit $EH \times FA$ et propterea datur, estqꝫ $EB \times FD = EH \times FA$ et inde $EB.EH::FA.FD$.(24)

Casum excipe ubi Figura GCH Parabola est. Nam puncta B, D in hoc casu simul abeunt in infinitum. Ideoqꝫ quæro puncta duo quævis determinata in quæ B ac D simul incidunt, puta m et n(25) et dico (per Theor 1) quod mB nD sunt in data ratione; nimirum in ratione mA ad nG propterea quod nD evadit nG ubi mB evadit mA. Si puncta m et H coincidunt, puncta n et A coincident, et nunc erit HB ad AD ut AH ad AG.(26)

Porisma 3. Inter rectas positione datas AE AF ducitur recta mobilis BCD quæ

(21) Which, since the point-correspondence $B \leftrightarrow D$ is 1, 1 (and hence projective), may be encapsulated in the equality of the cross-ratios $(\infty_{AE}\, EAB) = (F\, \infty_{AF}\, AD)$. This indeterminate case of the Apollonian *sectio spatii* (on which see §1.1: note (30) preceding) is listed by Pappus among the propositions of Euclid's first book *On porisms* under the head 'Quod hæc [*sc.* recta à puncto dato ducta] abscindit a positione datis [rectis spatium] datum continentes' (*Mathematicæ Collectiones:* 246).

(22) According to Pappus, Euclid himself (in the third book of his lost work where he passes to consider the porismatic properties of other than straight lines) would—if at all—have considered only the circular case.

(23) Initially more narrowly specified to be 'punctum contactus' (point of contact).

(24) That is, in equivalent cross-ratio form, $(\infty_{AE}\, EHB) = (F\, \infty_{AF}\, AD)$. The result is, of course, that obtained by Newton in a straightforwardly Apollonian form in Corollary 2 to Lemma XXV of the first book of his *Principia*; compare VI: 280–1, note (87).

(25) Newton first wrote 'qualia sunt H et A' (such as H and A)—with the justification in parenthesis 'nam si punctum B transit in H punctum D transit in A' (for if point B passes into

B fall at the point E; now let the point B pass to infinity and in that case let the point D fall at the point F, and then the rectangle $EB \times FD$ will be determined by Theorem 2. Let the points B and D fall at A, and that rectangle will prove to be $EA \times FA$, and is consequently given. There is therefore $EB \times FD = EA \times FA$ and thence $EB:EA = FA:FD$.(21)

Porism 2. Let any circle or conic(22) GCH be touched by three straight lines AG, AH, BCD in G, H, C, and let two of them, AG and AH, be given in position while the third, BD, remains mobile: for the resolution of some problem there is sought the simplest relationship of the lines AB, AD.

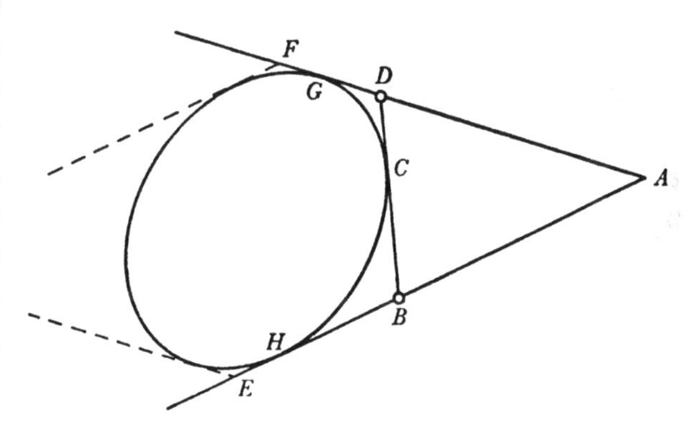

Seeing that the points B, D mutually determine one another simply, let the point D pass in A's direction to infinity and in that case let point B fall at E; now let B pass in A's direction to infinity and in that case let the point D fall at F, and then the rectangle $FD \times EB$ will be determined by Theorem 2. Let point D fall at the point A and point B will fall at the contact(23) H, and consequently that rectangle will now be $EH \times FA$ and accordingly given, and so there is $EB \times FD = EH \times FA$ and thence $EB:EH = FA:FD$.(24)

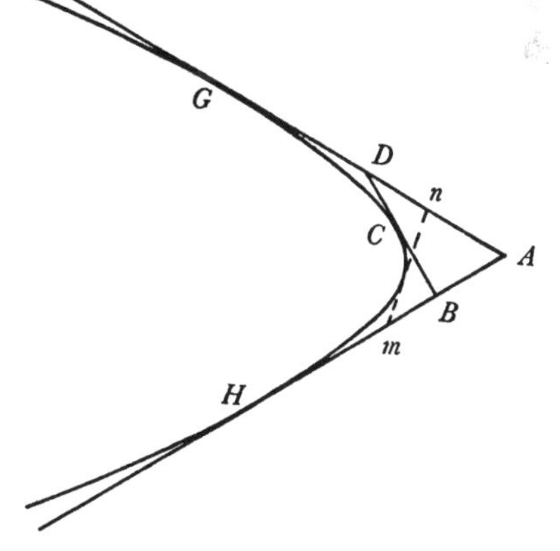

Exclude the case in which the figure GCH is a parabola. For the points B and D in this case pass simultaneously to infinity. Consequently I seek any two determinate points, m and n say,(25) at which B and D simultaneously fall, and state (by Theorem 1) that mB and nD are in a given ratio; namely that of mA to nG, inasmuch as nD comes to be nG when mB comes to be mA. If the points m and H coincide, the points n and A will coincide and there will now be HB to AD as AH to AG.(26)

Porism 3. Between straight lines AE, AF given in position there is drawn the mobile straight line BCD touching a circle or ellipse or hyperbola(22) GCH in

H, point D passes into A)—and then asserted ‘quod HB et AD sunt in ratione data, nimirum in ratione HA ad AG propterea quod AD evadit AG ubi HB evadit HA’ (that HB and AD are in a given ratio; namely that of HA to AG, inasmuch as AD comes to be AG when HB comes to be HA).

tangit circulum vel Ellipsin vel Hyperbolam(22) GCH in puncto aliquo C: invenienda est simplicissima relatio rectarum AB AD.

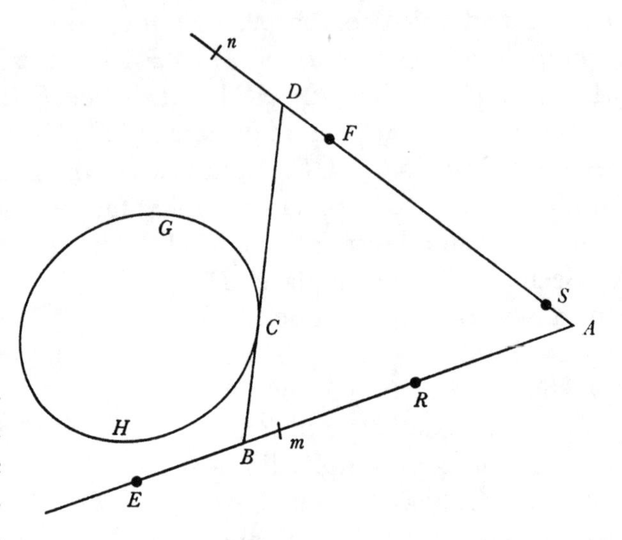

Quoniam puncta B, D se mutuò determinant dupliciter (hoc est si detur punctorum alterutrum B, ab eo duæ duci possunt rectæ curvam tangentes, eæq; duo dabunt puncta D) ideò efficiendo ut punctū B secundum determinationem utramq; abeat in infinitum invenio puncta F, S in quæ punctum D tunc incidit[,] et vicissim efficiendo ut punctum D abeat in infinitum invenio puncta duo E, R in quæ punctum B tunc incidit et concludo per Theor 4 quod rectangula EBR[,] FDS sunt determinata reciproce proportionalia.(27) Incidant simul puncta B, D in data quævis puncta m, n et rectangula illa evadent EmR, FnS. Incidant in punctum A et evadent EAR, FAS. [Sunt] igitur [rectangula] EBR, FDS [et] EmR, FnS nec non EAR, FAS [reciproce proportionalia].(28)

(26) This tangent property of the conic parabola that $mB/mA = nD/nG$—or, in equivalent cross-ratio form, $(\infty_{AH}\, mAB) = (\infty_{AG}\, nGD)$— is an immediate corollary of Apollonius, *Conics* III, 41 (which here prescribes that $AB:BE = FD:DA$ and likewise $Am:mE = Fn:nA$, and indeed that the respective tangents BD and mn are similarly divided by their points of contact with the parabola). Notice that, should the four tangents AE, AF, BD, mn be given in position, the point G in which AF touches the parabola is at once fixed by $nG = nD \times mB/mA$, and *mutatis mutandis* for the three other tangents. When Edmond Halley some fifteen years afterwards independently discovered this theorem that 'Tangentes Parabolæ auferunt semper segmenta in datâ ratione', he enthusiastically proclaimed it as a 'Propositio perpulchra' which 'nova mihi visa est' (*Apollonii Pergæi de Sectione Rationis Libri Duo Ex Arabico MS*to. *Latine versi* (Oxford, 1706): 64), though of his several equivalent enunciations of it he added in a more subdued tone that '[h]æ omnia manifesta sunt, ex eo, quod rectas quatuor Parabolam contingentes, ita sese intersecare necesse sit, ut quælibet Tangens similiter divisa sit (sive in partes proportionales) ad puncta intersectionum & contactuum'.

(27) Newton fails to notice that the second tangents drawn from E, R and F, S to the conic are not also simultaneously parallel to AD and AB, and hence that it is spurious here to apply Theorem 4. We need not stress that the remainder of his present paragraph is irremediably false. Since A itself is a common double point in the 2, 2 correspondence $B \leftrightarrow D$, and so in the related correspondence $AB = x \leftrightarrow AD = y$ there is 0 (twice) $\leftrightarrow$ 0 (twice), we would expect the latter relation to be defined by an equation of the (irreducible) general form

$$ax^2y^2 + (bx + cy + d)\, xy + ex^2 + fy^2 = 0.$$

That this is so may readily be confirmed by regarding AB and AD as the axes of an oblique

some point C: we have to find the simplest relationship of the straight lines AB, AD.

Since the points B and D mutually determine one another doubly (that is, if either one of the points, B, be given, two straight lines can be drawn from it to touch the curve and these will yield two points D), by making the point B—and its dual determinations in its train—to pass away to infinity I find in consequence points F, S at which the point then falls, and *vice versa* making the point D to pass to infinity I find two points E, R at which point B thereupon falls, and I conclude by Theorem 4 that the rectangles $EB \times BR$ and $FD \times DS$ are determined as reciprocally proportional.(27) Let the points B and D fall simultaneously at any given points m and n, and those rectangles will prove to be $Em \times mR$ and $Fn \times nS$. Let them fall at the point A and they will come to be $EA \times AR$ and $FA \times AS$. Therefore the rectangles $EB \times BR$ and $FD \times DS$, $Em \times mR$ and $Fn \times nS$, and also $EA \times AR$ and $FA \times AS$ are reciprocally proportional.(28)

---

system of Cartesian coordinates: in this representation the transversal $Y = (X-x)\, y/x$ passing through $B(x, 0)$, $D(0, y)$ and the general point $C(X, Y)$ meets the conic GH, defined (say) by $AX^2+2HXY+BY^2+2GX+2FY+C = 0$, such that

$$(Ax^2+2Hxy+By^2)\, X^2-2(Hxy+By^2+Gx-Fy)\, Xx+(By^2-2Fy+C)\, x^2 = 0,$$

and the condition

$$(Ax^2+2Hxy+By^2)\, (By^2-2Fy+C) = (Hxy+By^2+Gx-Fy)^2$$

for it to be tangent at once reduces to the above form on setting

$$AB-H^2 = a, \quad -2(AF+GH) = b, \quad -2(BG+FH) = c, \quad 2(CH+FG) = d,$$
$$AC-G^2 = e \quad \text{and} \quad BC-F^2 = f.$$

Clearly (corresponding to $y = \infty$) the lengths AE, AR are the roots of $ax^2+cx+f = 0$ and (corresponding to $x = \infty$) the lengths AF, AS are the roots of $ay^2+by+e = 0$. Since, however, five conditions are needed to determine the ratios $a:b:c:d:e:f$, no geometrical condition uniting these four given line-lengths can (without introducing some further restriction upon it) b framed to mirror the 2, 2 correspondence between the points B and D.

(28) We make trivial correction of some verbal inconsistencies in the manuscript text—where Newton first erroneously deduced 'per Theor 4' that 'rectangula *EBR FDS* sunt in data aliqua ratione', concluding correspondingly in sequel that similarly 'ratio illa evadet *EmR* ad *FnS* ... et ... *EAR* ad *FAS*', before amending his argument to its present form—to make it a fully accurate, if (see the previous note) thoroughly spurious, consequence thereof. In a cancelled following paragraph he initially went on, in line with the equivalent proviso in his preceding Porism 2, to observe: 'Attamen excipiendus est casus ubi curva *GCH* Parabola est propterea quod puncta *B*, *D* in eo casu simul abeunt in infinitum idq; bis' (Exception needs, however, to be made of the case where the curve *GCH* is a parabola, seeing that the points B and D in that case pass simultaneously off to infinity, and that twice). The latter assumption is as erroneous as before. In the algebraic equivalent of the previous note, the condition for the conic CGH to be a parabola is that $AB = H^2$ and so $a = 0$, so that the 2, 2 correspondence $AB = x \leftrightarrow AD = y$ reduces to be of the form $(bx+cy+d)\, xy+ex^2+fy^2 = 0$; accordingly, to $x = \infty$ there corresponds not only $y = \infty$ but $y = -e/b$, and to $y = \infty$ not only $x = \infty$ but also $x = -f/c$.

Porisma 4. Rectarum AB AD alterutra AD tangit circulum Ellipsin vel Hyperbolam in G[,] altera vero AB non tangit, et invenienda sit simplicissima relatio ipsarum AB, AD.

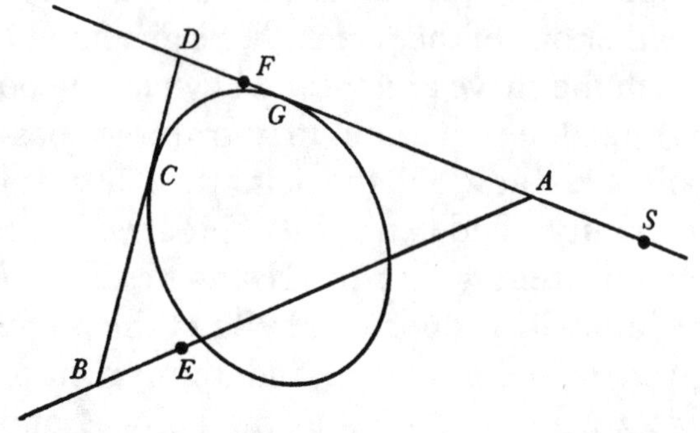

Quoniam punctum D determinat punctum B simpliciter, punctum vero B determinat punctum D dupliciter: abeat punctum D in infinitum et sit E locus in quem B jam incidit: abeat deinde punctum B in infinitum et sint F, S loca in quæ punctum D secundum geminam suam determinationem incidit et erit BE reciproce ut rectangulum FDS per Theor 5.(29)

Porisma 5. Invertendo præcedentia colligitur quod si recta aliqua mobilis BD secet(30) alias duas positione datas rectas AE, AF in quibus dantur aliquot puncta E, R, F, S: et si duo quævis EB, FD sint directe in proportione data tunc recta BD tangit Parabolam.(31) Sin duæ illæ sint reciproce ut EA ad FA tunc recta illa convergit ad punctum datum:(32) si verò sint reciproce in majore vel minore ratione quam est EA ad FA tunc recta illa circulum a[u]t Ellipsin aut Hyperbolam tangit et figura tacta tangit rectas positione datas AE, AF.(33) Porro si rectangulum EBR est ad rectang[ulum] FDS reciprocè ut rectangulum EAR ad rectangulum FAS,(34) tunc recta BD tangit circulum vel Ellipsin vel Hyperbolam et figura tacta non tangit rectas positione datas AE AF. Quod si proportio illa major sit vel minor quam EAR ad FAS tunc recta BD tanget curvam lineam uno

(29) Once more, much as in his previous Porism, Newton fails to notice that when D is at F or S both corresponding points B are, in each instance, not simultaneously at infinity in AE, so that his attempted present application of Theorem 5 is fallacious. To continue the algebraic equivalent of note (27), since the condition (necessary and sufficient) for the conic CG to touch AD is that $BC = F^2$ and so $f = 0$, in this case the relationship $AB = x \leftrightarrow AD = y$ is the pair of the constancy $x = 0$ and the general 1, 2 correspondence $(ay^2+by+e)\ x+cy^2+dy = 0$. On choosing as the four given conditions necessary to determine the ratios $a:b:c:d:e$ the distances from A of E (where $y = \infty$), F, S (at both which $x = \infty$) and the contact point G (the correspondent of $x = 0$ other than A)—that is, $AE = -c/a$, $AF = \gamma$, $AS = \delta$ (where γ and δ are the roots of $ay^2+by+e = 0$) and $AG = -d/c$ respectively—the correspondence is

$$(y-AF)\ (y-AS)\ x = AE.y^2 - AE \times AG.y \quad \text{or} \quad FD \times SD \times AB = AE \times AD \times GD.$$

And so Newton ought here to have found. We may add that when the conic touches both AB and AD (so that $AC = G^2$ and $BC = F^2$, and therefore $e = f = 0$) the relationship proves, on excluding the constancies $x = 0$ and $y = 0$, to be the general 1, 1 correspondence

$$axy+bx+cy+d = 0;$$

in which case S coincides with A, AF is equal to $-c/a$ and the relationship may be expressed geometrically in either of the equivalent forms $AB \times FD = AE \times GD$ or $EB \times FD = AE \times GF$, constant (the latter being *mutatis mutandis* that of Newton's Porism 2 above).

Porism 4. Of the straight lines AB, AD either one, AD, touches the circle, ellipse or hyperbola at G, but the other, AB, is not tangent: we need to find the simplest relationship of AB and AD.

Seeing that the point D determines the point B simply, while point B determines point D doubly, let point D pass away to infinity and E be the place at which B now falls; next, let point B pass to infinity and F, S be the places at which the point D in accordance with its twin determination falls, and there will then be BE reciprocally as the rectangle $FD \times DS$ by Theorem 5.[29]

Porism 5. By inverting the preceding we gather that, should some mobile straight line BD intersect[30] another two straight lines AE, AF, given in position, wherein there are a number of points E, R, F, S given, then, if any two lengths EB, FD are directly in a given proportion, the line BD touches a parabola;[31] but if those two be reciprocally as EA to FA, that line is directed through a given point;[32] while if they be reciprocally in a ratio greater or less than that of EA to FA, then that line touches a circle, ellipse or hyperbola and the figure so touched touches the straight lines AE, AF given in position.[33] Further, if the rectangle $EB \times BR$ is to the rectangle $FD \times DS$ reciprocally as the rectangle $EA \times AR$ to the rectangle $FA \times AS$,[34] then the straight line BD touches a circle, ellipse or hyperbola, while the figure so touched does not itself touch the straight lines AE, AF given in position; but if the former ratio be greater or less than that of $EA \times AR$ to $FA \times AS$, then the line BD will touch a

(30) Continue to understand in B and D respectively, of course.

(31) For in this converse of the second part of Porism 2, the constancy of the ratio EB/FD determines that the correspondence $B \leftrightarrow D$ is 1, 1 such that $E \leftrightarrow F$ and also $\infty_{AE} \leftrightarrow \infty_{AF}$; whence the transversal BD envelops a conic which is touched by the line at infinity, and consequently must be a parabola.

(32) Here the resulting equality $EB/EA = FA/FD$, that is $(\infty_{AE}\, EAB) = (F\, \infty_{AF} AD)$ in equivalent cross-ratio form, determines the correspondence $B \leftrightarrow D$ to be 1, 1 such that the intersection A is a self-corresponding point; it follows therefore, exactly as in Porism 1 above, that the general transversal passes always through the meet, C, of the parallel through E to AF and of the parallel through F to AE.

(33) If, to continue the preceding note, A is not a self-correspondent point where $B \leftrightarrow D$ such that EB/FD is constant, let there be $H \leftrightarrow A$ and the ensuing cross-ratio equality

$$(\infty_{AE}\, EHB) = (F\, \infty_{AF}\, AD)$$

then determines more generally that BD envelops a central conic (but not a parabola since ∞_{AE} does not correspond to ∞_{AF}) which touches AE in H, and is touched by the parallels through E to AF and through F to AE, exactly as in the first part of Porism 2 above (compare note (24)).

(34) That is, if $EB \times BR/EA \times AR = FA \times AS/FD \times DS$ and so

$$EB \times BR \times FD \times DS = EA \times AR \times FA \times AS$$

exactly as in Porism 3 preceding. The present converse is, we need scarcely add, no less spurious than its original (on which see note (28)).

gradu magis compositam quam sunt sectiones Conicæ, et linea tacta tanget rectas AE, AF.(35)

Porisma 6. A datis punctis A, B ad datam rectam CG concurrant rectæ duæ mobiles ACD, BCE et ulterius productæ incidant in rectas positione datas FD FE[,] prior AC in FD ad D et posterior BC in FE ad E, et requiratur relatio simplicissima rectarum FD FE ad invicem.

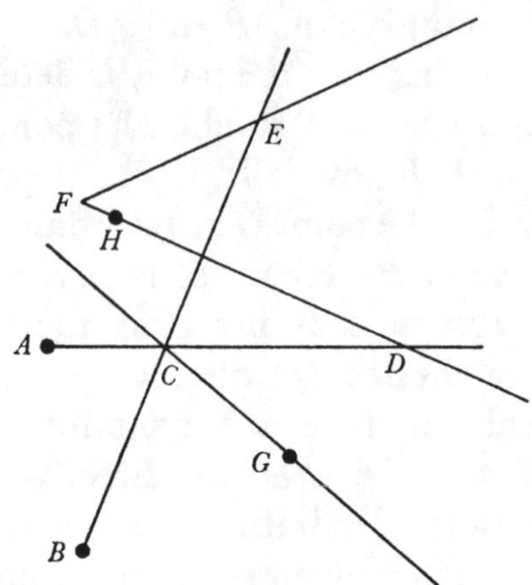

Quoniam puncta D, E se mutuo simpliciter determinant abeat punctum D in infinitum et sit ea ipsius FE positio ut punctum E simul abeat in infinitum. Incidat jam punctum E in datum aliquod puntum F(36) et sit H punctum in quod D tunc incidit et erit FE ad HD in data ratione per Theor 1.(37)

At si puncta D et E non simul abeant in infinitum, sit K punctum in quod E incidit ubi D abit in infinitum et L punctum in quod D incidit ubi E abit in infinitum et KE erit reciproce ut LD per Theor. 2.(38)

(35) Newton argues by analogy that, just as freeing the *reciproce proportionalia* $EB \times FD$ from the requirement that this product be equal to $EA \times FA$ raised the degree of the curve (C) touched by BD (namely, from a dimensionless point to a conic), so likewise here freeing $EB \times BR \times FD \times DS$ from the condition that it be equal to $EA \times AR \times FA \times AS$ will similarly elevate the locus (C) touched by BD to be a 'degree more compound' than the conic which, in the immediately preceding converse to Porism 3, he has fallaciously inferred it to be. We can, unfortunately, give no credence to this dubious 'induction'. For if, further to continue the algebraic equivalent of note (27), we again set $AB = x$ and $AD = y$, and suppose $AE = \alpha$, $AR = \beta$ to be the roots of $x^2+px+q = 0$ and $AF = \gamma$, $AS = \delta$ those of $y^2+ry+s = 0$—in other words, that $x^2+px+q \equiv (x-AE)(x-AR)$ and $y^2+ry+s \equiv (y-AF)(y-AS)$—then the 2, 2 correspondence $B \leftrightarrow D$ defined by $BE \times BR \times DF \times DS = k.AE \times AR \times AF \times AS$ will determine the equivalent correspondence $x \leftrightarrow y$ given by $(x^2+px+q)(y^2+ry+s) = kqs$, that is, $x^2y^2+(rx+py+pr)xy+sx^2+qy^2+psx+qry+(1-k)qs = 0$. The 'simplifying' supposition that k is unity removes the constant term here, it is true, but—as we may leave the reader to demonstrate—fails to reduce the (quartic) degree of the curve (C) enveloped by BD. (We exclude the degenerate instances when also $p, r = 0$ or $q, s = 0$ and the correspondence reduces to be $XY+sX+qY = 0$, $X \leftrightarrow x^2$, $Y \leftrightarrow y^2$ and $xy(x+p)(y+r) = 0$ respectively.) Further to pursue the consequent ramifications would here have little purpose. It is enough merely to point to a rare lack of geometrical insight on Newton's part.

(36) For simplicity Newton in his accompanying figure sets F to be the meet of the straight lines (D) and (E); any other fixed point in the latter line would serve equally well, of course.

(37) Notice that here (as in his following variations upon the present configuration) Newton takes it to be self-evident that the 1, 1 correspondence between quantities 'mutually related by simple geometry' is transitive; specifically, that because the points E, C and D are related in this manner by transversals defining $E \leftrightarrow C$ and $C \leftrightarrow D$, then without more ado we may assume that so too $E \leftrightarrow D$. No explicit demonstration of this transitivity was given—or, to be sure, thought necessary—till half a dozen years after Newton's death (as we have already

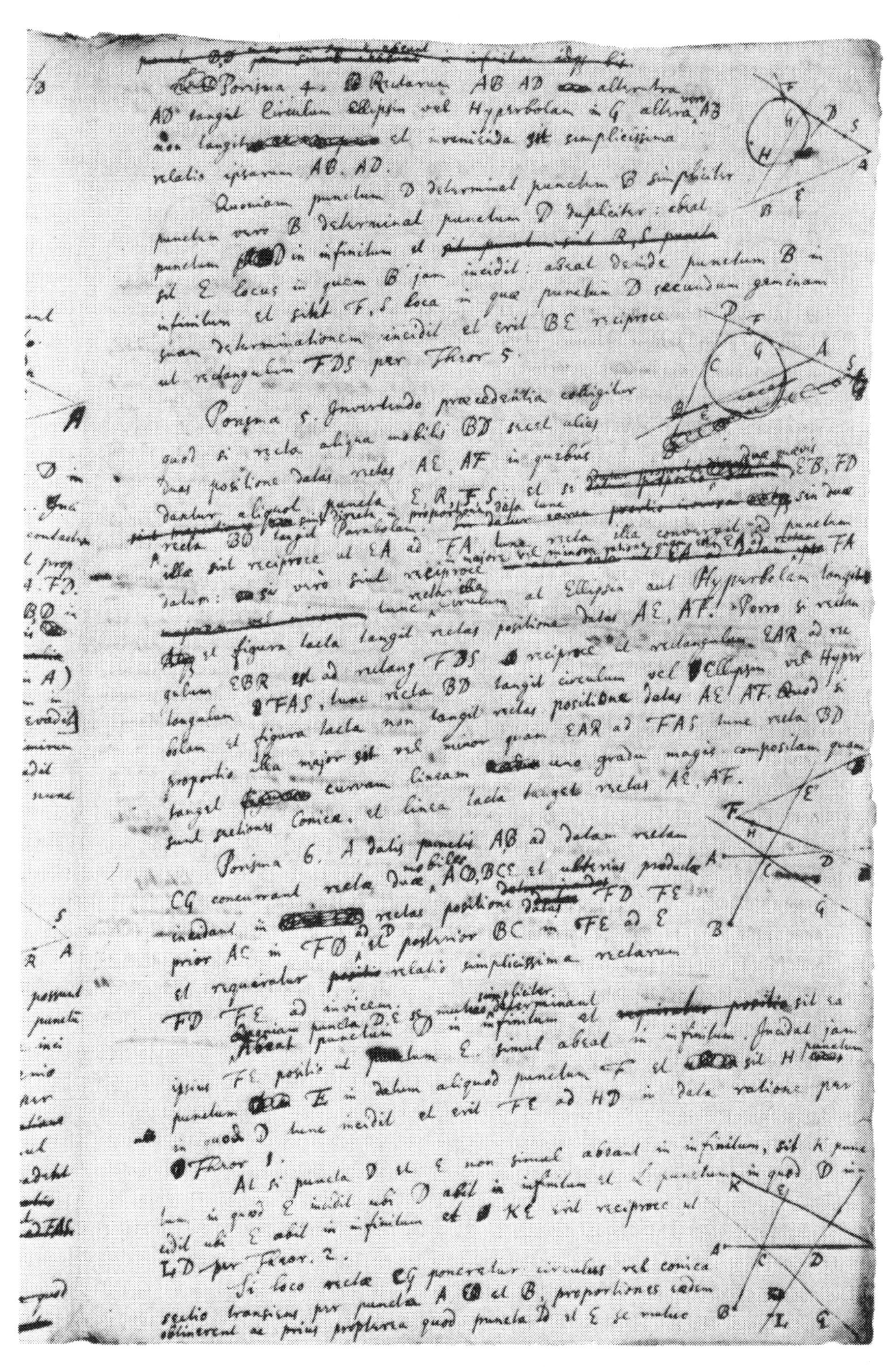

Plate II. 'The Finding of Porisms': 1, 2 and 2, 2 point-correspondences determined by tangents to a general conic (**2**, 1, §2).

curve one degree more compound than conics are, and the curve so touched will touch the straight lines AE, AF.(35)

Porism 6. From the given points A, B let there concur at the given straight line CG two mobile straight lines ACD, BCE meeting, when further produced, the straight lines FD, FE, given in position, the former, AC, intersecting FD at D and the latter, BC, FE at E: let there then be required the simplest relationship of the lines FD, FE to each other.

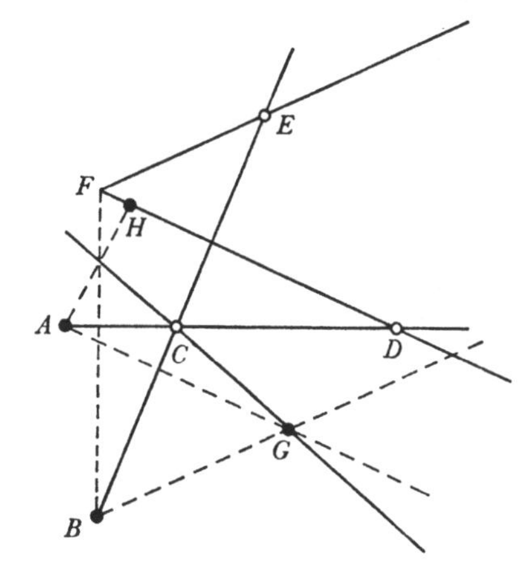

Seeing that the points D and E mutually determine one another simply, let point D pass away to infinity and let FE's position be such as to let point E pass simultaneously to infinity. Now let point E fall at some given point F(36) and H be the point at which D then falls, and there will then be FE to HD in a given ratio by Theorem 1.(37)

If, however, the points D and E should not pass simultaneously to infinity, let K be the point at which E falls when D passes to infinity and L the point at which D falls when E passes to infinity, and KE will then be reciprocally as LD by Theorem 2.(38)

---

mentioned in IV: 309, note (55)) William Braikenridge rediscovered the equation

$$axy+bx+cy+d = 0$$

defining the general 1, 1 correspondence $x \leftrightarrow y$ and applied it to the purpose in his *Exercitatio Geometrica De Descriptione Linearum Curvarum* (London, 1733): 67–8. The porism now restored is, in fact, the one proposition in Euclid's lost work *De Porismatibus* of which Pappus gives a wholly unambiguous account, namely: 'Si à duobus datis punctis ad rectam lineam positione datam rectæ lineæ inflectantur, abscindat autem vna à recta linea positione data ad datum in ipsa punctum, abscindet & altera proportionem habens datam' (*Mathematicæ Collectiones:* 246). Euclid in his formal proof of the proposition would here no doubt, having established that the correspondence $E \leftrightarrow D$ is such that $\infty_{FE} \leftrightarrow \infty_{HD}$, and that the perspectivities from B and A respectively entail the cross-ratio equalities $(E) = (C) = (D)$, have gone on to invoke a second corresponding pair of points E', D' to establish the conclusion $FE/FE' = HD/HD'$ in the form of the equivalent equality $(F\,\infty_{FE}\,EE') = (H\,\infty_{HD}\,DD')$. (Lest this seem too modern a restoration we would remind that full proof of the invariance of the cross-ratio of a line-pencil is contained in the related lemmas presented by Pappus in his *Mathematical Collection* VII, 129, 137 and 142—as indeed Newton well knew.)

(38) Though Pappus in his summary of Euclid's work does not cite this variant on the preceding porism—which is not to say, of course, that Euclid did not know it—the particular case in which the lines FK and FL coincide was enunciated by Pierre Fermat as the 'Porisma primum' of the *Porismatum Euclidæorum Renovata Doctrina, & sub formâ Isagoges recentioribus Geometris exhibita* which he composed in the late 1650's but which was published only after his death in his son Samuel's edition of his *Varia Opera Mathematica* (Toulouse, 1679): 116–19; see especially 117. Fermat gave no proof of his particular *porisma* (which, however, two applica-

Si loco rectæ CG poneretur circulus vel conica sectio transiens per puncta A et B, proportiones eædem obtinerent ac prius propterea quod puncta D et E se mutuo determinarent simpliciter.(39)

At si circulus ille vel sectio Conica non transiret per puncta A, B, tum puncta D, E se mutuò determinarent dupliciter, et inversa rectangulorum proportio per Theorema quartum prodiret.(40)

Porisma 7. In circuli dati AFB diametro DAB invenienda sunt puncta D, E a quibus rectæ DF EF ad circulum concurrentes simplicissimam habeant relationem ad invicem.

Jaceant puncta D, E ad easdem partes centri C, et quoniam rectæ DF, EF simul

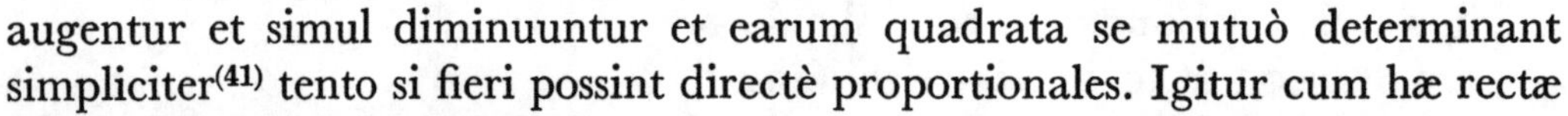

augentur et simul diminuuntur et earum quadrata se mutuò determinant simpliciter(41) tento si fieri possint directè proportionales. Igitur cum hæ rectæ nec in infinitum augeri possint nec evanescere: facio ut in casibus insignioribus redeant ad eandem rationem, veluti in casibus ubi punctum F incidit in rectam DE ad A et B, et DF tangit circulum vel FE ipsi AB perpendicularis est. Primum igitur rectam AB

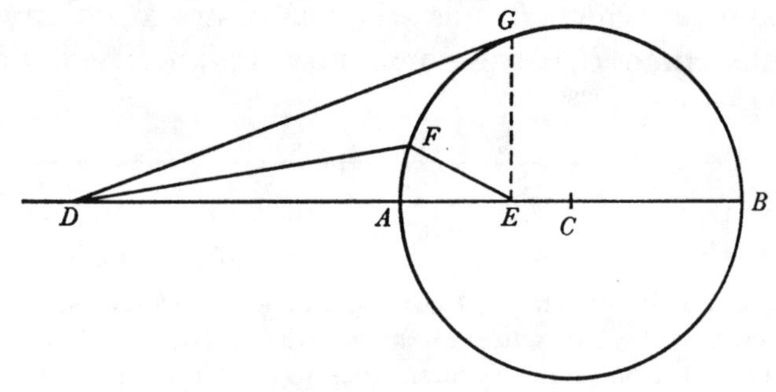

seco in D et E ut sit DA ad EA sicut DB ad $E[B]$. Dein faciendo ut DF tangat circulū, puta in G atq; adeo ut EF coincidat cum perpendiculari EG,(42) animadverto quod DA, DG, DB et EA, EG, EB sint continuè proportionales adeoq; DF EF ter sint in eadem ratione (existente $DA.EA::DB.E[B]::DG.EG$) hoc

tions of Menelaus' theorem readily provide) but, much as in the previous footnote, a demonstration accessible to Euclid in some equivalent form ensues straightforwardly from the premiss that in the present 1, 1 correspondence of $E \leftrightarrow D$ there is $K \leftrightarrow \infty_{FD}$ and $\infty_{FE} \leftrightarrow L$; whence at once $(K\,\infty_{FE}\,EE') = (\infty_{FD}\,LDD')$, that is, $KE/KE' = LD'/LD$. (We may add that, of the four other Euclidean porisms restored by Fermat in his *Renovata Doctrina*, only the 'Porisma tertium' in which (see IV: 311, note (59)) he stated the invariance of the cross-ratio of the pencil of lines drawn from four given points on a conic through any fifth point of it begins to make good his introductory boast that, whereas previous 'more recent' geometers had not even come to suspect what Euclid's lost treatise might have been about, 'Nobis tamen in tantis tenebris dudum cæcutientibus, & quâ ratione in hac materiâ Geometriæ opitularemur elaborantibus, tandem se clar[e] videndam obtulit [porismatum Euclidæorum doctrina], & purâ per noctem in luce refulsit'.)

If the straight line CG were to be replaced by a circle or conic passing through the points A and B, the same proportions would hold as before inasmuch as the points D and E would continue mutually to determine one another simply.(39)

But should that circle or conic not pass through the points A and B, then the points D and E would mutually determine one another doubly, and an inverse proportion of rectangles would ensue by Theorem 4.(40)

Porism 7. In the diameter AB of a given circle AFB there need to be found points D, E such that the straight lines DF, EF from them concurrent at the circle shall have the simplest relationship to each other.

Let the points D and E lie on the same side of the centre C, and then, because the lines DF and EF simultaneously increase and diminish and their squares mutually determine one another simply,(41) I test if they could be made directly proportional. Accordingly, since these lines can neither increase to infinity nor vanish, I contrive that in their more outstanding cases they return to the same ratio—in those, for instance, where F falls in the line DE at A and B, and where DF touches the circle, or FE is perpendicular to AB. First, therefore, I cut the line AB in D and E so that DA is to EA as DB to EB; and then, making DF tangent to the circle, at G say, and hence EF to coincide with the perpendicular EG,(42) I discern that DA, DG, DB and EA, EG, EB are in continued proportion, and hence (there being $DA:EA = DB:EB = DG:EG$) that DF and EF are

(39) For it is the anharmonic property of a conic that the equality $B(C) = A(C)$ of the cross-ratios of the line-pencils through the fixed poles A, B and the general point C is here preserved; whence, since by simple perspectivity from A and B there is again $B(C) = B(E)$ and $A(C) = A(D)$, the necessary equality of the cross-ratios $(E) = (D)$ of the points cut out by BC and AC in the lines FK and FL is also maintained. In particular, when (compare note (37)) there is $(K \infty_{FE} EE') = (L \infty_{FD} DD')$, then $KE/KE' = LD/LD'$ and so KE is directly proportional to LD; but when (see note (38)) there is $(K \infty_{FE} EE') = (\infty_{FD} LDD')$, then $KE/KE' = LD'/LD$ and therefore KE is reciprocally proportional to LD.

(40) While, to be sure, on taking $FE = x$ and $FD = y$ the ‘doubly mutual’ relationship of the points E and D will be determined by a general 2, 2 algebraic correspondence $x \leftrightarrow y$ of the form $ax^2y^2 + (bx+cy+d)\,xy + ex^2 + fy^2 + gx + hy + i = 0$, this will here be irreducible; in particular, since no single point E in Newton's configuration can (except in trivial degenerate cases) correspond to two points D lying simultaneously at infinity, or *vice versa* any single point D to two points E simultaneously at infinity—indeed, when the conic CG is an ellipse and the points A, B are placed far enough outside it, no real point C corresponds to the infinite points of either FK or FL—this fresh attempt geometrically to apply Theorem 4 is as spurious as those previously made in Porisms 3 and 4 above (compare notes (27) and (29) preceding).

(41) For, on drawing the radius FC, there is at once

$$(DC^2+FC^2-DF^2)/DC(=2FC.\cos \widehat{ACF}) = (EC^2+FC^2-EF^2)/EC,$$

where DC, EC and FC are given quantities.

(42) Since in this case (on drawing the radius GC) there is

$$DG:GE = DA:AE = (DA+DB):AB = DC:(AC \text{ or}) CG,$$

the triangles DGE and DCG sharing a common angle at D are similar, and in consequence $\widehat{DEG} = \widehat{DGC}$, right.

est bis redeunt ad eandem rationem et propterea semper sunt proportionales per Theor 1.(43)

Atq; hac methodo proportiones quantitatum eruere soleo.(44) Quinetiam in omni Problematum resolutione casus apprime(45) considero ubi puncta indeterminata abeunt in infinitum, et rectarum positiones et intersectiones in his casibus noto. Et si curva quam punctum indeterminatum percurrit, in orbem redit et seipsam decussat, casum etiam considero ubi punctum indeterminatum in intersectionem illam incidit, et noto positiones et intersectiones linearum in hoc casu. (46)Proprietates etiam noto linearum quæ vel pergunt in plagas illas in quas puncta indeterminata abeunt in infinitum vel convergant ad intersectiones prædictas curvilineas vel transeunt per puncta germana.(47) Relationes etiam noto quantitatum germanarum. Et si cupiam ut ignotæ quantitates evadant directè vel reciprocè proportionales aut aliam quamvis desideratam relationem ad invicem habeant, efficio ut in casibus insignioribus(48) ad eandem relationem semel vel sæpius redeant. Atq; hac ratione vel proportiones colligo vel alia Porismata quæ ad Problematis determinationem conducunt. Nam si proportiones hac methodo non prodeant prodibunt tamen aliæ ignotarum quantitatum relationes quæ ad resolutionem Problematis non inutiles esse possint.(49)

---

(43) Compare Newton's converse derivation of this defining property of the Apollonian circle in his earlier 'Solutio Problematis Veterum de Loco solido' (IV: 314–16). The subtleties of his present argument perhaps appear more nakedly to modern eyes in the Cartesian equivalent in which the general point $F(x, y)$ on the circle of centre $C(0, 0)$ and radius $CA = r$ is defined in perpendicular coordinates by $x^2+y^2 = r^2$, and the distances of F from the fixed points $D(a, 0)$ and $E(b, 0)$ are respectively

$$FD = \surd[(x-a)^2+y^2] = \surd[r^2+a^2-2ax] \quad \text{and} \quad FE = \surd[r^2+b^2-2bx],$$

so that $(r^2+a^2-FD^2)/a = (r^2+b^2-FE^2)/b$. If now for three distinct positions of F (excluding mirror-images in DAB)—corresponding to $x = x_1, x_2, x_3$ say—there is $FD/FE = k$, then for $i = 1, 2, 3$ there ensues $r^2+a^2-2ax_i = k^2(r^2+b^2-2bx_i)$ and hence

$$r^2(1-k^2)+a^2-k^2b^2-2(a-k^2b)\,x_i = 0;$$

which equation is therefore an identity, so that $a = k^2b$ and $(k \neq 1)$ $k^2b^2 = r^2$, and in consequence $ab = r^2$. Accordingly $FD/FE = k$, constant, for all points F provided that D and E are inverse points with respect to the circle.

(44) Newton initially began this last paragraph by remarking 'Et alia Porismata eadem methodo prodeunt' (And other porisms ensue by the same procedure). What he here had in mind is amply revealed in the extended lists of such propositions presented by him in the

thrice in the same ratio, that is, they twice return to the same ratio and consequently are ever proportional by Theorem 1.(43)

And this is my usual method for deriving the proportions of quantities.(44) To be sure, in all resolution of problems I chiefly(45) consider the cases where indeterminate points pass away to infinity, noting in these the positions of straight lines and their intersections. And if the curve traced by an indeterminate point loops back, crossing itself, then I also consider the case where the indeterminate point falls at that intersection, noting in this the positions and intersections of lines. (46)I also note the properties of lines which proceed in the same directions as those in which indeterminate points go off to infinity, or converge on the previously cited curvilinear intersections, or pass through points which are twin.(47) I also note the relationships of twin quantities. And should I desire unknown quantities to come to be directly or reciprocally proportional or have any other desired relationship one to another, I arrange that in the more outstanding cases(48) they should return once or more times to the same relationship. And by this means I gather either proportions or other porisms which are of service in the determination of a problem. For, though proportions should not ensue by this method, there will yet result other relationships of unknown quantities which may not be without avail(49) in the resolution of a problem.

---

several drafts of his contemporary treatise on classical 'Geometria' which we reproduce in Section 2 following.

(45) A more cautious replacement in afterthought for an ill-considered original 'semper' (ever).

(46) Newton here first went on: 'Computum verò quam maxime refero [? ad puncta germana]' (I for the greatest part, however, refer the computation [to brother-points]).

(47) Literally, of course, '[brothers] german'. Newton here briefly resumes his earlier 'Regula Fratrum' (see v: 612), where such 'quantitates gemellæ' were defined as ones 'quæ eodem modo se habeant ad conditiones Problematis quæq; cognitam aliquam habeant relationem ad invicem'. There, however, he had rejected the present terminology when it momentarily made its bow in his 'Exempl. 3'; see v: 617, note (162).

(48) These were initially specified to be 'ubi [ignotæ] maximæ sunt vel minimæ' (where the unknowns are greatest or least).

(49) Originally 'non parùm conferant' (may profit not a little). Whether Newton intended to go on to illustrate the point with examples is not clear: the manuscript text here terminates almost at the foot of the sheet's last page (f. 160v) leaving only a small blank space in sequel, but we can trace no continuation of the present piece in his surviving papers.

2

FIRST ESSAYS AT A MULTI-PARTITE TREATISE ON 'GEOMETRY'[(1)]

§1. THE INITIAL PREAMBLE.[(2)]

From the autograph[(3)] in the University Library, Cambridge

GEOMETRIÆ LIBRI TRES.

PROŒMIUM.

Quid est problema resolvere et quid solvere.

Synthesin et Analysin[(4)] Pappus ex veterum Geometrarum doctrina sic definit.[(5)] *Resolutio*, ait, sive Analysis *est via a quæsito tanquam concesso per ea quæ deinceps consequuntur ad aliquod concessum in compositione. In resolutione enim id quod quæritur tanquam factum ponentes quid ex hoc contingat consideramus: et rursum illius antecedens, quousq ita progredientes incidimus in aliquod jam cognitum vel quod sit e numero principiorum. Et hujusmodi processum* RESOLUTIONEM *vocamus, veluti ex contrario factam* SOLUTIONEM. *In compositione*[(6)] *autem per conversionem ponentes tanquam jam factam id quod extremum in resolutione sumpsimus, atq hic ordinantes secundum naturam ea antecedentia quæ illic consequentia erant, et mutua illorum facta compositione ad quæsiti finem pervenimus. Et hic modus vocatur* COMPOSITIO. Quod Pappus hic describit id ipsum est quod nos facimus ubi assumendo incognitum ut cognitum, et inde per debitam argumentationem colligendo aliquod cognitum ut incognitum, problema ad æquationem deducimus: dein ope

---

(1) Making good use of his preliminary surveys (see 1, §§1/2 preceding) of the manner of classical analysis *ex datis* and related porismatic techniques of discovery, Newton here leans yet more heavily on the account of Greek *analysis geometrica* uniquely afforded by Pappus in the seventh book of his *Mathematical Collection*—as delivered in Manolessi's second edition of Commandino's Latin *editio princeps* (see 1, §1: note (1) above), we continue to suppose—in drafting several successive openings of an intended major treatise on the principles and elements of plane geometry. This momentary preoccupation with the aims and purposes of classical Greek geometry all too quickly yielded to the insistent pressure of Newton's own day, and his planned treatise on 'Geometria' was altered radically in its content to be the advanced contemporary monograph on the Cartesian, projective and fluxional properties of algebraic curves and their quadrature which we reproduce in the next section. Though on many later occasions he was fulsomely to express his admiration of the neatness of its analytical resolutions and the elegance of its formal syntheses, he was never again to be as close in mood and style to the doctrine of 'the ancients' as in these present roughly penned and unfinished draft chapters of his 'Geometry'.

Translation

THREE BOOKS OF GEOMETRY.

INTRODUCTION

What it is to resolve a problem, and what to solve it.

Synthesis and analysis(4) are, in line with the teachings of the ancient geometers, defined by Pappus in this manner.(5)

> 'Resolution'—that is, analysis—'is the route from the required regarded as granted through what thereupon follows to something granted in the composition. For in resolution, putting what is sought as done, we consider what chances to ensue, and then again its antecedent, proceeding in this way till we alight upon something already known or numbered among the principles. And this type of procedure we call *resolution*, it being as it were a reverse *solution*. In composition,(6) however, putting as now done what we last assumed in the resolution and here, according to their nature, ordering as antecedents what were before consequences, we in the end, by mutually compounding them, attain what is required. And this method is called *composition*'.

What Pappus here describes is the very thing we do when, by assuming the unknown as known and therefrom by an appropriate argument gathering something known as unknown, we reduce a problem to an equation; and then

(2) Or so we infer (compare note (47) below) on juxtaposing its text with parallel passages in the not dissimilar proem which here follows in § 2, though perhaps no irrefragable case for its priority in sequence of composition is to be made. Nothing is known of the immediate historical background to these pieces. We may notice, however, that during his cataloguing visit to Hurstbourne Park in 1777 (see I: xxv–xxvii) Samuel Horsley wrote down on the backsheet (f. 112^{v}) of the manuscript's opening pages his judgement that it is 'A very curious fragment about the advantage of the Geometrical over the Algebraical analysis with [see Appendix 1 below] a solution of a most difficult Problem by way of Example', adding below in a smaller hand that 'This should be publishd but I think the Title *Procemium* ought not to stand as a more finishd Proem [*viz.* §2 following] is found. S. Horsley. Octr. 26th 1777'.

(3) Add. 3963.11: 109^{r}–110^{r}[+111^{r}–112^{r}]/ 3963.2: 9^{r}/ 3963.14: 157^{r}–158^{r}/ 3963.4: 29^{r}–32^{r}. The superseded (but uncancelled) ff. 111^{r}–112^{r} are reproduced in Appendix 1.

(4) 'sive Compositionem et Resolutionem' (that is, composition and resolution) is deleted in sequel.

(5) *Mathematicæ Collectiones* (see 1, §1: note (1)): 240. For a modern English rendering of Pappus' considerably less ambiguous original Greek see T. L. Heath, *A History of Greek Mathematics*, **2** (Oxford, 1921): 400 or Ivor Thomas, *Greek Mathematical Works*, **2** (London, 1941): 597–9. We would remind, however, that no portion of the original text of Pappus' work appeared in print till Edmond Halley set the full Greek of the preamble to the seventh book along with his fresh Latin translation of its text—insofar at least as he could make out its sense (see note (46) below)—in introduction (pages I–XVII/XXVIII–XLIV) to his Latin *editio princeps* of *Apollonii Pergæi De Sectione Rationis Libri Duo* (Oxford, 1706).

(6) Understand 'sive Synthesi' (that is, synthesis) to parallel Newton's preceding insert in Commandino's Latin version.

æquationis illius ex verè cognitis inverso ordine colligimus verè incognitum. Nec differre videtur Algebra nostra ab illorum[7] Analysi nisi in modulo expressionis.[8] Sed illi resolutum componendo formabant ejusmodi demonstrationes inventorum quæ forent ad captum vulgi accommodatæ: nos analyticam inventionem exponimus et de compositione minus sumus solliciti. Et tamen solutio quæstionis in compositione consistit. Nam cùm vulgi sit proponere & Geometrarum solvere, & cum quærenti non satisfaciat donec[9] ita respondeatur ut is responsum rectum esse sentiat:[10] ideo veteres in solutione problematum se semper accommodabant ad vulgus, et siquando ad inventionem solutionis opus esset Analysi, problema tamen solutum esse non putabant donec per compositionem ad responsum pervenerant quod sine Analysi (quam vulgus non intelligit) et enunciari et probari posset. Qua ratione Resolutionem et Solutionem ut contraria duo ab invicem distinguebant, ut ex verbis Pappi constat, existimantes Problema resolutum esse quando Geometra apud se absolverat Analysin, solutum quando sine Analysi componere didicerat. Unde solutio problematum per constructionem æquationis e Geometria pura, ex veterum sententia, excludenda videtur: nisi forte quatenus Algebraista qui Geometriam minus intelligit proponat hoc ipsum problema, *Radicem propositæ æquationis Geometricè designare*; aut quatenus Geometra ex constructione æquationis colligat ejusmodi solutionem quæ sine æquationis notitia proponi ac demonstrari potest.

Resolutio non nimis affectanda.[11]

Quinetiam circuitu utitur qui ad Analysin recurrit ubi quæstio solvi potest sine Analysi. Et quamvis ea sit imbecillitas humana ut sic per ambages sæpe tendamus[12] ad metam, Geometra tamen peritior[12] est, et in conclusiones sæpe simpliciores incidit idq͡ʒ minori cum labore [quam] qui m[agi]s aberrat a recto tramite. Nam problemata ferè omnia naturalem aliquem habent solvendi modum, quem qui invenerit nullo negotio solutionem assequetur, qui ab hac via deviaverit vim inferret Problemati, et in æquationem perplexam sæpius incidens, ægerrime ex æquatione illa perveniet ad simplicissimam conclusionem, quam natura ipsa secus nullo negotio ministrare potuisset. Unde factum puto quod Veteres qui ad Compositionem collimabant simpliciores conclusiones assequi solerent, quam recentes qui magis colunt Algebram.

Vis compositionis.

Sed et Problemata innumera sunt quæ per Algebram juxta methodum usitatum ægerrime perducantur ad æquationem, innumera quæ perduci

(7) 'Veterum' (the ancients'), that is.

(8) We have already (IV: 277, note (9)) cited Descartes' similar conviction in Rule 4 of his *Regulæ ad Directionem Ingenii* that 'nihil aliud esse videtur ars illa [inveniendi], quam barbaro nomine Algebram vocant'; see also IV: 222, note (20).

(9) 'quæstioni' (to the question) is deleted.

(10) Originally 'intelligat' (understand).

(11) This replaces an initial marginal head 'Solutionis modus optim^{s}' (The best mode of solution).

by aid of that equation we in inverse sequence gather from really knowns what is really unknown. Nor does our algebra seem to differ from their[7] analysis except in the mere manner of its expression.[8] But they in composing the resolved proof used to shape demonstrations of their findings in a form adapted to the common capacity to comprehend; whereas we exhibit the analysis of the finding and are less solicitous about its composition. And yet the solution of a question consists in its composition. For, since it is for the common person to propose and for geometers to solve, and since there is no satisfying an inquirer till reply be made[9] in such a way that he perceive[10] the reply to be right, the ancients in consequence always in the solution of problems adapted themselves to the common man, and wherever there was need of analysis in finding a solution to a problem they still considered it as unsolved till by composition they had attained an answer which might, without analysis (which the common person does not understand), be both enunciated and proved. For this reason they distinguished resolution and solution from one another as dual converses (so much is settled from Pappus' words), regarding a problem as resolved when a geometer had in his own view completed its analysis, and as solved once he had without analysis learned how to compose it. Whence the solution of problems by the construction of an equation would, to the ancients' mind, seem to be excluded from pure geometry, unless perhaps insofar as an algebraist who is less cognizant of geometry should propose this particular problem: *To denote the root of a proposed equation geometrically*, or insofar as a geometer should gather from the construction of an equation a solution of a kind propoundable and demonstrable without knowledge of the equation.

Resolution is not to be striven after too much.[11]

To be sure, he who has recourse to analysis when a question is solvable without it uses a circuitous route. Even though it be a human failing often to attain[12] our goal by digressing in this way, the geometer, however, is better trained[12] and comes upon conclusions that are often simpler—and gained with less effort—than (those of) a man who strays more from the direct path. For almost all problems have a natural way of being solved, and its discoverer will attain the solution with no trouble, whereas one who deviates from the path would do violence to a problem and, too often meeting up with a complicated equation, will pass with the greatest of difficulty from that equation on to the simplest of conclusions which its very nature could otherwise have furnished with a minimum of trouble. Whence happens it, I think, that the ancients, whose aim was composition, frequently arrived at simpler conclusions than the moderns, who are more devoted to algebra.

The power of composition.

But there are also innumerable problems which are by algebra, following the conventional method, brought only with extreme difficulty to an equation, and

(12) Originally 'collimemus' (aim at) and 'perfectior' (better accomplished) respectively.

nequeunt,[13] et quorum tamen solutio siquis recte procedat, satis facilis est. Ut si circulus ducendus esset qui tangeret tres datos circulos,[14] vel triangulum constituendum aut specie aut areæ magnitudine datum cujus anguli forent ad tres datos circulos & latus dato angulo conterminu[m] transiret per punctum datum in circumferentia circuli quem datus ille angulus contingit:[15] Per Algebræ methodum receptam demittendum esset perpendiculum a centro circuli quæsiti vel a puncto contactus aut ab aliquo angulo trianguli ad lineam aliquam positione datam; et perpendiculi hujus vel segmenti lineæ ad quam demittitur, longitudo quærenda, aut forte longitudo lateris trianguli aut partis ejus ad datum punctum terminatæ. Quo pacto in priori problemate difficulter incideretur in æquationem quatuor dimensionum,[16] in duobus autem posterioribus in æquationem octo vel potius sexdecim aut triginta duarum dimen-

(13) Again understand 'juxta methodum usitatum', that is, employing the algebraic approach of Cartesian analytical geometry. As his following examples show, Newton does not here mean that it would be logically impossible to reduce such problems by a conventional algebraic analysis, but rather asserts that it would be 'impossibly' difficult successfully to perform such a reduction.

(14) The classical Apollonian problem 'of [circle-]contacts'; see 1, §1: note (32) above. Despite his present citation of this as one 'impossible' manageably to reduce by an algebraic analysis, Newton himself had earlier (see IV: 254–8; V: 254–62, 614–18) given tolerably neat Cartesian solutions of a number of its particular cases, and its general resolution by an algebraic equivalent to the geometrical construction which he had some five years before published as Lemma XVI of Book 1 of his *Principia* is not (see note (16) below) qualitatively more difficult. He could not have known that Descartes himself, in the concluding 'Exemple 5e' of the *Introduction à la Geometrie* which he wrote (probably for Godefroid de Haestrecht) in the early summer of 1638, but which was to remain unprinted till the present century, had already successfully attacked by a straightforward algebraic method the analogous three-dimensional problem 'Estant donné quatre globes qui s'entretouchent, trouver le 5e qui les enferme et touche tous quatre' (see C. Adam and G. Milhaud, *Descartes: Correspondance*, **3** (Paris, 1941): 322–52, especially 346–52; the manuscript text is more faithfully reproduced by C. de Waard in *Correspondance du P. Marin Mersenne*, **7** (Paris, 1962): 453–62).

(15) By which understand that the given angle is free to rotate round the given point (on one of the circles) at its vertex. Newton's stated conditions that the triangle shall be given in species or area then manifestly determine that the sides adjacent to the given angle shall be respectively in direct or inverse ratio; whence, if one of the other vertices be constrained to move freely round its circle, the third vertex (by a corresponding homothety or inversion and then a rotation through the given angle) will describe a fourth circle whose (two) intersections with the given circle on which it is decreed to lie will fix its position and consequently that of the triangle itself. Newton will in sequel (compare note (17) below) grossly over-estimate the difficulty in pursuing the algebraic equivalent to this reduction. Initially, we may add, he concluded his present sentence with the restriction '& latera duo quævis transirent per puncta duo in circuli circumferentia ad quam concurrent data' (while any two sides should pass through two [*sc.* fixed] points in the circle-circumference at which these specified [sides] concur), but quickly replaced it, doubtless realizing that these conditions will not yield the 'plane' problem constructible by ruler and compass which his further argument requires.

(16) Not so, of course. Since, as Apollonius and Viète had long before shown (compare IV:

innumerable ones which cannot be so reduced at all,(13) and yet their solution, if anyone should go about it the right way, is easy enough. As, for instance, were a circle to be drawn to touch three given circles;(14) or were a triangle to be constructed given in species or area size, whose corners should be on three given circles, while a side adjacent to a given angle should pass through a given point in the circumference of the circle which that given angle touches:(15) by the accepted method of algebra a perpendicular would need to be let fall from the centre of the required circle, a point of contact or some corner of the triangle to some line given in position, and then the length of this perpendicular or a segment of the line to which it is let fall has to be sought, or perhaps the length of a side of the triangle or its portion ending at a given point. By this means you would in the first problem come with difficulty upon an equation of four dimensions,(16) while in the two latter you would meet with one of eight or rather sixteen

---

254, note (17) and VI: 164–5, note (154)), the 3-circles tangency problem is constructible by straight line and circle, its algebraic resolution will, if rightly ordered, arrive 'in æquationem duarum dimensionum' (...an equation of two dimensions). Indeed, given three circles of radii r, s, t and (with respect to a system of perpendicular Cartesian coordinates) of respective centres (a, b), (c, d) and (e, f), we may reduce the problem of constructing the centre (x, y) of the circle which touches them all externally or all internally to solving the simultaneous equations $\sqrt{[(x-a)^2+(y-b^2)]}\pm r = \sqrt{[(x-c)^2+(y-d)^2]}\pm s = \sqrt{[(x-e)^2+(y-f)^2]}\pm t$. (The other six cases where the circle touches two of the given circles externally and one internally or *vice versa*, are dealt with by appropriately altering the signs of r and s.) In making his present evaluation of the dimension of the equation in x or y which results on here eliminating y or x Newton would seem, by twice appropriately grouping terms, squaring and cancelling, to have entirely removed the radicals from a pair of these equations to attain two simultaneous quadratics in x and y. By arguing in exact algebraic parallel, however, to his earlier geometrical reduction of the problem in Lemma XV of his 'De motu Corporum Liber primus' (see VI: 162–4)—and indeed proceeding much as Descartes had done half a century before in his resolution of the analogous 4-spheres problem—we may, on squaring the equivalent equalities

$$\sqrt{[(x-a)^2+(y-b)^2]}\pm(r-s) = \sqrt{[(x-c)^2+(y-d)^2]}$$

and

$$\sqrt{[(x-a)^2+(y-b)^2]}\pm(r-t) = \sqrt{[(x-e)^2+(y-f)^2]},$$

deduce that

$$\pm\sqrt{[(x-a)^2+(y-b)^2]} = \tfrac{1}{2}(r-s)-((a-c)\,x+(b-d)\,y-\tfrac{1}{2}(a^2+b^2-c^2-d^2))/(r-s)$$
$$= \tfrac{1}{2}(r-t)-((a-e)\,x+(b-f)\,y-\tfrac{1}{2}(a^2+b^2-e^2-f^2))/(r-t)$$

and hence

$$(ax+by-\tfrac{1}{2}(a^2+b^2))\,(s-t)+(cx+dy-\tfrac{1}{2}(c^2+d^2))\,(t-r)+(ex+fy-\tfrac{1}{2}(e^2+f^2))\,(r-s) = \tfrac{1}{2}(s-t)\,(t-r)\,(r-s);$$

when, accordingly, x or y is eliminated between this linear equation and one or other of the quadratics in x and y which ensue on squaring the preceding, there results but a quadratic equation accurately defining the two centres (x, y).

sionum.(17) Et quid faceret Analysta cum tantis æquationibus?(18) Verùm insistendo vestigijs eorum quæ veteres tradiderunt problemata hæc, etiam sine Analysis subsidio, solvi, et per regulam et circinum expedite construi possunt.(19) Quod si in Ellipsi (aut alia Conica sectione) inscribenda esset figura rectilinea quatuor, sex, octo, vel plurium numeroq; parium laterum, ea lege ut anguli alterni dati forent & ad Ellipseos data puncta consisterent: ad solutionem problematis per solam Algebram, ubi plura sunt latera, vita hominis non sufficeret.(20). Et tamen Problema semper per regulam et circinum construi potest, et constructio sine Analysi inveniri. Neq; facilius foret ad curvam lineam e dato in ea puncto perpendiculum erigere, si modo curvæ illius lineæ hæc esset proprietas ut a quovis ejus puncto ad puncta quotcunq; data ductæ rectæ datam semper haberent summam.(21) Et tamen perpendiculum ad curvam non tantum

(17) An even grosser over-estimate! Read correctly 'in duobus...posterioribus item in æquationem duarum dimensionum' (...likewise one of two dimensions). Without passing to a detailed justification, we may in the 'plane' construction outlined in note (15) above take the fixed vertex to be the origin (0, 0), in which case the homothety $(x, y) \to (kx, ky)$ and the inversion $(x, y) \to (kx/(x^2+y^2), ky/(x^2+y^2))$ will transform the second given circle defined, say, by $x^2+y^2+2gx+2fy+c = 0$ into a further circle, say $x^2+y^2+2g'x+2f'y+c' = 0$, and the coupled transform $\begin{cases} x \to x\cos\alpha - y\sin\alpha \\ y \to x\sin\alpha + y\cos\alpha \end{cases}$ will rotate this clockwise through the given angle α into the congruent circular locus $x^2+y^2+2g''x+2f''y+c'' = 0$ of the triangle's third corner; since this must also lie on the third given circle, say $x^2+y^2+2Gx+2Fy+C = 0$, its two possible sites will be the pair of their intersections, readily determinable by means of the linear eliminant $2(g''-G)x+2(f''-F)y+c''-C = 0$.

(18) Newton was initially tempted to add a rhetorical flourish 'haud scio' (I know not) in sequel in the manuscript, but at once discarded it.

(19) See notes (14) and (15) above.

(20) Originally 'nec Herculis patientia nec anni Methusalem sufficerent' (neither Hercules' patience nor Methuselah's years would...suffice). Even the slightly more subdued replacement in the text above is vastly exaggerated. On assuming a standard system of perpendicular Cartesian coordinates, the locus (x, y) which subtends a constant angle $\tan^{-1}k$ at the fixed points (a, b), (c, d) evidently satisfies

$$k = \tan\left(\tan^{-1}\frac{y-b}{x-a} - \tan^{-1}\frac{y-d}{x-c}\right) = \frac{(x-c)(y-b)-(x-a)(y-d)}{(x-a)(x-c)+(y-b)(y-d)},$$

and is therefore a circle $x^2+y^2+Ax+By+C = 0$, where

$$A = b-d-k(a+c), \quad B = -a+c-k(b+d) \quad \text{and} \quad C = ad-bc+k(ac+bd).$$

In the present problem—already broached on IV: 268 in the simplest instance where a quadrilateral is to be inscribed in an ellipse—this circle will meet the given conic, which without loss of generality may be taken to be defined by $y^2 = rx \pm (r/q)x^2$, in four points, two of which are (a, b) and (c, d) (whence $b^2 = ra \pm (r/q)a^2$ and $d^2 = rc \pm (r/q)c^2$) and the other two determine the possible sites of the intervening vertex of the inscribed figure. The problem is therefore 'plane', by algebraic argument as by Euclidean postulate, and hence (as Newton proceeds to remark) constructible by straight-edge and circle.

(21) That is, $\sum_{1 \leqslant i \leqslant n} (r_i) = k$, constant, where the straight line joining the locus-point to the

or thirty-two dimensions.(17) And what were the algebraist to do with equations of such great degree?(18) However, by following in the steps of what the ancients have handed down these problems can, even without the aid of analysis, be solved and speedily constructed by ruler and compasses.(19) While if in an ellipse (or another conic) you needed to inscribe a rectilinear figure of four, six, eight or greater even number of sides with the restriction that alternate corners were given and positioned at given points of the ellipse, to solve the problem by algebra alone a man's life would, when there are a great many sides, not be long enough(20)—and yet the problem can always be constructed by ruler and compasses, and the construction ascertained without analysis. Nor would it be easier to erect the perpendicular to a curve at a given point in it if the curve's defining property be that the straight lines drawn from any point of it to any number of given points should always have a given sum(21)—and yet the per-

---

i-th fixed pole is of length r_i. While Newton had briefly considered an instance of this multi-polar locus some eight years before in his 'Matheseos Universalis Specimina' (see IV: 556–8), the present problem of constructing the perpendicular to the curve at any point owes its origin to Tschirnhaus who, in his *Medicina Mentis, sive Tentamen genuinæ Logicæ, in quâ disseritur de Methodo detegendi incognitas veritates* (Amsterdam, $_1$1687 [= Leipzig, $_2$1695: 92–101]), had sought to construct the normal to the slightly more general locus $\sum_{1\leqslant i\leqslant n} (\pm n_i r_i) = k$, in which the component focal radii r_i are severally multiplied by given constants $\pm n_i$. Tschirnhaus' own attempted solution—which none too clearly invoked an erroneous generalization of thc simplest bipolar case $r_1 + r_2 = k$, when the locus is an ellipse and the normal (bisecting the angle between them) cuts the line joining the foci instantaneously in the ratio $r_1 : r_2$ of the radii—was rightly dismissed by Nicolas Fatio de Duillier in his speedily published 'Réflexions ...sur une Methode de trouver les Tangentes de certaines Lignes courbes, laquelle vient d'être publiée dans un Livre intitulé *Medicina Mentis*' (*Bibliotheque Universelle et Historique*, **5** (Amsterdam, April 1687): 25–33). Fatio not only disproved Tschirnhaus' method in the case of the Cartesian oval $n_1 r_1 \pm n_2 r_2 = k$ by demonstrating *à l'ancienne* the accuracy of his own construction of its normal (*ibid.*: 26–9; compare M. Cantor, *Vorlesungen über Geschichte der Mathematik*, **3** (Leipzig, $_1$1898): 147–9) but also (*ibid.*: 31) accurately stated its extension to the multi-polar locus $\sum_{1\leqslant i\leqslant n} (\pm n_i r_i) = k$: the proof of the last he added two years later in a further article in the *Bibliotheque Universelle*, **13** (April 1689): 46–76, there extending it to embrace the yet more general locus defined by $\sum_{1\leqslant i\leqslant n} (\pm n_i r_i^p) = k$. Equivalent solutions were not long in forthcoming from Huygens (see his *Œuvres complètes*, **20**, 1940: 491–501), Leibniz (see P. Costabel, *Leibniz et la Dynamique. Les Textes de 1692* (Paris, 1960): 74–83 [in English as *Leibniz and Dynamics* (London, 1973): 73–83], where there is also given *ibid.*: 60–73 [= 69–83] a useful conspectus of the history of the problem and its statistical solutions) and L'Hospital (see his letter of 21 September 1693 to Johann Bernoulli in the latter's *Briefwechsel*, **1** (Basel, 1955): 189, and also his *Analyse des Infiniment Petits* (Paris, $_1$1696): § 32: 27–30). It was doubtless Fatio himself who first brought the problem to Newton's attention in a private conversation of which no record has survived. While Newton's own constructions (see note (23) and Appendix 1) of the normals to this and the other Tschirnhausian curves which he instances in sequel are in no way difficult to effect, it is, we would observe, impossible rigorously to justify them without a subtle appeal to the algebraic analysis which in his next sentence Newton claims here to avoid. Yet once more, accordingly, his appeal to the relative ease and neatness of a geometrical 'constructio sine Analysi inventa' is essentially spurious.

in hoc casu sine Algebra duci potest sed etiam in innumeris difficilioribus. Ut si summa aliquarum e ductis datam haberet rationem ad summam reliquarum, vel si summa rectangulorum sub singulis binis daretur[,] aut datam haberet rationem ad summam rectangulorum sub alijs.(22) Vel deniq; si harum linearum relatio quævis ad invicem statuatur quæ per æquationem exprimi potest, ut casus adhuc difficiliores non commemorem.(23)

Cum igitur hæc ita sint, nemini mirum esse debet quod Veteres non obstante eâ quam colebant Analysi ad varia subsidia se conferebant, et varios libros ad minuendam resolutionis difficultatem(24) spectantes edebant, ut Euclides libros tres *Porismatum* quos Pappus vocat *opus artificiosissimum ac perutile ad resolutionem obscuriorum problematum*;(25) Apollonius plures libros de sectionibus linearum, quæ sunt problematum simplex quoddam genus ad quod innumera difficiliora problemata reduci possunt; & Euclides alijq;(26) libros plures de locis tum planis tum solidis[,] de quibus Pappus hoc perhibet testimonium: *Locus resolutus*, inquit,(27) *propria quædam materia est post communem Elementorum constitutionem ijs parata, qui in Geometricis sibi comparare volunt vim ac facultatem inveniendi problemata quæ ipsis proponuntur; atq; hujus tantummodo utilitatis gratia inventa est. Scripserunt autem hac de re tum Euclides qui elementa tradit, tum Apollonius Pergæus, tum Aristæus senior.* Pergit autem Pappus in sequentibus libros prædictos describere.(28) Et Euclidis quidem *data* usq; nunc extant,(29) ut et libri *Conicorum* Apollonij partim græcè partim versione Arabica.(30) Cæterorum memoria solis Pappi descrip-

(22) Originally 'sub reliquis binis' (beneath the remaining pairs); Newton has made a similar change in a cancelled following alternative which reads in final form 'Vel si summa contentorum sub aliquibus ternis daretur aut datam haberet rationem ad summam contentorum sub alijs' (or if the sum of the products contained by some trios be given or have a given ratio to the sum of those contained by others).

(23) As the superseded following manuscript sheet (ff. 111–112, reproduced in Appendix 1 below) shows, Newton initially here had it in mind to go on to specify his mode of construction of the normal to the curve for each of the three types of Tschirnhausian multi-polar locus which he has instanced above. We may readily restore his approach on setting

$$f \equiv f(r_1, r_2, r_3, \dots, r_n) = k$$

to be his present most general 'linearum [r_i, $i = 1, 2, 3, \dots, n$] relatio quævis ad invicem' which is expressible 'per æquationem': it at once follows that $\sum_{1\leqslant i\leqslant n} (f_i . dr_i) = 0$ where $f_i = \partial f/\partial r_i$, and so the normal to the curve at any point of the Tschirnhausian locus thus defined is fixed by the condition $\sum_{1\leqslant i\leqslant n} (p_i) = 0$ where the $p_i = f_i . dr_i/ds$ are the perpendicular distances from it of points stationed in the individual radii r_i at a distance f_i from the locus-point; whence the normal passes instantaneously through the centre of gravity of these points. And such, in essence, are the prior constructions by Fatio, Huygens and Leibniz (see note (21)). In particular, when $f \equiv \sum_{1\leqslant i\leqslant n} (\pm n_i r_i) = k$, then $f_i = \pm n_i$ (which Fatio elegantly constructs by taking all the $f_i = \pm 1$ and 'weighting' their end-points individually by the factors n_i); and when $f = \sum_{\substack{1\leqslant i,j\leqslant n\\ [j\neq i]}} (\pm n_{ij} r_i r_j) = k$, then $f_i = \sum_{\substack{1\leqslant j\leqslant n\\ [j\neq i]}} (\pm n_{ij} r_j)$. We need not here further elaborate; but compare Appendix 1: note (5).

pendicular to the curve can be drawn without algebra not merely in this case but also in innumerable more difficult ones; as, for instance, if some of the lines drawn should have a given ratio to the sum of the remainder, or if the sum of the rectangles contained by individual pairs be given or have a given ratio to the sum of the rectangles beneath others,[22] or finally if any relationship of these lines to one another be decreed which is expressible by an equation, not to mention still more difficult cases.[23]

Such being so, therefore, no one ought to marvel that the ancients, notwithstanding their cultivation of analysis, had recourse to a variety of aids and put out various books regarding the way to diminish the difficulty of resolution:[24] Euclid's three books of *Porisms*, for instance, which Pappus calls 'a most skilful work, one extremely useful in the resolution of the obscurer problems';[25] Apollonius' several books on linear sections, which are a certain simple type of problem to which innumerable more difficult ones can be reduced; and the several books by Euclid and others[26] on both plane and solid loci. On these Pappus proffers the following information: 'The resolved locus', he says,[27] 'is a special subject-matter prepared, after the common groundwork of the *Elements*, for those who in things geometrical wish to acquire power and facility in finding (resolutions of) problems which are proposed to them, and with this use alone in view was it devised. Writers on this topic include Euclid, the author of the *Elements*; Apollonius of Perga; and Aristæus the elder'. In sequel Pappus proceeds to describe the above-mentioned books.[28] Of Euclid, indeed, the *Data* are still extant nowadays,[29] as are also—partly in Greek and partly in Arabic translation—Apollonius' books of *Conics*.[30] Memory of the rest is conserved in

(24) Originally 'ad inventionem et compositionem' (relating to discovery and [its] composition).

(25) *Mathematicæ Collectiones*: 244.

(26) Initially specified by Newton in the manuscript as 'Aristæus & Apollonius', doubtless in line with Pappus' survey of Euclid's successors in this field of the τόπος στερεός (*Mathematicæ Collectiones*: 251, already quoted in 1, § 1: note (33) preceding).

(27) These are in fact the opening sentences of Pappus' seventh book (*Mathematicæ Collectiones:* 240). On Commandino's rendering of Ὁ [τόπος] ἀναλυόμενος as 'Locus resolutus' see IV: 223–4, note (22).

(28) See *Mathematicæ Collectiones*: 241–52. In his introductory list of the twelve works which made up this classical 'treasury of analysis' (*ibid.*: 241, cited in full in 1, § 1: note (10) preceding) Pappus also included one, the two books of Eratosthenes *On means*, which Newton here ignores.

(29) This replaces a curious uncompleted first phrase 'vel nunc in omnium manibus ha[bentur]' (are even now in everyone's hands). Perhaps Newton here intends an oblique reference to Barrow's student edition of the *Data* (Cambridge, $_{1}$1657) which did indeed go through a number of printings in its original Latin and in English translation (again see 1, § 1: note (10)).

(30) It is an interesting index of his lack of close familiarity with Apollonius' *magnum opus* that Newton here first wrote 'ut et Apollonij *Conicorum* libri priores noti græcè & reliqui

tionibus conservatur præterquam quod Apollonij libri sex de triplici sectione[31] per Willebrordum Snellium ex Pappo restaurati sunt, *Tactionum* per Fr. Vietam, *Inclinationum* per Marinum Getaldum et Alexandrum Andersonum, & *locorum Planorum* per Franciscum a Schooten. Sed et Andersonus libros duos *de determinata sectione* aliter quam Snellius demonstravit.[32] Cæteris restituendis non sufficit Pappi descriptio nisi forte inde *Porismata* Euclidis ex parte aliquando revixerint.[33] Reliquum est ut scopo et usu horum librorum patefacto restituatur artificium loci resoluti quoad solutionem Problematum.[34]

Consistit autem artificium illud in prompta cognitione eorum quæ ex habitis & cognitis[35] dantur vel colligi possunt per Porismata, et maximè locorum. Sunt autem loci nihil aliud quam lineæ in quibus puncta quæsita locantur et quarum descriptione construuntur problemata.[36] Hujusmodi loci duo in omni Problemate investigandi sunt ac describendi ut in eorum intersectione inveniatur punctū quæsitum. Porismata sunt propositiones[37] quarum subsidio aliquid inter resolvendum et componendum ex præmissis colligitur. Quod colligitur vel

ex Arabico versi' (as also are of Apollonius' *Conics* the first books known in Greek and the remaining in a translation from the Arabic) before opting for an equivalent turn of phrase which does not require him to be so specific: not only can he not at once recall that it is the first four books which exist in their original Greek and the next three in Arabic versions alone, but he fails to mention that Apollonius' eighth book has not survived in any form (except for Pappus' brief remark (*Mathematicæ Collectiones:* 251) that it contained 'problemata conica determinata' and what little may be gathered from his few following lemmas (VII, 221–34) upon it). Though an inferior Latin translation of the first four books by G. B. Memo was published at Venice in 1537, their Latin *editio princeps* is that—based directly on several Greek manuscripts—which Commandino published at Bologna in 1566. As for the remaining three extant, while G. A. Borelli had published his Latin rendering of a broad Arabic paraphrase of their content at Florence in 1661, and Christian Rau had similarly brought out his Latin translation of an independent Arabic précis of these at Kiel six years later, it is fairly clear (compare VI: 34, note (14) for instance) that Newton had never looked at either—though he did later acquire, as a presentation from its editor, a copy (now Trinity College, Cambridge. NQ. 17.32) of Edmond Halley's monumental Greek/Latin *editio princeps* (Oxford, 1710) of the full surviving text.

(31) Two books each, that is, *De sectione rationis*, *De sectione spatii* and *De sectione determinata*. Newton is here, of course, yet unaware of the existence of the Arabic version of the first which Edward Bernard was soon to discover in manuscript at Oxford and whose Latin version Halley was eventually in 1706 to publish along with his restoration on like lines of the complementary *De sectione spatii* (see, 1, § 1: note (30)).

(32) Our respect for Newton's apparently expert knowledge of these previous restorations of works in the classical 'treasury of analysis' will somewhat diminish when we notice that this summary is taken virtually word for word from Schooten's preface to his *Apollonii Pergæi Loca Plana Restituta* [= Liber III: 191–292 of his *Exercitationum Mathematicarum Libri Quinque* (Leyden, 1657)] where he had observed (page 198) that 'paucorum annorum decursu duo Apollonii libri de rationis sectione, duo de spatii sectione, duo determinatæ sectionis nobis per Willebrordum Snellium restaurati sunt, [illi] Tactionum per Franciscum Vietam, Inclinationum per Marinum Getaldum, & Alexandrum Andersonum, qui duos de determinatâ sectione postea aliter quoque quàm Snellius demonstravit'. The citations here are respectively

Pappus' descriptions alone, except that Apollonius' six books on the triple section(31) have been restored by Willebrord Snell from Pappus, those on *Contacts* similarly by François Viète, those on *Inclinations* by Marin Ghetaldi and Alexander Anderson, and those on *Plane loci* by Frans van Schooten. Anderson has, I may add, also given demonstration of the two books *On determinate section* in a way differing from Snell's.(32) Pappus' description does not suffice to restore the remainder unless perchance Euclid's *Porisms* be one day in part revived therefrom.(33) It remains, once the scope and use of these books has been disclosed, to restore the technique of the resolved locus in regard to the solution of problems.(34)

That technique consists in the ready cognizance of what is given from what is had or known(35) or can be derived therefrom by porisms, and above all of loci. Loci, however, are none other than lines in which required points are located and by whose description problems(36) are constructed. Two loci of this sort are to be ascertained in every problem and described so that a required point may be found at their intersection. Porisms are propositions(37) by whose aid some mean between what is to be resolved and what to be composed is gathered from

---

of Snell's *Apollonius Batavus, seu Exsuscitata Apollonii Pergæi Geometrica* (Leyden, 1608); Viète's *Apollonius Gallus, seu Exsuscitata Apollonii Pergæi περὶ ἐπαφῶν Geometria* (Paris, 1600 [= (ed.) F. van Schooten, *Opera Mathematica* (Leyden, 1646): 325–38]; Ghetaldi's *Apollonius Redivivus. Seu, Restituta Apollonii Pergæi Inclinationum Geometria. Liber [primus/] secundus* (Venice, 1607/1613 [= *Opera Omnia* (Zagreb, 1968): 191–221/223–351]); and Anderson's *Variorum Problematum Practice* [appended to his *Supplementum Apollonii Rediuiui...περὶ νεύσεων* (Paris, 1612)]:7–35: Problemata V–XIII. Ghetaldi, we may add, also composed a *Supplementum Apollonij Galli. Seu exsuscitata Apollonii Pergæi Tactionum Geometriæ pars reliqua* (Venice, 1607 [= *Opera Omnia*: 173–90]).

(33) As indeed Pierre Fermat had attempted to do in the five propositions 'renewed' by him in the short 'Porismatum Euclidæorum Renovata Doctrina...' which his son had published in his father's posthumous *Varia Opera Mathematica* (Toulouse, 1679): 116–19 some dozen years earlier—much too late, of course, to be included in Schooten's list (see the previous note). While it is tempting to guess that Newton may have studied Fermat's tract before composing his own 'Inventio Porismatum' (see our note (38) in 1, § 2 above), his present omission to mention its existence must be strong negative evidence that he had not.

(34) In an equivalent cancelled preceding sentence Newton initially specified rather more fully that 'Antequam horum librorum usus exponam prænoscenda sunt quædam generalia, nimirum quid sit locus, quid sit Porisma et quænam sit mens *Datorum*' (Before I explain the uses of these books, certain generalities need first to be grasped—namely, what a locus is, what a porism is, and what is the purpose of the *Data*), thereafter defining much as in his revised sequel 'Cæterum per locos intelliguntur lineæ omnes quarum intersectione problemata solvuntur, per Porisma' (However, by loci are understood lines through whose intersection problems are solved, by porisms...) before abruptly breaking off.

(35) 'in omni propositorum investigatione' (in every exploration of the things proposed) is deleted.

(36) Originally 'omnia problemata' (all problems).

(37) Newton here initially inserted 'ex quibus pars Geometriæ dicta locus resolutus constat et' (out of which the branch of geometry called the resolved locus is established and).

datus est locus, vel datum aliud quod ad loci inventionem ac determinationem spectat[,] et nonnunquam nuda veritas seu theorema.[38]

[39]Porismata respectu eorum quæ in statu quæstionum ponuntur ac dantur sunt corollaria, et in componendis ac demonstrandis resolutis sæpe migrant in Lemmata. Unde fit ut Resolutio quæ per debita Porismata procedit sit aptior componendis demonstrationibus quam Algebra vulgi.[40] Per Algebram facile pervenitur ad æquationes sed inde sẽper ad elegantes illas constructiones ac demonstrationes pergere quæ per methodum Porismatũ prodire solent, non est adeò facile, sed nec ingenium et inventionis vis in hac Analysi tantoperè exercetur & excolitur.

Sunt autem Porismata duplicia. Quædam sunt in forma consequentiarum sine demonstrationibus[,] alia in forma propositionum demonstratarum. Illa inter operandũ investigantur[,] hæc asservantur in loco resoluto. Prioris generis Archimedes lib. 1 *De Sphæra & Cylindro* dedit exempla duo,[41] posteriora locum resolutum magna ex parte constituunt. *Horum autem* ait Pappus[42] *species omnes neq theorematum sunt neq problematum sed mediam quodammodo inter hæc formam ac naturam habent, ita ut eorum propositiones formari possint ut theorematum vel ut problematum: quo factum est ut ex multis Geometris alij quidem ea genere esse theoremata*[,] *alij vero Problemata opinati sint dum ad solam tantum propositionis formam respicerent. Horum autem trium differentiam Veteres multò melius cognovisse ex definitionibus perspicuum est. Dixerunt enim theorema esse quod proponitur in ipsius propositi demonstrationem, problema quod affertur in constructionem propositi*; *Porisma verò quod proponitur in porismum hoc est in inventionem et investigationem propositi*. Hæc Pappus. Differt ergo Porisma a Theoremate et Problemate tum ratione finis, tum quoad formam

(38) A deleted final clause originally went on to specify 'quo aliquid tandem vel detur si problema solvendum esset vel innotescat si theorema discutiendum eadem spectans' (whereby something shall at length either be given were there a problem to be solved or be made known were a theorem to be argued regarding the same data).

(39) A cancelled first version of the next three paragraphs (on ff. $157^r/158^r$ of the manuscript) is reproduced in Appendix 2.1 below. Even in present redraft Newton still found great difficulty in precisely defining the nature and purpose of the classical porism. He first began: 'Porismata sunt propositiones quæ ex positis consectantur et ad ulteriorem inventionem conducunt' (Porisms are propositions which are consectaries of what is posited and contribute to further discovery); and then specified alternatively that 'Porismata in resolutione et compositione eundem ferè locum obtinent quæ Lemmata & Lemmatum corrolaria in Solutione et in demonstrationibus componendis sunt instar Lemmatum' (Porisms hold in resolution and composition virtually the same place as lemmas and corollaries of lemmas in the [finished] solution, and in composing demonstrations are tantamount to lemmas).

(40) Originally 'longè aptior eleganti demonstrationum compositioni quam Analysis hodierna quâ problemata semper reducuntur ad æquationes et ex æquationibus sine porismatibus proceditur' (far more suited to the elegant composition of proofs than is present-day (algebraic) analysis whereby problems are always reduced to equations and we proceed from the equations without using porisms), after which Newton continued in a first version of the sequel: 'Hac analysi prompte pervenitur ad æquationes, sed ex æquationibus constructiones et

the premisses. What is gathered is either a given locus or another given thing which regards the finding and determination of a locus, and not infrequently is the bare truth, that is, a theorem.[38]

[39]Porisms in respect of what in the circumstances of a question is posited and given are corollaries, and in composing and demonstrating things resolved often pass over into being lemmas. Whence it comes that a resolution which proceeds by means of appropriate porisms is more suited to composing demonstrations than is common algebra.[40] Through algebra you easily arrive at equations, but always to pass therefrom to the elegant constructions and demonstrations which usually result by means of the method of porisms is not so easy, nor is one's ingenuity and power of invention so greatly exercised and refined in this analysis.

Porisms are, however, twofold. Certain are in the form of consequences without demonstrations, others in the form of demonstrated propositions. The first are discovered during the working, the latter are stored in the resolved locus. Of the former kind Archimedes has given two examples in Book 1 *Of the sphere and cylinder*,[41] the latter in great part constitutes the resolved locus. 'Of these', says Pappus,[42]

> 'all species are of the quality neither of theorems nor of problems, but keep in some manner a mean form and nature between these, so that their propositions can be formulated either as ones of theorems or as ones of problems; in consequence, of the mass of geometers some are of opinion that they are of the character of theorems, others in fact believe they are problems, so long as regard is had merely to the sole form of their proposition. It is clear, however, from their definitions that the ancients had a much better insight into the difference between these three. For they stated that a theorem is what is propounded in demonstration of a thing proposed, a problem what is adduced in construction of the proposed, whereas a porism is what is propounded poristically, that is, in the discovery and investigation of the proposed'.

So Pappus. A porism therefore differs from a theorem and a problem both by

---

demonstrationes melioris notæ derivare utplurimum difficilius est quam per resolutionem et compositionem totam feliciter perficere' (By this analysis we arrive promptly at equations, but it is in most cases more difficult to derive constructions and demonstrations of the better quality from the equations than happily to accomplish them by means of a complete resolution and composition).

(41) Newton evidently intends some two of the corollaries, not there given explicit proof, which Archimedes appended to several of his propositions—most probably the celebrated pair added in the train of I, 34 where its author states that the solid content and surface area of a sphere are each two-thirds the corresponding content and surface of the circumscribing right cylinder.

(42) *Mathematicæ Collectiones*: 245.

Propositionis. Non erant aliqua theoremata[,] alia Problemata sed mediam tenebant naturam ita ut Geometrarum aliqui ea omnia theoremata esse[,] alij omnia problemata esse putarent. Nimirum propositiones sunt de Datis discernendis & colligendis. Tales non sunt merè theoremata quia docent inventionem datorum, nec merè problemata quia non afferrunt constructiones ad invenienda data præterquam in demonstrationibus; de specie tamen utraqȝ participant, et inde facilè mutantur in utramvis, in Problemata afferendo constructiones, in Theoremata omittendo demonstrationem Datorum et relationem solummodo ponendo. Talem autem formam affectabant Porismata eo quod quantitatibus inveniendis id est Datis colligendis designabantur.

Sunt igitur Euclidis *data* nihil aliud quam Porismata sed his ob simplicitatem inventionis nomen *Datorum* potius impositum est. Sequuntur libri tres *Porismatum* quæ nihil aliud sunt quam continuatio *Datorum*. Id liquet ex forma propositionum a Pappo recensitarum in prooemio libri septimi quarum specimen ex versione Commandini hic subjungo. *Principio*, ait,(43) *septimi* (*porismatis libri primi habetur*)(44) *diagramma hoc. Si a duobus datis punctis ad rectam lineam positione datam rectæ lineæ inflectantur, abscindat autem una a recta linea positione data ad datum in ipsa punctum, abscindet et altera proportionem habens datam. In ijs autem quæ sequuntur. Quod hoc punctum tangit positione datam rectam lineam. Quod proportio hujus ad hanc data est. Quod proportio hujus ad apotomen. Quod hæc positione data est. Quod hæc ad datum punctum vergit. Quod proportio hujus ad aliquam ab hoc ut dato.* **Quod proportio hujus ad aliquam ab hoc ductæ. Quod proportio hujus spatij ad id quod data recta linea et hac continetur.* &c. Quarum Propositionum, si Pappi verba brevia et nimium depravata(46) recte interpretatus sum, sensus est talis.

* Suspicor hanc propositionem esse prioris repetitionem corruptā.(45)

(43) *Mathematicæ Collectiones*: 246.

(44) This parenthesis is Newton's conjectural interpolation of Commandino's Latin rendering—'In principio quidem septimi, diagramma hoc' to be precise—of an original Greek sentence whose surviving text is manifestly foreshortened by its copyist, if not otherwise corrupt (though we need not follow Pappus' modern editors Hultsch and Ver Eecke in their gratuitous inference that the whole phrase is 'une interpolation probable', as the last dubs it in his *Pappus d'Alexandrie*: *La Collection Mathématique* (Brussels, 1933): 490, note 4).

(45) Hultsch—for once!—here chooses not to cast a like doubt on the authenticity of the sentence whose Latin version seems thus suspect to Newton. (Compare Ver Eecke's *Pappus d'Alexandrie:* 491.)

(46) Some fourteen years afterwards, in his fresh attempt to render the corrupt Greek original of Pappus' survey of the content of Euclid's work, Edmond Halley here went far too far in his appended observation—as impercipient as it was defeatist in its attitude—that 'Hactenus *Porismatum* descriptio, nec mihi intellecta nec lectori profutura. Neque aliter fieri potuit: tam ob defectum schematis cujus fit mentio [see note (44) above]; unde rectæ satis multæ, de quibus hic agitur, absque notis Alphabeticis, ullove alio distinctionis charactere, inter se confunduntur: quàm ob omissa quædam ac transposita vel aliter vitiata in propositionis generalis expositione; unde quid sibi velit *Pappus* haud mihi datum est conjicere. His adde dictionis modum nimis contractum, ac in re difficili, qualis hæc est, minime usur-

reason of its end-purpose and with respect to the form of its proposition. It is not that some were theorems and others problems, but that they maintained a mean nature, with the result that certain of the geometers considered them all to be theorems, and others all problems. They are, of course, propositions about discerning givens and gathering them. As such, they are not merely theorems because they tell how to discover givens, nor are they merely problems because they do not present constructions for ascertaining givens other than in demonstrations; rather, they share the quality of both species and are in consequence easily changed into either—into problems by adjoining constructions, and into theorems by omitting the demonstration of givens and only positing their relationship. Porisms assumed such a form, however, in that they were designed for discovering quantities, that is, gathering givens.

Euclid's *Data*, therefore, are nothing else than porisms, but by reason of the simplicity of their invention the name 'Givens' was by preference set upon them. There follow three books of *Porisms* which are nothing other than a continuation of the *Data*. That is clear from the form of their propositions as they are reviewed by Pappus in the introduction to his seventh book. I here append a sample of these in Commandino's rendering:(43)

> 'At the beginning of the seventh (porism of the first book there is had)(44 this sketch. If from two given points straight lines are inclined to a straight line given in position, let one cut off from a straight line given in position (a segment terminated) at a given point in it, then the other will cut off one having a given proportion. In the sequel, however (it is premised) that this point traces a straight line given in position; that the proportion of this (line) to this is given; that the proportion of this to a segment cut off is; that this (line) is given in position; that this one verges to a given point; that the proportion of this to some from this is as given; that the proportion of this drawn to some of this is;* that the proportion of this space to the content of a given straight line and this is; ...'.

* I suspect that this proposition is a corrupt repeat of the previous one.(45)

The sense of these propositions, if I rightly interpret Pappus' concise and excessively debased text,(46) is as follows.

---

pandum' (*Apollonii Pergæi De Sectione Rationis Libri Duo.... Præmittitur Pappi Alexandrini Præfatio ad VIImum Collectionis Mathematicæ, nunc primum Græce edita* (Oxford, 1706): xxxvii). It was unfortunate enough for Halley that Newton had already, unbeknown to him, exhibited by his present limited restoration of propositions in Euclid's treatise the manifest error in this global rejection of any possibility of making sense of Pappus' outline of the *Porisms*. Just how amiss Halley's judgement is was soon publicly to be demonstrated by Robert Simson in the *Philosophical Transactions* for 1723 (see note (55) following), and it was given its *coup de grâce* more than a century later by Michel Chasles in his magistral restoration of *Les Trois Livres de Porismes d'Euclide, rétablis pour la première fois, d'après la Notice et les Lemmes de Pappus, et conformément au sentiment de R. Simson sur la forme des énoncés de ces propositions* (Paris, 1860).

Por[isma] 1. Si a duobus datis punctis *A*, *B* ad rectam lineam positione datam *CD* rectæ lineæ *AD BD* inflectantur, abscindat autem una *AD* a recta linea positione data *EF* ad datum in ipsa punctum *E* segmentum *EF* ad alterā positione datam *CD* proportionem habens datam, abscindet et altera *BD* segmentū *EG* ad idem *CD* proportionem habens datam. Quippe parallelæ erunt *EG*, *CD* et puncta *E*, *C*, *A*, *B* in eadem recta.

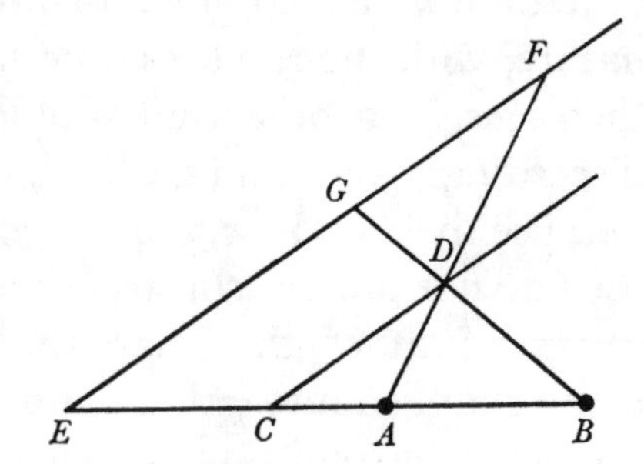

Porism. 2. In eodem Schemate[47] si detur ratio *EF* ad *EG*,[48] punctum *D* tanget positione datam rectam lineam *CD*.

Porism. 3. Et si punctum *D* tangit positione datam rectam *CD* proportio *EF* ad *EG* datur.

Porism. 4. Datur item proportio *EF* ad apotomen *FG*.

Porism. 5. Sed et recta *EF* positione data est.

Porism. 6.[49] Si a dato puncto *A* exiens recta *AD* secet parallelas duas positione datas *CD*, *EF*, detur autem proportio *EF* ad *EG*, recta *GD* ad datum punctum *B* converget.

Porism. 7. Et proportio hujus *GD* ad aliquam *DB* ab hoc dato puncto *B* ductam datur.

Porism. 8. Et proportio spatij *ECDG* (vel etiam spatij cujusvis *ECDF*, *GDF*, *DAB*, *CAD*, *EADG*) ad id quod data quavis recta linea et hac *EG* (vel *CD*) continetur, data est.

[50]Consimili Datorum forma pergebant propositiones reliquæ quantum ex hoc Pappi Procemio liquet, non obstante quod Pappus specimina quædam porismatū postea ponens mutavit formam propositionum.

Præcedentibus affinia sunt hæc.

Porism. 9. Si parallelæ *AB*, *CD* dentur positione et a datis punctis *A*, *B*

(47) That is, 'Diagrammate' (figure) to use Pappus' own word—and so, we may anticipate, Newton himself will alter it to be in the but minimally changed enunciation of the present restored *porismata Euclidæa* which he afterwards set down in his revised account 'De Compositione & Resolutione Veterum Geometrarum' which we reproduce in §2.2 below (see *ibid.*: note (84)).

(48) Of necessity, continue to understand 'et puncta data *E*, *A*, *B* in eadem recta sint' (and the given points *E*, *A*, *B* be in the same straight line) which Newton here initially specified but subsequently deleted in favour of his introductory reference to the circumstances of the preceding figure. Where *E* is not in line with *A* and *B* it will readily be seen that the locus of *D* is a hyperbola through the latter given points except when—exactly as in Porism 9 following on appropriately interchanging the designations of the points—*EGF* is parallel to *AB* and the locus degenerates to be the pair of the lines *AB* and *CD* (here not parallel to *EGF*).

(49) Newton initially went on to enunciate this in the equivalent form 'Si punctum *D* tangit positione datam rectam *CD* et in recta parallela et positione data *EF* datur proportio *EF* ad *EG* [, inclinetur autem *FD* per punctum *A* datum in EC, converget etiam *GD* ad punctum *B* in

Porism 1. If from the two given points A, B straight lines AD, BD are inclined to the straight line CD given in position, let one, AD, cut off from the straight line EF given in position a segment EF, terminated at the given point E in it, having a given proportion to the other, CD, given in position, then the other, BD, will cut off the segment EG having a given proportion to the same CD. Of course, EG and CD will be parallel, with the points E, C, A, B in the same straight line.

Porism 2. In the same scheme(47) if the ratio of EF to EG be given,(48) the point D will trace the straight line CD given in position.

Porism 3. And if the point D traces the straight line CD given in position, the proportion of EF to EG is given.

Porism 4. There is likewise given the proportion of EF to the segment FG cut off.

Porism 5. While the straight line EF is also given in position.

Porism 6.(49) If the straight line AD departing from the given point A shall intersect two parallels CD, EF given in position, let the proportion of EF to EG be given, and then the straight line GD will converge to the given point B.

Porism 7. And the proportion of this GD to some line DB drawn from this given point is given.

Porism 8. And the proportion of the space $ECDG$ (or indeed of any of the spaces $ECDF$, GDF, DAB, CAD, $EADG$) to that contained by any given line and the present EG (or CD) is given.

(50)The remaining propositions proceeded by a closely similar form of givens—so much is clear from this preamble of Pappus', notwithstanding that when he afterwards set out certain specimens of porisms Pappus altered their form of proposition.

Related to the preceding are these:

Porism 9. If the parallels AB, CD be given in position and two straight lines

---

hac datum]' (If the point D traces the straight line CD given in position and in the line EF parallel and given in position there is given the ratio of EF to EG, [let FD be inclined through the point A given in EC, and then GD will also pass through a point B given in it]).

(50) A cancelled sentence here immediately preceding reads 'His Porismatibus quantum sentio incipiebat liber primus' (With these porisms, as far as I perceive, the first book began). Newton thereafter continued in a first version of the following paragraph: 'Simili Datorum forma pergebant propositiones reliquæ quantum ex Pappo liquet, non obstante quod Pappus postea...mutavit formam propositionum, et more theorematum tradidit. Horum autem Porismatum ut defectus aliqua ex parte suppleatur visum est [? specimina quædam alia subjungere]' (The remaining propositions proceeded by a similar form of givens—so much is clear from Pappus, notwithstanding that he afterwards...altered their form of proposition, and presented them in the style of theorems. To make good the defects of these porisms in some part, however, it has seemed appropriate [to append a number of other examples]).

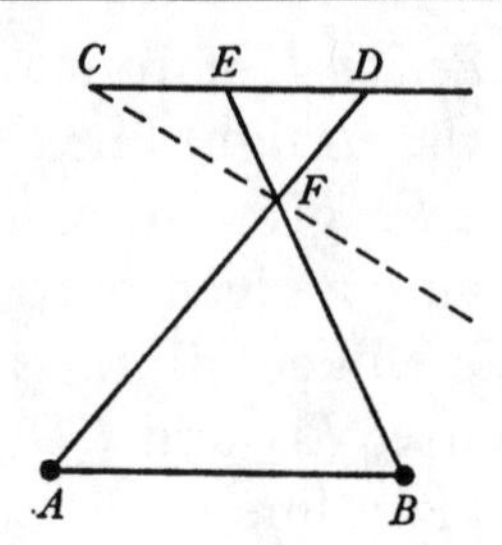

ducantur rectæ duæ AD, BE abscindentes CD CE in data ratione[,] ductarum concur[su]s F tanget rectam CF positione datam.

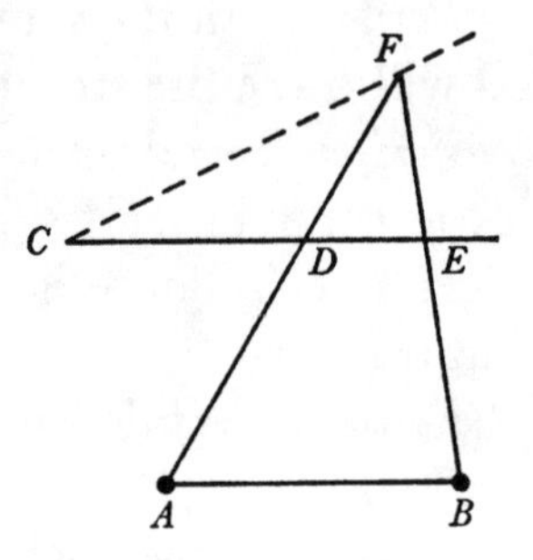

Porism. 10. Et si punctum concursus F tangit rectam CF positione datam, datur proportio CD ad CE.

Porism. 11. Et si a dato puncto A agatur recta AFD secans rectas positione datas CD CF in D et F[,] detur autem proportio CD ad CE, inclinat recta EF ad datum punctum B.

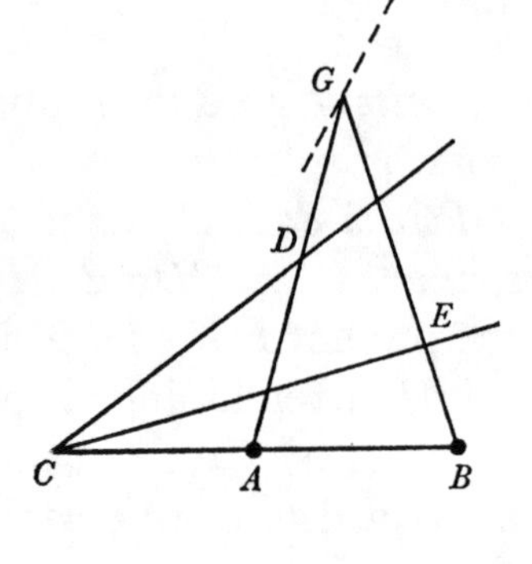

Porism. 12.[51] Si a duobus punctis A, B ad rectas duas positione datas CD CE ducantur rectæ duæ AD BE concurrentes in puncto G et sit CD ad CE in data ratione[,] puncta autem C, A, B in directum jaceant,[52] tanget punctum G rectam positione datam. Idem quocꝫ fiet si lineæ quævis CD FE ad rectam AB terminatæ datam habent rationem ad invicem.[53]

(51) This is a late insertion in the manuscript (on f. 29ᵛ). First versions of this and the Pappus–Desargues *propositio generalis* exist on Add. 3963.8: 62ʳ, reproduced in Appendix 2.2 below.

(52) A very necessary proviso yet again (compare note (48) above). When this does not hold, the locus of G will in general be a conic passing through each of the three points A, B and C, as Newton was well aware (see IV: 292; VI: 294–6).

AD, BE be drawn from the the given points A, B cutting off CD, CE in a given ratio, the meet F of the drawn lines will trace the straight line CF given in position.

Porism 10. While if the meeting-point F traces the straight line CF given in position, the proportion of CD to CE is given.

Porism 11. And if from the given point A there be drawn the straight line AFD cutting the straight lines CD, CF given in position in D and F, let the proportion of CD to CE be given, and the straight line EF is inclined through the given point B.

Porism 12.(51) If from the two points A, B there be drawn to the two straight lines CD, CE given in position two straight lines AD, BE concurrent at the point G, and CD be to CE in a given ratio, then—the points C, A, B lying in a straight line(52)—the point G will trace a straight line given in position. The same will also happen if any lines CD, FE terminated at the straight line AB have a given ratio to one another.(53)

---

(53) In general, where the two given lines CD, FE do not meet AB in corresponding points, because their given ratio determines the rotating lines AD and BE to be in a 1, 1 relationship *per simplicem geometriam* the meet G of these will (see IV: 308–10) trace a conic passing through A and B; and this, when (as here) a pair of corresponding points C, F are in line with AB, will break into the pair of the line AB and a rectilinear locus (G) (Compare 1, §2: note (37) preceding.) The same holds true, of course, for a general 1, 1 correspondence

$$a.CD \times FE + b.CD + c.FE = 0$$

between the points D and E of the two lines.

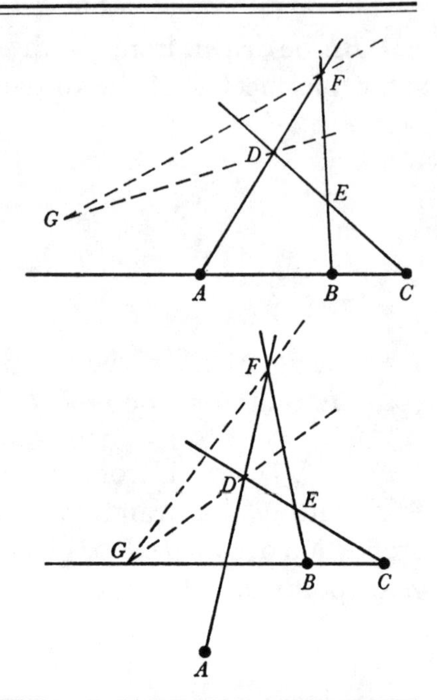

Hujusmodi quidem Porismata habebantur in principio libri primi, quæ omnia comprehendit Pappus hac propositione generali.(54) Si a tribus punctis datis A, B, C in directum jacentibus ducantur tres rectæ se mutuo secantes in punctis D, E, F et intersectionum duæ quævis D, F tangant rectas positione datas DG, FG, tanget etiam tertia E rectam positione datam.(55) Eadem propositio valet ubi puncta A, B, C non sunt in directum si modo puncta G, B, C in directum ponantur.(56)

Dixi constructiones Porismatum indicari(57) in demonstrationibus, Ejus rei exempla sunt plana in *Datis* Euclidis. Suffecerit exemplū unum et alterum hic subjungere.(58)

(54) On repairing a hiatus in Commandino's Latin version of a Greek original which is at this point highly corrupt, Pappus' own statement of this *propositio generalis* reads: 'Euclides... pauca ad principium primi libri posuit[,] consimilia ab vberima illa specie locorum, vt numero decem. Quare cum has vna propositione comprehendi posse intelligeremus, ita descripsimus[:] Si [quattuor rectæ lineæ se mutuo secent, quarum non plures quam duæ per idem punctum, data autem sint] tria puncta [*sc.* intersectionis] in vna recta linea, ... reliqua [*sc.* tria] vero præter vnum tangant positione datam rectam lineam, & hoc positione datam rectam lineam tanget' (*Mathematicæ Collectiones*: 245).

(55) Passing through the intersection G of the other two rectilinear loci (D) and (E), in fact, though whether Newton had here yet come explicitly to notice this corollary is doubtful. (With his preceding draft enunciations of this porism, certainly, he had failed to include any figure in which the property is even approximately observed; see IV: 239, note (28) and compare Appendix 2.2: note (16) below.) Once this consequence is specified, of course, we straightaway obtain Girard Desargues' familiar theorem—or, to be strict, its converse—which asserts that if a pair of triangles DEF are in perspective through a point G, then their pairs of corresponding sides DF, FE, ED meet in points A, B, C which are collinear. (This equivalent form of Pappus' proposition—which there is no evidence that Newton himself ever knew—had been discovered by Desargues half a century earlier and published by his disciple Abraham Bosse in his *Maniere Universelle de Mr Desargues pour Pratiquer la Perspective par Petit-pied, comme le Geometral* (Paris, 1648): 340–3; see R. Taton, *L'Œuvre Mathématique de G. Desargues: Textes publiés et commentés avec une introduction biographique et historique* (Paris, 1951): 201–12. As Taton remarks (*ibid.*: 204), 'Quoique la *Collection Mathématique* de Pappus ait éditée dès 1588 par Commandin, il est cependant peu probable que Desargues y ait puisé directement son inspiration; il semble plutôt que ce soient des considérations de perspective qui l'aient conduit à concevoir son théorème sous sa forme spatiale [*sc.* where the perspective triangles do not lie in the same plane, and only simple incidence axioms are required in its demonstration], et qu'ensuite, grâce à la relation de Ménélaüs dont il faisait fréquemment usage, il l'ait démontré pour le plan'. Despite the several efforts of such detractors of his originality as Jean Beaugrand and Grégoire Huret elsewhere to establish his debt to the *Mathematical Collection* (see Taton, *Desargues:* 37, 46) there

Porisms of this sort, indeed, were had at the beginning of the first book, and Pappus embraces all these in the following general proposition:(54) If from three given points A, B, C lying in a straight line there are drawn three straight lines mutually cutting one another in the points D, E and F, and any two of the intersections D and F should trace straight lines DG, FG given in position, then the third one E also will trace a straight line given in position.(55) The same proposition is valid when the points A, B, C are not in a straight line provided that the points G, B and C be set in a straight line.(56)

I have said that the constructions of porisms are indicated(57) in their demonstrations. There are 'plane' instances of this matter in Euclid's *Data*. It will suffice here to adjoin a couple of examples.(58)

---

is, indeed, only reasonable circumstantial surmise (compare *ibid.*: 125, note (33) and 143, note (55), for instance) that Desargues ever profited from reading Pappus' work.) On the other hand, however, we would insist that until Newton consciously made this 'evident' deduction he could not have untangled Pappus' further extension of his *propositio generalis*, according to which it is stated (*Mathematicæ Collectiones*: 245–6): 'Si quotcumque rectæ lineæ sese mutuo secent, non plures quam duæ per idem punctum, omnia autem in vna ipsarum data sint, & vnumquodq; eorum, quæ sunt in altera tangat positione datam rectam lineam[;] vel vniversalius hoc modo. Si quotcumque rectæ lineæ sese mutuo secent, non plures quàm duæ per idem punctum, omnia autem in vna ipsarum data sint, & reliquorum multitudinem habentium triangulum numerum, huius latus singula habet puncta tangentia rectam lineam positione datam, quorum trium non ad angulum existens trianguli spacij[:] vnumquodq; reliquum punctum rectam lineam positione datam tanget.' (With an agreeably modest show of deference to his predecessor Pappus added that 'Euclidem vero non verisimile est ignorare hoc, sed principium dumtaxat statuere'.) The meaning of this involved statement is not a little subtle, even in modern paraphrase: if in any configuration of $n+1$ straight lines (no more than two of which pass through the same point) the n points of intersection upon one of these be given, and it is further postulated that, out of the remaining 'triangular number' $\frac{1}{2}(n-1)n$ of these, the 'side' $n-1$ of the other intersections shall each be constrained to trace a rectilinear locus—but no three of which shall in the figure be located at the corners of a triangle (or, more generally, any i-th subgroup of them at the corners of a complete i-gon, $i = 4, 5, \ldots$) since, when all but one of the vertices trace given straight lines, so must the last—, then all the remaining (free) intersections in the configuration will also each trace a rectilinear locus. The significance of all this completely escaped Edmond Halley—in his *Apollonii Pergæi De Sectione Rationis*... (note (46) above): xxxv his sole contribution to the exegesis of the passage was erroneously to assert that Pappus' requirement that (in his own preferred Latin version) 'tres...intersectiones...non reperiantur ad angulos trianguli' should be taken to mean that these be 'in rectâ lineâ'—and a correct enodation of this cryptic *propositio generalissima* was achieved only at the end of Newton's life by Robert Simson in the first of 'Pappi Alexandrini Propositiones duæ generales, quibus plura ex Euclidis *Porismatis* complexus est, restitutæ...' (*Philosophical Transactions*, **32**, No. 377 [for May/June 1723]: §VI: 330–40, especially 330–7; the second restoration (*ibid.*: 337–40)—actually two mutually converse 'Porismata primum/secundum'—is but the particular case of the Pappus–Desargues porism where one of the fixed poles is at infinity).

Let us mention, finally, that when in July 1702 David Gregory met Newton in London to talk over what he should put in his *Præfatio* to the Greek and Latin edition of *Euclidis quæ supersunt omnia* (see 1, §1, Appendix 1: note (10)) whose proofs were then issuing from the press

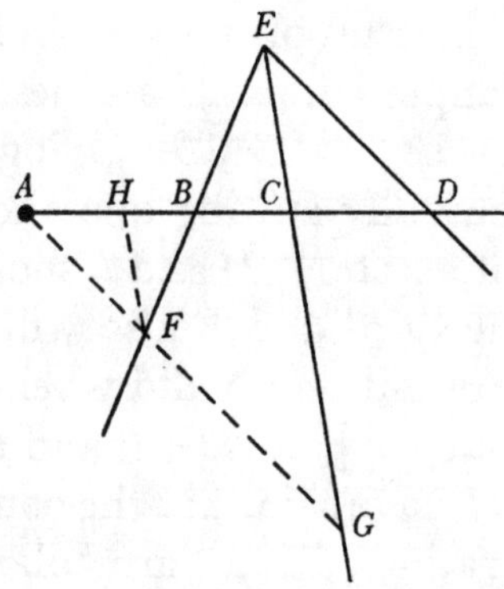

Porisma [1′]

Si a dato puncto A educta recta $ABCD$ secet rectas tres positione datas EB, EC, ED concurrentes in puncto communi E, fient rectangula

$$AB\times CD,\ AC\times BD\ \&\ AD\times BC$$

ad invicem in datis rationibus.(59)

Demonstratio.

A puncto A ipsi ED [parallela] agatur AFG occurrens ipsis EB et EC in punctis F et G, et ab F ipsi EG parallela agatur FH occurrens AB in H. Ob similes figuras $AFBH$ $DEBC$ sunt $AB.BD(::AF.ED)::AH.CD$. Ergo rectangula $AB\times CD$ et

in Oxford—and was told *inter alia* that 'In that præface the *Porismata Euclidi[s]* must be insisted upon. they were of the Nature of the *Data* but more difficult'—he made a rough note of what his mentor urged to be 'the chiefe proposition of them (which is obscurely spoken of by Pappus)', thereafter enunciating the present *porisma generale* in the little differing form 'Si circa Polos A, B, C rotentur rectæ AD, BE, CF ita ut intersectio ipsarum AD, BE sit ad rectam GH; et intersectio ipsarum AD, CF sit ad rectam KL: erit intersectio ipsarum BE, CF etiam ad rectam puta ad MN' (quoted from Gregory's memorandum C 36, now Royal Society MS 247.63). It is all too clear, we may add, that Gregory had no understanding of the theorem which he here parrots, because in his manuscript figure (which we see no need here to reproduce) he has not attempted either to make the rectilinear loci GH, KL and MN pass through a common point or to place the poles A, B and C even approximately in line. In the event his published preface, dated 'Oxoniæ, Junii 10. 1703', contained little more than the bare statement that 'Tres item *Porismatum* Libros edidit Euclides, teste Pappo & Proclo' along with his opinion that 'Perierunt hi Libri magno rei Geometricæ, Loci præsertim Resoluti, damno. Ex iis vero quæ in 7mo lib. Pappi (tum in Præf. tum à Prop. 127. ad 165.) supersunt, non difficile futurum reor illos quodammodo restituere, ubi Textus Græcus lucem aspexerit: Commandini namque versio Præfationis latina mentem Pappi minime assequitur' (*ΕΥΚΛΕΙΔΟΥ ΤΑ ΣΩΖΟΜΕΝΑ*...(Oxford, 1703): signature c2r). His fellow Savilian professor was to be markedly less optimistic regarding the possibility of effecting even a partial restoration of Euclid's work: we have already (see note (46) above) cited Halley's despairing aside that he could not begin to understand Pappus' terse and corrupt text at this point when just three years later he edited the hitherto unprinted Greek of the preamble to the seventh book of the *Mathematical Collection* along with a fresh translation of it into Latin.

(56) Newton here initially spelled out that 'Huic affinis est etiam hæc Propositio. Si a tribus punctis quibuscunqꝫ A, B, C agantur tres rectæ se mutuò secantes in punctis tribus D, E, F, et intersectionum duæ D, F tangunt rectas positione datas DG, FG concurrentes ad trianguli ABC latus illud infinitè productum BC quo angulus ad intersectionem tertiam E subtenditur, intersectio illa tertia tanget rectam positione datam' (Akin to this is also this proposition. If from any three points A, B, C there be drawn three straight lines mutually cutting one another in three points D, E, F, and two of the intersections trace straight lines DG, FG given in position and concurrent at the indefinitely extended side BC of the triangle ABC by which the angle at the third intersection E is subtended, that third intersection will trace a straight line given in position). In this variant case the porism reduces, much as we have already specified in IV: 239–40, note (29), to be an instance of Pappus' theorem (*Mathematical Collection* VII,

Porism 1'.

If when a straight line $ABCD$ is drawn out from the given point A it intersects three straight lines EB, EC, ED given in position and concurrent at the common point E, then will the rectangles $AB \times CD$, $AC \times BD$ and $AD \times BC$ prove to be to one another in given ratios.(59)

Demonstration.

From the point A parallel to ED draw AFG meeting EB and EC in the points F and G, and from F parallel to EG draw FH meeting AB in H. Because the figures $AFBH$ and $DEBC$ are similar, there is $AB:BD(=AF:ED) = AH:CD$. Therefore the rectangles $AB \times CD$ and $AH \times BD$ are equal. But the rectangle

139/143) on the collinear meets of opposite sides of a hexagon inscribed in a line-pair: here, namely, since the hexagon $ABFGDC$ is set with its alternate vertices on one and the other of the lines GBC and ADF, the intersection E of BF with CD is ever on the straight line through the fixed meets of AB with GD and of AC with GF. From a more general viewpoint, alternatively, when the poles A, B, C are not in line it follows by the 'Maclaurin–Braikenridge' theorem or Pascal's earlier equivalent generalization of Pappus' theorem to the *hexagrammum mysticum* inscribed in a conic (on both which see IV: 321–2, note (90)) that the intersection E will trace a conic passing through G, B and C which, when these are (as here) in line, degenerates to be the pair of GBC and a rectilinear locus (E).

(57) This replaces the considerably less tenable initial phrasing 'contineri' (are contained).

(58) Newton first introduced his remaining porisms with the balder statement that 'Continebantur præterea in his libris varia Porismatum genera qualia fere sunt hæc et quæ ex his derivare liceat' (There were contained in these books, furthermore, various kinds of porisms chiefly of the following sort and what may permissibly be derived from them). Superseded preliminary enunciations (on ff. 29r/30r) of the first two—unlike their revises they are given no demonstration—are quoted separately at their pertinent places in subsequent footnotes.

(59) This enunciation of the constancy of the cross-ratios of the intercepts made by three concurrent lines on a transversal constrained to pass through a fixed point is, like the proof which follows, taken over without essential change from Pappus, *Mathematical Collection*, VII, 129. In his discarded preliminary version of this (see the previous note) Newton also stated the converse proposition (*Mathematical Collection*, VII, 142), there announcing that 'Si a dato puncto A educta recta $ABCD$ secet rectas tres positione datas EB, EC, ED concurrentes in puncto communi E, erit rectangulum AB, CD ad rectangulum AD, BC in data ratione[;] et si rectangula illa sunt in data ratione et punctorum B, C, D duo quævis tangunt rectas positione datas, tanget etiam tertium rectam positione datam' (If from a given point A a straight line $ABCD$ is drawn out to cut three straight lines EB, EC, ED given in position and concurrent in a common point E, the rectangle $AB \times CD$ will be to the rectangle $AD \times BC$ in a given ratio; while if those rectangles are in a given ratio and any two of the points B, C, D trace straight lines given in position, the third one also will trace a straight line given in position)—understand, of course, that the last line passes through the meet of the two preceding ones. In sequel Newton stumbled trivially in restating this equivalence between the constancy of the cross-ratio $AB \times CD/AD \times CB$ and the concurrence of the pencil of lines through E, asserting that 'Idem fiet si fuerit AB ad AD ut BC ad CD[,] quippe rectangula illa tunc sunt æqualia atq; adeo in data ratione' (The same will happen if there were AB to AD as BC to CD, seeing that those rectangles will then be [proportional] and hence in a given ratio).

$AH \times BD$ æquantur. Sed rect. $AH \times BD$ est ad rect. $AC \times BD$ ut AH ad AC hoc est ut AF ad AG. Ergo $AB \times CD$ est ad $AC \times BD$ ut AF ad AG. Atqui ratio illa ob datas AF AG datur. Q.E.D. Simili argumento si ipsis EC, ED a puncto A ducantur parallelæ, demonstrabitur rectangula $AB \times CD$, $AD \times BC$ ad invicem & $AC \times BD$, $AD \times BC$ ad invicem esse in data ratione.(60). Q.E.D.

In hac Demonstratione constructio est ut a puncto A agatur AFG ipsi ED parallela et ponatur ratio $\frac{AB \times CD}{AC \times BD}$ eadem datæ rationi $\frac{AF}{AG}$.(61) Et sic de rationibus cæteris.

Porisma [2′]

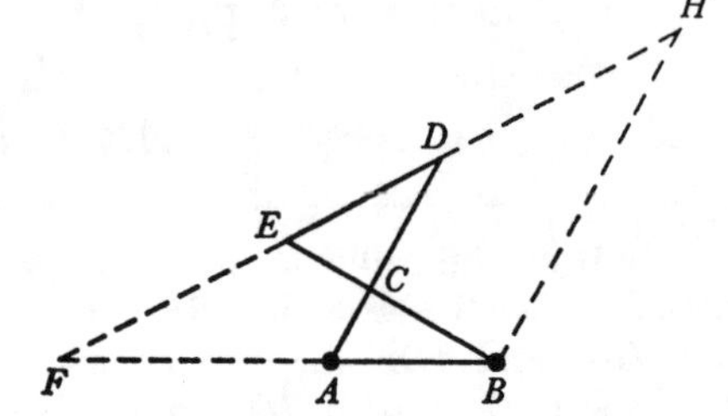

Si a datis punctis A, B ducantur rectæ duæ AD, BE se secantes in C ea lege ut rectangula $AD \times CE$, $BE \times CD$ inæqualia sint ac datam rationem ad invicem habeant, converget linea ED ad punctum datum.(62)

Demonstratio.

Ipsi AD parallela agatur BH et producta utrinq; ED occurrat ipsis AB, BH in F et H. Ob proportionales $EC.CD::EB.BH$ æqualia sunt rectangula $BH \times CE$ et $BE \times CD$[,] sed est AF [ad] BF ut AD ad BH hoc est ut $AD \times CE$ ad $BH \times CE$, hoc est ad $BE \times CD$ id est in data ratione[,] ergo *divisim AF est ad AB in data ratione adeoq; ob datam AB datur AF et inde datur punctum F ad quod recta ED convergit. Q.E.D.

* 5. *Dat.* 27. *Dat.*

Constructio hic est ut producatur BA ad F sic ut AF sit ad BF in data ratione rectangulorum $AD \times CE$, $BE \times CD$ et habebitur punctum datum F ad quod recta ED convergit.

Porisma [3′]

Si a datis duobus punctis A, B ad punctum tertium C inflectantur duæ rectæ

(60) We need scarcely remark that this trio of constant cross-ratios

$$\lambda = AB \times CD/AC \times BD \equiv (ADBC),$$
$$\mu = AB \times CD/AD \times BC \equiv -(ACBD)$$
$$\text{and } \nu = AC \times BD/AD \times BC \equiv (ABCD)$$

are inter-connected by the relation $1/(\lambda - 1) = \mu - 1 = \nu$.

(61) In modern terminology the construction is to 'project' the point D to infinity in ED (by rotating the transversal round A till it comes to be parallel to ED) and then by elementary Euclidean means—in exact pattern of Pappus' resolution (see note (59))—to determine the equality of the cross-ratios $(ADBC)$ and $(A\, \infty_{ED}\, FG)$.

(62) Much as before, Newton evidently borrows this porismatic equivalent to Menelaus' theorem—and also his following proof—from Pappus' enunciation of it in his *Mathematical Collection*, VIII, 3. (He was almost certainly unaware of Menelaus' own statement of it in its generalized spherical case in his *Sphærica*, III, 1 or Ptolemy's subsequent employment of it in his

$AH \times BD$ is to the rectangle $AC \times BD$ as AH to AC, that is, as AF to AG. Therefore $AB \times CD$ is to $AC \times BD$ as AF to AG. While, because AF and AG are given, that ratio is given. As was to be proved. By a similar argument, if parallels to EC and ED be drawn from the point A it will be demonstrated that the rectangles $AB \times CD$ and $AD \times BC$ are to one another, and so also $AC \times BD$ and $AD \times BC$ to one another, in a given ratio.(60) As was to be proved.

In this demonstration the construction is to draw AFG from the point A parallel to ED and to put the ratio $AB \times CD/AC \times BD$ the same as the given ratio AF/AG.(61) And the like for the remaining ratios.

Porism 2'.

If from the given points A, B there be drawn two straight lines AD, BE intersecting each other at C with the restriction that the rectangles $AD \times CE$ and $BE \times CD$ be unequal but have a given ratio one to the other, the line ED will converge on a given point.(62)

Demonstration.

Parallel to AD draw BH and let ED extended either way meet AB and BH in F and H. Because of the proportion $EC:CD = EB:BH$ the rectangles $BH \times CE$ and $BE \times CD$ are equal. But AF is to BF as AD to BH, that is, as $AD \times CE$ to $BH \times CE$ or $BE \times CD$, and so in a given ratio. Therefore *dividendo** AF is to AB in a given ratio, and so, because AB is given, AF is given and thence is given the point F on which the line ED converges. As was to be proved.

* *Data*, 5 and 27.

The construction here is to produce BA to F so that AF be to BF in the given ratio of the rectangles $AD \times CE$ and $BE \times CD$, and there will then be had the given point F on which the line ED converges.

Porism 3'.

If from the two given points A, B to a third point C there be inclined two

Almagest, I, Chapter 13. Even after Giovanni Ceva made extensive use of the theorem in his *De Lineis se invicem secantibus, Statica Constructio* (Milan, 1678)—a work of which Newton himself seemingly remained ignorant throughout his life—the theorem remained virtually unknown by mathematicians outside France and Italy; see Michel Chasles' 'Note VI' to his *Aperçu Historique sur L'Origine et le Développement des Méthodes en Géométrie* (Paris, ${}_3$1889: 291–3), and also J. Tropfke, *Geschichte der Elementar-Mathematik*, **4** (Berlin/Leipzig, ${}_2$1923): 171–3.) Newton's preliminary enunciation of this *porisma* on f. 30ʳ (see note (58) above) announces little differently that 'Si a datis punctis A, B ducantur rectæ AD, BE se secantes in C ea lege ut rectangula inæqualia $AD \times CE$, $BE \times CD$ sint in data ratione, converget linea ED ad datum punctum F [*sc.* in AB situm]', but thereafter adds: 'Idem fiet si CD, CE æquales sunt et AD, BE in data ratione vel AD, BE æquales et CD, CE in data ratione. Quod si rectangula illa sint ut AF, BF et detur solummodo punctum F, converget recta DE ad datum illud F' (The same will occur if CD, CE are equal and AD, BE in given ratio, or AD, BD equal and CD, CE in given ratio. While if those rectangles be as AF to BF and only the point F be given, the straight line DE will converge on that given F).

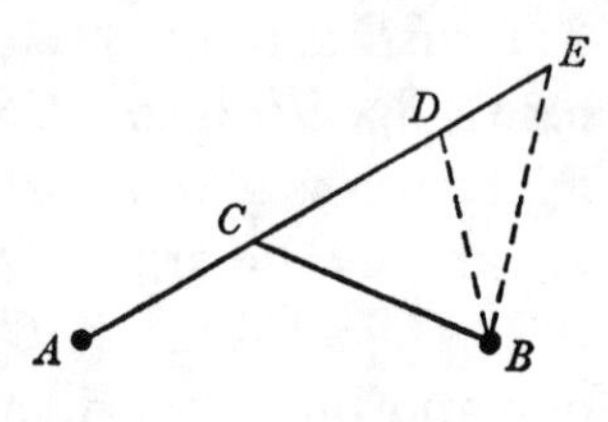

AC BC in dato ang[ulo] *ACB*, punctum illud *C* tanget circulū datum[,] et si producatur *AC* ad *D* ut sit *CD* ad *CB* in data ratione, tanget punctū *D* circulum datum[,] et si a dato puncto *A* agatur recta *ACDE* secans circulos duos datos(63) per punctū *A* transeuntes et habeat *DE* ad *CD* rationem datam, punctum *E* tanget circulum datum.(64)

Demonstratio.

Cas. 1. Ob datam rectam *AB* et datum angulum *ACB* datur(65) segmentum circuli capientis angulum *ACB*, hoc est quem punctum *C* tangit. Q.E.D.

Cas. 2. Ob datum angulum *ACB* vel *BCD* et datam rationem *BC* ad *BD*, triangulum *BCD* datur specie. Datur ergo angulus *CDB* adeoq̃ per Cas. 1 punctum *D* tangit circulum datum.(66) Q.E.D.

Cas. 3. Quoniam puncta *C*, *D* tangunt circumferentias datas dantur anguli *ACB*, *ADB* adeoq̃ datur specie triangulum *BCD*. Datur ergo ratio *BD* ad *DC*. Adde datam rationem *CD* ad *DE* et dabitur ratio *BD* ad *DE*. Ergo (per cas. 1) punctum *E* tangit circumferentiam datam. Q.E.D.

Constructio in his tribus casibus est, ut per puncta *A*, *B* describantur circuli qui capiant angulos *ACB*, *ADB*, *AEB*.

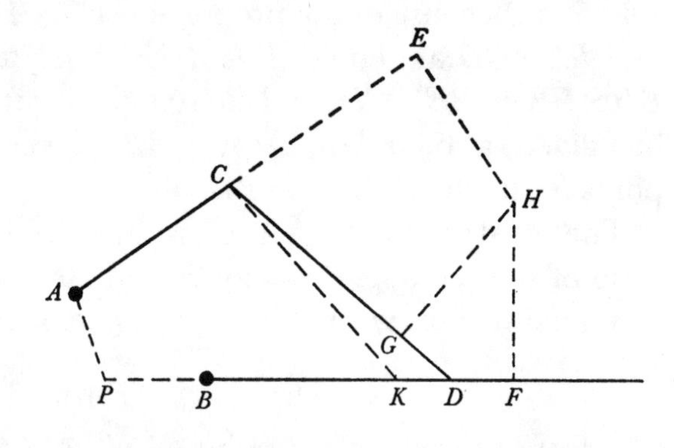

Porisma [4'](67)

Si a datis punctis *A*, *B* ducantur in datis positionibus rectæ duæ *AE* *BF* et utraq̃ secetur tertia *CD*[,] tangant autem omnes tres circulum

(63) Those through the given points *A*, *B* and also *C* and *D* respectively.

(64) At its first occurrence in Newton's papers (as *locus planus* 19 on IV: 240) we stated our belief that, as so defined, this porismatic theorem is his own original discovery (see *ibid.*: 241, note (39)). An extension of it—to the generalized case where the segments *AC*, *CD*, *DE* are no longer in line but form three sides of a quadrilateral given in species—had subsequently been introduced by him into his 'De motu Corporum Liber primus' as its Lemma XXVII (see VI: 292), the present proposition being thereto appended in corollary.

(65) Understand 'per 33. III. *Elem.*' (by *Elements* III, 33), as Newton indeed justified his equivalent initial requirement in a cancellation at this point that 'Per puncta *A*, *B* describatur circumferentia circuli capiens angulum datum *ACB* et dabitur centrū et intervallum, ergo et ipse circulus' (Let there through the points *A*, *B* be described a circle-circumference taking the given angle $\widehat{ACB}$, and there will be given its centre and radius, and therefore also the circle itself). The three last 'givens' are referred to '25. *Dat*', '26. *Dat*' and 'Def 8. *Dat*' respectively.

(66) Newton first concluded: 'Super data autem recta *AB* ([per] 33. III. *Elem.*) describantur segmenta circulorum quæ capiant angulos datos *ACB*, *ADB* et puncta *C*, *D* tangent horum circumferentias datas' (Upon the given straight line *AB*, however, let there (by *Elements* III, 33)

straight lines AC, BC at the given angle $A\widehat{C}B$, that point C will lie in a given circle; and if AC be produced to D so that CD be to CB in a given ratio, the point D will lie in a given circle; and if from the given point A there be drawn the straight line $ACDE$ intersecting the two given circles[63] passing through the point A, and DE shall have a given ratio to CD, the point E will lie in a given circle.[64]

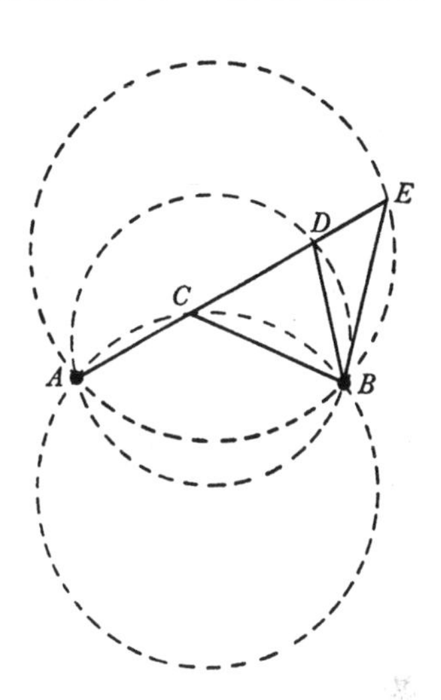

Demonstration.

Case 1. Because the straight line AB is given and the angle $A\widehat{C}B$ also, there is given[65] the circle segment taking the angle $A\widehat{C}B$, and this is that wherein the point C lies. As was to be proved.

Case 2. Because the angle $A\widehat{C}B$—and so $B\widehat{C}D$—is given and the ratio of BC to BD also, the triangle BCD is given in species. There is therefore given the angle $C\widehat{D}B$, and hence by Case 1 the point D lies in a given circle. [66] As was to be proved.

Case 3. Seeing that the points C and D lie in given circumferences, the angles $A\widehat{C}B$, $A\widehat{D}B$ are given, and hence the triangle BCD is given in species. There is therefore given the ratio of BD to CD. Adjoin the given ratio CD to DE, and there will be given the ratio of BD to DE. Therefore (by Case 1) the point E lies in a given circumference. As was to be proved.

The construction in these three cases is to describe circles through the points A, B which shall take the angles $A\widehat{C}B$, $A\widehat{D}B$ and $A\widehat{E}B$.

Porism 4'.[67]

If from the given points A, B two straight lines AE, BF be drawn in given positions and each be intersected by a third, CD, let all three touch a given circle which lies outside the

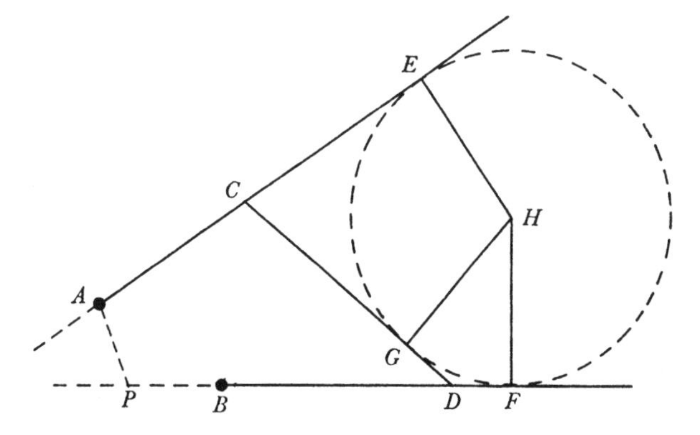

be described segments of circles which shall take the given angles $A\widehat{C}B$, $A\widehat{D}B$, and the points C, D will then trace the given circumferences of these).

(67) A first version (on ff. $31^r/32^r$) of this final 'plane' porism is reproduced in Appendix 3 below. We know no antecedents to it, and that Newton worked so hard to amend and polish both its enunciation and demonstration would seem to guarantee its originality.

datum qui jaceat extra figuram $ACD[B]$, dabitur summa trium $AC+CD+DB$. Et e contra si datur harum summa[,] tanget recta CD circulum datum.

Cas 1. Circulus centro H descriptus tangatur a tribus AE BF CD in E, F, G et ob æquales HE, HG, HF, æquales erunt CE CG et DF DG, adeoq; conjunctim $CE+DF$ et CD. Addatur utrobiq; $AC+BD$ et æquales erunt $AE+BF$ et $AC+CD+DB$. Sed $AE+BF$ datur, ergo et $AC+CD+DB$. Q.E.D.

Cas 2. Age AP constituens angulos ad A et P æquales et fiant AE PF æquales dimidio datæ summæ $AC+CD+D[P]$.(68) Ad E et F erigantur perpendicula concurrentia in H et centro H intervallo HE describatur circulus. Tanget recta CD hunc circulum. Si negas tangat recta CK eundem. Ergo

$$AC+CK+KP = AE+PF = AC+CD+DP.$$

Et ablatis utrinq; AC, PK restat $CK = CD+DK$. Quod est absurdum.(69) Simile consequetur absurdum si ducatur CK ad alteras partes rectæ CD. Tanget ergo recta CD circulum illum datum. Q.E.D.(70)

APPENDIX 1. CONSTRUCTION OF THE NORMAL AT A GENERAL POINT ON AN ARBITRARY MULTI-POLAR LOCUS.(1)

From a discarded draft(2) in the University Library, Cambridge

[Et tamen perpendiculum ad curvam non tantum in hoc casu(3) sine Algebra duci potest sed etiam in innumeris difficilioribus. Ut si summa aliquarum e ductis datam haberet rationem ad summam reliquarum, vel si...sub alijs.]

---

(68) Newton carelessly wrote 'DB' in the manuscript.

(69) This replaces the equivalent Barrovian stylization 'Quod fieri nequit' (Which cannot be); compare III: 410, note (12).

(70) On f. 30r of the manuscript, we may add—in immediate sequel to the cancelled preliminary enunciation of Porism 3′ quoted in note (62) above—Newton has deleted the bare statement of a fifth *porisma*: 'Si rectæ quævis AB, CD positione datæ terminantur ad data puncta A, C et datum continent rectangulum AB, CD, converget recta BD ad punctum datum' (If any straight lines AB, CD given in position terminate at given points A, C and contain a given product $AB\times CD$, the straight line BD will converge on a given point). This is, unfortunately, only true when the given 'rectangle' is equal to $AE\times CE$ where E is the point at which AB, CD intersect (see Porism 1 on pages 234–6 above); in all other cases, as we have already more than once observed (see 1, §2: note (24) in particular), the transversal BD will envelop a central conic touching AB and CD—and indeed Newton was (see Appendix 3: note (8) below) to add a circular example of the latter *sectio spatii* as 'Cas. 3' in an initial version of the present Porism 4′. Why he elected to discard this simplest rectilinear case of such a generalized Apollonian 'section'—one which is somewhat obscurely cited by Pappus in his recension of the content of Euclid's first book of *Porisms* (compare 1, §2: note (21)), though the pertinent sentence is engulfed in the terminal '&c' in Newton's preceding quotation of it—is not at all

figure $ACDB$, and there will be given the sum of the three $AC+CD+DB$. Conversely, if their sum is given, the straight line CD will touch a given circle.

Case 1. Let the circle described with centre H be touched by the three AE, BF, CD in E, F and G, and, because of the equals HE, HG and HF, both CE and CG and also DF and DG will be equal, and hence jointly so will $CE+DF$ and CD. Add $AC+BD$ to each member, and then $AE+BF$ and $AC+CD+DB$ will be equal. But $AE+BF$ is given, and so therefore is $AC+CD+DB$. As was to be proved.

Case 2. Draw AP forming equal angles at A and P, and let AE, PF be made equal to half the given sum $AC+CD+DP$.(68) At E and F erect perpendiculars meeting in H, and then with centre H and radius HE describe a circle. The straight line CD will touch this. If you deny it, let the straight line CK touch it. In consequence $AC+CK+KP = AE+PF = AC+CD+DP$. And so, when AC and PK are taken away from either side, there remains $CK = CD+DK$. Which is absurd.(69) A similar absurd consequence will follow if CK be drawn on the other side of the line CD. The straight line CD will therefore touch the given circle. As was to be proved.(70)

Vel deniq; si harum linearum relatio quævis ad invicem statuatur quæ per æquationem exprimi potest. Exempli gratia sit XY curva proposita ad cujus punctum quodvis Z erigendum est perpendiculum ZP, sintq; A, B, C, D, E

---

clear. Maybe he had it in mind to re-introduce it at some later stage in his current manuscript before he broke off to pen the revised draft of his proem to the 'Geometria' which we now print in 2, § 2 following.

(1) The historical background to this generalized Tschirnhausian problem is outlined in § 1: note (21) preceding. From the pattern of the way in which he constructs the normal in the three instances which he here treats it will be clear (compare note (5) below) that he had already attained its general solution on the lines of that pioneered in particular cases by Fatio de Duillier—who was, we may confidently surmise for want of any exact evidence, here Newton's stimulus and guide, if not his mentor. Some time later, we may add, when David Gregory visited Newton at Cambridge in 1694, he noted down the gist of the construction of the normal to the first species of Tschirnhausian curve here presented and preserved it along with other 'Adnotata Math: ex Neutono. 1694 Maio' which he then took away with him; see note (4) below.

(2) Add 3963.11: $111^r/112^r$. Why Newton chose ultimately to reject this sheet from the main text of his 'Procemium' above is not clear. Perhaps he came to feel that its detailed discussion of a particular problem—elegant and instructive though it might be—overly distracted the reader's attention from the general message of the relative ease and power of a geometrical resolution, as against the difficulty and inadequacy of an equivalent algebraic analysis, which it was intended to convey.

(3) Namely, that (see § 1: note (21)) of the Tschirnhausian locus defined by $\sum_i (r_i) = k$, constant, where the r_i are the individual *radii vectores* drawn from the fixed poles to the locus-point.

puncta quotcunq; data. Et si ea sit natura Curvæ ut rectarum ZA, ZB, ZC, ZD, ZE a quovis ejus puncto Z ad data illa puncta A, B, C, D, E ductarum summa maneat eadem, centro Z, radio quovis GZ describatur circulus secans ductas rectas & si opus est productas, in punctis G, H, I, K, L. Ab horum quovis G ad quodvis H age GH & biseca eam in M. Ab M ad tertium quodvis punctum I age MI & in ea sume tertiam ejus partem MN. Ab N ad quartum punctum K age NK et in ea sume quartam ejus partem NO. Ab O ad quintum punctum L age OL et in ea sume OP quintam ejus partem. Et sic perge donec ad ultimum punctum perveneris. Sit illud L et in actarum ultima OL sumpta pars OP. Junge PZ et hæc erit perpendicularis ad curvam in puncto Z. Vel brevius: Quære centrum gravitatis punctorum omnium G, H, I, K, L. Sit illud P et erit PZ perpendiculum.(4) Quod si Curvæ XY hæc sit natura ut quod sub omnibus rectis ZA, ZB, ZC, ZD, ZE &c in se continuo ductis continetur sit dato æquale: in ZA cape Zg quæ sit ad ZG ut ZG ad ZA, in ZB cape Zh ad ZH ut ZH ad ZB & sic in cæteris ZC, ZD, ZE &c cape Zi ad ZI ut ZI ad ZC, & Zk ad ZK ut ZK ad ZD, et Zl ad ZL ut ZL ad ZE. Dein omnium punctorum g, h, i, k, l quære centrum gravitatis & recta a centro illo ad punctū Z acta perpendicularis erit ad curvam XY. Rursus si curvæ XY hæc sit natura ut rectangulorum sub singulis binis rectis nempe horum ZA in ZB.

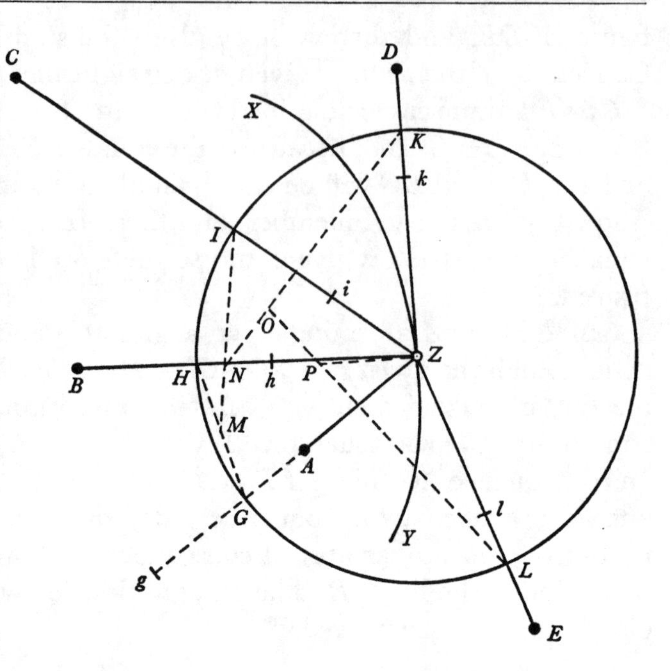

(4) David Gregory made a note of this construction among other 'Adnotata Mathematica' which he jotted down 'ex Neutono' during his visit to Cambridge early in May 1694: 'Si curvæ [XZY] ea sit proprietas ut [detur summa] rectarum A[Z], B[Z], C[Z], D[Z &c] a datis quotcunque punctis A, B, C, D [&c] ductarum ad assumptum utcunque punctum [Z]: si centro [Z] ducatur utcunque circulus secans predictas rectas in [G, H, I, K &c] respective, recta [ZP] jungens punctum [Z] cum communi centro gravitatis [P] punctorum [G, H, I, K] et [L] erit ad Curvam in [Z]...perpendicularis' (ULE. Gregory C43, poorly transcribed in *The Correspondence of Isaac Newton*, **3**, 1961: 332 along with (*ibid.*: 331) a woefully distorted reproduction of Gregory's accompanying sketch, whose designations of points we have here altered to accord with those of Newton's original figure). In even more rapid summary of what follows he further noticed that 'Similiter si loco summæ datæ, summa differentiarum, productorum etc. daretur, conficeretur facile problema'.

ZA in *ZC*. *ZA* in *ZD*. *ZA* in *ZE*. *ZB* in *ZC*. *ZB* in *ZD*. *ZB* in *ZE*. *ZC* in *ZD*. *ZC* in *ZE*. *ZD* in *ZE* summa sit dato æqualis; in *AZ* versus *Z* cape *AQ* æqualem summæ omnium linearum *AZ*, *BZ*, *CZ*, *DZ* & *EZ*[,] & similiter in *BZ*, *CZ*, *DZ*, *EZ* &c versus *Z* cape *BR*, *CS*, *DT*, *EV* sigillatim æquales eidem summæ, & punctorum omnium *Q*, *R*, *S*, *T*, *V* quære centrum gravitatis, & a centro illo ad punctum *Z* acta recta perpendicularis erit ad curvam. Et sic in alijs casibus semper inveniri possunt in lineis rectis puncta quædam a quorum communi centro gravitatis acta recta perpendicularis erit ad curvam.(5) Idem quoqȝ fieri potest ubi rectæ illæ lineæ ab eodem curvæ puncto ductæ sunt ad alias lineas seu rectas seu curvas versus data puncta, vel versus plagas assignatas aut sub alijs quibusvis legibus quæ in Geometria supponi solent, ac ductæ illæ quamvis habent ad invicem relationem quæ per æquationem exprimi potest.(6) Cujus rei perficiendæ Lector methodum generalem ex sequentibus(7) haud difficulter colliget. Sed et in curvis lineis quas vocant Mechanicas idem præstari potest. Hæc autem eo fusius commemoro quod Analyseos usus et vis in nihilo latius patuit quam in ducendis ad curvas perpendicu[li]s et tangentibus, et tamen

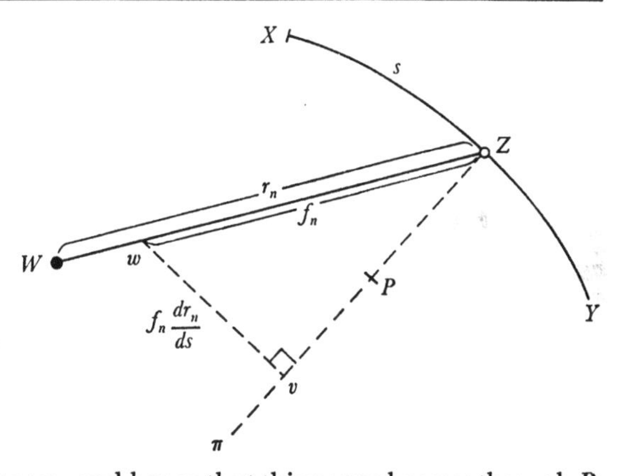

(5) If there are n fixed poles $A, B, C, \ldots, W$, from each of which to the locus-point Z there are drawn corresponding *radii vectores* $AZ = r_1, BZ = r_2, CZ = r_3, \ldots, WZ = r_n$, and the curve XZY—of arc-length $XZ = s$, say—is defined by some general *relatio ad invicem* of these $f \equiv f(r_1, r_2, \ldots, r_n) = k$, constant, then at once $\sum_{1\leqslant i\leqslant n} (f_i . dr_i/ds) = 0$ where $f_i = \partial f/\partial r_i$; accordingly, if in the individual radii $ZA, ZB, \ldots, ZW$ we set off $Zg = f_1, Zh = f_2, \ldots, Zw = f_n$, it follows that the sum of the perpendicular distances $Zg . dr_1/ds$, $Zh . dr_2/ds$, $\ldots$, $Zw . dr_n/ds$ of the points $g, h, \ldots, w$ from the normal $Z\pi$ at Z is zero, and hence that this normal passes through P, the centre of gravity of these points. In particular, if $f \equiv \sum_{1\leqslant i\leqslant n} (r_i) = k$, there is $f_i = 1$ and Newton chooses the points $g, h, \ldots$ to be the intersections $G, H, \ldots$ of $AZ, BZ, \ldots$ with a circle of some arbitrary unit-radius of centre Z; if, however, $f \equiv \prod_{1\leqslant i\leqslant n} (r_i) = k$, then $f_i = k/r_i \propto 1/r_i$; while if $f \equiv \sum_{\substack{1\leqslant i, j\leqslant n\\ [j\neq i]}} (r_i r_j) = k$, then $f_i = \sum_{\substack{1\leqslant j\leqslant n\\ [j\neq i]}} (r_j) = \sum_{1\leqslant j\leqslant n} (r_j) - r_i$.

(6) Here, of course, the points $A, B, C, \ldots$ are to be chosen as the instantaneous poles round which the *radii vectores* $AZ, BZ, CZ, \ldots$ effectively revolve. Since problems of the uniqueness of these radii raise themselves in these extensions, Newton is wise not to venture into a more detailed account of the various possibilities.

(7) Understand in the immediate sequel in his proem to the 'Geometria'. Newton in fact afterwards abandoned his intention of so discussing the 'general method' of comparing limit-motions of lines which he here presupposes, delaying it in the final draft of his 'Liber 1' (3, § 1 following) till he had explained what he meant by the fluxional increase of quantities; see pages 456–68 below.

nihil hactenus ea ratione effectum est præterquam in casu omnium simplicissimo, ubi duæ tantum lineæ rectæ ab eodem curvæ puncto ad alias duas positione datas ita ducuntur ut cum eis parallelogrammum constituant: aut quod perinde est, ubi una linea ad aliam in dato angulo ordinatur.[8]

APPENDIX 2. PRELIMINARY ATTEMPTS TO DEFINE THE NATURE OF PORISMS IN CONFORMITY WITH PAPPUS' REMARKS.[1]

From originals in the University Library, Cambridge

[1][2] Porismata, ut liquet ex Pappo, si formam propositionum spectes sunt partim Theoremata partim Problemata,[3] sed ratione finis distinguuntur ab utrisq. Hanc distinctionem veteres exposuerunt definiendo *Theorema esse quod proponitur in ipsius propositi demonstrationem, Problema quod affertur in constructionem propositi, Porisma verò quod proponitur in porismum hoc est in inventionem et investigationem propositi.*[4] Theoremata igitur et Problemata absolutè sunt quæ proponuntur sui ipsorum gratia, Porismata speculativa & problematica quæ proponuntur gratia inventionis. Est etiam Porismatum natura mutabilis et applicabilis ad omne subjectum sic ut quæ sunt problematica, accipiendo constructiones hypotheticè vertantur in Theoremata quæq Theorematum habent formam, determinando quantitatem aliquam quæ in ipsis est vertantur in genus Problematicum. Unde est quod Pappus dicit *horum species omnes neq*

---

(8) This is massively unappreciative of the wealth of contemporary research into general methods of tangents, of course. Even half a century before in his *Geometrie* Descartes had, as Newton was well aware, analytically determined the subnormal at a given point of a Cartesian oval defined in bipolars, admittedly by reducing the problem to its equivalent in standard Cartesian coordinates (see I: 555, note (15) and III: 138–9, note (228)). He was not to know that even as he penned these words Johann Bernoulli was composing a trail-blazing tract 'De Calculo Differentialium' which soon after formed the basis for L'Hospital's extensive 1696 monograph on the *Analyse des Infiniment Petits* (see P. Schafheitlin, 'Johannis (I) Bernoullii Lectiones de calculo differentialium', *Verhandlungen der Naturforschenden Gesellschaft in Basel*, **34**, 1922–3: 1–32; and also Otto Spiess' editorial 'Einleitung' to the Bernoulli–L'Hospital correspondence in the former's *Briefwechsel*, **1** (Basel, 1955): 123–57, especially 130–53).

(1) These first versions of Newton's attempts in § 1 preceding to clarify the logical form and purpose of the classical porism on the basis of Pappus' all too obscure statements regarding its nature—rendered yet further confusing in Commandino's Latin rendering of a manuscript original here excessively corrupt—and thence to construct simple rectilinear instances of individual *porismata* in Euclidean style are not in themselves of great significance, but their text is here reproduced to make the record complete.

(2) Add. 3963.14: 157r/158r; see § 1: note (39) above.

(3) Newton initially began with the equivalent statement that 'Porismata, ut refert Pappus, aliqua formam theorematum alia formam problematum induunt'.

theorematum esse neq Problematum sed mediam quodammodo inter hæc formam ac naturam habere, ita ut eorum propositiones formari possint ut theorematum vel ut Problematum.(4) Porismatum una species sunt propositiones quæ ad locorum inventionem ac determinationem spectant. Unde aliqui[,] apud Pappum[,] dixerunt *Porisma esse quod hypothesi deficit a locali theoremate:*(4) quasi illud hypotheticè in loco resoluto[,] hoc positive in Theorematum et Problematum resolutione et compositione proponeretur neq alia interesset differentia. Cæterum hæc species, ait Pappus,(4) *seorsum a Porismatibus collecta inscriptaq ac tradita est, quod magis diffusa et copiosa sit cæteris speciebus.* (5)Generaliter propositiones omnes in loco resoluto quæ ad inventionem spectant Porismata sunt, & eæ præsertim quæ vi determinandi magis pollent inventu sunt difficiliores, usu latius patent et inventionis facultatem magis exercent. Tot igitur sunt species Porismatum quot subjectorum in loco resoluto quamvis subjectis particularibus particularia dentur nomina, & Porismatū nomen Euclideis solis ob materiæ varietatem, inventionem minus obviam & usum frequentem in(6) difficilioribus enodandis impositum sit.

Porismatum pro varietate subjecti sunt species variæ. De ijs agitur promiscue in *Datis* et *Porismatibus* Euclideis, separatim in libris reliquis a Pappo memoratis. Unde his dabantur nomina particularia a subjectis quæ tractant, duobus prioribus nomina illa generalia *Datorum* et *Porismatum*: ac *Data* quæ magis obvia, *Porismata* in quibus inventio difficilior exercetur,(7) commodè dicta sunt.(8) Sunt enim libri *Porismatū* nihil aliud quam continuatio *Datorum* ut ex forma Propositionū constet quas Pappus juxta versionem Commandini sic obscurè(9) commemorat. *Principio,*(10) ait, habetur *diagramma hoc. Si a duobus datis punctis ad rectam lineam positione datam rectæ lineæ inflectantur, abscindat autem una a recta linea*

(4) All these quotations are from Pappus, *Mathematicæ Collectiones* (1, §1: note (1)): 245.

(5) Newton first continued: 'De genere Poristico sunt etiam *Data* Euclidis sed et hæc species ob simplicitatem inventionis & formam propositionum seorsim tradita, et potiori jure inscripta sunt ab Euclide *Porismata* quæ inventioni difficili juvandæ parabantur'.

(6) 'quæstionibus' is deleted.

(7) Initially 'quæ difficilioris sunt investigationis et inventioni difficiliori inserviunt'.

(8) Newton first wrote less cautiously 'optimè dicebantur', adding thereafter that 'Aliter vix differunt hæc duo, nisi quod *Porismatis* nomen generalius est. Quinetiam Data ipsa sunt una species porismatum, quanquam cæteris speciebus latius[!] patens' and then hastily inverting this to assert that 'Latius autem patet Porisma quam Datum, quatenus Porismatum aliqua sunt de datis alia de indeterminatis [afterwards changed to be 'veris'] relationem aliqualem habentibus, sed formam datorum magis amant [induere], ut quæ in resolutione Problematum aperti[or] et datis novis indigitandis et colligendis aptior sit'.

(9) In an immediately preceding cancelled first version of this clause Newton wrote 'breviter et perobscurè', adding in attenuation of this much harsher assessment that 'Commandinus ex MSS depravatis sic vertit'.

(10) That is, at the beginning of Euclid's *Porismata.* Commandino's Latin version here more fully renders the corrupt Greek manuscript as 'In principio septimi' (*Mathematicæ Collectiones*: 246), and in his revised draft on page 262 above Newton interpolates this to be 'In principio septimi porismatis libri primi'; see §1: note (44).

positione data ad datum in ipsa punctum, abscindet et altera proportionem habens datum. In ijs autem quæ sequuntur... &c.[(11)]...

---

[2][(12)] 1. Si a duobus punctis datis A, B inflectantur duæ rectæ ACF, BCG ad rectam positione datam CK: abscindat autem earum una ACF a recta positione data DK segmentum DF dato in ea puncto D adjacens, auferet etiam altera BCG ab alia recta EG[(13)] segmentum EG datam habens rationem ad abscissam priorem DF: puta si DF et EG simul evanescant.[(14)]

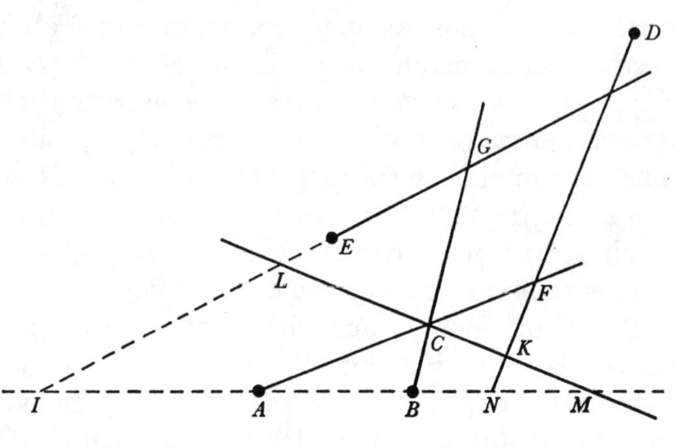

2. Si ratio DF ad EG datur[,] punctum C rectam CK positione datam tangit.[(15)]

3.

4.

---

(11) We here considerably foreshorten Newton's quotation of Pappus since we have reproduced Commandino's text in full at the equivalent place in the revised text in §1 preceding. Exactly as before, with the linking sentence 'Quarum Propositionum, si Pappi verba brevia et nimium depravata recte interpretatus sum, sensus est talis', he now passes to list his preferred restorations of the opening porisms of Euclid's treatise in line with Pappus' elliptical summary of their essence.

(12) Add. 3963.8: 62r: a loose scrap on which Newton has roughly penned out his first versions of what he subsequently incorporated in §1 preceding as Porism 12 and of the following *propositio generalis* of Pappus which we know today as Desargues' theorem (on which see §1: note (55)).

(13) Shortage of space in the manuscript, whose accompanying figure is drawn by Newton close to its left-hand edge, there dictated that he considerably curve the portion LI of this line in extending it below E (in broken line) to its meet with AB: this trivial blemish is here removed by slightly increasing the slope of EG to AB as well as by straightening the former over its whole length. (The remaining lines are accurately reproduced in relative proportion and direction.)

(14) Newton neglects also to add some equivalent to 'et simul abeant in infinitum' (compare Theorem 1 of his 'Inventio Porismatum' on page 232 above). It is, in other words, here necessary that the meet of the parallel through A to FD with the parallel through B to EG shall lie in the given straight line LCM (passing through the intersection of AD and BC, of course).

(15) A further proviso is here needed that the points N and I in which DF and EG meet AB shall mutually correspond (see §1: note (53) preceding); that is, that there shall in Newton's figure be $DF:EG = DN:EI$. When this is not so the locus of C is a conic through A and B.

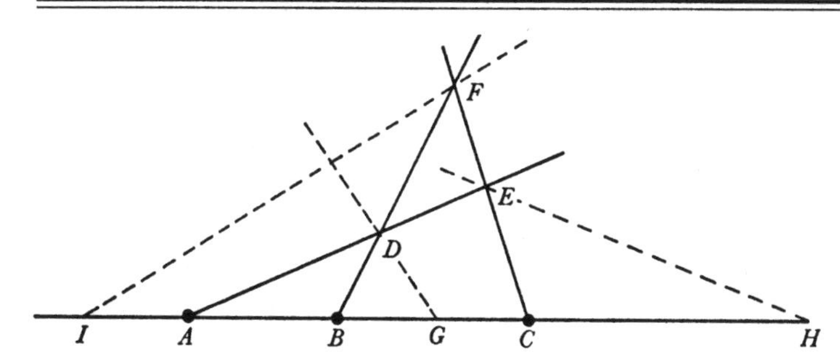

Duabus rectis AC, DG in eodem plano positione datis vel occurrentibus inter se vel parallelis: si dentur in una earum AC tria intersec[t]ionū puncta A, B, C, cætera vero intersectionum puncta D, E, F præter unum F tangant rectas DG, EH positione datas, etiam hoc quoq; F tanget rectam FI positione datam.(16)

APPENDIX 3. NEWTON'S SUPERSEDED 'PORISMA 4'.(1)

From the original draft(2) in the University Library, Cambridge

Porisma

Si detur ambitus trianguli ABC & positio duorum laterū AB, AC,(3) tertium latus BC tanget circulum datum. Idem fiet si datum rectangulum $EFGH$ recta BC secetur sic, ut vel summa trium GB, BC, CH detur vel rectangulum sub duobus CE, BF æquetur quartæ parti EF quadrati.

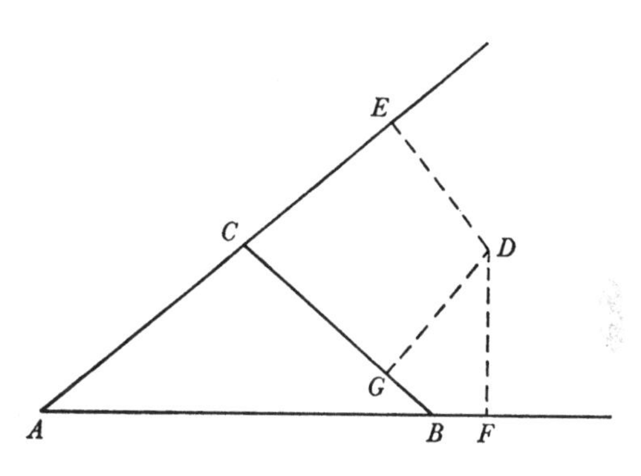

Demonstratio.

Cas. 1.(4) Capiantur AE AF æquales dimidio ambitus. Erigantur perp[endicula] ED FD concurrentia in D, et inde ad BC demittatur perpendiculum [D]G. Ob angulos ad E et G rectos est $CE^q+ED^q = CD^q = CG^q+GD^q$. Ergo

$$ED^q-GD^q = CG^q-CE^q.$$

Et simili argumento

$$FD^q-GD^q \text{ (hoc est } ED^q-GD^q) = BG^q-BF^q.$$

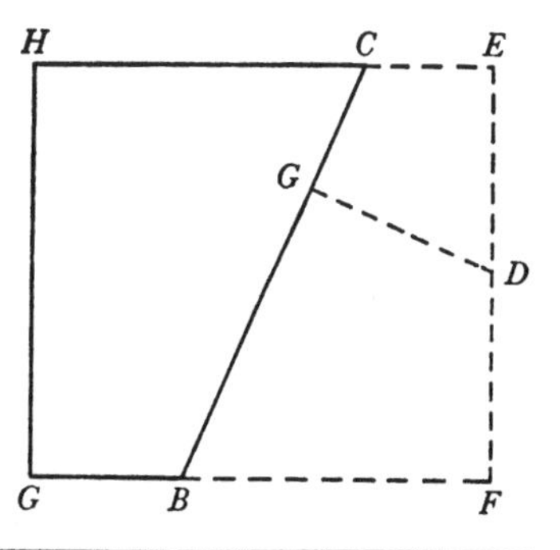

(16) The Pappus–Desargues theorem. Since in Newton's figure (here accurately reproduced) the line HE, when extended, comes nowhere near to passing through the meet of GD and FI, if would seem that Newton had not yet come to appreciate this important corollary of Pappus' *propositio generalis* (compare §1: note (55) preceding).

(1) A discarded version of the final *porisma* in §1 preceding (see *ibid.*: note (67)) superseded

Adeoq; si $ED \sqsubset = \sqsupset GD$, erunt una $CG \sqsubset = \sqsupset CE$ et $BG \sqsubset = \sqsupset BF$ et conjunctim $BC \sqsubset = \sqsupset CE+BF$.(5) Et e contra. Sed ex hypothesi $BC = CE+BF$[,] ergo $ED = GD$. Quare circulus centro D radio DE descriptus tangitur a recta CB.(6) Atqui circulus ille ob datum centrum et intervallum datur, ergo recta CB tangit circulum datum. Q.E.D.

Cas. 2.(7) Capiantur HE GF æquales dimidio trium GB, BC, CH et bisecta EF in D demittatur DG perpendicularis ad BC. Probabitur ut prius DG æqualem esse DE &c.

Cas. 3.(8) Junge CD, DB et ob æqualia rectangula CE, BF et EDF et angulos rectos ad E et F similia erunt triangula CED, DFB. Et inde angul[us] CDE æqualis est angulo $D[BF]$, et anguli externi EDB pars reliqua CDB æqualis angulo altero interno $[D]\,F[B]$ qui rectus est, et inde rectangulum $CGB = GD^q$. Jam si $GD \sqsubset$ vel $\sqsupset DE$ erit (ut supra) $GC \sqsupset$ vel $\sqsubset CE$ ct $GB \sqsupset$ vel $\sqsubset BF$[,]

by Newton's instruction at the head of its replacement on f. 31v: 'Put this Porisma in ye 4th place'. Except that he here treats in his 'Cas. 3' of a minor variant on the basic porism which was to be omitted from the revise, there is little essential difference between the two versions, however much they may seem superficially to differ in their verbal detail and sequence of argument.

(2) Add. 3963.4: 31r/32r.

(3) Newton first opened by enunciating 'Si detur summa laterum trianguli ABC et præterea latera duo AB, AC dentur positione' and then altered this to state 'Si angulus CAB positione et magnitudine datus subtendatur recta quavis BC et datur ambitus trianguli ABC' before changing his mind once more as to how to specify the circumstances of his porism.

(4) Where, namely (in the first figure) the *ambitus* $AB+BC+CA$ of the triangle ABC is given, whence by Newton's following construction of $AE = AF$ equal to half this perimeter there ensues 'ex hypothesi', his further argument requires, the equivalent condition

$$BC = (AE+AF)-(AB+CA) = CE+BF.$$

(5) The symbols $\sqsubset$ and $\sqsupset$ are the modified Oughtredian equivalents of the more familiar Harriotian signs $>$ and $<$ for greater and lesser inequality which Barrow prefixed (on signature a8v, facing page 1) to his *Euclidis Elementorum Libri XV. breviter demonstrati* (Cambridge, 1655) and with which Newton had been familiar since his earliest years at Cambridge (compare I: 578–85).

(6) The accompanying augmented figure, in which we have entered the constructed circle in broken line, may perhaps serve to clarify Newton's argument.

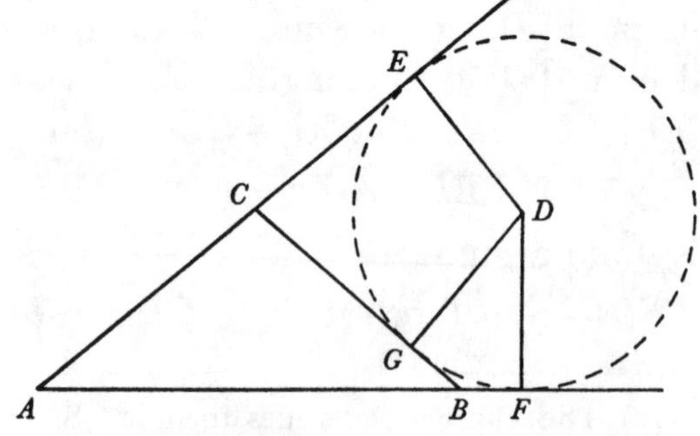

(7) That where (in the second figure) the sum $GB+BC+CH$ of the total length of the 'open trapezium' $GBCH$ is given. In his revised *Porisma* 4' on page 276 above Newton will neatly combine this case with that preceding by fixing like points G, H in AB and AC respectively and making the same equivalent stipulation (with distinction no longer kept between positions of A at a finite distance or at infinity in the—now parallel—lines BF and CE).

(8) That, namely, where $BF \times CE = \frac{1}{4}EF^2$ (and it is understood that EF is perpendicular to the parallels BF and CE). Since in the more general case of this species of Apollonian *sectio*

adeoqꝫ $CGB \sqsupset$ vel $\sqsubset$ $CE \times BF$, $GD^q \sqsupset$ vel $\sqsubset$ DE^q et $GD \sqsupset$ vel $\sqsubset$ DE[5] contra suppositionem. Ergo $GD = DE$. Adeoqꝫ recta BC circulum centro D intervallo DE descriptum[9] hoc est circulum datum tangit. Q.E.D.

---

spatii in which the lines *BF* and *CE* are no longer taken parallel the locus enveloped by *BC* is no longer a circle but a central conic, we will appreciate why Newton omitted any equivalent to this 'Cas. 3' in his revised porism (compare §1: note (70)).

(9) Here again an appropriate amplification of Newton's figure may aid the full comprehension of his argument.

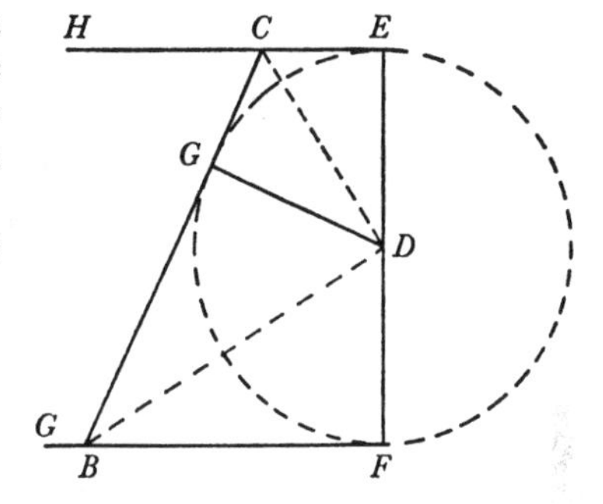

§2. THE FIRST BOOK: AN INCOMPLETE OPENING AND A 'CHAPTER 2'.(1)

From the originals in the University Library, Cambridge

[1](2) GEOMETRIÆ

LIBER PRIMUS.(3)

Geometria idem sonat quod Ars mensurandi terram. Inde Geometra Latinis a calamo mensuræ dictus est Decempedator.(4) Nec sane latius patere potuit hæc ars ante inventum Pythagoræ(5) quam Geodæsia hodierna. Opinio(6) verò est ipsam a mensurandis annuatim Ægyptiorū terris originem duxisse. Eò spectantia problemata expeditissimè et exactissimè per lineæ rectæ et circularis descriptiones construebantur. Rectis distinguebantur agri, circulis aliquando opus erat ad determinandas positiones et longitudines rectarum. Aliæ lineæ nec faciles erant descriptu nec istis usibus necessariæ. Ideo fundata est ars mensurandi in harum solarum descriptione. Mensurabat Artifex figuras planas universas obviam factas sed nullas formabat mensurando præter rectilineas et circulares. Sunt igitur hæ solæ instrumenta Geometriæ ac figuræ planæ universæ subjectum.

Planum describere Geometria nec docet nec postulat quamvis hoc sit Geometriæ totius fundamentum. Quippe plana agrorum non formantur ab artifice

(1) In this slightly overlapping but otherwise consecutive pair of drafts for a revised 'First Book of Geometry' Newton gives a general account of what in the received opinion of the 'ancients' it is permissible to postulate as an acceptable basis for ensuing geometrical argument and curvilinear constructions using 'plane' and 'solid' loci, and then in its 'Chapter 2' comes more particularly to elaborate (in a close revision and amplification of §1 preceding) the Greek notion of the 'resolved locus' and to exemplify it in the propositions of Euclid's lost work on *Porisms*, of which he here gives an improved restoration 'ex Pappo'. No succeeding 'Second Book' is found in his surviving papers but—had he gone on to adumbrate one—we may presume that in its essence it would not greatly have differed from 'Lib. 2' in §3 following, where Newton stresses the wider range of fundamental postulates allowed into the geometry of his own day and how these give the lead to the free employment of an extensive range of cissoidal, conchoidal, bipolar and yet more complicated curves in the geometrical construction of problems.

(2) Add. 3963.11: 119r–126r. When Samuel Horsley came in October 1777 to inspect Newton's mathematical papers at Hurstbourne he accurately specified on an insert— now placed with the wrong manuscript sheets by an errant modern cataloguer—that this is a 'Procemium Libri Primi Geometriæ', further initially noting in a now cancelled subtitle that it dealt with the theme 'De Geometriæ Veterum Materiâ et Instrumentis'. We have already (§1: note (2)) quoted his allied recommendation there that this 'more finishd Proem' should be published along with the preceding 'curious fragment about the advantage of the Geometrical analysis'.

(3) The much cancelled opening (Add. 3963.15: 177r/177v) of what would seem to have been a preliminary revise of the present introduction to Book 1 of the 'Geometry'—it bears

Translation

[1][2] GEOMETRY

BOOK ONE.[3]

'Geometry' means the art of 'earth-measure'. For this reason a geometer was called by the Romans from his measuring stick a 'ten-footer'.[4] Nor before Pythagoras' discovery,[5] certainly, could this art have had wider extent than modern land-surveying. There is, in fact, a belief[6] that it drew its origin from the Egyptians' annual measuring of their terrain. Problems regarding this were very speedily and accurately constructed by descriptions of straight and circular lines. With straight lines they made division between fields, and there was occasional need of circles to determine the positions and lengths of straight lines. Other types of line were neither easy to draw nor necessary for their uses. In consequence the art of measurement was founded on these alone. The practitioner measured plane figures universally as he met with them, but formed none by measuring other than rectilinear and circular ones. These, then, are the sole instruments of geometry and the subject of the plane figure universally.

Geometry neither teaches how to describe a plane nor postulates its description, though this is its whole foundation. To be sure, the planes of fields are not

an identical title—is reproduced in Appendix 1.1 below. In it Newton stresses the ties of geometry not only, as here, with *mechanica rationalis*, but also with arithmetic in so far as it 'accurately considers and demonstratively determines the quantities and ratios of all things'.

(4) The Roman wielder of the *decempeda* was in strict truth but a workaday surveyor who toiled exclusively in the field according to established rules of thumb. In Newton's present sense the title 'geometer' ought perhaps to be reserved for the rare few who formulated and modified such empirically founded principles.

(5) Presumably of the theorem (*Elements* I, 47) determining the equality of the sum of the squares of the sides of a right triangle to the square of the hypotenuse—and so, in modern terms, establishing the Euclidean metric—which traditionally bears Pythagoras' name, but which we now know was familiar (in its result at least) to Babylonian mathematicians more than a thousand years before. (See O. Neugebauer, *The Exact Sciences in Antiquity* (Providence, Rhode Island, $_2$1957): 34 ff.) It is curious that Newton should here ignore the 'father of Greek geometry', Thales, who a century earlier (according to Proclus) not only 'went to Egypt and thence introduced the study of geometry into Greece' but also 'discovered many propositions himself, and instructed his successors in the principles underlying many others' (see T. L. Heath, *A History of Greek Mathematics*, **1** (Oxford, 1921): 128).

(6) This widespread tradition—recounted in little differing form by writers as diverse as Heron of Alexandria, Strabo and Proclus—stems from Herodotus, who in his *History* (**2**. 109) reports how, in about 1300 B.C., the pharaoh Sesostris (Rameses II) established an annual levy on land in proportion to its size; when after each annual flooding of the Nile the individual plots had to be measured anew by official surveyors, the celebrated 'rope-stretchers' of legend and myth. (See Heath's *Greek Mathematics*, **1**: 120–2.) Whatever its historical truth, this unproven origin of 'land-measuring' was disputed by no classical author and it makes a good story for those ignorant of equivalent Babylonian achievements over the preceding millennium.

sed mensurantur tantū. Lineam rectam et circulum describere Geometria non docet sed postulat, hoc est[7] postulat Artificem has operationes prius didicisse quam attingit limen Geometriæ. His vero præcognitis et concessis ipsa cæteras omnes mensurandi operationes sub titulo Problematum docet, nec non sub titulo Theorematum comparat quantitates inter se ut ex mensuratis et inventis non mensuratæ deriventur.[8]

Pertinet igitur ad Mechanicam tum genesis subjecti Geometrici tum Postulatorum effectio. Figuras quasvis planas a Deo Natura Artifice quovis confectas Geometra ex hypothesi quod sunt exactè fabricatæ mensurat. Rectas et circulos describere Artifex prius didicisse requiritur et postulatur quam incipit esse Geometra. Ideo nil refert quomodo[9] describantur. Geometria modos descriptionum non ponit. Regulis admotis, radijs opticis, funiculis tensis, circino, angulo dato in perimetro, punctis discretim inventis, manus exquisitæ motu libero, ratione deniq; quacunq; mechanica eas[10] describere permittimur. Id solum postulat Geometria ut describantur exactè. Usu tamen jam venit Geometricum censere id omne quod [11]exactū est & mechanicum quod ejusmodi non existit, quasi nihil mechanicum et simul exactum esse posset. Crassa verò est hæc vulgi opinio et non aliunde orta quam quod Geometria postulat exactam praxin mechanicam in descriptione rectæ et circuli[,] et præterea in omnibus suis operationibus exacta est, Mechanica verò imperfectè et absq; legibus exactis vulgò exercetur. Ex imperitia et imperfectione Mechanicorum Vulgus definit Mechanicam. Qua ratione id magis mechanicum foret quod esset imperfectius. Pone Mechanicum [esse] perfectum et errorem corriges. Certe enim id magis mechanicum seu artificiosum est quod est exactius & is perfectior mechanicus qui perfectius et exactius operatur, is vero solus perfectus qui operatur exactè. Veteres[12] ut memorat Pappus Mechanicam distinxerunt in duas partes, et unam rationalem esse[,] alteram indigere manuum operatione constituerunt. Rationalis non minus exacta est quam ipsa Geometria. In manuali is solus perfectus est Mechanicus qui omnia ad normam rationalis exactissimè operari potest. Talem discipulum quoad descriptionem rectæ et

(7) 'præsumit et' (presumes and) is deleted. The present sentence is borrowed by Newton—consciously or no—from the first paragraph of his 1687 *Principia* preface (whose opening we quote in Appendix 1.2: note (14) below) and a number of other sharp echoes of this may here be heard in sequel.

(8) Newton initially terminated by writing '...ut aliæ ex alijs absq; novis mensurandi operationibus colligentur' (in order to gather some from others without recourse to new operations of mensuration).

(9) Originally 'qua ratione mechanica' (by what mechanical means).

(10) Understand 'figuras planas' (plane figures).

(11) 'omnino perfectū atq;' (entirely perfect and) was here momentarily inserted and then straightaway discarded.

(12) So likewise Newton identified the source of this distinction between rational and practical mechanics at the equivalent point in the stray sheet (ULC. Add. 3963.3: 19)—an

formed by the practitioner but merely measured. Geometry does not teach how to describe a straight line and a circle but postulates them; in other words, it[7] postulates that the practitioner has learnt these operations before he attains the threshold of geometry. Once, however, these are previously understood and granted it teaches all the other operations of mensuration under the head of 'problems', and likewise under the head of 'theorems' compares quantities with one another in order from the measured and the found to derive things which have not been measured.[8]

Both the genesis of the subject-matter of geometry, therefore, and the fabrication of its postulates pertain to mechanics. Any plane figures executed by God, nature or any technician you will are measured by geometry in the hypothesis that they are exactly constructed. A technician is required and postulated to have learnt how to describe straight lines and circles before he may begin to be a geometer. And it consequently does not matter how[9] they shall be described. Geometry does not posit modes of description: we are free to describe them[10] by moving rulers around, using optical rays, taut threads, compasses, the angle given in a circumference, points separately ascertained, the unfettered motion of a careful hand, or finally any mechanical means whatever. Geometry makes the unique demand that they be described exactly. It has now, however, come to be usual to regard as geometrical everything which is[11] exact, and as mechanical all that proves not to be of the kind, as though nothing could possibly be mechanical and at the same time exact. But this common belief is a stupid one, and has its origin in nothing else than that geometry postulates an exact mechanical practice in the description of a straight line and a circle, and moreover is exact in all its operations, while mechanics as it is commonly exercised is imperfect and without exact laws. It is from the ignorance and imperfection of mechanicians that the common opinion defines mechanics. On this reasoning a thing would be the more mechanical the more imperfect it was. Posit a mechanical thing to be perfect and you will correct the error. For assuredly the more mechanical—that is, skilfully wrought—a thing is, the more exact it is, and he the more perfect mechanic who works the more perfectly and exactly, while he alone is perfect who works exactly. The ancients,[12] as Pappus reminds us, distinguished mechanics into two parts, constituting one to be rational, whereas the other is reliant on manual working. The rational one is no less exact than geometry itself, while in the manual division he alone is a perfect mechanician who can with extreme accuracy work everything to the pattern of the rational.

---

abortive preliminary revise of the present passage?—which we reproduce in Appendix 1.2 following, though Pappus himself in the opening page of the eighth book of his *Mathematical Collection* here pointed to by Newton specifies him, in Commandino's Latin version, to be 'Hero mechanicus' (*Mathematicæ Collectiones*, $_2$1660: 447; see also Appendix 1.2: note (12) below).

circuli Geometria postulat et quasi concessū sua imbuit disciplina. Non distinguuntur Geometria et Mechanica quoad magis et minus exactum sed ex usu & fine disciplinarum. Intentio Mechanicæ est magnitudines imperatis figuris et motibus formare et movere. Intentio Geometriæ est magnitudines nec formare nec movere sed mensurare tantum. Nihil format Geometria præter mensurandi modos. Postulat Artificem qui formare novit rectas et circulos & hunc docet quomodo per illorum efformationem mensurandæ sint magnitudes imperatæ.(13)

Et hinc constet(14) Postulatorum conditio. Debent esse paucæ simplices utiles et abq; modis operationum expressæ. *Paucæ* quia sunt principia hujus scientiæ et de Mechanica petuntur. Ignominia est Geometriæ quod aliquorum Problematum solutiones postulare cogatur, gloria quod tam paucis aliunde solutis et concessis(15) tam multa suo marte præstet. *Simplices* ni fuerint, erunt instar multarum, præter indecoram perplexitatem quæ principijs non congruit. *Utiles* esse debent hoc est quibus Artifex quilibet in mensurando lubenter uteretur. Nam Geometria excogitata est, non in nudas speculationes, sed in usum vitæ: Et servanda est ratio institutionis primæ. Scientijs ex usu conciliatur gratia.(16) Et quamvis ferendæ fuerint speculationes quædam inutiles quæ ingenij excolendi gratia ex utilibus principijs deriventur[,] ipsa tamen principia inutilia ponere alienum est a ratione et scopo artium et scientiarum omnium(17) & scientiam totam vel nomine tenus inanem et futilem redderet. *Modi* deniq; operationum in postulatis exprimi non debent. Postulat Geometra quia modum effectionis docere nescit. Postulatū Geometria non construit: sola Mechanica hoc facere potis est.(18) Sunt enim Postulata de genere Problematum sed non de genere Problematum Geometricorum, et ideo modos ponere solutionum alienum est a Geometria. In definitionibus ponere licet rationem geneseos Mechanicæ, eò quod species magnitudinū ex ratione geneseos optimè intelligitur. Sic fecerunt Veteres in definitionibus Sphæræ coni et Cylindri. Sed

(13) A cancelled further sentence in the manuscript goes on: 'Et quoniam postulat in illis enchiriam seu praxin mechanicam exactam & præterea in omnibus suis operationibus exacta est, Mechanica vero imperfecte abq; legibus exactis vulgò exercetur: hinc fit ut perfectum omne et exactū ad Geometriam & imperfectum ad Mechanicam populari et parum accurato sermone referatur' (And because it demands manual skill—that is, an exact mechanical technique—in them and is, moreover, exact in all its operations, whereas mechanics is commonly practised in an imperfect manner and in divorce from exact laws, it hence comes that all which is perfect and exact is in popular—and too little accurate—speech referred to geometry, and the imperfect to mechanics).

(14) 'qualis debet esse' (the due quality of) is deleted in sequel.

(15) Originally 'tam paucis operationibus mechanicis postulatis' (with so few mechanical operations postulated).

(16) Newton first went on to write: 'Sunto principia inutilia et totum futilis fiet speculatio' (Let the principles be useless and their exploration will prove utterly futile), initially altering the latter pronouncement to state '...et nihil utile sequetur' (and nothing useful will ensue).

In regard to the description of a straight line and a circle geometry postulates just such a pupil and, when granted one as it were, imbues him with its own discipline. Geometry and mechanics are distinguished not inasmuch as they are more and less exact, but in the use and end of their disciplines. The purpose of mechanics is to form and move magnitudes in appointed figures and motions: that of geometry is neither to form nor move magnitudes, but merely to measure them. Geometry forms nothing except modes of measuring. It postulates a technician who knows how to form straight lines and circles, and teaches him how through their formation appointed magnitudes are to be measured.(13)

And from this(14) the character of its postulates should be agreed. They ought to be few, simple, useful and expressed without their modes of operation. *Few* because they are the principles of this science and are elicited from mechanics. It is geometry's disgrace that it should be compelled to postulate the solution of certain of its problems, its glory that it should with so few things solved and granted from outside it(15) by itself perform so many. *Simple* were they not, they would be the like of many, to ignore that an immodest intricacy is ill suited to principles. *Useful* they ought to be in that any practitioner should find them readily applicable in his measuring. For geometry was devised, not for the purposes of bare speculation, but for workaday use, and the reason for its first institution must be preserved. Science wins our gratitude in consequence of its usefulness.(16) Even though certain useless speculations which may be derived from useful principles with the motive of developing our ability were to be tolerated, it is none the less foreign to reason and the scope of every art and science(17) to set the very principles useless, and to do so would render the whole of science—and even its very name—empty and futile. The *modes*, finally, of operation ought not to be expressed in the postulates. Geometry postulates because it knows not how to teach the mode of effection. It does not construct a postulate: mechanics alone is capable of doing this.(18) Postulates are, of course, of the nature of problems, but not of geometrical problems, and it is accordingly foreign to geometry to posit modes of solution. In definitions it is allowable to posit the reason for a mechanical genesis, in that the species of magnitude is best understood from the reason for its genesis. The ancients so did in their definitions of the sphere, cone and cylinder. But the intention of postulates is neither to

What, we may wonder, would he now have said of his earlier speculations on 'Problemata numeralia' (IV: 74–114)?

(17) This was just 'foret...durum' (it would be harsh) in a cancelled first version.

(18) In a first continuation Newton specified: 'In eo differt Postulatum a Problemate quod hoc sit Problema Geometricum illud Mechanicum, et modos construendi problemata Me[chanicum sit ponere]' (A postulate differs from a problem in that the latter is a geometrical problem, the former a mechanical one, and [to posit] modes of constructing problems is a [mechanical requirement]).

Postulatorum intentio est nec magnitudines definire nec docere genesin sed hanc assumere tanquam præcognitam.(19)

Ortum habuit Geometria in plano ut supra diximus[,] et inde planis mensurandis accommodabantur postulata prima. Ea sunt Euclidea describendi rectas et circulos. Ante tempora Pythagoræ solidorum quoqȝ mensuras contemplati sunt Artifices[,] ut parallelipipidorum, prismatum, pyramidum, sphærarum, cylindrorum, conorum. Et sic Geometria duplex condita est, planorum et solidorum. At in ea solidorum nihildum postulabatur quantum ex *Elementis* ab Euclide collectis conjectare licet. Sphæra cylindrus & conus non ut instrumenta mensurandi sed ut magnitudines mensurandæ considerabantur et ideo non postulabatur eorum descriptio. Ars tota mensurandi hactenus in Postulatis Geometriæ Planorum fundabatur. Tandem verò ubi Problema Deliacum(20) agitari cœperat et una cum trisectione anguli diu multumqȝ per regulam et circinum tentatum fuerat, considerari cœperunt occursus linearum cum superficiebus solidorum perqȝ circulum et superficiem conicam(21) soluta sunt Problemata. Unde nomina Problematum planorum et solidorum. Plana dicebantur quæ in plano absqȝ consideratione solidi[,] solida quæ non nisi per solidum construere licebat. Defectu tamen postulatorum describendi solida, constructiones hujus generis, quantum sentio, non recipiebantur in Geometriam.(22) Considerari jam vero cœperunt figuræ quas plana circulorum generant secando conos. Dictæ sunt hæ Sectiones Conicæ. Utqȝ earum in plano fieret copia postularunt Scriptores, *Datum conum dato plano secare*. Hoc fuit primum atqȝ unicum postulatum Veterum in Geometria Solidorum, quantum prodit memoria.(23) Subinde ex varijs

(19) Newton initially added 'Hæc autem exposui ut Geometriam quæ labentibus annis immutata fuit, ad rationem institutionis quoad fieri potest reducam' (These observations I have delivered, however, so that I may return geometry, which has altered with the passing years, as far as may be to the manner of its first institution) and then proceeded in a superseded following paragraph to state: 'Talia debent esse Postulata. Problematum vero hæc est conditio. Geometricè soluta sunt omnia quæ per operationes postulatas in planis datis exercitas construi possunt. Cæterorū solutiones ad Mechanicam pertinent. Ut si concedantur sola Postulata Euclidea et inveniendum sit centrum datæ Ellipseos: licebit sic operari' (Such ought the postulates to be. The characteristic of problems, however, is this. All are solved geometrically which can be constructed by means of postulated operations performed in given planes; the solutions of the rest belong to mechanics. If, for instance, the Euclidean postulates alone be granted and you need to find the centre of a given ellipse, you will be free to work in this way). Newton's problem is, of course, solved at once *modo geometrico* by determining the common intersection of straight lines drawn through the mid-points of pairs of parallel chords—indeed, as Newton had long been aware (see IV: 276–8, and compare V: 308–10), the Euclidean construction of its centre is readily effectable by Pappus, *Mathematical Collection* VIII, 13 when only five points of the conic are given.

(20) That of constructing a cube which has twice the volume of a given one; see 1, §1.1: note (8) preceding.

(21) A cancelled parenthesis 'ut memorat Pappus' (as Pappus recounts) at this point in the manuscript is one more reminder of Newton's present indebtedness to his Greek forebear. In

define magnitudes nor to teach their genesis, but to assume this as previously ascertained.[19]

Geometry had its origin in the plane, as we said above, and its first postulates were in consequence adapted to measuring planes. These are the Euclidean ones for describing straight lines and circles. Before the time of Pythagoras practitioners considered also the measures of solids such as parallelepipeds, prisms, pyramids, spheres, cylinders and cones. And so a two-fold geometry was established, of planes and of solids. But in that of solids there were as yet no postulates as far as we may permissibly conjecture from the *Elements* collected by Euclid. The sphere, the cylinder and the cone were regarded not as measuring instruments but as magnitudes to be measured, and their description was accordingly not postulated. The whole art of measuring was founded thus far in the postulates of plane geometry. At length, however, when the Delian problem[20] began to make a stir and together with angle-trisection was long and extensively attempted by means of ruler and compasses, the meets of lines with the surfaces of solids began to be considered and problems were[21] solved by means of a circle and a cone surface. Whence the names 'plane' and 'solid' problems. Plane ones were stated to be those which it was allowable to construct in a plane in divorce from consideration of a solid; solid ones were those constructible only by aid of a solid. Through the lack, however, of postulates for describing solids constructions of this kind were not, as far as I may discern, received into geometry.[22] Figures now, however, began to be considered which are generated by the planes of circles intersecting cones: these were called 'conic sections'. And so that there might be a ready supply of these in the plane writers adopted the postulate *To cut a given cone by a given plane*. This was the first and only postulate of the ancients in the geometry of solids as far as the record[23]

---

a preamble to Proposition 31 of the fourth book of his *Mathematical Collection* Pappus had stated, in Commandino's Latin translation: 'Antiqui Geometræ datum angulum rectilineum tripartito secare volentes ob hanc caussam hæsitarunt.... Quæ [problemata] per rectas lineas, & circuli circumferentiam solui possunt, merito dicuntur plana; lineæ enim per quas talia problemata inueniuntur, in plano ortum habent. Quæcumque vero soluuntur, assumpta in constructionem aliqua coni sectione, vel etiam pluribus, solida appellata sunt, quoniam ad constructionem solidarum figurarum superficiebus videlicet conicis vti necessarium est' (*Mathematicæ Collectiones*: 95; we have given an English rendering in v: 423, note (617)).

(22) These last three sentences are a late insertion in the manuscript; there Newton initially went on to observe: 'Et quoniam Sphæra Cylindrus Conus et figuræ planis superficiebus terminatæ eundē habere locum viderentur in Geometria solidorū atqꝫ circulus et figuræ rectilineæ in planorū, solutiones illæ pro Geometricis a plerisqꝫ sunt agnitæ' (And seeing that the sphere, cylinder, cone and (polyhedral) figures bounded by plane surfaces might seem to have the same place in the geometry of solids as the circular and rectilinear figures in that of planes, those solutions are recognized by most as geometrical).

(23) Newton would appear to be ignorant of Eudoxus' hippopede (see O. Neugebauer in *Scripta Mathematica*, **19**, 1953: 225–9) or the general 'spiric' (torus) sections investigated by Perseus (on which see Heath's *Greek Mathematics*, **2**: 203–6).

novarum superficierum curvarum et planorum intersectionibus ut et ex linearum in plano motibus complicatis excogitatæ sunt lineæ novæ complures[,] ut Conchoides, Cissoides, Helices, Quadratrices, quarum intersectionibus Problemata difficiliora solverentur. Et ab his lineis dicta sunt talia Problemata linearia.(24) Sic nata sunt tria genera Problematum a triplici magnitudinum dimensione denominata. Hæc liquent ex Pappo.(25) Nostris autem temporibus postulant aliqui(26) lineas quasvis motibus quibuscunq; complicatis movere et sic lineas omnes quæ per earum intersectiones describi possunt introducunt in Geometriam ad constructionem problematum Solidorum et linearium. Alij(27) verò lineas omnes quæ exactæ sunt pro Geometricis haberi volunt sed nondum postulant omnes describere.(28) In tanta sententiarum diversitate videamus quid tandem sequendum sit.

Postulatum illud movendi lineas quascunq; secundum leges imperatas et earum intersectionibus describendi lineas novas, multis rationibus improbatur. Complexissimum est et instar infinitorum Postulatorum cum tamen deberet esse simplicissimum. Durum nimis est in Tyrones. Requirit enim ut Tyro antequam is incipit esse Geometra lineas quasvis rationibus quibuscunq; et Geometricis et in posterum inveniendis compingere et movere didicerit. Fundamentum est totius Mechanicæ exactæ et sic in Geometriam introducta confundit scientias inter se. Quid enim aliud potest Mechanicus exactus quam figuras modis imperatis movere et formare? Superflua reddit Postulata Euclidea, et linearum descriptiones quæ solum postulari debent, convertit in Problemata Geometrica. Nam Postulatum secundum producendi lineam quamvis continuò, construetur problematicè faciendo ut linea illa moveatur de loco in locum ex parte coincidentem cum loco priore. Postulatum primum, datâ vel unâ rectâ, construetur producendo rectam illam ubi opus est et applicando eandem ad data duo puncta. Postulatum tertium describendi circulum solvetur volvendo rectam circa terminum ejus. Cæteras lineas describere(29) tot administrabit Problemata quot excogitari possunt lineæ, totq; singulorum solutiones quot modi descriptionum. Cum tamen descriptiones linearum sint operationes Mechanicæ[,] postulari debent non doceri. Postulat Geometria Tyronem has didicisse prius. Mechanicas esse docent vel modi descriptionum. Regulis ad invicem applicatis et filis tensis Curvas describere idem est quod earum puncta

(24) In a preceding cancelled sentence, otherwise appropriately absorbed into his revised text, Newton here added 'a lineis illis recens inventis difficilioribus ad eorum solutionem requisitis' (from the more complicated lines recently invented which are required for their solution). Pappus had stated to the same effect that 'Relinquitur tertium genus problematum, quod lineare appellatur; lineæ enim aliæ præter iam dictas in constructionem assumuntur, quæ varium, & difficilem ortum habent, ex inordinatis superficiebus, & motibus implicatis factæ. Eiusmodi vero sunt etiam lineæ, quæ in locis ad superficiem dictis inueniuntur, & aliæ quædam magis variæ.... Ex hoc genere sunt lineæ helices, & quadrantes, & conchoides, & cissoides' (*Mathematicæ Collectiones*: 95; again see v: 423, note (617) for a modern English rendering).

relates. Subsequently, from a variety of intersections of fresh curved surfaces and planes and also from complicated motions of lines in a plane a great many new lines were devised—conchoids, cissoids, helixes and quadratrixes, for instance—so that by their intersections more difficult problems might be solved. And by reason of these lines such problems were called 'linear'.(24) So were born three classes of problem named after a three-fold division of magnitudes. These things are clear from Pappus.(25) In our own day some(26) postulate that any lines may move with any complicated motions whatever, and in this way introduce into geometry all lines which are describable by their intersections in order to construct solid and linear problems. Others,(27) however, want all lines which are exact to be considered as geometrical, but are as yet unwilling to postulate the description of all.(28) In such a diversity of opinion let us see what lead we need at length follow.

The postulate of moving any lines whatever according to appointed laws and by their intersections describing new lines is to be rejected for many reasons. It is exceedingly complex and akin to an infinity of postulates when, however, it ought to be very simple. It is too hard for the novice; for it requires him before he starts to be a geometer to frame any arbitrary lines and move them in any manner whatever, both geometrical and in ways later to be ascertained. This is the foundation of all exact mechanics and when it is thus introduced into geometry it confuses these sciences one with the other. For what else is an exact mechanician able to do than move and fashion figures in appointed ways? Its introduction renders Euclid's postulates superfluous, and converts descriptions of lines which ought only to be postulated into geometrical problems. For his second postulate of extending any line without break will be constructed in the style of a problem by making that line move position to a position coinciding in part with its previous one. The first postulate, given even one straight line, will be constructed by extending that line as need be and applying it to the two given points. The third postulate of describing a circle will be solved by revolving a straight line round its end-point. The description of other lines(29) will cater for as many problems as there are lines which can be devised, and furnish as many solutions for each singly as there are modes of description. To describe curves by applying rulers to one another and stretching threads is the same as to

(25) See notes (21) and (24) above.

(26) Newton would here have Frans van Schooten and maybe Jan de Witt in mind; compare I: 29–45 and II: 7–8.

(27) Above all, of course, Descartes in his *Geometrie*.

(28) This replaces the equivalent 'omnium descriptiones' (their universal description).

(29) Originally 'Cæterarum linearum descriptiones invenire' (Finding the descriptions of other lines).

omnia invenire successivè. Si puncta illa omnia inveniuntur geometricè[,] tum etiam singula. Atqui puncti singuli inventionem per regulas applicatas et fila tensa absq; postulatis Euclideis institutam referunt omnes ad Mechanicā. Porrò hoc postulato introducentur in Geometriam solutiones Problematum planorum absq; postulatis Euclideis. Sic recta *CD* alteri rectæ *AB* parallela et æqualis a dato puncto *C* ducetur jungendo *CA* et in angulo dato [*BAC*] movendo junctam per longitudinem *AB*. Nam punctum *C* eo motu describet longitudinem ducendam. Introducentur etiam solutiones Problematum absq; descriptione Curvarum. Ut si angulum datum *CBH* recta *EF* quæ datæ *GH* æqualis sit et ad datum punctum *D* inclinet[30] subtendere oportet: a puncto *D* agatur recta quævis *DI* occurrens rectæ *GH* in *I* et inter anguli latera *CB*, *BH* labatur *GH* usq; dum intersectionis punctum *I* (per Post. 1 lib. 1 *Elem*[31]) describat longitudinem *ID*. Tales admitte constructiones et Geometria omnis antiqua luxabitur. Deniq; inutile est hoc Postulatum. Quippe in praxi mensurandi non solent Artifices[32] alias describere lineas quam rectas & circulares.[33] Nondum efficit Geometria ut vel Coni sectiones in solutionem solidorum problematum ab hominibus operarijs describantur. Quod si postulato illo non praxin Geometricam augere sed speculationi subjectum solummodo comparare volunt, suffecerit lineas definire. Geometria postulat facultatem[34] non cogitandi sed operandi.[35] Subjectum in quo agat non postulat sed definit tantum. Adinventa est enim non ut conderet magnitudines mensurandas sed ut mensuraret conditas. Ideo primi Geometræ non postulabant Planum, Sphæram, Cylindrum et Conum describere. Vi definitionum his dabatur locus in Geometria primitiva. Et par est ratio linearum superficierum et solidorum omnium quæ Geometra definierit.

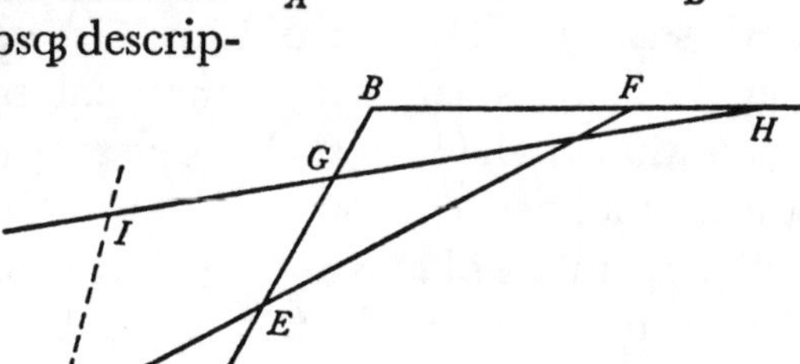

Ex his liquet, credo, quid de Postulato altero sentiendum sit Datum conum dato plano secandi. In praxi inutile est, ad speculationem non requiritur. Scripsit Archimedes geometrice de sectionibus Conoidum et Sphæroidum[36]

(30) 'convergat' (be convergent) is replaced.

(31) Namely, in Barrow's Latin version, 'Postuletur, ut à quovis puncto ad quodvis punctum rectam lineam ducere concedatur' (*Euclidis Elementorum Libri XV. breviter demonstrati*, Cambridge, $_1$1655: 6).

(32) Practical working geometers, that is. In immediate sequel Newton has deleted 'ad constructionem Problematum' (in order to construct problems).

(33) A cancelled following sentence continues in the same vein: 'Geometræ practici ad invenienda media proportionalia, angulumve in partes imperatas secandum non solent Coni sectiones lineasq; magis compositas describere' (Practical geometers, to find mean propor-

find all their points successively. If all those points are found geometrically, then so is each individual one. And yet the finding of an individual point, when undertaken by means of applied rulers and stretched threads in divorce from the Euclidean postulates, is referred by all to mechanics. Furthermore, acceptance of this postulate will introduce into geometry solutions of plane problems independently of the Euclidean postulates. The straight line *CD* will thus be drawn from the given point *C* parallel and equal to a second straight line *AB* by joining *CA* and moving the joined line at the given angle $B\widehat{A}C$ through the length *AB*; for the point *C* will by that motion describe the length to be drawn. Solutions of problems will also be introduced independently of the description of curves. If, for instance, it be required to subtend the given angle $C\widehat{B}H$ by a straight line *EF* which shall be equal to the given one *GH* and inclined(30) through the given point *D*: from the point *D* draw any straight line *DI* meeting the line *GH* in *I*, and let *GH* slide between the sides *CB*, *BH* of the angle till the point *I* of intersection describe (by Postulate 1 of *Elements*, Book 1(31)) the length *ID*. Admit such constructions and all ancient geometry will be put out of joint. Lastly, this postulate is useless. To be sure, in their mensurational practice technicians(32) do not usually describe lines other than straight and circular ones.(33) Geometry has not yet achieved that even conics be described by working men to effect the solution of solid problems. But if they do not wish to augment geometrical practice by that postulate, but want merely thereby to provide a subject for speculation, it will be enough to define lines. Geometry postulates the faculty(34) not of pondering but of performing work:(35) it does not postulate the subject on which it shall act but merely defines it. It was, of course, devised not to establish magnitudes to be measured, but to measure them once established. The first geometers accordingly did not postulate the description of the plane, sphere, cylinder and cone—a place was given to these in primitive geometry by force of definition, and the case is the same for all the lines, surfaces and solids which the geometer shall define.

From these considerations it is clear, I believe, what our feelings must be regarding the other postulate, that of intersecting a given cone by a given plane. In practice it is useless, while it is not requisite to speculation. Archimedes wrote geometrically on the sections of conoids and spheroids(36) and yet postulated

tionals or cut an angle into ordained parts, do not usually describe conics or more compound lines).

(34) That is, manner; 'modum' (mode) was first written.

(35) 'Postulat instrumenta ad agendum' (It postulates the instruments to do it with) is cancelled in sequel.

(36) In the treatise in thirty-two propositions which he composed on the subject. Archimedes there considers only single ellipsoids, paraboloids and hyperboloids of revolution, and their oblique-angled equivalents, defining each of the former to be the locus of the pertinent species of conic as it rotates round an axis.

nec tamen solida illa vel secare vel describere postulavit.(37) Sed nec Veteres Postulato secandi Conum acquieverunt. Hi tantummodo constructiones illas appellabant Geometricas quæ per lineas rectas et circulares perficiebantur: reliquas omnes tam lineas quam constructiones referebant ad Mechanicā. Id ex Pappo agnoscit et fuse docet Franciscus a Schooten Præfat *Comment. in lib. 2 Geom. Cartesij.*(38) Idem ex scriptis Archimedis colligitur. Is Problemata solida sæpe tractat[,] nunquam verò solvit per conica. Constructiones eorum aut prætermittit aut exponit per inclinationes rectarum(39) pro more veterum apud Pappum. Ex conicis Theoriam solidorum Problematum derivabant veteres: in praxi utebantur inclinationibus et operationibus quibuscunq mechanicis quæ faciliores viderentur[,] neglectis fere conicis ob difficilem, ut ait Pappus,(40) descriptionem in plano.(41) Sed nec e recentioribus inventus est quisquam qui secet conos ad constructionem Problematum. Sed videamus quam late patuit Geometria Veterum absq his Postulatis, nequis talia ad ampliandam hanc scientiam necessaria esse credat.

---

(37) Two consecutive preliminary versions of the sequel which in the manuscript here immediately follow (on ff. $123^r/124^r$) are, because of their textual complexity, separately reproduced in Appendix 2 below.

(38) See his introductory 'Argumentum Secundi Libri' in his *In Geometriam Renati Des Cartes Commentarii* [= *Geometria à Renato Des Cartes Anno 1637 Gallicè edita; postea...in Latinam linguam versa, & Commentariis illustrata, Operâ atque studio Francisci à Schooten* (Amsterdam, $_2$1659): 143–344]: 167–70. Schooten there (*ibid.*: 168) paraphrases 'quæ afferuntur à Pappo ad propositionem 4^{tam} libri tertii, ut & ad prop^{nem} 30 libri quarti *Collectionum Mathematicarum*.... Vbi apparet, quòd tantummodo constructiones illas Geometricas appellaverint, quæ per rectas lineas & circulorum circumferentias perficiebantur; quodque constructiones in genere non aliter respexerint, quàm quatenus ipsarum perfectio à manuum dexteritate & instrumentorum perfectione proficisceretur. Vnde cum ad planorum Problematum constructiones non nisi rectas lineas & circulorum circumferentias adhibendas esse viderent, quæ omnium facillimè atque expeditissimè regulæ & circini beneficio (utpote per instrumenta omnino simplicia) in plano describuntur, & sectiones Conicas reliquasque curvas lineas, varium & difficilem ortum habentes, in plano designare difficile existimarent, ideoque descriptionem earum minus certam statuerent; factum inde quoque, ut solam Planorum constructionem, Geometricam pronuntiarent: adeoque non nisi rectas lineas & circulares, reliquas vero non item, pro Geometricis agnoscerent'. His own reaction (*ibid.*: 168–9) was forcefully sharp: 'Quod quare ita dixerint, non video. Quandoquidem rectas lineas & Circulos perinde atque Parabolas, Hyperbolas, & Ellipses ex Cono secari posse ab Apollonio scio ostensum. Qui porrò postquam plurimas proprietates tribus hisce sectionibus pariter atque Circulo convenire ostendit, & quidem propter mirificas Conicorum Theorematum demonstrationes, cum non solùm illâ tempestate, verùm etiam sequentibus sæculis, magnus Geometra sit appellatus, non apparet quam ob causam prædictæ lineæ non æquè ac rectæ & circulares pro Geometricis fuerint habitæ....Cæterùm si afferatur...hasce lineas Geometricas non fuisse dictas, eò quòd instrumentis describi viderent[ur]; Annon ob eandem rationem linea recta & circularis non Geometricæ fuissent dicendæ, cum ad illas in plano describendas regulâ atque circino sit opus?...Annon pari jure artificiosam atque scientificam appellare licebit constructionem illam, quæ non nisi instrumentis perfici potest, quæ majorem industriam atque artificium in sui compositionem requirunt, cujusque demonstratio simul ex penitiori Geometriæ penu est depromenda?' It was, we have

neither the intersection nor description of those solids.(37) But then the ancients did not acquiesce in the postulate of intersecting a cone. They merely called those constructions geometrical which were accomplished by means of straight and circular lines: all the rest, lines and constructions alike, they referred to mechanics. That Frans van Schooten acknowledges from Pappus and profusely explains in the preface to his *Commentary on Book 2 of Descartes' Geometry*.(38) The same is gathered from the writings of Archimedes. He often treats solid problems, but never in fact solves them by conics; their constructions he either passes over or exhibits by inclinations of straight lines in the ancients' manner,(39) as in Pappus. While the ancients derived the theory of solid problems from conics, in practice they employed inclinations and any other mechanical operations which might seem easier, virtually neglecting conics on account, as Pappus remarks,(40) of their difficult description in a plane.(41) But nor has anyone of those more recent been found to intersect cones for the construction of problems. Let us see, however, how wide the extent of the ancients' geometry was without these postulates, lest anyone should believe such demands necessary to develop this science.

---

earlier remarked (II: 8), Schooten's life-long aim to contrive a plane construction of all species of conic 'organicè & uno ductu'—and Newton who had made this ideal of a uniform *constructio organica* a reality not merely for the Apollonian conic but also for all higher algebraic curves which are possessed of the requisite number and type of (real) multiple points; see II: 106–50. In the latter of the preliminary versions which Newton drafted of his present text—that here reproduced in Appendix 2.2—he made lengthy quotation (in Commandino's Latin version) of the pertinent passage in the preamble to Pappus, *Mathematical Collection* III, 5 which is here at issue; see page 344 below.

(39) At this point in his initial draft (see Appendix 2.2: note (15) below) Newton loosely specified the neusis constructions variously employed by Archimedes in his *Lemmas* and *On Spirals*, and also referred obliquely to the latter's omission in *On the Sphere and Cylinder* II, 4 to give exact construction of the 'solid' problem—a cubic equation in algebraic equivalent—to which he there reduced its requirement that a sphere be cut by a plane into segments whose volumes are in given ratio.

(40) Namely in the preamble to *Mathematical Collection* III, 5 where in elaborating modes of solving the problem of inserting two mean proportionals between two given quantities—'quod natura solidum est' (inasmuch as it leads at once to an irreducible cubic equation whose roots are not definable by the intersections of 'plane' rectilinear and circular loci)—Pappus states that 'antiqui geometræ problema...geometrica ratione innixi construere non potuerant, quoniam neque coni sectiones facile est in plano designare. Instrumentis autem ipsum in operationem manualem, & commodam, aptamque constructionem mirabiliter traduxerunt....' (*Mathematicæ Collectiones:* 7).

(41) A stray cancelled phrase on the facing page (f. 123v) reads at this point: 'Absq speculatione locorum solidorum genus Problematis sæpe cognosci non posset, eo cognito traducebatur constructio ad simplicissimam quamvis operationem manualem' (While in divorce from the investigation of 'solid' loci the class of a problem might not often have been able to be ascertained, once that was known a construction was converted into any simplest manual operation). This is pure speculation on Newton's part, one for which Pappus offers no support, and we may suppose that he straightaway realized as much in so quickly discarding it.

Videntes igitur Veteres quod in problematum constructionibus puncta quæsita semper invenirentur in binarum linearum intersectionibus, nominabant lineas illas loca punctorum. *Locorū* autem *alij* ab ipsis dicebantur *ἐφεκτικὸι hoc est in seipsis tantum consistentes* quo sensu(42) *puncti locus punctū est, lineæ linea, superficiei superficies et solidi solidum: alij* autem *διεξοδικὸι*(43) *hoc est sese extra tendentes*, et sic *puncti locus linea est, lineæ superficies, superficiei, solidum.* Hi loci quatenus tractantur ea ratione quæ conducit ad resolutionem Problematum,(44) constituunt partem Geometriæ quæ locus resolutus olim appellabatur. *Locorum autem, qui* (*sunt*) *in resoluto loco, alij quidem positione dati ἐφεκτικὸι sunt* (*punctorum*), *alij autem dicti plani solidi et lineares sunt punctorum διεξοδικὸι; et alij ad superficies ἀνασ[τ]ροφικὸι quidem punctorum, διεξοδικὸι autem linearum. Lineares loci ex ijs quæ sunt ad superficies demonstrantur*, quippe qui considerantur ut superficierum intersectiones. *Dicuntur autem plani loci quicunꝗ sunt rectæ lineæ vel circulares:*(45) *solidi loci quicunꝗ sunt conorum sectiones, parabolæ* [*vel*] *Ellipses, vel Hyperbolæ; lineares loci quicunꝗ lineæ sunt neꝗ rectæ neꝗ circuli neꝗ aliqua dictarum coni sectionum.*(46) Sic veterum Geometrarum Locus resolutus in se lineas universas complexus est nec indiget postulatis novis ut Geometria, quoad hanc speculationem, auctior evadat.

Est autem *locus resolutus* inquit Pappus,(47) *propria quædam materia post communem Elementorum constitutionem, ijs comparata qui in Geometricis sibi comparare volunt vim ac facultatem inveniendi Problemata quæ ipsis proponuntur: atꝗ hujus tantummodo utilitatis gratia adinventa est.* (48)Tractabat igitur Geometria Veterum problemata omnia quæ ad locorum speculationem deduci possent, dein cujus sit speciei locus quivis qui obvenerit, tum qua ratione perveniendum sit ad locorum speciem simpliciorem ubi id fieri possit.(49) Hoc pacto innotescit genus Problematis. Si id planum fuerit, habebitur ejus constructio ex locis planis jam inventis: si solidum est vel lineare traducitur ad quamcunꝗ operationem manualem quæ facillima videatur. Sed operationē illam in compositione Geometra non imperat. Eam

(42) Or so Apollonius stated in his lost treatise *De locis planis* according to Pappus' more specific phrase 'de quibus et Apollonius ante propria elementa dicit' (*Mathematicæ Collectiones*: 247) which Newton—having indeed first started to transcribe it in the manuscript along with the rest of his present quotation from the *Collection*'s seventh book—here discards in favour of a vaguer connective.

(43) Here and in sequel Newton miscopies the more correct termination *-οὶ* from the 1660 edition of Commandino's version of Pappus' text which was before him.

(44) Originally 'quatenus per Analysin investigantur et analyticè traduntur' (in so far as they are investigated by analysis and analytically delivered).

(45) Another careless slip on Newton's part. Commandino's Latin version here reads 'circuli' and so we translate on the facing page.

(46) In sequel Newton has deleted a dubious paraphrase of Pappus' next sentence 'Loci autem ab Eratosthene inscripti ad medietates ex prædictis genere sunt a proprietate hypotheseon in illis' (*Collectiones:* 247), which he understood to signify 'Ex his genere sunt loci dicti ad medietates seu ad media puncta duorum id est ad intersectionum puncta' (Of this type in general are the loci called 'to means', that is, to mid-points of two, or to points of intersection).

Seeing, then, that points required in the constructions of problems might always be found at the intersections of pairs of lines, the ancients named those lines the 'loci' of the points. 'Of these loci some', however, were called by them 'ἐφεκτικοί, that is, consisting merely in themselves' in the sense[42] in which 'a point is the locus of a point, a line that of a line, a surface that of a surface, and a solid that of a solid; others are διεξοδικοί, that is, tending outside themselves', and in this way 'a line is the locus of a point, a surface one of a line, a solid one of a surface'. These loci, insofar as they are treated in a manner contributing to the resolution of problems,[44] constitute the part of geometry which was once called the resolved locus. 'Of the loci, however, treated in the resolved locus some indeed, given in position, are ἐφεκτικοί of points, while others called plane, solid and linear are διεξοδικοί thereof, and others relating to surfaces are ἀναστροφικοί of points, indeed, but διεξοδικοί of lines. Linear loci are demonstrated from those which relate to surfaces'—those considered, of course, as the intersections of surfaces. 'Loci, however, are called plane whenever they are straight lines or circles; solid loci are any (plane) sections of cones—parabolas, ellipses or hyperbolas—whatever; linear loci are any lines at all which are neither straight lines or circles, nor any one of the said conic sections'.[46] In this way the resolved locus of the ancient geometers embraced within itself the entirety of lines, nor does it stand in need of new postulates to make geometry, in regard to this line of investigation, yet more extensive.

Now, says Pappus,[47] 'the resolved locus is a special subject-matter prepared, after the common groundwork of the *Elements*, for those who in things geometrical wish to acquire power and facility in resolving problems which are proposed to them, and with this use alone in view was it devised'. [48]The geometry of the ancients therefore treated all problems which might possibly be reduced to the investigation of loci, inquiring both the species of any locus which is met with, and what means are needed to attain a simpler species of loci when that may be.[49] In this manner the problem's kind is ascertained. If it be plane, its construction will be obtained in terms of plane loci already found; if it is solid or linear, that is made over to any manual operation which shall seem simplest. But the geometer does not order that operation in composition—he

(47) In the opening sentence of his seventh book (*Mathematicæ Collectiones*: 240); on Commandino's rendering of τόπος ἀναλυόμενος as 'locus resolutus' again (compare §1: note (27)) see IV: 223–4.

(48) Newton first went on: 'Nimirum proposito aliquo Problemate scire debet Geometra quomodo Problema idem ad locos traduci possit' (In other words, where some problem is proposed, the geometer ought to know how that problem can be converted to loci).

(49) A cancelled final clause here concludes: 'et ultimo quibus gradibus perveniendum sit ad ejusdem loci compositionem simplicissimam qua Problema commodissimè construatur' (and lastly by what steps one must attain the simplest composition of this same locus whereby the problem may most conveniently be constructed).

insinuat tantum ut possibilem vel proponit hypothetice et sub specie Theorematis vel ab assumptione quæsiti(50) deducit invertendo propositionem vel deniqȝ ut concessam assumit in statu problematis. Primi casus exempla aliquot dedit Archimedes in libro *de spiralibus* Prob 3, 4, 5, 6, 7, 8, 10, secundi casus exempla præbent hypotheticæ solidorum sectiones in ejusdem libro *de Conoidibus et Sphæroidibus* nec non hypotheticus ductus lineæ spiralem tangentis ad inveniendam longitudinem perimetri circuli Theor 11, 12, 13(51) libri *de Spiral.* Tertij casus exemplum extat in *Lemmatis* Archimedis ubi angulum per inclinationem rectæ trifariam secare docetur(52) invertendo propositionem et quærendo triplum trientis. Ex quibus omnibus videre licet quam caute Archimedes se gessit. Constructiones per plana non dubitat directe tradere, eas solidorum Problematum semper tradit indirectè ne operationem aliquam mechanicam in Geometria fieri juberet: *Tanto* enim (ut ait Carpus Antiochenus(53)) *Geometriæ et Arithmeticæ amore inflammatus erat ut nihil extrinsecus in eas introducendum statuerit.* Sequamur itaqȝ Archimedem(54) imo et veteres Geometras et Geometriam servabimus incorruptam et simul universalem exercebimus. Nulla definiri potest figura quā hæc non speculetur, nullum proponi Problema quod hæc non conficit. Operationes ad constructionem faciliorum Problematum penes omnes suos discipulos esse postulat et propterea fieri imperat pro lubitu: facultatem verò operationū difficiliorū in Tyronibus postulare absurdum esse putat et proinde has non imperat omnibus(55) sed speculatur(56) tantum inqȝ eorum usum qui peritiores fuerint & sibi ipsis facultatem earundem concesserint proponit oblique. Solvit omnia problemata, quædam practicè cætera speculativè quædam præcipiendo cætera docendo sine præceptis. In eo solo differt constructio Geometrica et Mechanica quod illam Geometria imperat atqȝ adeo habet penes se: alteram Geometria non habet penes se sed tantum speculatur.

Sunt igitur constructiones Geometricæ quæ per rectas et circulos et figuras quascunqȝ in plano datas perficiuntur. Nam figuræ quæ in statu quæstionum dantur(57) sunt penes Geometram. Intelligo figuras non solū cognitione sed etiam descriptione datas. Sic dato Ellipseos segmento problema planum est invenire centrum et axes Ellipseos totius.(58) Nec tamen aliter construitur quam

(50) Originally 'ab alijs positis' (from other suppositions).

(51) For some reason Newton first cited '9 & 10' here.

(52) See v: 464, note (678).

(53) As cited by Pappus in the introduction to his eighth book (*Mathematicæ Collectiones*: 448). On Carpus see also Appendix 1.2: note (21) below.

(54) Newton initially began to add the epithet 'Geometrarum facile [? principem]' (easily the [prince] of geometers), and then changed this to be 'qui unus est instar omnium' (who singly is the like of all) in an even more lavish Ciceronian compliment.

(55) 'suis discipulis' (its disciples) is deleted.

(56) Originally 'insinuat' (hints at them).

(57) Newton initially added 'quoad descriptionem' (in regard to their description).

merely hints it as a possibility, or proposes it hypothetically and as a species of theorem, or deduces it from the assumption of what is required,[50] or finally assumes it as granted in the circumstances of the problem. Archimedes has given a number of examples of the first event in Problems 3, 4, 5, 6, 7, 8 and 10 of his book *On spirals*; instances of the second are afforded by the hypothetical sections of solids in the same author's book *On conoids and spheroids*, and also in the hypothetical drawing of a line tangent to his spiral to find the length of a circle's circumference in Theorems 11, 12 and 13[51] of his book *On spirals*; while an example of the third case exists in Archimedes' *Lemmas* where we are by the inclining of a straight line taught how to cut an angle into three parts[52] by inverting the proposition and seeking triple its third. From all these you may well appreciate how prudently Archimedes deported himself. Constructions by planes he does not hesitate to present in direct fashion, those of solid problems he always presents indirectly lest he bid some mechanical construction to be done in geometry; for 'he was', says Carpus of Antioch,[53] 'fired with so great a love of geometry and arithmetic that he decreed nothing foreign must be introduced into them'. Let us accordingly follow Archimedes[54]—indeed, the ancient geometers generally—and we shall keep geometry unspoiled and at the same time practise its whole. No figure can be defined which this shall not regard, no problem proposed which this does not accomplish. It demands that operations for the construction of easier problems be in the power of all its disciples and consequently commands at will that they be performed; it thinks it absurd, however, to postulate that novices should control the more difficult operations, and accordingly does not dictate these for all[55] but merely takes them within its scope[56] and obliquely propounds them for the use of those who are more proficient and have acquired in themselves a mastery over them. It solves all problems, some in a practical manner, the others speculatively, some by laying down precepts and the rest by instruction without precept. Geometrical and mechanical construction differ solely in this, that geometry commands the former and hence has it in its power, while it has no power over the other but merely speculates upon it.

Geometrical constructions, then, are ones which are achieved by means of straight lines and circles and any figures whatever given in the plane; for, of course, figures given[57] in the circumstances of a question are in a geometer's power. I understand figures given not only in notion but also in their description. Thus, given an ellipse's segment it is a plane problem to find the centre and axes of the whole ellipse[58]—and no other means is needed for its construction

(58) For, given the segment, we may readily fix five suitable points on the ellipse, and thence by Pappus, *Mathematical Collection* VIII, 13/14 (compare note (19) above) successively determine a pair of conjugate diameters and therefrom the ellipse's centre (in which they intersect) and its axes—all by 'plane' constructions involving but straight lines and circles.

ducendo rectas occurrentes Ellipsi. A dato puncto ad datam Ellipsin perpendiculū demittere est etiam problema planum eo quod per Ellipsin datam et operationes Geometriæ planæ construi potest.(59)

---

[2](60)

Cap 2(61)

De Compositione & Resolutione Veterum Geometrarum.

Scripta Veterum Geometrarum quæ ad nostram notitiam pervenerunt partim exhibent inventa, ut libri Archimedis, partim spectant ad methodum inveniendi, ut Euclidis *data* et opera Apollonij. Ea posterioris generis non docent methodum aliquam computandi sed argumenta continent jam resoluta et præparata quæ tanquam loci communes inserviant inventioni, eamq reddant facilem simplicem compendiosam et elegantem. Quicquid enim jam ante confectum reperiebatur non opus erat iterum conficere. Id omne ut datum & concessū(62) assumebatur.

---

(59) This is untrue. A quarter of a century earlier, in Problem 3′ of his 'Problems of Curves' (see II: 179–80) Newton had accurately inferred that, if the (four) feet of the normals from the given point to the given ellipse 'may bee found by plaine Geometry', then these must lie on some circle which 'will cut the curve in the desired points. This I say might be tryed but it would bee found impossible (unless ye conick bee a parabola) because there are foure given points through wch ye circle must passe wch make ye problem contradictious since 3 are enough to determine a circle' (*ibid.*: 180). As an instance, where the given point is the ellipse's centre the feet of the normals from it to the ellipse are the four (non-concyclic) principal vertices—which here, however, may be determined by constructing a circle of arbitrary radius concentric with the ellipse and then bisecting the angle formed by joining its opposite meets with the ellipse to determine the axis on which these feet lie. More generally, since at its meet (x, y) with the ellipse $y^2 = rx-(r/q)\,x^2$ the latter's slope must be perpendicular to the normal from the given point (a, b), then this must lie also on the hyperbola defined (in perpendicular Cartesian coordinates) by

$$(\tfrac{1}{2}r-(r/q)\,x)/y = -(x-a)/(y-b) \quad \text{or} \quad xy(1-r/q)+(br/q)\,x+(\tfrac{1}{2}r-a)\,y = \tfrac{1}{2}br,$$

and the condition $(a \neq \frac{1}{2}r, b \neq 0)$ for the intersections of this with the ellipse to lie on some circle $x^2+y^2+cx+dy+e = 0$ is that $r/q = 0$, that is, (since $r \neq 0$) $q \to \infty$: in which case the given ellipse becomes the parabola $y^2 = rx$, meeting its 'normal' hyperbola $xy+(\frac{1}{2}r-a)\,y = \frac{1}{2}br$ in the three feet of the perpendiculars from (a, b) to it, which intersections are readily proved to lie on the circle $x^2+y^2-(\frac{1}{2}r+a)\,x-\frac{1}{2}by = 0$ which also passes through the parabola's vertex $(0, 0)$. The 'solid' constructions of the normals by means of the given conic's intersections with the appropriate auxiliary rectangular hyperbola are given by Apollonius in his *Conics* v, 58/59 (which treat the cases of a parabola and central conic respectively) without, in the first, specifying the further reduction to a 'plane' construction by an auxiliary circle which is there possible. While T. L. Heath has perceptively conjectured in his *History of Greek Mathematics*, 2: 166 that Pappus had the latter deficiency in mind when, in the preamble to *Mathematical Collection* IV, 31, as an instance of the 'peccatum non paruum...apud geometras, cum problema planum per conica, vel linearia...inuenitur' he cited 'quale est in quinto libro *Conicorum* Apollonij problema in parabola' (*Collectiones*: 95), no full enunciation of the construction *per circulum* was published till Edmond Halley set it out in a scholium to his 1710 *editio*

than drawing straight lines meeting the ellipse. To let fall a perpendicular from a given point to a given ellipse is likewise a plane problem in that it is constructible with the aid of the given ellipse and the operations of geometry.(59)

---

[2](60)

CHAPTER 2.(61)

THE COMPOSITION AND RESOLUTION OF THE ANCIENT GEOMETERS.

The writings of the ancient geometers which have reached our (modern) notice partly—Archimedes' books, for instance—display discoveries and partly—Euclid's *Data*, for example, and the works of Apollonius—have regard to the method of invention. Those of the latter kind do not explain some particular method of computation, but contain arguments already resolved and made ready beforehand to serve the purpose, as it were, of discovering a locus and making it easy, simple, compendious and elegant. For to be sure there was no need to accomplish anew anything that was already before found accomplished. All that was assumed as given and (62)granted. And in consequence the ancients

---

princeps of Serenus' *De Sectione Cylindri et Coni*. (See II: 180, note (28). We ought there to have stated, however, that this 'plane' Cartesian construction of the normals to a parabola from a given point was already attained by Christiaan Huygens in September 1653 (see his *Œuvres complètes*, **12**, 1910: 81–2) and quickly imparted by him to several of his correspondents. Newton himself in the late 1660's had probably seen only the brief report on the *problema* which Schooten inserted shortly afterwards in his second Latin edition of Descartes' *Geometrie:* '...Cujus porrò demonstrationem universalem, quam sibi vulgari modo Geometrarum, continuæ contemplationi figuræ obnoxiam, acutissimus pariter atque eruditissimus noster Chr. Hugenius concinnavit, cum ipsa jam pridem nobis aliisque ab eo communicata fuerit, nec illa etiam hujus loci existat, eandem hîc prætereundum duximus' (*Geometria*, $_2$1659: 322).)

Newton here (at the top of f. 126r) breaks off, leaving the remainder of his manuscript page blank. Though we may only guess how he intended to proceed, these last few paragraphs of his discussing the ancients' notions of point-loci and how they applied these in the 'linear' and 'organic' construction of more difficult problems lead naturally and with little overlap—we need suppress little more than a couple of duplicated quotations from Pappus—into his elaborate ensuing discussion of the classical methods of analysis and synthesis, and of the dominating rôle played in the former by the several varieties of 'resolved locus' developed by Euclid, Apollonius and other Greek geometers. The reader may judge whether the sequence we here impose is more than an editorial fancy.

(60) Add. 3963.14: 161r–170r. On a loose slip of paper inserted with them in 1777 (compare note (2) above) Horsley contributed his needlessly tentative and scarcely enlightening opinion that these folios are 'Portions apparently of an intended treatise on geometry'.

(61) It is our surmise (see notes (1) and (59) above) that Newton's opening 'CAP. 1'—of which no trace now survives in the Portsmouth papers—would have been (if indeed it was ever written) some appropriately revised version of the draft introduction to the 'Geometria' which is reproduced in [1] preceding.

(62) 'gratis' (freely) is deleted.

Et hinc veteres omnia hujus generis edita argumenta[63] collectivè appellabant Locum resolutum. Hunc autem locum una cum resolutione et compositione cui is inserviebat, Pappus procemio libri septimi *collect. Math.* sic describit.[64]

Locus, inquit, *qui vocatur ἀναλυόμενος, hoc est resolutus, ut summatim dicam, propria quædem est materia post communem Elementorum constitutionem, ijs parata, qui in Geometricis sibi comparare volunt vim ac facultatem inveniendi problemata quæ ipsis proponuntur: atq hujus tantummodo utilitatis gratia inventa est. Scripserunt hac de re tum Euclides qui Elementa tradit, tum Apollonius Pergæus, tum Aristæus senior: Quæ quidem per resolutionem et compositionem procedit. Resolutio igitur est via a quæsito tanquam concesso per ea quæ deinceps consequuntur ad aliquod concessum in compositione: in resolutione enim id quod quæritur tanquam factum ponentes quid ex hoc contingat consideramus: et rursum illius antecedens, quousq ita progredientes incidamus in aliquod jam cognitum vel quod sit e numero principiorum. Et hujusmodi processum* RESOLUTIONEM *appellamus, veluti ex contrario factam* SOLUTIONEM. *In compositione autem per conversionem ponentes tanquam jam factum id quod postremum in resolutione sumpsimus: atq hic*[65] *ordinantes secundū naturam ea antecedentia quæ illic consequentia erant et mutua illorum facta compositione, ad quæsiti finem pervenimus; & hic modus vocatur* COMPOSITIO. Quibus verbis intelligas methodum generalem solvendi problemata Veteribus innotuisse, hanc methodum maximè constitisse in Loco resoluto et procedere per resolutionem & compositionem conjunctim, resolutione perveniri ad compositionem, in compositione plena Geometricum omne perfici, solutionem verò ita contrariam esse resolutioni ut ea non prius habeatur quā resolutio omnis a principio ad finem per compositionem plenam et perfectā excludatur. Verbi gratia si quæstioni per constructionem æquationis alicujus respondeatur, quæstio illa resolvitur per inventionem æquationis, componitur per constructionem ejusdem, sed non priùs solvitur quàm constructionis enunciatio ac demonstratio tota[66] componitur, æquatione neglecta. Hinc est quod resolutio in veterum scriptis extra Pappi *collectanea* tam rarò occurrat. Ea solus Archimedes usus est idq non nisi in propositione una et altera ubi hærebatur. Quippe Geometrarum est homines imperitos et mechanicos docere,[67] & multitudini docendæ inepta est resolutio. Ea nec intellectu facilis est nec demonstrando clara neq ordine rerum naturalis.[68] Qui quæstionem proponit profitetur se solutionem nescire. Hoc non decet artis analyticæ gnaros. Cæteris non prius

(63) Originally 'scripta' (writings).

(64) Newton once more quotes Commandino's version of these opening sentences of Pappus' from *Mathematicæ Collectiones*: 240; compare note (47) above and §1: note (27).

(65) We correct a mistranscription 'hinc' in Newton's manuscript.

(66) A superfluous 'a principio ad finem exclusa omni analysi perfectè' (from beginning to end excluding all analysis perfectly) is here deleted.

(67) Well spoken by a university professor of mathematics who never did so in his life! Or did Newton roundly despair of the skill and intelligence of his Cambridge audience?

called the arguments(63) of this kind which they produced collectively the 'resolved locus'. In the introduction to the seventh book of his *Mathematical Collection*, however, Pappus describes this locus, along with the resolution and composition to which it was servant, as follows:(64)

'The locus called ἀναλυόμενος, that is, resolved, is in summary a special subject-matter prepared, after the common ground-work of the *Elements*, for those who in things geometrical wish to acquire power and facility in resolving problems which are proposed to them, and with this use alone in view was it devised. Writers on this topic include Euclid, the author of the *Elements*; Apollonius of Perga; and Aristæus the elder. Its development, indeed, is by way of resolution and composition. Resolution, accordingly, is the route from the required as it were granted through what thereupon follows in consequence to something granted in the composition. For in resolution, putting what is sought as done, we consider what chances to ensue, and then again its antecedent, proceeding in this way till we alight upon something already known or numbered among the principles. And this type of procedure we call *resolution*, it being as it were a reverse *solution*. In composition, however, putting as now done what we last assumed in the resolution and here, according to their nature, ordering as antecedents what were before consequences, we in the end, by mutually compounding them, attain what is required. And this method is called *composition*'.

By these words you should understand that a general method for solving problems was known to the ancients, and that this method consisted in its greatest part in the resolved locus, proceeding by means of resolution and composition jointly: by resolution composition is attained and in the fullness of composition all that is geometrical is accomplished; solution is, however, the opposite of resolution in that it may not be had till all trace of resolution be removed from start to finish by means of a full and perfect composition. For example, if a question be answered by the construction of some equation, that question is resolved by the discovery of the equation and composed by its construction, but it is not solved before the construction's enunciation and its complete demonstration is, with the equation now neglected,(66) composed. Hence it is that resolution so rarely occurs in the ancients' writings outside of Pappus' *Collection*. Archimedes alone employed it and then only in a couple of propositions where he came to be stuck. To be sure, it is the duty of geometers to teach unskilled men and mechanics,(67) and resolution is ill-suited to be taught to the masses. It is neither easy to understand nor plain in its demonstration nor natural in its sequence of content.(68) One who propounds a question confesses that he is ignorant of its solution: this does not befit those practised in the analytical art, and to the rest no answer may be made before the reply is

(68) In sequel Newton initially continued: 'Indoctorum est æque ac doctorum quæstiones proponere, artificis est responsa [dare]' (It is the rôle of the unlearned equally as of the learned to propose questions, that of the expert to give replies).

respondetur quam aptatur responsum eorum captui. Et absq; facultate respondendi quæstio non solvitur quærentibus. Differt ergo solutio a resolutione et compositione quatenus hæ sunt media[69] et operationes inveniendi per locū resolutū, illa finis et inventum in quo cessat omnis operatio. Extrema duo per totum contraria sunt resolutio et solutio[,] et ab uno ad alterum per compositionem gradatim pergit artifex. Absq; demonstratione non fit solutio scientifica, et demonstratio resolutioni ex contraria parte respondet[,] ut etiam docet Pappus in sequentibus.[70]

Duplex autem, ait, *est resolutionis genus*[,] *alterum quidem quod veritatem perquirit et contemplativum appellatur: alterum verò quo investigatur id quod dicere proposuimus, vocaturq; problematicum. In contemplativo igitur genere quod quæritur ut jam existens,*[71] *et ut verum ponentes*[,] *per ea quæ deinceps consequuntur tanquam vera, & quæ ex positione sunt, procedimus ad aliquod concessum, quod quidem si verum sit, verum erit & quæsitum & demonstratio quæ resolutioni ex contraria parte respondet. Si vero falso evidenti occurramus, falsum erit quæsitum. In problematico autem genere, quod propositum est ut cognitum ponentes, per ea quæ deinceps consequuntur ta[m]quam vera procedimus ad aliquod concessum, quod quidem si fieri comp[er]iriq; possit (quod datum vocant Mathematici)*[,] *etiam illud quod propositum est fieri poterit, & rursus demonstratio resolutioni ex contraria parte respondens. At si evidenti quod fieri non possit occurramus: et problema itidem fieri non poterit. Determinatio autem est quæ declarat quando, et qua ratione, et quot modis problema fieri possit. Hæc igitur de resolutione et compositione dicta sint.*

Dictorum autem librorum, pergit Pappus,[72] *qui ad locum resolutum pertinent ordo talis est. Euclidis datorum liber unus, Apollonij de proportionis sectione libri duo, de spatij sectione libri duo, (determinatæ sectionis duo,)*[73] *tactionum duo, Euclidis porismatum tres, Apollonij inclinationum duo, planorum locorum duo, Conicorum octo, Aristæi locorum solidorum quinq;, Euclidis ad superficiem duo, et Eratosthenis de medietatibus duo.*[74] Tot libris illam Geometriæ partem quæ locus resolutus dicebatur, Veteres prosecuti sunt, et ni fallor pluribus qui ad Pappi notitiam non pervenerant. Quippe Arist[æus][75] Euclide et Apollonio antiquior, non prius tractabat locos solidos

(69) Newton originally here added 'perveniendi ad illam' (of attaining it), continuing thereafter: 'Resolutioni contraria est solutio[:] a principio ad finem et a resolutione ad solutionem per compositionem pergit artifex. Cum compositione permisceri potest resolutio, cum solutione minimè' (Solution is contrary to resolution: the expert proceeds from beginning to end and from resolution to solution by means of composition. Resolution can be intermingled with composition, but with solution not at all).

(70) In immediate continuation of the previously quoted passage (*Mathematicæ Collectiones*: 240–1).

(71) In a manner unprecedented elsewhere in Newton's private mathematical papers the remainder of this paragraph is copied from Commandino's printed text in an alien hand which we have been unable to identify. We can think only that, when at this point his attention was distracted for one reason or another by some visitor to his room, Newton requested the latter to carry on with transcribing this quotation from Pappus till be himself could return to take

adapted to their capacity for understanding—without their comprehending the answer a question is not solved for those who ask it. Solution therefore differs from resolution and composition inasmuch as the latter are means[69] and procedures for discovering through the resolved locus, the former is the end-discovery at which the whole process terminates. Resolution and solution are two totally opposite extremes, and the expert makes his way in stages from one to the other by composition. Without demonstration a solution does not become scientific, and demonstration corresponds to the reverse of resolution, as Pappus also explains in sequel.[70]

'There exist two kinds of resolution. One pursues the truth and is named contemplative, while the other whereby what we proposed to state is investigated is called problematic. In the contemplative kind, accordingly, on positing what is sought already to exist as true, by way of what follows thereon as true and what is so by supposition we proceed to something granted: if this indeed be true, then both what is sought and the demonstration corresponding to the reverse of the resolution will be true; but if we meet with an evident falsehood, then what is sought will be false. In the problematic kind however, on positing what is proposed as known, by what follows thereon as true we proceed to something granted: if this may indeed be done and ascertained (what mathematicians call given), then what is proposed will also be able to be done, and in turn the demonstration answering to the resolution as its reverse; while if we meet with what evidently cannot be done, then the problem will be likewise impossible. Its determination, however, is what proclaims when, by what method and in how many ways a problem is possible. So much, then, for what may be said on resolution and composition.'

Pappus goes on:[72]

'Of the books mentioned, however, which relate to the resolved locus the order is as follows: Euclid's *Data*, one book; Apollonius' *On cutting off a ratio*, two books; his *On cutting off a space*, two books; his *On determinate section*, two;[73] *On contacts*, two; Euclid's *Porisms*, three; Apollonius' *On inclinations*, two; *Plane loci*, two; *Conics*, eight; Aristæus' *Solid loci*, five; Euclid's (*Loci*) *to surfaces*, two; and Eratosthenes' *On means*, two.'[74]

In so many books was the part of geometry called the resolved locus pursued by the ancients—and in yet more, if I am not mistaken, which did not reach Pappus' notice; for we may be sure that Aristæus,[75] earlier than Euclid and

over. With the next paragraph the manuscript hand reverts to be his own characteristic autograph and so continues to the end.

(72) *Mathematicæ Collectiones*: 241.

(73) Newton here rightly makes good the accidental omission of this item from the 1660 printing of Commandino's Latin version.

(74) An unnecessary termination 'Hæc Pappus' (So Pappus) to this quotation is deleted in the manuscript.

(75) By a slip of his pen Newton himself conflated this minor Greek geometer with 'Euclides' following to beget the *nom de plume* 'Aristides'. Bourbaki had, we see, his ancient precursor!

quam alij hic non recitati tractarent planos. Sed et aliqua intelligendis Apollonij libris necessaria deesse videntur. Quo factum est ut Pappus non modò scopum librorum de sectione triplici(76) exponere nesciret sed etiam negaret compositionem quandam loci(77) a Veteribus perfectam esse[,] quam tamen ab Apollonio perfectam esse demonstrant libri *sectionum* cum *conicis* collati, ut posthac ostendam.(78)

Principio, ait,(79) *septimi* (*Porismatis libri primi habetur*)(80) *diagramma hoc. Si a duobus datis punctis ad rectam lineam positione datam rectæ lineæ inflectantur, abscindat autem una a recta linea positione data ad datum in ipsa punctum, abscindet et altera proportionem habens datā. In ijs autem quæ sequuntur. Quod hoc punctum tangit positione datam rectam lineam. Quod proportio hujus ad hanc data est. Quod proportio hujus ad apotomen. Quod hæc positione data est. Quod hæc ad datum punctū vergit. Quod proportio hujus ad aliquam ab hoc ut dato. Quod proportio hujus ad aliquam ab hoc ductæ.*(81) *Quod proportio hujus spatij ad id quod data recta linea et hac continetur. Quod hujus spatij alterum quidem est datum*[,] *alterum proportionem habet ad apotomen. Quod hoc spatium vel hoc una cum aliquo spa*[*t*]*io dat*[*um*](82) *est*[,] *illud autem proportionem habet ad apotomen. Quod hæc cum qua ad quam hæc proportionem habet datam, proportionem habet ad aliquam ab hoc ut dato. Quod id quod sub dato et hac æquale est ei quod sub dato et ab hoc ut dato. Quod proportio hujus et hujus ad aliquā ab hoc ut dato. Quod hæc abscindit a positione datis datum continentes.*(83) Quarum propositionum si Pappi verba brevia et nimiùm depravata rectè interpretatus sum, sensus est talis.

[*Porisma 1.*] Si a duobus datis punctis A, B ad rectam lineam positione datam CZ rectæ lineæ AZ, BZ inflectantur, abscindat autem una AZ a recta linea

(76) Apollonius' trio of works *De sectione rationis/spatii/determinata*, of course.

(77) That, namely, which shall—as Euclid himself, according to Apollonius (see IV: 220), had been unable fully to do—rigorously establish the identity of the Euclidean 3/4-line locus with the general conic. As we have earlier remarked (IV: 221), Pappus had asserted that 'Quem...dicit [Apollonius] locum ad tres, & quatuor lineas ab Euclide perfectum non esse, neque ipse perficere poterat, neque aliquis alius: sed neque paululum quid addere ijs, quæ Euclides scripsit per ea tantum conica, quæ usque ad Euclidis tempora præmonstrata sunt...' (*Mathematicæ Collectiones*: 251; compare 1, §1.1: note (29) preceding).

(78) Though in the present unfinished draft of the 'Geometria' Newton did get so far as to record in his final paragraph below both Pappus' enunciation of the 3/4-line locus and his statement of its identity with the general conic, he never went on here to formulate his long-standing conviction (on which see 1, §1.1: note (28) above) that Apollonius must have attained the requisite 'perfect composition' of the proof of this identity which ensues from enumerating all the possibilities of the related quadratic 'determinate section' which the locus cuts out in a straight line given in general position in its plane. Had he done so, his ensuing 'collation' of Apollonius' *sectio determinata* with the Euclidean 'solid' locus would not, we may be certain, have differed in its essentials from that set out by him (see pages 206–10 above) in his preceding survey of the 'Analysis Geometrica' of the ancients.

(79) A few pages later when, in reviewing the content of Euclid's lost teatise *De Porismatibus*, he comes to specify the main varieties of proposition in its first book (see *Mathematicæ Collectiones*: 246).

Apollonius, did not treat of solid loci before others not here mentioned treated plane ones, while certain other items necessary for comprehending Apollonius' books seem to be missing. The result was that Pappus was not only ignorant of how to display the scope of the books on the triple section,(76) but also denied that the ancients had accomplished a certain locus-composition(77) even though Apollonius' books on the sections when compared with his *Conics* prove that he had achieved it, as I shall afterwards show.(78)

He further states:(79)

'At the beginning of the seventh (porism of the first book there is had)(80) this sketch. If from two given points straight lines are inclined to a straight line given in position, let one cut off from a straight line given in position (a segment terminated) at a given point in it, then the other will cut off one having a given proportion. In the sequel, however, (it is premised) that this point traces a straight line given in position; that the proportion of this (line) to this is given; that the proportion of this to a segment cut off is; that this (line) is given in position; that this one verges to a given point; that the proportion of this to some from this is as given; that the proportion of this drawn to some of this is;(81) that the proportion of this space to the content of a given straight line and this is; that of this space one member indeed is given, while the second has a ratio to a segment cut off; that the latter space, or this together with some space, is given, while the former has a proportion to a segment cut off, that this (line) with that to which this one has a given proportion has a proportion to some (line extended) from this (point) as given; that the content beneath a given and this (line) is equal to that beneath a given and (a line extended) from this (point) as given; that the proportion of this and this (line) to some (line extended) from this (point) as given is; that this (line) cuts off from ones given in position (segments) containing a given.'(83)

The sense of these propositions, if I rightly interpret Pappus' concise and excessively corrupt text, is as follows.

Porism 1. If from the two given points A, B straight lines AZ, BZ be inclined to the straight line CZ given in position, let one, AZ, cut off from the straight

(80) On this interpolation in Commandino's Latin version—one embodying a conjecture on Newton's part which is at least as plausible as Hultsch's dogmatic contention that the whole phrase is probably a scholiast's intercalation—compare §1: note (44) preceding.

(81) In his prior draft at this point (see §1: note (45)) Newton in a marginal postil passed on his suspicion (not supported by Pappus' more recent editors) that 'the proposition is a corrupt repetition of the previous one'.

(82) We take it upon ourselves to amend the illogical case of the printed text's 'dato' (*Collectiones*: 246, l. 22), here faithfully copied by Newton in the manuscript.

(83) With this repeated quotation—here in full—of Pappus' cryptic outline of the main types of Euclidean porism Newton now passes to revise (and partially to augment) his previous restorations in §1 preceding (see pages 262–8 above) of their rectilinear instances. The several roughnesses in enunciation and sequence in these listed *Porismata* to which we point in following footnotes were as evident to their author as to his modern reader, and it was not long before he began to prepare the further amplification of their content which we reproduce in Appendix 3.1 below, but which all too quickly he left off.

positione data EX ad datum in ipsa punctum E segmentum EX ad alteram positione datam CZ proportionem habens datam, abscindet et altera BZ segmentum EY ad eandem CZ proportionem habens datam. Quippe parallelæ erunt EX, CZ, puta si puncta E, C, A, B jaceant in directum.

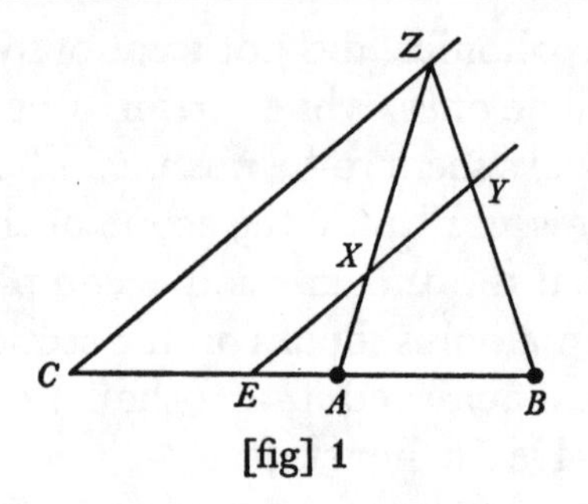

[fig] 1

Porisma [2]. In eodem Diagrammate,[84] si detur ratio EX ad EY, punctū Z tanget positione datam rectam lineam CZ.

Porisma [3]. Et si punctum Z tangit positione datam rectam CZ[,] proportio EX ad EY datur.[85]

Porisma [4]. Datur item proportio EX ad apotomen XY.

Porisma [5]. Sed et recta EX positione data est.

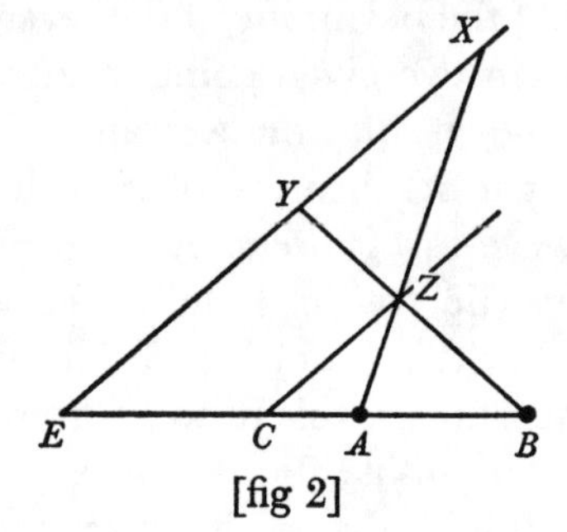

[fig 2]

Porisma [6]. Si a dato puncto A exiens recta AZ secet parallelas duas positione datas CZ, EX[,] detur autem proportio EX ad EY, recta ZY ad datum punctum B vergit.

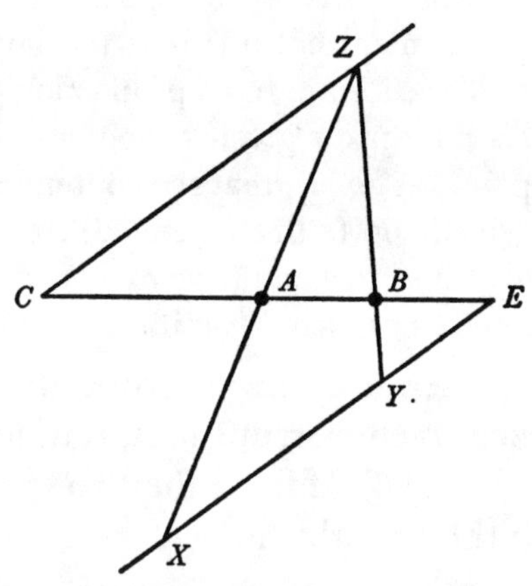

[fig 3]

Porisma [7]. Si[86] recta positione data EX non sit ipsi CZ parallela, datur proportio hujus ad aliquam[87] positione datam EY ductam a puncto communi E in recta AB dato. Fig. 4.

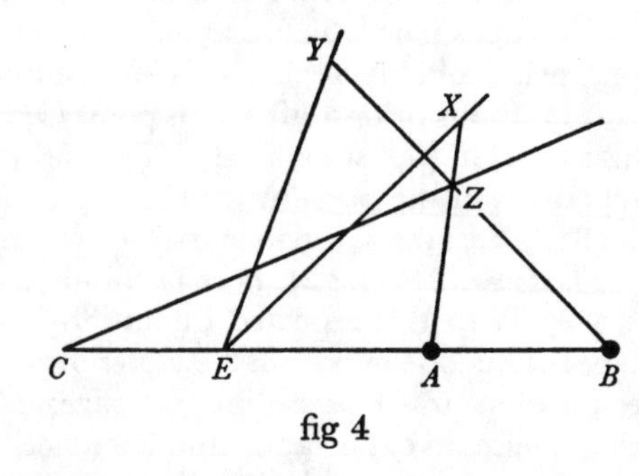

fig 4

(84) That whose several possibilities are exhibited in Newton's figures 1–3 above, in all of which CZ is parallel to EXY. Here, as in Porisms 3–6 following which display minor variations

line EX given in position a segment EX, terminated at the given point E in it, having a given proportion to the other, CZ, given in position, then the other, BZ, will cut off the segment EY having a given proportion to the same CZ. Of course EX and CZ will be parallel—understanding that the points E, C, A, B lie in a straight line.

Porism 2. In the same scheme[84] if the ratio of EX to EY be given, the point Z will trace the straight line CZ given in position.

Porism 3. And if the point Z traces the straight line CZ given in position, the proportion of EX to EY is given.[85]

Porism 4. There is likewise given the proportion of EX to the segment XY cut off.

Porism 5. While the straight line EX is also given in position.

Porism 6. If the straight line AZ departing from the given point A shall intersect two parallels CZ, EX given in position, let the proportion of EX to EY be given, and the straight line ZY then verges to the given point B.

Porism 7. If[86] the straight line EX given in position should not be parallel to CZ, there is given the proportion of this to some line[87] EY which is drawn from the common point E given in the straight line AB. Figure 4.

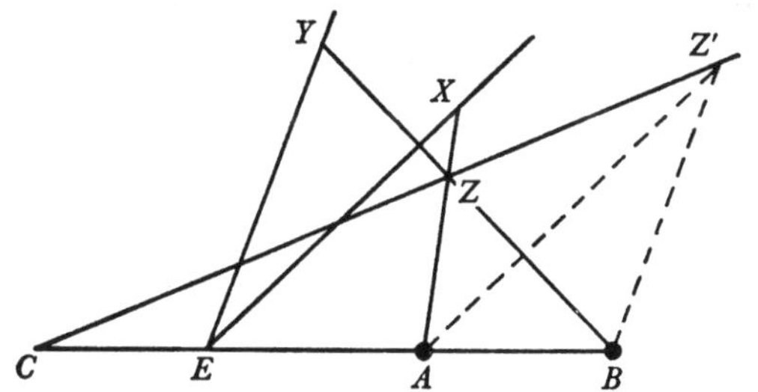

upon the same basic configuration, the constancy of the ratio EX/EY determines a 1, 1 correspondence between the points X and Y, and hence the meet Z of AX and BY defined therefrom *per simplicem geometriam*, such that X and Y come (at E) simultaneously to lie in AB and simultaneously to pass to infinity in EXY, whence correspondingly Z comes to be at C (in AB but distinct from either A or B) and to pass to infinity in the direction of EXY. The first consequence entails that the locus (Z)—in general (see IV: 308–10) a conic passing through A and B—degenerates into the pair of AB and some other straight line CZ, while the latter decrees that CZ shall be parallel to EXY.

(85) For at once, in modern equivalent, the perspective correspondences $Z \leftrightarrow X$ and $Z \leftrightarrow Y$ are such that $C \leftrightarrow E$ and $\infty_{CZ} \leftrightarrow \infty_{EXY}$ in both cases, whence there ensue the cross-ratio equalities $(E\, \infty_{EXY}\, X) = (C\, \infty_{CZ}\, Z) = (E\, \infty_{EXY}\, Y)$.

(86) In a cancelled preceding enunciation Newton here inserted—to no point as he here in redraft realizes—the repeated specification 'a datis duobus punctis A, B ad rectam positione datam CZ inflectantur duæ rectæ AZ BZ et' (from the two given points A, B to the straight line CZ given in position there be inclined the two straight lines AZ, BZ and).

(87) So angled through E, namely, that corresponding to some point Z' in CZ the points X and Y shall pass simultaneously off to infinity in their respective lines—in other words (see our augmented figure in the English version), such that the parallels through A and B to EX and EY respectively shall meet (in Z') on the base line CZ. This present porism is considerably elaborated—and given explicit proof—in the equivalent Porism 11 in Appendix 3.1 below, there being extended (in its Case 3) to hold *mutatis mutandis* for a like configuration in which the rectilinear loci (X) and (Y) meet AB in distinct points E.

Porisma [8]. Datur etiam proportio hujus EX ad aliquam(88) EY ductam a puncto communi E in recta CZ dato. Fig 5.

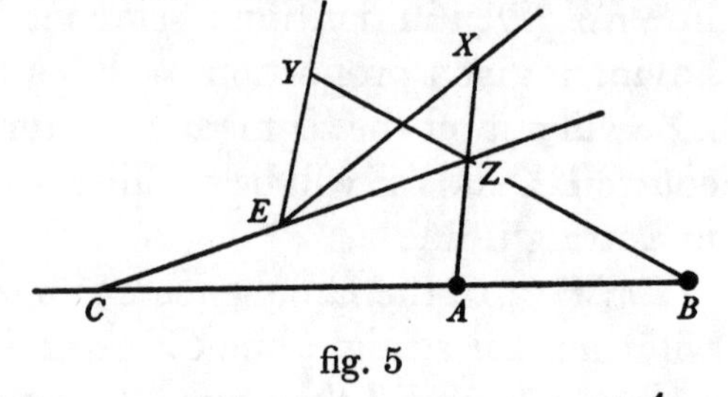

fig. 5

Porisma. Si rectæ EX eY utcun[q;] dentur positione secantes AB in E et e, ipsi EX parallelam age BK secantem [e]Y in G et proportio spatij $EX \times GY$ ad spatium quod data recta linea et hac ipsa [e]Y continetur data est.(89)

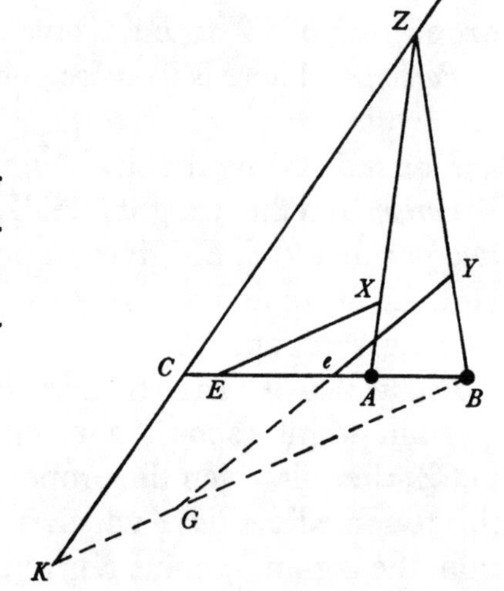

Porisma [9]. Si recta EX secet lineam CF in puncto F, proportio spatij $EX \times FY$ ad id quod data recta linea et ipsa XY continetur data est, ut et proportio spatij $EY \times FX$ ad utrumq;.(90)

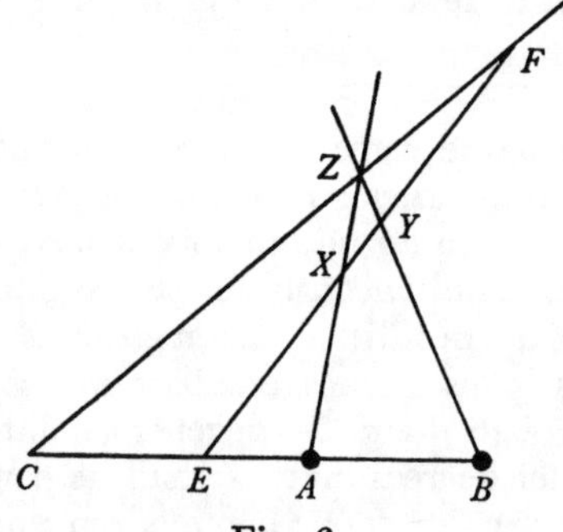

Fig. 6

Porisma [10]. Si recta AZ secet rectas duas quasvis positione datas CZ, EX,(91) comple parallelogrammum $A\upsilon F\xi$ et rectangulum $\xi X \times FZ$ est ad apotomen AFX (ut et rectangulum $\upsilon Z \times FX$ ad apotomen AFZ, nec non rectangulum $\xi X \times CZ$ ad apotomen AEX) in ratione data, rectangulum verò $\xi X \times \upsilon Z$ datur.(92)

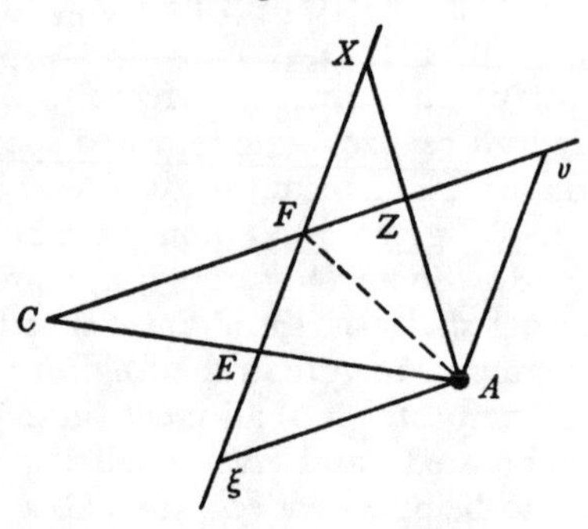

Fig 7

Porism 8. There is also given the proportion of this EX to some line[88] EY drawn from the common point E given in the straight line CZ. Figure 5.

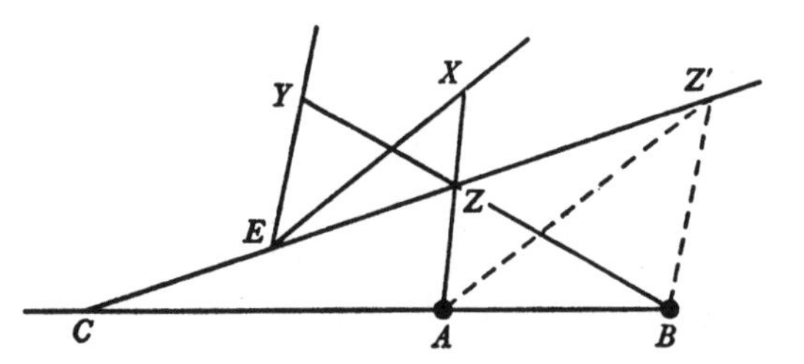

Porism. If the straight lines EX, eY given arbitrarily in position intersect AB in E and e, parallel to EX draw BK cutting eY in G, and the proportion of the space $EX \times GY$ to the space contained by a given straight line and the present one eY is given.[89]

Porism 9. Should the straight line EX intersect the line CF in the point F, the proportion of the space $EX \times FY$ to that contained by a given straight line and by XY is given, as also the proportion of the space $EY \times FX$ to each.[90]

Porism 10. Should the straight line AZ intersect any two straight lines CZ, EX[91] given in position, complete the parallelogram $A\upsilon F\xi$, and the rectangle $\xi X \times FZ$ is then to the space $AF \times FX$ cut off (as also the rectangle $\upsilon Z \times FX$ to the space $AF \times FZ$ cut off, along with the rectangle $\xi X \times CZ$ to $AE \times EX$ cut off) in a given ratio, while the rectangle $\xi X \times \upsilon Z$ is given.[92]

(88) Here again (compare the previous note) this needs to be drawn parallel to BZ' where Z' is the meet with CZ of the parallel through A to EX.

(89) Newton has cancelled this porism only because its proof anticipates—directly so or in explicit equivalent—the fundamental theorem of the invariance of the cross-ratio of the intercepts of a general transversal with a pencil of lines which he presents in its train. Here the perspectivities $X \leftrightarrow Z$ and $Z \leftrightarrow Y$ from the respective poles A and B define $E \leftrightarrow C \leftrightarrow e$ and $\infty_{EX} \leftrightarrow K \leftrightarrow G$, so that if also $x \leftrightarrow z \leftrightarrow y$ there is $(E\,\infty_{EX}\,Xx) = (eGYy)$, or

$$EX/Ex = eY \times Gy/ey \times GY,$$

whence $EX \times GY/k.eY = Ex \times Gy/k.ey$, constant, where k is the *data recta*.

(90) The 1, 1 perspective correspondences $X \leftrightarrow Y$ and $Z \leftrightarrow Y$ defined by AXZ and BYZ as they rotate round their respective poles A and B relate $E \leftrightarrow C \leftrightarrow E$ and have F as a common point. On restating Newton's constant proportion in the form

$$EX \times FY/EY \times FX = 1 + EF \times XY/EY \times FX$$

it will be seen that he merely enunciates the constancy of the cross-ratio $(EFXY)$ of the involution cut out in the fixed straight line EF by its meets X, Y with the pair of lines ZA, ZB drawn from an arbitrary point Z in CF to the fixed points A, B collinear with E. His demonstration of the constancy of the proportion would doubtless have proceeded, by appealing to Pappus, *Mathematical Collection*, VII, 129 or equivalently to Porism 1' in §1 preceding (see *ibid.*: note (59)), to deduce that for all line-pencils ZE, ZC, ZA, ZB through the variable point Z there is $(EFXY) = (ECAB)$, constant.

(91) Newton understands, of course, that X lies ever in AZ.

(92) In this generalized Apollonian *sectio spatii* (compare Porism 1 of his 'Inventio Porismatum' on page 234 above) by the 1, 1 perspective correspondence $X \leftrightarrow Z$ of points in EX and CZ which are in line through the fixed pole A Newton successively relates $E \leftrightarrow C$, $F \leftrightarrow F$, $\xi \leftrightarrow \infty_{CZ}$ and $\infty_{EX} \leftrightarrow \upsilon$. His stated constant proportions and rectangles are, when reduced to the equivalent forms

$$\xi X/FX = F\upsilon/FZ\,(\propto AF/FZ),\quad (FX/AF \propto)\; FX/F\xi = FZ/\upsilon Z$$
$$\text{and}\quad \xi X/\xi F = \upsilon F/\upsilon Z,$$

Porisma [*11*]. Si rectæ AZ BZ a datis punctis A[,] B exeuntes et ad rectam positione datam CZ concurrentes secuerint rectas duas positione datas EX eY in X et Y: age $A[\eta]$(93) parallelam EX et $B\zeta$ parallelam eY, jungeq $B[\eta]$, $A\zeta$ secantes eY, EX in v et ξ, et spatium rectangulum $\xi X \times [eY]$ est ad apotomen AEX (ut et spatium $vY \times [EX]$ ad apotomen BeY) in ratione data[,] et spatium rectangulum $\xi X \times v[Y]$ datur.(94) Hujus autem generationis Porismatis casus sunt varij propositionibus totidem tractandi, prout lineæ EX, eY coincidunt vel non coincidunt aut punctorum A, B alterutrum infinitè distat, vel utrumq finitè.

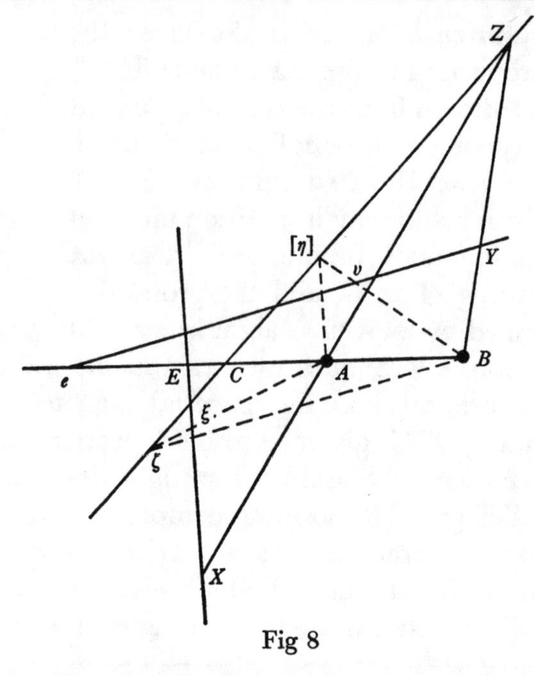

Fig 8

His adde propositiones inversas quibus affirmatur vel punctum Z tangere rectam positione datam vel lineam EY(95) convergere ad punctum datum et habebuntur fere propositiones libri primi *Porismatum*: quas quidem dicit Pappus(96) hoc generali diagrammate contineri. Si a tribus punctis A, B, C in recta aliqua datis exeuntes tres rectæ se secuerint in punctis X Y et Z[,] intersectionum verò duæ quævis ut Z et Y tangant rectas positione datas, tanget etiam tertia X rectam positione datam.(97) Idem obtinet ubi puncta A, B, C non jacent in directum, si modo rectæ positione datæ quas puncta Z et Y tangunt concurrant ad rectam $A[C]$ infinite productam.(98)

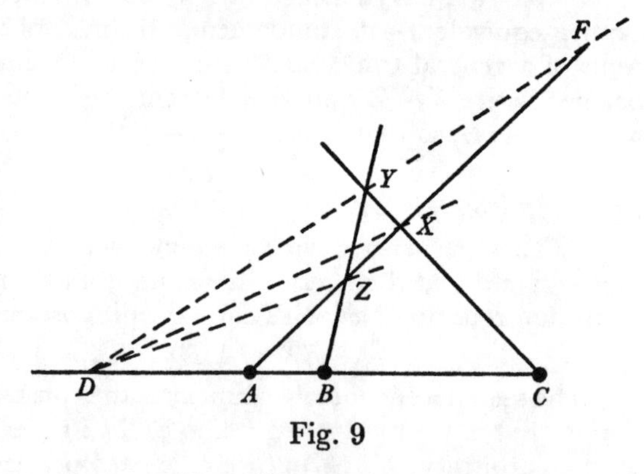

Fig. 9

Eodem diagrammate continentur hæc porismata.(99) Si a datis duobus punctis

---

readily seen to be but variations

$$(F\xi \infty_{EX} X) = (F \infty_{CZ} vZ), \quad (F \infty_{EX} X\xi) = (FvZ \infty_{CZ}) \quad \text{and} \quad (\xi \infty_{EX} XF) = (\infty_{CZ} vZF)$$

on the same basic cross-ratio equality.

(93) Here and in his accompanying manuscript figure Newton employs a variant lower-case zeta to designate the point which we, for distinction's sake, denote by η. In sequel we

Porism 11. If the straight lines AZ, BZ departing from the given points A, B and concurring at the straight line CZ given in position should intersect two straight lines EX, eY given in position in X and Y, draw $A\eta$(93) parallel to EX and $B\zeta$ parallel to eY, and join $B\eta$, $A\zeta$ cutting eY, EX in υ and ξ: the rectangle $\xi X \times eY$ is then to the space $AE \times EX$ cut off (as also the space $\upsilon Y \times EX$ to $Be \times eY$ cut off) in a given ratio, and the rectangle $\xi X \times \upsilon Y$ is given.(94) Of this generation of a porism, however, there are various cases—to be treated in as many propositions—according as the lines EX, eY coincide or do not, or whether one or other of the points A, B is at an infinite distance or each is at a finite one.

Add to these the inverse propositions whereby it is affirmed either that the point Z traces a straight line given in position or that the straight line EY(95) verges to a given point, and there will be obtained virtually all the propositions in the first book of the *Porisms*. These indeed Pappus states(96) to be contained in this general scheme: If three straight lines departing from the three points A, B, C given in some straight line should cut one another in the points X, Y and Z, while any two of the intersections such as Z and Y trace straight lines given in position, then the third one X will also trace a straight line given in position.(97) The same holds when the points A, B, C do not lie in a straight line, provided that the straight lines given in position which the points Z and Y trace concur at the straight line AC when indefinitely extended.(98)

In the same scheme are contained these porisms.(99) If two straight lines AZ,

---

correct a couple of inessential slips of his pen where, by an over-hasty glance at his diagram, he confused Z with both X and Y, and erroneously took both E and e to be C.

(94) In extension of the preceding (in tune with Fermat; compare 1, §2: note (38) above), where now twin perspectivities through a pair of poles A and B determine the coupled 1, 1 correspondences $X \leftrightarrow Z \leftrightarrow Y$ between single points in the given straight lines EX, CZ and eY there is $E \leftrightarrow C \leftrightarrow e$, $\xi \leftrightarrow \zeta \leftrightarrow \infty_{eY}$ and $\infty_{EX} \leftrightarrow \eta \leftrightarrow \upsilon$. On reducing Newton's present proportions to the equivalent forms

$$\xi X/EX = e\upsilon/eY\,(\propto AE/eY), \quad (Be/EX \propto)\; E\xi/EX = \upsilon Y/eY \quad \text{and} \quad \xi X/\xi E = \upsilon e/\upsilon Y$$

these likewise prove to be the variants

$$(\xi EX\,\infty_{EX}) = (\infty_{eY}\,eY\upsilon), \quad (\infty_{EX}\,EX\xi) = (\upsilon eY\,\infty_{eY}) \quad \text{and} \quad (\xi\,\infty_{EX}\,XE) = (\infty_{eY}\,\upsilon Ye)$$

on one and the same basic cross-ratio equality.

(95) Or of course EX or eY, whichever is appropriate.

(96) *Mathematicæ Collectiones*: 245; see §1: note (54), where we give a slightly edited quotation of Pappus' enunciation of the following Euclid–Desargues theorem.

(97) Here for the first time (compare §1: note (55) above) Newton draws the three rectilinear loci (X), (Y) and (Z) to be concurrent in a single point D, though—out of a false sense of economy which makes him set a single figure to illustrate both this and the following variant porism—their confluence is needlessly (and in a manner that is potentially misleading) now set in line with the poles A, B, C.

(98) On this variant theorem of Pappus (*Mathematical Collection* VII, 139/143) see §1: note (56) above.

(99) These are reworked (and given full synthetic proof) as Porisms 1–4 in Appendix 3.2 below.

A, B ad datam positione rectam EZ concurrentes duæ rectæ AZ BZ secuerint in X et Y rectam EXY ipsi AB parallelam & positione datam[,] datur proportio EX ad EY. Et si datur proportio illa punctum Z tangit rectam CZ positione datam. Et si recta AZ exiens a dato puncto A secuerit rectas duas positione datas EZ EY[,] fuerit autem EX ad EY in data ratione[,] recta ZY convergit ad datum punctum B. Similia obtinent ubi lineæ EX et EY non coincidunt sed sunt parallelæ ut in annexo schemate. Et hæcce sex Porismata initio libri primi posita fuisse suspicor eo quod sunt cæteris simpliciora.

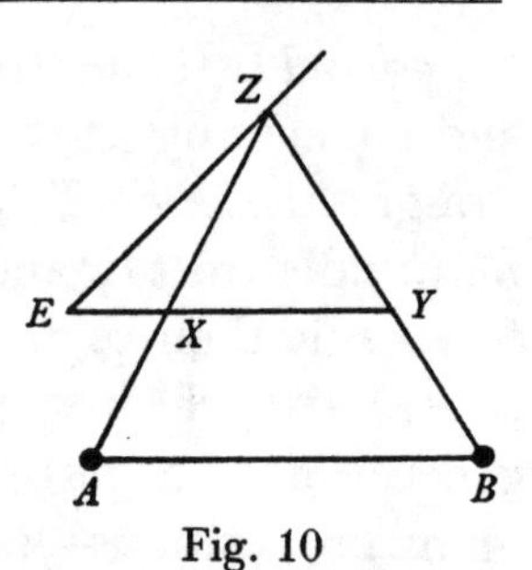

Fig. 10

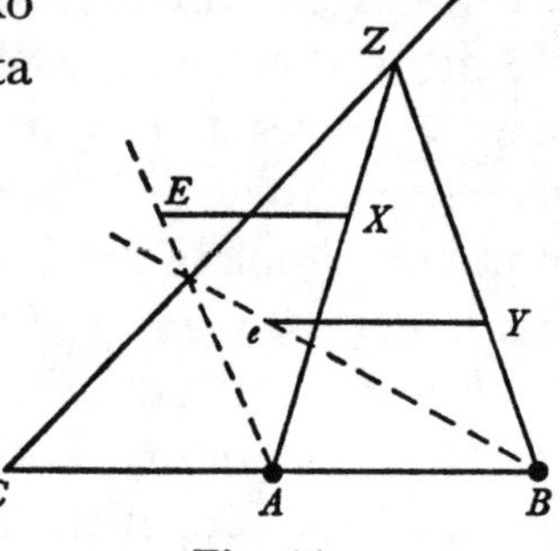

Fig. 11

Ejusdem diagrammatis casus singulos evolvendo colliguntur adhuc plura Porismata. Nam in Fig 7 contenta tria $AX \times FZ$, $AZ \times FX$ et $AF \times XZ$ sunt ad invicem in datis rationibus.(100) In figura 9 si quatuor rectæ DA, DZ, DX, DY positione datæ coeant ad idem punctum D et secentur a recta quinta in punctis A, Z, X, F contenta $AX \times FZ$, $AZ \times FX$ & $AF \times XZ$ sunt ad invicem in datis rationibus.(101) Rursus in eadem fig 9 si puncta A, B, C jaceant in directum, ratio contentorum $XA \times ZY$, $BY \times ZX$ eadem est rationi linearū AC, BC,(102) adeoq; data ratione alterutra datur altera. Porrò si lineæ AX, BY cum lineis AZ BZ ad rectam positione datam DZ concurrentibus, datos contineant angulos ZAX, ZBY et secent rectas positione datas EX EY,(103) multæ hinc emanabunt propositiones prioribus affines, adeò ut Euclidi

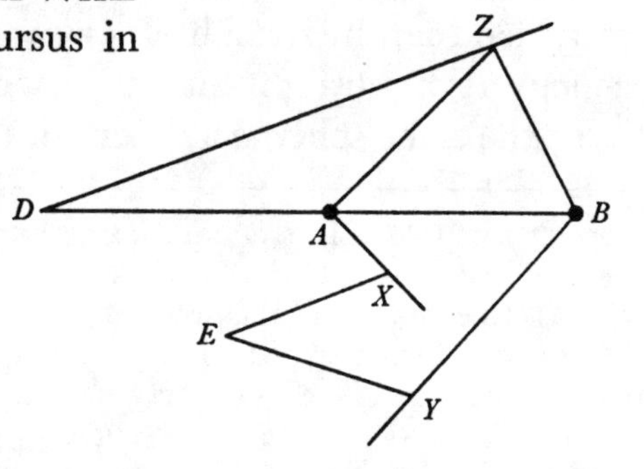

(100) An immediate extension of Porism 10 above since, in its Figure 7, there ensues by similar triangles $\xi X:FX = AX:ZX$ and $\upsilon Z:FZ = AZ:XZ$.

(101) Because for all collinear points A, F, Z and X there is

$$AZ \times FX = AX \times FZ + AF \times ZX,$$

this is essentially but a restatement (see note (90) above) of the invariance of the cross-ratio $(AFZX) = ZA \times FX / AX \times FZ$ of the intersections A, F, Z and X—or of the equivalent ratios arising when the latter are taken in a different order—cut out in any rectilinear transversal AF by a given pencil of straight lines DA, DF, DZ, DX.

(102) A rather more explicit phrasing of the theorem, due historically to Menelaus (see §1: note (62)) but doubtless here again adapted from Pappus' restatement of it in his *Mathematical Collection* VIII, 3, which made its present Newtonian bow as Porism 2′ on page 272 above.

BZ concurring from the two given points A, B at the straight line EZ given in position should cut in X and Y the straight line EXY given in position parallel to AB, the proportion of EX to EY is given. While if that ratio is given, the point Z traces the straight line CZ given in position. And if the straight line AZ departing from the given point A shall interesect two straight lines EZ, EY given in position, where now EX is to EY in a given ratio the straight line ZY converges to the given point B. The like obtains when the lines EX and EY are not coincident but parallel, as in the adjoining figure. And I suspect that these six porisms were set at the beginning of the first book, they being simpler than the rest.

By developing the individual cases of this same scheme still more porisms are gathered. For in Figure 7 the three rectangles $AX \times FZ$, $AZ \times FX$ and $AF \times XZ$ are to one another in given ratios.(100) In Figure 9, if the four straight lines DA DZ, DX, DY be concurrent at the same point D and intersected by a fifth straight line in the points A, Z, X and F, the rectangles $AX \times FZ$, $AZ \times FX$ and $AF \times XZ$ are to one another in given ratios.(101) In the same Figure 9 again, if the points A, B, C lie in a straight line, the ratio of the rectangles $AX \times ZY$ and $BY \times ZX$ is the same as that of the lines AC and BC,(102) and hence when either ratio is given so is the other. Furthermore, if the lines AX, BY should contain given angles $Z\widehat{A}X$, $Z\widehat{B}Y$ with the lines AZ, BZ concurrent at the straight line DZ given in position and should also intersect the straight lines EX, EY given in position,(103) many propositions akin to the preceding ones will issue therefrom.

As before, this relationship between the intercepts made (externally as illustrated in Figure 9) by a rectilinear transversal CXY in the sides of a general triangle ABZ would presumably be established by drawing a parallel to AZ through B to meet the transversal in V and thereby, from pairs of similar triangles, deducing that $AC/BC = AX/BV$ where $BV/BY = ZX/ZY$.

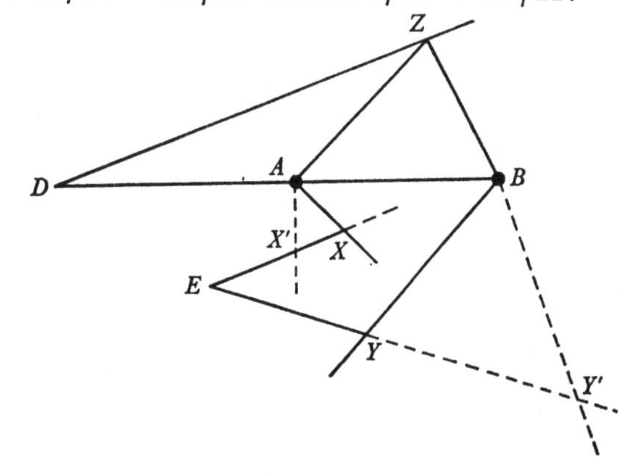

(103) As in the analogous Proposition 12 of his antecedent 'Solutio Problematis Veterum de Loco solido' (see IV: 308–10), Newton here understands that the points X and Y are mutually determinable *per simplicem geometriam*—in other words, that the segments EX, EY are inter-related by some 1, 1 correspondence of the form $a.EX \times EY + b.EX + c.EY + d = 0$. In order that the conic locus (Z) thence traced by the intersecting arms AZ, BZ of the fixed angles $X\widehat{A}Z$, $Y\widehat{B}Z$ as they rotate round the poles A and B shall, as Newton here requires, degenerate to be the pair of AB and a second straight line DZ, it is manifestly further necessary that these arms should come simultaneously to coincide with AB; that is, to the point X' of EX positioned such that $X'\widehat{A}B = X\widehat{A}Z$ there should correspond the point Y' of EY for which $Y'\widehat{B}A = Y\widehat{B}Z$.

non defuerit copia scribendi librum secundum *Porismatum* de figuris rectilineis.(104)

In Libro tertio, ait Pappus,(105) *plures erant propositiones de semicirculis, paucæ autem de circulis et circulorum portionibus: Quæsitorum autem multa similia erant antedictis*. Nam Propositiones in his libris erant aliæ atq; aliæ sed quæsita et symptomata genere affinia et ferè eadem.(106) Igitur loco rectæ quam punctum *Z* tangit substitue circumferentiam circuli et incides in Porismata libri tertij, qualia sunt hæc.(107)

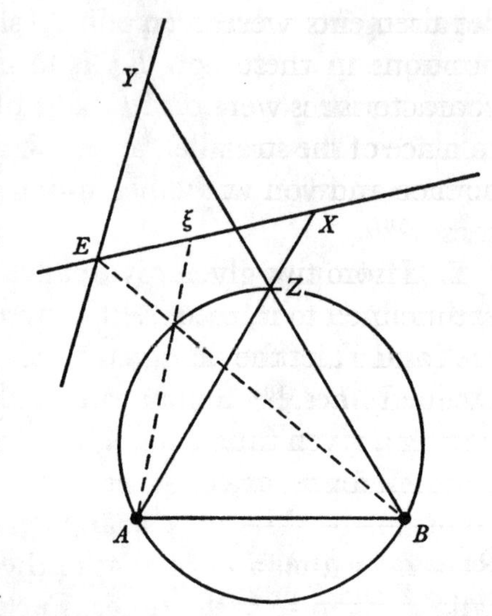

[*1*.] Si a datis duobus in circumferentia punctis *A*, *B* inflectantur ad circumferentiam eandem rectæ duæ *AZ BZ* secantes alias duas positione datas rectas *EX* et *EY* in *X* et *Y*, sit autem angulus [*XE*]*Y* æqualis angulo dato *AZB* quem inflexæ continent,(108) inflexæ illæ abscindent a positione datis partes datam rationem habentes ad invicem, nimirum *EX EY* si punctum *E* est in circumferentia[,] aut alias quasdam *ξX*, *EY* si punctum *E* extra circumferentiam positum est.(109)

(104) While (in the absence of any known evidence on the matter) it is no great strain on plausibility to conjecture that Euclid was as aware as Pappus long afterwards (see notes (101) and (102)) both of the anharmonic property of a pencil of lines and of Menelaus' theorem for a plane triangle, it would seem far-fetched to suppose that he could have had any prior awareness of the complicated mechanism of the *constructio organica* which it had cost Newton himself (see II: 108 ff.) a deal of trouble to fashion and refine, and in whose invention, even in his own day, he had no recorded precursor.

(105) *Mathematicæ Collectiones*: 246.

(106) A cancelled preliminary version of the sequel which is found in the manuscript at this point (f. 165ʳ) is, because of its length, reproduced separately in Appendix 3.2 below.

(107) It will be evident that these porismatic relationships—all which are, in Poncelet's sense, projective—hold equally true *mutatis mutandis* when the circle *AZB* is replaced by any conic, as Newton knew as well as anyone. Here he obeys the same restriction as Euclid in the third book whose 'quality' he now epitomizes, treating the properties of 'plane' loci alone without extending his scope to specify their equivalents in analogous 'solid' cases.

(108) Newton initially began to specify 'ad circu[mferentiam]' (at the circumference) but changed his mind. From this proviso it ensues that, when *Z* attains the point *Z′* at which *AZ*

As a result Euclid had no lack of ample material for writing the second book of *Porisms* on rectilinear figures.(104)

'In the third book', says Pappus,(105) 'there were a great many propositions on semicircles, with a few on circles and portions of circles. But many of the requirements were similar to those before mentioned.' To be sure, the propositions in these books were of different sorts, but their requirements and characteristics were related in their kind and almost the same.(106) Accordingly, in place of the straight line which the point Z traces substitute a circle's circumference and you will meet with the porisms of the third book, of the quality of these:(107)

1. If from two given points A, B in a circumference two straight lines AZ, BZ are inclined to it, meeting two other straight lines EX and EY given in position in X and Y, let the angle $X\widehat{E}Y$ be equal to the given angle $A\widehat{Z}B$ contained by the inclined lines,(108) and these will cut off from the ones given in position parts having a given ratio to one another—namely, EX and EY if the point E is in the circumference, or certain others ξX, EY if the point E is set outside it.(109)

---

comes to be parallel to EX, then simultaneously BZ comes to be parallel to EY, so that the points X and Y pass away to infinity at the same instant.

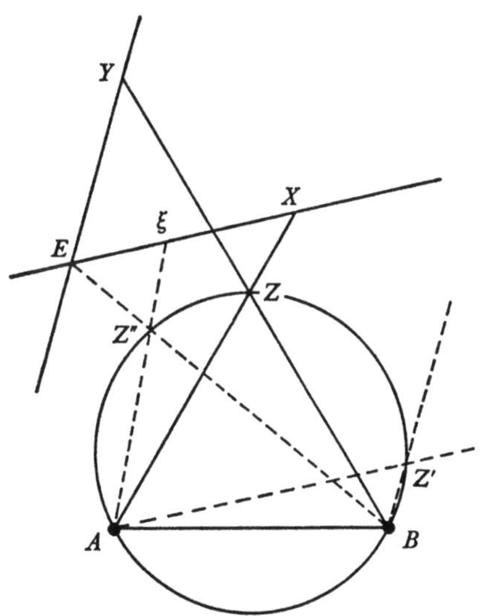

(109) The particular case where E is sited in the circle's perimeter (and AB is needlessly specified, following Euclid, to be its diameter) is that of Porism 3 in Appendix 3.2 below. In general, since Newton's configuration determines the points X and Y (each in perspective from Z through the poles A and B respectively) to be in a 1, 1 correspondence $X \leftrightarrow Y$ such that $\xi \leftrightarrow E$ and (see the previous note) $\infty_{EX} \leftrightarrow \infty_{EY}$, at once by Rule 1 of his 'Inventio Porismatum' (1, §2) above) there is given the ratio $\xi X/EY$. Equivalently, since the line-pencils $A(Z)$ and $B(Z)$ issuing from A and B are equiangular, these have the same cross-ratio; whence, if also $x \leftrightarrow y$, there is $(\xi\, \infty_{EX}\, Xx) = (E\, \infty_{EY}\, Yy)$ or $\xi X/\xi x = EY/Ey$ and therefore $\xi X/EY = \xi x/Ey$, constant, for all $X \leftrightarrow Y$.

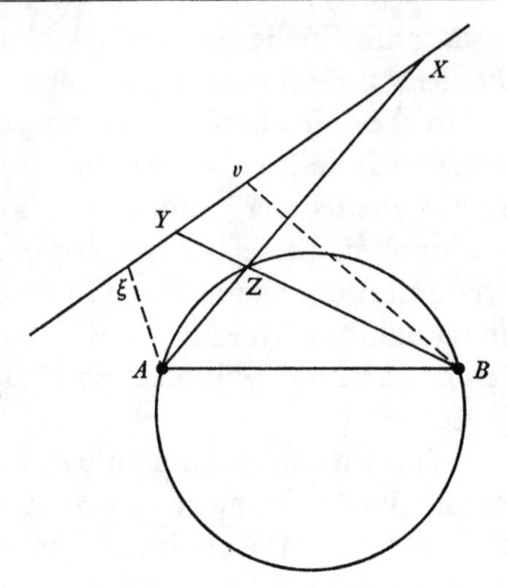

[2.] Si inflexæ AZ BZ secuerint rectam quamvis positione datam XY in X et Y[,] age $A\xi$, $B\upsilon$(110) quae constituant angulos $A\xi\upsilon$, $B\upsilon\xi$ æquales angulo dato AZB(111) et rectangulum $\xi X \times \upsilon Y$ dabitur. Simile quid fit ubi loco unius XY adhibentur duæ positione datæ rectæ.(112)

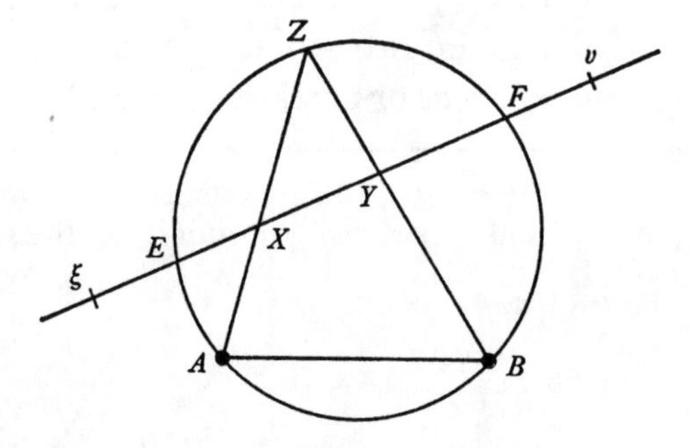

[3.] Si datam positione EF secuerint inflexæ AZ, BZ in X et Y, sunt rectangula tria $EX \times FY$, $EY \times FX$, $EF \times XY$ in datis rationibus ad invicem. Inveniri etiam possunt puncta ξ, υ in linea EF(113) sic ut detur tum ratio $\xi X \times EY$ ad EX in datam, tum rectangulum $\xi X \times \upsilon Y$.(114)

Posuit etiam Euclides(115) propositiones quasdem de circulis binis deq; rectis datum angulum ad datum punctum continentibus, ut his expressit Pappus:(116) *Quod est aliquod datum punctum a quo junctæ ad datum*(117) *continent specie triangulum.*

---

(110) By a careless slip Newton has here and in his next clause erroneously interchanged ξ and υ.

(111) So determining that, corresponding to ξ and υ, the related lines BZ and AZ (when, in our figure, Z is at Z' and Z'' respectively) shall be parallel to XY.

(112) In either case, since the configuration determines the points X and Y to be in a 1, 1 correspondence $X \leftrightarrow Y$ such that (see the previous note) $\xi \leftrightarrow \infty_{\upsilon Y}$ and $\infty_{\xi X} \leftrightarrow \upsilon$, it ensues immediately by Rule 2 of Newton's preceding 'Inventio Porismatum' (1, §2) that the product $\xi X \times \upsilon Y$ is given. From first principles, alternatively, because (as before) the line-pencils $A(Z)$ and $B(Z)$ issuing from A and B are equiangular and hence have the same cross-ratio, on setting also $x \leftrightarrow y$ there results $(\xi\, \infty_{\xi X}\, Xx) = (\infty_{\upsilon Y}\, \upsilon Yy)$ or $\xi X/\xi x = \upsilon y/\upsilon Y$ and in consequence

$$\xi X \times \upsilon Y = \xi x \times \upsilon y,$$

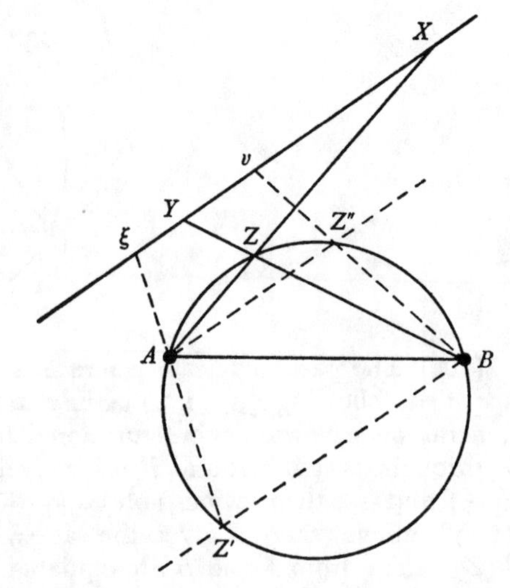

constant. Porism 4 in Appendix 3.2 is the Euclidean semicircular instance in which AB is the diameter of the arc $\widehat{AB}$.

2. Should the inclined lines AZ, BZ intersect any straight line XY given in position in X and Y, draw $A\xi$, $B\upsilon$(110) so as to form angles $\widehat{A\xi\upsilon}$, $\widehat{B\upsilon\xi}$ equal to the given $\widehat{AZB}$(111) and the rectangle $\xi X \times \upsilon Y$ will be given. Something of the like occurs when in place of the single XY two straight lines given in position are employed.(112)

3. Should the inclined lines AZ, BZ intersect EF given in position in X and Y, the three rectangles $EX \times FY$, $EY \times FX$ and $EF \times XY$ are in given ratios to one another. Points ξ and υ can also be found in the line EF(113) such that there is given both the ratio of $\xi X \times EY$ to EX times a given, and also the rectangle $\xi X \times \upsilon Y$.(114)

Euclid also set(115) certain propositions on pairs of circles and on straight lines containing a given angle at a given point, premissing—as Pappus expressed it(116)—'that there is some given point such that when lines are joined from it (each) to a given(117) they contain a triangle (given) in species; that there is

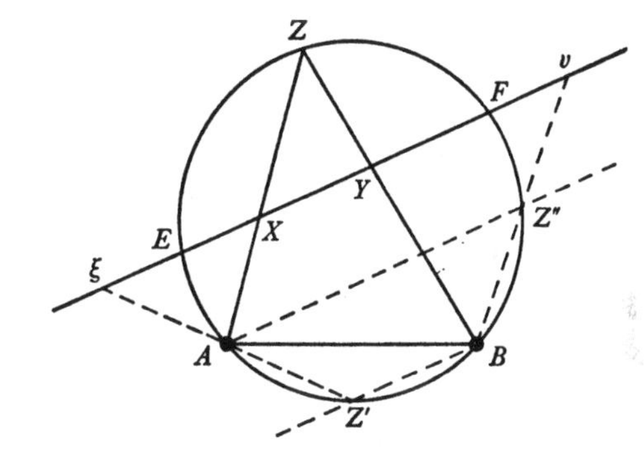

(113) Corresponding to positions of BZ and AZ (when, in our amplified figure, Z is at Z' and Z'' respectively) at which these are parallel to EF.

(114) In this extended Fermatian porism (on whose contemporary antecedents see IV: 310–11, note (60)) the configuration evidently, to use Desargues' parlance, determines a 1, 1 involutory correspondence $X \leftrightarrow Y$ in which E and F are double points. In proof of Newton's stated relationships, firstly, since the line-pencils ZE, ZF, ZA, ZB remain, as Z traverses the circle, equi-angular and hence equi-cross, at once the cross-ratio

$$(EFXY) = EX \times FY / EY \times FX = 1 + EF \times XY / EY \times FX$$

is given in value. Again, because the correspondence $X \leftrightarrow Y$ is also defined by the meet with EF of the equiangular line-pencils $A(Z)$ and $B(Z)$ issuing from the poles A and B, and since Newton's points ξ and υ are to be sited (see the previous note) such that $\xi \leftrightarrow \infty_{EF}$ and $\infty_{EF} \leftrightarrow \upsilon$, there further ensues $(\xi EXF) = (\infty_{EF}\, EYF)$ or $\xi X \times EF / \xi F \times EX = EF/EY$, whence

$$\xi X \times EY / EX = \xi F,$$

given; and similarly $(\xi\, \infty_{EF}\, XE) = (\infty_{EF}\, \upsilon YE)$ or $\xi X / \xi E = \upsilon E / \upsilon Y$, and consequently

$$\xi X \times \upsilon Y = \xi E \times \upsilon E = \xi F \times \upsilon F,$$

given.

(115) Continue to understand 'in Libro tertio *Porismatum*' (in Book 3 of his *Porisms*).

(116) *Mathematicæ Collectiones*: 247.

(117) Namely a 'circulum' (circle) rather than the 'puncta quævis' which Edmond Halley—and after him Chasles—would have it be in his *Apollonii Pergæi De Sectione Rationis Libri Duo*... (Oxford, 1706): xxxvi. The corrected emendation was first made at these points by Robert Simson in his posthumously printed 'De Porismatibus Tractatus' [= *Opera Quædam Reliqua* (Glasgow, 1776): 315–594]: 455, 463.

Quod est aliquod datum punctum a quo junctæ rectæ lineæ ad hoc(117) *æquales assumunt circumferentias. Quod quæ in parathesi*(118) *erit vel una cum aliqua recta linea ad datum punctum vergente datum continet angulum.* Quorum sensus si de circulorum non tantum æqualium circumferentijs æqualibus sed etiam de inæqualiū proportionalibus intelligantur, ita se habebit.

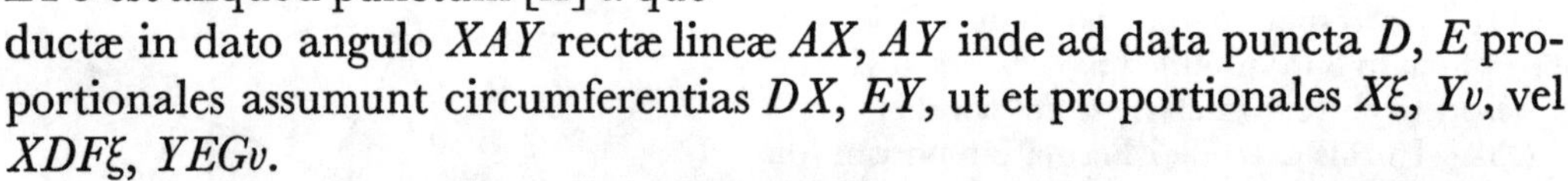

[*1.*] In circulis duobus datis *DXF*, *EYG* a punctis datis *D*, *E* captis circumferentijs *DXEY* quæ circulorum diametris proportionales sint, seu angulos æquales *DPX*, *EQY* subtendant ad centra: est aliquod datum punctum *A*(119) a quo junctæ *AX AY* datum contineant specie triangulum *AXY*.

[*2.*] Datis duobus circulis *DXF EYG* est aliquod punctum [*A*] a quo ductæ in dato angulo *XAY* rectæ lineæ *AX*, *AY* inde ad data puncta *D*, *E* proportionales assumunt circumferentias *DX*, *EY*, ut et proportionales *Xξ*, *Yυ*, vel *XDFξ*, *YEGυ*.

[*3.*] Si ductæ illæ *AX*, *AY* proportionales assumunt circumferentias[,] recta *XY* constructione ulteriore adjecta, una cum alterutra linea *XA*, *YA* ad datum punctū vergente datum continet angulum *AXY* vel *AYX*.

Horum reciprocum(120) est quod si proportio *AX* ad *AY* et angulus *XAY* vel proportio *AX* ad *XY* et angulus *AXY* detur, punctorum autem *X* vel *Y* alterutrum tangit circulum datum, tanget et alterum circulum datum. Quinetiam quia rectangulum *XAξ*(121) datur et *XA* datam habet rationem ad *AY*[,] dabitur rectangulum *YAξ*: Et vicissim si detur rectangulum illud *YAξ*[,] punctorum autem *ξ*, *Y* alterutrum tangat circulum datum[,] tanget et alterum circulum datum præterquam ubi punctum *A* reperitur in circuli alterutrius circumferentia: quo casu loco circuli alterius habebitur linea recta.(122)

Cæterum datis circulis et punctis *D*, *E* inventio puncti *A* pendet ab hoc Porismate.(123) Si fuerint *AC DC BC* continue proportionales et ad circum-

(118) That is, parallel to some other straight line given in position.

(119) Most simply constructed as the meet of two circles drawn respectively through *D*, *E* and *X*, *Y* to subtend at *A* an angle equal to the inclination of *DP* to *EQ*; for, when the circle of centre *Q* is rotated anti-clockwise round *A* through this angle, the pairs *D*, *E*; *P*, *Q*; *X*, *Y* come manifestly to be in homothety from the pole *A*. Newton himself in sequel (see note (123) below) adduces a somewhat more complicated equivalent construction by two Apollonius circles, determining *A* as the meet of the loci whose distances from *D* and *E* and also from *X* and *Y* are in the constant ratio $AD:AE = AX:AY = DP:EQ$. The two following porisms are but simple variations upon this basic proposition.

some given point such that when straight lines are joined from it to this[117] they take to themselves equal circumferences; that a line in juxtaposition[118] together with some straight line verging to a given point contains a given angle'. The sense of these words, if they be understood not merely of the equal circumferences of equal circles but also of the proportional perimeters of unequal ones, will be thus exhibited:

1. When in two given circles DXF, EYG circumferences $\widehat{DX}$, $\widehat{EY}$ are taken from given points D, E such as to be proportional to the diameters of their circles or to subtend equal angles $D\hat{P}X$, $E\hat{Q}Y$ at the centres, then there is some given point A[119] such that the joined lines AX, AY contain a triangle AXY given in species.

2. Given the two circles DXF, EYG there is some point A such that when the straight lines AX, AY are drawn from it at the given angle $X\hat{A}Y$ they thence take away, terminated at the given points D and E, proportional circumferences $\widehat{DX}$, $\widehat{EY}$, as also proportional ones $\widehat{X\xi}$, $\widehat{Yv}$ or $\widehat{XDF\xi}$, $\widehat{YEGv}$.

3. If those drawn lines AX, AY take away proportional circumferences, the straight line XY, on adjoining the latter construction, contains together with one or other of the lines XA, YA verging to the given point the given angle $A\hat{X}Y$ or $A\hat{Y}X$.

The reverse of these[120] is that if the ratio of AX to AY and the angle $X\hat{A}Y$ or the ratio of AX to AY and the angle $X\hat{A}Y$ be given, while one of the points X or Y traces a given circle, then the second also will trace a given circle. To be sure, because the rectangle $AX \times A\xi$[121] is given and AX has a given ratio to AY, the rectangle $AY \times A\xi$ will be given; and conversely, if that rectangle $AY \times A\xi$ be given, while one of the points ξ, Y shall trace a given circle, then the second also will trace a given circle except when the point A is found in the circumference of either circle, in which case instead of the second circle a straight line will be had.[122]

With the two circles and the points D, E given, however, the discovery of the point A depends on the following porism.[123] If AC, DC, BC be in continued

(120) Originally 'His affine' (Related to these).

(121) Equal to $AP^2 - DP^2$, that is.

(122) In other words, the inverse circle whose centre lies at infinity in the perpendicular at A.

(123) Yet one more Newtonian statement of the circular Apollonian locus (Z) defining or defined by the related condition that the ratio AZ/BZ of its distances from two fixed points A, B (mutually inverse with respect to the circle) be constant; compare Porism 7 of the preceding 'Inventio Porismatum' (1, §2) and Porism 1 in Appendix 3.2 below. While Newton's suggested construction of A by way of the given ratios $AD:AE = AX:AY (= DP:EQ)$ in this manner is not inelegant, we have in note (119) above outlined an equivalent Euclidean construction *per circulos* which assumes only *Elements* III, 21.

ferentiam centro C intervallo CD descriptam inflectantur a datis punctis A, B rectæ duæ AZ BZ[,] datur inflexarum AZ, BZ ratio ad invicem: Et vicissim si datur ratio illa[,] punctum Z tangit circumferentiam datam. Hoc Porisma ex eorum numero fuisse quæ Euclides posuit ad semicirculos verisimile est.(124) Nonnullis autem casibus punctum A(125) determinatur absq; hoc porismate. Ut si duo circuli se secuerint in A, B et recta BZ secentur in X et Y[,] junge AX AY et triangulum(126) dabitur specie. Unde et si fiat YZ ad XY in data ratione[,] punctum Z tanget circulum tertium per puncta A, B transeuntem eo quod triangulum AYZ datur specie: Et vicissim si tres circuli se secuerint in ijsdem punctis A, B[,] a recta BZ secentur in X, Y et Z [:] datur proportio XY ad YZ et figura rectilinea $XAYZ$ datur specie.(127)

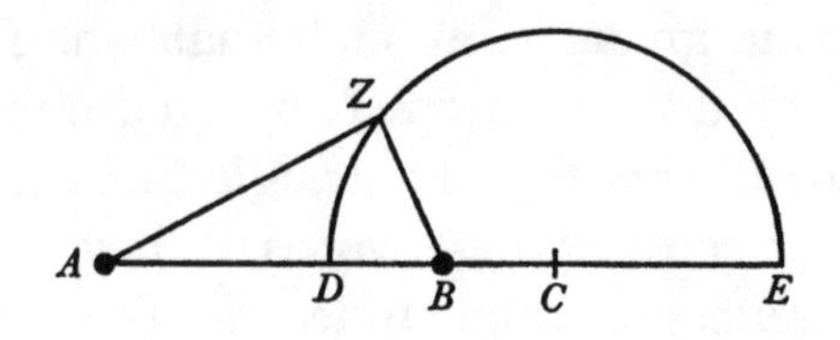

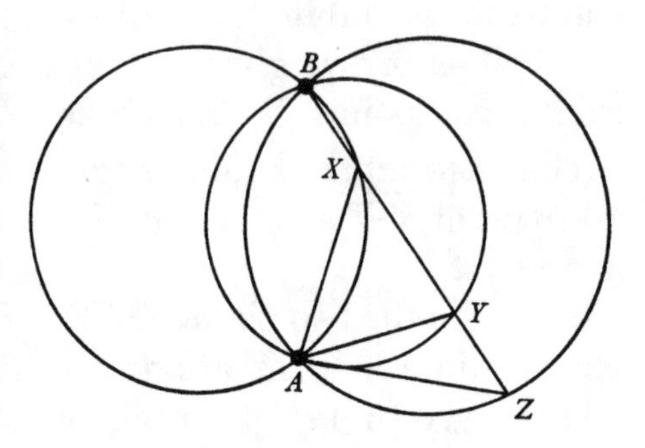

Quæ omnia fusiùs enarravi, pártim ut restituendis *Porismatum* libris facem accenderem, partim ut Loci resoluti vim quæ in his *Porismatibus* maximè constitit pleniùs exponerem in sequentibus[,] Pappo appellante hos libros E[u]clideos *opus artificiosissimum ac perutile ad resolutionem obsc[u]riorum Problematum*, habens in se *contemplationem subtilem, necessariam, et admodū universalem, et ijs qui hæc valent perspicere atq; investigare suavem.*(128)

(129)Post Euclidis *Porismata* considerandi veniunt libri duo Apollonij *de locis planis. Locorum* verò *omnium*(130) *alij sunt ἐφεκτικòι hoc est in seipsis tantum consistentes*, quo sensu *puncti locus est punctū, lineæ linea, superficiei superficies & solidi solidum: alij autem διεξοδικòι hoc est sese extra tendentes*, et sic *puncti locus linea* (*est*), *lineæ superficies, superficiei solidum. Locorum autem qui in resoluto loco* (*tractantur*) *alij quidem positione dati ἐφεκτικòι sunt* (*punctorum*), *alij autem plani dicti et solidi et lineares sunt punctorum διεξοδικòι, & alij ad superficies ἀνασ[τ]ροφικòι quidem* (*sunt*) *punctorum, διεξοδικòι autem linearum. Lineares loci ex ijs qui sunt ad superficies demonstrantur.*

(124) While this conjecture is eminently plausible—the Apollonian locus is employed (and given vigorous demonstration) yet earlier in Aristotle, *Meteorologica* III, v, 376a (see IV: 315–16, note (76))—it still today remains wholly undocumented.

(125) Newton has deleted the dubious qualification 'facilius' (more easily).

(126) Namely AXY.

(127) This trivially rephrases Porism 3′ on page 274 above, where (note (64)) its Newtonian antecedents are specified.

(128) *Mathematicæ Collectiones*: 244/245.

(129) Newton initially went on in the manuscript: 'Cum his *Porismatibus* conjungenda sunt *loca plana* Apollonij. Nam *Porismatum*, ait Pappus, *loci ipsi sunt una species de qua tractatur in loco*

proportion and two straight lines AZ, BZ are inclined from the given points A, B to the circumference drawn with centre C and radius CD, then the ratio of the inclined lines to one another is given; conversely, if that ratio is given, the point Z traces a given circumference. This porism was likely one of those which Euclid set on semicircles.(124) But in some cases the point A is (125) determined without appealing to this porism. Should, for instance, the two circles intersect in A and B and be cut by a straight line BZ in X and Y, join AX and AY and their triangle(126) will be given in species; whence, if also YZ be made to be to XY in a given ratio, the point Z shall trace a third circle passing through the points A and B, seeing that the triangle AYZ is given in species. And conversely, should three circles intersect in the same points A and B, let them be cut by a straight line BZ in X, Y and Z, and there is then given the ratio of XY to YZ, while the rectilinear figure $XAYZ$ is given in species.(127)

All this I have rather elaborately recounted partly to light a torch for restoring the books of *Porisms*, and partly that I might more fully display in sequel the power of the resolved locus which found its strongest footing in them, Pappus calling these Euclidean books 'a most skilful work, one extremely useful in the resolution of the obscurer problems' and comprehending 'a subtle, indispensable and entirely universal survey which for those so capable is pleasant to peruse and explore'.(128)

(129)After Euclid's *Porisms* Apollonius' two books *On plane loci* come into consideration. Now(130) 'of the totality of loci some are ἐφεκτικοί, that is, consisting merely in themselves' in the sense in which 'a point is the locus of a point, a line that of a line, a surface that of a surface, and a solid that of a solid; others are διεξοδικοί that is, tending outside themselves', and in this way 'a line is the locus of a point, a surface one of a line, a solid one of a surface. Of the loci, however, which are treated in the resolved locus some indeed, given in position, are ἐφεκτικοί of points, while others called plane, solid and linear are διεξοδικοί thereof, and others relating to surfaces are ἀναστροφικοί of points, indeed, but διεξοδικοί of lines. Linear loci are demonstrated from those which relate to

resoluto. Seorsim verò ab alijs Porismatum collecta inscriptaq; ac tradita est hæc species quod magis diffusa et copiosa sit cæteris speciebus. Propositiones autem Apollonij de his locis omninò formantur ad modum Porismatum, de Datis passim agentes' (With these *Porisms* are to be conjoined Apollonius' *Plane loci*. For, says Pappus, 'Loci are themselves one species of porism, and discussion of it is given in the resolved locus. This, however, is separately collected, written down and delivered because it is more diffuse and abundant than the remaining species'. Apollonius' propositions on these loci are, we may add, fashioned entirely in the style of porisms, having everywhere to do with givens.) The quotation (from *Mathematicæ Collectiones*: 245) is, along with the gist of his surrounding remarks, re-introduced by Newton in his next paragraph.

(130) A parenthetic 'ait Pappus' (says Pappus), subsequently rendered superfluous by a following reminder (see the next note) that the present citation of Commandino's text is not quite word for word, is here deleted in the manuscript.

Nimirum ut puncta investigamus per intersectiones linearum[,] sic naturalis est inventio linearum per intersectionem superficierum. *Dicuntur autem plani loci quicunq; sunt rectæ lineæ vel circuli: solidi loci quicunq; sunt conorum sectiones, Parabolæ vel Ellipses vel Hyperbolæ. Lineares loci quicunq; lineæ sunt neq; rectæ neq; circuli neq; aliqua dictarum coni sectionum.* Ita fere Pappus(131) ex mente veterum.

Locorum ad punctum exempla habemus in porismatibus rectis ad datum punctum convergentibus, ut et in omni propositione qua punctum dari affirmatur. Lo[co]rum planorum exempla quædam posuit Euclides in *Porismatibus* ubi affirmat puncta datas rectas vel datas circumferentias tangere. *Sunt* enim *loci ipsi* (affirmante Pappo(132)) *una species Porismatum*, at maxima ex parte *seorsim collecta inscriptaq; ac tradita, quod magis diffusa et copiosa sit cæteris speciebus.* Apollonij igitur *de locis planis* libri duo omninò poristici sunt, non obstante alio nomine inscribantur. Id docet vel ipsa propositionum forma, de Datis passim agentium.

Primo autem libro agit Apollonius de locis rectilineis, ea supplens quæ Euclides omiserat. Euclides enim in primis duobus *Porismatum* libris tradiderat loca rectilinea quæ rectæ a datis punctis exeuntes terminis suis contingunt: hic tradit Apollonius loca rectilinea ad quas rectæ inclinatione datæ seu datis parallelæ terminantur. Propositiones ad capita generalia a Pappo(133) reductæ ita fere sonant.

[*1.*] Si rectæ lineæ magnitudine datæ *DZ* et cuipiam positione datæ *M* æquidistantis unus terminus *D* contingat rectam lineam positione datam *AD*[,] et alius terminus *Z* rectam lineam positione datam(134) continget.

[*2.*] Si a puncto quodam *Z* ad positione datas duas rectas lineas vel parallelas vel inter se convenientes *AD AE* ducantur rectæ lineæ *ZD ZE* in datis angulis *ADZ AEZ*, vel datam habentes proportionem *ZD* ad *ZE*, vel quarum una *ZD* simul cum ea ad quam altera *ZE* proportionem habet datam, data fuerit; continget punctum *Z* rectam lineam positione datam. Idem fit ubi datur differentia inter unam *ZD* et eam ad quam altera *ZE* proportionem habet datam.(135)

[*3.*] Si sunt quotcunq; rectæ lineæ positione datæ *AD*, *AE*, *BF*, *BG* atq; ad ipsas a quodam puncto *Z* agantur rectæ totidem lineæ *ZD*, *ZE*, *ZF*, *ZG* in datis

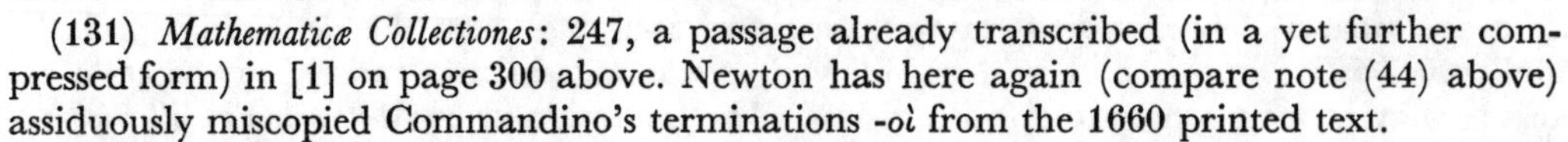

(131) *Mathematicæ Collectiones*: 247, a passage already transcribed (in a yet further compressed form) in [1] on page 300 above. Newton has here again (compare note (44) above) assiduously miscopied Commandino's terminations *-oì* from the 1660 printed text.

(132) *Mathematicæ Collectiones*: 245. Newton's earlier version of the present paragraph is reproduced in note (129) above.

(133) *Mathematicæ Collectiones*: 248. In sequel Newton fills out Pappus' verbal enunciations of the *proposita generalia* in Apollonius' first book by interpolating references to an accompanying (restored) figure where their import is visually evident.

surfaces'—and of course, just as we ascertain points by means of the intersections of lines, the discovery of lines by the intersection of surfaces is equally natural. 'Loci, however, are called plane whenever they are straight lines or circles; solid loci are any (plane) sections of cones—parabolas, ellipses or hyperbolas—whatever; linear loci are any lines at all which are neither straight lines or circles, nor any of the said conic sections'. So, virtually, states Pappus(131) in the spirit of the ancients.

Of loci to a point we have examples in porisms where straight lines converge on a given point, as also in every proposition whereby it is asserted that a point is given. Of plane loci Euclid set certain instances in his *Porisms* where he asserts that points trace given straight lines or given circumferences. To be sure, affirms Pappus,(132) 'loci are themselves one species of porism', though for the most part this 'is separately gathered, written down and recounted since it is more diffuse and abundant than the remaining species'. Accordingly, Apollonius' two books *On plane loci* are entirely porismatic, notwithstanding that they are listed under a different title. Even the very form of their propositions tells us that, universally concerned as they are with givens.

In his first book specifically, Apollonius concerns himself with rectilinear loci, supplying what Euclid had omitted. Where Euclid in the first two books of his *Porisms* had, as we have seen, given account of rectilinear loci traced by the end-points of straight lines departing from given points, Apollonius here communicates rectilinear loci at which straight lines given in inclination—that is, parallel to given ones—terminate. Its propositions, as reduced to general heads by Pappus,(133) have effectively the following meaning:

1. If one end-point D of a straight line DZ given in size and equidistant from any other, M, should trace the straight line AD given in position, then its other end-point Z will also trace a straight line given in position.(134)

2. If from a point Z to two straight lines AD, AE given in position, either parallel or meeting each other, there be drawn straight lines ZD, ZE at given angles $A\widehat{D}Z$, $A\widehat{E}Z$ and having either the ratio of ZD to ZE given or the sum of one, ZD, and a given proportionate multiple of the other, ZE, given, then the point Z will trace a straight line given in position. The same happens when the difference between one, ZD, and a given proportionate multiple of the other, ZE, is given.(135)

3. If there are any number of straight lines AD, AE, BF, BG given in position, and to these from a point Z there be drawn an equal number of straight lines

(134) Parallel to AD at the given oblique distance DZ, of course.

(135) For if the given defining relationship be $a.ZD \pm b.ZE = c$, the locus of Z is manifestly a straight line meeting AD and AE in points (constructible by the preceding porism) such that $ZE = \pm c/b$ and $ZD = c/a$ respectively.

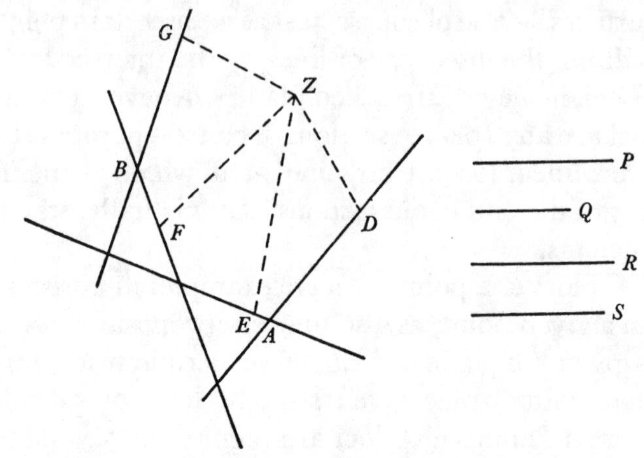

angulis D, E, F, G, et actæ singulæ seorsim ducantur in datas totidem P, Q, R, S, ZD in P, ZE in Q, ZF in R, ZG in S; sit autem summa aliquorum contentorum ZD in P et ZE in Q æqualis summæ reliquorum ZF in R et ZG in S, punctū Z rectam lineam positione datam continget. Idem fit ubi summa aliquorum contentorum æqualis est summæ reliquorum una cum spatio dato.(136)

[4.] Si ab aliquo puncto Z ad positione datas parallelas AD, BE ducantur rectæ lineæ ZD ZE in datis angulis D, E quæ ad puncta in ipsis data A, B abscindant rectas AD, BE datam proportionem habentes, quæve spatium contineant datum $ZD \times ZE$[,] vel ita ut specierum ab ipsis ductis summa $ZD^{\text{quad.}}$ & $ZE^{\text{quad.}}$ vel excessus specierum $ZD^q - ZE^q$ æqualis sit dato spatio, punctum Z continget positione datam rectam lineam.(137) Idem intellige de pluribus parallelis deq; alia quavis relatione ductarum ad invicem qua longitudines earundem ductarum determinantur.(138)

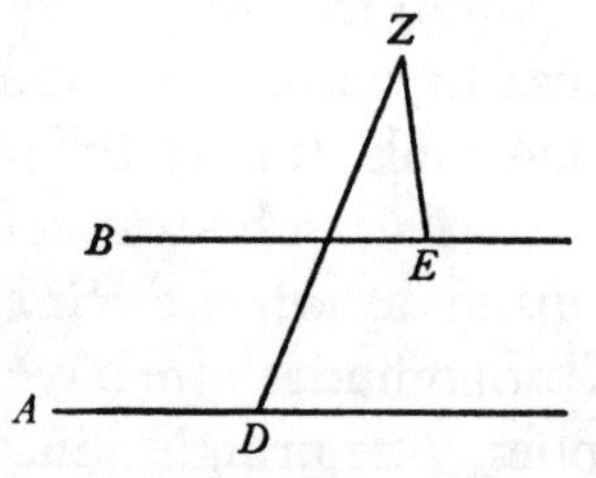

Talia igitur exposuit ac demonstravit Apollonius in libro primo, de lineis solis inclinatione datis scribendo. In libro secundo de lineis agit quæ a datis punctis exeunt. Et quoniam Euclides hujus generis diagrammata(139) purè rectilinea duobus *porismatum* libris copiosè satis evolverat, parcius autem scripserat de circularibus in tertio, Apollonius ferè totus est in circularibus enucleandis. Propositiones autem ita se habent.(140)

[1.] Si a datis punctis A, B rectæ lineæ inflectantur(141) & sunt quæ ab ipsis fiunt AZ^q & BZ^q dato spatio N^q differentia (id est si fuerit $AZ^q - BZ^q$ æquale N^q)[,] punctum Z positione datam rectam continget. Sin AZ & BZ sint in proportione data

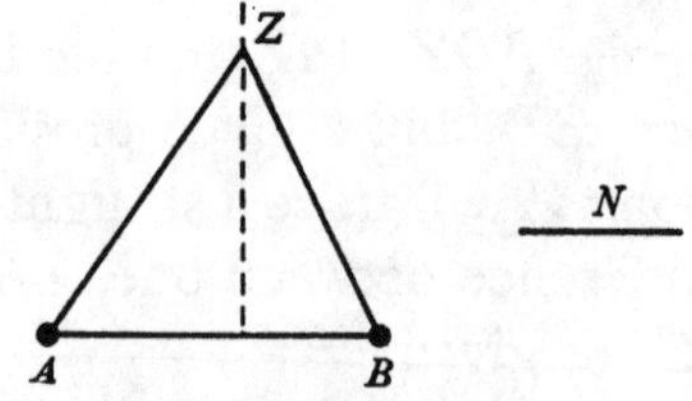

(136) In this further generalization, as Descartes definitively established in the first book of his *Geometrie* in 1637 (see 1, §1: note (19) above, and compare D. T. Whiteside, 'Patterns of Mathematical Thought in the later Seventeenth Century', *Archive for History of Exact Sciences* **1**, 1961: 179–388, especially 291), since in any system of Cartesian coordinates in which Z is the point (x, y) each of its oblique distances ZD, ZE, ZF, ZG from the given lines has the form $a_i x + b_i y + c_i = 0$, the locus will have the linear defining equation $\sum_i k_i(a_i x + b_i y + c_i) = 0$, where $P = k_1$, $Q = k_2$, $\pm R = k_3$, $\pm S = k_4$.

ZD, ZE, ZF, ZG at given angles $\hat{D}$, $\hat{E}$, $\hat{F}$, $\hat{G}$, let each drawn line be multiplied individually by (one of) the corresponding number of givens P, Q, R, S: should the sum of certain of the rectangles $ZD \times P$ and $ZE \times Q$ now be equal to the sum of the rest $ZF \times R$ and $ZG \times S$, then the point Z will trace a straight line given in position. The same happens when the sum of some of the rectangles is equal to the sum of the rest together with a given space.(136)

4. If from some point Z to parallels AD, BE given in position there be drawn at given angles $\hat{D}$, $\hat{E}$ straight lines which shall cut off, terminating at points A and B given in these, straight lines AD, BE having a given ratio, or to contain a given space $ZD \times ZE$, or so that the sum $ZD^2 + ZE^2$ or excess $ZD^2 - ZE^2$ of their square powers be equal to a given space, then the point Z will trace a straight line given in position.(137) Understand the same of several parallels and for any other relationship of the drawn lines to one another whereby their lengths are determined.(138)

Such, then, were the things expounded and demonstrated by Apollonius in his first book, writing on lines given in inclination alone. In the second book he dwells on lines which depart from given points. And because Euclid in two books of his *Porisms* had sufficiently copiously developed purely rectilinear schemes(139) of this kind, but had written more sparingly on circular ones in the third, Apollonius is almost wholly concerned to give a detailed account of circular ones. The propositions are, however, of this nature:(140)

1. If from the given points A, B straight lines are inclined,(141) and their powers AZ^2 and BZ^2 differ by a given space N^2 (that is, if $AZ^2 - BZ^2$ be equal to N^2), then the point Z will trace a straight line given in position. But if AZ and BZ be in a given ratio, the point Z will trace either a straight line or a circle—namely,

(137) If, for simplicity, we choose $AD = x$ and $DZ = y$ to be the abscissa and ordinate of the point Z, then the lengths of BE and ZE will be respectively of the form $x + ay + b$ and $m(y-c)$, where c is the oblique distance (measured along ZE) between AD and BE. Whence, first, if $AD/BE = k$, the defining equation of Z is at once $x + (a-km)y + b + kmc = 0$. Alternatively, if there is either $ZD \times ZE = k$ or $ZD^2 \pm ZE^2 = l$, the locus of Z is the pair of parallels $y = Y_1$, $y = Y_2$ where Y_1 and Y_2 are the roots of the ensuing quadratics $y(y-c) = k/m$ or $y^2 \pm m^2(y-c)^2 = l$ respectively.

(138) To continue the previous note, if there are any number of parallels AD, BE, ... defined by the respective Cartesian equations $y = 0$, $y = c$, ..., the corresponding oblique distances of Z from these will be $ZD = y$, $ZE = m(y-c)$, ...; if, therefore, the relationship defining Z be of any general (finite algebraic) form $f(ZD, ZE, ...) = k$, its locus will be a multiple of parallels $y = Y_1, Y_2, ...$ where $Y_1, Y_2, ...$ are the roots of the equation $f(y, m(y-c), ...) - k = 0$.

(139) Newton originally wrote 'loca' (loci).

(140) Much as before in his preceding amplification of Pappus' verbal précis of the first book *De locis planis* (compare note (133) above) Newton fills out slightly paraphrased repeats of the *capita generalia* of Apollonius' second book (*Mathematicæ Collectiones*: 248–9) with references to accompanying figures of his own which exemplify and explain the meaning of Pappus' words.

(141) Understand 'ad punctum Z' (to the point Z).

punctum Z continget vel rectam lineam vel circumferentiam: nempe rectam lineam ubi AZ et BZ æquales sunt, alijs in casibus circumferentiam.(142)

[2.] Si sit positione data recta linea EF et in ipsa datum punctum A a quo ducatur quædam linea terminata AZ, a termino autem ipsius Z ducatur etiam ZV (in dato angulo AVZ)(143) ad positione datam EF et sit quod fit a ducta $AZ^{\text{quad.}}$ æquale ei quod a data M et abscissa vel ad datum punctum A vel ad alterum punctum datum E vel ad alterum datum F in linea data positione, hoc est ipsi $M \times AV$ vel $M \times EV$ vel $M \times FV$, terminus Z ipsius AZ circumferentiam positione datam continget.(144)

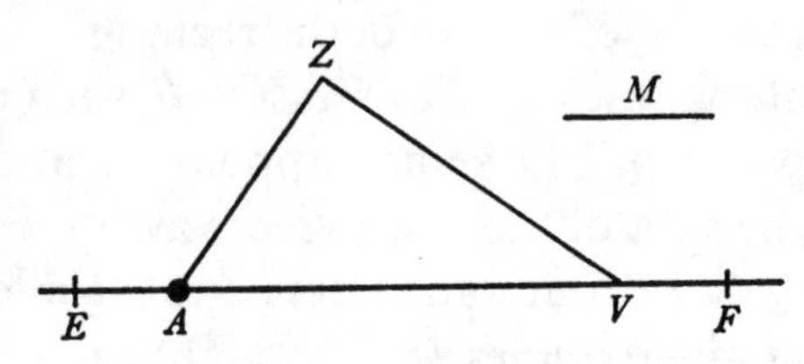

[3.] Si a duobus punctis datis A, B rectæ lineæ AZ, BZ inflectantur & sit quod ab una efficitur $AZ^{\text{quad.}}$ eo quod ab altera $BZ^{\text{quad.}}$ dato majus quam in ratione,(145) punctum Z positione datam circumferentiam continget.(146)

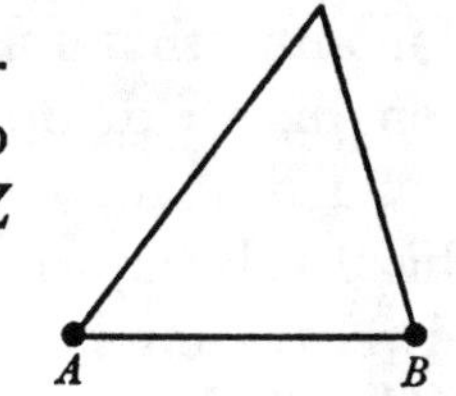

[4.] Si a quotcunq; datis punctis A, B, C ad punctum unum Z inflectantur rectæ AZ, BZ, CZ et sint species quæ ab omnibus fiunt $AZ^q + BZ^q + CZ^q$ dato æquales[,] punctum Z continget positione datam circumferentiam.(147)

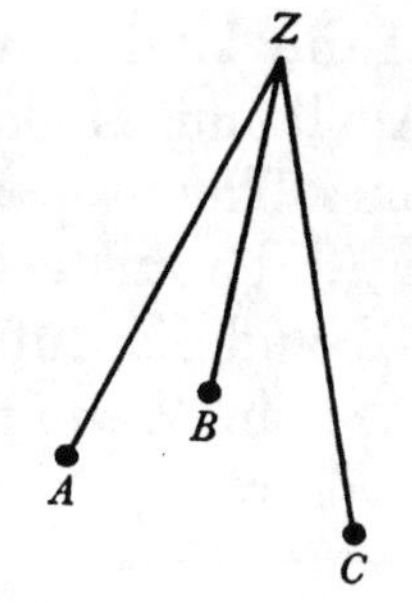

[5.] Si a duobus punctis datis A, B inflectantur rectæ lineæ AZ BZ, a puncto autem Z ad positione datam lineam EF agatur ZV in dato angulo V abscindens EV a recta linea positione data ad datum punctum E et sint species ab inflexis $AZ^q + BZ^q$ æquales ei quod a data N et

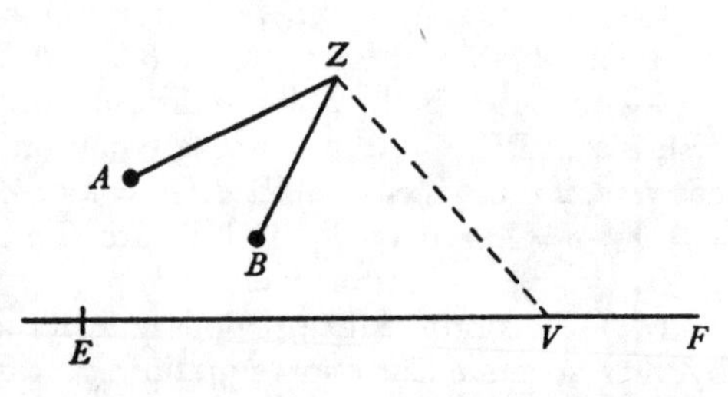

(142) If in a system of perpendicular Cartesian coordinates in which the poles A and B are $(a, 0)$ and $(b, 0)$ respectively the general point Z is (x, y), the lengths of the 'inflected' lines are readily determined to be $AZ = \sqrt{[(x-a)^2+y^2]}$ and $BZ = \sqrt{[(x-b)^2+y^2]}$. If the defining relationship of Z be $AZ^2 - BZ^2 = N^2$, then its locus is the straight line $x = \frac{1}{2}(a+b-N^2/(a-b))$ through Z perpendicular to AB. (An equivalent Euclidean demonstration departs from *Elements* I, 47 to establish that this perpendicular from Z meets AB, in C say, in the unique point satisfying $AC^2 - BC^2 = N^2$.) Where, alternatively, there is $AZ/BZ = k$, the locus of Z is the Apollonian circle $(1-k^2)(x^2+y^2) - 2(a-k^2b)x + a^2 - k^2b^2 = 0$ meeting AB $(y = 0)$ in the diametral end-points $((a \pm kb)/(1 \pm k), 0)$ which divide A and B internally/externally in the

a straight line when AZ and BZ are equal, and in other cases a circumference.(142)

2. If the straight line EF be given in position and in it be a given point A, from this draw a terminated line AZ and from its end-point Z draw also ZV, at a given angle $A\widehat{V}Z$,(143) to EF given in position, and let the power AZ^2 of the drawn line be equal to the product of a given one, M, and a segment ending either at the given point A or one or other of the given points E or F in the line given in position—that is, to either $M \times AV$ or $M \times EV$ or $M \times FV$: the end-point Z of AZ will then trace a circumference given in position.(144)

3. If from the two given points A, B straight lines AZ, BZ are inclined, and the power, AZ^2, of one be proportionately greater than that, BZ^2, of the other by a given (space),(145) the point Z will trace a circumference given in position.(146)

4. If from any number of given points A, B, C straight lines AZ, BZ, CZ are inclined to a single point Z, and the sum-total $AZ^2+BZ^2+CZ^2$ of their powers be equal to a given space, the point Z will trace a circumference given in position.(147)

5. If from the two given points A, B straight lines AZ, BZ are inclined, from the point Z to the line EF given in position draw ZV at the given angle $\widehat{V}$ cutting off from the straight line given in position EV terminating at the given point E, and let the (joint) powers AZ^2+BZ^2 of the lines inclined be equal to the

given ratio k. On page 326 above, of course, Newton has already taken cognizance of the latter locus as a porism 'probably among those which Euclid set out for semicircles' (on which conjecture see note (124)).

(143) A very necessary addition by Newton to Pappus' text.

(144) If A is the origin of a system of perpendicular Cartesian coordinates in which E and F are the points $(e, 0)$ and $(f, 0)$ of the abscissa and Z is the general point (x, y), on setting $\cot A\widehat{V}Z = k$ it follows (as Newton's figure has it) that $AV = x+ky$ and so $EV = x+ky+e$, $VF = f-x-ky$, while $AZ = \sqrt{[x^2+y^2]}$. The 'plane' loci (Z) defined by Newton's respective relationships readily prove to have the somewhat uninteresting corresponding Cartesian equations $x^2+y^2 = M(x+ky)$, $x^2+y^2 = M(x+ky+e)$ and $x^2+y^2 = M(x+ky-f)$.

(145) That is, if there be $AZ^2-c^2 = k.BZ^2$. Where the *spatium* c^2 is zero, of course, the locus reduces to be the simple Apollonius circle of Porism 1 preceding.

(146) Much as in note (142) above—and to continue its Cartesian equivalent—the present generalized Apollonius circle is defined by $(x-a)^2+y^2-c^2 = k^2((x-b)^2+y^2)$, that is,

$$(1-k^2)\,(x^2+y^2)-2(a-k^2b)\,x+a^2-k^2b^2-c^2 = 0.$$

(147) This familiar 'plane' locus had earlier been restored—and given virtually identical proof—by Frans van Schooten in his *Apollonii Pergæi Loca Plana Restituta* [= Liber III of his *Exercitationum Mathematicarum Libri Quinque* (Leyden, 1657): 191–292]: Liber II, Problema X: 273–6; by Pierre Fermat in his posthumously printed *Apollonii Pergæi Libri Duo De Locis Planis Restituti* [= *Varia Opera Mathematica* (Toulouse, 1679): 12–43]: Liber II, Propositio Altera [Generalis]: 38–41; and not least, as we have seen (IV: 318–20), by Newton himself some dozen years before in his 'Solutio Problematis Veterum de Loco solido'. In Porism 2 of the preliminary draft reproduced in Appendix 3.2 below (see p. 351: note (17)) Newton had adumbrated the extension, there considering only the bipolar instance, in which a *locus planus* (Z) is defined by the more general condition that $k.AZ^2+l.BZ^2+\ldots$ be equal to a *datum spatium*.

abscissa EV continetur, punctum Z ad inflexionem positione datam circumferentiam continget.(148)

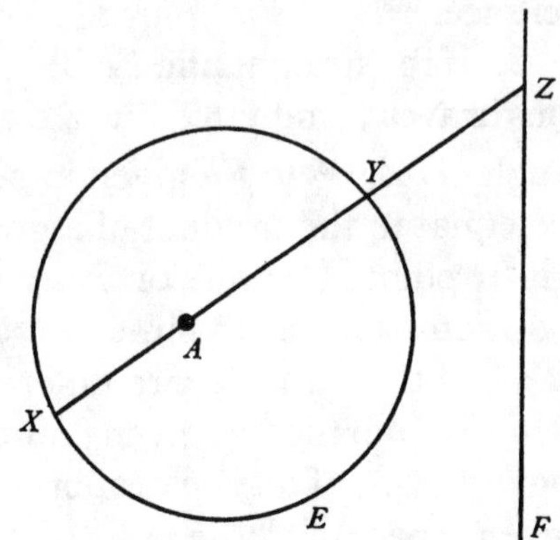

[6.] Si in circulo positione dato EXY sit datum punctum A perq; ipsum agatur quædam recta linea XYZ, inq; ipsa punctum Z extra sumatur, si autem $AZ^{quad.}$ quod fit a linea ducta usq; ad punctum A intra datum circulum [sit] æquale ei quod a tota XZ et extra sumpta YZ vel soli vel una cum contento XAY quod duabus quæ intra circulum portionibus AX AY continetur: punctū extra sumptum Z positione datam rectam lineam continget.(149)

[7.] Si duæ rectæ lineæ agantur (AY, AZ fig 1.2 vel AY, aZ fig. 3, 4) vel ab uno dato puncto (A fig 1, 2) vel a duobus (A, a fig 3, 4) vel in rectam lineam (AYZ fig 1) vel parallelæ (AY, aZ fig 3) vel datum continentes angulum (YAZ fig: 2 aut YEZ fig 4) eæq; vel inter se datam proportionem habentes (AY ad AZ vel AY ad aZ) vel datum comprehen[den]tes spatium ($AY \times AZ$ vel $AY \times aZ$)[,] contingat autem terminus unius Y locum planum positione datum et alterius terminus Z locum planum positione datum continget, interdum quidem ejusdem generis(150) interdum vero diversi(150) interdum simi[li]ter positum ad rectam lineam AY et AZ interdū contrario modo. Nimirum ubi datur proportio AY ad AZ vel AY ad aZ[,] si punctum Y tangit rectam lineam punctum Z tanget rectam lineam[,] sin punctum Y tangit circulum punctum Z tanget circulum similiter positum ad lineam AZ vel aZ atq; circulus alter ad lineam AY. Ubi vero datur rectangulum $AY \times AZ$

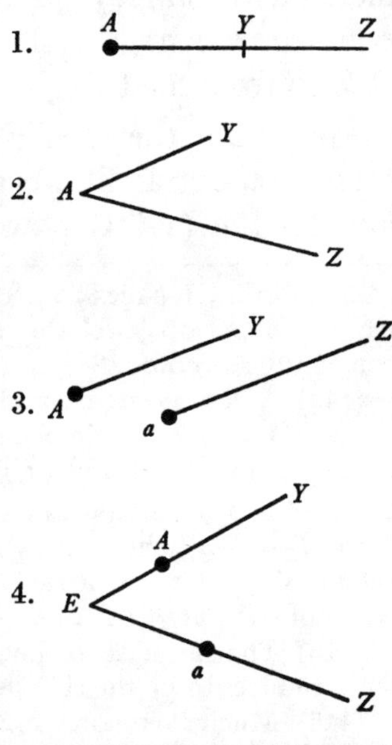

(148) Much as in note (144), but where the fixed poles $A(a, r)$ and $B(b, s)$ are in general position in the plane, the distances $AZ = \sqrt{[(x-a)^2+(y-r)^2]}$ and $BZ = \sqrt{[(x-b)^2+(y-s)^2]}$ of $Z(x, y)$ from these yield, when related to the abscissa $EV = x+ky$ (where, as before, $k = \cot \widehat{V}$) by the defining relationship $AZ^2+BZ^2 = N \times EV$, the corresponding Cartesian equation $(x-a)^2+(y-r)^2+(x-b)^2+(y-s)^2 = N(x+ky)$ or

$$x^2+y^2-(a+b+\tfrac{1}{2}N)\,x-(r+s+\tfrac{1}{2}kN)\,y+\tfrac{1}{2}(a^2+b^2+r^2+s^2) = 0.$$

(149) For, on setting O to be the centre of the given circle and drawing ZB instantaneously tangent to it at B, there at once comes to be $XZ \times YZ = BZ^2 = OZ^2-(OB^2$ or) OC^2 and $XA \times AY = (OY^2$ or) OC^2-OA^2. If therefore the locus (Z) is defined by the condition $XZ \times YZ = AZ^2$, that is, $OZ^2-AZ^2 = OC^2$, it is a corollary of Newton's preceding Locus 1 that Z is placed in a straight line ZF perpendicular to OA and meeting it in D where $OD^2-AD^2 = OC^2$ and hence $OD = \tfrac{1}{2}(OC^2/OA-OA)$. From a more modern viewpoint,

rectangle contained by a given line, N, and the segment EV: the point Z will at the point of inclination trace a given circumference.(148)

6. If in the circle EXY given in position the point A be given and through it be drawn a straight line XYZ with an external point Z assumed in it, let the power AZ^2 of the line drawn therefrom up to the point A within the given circle be equal to the product of the whole line XZ and its exterior portion YZ, either by itself or together with the rectangle $XA \times AY$ contained by the two portions XA, AY within the circle: the externally taken point Z will trace a straight line given in position.(149)

7. If two straight lines AY, AZ (Figures 1 and 2) or AY, aZ (Figures 3 and 4) be drawn either from one given point A (Figures 1 and 2) or from two A, a (Figures 3 and 4) either in a straight line AYZ (Figure 1) or as parallels AY, aZ (Figure 3) or containing a given angle $Y\hat{A}Z$ (Figure 2) or $Y\hat{E}Z$ (Figure 4), and let these either have a given ratio AY to AZ or aZ one to the other or embrace a given space $AY \times AZ$ or $AY \times aZ$: then if the end-point Y of one should trace a plane locus given in position, the end-point Z of the other will also trace a plane locus given in position, sometimes indeed of the same kind(150) and sometimes to be sure of a different one,(150) sometimes similarly placed with regard to the straight lines AY and AZ and sometimes set the inverse way. Namely, when the ratio of AY to AZ or aZ is given, if the point Y traces a straight line, the point Z will trace a straight line; but if the point Y traces a circle, the point Z will trace a circle set with regard to the line AZ or aZ the like way to that in which the other circle regards the line AY. When, however, the rectangle $AY \times AZ$ or $AY \times aZ$ is given,

---

equivalently, the line ZF is the radical axis of the given circle and the point-circle A. Where, alternatively, the locus (Z) is defined by $XZ \times YZ = AZ^2 + XA \times AY$, on reducing this to be $OZ^2 - OC^2 = AZ^2 + OC^2 - OA^2$ or $OZ^2 - AZ^2 = 2OC^2 - OA^2$, constant, it again likewise ensues that Z is found in a straight line perpendicular to OA, now meeting it in D such that

$$OC^2 = \tfrac{1}{2}(OA^2 + OD^2 - AD^2)$$

and so $OD = OC^2/OA$, whence the rectilinear locus (Z) is the polar of A with respect to the given circle. More directly, since

$$XZ \times YZ - AZ^2 - XA \times AY = XA \times YZ - XZ \times AY$$

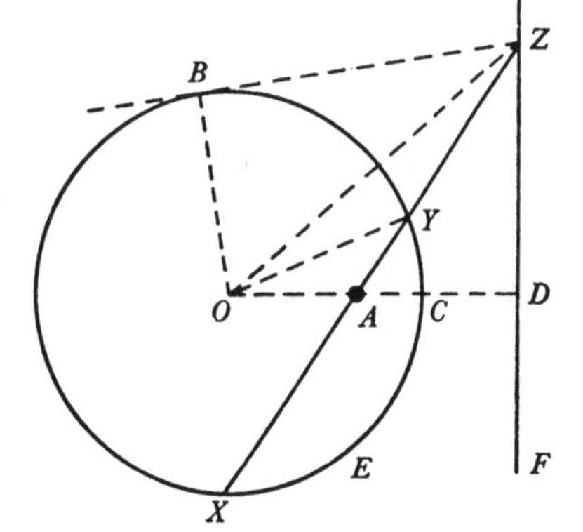

and consequently the cross-ratio

$$(AZXY) = AX \times ZY/AY \times ZX = -1,$$

the line joining all points Z on the required locus to the fixed pole A meets the circle in points X, Y which divide A, Z harmonically. These converses of Pappus, *Mathematical Collection* VII, 159/161 (which are there listed as lemmas on Euclid's *Porisms*) are already entered jointly as Locus 21 of an analogous tabulation of Apollonian *loca plana* on IV: 242. Notice that—con-

vel $AY \times aZ$ si punctum Y tangit rectam punctum Z tanget circulum[,] sin punctum Y tangit circulum & circulus ille transit per punctum A, punctum Z tanget rectam lineam, quod si circulus ille non transit per punctum A, punctum Z tanget circulum aliter positum ad rectam AZ vel aZ quam circulus alter ad lineam AY.(151) Hanc propositionem seu potius propositionum complexionem Pappus primo loco recenset, dicens eam propositi posteriorem esse ordine autem priorem.(152) id est in propositionibus Apollonij ultimum locum tenere sed ordine naturæ priorem esse. Ordinem igitur restitui quem Pappus immutaverat. Primo libro locos rectilineos, priore parte secundi circulum unicum, posteriore circulos duos vel locum circularem et rectilineũ conjunctim considerare, pace Pappi, ordo erat optimus.(153)

Post locos planos considerandi sunt loci solidi. At *Conica* Apollonij hic in censum non veniunt. Ea ita se habent ad hos locos ut Euclidis *Elementa* ad locos planos,(154) proinde a Pappo minus rectè ponuntur in loco resoluto. (155)E reliquis autem libris qui huc spectabant Pappus nihil ferè descripsit præter pauca de locis universis porismata quæ ita fere se habent.(156) Si ducantur a puncto quovis ad datas positione quotcunq; lineas aliæ totidem lineæ in datis angulis, et si datur ductarum proportio ad invicem ubi duæ sunt[,] punctum illud contingit rectam lineam positione datam, sin datur proportio contenti sub duabus ad contentum sub tertia et data recta aut ad quadratum tertij si tres sint[,] vel ad

trary to its position in Newton's figure (here faithfully reproduced in its proportions)—the point A can in neither case lie at a greater distance from ZF than the circle-centre O.

(150) Newton here initially added in parenthesis 'puta si detur linearum AY AZ vel AY aZ proportio' (if, say, there be given the ratio of the lines AY, AZ or AY, aZ) and 'puta si detur earundem rectangulum' (if, say, there be given the product of these same lines) respectively.

(151) It will be clear that neither a simple rotation of AZ round A through an angle $\widehat{YAZ}$, nor its translation to a new parallel position aZ, nor, by suitably compounding these, its removal to any arbitrary position in the plane will distort the basic homothety $AZ = k.AY$ or inversion $AZ = k/AY$ by which any given *locus planus* (Y) is transformed into a new plane locus (Z): for even in the last case there exists a point O (the meet of the perpendicular bisectors of Aa and ZZ', namely) round which aZ' may be rotated into line with AY such that a comes to coincide with A. In §1 preceding, as we have seen (compare *ibid.*: notes (15) and (17)), Newton adduces an equivalent generation of *loci plani* from given ones in a much overdone arrogation of the 'vis compositionis' in resolving difficult geometrical problems.

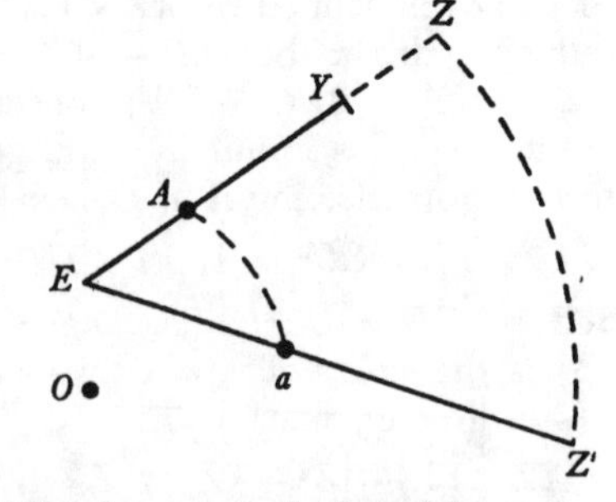

(152) *Mathematicæ Collectiones*: 247: 'Antiqui... horum planorum [locorum] ordinem respicientes elementa tradiderunt, quem cum negligerent posteriores, alios apposuerunt.... [Ego] ponam propositi quidem posteriora, ordine autem priora, vna propositione comprehendens hoc modo. Si duæ lineæ agantur, vel ab vno dato puncto, vel à duobus, & vel in rectam lineam, vel parallelæ, vel datum continentes angulum, vel inter se datam proportionem habentes, vel datum comprehendentes spacium: contingat autem terminus vnius locum

if the point Y traces a straight line, the point Z will trace a circle; but if the point Y traces a circle and that circle passes through the point A, the point Z will trace a straight line; while if that circle does not pass through the point A, the point Z will trace a circle set with regard to the straight line AZ or aZ in the inverse way to that in which the other circle regards the line AY.(151) This proposition, or rather complex of propositions, Pappus lists in primary place, stating however that the latter one propounded is the prior one in sequence(152)—that is, it occupies the end-place in Apollonius' propositions though it is prior in natural order. I have therefore restored the sequence which Pappus had altered. It was, despite Pappus, the best order to consider(153) rectilinear loci in the first book, a single circle in the former part of the second, and two circles or a circular locus and a rectilinear one jointly in the latter.

After plane ones solid loci need to be considered. But Apollonius' *Conics* do not here enter the reckoning. That is related to these loci the same way as Euclid's *Elements* is to plane loci, (154) and is accordingly not at all correctly put in the resolved locus. (155)Of the remaining books, however, which regard this Pappus has given virtually no description except for presenting a few porisms on universal loci which are effectively as follows.(156) If there be drawn from any point to any number of lines given in position an equal number of other lines at given angles, then if there is given the ratio of the lines drawn to one another when there are two, that point traces a straight line given in position; but if there is given the ratio of the rectangle contained beneath two of those drawn to that beneath the third and a given straight line or to the square of the third if there

---

planum positione datum, & alterius terminus locum planum positione datum continget, interdum quidem eiusdem generis, interdum vero diuersum, & interdum similiter positum ad rectam lineam, interdum contrario modo.'

(153) Momentarily forgetting that Apollonius would be classed by Pappus among the *recentiores* who had (see the previous note) changed the pristine sequence of the *elementa* of the classical plane locus, Newton here initially wrote 'Apollonius ordine optimo consideravit' (It was Apollonius who in best order considered) but quickly rephrased it into the vaguer equivalent form here printed.

(154) Originally 'Ea sunt harum figurarum elementa' (That is to do with the elements of these figures).

(155) Newton first went on: 'Sic et in Aristæi *locis solidis* magna ex parte Elementa horum locorum tradebantur. Quicquid amplius erat, pertinebat forte ad locum resolutum sed oblivioni traditum est. In poristico scribendi genere Pappus nihil omninò de his locis ex Veterũ libris recensuit præter secundum Porisma eorum quæ sequuntur' (So also in Aristæus' *Solid loci* in great part were the elements of these loci recounted. What more there was had perhaps to do with the resolved locus, but it is consigned to oblivion. In his poristic *genre* of writing Pappus reviewed nothing at all of these loci in the ancients' books except in the porismatic manner of the following).

(156) What follows is indeed a close paraphrase of Commandino's Latin version of Pappus' enunciation of the 3/4-line 'solid' locus and its 5/6-line cubic extension (*Mathematicæ Collectiones*: 251–2), extracts from which are quoted in our two remaining footnotes.

contentum sub reliquis duobus si quatuor sint, punctū contingit coni sectionem positione datam et aliquando locum planum:(157) quod si data sit proportio contenti sub tribus ductis ad contentum sub reliquis du[a]bus et data si quinqꝫ sint aut sub reliquis tribus si sex sint, punctum continget locos nondum [cognitos](158)

APPENDIX 1. PRELIMINARY OBSERVATIONS ON THE PLACE OF ARITHMETIC AND MECHANICS IN GEOMETRY(1)

From the original drafts in the University Library, Cambridge

[1](2) GEOMETRIÆ LIBER PRIMUS.

Arithmetica rerum omnium quantitates et rationes abstractas, Geometria figuras superficierum et solidorum, Mechanica motus locales figurarum accuratè considerat et demonstrativè determinat.

Antecedentes rationum ubi consequens est unitas sunt numeri, ijqꝫ vel integri si consequens metitur antecedentes, vel fracti si consequentis pars submultiplex metitur antecedentem, vel surdi si ante[ce]dens et consequens sunt incommensurabiles. Quatenus hæ sunt rationes linearum superficierum et solidorum(3) Arithmetica locum habet in Geometria ideoqꝫ prius addiscenda est quam Tyro ad Geometriam accedat. (4)Arithmetica per computationem quantitatum et rationum abstractarū æquationes colligit et extrahendo eorum latera problemata solvit, ideoqꝫ Problemata in genera distinguit secundum rationes complicatas seu dimensiones æquationum. Geometria vero Problemata solvit per descriptiones linearum ideoqꝫ descriptiones prius postulat quæ faciliores sunt et in mensuris determinandis utiliores, et problemata perinde in genera distinguit.

---

(157) See IV: 236 and VI: 248. This last proviso is not found in Pappus, who (in Commandino's Latin rendering) merely states that 'locus ad tres, & quattuor lineas...est huiusmodi. Si positione datis tribus rectis lineis ab vno, & eodem puncto ad tres lineas in datis angulis rectæ lineæ ducantur, & data sit proportio rectanguli contenti duabus ductis ad quadratum reliquæ: punctum contingit positione datum solidum locum, hoc est vnam ex tribus conicis sectionibus. Et si ad quattuor rectas lineas positione datas in datis angulis lineæ ducantur; & rectanguli duabus ductis contenti ad contentum duabus reliquis proportio data sit: similiter punctum datam coni sectionem positione continget' (*Collectiones*: 251).

(158) Newton breaks off in mid-sentence: we complete his sense by borrowing a participle from the corresponding passage in the *Mathematical Collection* where Pappus more briefly remarks (in Commandino's Latin rendering) that 'si ad plures [rectas] quam quattuor [lineæ ducantur] punctum continget locos non adhuc cognitos, sed lineas tantum dictas; quales autem sint, vel quam habeant proprietatem, non constat' (*Mathematicæ Collectiones*: 251). We will appreciate how a man who had over the previous quarter of a century (see II: 10–88; IV: 354–80) not only given detailed enumeration of the principal species of cubic but also established several universal properties of algebraic curves in general should find it

be three, or to that contained beneath the remaining two if there be four, then the point traces a conic section given in position, and sometimes a plane locus;[157] while if there be given the ratio of the content beneath three of those drawn to that beneath the remaining two and a given line if there be five, or beneath the remaining three if there be six, the point will trace loci not yet (investigated).[158]

Unde primi sunt generis geometrici quæ per descriptionem rectarum et circulorum solvuntur quamvis æquatio ad quam arithmetice deducuntur sit duarum dimensionum vel etiam quatuor dimensionum sine terminis imparibus & nonnunquam plurium.

Sed et Mechanicam prius degustandum esse Geometræ postulant. [5]Figuras quidem omnes oblatas (uti Sphæram, Cylindrum, Conum, Cissoidem, Conchoidem, Spiralem Archimedeam) considerant et mensurant arte propria, etiam absq harum descriptionibus postulatis.[6] Sed ars illa in debita appli-

---

pointless to continue transcribing this now badly obsolete observation. Newton's present statement of the 5/6-line locus is, we may add, a deal less cumbrous than Pappus' equivalent eunUnciation, which reads in sequel (*ibid.*: 251–2): 'Si ab aliquo punct[o] ad positione datas rectas lineas quinque ducantur rectæ lineæ in datis angulis, & data sit proportio solidi parallelepipedi rectanguli, quod tribus ductis lineis continetur ad solidum parallelepipedum rectangulum, quod continetur reliquis duabus, & data quapiam linea, punctum positione datam lineam continget. Si autem ad sex, & sit data proportio solidi tribus lineis contenti ad solidum, quod tribus reliquis continetur, rursus punctum continget positione datam lineam'. Had he gone on to consider Pappus' general $(2n-1)/2n$-line locus (*ibid.*: 252) Newton would doubtless, as before (see IV: 240–2), have drawn attention to Descartes' error in his *Geometrie* of supposing that it would 'usefully' define all possible 'geometrical' curves.

(1) Two discarded scraps, roughly consecutive in their content, which would appear to be half-hearted attempts on Newton's part to improve on corresponding portions (see §2.1: notes (3) and (12) above) of the preceding proem to Book 1 of the 'Geometria'. The first piece is fairly trite, but the second presents an interesting *arrière-pensée* on the first paragraph of the 'Præfatio ad Lectorem' with which (see note (14) below) he had in 1687 introduced his *Principia* to the world.

(2) Add. 3963.15: 177r–178v.

(3) Originally 'quantitatum Geometrica[rum] ad lineas superficies et solida [pertinentium]'.

(4) These following sentences (to the end of the paragraph) exist in isolation overleaf (on f. 177v) in the manuscript, but we here insert them at an appropriate place in Newton's main text.

(5) Newton initially observed at this point in a preceding cancellation that 'Figuras accuratè describere mechanicum est sed [talium problematum solutio] vi postulatorum locum habet in Geometria' before beginning to continue much as in his revise 'Figuras oblatas G[eometræ considerant...]'.

(6) Originally 'Figuras omnes oblatas considerant et mensurāt arte propria, non oblatas describunt arte mechanica vi Postulatorū ideoq necesse est ut omnis Geometra sit etiam Mechanicus. Describunt autem non ut has consid[er]ent et mensurent sed ut harum beneficio mensurent oblatas et mensuras demonstrent. Ideoq non opus est ut figuræ omnes a Geometris describi possint. Sufficit earum descriptiones postulare quæ ad constructionem Problematum

catione operationum manualium describendi rectas et circulos ut horum intersectionibus mensuræ determinentur. Ideocꝫ postulatur in limine Geometriæ ut Tyro problemata has accurate describendi[7] in Mechanica prius didicerit.

Et in his tribus fundantur scientiæ omnes mathematicæ. Arithmetica cæteris generalior est et facilior et primo addisci debet. Geometria ad Mechanicam requiritur et vicissim in operationibus Mechanicis fundatur.[8]

---

[2][9] [Ait Pappus Mechanicam][10] *a Philosophis maxima laude dignam existimatam esse et omnes Mathematicos non mediocri studio in eam incubuisse.*[11] Ejus vero partes duas a Veteribus[12] constitutas esse, *unam rationalem, alteram (quæ) manuum opera indiget, & rationalem quidem partem ex Geometria et Arithmetica et Astronomia et ex physicis rationibus constare, eam vero quæ manuum opera indiget ex æraria et ædificatoria et te[c]tonica et pictura*[13] *et in om[n]ibus manuum exercitatione.* Ita ille. Est igitur

---

sunt utiles. Nam et Geometræ antiqui Sphæram Cylindrum & Conum ceu figuras oblatas in Geometriam absqꝫ Postulatis admiserunt'. In his present revise Newton first here similarly went on: 'at non oblatas ubi opus est describunt arte mechanica vi postulatorum ut harum beneficio mensurent oblatas. Affinitate igitur quam-maxima junguntur Geometria et Mechanica, et ut verius dicam nihil aliud sunt quam duæ partes unius et ejusdem scientiæ. Quid enim aliud est Geometria quam ars mensurandi longitudines & figuras oblatas per concessas operationes manuales. Et quid aliud Mechanica quam ars determinandi et mensurandi longitudines & figuras quas corpora mota datis temporibus describunt'.

(7) This replaces 'praxin accuratam describendi rectas et circulos'.

(8) Newton first began this final paragraph by concluding 'Unde Arithm[etica] in scientijs omnibus Mathemat[icis] locum habet et primò addiscenda est: Geometria et Mechanica locum proximum tenent', and then initially amended this to assert that 'Geometria in Mechanica et Arithmetica in utracꝫ locum habet, estcꝫ [Arithmetica] scientiarum generalissima ...et prius addisci debet, deinde Geometria et ultimo Mechanica', leaving unfinished a last sentence 'Geometria et Mechanica affinitate maxima junguntur cùm motus [? nihil aliud sunt quam longitudines et figuræ quas corpora mota datis temporibus describunt]' (compare the equivalent discarded phrase quoted in note (6) above).

(9) Add. 3963.3: $19^r/19^v$; compare §2.1: note (12) preceding.

(10) As it survives today in the Portsmouth papers, the manuscript sheet begins abruptly in mid-quotation. To make better sense of what follows we interpolate a few opening words. Here and below, we may add, Newton's citations of Pappus are from the latter's preamble to the eighth book of his *Mathematical Collection* (= *Mathematicæ Collectiones*, $_2$1660: 447–9).

(11) In Pappus' own direct speech (*Mathematicæ Collectiones*: 447) this opening to his eighth book reads more fully: 'Cum mechanica contemplatio...multis, & magnis vitæ nostræ rationibus conducat, iure optimo à philosophis maxima laude digna existimata est: ...'.

(12) In Commandino's Latin version (*ibid.*) Pappus attributes the following distinction between 'rational' and 'practical' mechanics to 'Hero mechanicus', doubtless in the lost Greek text of his work on the subject. Pappus' modern editors Hultsch and Ver Eecke—ever to be different—here prefer to understand a vaguer reference to Heron's 'mechanical' adherents.

(13) That is, the arts of *χαλκευτικὴ* (working in brass or, more generally, metal), *οἰκοδομικὴ* (house-building), *τεκτονικὴ* (working in wood, carpentry) and *ζωγραφικὴ* (painting) to cite Pappus' original Greek text.

Mechanica tam rationalis seu speculativa et eo nomine accurata quàm practica et non accurata: Et similiter Geometria duabus partibus constat, rationali ubi proprietates quantitatum perscrutamur et Problemata imaginamur per accuratam Postulatorum executionem solvi, & manualem ubi Postulata per manuales operationes exequimur imperfectè. Futilis est ergo distinctio illa inter Mechanicam et Geometriam quod hæc accurata sit[,] illa secus.(14)

Sed ut Geometriam cum Mechanica conferam dico hanc ex natura sua nihil aliud esse quam partem selectiorem Mechanicæ, imò scientias pene omnes Mathematicas ex Mechanica ortum duxisse. Res diversas coram positas et ejusdem rei partes continuas numerare et operationes arithmeticas exercere in chartis Mechanicum est. Observare motus Astrorum Mechanicum est. Lucis refractiones mensurare, et fabricare perspicilla Mechanicum est. Instrumenta musica tractare(15) Mechanicum est, et sic in alijs. Hæc sunt mechanica practica et speculationes in his fundatæ sunt Mechanica rationalis. Quoties pars aliqua Mechanicæ digna visa est quæ seorsim tractaretur, datum est huic nomen aliud et reliquæ tantum Mechanicæ mansit nomen proprium. Et quoniam ejus partes principales abierunt in diversas scientias et inter has Geometria eminuit, ideo Geometria inter omnes scientias mathematicas principem locum obtinere

(14) Newton here echoes the opening words of his *Principia* half a dozen years before: 'Cum Veteres [originally 'Philosophi antiqui'] Mechanicam (uti Author est Pappus) in rerum naturalium investigatione maximi fecerint, et recentiores, missis formis substantialibus et qualitatibus occultis, phænomena Naturæ ad leges Mathematicas revocare aggressi sint: visum est in hoc Tractatu *Mathesin* excolere quatenus ea ad *Philosophiam* spectat. *Mechanicam* verò duplicem Veteres constituerunt: *rationalem* quæ per Demonstrationes accuratè procedit, et *practicam*. Ad *practicam* spectant Artes omnes manuales a quibus utiq; *Mechanica* nomen mutuata est. Cùm autem Artifices parum accuratè operari soleant, fit ut *Mechanica* omnis a *Geometria* ita distinguatur ut quicquid accuratum sit ad *Geometriam* referatur, quicquid minus accuratum ad *Mechanicam*. Attamen errores non sunt Artis sed Artificum. Qui minus accuratè operatur, imperfectior est Mechanicus, et siquis accuratissimè operari posset, hic foret Mechanicus omnium perfectissimus. Nam et linearum, rectarum et circulorum descriptiones, in quibus *Geometria* fundatur, ad *Mechanicam* pertinent. Has lineas describere *Geometria* non docet sed postulat. Postulat enim ut Tyro easdem accuratè describere prius didicerit quam limen attingat *Geometriæ*; dein, quomodo per has operationes Problemata solvantur, docet. Rectas et circulos describere problemata sunt sed non Geometrica. Ex *Mechanica* postulatur horum solutio, in *Geometria* docetur solutorum usus. Ac gloriatur *Geometria* quod tam paucis principijs aliunde petitis tam multa præstet. Fundatur igitur *Geometria* in praxi Mechanica, et nihil aliud est quam *Mechanicæ* universalis pars illa quæ artem mensurandi accuratè proponit ac demonstrat. Cùm autem artes manuales in corporibus movendis præcipuè versentur, fit ut *Geometria* ad magnitudinem *Mechanica* ad motum vulgo referatur. Quo sensu *Mechanica rationalis* erit scientia motuum qui ex viribus quibuscunq; resultant... accuratè proposita ac demonstrata...' (quoted from the corrected partial secretarial transcript of Newton's 'Præfatio ad Lectorem' in ULC. Add. 3963.15: 182^r, afterwards printed with only trivial change in his *Philosophiæ Naturalis Principia Mathematica* (London, $_1$1687): signature A3^r).

(15) Newton perhaps has in mind Mersenne's lavish treatise on the theory and practice of *Harmonie Universelle* (Paris, 1636)?

censetur, Mechanica verò quæ omnium mater est habetur omnibus inferior.(16) Contumelia Mechanicæ est quod ejus pars manualis tam late exercetur & pars rationalis si cum altera parte conferatur tam angusta est. E contra Geometriæ laus et perfectio est quod ejus pars manualis in Postulatis contenta tam simplex et angusta est et pars rationalis tam latè patet[,] imò quod Mechanicæ manualis operationes omnium simplicissimas et facillimas selexit easq non nisi duas, Rectam et Circulum describere ubi in Planis res est: at Mechanicæ omnis rationalis quam homines hactenus assequi valuerunt, assumsit sibi partem tam majorem quam potiorem(17) quæq reliquam Mechanicā sustinet. Nu[m]quid ergo in Geometriam per novum Postulatum introducemus innumeras operationes manuales easq compositissimas ita ut nihil fere circa magnitudines regulariter formandas(18) et movendas in Mechanica exerceri possit quod hoc non in posterum censeatur operatio Geometrica[?] Viderint Geometræ si hoc non sit Geometriam contaminare & similiorem reddere Mechanicæ. Siquas Mechanica figuras habet quæ exacte mensurari possint, licet hic exercere Geometriam. Nam Geometria se diffundit per omnes scientias, sed in se nihil recipit aliunde. Et si authoritas novorum Geometrarum(19) contra nos facit[,] tamen major est authoritas Veterum. Audi enim quod Pappus initio libri Mechanices(20) de Archimede scribit. *Carpus Antiochensis*, ait,(21) *quodam in loco dicit Archimedem Syracusanum unum duntaxat librum mechanicum co[m]posuiss[e] de Sphæropœia,*(22) *de alijs verò sibi scribendum non existimasse quamvis apud multos ob mechanicam facultatem summo in honore semper fuerit et admirabilis magno quodam ingenio habitus sit adeo ut adhuc apud omnes homines ejus fama mirandū in modum celebretur. Sed de ijs quæ præcipua sunt & Geometricam Arithmeticamq contemplationem*

---

(16) In sequel Newton went on initially to adjoin the yet more biting phrase 'et eò tandem ignominiæ processit ut Geometriæ laus et perfectio [sit] quod [? ejus pars manualis tam angusta est]'.

(17) Originally 'partem maximam'.

(18) 'figurandas' was first written.

(19) Yet once more understand the 'arch enemy' Descartes and such adherents as Frans van Schooten.

(20) The eighth book of Pappus' *Mathematical Collection*, that is. While by no means all of this pertains to mechanics, its last portion (*Mathematicæ Collectiones*: 482–90) presents a detailed account of the five classical 'powers' exemplified (compare VI: 102, note (24)) in the windlass, lever, pulley, wedge and screw—the only significant 'Pars *Mechanicæ*...a Veteribus exculta', to quote Newton's *Principia* preface (see note (14)) once more.

(21) *Mathematicæ Collectiones*: 448. Apart from the present passage, an obscure citation by Iamblichus of his quadratrix curve 'of double motion'—which Tannery for no good reason guessed to be a cycloid—and Proclus' identification of him as μηχανικός (see T. L. Heath, *Greek Mathematics*, **1**: 225, 232; and P. Ver Eecke, *Pappus d'Alexandrie:* 813, note 3) nothing is known of Carpus of Antioch or his mathematical work.

(22) Archimedes' work *On the Armillary Sphere* has not survived, though the planetarium which he constructed to embody its Eudoxian principles of spherical celestial motion was much admired in antiquity (see E. J. Dijksterhuis, *Archimedes* (Copenhagen, 1956): 23–5).

continent, quanquam ea brevissima videantur esse, diligenter conscripsit, tanto ut apparet prædictarum scientiarum amore inflammatus ut nihil extrinsecus in eas introducendum statuerit. Ipse autem Carpus et alij quidam jure optimo usi sunt Geometria etiam ad aliquas artes. Geometria enim nihil læditur quæ multas artes stabilire consuevit quando ejs adjungitur. Itaq; cum sit tanquam mater Artium[,] *non læditur quod curam habet Organicæ et Architectonicæ:*(23) *Neq; enim propterea quod simul sit cum ea quæ terras dimetitur & cum Gnomonica*(24) *& Mechanica & Scenographia*(24) *aliqua ex parte læditur; sed contra potius videtur eas promovere quod et honoretur & ab ipsis pro dignitate ornetur.* Hæc [Pappus.]

APPENDIX 2. SUPERSEDED REMARKS ON THE ANCIENTS' RELUCTANCE TO ACCEPT 'SOLID' SECTIONS INTO GEOMETRY.(1)

From the original drafts(2) in the University Library, Cambridge

[1] (3)[Ex his liquet, credo, quid de Postulato altero sentiendum sit Datum conum dato plano secandi....Scripsit Archimedes geometrice de sectionibus Conoidum et Sphæroidum,(4) nec tamen solida illa vel secare vel describere postulavit.] Porrò fundatum videtur hoc Postulatum in hypothesi incerta, quasi Conus omnis datus sit etiam descriptus. Ambigua est vox dati. Primò significat id quod actu habetur seu quod descriptum est, deinde determinatū et cognitum omne cujus descriptio est penes Geometram etsi id nondum descriptum sit, tertiò determinatum et cognitum omne cujus descriptio non est penes Geometram. Hoc tertio sensu Euclides lib[ro] 13(5) *Elementorum* datam vocat sphæram cujus centrum et intervallum datur. Simili de his solidis locutione priores Geometras in libris suis usos esse verisimile est[,] et e posterioribus nonnullos ista legentes

(23) That is, the theory of machines and architecture. Hultsch, we will not be surprised to learn, writes this and the following phrase off as one more scholiast's interpolation.

(24) 'Dialling' and 'scenic design' respectively. (The latter is probably to be understood in its theatrical sense, much as in Vitruvius' *De Architectura*.)

(1) Two consecutive preliminary castings of Newton's arguments and comments to the same effect on pages 296–302 above. While of no great intrinsic importance, these contain a few interesting minor originalities in emphasis and detail which will justify our reproducing their text here in full.

(2) Add. 3963.11: 123r/124r, extracted from the sequence of the main text in §2.1 preceding (see *ibid.*: note (37)).

(3) We repeat a few preceding sentences (on f. 123r) to establish the context.

(4) See §2.1: note (36).

(5) Namely in Propositions 13–17, where Euclid takes the sphere to be the locus of a semicircle which rotates round its base diameter. The construction of the latter is given by his opening Postulate 3 'quovis centro & intervallo circulum describere' (Barrow, *Euclidis Elementorum Libri XV...* (Cambridge, 11655): 6).

credidisse conos sic datos esse in Geometria ut secari possent et sectiones ad solutionem problematum adhiberi. Has sectiones liceret speculari sed non prius postulare quam Conum describere possent et planis imperatis secar[e].

---

[2] (6)Sed nec Veteres postulatum secandi conos receperunt(7) in geometricam constructionem problematum, neq; prætulerunt operationibus Mechanicis. *Antiqui Geometræ*, inquit Pappus,(8) *problema (Deliacum)*(9) *quod natura solidum est Geometrica ratione innixi construere non potuerunt; quoniam neq; coni sectiones facile est in plano designare: instrumentis autem ipsum in operationem manualem & commodā aptamq; contructionem mirabiliter traduxerunt ut videre licet in eorum voluminibus.*(10) *Asserentes enim Problema solidum esse, ipsius constructionem instrumentis tantum perfecerunt, congruenter Apollonio Pergæo qui et resolutionem ejus fecit per Coni sectiones; alij per locos solidos Aristæi; nullus autem per ea quæ propriè plana appellantur.* (11) Quibus verbis docere videtur Pappus quod etsi Problemata solida nomen haberent a solidis sectis, tamen istæ sectiones ob difficilem descriptionem in plano non sunt receptæ ad Geometricam constructionem talium Problematum. Veteres ea construxerunt per operationes quascunq; manuales et nonnulli per conica, nemo autem per ea quæ proprie plana appellantur, atq; adeo Problema Deliacum quod natura solidum est ratione Geometrica innixi construere non potuerunt. Alibi(12) docet

(6) Newton originally began this redraft: 'Verum tamen id non effecit ut hujusmodi operationes in Geometriam omnino reciperentur, nedum ut a plerisq; aliter ponerentur operationibus mechanicis. Si Apollonius et Menæchmus ad solutionem problematū solidorum usi fuerint sectionibus conicis, alij rursus ut Nic[omedes illas rejecerunt?]'.

(7) Initially 'admiserunt'.

(8) In his preamble to *Mathematical Collection* III, 5 (= *Mathematicæ Collectiones*: 7).

(9) On which see 1, §1.1: note (8) preceding. However, Newton's discarded initial parenthesis 'inveniendi duo media proportionalia' here more accurately specified Pappus' general 'problema ante dictum', *viz:* 'datis duabus rectis lineis duas medias proportionales in continua analogia inuenire' (*Collectiones*: 1). Perhaps he conflates the present preamble with a closely similar passage in *Mathematical Collection* VIII, 11 where Pappus states: 'problema, quod deliacum appellatur, cum natura solidum sit, fieri non potest, vt geometricis rationibus innixi construamus; quoniam neque coni sectiones facile est in plano describere. Instrumentis autem mutat[is] in manuum operationem, & constructionem magis idoneam...sic reducetur propositum...' (*ibid.*: 463).

(10) In a following clause, here omitted by Newton, Pappus in fact gives instances of these 'in Eratosthenis *mesolabo*, [et] in Philonis & Heronis *mechanicis* & *catapulticis*'. We will appreciate Newton's reluctance to introduce such works—fragments only of whose Greek texts survive—into his present argument, which their citation would needlessly complicate to no good purpose.

(11) In a cancelled sentence immediately preceding, Newton asserted to much the same effect as in his sequel that 'Talibus verbis significare videtur Pappus, quod nemo problema illud construere posset geometrice seu per ea quæ proprie plana appellantur eò quod solidum erat: quodq; veteres in solidorum problematum constructione ob difficilem descriptionem sectionum conicarum in plano instrumentis potius usi sunt'.

(12) In the preamble to *Mathematical Collection* IV, 31 (= *Collectiones*: 95).

Pappus *peccatum non parvum esse cum problema planum per conica vel linearia ab aliquo invenitur*, quippe quod *ex improprio solvitur genere:* at interea Veteres problemata solida liberè per linearia & Organa solvebant.(13) Idem factitabat ipse Pappus(14) ad exemplar veterum, quamvis hic sectiones coni nonnunquam præferat. Magnus ille Archimedes problemata solida sæpe tractat, nunquam verò construit per conica. Solutiones aut prætermittit aut exponit per inclinationes rectarum[,] ut videre licet in *Lemmatis* & in libris *de Spiralibus deq; Sphæra et Cylindro*.(15) Quod si Veteres pro Geometricis agnovissent constructiones per conica, hi earum loco neutiquam substituissent operationes mechanicas: certè non Archimedes, qui, ut author est Carpus Antiochenus, *tanto Geometriæ et Arithmeticæ amore inflammatus est ut nihil extrinsecus in eas introducendū statuerit*. Mitto recentiores Vietam, Rivaltum, Oughtredum(17) aliosq; vestigijs Veterum insistentes.

Apud Pappū in P[r]æfat, lib. 8(16)

---

(13) Pappus had observed in Proposition 11 of his eighth book that 'Organicæ...multæ sunt species, & partes. alia enim a mechanica, & gnomonica...per instrumenta ab ipsa confecta demonstrantur. multa vero, & seorsum a mechanicis extrinsecus ab ea perficiuntur. & nonnulla, quæ Geometricis rationibus non facile tractantur, assumens, instrumentis ad faciliorem constructionem [Archimedes] perduxit' (*Mathematicæ Collectiones*: 463).

(14) See especially *Mathematical Collection* III, 5; IV, 31, 33, 34; and VIII, 11 (on which compare T. L. Heath, *History of Greek Mathematics*, **1**: 226–7, 241–4).

(15) In both Lemma 8 of the Archimedean *Lemmas* (see V: 464, note (678)) and in Propositions 5–9 of *On spirals* (see E. J. Dijksterhuis, *Archimedes* (Copenhagen, 1956): 134–8) a straight line satisfying a stated condition has by neusis to be inclined through a given point in a circle so as to lie between its circumference and a straight line given in position; in the celebrated Proposition 4 of Book 2 of his *On the Sphere and Cylinder* Archimedes broaches the problem of cutting a given sphere by a plane so that the two ensuing segments shall be in a given ratio, reducing it to the geometrical equivalent of a cubic equation which, despite his opening promise, he does not—in the text as it survives anyway—attempt to solve. (In the classical period, however, variant constructions of it by means of intersecting conics were given by Diocles and Eutocius; see Dijksterhuis, *Archimedes:* 193–204.)

(16) See *Mathematicæ Collectiones*: 448, lines 27–9.

(17) A motley trio of promoters of classical Archimedean geometry to be sure! Among his many other fertile mathematical pursuits François Viète had in Capita XIV–XVII of his *Variorum de Rebus Mathematicis Responsorum Liber* VIII (Tours, 1593 [= (ed. Frans van Schooten) *Francisci Vietæ Opera Mathematica* (Leyden, 1646): 347–435]) 'renewed' Archimedes' tract 'On Spirals' in several ingenious ways; see I: 84–7 and compare J. E. Hofmann, 'François Viète und die Archimedische Spirale' (*Archiv der Mathematik*, **5**, 1954:138–47). David Rivault, sometime mathematician at the court of Louis XIII, is today remembered only for his scholarly but not acutely percipient edition of *Archimedis Opera quæ exstant Novis Demonstrationibus Commentariisque illustrata* (Paris, 1615). The gifted 'philomath' William Oughtred compiled an elementary 12-page digest, *Theorematum in Libris Archimedis De Sphæra & Cylindro Declaratio*, which is found appended to all editions of his *Clavis Mathematicæ* from the third (Oxford, 1652) which Newton had studied in his youth (see I: 22), but nowhere, so far as we are aware, did he attempt any geometrical resolution of problems of order higher than 'plane'.

APPENDIX 3. VARIANT LISTINGS OF RECTILINEAR AND CIRCULAR PORISMS.(1)

From the originals in the University Library, Cambridge

[1](2)

PORISMATA

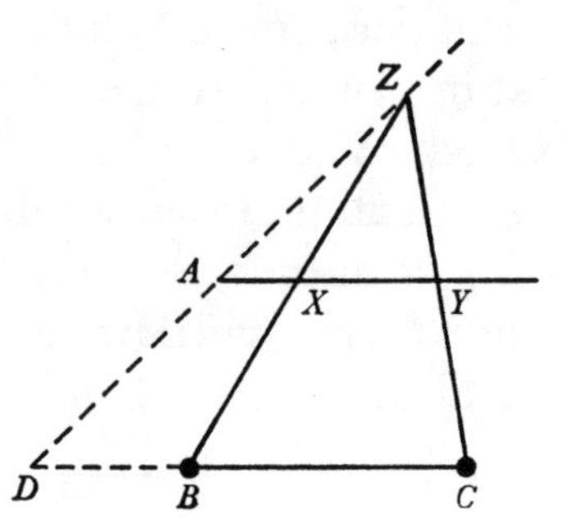

Porism. 1.(3) Si a datis punctis B, C ad rectam AZ positione datam concurrentes rectæ secent in punctis X, Y rectam AY a dato puncto A ipsi BC parallelam,(4) erit AX ad AY in data ratione.

Nam si AZ producta occurrat BC in D erit $AX.XY::DB.DC::$dat.dat. Q.E.D.

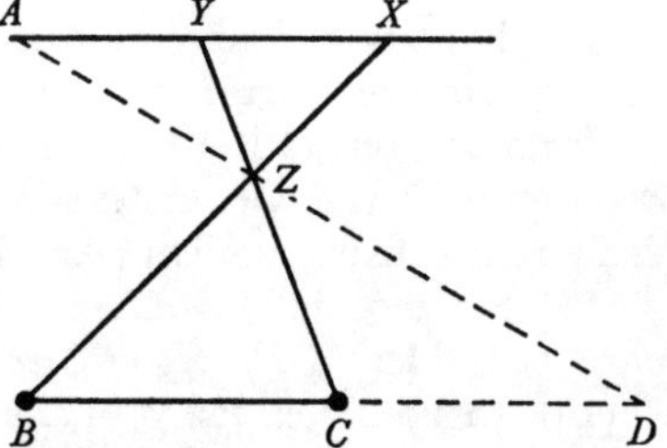

Porism. 2. Et si a datis duobus punctis B, C ductæ rectæ BZ, CZ secent AY in data ratione, punctum Z tanget rectam positione datam.

Age rectam ZAD occurrentem BC in D et erit $AX.XY::DB.BC$. Ergo datur ratio DB ad BC. Ergo datur DB. Ergo datur punctũ D.(5)

---

(1) In these minor preliminary elaborations and extensions of two of the groups of Euclidean *porismata* restored by him, to accord with Pappus' schematic outline, in §2.2 above Newton has, it will be evident, no radical departures to offer from their text in the way of novel configurations or freshly restyled supporting sequences of argument. For all their essential repetitiveness, however, they do possess some small interest in their own right, if only as superseded variants revealing Newton's preliminary groping towards an end which he later achieved more neatly and in more general form, and they will corporately emphasize yet once more how deeply engrossed he became in giving precise and detailed expression to his preferred notions of how Euclid's lost treatise on this topic in Greek geometrical analysis should be 'renewed'.

(2) Add. 4004: 183r–185r; compare §2.2: note (83) above. In these discarded folios from his 'Waste Book'—conscientiously, if imperceptively, returned to their parent by J. C. Adams in the late nineteenth century when (see I: xxxi ff.) he came roughly to sort these geometrical papers—Newton lightly reworked the six opening *porismata* of his preceding 'Cap. 2' (see pages 310–12 above) into an equal number of revised propositions (on f. 183r), and then in preliminary (on f. 184r) inserted recastings of Porisms 9–11 of his first draft (§1 above) together with an extension of these to form a quartet treating special cases of parallelism of the generating lines; in sequel (on ff. 184r/185r) he elaborated the following Porism 7 on page 312 into a new eleventh porism, distinguishing three successively more general cases which he adroitly reduced to the first (and simplest), and finally appended the unproved enunciation of a yet more general instance. These we present in the order of their numbering.

(3) This and the two following propositions repeat without essential change the previous Porisms 9–11 on pages 314–18 above.

(4) It is not of course necessary that AXY should, as Newton places it in both of his accompanying figures, be stationed above the poles B and C.

Porism. 3. Et si a dato puncto B agatur BXZ occurrens rectis positione datis AX, AZ in X et Z[,] detur autem ratio AX ad XY, inclinabit ZY ad datum punctum C.

Per B ipsi AY parallela, agatur DBC occurrens rectis ZA, ZY in D et C. Ergo [a]datur linea DB. Est $DB.BC::AX.XY$. Ergo datur BC. Ergo datur punctum C. Q.E.D.

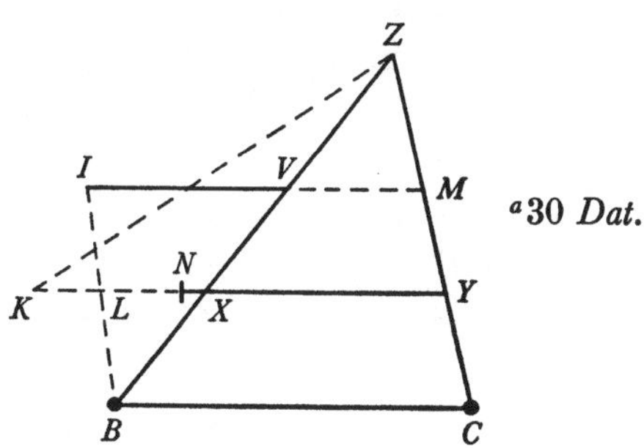

[a] 30 *Dat.*

Porism. 4. Si a datis punctis B, C concurrentes rectæ BZ CZ secent in V et Y rectas a datis punctis I, N ipsi BC parallelas ductas IV, NY sitq IV ad NY in data ratione[,] tanget punctum Z rectam positione datam.(6)

Age BI occurrentem NY in L et erit IV ad LX ut IB ad LB, hoc est in data ratione. Ergo LX est ad NY in data ratione. In eadem ratione capiatur KL ad KN et erit KX ad KY in eadem data ratione. Ergo (per Porism 2) punctum Z tangit rectam ZK positione datam. Q.E.D.

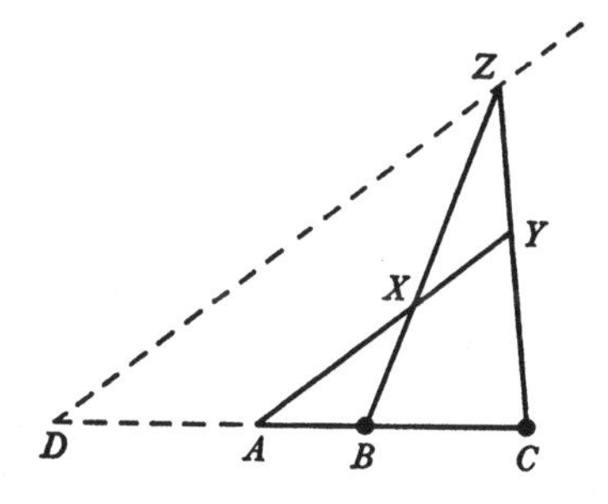

[*Porisma*] *5.*(7) Si a duobus datis punctis B, C ad rectam DZ positione datam inflectantur duæ rectæ BZ, CZ secantes rectam AY positione datam sitq AY parallela DZ, habebunt AX, AY, XY datas rationes ad invicem.

Est enim $AX.DZ::AB.DB::$dat.dat et

$$DZ.AY::DC.AC::\text{dat.dat.}$$

Ergo [a]$AX.AY::$dat.dat et [b]$AY.XY::$dat.dat. Q.E.D.

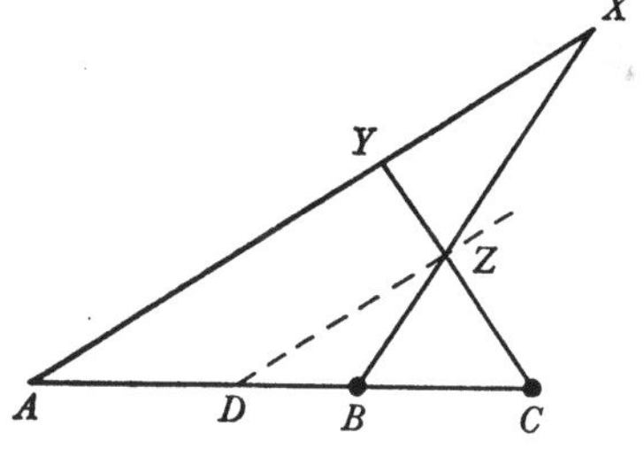

[a] 8 *Dat.*
[b] 5 *Dat.*

(5) In general, we may note, the given ratio AX/AY determines the rotating lines BX, CY to be in 1, 1 correspondence *per simplicem geometriam* and hence that the locus (Z) of their intersection is (compare IV: 308–10) a conic passing through B and C; but when X and Y pass (simultaneously) to infinity in AXY both BX and CY coincide with the parallel DBC, so that the locus (Z) is the pair of this and some second straight line.

(6) Here, much as before (see the previous note), the consequent simultaneous passage of V and Y to infinity in the parallels IV and NY again determines that the lines BV and CY in this position each coincide with BC, so that their intersection (Z) is the pair of BC and some second straight line. Newton in sequel makes use of the corollary that the latter will meet NY in K such that (by Porism 1) the ratio KX/KY is given to contrive an elegant reduction to Porism 2.

(7) Originally '1'; see note (2) above. Similarly, Porisms 6–10 following were initially numbered 2–6 by Newton.

[*Porisma*] *6*. Si a duobus datis punctis B, C ad punctum tertium Z concurrant duæ rectæ BZ, CZ secantes rectam AY positione datam in X et Y et habeat AX ad AY datam rationem, tanget punctum Z rectam positione datam.(8)

Agatur enim DZ ipsi AY parallela. Et quia

$$AX.DZ::AB.DB \quad \text{et} \quad DZ.AY::DC.AC$$

et conjunctis rationibus $AX.AY::AB \times DC.DB \times AC$, datur ratio $AB \times DC$ ad $DB \times AC$. Sed datur etiam ratio AB ad AC. Ergo datur ratio DC ad DB et divisim ratio DC ad datam BC atq; adeo datur punctum D. Datur etiam angulus D et proinde recta DZ quam punctum Z tangit datur positione. Q.E.D.

[*Porisma*] *7*. Si a duobus datis punctis B, C ad rectam positione datam DZ inflectantur rectæ duæ BZ, CZ secantes rectam AY ipsi DZ parallelam in punctis X et Y et detur ratio AX ad AY, datur AY positione.

Nam ob parallelas AY, DZ fit $AX.AY::AB \times DC.DB \times AC$ ut supra. Ergo datur ratio $AB \times DC$ [ad] $DB \times AC$. Sed datur etiam ratio DC ad DB[,] ergo et ratio AB ad AC, ut et AB ad BC. Et inde ob datam BC datur AB. Dato autem tum puncto A tum angulo BAY datur positione AY. Q.E.D.

[*Porisma*] *8*. Si a dato puncto B agatur recta BZ secans parallelas duas positione datas in X et Z[,] capiatur autem AY ad AX in data ratione et jungatur ZY[,] converget ZY ad datum punctum C.

Est enim AX ad DZ ut AB ad DB hoc est in data ratione et AX ad AY in data ratione[,] adeoq; [a]AZ ad AY in data ratione. Sed est DC ad AC in eadem ratione. Ergo divisim ratio DC ad AD datur adeoq; DC etiam datur. Et inde datur punctum C. Q.E.D.

[a] 8 *Dat.*

[*Porisma*] *9*. Ijsdem positis, datur ratio ZY ad YC.

Nam $ZY.YC::DA.AC::$dat.dat.

[*Porisma*] *10*. Ijsdem positis dantur rationes arearum omnium YXZ, AXB, DZB, $DAXZ$, $DAYZ$, BZC, $B[X]YC$, ACY et DZC in datam.

Porism. 11.(9) Si a datis punctis B, C concurrant rectæ duæ BZ, CZ secantes rectas positione datas AX AY in data ratione, jaceant autem puncta A, B, C in directum, punctum Z tangit rectam positione datā.

Cas. 1.(10) Junge XY et triangulum AXY dabitur specie. Jam si XY parallela

(8) Much as before (see note (5)) the given ratio AX/AY again determines the lines BX and CY to be in 1, 1 correspondence *per simplicem geometriam;* but the necessary degeneration of the conic locus (Z) of their meets into the pair of the straight lines DZ and BC is now effected by setting the point A (at which X and Y come simultaneously to be) in the latter line, with which in consequence BX and CY are constrained likewise to coincide.

(9) As we have stated in note (2), this elaborates Porism 7 on page 312 above into three progressively more general cases.

(10) Here, though they are now located in separate lines AX and AY, the points X and Y continue at A (by reason of the constant ratio AX/AY) simultaneously to coincide with BC, and so also therefore do the lines BX and CY, thus decreeing that their conic meet (Z) shall,

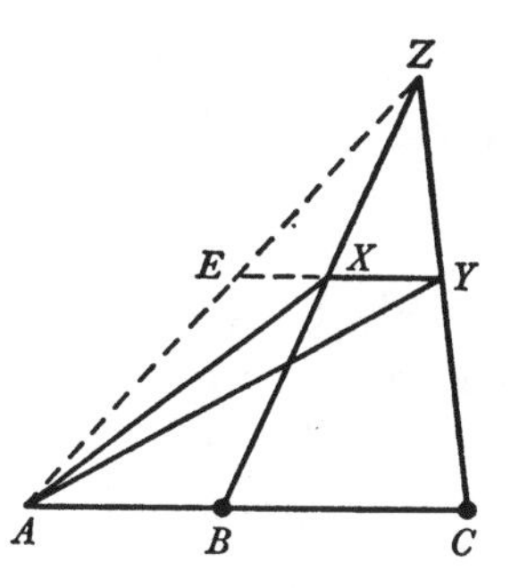

sit ipsi BC, produc XY ad E ut sit EX ad XY ut AB ad BC et concurrent tres rectæ AE, BX, CY in eodem puncto Z. Atqui ob datas rationes AX ad XY et EX ad XY datur ratio EX ad AX. Ergo triangulum AXE datur specie et recta AE positione. Ergo punctum Z tangit rectam positione datā. Q.E.D.

Cas. 2.(10) Sin XY parallela non sit ipsi BC, duc VY ipsi BC parallelam et ob datam specie figuram $AVXY$ dabitur ratio AV ad AX. In ista ratione fac ut sit AC ad EC[,] nec non in ratione EC ad AB ut sit DC ad DB[,] et dabitur punctum D.(11) Ipsis AX AY age parallelas DO DN occurrentes BZ CZ in O et N. Converte rationem novissimam et fiet $EC.DC::AB.DB::AX.DO$. Ergo cum sit $AV.AX::AC.EC$ et

$$AX.DO::EC.DC$$

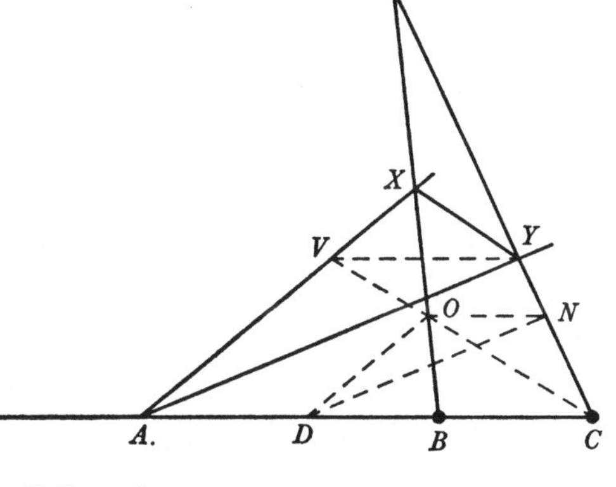

erit ex æquo $AV.DO::AC.DC$. Sed in eadem ratione est $AY.DN$. Ergo simili[a sunt] triangula AVY, DON. Ergo ON parallela est BC et ratio DO ad DN datur. Ergo (per cas. 1) punctum Z tangit rectam positione datam. Q.E.D.

Cas. 3.(12) Si datur ratio AX ad aY[,] age AT ipsi aY parallelam et dabitur ratio AT ad aY: quippe quæ eadem sit rationi AC ad aC. Ergo datur ratio AX ad AT. Ergo per cas 2 punctum Z tangit rectam positione datam. Q.E.D.

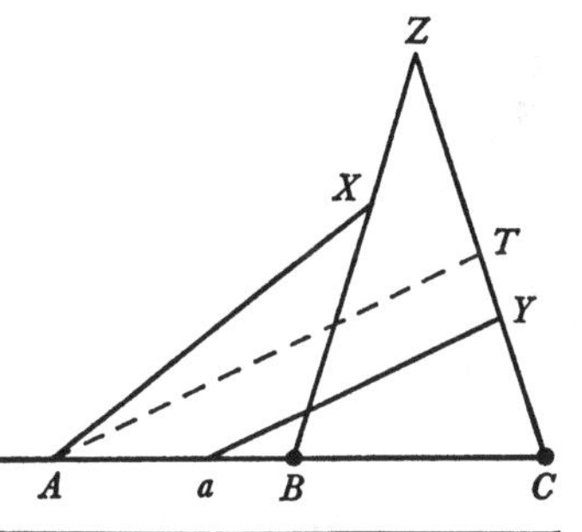

as ever, degenerate into the pair of BC and some other straight line. With his usual efficiency Newton gives elegant demonstration *ab initio* of the rectilinearity of the locus (Z) in each case, reducing the second (where XY is not parallel to BC) neatly—see our analysis in the next note —to the first (where it is).

(11) That, namely, in which the rectilinear locus (Z) meets BC. The preliminary analysis underlying this by no means evident construction is not difficult: for, if the parallel through X to BC meets CY in W, then at once $ZV:AX = CY:CW$ where $CW:BX = CZ:BZ$; in the limit as X and Y come to coincide with A, therefore, W will pass into E in BC such that $CA:CE = AV:AX$ and so Z into D such that $CD:BD = CE:BA$, exactly as Newton prescribes. The sequel is immediate. Since, as X and Y pass (simultaneously) to infinity, BZ and CZ come to be parallel to AX and AY respectively; and hence, where DO and DN are drawn in these same directions through D, it at once follows that ON is parallel to BC and also that DO is in a constant ratio to DN. Newton proceeds to give the straightforward synthetic proof that this is indeed so.

(12) In this further extension, though X and Y are now located on lines which do not

(13)*Porisma* [*12*]. Ijsdem positis si AX est ad aY ut datum data parte ipsius AX auctum vel diminutum ad datum, tangit punctum Z rectam positione datam.(14)

---

[2](15) [*In Libro tertio*, ait Pappus,(16) *plures erant propositiones de semicirculis, paucæ autem de circulis et circulorum portionibus: Quæsitorum autem multa similia erant antedictis....*] De semicirculis igitur verisimile est Euclidem hujusmodi propositiones posuisse.

[*1.*] In semicirculi PQ semidiametro PQ bisecta in C si sint AC, PC, BC continuè proportionales[,] rectæ AZ, BZ ad circumferentiam inflexæ datam habeant proportionē. Et vicissim si data est inflexarum proportio[,] punctum Z tangit circumferentiam positione datam.(17)

[*2.*] Si puncta a, B hinc inde æqualiter distant a centro C[,] datur summa quadratorum $[a]Z^q+BZ^q$. Sin distantiæ Ca CB sint inæquales[,] applicetur aZ^{quadr} ad distantiam aC et BZ^{quad} ad distantiam BC et laterum genitorum summa si puncta a, B cadunt ad diversas partes centri C, differentia si ad easdem, dabitur.(18)

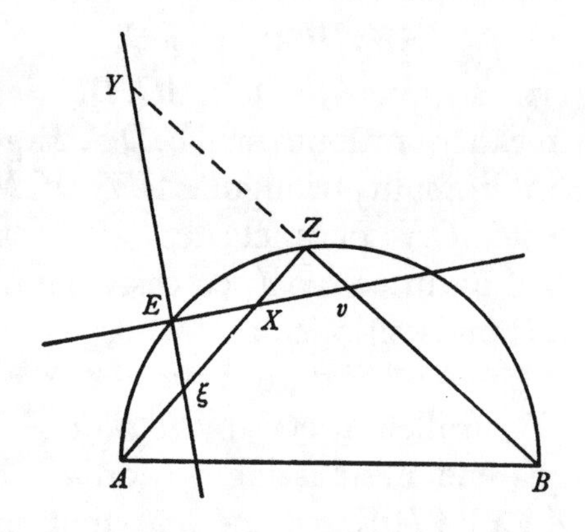

[*3.*] A diametri terminis A, B ad circumferentiam inflectantur rectæ AZ, BZ quæ secent in X et Y rectas EX EY positione datas, et ad circumferentiam rectum(19) continentes angulum XEY, data est proportio abscissarum EX, EY, ut et proportio interjectarum ξX, vY.(20)

---

intersect in a point of BC, the constant ratio AX/aY yet again furnishes the crucial restriction for the conic locus (Z) to break into the pair of BC and some other straight line that they should (at A and a respectively) come simultaneously to lie in BC at points other than the poles B and C. In sequel Newton states the obvious reduction to the preceding case.

(13) Immediately before this Newton has cancelled a '*Definitio.* Magnitudo P magnitudine [Q] major est sui parte [R] quam in ratione $\left[\frac{d}{e}\right]$ quando ablata sui parte, reliqua ad eandem habet rationem datam'; in other words, there is $P = (d/e)\,Q+R$ if $(P-R):Q = d:e$, constant. In present context it will be clear that he intended to apply it in a variant on the preceding porism in which $AX = (d/e).aY+R$ was to define the 1, 1 relationship between X and Y. If this be so, Newton would at once have realized that in this case, since X and Y cannot (when $R \neq 0$) simultaneously be located at the intersections A and a of their lines with BC, the locus of Z defined by the preceding configuration will be a non-degenerate conic through the poles B and C.

(14) Newton's stated relationship between the general segments AX and aY in the preceding configuration, namely $AX/aY = (c\pm(d/e).AX)/b$ or $b.AX-c.aY = \pm(d/e).AX\times AY$, mani-

[*4.*] Ad rectam quamvis positione datam XY demissis perpendiculis $A\xi$ Bv datur rectangulum $\xi X \times vY$.(21)

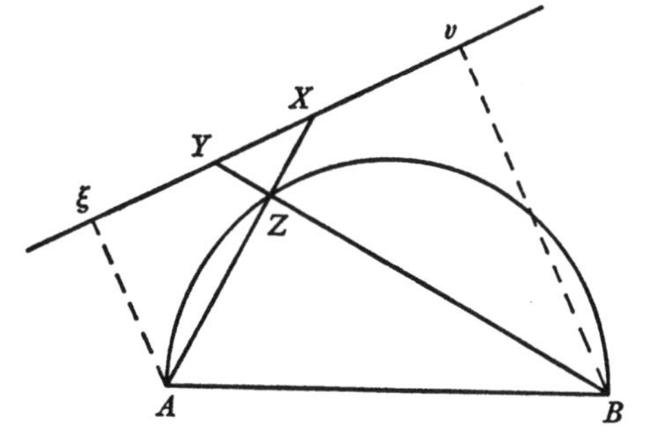

festly defines the most general 1, 1 correspondence between the points X and Y which concludes that (when $AX = aY = 0$) these may simultaneously be located at A and a respectively in BC, thus determining that the lines BX and CY may likewise simultaneously coincide with BC, and consequently that the conic locus (Z) constructed by the configuration may reduce to be the pair of $AaBC$ and some other definable straight line. The problem of finding an effective synthetic demonstration of this porism on a par with the elegant compositions given above of its particular cases is well-nigh insoluble, and Newton very wisely does not here attempt it.

(15) Add. 3963.14: 165^r, a cancelled first version of pages 320–2 above where (compare §2.2: note (106)) Newton has generalized his following Propositions 3 and 4 to hold for an arbitrary circle-arc $\widehat{AZB}$.

(16) *Mathematicæ Collectiones*: 246. We repeat this quotation from Pappus' survey of Euclid's lost work on porisms to afford a smooth lead into Newton's next sentence.

(17) One more enunciation of Apollonius' circular locus (on which see IV: 314–16 and compare 1, §2: note (43) above). The porism reappears in minimally revised form on pages 324–6 above.

(18) Because

$$BZ^2 = CZ^2+BC^2-2BC\times CZ.\cos \widehat{BCZ} \quad \text{and} \quad aZ^2 = CZ^2+Ca^2-2Ca\times CZ.\cos \widehat{aCZ},$$

at once, since the angles $\widehat{BCZ}$ and $\widehat{aCZ}$ are supplementary, there ensues

$$BZ^2/BC+aZ^2/Ca = CZ^2(1/BC+1/Ca)+Ba,$$

constant; in particular, where $BC = Ca$, then is $BZ^2+aZ^2 = 2CZ^2+\frac{1}{2}Ba^2$, constant. Here BC and Ca have the same or opposite signs according as B and a are located on opposite sides of C or the same one. The converse proposition—namely, if the locus (Z) is defined by the symptom $BZ^2/k+aZ^2/l = m$, then it is a circle—also holds true, of course; compare our note (147) on the generalized Apollonius *locus planus* restored by Newton on page 332 above.

(19) Equal to $\widehat{AZB}$, that is, whence it is prescribed that BZ shall be parallel to ξEY when AZ is parallel to EXv.

(20) For the configuration determines the points X and Y to be in 1, 1 correspondence such that they simultaneously coincide at E and also (see the previous note) simultaneously pass off to infinity in their lines Ev and ξE, and consequently by Rule 1 of Newton's preceding 'Inventio Porismatum' (1, §2 above) the ratio EX/EY is given. Because the right triangles $EX\xi$ and EYv are similar, this constant ratio is also that of $X\xi$ to Yv. (The latter property evidently no longer—unlike the first—holds true when AB is not the diameter of the circle-arc $\widehat{AB}$, but the equality $\widehat{AZB} = \widehat{XEY}$ is maintained.)

(21) Here Newton's configuration determines a 1, 1 relationship $X \leftrightarrow Y$ between points of the same straight line such that $\xi \leftrightarrow \infty$ and $\infty \leftrightarrow v$; by Rule 2 of the 'Inventio Porismatum', accordingly, there is given the product $\xi X \times vY$.

§3. THE ELEMENTS OF PLANE GEOMETRY ELABORATED IN TWO BOOKS.[1]

From the incomplete original[2] in the University Library, Cambridge

GEOMETRIÆ LIBER 1[3]

... ...

[PRINCIPIA RECTILINEA.]

... ...[4]

4. In quatuor proportionalibus rectangulum sub medijs æquatur rectangulo sub extremis. Eucl. 16, 17. VI. Et vice versa.

5. Summa[5] rationum e[s]t ratio rectangulorum $A.B+C.D=AC.BD$. Unde duplicata ratio est ratio quadratorū $A.B+A.B=A^q.B^q$. triplicata cuborum &c.

6. Differentia[5] rationum e[s]t ratio reciprocorum rectangulorum

$$A.B-C.D=AD.BC.$$

... ...[4]

9. In omni triangulo ABC circa verticem C [radio] latere alterutro BC descripto circulo basem si opus est productam secante in D, erit Basis ad summam laterum ut differentia laterum ad AD. Vel

$$AC^q-BC^q=AB\times AD.$$

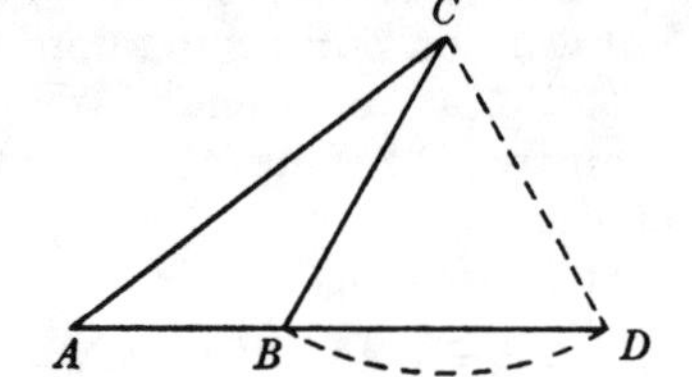

(1) In these roughly penned drafts of Books 1 and 2 of yet one more version of his projected treatise on 'Geometria' Newton leaves off his previous concentrated efforts to 'renew' the essence of ancient geometrical lore by a minute exegesis of the (Latin text of the) *Mathematical Collection.* While not entirely spurning the unique insights into the lost reaches of Greek higher analysis which Pappus' preamble to his seventh book affords, he now launches into a broader exposition of the principles of plane geometry, founding his account on the postulation of a few select theorems from the *Elements* coupled with the Euclidean notion of a 'given' and the assumption of the standard compositions of ratios, and appealing to the Apollonian determinate and spatial sections only as derivable, secondary techniques. In Book 1 he passes on to lay down seven rules for the effective application of these general precepts to the construction of geometrical problems, instancing their employment in the solution of questions involving but straight lines and circles (giving *inter alia* an elegant composition of the rhombic case of the Apollonian neusis). Thereafter, in Book 2, he lays down a trio of postulates whereby a wide class of 'organic' constructions of curves—not merely of such 'geometrical' ones as conics, cissoid and Cartesian ovals, but also of such 'mechanical' loci as spirals and trochoids—may be received into geometry and so make possible the general 'linear' construction of problems.

(2) Add. 3963.3: 33^r–39^v/15^r–16^v/18^v. The manuscript's opening sheet(s) would appear to be irretrievably lost. Therein, no doubt, Newton once more introduced his reader to what he

Translation

GEOMETRY. BOOK 1[3]

... ...

[RECTILINEAR PRINCIPLES.]

... ...[4]

4. In four (continued) proportionals the rectangle beneath the middles equals the rectangle beneath the extremes: Euclid, [*Elements*] VI, 16/17. And conversely so.

5. The 'sum'[5] of ratios is the ratio of the rectangles:

$$(A:B)\times(C:D) = AC:BD.$$

Whence the doubled ratio is the ratio of the squares: $(A:B)\times(A:B) = A^2:B^2$; the tripled that of the cubes; and so on.

6. The 'difference'[5] of ratios is the ratio of the reciprocal rectangles: $(A:B)\div(C:D) = AD:BC$.

... ...[4]

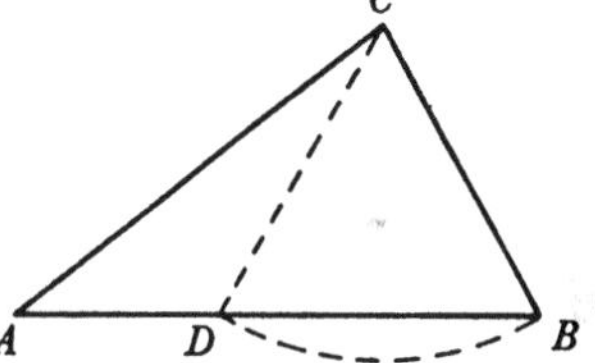

9. In every triangle *ABC* where a circle described round the vertex *C* and with one or other of the sides *BC* as radius cuts the base (extended if need be) in *D*, the base will be to the sum of the sides as the difference of

conceived to be the purpose and use of geometry before passing to specify the *principia rectilinea* with which the surviving text opens.

(3) The precise title which Newton gave to the following piece has, along with its opening paragraphs (see the previous note), disappeared into oblivion if not perdition. The one we here propose is conjecturally restored to be on a par with that of 'Lib. 2' on page 382 below.

(4) From Newton's subsequent appeals 'per Princip. ' which do not relate immediately to the surviving five it would seem that these missing Principles 1–3 and 7/8 included both the postulate that if $A:B = C:B$ then $A = C$ (see the second of the cancelled examples, originally set to illustrate *Datum* 3, which are reproduced in note (11)) and also a statement that if $A:B = C:D$ then this ratio is that of $(A \pm mC):(B \pm mD)$ (see the revised Case 1 of the same *Datum* 3), though these are of course straightforwardly derivable from his present *Principium* 4. The remaining lost principles, we suppose, dealt similarly with the standard combinations of ratios *componendo* and *dividendo*.

(5) In modern non-logarithmic parlance, the product and quotient respectively. As earlier in his 'De motu Corporum' (see VI: 47–8, note (47)), Newton derives this medieval nomenclature and following notation from Isaac Barrow, who initially employed it in his *Euclidis Elementorum Libri XV. breviter demonstrati* (Cambridge, $_1$1655): 165, 167–8 to epitomize the Euclidean propositions (*Elements* VIII, 5/11/12) which Newton here encapsulates in Principle 5, but with a late added note to the 'candidus lector'—evidently ignored by Newton—that 'in octavo libro:: [legendum est] pro =, sed locum non memini' (*ibid.:* signature A8^v, *lin. ult.*).

Eucl Prop 12 & 13 lib 2.(6) Eodem recedunt Prop 47 lib 1 & Prop 35, 36, & 37 lib 3.(7)

10. Si bisecatur angulus secabitur latus oppositum in ratione laterum conterminorum et vice versa. Eucl 3. VI.

De Datis.

Proposito Problemate(8) videndum est semper quid ex datis consequatur ut ex pluribus datis(9) quod quærimus facilius tandem eliciatur. Huic inventionis generi inserviunt Theoremata sequentia.

De rationibus datis.

1. Quorum dantur rationes ad tert[i]um, datur ratio ad invicem.

2. Si duarū vel plurium quantitatum dantur rationes inter se, dantur quoqȝ rationes homogeneorum compositorum ex ipsis. Ut si dentur rationes A, B, C ad invicem, datur quoqȝ ratio A ad $2A+3B$ & ratio $A+2B$ ad $A-B+C$ et ratio A^q ad $BC+2C^q$. Homogeneas verò quantitates voco quæ ad dimensiones æquè altas ascendunt[,] ut A, B, C vel A^q, BC, C^q.

De sectione determinata.

3. Si in linea aliqua AE dentur puncta quotcunqȝ A, B, C, D, E et horum ab alio puncto(10) X in eadem recta sito distantiæ AX, BX, CX, DX, EX vel inter se vel una cum datis quibusvis distantijs M, N[, O, P, Q] utcunqȝ componantur; detur vero vel compositum illud vel duorum ejusmodi compositorum ratio ad invicem, dabitur etiam illud punctum commune idqȝ per geometriam planam ubi in compositis illis distantiæ ad commune punctū illud terminatæ non nisi ad duas dimensiones ascendunt.(11)

A B C X D E

(6) In these familiar extensions of I, 47 it is equivalently demonstrated (compare III: 406) that there is $AC^2-BC^2 = AB^2+2AB\times BP$, where P is the foot of the perpendicular let fall from C onto BD (and BP is to be taken negative when P falls between A and B).

(7) These latter propositions need to be taken jointly: for on drawing AE tangent to the circle-arc BD at E, and hence perpendicular to the corresponding radius EC, by I, 47 (Pythagoras' theorem) there is AC^2-EC^2 (or BC^2) $= AE^2$ where, by III, 35–7, $AE^2 = AB\times AD$.

(8) Initially 'In omni problemate' (In every problem).

(9) Newton first specified more broadly 'ex datis et consequentijs illis simul sumptis' (from those givens and consequences jointly assumed).

(10) Originally 'commun[i]' (a common) in an equivalent preliminary version immediately preceding.

(11) A dubious final phrase 'vel etiam ubi ad tres ascendunt' (or even when they rise to three)—true only if the ratio is reducible to a simpler quadratic form—is cancelled in the manuscript. There Newton first went on to pen a first version of the sequel whose content will be readily understood without our rendering it also in a parallel English version:

'Ut si in recta AE dentur puncta tria A, B, E sitqȝ X punctum ignotum: ex dato

$$AX\times BX-MN\times EX$$

dabitur punctum X. Nam BX est $AX-AB$ et EX est $AE-AX$, adeoqȝ $AX\times BX-MN\times EX$ est $AX^q-AB\times AX-MN\times AE+MN\times AX$, seu $AX^q+V\times AX-MN\times AE$ si modò pro dato $MN-AB$ scribas V. Datur igitur $AX^q+V\times AX-MN\times AE$. Huic adde data $MN\times AE$ et $\frac{1}{4}V^q$

the sides to AD; that is, $AC^2 - BC^2 = AB \times AD$: Euclid, [*Elements*] II, 12 and 13.[6] And [*Elements*] I, 47 and III, 35, 36 and 37 come to the same.[7]

10. If an angle is bisected, the opposite side will be cut in the ratio of the adjacent sides; and conversely so: Euclid, [*Elements*] VI, 3.

Givens

When a problem is proposed,[8] we need always to see what shall be the consequence of the givens in order that from several givens[9] what we seek may the more easily at length be derived. To this kind of investigation the following theorems are of service:

Given ratios.

1. Magnitudes whose ratios to a third are given have a given ratio one to the other.

2. If the ratios of two or more quantities are given to each other, there are also given the ratios of homogeneous ones compounded from these. If, for instance, there be given the ratios of A, B, C to each other, there is also given the ratio of A to $2A+3B$, and that of $A+2B$ to $A-B+C$, and that of A^2 to $BC+2C^2$. I call homogeneous quantities, however, ones which rise to equally high dimensions, as A, B and C; or A^2, BC and C^2.

Determinate section.

3. If in some line AE there be given any number of points A, B, C, D, E and the distances AX, BX, CX, DX, EX of these from any[10] point X situated in the same straight line be arbitrarily compounded either with one another or together with any given distances M, N, O, P, Q whatever, let there be given, however, either that compound or the ratio of two compounds of such a type to one another, and there will also be given that common point—and this by plane geometry when in those compounds the distances terminating at the common point rise but to two dimensions.[11]. Should, for instance, there be given the

---

et dabitur summa $AX^q + V \times AX + \frac{1}{4}V^q$ ejusq; latus quadratum $AX + \frac{1}{2}V$, a quo si auferas datū $\frac{1}{2}V$ restabit datum AX[,] unde datur punctū X.

'Rursus si detur ratio $AX \times BX$ ad $MN \times EX$, cape T ad MN in ea ratione et erit $T \times EX$ ad $MN \times EX$ in eadem ratione adeoq; $AX \times BX = T \times EX$, per princip. [?; see note (4) above]. Ergo $AX \times BX - T \times EX$ datur, utpote nihilo æquale, et proinde dabitur punctum X ut in casu superiori.

'Adhæc si detur ratio $AX \times BX$ ad EX^q. cum sit $AX = AE - EX$ & $BX = BE - EX$ erit

$$AX \times BX = \overline{AE-EX} \times \overline{BE-EX} = AE \times BE - AE \times EX - BE \times EX + EX^q,$$

cujus itaq; ratio ad EX^q datur. Aufer posteriorem de priori et dabitur ratio

$$AE \times BE - EX \times \overline{AE+BE} \text{ ad } EX^q.$$

In ea ratione fac ut sit AE ad M et $AE+BE$ ad N et erit $M \times BE - N \times EX = EX^q$. Æqualibus adde $N \times EX$ et fiet $M \times BE = EX^q + N \times EX$, huic dato adde $\frac{1}{4}N^q$ et dabitur $EX^q + N \times EX + \frac{1}{4}N^q$ una cum ejus latere quadrato $EX + \frac{1}{2}N$. Aufer datum $\frac{1}{2}N$ et restabit datum EX'.

Ut si detur ratio $AX \times BX - M \times CX$ ad $DX \times EX + 2BX \times CX$[(12)] dabitur punctum X. Sed qua ratione ex hujusmodi datis perveniatur ad hoc punctū lubet paucis exponere.[(13)]

Cas. 1. Data ratione AX^q ad $BX \times M$, datur punctum X. Est enim

$$BX = AX - AB,$$

ergo datur ratio AX^q ad $AX \times M - AB \times M$. Capiatur R ad M in ea ratione et per Princip. [(14)] dabitur ratio $AX^q - AX \times R$ ad $-AB \times M$ seu $AX \times R - AX^q$ ad $AB \times M$. Ergo datur $AX \times R - AX^q$. Aufer hoc de dato $\frac{1}{4}R^q$, et dabitur residuum $\frac{1}{4}R^q - AX \times R + AX^q$ una cum ejus latere quadrato $\frac{1}{2}R - AX$, quod ablatū de dato $\frac{1}{2}R$ relinquit datum AX.[(15)] Datur ergo punctum X.

Cas. 2. Data ratione AX^q ad $BX \times N + P, Q$ datur X. Nam fac $N.P::Q.GB$ et $BX \times N + P, Q$ valebit $BX \times N + GB \times N$ seu $GX \times N$. Datur ergo ratio AX^q ad $GX \times N$ adeoq; datur punctū X ut in casu priori.[(16)]

Cas. 3. Data ratione AX^q ad $BX^q + M \times BX + N, O$ datur X. Nam auferendo prius de posteriore datur ratio AX^q ad residuum[,] quod residuum si pro AX scribas $AB + BX$ erit $N, O + M \times BX - 2ABX - AB^q$ seu $P \times BX + Q, R$ si modo scribas insuper P pro dato $M - 2AB$ et Q, R pro dato $N \times O - AB^q$. Datur ergo ratio AX^q ad $P \times BX + Q, R$, ut in casu 2.[(17)]

Cas. 4. Data ratione AX^q ad $BX^q + \frac{d}{e} CX^q + M, CX + P, Q$ datur X. Nam BX

---

(12) Since this is intended to be a canonical form of the general 'determinate section' we would perhaps have expected Newton to set the second member of his ratio to be

$$\text{`}DX \times EX + BX \times N + O, P\text{'}.$$

As enunciated by Pappus in his report on Apollonius' lost work upon it (*Mathematicæ Collectiones*, $_2$1660: 243, quoted in 1, §1: note (28) above) the classical *sectio determinata* posed the restricted problem of defining the point X in the line $ABCD$ such that the ratio of $AX \times BX$ to $CX \times DX$ is given. Newton's present extension of this is (see the next note) narrowly in the spirit of his equivalent elaborations of Rule 1 in 1, §1.3 preceding.

(13) Exactly as in his parallel progress to increasingly more complex cases of a quadratic *sectio determinata* on pages 216–20 above, given here in general the ratio of $AX \times BX - M.CX$ to $(DX \times EX + N.FX + O.P$ and hence *dividendo* to)

$$AX \times BX - M.CX - DX \times EX - N.FX - O.P,$$

Newton's technique of reduction is to express all the remaining variable line-segments $BX (= AX - AB)$, $CX (= AX - AC)$, $DX (= AX - AD)$, ... in terms of a single one of these, AX, and then, on putting the given ratio to be k and collecting the constant segments

$$l = (k-1).(AB+M) - k.(AD+AE-N)$$

and $$m = -(k-1)\, M.AC + k.(AD \times AE - N.AF + O.P),$$

to solve the simple quadratic equation $AX^2 + l.AX + m = 0$ which ensues; namely, as in Case 1, by completing the square $(AX + \frac{1}{2}l)^2 = \frac{1}{4}l^2 - m$ and thence extracting the square root to obtain $AX = -\frac{1}{2}l \pm \sqrt{[\frac{1}{4}l^2 - m]}$. In our reproduction of the following enumerated cases of the general problem we have silently standardized a few trivial inconsistencies of mathematical detail in the manuscript text.

ratio of $AX\times BX-M\times CX$ to $DX\times EX+2BX\times CX$,[12] then the point X will be given. But it is agreeable to disclose in a few instances the method by which we may from givens of this sort attain this point.[13]

Case 1. Given the ratio of AX^2 to $BX\times M$, there is given the point X. For $BX=AX-AB$, and consequently the ratio of AX^2 to $AX\times M-AB\times M$ is given. Take R to M in that ratio, and by Principle [14] there will be given the ratio of $AX^2-AX\times R$ to $-AB\times M$, that is, $AX\times R-AX^2$ to $AB\times M$. Therefore $AX\times R-AX^2$ is given. Take this away from the given $\frac{1}{4}R^2$, and the remainder $\frac{1}{4}R^2-AX\times R+AX^2$ will be given, along with its square root $\frac{1}{2}R-AX$; and when this is taken from the given $\frac{1}{2}R$ there remains the given AX.[15] The point X is therefore given.

Case 2. Given the ratio of AX^2 to $BX\times N+P\times Q$, there is given X. For make $N:P=Q:GB$, and $BX\times N+P\times Q$ will then be equivalent to $BX\times N+GB\times N$, that is, $GX\times N$. There is therefore given the ratio of AX^2 to $GX\times N$, and hence the point X is given as in the previous case.[16]

Case 3. Given the ratio of AX^2 to $BX^2+M\times BX+N\times O$, there is given X. For on taking the former away from the latter there is given the ratio of AX^2 to the residue, and if in place of AX you write $AB+BX$ this remainder will be $N\times O+M\times BX-2AB\times BX-AB^2$, that is, $P\times BX+Q\times R$ if you now in addition write P for the given $M-2AB$ and $Q\times R$ for the given $N\times O-AB^2$. There is therefore given the ratio of AX^2 to $P\times BX+Q\times R$, as in Case 2.[17]

Case 4. Given the ratio of AX^2 to $BX^2+(d/e)\,CX^2+M\times CX+P\times Q$, there is

---

(14) Some one of the missing Principles 1–3 and 7/8, we presume; compare note (4) above.

(15) Namely, $\frac{1}{2}R-\sqrt{[\frac{1}{4}R^2-S^2]}$ where $S^2=R.AX-AX^2$. Newton here makes use of the standard mode of solution of a quadratic equation which he had expounded as 'Reg: 7' of his *Arithmetica Universalis* (see v: 116–18) .

(16) Initially this was followed (on f. 34r) by:

'[*Cas.*] *3.* Data ratione AX^q ad BX^q+P,Q datur X. Nam auferendo prius de posteriori datur ratio AX^q ad BX^q-AX^q+P,Q. Sed AX^q valet $BX^q+2ABX+AB^q$. ergo datur ratio AX^q ad $P,Q-AB^q-2ABX$, et inde datur punctum X ut in casu [2]'. As before (compare note (11)) the sense of this will be clear without our tiresomely rendering it word by word into its English equivalent.

(17) This in turn was followed by two separate further cases (whose Latin text we again omit to translate):

'*4.* Data ratione AX^q ad $BX\times CX+P\times BX+N,O$ datur punctum X. Nam $BX\times CX$ est $BX\times\overline{BX-BC}$ seu $BX^q-BC\times BX$. Ergo datur ratio AX^q ad $BX^q\begin{smallmatrix}-BC\\+P\end{smallmatrix}\times BX+Q,R$ et inde punctū X ut in casu 3.

'*5.* Data ratione $AX\times DX$ ad $BX\times CX+P\times BX+Q,R$, datur X. Nam DX est $AD-AX$, adeoq; $AX\times DX$ est $AX\times AD-AX^q$. Datur ergo ratio hujus ad $BX\times CX+P\times BX+Q,R$. In ea ratione fac esse AD ad S et per Princip. [?; see note (4)] auferendo hinc $AX\times AD$ illinc $AX\times S$ seu $S\times\overline{AB+BX}$ restabit $-AX^q$ ad $BX\times CX\begin{smallmatrix}-S\\+P\end{smallmatrix}\times BX\begin{smallmatrix}+Q,R\\-S,AB\end{smallmatrix}$ in data rationc. Unde datur [X] per cas. 4.'

est $BC+CX_{[,]}$ cujus quadrato substituto dabitur ratio AX^q ad

$$BC^q+2BCX+\frac{d+e}{e}CX^q[+M,CX]+P,Q.$$

Applica posteriorem rationis terminū ad $\frac{d+e}{e}$ et dabitur ratio AX^q ad

$$\frac{e}{d+e}\times\overline{BC^q+P,Q}+\frac{2e}{d+e}\overline{B[C]+M},CX+CX^q.$$

Unde per cas. 3 datur punctum X.

Ad hosce vero casus cæteri omnes facile reducuntur. Nam ubi in rationis terminis habetur duarum incognitarum rectangulum,(18) reductio fit ad modum exempli hujus. Sit $AX\times BX$ ad $CX\times DX$ in data ratione. Biseca AB et CD in M et N et fiet $\overline{MX+AM}\times\overline{MX-AM}$ ad $\overline{CN-NX}\times\overline{CN+NX}_{[,]}$ hoc est

$$MX^q-AM^q \quad \text{ad} \quad CN^q-NX^q$$

in data ratione. Fac AM ad P in ea ratione et erit MX^q ad $CN^q+AM\times P-NX^q$ in eadem ratione. Ergo datur X ut in casu 3. Porro ubi linea incognita in terminis rationis tam unius quam duarum est dimensionum reductio fit hoc modo. Sit AX^q-BAX ad $CX\times DX$ in data ratione. Pro $AX-AB$ scribe BX et fiet $AX\times BX$ ad $CX\times DX$ in data ratione. Unde per reductionem in casu priore dabitur X. Deniq; ubi linea aliqua incognita non est duarum dimensionū sed unius tantum, reductio magis in pro[m]ptu est. Sed his immorari tædet.(19) Subnoto tantum veteres sectionem lineæ in pun[c]to X vocasse sectionē determinatam et [Apollonius] de sectionis hujus casibus simplicioribus et elegantioribus librum scripsisse, quem [Willebrordus Snellius] ex Pappi *collectionibus* restituere conatus est.(20) Sufficit hic methodum generalem determinandi hanc sectionem breviter indicasse ut sectio quoties in aliquo Problemate determinanda occurrat jam determinata habeatur.

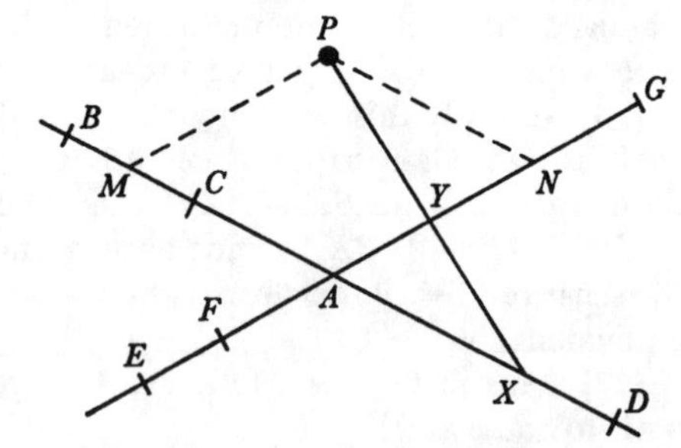

De sectione spatij.

4. Si a dato puncto P in rectas duas positione datas AX, AY recta tertia utcunq; incidat ad X et Y, in rectis vero prioribus dentur puncta quotcunq; A, B, C, D, E,[F, G] et partes unius rectæ AX, BX, CX, DX a

(18) Newton originally continued: 'ut $AX\times BX$: bisecando distantiam datorum punctorū A, B, puta in Q, et pro AX et BX scribendo $QX+AQ$ & $QX-AQ$, transmutatur rectangulū in unius incognitæ et unius datæ differentiam quadratorum QX^q-AQ^q. Ubi habetur lineæ incognitæ ut AX tum quadratum AX^q tum rectangulum cum dato $AX\times[AB]$ utrumq; pro uno rectangulo sub duabus incognitis habendum est, et ut in priori casu reducendum.

$$AX^q-AX\times AB = AX\times BX = [Q]X^q-[Q]A^q.$$

Ubi incognitæ non habetur quadratum sed contentū rectangulum tantum cum cognita, pro incognita substituenda est alia incognita aucta vel diminuta excessu dato quo una superat alteram' (as $AX\times BX$, by bisecting the distance between the given points A and B, say in Q,

given X. For BX is $BC+CX$, and when its square is substituted there will be given the ratio of AX^2 to $BC^2+2BC\times CX+((d+e)/e)\,CX^2+M\times CX+P\times Q$. Divide the ratio's latter member by $(d+e)/e$ and there will be given the ratio of AX^2 to $(e/(d+e))\,(BC^2+P\times Q)+2(e/(d+e))\,(BC+M)\,CX+CX^2$. Whence by Case 3 there is given the point X.

To these cases to be sure, all the rest are easily reduced. For when in the members of a ratio there is had a rectangle of two unknowns,(18) reduction is made in the manner of the following example. Let $AX\times BX$ be to $CX\times DX$ in a given ratio: bisect AB and CD in M and N, and there will come to be

$$(MX+AM)\,(MX-AM) \quad \text{to} \quad (CN-NX)\,(CN+NX),$$

that is, MX^2-AM^2 to CN^2-NX^2, in the given ratio; make AM to P in that ratio, and then will MX^2 be to $CN^2+AM\times P-NX^2$ in the same ratio; therefore X is given as in Case 3. Moreover, when the unknown line in the ratio's members is of one as well as two dimensions reduction is made in this way. Let

$$AX^2-AB\times AX$$

be to $CX\times DX$ in a given ratio: in place of $AX-AB$ write BX and there will come to be $AX\times BX$ to $CX\times DX$ in a given ratio; thence by the reduction in the previous case there will be given X. Lastly, when some unknown line is not of two dimensions but merely of one, the reduction is more immediate. But it is tiresome(19) to linger over these things. I merely note in passing that the ancients called the cutting of the line in the point X 'determinate section', and that Apollonius wrote a book on the simpler and choicer cases of this section—one which Willebrord Snell has attempted to restore out of Pappus' *Collection*.(20) It is here sufficient to have briefly indicated a general method for determining this section, so that each time it occurs to be determined in some problem it may be regarded as determined.

4. If two straight lines AX, AY given in position should be arbitrarily met in X and Y by a third straight line from P, let there be given in the former lines any number of points A, B, C, D, E, F, G and individually multiply the parts AX, BX, CX, DX of one line, extending from the given points to the cut-point X,

Cutting off a space.

and in place of AX and BX writing $QX+AQ$ and $QX-AQ$ the rectangle is transformed into the difference, QX^2-AQ^2, of the squares of one unknown and one given. Where of an unknown line such as AX there is had both its square AX^2 and its rectangle $AX\times AB$ with a given, both need jointly to be considered as one rectangle beneath two unknowns and reduced as in the former case: $AX^2-AX\times AB = AX\times BX = QX^2-AQ^2$. Where of the unknown there is had not the square but merely the rectangle contained by it with a known, in place of the unknown you must substitute another unknown augmented or diminished by the given excess by which one surpasses the other). In the last case understand that, given $k.AX+l.BX$, reduction is to be made by substituting $BX = AX-AB$.

(19) Originally 'non vacat' (there is no time) and then 'pertæsum est' (it is wearisome).

(20) Namely, in his *Apollonius Batavus, seu Exsuscitata Apollonii Pergæi Geometria* (Leyden, 1608); compare §1: note (32) preceding.

datis punctis ad sectionem X extensæ, ducantur in partes alterius AY, EY, [FY, GY] a datis ad punctum sectionis Y extensas singulæ in singulas: ex dato rectangulorū uno[21] vel duorum summa differentia vel ratione ad invicem vel plurium per additionem aut subductionem aggregato, dantur puncta sectionū X et Y.

Ipsis AX, [A]Y parallelæ [agantur] PN, PM complentes parallelogrammū AP. Et propter sim[ilia] triangula PNY, XMP erit $MX.MP::PN.NY$. Ergo $NY = \frac{MPN}{MX}$. Proponatur jam rectangulū $BX \times EY$ et cum sit $EY = EN - NY$, rectangulum illud erit $BX \times \overline{EN - \frac{MPN}{MX}}$ seu $BX \times EN - BX \times \frac{MPN}{MX}$. Eodem modo rectangulum $CX \times FY$ est $CX \times FN - CX \times \frac{MPN}{MX}$ et sic in cæteris. Si jam detur hujusmodi rectangulum aliquo[d] vel rectangulorum summa vel differentia vel ratio ad invicem, ductis terminis in MX dabitur punctum X per sectionem determinatam.[22] Hanc autem sectionem Veteres vocabant sectionem spatij eo quod pars lineæ unius cum parte lineæ alterius spatiū comprehendat. Casus vero sectionis hujus simpliciores exposuit olim [Apollonius] cujus opus ex Pappi *collectaneis* restituit [Snellius].[20] Sed ad inventionem minus confert hæc sectio quam sectio determinata.

De Triangulis specie datis.

5.[23] In triangulo aliquo ABC demisso perpendiculo CD, si ex angulis omnibus A, B, ACB, ACD, BCD et rationibus duarū quarumvis ex lineis AB, BC, CD, AD, BD linearumq; summis ac differentijs dentur duo quævis quarum unum non dat alterū, datur triangulum specie.

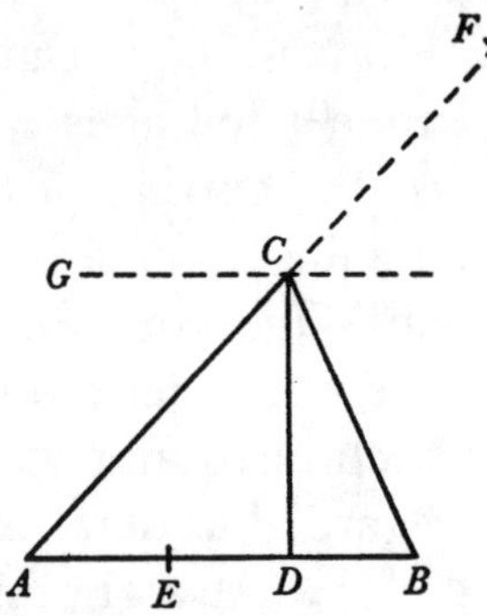

F G C A E D B

Ut si detur angulus ACD et ratio $AD - BD$ ad BC. Sit $DE = DB$ et erit $AE = AD - BD$[24] et $EC = BC$[,] ergo datur ratio AE ad EC. Assume AE et dabitur EC[,] et centro E radio EC describendo circulū secantem

(21) That is, $AX \times AY$, $BX \times EY$, $CX \times FY$ or $DX \times GY$ and so on.

(22) Other than in the porismatic instance in which the rectangle $MX \times NY$, constant for all rectilinear transversals XY through the pole P, is given; on this indeterminate Apollonian case see Porism 1 of Newton's preceding 'Inventio Porismatum' (1, §2 above).

(23) In the manuscript this section is entered after the one next following (which was itself originally numbered '5').

(24) Newton first went on to argue: 'Assumatur ergo AE et ob datam ejus rationem ad BC dabitur BC. Est autem (per Princip [9]) $AE \times AB = AC^q - BC^q$ seu $AC^q - AE \times AB = BC^q$. Ob datum angulum ACD datur ratio AC ad AD adeoq; ratio AC^q ad AD^q. In ea ratione fac esse AE ad N et dabitur $AD^q - N \times AB$. Est $AB = 2AD - AE$. ergo datur $AD^q - 2N \times AD + N \times AE$. Aufer datum $N \times AE$ & adde datum N^q et dabitur $AD^q - 2N \times AD + N^q$ ut et latus ejus $AD - N$ vel $N - AD$. Unde cum detur N dabitur AD. Quo invento cætera simul dantur. Ex una igitur

by one each of the parts AY, EY, FY, GY of the other, extending from the givens to the cut-point Y: then, given one of the rectangles,[21] or the sum, difference or mutual ratio of two, or the aggregate of several by addition or subtraction, therefrom the cut-points X and Y are given.

Parallel to AX, AY draw PN, PM completing the parallelogram AP. Then because of the similar triangles PNY, XMP there will be $MX:MP = PN:NY$, and therefore $NY = MP \times PN/MX$. Let the rectangle $BX \times EY$ be propounded, and then, since $EY = EN-YN$, that rectangle will be

$$BX \times (EN - MP \times PN/MX),$$

that is, $BX \times EN - BX \times MP \times PN/MX$. In the same way the rectangle $CX \times FY$ is $CX \times FN - CX \times MP \times PN/MX$, and the like in other instances. If there now be given some rectangle of this sort, or the sum, difference or mutual ratio of such rectangles, when the terms are multiplied by MX the point X will be given by determinate section.[22] The ancients called this section, I may add, cutting off a space seeing that part of one line together with part of the other is required to comprehend a space. The simpler cases of this section were, in fact, exhibited by Apollonius in times long past, and his work has been restored, from Pappus' *Collection*, by Snell.[20] But this section has less to contribute to discovery than determinate section.

Triangles given in species.

5.[23] When in some triangle ABC the perpendicular CD is let fall, if out of all the angles, $\hat{A}$, $\hat{B}$, $A\widehat{C}B$, $A\widehat{C}D$, $B\widehat{C}D$ and the ratios of any two of the lines AB, BC, CD, AD, BD and the sums and differences of these lines there be given any two, provided that one does not give the other the triangle is given in species.

If, for instance, there be given $A\widehat{C}D$ and the ratio of $AD-BD$ to BC, let $DE = DB$ and there will be $AE = AD-DB$[24] and $EC = BC$, and therefore the ratio of AE to EC is given; assume AE and there will be given EC, and then, on describing a circle with centre E and radius EC cutting the side AC of the given

---

linea AE assumpta triangulum determinatur magnitudine et proinde ante lineæ illius assumptionem dabatur specie' (Let AE be assumed, therefore, and because of its given ratio to BC there will be given BC. However, by Principle 9 there is $AE \times AB = AC^2 - BC^2$, that is, $AC^2 - AE \times AB = BC^2$. Because the angle $A\widehat{C}D$ is given, there is given the ratio of AC to AD and hence that of AC^2 to AD^2. In that ratio make there to be AE to N, and there will be given $AD^2 - N.AB$. Now $AB = 2AD - AE$, therefore $AD^2 - 2N.AD + N.AE$ is given. Take away the given $N.AE$ and add the given N^2, and there will be given $AD^2 - 2N.AD + N^2$, as also its 'side' (root) $AD-N$ or $N-AD$. Whence, since N is given, AD will be given. From one line AE assumed, therefore, the triangle is determined in size, and consequently before the assumption of that line it was given in species). The cogency of Newton's argument will appear to better effect, perhaps, if we look upon AE as having unit-length: since the linear elements of the figure (and also N, in constant ratio to AE) are then given in size relatively to AE and so to each other, it is—independently of the absolute size of the unit AE— given in species.

Anguli dati EAC(25) latus AC in C dabitur AC una cum angulo CEB vel CBA, quibus determinatur triangulū. Triangulū igitur cum ex unius lineæ AE assumptione determinetur, ante lineæ illius assumptionem dabatur saltem specie.(26)

Detur angulus A et ratio $AD-B[D]$ ad $AC+BC$. Cape $DE=DB$, produc AC ad F ut sit $CF=CB$ et dabitur ratio AE ad AF. Ergo dantur anguli AEF, AFE quorum differentia est AEC.(27) Unde datur triangulum specie ut ante.

Dētur ratio AC ad CB et ratio AB ad CD, & dabitur specie triangulū. Nam si assumatur AB, dabitur CD. Ipsi AB ad distantiam CD parallelam age CG. Et cum detur ratio AC ad BC dabitur punctum C per sectionem incidentiæ.(28) Quo invento datur triangulum ABC. Cognitis igitur angulis ejus et linearum rationibus(29) cognoscitur species trianguli ubi AB non assumitur.

Sed casus omnes percurrere non vacat. Neq; his difficiliores usui esse(30) solent.

De triangulis datis.

6.(31) Si in triangulo aliquo ABC demisso perpendiculo CD, dentur ex his omnibus, nimirum lineis AC, BC, AB, AD, BD, DC, $AC\pm BC$, $AC\pm AB$, $AD-BD$, $AC+BC+AB$, angulis A, B, C, areis ABC, ADC, BDC, tria quævis quorum duo sola non dant tertium; ex datis illis tribus datur triangulum. Casus sing[u]los percurrere nimium esset. Nos unum et alterum in specimen exhibemus.

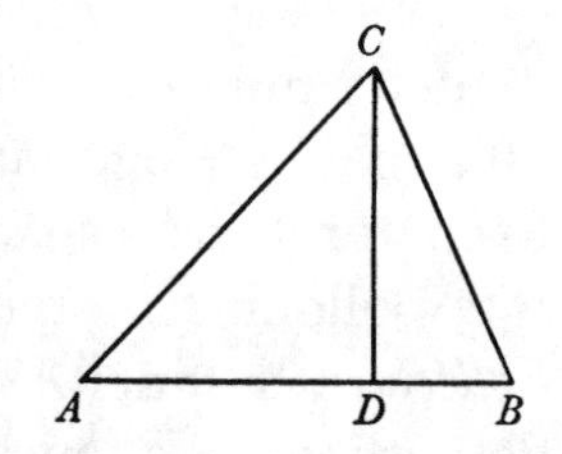

Dato trianguli cujusvis ABC perimetro $AC+BC+AB$, area ABC & angulo uno A, invenire triangulum.(32) Ob datā areā datur $CD\times AB$ [et] ob datum ang. A datur ratio DC ad AD et $AC_{[,]}$ ergo datur DAB et $AC\times AB$. At

$$2DAB = AC^q + AB^q - BC^q.$$

Huic dato adde datum $2AC\times AB$ et summa quæ est $\overline{AC+AB}^q - BC^q$, seu $\overline{AC+AB+BC}\times\overline{AC+AB-BC}$ dabitur. Ergo cum latus $AC+AB+BC$ detur, dabitur et latus $AC+AB-BC$ et laterum semisumma $AC+AB$, semidifferentia BC. A dati $AC+AB$ quadrato $AC^q+2AC\times AB+AB^q$ aufer datum $4AC\times AB$ et

(25) The complement of the given angle $\widehat{ACD}$.

(26) By virtue, of course, of the homogeneity of the given relationship (of constant ratio) between the linear elements $AB-BD = AE$ and $BC = EC$. (If, for instance, the ratio of AE^2 to EC had been given, the argument would fail.) More straightforwardly, since both the angle $\widehat{EAC}$ and the ratio $AE/EC = \sin\widehat{ACE}/\sin\widehat{EAC}$ are given, so too is $\widehat{ACE}$ and hence also is $\widehat{EBC} = (\widehat{BEC}$ or) $\widehat{EAC}+\widehat{ACE}$, whence the triangle ABC is determined in its angles and so in species.

(27) Newton first added 'et hujus complementum' (and the supplement of this), *viz.* $\widehat{BEC}$.

(28) Namely, by CG's meet with the Apollonius circle (C) defined by the constancy of the ratio AC/BC.

(29) So rephrased from an initial choice of 'proportionibus' (proportions). The distinction is not here trivial; compare our comment in note (26) above.

(30) 'occurrere' (met with) was first written. Newton has made the change solely, we

angle $\widehat{EAC}$(25) in C, there will be given AC along with $\widehat{CEB}$, that is, $\widehat{CBA}$, and by these the triangle is determined. Since, then, the triangle is determined from assuming the single line AE, before the assumption of that line it was given at least in species.(26)

Let there be given the angle $\hat{A}$ and the ratio of $AD-DB$ to $AC+BC$. Take $DE = DB$ and extend AC to F so that $CF = CB$, and the ratio of AE to AF will be given; therefore the angles $\widehat{AEF}$ and $\widehat{AFE}$ are given, and so too their difference $\widehat{AEC}$.(27) Whence the triangle is given in species as before.

Let the ratio of AC to CB and that of AB to CD be given, and the triangle will be given in species. For, if AB be assumed, CD will be given; draw CG parallel to AB at the distance CD and then, since the ratio of AC to BC is given, the point C will be given by the intersection of an incidence.(28) Once this is found, the triangle ABC is given. With its angles and ratios(29) of lines consequently known, the species of the triangle is known when AB is not assumed.

But time does not allow us to run through all the cases. Nor are ones more difficult than these usually of use.(30)

Given triangles.

6.(31) If in some triangle ABC where the perpendicular CD is let fall there be given of the following total—namely, the lines AC, BC, AB, AD, BD, DC, $AC \pm BC$, $AC \pm AB$, $AD-BD$ and $AC+BC+AB$; the angles $\hat{A}, \hat{B}, \hat{C}$; and the areas ABC, ADC and BDC—any three, no two of which by themselves yield the third, the triangle is given therefrom. It would be tedious to run through the separate cases. We display a couple as a sample.

Given in any triangle ABC the perimeter $AC+BC+AB$, the area ABC and one angle $\hat{A}$, to find the triangle.(32) Because of the given area there is given $CD \times AB$, and because of the given angle $\hat{A}$ there is given the ratio of CD to AD and to AC, so consequently there is given $AD \times AB$ and $AC \times AB$. But

$$2AD \times AB = AC^2 + AB^2 - BC^2:$$

to this given add the given $2AC \times AB$ and the sum, that is, $(AC+AB)^2 - BC^2$ or $(AC+AB+BC)(AC+AB-BC)$ will be given. Therefore, since the component $AC+AB+BC$ is given, so also will be the component $AC+AB-BC$, along with their half-sum $AC+AB$ and half-difference BC. From the square,

$$AC^2 + 2AC \times AB + AB^2,$$

of the given $AC+AB$ take away the given $4AC \times AB$, and there will be given the

presume, to avoid the harsh assonance thereby resulting with the parallel infinitive 'percurrere' in the previous sentence.

(31) Originally—when the present section was set to follow immediately after that 'De sectione spatij'—Newton wrote '5'; see note (23).

(32) This was the fourth geometrical problem of Newton's 'Arithmetica Universalis' some fifteen years before; see v: 188. The present analysis by 'givens' is essentially that already set out on v: 510.

dabitur reliquum $AC^q - 2AC \times AB + AB^q$ una cum ejus latere $AC - AB$. Hujus et $AC + AB$ semisumma est AC[,] semidifferentia AB.

Dato angulo ACD, ratione laterum AC, CB et segmento basis BD, datur triangulum. Nam duobus prioribus datis datur specie, et ex tertio datur magnitudine.

7. In figuris quadrilateris(33) ductis diagonalibus & demissis perpendiculis et lateribus oppositis productis donec concurrant: ubi ex angulis et rationibus linearū quæ in figura sunt dantur quævis quatuor quorum tria non dant quartum, determinatur figura specie.(34)

8. Et ubi ex angulis, lineis et linearum rationibus dantur quinqꝫ quarum quatuor non dant quintum, determinatur figura tam magnitudine quam specie.

Eadem obtinent ubi alia quæcunqꝫ quatuor vel quinqꝫ dātur. Ex quibusvis quatuor datis ubi lineæ nullius earum quæ in figura sunt longitudo aut immediate aut per consequentiam datur, figura determinatur specie: ex quinqꝫ ubi linea una vel plures earum quæ in figura sunt datur, figura determinatur specie et magnitudine. Ut si in datis est perpendiculum aliquod, vel ratio perpendiculi ad aliam lineam earum quæ in figura sunt, vel quod figura in circulo inscribitur, vel præterea quod circulus ille datur. Nam duorum datorum vicem habet si figura quadrilatera in circulo dato(35) inscribitur. Cæterum dum ex hujusmodi datis figuram specie vel specie et magnitudine determina[r]i affirmem, hoc non dico ut figura quoties ejusmodi data occurrant pro jam cognita habeatur, nisi forte in casibus quibusdam simplicioribus, sed tantum ut inde innotescat difficultates Problematis propositi ad inventionem determinatæ speciei vel determinatæ speciei et magnitudinis figuræ quadrilateræ reduci. Ubi vero ejusmodi data occurrerint, inventioni speciei vel speciei et magnitudinis figuræ sæpe inservient hæc Theoremata.

1. Angulus ad circumferentiam dimidium est anguli ad centrum, adeoqꝫ in dato circuli segmento datur. Eucl [20, 21. III].(36) Hoc theorema ad figuras omnes in circulo inscriptas spectat.

2. In Figura quadrilatera in circulum inscripta, anguli oppositi simul sumpti æquantur duobus rectis, adeoqꝫ ex uno dato datur alter. Eucl [22. III].

3. In Figura quadrilatera in circulum inscripta, summa rectangulorū sub lateribus oppositis æquatur rectangulo sub diagonijs.(37)

4. In omni Figura quadrilatera $ABCD$, lateribus oppositis AB, DC & AD, BC

(33) Originally 'æquilateris' (equilateral) by a slip of the pen. Newton would evidently understand these to be convex, though it would not here matter to include re-entrant figures.

(34) This is not quite true. If in the quadrilateral $ABCD$ we are given the ratios of the diagonal AC to each of the sides AB, BC and CD, and also the angle $\widehat{CAD}$, then it is not uniquely determined in species. And the same is true when the angles $\widehat{BAC}$, $\widehat{BCA}$ and $\widehat{CAD}$ are given along with the ratio of AC to CD.

remainder $AC^2 - 2AC \times AB + AB^2$ together with its 'side' $AC - AB$. The half-sum of this and $AC + AB$ is AC, the half-difference AB.

Given the angle $A\widehat{C}D$, the ratio of the sides AC and CB, and the base segment BD, the triangle is given. For, given the two first, it is given in species, and so from the third it is given in size.

7. In four-sided figures(23) where diagonals are drawn, perpendiculars let fall and opposite sides produced till they meet, when from the angles and ratios of lines in the figure any four are given, no three of which yield the fourth, the figure is determined in species.(34)

8. And when from the angles, lines and ratios of lines five are given, no four of which yield the fifth, then the figure is determined both in size and species.

The same holds when any four or five elements whatever are given. From any four given, when the length of none of the figure's lines is given either directly or by inference, a figure is determined in species; from five, when one or more of the lines in the figure are given, a figure is determined in species and size—if among the givens, for instance, there is some perpendicular or the ratio of that perpendicular to another of the lines in the figure, or it is given that the figure is inscribable in a circle or, further, that this circle is given. (If the figure is inscribed in a given circle,(35) that of course does duty for two givens.) While I may, however, affirm that from givens of this sort a figure is determined in species or in species and size, I say this not that a figure should, except maybe in certain simpler cases, be regarded as now known each time that givens of the sort occur, but merely so that it should thereby be understood that the difficulties of a problem propounded are reduced to finding a quadrilateral of determinate species or determinate species and size. When, to be sure, givens of the sort do occur, the following theorems are often of service in ascertaining a figure's species or species and size.

1. The angle at a circumference is half the angle at the centre, and hence in a given circle segment it is given. Euclid III, 20/21.(36) This theorem has regard to all figures inscribed in a circle.

2. In a quadrilateral inscribed in a circle the opposite angles taken together equal two right angles, and hence when one is given the other is given from it. Euclid III, 22.

3. In a quadrilateral inscribed in a circle the sum of the rectangles contained by opposite sides is equal to the rectangle beneath the diagonals.(37)

4. In every quadrilateral $ABCD$ in which the opposite sides AB, DC and

(35) Of given centre and radius, that is.

(36) The manuscript erroneously reads 'Eucl IV'.

(37) Ptolemy's theorem relating the sides and diagonals of a cyclic quadrilateral; compare IV: 165, note (67).

concurrentibus in E et F, solidum $AB \times FD \times EC$ æquatur solido

$FA \times E[B] \times DC$.(38)

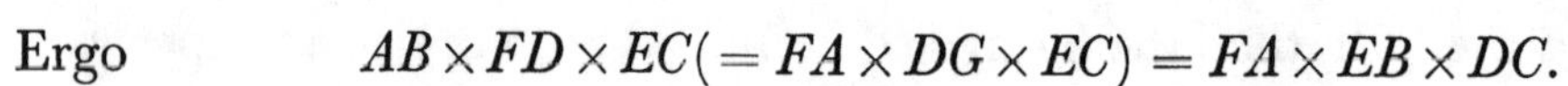

Nam duc DG parallelam AB, et erit

$$FA.AB(::FD.DG):: FD \times EC.DG \times EC$$

et $$EC.EB(::DC.DG):: FA \times DC.FA \times DG.$$

Ergo $$AB \times FD \times EC (= FA \times DG \times EC) = FA \times EB \times DC.$$

Hæc sunt ferè Geometriæ principia rectilinea. Quorum usus e præceptis sequentibus magis elucescet.(39)

Reg[*ula*] *1.* Proposito aliquo problemate de schematis constructione videndum est qua(40) ratiocinia p[r]ocedere possint. Ea fieri solet producendo lineas figuræ donec aliquæ cum alijs(41) concurrant, jungendo puncta ac ducendo novas lineas prioribus aut parallelas aut perpendiculares ut figura in triangula aut specie data aut similia aut rectangula vel in alias forte figuras notam aliquam proprietatem habentes resolvatur. Qua de causa tangentes etiam circulorum aut aliarum curvarum ubi de duarum curvarū contactu vel intersectione agitur non raro ducimus. Aliquando tamen sufficit schema primo exhi[bi]tum sine u[l]teriori constructione. Ubi verò non sufficit id agendum est ut constructio sit, quam fieri potest(42) simplex, et ad Regulas sequentes in usum deducendas apta.

Reg. 2.(43) Quamvis per constructionem lineæ multiplic[e]ntur, eo tamen semper collimemus ut omnes Problematis conditiones ad lineas paucissimas & si commodè fieri potest ad unam aliquam per debita ratiocinia tandem reducantur. Hoc enim pacto deveniri solet ad sectionem determinatam & aliquando ad sectionem angularē vel forte ad sectionem spatij. Quod ubi contigit Problema ob notas illas sectiones solutum est.

Ut si dentur positione quatuor rectæ AE, BD, BF, AG et quinta DG ducenda est quæ ab his secabitur in partes DE, EF, FG datam inter se rationem

(38) Menelaus' theorem for the plane triangle. The following proof is that given to the equivalent Porism on page 272 above (see §1: note (62)).

(39) The ensuing rules are in the main a revision of the draft set in 1, §1.2 preceding, filled out from the equivalent passages in Newton's earlier 'Arithmetica' (see especially v: 168–70 and 612–20) where he lists similar precepts for reducing a given problem to a more straightforwardly analysable form. From his 'Arithmetica', too, he borrows several of the worked examples which he now inserts in illustration.

AD, BC meet in E and F the 'solid' $AB \times FD \times EC$ is equal to the 'solid' $FA \times EB \times DC$.[38] For draw DG parallel to AB and there will then be

$$FA:AB = (FD:DG \text{ or}) \, FD \times EC:DG \times EC$$

and also $$EC:EB = (DC:DG \text{ or}) \, FA \times DC:FA \times DG.$$

In consequence

$$AB \times FD \times EC = (FA \times DG \times EC =) \, FA \times EB \times DC.$$

These, virtually, are the rectilinear principles of geometry. Their use will emerge more lucidly from the following precepts.[39]

Rule 1. When some problem relating to the construction of a scheme is proprosed, we need to see by what path our reasoning can proceed. That is usually done by extending the lines in the figure till some meet others,[41] joining points and drawing new lines either parallel or perpendicular to the previous ones so that the figure may be resolved into triangles either given in species or similar or right-angled or, it may be, into other figures possessing some known property. For this reason we not infrequently draw tangents to circles or other curves as well when it is a matter of the contact or intersection of two curves. Sometimes, however, the scheme initially presented suffices without further construction. But when it is insufficient we need to contrive the construction to be as simple as possible[42] and appropriate for bringing the following rules into play.

Rule 2.[43] Though by construction lines are multiplied, we should always aim nonetheless ultimately to reduce all the conditions of a problem to the fewest number of lines and, if it can conveniently be done, just to some particular one. In this manner, to be sure, we usually arrive at a determinate section or sometimes an angular one or maybe the cutting off of a space. And when this happens the problem is, because those sections are known, then solved.

If, for instance, four straight lines AE, BD, BF, AG be given in position and a fifth, DG, is to be drawn which shall be cut by these into parts DE, EF, FG

(40) Understand 'viâ', and so we render the sense in our facing English translation.

(41) 'si opus est etiam productis' (if need be, also extended) is deleted in the manuscript.

(42) Originally 'quantum liceat' (... as permissible).

(43) This was initially combined with the preceding rule to be a 'Reg. 1' prescribing: 'Proposito aliquo Problemate circa figuram plurium linearū, id apprimè agendum est ut quæstionis conditiones ad lineas paucissimas ac tandem si commodè fieri potest ad unam transferamus. Hoc pacto Problema reduci solet ad sectionem determinatam, & nonnunquam ad sectionem angularem vel sectionem spatij' (Where some problem is proposed about a figure of several lines, what needs first of all to be done is that we should translate the conditions of the question to the fewest lines and ultimately, if it can conveniently be so, to one. In this manner a problem is usually reduced to a determinate section, and on the odd occasion to an angular section or the section of a space).

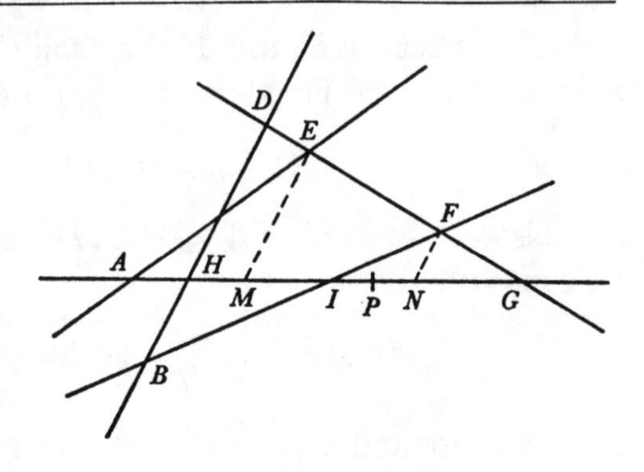

habentes.(44) E lineis seligatur aliqua ut AG ad quam problematis conditiones omnes reducantur.(45) Igitur hanc secet DB in H et FB in $I_{[,]}$ et ad eam a punctis E et F si agas EM FN parallelas DH, ob datas rationes DE, EF, FG ad invicem dabuntur rationes HM, MN, NG ad invicem. In triangulis EMA, FNI specie datis dantur rationes EM ad AM et FN ad IN; datur et ratio $[E]M$ ad $[F]N$, quippe quæ est EG ad $FG_{[,]}$ ergo datur ratio AM ad IN. Fac ut sit AH ad IP in ea ratione et erit HM ad PN in eadem ratione. Datur ergo HM tam ad PN quam ad $HN_{[,]}$ ergo datur punctum N per sectionem determinatam.(46) Et agendo NF parallelam HD et capiendo NG in data ratione(47) ad HN dabuntur puncta F, G per quæ recta DG agenda est.

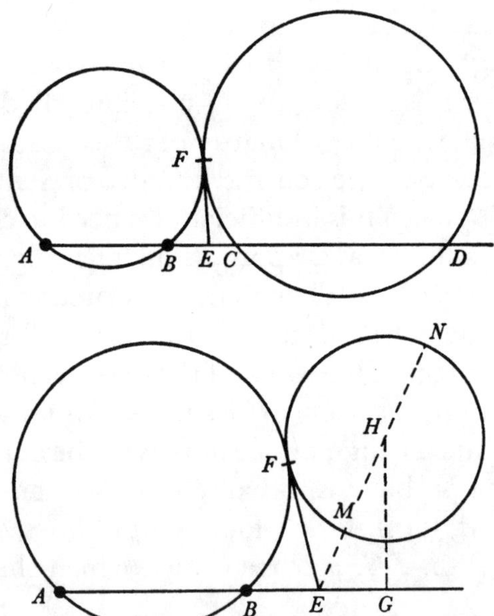

Rursus si per data duo puncta A, B describendus est circulus ABF qui tangat alium circulum positione datū FCD. Junge AB, et ei productæ occurrat tangen[s] FE in E. Si AB producta secet circulū datum puta in C, D erit $EA\times E[B]$ $(=EF^q)=EC\times ED$, adeoq; datur punctum E per sectionem determinatam. Sin AB non secet circulū illum, a centro ejus H ad AB demitte perpendiculum HG et ab EH ducta et producta secetur circulus ille in M et N. Hoc facto erit $AEB(=EF^q=MEN$

$$=\overline{EH-HM}\text{ in }\overline{EH+HN}=EH^q-HM^q)$$
$$=EG^q+HG^q-HM^q,$$

unde $AEB-EG^q=$ dato: adeoq; datur punctum E per sectionem determinatam.(48) Ex E igitur age EF tangentem circulū datum in F et habebuntur tria puncta A, B, F per quæ circulus ABF describi debet.(49)

(44) The bare geometrical bones of the Wrennian problem of fixing the path $DEFG$ of a comet from four timed sightings BD, AE, BF, AG in the hypothesis that this is a uniformly traversed straight line. Newton had, as we have seen, originally introduced it in its full astronomical dress as Problem 52 of his 'Arithmetica' (see v: 298–302) and more recently set its denuded mathematical skeleton as Example 2 of the draft 'Analysis Geometrica' reproduced in 1, §1, Appendix 1 above. His present construction of the transversal DG is, however, as novel in its detail as it is elegant in its mode of solution.

having a given ratio to one another,[44] from the lines choose a particular one, such as AG, to which all the conditions of the problem are to be reduced:[45] accordingly, let DB and FB cut this in H and I, and if from the points E and F you then draw to it EM, FN parallel to DH, because the ratios of DE, EF, FG to each other are given, so too will the ratios of HM, MN, NG to each other be given. Now in the triangles EMA, FNI given in species the ratios of EM to AM and FN to IN are given; there is also given the ratio of EM to FN seeing that this is that of EG to FG, and hence there is given the ratio of AM to IN. Construct AH to IP in that ratio, and HM will then be to PN in the same ratio. The ratio of HM to both PN and HN is therefore given, and consequently the point N is given by determinate section.[46] And by then drawing NF parallel to HD and taking NG in its given ratio[47] to HN there will be given the points F and G through which the straight line DG is to be drawn.

Again, if through two given points A, B there has to be drawn a circle ABF to touch another circle FCD given in position: join AB and let the tangent FE meet its extension in E, and then, should AB produced intersect the circle, say in C and D, there will be $EA \times EB (= EF^2) = EC \times ED$, and so the point E is given by determinate section; but if AB does not intersect that circle, from its centre H let fall to AB the perpendicular HG and let it also be cut by EH, when drawn and extended, in M and N, whereupon there will be

$$EA \times EB (= EF^2 = EM \times EN = (EH-MH)(EH+HN) = EH^2 - MH^2) = EG^2 + GH^2 - MH^2,$$

so that $EA \times EB - EG^2$ equals a given, and thence the point E is given by determinate section.[48] From E, therefore, draw EF tangent to the given circle at F, and there will be had three points A, B, F through which the circle ABF ought to be described.[49]

(45) Newton initially wrote 'transferantur' (are to be translated).

(46) Except where the constructed point P coincides with H and the problem proves to be indeterminate. For this porismatic configuration the necessary and sufficient Boscovichean conditions (see v: 300–3, note (405)) are readily shown to be $HA:HI = DE:DF$ and, on taking K to be the point in which BD and AE intersect, also $HB:HK = FG:EG$.

(47) Namely that of FG to DF.

(48) The former case in which AB intersects the given circle in real points is already treated by an analogous geometrical argument in Problem 3 of Newton's 'Quæstionum Solutio Geometrica' on IV: 256, and is accorded an equivalent algebraical analysis in Problem 39 of his contemporary 'Arithmetica Universalis' (see v: 256–8). A near-identical algebraic reduction of the latter case in which it does not—one immediately applicable to that in which it does—is expounded in the superseded 'Exempl. 1' of his 'Regula Fratrum' on v: 614–16.

(49) By *Elements* III, 25, namely. This perfectly acceptable alternative worked exemplification of Newton's present rule had the misfortune to be entered in the manuscript at the top of a new sheet (f. 36^r), where it is immediately followed by the likewise cancelled first version of an ensuing 'Reg. 2' (so numbered to follow the initial combined Rule 1/2; compare note (43)

Reg. 3. Si in quæstione aliqua non nisi una magnitudo detur sive ea linea sit sive superficies: potest hæc magnitudo pro ignota haberi et alia quævis ex qua ignotæ quantitates facilius colligantur pro nota assumi. Dein ubi figura juxta assumptionem determinata est, si alia huic similis ad lineam illā areamve verè datam construatur, solvetur problema.(50)

Ut in triangulo ABC si detur summa laterum $AC+BC$[,] angulus(51) sub ipsis ACB et ratio basis [AB] ad perpendiculū [CD]. Fingo summa[m] laterum ignotam esse et ejus vice assumo Basem. Inde propter datam rationem basis ad perpendiculum datur perpendiculum. Basi igitur AB ad distantiam perpendiculo æqualem duco lineā parallelam, describoq per puncta A, B circuli circumferentiā quæ datū angulum ACB comprehendat, et ad horum intersectionē duco lineas AC BC. Dein triangulo huic(52)[simile] aliu[d] constituo cujus basis et latera ad basem et latera hujus eam habebit rationem qu[am] habet data summa laterum illius ad inventam summā laterum hujus.(53)

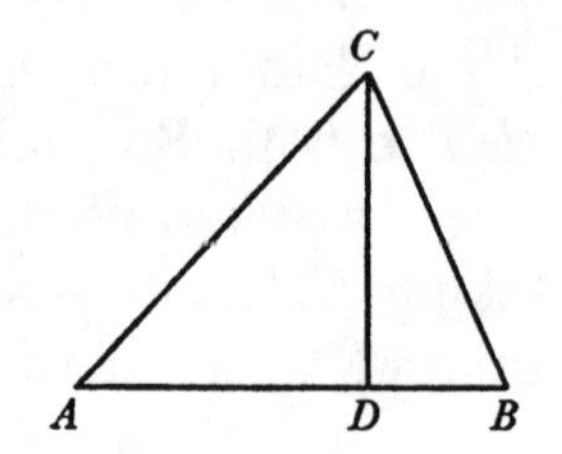

Eodem modo ubi plures magnitudines datæ sunt possunt omnes pro ignotis haberi et earum vice tum rationes earum ad invicem collatis lineis cum arearum lateribus quadratis, tum alia quævis magnitudo quam libuerit assumere pro datis haberi, dein figuræ ad magnitudinem assumptam determinatæ alia similis ad magnitudines vere datas construi. Sed hoc rarò usui erit.

Reg. 4. Diximus(54) in omni problemate id apprime agendum esse ut ex datis colligamus quicquid possumus colligere. Nam ex pluribus jam cognitis cætera facilius colligentur, et conclusiones simpliciores esse solent quæ sic prodeunt. Cæterum ubi in quæstione aliqua hæretur nec possumus hac ratione quæ desiderantur colligere: fingendum est(55) omnes quantitates cognitas esse ita ut

above) which Newton suppressed subsequently to reincarnate as 'Reg. 5' below without thinking similarly to restore the present worked illustration which originally preceded it. So fickle is fortune.

(50) Newton first carried straight on into his terminal paragraph with the words 'Atq ita si plures magnitudes datæ sunt...' (And thus if there are several given magnitudes...) before determining to insert the illustrative *exemplum* of this precept which immediately follows.

(51) In our English version we render a participle 'comprehensus' which Newton here chose to delete in his manuscript text.

(52) ABC, that is.

(53) A tedious example, to be sure! Newton could have illustrated his precept so much more interestingly for his reader—for instance, by adapting the construction which he had earlier presented as Lemma XXVII of his 'De motu Corporum Liber primus' (see VI: 290–2).

(54) In the preceding Rule 1, to be precise. Newton found some difficulty in formulating the following precept, making a number of hesitant prior attempts to shape his words into best form before attaining the version which we reproduce above: of these we record in following footnotes only those variants whose content is not fully absorbed *mutatis mutandis* into the final text.

Rule 3. If in some question only a single magnitude—either a line or a surface—be given, this can be considered as unknown, and any other from which the unknown quantities shall be more readily gatherable assumed as the known. Then, once the figure is determined according to this assumption, if another one similar to it be constructed to the scale of that really given line or area, the problem will be solved.(50)

So in the triangle ABC, if there be given the sum of the (inclined) sides $AC+BC$, the angle $A\widehat{C}B$ contained by them, and the ratio of the base AB to the perpendicular CD, I imagine that the sum of the sides is unknown and in its place I assume the base; in consequence, because of the given ratio of the base to the perpendicular, this also is given, and therefore parallel to the base AB I draw a line at a distance equal to the perpendicular, and then through the points A, B describe a circle circumference which shall embrace the given angle $A\widehat{C}B$: to the intersection of these I draw the lines AC, BC, and then similar to this triangle(52) set up another whose base and sides shall have to the base and sides of this one the same ratio as that of the given sum of sides to the sum of the sides now ascertained of this.(53)

In the same way, where there are several given magnitudes they can all be regarded as unknown and in their stead not only their ratios one to another, comparing lines with the squared sides of areas, but also any other magnitude which it is permissible to assume may be had as given, and then, similar to the figure determined to the size assumed, another can be constructed to the scale of the magnitudes really given. But this will rarely prove of use.

Rule 4. We have said(54) that in every problem we need first and foremost to contrive to gather from the givens anything we can collect. For, once several things are already known, the rest will easily be gathered from them, while the conclusions thus resulting are usually simpler. When, however, we stick in some question and cannot by this procedure deduce what is required, we must imagine (55)that all quantities are known, making in consequence no distinction

(55) Newton initially went on to write 'quæstionem jam solutam esse et quantitates omnes cognosci, et sic nullo inter cognitas & incognitas habito discrimine seligendæ sunt [quantitatum relationes simplicissimæ]' (that the question is already solved and that all the quantities are known, and in this way without making distinction between knowns and unknowns the simplest relationships of the quantities are to be selected), thereafter provisionally replacing this with the equivalent prescription 'quantitates nullas dari sed liberum esse quaslibet assumere, deinde nunc has nunc illas assumendo videndum quid ex assumptis si revera darentur colligi posset et conclusiones simplicissimæ tanquam theoremata notandæ sunt. Nam harum beneficio solutio Problematis aliquando invenietur, aliquando reddetur concinnior' (that no quantities are given but that we are free to assume any we like, and then by assuming now these and now those we must see what might be gatherable from the ones assumed were they in fact given and note the simplest conclusions as though they were theorems. For by their aid the solution of a problem will sometimes be found and at others be made tidier).

nullum inter cognitas et incognitas discrimen habeatur. Et incipiendo a duarum vel trium quantitatum relatione aliqua, illic pro una quantitatum substituenda est quantitas aliqua æquipollens et habebitur nova quantitatū[56] relatio. Dein pro una harum substituenda alia quævis quantitas æquipollens et habebitur alia quantitatum relatio. Hoc pacto pergendo relationes quantitatum simplicissimæ quæ colligi possunt tanquam Theoremata notandæ sunt. Nam harum beneficio solutio Problematis aliquando invenietur, aliquando jam inventa reddetur concinnior.[57] Hac methodo solvuntur etiā Problemata Algebraice:[58] sed ad eos jam loquor qui Algebram aut non intelligunt aut propter ejus complexas atq; adeo ineligantes operationes minus colunt.

Reg. 5.[59] Et ubi quæstionis duo vel forte plura sunt responsa ut fere fit in difficilioribus, apprimè quærenda est summa vel semisumma quantitatum quibus respondebitur aut alia quantitas punctumve quod eodem modo se habeat ad utram. Nam hujus determinatio simplicior esse solet, & manuducere ad quantitatis utriusq; quæsitæ determinationem. Quamobrem ubi Regula quarta opus est, debemus ad omnem quantitatem quæ eodem modo se habet ad ambas illas quantitates[60] quibus quæstioni respondendum est animum advertere, sedulo colligentes relationes quas habent illæ ad alias quantitates.

[61]Ut si a quadrati *ABCD* angulo *A* recta ducenda est cujus pars *GE* inter

(56) This replaces the more explicit phrase 'inter quantitatem substitutam et quantitates quæ manent' (between the quantity substituted and the quantities which remain).

(57) Initially 'simplicior' (simpler).

(58) Compare v: 170–2, 604–6.

(59) A new statement of the 'Regula Fratrum' which Newton had earlier adumbrated in his 'Arithmetica' (see v: 204) and more fully in his separate piece (v: 612–20) of that title. In a preliminary equivalent 'Reg. 2' (on which see note (49) above) he had here first enunciated: 'Ubi quantitas incognita anceps & ambigua est, existentibus ejusmodi duabus quæ eodem modo se habent ad omnes conditiones quæstionis, præstat aliam aliquā quærere quæ eodem modo se habet ad ambiguas, ut ambiguarum summā, differentiam, duarum rectangulum, vel summam quadratorum, vel trianguli basem, angulū verticalem, perpendiculum aut aream ubi ambigua sunt latera angulive ad basem, maxime vero quantitatum duarū summam vel semisummam. Nam hujus determinatio in quæstionibus quæ per regulam et circinū solvi possunt, semper est simplicior et plerumq; facilius investigatur, præsertim in quæstionibus arduis' (When an unknown quantity is twosome and ambiguous, there existing two of the kind which are involved in the same relationship to all the conditions of the question, it is better to seek some other which is in the same relationship to the ambiguous ones, such as their sum or difference, the product of the two or the sum of their squares, or the base, vertical angle, altitude or area of a triangle in which the ambiguous quantities are its (inclined) sides or the angles at its base—but above all the sum or half-sum of the quantities. For the determination of this in questions solvable by ruler and compasses is always simpler and for the most part more easily explored, especially in the case of harder questions).

(60) Initially 'ad summam vel punctum medium quantitatum quæ eodem modo se habent ad utramq; quantitatem' (to the sum or mid-point of quantities which stand in the same relationship to each quantity). Compare the superseded preliminary draft cited in the previous note.

between knowns and unknowns. Then, beginning from some relationship of two or three quantities, some equivalent quantity is to be substituted in this in one of those quantities' place, and there will be obtained a new relationship of quantities.(56) Thereafter, any other equivalent quantity is to be substituted for one of these and a further relationship of quantities will be had. By proceeding in this manner the simplest relationships of quantities which can be gathered are to be noted as though they were theorems; for by their aid the solution of a problem may sometimes be discovered, and sometimes, when it is already discovered, made neater.(57) In this manner algebraic problems, too, are solved(58) —but I now speak to those who either do not understand algebra or are, on account of its complex and accordingly inelegant operations, less than well up in it.

Rule 5.(59) And when a question has two or maybe more answers, as almost always happens with the more difficult ones, we need principally to seek the sum or half-sum of the quantities by which it will be answered, or another quantity or point which is related in the same way to each. For the determination of this is usually simpler and a guide to the determination of each of the individual quantities required. Consequently, when we have need of the fourth Rule, we ought to turn our attention to every quantity which stands in the same manner to both those quantities(60) by which reply is to be given to the question, painstakingly collecting the relationships which they bear to other quantities.

(61)If, for instance, you have from the corner A of a square $ABCD$ to draw a

(61) In a first illustration of his precept on f. 36^r in the manuscript—there straightaway abandoned in favour of a preliminary enunciation (see note (63)) of the various possibilities of 'ambiguous' twin 'responses' to the Apollonian neusis problem which here follows—Newton initially instanced: 'Ut si datis positione rectis *AB*, *CD* super data basi *AB* triangulū constituendum sit cujus vertex sit ad datam lineam *CD*, latera verò *AC*, *BC* datam habeant different[iam]. Hic latera trianguli eodem modo se habent ad conditiones problematis ita ut ratio nulla sit cur unum potius quæratur quam alterum' (If, for instance, where the straight lines *AB* and *CD* are given in position we need to set up on the given base *AB* a triangle whose vertex shall be in the line *CD* while its sides *AC*, *BC* shall have a given difference: the sides of the triangle here stand related in the same way to the conditions of the problem and there is in consequence no reason why one should be sought rather than the other...). We may presume that he had it in mind to go on to refer the 'conditions' of this problem to the mid-point, F say, of AB and thence on dropping the perpendicular CG from C onto AB ascertain the horizontal distance FG of C away from it: in which case, since $AC^2-BC^2 = (AG^2-GB^2 \text{ or}) \; 2AB\times FG$ and consequently $(AC+BC \text{ or}) \; 2BC\pm k = (2/k).AB\times FG$ where, on setting $\hat{E} = \tan^{-1} l$, there is $BC^2 = (FB-FG)^2+l^2.(FE-FG)^2$, the segment FG is determined *dupliciter* as a root of the quadratic equation

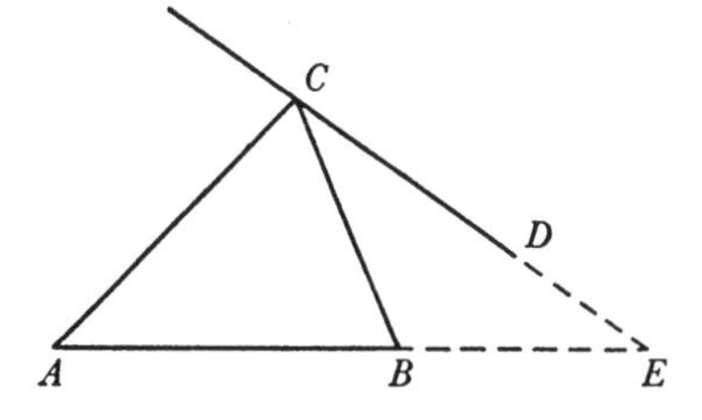

$$(l^2+1).FG^2-2(FB+l^2.FE)FG+FB^2+l^2.FE^2 = (k^{-1}.AB\times FG\mp \tfrac{1}{2}k)^2.$$

The ensuing construction *per sectionem determinatam* of the point G in AB (and thereby of C in

latera opposita producta DC, BC datæ sit longitudinis:[62] potest linea GE æque in angulo DCg ac in angulo BCE duci et perinde duplex[63] erit responsum. Si quæritur linea DE quæ a puncto D ad intersectionem linearum DC et AE extenditur[,] duæ erunt ejusmodi lineæ DE & De. Si linea AE quæ a puncto A ad intersectionem eandem extenditur[,] duæ sunt etiam ejusmodi lineæ AE et Ae. Quare neutras quæro sed quæ ad ambas eodem modo referuntur[,] duarum summam, differentiam, rectangulum, summā quadratorū, maximè verò summam. Sumendo igitur $EH = De$ erit DH summa illa. Ad E et H ipsis AE, DH perpendiculares erigendo EF, HF constituetur triangulum EHF simile et æquale triangulo ADe vel ABG et linea AB producta transibit per punctum F. Quærendæ sunt itaq; relationes AF vel BF ad alias lineas. Est itaq; $AF^q = AE^q + EF^q$. Pro EF sc[r]ibe AG. Fit $AF^q = AE^q + AG^q$. Biseca GE in K. Et [pro] AG et AE scribe $AK - GK$ et $AK + GK$. Fit

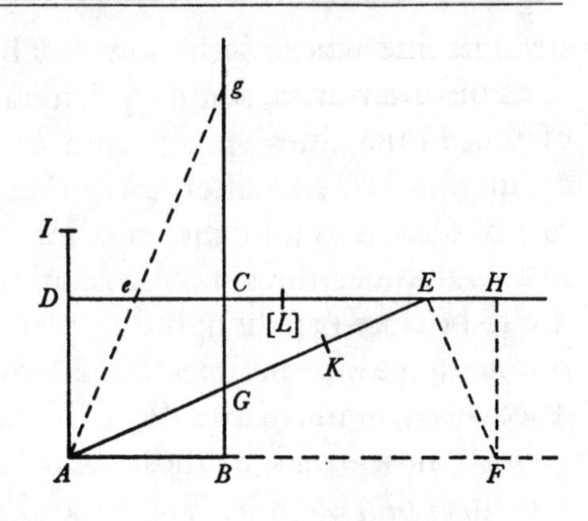

$$AF^q = 2AK^q + 2GK^q,$$

Theorema notandum. Rursus ob sim[ilia] triangula ABG AEF est

$$AB.AG::AE.AF_{[,]}$$

ergo $BAF = GAE$. Pro GA et AE scribe $AK - GK$ et $AK + GK$ et fiet

$$BAF = AK^q - GK^q,$$

aliud Theorema. Jam de AF^q aufer $2BAF$ et residuū erit $4GK^q$ seu GE^q. Ergo

---

CD) is straightforward enough, but no simpler than any equivalent reduction from first principles—for instance, by computing (geometrically or algebraically) the meet with CD of the hyperbolic locus (C) defined by the condition that the focal radii AC, BC shall have a constant difference—and we will appreciate Newton's reluctance further to proceed with a tentative worked example which proves in the event so poorly to evidence the claimed gain in simplicity and neatness of demonstration attainable (in the case of problems appropriately possessed of multiple pairs of solutions) by appeal to the present 'rule of mates'.

(62) Yet one more exposition by Newton of this familiar Apollonian neusis, Heraclitus' elegant 'plane' construction of which (as presented by Pappus in his *Mathematical Collection* VII, 71/72) Newton here essentially repeats but whose formal justification he considerably shortens by aptly choosing a sum of 'brother' quantities as base in his following concise synthesis. He had initially come upon the problem in his formative years (see I: 509–11) as cited from Pappus—and given a rather clumsy algebraic solution—by Descartes in the third book of his *Geometrie* [= *Geometria*, $_2$1659: 82–4] and had subsequently introduced his own improved Cartesian reduction of it into his 1670 'Observationes' on Kinckhuysen's *Algebra* (see II: 434–8), afterwards incorporating this as Problem 13 of his reworked 'Arithmetica' (see

straight line whose part GE between the extended opposing sides DC and BC shall be of given length,[62] the line GE can be drawn equally as well in the angle $D\widehat{C}g$ as in the angle $B\widehat{C}E$, and in consequence there will be a double[63] answer. If the line DE which extends from the point D to the intersection of the lines DC and AE is sought, there will be two lines of the kind, DE and De; if the line AE extending from the point A to the same intersection, there are likewise two lines of the sort, AE and Ae. I therefore seek neither, but rather ones related in the same manner to both—the sum, difference or rectangle of the two, or the sum of their squares, but most of all their sum. On taking $EH = De$, accordingly, DH will be that sum. Now by erecting EF, HF perpendicular at E and H to AE, DH there will be formed a triangle EHF congruent to the triangle ADe and so to ABG, while the line AB when extended will pass through the point F. The relationship of AF or BF to the other lines is consequently to be sought. There is, accordingly, $AF^2 = AE^2+EF^2$; in place of EF write AG and there comes to be $AF^2 = AE^2+AG^2$; bisect GE in K and in place of AG and AE write $AK-GK$ and $AK+GK$; there comes $AF^2 = 2AK^2+2GK^2$, a theorem to be noted. Again, because the triangles ABG and AEF are similar, there is $AB:AG = AE:AF$ and therefore $AB\times AF = AG\times AE$; in place of AG and AE write $AK-GK$ and $AK+GK$, and there will come to be $AB\times AF = AK^2-GK^2$, another theorem. Now from AF^2 take away $2AB\times AF$ and the residue will be $4GK^2$, that is, GE^2;

---

v: 202–4) with stress (*ibid.*: 204; compare also v: 614) on the further simplifications possible by effective prior employment of the present *regula fratrum*. His equivalent geometrical synthesis of the same Problem 13 on v: 512 is here but trivially refashioned.

(63) The two other real solutions possible when the given line-length GE is large enough for it also to lie in the angle $B\widehat{C}D$ (as *mutatis mutandis* it does in Newton's figures on I: 509 and v: 610) are here implicitly excluded by his present insistence that it shall be placed between one of the sides BC, DC of the given square and the respective extension CE, Cg of the other. In a cancelled preliminary version of this *exemplum* on f. 36r of the manuscript (see note (61) above) Newton stated equivalently—on trivially replacing the designations F, p and q in his draft figure by G, H and K to correspond with his revised diagram—that 'Potest linea illa duci a parte puncti C versus B ut ad AGE vel ab altera parte ut ad Aeg' (that line can be drawn on the side of the point C towards B, as at AGE, or on the other side, as at Aeg), continuing thereafter: 'adeo ut puncta E & e, G et g lineæq DE ac De, CE ac Ce, BG ac Bg, CG ac Cg, AE et Ae, AG et Ag ambigua sint[,] proinde horum nullum quæro sed medium punctum inter bina ambigua ut L quod medium inter E & e vel K quod medium est inter G et E vel DL aut AK, vel summam duarum $DE+De$, $AG+AE$, $CE-Ce$; nam quoniam CE et Ce a puncto dato C ad plagas oppositas tendunt, si una earum affirmativa statuatur altera negativa erit adeoq cum diversis signis connecti debent ad summam constituendam' (and hence the points E and e, G and g, and lines DE, De; CE, Ce; BG, Bg; CG, Cg; AE, Ae; and AG, Ag are ambiguous. I consequently seek none of these but a mid-point between ambiguous pairs, such as L which is the middle between E and e or K which is middle between G and E, or either DL or AK, or the sum of the pair $DE+De$, $AG+AE$, or $CE-Ce$; for because CE and Ce extend opposite ways from the given point C, if one of them be set positive the other will be negative, and hence they must be connected with opposite signs to form their (geometrical) sum).

datur punctum F per sectionē determinatam. Quam sectionem si concinnare velis adde utrobiq(64) AB^q et fiet

$$AF^2-2BAF+AB^q \text{ seu } BF^q = AB^q+GE^q = BI^q \quad \text{et} \quad BF = BI.^{(65)}$$

Quod si non AF sed BF prima vice quæsivissem, ducta GF, fuisset

$$BF^q = GF^q-BG^q.$$

Pro $GF^{[q]}$ scribe GE^q+EF^q, fiet $BF^q = GE^q+EF^q-BG^q$. Pro EF^q seu AG^q scribe AB^q+BG^q, fiet $BF^q = GE^q+AB^q$ ut prius.(66)

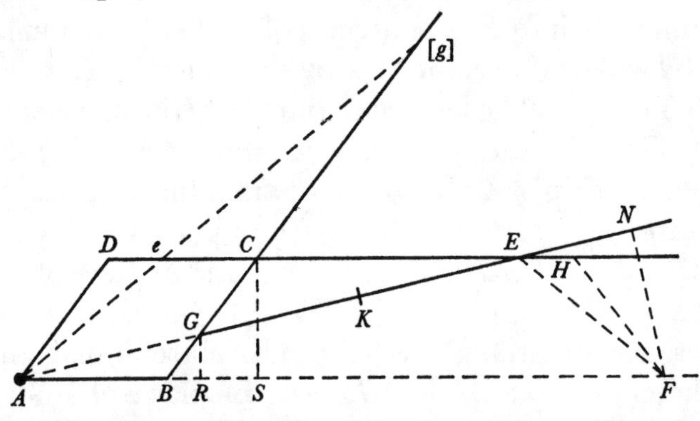

Reg 6. Ubi his non obstantibus in problemate aliquo difficiliori hæretur, evolvendi sunt casus ejus faciliores. Nam ex horum determinatione aditus patefieri solet ad determinationem difficiliorum. Ut si a Rhombi $ABCD$ angulo A ducenda esset linea cujus pars GE inter Rhombi latera opposita producta DC BC datæ sit longitudinis.(67) Primo considero quomodo solvendum sit Problema ubi anguli figuræ recti sunt. Dein postquam Problema in hoc casu solui ut supra, insistendo vestigijs illius solutionis sic argumentatur. Capta $EH = BG$ et constituto triangulo EHF simili et æquali triangulo ABG: recta AB producta transibit per punctum F et erit (juxta priorem e præcedentibus solutionibus casus illius simplicioris) $AF^q = AE^q+EF^q+2AE\times EN$ demisso scilicet ad AE productam perpendiculo FN. Hic pro EF scribe AG, et fiet

$$AF^q = AE^q+AG^q+2AE\times EN.$$

Demissis ad AB perpendiculis GR, CS propter similes figuras $AFEN$, $AGBR$ est $AF.EN::AG.BR::AE.BS$. Nam $AG.AE::BG.BC::BR.BS$. Ergo

$$AF\times BS = AE\times EN.$$

Quare pro $2AE\times EN$ scribe $2AF\times BS$ et fiet $AF^q = AE^q+AG^q+2AF\times BS$. seu $AF^q-2BS\times AF = AG^q+AE^q$. Porro ob similia triangula ABG, AEF est

(64) Of the equation $AF^2-2AB\times AF = GE^2$, that is.

(65) Understanding that I is taken (in AD perpendicular to AB) such that $AI = GE$. In a cancelled preliminary draft (see the next note) Newton at this point went on to specify Heraclitus' ensuing construction of the neusis (as given by Pappus in his *Mathematical Collection* VII, 72) in an added sentence: 'Invento autem BF, super diametrum AF descriptus semicirculus secabit DH in duobus punctis quæsitis E et e' (Once BF is found, however, when a semicircle is described on the diameter AF it will cut DH in the two required points E and e) —evidently so, since the angles $\widehat{AEF}$ and $\widehat{AeF}$ are each right.

(66) Initially, in a yet closer imitation of Pappus, Newton set this *idem aliter* immediately

consequently the point F is given by determinate section. Should you wish to refine this section, add AB^2 to each side(64) and there will ensue

$$(AF^2 - 2AB \times AF + AB^2 \text{ or}) \; BF^2 = AB^2 + GE^2 = BI^2$$

and so $BF = BI$.(65)

But had I in the first instance sought not AF but BF, when GF is drawn there would have been $BF^2 = GF^2 - BG^2$. In place of GF^2 write $GE^2 + EF^2$ and there will come to be $BF^2 = GE^2 + EF^2 - BG^2$; in place of EF^2, that is, AG^2, write $AB^2 + BG^2$ and there will come $BF^2 = GE^2 + AB^2$ as before.(66)

Rule 6. When, notwithstanding this, you stick in some more difficult problem, its easier cases must be unwrapped. For once these are determined an avenue to the more difficult ones is usually opened up therefrom. If, for example, you had from the corner A of a rhombus $ABCD$ to draw a line whose portion GE between the extended opposing sides DC, BC of the rhombus shall be of given length,(67) I first consider how the problem is to be solved when the angles in the figure are right; then, after having in this case solved the problem as above, by treading in the steps of that solution I would reason as follows. With EH taken equal to BG and the triangle EHF constructed congruent to the triangle ABG, the straight line AB will, when produced, pass through the point F and there will then (according to the former of the preceding solutions of the simpler case) be $AF^2 = AE^2 + EF^2 + 2AE \times EN$, where FN is, of course, the perpendicular left fall to AE produced; here in place of EF write AG and there will come to be $AF^2 = AE^2 + AG^2 + 2AE \times EN$. With the perpendiculars GR, CS let fall to AB there is again, because of the similar figures $AFEN$ and $AGBR$, $AF:EN = AG:BR = AE:BS$ (for $AG:AE = BG:BC = BR:BS$) and therefore $AF \times BS = AE \times EN$; consequently, in place of $2AE \times EN$ write $2AF \times BS$ and there will come $AF^2 = AE^2 + AG^2 + 2AF \times BS$, that is,

$$AF^2 - 2BS \times AF = AG^2 + AE^2.$$

after his preceding derivation of the lemmatical 'Theorema notandum' $AF^2 = 2AK^2 + 2GK^2$ (whose result is *Mathematical Collection* VII, 71) there to deduce equivalently that 'Rursus $BF^q = GF^q - BG^q$. Pro GF^q scribe $GE^q + EF^q$. Fit $BF^q = GE^q + EF^q - BG^q$. Pro EF^q substitue AG^q seu $AB^q + BG^q$. Fit $BF^q = GE^q + AB^q$. Ergo datur BF^q' before passing to specify the Heraclitan construction which ensues 'invento puncto F' as we have quoted it in the previous note.

(67) 'Rhombo dato, & vno latere producto aptare sub angulo exteriori magnitudine datam rectam lineam, quæ ad oppositum angulum pertingat' as Commandino (*Mathematicæ Collectiones:* 249) rendered Pappus' enunciation of this more general neusis in his summary of the content of Apollonius' lost work *On Inclinations*. Apollonius' own 'plane' construction of this problem is (see note (70) straightforwardly restorable from a following lemma by Pappus upon it, but Newton—despite his previous citation of Marin Ghetaldi (see §1: note (32)) for his restoration of just such a solution in (Problems II–IIII of) his *Apollonius Rediuiuus. Seu, Restituta Apollonii Pergæi Inclinationum Geometria* (Venice, 1607): 4–22 [= *Opera Omnia* (Zagreb, 1968): 202–20]—is here seemingly ignorant of the fact.

$AB.AG::AE.AF$ adeoq; $AB \times AF = AG \times AE$. Aufer hujus duplum et manebit $AF^q - 2SAF = AG^q + AE^q - 2AG \times AE$ hoc est = quadrato differentiæ AG & AE seu GE^q. Unde datur punctū S per sectionē determinatam. Quæ sectio concinnabitur addendo AS^q utrobiq;[,] fiet enim

$$AF^{[q]} - 2SAF + AS^q \text{ seu } FS^q = GE^q + AS^q.$$

Sic et juxta posteriorem solutionem casus simplicioris, est

$$BF^q - 2RBF = GF^q - GB^q = GE^q + EF^q - GB^q + 2G[EN].$$

Pro EF^q seu AG^q scribe $AB^q + BG^q + 2ABR$ et fit

$$BF^q - 2RBF = GE^q + AB^q + 2ABR + 2G[EN].$$

Est [et] $AF.EN::AG.BR::GE.RS$. Scribe ergo $2AF \times RS$ pro $2GEN$ et fiet $BF^q = GE^q + AB^q + 2RBF + 2RBA + 2RS \times AF$ seu $= GE^q + AB^q + 2BS \times AF$. Unde datur [punctum] F per sectionem determinatam. Quæ sectio ad terminos simplicissimos reducta tan[tum] fiet $SF^q = AS^q + GE^q$ ut ante.(68) Invento autem puncto F per A et F describendus est circulus qui capiat angulum AEF.(69) Hic enim rectā DC secabit in punctis quæsitis E et e.(70)

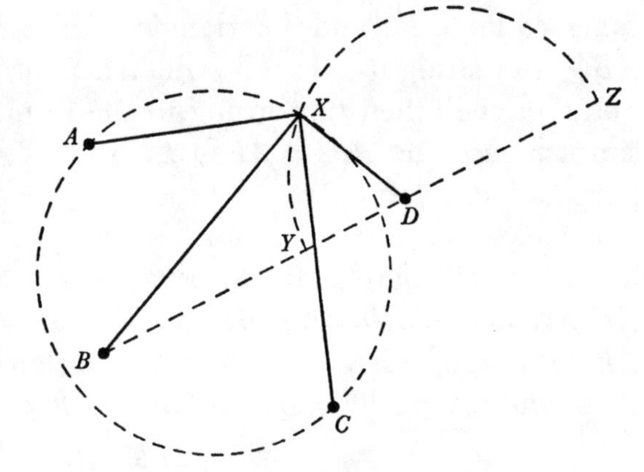

Reg. 7. Quoniam lineæ alicujus tam curvæ quam rectæ in qua punctum quæsitum existit(71) inventio ac determinatio facilior esse solet quam puncti ipsius determinatio omnimoda: ideo quærendo duas lineas in quibus punctum illud existit, per eorum communem intersectionem problema sæpissimè solvetur. Ut si

---

(68) For on reordering Newton's previous equation there is

$$GE^2 = BF^2 - AB^2 - 2BS \times AF = AF(BF - AB - 2BS) = (SF + AS)(SF - AS) = SF^2 - AS^2.$$

(69) Equal to $\widehat{AeF}$ since the trapezia $AFED$ and $FAeH$ are congruent. Newton here exactly imitates his suppressed Heraclitan construction (see note (65) above) of the previous particular neusis in which the rhombus $ABCD$ is a square.

(70) Notwithstanding the deftness of this resolution by Newton of Apollonius' rhombic neusis problem—now elegantly succeeding where he had earlier (see v: 513, note (19)) failed—we may yet be surprised (compare note (67) preceding) at his apparent ignorance that Pappus' lemma upon it (*Mathematical Collection* VII, 70, where it immediately precedes the construction by Heraclitus of the square case which he here generalizes) straightforwardly yields a solution which is not only neater and more compact still but is a signal instance of the effective application of his favourite *Regula fratrum* (Rule 5). If, namely, at the mid-point K of GE we erect a perpendicular to meet the diagonal AC produced in I, it is readily shown that the triangles

Further, because of the similar triangles ABG and AEF there is

$$AB:AG = AE:AF$$

and hence $AB \times AF = AG \times AE$. Take twice this away and there will remain $AF^2 - 2AS \times AF = AG^2 + AE^2 - 2AG \times AE$, that is, the square of the difference of AG and AE, or GE^2. Whence the point S is given by determinate section. This section will be refined by adding AS^2 to each side, for there will then come to be $(AF^2 - 2AS \times AF + AS^2$ or) $FS^2 = GE^2 + AS^2$.

So also, according to the latter solution of the simpler case, there is

$$BF^2 - 2BR \times BF = GF^2 - GB^2 = GE^2 + EF^2 - GB^2 + 2GE \times EN.$$

In place of EF^2, that is, AG^2, write $AB^2 + BG^2 + 2AB \times BR$ and there comes to be $BF^2 - 2BR \times BF = GE^2 + AB^2 + 2AB \times BR + 2GE \times EN$. There is also

$$AF:EN = AG:BR = GE:RS,$$

so write $2AF \times RS$ in place of $2GE \times EN$ and there will come

$$BF^2 = GE^2 + AB^2 + 2BR \times BF + 2BR \times AB + 2RS \times AF$$

that is $$= GE^2 + AB^2 + 2BS \times AF.$$

Whence the point F is given by determinate section. And when this section is reduced to simplest terms it will come but to be $FS^2 = AS^2 + GE^2$, as before.(68) Once the point F is found, however, you need to describe through A and F a circle which shall take the angle $A\widehat{E}F$,(69) for this will cut the straight line DC in the required points E and e.(70)

Rule 7. Seeing that the discovery and determination of a particular line, both curve and straight, in which a required point is located (71) is usually easier than the universal determination of the point itself, by seeking in consequence two lines in which that point exists a problem will very often be solved through their

---

IGE, DAC and also AGI, GCI are similar, so that

$$IG:GE = DA:AC$$

and $$AI:IG = IG:CI;$$

whence, since GE is fixed in length by the given neusis condition, on constructing I in AC *per sectionem determinatam* to satisfy $AI \times CI = IG^2$ where

$$IG = GE \times DA/AC,$$

constant, a circle of centre I and radius IG will meet BC and DC in the pairs of required end-points G_1, G_2 and E_1, E_2. (The second point I' constructed in AC by the determinate section will—when its distance from

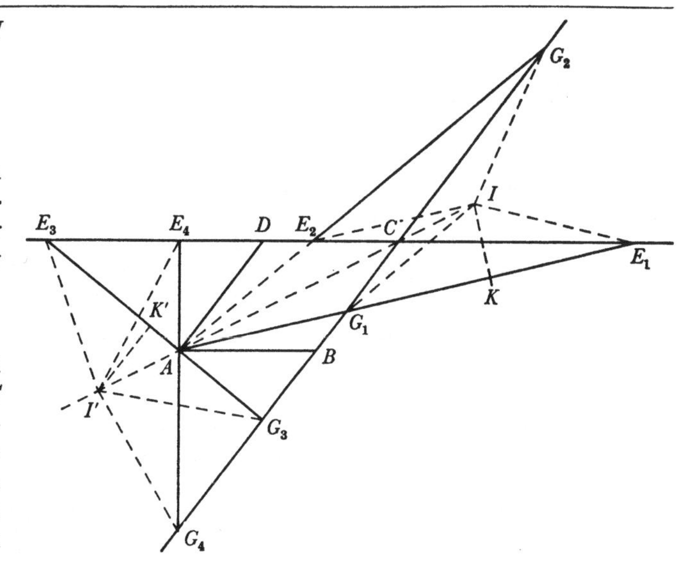

punctum X quæratur a quo si ad data quatuor puncta A, B, C, D ducantur totidem rectæ quarum duæ AX CX comprehendant datum angulum AXC, aliæ duæ [BX, DX datam habeant rationem ad invicem].(72) Quoniam puncta A, C et angulus AXC da[n]tur, punctum X erit in circumferentia circuli qui per puncta A et C transibit et in segmento suo ACX angulũ illum datum AXC capiat. Et quoniam puncta B ac D et [ratio BX ad DX](72) dantur[,] idem punctum X erit in circumferentia alterius circuli cujus centrum erit in linea BD, diameter vero YZ invenietur jungendo et producendo BD et secando eam in Y et Z ita ut sit BY ad DY et BZ ad DZ ut BX ad DX.(73) Descriptis igitur duobus illis circulis, in earum intersectione habebitur punctum quæsitum X.

Cæterum linea in qua punctum aliquod existit illius Locus dici solet. Et in Loci speculatione veteres plurimum posuerunt. *Locus*, inquit Pappus,(74) *propria quædam materia est, post communium elementorum constitutionem*[,] *ijs parata qui in geometricis sibi comparare volunt vim ac facultatem inveniendi problemata quæ ipsis proponuntur: atq; hujus tantummodo gratia inventa est.* Locus vero apud Veter[e]s duplex erat, *adæquatus*,(75) quo sensu *puncti locus punctum est, lineæ linea, superficiei superficies*; et *excurrens*,(75) ut *linea puncti, superficies lineæ*.(76) Punctorum vero omniũ quæ in linea sunt simul sumptorum linea illa locus adæquatus est et linearũ omnium punctorum quæ in superficie sunt superficies est locus adæquatus. Pappus itaq; de utroq; locorum genere locutus est. Nam definit utrumq; & statim recensendo quæ veteres de loco scripserunt,(77) non tantum ponit Euclidis libros *de Porismatibus*, Apollonij libros *de locis planis deq; conicis sectionibus* et Aristæi libros *de locis solidis*, sed et Euclidis libros *datorum*, & Apollonij libros *de sectione Proportionis, de sectione determinata deq; sectione spatij* ut prætereram ejusdem libros *tactionum* et *inclinationum*. Hæc igitur Veteres non propositionum gratia scripserunt sed ut inventionis fundamenta sternerent. Quod genus inventionis diu neglectum his

BC, BD is not greater than IG—yield a further pair of solutions of the neusis in which GE is placed in the angle $\widehat{BCD}$. In the particular case where the rhombus $ABCD$ is square, and hence $AC^2 = 2AD^2$, the radius of the constructing circle reduces to be $GE/\sqrt{2}$.)

(71) Originally 'reperitur' (is found).

(72) We mend a slip of Newton's pen which at these points in the manuscript wrote 'aliæ duæ si in se ducantur conficiant datum rectangulum $BX \times DX$' (while the other two, if they be multiplied together, shall make a given rectangle $BX \times DX$) and 'rectangulum $BX \times DX$' (the rectangle $BX \times DX$) respectively, in manifest contradiction with his further argument where (see the next note) he takes the locus (X) to be an Apollonius circle— and not the Cassini oval of foci B and D which the constancy of the product $BX \times DX$ would define it to be.

(73) Whence XY and XZ are the internal and external bisectors of $\widehat{BXD}$, so that the angle $\widehat{YXZ}$ is right and hence YZ is a diameter of the Apollonius circle (X) defined by the constancy of the ratio of BX to DX.

(74) In the opening sentence of the seventh book of his *Mathematical Collection*, here quoted yet once more (compare pages 256 and 306 above) from Commandino's Latin version (*Mathematicæ Collectiones*, ${}_2$1660: 240).

(75) Notice that Newton here rejects Commandino's Latin paraphrases 'in se ips[o] consisten[s]' and 'sese extra tenden[s]' (see *Mathematicæ Collectiones*: 247) in favour of his own

common intersection. Should, for instance, a point X be required such that if from it to the four given points A, B, C, D there be drawn an equal number of straight lines, two of which, AX and CX, shall comprehend a given angle $A\hat{X}C$, while the other two, BX and DX, have a given ratio one to the other:(72) because the points A, C and the angle $A\hat{X}C$ are given, the point X will be in the circumference of a circle which will pass through the points A and B and shall in its segment $\widehat{AXC}$ take the given angle $A\hat{X}C$; and, because the points B and D and the ratio of BX to DX(72) are given, the same point X will be in the circumference of a second circle whose centre will be in the line BD, while its diameter YZ will be found by joining and extending BD and then cutting it in Y and Z such that BY shall be to DY and BZ to DZ as BX to DX.(73) When, therefore, those two circles have been described, the required point X will be had at their intersection.

For the rest, a line wherein some point is located is usually called its 'locus'. On the investigation of loci the ancients set very great store. 'The locus', says Pappus,(74) 'is a special subject-matter prepared, after the common groundwork of the *Elements*, for those who in things geometrical wish to acquire power and facility in exploring problems which are proposed to them, and with this intention alone was it devised'. In the ancients' understanding, however, the locus had a twofold character, being 'equivalent'(75) in the sense in which 'a point is the locus of a point, a line that of a line, a surface that of a surface', and also 'surpassing'(75) as 'a line is one of a point, a surface one of a line'—though, indeed, of the joint set of all points which are in a line that line is the 'equivalent' locus and of the points in all lines which are in a surface the surface is the 'equivalent' locus.(76) Pappus accordingly spoke of both kinds of locus. To be sure, he defines both and in straightaway reviewing what the ancients wrote on the locus(77) he cites not merely Euclid's books *On porisms* and Apollonius' *On plane loci* and *On conic sections* and Aristæus' *On solid loci*, but also Euclid's books of *Data* and Apollonius' *On cutting off a ratio*, *On determinate section* and *On cutting off a space*—to pass by the same author's books *Of contacts* and *On inclinations*. These, then, the ancients wrote not for the (intrinsic) sake of their propositions, but to lay

terser renderings of Pappus' adjectives ἐφεκτικὸς and διεξοδικὸς. Initially in the manuscript he here set these in still sharper contrast as being 'definitus' (definite) and 'indefinitus' (indefinite) respectively.

(76) A cancelled first continuation reads: 'Ad prius genus loci referenda sunt quæ de datis scripsimus nisi mavis data specie ad secundum genus referre' (To the first type of locus are to be referred what we have written on givens, unless you prefer to refer those given in species to the second kind).

(77) Namely in the surrounding pages (*Mathematicæ Collectiones*: 241–9) of the preamble to his seventh book. Since Newton in the geometrical papers reproduced in preceding pages (see especially 1, §1.1 and 2, §§1/2) has commented variously and at length on Pappus' summaries of the following works of Greek higher geometrical analysis, we need not here pause to specify either their detail or Newton's interpretations and restorations of their content.

scriptis restituere conatus sum. Quamobrem cum punctorum quæsitorū loca adæquata quæ ad inventionem magis conferunt, in præcedentibus Datis complexus fuerim,[78] restat jam ut de eorundem locis excurrentibus agam. Sed hæc ut linearum vel punctorum omnium quæ in lineis sunt loca adæquata primò considerabo, tradendo naturam et proprietates Curvarum in quibus puncta communem habentes proprieta[te]m locantur, et quomodo et datis proprietatibus curvæ cognosci ac describi possint.

GEOM. LIB 2.

Veteres in Geometriam lineas solas receperunt quæ per Postu[lat]a Geometrica[79] describi possunt, quorum tria Euclides in *Elementis* posuit,[80] quartum quod est *Datum Conū dato Plano secare*, addiderunt qui de Conicis scripserunt sectionibus. Hinc rectas, circulos & Conicas sectiones in Geometriam receperū[t], cæteras curvas eo quod non nisi per instrumenta describuntur appellarunt mechanicas. Nam curvarum per instrumenta quævis, etiam circuli ipsius per circinum et rectæ lineæ per regulam descriptio mechanica est. Et ideo postularunt earū descriptiones quæ in Geometriam receperunt. Non quod hæ ab hominibus quatenus geometricæ sunt [81]describi possint (quis enim vidit lineam sine latitudine?) sed quod ex concessa earum descriptione, cætera omnia quæ Geometræ inde derivant accuratè consequentur. Quoniam tamen Geometria in usus humanos condita est, ideo linearum quæ omnium facillimè & accuratissimè per instrumenta describi possunt, rectæ nimirum et circuli descriptiones primo loco postularunt. Attamen qui postea postulatum addiderunt de generatione lineæ per sectionem Coni haud usibus humanis consulerunt.[82] Nam sectio illa ob difficilem praxin mechanicam nuda est speculatio, nihil habens cum usibus humanis conjunctum. Qua de causa non male de Geometria forte mereatur qui ejus vice aliud Postulatum substituat cujus effectio mechanica facilior sit quodꝗ ad descriptionem plurium curvarum extendat. Nam et veteres ipsi dum curvam quandam[83] conicis sectionibus magis com-

(78) Originally 'cum locorum casus simpliciores in præcedentibus Datis absolverim' (since I have dealt with the simpler cases of loci in the preceding (section on) givens) *tout court*.

(79) Newton initially wrote 'per principia Geometriæ ab ijs assignata' (through the principles of geometry assigned by them).

(80) As listed by Isaac Barrow in his *Euclidis Elementorum Libri XV* (Cambridge, $_1$1655): 6 these postulate respectively:

'*1*. ...à quovis puncto ad quodvis punctum rectam lineam ducere....

'*2*. Et rectam lineam terminatam in continuum rectà producere.

'*3*. Item, quovis centro, & intervallo circulum describere.'

For Newton's preferred trio of 'instrumental' axioms see page 388 below.

(81) 'accurate' (accurately) was here initially added and afterwards deleted solely, we would suppose, because the adverb is twice repeated at equivalent points in the immediate sequel.

foundations for invention. This *genre* of discovery, after long years of neglect, I have attempted to restore in this present script. Now, therefore, since I have embraced the 'equivalent' loci of required points, as having a greater contribution to make to invention,[78] in the preceding (account of) givens, it remains for me to discuss their 'surpassing' loci. But I shall in the first instance consider these as the 'equivalent' loci of lines or the totality of points in them, recounting the nature and properties of curves in which points sharing a common property are located, and explaining how also, when properties of curves are given, these may then be identified and described.

Geometry. Book 2

The ancients received into geometry those lines alone which are describable by means of geometrical postulates:[79] three of these Euclid set down in his *Elements*,[80] while a fourth—positing 'To cut a given cone by a given plane'—was added by those who wrote on conic sections. On this basis they received straight lines, circles and conics into geometry, while the remaining curves they called mechanical from the fact that they are described but by means of instruments. Of course, any description of curves by instruments, even that of the circle itself by compasses and of the straight line by a ruler, is mechanical, and they consequently postulated the descriptions of those they received into geometry—not that these might, insofar as they are geometrical, be[81] described by men (for who has seen a line without breadth?) but that, once their description is granted, all the rest of what geometers derive therefrom shall accurately follow from it. Seeing, however, that geometry was established for human uses, they accordingly in the first place postulated the descriptions of those lines which are describable by instruments most easily and accurately of all, namely the straight line and the circle. But those who afterwards added the postulate on the generating of a line by the section of a cone were not so mindful of human utility;[82] for, because of its difficult mechanical accomplishment, that section is a barren speculation having nothing in joint with the uses of men. For this reason, geometry would perhaps not be ill-served by any one who should substitute in its stead another postulate easier in its mechanical execution and which would extend to the description of further curves. To be sure, even the ancients themselves all the while they determined a certain curve[83] more

(82) In an equivalent preceding cancellation Newton's original phrase was 'non recte f[e]c[er]unt' (did not do right).

(83) Newton evidently here refers to the (upper shell of the) quartic conchoid, whose familiar construction by means of a pivoting ruler is elaborated by Pappus in the preamble to *Mathematical Collection* IV, 23 (= *Collectiones*: 86); compare T. L. Heath, *Greek Geometry*, **1**: 238–40, and see also note (156) on page 469 below.

positam determinabant et in solutione problematum utilem esse docebant, ut commemorat Pappus; certe non abhorrebant a plurium curvarū receptione.

Sed ne quid non legitimum postulemus, prius excludenda sunt postulata spuria. Ut *Curvam describere per inventionem punctorum.* Nam si millena puncta per quæ curva transibit inveneris, nunquam tamen omnia sigillatim invenire potes sed curva tandem de puncto ad punctum intervalla singula fortuito manus ductu describenda est, quæ ideo mechanica erit. Similiter postulata vix admittenda sunt quæ alludunt ad extensa fila. Ut si circa datum punctū *B* volvatur baculus *BC*, et interea Filum *ADC* datæ longitudinis una extremitate semper desinens ad punctum *A* paxillo *A* affixū, altera extremitate ligatum ad baculi terminum *C*, trahatur a paxillo *D* versus baculum ita ut pars *CD* per totam longitudinē tangat baculū, pars vero *AD* in directum extensa tendendo recta fiat, paxillo *D* juxta latus baculi movente describatur curva *ED*: descriptionis hujus eadem erit lex ubi baculus *BC* rectus est et ubi curvus est si modo convexo sui latere filum tangat. Ubi vero curv[us] est, descriptio lineæ *ED* nihil habet commune cum Geometria.(84) Rursus descriptiones rejiciendæ sunt quibus curvarum puncta non determinantur ratione vere Geometrica. Nam curva non est geometricè descripta cujus puncta omnia non sunt geometrice inventa. Ut si baculus *AE* circa polū *A* volvatur et a termino ejus *E* circa punctū in eo mobile *C* et punctum datum *B* et rursus circa punctum *C* ad punctum datum *D* tendatur filum cujus tota longitudo *ECBCD* datur, et puncto mobili *C* circa quod filum bis flectitur describatur curva linea:(85) hæc descriptio a Geometria

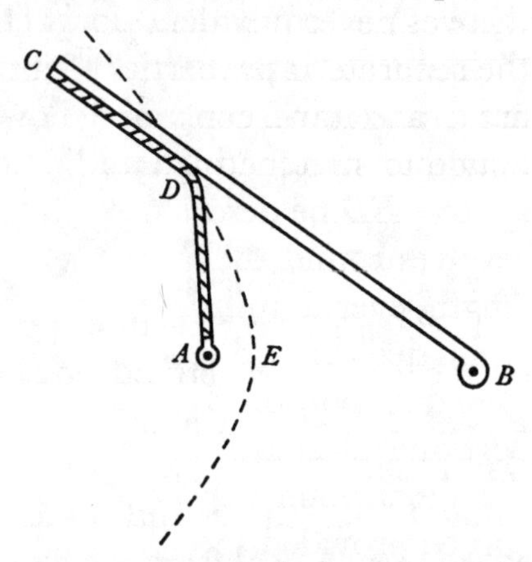

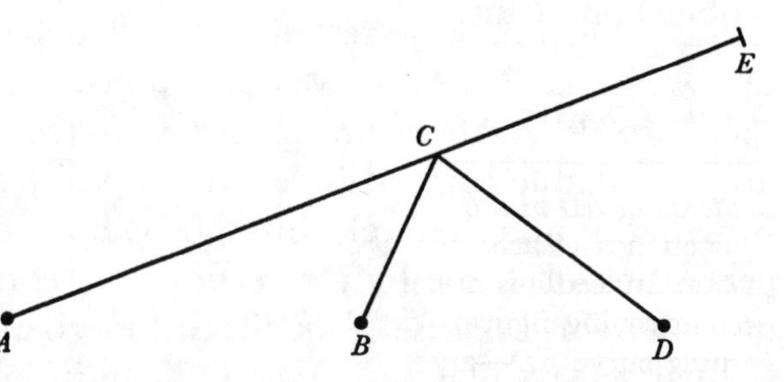

(84) Not even where the general arc $\widehat{BD}$ of the *baculus* is rectifiable and the described curve *ED* is in consequence geometrical? If for instance— in a system of perpendicular Cartesian coordinates in its rotating plane—the *baculus* is (compare our analogous counter-example in v: 470–1, note (689)) a semicubical parabola of fixed cusp $B(0, 0)$ and general point $D(x, y)$ defined by $y^2 = x^3$, so that the length of the arc $\widehat{BD}$ and its chord are respectively

$$\sigma = \tfrac{8}{27}((1+\tfrac{9}{4}x)^{\frac{3}{2}}-1) \quad \text{and} \quad s = \sqrt{[x^2+y^2]} = x(1+x)^{\frac{1}{2}},$$

then Newton's instrument determines the curve *ED* by the general condition $r+k = \sigma \equiv f(s)$, where $AD = r$ and $f(s)$ is the algebraic expression which ensues for σ on eliminating the parameter x in favour of s by means of the preceding equations; that is, by a bipolar equation

compound than the conics and explained its use in the solution of problems, as Pappus recounts, assuredly did not shrink from admitting further curves.

But, lest we postulate anything which is not legitimate, we need first to exclude spurious postulates. Such as 'To describe a curve by the finding of points': for, even if you have found a thousand points through which a curve shall pass, you will yet never be able so to find them all individually but must at length trace the separate intervals from one point to the next by a chance drawing of the hand, and it will consequently be a mechanical line. Similarly, postulates which allude to stretched threads are scarcely to be admitted. Should, for instance, a curve *ED* be described by the peg *D* moving along the edge of a rod *BC* as it revolves round the given point *B*, while all the while a thread *ADC* of given length, ever terminating in its one extremity at the peg *A* attached at the point *A* and tied at its other to the rod's end-point *C*, is pulled taut in the rod's direction so that its portion *CD* shall touch the rod along the whole of its length while its portion *AD* shall stretch out straight to become a straight line: the principle of this description will be the same when the rod *BC* is straight and when it is curved provided the thread touch it along its convex side; when, however, it is curved, the description of the line *ED* has nothing in common with geometry.(84) Again, descriptions are to be rejected when by them the points of curves are not determined by a truly geometrical method. Of course, a curve is not geometrically described when all its points are not geometrically found. If, for instance, the rod *AE* should revolve round the pole *A*, and from its end-point *E* round a mobile point *C* in it and on round the given point *B* and again round the point *C* and as far as the given point *D* there be stretched a thread whose total length *ECBCD* is given, let a curved line(85) then be described by the mobile point *C*

---

connecting *AD* and *BD* which is, in Descartes' sense, 'exact' and so admissibly 'geometrical'. Newton has clearly not pondered such paradoxical consequences of his present exclusive dictate. Indeed it is difficult to see why he at all tolerates the primary case (illustrated in his accompanying figure) where the *baculus BDC* is a straight line since the description of the ensuing curve *ED*—an Apollonian hyperbola of foci *A* and *B*, of course—is equally 'ungeometrical', but presumably he is reluctant to concede that when all the points of so basic a curve are thus 'mechanically' generated they are not thereby 'geometrically found' (to anticipate his phrase in the next sentence but one). This 'organic' description of a hyperbola by a string manually pressed taut along a straight rotating stick is exactly that which Frans van Schooten had presented half a century earlier in his *Organica Conicarum Sectionum in Plano Descriptione* (Leyden, 1646), reissued as Liber IV (pages 293–368) of his *Exercitationum Mathematicarum Libri Quinque* (Leyden, 1657) where the construction is twice presented in little differing forms (see *ibid.*: 337–8, 352–3) and referred for its proof to Apollonius *Conics* III, 51; and Newton himself had duly noted it down in one of his student notebooks (see I: 36) when he came to read Schooten's *Exercitationes* in his last undergraduate year.

(85) Namely—when (as Newton so sets them in his manuscript figure, here accurately reproduced) the poles *A*, *B* and *D* are in line—a Cartesian oval defined parametrically with respect to the pair of poles *A* and *D* by $AC = c+z$, $DC = c-nz$, z free; and so indeed Descartes himself constructs it in the second book of his *Geometrie* (*Discours de la Methode* (Leyden, 1637):

aliena est non tantum quod filum adhibetur sed etiam quod ex assumpta ali[qua] e lineis quæ in figura sunt aut positione aut longitudine, inventio reliquarum problema est quod per extensionem filarem solvere a Geometria abhorreret.(86) Atq; ita si lineæ datarum longitudinum *CD EF* angulos datos *CHD EHF* subtendentes convergant ad punctum *K* quod ubivis in linea *AB* situm sit, et recta *DG* datum angulum cum *CD* continens secet rectam *EF* productam in *G*, dum

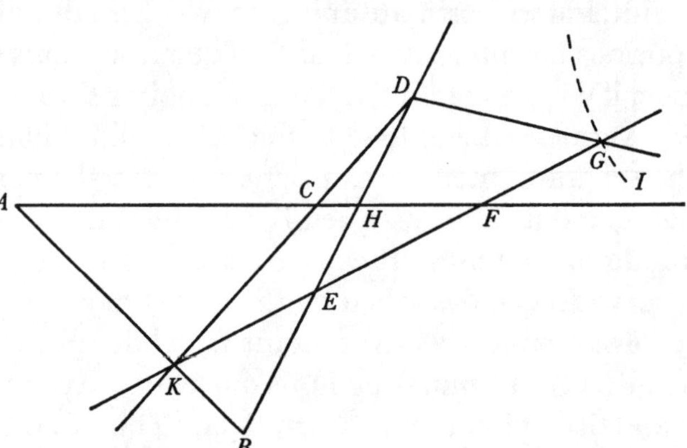

punctum *K* per omnia puncta lineæ *AB* vagatur, punctum *G* motu suo describet curvam quandam *GI*[:] sed hæc descriptio non erit Geometrica eo quod ex assumpta alterutrius lineæ *CD* vel *EF* positione qua punctum *K* determinetur positio alterius lineæ determinatur per applicationem ejus a puncto *K* ad angulum *CHD* vel *EHF* quem subtendit, quæ applicatio nulli operationi Geometricæ respondet,(87) et proinde punctorum *G* hac methodo inventio non est geometrica. (88)Jā vero siquis, ne inventio punctorū *C* vel *G* mechanica habeatur, postulet *lineas datas quacunq; ratione assignata movere ut earum intersectionibus novæ lineæ describantur;* hic postulabit puncta ignota in casibus

356 = *Geometria* (Amsterdam, $_2$1659): 54) but without adding any demonstration. In proof, let the described locus (*C*) meet *ABD* in *F*, and take $AF = c$, $FD = d$, $FB = e$; suppose, further, that there exist constants m and n such that, where $AC = c+z$, then correspondingly $BC = e+mz$ and $DC = d-nz$. Because by construction CE(or $AE-AC)+2BC+DC$ is fixed,

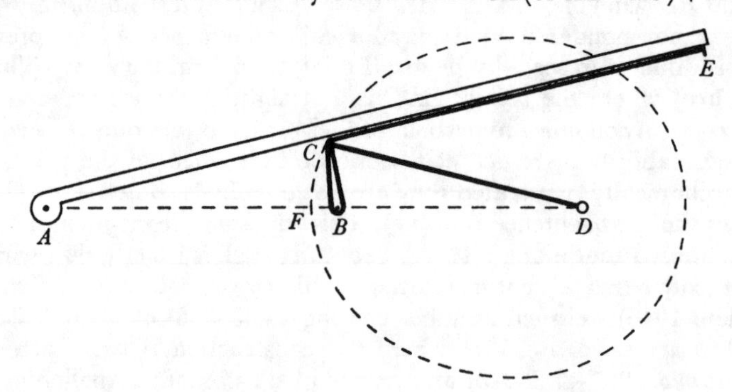

there is $-AC+2BC+DC = -(c+z)+2(e+mz)+d-nz = -c+2e+d$, constant, and hence $m = \frac{1}{2}(1+n)$, so that $BC = e+\frac{1}{2}(1+n)\,z$. Again, since the angles

$$\widehat{ABC} = \cos^{-1}[\tfrac{1}{2}((c+e)^2+(e+\tfrac{1}{2}(1+n)\,z)^2-(c+z)^2)/(c+e)\,(e+\tfrac{1}{2}(1+n)\,z)]$$

and $$\widehat{DBC} = \cos^{-1}[\tfrac{1}{2}((d-e)^2+(e+\tfrac{1}{2}(1+n)\,z)^2-(d-nz)^2)/(d-e)\,(e+\tfrac{1}{2}(1+n)\,z)]$$

are supplementary, and hence $\cos\widehat{ABC} = -\cos\widehat{DBC}$, on substituting their values, cancelling

round which the thread is twice bent: this description is foreign to geometry not merely because a thread is employed, but also because, when some one of the lines in the figure is assumed either in position or length, the finding of the rest from it is a problem whose solution by a thread-stretching would shrink away from geometry.(86) And likewise, if lines CD, EF of given lengths and subtending given angles $C\widehat{H}D$, $E\widehat{H}F$ should converge on the point K situated anywhere in the line AB, and the straight line DG containing a given angle with CD should cut the line EF produced in G, while the point K roams through all the points of the line AB the point G shall in its motion describe a certain curve GI: but this description will not be geometrical inasmuch as, when either of the lines CD or EF is assumed in position, so thereby determining the point K, the position of the other line is determined by its application (in direction) from the point K to the angle $C\widehat{H}D$ or $E\widehat{H}F$ which it subtends; since this application corresponds to no geometrical operation,(87) the finding of the points G by this method is not geometrical. (88)Should anyone now, however, to prevent the finding of the points C or G being regarded as mechanical, postulate that 'given lines may move in any assigned manner whatever so that by their intersections fresh lines are described', he will postulate that in innumerable cases unknown points are

---

the common factor $e+\frac{1}{2}(1+n)\,z$ ($\neq 0$ where F is distinct from B) and reordering the resulting equation, there ensues true for all z

$$(d-e)\,(2(c+e)\,e+(e(1+n)-2c)\,z-\tfrac{1}{4}(3+n)\,(1-n)\,z^2)$$
$$+(c+e)\,(-2(d-e)\,e+(e(1+n)+2dn)\,z+\tfrac{1}{4}(1+3n)\,(1-n)\,z^2) = 0;$$

accordingly, after deleting the constant terms and equating the coefficients of z and z^2 to zero, there comes $e = 2cd(1-n)/(c(3+n)+d(1+3n)) = \frac{1}{4}(-c(1+3n)+d(3+n))/(1+n)$, whence $c^2 = d^2$ and so (since D is taken distinct from A and therefore $c \neq -d$) $c = d$, that is, $AF = FD$, and in consequence $FB = e = \frac{1}{2}c(1-n)/(1+n)$. It follows that ($C$) is a Cartesian oval defined parametrically by any pair of its respective distances $c+z$, $\frac{1}{2}c(1-n)/(1+n)+\frac{1}{2}(1+n)\,z$ and $c-nz$ from its three poles A, B and D. However unacceptable in his modified Euclidean plane geometry Newton may—as in sequel he does—regard this 'mechanical' description *per extensionem filarem*, there is no denying that the curve thus constructed (which is readily shown to be of fourth degree) is itself inalienably 'geometrical'. He stands once more (compare note (84) above) badly in need of some modified Cartesian criterion of 'exactness' which will helpfully distinguish those instrumental descriptions which produce an admissibly 'geometrical' (algebraic) curve from those which yield a 'mechanical' (transcendental) one.

(86) Originally 'nulli operationi Geometricæ respondet' (corresponds to no geometrical operation). This phrase, unsatisfactory as it is (compare notes (84) and (85) preceding), reappears uncancelled a few lines below in an equivalent context; see the next note.

(87) This is, as we have already observed in the case of Newton's preceding 'mechanical' descriptions (see notes (84) and (85) above), not a sufficient criterion for the constructed curve to be itself non-'geometrical'.

(88) Newton initially went on: 'Adhæc descriptiones illæ rejiciendæ sunt quibus tolluntur Problemata Geometrica. Ut si quis postulaverit *lineas datas...describantur:* hoc postulatum tollit plurima problemata' (Furthermore, those descriptions must be rejected whereby geometrical problems are eliminated. If, for instance, anyone should postulate that 'given lines...are described', this postulate eliminates the vast majority of problems).

innumeris dari quorum inventionem Geometræ omnes inter problemata numerant, adeoq problemata tollet. Ut si angulus datus recta data subtendendus esset quæ ad datum punctum vergeret: Quoniam in figura præcedente lineæ *CD*, *EF* in angulis datis *CHD* *EH*[*F*] juxta postulatū illud ita moveri possint ut ad commune punctum *K* in recta *AB* situm semper convergant, adeoq ex dato puncto *K* positiones earum vi postulati illius dantur, potest angulus quivis datus *CHD* recta data *CD* quæ ad datum punctum *K* converget, vi postulati illius subtendi ideoq ut Problema solvendum inepte proponitur[:] nam problema non est quod fieri postulatur. Cavendū est igitur ne dum Geometriam studemus augere, [eam] contaminemus hujusmodi postulatis.

Quoniam igitur postulata debent esse pauca facillima et ab invicem distincta ita ut quod in uno postuletur non includatur in altero: Quid si igitur postulemus[(89)]

1. A dato puncto ad datū punctum rectam lineam ducere.

2. Per data duo puncta lineam[(90)] datæ lineæ distantia punctorum non minori similem et æqualem ducere, cujus punctum imperatum in alterut[r]um punctum datum incidet.

3.[(91)] Ducere lineam quamvis in quam semper incidet punctum quod certa lege[(92)] ducendo a punctis per puncta lineas datis similes et æquales datur.

His tribus postulatis Geometriam om[n]em legitimam comprehendi puto. Secundo includitur secundum Euclidis[(93)] nam data linea *AB* continuo[(94)] produci potest si ab aliquo ejus puncto *C* per punctum *D* agatur *CD* similis et æqualis *AB* et sic in infinitum. Tertio includitur tertium.[(93)] Nam si a dato puncto *C* per punctum quodvis *E* agatur linea *CD* similis et æqualis rectæ *CB*[,] circuli erit periph[e]ria in quam datum punctum *D* incidet, proinde[q] centro quovis *C* et intervallo *CB* circulus describi potest. Hæc autem non trado ut

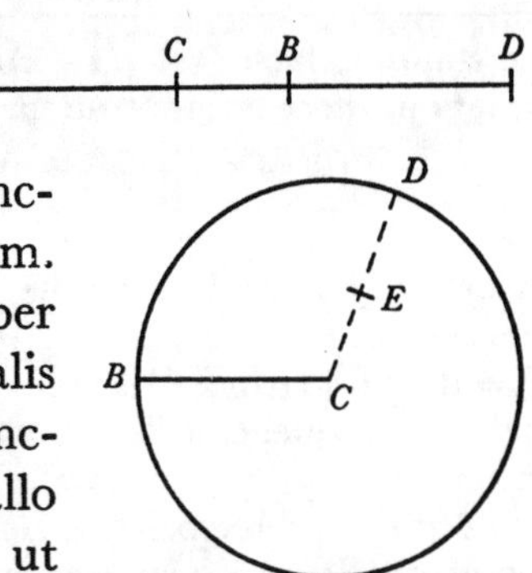

(89) Other than that Newton stresses that the points in question are *data*, his first *postulatum* following is exactly Euclid's (compare Barrow's Latin version, which we have cited in note (80) above), while—as he himself goes on to make clear—the second and third comprehend Euclid's similar *postulata* as special cases.

(90) Initially 'A dato puncto per datum punctum rectam lineam alteræ' (from a given point through a given point a straight line...to a second).

(91) Newton found great difficulty in enunciating this third postulate, first requiring (f. 39ᵛ) 'Datis aliquot lineis alias ducere quasvis in quas incident puncta assignata et cognita linearum omnium quæ ad datarum normam secundū postulatum prius communi aliqua lege ducuntur' (Given a number of lines, to draw any others in which shall fall assigned and known points of all lines which are drawn, following the previous postulate, to the pattern of the given ones by some common law), and then demanding in turn (on f. 15ʳ) 'Datis aliquot lineis et ad harum normam per postulatum prius juxta legem aliquam assignatam ductis alijs, invenire aliam quamvis in quam incidet punctū datum' (Given a number of lines and, by means of the previous postulate, drawing others to their pattern according to some assigned

given the discovery of which all geometers count as problems, and hence he will do away with problems. Thus, if a given angle were to be subtended by a given straight line so as to verge on a given point, because the lines CD, EF in the preceding figure might in accord with that postulate be moved in the given angles $C\widehat{H}D$, $E\widehat{H}F$ so as always to converge on a common point K sited in the straight line AB, and hence, given the point K, their positions are by dint of that postulate given therefrom, any given angle $C\widehat{H}D$ you wish can by force of that postulate be subtended by a given straight line CD which shall converge to the given point K, and in consequence it is ineptly proposed as a problem to be solved; for, of course, a problem is not something which is postulated to be done. We must therefore take care lest, in our zeal to augment geometry, we at the same time pollute it with postulates of this sort.

Seeing, then, that postulates ought to be few, very easy and distinct one from another, so that what shall be postulated in one be not included in another, what therefore of it if we should postulate these?[89]

1. To draw a straight line from a given point to a given point.

2. To draw through two given points a line congruent to a[90] given line at a distance not less than that of the points, an ordained point of which shall coincide with one or other given point.

3.[91] To draw any line on which there shall always fall a point which is given according to a precise rule[92] by drawing from points through points lines congruent to given ones.

In these three postulates all legitimate geometry is, I think, embraced. In the second is included Euclid's second,[93] for the given line AB can be continuously extended[94] if from some point C of it through the point D there be drawn CD congruent to AB, and so on indefinitely. In the third is included his third,[93] for if from the given point C through any point E there be drawn the straight line CD congruent to the straight line CB, the periphery on which the given point D shall fall will be that of a circle, and consequently a circle can be described

law, to find any other in which shall fall a given point) and yet again requiring 'Ducere lineam quamvis in quam incidet punctum quod ducendo solummodo datis similes et æquales constanti lege datum est' (To draw any line in which shall fall a point which is given by drawing only lines congruent to given ones according to a constant rule) before arriving at the final version here reproduced.

(92) By which 'certain law' Newton evidently understands—or so we presume—the *simplex geometria* to which he has so often previously appealed in like circumstances (and on which see especially IV: 302–4, where he defined the latter to comprehend 'quæ per ductum solarum rectarum...determinari possunt'). Initially he here wrote 'utcunq juxta legem assignatam per postulatum prius' (howsoever according to an assigned law by means of the previous postulate); compare the equivalent prior versions reproduced in the preceding note.

(93) These are quoted, in Barrow's 1655 Latin version, in note (80) above.

(94) Newton first began to write 'ad ar[bitrium]' (arbitrarily).

Postulata Euclidis submoveam. Illis Geometria plana[95] optime innititur. His uti licebit quoties in Geometria solida et sursolida[95] res est. Nam cùm Geometræ jactitent se tribus tantum postulatis ad omnium planorum problematum solutionem usos esse, insinuare volui tria etiam ad altiorum solutionem sufficere.

Postulatum secundum nil difficilius est quam primū et secundū Euclidis. Illa ad usum normæ alludunt. Et linea ad normam quamvis sive rectam sive curvam æque facile et accuratè[96] ducitur. Hoc tamen postulato tollitur problema unum et alterum Euclidis, ut illud secundæ[97] propositionis primi *Elementorum*. Sed ea tam simplicia sunt ut in postulatis includi melius sit quam per nimias ambages solvi. [98]Postulatum tertiū affinitatem habet cum descriptione mechanica per motas regulas. Ut si a rectæ interminatæ positione datæ AB puncto quolibet B per punctum datum P ducatur linea BCP similis et æqualis lineæ bcp quæ ex duabus rectis, data bc & interminata cp, angulum rectum bcp continentibus constat, dabitur punctum C et si sumatur BD datæ longitudinis dabitur etiam punctū D. Unde per postulatum tertium describi possunt lineæ illæ duæ in quibus lineæ normalis BCP sic ductæ[99] puncta C ac D semper reperiuntur. Organicè vero describentur si norma materialis BCP ita moveatur ut crus ejus CP semper transeat per polum P, crus vero CB termino suo B dilabatur super regulam AB et puncta ejus data C, D in plano subjecto & immoto ducant lineas. Et linea quidem quæ puncto D describitur ubi BC bisecatur in D erit Cissoides Veterum.[100] Rursus si a datæ positione rectæ AB puncto quovis C per puncta data D, E

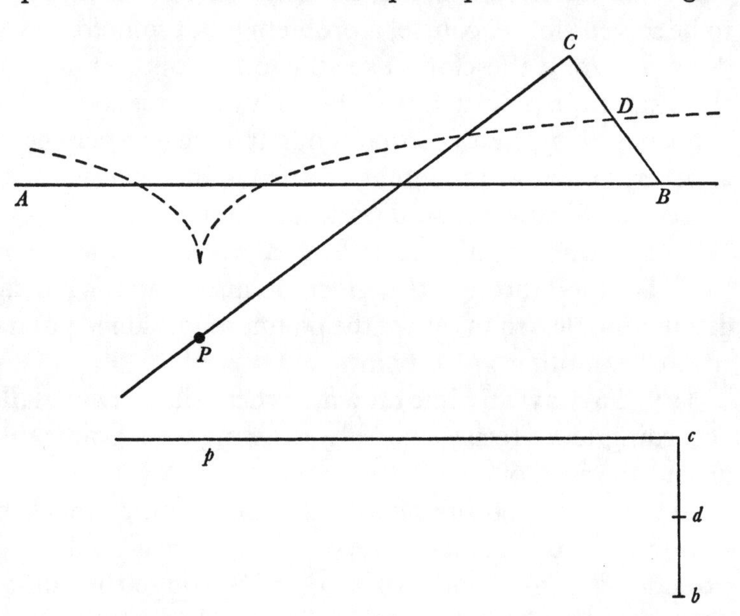

(95) As ever, in the classical sense of the geometry of straight-lines/circles and that also invoking conics and higher curves respectively.

(96) 'eadem ακριβεια' (with the same precision) was initially written by Newton in a rare appeal to the nuances of Greek to supply a *mot juste* lacking in Latin.

(97) Originally 'secundæ & tertiæ' (...second and third).

(98) Newton first went on: 'Postulato tertio alludimus ad descriptiones curvarum per motas regulas. Nam quemadmodū Euclides lineas quæ facilius et accuratius per instrumenta

with any centre C and radius CB. I do not recount these, however, in order to displace Euclid's postulates. On those 'plane' geometry[95] best relies. The present ones we will be free to employ each time the topic is one of 'solid' and 'higher than solid' geometry.[95] Though geometers may, to be sure, boast that they make use of but three postulates to attain the solution of 'plane' problems, I did wish to suggest that three are also sufficient for the solution of higher ones.

The second postulate is no more difficult than Euclid's first and second. The latter allude to the use of a rule; and a line is drawn along any rule, be it straight or curved, equally easily and accurately.[96] By the present postulate, however, a problem or two of Euclid's is done away with, such as that of the second[97] proposition of the *Elements*' first book. But these are so simple that they are better included in the postulates than solved by excessive digression. [98]The third postulate has a kinship with mechanical description by moving rulers. Thus, if from any point B you please of an unterminated straight line AB given in position you draw through the given point P a line BCP congruent to the line bcp made up of two straight lines, a given one bc and an unterminated one cp, containing the right angle $\widehat{bcp}$, then the point C will be given, and if BD be taken of given length the point D will also be given; whence by the third postulate the two lines can be described in which the points C and D of the right-angled line BCP so drawn[99] are always located. They will, indeed, be organically described if the solid right-angle BCP shall move so that its leg CP shall always pass through the pole P, while its leg CB at its end-point B slides along the ruler AB, its given points C, D tracing the lines on the stationary plane below. And when, to be sure, BC is bisected at D, the line described by the point D will be the ancients' cissoid.[100] Again, if from any point C of a straight line AB given in position

---

mechanica ducebantur postulavit Geometricè duci posse, sic æquum est Geometricam descriptionem aliarū linearum postulari quæ per instrumenta facilior et accuratior est' (In the third postulate we allude to the descriptions of curves by moved rulers. For, just as Euclid postulated lines which were drawn more easily and accurately by mechanical instruments to be geometrically drawable, so it is fair for there to be postulated the geometrical description of other lines where it is easier and more accurate by means of instruments). In a preceding preliminary draft he had started to write to the same effect that 'Tertio postulato alludimus ad mechanicam descriptionem curvarum per puncta vel intersectiones motarum regularum. Ut si' (In the third postulate we allude to the mechanical description of curves by points or the intersections of moved rulers. If, for instance, ...).

(99) 'ab datæ lineæ AB alio atq; alio puncto B' (from one or another point B of the given line AB) was originally added.

(100) See Problem 20 of Newton's 'Arithmetica Universalis' (v: 218–20) *mutatis mutandis*. In these Cambridge '*lectiones*' he subsequently (*ibid.*: 464–6) sought to apply the more general cubic curve described by a point which does not bisect the travelling arm in an abortive attempt geometrically to construct a cubic equation, noticing—and correcting—his error (see *ibid.*: 466–9, note (686)) only at a late stage after the pages of his 1722 revised Latin edition of the *Arithmetica* had been initially printed off.

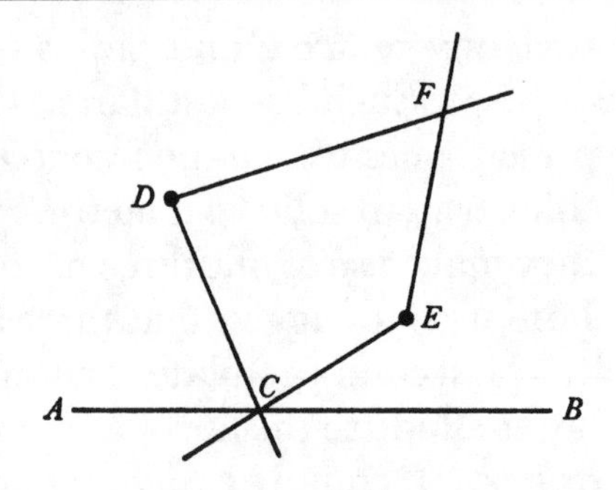

ducantur lineæ *CDF*, *CEF* quarum partes *CD*, *DF*, *CE*, *EF* rectæ sunt et angulos *CDF CEF* datis æquales continent, dabitur punctum *F* adeoq per postulatum tertium linea duci potest in quam punctum *F* semper incidit. Organicè[101] verò ducetur linea illa si regulæ angulares *CDF CEF* circa polos *D*, *E* ita volvantur ut crura *DC*, *EC* sese semper secent ad[102] lineam *AB*, & crurum alterorum *DF*, *EF* intersectio *F* in plano subjecto[103] describat lineam. Et hoc modo pro diverso situ lineæ *AB* diversaq magnitudine angulorum *D*, *E* innumeræ lineæ describi possunt, quæ tamen omnes erunt Conicæ sectiones.[104] Et si loco rectæ *AB* adhibeantur jam descriptæ Conicæ sectiones, punctum *F* describet alias curvas innumeras[105] quarum aliquæ rursus loco conicarum s[ectionum] usurpari possint, et sic in infinitum. Et sic vice rectarum *DC*, *EC*, *DF*, *EF* aliquæ ex curvis jam descriptis adhiberi possunt.[106] Hæ autē descriptiones quatenus per Organa ma[n]ufacta perficiuntur mechanicæ sunt: quatenus verò per lineas geometricas quas organorum regulæ representant subintelliguntur fieri, eæ ipsæ sunt quas postulato tertio ut geometricas amplectimur.[107]

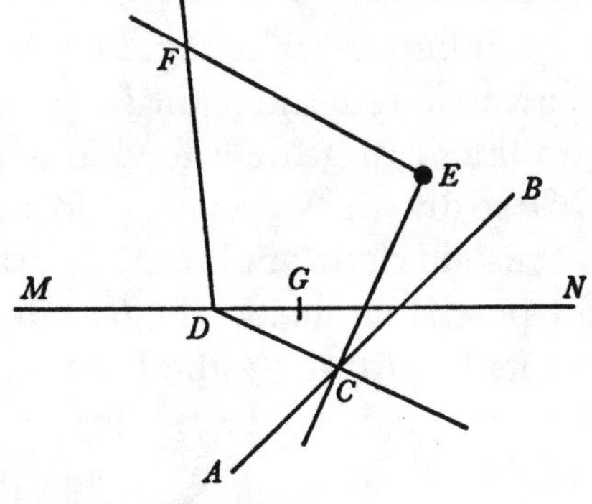

Rursus si per datæ positione rectæ *MN* puncta quævis *G*, *D* agantur lineæ *GDF GDC* similes et æquales lineis *gdf*, *gdc* et per punctū *C* ubi *DC* secat rectam *AB* positione datam & punctum datum *E* agatur linea *CEF* alteri datæ similis et æqualis, dabitur linearum *DF EF* intersectio *F*.[108] Ergo curva in quam *F* semper incidet duci potest per postulatum tertium. Nempe si linea composita *GDFC* dilabatur super lineam *MN* & linea *CEF* vertatur circa polum *E* ita ut earum crura se semper secent ad lineam *AB*, aliorum crurū intersectio *F* in plano subjecto describet curvam desideratam. Quæ descriptio Mechanica est quatenus instrumentis manufactis utimur,[109] Geometrica vero quatenus juxta postulatū tertiū conceditur

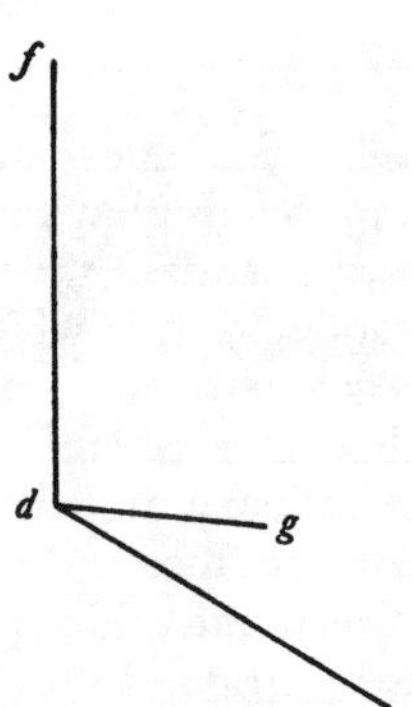

(101) Originally 'Mechanicè' (mechanically).

(102) This replaces 'alicubi in' (somewhere in).

(103) Initially 'in plano super quod moventur' (in the plane upon which they move).

(104) Passing through the poles *D* and *E*, of course. Compare Newton's previous discussions of the conic case of this his favourite *organica descriptio* of curves, geometrically in the two versions of his late 1660's treatise 'De Modo describendi Conicas sectiones...' (see II: 118–28, 148) and algebraically a decade later in Problem 53 of his 'Arithmetica Universalis' (v: 304–6).

through given points D, E there be drawn lines CDF, CEF whose portions CD, DF, CE, EF are straight lines and contain angles $C\widehat{D}F$, $C\widehat{E}F$ equal to givens, then the point F will be given, and hence by the third postulate the line can be drawn in which the point F always falls. That line will, of course, be drawn organically[101] if the angular rulers CDF, CEF should revolve round the poles D, E such that the legs DC, EC shall always intersect each other at [102] the line AB, and the intersection F of the other two legs DF, EF shall then describe the line in the plane below.[103] And in this way, according to the differing situation of the line AB and the differing size of the angles $\widehat{D}$ and $\widehat{E}$, innumerable lines can be described, all of which, however, will be conics.[104] And if in place of the straight line AB you employ the conics just now described, the point F will describe innumerable other curves,[105] some of which in turn may be used to replace the conics, and so on indefinitely. And, likewise, instead of the straight lines DC, EC, DF, EF certain of the curves already described can be employed.[106] But these descriptions, insofar as they are achieved by manufactured instruments, are mechanical; insofar, however, as they are understood to be accomplished by the geometrical lines which the rulers in the instruments represent, they are exactly those which we embrace[107] in the third postulate as geometrical.

Yet again, if through any points G, D in the line MN given in position there be drawn lines GDF, GDC congruent to the lines gdf, gdc, and through the point C in which DC cuts the straight line AB given in position and the given point E there be drawn the line CEF congruent to a second given one, then the intersection F of the lines DF, EF will be given.[108] Therefore, by the third postulate the curve in which F shall always fall can be drawn—specifically, if the compound line $GDFC$ should slide away along the line MN and the line CEF turn round the pole E such that their legs always intersect one another at the line AB, the intersection F of the other legs will describe the desired curve in the plane beneath. This description is mechanical inasmuch as we employ manufactured instruments,[109] but geometrical insofar as it is, in accordance with the third postulate, granted that those instruments can be accurately constructed and

(105) Namely, cubics or quartics with double points at one or both respectively of the poles D and E. Compare II: 128–32, 148–50.

(106) At the expense of destroying the basic 1, 1 correspondence *per simplicem geometriam* which is continuously constructed between the points C and F of the describing and described curves. That way lies trouble ahead!

(107) Originally 'indigitamus' (invoke).

(108) This *descriptio mechanica* has no precedent that we have been able to discover, either in Newton's own earlier papers (though it has some echoes of his 'Engines' on I: 264 and 462) or in the works of others whom he might have read. Nor, to be honest, is it very practical.

(109) Newton first began to add 'quæ non sunt accurate [construendæ ac tractandæ]' (which are not accurately to be constructed and handled) before subsuming its essence in the following clause.

instrumenta illa accurate construi ac tractari posse. Et eodem modo descriptio organica semper comitatur postulatum.

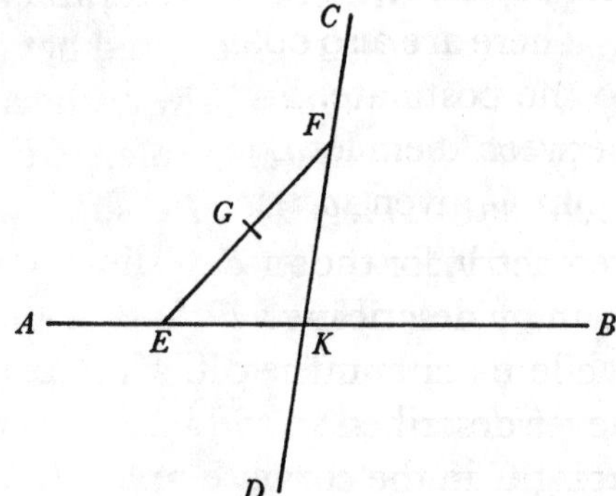

Sunt et aliæ descriptiones organicæ satis faciles quæ Postulato non sunt affines. Ut si dentur positione lineæ *AB*, *CD* et inter has labatur recta *EF* longitudine data quæ puncto suo quovis dato *G* Ellipsin describat.(110) Sed hæc descriptio aliena est a geometria[,] nam recta *EF* non ponitur geometrice in angulo *EKF* nisi per descriptionem circuli cujus centrum sit in alterutro punctorum *E*, *F* [et] circumferentia transeat per alterum. Hic autem circulus in motu continuo lineæ *EF* nunquā describitur et proinde nullum curvæ descriptæ punctum *G* invenitur Geometricè.

Fateor tamen hujusmodi postulata fieri posse: *A dato puncto ad datam positione lineam ducere rectam quæ datæ erit longitudinis:* Et *a dato puncto rectam ducere cujus pars inter duas positione datas lineas datæ erit longitudinis.*(111) Atq hæc ad omnium planorum et solidorum problematum(112) solutionem sufficiunt. Sed duo obstant quo minus recipiantur. Ijs tolleretur usus circuli et conicæ sectionis(113) in solutione problematum, neq possunt eorum subsidio figuræ illæ curvilineæ quæ Geometriæ objecta sunt describi. Ad horum igitur descriptionem opus est postulatis. Quibus admissis superfluum erit aliquid amplius postulare. Atq hæc de postulatis scripsi ut eorum natura intellecta constaret legitima esse quæ postulavimus, neq aliud amplius postulandum esse.

Superest ut de curvis agamus quæ per hæc postulata describi possunt. Sed ne immorer in distinguendis his ab alijs quæ non liceat sic describere, imprimis considerabo indistincte(114) lineas omnes regulares qu[as] licebit imaginari. Lineam regularē voco cujus omnia puncta eadem(115) lege determinantur. Hæ lineæ eo magis intricatæ et compositæ sunt quo plures habent sinus aut sæpius

(110) As we have recorded (I: 31–2), Newton had encountered this trammel construction of the ellipse in his undergraduate reading of Liber IV of Schooten's *Exercitationum Mathematicarum Libri Quinque* (Leyden, 1657): 314–21, and he subsequently made elegant application of it both in Problem 4 of his 'Problems for construing æquations' (see II: 478–84) and in the corresponding portion (V: 474–84) of the appendix to his 'Arithmetica Universalis' to construct the real roots of general cubic and quartic equations by determining the intersections of the described ellipse with a given circle.

(111) The classical problem of 'verging' (νεῦσις), explored in its simplest cases where the given curves are straight lines and circles by Apollonius in his lost work *De Inclinationibus* and (compare §2, Appendix 2: note (15) above) by Archimedes.

(112) As ever, those resolvable geometrically by the intersections of 'plane' and 'solid' loci, and hence reducible algebraically to equations of first/second and third/fourth degrees respectively. As Newton had shown long before in a variety of ways in his 'Problems for

manipulated. And in the same way the organic description always attends upon the postulate.

There are also other organic descriptions, easy enough, which are not related to the postulate. For instance, let the lines AB, CD be given in position, and between them let there slide the straight line EF given in length so as by any point G given in it to describe an ellipse.(110) But this description is foreign to geometry, for the straight line EF is not set geometrically in the angle $E\widehat{K}F$ other than by describing a circle whose centre shall be at either one of the points E, F while its circumference shall pass through the other. This circle, however, is never described in the continuous motion of the line EF, and consequently no point G in the curve described is found geometrically.

I admit nonetheless that postulates of the following sort can be made: 'From a given point to draw to a line given in position a straight line which shall be of given length'; and 'From a given point to draw a straight line whose portion between two lines given in position shall be of given length'.(111) And these are sufficient for the solution of all plane and solid problems.(112) But there are two obstacles to their being received. By them the employment of a circle and conic(113) in the solution of problems would be done away with, nor can by their aid the curvilinear figures which are the objects of geometry be described. For the latter's description, accordingly, we have need of postulates; and, once these are accepted, it will be superfluous to make further postulate of something. I have penned these observations on postulates so that, with their nature understood, what we have postulated might be agreed to be legitimate and not stand in need of further postulation.

It remains for us to discuss what is describable by means of these postulates. But, so as not to dwell on distinguishing this from what else it is impermissible thus to describe, I shall in the first instance consider indiscriminately(114) all regular lines which it is allowable to conceive. I call a line 'regular' when its every point is determined by the same(115) principle. Such lines are the more intricate and compound the more folds they possess or the more often they are bent into loops.(116) In consequence of this, from the number of points in which

---

construing æquations' (see II: 460–76, 508–12) and afterwards in the corresponding portion of his 'Arithmetica' (see V: 428–64), all such 'plane' and 'solid' problems are constructible *infinitis modis* by conchoidal neuses setting a line-segment of given length to lie with its end-points one each on a pair of given straight lines or on a given straight line and a given circle.

(113) As, for example, Newton himself has applied these in constructing the roots of general cubic/quartic equations— and hence, by reduction, all 'solid' problems—on II: 484–98 and equivalently on V: 484–90.

(114) 'conjunctim' (jointly) was first written.

(115) This replaces 'communi' (a common).

flectuntur in gyros.[116] Unde fit ut a numero punctorum in quibus secari possint a linea recta, distinguantur in genera. Lineā primi generis definire licet quam recta in unico tantum puncto secare potest nimirum ut rectam lineam, secundi quam secare potest in duobus et non in pluribus ut circulum et conicas sectiones, tertij quam secare potest in tribus tantum ut Cissoidem Veterum, & sic in infinitum: ultimi vero generis est curva quam recta secare potest in punctis infinitis ut Spiralis Archimedea, Quadratrix, Trochoides. Verum quot punctis recta secare potest curvam optime noscitur per dimensiones æquationis cujus radicibus intersectiones illæ determinantur. Nam tot habetur dimensiones quot sunt intersection[e]s[,] at possunt intersectionum aliquæ nonnunquam impossibiles esse (ut intersectio circuli [*A*] et rectæ *B*) et tunc supplebuntur per imaginarias radices æquationis. Linea igitur primi generis est si intersectio[ne]m ejus et rectæ lineæ generaliter[117] determinare sit problema simplex seu primi generis; Secundi generis est linea si problema illud planum est seu secundi generis; Tertij generis est si problema illud solidum est; quarti generis, si sursolidum;[118] & sic in infinitum.

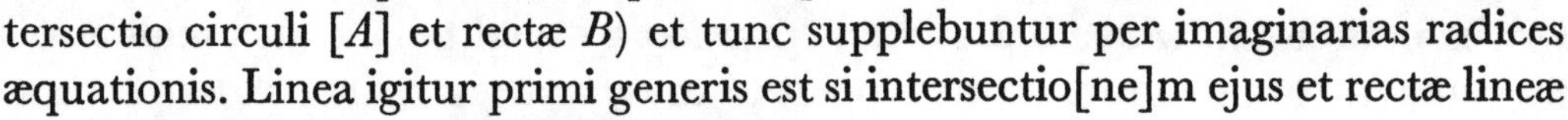

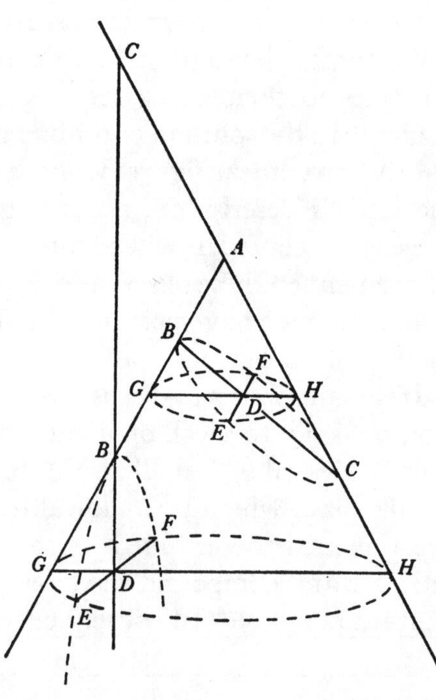

Ut si cujusnam generis sint conicæ sectiones scire vellem, secetur conus *BAC* plano determinato *BEF*. Concipe tum hoc planū ab alio quovis plano *GEHF* secari quod sit coni basi circulari parallelum atq; adeo et ipsum circulare. Ad plani utriusq; intersectionem [*F*]*E* bisectam in [*D*] erige in planis *FEB* [*FHEG*] normales *DB* *DG*. Quæ productæ occurrant lateribus Coni in *B*, *C* ac *G*, *H*. Et erit *BC* diameter sectionis ac *GH* diameter circuli. Ergo

$$GD \times DH = DE^q.$$

At *DH* ad *DC* et *DG* ad *DB* datam habent rationem. Ergo *GDH* hoc est $D[E]^q$ ad *BDC* datam habet rationem.[119]

(116) We are here strongly reminded of James Gregory's earlier restriction of his 'Methodus Universalis transmutandi, & mensurandi quantitates curvas', as he expounded it in his *Geometriæ Pars Universalis* (Padua, 1668), to a 'curva quæcunque...simplex & non sinuosa' (see *ibid.*: 1, 3, 13, 17).

(117) Newton initially began to pen in sequel a superfluous 'per geo[metriam]' (by geometry).

they can be cut by a straight line they may be separated into kinds. A line of first kind is permissibly defined to be one intersectable by a straight line in but a single point, a straight line itself to be sure; of second, one so intersectable in two and no more, namely the circle and the conics; of third, one so intersectable in but three, such as the ancients' cissoid; and so on indefinitely: of ultimate kind, however, is a curve which a straight line can cut in an infinity of points, such as the Archimedean spiral, quadratrix and cycloid. But the number of points in which a straight line can cut a curve is best ascertained through the dimensions of the equation by whose roots those intersections are determined; for it has as many dimensions as there are intersections, though some of the intersections can on occasion be impossible (as here the intersection of the circle A and straight line B) and their place will then be filled by imaginary roots of the equation. A line is therefore of first kind if the general determination(117) of its intersection with a straight line be a simple problem, that is, one of first kind; a line is of the second kind if that problem is plane, or of second kind; it is of a third kind if that problem is solid; of a fourth kind if sursolid;(118) and so on indefinitely.

Should I, for instance, wish to know of what kind conics are, let the cone BAC be cut by the determinate plane BEF; then conceive that this plane is cut by any other plane $GEHF$ which shall be parallel to the cone's circular base and hence itself circular. At the bisection point D of their common intersection EF erect in the planes FEB, $FHEG$ the normals DB, DG and extend them to meet the cone's sides in B, C and G, H. Then BC will be the section's diameter and GH the circle's diameter, and therefore $GD \times DH = DE^2$. But DH and GD have given ratios to DC and DB. Therefore ($GD \times DH$ or) DE^2 has a given ratio to $BD \times DC$.(119)

(118) Compare note (95) above.

(119) Except where the cutting plane $BEDF$ is parallel to the corresponding cone-generator AH, so that the diametral end-point C is at infinity and the resulting section is a parabola defined by the constancy of the ratio of DE^2 to BD. In the general case where C is finite and the (second-degree) section determined by Newton's classical Apollonian symptom is a central conic, this will evidently be an ellipse or hyperbola according as the abscissal point D lies within or outside the diameter BC.

In cancelling this intended exemplification of his preceding rule for classifying the 'kind' of a given or constructed geometrical *linea* by the number of its possible intersections with a fixed straight line, Newton began tentatively to replace it with an analogous conic instance before breaking abruptly off to discard this too, initially continuing in the manuscript (f. 16v): 'Ut si curva CE ejusmodi sit ut rectarum AC BC a quovis ejus puncto C ad data duo puncta A, B ductarum summa detur et cujus generis sit hæc curva scire cupiã: fingo rectam aliquam CD positione

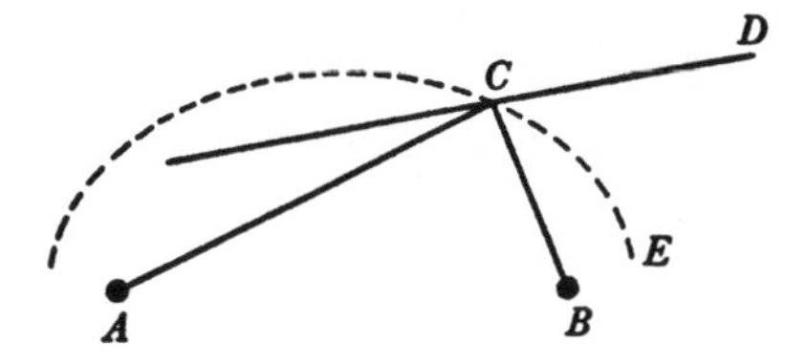

datam esse et [? determino puncta in quibus ea curvam secare potest]' (If, for instance, the curve CE be of the sort that when straight lines AC, BC are drawn from any point C of it

[120]Et ex cognitis linearum generibus vicissim innotescunt genera problematum quæ per earum intersectiones solvuntur. Numerus enim punctorū in quibus duæ lineæ se possunt secare confit ex multiplicatione numerorum ordinis utriusꝗ[:][121] ut si una linea sit secundi ordinis altera tertij, numerus intersectionum erit 2×3 seu 6, nisi ubi ex intersectionibus illis duæ vel quatuor vel omnes sex evaserint impossibiles adeoꝗ Problema per eas solutum erit sexti ordinis. Aut si intersectionū una vel plures datæ sint vel ratione aliqua quæ non extendit ad intersectiones omnes inveniri possint[,] Problema erit quinti vel inferioris ordinis.

Veteres[122] ut omnia Problemata ad intersectiones linearum deducerent proposuerunt hujusmodi Porisma:[123] *Invenire lineam a cujus puncto quovis ad datas positione duas tres vel plures rectas aliæ totidem rectæ singulæ ad singulas in datis angulis ducantur ea lege ut ex ductis illis vel solis vel una cum alijs quibusvis datis rectis conflentur contenta duo plana vel duo solida quæ æqualia sint vel aliam quamvis datam habeant*

---

to two given points A, B their sum is given, and I should desire to know of what kind this curve is: I imagine there to be some straight line CD given in position and [I determine the points in which it can cut the curve]). The following argument requisite to show that there are just two such possible meets is easily restored in a variety of ways: where, for example, a and b are the feet of the perpendiculars let fall from A and B onto CD, the twin intersections of CD with the locus (C) defined by $AC+BC = k$, constant, are straightforwardly constructible by the quadratic *sectio determinata* $4(ab^2-k^2).aC^2+2l.aC+m = 0$, $l = Aa^2-Bb^2-ab^2+k^2$ and $m = (Aa^2-Bb^2-ab^2)^2-2k^2.(Aa^2+Bb^2+ab^2)+k^4$, which ensues on rationalizing the equation $\sqrt{[aC^2+Aa^2]}+\sqrt{[(ab-aC)^2+Bb^2]} = k$ by which the indeterminate line-length aC is fixed in aD. As a corollary, because the conic locus (C) so ordered to be 'of second kind' can have no real points at a distance from A or B greater than the constant sum of AC and BC, it must perforce be an ellipse—as indeed Newton had known since he studied Schooten's *Exercitationes Mathematicæ* as an undergraduate (see I: 29, note (15)) if he had not encountered this 'gardener's description' yet earlier still.

(120) We omit an introductory 'VI' which Newton afterwards interposed to number the ensuing paragraph, evidently then planning to insert it as the sixth section of the following final version of his 'Geometriæ Liber primus', but where it does not in fact—maybe through simple oversight rather than conscious intent—reappear, its place there being usurped (see 3, §1: note (24) below) by a long inserted discussion of the projectively invariant 'order' of a curve.

(121) 'Cramer's rule', as we now know it, and 'Euler's' as it ought to be named by virtue of its prior publication and proof; see our comment (II: 117, note (15)) on Newton's first enunciation of it some quarter of a century before as Problem 1—or 'rather a principle' in his own words—of his 'Problems of Curves'.

(122) Newton initially changed this vague term to cite 'Euclides' (Euclid) as the author of the following 'porism' (on which see the next note) but speedily restored his earlier word, doubtless aware that Pappus—on whom, of course, he here uniquely relies—supports no such surmise.

(123) Originally 'hoc Problema' (this problem). As on page 338 above (compare §2: notes (157) and (158)) Newton here too found great difficulty in achieving a satisfactory précis of Pappus' long and rambling enunciation of the general $(2n-1)/2n$-line locus (as rendered by Commandino in *Mathematicæ Collectiones*, ${}_{2}1660$: 251–2). While it would in our opinion be both

(120)And, once the kinds of lines have been ascertained, therefrom in turn are known the kinds of problems which are solved by their intersections. For the number of points in which two lines can cut one another is the product of the multiplication of the numbers of their individual orders(121) for example, if one line be of second order and the other of third, the number of their intersections will be 2×3 or 6, except when two or four or all six of those turn out to be impossible, and hence a problem solved by them will be of sixth order. Or if one or more of the intersections be given or can be found by some method which does not extend to all the intersections, then the problem will be of fifth or lower order.

The ancients(122) in order to reduce all problems to the intersections of lines proposed a porism of this nature:(123) 'To find a line so restricted that, when from any point in it to two, three or more straight lines given in position there be drawn at given angles, one to each, an equal number of other straight lines, there shall be produced out of those drawn lines, either by themselves or together with any other given straight lines, two plane or solid contents which shall either be

---

tedious and pointless here to cite every fine detail of the many false starts and discarded provisional improvements in phrasing which are found at this point in the manuscript, we may with some slight profit quote an initial version of the sequel wherein Newton required: 'Si a puncto aliquo ad datas positione duas tres vel plures rectas aliæ totidem rectæ singulæ ad singulas in datis angulis agantur et ex actis illis vel solis vel simul cum una vel pluribus datis rectis conflentur utcunꝗ contenta duo plana vel duo solida quæ æqualia sint vel datam habeant rationem ad invicem: componere locum quem punctum illud continget. Locum vero [veteres] dixerunt lineā in quo punctum illud indeterminate locatur, sive cujus puncta omnia et sola pro puncto illo accipi possint. Sunt utiꝗ infinita puncta conditionem assignatā habentia quæ omnia lineam aliquam constituunt' (If from some point to two, three or more straight lines given in position there be drawn an equal number of other straight lines, one to each, at given angles, and from those drawn either by themselves or together with further given straight lines there be formed products in any manner—two 'plane' ones, or two 'solid' ones (and so on)—which shall be equal or have a given ratio to each other: to construct the locus which that point shall trace. The 'locus', however, the ancients called a line in which that point is indeterminately located, that is, all of whose points, and they alone, can be accepted in that point's place. There are, of course, an infinity of points having an assigned [*sc.* indeterminate] condition, and the totality of these constitutes some line). A cancelled example in sequel reads: 'Ut si a puncto aliquo D ad rectas duas AB, AC positione datas ducendæ sint in datis angulis aliæ totidem DB, DC ea lege ut rectangulum sub ducta BD et data aliqua E æquale sit rectangulo sub altera ducta DC et alia data F: locus puncti D erit recta linea transiens per punctum [A]' (If, for instance, from some point D to two straight lines AB, AC given in position there need to be drawn at given angles an equal number of others DB, DC with the restriction that the rectangle contained by one of the drawn lines DB and some given one E be equal to the rectangle beneath the other drawn line DC and another given one F: the locus of the point D will be a straight line passing through the point A).

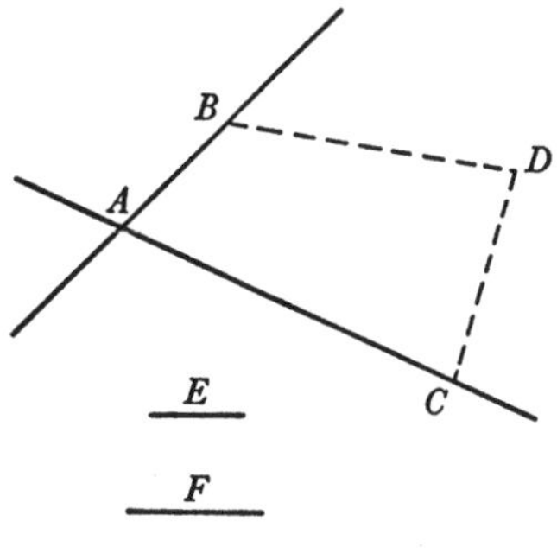

rationem ad invicem. Vel quod perinde est ut proportio conjuncta aliquarum ductarum ad alias ductas vel etiam ad datas quasvis detur. Et quoniam punctum illud a quo rectæ ducendæ erant indeterminate locabitur in linea illa invenienda,(124) nominabant lineam illam locum puncti, et quæstionem proponebant de inventione Loci quem punctum illud contingeret.

Duplex autem erat locorum inventio: prior Analytica, posterior Synthetica. Resolvendo quæstionem propositam perveniebant Veteres ad contenta illa quæ datam haberent rationem ad invicē, deinde ex contentis componebant loca punctorum et per locorū duorum intersectiones respondebant Quæstioni.(125)

---

(124) 'extra eam nullibi' (and outside it nowhere) was initially added at this point in a cancelled preceding version. In place of 'indeterminate' (indeterminately) Newton first wrote 'semper' (ever).

(125) Newton breaks off, not quite at the bottom of his manuscript page (f. 18v), to begin—or so our interpretation of the sequence of these present geometrical papers has it—the final revise of the 'Geometria' which now follows immediately on.

equal or have any other given ratio one to the other'—or, what is exactly the same, that the joint ratio of some of the drawn lines to others so drawn, or also to any given lines, be given. And, because that point from which the straight lines are to be drawn will be located in indeterminate position in the line to be found,[124] they named that line the locus of the point, and proposed the question in reference to finding the locus which that point should trace.

The discovery of a locus was, however, in two parts: the first analytic, and the latter synthetic. By resolving the question proposed the ancients attained the contents which should have a given ratio to one another, and then from those contents they composed loci of points, and so through the intersections of two loci answered the question.[125]

3

THE FINAL 'GEOMETRIÆ LIBRI DUO'(1)

§1. THE FIRST BOOK: ON THE GENERATION OF GEOMETRICAL CURVES, THEIR PROJECTIVE CLASSIFICATION, ASYMPTOTIC PROPERTIES AND TANGENTS.(2)

From the original(3) in the University Library, Cambridge

Geometriæ

Lib. i.(4)

I Gradus Problematum.

Problemata pro numero solutionum quas admittunt(5) distingui possunt in gradus. Quæ unicam tantum admittunt solutionem sunt primi gradus, quæ duas secundi, quæ tres tertij, & sic in reliquis. Ut si data recta AB producenda est ad D ita ut punctum D dato intervallo distet a puncto aliquo C quod in sublimi datur: solvetur Problema si

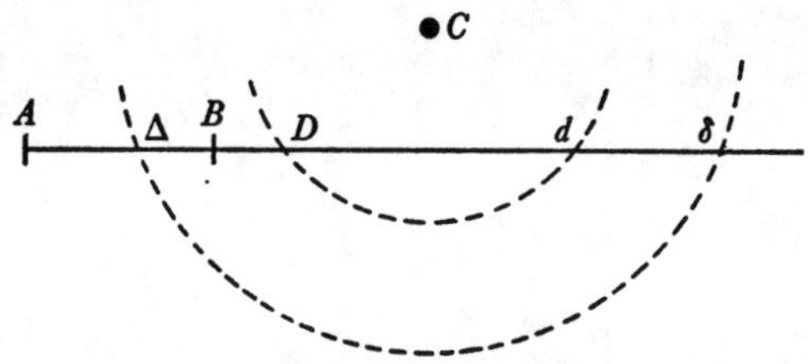

(1) Or so it appears if we trace aright the internal sequence of the corpus of surviving autograph drafts of Newton's projected treatise on 'Geometry', the other—and, we would strongly argue, earlier—fragments and versions of which are reproduced in preceding pages. While there is no explicit documentary evidence which we may cite to support our conjoining in this neat fashion the separate texts of the following 'Book 1' and 'Book 2', in the common modernity of their subject-matter these certainly go together far better than any desultory adjunction of the present Liber 2 'De quadratura Curvarum' to a Liber 1 subsuming the gist of the above drafts 'de Veterum Geometria' to form, as David Gregory in May 1694 reported it to be Newton's intention (see our preceding introduction), a dual appendix to an extensively reworked revision of the *Principia*.

(2) Although in expounding these topics—most taken over without essential change from his earlier private research papers on geometry and calculus, but the application of perspective projection to enumerate the component species of the general cubic curve totally without precedent—Newton makes some small use (compare note (20) below) of his immediately antecedent 'Geometriæ Liber 2', it will be clear that he here eschews all his previous elaborate efforts to penetrate and probe the purpose and methods of classical geometrical analysis as it was possible to understand and reconstitute its aims and work techniques from the surviving works of Euclid and Archimedes and above all through the distortions of Pappus' survey of

Translation

Geometry

Book 1.(4)

I The grades of problems.

Problems according to the number of solutions which they admit (5) are distinguishable into grades. Those allowing but a single solution are of first grade, those admitting two are of the second, those three of the third, and so on for the rest. If, for instance, a straight line AB is to be extended to D so that the point D shall be at a given distance away from some point C which is given up

---

the lost Greek writings on the 'resolved locus'. For this remarkable *volte-face* on Newton's part we have no firm explanation to offer. Maybe, as so often happened with him, he grew suddenly tired and impatient to be finished with reviving the low-level insights, wordinesses of argument and relative inferiority of technique which except in rare instances characterized and delimited the range and quality of the classical heritage in geometry. As he here slips again into his accustomed rôle of creative innovator and exponent of current mathematical notions and methods one may all but tangibly sense his relief, after so long a backwards glance to a past whose fixed patterns he could but restore and interpret, in being once more at grips with problems of contemporary geometry whose future development he could determine and shape.

(3) That formed by combining Add. 3963.2: 3^r–5^r with (in sequence) Add. 4004: 128^r–130^r/153^r–159^r/130^r–138^v/145^r–150^r. (In the case of the latter Waste Book folios we would again warn against the misguided zeal of the cataloguers—notably Adams and Luard—who in the late nineteenth century religiously and thoughtlessly returned all such scattered leaves to their pristine matrix.) A number of Newton's initial workings and preliminary drafts of sections of this final text are now found in various other places in Add. 3963 and Add. 4004; the non-trivially variant parts of these are reproduced in the Appendix which follows. As with all folios from the Waste Book the bottom edge of each of the manuscript leaves from Add. 4004 is irregularly eroded and the immediately adjacent portion badly damp-stained. Where any loss of Newton's written text has thereby occurred we have (as it is here everywhere possible uniquely to do) made good the defect, setting our restorations within square brackets according to our editorial habit.

(4) In Newton's preliminary draft of the three opening sections which is now Add. 4004: 128^r–130^r his original title—when, we may presume, he had no plan to attach a complementary 'LIBER 2'—was simply 'GEOMETRIA' (GEOMETRY). From this initial version we have taken over the marginal subheads which are lacking in the revise and have retained a couple of phrases in the main text whose deletion leads to a slight loss in clarity; other significant variants are recorded in following footnotes and a longer, superseded passage regarding the geometrical representation of multiplication and division is reproduced in section [1] of the Appendix. For what minimal interest it affords, we may add that when on 26 October 1777 Samuel Horsley came to examine the revised opening of the 'Geometria' (Add. 3963.2: 3^r–5^r) he with his usual impercipience on a loose inserted sheet passed his judgement of this 'Geometriæ Lib 1^s. Cap I' that 'This is imperfect & contains nothing but what is very commonly known. Not to be published'.

(5) As a general possibility, that is. Though deleted by Newton on f. 3^r simply because it is repeated in equivalent form in the next sentence, we here retain this clarifying clause to avoid an uncomfortable lacuna in his intended sense. A more emphatic 'distinguuntur' (are distinguished) was first written in sequel.

centro C intervallo isto dato describatur circulus et producatur recta illa usq; dum circumferentiæ[6] circuli hujus occurrat: et duplici occursu in D et in d fiet duplex solutio[,] una per lineam AD altera per lineam Ad: quod ostendit Problema secundi gradus esse.

II Quantitates positivæ et subductitiæ.

Quantitates autem per quas Quæstioni respondetur aliquando directæ et positivæ sunt, aliquando retrorsæ vel subductitiæ quas et negativas vocant. U si datum illud intervallum DC majus sit quam distantia BC ita ut recta illa occurrat circulo in Δ et δ, respondebitur quæstioni directè producendo AB ad δ et contrario modo ducendo $B\Delta$ retrorsum. Directas quantitates notamus[7] præfigendo signum $+$ retrorsas præfigendo signum $-$: ut in his $+B\delta$ & $-B\Delta$. Et ubi neutrum signum præfigitur quantitas directa est. His signis etiam additionem & subductionem significamus. Ut in $AB+B\delta$ & $AB-B\Delta$, ubi $B\delta$ addi $B\Delta$ subduci intelligitur. Signum incertum sic notamus $\pm$ vel sic $\perp$[8] & huic contrarium signum sic $\mp$, $\top$.[8]

III Quantitates impossibiles.

Hæ quantitates per quas respondemus quæstioni aliquando etiam impossibiles evadunt; ut in hoc casu quantitates BD et Bd ubi intervallum CD minus assignatur quam ut circulus rectam productam secare possit. Et quando duæ vel forte quatuor aut plures sunt impossibiles (nam numerus impossibilium semper est par) gradus Problematis non æstimabitur ex numero solarum realium sed ex numero omnium, id est omnium qui in quocunq; casu Problematis generaliter propositi reales evadere possunt. Problema verò generaliter proponi dico in qua quantitates nullæ ita limitantur quin possint additis vel subductis datis majores vel minores sumi. Ut si inter duas quasvis quantitates A et B inveniendæ sint duæ mediæ proportionales X et Y, unica tantum est hujus solutio realis, nec tamen ideo primi erit gradus Problema. Nam si omnes ejus termini exprimantur et quæ modo quovis[9] limitantur, datis quibusvis C, D, E augeantur vel diminuantur, Problema generalius enunciabitur hoc modo[:] Invenire quantitates X et Y ita ut sint A ad X, $X\pm C$ ad Y & $Y\pm D$ ad B in eadem ratione. Hujus generalis Problematis tres possunt esse solutiones reales,[10] adeoq; cas[u] ill[o] ubi C et D nulla sunt id est ubi X et Y mediæ sunt proportionales inter A et B, problema erit tertij gradus quamvis duæ ex solutionibus in hoc casu evaserint impossibiles. Casus enim omnes ejusdem sunt gradus cum

(6) Initially, much as in Newton's preliminary draft on Add. 4004: 128r, 'circulus datum illam rectam infinite [productam secans]' (...a circle cutting the given line when indefinitely produced).

(7) Originally 'signamus' (designate) in first draft.

(8) Of these less immediately comprehensible variant symbols—earlier introduced by him in passing in Problem 42 of his 'Arithmetica Universalis' (see v: 268 and compare *ibid.*: 269, note (349))—Newton makes no use in sequel. In his initial draft (on Add. 4004: 127v) he had stated more succinctly that 'Signo verò $\pm$ additionem et subductionem a[m]biguè denoto' (By the symbol $\pm$, however, I denote addition and subtraction ambiguously).

above, the problem will be solved if with centre C and that given distance as radius there be described a circle and the straight line be produced till it shall meet its circumference:(6) from the double meet at D and d there will prove to be a double solution, one by means of the line AD and the other by the line Ad—which shows that the problem is of second grade.

II Positive and subtractive quantities.

The quantities by means of which a question is answered, however, are sometimes direct and positive, and sometimes reverse or subtractive (these are also called negative). Should, for instance, the given interval DC be greater than the distance BC, so that the previous straight line meets the circle in Δ and δ, the question will be answered in a direct sense by extending AB to δ, and in the contrary way by drawing $B\Delta$ backwards. We mark(7) direct quantities by prefixing the sign $+$, reverse ones by prefixing the sign $-$; so in the present case we set down $+B\delta$ and $-B\Delta$. And when neither sign is prefixed the quantity is direct. By these signs, also, we signify addition and subtraction; as in $AB+B\delta$ and $AB-B\Delta$, where $B\delta$ is understood to be added and $B\Delta$ subtracted. An unknown sign we mark thus $\pm$ or $\perp$(8) and the contrary sign $\mp$ or $\top$.(8)

III Impossible quantities.

These quantities by means of which we answer a question may sometimes also prove to be impossible: as in the present case the quantities BD and Bd when the interval CD is assigned too small for the circle to be able to cut the extended straight line. And whenever two or maybe four or more solutions are impossible (for the number of impossibles is always even) the grade of a problem will be reckoned not from the number of reals alone but from the total number, that is, of all which in any case whatever of the problem generally proposed can come to be real. I say that a problem is generally proposed, however, when in it there are no quantities so limited as to be unable to be taken greater or less by the addition or subtraction of givens. If, for instance, between any two quantities A and B two mean proportionals X and Y were to be found, this has but a single real solution, but yet the problem will not in consequence be of first grade. For, if all its terms be expressed and those limited in any way(9) be increased or diminished by any givens C, D, E you wish, the problem will be more generally enunciated in this manner: to find quantities X and Y such that A be to X, $X\pm C$ to Y and $Y\pm D$ to B in the same ratio. This general problem can have three real solutions,(10) and hence in the case where C and D are nil—that is, when X and Y are mean proportionals between A and B—the problem will be of third grade even though two of the solutions in this case prove impossible. For all cases are of the same grade as the general problem unless perchance by

(9) In first draft 'quotquot aliquo modo' (however many they are, in some way) was written.

(10) Namely the roots of the cubic which ensues on eliminating X or Y from the pair of equations $A/X = (X+C)/Y = (Y+D)/B$. It is readily determined that all three are real when $(-\frac{1}{3}AD+\frac{1}{9}C^2)^3 > (\frac{1}{2}A^2B+\frac{1}{6}ACD-\frac{1}{27}C^3)^2$, but otherwise two are conjugate 'impossibles'.

generali Problemate nisi forte per conditiones quasdam de quibus post agetur deprimantur ad gradum inferiorem.

[Distinctio Problematum per æquationes.]

(11)Obtinet autem illa Problematum distinctio tam in Arithmetica quam in Geometria. Nam simile est utriusq; computum et ad similes conclusiones tendit. Si Arithmeticus tractat quæstionem Geometricam, substituit hic numeros pro lineis, multiplicationem pro ductu in lineam, Divisionem pro applicatione ad lineam et extractionem radicis pro inventione mediæ proportionalis: et vice versa Geometra ubi quæstio proponitur in numeris pro operationibus Arithmeticis substituit respondentes Geometricas. Unde fit ut eædem notæ in utroq; computo nunc usurpentur. Designent A, B, C &c vel numeros vel lineas pro Artificis arbitrio, et AB, ABC &c designabunt illorum factos planos et solidos, harum contenta plana et solida; AA vel A^2, AAA vel A^3 &c illorum factos quadratos et cubicos, harum quadrata et cubos; $\frac{AB}{C}$ latus tam divisione numeri AB per numerum C quam applicatione areæ AB ad lineam C genitum; (12) $\frac{AA}{C}$, $\frac{A^3}{CC}$, $\frac{A^4}{C^3}$, $\frac{A^5}{C^4}$ &c tam numeros continua multiplicatione et divisione quam lineas continuo ductu et applicatione gradatim factas: Pro quibus si C sit unitas in Arithmetica vel subintelligatur ut universale quoddam latus vel mensura in Geometria, scribere licet AA, A^3, A^4, A^5 &c. Item $\overline{AB}^{\frac{1}{2}}$ latus quadratum est tam numeri AB quam areæ AB; $\overline{ABC}^{\frac{1}{3}}$ latus cubicum tam numeri quā solidi ABC; & $\overline{ABC}^{\frac{2}{3}}$ quadratum illius lateris in utroq; casu, et sic in reliquis. A qua notarū Analogia fit ut qui computum symbolicum in utravis scientia intellexerit intelliget in utraq;. (13)Repetamus Problema superius ubi erat A ad X ut $X \pm C$ ad Y & $Y \pm D$ ad B, et ob proportionales A ad X ut $X \pm C$ ad Y si applicetur contentum sub medijs ad extremum prius fiet $\frac{XX \pm CX}{A}$(14) extremum posterius dictum Y. Deinde ob proportionales A ad X ut $Y \pm D$ ad B, si scribatur $\frac{XX \pm CX}{A}$

(11) To fill a manifest lacuna at this point in the sequence of Newton's explanatory postils, in the margin alongside we interpose an appropriate subhead epitomizing the content of the next two paragraphs. A widely variant initial draft (on Add. 4004: 129r/129v) of the ensuing statement of the close correspondence—but not absolute identity—which exists between arithmetic and geometry is, along with a preliminary amplified recasting (on Add. 3963.2:11r), reproduced in sections [1] and [2] of the Appendix; in note (12) thereto, furthermore, we quote a first version (on Add. 3963.2: 7r) of the present sequel in which Newton momentarily set his trust in an erroneous rule establishing an exact correspondence between the geometrical 'grade' of a problem and the arithmetical degree of the root needing to be extracted in its solution.

(12) The remainder of this sentence is a late insertion on f. 3v of the manuscript, there added to fill a small gap left when (see the next note) Newton refashioned the sequel.

(13) In the manuscript the rest of the paragraph here following is inserted in afterthought

means of certain conditions which we shall subsequently discuss they may be reduced to a lower grade.

[Distinguishing problems by means of equations.]

(11)That distinction between problems obtains both in arithmetic and geometry. To be sure, in each the computation is similar and it reaches similar conclusions. If the arithmetician handles a geometrical question, he substitutes therein numbers for lines, multiplication for 'drawing' into a line, division for 'application' to a line, and the extraction of a root for the finding of a mean proportional; and, *vice versa*, where a question is proposed in numbers the geometer in place of its arithmetical operations substitutes corresponding geometrical ones. Whence it happens that the same symbols are nowadays employed in both schemes of computation. Let $A, B, C, \ldots$ denote either numbers or lines at the expert's will, and AB, ABC and so on will then denote the 'plane' and 'solid' products of the former, and the plane and solid contents of the latter; AA or A^2, AAA or A^3 and so on the former's square and cubic products, and the latter's squares and cubes; AB/C the 'side' generated both through division of the number AB by the number C, and through application of the area AB to the line C; (12)A^2/C, A^3/C, A^4/C^3, A^5/C^4, and so on—in place of which, if C be the unit in arithmetic or understood to be a universal side or measure in geometry, we are free to write $A^2, A^3, A^4, A^5, \ldots$—both numbers produced step by step by continued multiplication and division, and also lines equivalently attained by continued 'drawing into' and 'application'. Likewise, $(AB)^{\frac{1}{2}}$ is the square root both of the number AB and the area AB; $(ABC)^{\frac{1}{3}}$ the cube root both of the number and the solid ABC; and $(ABC)^{\frac{2}{3}}$ the square of that side in either case; and so forth in the rest. The consequence of this notational analogy is that anyone who understands symbolic computation in either science shall understand it in both. (13) Let us turn again to the preceding problem where A was to X as $X \pm C$ to Y and as $Y \pm D$ to B: because of the proportion A to X as $X \pm C$ to Y, if the content beneath the middles be divided by the first extreme, $(X^2 \pm CX)/A$ will come to be the latter extreme, namely Y; next, because of the proportion A to X as $Y \pm D$ to B, if $(X^2 \pm CX)/A$ be written in place of Y, and the extremes and middles then be multiplied together, there will result $(X^3 \pm CX^2)/A \pm DX$

on the facing page (f. 3v). Newton initially went on to register (on ff. 4r/5r, reproduced in section [3] of the Appendix below) yet one more version of his earlier *caveat* against confusing the essentially separate sciences of arithmetic and geometry, distinct in their concepts and subject-matter if analogous in much of their working terminology. (Compare sections [1] and [2] of the Appendix.) As we remarked in the previous note, the suppression of this original continuation necessitated some small adjustment of the preceding text.

(14) At this point Newton in the manuscript slips into designating A, B, C, D, X, Y, by their lower-case equivalents, maintaining the change (though not consistently so) through the rest of his paragraph. For uniformity's sake we here retain his original capitalization of constants and variables, but in the next paragraph (where the transition is less harshly made) reproduce his lower-case letters exactly as he there wrote them.

pro Y, et extrema et media ducantur in se, fiet $\dfrac{X^3 \pm CXX}{A} \pm DX$ æquale AB. Et omnibus in A ductis et æqualitate per notam $=$ significata fiet

$$X^3 \pm CXX \pm ADX = AAB.$$

Hic perinde est sive lineas per symbola A, B, C, D, X sive numeros intelligas. In utroq; casu computu[s] idem est et eadem conclusio.

Ut in hoc(15) Problemate assumendo ignotam quantitatem X et inde per analogiarū unam inveniendo ignotam alteram Y pervenimus per analogiam alteram ad conclusionem: sic in quæstione quavis legitima assumendo quantitatem aliquam ignotam a qua et quantitatibus datis cæteræ ignotæ possunt facillimè colligi, et nomina datis et ignotæ illi imponendo, postquam cæteræ ignotæ quotquot in Analogia vel Propositione quavis continentur, ex nominatis collectæ fuerint, Analogia vel Propositio illa Conclusionem dabit, quæ Æquatio dicitur. Vel si duæ aut plures ignotæ ad cæteras colligendas assumantur, debent tot æquationes inveniri quot assumuntur ignotæ et inventæ omnes ad unam tandē reduci in qua non nisi una ignota quantitas manebit. Et æquatio illa postquam termini ritè dispositi fuerint erit alicujus ex his(16) formis & gradibus

$$\begin{aligned} x &= a \\ xx &= -ax+bb \\ x^3 &= -axx+bbx-c^3 \\ x^4 &= ax^3-bbxx-c^3x+d^4 \\ &\&c \end{aligned}$$

ubi x ignotam quantitatem designat ab altissimo suo gradu per singulos terminos dextrorsum descendentem & a, b, c, d &c quantitates quasvis notas sub signis quibusvis $+$ & $-$.

Ex altissimo ignotæ quantitatis gradu innotescit gradus æquationis et ex gradu æquationis gradus Problematis quod ad æquationem deducitur.(17) Nam si ponantur æquationes quotcunq; primi gradus $x=a$, $x=b$, $x=-c$ seu quod perinde est $x-a=0$, $x-b=0$, $x+c=0$ et una $x-a=0$ ducatur in aliā $x-b=0$, fiet æquatio secundi gradus $xx \begin{smallmatrix}-a\\-b\end{smallmatrix} x+ab=0$, ubi ignota quantitas x duplex est a et b. Et si hæc æquatio ducatur in tertiam $x+c=0$ fiet æquatio tertij gradus $x^3 \begin{smallmatrix}-a\\-b\\+c\end{smallmatrix} xx \begin{smallmatrix}+ab\\-ac\\-bc\end{smallmatrix} x+abc=0$ ubi x triplex est a b et $-c$. Tot sunt quantitates x quot latera æquationis (nisi ubi aliquæ earum absurdæ(18) fuerint) & quantitates illæ responsa sunt quæstionum.(19)

---

(15) Initially—before (see note (13)) he inserted the preceding lines where this earlier problem is restated and reduced to an equation—Newton wrote 'superiore' (previous).

(16) In the manuscript this was momentarily replaced by the more explicit 'sequentibus' (the following).

(17) A superfluous 'Idem enim est gradus omnium' (Of all, namely, the degree is the same) is cancelled in sequel.

equal to AB; and so, when everything is multiplied by A and the equality is signified by the notation $=$, there will come $X^3 \pm CX^2 \pm ADX = A^2B$. Here it is exactly the same whether by the symbols A, B, C, D, X you understand lines or numbers. In either case the computation is identical and the conclusion the same.

Just as in this(15) problem, by assuming the unknown quantity X and by means of one of the proportions finding the other unknown Y in its terms, we arrived by way of the second proportion at the conclusion, so in any legitimate question, by assuming as unknown some quantity from which, along with given quantities, the other unknowns can most easily be gathered and then setting names on the givens and that unknown, after as many other unknowns as are contained in any proportion or proposition have been gathered in terms of the named quantities that proportion or proposition will yield a conclusion which is called an 'equation'. Or if two or more unknowns be assumed in order to gather the rest, as many equations ought to be found as there are unknowns assumed, and when all are found they should ultimately be reduced to one in which there shall remain but a single unknown quantity. And that equation, after its terms have been duly arranged, will be of one of these(16) forms and degrees:

$$x = a,$$
$$x^2 = -ax + b^2,$$
$$x^3 = -ax^2 + b^2x - c^3,$$
$$x^4 = ax^3 - b^2x^2 - c^3x + d^4, \ \ldots$$

where x denotes the unknown quantity descending rightwards from its highest degree through its separate terms, and $a, b, c, d, \ldots$ any known quantities under any signs $+$ and $-$. From the highest degree of the unknown the degree of the equation is established, and from the degree of the equation the degree of the problem which is reduced to the equation.(17) For if there be supposed any number of equations of first degree $x = a$, $x = b$, $x = -c$, or what is exactly the same $x - a = 0$, $x - b = 0$, $x + c = 0$, and one, $x - a = 0$, be multiplied by another, $x - b = 0$, there will ensue an equation of second degree, $x^2 - (a+b)\,x + ab = 0$, in which the unknown x has the double value a and b. And if this equation be multiplied by the third, $x + c = 0$, there will come an equation of third degree, $x^3 - (a+b-c)\,x^2 + (ab - ac - bc)\,x + abc = 0$, in which x has the triple value a, b and $-c$. There are as many quantities x as the equation has roots (except when some of these are absurd(18)) and those quantities are the answers to the questions.(19)

(18) That is, 'imaginariæ' (unreal), which Newton will not permit the 'responses' to geometrical 'questions' to be. A more pithy equivalent 'demptis absurdis' (excepting the absurds) was first written in parenthesis.

(19) Originally '...responsa sunt per quæ quæstiones solvuntur' (the answers by which the questions are solved).

IV Quibus lineis problemata solvuntur.

Geometria per intersectiones linearum solvit omnia Problemata, singulas ejusdem problematis solutiones per totidem intersectiones una vice exhibens. Nam solutionum omnium eadem est lex et natura ita ut una Geometricè exhiberi non possit quin reliquæ eadem constructione simul prodeant. Unde fit ut ad constructionem cujusq Problematis lineæ duæ adhiberi debe[a]nt quæ se mutuò in tot punctis secare possunt quot Problema admittit solutiones. Ad constructiones omnium problematum primi gradus sufficiunt lineæ rectæ, ad eas secundi requiruntur recta et circulus vel duo circuli, ad eas tertij requiritur linea magis complexa quæ rectam aut circulum in tribus punctis ad minimum secare possit[,] et sic in infinitum.[20]

V Ordines[21] Linearum.

Et hinc pro numero punctorum in quibus linea quævis a linea recta secari potest, oritur distinctio linearum in gradus. Primi gradus vel ordinis est linea quam recta in unico tantum puncto secare potest vel cujus intersectionē cum imperata quavis recta determinare Problema est primi gradus. Et hujusmodi sunt solæ rectæ lineæ. Secundi gradus linea est cujus intersectionem cum recta quavis determinare Problema est secundi gradus et hujusmodi sunt circulus et lineæ illæ omnes quas Conicas Sectiones appellant. Tertij verò gradus linea est cujus intersectionem cum recta determinare problema est tertij gradus, ut Cissoides Veterum. Et sic in infinitum. Lineæ vero quas recta in punctis infinitis secare potest (quales sunt Spiralis,[22] Quadratrix, Trochoides & similes) meritò dicētur ordinis ultimi.[23]

VI[24] Genera Linearum ejusdem Ordinis.

Si linea aliqua oculo extra planum ejus sito spectetur per planum translucidum, et in plano illo locus ejus apparens vel (ut voce mathematica utamur) projectio notetur, erit linea projecta ejusdem ordinis cum projiciente. Si projiciens est recta projectio erit recta, si curva est quæ rectam secare potest in duobus vel pluribus punctis, projectio ejus projectionem rectæ in totidem punctis secare potest. Et hinc habita linea aliqua cujusvis ordinis possunt aliæ plures ejusdem ordinis inde derivari. Sic Veteres ex circulo derivarunt omnes secundi ordinis figuras et inde Conicas Sectiones nominarunt, considerantes spatium illud solidum quod radijs per circuli spectati perimetrum transeuntibus terminatur ut conum quem planum figuræ projectæ secat. Sic et figuræ superio-

(20) It will be clear that this and the following paragraphs are but a light recasting of the equivalent passage on pages 394–8 above in Newton's draft 'Geometriæ Lib. 2'.

(21) 'Genera' (kinds) was first written. Newton has made similar replacements of 'genus' by 'ordo' several times in his ensuing Latin text.

(22) Understand 'Archimedea' (Archimedean), of course.

(23) By way of a transitional sentence 'Quas quidem lineas omnes sic licebit exprimere' (All these lines, indeed, we will be at liberty to express in this way), afterwards suppressed when its sense was gainsaid in the reordered manuscript text, Newton initially at this point (on Add. 4004: 130r) passed without break to expound the 'Modus exprimendi lineas' which here followed in natural sequence but was subsequently postponed to make way for the long interpolated passage on the projective classification of curves which now ensues.

IV By what lines problems are solved.

Geometry solves all problems by the intersections of lines, exhibiting the individual solutions of the same problem at one go by means of an equal number of intersections. For of all solutions the law and nature is the same, so that one cannot be displayed geometrically without the rest simultaneously resulting by the same construction. In consequence, to construct each problem two lines ought to be adduced which are able mutually to cut one another in as many points as the problem admits solutions. For the constructions of all problems of first degree straight lines suffice, for those of second there is required a straight line and a circle or two circles, for those of third there is needed a more complicated line which a straight line or a circle can cut in a minimum of three points, and so on indefinitely.(20)

V Orders(21) of lines.

And hence, according to the number of points in which any line can be cut by a straight line, there arises the distinction of lines into degrees. Of first degree or order is a line which a straight line can cut in but a single point, or the determination of whose intersection with any ordained straight line is a problem of first degree: of this sort there are straight lines alone. Of second degree is a line the determination of whose intersection with any straight line is a problem of second degree: of this sort are circles and all those lines called conics. Of third degree, however, is a line the determination of whose intersection with a straight line is a problem of third degree: the ancients' cissoid, for instance. And so on indefinitely. Lines, however, which a straight line can cut in an infinity of points—such as the(22) spiral, quadratrix, cycloid and like curves—will deservedly be said to be of ultimate order.(23)

VI(24) Kinds of lines of the same order.

If some line be looked at through a translucent plane by an eye situated outside its plane, and in that plane its apparent place or (to use the mathematical jargon) projection be marked, the projected line will be of the same order as the projecting one. If the projecting line is straight, the projection will be a straight line; if it is a curve which a straight line can cut in two or more points, its projection can cut the projection of a straight line in an equal number of points. And accordingly, when some line of any order is had, many others of the same order can be derived from it. In this way the ancients derived from the circle all figures of the second order and thence named them conic sections, considering the solid 'space' which is terminated by rays passing through the perimeter of the observed circle as a cone which the plane of the projected figure cuts. So, too, all figures of higher orders can be derived from certain

(24) This long section is a late interpolation at this point in the Waste Book manuscript (from ff. 153r–159r at its very end) in line with Newton's instruction at its head (f. 153r) that it be added 'pag 13; post verba—*meritò dicentur ordinis ultimi*'. We have previously suggested (2, §3: note (120)) that he originally had it in mind here to outline the rule—'Cramer's'—specifying the number of possible intersections of two algebraic curves which (see *ibid.*: note (121)) he had long before set down as a primary theorem in his 'Problems of Curves'.

rum ordinum possunt omnes a simplicioribus quibusdam ejusdem ordinis figuris per successivas projectiones derivari, et inde distingui in genera coordinata[(25)] positis illis ejusdem esse generis quæ ab eadem figura derivantur. Nam hæ omnes & solæ in se mutuò per projectiones transeunt et ea ratione cognatæ sunt, a cæteris verò in quas non transeunt alienæ.[(26)] Hac lege unicum tantum est genus linearum secundi ordinis, eo quod omnes derivantur a circulo: at ordinis tertij genera sunt quinꝗ.[(27)]

Exemplum in lineis tertij ordinis.

In recta infinita EAB dentur puncta duo A, E et ad tertium quodvis ejus punctum B in dato angulo erigatur ordinata BC cujus quadratum, si præterea dentur rectæ duæ M et N æquale fuerit $\dfrac{AB^{\text{cub}}}{M}+N\times EB$.[(28)] Et curva linea ad

(25) An unwanted 'pro numero figurarum quæ ad omnium projectionem sufficiunt' (proportionate to the number of figures which suffice for the projection of all) is deleted in sequel.

(26) In a first draft of this passage on Add. 3963.3: 13r Newton initially wrote: 'Ejusdem ordinis Lineæ sic distinguuntur in genera coordinata [per projectionem]. Oculo O immoto spectētur plana duo quævis translucida et immota $ACBD$ $acbd$ in quorum alterutro figura aliqua [descripta sit]' (Lines of the same order are distinguished in this way into coordinate kinds by projection. With the eye O stationary let any two translucent, stationary planes $ACBD$, $acbd$ be looked at, and in one or other of these let some figure be described). In Lemma XXII of his 'De motu Corporum Liber primus' (see VI: 268–72) he had earlier specified the linear transformation of one plane into another which is embodied in such an 'ocular' (perspective) projection. Here, little differently, on drawing through O the lines OE, Oe parallel to acb and ACB respectively, we may define the projection d (in the plane $acbd$) of any point D (in the plane $ACBD$) by drawing DC at some fixed angle $\widehat{ECD}$ to the base line ACB in its plane, joining OC to intersect ab in c and then (in the plane $acbd$) erecting cd parallel to CD—and hence in a fixed angle $\widehat{ecd}$ to the base line bca—to meet OD in d; whence, if the curve ADB is defined by some given relationship $C_n(X, Y) = 0$ of n-th algebraic degree between the oblique coordinates $EC = X$ and $CD = Y$, the relationship between the coordinates $ec = x$ and $cd = y$ defining the perspective curve adb in the projection plane will be some corresponding $C'_n(x, y) = 0$ derivable therefrom by way of the linear transformation ($x/k = l/X = y/Y$ and so) $X = kl/x$, $Y = ky/x$ where $k = OE$ and $l = Oe$. It is an immediate corollary that, since the degrees of the equations determining $D(X, Y)$ and $d(x, y)$ are the same, so (in Newton's present sense) are the 'orders' of the curves ADB, adb which these define.

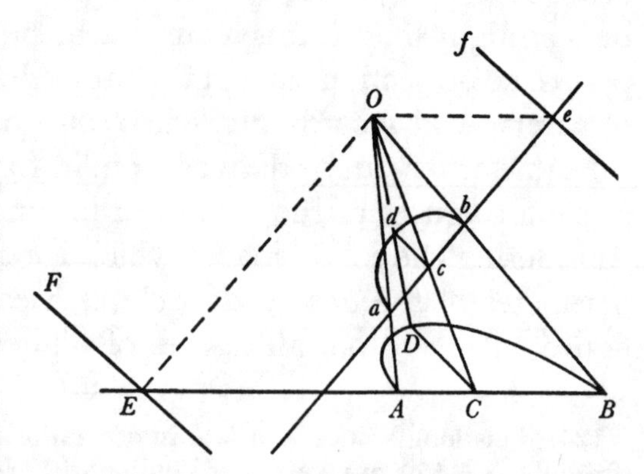

(27) The individual species of which are, as Newton passes straightaway to assert, likewise derivable—according to their projective 'kind'—from one or other of the five divergent parabolas. In first draft on Add. 3963.3: 13r he had written to the same effect that 'Qua

simpler figures of the same order by means of successive projections and thereby be distinguished into coordinate kinds(25) on setting those to be of the same kind which are derived from the same figure: for, to be sure, all these both by themselves pass mutually one into another by projection and for that reason are cognate, while they are unrelated to the rest into which they do not pass.(26) On this principle there is but a single class of lines of second order, in that all may be derived from the circle; but of the third order there are five kinds.(27)

An example in lines of the third order.

In the unbounded straight line EAB let there be given the two points A, E and at any third point B in it at a given angle erect the ordinate BC whose square—given in addition the two straight lines M and N—shall be equal to $AB^3/M+N\times EB$,(28) and the curve at which every straight line BC of this sort

ratione unicum tantū est genus figurarum secundi ordinis, eo quod circulus projicit has omnes et vice versa omnes projiciunt circulum cæteræqȝ omnes illius ordinis lineæ se mutuo projiciunt. At ordinis tertij genera sunt quinqȝ quæ sic enumero' (On this ground there is but a single kind of figures of second order, since a circle projects all these and they *vice versa* all project a circle, and all the rest of the lines of the order mutually project one another. But of the third order there are five kinds, which I enumerate in this manner). How and when Newton first achieved this insight that the numerous species of cubic curve fall into one or other of five projectively distinct 'kinds' his extant papers do not reveal. We would guess that it came to him as an unexpected bonus while mulling over the series of sketches of the main individual species of cubic which he had prepared long before for his initial, algebraically based 'Enumeratio Curvarum trium Dimensionum' (see II: 10–85, especially 18, 28/29, 32–5 and the revised equivalents on 48/49, 72–84). Once conceived, however, the notion is readily tested by mentally projecting to infinity any arbitrary straight line in the plane of a given cubic and identifying the species of like 'kind' into which this is transformed; and from this it is but a small step to recognize that the family of divergent parabolas form—much more obviously so for Newton (who had already given them separate status in his cubic enumeration) than the five central cubics which, as Chasles was to notice only more than a century and a quarter afterwards, fill the rôle equally well—a quintuple of 'simpler figures of the same order' here characterized, members one each of the five projectively distinct 'coordinate kinds' of cubic, which are collectively able (when 'lit' from a given ocular point-centre) to cast all other species of cubic as their plane perspective 'shadow'. General analytical proof that this is universally true is, as Nicole found to his cost in 1731, formidably complicated until one attains the converse insight that all cubics may be transmuted in this way into their correspondent 'kind' of divergent parabola by projecting to infinity the tangent at an inflection point (of which they all possess at least one which is real, albeit—as here makes no matter—not finite). However evident the truth of this may now appear to us—and certainly was so for Newton, who in the ensuing (hitherto unprinted) projective generation of the component species of the general cubic off-handedly assumes it (see note (60) below) in his listing of the perspective 'shadows' of the punctate diverging parabola—its very enunciation lay hidden to the geometers of the early eighteenth century before Clairaut. See our more detailed comment in note (30) following.

(28) Newton originally wrote out in full that 'BC...quadratum...æquale fuerit rectangulo sub N et $[E]B$ una cum cubo ex AB applicato ad M' (BC...square...shall be equal to the rectangle of N and EB together with the cube of AB divided by M). Whence, on setting $AB=x$, $BC=y$, $EA=a$ and replacing M and N by their lower-case equivalents, there ensues $y^2=x^3/m+n(x+a)$, that is (where $mn=k$ and $amn=l$) $my^2=x^3+kx+l$, as the Cartesian equation defining the curve here geometrically specified. This is, of course, given (see IV: 366) in its most fully reduced canonical form.

quam hujusmodi recta omnis BC terminetur erit Parabola alata[29] quæ per projectionem dabit omnes lineas tertij ordinis.[30] Hujus autem Parabolæ casus sunt quinq; principales; primus et simplicissimus ubi linea N nulla est: Secundus ac tertius ubi N negativè ponitur et præterea AE est $\frac{2}{3}\sqrt{\frac{MN}{[3]}}$, et secundus quidem ubi AE capitur ab A versus D seu versus alas figuræ, tertius verò ubi AE capitur ad contrarias partes ipsius A: Quartus et quintus sunt ubi AE est alterius cujusvis longitudinis, & quartus quidem ubi Parabola illa secat lineam

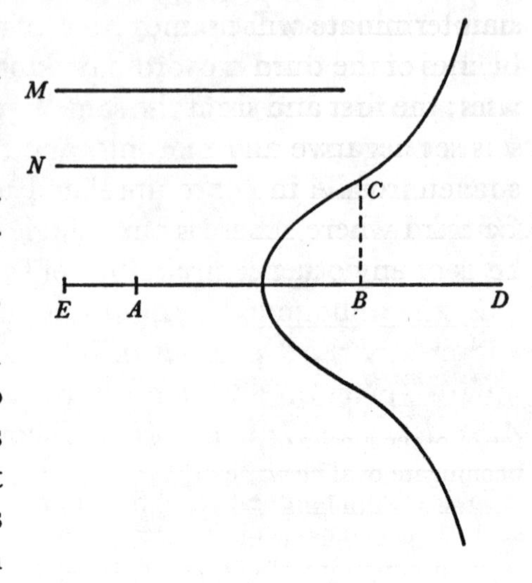

(29) An imaginative variant on the dull technical appellation 'parabola divergens' (divergent parabola) which Newton had earlier (see IV: 374) tentatively set on this class of cubic and was soon to standardize as its name in the final version of his 'Enumeratio' (reproduced in **3**, **1**, §2 below), though this is perhaps more accurately evocative of its pure and cusped species than of those possessed of a node, conjugate oval or double point. We may add the somewhat irrelevant aside—since Newton himself could scarcely have been aware of this forerunner of his own 'winged' cubic—that François du Verdus had half a century before similarly described Roberval's 'pteroid' (namely the analogous hyperbolic *alæ* of a strophoid) in a letter to Torricelli in April 1645; see *Opere di Evangelista Torricelli*, **3** (Faenza, 1919): 314.

(30) There is, as we stated in note (27), nothing in his surviving papers to show that Newton ever formulated a proof of this assertion, beyond at least the qualitative enumeration of the component species of the five projectively distinct 'kinds' of cubic curve which the divergent parabolas cast as their 'shadows', as he proceeds to outline it in sequel. Given this dearth of evidence, we can place little faith in Rouse Ball's confident surmise in his otherwise valuable article 'On Newton's Classification of Cubic Curves' (*Proceedings of the London Mathematical Society*, **22**, 1891: 104–43) that 'Newton had arrived at this remarkable result...by the method of projection indicated in the *Principia*, Bk. I, ...Lemma XXII' (*ibid.*: 123). In the simplified summary of the three-dimensional analogue of this projective 'transmutation' given in note (26), where EF is the line $X = 0$—the 'horizontal' as Newton will name it below—in the plane ABD which is cast to infinity in the plane abd, and likewise ef is the line $x = 0$ in the latter plane cast to infinity by the converse projection of the point-loci $(d) \rightarrow (D)$, we need to prove that where the five species of divergent cubic (D) are severally set in given position in the projecting plane ABD they jointly cast as their perspective shadows from the point O in the freely mobile projected plane abd all other species of cubic (d). While in particular instances—that, for example, where the projecting parabola, set with its diameter along EF, is defined by the equation $X^2 = aY^3+bY^2+cY+d$ and the projected curves, of Cartesian equation $x = a'y^3+b'xy^2+c'x^2y+d'x^3$ are the related family of the five distinct 'kinds' of central cubic—the species of the resulting lines of third order are easily catalogued, the equations of the projected cubics have in general their full complement of terms and their reduction in each case to the standard Newtonian form by which their type may be identified is for the most part tediously complicated, if it poses no great theoretical difficulty. And so Nicole found it when, in the 'Maniére d'engendrer dans un Corps solide toutes les lignes du troisième ordre' which he presented to the French *Académie des Sciences* on 1 December 1731 (and afterwards published

shall terminate will be a 'winged' parabola[29] which shall by projection yield all lines of the third order.[30] Of this parabola, however, there are five principal cases: the first and simplest is when the line N is nil; the second and third when N is set negative and moreover AE is $\frac{2}{3}\sqrt{(\frac{1}{3}MN)}$, the second indeed where AE is taken from A in D's direction—that is, towards the wings of the figure—but the third where AE is taken to A's opposite side; the fourth and fifth are when AE is of any other length, the fourth indeed where the parabola cuts the line

---

in its *Mémoires.... Année MDCCXXXI* (Paris, 1733): 494—510), he sought equivalently to do so by geometrically determining the curve of the arbitrary plane section of the cubic cone which has one or other species of divergent parabola as its base; though in the case of the parabola having a conjugate oval he was easily enough able to determine the nine cubic sections (species 2–10 of Newton's 'Fifth kind' below) produced when the 'horizontal' is parallel to the ordinate of the parabola which lies with its diameter along the abscissa EC, its more general sections produced defining equations of cubics which were half a page long, and Nicole weakly promised to remit 'à un autre Mémoire' the identification of their individual species. The breakthrough came only days later in a report 'Sur les courbes que l'on forme en coupant une surface courbe quelconque par un plan donné de position' (*ibid.*: 483–93) which Clairaut independently submitted to the *Académie* on 12 December, and where he argued that 'un Cone fait sur une ligne qui a des points d'inflexions a des côtés...qui séparent des parties convexes d'avec des parties concaves, &...la section faite par un plan aura autant de points d'inflexions que ce plan coupera de côtés d'inflexion; de façon que pour que cette section perde une ou plusieurs inflexions, il faut que le plan qui la donne soit parallèle à un ou plusieurs côtés d'inflexions' (*ibid.*: 491–2); in plainer words (and as De Gua was soon to restate it in his *Usages de l'Analyse de Descartes* (Paris, 1740): 222–4), because the family of divergent parabolas are uniquely characterized among cubic curves by being touched by the line at infinity in an inflection point, all other cubics may be projected into the like 'kind' of divergent parabola by casting to infinity the tangents at their inflections (one or all three of which will be real). To return to Newton's method of projective 'transmutation', we need in consequence (in the second figure of note (26) above) only linearly to transform any given general cubic (d) so that it is touched by the 'horizontal' ef in a point of inflection, and then the projecting cubic (D) will be a divergent parabola. In analytical equivalent, since the simultaneous conditions $x = y = dx/dy = d^2x/dy^2 = 0$ readily determine $ax^3+bx^2y+cxy^2+dy^3+ex^2+fxy+hx = 0$ to be the most general Cartesian equation of the cubic (d) which touches ef ($x = 0$) at its inflection $(0, 0)$, the Newtonian 'transmutation' $x = kl/X$, $y = lY/X$ determines that the locus (D) of the corresponding projecting cubic has a defining equation of form

$$X^2+AXY+BX+CY^3+DY^2+EY+F = 0,$$

which reduces to the canonical equation $X^2+C'Y^3+D'Y^2+E'Y+F' = 0$ of the general divergent parabola on maintaining the Y-axis in given direction but transforming the coordinates linearly so that the new X-axis is that determined (in the old coordinates) by $X+\frac{1}{2}AY+\frac{1}{2}B = 0$. Yet more elegantly, we could equivalently transform the coordinates x, y of the projected curve (d), maintaining the y-axis in direction but taking as new abscissa the harmonic polar $\frac{1}{2}ex+\frac{1}{2}fy+h = 0$ of the inflection point $(0, 0)$, and then the equation of the projecting divergent parabola (D) would be at once in standard form. If instead we thereafter cast this harmonic polar of the inflection point to infinity, the new projecting cubic would be the transmutation of the latter parabola in which its diameter is sent to infinity and hence one of the five 'kinds' of central cubic $X = \alpha Y^3+\beta XY^2+\gamma X^2Y+\delta X^3$ whose centre $E(0, 0)$ is the transform of the infinite point of inflection in the divergent parabola. This alternative pro-

utrinq; productam AD in unico tantum puncto, quintus verò ubi secat in tribus.(31) Primo casu habetur Parabola *Neiliana* cujus utiq; longitudinem ubi angulus ABC rectus est Neilius noster primus invenit:(32) Secundo habetur Parabola alata se decussans quam inde *nodosam* nominabimus: tertio Parabola campaniformis punctum habens conjugatum: quarto Parabola campaniformis solitaria: quinto Parabola campaniformis cum Ellipsi conjugata, quæ si in punctum contrahitur fit punctum illud conjugatum in casu tertio. Et hæ quinq; figuræ cum projectionibus suis constituunt quinq; genera curvarum tertij ordinis, quarum nulla unius generis projicit aliquam alterius, omnes verò quæ sunt ejusdem generis per successivas projectiones in se mutuò transeunt.(33) Et eadem ratione curvæ superiorum ordinum distinguuntur in genera.

Species Linearum ejusdem Generis.

Quinetiam per casus Projectionum distinguuntur genera linearum in species. Nominemus planum illud *Horizontem* quod per oculum transit et plano lineæ projectæ parallelum est, et lineam illam *Horizontalem* in qua Horizon secat planum lineæ projicientis: et linea omnis projiciens dabit tot projectionum species quot sunt casus positionum lineæ Horizontalis.(34) Si linea Horizontalis alicubi secat projicientem intersectio illa generabit in projectione crura duo Hyperbolici generis circa eandem Asymptoton ad oppositas plagas in infinitum tendentia, idq; ex eodem Asymptoti latere si intersectio sit in puncto flexus contrarij, aliter ex latere diverso, et Asymptotos erit projectio rectæ quæ curvam projicientem tangit in puncto intersectionis, totq; ejusmodi crurum paria in projectione quot sunt intersectiones lineæ Horizontalis cum projiciente. Unde Linea secundi ordinis non nisi duo paria crurum Hyperbolicorum habere potest, Linea tertij ordinis non nisi tria paria, Linea quarti quatuor &c; et earum Asymptoti tres vel plures se secabunt in uno puncto si tangentes(35) se secant in uno, aliter in tribus vel pluribus; & si Projiciens se semel vel sæpius in eodem

---

jective generation of the cubic's species is the discovery of Michel Chasles, who initially set it out in his ever graceful prose in Note XX 'Sur la génération des courbes du troisième degré, par les cinq paraboles divergentes, et par les cinq courbes à centre' of his *Aperçu Historique sur l'Origine et le Développement des Méthodes en Géométrie* ([Brussels, $_1$1837 →] Paris, $_2$1875): 348–9. Of Clairaut's guiding insight Newton himself in his ensuing catalogue of the perspective *umbræ* of the divergent parabolas is, as we have stated, explicitly aware, but we know no evidence which begins to suggest that he ever applied it to give rigorous proof of his present assertion, being no doubt content to let that listing manifest its truth *per se*.

(31) In the algebraic terms of note (28) preceding, the given parabola $my^2 = x^3+mnx+amn$ meets the abscissa AD ($y = 0$) in one or three real points whose distance x from A is fixed by the equation $x^3+mnx+amn = 0$. The condition for two of these roots to coincide—either at a node or conjugate double point—is manifestly $3x^2+mn = 0 = 2mnx+3amn$, whence $x = \pm\surd(-\frac{1}{3}mn) = -\frac{3}{2}a$, and so, on returning to Newton's geometrical expression,

$$AE \text{ (or } -a) = \pm\tfrac{2}{3}\surd(-\tfrac{1}{3}MN).$$

(32) In thus firmly specifying (unlike on IV: 374) the 'prime inventor' of the algebraic rectification of the general arc of this semi-cubical parabola, Newton here patriotically ignores

AD extended either way in but a single point, the fifth, however, where it intersects it in three.[31] In the first case there is had the 'Neilian' parabola whose length, when the angle $A\widehat{B}C$ is right, our own Neil of course first discovered;[32] in the second is had a winged parabola which crosses itself and which we consequently shall name 'knotty'; in the third, a bell-shaped parabola having a conjugate point; in the fourth, a bell-shaped parabola by itself; in the fifth a bell-shaped parabola with a conjugate oval which, if it contracts into a point, becomes the conjugate point in the third case. These five figures together with their projections constitute five kinds of curves of the third order, none of one kind of which projects some one of another, but all which are of the same kind pass mutually into one another by successive projections.[33] And on the same basis curves of higher orders are distinguished into kinds.

Species of lines of the same kind.

To be sure, by the cases of their projections classes of lines are distinguished into species. Let us name the 'horizon' that plane which passes through the eye and is parallel to the plane of the projected line, and the 'horizontal' that line in which the horizon cuts the plane of the projecting line; and then every projecting line will yield as many species of projections as there are cases of position of the horizontal.[34] Should the horizontal cut the projecting line somewhere, that intersection will generate in the projection two branches of hyperbolic kind stretching round the same asymptote in opposite directions to infinity, this on the same side of the asymptote if the intersection be at a point of inflection, but otherwise on opposite sides, while the asymptote will be the projection of the straight line touching the projecting curve at the point of intersection; and there will be as many pairs of branches of this sort in the projection as there are intersections of the horizontal with the projecting curve. Consequently, a line of second order can have only two pairs of hyperbolic branches, a line of third order but three pairs, a line of fourth order but four, and so on; and three or more asymptotes of these shall cut one another in a single point if the tangents[35] intersect in one point, but otherwise in three or more; while if the projecting

Heuraet's valid claim to its independent—if not indeed its prior—discovery, one whose justice he had freely allowed in his youthful notes on the conic parabola's evolute (see 1: 68).

(33) In sequel Newton has deleted 'Et in his quinqꝫ generibus curvæ omnes tertij ordinis comprehenduntur' (And in these five kinds all curves of third order are comprised). The remark does no more, of course, than repeat what has gone before.

(34) An unwanted termination 'respectu lineæ projicientis' (in regard to the projecting line) is deleted. It may help to clarify Newton's cramped definition of these *termini technici* if we observe that in our figure in note (26) above, when *O* is the 'eye', *ADB* the projecting curve and *adb* the projection, the 'horizon' is the plane *OEF* drawn parallel to the plane *ebd* of projection, while *EF* is the 'horizontal (straight) line' in which this intersects the plane of the projecting curve.

(35) Understand 'in projiciente' (in the projecting curve).

puncto decussat et Horizon secat ipsam in puncto decussationis, Asymptoti duæ vel plures parallelæ erunt.

Si Linea Horizontalis tangit Projicientem, contactus omnis generabit crura duo Parabolici generis ad eandem plagam in infinitum tendentia et concavis partibus se mutuò respicientia, nisi ubi contactus ille aut est in puncto flexus contrarij, quo casu crura Parabolica ad modum Alarum tendent ad plagas oppositas et ex eodem latere concavæ erunt, aut in cuspide seu vertice acuto curvæ alicujus Cissoidalis, quo casu si Horizon quam obliquissimè tangit Projicientem seu (ut propriè loquar) secat ipsam in angulo contactus, crura Parabolica tendent ad plagas oppositas et ex diverso latere concavæ erunt, at si tangit ipsam in angulo qui rectilineo æqualis sit contactus ille generabit crura duo Hyperbolica ex eodem latere ejusdem Asymptoti ad eandem plagam in infinitum tendentia.

Si linea Horizontalis e[s]t Asymptotos Projicientis[,] crura Hyperbolica quæ circa Asymptoton illam sunt, convertentur in Parabolica: Et vice versa si linea Horizontalis tendit ad plagam crurum Parabolicorum[,] crura illa convertentur in Hyperbolica. Omnia vero crura infinita quæ non tendunt ad plagam lineæ horizontalis in omni casu evanescent.

Si deniq; Linea Horizontalis transit per punctum conjugatum, generabitur curva linea cujus punctum conjugatum in infinitum abijt. Et ne punctum conjugatum infinite distans absurdum videatur scias projectiones hujus curvæ habere puncta conjugata finitè distantia quæ sunt puncti illius infinitè distantis projectiones.

Atq; hæ sunt mutationes curvarum linearum quæ projectione fiunt: quarum casus omnes et eorum complexiones siquis ad curvam aliquam (36)projicientem enumeravit, is simul enumerabit linearum species omnes quæ sunt ejusdem generis cum projiciente: saltem si in lineis altiorum ordinum Projiciens satis latè sumitur.

Exemplum in lineis secundi ac tertij ordinis.

Sic ubi Projiciens est circulus, Linea Horizontalis hunc circulum aut secabit in duobus punctis aut tanget in uno aut tota cadet extra circulum, et perinde Projectio aut quatuor habebit infinita crura Hyperbolica aut duo Parabolica aut nullum. Unde hujus ordinis tres erunt species Hyperbola Parabola et Ellipsis præter Circulum. At in generibus linearum tertij ordinis casus sunt plures.(37)

(36) The significant variants (mainly in terminology and the detailed sequence of enumeration) occurring in a preliminary draft, now Add. 3961.3: 17^r–18^r, of the following generation of the cubic's species which begins at this point *in medio* are specified at their pertinent place in following footnotes. This, wrongly listed by J. C. Adams on its present foolscap wrapper as being but 'On the curves of the 3rd Order produced by the projections of the *Parabola Neiliana*', is briefly but accurately described by Rouse Ball—in ignorance of the revised text which we here reproduce—as 'a sheet...containing rough holograph notes, enumerating the various species which arise from the section by a plane in some defined position of one of the five cones

curve crosses itself once or more times at the same point and the horizon intersects it at the cross-point, two or more of the asymptotes will be parallel.

If the horizontal touches the projecting curve, every contact will generate two branches of parabolic kind stretching the same way to infinity and with their concavities facing one another, except when the contact is either at a point of inflection, in which case the parabolic branches will stretch wing-like in opposite directions and be concave on the same side, or at a cusp—that is, the sharp vertex of some cissoidal curve—in which case if the horizon touches it at an extreme slant (or, to use the proper phrase, intersects it at an angle of contact) the parabolic branches will stretch in opposite directions and be concave on different sides, but if it touches it at an angle equal to a rectilinear one then the contact will generate two hyperbolic branches stretching to infinity the same way on the same side of the same asymptote.

If the horizontal is an asymptote of the projecting curve, the hyperbolic branches round that asymptote will be changed into parabolic ones; and *vice versa* if the horizontal tends in the direction of parabolic branches, those branches will be changed into hyperbolic ones. All infinite branches, however, which do not tend in the horizontal's direction will in every case vanish.

If, finally, the horizontal passes through a conjugate point, a curve will be generated whose conjugate point has gone off to infinity. And lest you regard an infinitely distant conjugate point as absurd, you should know that the projections of this curve have finitely distant conjugate points which are the projections of that infinitely distant point.

And these are the transformations of curves which take place by projection. If anyone lists all the cases and their combinations with respect to some projecting curve,[36] he will simultaneously enumerate all species which are of the same kind as the projecting curve—at least if in lines of higher orders the projecting curve is sufficiently broadly taken.

An example in lines of second and third order.

Thus, when the projecting curve is a circle, the horizontal will either intersect this circle in two points, or touch it in one, or fall wholly outside the circle; and correspondingly the projection will have either four hyperbolic infinite branches, or two parabolic ones, or none at all. Consequently, in this order there will be three species—the hyperbola, parabola and ellipse—apart from the circle. But among the kinds of lines of third order there are several cases.[37]

having the diverging parabolas for bases....But no demonstrations are given, and the numerous erasures, corrections and interlineations make the manuscript somewhat difficult to read' ('On Newton's Classification of Cubic Curves' (see note (30) above): 123).

(37) Newton passes to give a detailed listing of the species of cubic which are severally comprehended in these five projectively distinct 'kinds' of curve of cubic order—those typified, namely, by the five respective *genera* of divergent parabola: the semicubical (Neil) parabola, that possessed of a node, that having a conjugate double point, the pure divergent

IN PRIMO GENERE.(38)

1. Si oculus infinitè distat vel si planum projectionis plano projicientis parallelum est, projectio erit Parabola ejusdem speciei cum projicente[,] id est *Parabola cuspidata* quam *Neilianam* nominavimus.

2. Si Linea Horizontalis transit per verticem cuspidatum Projicientis idq in angulo contactus, Projectio erit *Parabola Wallisiana*,(39) habens crura duo Parabolica ad oppositas plagas in infinitum tendentia et ex diverso latere concava, et centrum in puncto flexus contrarij.

3. Si linea illa Horizontalis transit per verticem cuspidatum et tendit ad plagam infinitorum crurum Projicientis, Projectio erit *Crux Hyperbolica librata*,(40) habens duas Hyperbolas in eodem latere unius Asymptoti ex diverso alterius. Libratam vero voco curvam quæ diametrum rectilineā habet ordinatas hinc inde æquales terminantem: non libratam quæ talem diametrum non habet.(41)

4. Si linea illa transiens per verticem cuspidatum tendit ad aliam quamvis plagam: Projectio erit *Crux hyperbolica non librata*, habens Hyperbolas duas quarum duo crura ex diverso latere alterius Asymptoti ad plagas oppositas tendunt.

---

parabola, and that, lastly, having a conjugate oval. While it will be clear that this projective generation is, in the case of the four latter 'kinds', not entirely complete and does not elsewhere point uniquely to one or other of the 72 derived (in the fullest version of this algebraic enumeration which is reproduced in **3**, 1, §2 below and was published by Newton in 1704 in appendix to his *Opticks*) by considering all conceivable non-degenerate particular varieties of the four reduced canonical forms of the cubic's general Cartesian defining equation, it does in compensation—and in refutation of a tradition which has for a quarter of a millennium cited Newton's deficiency in this respect—accurately list (see notes (54) and (65) below) the six species omitted by him in that algebraically based 'Enumeratio Linearum tertij Ordinis'. A full (if not quite definitively evaluated) equivalent genesis of the species of perspective 'shadows' of the five divergent parabolas was afterwards carefully set out by Patrick Murdoch in Sectio IV (pages 74–126) of his *Neutoni Genesis Curvarum per Umbras. Seu Perspectivæ Universalis Elementa; Exemplis Coni Sectionum et Linearum Tertii Ordinis illustrata* (London, 1746), Newton's five present projective 'kinds' of cubic there being catalogued in a different permutation as 'classes' of component species (*ibid.*: 118—21/111–18/86–90/92–7/74–86); compare H. Wieleitner, 'Die Behandlung der Perspektive durch Murdoch', *Bibliotheca Mathematica*, $_3$**14** (1913/14): 320–35, especially 333–5. (The like enumeration given by C. R. M. Talbot in his more accessible *Sir Isaac Newton's Enumeration of Lines of the Third Order.... Translated from the Latin with Notes and Examples* (London, 1860): 77–82 is, we would warn, somewhat faulty in its detail: projection 5 of Case 1 of the punctate parabola (*ibid.*: 80) does not, for example, yield—as it is there stated to do—Newton's published species 13 and 25, but only the former one together with two of the four additional species adjoined by Stirling in 1717.)

(38) That of the perspective 'shadows' cast by the cusped Neilian parabola, namely. We may visually display Newton's projections of this cubic in compact form in the accompanying figure, where the thin, numbered lines—the first being conceived to lie at infinity 'off to the left'—mark the various sites of the 'horizontal' which passes to infinity in the projected figure. In the draft enumeration on Add. 3961.3: 17r (on which see note (36) preceding) projections 3, 4, 5, 6, 7, 8, 9 were originally ordered somewhat more logically in the sequence 4, 7, 3, 5, 6, 9, 8. As they are listed by Newton in his final algebraic 'Enumeratio' (**3**, 1, §2 below) the

IN THE FIRST KIND.[38]

1. If the eye is infinitely distant, or if the plane of the projection is parallel to the plane of the projecting cubic, the projection will be a parabola of the same species as the projecting cubic, namely, the *cusped parabola* which we have named *Neilian*.

2. If the horizontal passes through the cusped vertex of the projecting cubic, and this at an angle of contact, the projection will be a *Wallisian*[39] *parabola*, having two parabolic branches extending in opposite directions to infinity, concave on their different sides, and with a centre at the point of inflection.

3. If the horizontal passes through the cusped vertex and tends in the direction of the infinite branches of the projecting curve, the projection will be a *balanced*[40] *hyperbolic cross*, having two hyperbolas (lying) on the same side of one asymptote and on different sides of the other. I call a curve 'balanced', however, when it has a rectilinear diameter terminating ordinates equal on either hand; 'unbalanced' when it lacks such a diameter.[41]

4. If that line passing through the cusped vertex tends in any other direction, the projection will be an *unbalanced hyperbolic cross*, having two hyperbolas, two branches of which stretch in opposite directions on different sides of either asymptote.

projected cubics are of respective species 70, 72, 65, 64, 42, 12, 35, 48 and 3. There are, we may add, no further projections of this 'kind' possible.

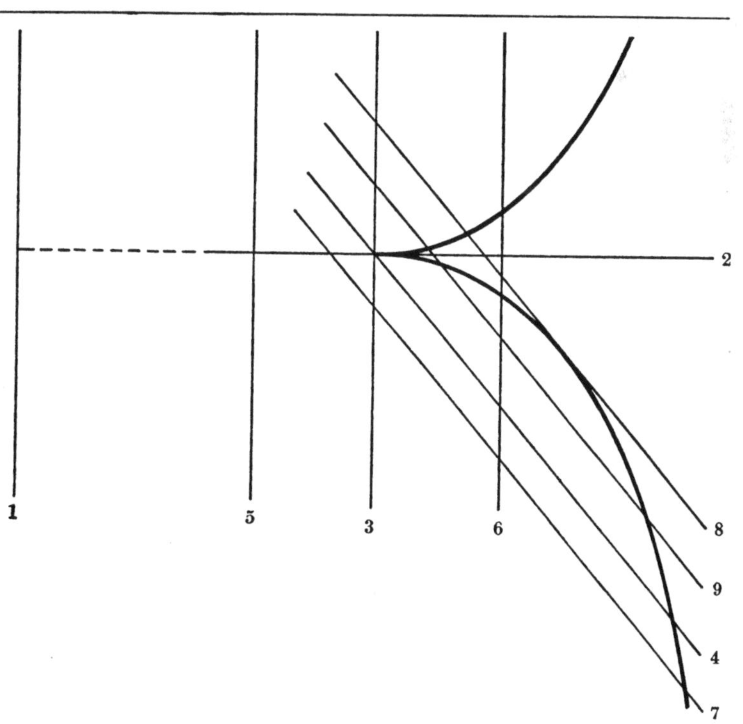

(39) Borrowing in each case from his earlier designation of this cubic on II: 86, Newton originally—on Add. 3961.3: 17ʳ—named this parabola '*simplex cubica*' (*simple cubic*) and then—in the present manuscript—initially changed this to be '*circumflexa*' (*around-bent*).

(40) First—as on IV: 370 ff.—'*bifida*' (*cleft*), then '*par ad diametrum*' (*diametrally matching*), and yet again thereafter '*bijuga*' (*double-yoked*). Like transitions to the final choice of '*librata*' occur in following places in the manuscript, but these we will not cite piecemeal.

(41) Referring back to his original designation of curvilinear symmetry round an axis (see the previous note), Newton here first equivalently explained that '*Bifariam* vero voco lineam

5. Si tendit ad plagam crurum infinitorum et Projicientem nec secat nec tangit Projectio erit *Cissois librata*, et uno casu *Cissois Veterum.*

6. Si tendit ad plagam crurum infinitorum et secat Projicientem in duobus punctis Projectio erit *Hyperbola triplex*(42) *librata cuspidata*. Hyperbolarum una quæ cuspidata erit jacebit extra angulum Asymptotorum, alteræ duæ non cuspidatæ jacebunt intra.

7. Si secat Projicientem in unico tantum puncto et non transit per cuspidem ejus Projectio erit *Cissois circa Asymptoton torta.*

8. Si tangit Projicientem extra cuspidem, atq; adeo in alio etiam puncto secat[,] Projectio erit *Crux Parabolica cuspidata*. Ejus crura duo Parabolica tendunt ad eandem plagam et concavitate se mutuò respiciunt, in vertice verò non junguntur sed postquam convergendo unum eorum processit in cuspidem, divergunt denuò et ad plagas oppositas cruribus Hyperbolicis ex diverso latere ejusdem Asymptoti in infinitum tendunt.

9. Si secat Projicientem in tribus punctis Projectio erit *Hyperbola triplex cuspidata non librata.*

IN SECUNDO GENERE.(43)

1. Si oculus infinitè distat vel si plana Projectionis et Projicientis parallela sint, Projectio erit ejusdem speciei cum Projiciente id est *Parabola nodosa*.(44)

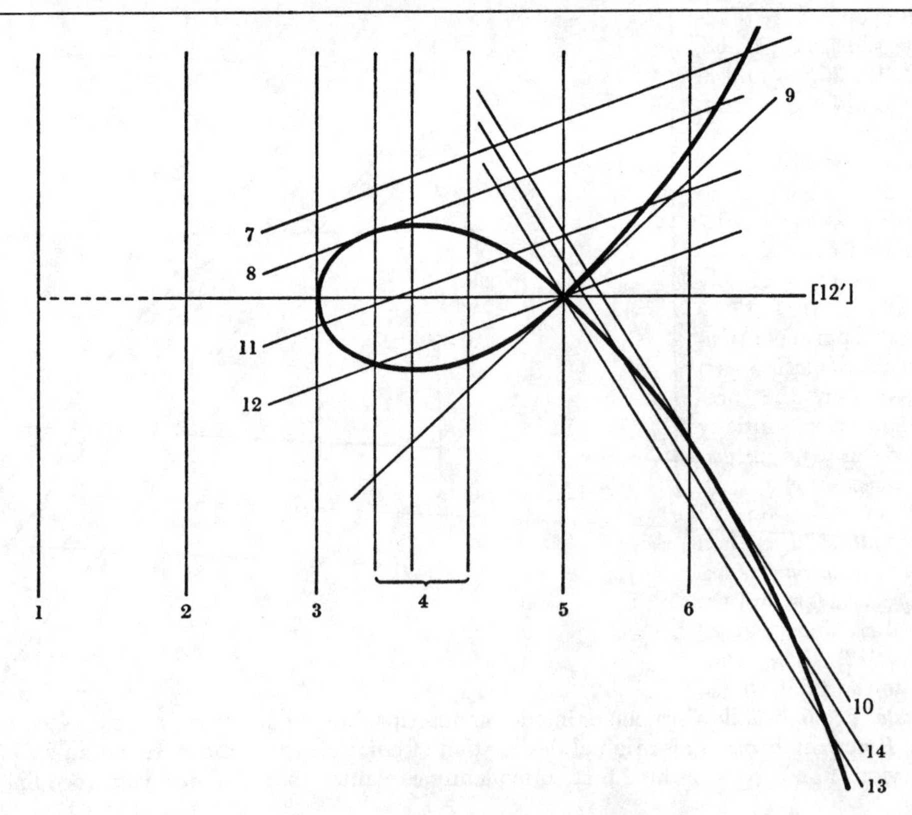

5. If it tends in the direction of the infinite branches and neither cuts nor touches the projecting cubic, the projection will be a *balanced cissoid*, and in one case the *ancients' cissoid*.

6. If it tends in the direction of the infinite branches and cuts the projecting cubic in two points, the projection will be a *triple*(42) *balanced and cusped hyperbola*. One of the hyperbolas—the cusped one—will lie outside the angle of the asymptotes, the other two (uncusped) ones will lie within it.

7. If it cuts the projecting cubic in but a single point and does not pass through its cusp, the projection will be a *cissoid twisted round its asymptote*.

8. If it touches the projecting cubic outside of its cusp, and hence cuts it in yet a further point, the projection will be a *cusped parabolic cross*. Its two parabolic branches stretch the same way and face one another in their concavity; at the vertex, however, they are not joined but after, in its converging path, one of them has proceeded to the cusp they diverge anew and tend in opposite directions to infinity in hyperbolic branches on different sides of the same asymptote.

9. If it cuts the projecting cubic in three points, the projection will be a *triple cusped, unbalanced hyperbola*.

IN THE SECOND KIND.(43)

1. If the eye is infinitely distant, or if the planes of the projection and projecting curve be parallel, the projection will be of the same species as the projecting cubic, that is, a *knotty parabola*.(44)

---

cujus area comprehensa ita in æquales partes a recta aliqua secari potest ut ordinatæ ad idem punctum ex utroq bisecantis latere æquales sint: non-bifidam cujus area non potest ita secari' (I call a line 'cleft', however, when the area it comprehends can so be cut by some straight line into equal parts that the ordinates at the same point of it on either side of the bisecting line shall be equal; 'uncleft' when its area cannot be cut in this way).

(42) Initially '*triformis*' (*triple-figured*) on Add. 3961.3: 17^r.

(43) That of the divergent parabola possessing a node. Much as before (in note (38)) we can summarize Newton's following projections of this base cubic in the accompanying scheme, where the first numbered thin line again denotes the position of the 'horizontal' when it lies off at infinity. In order, projections 1–14 (which repeat those of the draft on Add. 3961.3: $17^r/17^v$ except that numbers 9–14 are there listed in the sequence 11, 14, 9, 10, 12, 13) yield respectively species 68, 41, 54, 19/30/18 [see note (46)], 60, 11, 34, 51, 66, 47, 7/8/25 [see note (49)], 57/59, 58 and 2 of the related algebraic enumeration in **3**, 1, §2 below. It would, we may add, have been more consistent if Newton had here, in line with Case 2 of Genus 1 preceding, separately specified the projection of the parabola's axis rather than obscurely included it as a particular instance of his present case 12.

(44) Originally '*Parabola se decussans* quam et *nodosam* nominabimus' (a *parabola crossing itself*, which we shall also name *knotty*) on Add. 3961.3: 17^r.

2. Si linea Horizontalis tendit ad plagam crurum infinitorum Projicientis et Projicientem nec[45] secat nec tangit Projectio erit *Cissois nodosa librata.*

3. Sin Projicientem tangit in vertice Projectio erit *Crux Parabolica nodosa librata.*

4. Si secat eam inter verticem et nodum projicietur *Hyperbola triplex librata cum nodo in pari Hyperbolarum.*[46]

5. Si secat in ipso nodo, Projectio erit *Hyperbola triplex librata duas ex tribus asymptotis parallelas habens.*

6. Si secat ultra nodum versus crura infinita Projectio erit *Hyperbola triplex librata cum nodo in impari Hyperbola.*

7. Quod si linea Horizontalis non tendit ad plagam crurum infinitorum et occurrit Projicienti in unico tantum puncto, Projectio erit *Cissois nodosa*[47] *circa Asymptoton torta.*

8. Si præterea tangit Projicientem inter verticem et nodum, projicietur *Crux Parabolica nodosa, non librata, clausa in vertice.*

9. Si tangit eam in ipso nodo Projectio erit *Parabola Cartesiana.*[48]

10. Si tangit eam ultra nodum in crure infinito projectio erit *Crux Parabolica nodosa, non librata, aperta in vertice.*

11. Si secat eam bis ad partes nodi versus verticem et semel alicubi projicietur *Hyperbola triplex non librata cum nodo in pari Hyperbolarum.*[49]

12. Si secat semel ad partes nodi versus verticem et bis in nodo projicietur *Hyperbola triplex non librata duas ex Asymptotis parallelas habens et inter eas Hyperbolam concavo-convexam,*[50] et præterea *centrum habens in flexu contrario* si modo linea Horizontalis secat Projicientem in ipso Vertice.

13. Si secat eam bis in nodo et semel extra eum versus crura infinita Projectio erit *Hyperbola triplex non librata duas ex Asymptotis parallelas habens et inter eas Hyperbolam ad easdem partes omnino concavam.*[51]

14. Si secat Projicientem in tribus punctis extra nodū versus crura infinita, Projectio erit *Hyperbola triplex non* [*librata*][52] *nodum habens in impari Hyperbola.*[53]

---

(45) 'nullibi' (nowhere) was first written.

(46) Initially '*in duarum Hyperbolarum intersectione*' (*at the intersection of two of the hyperbolic branches*). In fact, according as the 'horizontal' meets the node at right angles—the tangents to the parabola at these twin points of intersection being parallel—or to either side of this site (see our figure in note (43) above), the projection will yield either species 30 or species 18 and 19 of the algebraic enumeration.

(47) Newton presumably here meant also to add '*non librata*' (*unbalanced*).

(48) Originally '*Tridens Parabolica* sive Parabola illa cujus proprietates Cartesius in Geometria explicuit' (a *parabolic trident*, that is, the parabola whose properties Descartes explained in his *Geometrie*); on which see IV: 478, note (155).

(49) Much as in case 4 above (see note (46)) Newton here initially specified the node to be '*in intersectione duarum Hyperbolarum*' (*at the intersection of two of the hyperbolic branches*). According to the relative position of the tangents at the three meets of the 'horizontal' with the parabola, the projected cubics will be one or other of species 7, 8 and 25 in the algebraic enumeration, the last case being that where the tangents in the projecting parabola are concurrent, so

2. If the horizontal tends in the direction of the infinite branches of the projecting curve and neither(45) intersects nor touches it, the projection will be a *knotty, balanced cissoid*.

3. But if it touches the projecting cubic at its vertex, the projection will be a *knotty, balanced parabolic cross*.

4. If it cuts it between its vertex and its node, there will be projected a *triple balanced hyperbola with a node in a pair of its hyperbolas*.(46)

5. If it cuts it in the node itself, the projection will be a *triple balanced hyperbola having two of its three asymptotes parallel*.

6. If it cuts it beyond the node towards its infinite branches, the projection will be a *triple balanced hyperbola with a node in the odd one of the hyperbolas*.

7. But if the horizontal does not tend in the direction of the infinite branches and meets the projecting cubic in but a single point, the projection will be a *knotty*(47) *cissoid twisted round its asymptote*.

8. If, moreover, it touches the projecting cubic between its vertex and node, there will be projected a *knotty, unbalanced parabolic cross, closed at its vertex*.

9. If it touches it in the node itself, the projection will be a *Cartesian parabola*.(48)

10. If it touches it beyond the node in an infinite branch, the projection will be a *knotty parabolic cross, unbalanced and open at its vertex*.

11. If it cuts it twice in the vicinity of the node towards the vertex and once elsewhere, there will be projected a *triple unbalanced hyperbola with a node in a pair of its hyperbolas*.(49)

12. If it cuts once in the region of the node towards the vertex and twice in the node, there will be projected a *triple unbalanced hyperbola having two of its asymptotes parallel and a concavo-convex hyperbola between them*,(50) and moreover *having a centre at its inflection point* if, further, the horizontal cuts the projecting cubic in the vertex itself.

13. If it cuts it twice in the node and once outside it towards the infinite branches, the projection will be a *triple unbalanced hyperbola having two of its asymptotes parallel and between these a hyperbola wholly concave the same way*.(51)

14. If it cuts the projecting cubic in three points outside the node towards the infinite branches, the projection will be a *triple unbalanced hyperbola having a node in its odd hyperbola*.(53)

---

determining that the corresponding asymptotes in the projected cubic pass through the same point.

(50) On Add. 3961.3:17ᵛ Newton initially specified this to be '*Hyperbola triformis duas ex asymptotis parallelas habens et ex Hyperbolis unam inter parallelas asymptotos contortam*' (*a triple-figured hyperbola having two of its asymptotes parallel and one of its hyperbolas twisted round between the parallel asymptotes*).

(51) Originally '*cum una ex Hyperbolis interjecta*' (*with one of the hyperbolas interposed*).

(52) We replace an equivalent '*bifida*' in the manuscript which Newton himself surely meant thus to alter (compare note (40) above).

(53) Initially '*...in una triũ Hyperbolarum*' (*in one of the three hyperbolas*).

IN TERTIO GENERE.(54)

1. Si oculus infinite distat vel si plana projectionis et projicientis parallela sint,(55) projectio est ejusdem speciei cum projiciente id est *Parabola campaniformis cum puncto conjugato*.(56)

2. Si Linea Horizontalis vel tendit ad plagam crurum infinitorum vel transit per flexum contrarium Projicientis et præterea transit ultra punctum conjugatum Projectio erit *Concha librata punctum habens conjugatum ad convexitatem verticis*.

3. Sin transit per punctum conjugatum, orietur *Concha librata cum puncto conjugato ad infinitam distantiam*.

4. Si transit inter punctum conjugatum et Projicientem, Projectio est *Concha librata punctum conjugatum habens ad concavitatem verticis*.

(54) That of the divergent parabola having a conjugate double point. Of Newton's listed projections (identical with those of the draft on Add. 3961.3: 17v/18r except that numbers 9 and 10 are here interchanged) 1–4 and 8–12 yield species 69, 43, 63, 44 and 8, 61/62, 36, 49, 4 of the corresponding algebraic enumeration in **3**, 1, §2 below, while projections 5–7 produce species of cubic there omitted by Newton. Of these last the two defined by projections 6 and 7 —a monodiametral and a tridiametral 'hyperbolic' (redundant) hyperbola respectively— were first publicly identified by James Stirling in the augmented algebraic enumeration of cubic curves which he set out in his *Lineæ Tertii Ordinis Neutonianæ, sive Illustratio Tractatus*

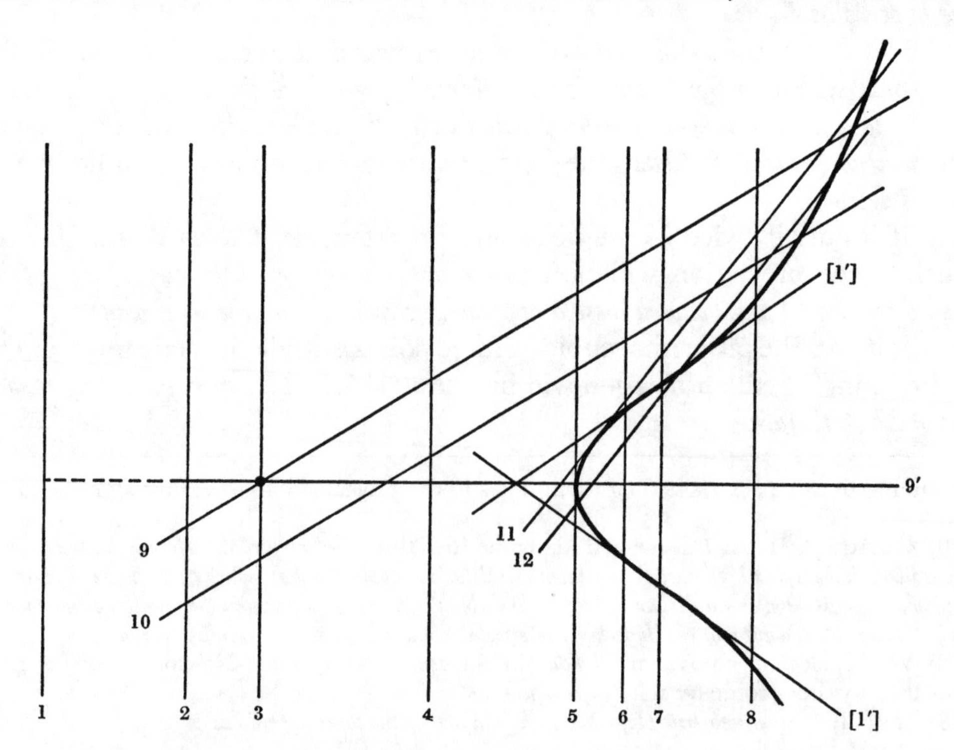

IN THE THIRD KIND.(54)

1. If the eye is infinitely distant, or if the planes of the projection and the projecting curve be parallel,(55) the projection is of the same species as the projecting cubic, namely a *bell-shaped parabola with a conjugate point.*(56)

2. If the horizontal either tends in the direction of the infinite branches or passes through an inflection in the projecting cubic and, furthermore, passes beyond the conjugate point, the projection will be a *balanced shell having a conjugate point on the convex side of its vertex.*

3. But if it passes through the conjugate point, there will ensue a *balanced shell with a conjugate point at an infinite distance.*

4. If it passes between the conjugate point and the projecting cubic, the projection is a *balanced shell having a conjugate point on the concave side of its vertex.*

D. Neutoni De Enumeratione Linearum Tertii Ordinis (Oxford, 1717): 83–120, where they form his species 15 and 25. The parabolic hyperbola determined by projection 5 was afterwards privately brought to Stirling's notice—as an addition to page 112 of his book where he treated its fellows—by Niklaus I Bernoulli in a letter of 1 April 1733. (See C. Tweedie, *James Stirling: A Sketch of his Life and Works, along with his Scientific Correspondence* (Oxford, 1922): 149. Bernoulli also communicated his discovery to Gabriel Cramer, for Patrick Murdoch subsequently remarked in his *Neutoni Genesis Curvarum per Umbras* (note (37)): 87–8 that 'Speciem hanc... apud *Neutonum* animadverterat D. *Nic. Bernoulli*; quod me olim monuit D. *Cramer, Phil.* et *Math.* apud *Genevenses* celebris Professor'.) First public announcement of its existence was made by Edmund Stone in an open letter, dated 31 July 1736, 'concerning two Species of Lines of the Third Order, not mentioned by Sir *Isaac Newton*, nor Mr. *Sterling*' which was printed four years later in *Philosophical Transactions* **41**, No. 456 [for January–June 1740]: 318–20, where Bernoulli's cubic species appears (*ibid.*: 320) as the punctate instance of Stone's two 'Hyperbolo-parabolical Curves', the other being the analogous cubic with an oval which had—though Stone seems unaware of this—been described in 1731 by Nicole (see note (30) above). Simultaneous publication of both these 'new' species was given by De Gua, with equal confidence in being their discoverer, in a 'Remarque' on the inadequacies of Newton's published enumeration in his *Usages de l'Analyse de Descartes* (Paris, 1740): 367–8. To return to Newton's present listing, we may point to his lack of forethought in here not separately specifying the case (marked 9′ in our scheme) where the 'horizontal' coincides with the axis of the projecting parabola. By comprehending it as a particular instance of projection 9 he will forget to carry this generation of the like 'kind' of central cubic over into Genus 4 following where the latter is rightly discarded.

(55) Newton might well for completeness' sake have here adjoined 'vel si Linea Horizontalis tangit in puncto flexûs contrarij' (or if the horizontal line is tangent at an inflection point)—in one or other of the positions, that is, which we have numbered 1′ in the figure of our previous note. In first draft he had (see note (60) below) initially set this variant generation of the punctate diverging parabola as a separate case of projection before coming to appreciate its equivalence with the present case 1.

(56) Originally, on Add. 3961.3: 17^v, Newton equivalently designated this a '*Parabola... punctata*' (*punctate...parabola*). We omit individually to notice several analogous replacements of 'punctata' by 'cum puncto conjugato' in sequel.

5. Si tangit Projicientem[57] fit *Crux Parabolica librata cum vertice aperto et puncto conjugato ultra verticem.*

6. Si secat Projicientem[57] inter verticem et puncta flexus contrarij fit *Hyperbola triplex librata cum flexibus contrarijs in pari Hyperbolarum et puncto conjugato inter tres Asymptotos.*

7. Si secat Projicientem[57] in utroq flexu contrario fit *Hyperbola triplex, trifariam librata, cum puncto conjugato in centro trianguli Asymptotis inclusi*, quod *centrum* est projectionis.[58]

8. Si secat Projicientem[57] ad alteras partes alterutrius vel utriusq flexus contrarij fit *Hyperbola triplex librata cum flexibus contrarijs in impari Hyperbola et puncto conjugato inter tres Asymptotos.*

9. Quod si linea Horizontalis nec tendit ad plagam crurum infinitorum nec transit per flexum contrariũ, transit verò per punctum conjugatum, fit *Concha flexu contrario circa Asymptoton torta* cujus *punctum conjugatum in infinitum abijt* quæq insuper *centrum* habebit in flexu contrario si modò linea Horizontalis transit per verticem Projicientis.[59]

10. Sin transit ultra vel citra punctum conjugatum et Projicientem secat in unico tantum puncto extra flexus contrarios, Projectio erit *Concha flexu contrario circa Asymptoton torta cum puncto conjugato ad finitam distantiam.*

11. Quod si transiens ultra vel citra punctum conjugatum tangit Projicientem[60] habebitur *Crux Parabolica non librata aperta in vertice cum puncto conjugato ultra verticem.*

12. Si deniq secat Projicientem in tribus punctis projicitur *Hyperbola triplex non librata cum puncto conjugato inter tres Asymptotos.* Et una Hyperbolarum jacet intra Asymptotos suas, altera jacet uno crure intra alio extra, tertia utroq crure extra.[61]

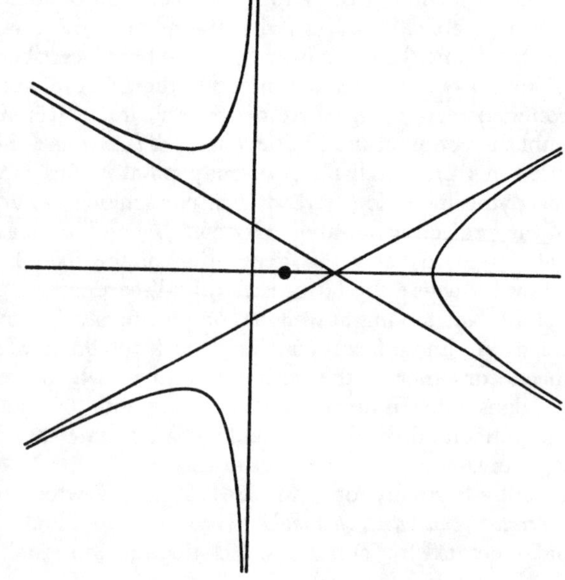

(57) So changed from 'Parabolam' (parabola) which Newton first wrote at these places in his preliminary draft.

(58) A thoroughly misleading observation. The double point in this projected cubic—Stirling's species 25 (see note (54))—lies at its 'centre' only in the loose sense that it is located (on its axis) roughly at its middle, within the triangle formed by the asymptotes. It is of course impossible that a cubic's double point should be its centre of bilateral symmetry (because a straight line drawn through it might then intersect the cubic in four points: *quod fieri nequit*).

(59) If, that is, it coincides with the projecting parabola's axis (as in position 9′ in our figure in note (54) above).

(60) For no very cogent reason—unless

5. If it touches the projecting curve,[57] there comes a *balanced parabolic cross with an open vertex and a conjugate point beyond the vertex.*

6. If it cuts the projecting curve[57] between its vertex and inflection points, there comes a *triple balanced hyperbola with inflections in a pair of hyperbolas and a conjugate point between its three asymptotes.*

7. If it cuts the projecting curve[57] in either inflection, there comes a *triple, three-way balanced hyperbola with a conjugate point in the centre of the triangle shut off by the asymptotes:* this is the projection's *centre.*[58]

8. If it cuts the projecting curve[57] on the other side of either inflection or both, there comes a *triple balanced hyperbola with inflections in its odd hyperbola and a conjugate point between its three asymptotes.*

9. But if the horizontal neither tends in the direction of the infinite branches nor passes through an inflection, if however it passes through the conjugate point there comes a *shell with an inflection, twisted round an asymptote,* whose *conjugate point has gone off to infinity*; this will in addition have a *centre* at the inflection if, further, the horizontal passes through the projecting cubic's vertex.[59]

10. While if it passes to one side or other of the conjugate point and cuts the projecting cubic in but a single point outside of the inflections, the projection will be a *shell with an inflection, twisted round an asymptote and with a conjugate point at a finite distance.*

11. But if in passing on one side or other of the conjugate point it touches the projecting cubic,[60] there will be had an *unbalanced parabolic cross, closed at its vertex and with a conjugate point beyond the vertex.*

12. If, finally, it cuts the projecting cubic in three points, there is projected a *triple unbalanced hyperbola with a conjugate point between its three asymptotes. And one hyperbola lies within its asymptotes, a second lies with one branch within and one without, the third with both branches outside.*[61]

---

he believed that both exceptions are trivially self-evident—Newton has in immediate sequel in his manuscript deleted 'extra verticem et flexum contrariũ' (other than at the vertex or an inflection point). The instance where the 'horizontal' is tangent at the vertex is, of course, case 5 preceding. In preliminary draft on Add. 3961.3; 18r he had initially determined to specify as a separate projection the latter excluded instance where 'Si [linea horizontalis] secat Parabolam in flexu contrario in angulo contactus[,] projicitur Parabola ejusdem speciei cum projiciente' (If the horizontal line cuts the parabola at an inflection in an angle of contact [*sc.* coincident with the tangent there], there is projected a parabola of the same species as the projecting one), afterwards omitting this as being 'idem cum...cas. 1' (identical with case 1).

(61) Apart from the enunciation of the variant generation of the punctate diverging parabola cited in the previous note, Newton in his preliminary draft on Add. 3961.3: 18r tentatively set out a number of other would-be distinct projections which he subsequently rejected as identical with one or other of cases 5, 8, 11 and 12. Since their nuances were afterwards incorporated by him in his revised statements of these, we see no need here to set out their texts in full, but may instance their quality by quoting the most elaborate of these,

In quarto genere.[62]

Species 1. 2. 3. 4. 5. 6. 7. 8. 9 ijsdem ferè verbis describuntur ac in Genere tertio species 1. 2. 5. 6. 7. 8. 10. 11. 12 respectivè, nisi quod projectiones hic non habent punctum conjugatum.[63] Et speciei 4. 5. 6. 9 casus sunt simplicissimi ubi tres Asymptoti in unico puncto concurrunt.[64]

In quinto Genere.[65]

Species 1, 2, 6, 7, 8, 9, 10. 11. 14. 15 ijsdem verbis describuntur ac in Genere

where—before recognizing that 'Est casus idem cum casu 8'—he announced: 'Si [linea horizontalis] secat Parabolam in flexu contrario et alibi[,] fit *Hyperbola triformis bifida cum puncto conjugato inter tres asymptotos*, et duæ hyperbolarum ja[c]ent intra asymptotos tertia jacet extra' (If the horizontal line cuts the parabola in an inflection and elsewhere, there comes a triple-figured, cleft [*sc.* 'balanced', diametral] hyperbola with a conjugate point, between three asymptotes, and two of the hyperbolas lie within the asymptotes, the third one outside).

(62) That of the 'pure' divergent parabola, to which Newton has above given the name *Parabola campaniformis solitaria*. Apart from its negligent omission of the case where the 'horizontal' coincides with the parabola's axis, however, his following allocation of its species is somewhat foreshortened by his failure to notice that in this projective genus the 'solitary bell-shaped parabola' may, because it is free of cusps, nodes, conjugate double points and ovals, have a pronounced vase-like 'neck'; see the figure in the next note.

(63) In preliminary draft on Add. 3961.3: 18r Newton concluded a little differently by writing 'nisi quod projectio non est punctata nec plagam habet puncti conjugati in infinitum

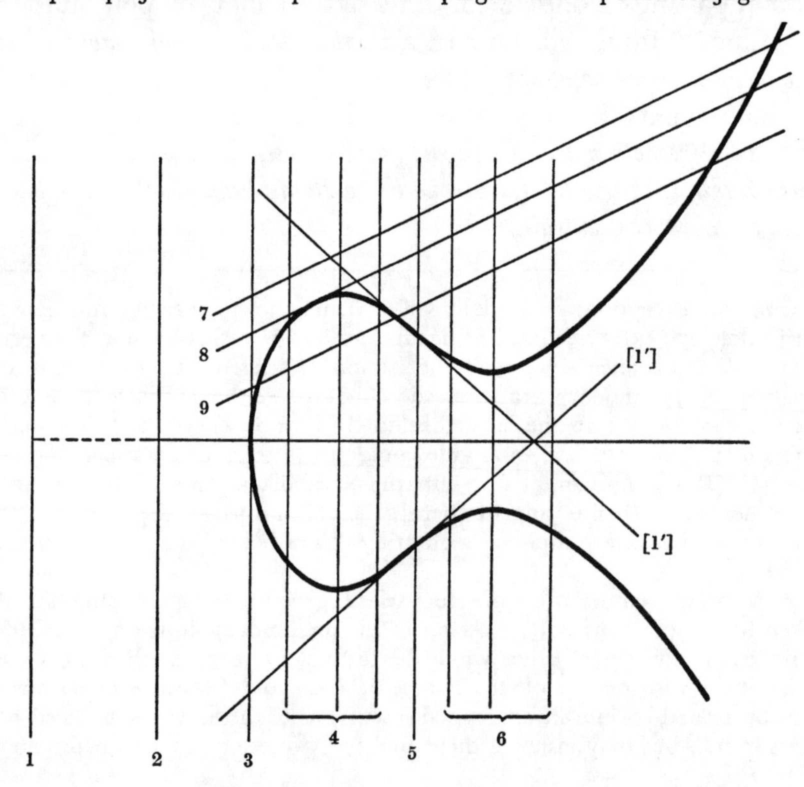

In the fourth kind.(62)

Species 1–9 are described in virtually the same words as species 1, 2, 5, 6, 7, 8, 10, 11, 12 respectively in the third kind except that the projections do not here have a conjugate point.(63) And the simplest cases of species 4, 5, 6 and 9 are those where the three asymptotes concur in a single point.(64)

In the fifth kind.(65)

Species 1, 2, 6–11 and 14/15 are described in the same words as species

translati' (except that the projection is not punctate, nor does it have the direction of a conjugate point translated to infinity)—the latter case being, of course, projection 3 in the preceding Genus, which yields species 63 of the algebraic enumeration in **3**, 1, §2 below. In that preliminary version he had first begun explicitly to list '1. Parabola campaniformis solitaria si oculus infinite distat vel plana projectionis et projicientis parallela sint, suæ speciei figuram projicit' (1. If the eye is infinitely distant or the planes of the projection and the projecting curve be parallel, the solitary bell-shaped parabola projects a figure of its own species) and thereafter '2. Si linea hori[zontalis vel tendit...]' (2. If the horizontal line [either tends...]) before breaking off to summarize this repeat *mutatis mutandis* of the projections of the previous punctate genus which do not involve its conjugate point. In thus tailoring his present catalogue to that which precedes, Newton has missed certain nuances, not only forgetfully discarding along with his previous projection 9 the particular instance which casts the parabola's axis to infinity (and would here yield the central cubic which is species 38 of his algebraic enumeration) but failing to see that the generating divergent parabola is in general 'ampullate' (to use Murdoch's word) with a jar-like 'neck' as here illustrated. As before, in the first projection the 'horizontal' is conceived to lie at infinity parallel to the ordinate direction, and we have inserted its variant generations where the 'horizontal' touches the parabola at an inflection as the twin projections 1′. It will be clear that projections 4, 5 and 6 split into differing cases according to the angle at which the 'horizontal' in these instances meets the parabola. In sum, Newton's projections 1–9 produce respectively species 71, 45, 53, 15/17/29, 22/23/32, 14/16/28, 37, 50 and 5/6/24 of the ensuing algebraic enumeration. The omission of the tenth projection where the 'horizontal' coincides with the parabola's axis (yielding species 38) is readily made good.

(64) In projections 4, 5 and 6 the 'horizontal' will in these special instances intersect the parabola at points which have parallel slope and hence whose tangents meet on the (vertical) line at infinity which touches the parabola at its infinitely distant point of inflection; in the case of projection 9, more simply, the tangents at the three intersections of the 'horizontal' with the parabola will here be concurrent in some finite point. The respective species of the algebraic enumeration are 29, 32, 28 and 24.

(65) That of the divergent parabola possessing an oval. The cases of projection, as Newton accurately specifies them in sequel, are encapsulated much as before in the accompanying scheme. Projections 1–6 and 10–15 yield respectively, when due consideration is given to their component cases, species 67, 39, 55, 20/21/31, 56, 40 and 10, 33, 62, 9/26/27, 46, 1 of the algebraic enumeration in **3**, 1, §2 below. The monodiametral and tridiametral 'hyperbolic' (redundant) hyperbolas corresponding to projections 8 and 9 were first catalogued by Stirling in 1717 in his *Lineæ Tertii Ordinis Neutonianæ* (see note (54) above) as his species 11 and 24 (*ibid.*: 99, 102). The parabolic hyperbola produced by projection 7 was first identified in 1731 by Nicole, who described it, along with the rest of Newton's projections 2–10, in his 'Manière d'engendrer dans un Corps solide toutes les lignes du troisième ordre' (see note (30) above): §XIII: 500–1. We pass comment in note (70) below on Newton's somewhat confused jumbling

tertio species: 1. 2. 4. 5. 6. 7. 8. 10. 11. 12 nisi quod loco puncti conjugati Ellipsis conjugata ponenda est.(66) [Insuper sunt]

3. Si linea Horizontalis vel ad plagam infinitorum crurum tendens vel per punctum flexus contrarij transiens tangit Ellipsin ad partem exteriorem Projectio erit *Parabola librata cum Concha quæ convexitate sua Parabolam respicit.*

4. Sin secat Ellipsin Projectio erit *Hyperbola duplex librata cum Concha interjecta*:(67) cujus casus est simplicissimus ubi tres Asymptoti concurrunt in eodem puncto.

5. Quod si tangit Ellipsin ad partes interiores seu versus Parabolam Campaniformem Projectio erit *Parabola librata cum concha quæ concavitate sua Parabolam respicit.*

12. Si tangit Ellipsin et non tendit ad plagam infinitorum crurū nec transit per flexum contrarium Projectio erit *Parabola non librata cum Concha flexu contrario circa Asymptoton torta.*

13. Si secat Ellipsin in duobus punctis et alibi in tertio extra flexum contrarium, Projectio erit *Hyperbola duplex non librata cum Concha flexu contrario circa Asymtoton torta*:(68) et præterea *centrum* habebit in flexu illo contrario si linea horizontalis per tres Projicientis vertices transit;(69) quo casu tres etiam asymptoti per centrum illud transibunt.(70)

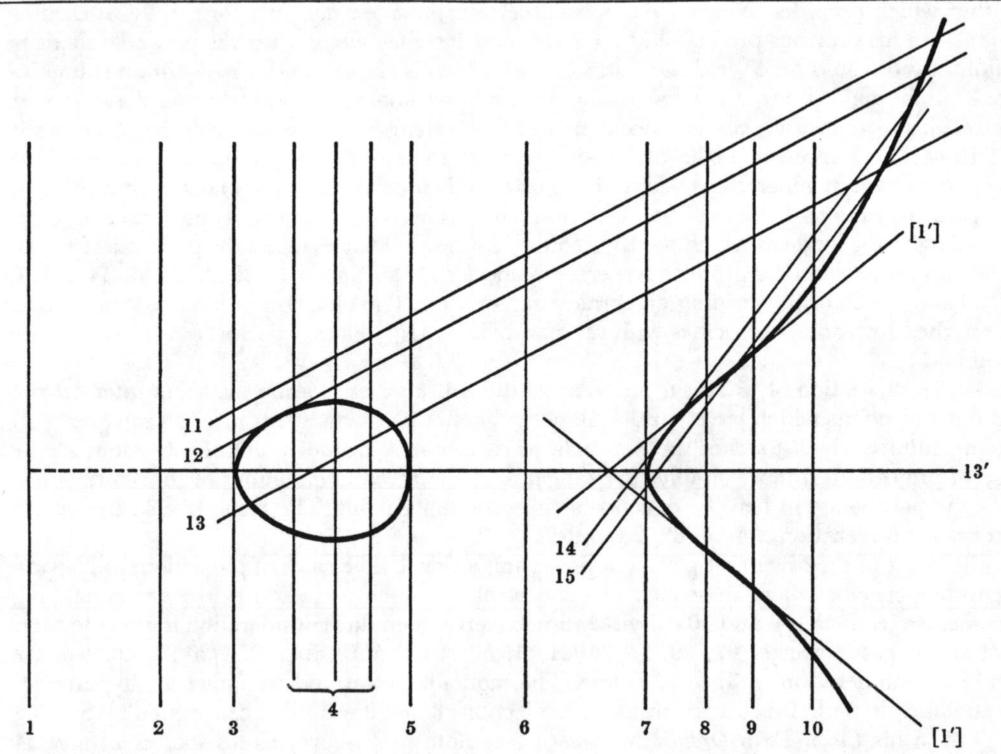

together in projection 13 not only of the cases where the 'horizontal' intersects the oval but also of the special instance—that marked 13' in our figure—where, in so doing, it comes to coincide with the parabola's axis (thus producing the central cubic which is species 27 in Newton's algebraic listing).

1, 2, 4–8 and 10–12 in the third kind except that in place of 'a conjugate point' you need to put 'a conjugate oval'.(66) In addition there are:

3. If the horizontal either in tending in the direction of the infinite branches or in passing through an inflection point touches the oval on its outer side, the projection will be a *balanced parabola with a shell which faces the parabola in its convexity*.

4. But if it cuts the oval, the projection will be a *double balanced hyperbola with an interposed shell*:(67) the simplest case of this is where the three asymptotes are concurrent in the same point.

5. While if it touches the oval on its inner side—that is, towards the bell-shaped parabola—the projection will be a *balanced parabola with a shell which faces the parabola in its concavity*.

12. If it touches the oval and neither tends in the direction of the infinite branches nor passes through an inflection, the projection will be an *unbalanced parabola with a shell twisted in an inflection round its asymptote*.

13. If it cuts the oval in two points and (the hyperbola) elsewhere in a third point beyond the inflection, the projection will be a *double unbalanced hyperbola with a shell twisted*(68) *in an inflection round an asymptote*; and will, furthermore, have a *centre* at that inflection if the horizontal passes through the three vertices of the projecting cubic(69)—in which case the three asymptotes too will pass through that centre.(70)

(66) In preliminary draft on Add. 3961.3: 18r Newton added at the end that 'Præterea in casu 4 & 13 tres asymptoti se in unico puncto secare possunt' (Furthermore, in cases 4 and 13 the three asymptotes can cut one another in a single point). In the former instance, namely, the 'horizontal' intersects the parabola's oval in points where it has parallel slope, and the tangents at which therefore meet on the line at infinity which touches the parabola in its infinite inflection; the latter arises where the 'horizontal' meets the parabola in three points, the tangents at which are either concurrent in a finite point or (case 13′ in our above figure) parallel, the points of intersection being the parabola's vertices. In the present revise, where this observation is separately appended to cases 4 and 13, Newton fails (see note (70) following) to make the generality of the latter occurrence clear.

(67) In his draft version on Add. 3961.3: 18r Newton specified little differently that 'Projectio erit *Hyperbola duplex bifida cum concha transversè interjecta*' (...a *double cleft* [*sc.* diametral] *hyperbola with a transversely interposed shell*), but in his present revise initially changed this to be '...*Hyperbola triplex librata, quarum Hyperbolarum una est Conchoidalis inter alias duas sita*' (a *triple balanced hyperbola, of whose hyperbolic branches one* which is *conchoidal is positioned between the other two*) before reverting to his prime designation.

(68) Originally '*flexa*' (*bent*).

(69) That is, when (in position 13′ in the figure of note (65) above) it coincides with the parabola's axis.

(70) For the tangents at the three vertices of the projecting parabola are parallel to the direction of its ordinates and so concur at its infinite point of inflection, which accordingly passes to be the centre of the projected cubic when the parabola's axis (the harmonic polar of that inflection; see note (30) above) is cast to infinity. Newton should here also, however, have mentioned that there are innumerable other projections of the generating parabola which

Atq; hæ sunt species linearum tertij ordinis quarum formas et particulares conditiones[71] fusius describere non operæ pretium duxi quoniam has ubi opus est Geometræ speculando formam situm et conditiones Lineæ Projicientis haud difficulter colligent.[72] Malui paucis[73] inventionem generaliorum proprietatum linearum aperire.[74]

Considero igitur quod quæ conveniunt duabus linearum speciebus convenire solent generi et quæ convenient duobus generibus convenire solent ordini et quæ conveniunt duobus ordinibus observato progressionis tenore convenire solent ordinibus universis: deinde quod combinatio [75]simpliciorum linearum quarū ordines conjuncti ascendunt ad ordinem superiorem[76] vicem obire potest lineæ illius ordinis superioris. Ut combinatio duarum linearum primi ordinis vicem lineæ secundi, combinatio trium, quatuor vel plurium vicem tertij quarti aut superioris ordinis, combinatio unius lineæ primi et unius secundi ordinis vicem lineæ tertij ordinis et sic in reliquis. Nam linea superioris ordinis sæpe transit in combinationem linearum simplicium et combinatio cujusvis ordinis secari potest a recta in tot punctis quot linea quævis ejusdem ordinis. Quæro igitur proprietates combinationum incipiendo a simplicioribus, & primum considero proprietates rectarum combinatarum in infinitum, deinde proprietates circuli vel alterius cujusvis notæ curvæ combinatæ cum rectis in infinitum. Nam quæ duobus combinationum generibus per ordines universos conveniunt, fieri vix potest quin conveniant lineis et linearum combinationibus universis.

Modus exprimendi lineas.

[77]Concipe rectam BC parallelo motu ad latus ferri et interea secare rectas quotcunq; positione datas AB in B, HP in P, IQ in Q, KR in R et punctis in ea mobilibus C, D, E, F alias lineas $c[C]$, $d[D]$, $e[E]$, $f[F]$ describere. Determinari

---

yield cubics having three concurrent asymptotes; namely those (species 26 in his following algebraic enumeration in **3**, 1, §2) which are its 'shadows' when the horizontal line meets the projecting parabola—and its companion oval—in separate triples of intersections, the tangents at which concur in a finite point. (Compare note (66) above.) In preliminary draft on Add. 3961.3: 18^r, we may add, Newton initially set down a further case of projection, premising 'Si [Linea Horizontalis] tangit Ellipsin & transit per flex[um] contr[arium]' (If the horizontal line touches the oval and passes through an inflection), but quickly discarded it with a jotted reminder that this is identical with 'cas. 3 & 5' above.

(71) 'proprietates' (properties) was first written.

(72) In a preliminary draft on Add. 3961.3: 13^r/13^v, reproduced in section [4] of the Appendix below, Newton initially at this point entered into just such a 'profuse' discussion of how the 'forms and particular conditions' of curves are transmuted by optical projection.

(73) Originally 'propositionibus quibusdam poristicis' (in certain poristic [*sc.* indeterminate] propositions).

(74) Newton first phrased this sentence a little differently to read 'Cæterum qua ratione generaliores linearum proprietates ex inventis particularibus eruantur non pigebit paucis insinuare' (However, by what method the more general properties of lines are to be dug out from particular ones discovered will not be irksome to suggest in a few words). A variety of preliminary versions of his ensuing considerations on the general properties of curves and their

And these are the species of lines of third order. Their shapes and particular circumstances(71) I have not considered it worth while to describe more elaborately, seeing that geometers will, when need be, arrive at them without much difficulty by observing the form, position and circumstances of the projecting line.(72) I have chosen rather to lay bare in a few words(73) the discovery of more general properties of curves.(74)

I consider, accordingly, that what obtains in common for two species of line usually obtains for their kind, and what obtains in common for two kinds usually obtains for their order, and—observing the sequence of progression—what obtains for two orders usually does so for all orders universally; and, again, that a combination of(75) simpler lines whose orders rise jointly to that of a higher one(76) can fill the place of that line of superior order: thus, for instance, a combination of two lines of first order can replace a line of second; a combination of three, four or more one of third, fourth or higher order; a combination of a line of first order and one of second can take the place of a line of third order; and the like in other cases. To be sure, a line of superior order often passes into a combination of simpler lines, while a combination of any order can be cut by a straight line in as many points as any line of the same order. I therefore seek out the properties of combinations, beginning with the simpler ones. And first I consider the properties of straight lines combined indefinitely, and thereafter the properties of a circle or any other known curve combined with straight lines indefinitely; for what obtains in common in two kinds of combinations through their entire orders can scarcely do other than obtain in the entirety of lines and combinations of lines.

A way of expressing lines.

(77) Conceive that the straight line BC is carried sideways in a parallel motion and in the meantime cuts any number of straight lines given in position—AB in B, HP in P, IQ in Q, KR in R—while by the mobile points C, D, E, F in it other lines cC, dD, eE, fF are described, the lengths BC, BD, BE and BF, however,

---

particular exemplifications are reproduced in sections [4]–[8] of the Appendix from the originals in Add. 3963.3: 13r–14v/ 3963.10: 107r/ 3963.3: 21r–26r/ 3963.15: 181r.

(75) 'duarum' (two) is deleted.

(76) Initially 'ad ordinem lineæ minus simplicis' (to the order of a less simple line).

(77) Newton here first began: 'Datis positione quotcunq rectis AB, $[f]F$, HP, IQ, KR concipe rectam BC super recta AB in dato angulo $A[B]C$ incedere et interea productam secare reliquas positione datas in F, P, Q, R, ac termino suo C lineam cC describere ea lege ut semper sint AB ad BC, PC ad BD, QD ad BE, RE ad BF in eadem ratione [...]' (Given in position any number of straight lines AB, fF, HP, IQ, KR, conceive that the straight line BC passes along on the straight line AB at the given angle $A\widehat{B}C$ so as all the while, when extended, to cut the remaining ones given in position in F, P, Q, R and at its end-point C to describe a line cC subject to the restriction that there shall always be AB to BC, PC to BD, QD to BE, RE to BF in the same ratio...). This clumsily phrased opening is inessentially, if a deal more smoothly, recast in its ensuing revise.

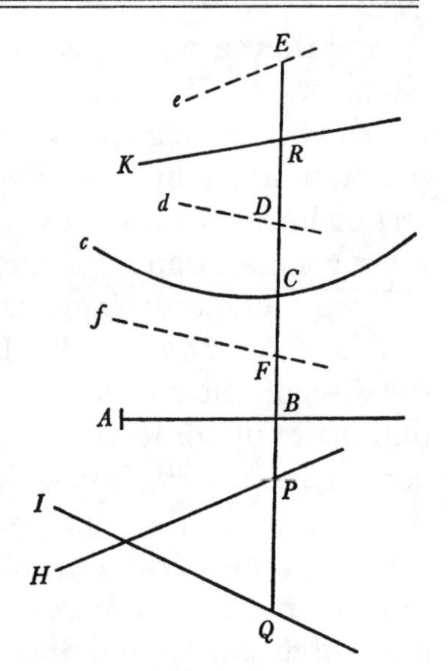

autem concipiantur longitudines BC, BD, BE, BF hac semper lege ut sint AB ad BC et PC ad BD et QD ad BE et RE ad BF in eadem ratione. Et si linea dD ad quam ratio secunda desinit recta est, tunc linea cC ad quam prima desinit erit secundi generis et aliquando primi nec ulla est linea secundi generis quæ non potest hoc modo exhiberi.(78) Sin linea eE ad quam ratio tertia desinit recta assumitur, linea cC ad quam prima desinit erit tertij generis et aliquando secundi vel primi.(79) Quod si linea fF ad quam quarta ratio desinit recta statuatur, tunc linea cC ad quam prima desinit erit aut quarti aut inferioris alicujus generis.(80) Et sic novas(81) in infinitum line[as desi]gnare licet, et numerus rationum gradum altissimum lineæ cC semp[er æqua]bit. Tot enim punctis & non pluribus possibile est Curvam illam cC a recta BC secari quot sunt rationes. Nam si verbi gratia tres sint rationes et linea BC detur positione dabuntur AB, BE, BP, BQ et BC invenienda erit ea lege ut sit $AB.BC::BC+BP.BD::BD+BQ.BE$[,] quod Problema triplicem admittere solutionem ex superioribus(82) constat adeoq linea BC triplex(83) est. Tria igitur et non plura possunt esse puncta C in quibus recta BC occurrat Curvæ cC, proinde Curva illa tertij est generis.

Facilius autem imaginamur has curvas ubi per motus locales linearum inter se cohærentium tanquam per organa quædam describi concipimus. Ut si(84) regulæ

(78) If we understand that the angle $A\widehat{B}C$ is allowed to vary in magnitude. For on specifying $AB = x$ and $BC = y$ to be the Cartesian coordinates of the locus-point $C(x, y)$, and thence assigning $BP = k(x-a)$ and $BD = l(b-x)$ as the respective equations of the straight lines HP and dD, Newton's stated proportion $AB:BC = PC:BD$ determines the defining relationship of the conic locus (C) to be

$$x/y = (y+k(x-a))/l(b-x) \quad \text{or} \quad y^2+ky(x-a)+lx(x-b) = 0.$$

By itself this equation contains only four independent constraints (the constants a, b, k, l) out of the five required to delimit an arbitrary conic. It is readily shown that when (most generally) there is $l = k^2a(b-a)/b^2$ the conic so defined degenerates to be the pair of the straight lines $y = k(a/b-1)\,x$ and $y = -k(a/b)\,(x-b)$.

(79) Newton has wisely deleted an initially following assertion—manifestly framed overhastily on the pattern of that which holds true in the preceding conic case—that 'neq ulla est linea tertij generis quæ non potest sic designari' (nor is there any line of third kind which cannot be thus denoted). Much as in the previous note, on assigning the equations of the straight lines IQ and eE to be respectively $BQ = m(x-c)$ and $BE = n(x-d)$ Newton's present proportion $AB:BC = PC:BD = QD:BE$ yields there to be

$$x/y = (y+k(x-a))/BD = (BD+m(x-c))/n(x-d),$$

whence on eliminating BD there comes

$$y^3+ky^2(x-a)+myx(x-c)-nx^2(x-d) = 0.$$

being determined by the perpetual restriction that AB shall be to BC, PC to BD, QD to BE and RE to BF ever in the same ratio. Then, if the line dD at which the second ratio ends up is a straight line, the line cC at which the first ceases will be of second kind and on occasion of first, nor is there any line of second kind which cannot be displayed in this way.(78) But if the line eE at which the third ratio terminates is assumed straight, the line cC at which the first one ends will be of third kind and sometimes of second or first.(79) While if the line fF at which the fourth ratio ceases be decreed straight, then the line cC at which the first one ends will be either of fourth or of some lower kind.(80) And in this manner we are free to designate fresh(81) lines indefinitely, the highest degree of the line cC being ever equal to the number of ratios—for it is possible for the curve cC to be cut by the straight line BC in as many points (and no more) as there are ratios. To illustrate, if for instance there be three ratios and the line BC be given in position, then AB, BE, BP, BQ will be given and we shall need to find BC subject to the restriction $AB: BC = (BC+PB): BD = (BD+QB): BE$; from the preceding(82) it is established that this problem admits a triple solution, and hence the line BC is triple.(83) There can therefore be three points C, and not more, in which the straight line BC may meet the curve cC, and consequently that curve is of third kind.

We may more easily picture these curves, however, when we conceive them to be described by means of the local motions of lines interconnected one with another as if by a kind of mechanism. For instance,(84) if the rulers PC, PD

The six constants a, c, d, k, m, n in this equation fail—even when the seventh freedom of varying $A\widehat{B}C$ is added—to furnish the nine conditions required to determine the general cubic curve. We need not bother to specify the condition for the cubic locus (C) to degenerate into the pair of a conic and a straight line—or indeed the line-triple which (as Newton omits to observe) it may also be.

(80) Much as in the two preceding notes, if we now take the equations of the straight lines KR and fF to be respectively $BR = p(x-e)$ and $BF = q(x-f)$ Newton's proportion $AB:BC = PC:BD = QD:BE = RE:BF$ defines a quartic locus (C) whose Cartesian equation $y^4+ky^3(x-a)+my^2x(x-c)+pyx^2(x-e)-qx^3(x-f) = 0$ readily ensues on eliminating BD and BE in the equivalent $x/y = (y+k(x-a))/BD = (BD+m(x-c))/BE = (BE+p(x-e))/q(x-f)$. The independent constants a, c, e, f, k, m, p, q taken with the ninth freedom of permitting $A\widehat{B}C$ to vary fall, of course, far short of supplying the fourteen conditions needed to determine a general curve of fourth algebraic 'kind'.

(81) So corrected by Newton from an over-confident 'omnes' (all) when he came in retrospect (compare note (79) above) to appreciate that his preceding generation of *lineæ curvæ* is far from adequate enough to construct the totality of algebraic curves.

(82) See pages 404–8 above.

(83) 'threefold' inasmuch as it is of third algebraic 'kind'.

(84) Newton yet once more outlines his favourite *descriptio organica* of curves from given ones of lower degree, setting fixed angles to rotate round the poles P and Q so as in one of their arms to intersect instantaneously at a point D of the describing curve, thence tracing at the meet C of their other arms the locus—of higher degree but with additional double points at one or

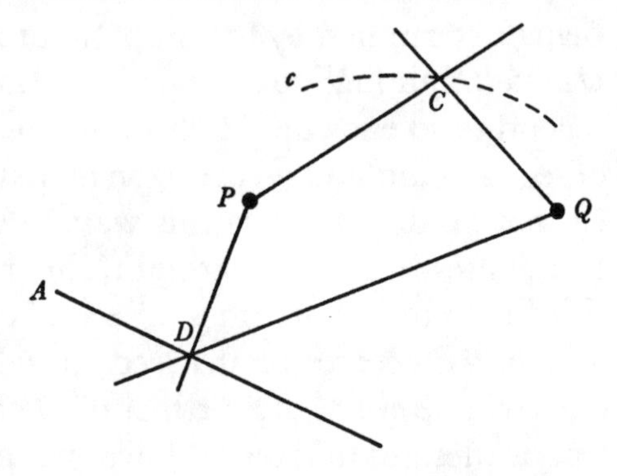

PC PD datum angulum *CPD* continentes volvantur circa datum punctum *P* quod in anguli illius vertice est et similiter regulæ *QC QD* datum angulum [*C*]*Q*[*D*] continentes circa punctū *Q* ea lege ut regulæ *PD*, *QD* se mutuo semper secent ad rectam aliquam lineam positione datam *AD* et interea reliquarum regularum *PC*, *QC* intersectio *C* motu suo lineam *cC* designet:(85) Erit hæc *cC* linea secundi gradus et aliquando primi.(86) Et hac ratione possunt omnes lineæ secundi gradus designari.(87) Deinde si loco rectæ *AD* substituatur linea aliqua secundi generis per neutrum punctorum *P*, *Q* transiens et regularum intersectio *D* in hac movere co[n]cipiatur, altera intersectio *C* designabit lineam quarti gradus aut etiam tertij.(88) Qua ratione et omnes tertij gradus lineas quarum commoda aliqua descriptio organica hactenus reperta fuit designare liceat.(89) Atq; ita ad lineas superiorum generum pergitur.(90) Quod idem fiet si regularum duæ *PC*, *PD* non volvantur circa polum *P* sed parallelo motu ferantur ita ut concursus earum *P* pergat in recta aliqua positione data vel etiam in curva jam ante descripta. Sed et ad majorem designandi copiam vice rectarum regularum adhiberi possunt curvæ.

Locus linearis puncti vagi.

Lineam vero ut *cC* in qua punctum aliquod indeterminatū ut *C* perpetuò reperitur Veteres dixerunt puncti illius locū, et quoniam problematum constructiones pendebant a descriptione(91) duorum locorum puncti quæsiti in quorum intersectione situm inveniretur, ideo Veteres ad hujusmodi locorum compositionem ut loquebantur[,] id est ad eorum inventionem ac determina-

both poles—to be described. He had given his earliest (and fullest) elaboration of the continuous correspondence thereby connecting the loci (*D*) and (*C*) some quarter of a century before. (See II: 106–50; compare also G. A. Shkolenok, 'Geometrical Constructions Equivalent to Non-linear Algebraic Transformations of the Plane in Newton's Early Papers', *Archive for History of Exact Sciences*, **9**, 1972: 22–44.)

(85) Originally 'describat' (shall describe). Several analogous replacements have been made by Newton in the immediate sequel in his manuscript text.

(86) Where, namely, the describing line *AD* passes through one or other of the poles *P*, *Q*.

(87) Initially 'describi' (...described); compare note (85). We will specify none of the further similar replacements made by Newton in sequel. His first rigorous proof of his present assertion—earlier attempted on II: 152–5 (see our comments thereon *ibid.*: 155–6)—was that set down as Problem 53 of his 'Arithmetica Universalis' (V: 304–6). In the preliminary draft reproduced in section [8] of the present Appendix he had momentarily been minded here to cite the particular constructions of conics to pass through five points or, equivalently, to touch given straight lines as he had earlier stated them on IV: 298–302.

(88) When, that is, for some point *D* of the describing conic the arms *PC*, *QC* come simultaneously to coincide with *PQ*.

containing the given angle $C\hat{P}D$ revolve round the given point P at the vertex of that angle, and the rulers QC, QD containing the given angle $C\hat{Q}D$ rotate similarly round the point Q, with the restriction that the rulers PD, QD shall mutually intersect at some straight line AD given in position, and as they do so the intersection C of the remaining rulers PC, QC shall trace(85) in its motion the line cC: this line cC will then be of second degree and, on occasion, of first.(86) And by this means all lines of second degree can be traced.(87) Next, if in place of the straight line AD there be substituted some line of second kind passing through neither of the points P and Q, and the rulers' intersection D be conceived to move in this, the other meet C will trace a line of fourth degree, or even one of third.(88) And by this means, also, we are at liberty to trace all lines of third degree of which some convenient 'organic' (instrumental) description or other has up to now been discovered.(89) In like manner we may pass to treat lines of higher kinds.(90) And the same will hold if the pair of rulers CP/PD no longer revolve round the pole P but should be carried in a parallel motion such that their meeting-point P travels along in some straight line given in position or, again, in a curve already previously described. Yet further to add to the rich profusion of this (mode of) tracing, instead of straight-line rulers curved ones can be employed.

The linear locus of a roving point.

A line such as cC, however, in which some indeterminate point like C is perpetually located was dubbed by the ancients the 'locus' of that point. And, seeing that their constructions of problems depended on the description(91) of a pair of loci of a required point at whose intersection this might be found situated, the ancients in consequence put their utmost effort into achieving the 'composition', as they spoke of it, of loci of this sort, that is, their discovery

(89) This is untrue unless Newton frames his criterion of convenience to exclude Descartes' variant 'organic' construction of his trident as the locus of the meets of a parabola sliding freely along its axis with a straight line rotating round a fixed point and constrained instantaneously to pass through a point in the former's (moving) axis (see pages 468–9 below, and II: 363, note (125)). Even where the describing conic is permitted to pass through one or other of the poles, the described cubic will in all cases have a real double point there, so that no 'pure' cubics such as the Cartesian 'hyperbole' are traceable by Newton's *descriptio organica*.

(90) A cancelled following qualification initially added the codicil 'licet omnes non possunt hoc modo describi' (though all cannot be described in this manner). Newton evidently thought better of here worrying his innocent reader by specifying the limitations of his *descriptio organica*, content now to let it appear that this technique is universally valid to construct all 'conveniently describable' higher curves. Let us not censure too harshly what was not necessarily mere deviousness on Newton's part: only too few of his contemporaries were capable or skilled enough to appreciate such subtleties.

(91) Originally 'solutiones pendebant ab inventione' (their solutions...depended on the discovery).

tionem summis viribus nitebantur. Duo autem hic requiruntur. Primum ut ex datis loci conditionibus sciamus qualis sit et quomodo describendus, deinde ut in quolibet problemate loca inveniamus quæ simplicissima sint et facillimè determinari ac describi possint. Sed antequam de his agamus, proprietates curvarum cognoscendæ sunt: quarum insigniores sunt hæ.

Curvarum proprietates generales.

Si parallelæ quotcunq quævis AC, DF, GI agantur seca[nte]s curvam quamvis in tot punctis A, B, C[;] D, E, F ac G, H, I quot curva e[jus] generis a rectis secari potest. Dein tertia agatur recta [dua]s e prioribus ita secans in K et L ut utriusq pars vel summa partium ad curvam extensarum ex uno latere æqualis sit parti vel summæ partium ad curvam extensarum ex altero latere, vizt

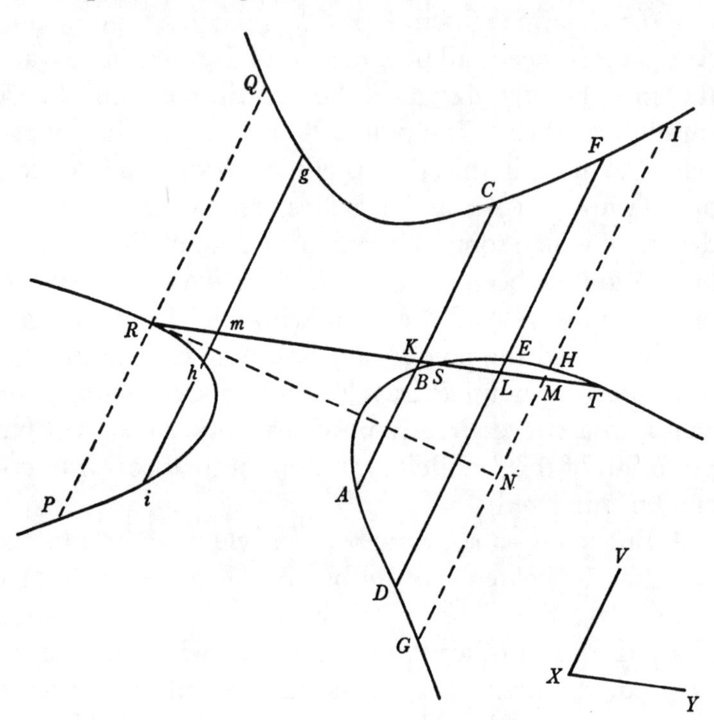

$$KA+KB = KC$$

et $LD = LE+LF$: tunc partes etiam reliquarum parallelarum hinc inde æquales erunt $MG = MH+MI$.(92) Lineā vero quæ parallelas ita secat in partes *Diametrum* Curvæ appellamus et partes illas *ordinatim applicatas* ad Diametrum.(93) Quinetiam si datis positione rectis VX, XY parallelæ duæ IG, RT utcunq ducantur secantes se mutuo in M[,] curvam vero in tot punctis quot recta curvam ejus generis secare potest puta in G, H, I et R, S T[:] contentum sub omnibus partibus unius rectæ inter curvam et alteram rectam sitis MG, MH,

(92) This is his Theorem 1 on II: 98–100. Much as there (see *ibid.*: 98–9, note (36)), if with respect to some fixed origin and abscissa we take the Cartesian equation of the given curve—of n-th algebraic degree, say—to be of the form $y^n-(ax+b)\,y^{n-1}+\ldots+lx^n+mx^{n-1}\ldots+p=0$ where the ordinates y are drawn parallel to ABC, then on setting $Y=\sum_{1\leqslant i\leqslant n}(y_i)/n$, that is, $(ax+b)/n$, it follows that the locus of the point (x, Y) is the rectilinear diameter $nY = ax+b$ whose property it is that $\sum_{1\leqslant i\leqslant n}(y_i-Y) = 0$. Manifestly, the intersections K and L of this with any two ordinates ABC and DEF will suffice to fix it uniquely in position.

(93) Newton initially went on to write: 'puncta item ubi Diameter secat curvam ut R, S, T

and determination. Two things, however, are here required: first, that from the given circumstances of a locus we should know what type it is and how it is to be described; and then that in any particular problem we should find out the loci which are simplest and can most readily be determined and described. But, before we discuss these matters, we need to be acquainted with the properties of curves: their more outstanding ones are the following.

General properties of curves.

If any number of any parallels AC, DF, GI be drawn cutting any curve in as many points A, B, C; D, E, F; and G, H, I as a curve of its kind can be cut by straight lines, and then a third straight line be drawn cutting two of the previous parallels in K and L such that the part or sum of the parts of each extending on one side as far as the curve be equal to the part or sum of the parts extending to the curve on the other, to wit (here) $KA+KB = KC$ and $LD = LE+LF$: the parts of the remaining parallels also will then be equal on either side, viz. $MG = MH+MI$.(92) The line, however, which cuts parallels into parts in this way we call the curve's *diameter*, and those parts *ordinates* to the diameter.(93) Furthermore, if parallel to the straight lines VX, XY given in position any two IG, RT be arbitrarily drawn intersecting one another in M and the curve in as many points—G, H, I and R, S, T say—as a straight line can cut a curve of its kind: the product contained by all parts MG, MH, MI of one straight line

---

Vertices ejus, et verticum duarum quarumvis intervallum *latus transversũ* et partem Diametri inter verticem quamvis et ordinatim applicatam *segmentum* Diametri. Quibus nominibus analogiam inter Conicas sectiones et superiores figuras insinuare volui. Porrò ubi Diameter curvam in tot punctis secat quot ipsam recta secare potest et ab aliqua ordinatim applicatarum ita dividitur ut segmentum vel summa segmentorum ex uno latere æqualis sit segmento vel summæ segmentorum ex altero latere, illa ordinatim applicata dicetur *Diameter conjugata*, Diametrorum verò intersectio *centrum* vel unum ex *centris*, et lineæ quævis *latera recta* quæ ita sunt ad latera transversa ut contentũ sub omnibus ordinatim applicatis ad contentũ sub omnibus segmentis diametri[;] verbi gratia ut contentum sub MG, MH, MI ad contentũ sub MR, MS, MT. Nam in omnibus figuris contentum sub ordinatim applicatis est ad contentum sub segmentis diametri in data ratione si modò applicatarum numerus pro genere figuræ plenus est' (Likewise, the points in which a diameter cuts the curve—such as R, S, T—are its *vertices*, the distance between any two vertices is a '*transverse side*'(*principal diameter*), and the portion of a diameter between any vertex and an ordinate is a *segment* of the diameter. By these names I wanted to convey the analogy (existing) between conics and higher figures. Furthermore, when a diameter cuts a curve in as many points as a straight line can cut it, and it is so divided by some ordinate that the segment or sum of segments on one side is equal to the segment or sum of segments on the other side, that ordinate will be said to be the *conjugate diameter*, while the meet of diameters is a *centre* or one of the *centres*, and *latera recta* are any (straight) lines which are to 'transverse sides' as the product contained by all the ordinates to that contained by all the segments of a diameter—for example, as the product contained by MG, MH and MI is to that contained by MR, MS and MT—since, to be sure, in all figures the product contained by the ordinates is to that contained by the segments of a diameter in a given ratio, provided the number of applicates [*sc.* ordinates/segments] is the full one for the kind of figure). Compare II: 94 and IV: 354–8.

MI erit ad contentum sub omnibus ejusmodi partibus alterius rectæ *MR*, *MS*, *MT* in data ratione.(94)

Tangentes ad Curvas descriptas ducere.

Et hinc recta duci potest quæ curvam quamvis descriptam in puncto imperato tanget, secetve in dato angulo. Sit illud punctum *R*. Per quod age duas quasvis rectas *RP*, *RT* et uni earum *RP* parallelā *IG* secantem alteram *RT* in *M*, quæ omnes etiam secent Curvam in pleno numero punctorum *P*, *R*, *Q*; *R*, *S*, *T*; *G*, *H*, *I*. In *IG* cape *MN* ita ut sit contentum sub *PR*, *QR*, *MN* ad contentum sub *GM*, *HM*, *IM* ut contentum sub *RS*, *RT* ad contentum sub *MS*, *MT*, et acta *RN* tanget curvam in *R*, si modo [*M*]*N* capiatur in eo angulo *PRM* quem curva secat. Nam concipe *ig* parallelam esse *PQ* et ad eam accedere interea dum secat curvam in *g*, *h*, *i* et evanescentium *Rm*, *mh* ultima ratio erit ea quæ est *RM* ad *MN* ubi *RN* tangit curvam. Est autem contentum sub *gm*, *hm*, *im* ad contentum sub *GM*, *HM*, *IM* ut contentum sub *mR*, *mS*, *mT* ad contentum sub *MR*, *MS*, *MT*. Hic pro ratione *hm* ad *mR* substitue æquipollentem rationem *NM* ad *MR* et coalescentibus lineis *PQ* et *ig* scribe *R* pro *m*, *Q* pro *g* et *P* pro *i* et incides in proportionē qua tangentem *RN* determinavimus.(95)

De cruribus infinitis et Asymptotis curvarum.

Cæterum hæ lineæ utplurimum crura habent in infinitū serpentia quæ aut Hyperbolici sunt generis aut Parabolici. Concipe punctum *B* secundum lineæ curvæ crus *AB* delatum abire in infinitum et interea curvam a linea mobili *BC* semper tangi. Incidat autem semper a puncto aliquo *G* in tangentem illam perpendiculum *GC*, et ubi punctum *B* in infinitū abit

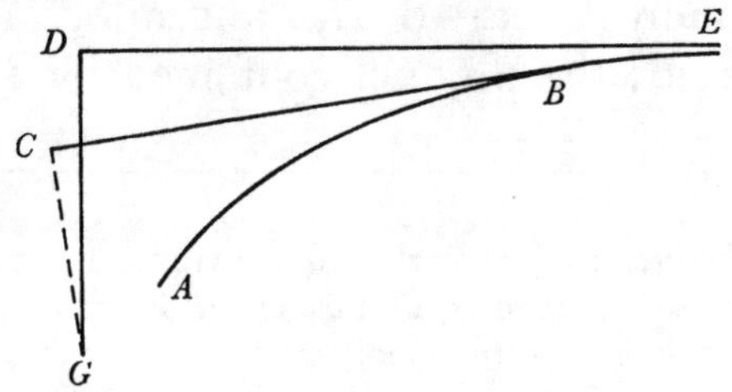

si *GC* fit infinitè longum tangente *BC* prorsus evanescente(96) crus illud *AB* Parabolicum est, sed si *GC* non fit infinitè longum, crus Hyperbolicum est et tangens in ultima positione seu recta illa (uti *DE*) quacum tangens ultimò convenit, cruris illius *Asymptotos* appellatur. Crus vero infinitum semper habet socium suum qui nunc ad eandem(97) nunc ad oppositam plagam tendit. Et paris Hyperbolici semper eadem est Asymptotos. Estq; hæc Assymptotorum insignis proprietas, quod si curva cujusvis generis plenè Hyperbolica(98) secetur a recta in pleno numero

(94) Newton had originally enunciated this general theorem some fifteen years earlier as the last of an analogous set of 'proprietates insigniores Conicarum sectionum [quæ] curvis etiam superioris ordinis conveniunt' (see IV: 358). Much as before (*ibid.*: note (18)) we may continue the terms of note (92) above straightforwardly to remark that, corresponding to any given x, there is $\prod_{1\leqslant i\leqslant n}(y_i) = lx^n+mx^{n-1}\ldots+p\equiv l\times\prod_{1\leqslant i\leqslant n}(x-x_i)$, where $(x_i, 0)$, $i = 1, 2, \ldots, n$, are the points in which the abscissa ($y = 0$) cuts the curve *in pleno numero punctorum*; whence at once there ensues $\prod_{1\leqslant i\leqslant n}(y_i)/\prod_{1\leqslant i\leqslant n}(x-x_i) = l$, constant.

(95) While the present application of his preceding theorem on the constancy of the ratio of the products of intercepts cut out in its ordinate and abscissa by a general algebraic curve is novel, Newton had earlier similarly applied its quadratic Apollonian instance (*Conics*, III,

between the curve and the other line will then be to the product by all parts *MR*, *MS*, *MT* of the sort in the other line in a given ratio.[94]

To draw tangents to described curves.

And hence a straight line can be drawn which shall touch any described curve in an ordained point, or cut it at a given angle. Let that point be *R*. Through it draw any two straight lines *RP*, *RT* and parallel to one of them, *RP*, draw *IG* cutting the other, *RT*, in *M*; and let all these also cut the curve in their full number of points *P*, *R*, *Q*; *R*, *S*, *T*; *G*, *H*, *I*. In *IG* take *MN* such that the product contained by *PR*, *QR*, *MN* shall be to that contained by *GM*, *HM*, *IM* as the product of *RS*, *RT* to that of *MS*, *MT*, and when *RN* is drawn it will touch the curve at *R* provided *MN* be taken in the angle $P\widehat{R}M$ which the curve cuts. For conceive *ig* to be parallel to *PQ* and to approach it while all the time it cuts the curve in *g*, *h* and *i*, and the last ratio of the vanishing segments *Rm*, *mh* will be that of *RM* to *MN*, where *RN* touches the curve. But the product contained by *gm*, *hm*, *im* is to that contained by *GM*, *HM*, *IM* as the product of *mR*, *mS*, *mT* to that of *MR*, *MS*, *MT*. Here in place of the ratio *hm* to *mR* substitute the equivalent ratio *NM* to *MR*, and with the lines *PQ* and *ig* coalescing write *R* in place of *m*, *Q* for *g* and *P* for *i*, and you will arrive at the proportion by which we determined the tangent *RN*.[95]

On the infinite branches and asymptotes of curves.

For the rest, these lines for the most part have branches snaking away to infinity which are either of hyperbolic or parabolic kind. Conceive the point *B* to go off to infinity following the branch *AB* of a curved line, and all the while the mobile line *BC* to touch the curve. To that tangent, however, from some point *G* let there ever fall the perpendicular *GC*; and then, as the point *B* goes off to infinity, should *GC* become infinitely long—the tangent *BC* entirely disappearing[96]—that branch *AB* is parabolic, but if *GC* does not become infinitely long the branch is hyperbolic and the tangent in its ultimate position, that is, the straight line (such as *DE*) with which the tangent ultimately coincides, is called the *asymptote* of that branch. An infinite branch, however, always has its partner tending now in the same direction,[97] now in the opposite one. And a hyperbolic pair always possess the same asymptote. To asymptotes belongs this outstanding property that, if any fully hyperbolic[98] curve of any

17/18) analogously to construct the tangent at a conic's general point in a variant resolution of Problem 56 of his 'Arithmetica Universalis' (see V: 318 and 319, note (438)). In the preliminary draft of his present text which is given in section [6] of the following Appendix he refines the particular application in the conic case to yield an elegant derivation of the basic pole-polar property (*Conics*, I, 34/36/37); see our note (44) thereto. Little before, we may remind, he had contrived the related application of his theorem on products of intercepts which equally neatly constructs the chord of the osculating circle at an arbitrary point of a general algebraic curve and thence determines its curvature there (see VI: 606–8).

(96) Understand 'in infinitum' (to infinity), whence it vanishes from our finite ken.

(97) When, namely, the point at infinity is an inflection or a cusp.

punctorū G, H, I et recta illa secet etiam omnes Asymptotos puta AB, AC, BC in D, E et F, pars vel summa partium rectæ ab Asymptoto vel Asymptotis ad crura totidem versus unam plagam extensarū æqualis est parti vel summæ partium similium a reliquis Asymptotis versus alteram plagam ad reliqua crura tendentium

$$DG = EH + FI,$$

vel $EG + FH = DI$.(99)

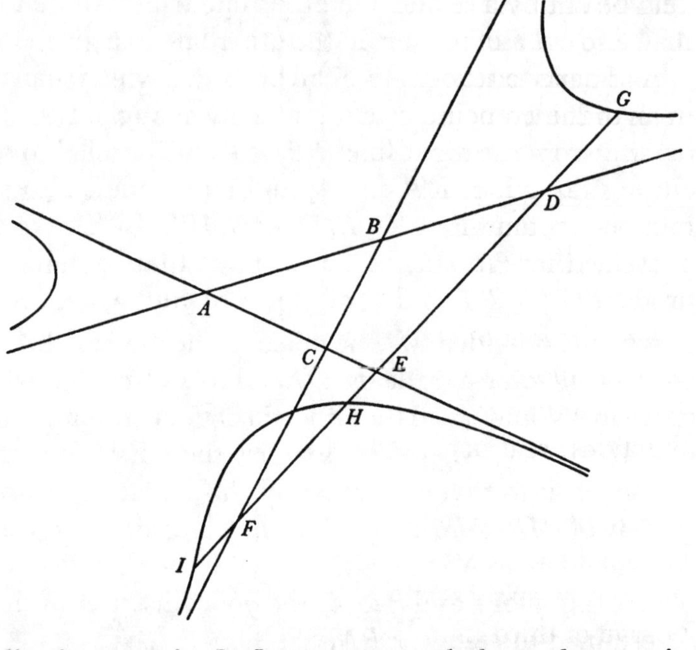

Curvā(100) verò plenè Hyperbolicam voco qua alia ejus generis c[ur]va non potest habere plura paria crurum Hyperbolicorum. Habet autem tria paria si sit tertij generis, quatuor si quarti et sic deinceps.(101)

Quomodo curvæ in species distinguendæ.

Ex crurum infinitorum numero et diversitate pendet distinctio curvarum in species(102) principales. Sunt autem alia crura conspirantia seu ad eandem plagam tendentia, alia divergentia seu vergentia in oppositas plagas[,] et utraꝗ rursus vel ad easdem partes convexa vel ad contrarias. Par crurum Parabolicorū et Hyperbolicorum conspirantium æquipollet duobus paribus Hyperbolicorum divergentium, et par Parabolicorum divergentium(103) æquipollet tribus: saltem in curvis tertij generis.(104) Unde sciri potest quot crura cujusvis generis curva quævis habere potest. Ut si curva tertij generis habeat par sive Parabolicum sive Hyperbolicum conspirans: non habebit nisi aliud par Hyperbolicum divergens. Sed et Ellipses conjugatæ considerandæ sunt, quæ et aliquando in puncta conjugata contrahuntur,(105) aliquando prorsus evanescunt: Asymptotorum item situs, an parallelæ sunt vel inclinatæ, et aliæ quædam differentiæ minoris

(98) That is, possessed of the full quota of hyperbolic infinite branches—and hence of the full number of related asymptotes—as Newton explains in sequel.

(99) This is Newton's Theorem 2 on II: 102, here trivially generalized from its cubic instance. The proof is immediate. Since, in the figure on page 440 above, KLM is evidently the Newtonian diameter not only of the given curve but of the line-multiple formed by the set of chords $AD/BE/CF/\ldots$ (because for each there is

$$KA + KB + KC + \ldots = 0 = LD + LE + LF + \ldots,$$

due account being taken of directed sense), in the limiting case where DEF comes to coincide

kind be cut by a straight line in the full number of points G, H, I and that line shall also cut all the asymptotes, AB, AC, BC say, in D, E and F, then the part or sum of parts extending from the asymptote or asymptotes an equal number of times to the branches in one direction is equal to the part or sum of similar parts tending from the remaining asymptotes to the remaining branches in the other: $DG = EH+FI$, or $EG+FH = DI$.(99) I call a curve(100) 'fully hyperbolic', however, when another curve of its kind cannot have more pairs of hyperbolic branches than it does. It has three pairs if it be of third kind, four if of fourth, and so on.(101)

How curves are to be distinguished into species.

The distinction of curves into their principal species(102) depends on the number and diversity of their infinite branches. Some branches are in unison, that is, tending in the same direction; others are divergent or verging in opposite directions; and both, again, may be convex the same way or in contrary ones. A pair of parabolic and hyperbolic branches in unison are equivalent to two pairs of divergent hyperbolic ones, and a pair of divergent(103) parabolic ones is equivalent to three—in curves of third kind at least.(104) From this it can be known how many branches of any kind any curve can have. Should, for instance, a curve of third kind have a pair, either parabolic or hyperbolic, in unison, it will have no other except a hyperbolic pair. But conjugate ovals, too, are to be considered, and these may also on occasion contract into conjugate points(105) and sometimes completely vanish: likewise, the lie of asymptotes, whether they are parallel or inclined to one another, and certain other differences of lesser

---

with ABC it is manifestly also the diameter of the n-ple of tangents at A, B, C, ...; whence, if these last meet GHI in g, h, i, ... respectively, there is in consequence

$$MG+MH+MI+\ldots = 0 = Mg+Mh+Mi+\ldots$$

and therefore $Gg+Hh+Ii+\ldots = 0$. Newton's theorem is the instance of this where ABC is transposed to lie at infinity, and hence the tangents at A, B, C, ... pass into an equal number of asymptotes to the curve.

(100) 'Figuram' (figure) was first written.

(101) Newton has cancelled a final sentence where he initially added 'Et par crurum Parabolicorum æquipollet duabus paribus crurum Hyperbolicorum' (And a pair of parabolic branches is equivalent to two pairs of hyperbolic ones)—true, of course, since each pair of hyperbolic branches corresponds to a single intersection of the curve with the line at infinity, while a parabolic pair occurs where the curve is touched by it, that is, has a 2-point contact with it there.

(102) Originally 'gradus' (grades).

(103) Newton first adjoined 'et ad contrarias partes convexorum' (and contrariwise convex).

(104) In the respective instances of the Cartesian trident and 'pure' Wallisian cubic, for example.

(105) 'cujusmodi est polus Conchoidis' (such as the conchoid's pole is) is deleted—rightly so since this instance of a conjugate point is formed from the contraction of a nodal loop which passes into a cusp and then splits off from the main curve. See our figures illustrating the three cases on I: 503–4.

notæ(106) quas enumerare non operæ pretium est. De lineis enim superiorum generum fuse disserere non est instituti. Qua de causa nec in demonstrandis quæ dicta sunt tempus conteram. Volui tantum quæ de lineis secundi generis ab Apollonio et alijs demonstrata habentur ita commemorare ut eadem lineis etiam superiorum generū competere insinuarem.

De curvarū tangentibus.

Cæterum quoniam cognitio ac determinatio curvarum maximè pendet a cruribus infinitis, hæc autem cum eorum Asymptotis noscuntur ex tangentibus; tum etiam quia tangentium inventio posthac alijs inserviet usibus: methodum jam subjungam ducendi rectas quæ curvas quasvis nondum descriptas postquam describuntur tangent. Sed notæ quibus in operationibus Geometricis(107) utimur sunt prius explicandæ.

Notarum quarundam explicatio.

Ubi linea aliqua AB ducitur in aliam lineam CD rectangulum genitum significamus scribendo AB' CD et si id rursus ducatur in tertiam lineam EF ad exprimendum parallelipipedum genitum scribimus AB' CD' EF.(108) Latus verò quod oritur applicando rectangulum illud ad lineam quamvis GH sic notamus $\frac{AB'CD}{GH}$. Atq; ita in alijs. Sed et exposita linea aliqua ad quam tanquam mensuram universalem aliæ omnes (ut fit in decimo *Elementorum*(109)) referantur scribimus AB' CD ad designandam quartam proportionalem ab hac linea ubi mediæ duæ sunt AB et CD, et AB' CD' EF ad designandam etiam quartam ab eadem linea ubi mediæ duæ AB' CD et EF, et sic in infinitum. Et si linea illa sit prima continuè proportionalium et alia quævis AB secunda, tertiam sic designamus AB^2, quartam sic AB^3, quintam sic AB^4 atq; ita deinceps. Et inter lineam illam et aliam quamvis AB notamus mediam proportionalem sic $AB^{\frac{1}{2}}$, primam e duabus medijs sic $AB^{\frac{1}{3}}$ secundam sic $AB^{\frac{2}{3}}$ [&c]. Similiter $\overline{AB'CD}^{\frac{1}{2}}$ denotat tum latus quadrati æqualis rectangulo AB' CD[,] tum mediam proportionalē inter mensuram illam universalem et AB' CD vel quod perinde est inter AB et CD. Sed $AB'CD^{\frac{1}{2}}$ est quarta proportionalis a mensura illa ubi mediæ sunt AB et $CD^{\frac{1}{2}}$. Et has quantitates nominibus usitatis significamus præterquam quod a vocabulis Arithmeticis certas ob rationes cum Veteribus abstinendum esse duximus. Porro quantitates compositæ eodem modo signantur Sic $\overline{A+B}'\overline{C-D}$ denotat rectangulum sub $A+B$ et $C-D$, et $\overline{A\pm B}^2$ quadratum ipsius $A\pm B$. Quæ quantitates

(106) Initially just 'minutiores' (more minute).

(107) Newton originally specified these to be 'tam Analyticis quam syntheticis' (both analytic and synthetic) and more accurately qualified his usage of the following symbols as being 'nonnunquam' (sometimes); compare the next note.

(108) Where, however, single elements X, Y, Z, ... are multiplied together, Newton usually continues in sequel the standard practice of denoting their products merely by conjoining them—as XY, XYZ, and so on—without inserting any intervening symbols of multiplication whatever.

(109) Whose Definition 1, for instance, declares in Barrow's Latin version (the one which

significance[106] which it is not worth while to list. On lines of higher kinds, of course, it is not my intention elaborately to discourse. Nor for this reason shall I waste time in demonstrating what has been asserted. I wanted merely to recall what is held demonstrated by Apollonius and others in regard to lines of second kind so as to suggest that the same holds for lines of higher kinds also.

On the tangents of curves.

For the rest, seeing that the ascertaining and determination of curves mostly depends on the infinite branches, while these together with their asymptotes are known from their tangents, and then, again, because the finding of tangents will subsequently serve other purposes, I will now subjoin a method of drawing straight lines to touch any curves not yet described after they are described. But the symbols we use in geometrical operations[107] need first to be explained.

The explanation of certain symbols.

When some line AB is drawn into (multiplied by) another line CD we denote the rectangle engendered by writing $AB'CD$ ($AB \times CD$), and if this should be again drawn into a third line EF, to express the 'parallelepiped' engendered we write $AB'CD'EF$ ($AB \times CD \times EF$).[108] The 'side', however, which ensues on applying the former rectangle to any line GH we designate thus: $AB \times CD/GH$. And the like in other cases. But also, when some line is exhibited to which, a universal measure as it were, all others may (as happens in the tenth book of the *Elements*[109]) be related, we write $AB \times CD$ to denote the fourth proportional to it where the two means are AB and CD, and $AB \times CD \times EF$ to denote the fourth one also to this line where the two means are $AB \times CD$ and EF, and so on indefinitely. And if that line be the first of a string of continued proportionals, and any other, AB, be the second, we denote the third one thus: AB^2, the fourth thus: AB^3, the fifth thus: AB^4, and so on. And between that line and any other, AB, we designate the mean proportional thus: $AB^{\frac{1}{2}}$, the first of two means thus: $AB^{\frac{1}{3}}$ and the second thus: $AB^{\frac{2}{3}}$, and so forth. Similarly, $(AB \times CD)^{\frac{1}{2}}$ denotes both the side of a square equal to the rectangle $AB \times CD$, and also the mean proportional between the universal measure and $AB \times CD$, or what is exactly the same between AB and CD. But $AB \times CD^{\frac{1}{2}}$ is the fourth proportional to that measure when the means are AB and $CD^{\frac{1}{2}}$. These quantities we indicate by their customary names except that for certain reasons we have, along with the ancients, thought fit to abstain from using arithmetical language. Further, compound quantities may be designated in the same manner. Thus,

$$(A+B)\,(C-D)$$

denotes the rectangle contained by $A+B$ and $C-D$, and $(A \pm B)^2$ the square of $A \pm B$. When their parts are drawn into one another according to the second

Newton himself would doubtless here cite) that 'Commensurabiles magnitudines dicuntur, quas eadem mensura metitur' (*Euclidis Elementorum Libri XV. breviter demonstrati* (Cambridge, ${}_1$1655): 190). In following illustrations (*ibid.*: 190, 191) a 'pes' (foot) is employed as the unit of linear measure.

etiam partibus juxta secundū *Elementorum*(110) in se ductis sic scribuntur $A'C+B'C-A'D-B'D$ et $A^2\pm 2A'B+B^2$. Ubi notes quod pars positiva ducta in subductitiam vel subductitia in positivam producit subductitiam[,] duæ vero subductitiæ in se ductæ producunt positivam. Secentur AB AD parallelis BD, EG ita ut sit AB ad AD ut AE ad AG, et posita AB mensura illa ad quam lineæ omnes referuntur AG erit $AD'AE$. Diminuatur AE donec evanescat et postea evadat

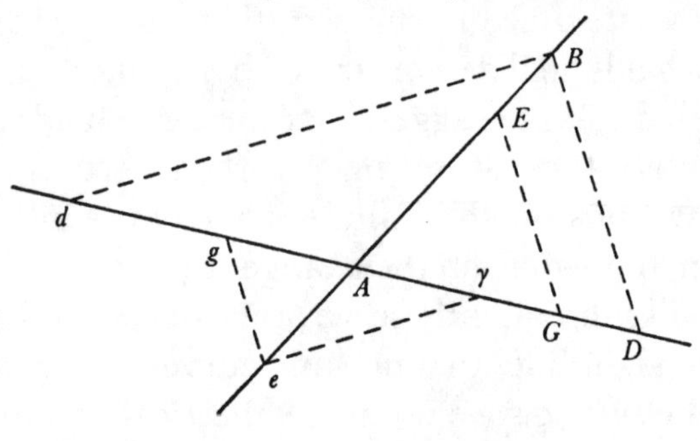

retrorsa Ae et AG simul diminuetur evanescet & convertetur in retrorsam Ag.(111) Diminuatur etiam AD donec evanescat et postea retrorsa evadat Ad, et retrorsa Ag simul diminuetur evanescet et convertetur in directam $A\gamma$. Duæ igitur retrorsæ Ae, Ad bis retrorsam id est directam $A\gamma$(112) efficiunt.

Fluxiones quantitatum.

Notis præcognitis, præmittenda est etiam methodus determinandi fluxiones linearum et fluxionum plagas. Per *Fluxionem* intelligo celeritatem incrementi vel decrementi lineæ cujusvis indeterminatæ ubi lineæ aliquæ super alias in descriptione curvarum moveri concipiantur et inter movendum (113)augeri vel diminui aut motu punctorum describi. Unde et indeterminatas illas quantitates *fluentes* nominare licebit. Proponantur datæ aliquot quantitates A, B, C, D et fluentes V, X, Y, Z quarum fluxiones respective designent minusculæ v, x, y, z.(114) et requiratur fluxio lineæ alicujus quæ ex his fit ut lineæ $X'Y$. Maneat primum X et fluat Y donec ipsa fiat Z et XY fiat XZ[,] et quia Y et XY fluendo non mutant rationem fluxiones earum erunt ut ipsæ hoc est ut 1 ad X. Unde cum fluxio Y sit y fluxio XY erit Xy. Maneat jam Z et fluat X donec ipsa fiat V et XZ fiat VZ et fluxio XZ erit Zx ut in casu priore. Fluant jam X et Y simul motu priore ut XY una vice fiat VZ et quia fluxio Xy sufficit ad mutandum XY in XZ et fluxio Zx ad mutandum XZ in VZ[,] Fluxio tota qua XY mutatur in VZ erit $Xy+Zx$. Pone V æqualem X et Z æqualem Y ut fit ipso fluendi initio et fluxio initialis ipsius XY erit $Xy+Yx$.

---

(110) *Elements* II, 1–9; compare Newton's earlier reworkings of these propositions on III: 402–4.

(111) That is, $-gA = (AD\times -eA$ or) $AD\times Ae$.

(112) Equal to $(AD\times eA$ or) $-Ad\times -Ae$, namely.

(113) 'quomodocunꝗ' (in any manner whatever) was initially added. As in his previous informal definitions of the fluxion of a continuously 'flowing' quantity (compare III: 72; IV: 422, 566) Newton yet again makes crucial appeal to his reader's intuitive understanding of 'instantaneous' speed.

(114) Much as in his earlier exposition on IV: 566 Newton originally specified '...fluentes x, y, z quarum fluxiones respectivè designentur r, s, t' (fluents x, y, z whose fluxions shall

book of the *Elements*,(110) these quantities may also be written thus:

$$A \times C + B \times C - A \times D - B \times D \quad \text{and} \quad A^2 \pm 2A \times B + B^2.$$

Here note that a positive part drawn into a subtractive one or a subtractive one drawn into a positive produces a subtractive, but two subtractive ones drawn into each other produce a positive. Let AB, AD be cut by the parallels BD, EG so that AB shall be to AD as AE to AG, and when AB is set as the measure to which all lines are related AG will be $AD \times AE$. Let AE diminish till it vanishes and thereafter comes to be the reverse magnitude Ae, and then AG will simultaneously vanish and be converted into the reverse one Ag.(111) Let also AD diminish till it vanishes and thereafter comes to be the reverse line Ad, and the reverse line Ag will simultaneously vanish and be converted into the direct one $A\gamma$. The two reverses Ae and Ad, therefore, achieve the twice reversed, that is, direct, line $A\gamma$.(112)

The fluxions of quantities.

Once these notations are presupposed, we need also to premise a method of determining the fluxions of lines and the directions of those fluxions. By a *fluxion* I understand the speed of increment or decrement of any indeterminate line where in the describing of curves some lines are to be conceived to move upon others and, all the while they move, to increase or diminish or to be described by the motion of points.(113) Whence also it will be permissible to name these indeterminate quantities *fluents*. Let there be proposed any number of given quantities A, B, C, D and fluents V, X, Y, Z, the fluxions of which shall respectively be denoted by lower-case v, x, y, z,(114) and let there be required the fluxion of some line made up out of these—for instance, the line $X \times Y$. Let the first magnitude X stay fast and Y flow till it becomes Z, and so XY becomes XZ; and then because Y and XY in flowing do not change their ratios, their fluxions will be as they themselves are, that is, as 1 to X. Hence, since the fluxion of Y is y, the fluxion of XY will be Xy. Now let Z stay fast and X flow till it becomes V, and so XZ becomes VZ, and the fluxion of XZ will be Zx, as in the previous case. Now let X and Y flow simultaneously with their previous motion, so that XY comes at one go to be VZ, and then, because the fluxion Xy suffices to change XY into XZ and the fluxion Zx is sufficient to change XZ into VZ, the total fluxion whereby XY is changed into VZ will be $Xy + Zx$. Set V equal to X and Z equal to Y as happens at the very start of the flowing, and the initial fluxion of XY will be $Xy + Yx$.

respectively be designated r, s, t), passing in sequel (on f. 135r) to give variant proof of the following algorithm for deriving the fluxion of a product by letting x, y, z 'flow' to become $x + or$, $y + os$, $z + ot$ and thereafter considering the 'last ratios' which ensue when the fluent increments or, os, ot flow 'back' to become vanishingly small. See section [9] of the Appendix, especially the preliminary draft cited in our note (99) thereto.

(115)Proponatur jam factum XYZ et ponendo $XY = V$ erit $XYZ = VZ$. Cujus fluxio juxta casū priorē est $Vz+Zv$. Sed et ob $XY = V$ est $Xy+Yx = v$. Pro V et v substitue æquipollentia et $Zz+Zv$ hoc est fluxio ipsius XYZ fiet

$$XYz+XZy+YZx.$$

Et progressionis modum observando colligitur universaliter quod facti cujuscunq ut $VXYZ$ fluxio invenietur substituendo sigillatim in facto illo pro unoquoq factore fluxionem ejus et sumendo resultantium terminorum aggregatum. Quæ regula etiam obtinet ubi aliqui factores æquales sunt. Ut si X et Y æquales sint ita ut XY valeat X^2, ejus fluxio $Xy+Yx$ fiet $2Xx$. Et similiter ipsius X^3 fluxio est $3X^2x$ et ipsius X^2Z fluxio $X^2z+2XZx$. Atq ita in compositis fluxio ipsius $AX-3X^2$ est $Ax-6Xx$. Nam fluxiones partium simul sumptæ sunt fluxio totius.

In lateribus applicatorum ad fluentia methodus hæc est. Proponatur latus $\frac{X^2}{Y}$. Pone ipsum æquale V et erit $XX=YV$ adeoq $2Xx=Yv+Vy$. Nam fluentium semper æqualium fluxiones æquales sunt. Aufer utrobiq Vy et reliquum divisum per Y nempe $\frac{2Xx-Vy}{Y}$ erit v. Est autem v ipsius V id est ipsius $\frac{X^2}{Y}$ fluxio quam invenire oportuit.

Similis est methodus in lateribus quadraticis, cubicis alijsq.(116) Proponatur latus(117) $\overline{AX-X^2}|^{\frac{1}{2}}$. Pone ipsum æquale V, et erit $AX-X^2 = V^2$, adeoq

$$Ax-2Xx = 2Vv \quad \text{et} \quad \frac{Ax-2Xx}{2V} = v.$$

[Fluxiones linearum et angulorum.]

(118)In figuris pro significanda fluxione lineæ alicujus pono lineam illam literis minusculis: ut *bc* pro significanda fluxione lineæ *BC*. Angulorum verò fluxiones expono per fluxiones arcuum quibus subtenduntur ad datam distantiam. Et distantiam illam quæcunq tandem assumatur designo per literam R; fluxionem arcus per angulum literis minusculis scriptum, oppositam ordinatim applicatam in hoc circulo, id est sinum hujus arcus per literam s angulo præfixam, distantiam ordinatæ a centro id est sinum complementi

(115) Content initially with a mere statement of the ensuing extension to triple and higher products, Newton first went on: 'Simili argumentatione fluxio ipsius XYZ invenietur $XYz+XZy+YZx$ et fluxio ipsius $VXYZ$ invenietur $VXYz+VXZy+VYZx+XYZv$ et sic in infinitum fluxio facti semper invenietur substituendo sigillatim pro unoquoq factore fluxionem ejus et sumendo resultantium terminorum aggregatum' (By a similar reasoning the fluxion of XYZ will be found to be $XYz+XZy+YZx$, and the fluxion of $VXYZ$ will be found to be $VXYz+VXZy+VYZx+XYZv$, and so indefinitely the fluxion of a product will ever be found by singly substituting for each individual factor its fluxion and taking the aggregate of the resulting terms).

(116) Originally just 'in medijs proportionalibus' (in mean proportionals), to which Newton adjoined as his following example: 'Proponatur $\overline{XY}^{\frac{1}{2}}$. Pone ipsum æquale V et erit

(115)Now let the product XYZ be proposed. On setting $XY = V$ there will be $XYZ = VZ$, the fluxion of which according to the previous case is $Vz+Zv$. But, also, because $XY = V$ there is $Xy+Yx = v$; so in place of V and v substitute their equivalents and the fluxion of XYZ, namely $Vz+Zv$, will come to be

$$XYz+XZy+YZx.$$

And by observing the mode of progression it is gathered universally that the fluxion of any product whatsoever, such as $VXYZ$, will be found by separately substituting in that product for each factor its fluxion and taking the aggregate of the resulting terms. This rule also holds good when some of the factors are equal. Should X and Y, for instance, be equal, so that XY is equivalent to X^2, its fluxion $Xy+Yx$ will come to be $2Xx$. Similarly, the fluxion of X^3 is $3X^2x$, and the fluxion of X^2Z is $X^2z+2XZx$. And the like in composite cases: the fluxion of $AX-3X^2$ is $Ax-6Xx$. For, of course, the fluxions of the parts are, taken together, the fluxion of the whole.

In the 'sides' of quantities 'applied' to fluents the method is this. Let there be proposed the 'side' X^2/Y. Set it equal to V and there will be $X^2 = YV$, and hence $2Xx = Yv+Vy$; for of fluents which are ever equal the fluxions are equal. Take Vy away from each side and the remainder divided by Y, namely $(2Xx-Vy)/Y$ will be v. Here, however, v is the fluxion of V, that is, X^2/Y, which it was necessary to find.

The method is similar in the case of quadratic, cubic and other 'sides'.(116) Let the root(117) $(AX-X^2)^{\frac{1}{2}}$ be proposed. Set it equal to V and there will be $AX-X^2 = V^2$, whence $Ax-2Xx = 2Vv$ and so $(Ax-2Xx)/2V = v$.

[The fluxions of lines and angles.]

(118) In figures in order to signify the fluxion of some line I set the line in lower-case letters: as bc to signify the fluxion of the line BC. Fluxions of angles, however, I represent by means of the fluxions of the arcs whereby they are subtended at a given distance. That distance, whatever it eventually be assumed to be, I denote by the letter R; the fluxion of an arc by the angle written in lower-case letters; the opposing ordinate in this circle—that is, the sine of this arc— by the letter s prefixed to the angle; the ordinate's distance from the centre—that is, the com-

---

$XY = V^2$. Adeoq; $Xy+Yx = 2Vv$ et $\frac{Xy+Yx}{2V} = v$' (Let there be proposed $(XY)^{\frac{1}{2}}$. Put it equal to V and there will be $XY = V^2$. Hence $Xy+Yx = 2Vv$ and so $(Xy+Yx)/2V = v$).

(117) A superfluous 'quadrati alicujus' (of some square) is deleted. Newton first wrote in vaguer geometrical equivalent 'medium...in lateribus æqui[are]orum' (the middle...in the sides of (figures of) equal area).

(118) The preliminary drafts of portions of the following which are reproduced (from ff. 136r[+135v]−144r of the manuscript) in sections [10] and [11] of the Appendix will reveal even more clearly than their ensuing revised texts how narrowly Newton in the following section is indebted to his earlier 'Geometria Curvilinea' (see IV: 420–84, particularly 428–64). In the margin alongside we interpolate a suitable heading in summary of its theme.

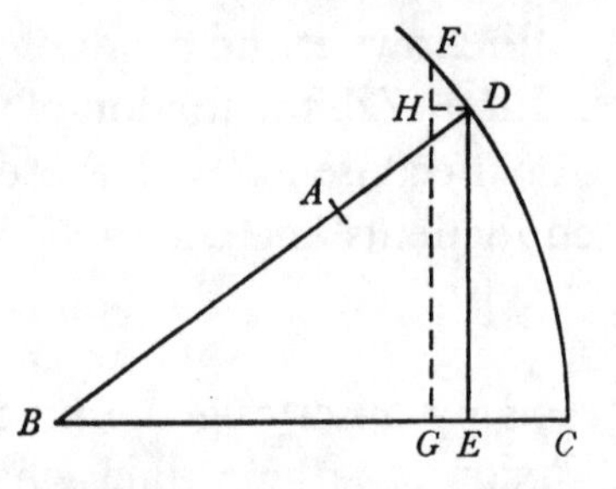

rectum per literam s' angulo præfixam & horum sinuum fluxiones per literas easdem s et s' angulo minusculis literis notato præfixas.(119) Sit ABC angulus quilibet fluens, DC arcus quo ad datam distantiam BC subtenditur & DE sinus ejus, et R significabit distantiam illam BC vel BD, b vel abc fluxio arcus CD,(120) sB lineam DE, s'B lineam BE & sb s'b earum fluxiones.(121)

Illud etiam præmittendum est, fluxionem arcus esse ad fluxionem sinus ejus ut Radius ad sinum complementi[,] et ad defluxionem sinus complementi ut radius ad sinum. b ad sb ut R ad s'B et b ad $-$s'b ut R ad sB. Fluant enim omnes aliquantulum donec CD fiat CF, DE fiat FG et CE fiat CG et ipso fluendi initio fluxiones erunt ut augmenta incipientia FD, FH, HD id est ut BD, BE, DE.(122)

His præmissis proponatur triangulum aliquod ABC et demissis ad latera singula perpendiculis AE, BF, CG erit(123) ut sA ad sB ita BF ad AE & ita BC ad AC, adeoq sA'AC = sB'BC. Ergo sA'ac+AC'sa = sB'bc+BC'sb. Sed

$$sa.a::s'A.R::AG.AC.$$

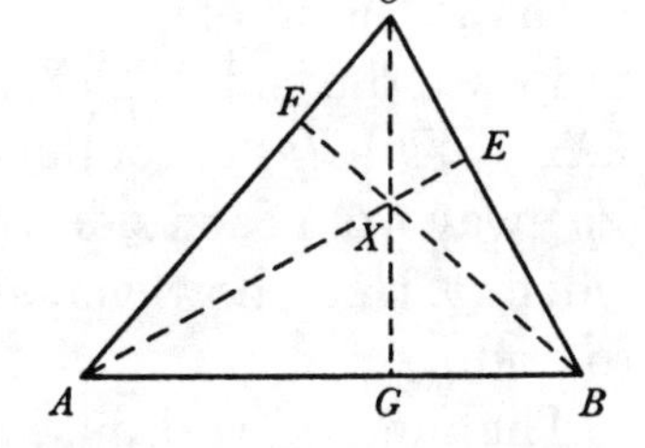

Ergo AC'sa = AG'a. Et eodem modo est BC'sb = BG'b. Quare sA'ac+AG'a = sB'bc+BG'b.(124)

Quo Theoremate conferre possumus fluxiones angulorum duorum et laterum oppositorum trianguli cujuscunq et ex tribus cognitis invenire quartam.

Et cum summa trium angulorum detur adeoq aggregatum fluxionum omnium nullum sit, vel quod perinde est duorum fluxio æqualis sit defluxioni tertij, si pro $+BG$'b scribas $-BG'\overline{a+c}$ & idem utrobiq auferas fiet

$$sA'ac+AB'a+BG'c = sB'bc.^{(125)}$$

(119) Newton returns to the Wardian notation for the sine and cosine functions which he had earlier employed in his 'Epitome Trigonometriæ' (see IV: 116). As before (compare *ibid.*: 117, note (4)), in our English version we render s and s' by Sin (= R sin) and Cos (= R cos).

(120) That is, $R.\hat{B}$.

(121) For clarity we render these in our English version by using the contraction 'fl' (for 'fluxio') which Newton had pioneered in the *addendum* to his 1671 tract (see III: 330–52) and afterwards employed ubiquitously in the later propositions of his 'Geometria Curvilinea' (see IV: 436–520, *passim*). Thus we translate his b (or $\widehat{abc}$) as 'fl($R\hat{B}$)', and his sb, s'b as 'fl(Sin $\hat{B}$)' and 'fl(Cos $\hat{B}$)' respectively, where (see the previous note)

$$\text{Sin } \hat{B} = R \sin \hat{B} \quad \text{and} \quad \text{Cos } \hat{B} = R \cos \hat{B}.$$

(122) Proposition 14 of the 'Geometria Curvilinea' (see IV: 440–2), here a little more shortly derived. In more modern form Newton's theorem demonstrates the fluxional relationships

plementary right sine (cosine)—by the letter s' prefixed to the angle; and the fluxion of this sine and cosine by the same letters s and s' prefixed to the angle denoted in lower-case letters.(119) Let $\widehat{ABC}$ be any fluent angle, $\widehat{DC}$ the arc whereby it is subtended at the given distance BC, and DE its sine: then R will signify the distance BC or BD, $\hat{b}$ or $\widehat{abc}$ the fluxion of arc $\widehat{CD}$,(120) sB the line DE (Sin $\hat{B}$), s'B the line BE (Cos $\hat{B}$) and sb, s'b their fluxions.(121)

It needs also to be premised that the fluxion of an arc is to the fluxion of its sine as the radius to its cosine, and to the negative fluxion of its cosine as the radius to its sine: fl($R\hat{B}$) to fl(Sin $\hat{B}$) as R to Cos $\hat{B}$, and fl($R\hat{B}$) to $-$fl(Cos $\hat{B}$) as R to Sin B. For let all flow ever so slightly till $\widehat{CD}$ becomes $\widehat{CF}$, DE becomes FG and CE comes to be CG, and at the very start of flowing their fluxions will be as the incipient increments DF, HF and DH, that is, as BD, BE and DE.(122)

With these premisses, let some triangle ABC be proposed. When the perpendiculars AE, BF, CG are let fall to the individual sides,(123) Sin $\hat{A}$ will be to Sin $\hat{B}$ as BF to AE and so as BC to AC, and hence Sin $\hat{A}.AC$ = Sin $\hat{B}.BC$; therefore

$$\text{Sin}\,\hat{A}.\text{fl}(AC)+AC.\text{fl}(\text{Sin}\,\hat{A}) = \text{Sin}\,\hat{B}.\text{fl}(BC)+BC.\text{fl}(\text{Sin}\,\hat{B}).$$

But
$$\text{fl}(\text{Sin}\,\hat{A}):\text{fl}(R\hat{A}) = \text{Cos}\,\hat{A}:R = AG:AC,$$

and therefore $AC.\text{fl}(\text{Sin}\,\hat{A}) = AG.\text{fl}(R\hat{A})$; and in the same way there is $BC.\text{fl}(\text{Sin}\,\hat{B}) = BG.\text{fl}(R\hat{B})$. Consequently

$$\text{Sin}\,\hat{A}.\text{fl}(AC)+AG.\text{fl}(R\hat{A}) = \text{Sin}\,\hat{B}.\text{fl}(BC)+BG.\text{fl}(R\hat{B}).\text{(124)}$$

By this theorem we can compare the fluxions of two angles and the opposing sides in any triangle whatever, and when three are ascertained find the fourth one therefrom.

And since the sum of the three angles is given, and hence the aggregate of all their fluxions is nil—or, what is exactly the same, the fluxion of two is equal to the negative fluxion of the third—if in place of $+BG.\text{fl}(R\hat{B})$ you write

$$-BG.\text{fl}(R\hat{A}+R\hat{C})$$

and you take this away on either side, there will come

$$\text{Sin}\,\hat{A}.\text{fl}(AC)+AB.\text{fl}(R\hat{A})+BG.\text{fl}(R\hat{C}) = \text{Sin}\,\hat{B}.\text{fl}(BC):\text{(125)}$$

$$sb:b = \text{fl}(\text{Sin}\,\hat{B}):\text{fl}(R\hat{B}) = (d(\sin\hat{B})/d\hat{B}\text{ or})\ \text{Cos}\,\hat{B}:R,$$

and
$$s'b:b = \text{fl}(\text{Cos}\,\hat{B}):\text{fl}(R\hat{B}) = (d(\cos\hat{B})/d\hat{B}\text{ or})\ -\text{Sin}\,\hat{B}:R.$$

(123) Newton has here deleted the none too illuminating parenthesis 'ut e demonstratis trigonometricis notum est' (as is known from what is demonstrated in trigonometry). From first principles, of course, sA = Sin $\hat{A}$ = $R.BF/AB$ and sB = Sin $\hat{B}$ = $R.AE/AB$, so that their ratio is indeed that of BF to AE.

(124) Proposition 23 of the 'Geometria Curvilinea' (see IV: 460), here freshly demonstrated.

(125) Proposition 24 on IV: 462, here minimally recast (and again derived as a straightforward extension of the previous theorem).

Theorema ad comparandas fluxiones duorum angulorum totidemq laterum quorum unum angulis illis interjicitur.

Rursus est $R.\text{s'}A::AC.AG$ seu $\text{s'}A'AC = R'AG$. Ergo $\text{s'}A'ac+AC'\text{s'}a = R'ag$. Est et (per præmissa) $a.-\text{s'}a$ vel $-a.\text{s'}a::R.\text{s}A::AC.CG$. Ergo pro $AC'\text{s'}a$ scribendo $-CG'a$, fit $\text{s'}A'ac-CG'a = R'ag$. Eodem modo est

$$\text{s'}B'bc-CG'b = R'bg.$$

Et æqualibus æqualia addendo fit $\text{s'}A'ac+\text{s'}B'bc-CG'\overline{a+b} = R'ab$. Ob datam summa[m] trium angulorum pro $a+b$ scribe $-c$ et fit

$$\text{s'}A'ac+\text{s'}B'bc+CG'c = R'ab.$$ (126)

Theorema ad comparandas fluxiones trium laterum et anguli cujusvis.

Simili argumentatione possunt alia(127) Theoremata colligi ubi perpendicula triangulorum et segmenta basium aliæve lineæ considerantur. Sic

$$AC'ac-BC'bc+BG'ab = AB'ag$$ (128)

Theorema est ubi latera tria et segmentum basis considerantur, et, posito X communi trium perpendiculorum intersectione,(129) est

$$BX'ac+AX'bc = GX'ab+AB'gc$$ (128)

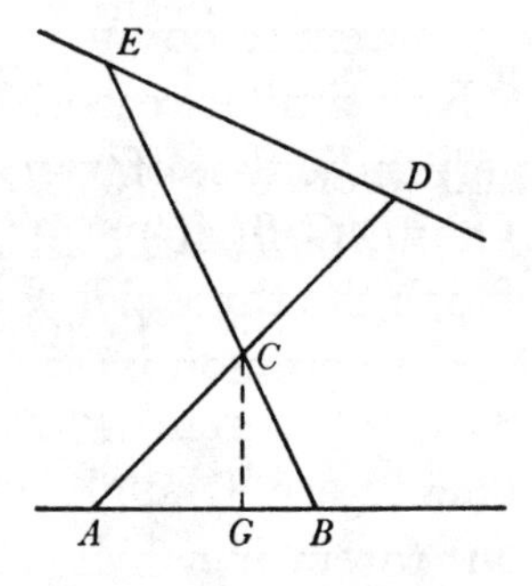

Theorema ubi agitur de lateribus et perpendiculo.(130) Sed hæc non prosequar.(131) Satis est investigandi methodum aperuisse.

Horum verò Theorematum beneficio possumus in propositis quibusvis figuris fluxiones linearum et angulorum haud secus ac in computo trigonometrico has ab alijs colligere donec ad quæsitam pervenimus. Ut si

(126) Proposition 22 of the 'Geometria Curvilinea' (see IV: 458), here given much the same proof as before. In his initial draft of the present section (on ff. 136r[+135v]/137r) it had, however, been Newton's original intention to present an elaborate demonstration of this theorem from first principles, computing the individual components of the total fluxion of AB step by step from the contemporaneous vanishing increments of AC, BC and $R\hat{C}$ to which *initio fluxionis* they are proportional; see section [10] of the following Appendix.

(127) Originally 'varia' (various) in a first, cancelled draft of the present paragraph on f. 138r; other slight divergences therein from the revised version here reproduced (from f. 137v) are cited in notes (129) and (130) following.

(128) Except that their fluxions are here represented in his now preferred lower-case notation, these equations are exactly those derived by Newton in Propositions 20 and 21 of his earlier 'Geometria' (see IV: 452–4 and 456 respectively).

(129) Initially 'si trium perpendiculorum communis intersectio sit X' (if the common meeting-point of the three perpendiculars [*sc.* altitudes] be X). In a fuller account Newton would here doubtless have been careful to add, on the pattern of his earlier Lemma on IV: 454, a rider explicitly demonstrating the uniqueness of a triangle's orthocentre which he here baldly asserts.

(130) This replaces an equivalent 'ubi latera et perpendiculum in quæstione sunt' (where the sides and a perpendicular are in question) on f. 138r.

a theorem to compare the fluxions of two angles and an equal number of sides, one of which is interposed between those angles.

Again, there is $R:\operatorname{Cos}\hat{A} = AC:AG$ or $\operatorname{Cos}\hat{A}.AC = R.AG$, and therefore $\operatorname{Cos}\hat{A}.\mathrm{fl}(AC)+AC.\mathrm{fl}(\operatorname{Cos}\hat{A}) = R.\mathrm{fl}(AG)$. Also (by premiss)

$$\mathrm{fl}(R\hat{A}): -\mathrm{fl}(\operatorname{Cos}\hat{A}), \quad \text{that is} \quad -\mathrm{fl}(R\hat{A}):\mathrm{fl}(\operatorname{Cos}\hat{A}) = R:\operatorname{Sin}\hat{A} = AC:CG.$$

Therefore on writing $-CG.\mathrm{fl}(R\hat{A})$ in place of $AC.\mathrm{fl}(\operatorname{Cos}\hat{A})$ there comes $\operatorname{Cos}\hat{A}.\mathrm{fl}(AC)-CG.\mathrm{fl}(R\hat{A}) = R.\mathrm{fl}(AG)$. In the same way there is

$$\operatorname{Cos}\hat{B}.\mathrm{fl}(BC)-CG.\mathrm{fl}(R\hat{B}) = R.\mathrm{fl}(BG).$$

And by adding equals to equals there comes

$$\operatorname{Cos}\hat{A}.\mathrm{fl}(AC)+\operatorname{Cos}\hat{B}.\mathrm{fl}(BC)-CG.\mathrm{fl}(R\hat{A}+R\hat{B}) = R.\mathrm{fl}(AB).$$

Because of the given sum of the three angles in place of $\mathrm{fl}(R\hat{A}+R\hat{B})$ write $-\mathrm{fl}(R\hat{C})$, and there comes to be

$$\operatorname{Cos}\hat{A}.\mathrm{fl}(AC)+\operatorname{Cos}\hat{B}.\mathrm{fl}(BC)+CG.\mathrm{fl}(R\hat{C}) = R.\mathrm{fl}(AB):^{(126)}$$

a theorem to compare the fluxions of the three sides and any angle.

By a similar reasoning there can be gathered other(127) theorems in which the perpendiculars of triangles and their base segments or other lines are considered. Thus $AC.\mathrm{fl}(AC)-BC.\mathrm{fl}(BC)+BG.\mathrm{fl}(AB) = AB.\mathrm{fl}(AG)$(128) is a theorem in which the three sides and a base segment are considered, and, when X is put to be the common intersection of the three perpendiculars,(129) there is

$$BX.\mathrm{fl}(AC)+AX.\mathrm{fl}(BC) = GX.\mathrm{fl}(AB)+AB.\mathrm{fl}(GC):^{(128)}$$

a theorem in which it is a matter of the three sides and a perpendicular.(130) But these things I shall not pursue.(131) It is enough to have disclosed the method of discovery.

With the benefit of these theorems, to be sure, we can in any propounded figures gather the fluxions of lines and angles, no differently than in a trigonometrical computation, these from others till we attain the one required. If, for

(131) Newton first affirmed more fully: 'Sed hæc et similia minus necessaria sunt et a Geometris ubi usus eorum inciderit, ex inventis haud difficulter colligentur' (But these and similar ones are less necessary and geometers will, when the need to use them arises, gather them with no great difficulty from ones already found). In the initial draft of the remainder of the 'Geometriæ Liber primus' which is found in immediate sequel on ff. 138^r–144^r (and which we reproduce in section [11] of the Appendix) he had originally at this point gone on to repeat from his earlier 'Geometria Curvilinea' whole blocks of analogous fluxional theorems 'De Proportionalibus' and 'De Triangulis' before cancelling their text. He did not, we may add, himself there find it as easy as he states to derive these 'Theoremata minus necessaria' from ones already established without making significant (if elementary) slips in his deduction; see note (115) of the Appendix.

dentur positione lineæ AB, AD, DE et BC datæ longitudinis moveatur perpetuo subtendens angulum A et producta secans rectam ED in E, et ex cogn[i]ta vel assumpta fluxione lineæ AC desideretur fluxio lineæ EC: primum in triangulo ABC per secundum Theorematum invenietur $sA'ac+BG'c = 0$[,] evanescunt enim termini [$AB'a$ et $sB'bc$ ob datum angulum A et latus BC].(132)

[Quomodo plagæ motuum inveniendæ.]

In figuris hæc est methodus.(133) Puncti mobilis considero semper motus diversos juxta diversas plagas quarum principalis sit via puncti. Et hos motus expono vel saltem exponi imaginor describendo per punctum illud circulum quemvis cujus centrum sit in via illa et in singulis plagis ducendo rectas usq; ad hunc circulum. Ut si punctum A moveatur in linea BA, per illud A de[s]cribo circulum quemvis cujus centrum sit in BA et cui illa BA aliæq; lineæ quævis CA, DA, EA occurrant in F, G, H, I, et linearum partes intra circulū AF, AG, AH, AI erunt inter se ut motus puncti A in illarum plagis. Adeo ut si motus puncti A a B exponitur per AF, motus ejus a C exponatur per AG et sic in reliquis: Aut quod perinde est si fluxio lineæ BA ex parte termini A exponitur per AF, aliarum linearum CA, DA ad idem mobile punctum A semper desinentium fluxiones ex parte termini illius A exponantur per AG AH, et lineæ EA defluxio per AI.(134) Unde ex cognitis motibus duorum punctorum ad quæ linea quævis utrinq; terminatur, cognoscetur et exponi potest ejus fluxio absoluta: quippe quæ summa est fluxionum ejus ad utrumq; terminum, vel excessus fluxionis ad unum terminum supra defluxionem ad alterum.

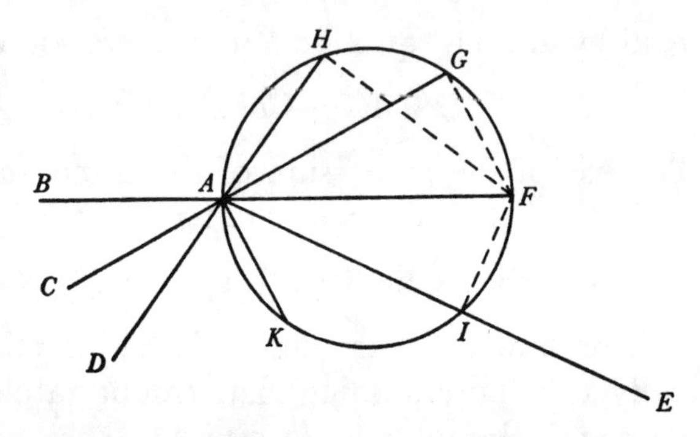

(132) We complete Newton's sentence by an appropriate editorial insertion at the point where he broke it off, leaving the rest of his manuscript page (f. 138v) blank. That he never filled it is evidence at once of his discontent with an approach which pointlessly invokes the complication of considering the fluxion of the angle $\widehat{ACB}$, but also of his inability— on the spur of the moment at least—to devise a more fruitful avenue of attack on the problem of expressing the fluxion of CE in terms of that of AC. The knot is, in fact, readily cut by extending AB and ED to their meet in H, say, and then applying Menelaus' theorem to the triangle ABC crossed (externally) by the transversal EDH to deduce that $BE \times CD = k.CE \times BH$, $k = AD/AH$, constant; whence (on employing Newton's present lower-case denotations for the fluxions of the line-segments defined by the corresponding upper-case end-points) there comes $BE \times cd + CD \times be = k.(CE \times bh + BH \times ce)$. Here, since AD, BC and AH are fixed in length, there is at once $cd = -ac$, $be = ce$ and $bh = -ab$, that is (because

$$AB^2 + AC^2 - 2AB \times AC.\cos \hat{A} = BC^2$$

and consequently, on taking fluxions, $AB \times ab + AC \times ac - (AB \times ac + AC \times ab).\cos \hat{A} = 0$) $bh = (CF/BG).ac$ where, as before, BF is let fall perpendicular to the side AC of the triangle

instance, the lines AB, AD, DE be given in position and BC of given length should move so as perpetually to subtend the angle $\hat{A}$ and, when produced, to cut the straight line ED in E, and were there from the known or assumed fluxion of the line AC desired the fluxion of the line CE: first, in the triangle ABC there will be found by the second theorem

$$\mathrm{Sin}\,\hat{A}.\mathrm{fl}(AC)+BG.\mathrm{fl}(R\hat{C})=0,$$

for the terms $[AB.\mathrm{fl}(R\hat{A})$ and $\mathrm{Sin}\,B.\mathrm{fl}(BC)]$ will vanish [because the angle $\hat{A}$ and the side BC are given.](132)

[How the directions of motions are to be found.]

In figures this is the method.(133) I consider always a mobile point's differing motions in line with differing directions, the chief one of which shall be the path of the point. And these motions I represent—or at least imagine to be represented—by describing through that point any circle whose centre shall be in that path and in each separate direction drawing straight lines as far as this circle. Should, for instance, the point A move in the line BA, through that A I describe any circle whose centre shall be in BA; let that BA and any other lines CA, DA, EA meet this in F, G, H, I, and the portions AF, AG, AH, AI of those lines within the circle will then be to one another as the motions of the point A in their respective directions. In consequence, if the motion of point A from B is represented by AF, its motion from C shall be represented by AG, and the like of the remainder; or, what is exactly the same, if the fluxion of the line BA in respect of its end-point A is represented by AF, the fluxions of other lines CA, DA ever terminating at the same mobile point A shall in respect of that end-point A be represented by AG, AH, and the negative fluxion of the line EA by AI.(134) When, therefore, the motions of two points at which any line is terminated at its either end are known, its absolute fluxion will be known therefrom and can be represented, since it is, of course, the sum of its fluxions at either end-point, or the excess of the fluxion at one end-point over the negative fluxion at its other.

ABC. When these values are substituted in the preceding equation there results straightforwardly $ce/ac = (BE+k.CE\times CF/BG)/(CD-k.BH)$ as the desired relation.

(133) Understand 'inveniendi fluxiones linearum' (of finding the fluxions of lines), as before, but in curvilinear figures we need of course to determine not only the speeds of fluxional increase of their elements but also their instantaneous, tangential directions of motion. In the margin alongside we again insert a caption which appropriately epitomizes the content of this final portion—as its manuscript has survived (and was seemingly penned by Newton)—of the 'Geometriæ Liber primus'. As in the related theorems 'De motuum plaga et celeritate' of the preliminary draft reproduced in section [11] of the Appendix, Newton here borrows heavily from the corresponding Propositions 26–30, and dependent Problems 3–9 of his earlier 'Geometria Curvilinea' (see IV: 466–74 and 476–82 respectively), though these are here much recast and improved in their verbal form.

(134) This is all much as in the cancelled Proposition 30 of Newton's 'Geometria Curvilinea' (see IV: 472 and compare *ibid.*: 473, note (140)).

Porro motus punctorum circa polos quosvis ijdem sunt et easdem habent exponentes ac motus in plagis perpendicularibus ad radios. Sic motus puncti *A* circa polum quemvis in linea *CA* situm exponens est normalis *AK* circulo occurrens in *K*. Expositis vero duorum punctorum rectæ cujusvis motibus circumpolaribus, recta alia per terminos exponentium acta secabit rectam illam in Polo suo. Et per harum exponentium rationem ad radios id est ad distantias suas a Polo, exponere licebit motum angularem hujus rectæ seu fluxionem angulorum quos ea cum rectis positione datis continet.

Et ut ex motibus punctorum inveniri et exponi possunt fluxiones linearum et angulorum, sic vice versa ex horum fluxionibus colligere licet motus punctorum. Nimirum considerando lineam *AF* in qua punctum quodvis *A* movetur ut exponentem motus ejus, et exponentis illius terminum ulteriorem *F* ut metam ad quam punctum illud *A* tendit, et lineas omnes *FA*, *FG*, *FH*, *FI* per metam transientes ut loca metæ ex inventione duorum ejusmodi locorum, meta quæ in utriusq intersectione est determinabitur. Loca verò sic inveniuntur. [135]Quando mobile punctum ex assumptione duarum quarumvis vel plurium fluentium quantitatum determinatum et stabile redditur, inveniendus est motus quem punctum illud haberet si una quantitatum assumeretur et alterius tantum vel reliquarum fluxio maneret, et motûs illius quoad plagam et quantitatem exponens[136] ducenda est. Cognoscenda est etiam plaga motus quem punctum idem haberet si vice versa illa una quantitas flueret et altera vel reliquæ assumerentur. Et in plaga illa per terminum exponentis acta recta erit unus e locis metæ.

Hoc modo a motibus punctorum ad fluxiones quantitatū et vicissim ab harum fluxionibus ad illorum motus pergere licebit donec quoadusq libuerit perventum sit. Et ubi exponens motus puncti curvam propositam describentis inventa est, hæc et curvam in puncto illo tanget et exponens erit fluxionis ejus. Sed res exemplis clarior fiet.

Exempla prima.[137]

[1.] *A mobili puncto A qua curva quævis EA describitur ad rectas duas positione datas*

(135) In a first version of the remainder of this present paragraph Newton (on ff. 145r/146r) initially went on: 'Quando motus puncti ex duabus vel pluribus quantitatum fluxionibus certus ac determinatus redditur, quarum una ignota est vel ut ignota spectatur, ex altera vel reliquis inveniendus est motus ejus qualis foret si fluxio illa ignota esset nulla. Et si motus inventi exponens ducatur & per terminum ejus in plaga qua punctum vi solius fluxionis ignotæ pergeret, recta agatur, erit hæc unus e locis metæ. Metâ vero ex duabus locis inventa, simul habetur exponens motus quæsiti' (When the motion of a point is rendered certain and determinate by way of two or more fluxions of quantities, one of which is unknown or regarded as unknown, from the other or the remaining ones we need to find its motion such as it were if that unknown fluxion had been nil. And if the exponent of the motion when found be drawn, and through its end-point in the direction which a point would proceed in by dint of the unknown fluxion alone there be drawn a straight line, this will be one of the loci of the goal-point. Once, however, the goal-point is found from (the meet of) two of its loci, there is at once had the exponent of the required motion).

Moreover, the motions of points round any poles are the same, and have the same exponents, as motions in directions perpendicular to the radii. Thus the motion of the point A round any pole situated in the line CA has as its exponent the normal AK meeting the circle in K. Where, however, the motions of two points of any straight line about its pole are so represented, any straight line drawn through the end-points of their exponents will cut that straight line at its pole. And by the ratio of these exponents to their radii, that is, their distances from the pole, we shall be free to represent the angular motion of this straight line, in other words, the fluxion of the angles which it contains with straight lines given in position.

And just as from the motions of points the fluxions of lines and angles can be ascertained and represented, so *vice versa* from the fluxions of these we are free to gather the motions of points. Specifically, by considering the line AF wherein any point A moves as the exponent of its motion, and the further end-point F of that exponent as the goal towards which that point A tends, and all lines FA, FG, FH, FI passing through that goal-point as loci of the goal, from finding two loci of this sort the goal-point which is at the intersection of both will be determined. Those loci, however, are found in this manner. (135)When a mobile point is rendered fixed and determinate in consequence of the assumption of any two or more quantities, we need to find the motion which that point would have were one of the quantities assumed while the fluxion of just the second or the remaining ones should remain, and then to draw the exponent(136) of that motion in regard to both its direction and quantity. We need also to know the direction of motion which the same point would have if *vice versa* that one quantity were to flow and the other or remaining ones should be assumed. And when a straight line is drawn in that direction through the end-point of the exponent it will be one of the loci of the goal-point.

In this way it will be permissible to proceed from the motions of points to the fluxions of quantities, and conversely from the latter's fluxions to the former's motion, till you have arrived at whatever you will. And when the exponent of the motion of a point describing a propounded curve has been found, this will both touch the curve at that point and be the exponent of its fluxion. But the matter will become clearer by examples.

First examples.(137)

1. *From the mobile point A whereby any curve EA is described two straight lines AB,*

(136) Newton first began to write 'signu[m]' (symbol).

(137) For convenience of reference we number the three instances of this primary group of examples which Newton sets down in conclusion to his 'Geometriæ Liber primus'. (The manuscript text in its surviving, unfinished state includes no *Exempla altera*, 'second examples', to complement these.) Though all three, as we shall specify in following footnotes, derive from correlative propositions in his previous 'Geometria Curvilinea', it is no mere coincidence that their reference systems—generalized oblique Cartesian coordinates, standard bipolars and the monopolar/translating-curve compound—are exactly those pioneered by Descartes in the

DB, DC in datis angulis(138) *ducuntur rectæ duæ AB AC et ductarum relatio ad invicem habetur. Ducenda est recta quæ curvam hanc tangat in A.*

Ut hoc fiat exponantur ductarum fluxiones per *AG* et *AH*. Jam quia punctum *A*, assumptione fluentium *DB*, *BA* determinatur, et ubi earum una *DB* assumitur et altera *BA* solummodo fluit, linea *AG* exponens est tam motus puncti(139) quam fluxionis lineæ *BA*, ubi vero vice versa altera *BA* assumitur et prior *DB* fluit punctum *A* movetur in plaga lineæ *DB*, recta *GF* quæ per exponentis terminum *G* in plaga lincæ *DB* parallela ducitur erit unus locus Metæ. Et simili argumento recta *HF* quæ per exponentis *AH* terminum *H* in plaga lineæ *DC* ducitur erit alius locus Metæ. Et locorum intersectio *F* metam dabit ad quam tangens quæsita *AF* ducenda est. Quam conclusionem sic concinnare licebit. Secet tangens rectam *DB* in *M* et ipsi *DC* parallela agatur *MN* occurrens *AC* in *N*: et *AB*, *AM*, *AN* erunt inter se ut *AG*, *AF*, *AH*, adeoq; vice exponentium *AG*, *AH* adhiberi possunt *AB*, *AN*: qua ratione longitudo *AN* atq; adeo punctum *M* ad quod tangens duci debet invenietur.

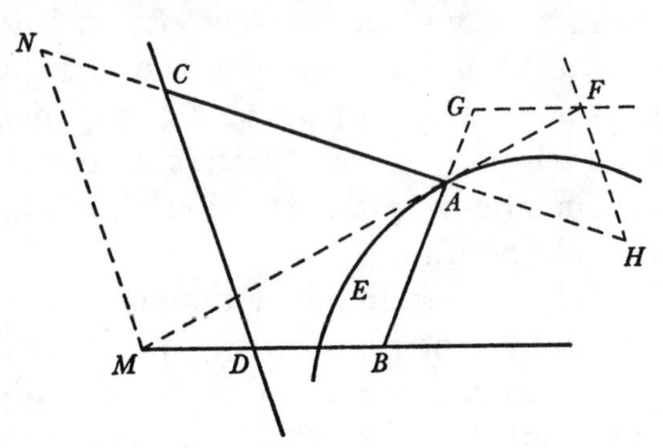

Ut si relatio inter *AB* et *AC* sit quod rectangulum sub *AC* et data quavis recta *R* æquale sit quadrato AB^2,(140) æquales erunt etiam harum fluxiones *R'ac* & *2AB'ab*. Hic pro fluxionibus *ac* et *ab* substitue earum exponentes *AH*, *A*[*G*], vel potius harum vice lineas *AN*, *AB*, et fiet $R'AN = 2AB^2$. Unde *R'AN* et *2R'AC* æquales sunt utpote eidem $2AB^2$ æquales; adeoq; $AN = 2AC$. Cape ergo $CN = AC$ & per *N* ipsi *CD* parallelam age *NM* occurrentem *DB* in *M* et recta *AM* curvam propositam tanget in *A*.

Haud secus si ad defin[i]endam relationem inter *AB* et *AC* ponatur *R'AC*−AC^2 esse ad AB^2 in data ratione,(141) colligentur horum fluxiones *R'ac*−*2AC'ac*

second book of his *Geometrie* (*Discours de La Methode* (Leyden, ₁1637): 315–69 = *Geometria à Renato Des Cartes*... (Amsterdam, ₂1659): 17–66) more than half a century before. If Newton did not here consciously borrow from the work of his predecessor, it could not have been far from the forefront of his mind.

(138) In the preliminary version of this prime instance of the preceding general tangent-rule which is reproduced in section [12] of the Appendix, Newton specified that *AB* and *AC* should be drawn parallel to the axes *DC* and *DB* respectively—in other words that

$$A\widehat{B}D = A\widehat{C}D = \text{(exterior angle) } B\widehat{D}C.$$

There is no real gain to be had by relaxing this restriction, since if *AB'* and *AC'* are the parallels so drawn, then $AB = k.AB'$ and $AC = l.AC'$ on setting $k = \sin \hat{D}/\sin \hat{B}$ and $l = \sin \hat{D}/\sin \hat{C}$, both constant; whence any given relationship $f(AB, AC) = 0$ defining the locus (*A*) reduces at once to the *relatio* $f(k.AB', l.AC') \equiv f'(AB', AC') = 0$ of analogous type

AC are drawn at given angles[138] *to two straight lines DB, DC given in position, and the relationship of the lines drawn to one another is had. A straight line has to be drawn to touch this curve at A.*

That this may be done, represent the fluxions of the lines drawn by AG and AH. Now because the point A is determined by the assumption of the fluents DB, BA, and when one of these, DB, is assumed and the other, BA, alone flows the line AG is the exponent both of the motion of the point[139] and of the fluxion of the line BA, while when *vice versa* the second, BA, is assumed and the first, DB, flows the point A moves in the direction the line DB does, the straight line GF which is drawn through the end-point G of the exponent parallel to the direction of the line DB will be one locus of the goal-point. And by a similar argument the straight line HF which is drawn through the end-point H of the exponent AH in the line DC's direction will be another locus of the goal-point. Then the intersection F of the loci will yield the goal-point towards which the required tangent AF is to be drawn. It will be allowable to refine this conclusion as follows. Let the tangent cut the straight line DB in M and parallel to DC draw MN meeting AC in N: AB, AM and AN will then be to one another as AG, AF and AH, and hence instead of the exponents AG and AH we can employ AB and AN; by this means the length of AN will be found, and hence the point M to which the tangent ought to be drawn.

If, for instance, the relationship between AB and AC should be that the rectangle contained by AC and any given straight line R is equal to the square AB^2,[140] the fluxions $R.\text{fl}(AC)$ and $2AB.\text{fl}(AB)$ of these will also be equal. Here in place of the fluxions $\text{fl}(AC)$ and $\text{fl}(AB)$ substitute their exponents AH, AG, or preferably in their stead the lines AN, AB, and there will come to be

$$R.AN = 2AB^2.$$

In consequence $R.AN$ and $2R.AC$ are equal (being equal, namely, to the same $2AB^2$), and hence $AN = 2AC$. Take $CN = AC$, therefore, and through N parallel to CD draw NM meeting DB in M, and the straight line AM touches the propounded curve at A.

No differently, if to define the relationship between AB and AC it should be supposed that $R.AC - AC^2$ is to AB^2 in a given ratio,[141] gather the fluxions

connecting the standard oblique Cartesian coordinates AB' and AC' of the locus-point A. In Newton's two ensuing worked instances, in particular, where the given relationships are $AB^2 = R \times AC$ and $AB^2 \propto R \times AC - AC^2$, the corresponding standard Cartesian equations of the loci (A) thereby defined—a parabola and an ellipse respectively—are $AB'^2 = R' \times AC'$, $R' = lR/k^2$, and $AB'^2 \propto R' \times AC' - AC'^2$, $R' = R/l$.

(139) Understand of A as it 'flows' in the moving ordinate BA.

(140) Thus (see note (138)) defining the locus (A) to be an Apollonian parabola.

(141) When (see note (138)) the curve (A) is an ellipse.

& $2AB'ab$, et inde [erunt] $R'AN-2AC'AN$ [et $2AB^2$] in eadem ratione. Unde $R'AC-AC^2$ & $\frac{1}{2}R'AN-AC'AN$ æqualia erunt, utpote eandem rationem ad AB^2 habentia. Capiatur ergo AN ad AC ut $R-AC$ ad $\frac{1}{2}R-AC$ et, actâ MN parallelâ CD, habebitur tangens AM.

Porrò si curvæ EA hæc sit prop[r]ietas ut si a dato circulo FK per data puncta P, Q ducantur rectæ duæ LI, LK concurrentes ad datum circulum EL, ponatur AB(142) æqualis LI et AC(142) æqualis LK: ducantur circulorum tangentes IM, KN, LR et fluxio arcus $E[L]$ exponatur per LR cujusvis longitudinis. Super diametro LR describatur circulus secans PL productam in S et QL in T, et erit LS exponens fluxionis rectæ PL et LT exponens fluxionis retrogradæ rectæ QL. Erigantur normales LV ad LP et LX ad LQ occurrentes circulo LTR in V et X et erunt hæ exponentes motuum puncti L circa polos P et Q.

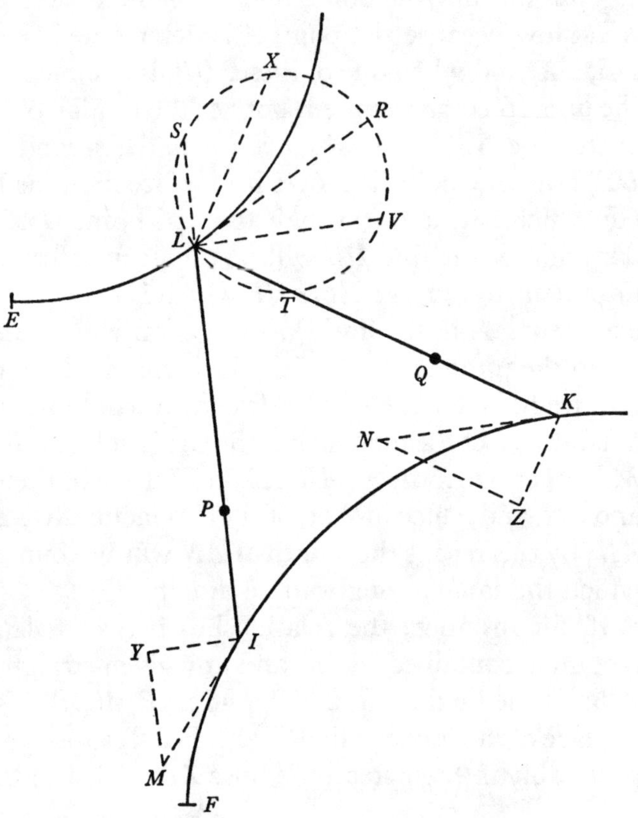

Erigantur etiam normales YI ad PI et ZK ad QK ita ut sit YI ad IP ut VL ad LP et ZK ad KQ ut XL ad LQ et erunt hæ(143) exponentes motuum punctorum I et K circa polos eosdem P et Q.(144) Concipe per puncta I et Y circulum describi cujus centrum sit in tangente IM et pariter per puncta K et Z alium circulū cujus centrum sit in tangente KN, et horum circulorum diametri IM KN exponentes erunt motuum punctorum I et K in circumferentia circuli IK: item YM æqualis erit exponenti fluxionis lineæ PI et ZN æqualis exponenti fluxionis retrogradæ ipsius QK. Quare $SL+YM$ exponens erit fluxionis totius IL et $TL+ZN$ exponens fluxionis retrogradæ totius KL. Cape ergo(145) $AG = SL+YM$ et $AH = [TL]+ZN$, sed ob fluxionem retrogradam ipsius LK vel AC cape AH ad partes ipsius A versus C, et HF acta parallela DC secabit GF actam parallelam DB in Meta F ad quam tangens quæsita AF duci debet.

(142) In the previous figure, that is.

$R.\mathrm{fl}(AC)$ and $2AB.\mathrm{fl}(AB)$ of these, and thence $R.AN-2AC.AN$ and $2AB^2$ will be in the same ratio. Consequently, $R.AC-AC^2$ and $\frac{1}{2}R.AN-AC.AN$ will be equal (having, namely, the same ratio to AB^2); so take AN to AC as $R-AC$ to $\frac{1}{2}R-AC$, and when MN is drawn parallel to CD the tangent AM will be obtained.

Furthermore, if the curve EA has this property that, if from the given circle FK there be drawn through the given points P, Q two straight lines LI, LK concurrent at the given circle EL, it be supposed that AB[142] is equal to LI and AC[142] equal to LK: draw the circles' tangents IM, KN, LR and represent the fluxion of the arc $\widehat{EL}$ by LR of any length. On LR as diameter describe a circle cutting PL produced in S and QL in T, and LS will then be the exponent of the fluxion of the line PL and LT the exponent of the backwards fluxion of the line QL. Erect the normals LV (to LP) and LX (to LQ) meeting the circle LTR in V and X, and these will then be the exponents of the motions of point L round P and Q. Erect also the normals YI to PI and ZK to QK such that YI be to IP as VL to LP, and ZK to KQ as XL to LQ, and these[143] will be the exponents of the motions of I and K round the same poles P and Q.[144] Conceive that through the points I and Y a circle is described whose centre shall be in the tangent IM, and correspondingly through the points K and Z another circle whose centre shall be in the tangent KN, and the diameters IM and KN of these circles will be exponents of the motions of the points I and K in the circumference of the circle IK; likewise YM will be equal to the exponent of the fluxion of the line PI, and ZN equal to the exponent of the backwards fluxion of QK. In consequence, $SL+YM$ will be the exponent of the total fluxion of IL, and $TL+ZN$ the exponent of the total backwards fluxion of KL. Therefore take[145]

$$AG = SL+YM \quad \text{and} \quad AH = TL+ZN,$$

but because of the backwards fluxion of LK or AC take AH in direction towards C: when HF is drawn parallel to DC, it will cut GF drawn parallel to DB in the goal-point F towards which the required tangent AF ought to be drawn.

(143) The lines YI and ZK respectively.

(144) Newton here first went on: 'Ad *IY* et *KZ* erige normales *YM*, & *ZN* occurrentes *IM* et *KN* in *M* et *N* et erit *YM* exponens fluxionis ipsius *PI* eo quod æqualis sit parti lineæ *PI* productæ quæ intra circulum per puncta *I* et *Y* circa centrum constitutum in *IM* descriptum caderet. Et simili argumento *NZ* exponens est fluxionis retrogradæ ipsius *QK*' (To IY and KZ erect the normals YM and ZN meeting IM and KN in M and N, and then YM will be the exponent of the fluxion of PI inasmuch as it is equal to the part of the line PI which, if extended, would fall within the circle described through the points I and Y about a centre stationed in IM; and by a similar reasoning NZ is the exponent of the backwards fluxion of QK).

(145) Again in the previous figure.

Quod si vice rectarum *LI*, *LK* adhibeantur circulorum arcus *EL*, *FK* ponendo *AB* æqualem arcui *EL* et *AC* æqualem arcui *FK*, tunc *AG* sumenda erit æqualis *LR* et *AH* æqualis *KN*, eo quod *LR* exponens sit fluxionis arcus *EL* et *KN* exponens defluxionis arcus *FK*, et actæ *GF*, *HF* ut prius tangentem determinabunt. Neqꝫ problema difficilius erit[(146)] si vice circulorum *EL FK* adhibeantur aliæ quævis curvæ lineæ quarum tangentes *LR*, *KN* ductæ habentur. Sed et alijs modis innumeris relatio inter *AB* et *AC* exprimi potest, imò et vice rectarum *DB*, *DC* curvæ quævis adhiberi ad quas *AB*, *AC* ducantur in datis plagis et quarum tangentes ad puncta *B* et *C* sint *DB* et *DC*.[(147)]

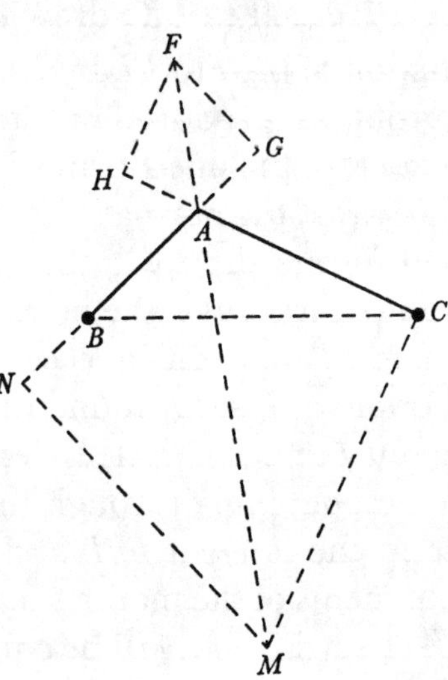

[2.] Ducantur verò jam lineæ *DB*, *DC* non in datis plagis sed *ad data puncta B et C*,[(148)] et earum fluxiones exponantur per *AG* et *AH*. Et quoniam assumptione anguli *ABC* et longitudinis *BA* determinatur punctum *A*, et ubi angulus ille solummodo assumitur exponens motus puncti *A* est linea *AG*, ubi vero e contra angulus ille fluit et longitudo *BA* assumitur plaga motus puncti *A* perpendicularis est ad *BA*, recta *GF* in plaga illa per exponentis terminum *G* ducta erit unus locus metæ. Et simili argumento recta *HF* per terminum exponentis *AH* in plaga perpendiculari ad *CA* ducta erit alius locus Metæ. Et meta in utroqꝫ loco consistens erit in eorum intersectione *F*, adeoqꝫ *AF* ad intersectionem illam ducta curvam motu puncti *A* descriptam tanget in *A*.

Ut si ea sit natura curvæ hujus ut summa vel differentia fluentis[(149)] *AB* et datæ cujusvis *R* sit ad fluentem *AC* in data ratione (qui casus est quatuor Ovalium Cartesij[(150)]) fluxiones illarum *AB* et *AC* erunt in eadem data ratione, adeoqꝫ si in plagis fluxionum illarum capiãtur *AG* & *AH*[,] vel quod perinde est si in plagis contrarijs capiantur *AN* et *AC* in illa ratione et ad terminos captarum erigantur perpendicula concurrentia in *F* vel *M*, acta *AF* vel *AM* curvam

(146) Originally 'esset' (would...have been).

(147) For all their elegance and individual interest, these elaborations and variations upon the primitive determination of the tangential *plaga motus* of a curve defined in standard Cartesian coordinates—as he initially intended here to present it in the discarded preliminary draft reproduced in section [12] of the Appendix—must appear to Newton's modern reader in the main as so much sterile and uselessly complicated embellishment of a simple (but universal) basic mode of construction which is an exact geometrical analogue of the operation of taking the derivative of the corresponding defining equation. (Compare note (157) of the Appendix, and see also I: 279–80/386–7.)

(148) Whence the locus (*A*) is referred to standard Cartesian bipolars *AB* and *AC*. Newton's ensuing determination of the tangential direction *AF* at *A* proceeds much as in the corresponding Proposition 29 of his earlier 'Geometria Curvilinea' (see IV: 472); compare also Mode 3 of Problem 4 of his 1671 tract (III: 136–8).

But if instead of the straight lines *LI*, *LK* the arcs $\widehat{EL}$, $\widehat{FK}$ of the circles be employed by setting *AB* equal to $\widehat{EL}$ and *AC* equal to $\widehat{FK}$, then *AG* will need to be taken equal to *LR* and *AH* equal to *KN*, seeing that *LR* is to be the exponent of the fluxion of the arc $\widehat{EL}$ and *KN* the exponent of the negative fluxion of $\widehat{FK}$; and then, when *GF* and *HF* are drawn as before, they will determine the tangent. Nor will the problem be[(146)] more difficult if in place of the circles *EL*, *FK* there be employed any other curve lines whose tangents *LR*, *KN* are had drawn. While, too, the relationship between *AB* and *AC* can be expressed in innumerable other ways, and indeed in place of the straight lines *DB* and *DC* any curves you wish may be employed, to which *AB* and *AC* are to be drawn in given directions, and whose tangents at the points *B* and *C* shall be *DB* and *DC*.[(147)]

2. Now, however, let the lines *DB* and *DC* be drawn not in given directions, but *to the given points B and C*,[(148)] and let their fluxions be represented by *AG* and *AH*. Then, since the point *A* is determined by the assumption of the angle $A\widehat{B}C$ and the length *BA*, and when that angle alone is assumed the exponent of the motion of the point *A* is the line *AG*, while conversely when that angle flows and the length *BA* is assumed the direction of motion of the point *A* is perpendicular to *BA*, the straight line *GF* drawn in that direction through the end-point *G* of the exponent will be one locus of the goal-point. And by a similar argument the straight line *HF* drawn through the end-point of the exponent *AH* in a direction perpendicular to *CA* will be another locus of the goal-point. So the goal-point, lying in each locus, will be at their intersection *F*, and hence *AF* drawn to that intersection touches at *A* the curve described by the motion of the point *A*.

For instance, if it should be the nature of this curve that the sum or difference of the fluent[(149)] *AB* and any given line *R* shall be to the fluent *AC* in a given ratio —which is the case of Descartes' four ovals[(150)]—, their fluxions *AB* and *AC* will be in the same given ratio, and hence if *AG* and *AH* be taken in the directions of those fluxions, or what is exactly the same if *AN* and *AC* be taken in the contrary directions and in that ratio, and then at the end-points of these lines so taken there be erected perpendiculars concurrent in *F* or *M*, when *AF* or *AM* is drawn

(149) 'lineæ' (line) is deleted.

(150) As in Problem 5 of the 'Geometria Curvilinea' (IV: 478). This simplest of the bipolar curves—unnecessarily distinguished by Descartes himself (*Geometrie*, $_1$1637: 252–7) into four component 'ellipses' according to the sign of the given ratio and the magnitude of the constant *R* (which determines whether or not the oval meets its diametral axis between the poles or outside them)—had still earlier been Newton's 'Exemplum' of Mode 3 of drawing tangents in his 1671 tract (see III: 138), and reappeared little afterwards in Proposition 34 of his 'Optica' (see III: 494–6) as the curvilinear interface possessed of the differential property (implicit in Newton's present construction of its tangent) of refracting all rays of white light incident from one pole to pass through the second—or more strictly (compare 2, §2: note (85) preceding) one of its two others.

propositam tanget in A. Unde si ratio illa est æqualitatis (qui casus est Hyperbolæ et Ellipsis) tangens bisecabit angulum $CA[B]$.(151)

[3.] Ponamus jam super plano immobili in quo puncta P et K et recta infinita KD positione data habentur, planum mobile BCA curva aliqua CA terminatum ita ferri(152) ut recta BC in eo data semper coincidat cum recta KD, et interea secum trahere regulam PB per punctum suum B perpetuò transeuntem et circa polum P rotantem, & ejus intersectione cum termino suo curvilineo CA describere curvam lineam PAL in plano immobili,(153) et requiratur hujus curvæ tangens ad punctum quodvis A. Quoniam assumptione rectæ KC et curvæ CA determinatur punctum A[,] assumatur solummodo curva AC et sit CQ exponens fluxionis lineæ KC et huic parallela et æqualis AG exponens erit motus puncti A, et GF ducta in plaga motus quem punctum A haberet si vice versa KC assumeretur et curva CA solummodo flueret id est ducta parallela rectæ AD quæ curvam AC tangit in A, erit unus locus metæ. Rursus quoniam punctum A assumptione longitudinis KB et proportionis PA ad PB determinatur, assumamus solummodo proportionem illam et punctum movebit in linea AG eritq; motus ejus ad motum puncti B ut PA ad PB. Exponatur ergo motus ejus per AH quæ sit ad alterius exponentem id est ad CQ vel AG ut PA ad PB, et per punctum H in plaga motus quem haberet punctum A si vice versa KB assumeretur et ratio PA ad PB flueret, id est parallela PB acta recta HF erit alter locus metæ. Habitis autem duobus metæ locis habetur Meta in eorum intersectione F una cum tangente AF quæ ad metam duci debet.(154) Quam conclusionem si concin[n]are animus est, produc tangentem donec secet BK in N, et ob similes figuras $AFGH$, $NADB$ erit BN ad BD ut AH ad HG hoc est ut AP ad AB.

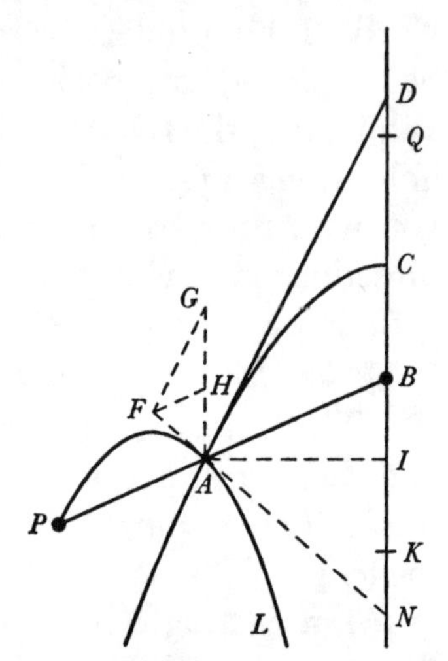

(151) This special case is dealt with separately in the 'Geometria Curvilinea' as its Problem 4 (see IV: 478).

(152) 'motu parallelo' (in a parallel motion) is deleted.

(153) This is the 'instrument' (as its author names it) originally devised by Descartes and employed by him in the second book of his *Geometrie* ($_1$1637: 319–22 [= *Geometria*, $_2$1659: 21–3]) successively to generate sequences of higher-order 'paraboles' from ones of lower genus (on which see II: 363, note (123)). Newton's two following instances, where (see notes (155) and (156) below) a Cartesian trident and a conchoid are described by the intersection of the rotating 'reigle' PAB with a translating parabola and a circle respectively, are the latter pair of Descartes' three examples (*ibid.*: 322)—the first being that of the Apollonian hyperbola described by an oblique straight line AC moving parallel to itself (see *ibid.*: 320–1).

(154) As we remarked in our note (155) on IV: 478 Roberval had already more than half a century before, in the particular instance where the curve generated is the 'Parabole de

it will touch the curve proposed at A. Consequently, if that ratio is one of equality—which is the case of the hyperbola and ellipse—the tangent will bisect the angle $C\widehat{A}B$.(151)

3. Now let us posit that on a stationary plane in which the points P and K and the unbounded straight line KD are had given in position the mobile plane BCA terminating in some curve CA is so carried(152) that the straight line BC given in it shall ever coincide with the straight line KD, and that all the while it pulls with it the ruler PB perpetually passing through its own point B and rotating round the pole P, and so by the intersection of this with its curvilinear boundary $\widehat{CA}$ describes the curved line PAL in the stationary plane;(153) and let there be required the tangent of this curve at any point A. Because the point A is determined by the assumption of the straight line KC and the curve CA, assume only the curve CA and let CQ be the exponent of the fluxion of the line KC, and then AG parallel and equal to this will be the exponent of the motion of the point A, and GF drawn in the direction of the motion which the point A would have if *vice versa* KC were to be assumed and the curve CA alone should flow, that is, drawn parallel to the straight line AD which touches the curve AC at A, will be one locus of the goal-point. Again, because the point A is determined by the assumption of the length KB and the ratio of PA to PB, let us assume that ratio alone and the point will move in the line AG, and so its motion will be to the motion of the point B as PA to PB; therefore represent its motion by AH which shall be to the other's exponent, namely CQ or AG, as PA to PB, and then through the point H in the direction of the motion which point A would have if *vice versa* KB were to be assumed and the ratio of PA to PB should flow, that is, parallel to PB, draw the straight line HF and it will be the other locus of the goal-point. Once the two loci of the point are obtained, however, that point is had at their intersection F, along with the tangent AF which ought to be drawn towards the goal-point.(154) If you have a mind to refine this conclusion, extend the tangent till it cuts BK in N, and there will then, because of the similarity of the figures $AFGH$ and $NADB$, be NB to BD as AH to HG, that is, as PA to AB.

M. Descartes', set down an analogous argument determining the *plaga motus* defined by this goal-point in the 'Treizième exemple' of the 'Observations sur la composition des Mouvemens, et sur le moyen de trouver les Touchantes des lignes courbes' (now Bibliothèque Nationale, Paris. MS fonds français 9119) which he delivered in the late 1630's from his professorial rostrum at the Collège Royal. But this treatise was, in (or around) 1693 when Newton penned his present generalized construction, still not to appear in print for another half dozen years—when Du Verdus edited it, together with such other of his 'Divers Ouvrages' as he could locate, in Tome 6 of the *Mémoires de l'Académie Royale des Sciences. Depuis 1666 jusqu'à 1699* (Paris, $_1$1699)—and its content remained unknown to the public at large. We do not know if Newton ever read Roberval's work at a later date.

Ut si Curva CA Parabola sit cujus vertex C diameter CK ordinatim applicata AI, (quo casu AL Parabola erit Cartesij(155)) imprimis ducenda erit AD quæ Parabolam CA tanget in A[,] quod fiet si capiatur CD æqualis CI. dein capienda est BN ad BD ut AP ad AB et acta AN tanget curvam AL in A.

Quod si AC circulus sit centro B descriptus, quo casu AL Conchoides erit Veterum,(156) erigenda est ad AP normalis AD occurrens BN in D, hæc enim circulum illum tanget. Dein capienda est BN ad BD ut AP ad AB. Vel brevius capienda est $BM = AP$ et erigenda normalis MN occurrens BD in N et acta AN figuram AL tanget in A.

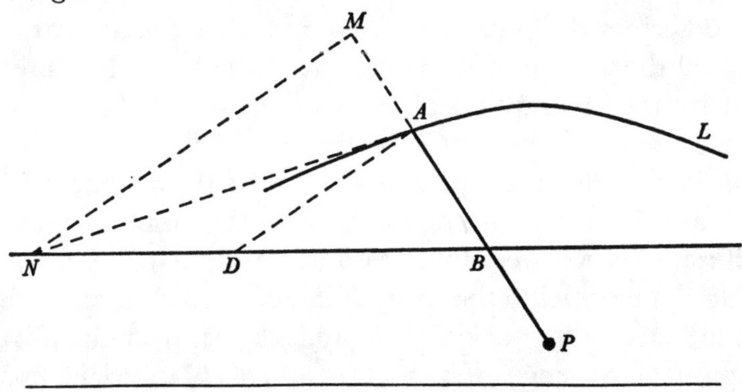

(155) The 'Parabole...que i'ay tantost dit estre la premiere, & la plus simple pour la question de Pappus, lorsqu'il n'y a que cinq lignes droites données de position', as Descartes himself specified it in his *Geometrie* ($_1$1637: 322; compare *ibid.*: 309, and see also IV: 344, note (30)). Though Descartes himself afterwards (*ibid.*: 338) drew a somewhat confusing diagram in which the full figure of the Cartesian trident was depicted along with its mirror-image in a horizontal axis, in his previous figure (which does double duty both for the preceding general case and its present instance) Newton has drawn only one of the curve's parabolo-hyperbolic branches. We have added the second branch in the expanded sketch set in our English version. This construction of the tangent to the 'Parabola Cartesiana 2^{di} generis' is no doubt identical with that which Newton earlier intended to present in Problem 6 of his 'Geometria Curvilinea' (IV: 478)—and so, indeed, we there restored it (see *ibid.*: 478–9, note (155)).

(156) Newton here follows Descartes (*Geometrie*, $_1$1637: 322) in taking into consideration only 'la première Conchoïde des anciens', that is, the upper shell of this classic 'mussel' curve. In illustration of the full curve here generated by the translating circle, in the augmented figure set alongside in our English text we have introduced the nodal instance of the omitted 'lower Conch' (as Newton described the conchoid's second branch on I: 503 in there commenting upon the possible locations of the 'point betwixt its convexity & concavity'). His present construction of the tangent to the conchoid is that which he had himself discovered nearly thirty years before in May 1666 (see I: 394).

If, for instance, the curve CA be a parabola whose vertex is C, diameter CK and ordinate AI—in which case AL will be Descartes' parabola,[155] you will need first to draw AD to touch the parabola CA at A (which will be done if you take CD equal to CI) and then take BN to BD as AP to AB: when AN is drawn, it touches the curve AL at A.

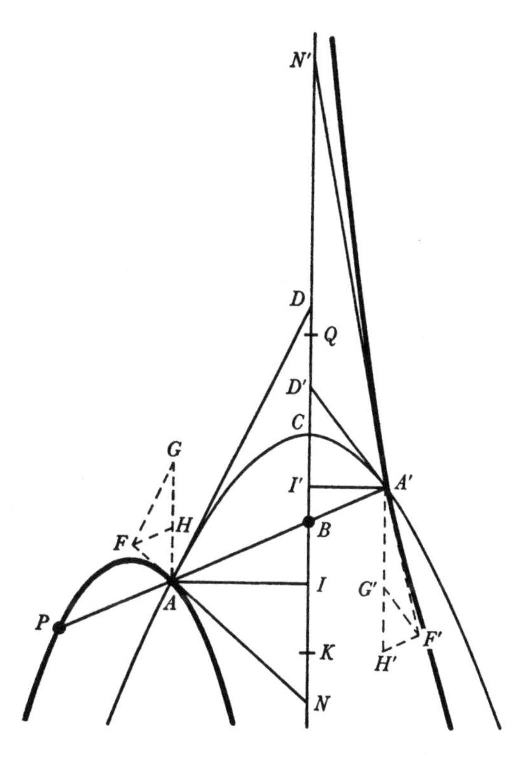

But if AC be a circle described with centre B—in which case AL will be an ancients' conchoid[156]—you need to erect to AP the normal AD meeting BN in D, for this of course will touch that circle, and then take BN to BD as AP to AB. Or, more briefly, you are to take BM equal to AP and erect the normal MN meeting BD in N. Then, when AN is drawn, it touches the figure AL at A.

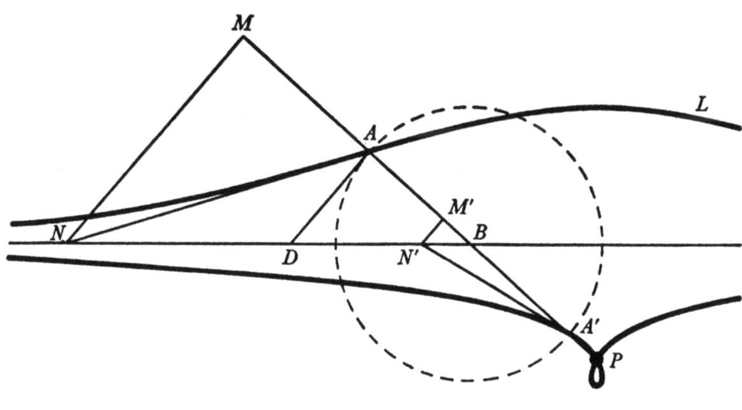

APPENDIX. MISCELLANEOUS PRELIMINARY DRAFTS FOR THE FINAL 'LIBER 1'.(1)

From the originals in the University Library, Cambridge

[1](2) Si linea A ducitur in lineam B rectangulum genitum signamus scribendo $A\times B$ vel AB et si id rursus ducatur in lineā C parallelipipedum genitum signamus scribendo $A\times B\times C$ vel ABC. Latus vero quod oritur applicando rectangulum(3) illud ad lineam quamvis D sic notamus $\frac{AB}{D}$. Et sic in reliquis. Sed et exposita linea aliqua ad quam tanquā mensuram universalem aliæ omnes lineæ referantur, scribimus $A\times B$ vel $A'B$ aut AB ad designandam quartam proportionalē ab hac linea ubi duæ mediæ sunt A et B[,] et $A'B'C$ ad designandam etiam quartam ab eadem linea ubi $A'B$ et C sunt duæ mediæ et sic in infinitum. Et si linea illa sit prima continuè proportionalium et alia quævis A secunda, tertiam sic designamus AA vel A^q vel A^2, quartam sic A^c vel A^3, quintam sic A^{qq} vel A^4[,] ubi A^2 A^3 A^4 A^5 dici possunt A duarum, trium, quatuor, quinꝗ rationum.(4) Et sic $A^{\frac{1}{2}}$ est A dimidiæ rationis seu media proportio inter mensuram universalem et A. Deniꝗ ad designandum tum latus quadrati æqualis areæ(5) $A'B$ tum medium proportionale inter mensuram universalem et $A'B$ scribo $\surd A'B$ vel $\overline{A'B}|^{\frac{1}{2}}$[,] et ad designandū tum latus cubicum solidi

(1) Out of the mass of Newton's numerous preliminary castings of the text of his final 'Geometriæ Liber primus' which (compare §1: note (3) preceding) survive in ULC. Add. 3963 and Add. 4004 we here reproduce in their full detail only those more connected drafts which have especial verbal interest or mathematical significance. Of the remainder we give much more fleeting account as and where it is pertinent to do so in following footnotes.

(2) Add. 4004: 129r/129v; see §1: note (11) above. Newton illustrates how, by fixing on some unit line-length as a 'universal measure', geometrical equivalents of the arithmetical operations of multiplication and division may be assigned, but appeals to the authority of Euclid and the other 'ancient' geometers in warning against untenably extending the 'affinity' between number and linear magnitude thus exemplified to be an expression of their essential identity.

(3) 'vel parallelipipedum' is deleted.

(4) Newton initially went on to specify 'per has notas A^q, A^c, A^{qq}, A^{qc}, A^{cc} intelligendo quadratum, cubum, quadrato-quadratum, quadrato-cubum, cubo-cubum de A et sic in infinitum. Nam et quadratum et cubum super latere A constitutum designamus ijsdem notis ac tertium quartumꝗ proportionalem. Unde et reliquis proportionalibus per analogiam nomina dantur. Quæ et analogicè etiam dicuntur dimensiones ac potestates lineæ A. Sic A^4 dicitur A quatuor dimen[s]ionū vel A potestatis quadrato-quadraticæ quamvis revera nihil u[l]tra trinam dimensionem et potestatem cubicam in Geometria reperiatur. Et simili analogia dicimus proportionales A^q, A^c, A^{qq} generari ducendo A in se et A^q in A et A^c in A. Et quartam proportionalem AB generari ducendo A in B. Et vicissī A^{qq} applicatum ad A producere A^c et AB applicatum ad B producere A, applicatum vero ad C producere $\frac{A'B}{C}$.'

(5) Newton first began to write 'rectangu[lo]'.

$A'B'C-B'C^q$ tum primum e duobus medijs proportionalibus inter mensuram universalem et $A'B'C-B'C^q$ scribo $\sqrt[3]{A'B'C-B'C^q}$ vel $\overline{A'B'C-B'C^q}|^{\frac{1}{3}}$[,] & ad hujus quadratum designandum scribo $\overline{A'B'C-B'C^q}|^{\frac{2}{3}}$. Eademq; notarum ratio in magis compositis tenenda est.

Terminis Arithmeticis multiplicandi dividendi et extrahendi radices non utor quod improprij sint et minimè necessarij et scientias diversi generis confundere nolui. Quantitas per aliam quantitatem multiplicari absurdè dicitur. Solus numerus est per quem possumus multiplicare. Tres homines multiplicari possunt per 4[,] non autem per quatuor homines, et linea trium pedum per quatuor[,] non autem per lineam quatuor pedum. Multiplicatio non fit per lineas nisi quatenus hæ per numeros exponi ac designari concipiantur id est de propria natura in quantitates Arithmeticas convertantur. Est autem Arithmetici exponere et speculari quanti[it]ates omnis generis per numeros, Geometræ verò per lineas superficies et solida. (6)Certe Veteres inter has scientias maximam esse affinitatem animadverterunt, ita ut ex analogia terminos Geometricos quadrati cubi et similium in Arithmeticam introducerent et Euclides libros Arithmeticos misceret Geometricis;(7) sed Geometriam tamen quæ scientiarum Mathematicarū regina est terminis exoticis contaminare noluerunt. Inventa est utiq; Geometria ut ejus succinctis operationibus in terris metiendis(8) effugeremus tædium computi Arithmetici. Proinde ut a computis quantum fieri potest vacare debet, sic etiam a computi nominibus ne horum usu ad rem significatam plus nimio invitemur et sic scientiam nobilissimam contra institutū ejus cum Arithmetica tandem confundamus. Hac igitur in re reprehendi non debeam si Veteres sequar.(9)

---

[2](10) [Certe Veteres inter has scientias maximam esse affinitatem agnoverunt...; sed Geometriam...terminis exoticis contaminare noluerunt ne]

---

(6) In a first version of the following phrase Newton added a sentence of dubious explanation to assert: 'Qua de causa Veteres ut Geometriam servarent incontaminatā a terminis illis exoticis maximè abstinuerunt. Tantam inter has scientias esse affinitatem animadverterunt ut....'

(7) In his *Elements*, Books v, vii–ix and x of which are essentially arithmetical in their basis, as opposed to the truly geometrical Books i–iv, vi and xi–xiii.

(8) See 2, §2: note (6) above for more detailed reference to the classical sources— all stemming ultimately from Herodotus, it would seem—that 'geo-metry' has not merely its etymological origin but also its historical source in the codified principles of land-measurement.

(9) Newton broke off at this point, initially (it would seem) to draft the partial amplification which we reproduce in [2] following and then—aware maybe that he had strayed rather far from his main point—to discard the whole passage.

(10) Add. 3963.2: 11r, a preliminary amplified recasting of the concluding portion of [1] above. The manuscript begins *in medio*, but we repeat the opening of a sentence in the preceding text to smooth the join between the two.

hæc non amplius pura et sincera manens de dignitate sua recederet. In pura Geometria nulla alia est operatio Arithmetica præter Additionem unius ad unum[,] subductionem unius ab uno & harum imperatam repetitionem quæ verius Numeratio est quam Multiplicatio: & res numerare commune est omnium scientiarum Postulatum. Numeri numerando inventi considerantur in Arithmetica, res numeratæ in scientijs reliquis. Debebit tamen Geometra Arithmeticam omnem probè cognitam habere ne forsan hæreat ubi inciderit Numerorum consideratio. Sic et Mechanicam, Opticam, Staticam, Trigonometricam, Astronomicam, Nauticā cæterasqꝫ id genus scientias debebit intelligere et quæstionibus quibusvis a Geometria pendentibus sciat aptè respondere: Geometriam tamen, si Veteres Geometras æmulari velit, servabit ab his omnibus puram. Et quamvis Arithmetica apud Vulgus notior est quam Geometria, qui tamen Geometriæ limen ingressus est, melius intelligit quid sit ducere in lineam, applicare ad lineam, & invenire medias proportionales inter lineas, quam quid sit multiplicare et dividere per lineas et extrahere radices linearum. Priores expressiones rem ipsam proprie significant(11) posteriores valdè impropriè. Possunt tres homines multiplicari per quatuor, non autem per quatuor homines et linea trium pedum per quatuor[,] non autem per lineam quatuor pedum. Multiplicatio non fit per res numeratas nisi quatenus pro his significandis substituuntur numeri. Est autem Arithmetici quantitates omnis generis exponere et speculari per numeros, Geometræ verò per lineas superficies & solida. Deniqꝫ inventa est Geometria ut per ejus succinctas operationes in terris metiendis effugeremus tædiū computi Arithmetici: proinde ut a computis quantum fieri potest vacare debet, sic etiam a computi nominibus, ne horum usu ad rem significatam plus nimio invitemur, et sic scientiam nobilissimam contra institutum ejus cum Arithmetica tandem confundamus. Siquis tamen aliter senserit, res non tanti est ut latius disputem.(12)

---

(11) 'exprimant' was first written.

(12) Finding this amplified revise of [1] no more to his satisfaction than before, Newton began anew, passing to pen (on Add. 3963.2: 7r) a rough draft of a replacement whose uncancelled portion—the deletions are insignificant—reads: 'Obtinet autem illa Problematum distinctio in Arithmetica verius quam in Geometria. Nam problemata omnia primi & secundi gradus et aliqua gradus quarti, octavi, decimi sexti &c solvuntur in Geometria per regulam et circinum. Et quæ ita solvuntur sunt primi gradus Geometrici. In Arithmetica vero hi gradus distinguuntur. Nam Problemata quæ primi sunt gradus solvuntur ibi per multiplicationē et divisionem et ad harum solutionem sufficit regula falsæ Positionis in Arithmetica vulgi. Quæ secundi sunt gradus solvuntur ibi per extractionem radicis quadraticæ. Quæ tertij sunt gradus per extractionem radicis cubicæ. Quæ quarti sunt gradus per extractionem radicis biquadraticæ[!] & aliquando per extractiones duarum radicum quadraticarum. Et sic in infinitum. Deducit Arithmetica Problemata omnia ad æquationes[,] dein radices æquationum

[3][13] Cavendum est tamen ne per hanc analogiam confundantur scientiæ. Utitur quidem Arithmeticus Geometricis numerorum planorum et solidorum vocabulis sed non concipit numeros per modum planorum et solidorum, neq; multiplicationem et divisionem repræsentat per ductū in lineam et applicationem ad lineam; aut si hoc facit jam incipit esse Geometra. Et vice versa Geometra qui concipit lineas per modum unitatum et numerorum & per operationes Arithmeticas repræsentat Geometricas jam induit personam Arithmetici. Geometræ est per extensiones[,] Arithmetici per numeros repræsentare & exponere omnis generis quantitates. Dices forte quantitates A^4, A^5, A^6 et similes in Geometria nullas fore si non licuerit exponere lineas per numeros. Imò numeri erunt per quos lineæ exponuntur, ipsæ lineæ non erunt. Verùm tamen $\frac{A^4}{V^3}$ lineam denotare potest quæ fit applicando ad V contentum sub lineis $\frac{A^3}{VV}$ et A[14], et $\frac{A^5}{V^4}$ lineam quæ fit applicando ad V contentum sub lineis $\frac{A^4}{V^3}$ et A et sic in reliquis. Eodem recidit si consideres A, $\frac{AA}{V}$, $\frac{A^3}{VV}$, $\frac{A^4}{V^3}$ &c ut lineas quæ sint ad V in ratione simpla, dupla, tripla, quadrupla ipsius A ad V. Et si V jam consideretur non ut unitas quidem sed ut universale latus applicationum aut ut universale consequens rationum, licebit illud V semper subintelligere & scribere tantum A, AA, A^3, A^4 vel pro his dicere A rationis unius duarum trium quatuor, ut et A dimidiæ vel sesquialteræ rationis pro $A^{\frac{1}{2}}$ vel $A^{\frac{3}{2}}$, respectu scilicet habito ad communem consequentem, quæ cujus sit longitudinis nil refert; vel etiam generali voce Graduum vel dignitatum indifferenter significare licebit contenta,

---

extrahit. Gradui verò problematis respondēt dimensiones æquationis, & dimensionib^s genus radicis. Ut si problema sit tertij gradus deducetur hoc ad æquationem trium dimensionum & talis æquationis radix semper cubica est.' His patent error here of supposing in particular that a problem of fourth grade (resolvable by an equation of fourth degree which is universally reducible in standard form to a cubic) requires for its solution the extraction of a quartic root, and more generally of positing an exact correspondence between the grade of a problem and the dimension of the root needing to be extracted in its enodation was, as soon as he appreciated its mistakenness, no doubt Newton's reason for speedily suppressing this preliminary version of what he more cautiously and accurately restated in the equivalent revised passage on pages 406–10 above (see §1: note (11)).

(13) Add. 3963.2: $4^r/5^r$. In this suppressed passage (whose initial place in the main text of the preceding 'Liber primus' is specified in §1: note (13)) Newton enters, in line with his previous *caveat* in [1] and [2] above, yet one more warning to his reader not to confuse the distinct mathematical sciences of arithmetic and geometry, despite their many analogies in descriptive terminology and operational structure.

(14) Originally 'ducendo lineam $\frac{A^3}{VV}$ in A et contentum applicando ad lineam V' in a more long-winded equivalent phrasing.

rationes & numeros factos, dicendo A tertij gradus pro A^3, A dimidij gradus pro $A^{\frac{1}{2}}$ et sic in reliquis.(15)

---

(15) In an abandoned recasting of this passage on Add. 3963.2: 3v Newton went on 'Longius adhuc progreditur Arithmetica et per notas a^4, a^5, a^6, a^7; a^3b, a^4b, a^5b, a^6b; $abbc$, ab^3c, ab^4c, ab^5c et similes intelligit numeros quadrato-quadratos, quadrato-cubos, cubo-cubos, quadrato-quadrato-cubos aliosq; plano-planos, plano-solidos, solido-solidos, plano-plano-solidos et ita deinceps. Unde et aliqui lineas aliasq; quantitates per modum numerorum considerando, usurpant easdem notas pro his omnibus significandis. Quod quidem in Arithmetica et Algebra ad subjectum Geometricum applicata proprie fit[,] in Geometria verò improprie. Utitur quidem Arithmeticus Geometricis— — — generis quantitates. Scribi tamen potest in Geometria $\frac{a^4}{b^3}$ pro linea quæ fit applicando contentum sub lineis $\frac{a^3}{bb}$ et a ad latus b, atq; $\frac{a^5}{b^4}$ pro linea quæ fit applicando contentum sub lineis $\frac{a^4}{b^3}$ et a ad latus b et ita deinceps. Et eodem recidit si consideres $\frac{aa}{b}$, $\frac{a^3}{bb}$, $\frac{a^4}{b^3}$ &c ut lineas quæ sint ad b in ratione simpla, dupla, tripla, quadrupla, quintupla ipsius a ad b. Sed aa, a^3, a^4, a^5 per se non sunt lineæ. Numeri esse possunt per quos lineæ significantur, sed non ipsæ lineæ. Aut si forte usurpentur in Geometria pro lineis, subintelligendum est earum aliquod latus vel consequens rationum perinde quasi scriberentur pro $\frac{aa}{b}$, $\frac{a^3}{bb}$, $\frac{a^4}{b^3}$, $\frac{a^5}{b^4}$ &c. Quod quidem satis aptè fieri potest si latus illud vel consequens rationum idem subintelligatur ubiq;, et non concipiatur per modum unitatis. Et quoniam in his gradatim ascenditur a in [? numeratoribus...]'. (He here first wrote '...Quod quidem in Arithmetica dandum est, in Geometria minime ferendum....Sed a^4 et a^5 per se lineæ non sunt. Lineas significare possunt. Numeri esse possunt per quos lineæ [repræsentantur, sed hoc modo locum non habent in Geometria]. Et satis est quod hac ratione in Algebra seu Arithmetica universali ad subjectum Geometricum applicata locum habent'. The interpolated words are supplied, to fill a lacuna in Newton's cancelled text, from his original rough jotting on Add. 3963.3: 20v, which contains a confused sequence of such preliminary drafts of the present passage. These we see no need to reproduce *in extenso* since the verbal changes they ring are by and large without significance, but we may take notice of the considerable difficulty Newton had in choosing words to deliver the core of his argument, there first affirming that 'ubi Geometra in solutione problematum solus cum solo ratiocinatur potest quidem sic more Arithmetico lineas considerare, personam Arithmetici indutus, sed ubi solutionem pro Geometrica venditat, abstinere debet ab hujusmodi conceptibus personam Arithmetici non amplius indutus', passing thereafter in its place to assert that 'aliqui lineas per modum numerorum concipiendo usurpant easdē notas [*sc.* Arithmeticas] pro significandis lineis: quod quidem in Arithmetica laudandum est[,] in Geometria minimè ferendum. Analystis tamen concessum est pro lubitu suo nunc Geometras agere nunc Arithmeticos, atq; adeo lineas repræsentare per numeros', and thereafter wavering between reshaping the last sentence as 'Analystis tamen qui personas Arithmeticorum et Geometrarum indifferenter induunt concessum est ejusmodi notas ad lineas significandas usurpare' or adding it to the previous one in the recasting 'in Analysta verò qui utriusq; personam induit tolerari potest' before deciding to omit it altogether in the revised version which precedes.) In the final text on page 406 above, of course, the reader's attention is nowhere drawn to any such potential confusion in ontological status between arithmetic and geometry.

[4][16] In quibus omnibus nota quod si ad punctum aliquod curvi cruris infiniti ducatur recta curvam illam tangens et punctum contactus in infinitum abire concipiatur: crus illud Parabolici erit generis si recta illa tangens utrinq; in infinitū producta simul abit tota in infinitum & evanescit: sin recta illa tota non abit in infinitum, crus illud Hyperbolici erit generis et ultimus locus rectæ illius erit Asymptotos ejus.

Cæterum qua ratione generaliores linearū proprietates ex inventis particularibus eruantur et quantum contemplatio projectionum ad id conducat non pigebit paucis insinuare.

Th. 1. Oculo ad infinitam distantiam sito[17] propositiones de partibus ejusdē rectæ vel parallelarum rectarum in figura Projiciente valent de partibus ejusdem rectæ vel parallelarum rectarū in Projectione, eo quod rectæ in projectione secantur ut in projiciente.

E.g. Quoniam in circulo recta bisecans duas parallelas bisecat omnes idem fiet in Ellipsi. Et quoniam hæ bisecantes a se mutuo bisecantur in circulo, bisecabuntur etiam in Ellipsi, adeoq; tam in Ellipsi quam in circulo transeunt omnes per idem punctum quod ideo centrum figuræ dicitur. Et quoniam in circulo partes semidiametri a centro ad tangentem, ad perimetrum et ad ordinatam a puncto contactus demissam sunt continue proportionales idem etiam fiet in Ellipsi.[18]

Th.1. Quæ de solis lineis inclinatione datis dicuntur, manent mutatis inclinationibus in alias inclinationes datas.

The. 2. Quæ de lineis quibusvis dicuntur, obtinent mutatis linearum longitudinibus in data ratione.

Th. 3. Linearum quæ communi lege exprimuntur communes sunt omnes conditiones mutatis mutandis nisi ubi quantitas aliqua a quibus hæ pendent in aliquo casu absurda est vel in infinitum abit.

---

(16) Add. 3963.3: 13r/13v. A lengthy discussion of how the particular forms and 'conditions' are altered by perspective projections, subsequently omitted by Newton from the revised text of his 'Geometriæ Liber primus' as being too 'profuse' a digression from his main theme; see §1: note (72) above.

(17) Evidently conceding that his reader's imagination might take the strain of conceiving such an occurrence without his apology, Newton first altered an initially following parenthesis 'des veniam duræ expressioni' to a formally polite 'ut loquar', and then cancelled this interjection too. He here of course understands the projection to be orthogonal.

(18) In other words, the tangent point and ordinate point are mutually inverse with respect to the ellipse: a standard pole-polar property which remains unchanged when the circle is optically projected into a general conic from a finite position of the 'eye'. Newton now passes— in the style of his youthful assignation of 'Conick propertys to bee examined in other curves' (II: 92)—to specify (compare §1: note (74) above) how such general properties of classes of curves may be identified, and also, in the case of curves which are of like perspective kind, what 'forms' and 'conditions' in the projecting curve produce given sorts of infinite branch in the projection.

Theor. 3B[*is*].(19) Quæ conveniunt omnibus lineis inferioris ordinis mutatis mutandis convenire solent omnibus superioris. Ut 1 si Con. Sect. habent diametros & ordinat[a]s[,] sic et reliquæ. Si in his rectæ duæ inclinatione datæ se secuerint contenta segmentorum sunt in ratione data[,] sic in superioribus. 2 Si summa quadratorum ordinatarum vel rectangulum sub ipsis applicata ad datam efficit ordinatam ad parabolam conicam[,] sic et in superioribus.

Theor. 4. Si manente linea curvâ mutetur data inclinatio rectæ alicujus donec intersectio ejus cum curva abeat in infinitum, et intersectio illa sit cruris Parabolici, pro recta illa ad intersectionem terminata substitui potest in Theoremate quovis recta data.

Th. 5. Sin ea cruris(20) Hyperbolici sit, pro recta intersecante substitui potest vel distantia ejus ab illius Asymptoto vel etiam recta aliqua quæ ad hanc dist[ant]iam sit in data ratione.

Th. 6. Et ubi distantia illa vel data est vel nulla, substitui potest linea aliqua data.

Th: 7, 8, 9. Si manente inclinatione rectæ, curva mutetur donec intersectio ejus &c. Eadem cum 4, 5, 6.(21)

Schol. Eadem obtinent ubi duo vel plura puncta simul abeunt in infinitū.

Theor. 10. Si data rectæ inclinatio mutetur donec duæ intersectiones absurdæ fiunt, et postea donec recta illa versus punctum conjugatū infinitè distans tendit, pro partibus rectæ ad duas illas intersectiones terminatis substitui possunt longitudines datæ.

Theor. 11. Si linea ad datum punctum convergit et punctum illud abit in infinitum[,] pro linea substitui potest longitudo data.(22)

---

[5](23) *Porisma 1.* Quæ de solis lineis inclinatione datis dicuntur, obtinent in alijs inclinationibus et rationibus datis. Ut ordinatis, diametris, inventione tangentis, data ratione rectangulorū, circumscripto pḡrō(24) &c ubi circulus in Ellipsin mutatur.

---

(19) Originally set as '*Theor. 12*' at the end. In thus advancing it, Newton has not taken the trouble to renumber his theorems.

(20) 'rectilinei vel' is deleted in sequel.

(21) On interchanging the 'linea curva' and 'recta', that is.

(22) Newton leaves off to recast these 'Theorems' as the equivalent 'Porisms' of the next section.

(23) The revised portions of Add. 3963.3: 13r–14r, altered and rearranged in sequence according to Newton's manuscript directions.

(24) Read 'parallelogrammo'. The theorem (first enunciated by Apollonius in his *Conics* VII, 31 but independently discovered by Grégoire de Saint-Vincent in the early seventeenth century) which is here referred to —asserting that any parallelogram circumscribed in general position to a given ellipse or hyperbola is constant in area—had been invoked by Newton a few years earlier as a crucial prop in his geometrical evaluation of the *vis centripeta* induced

Porisma 2. Linearum ejusdem ordinis quæ communi lege exprimuntur communes sunt conditiones mutatis mutandis, nisi ubi quantitas aliqua a quibus hæ dependent in aliquo casu absurda fit vel infinita. Ut in prædictis ubi Ellipsis in Hyperbolam vel Parabolam mutatur. Et in magna quæstione Veterum.(25)

Porisma 3. Quæ generaliter conveniunt lineis primi et secundi ordinis generaliter convenire solent lineis superiorum ordinum observato progressionis tenore. Ut in diametris & datis rationibus rectangulorū.

Porisma 4.(26) Duæ rectæ positione datæ vicem obeunt lineæ secundi ordinis, tres rectæ vel una et linea secundi ordinis vicem obeunt lineæ tertij, quatuor rectæ, vel duæ et linea secundi ordinis(27) vicem supplent lineæ quarti ordinis. Et sic in infinitum. Nimirum lineæ superiorum ordinum abire possunt in has inferiorum. Unde et superiorū proprietates generales (id est quæ ordini alicui generaliter conveniunt) convenient his inferioribus conjunctim sumptis, et vicissim quæ his generaliter conveniunt etiam illis generaliter convenire solent.

(28)Ponantur exempla in diametris vel summa diametrorum, summaꝗ quadra-

---

to its centre or a focus by constrained (but unresisted) motion in a central conic; see VI: 34, note (14) and 138, note (103). His present outline proof by orthogonal projection is novel.

(25) That seeking rigorous proof of the identity of the Greek 3/4-line locus with the general conic (on which see IV: 219–21) and, in extension, inquiring into the $(2n-1)/2n$-line locus, $n = 3, 4, \ldots$, analogously defined by Pappus in his *Mathematical Collection* (see 2, §3: note (123) preceding) whose like identity with the general 'geometrical' curve of n-th algebraic degree Descartes had, in particular instances, assumed—wrongly, as Newton in criticism of such 'Errores Cartesij' pointed out on IV: 340–4. Compare the initial version of Porism 4 following (reproduced in the next note), in whose 'Exempla quinta' is spelled out that the subject of this 'magna quæstio Veterum' is 'de lineis a loci puncto ad rectas positione datas ductis'; this reappears 'latissime sumpta' in note (28).

(26) On f. 13v of the manuscript Newton originally enunciated this to affirm: 'Si rectæ inclinatione datæ secuerint alias positione datas rectas[,] quæ de duabus positione datis dicuntur conveniunt omnibus conicis sectionibus, quæ de tribus dicuntur conveniunt omnibus lineis tertij ordinis, quæ de quatuor omnibus quarti & sic in infinitū. Et quæ de duabus dicuntur si mutentur mutanda convenire solent tribus, quatuor, quinꝗ aut pluribus'; after which he jotted down in outline a list of 'Exempla prim[a] in diametris ordinatarum. Exempla secunda in diametris quadratorum vel rectangulorum ex ordinatis. Exempla tertia in diametris cuborum vel parallelopipidorum ex ordinatis. &c. [*viz.* the 'Proprietatum genus primum' in section [6] following.] Exempla quarta in data ratione contenti sub omnibus ordinatis ad contenta sub segmentis basis. Coroll. de ductu tangentium. [*i.e.* the 'Proprietatum genus secundum' in [6] below.] Exempla quinta in quæstione veterum [see the previous note] de lineis a loci puncto ad rectas positione datas ductis'. He then altered this 'porism' to announce equivalently that 'Quæ de rectis curvas cujusvis ordinis in pleno punctorū numero secantibus generaliter dicuntur, conveniunt rectis alias positione datas rectas in tot punctis secantibus' before passing (on f. 14r) to the revised enunciation which here ensues.

(27) Here, of course, Newton could have added 'vel una et linea tertij ordinis'.

(28) Closely adhering to his prior list of such illustrations (given in note (26) above) Newton first went on to specify at this juncture five 'Exempla horum duorum Porismatum [3 & 4]', namely '1 In diametris, seu loco termini ordinatæ unius, semisummæ duarum, $\frac{1}{3}$ summæ trium &c. 2 Idem in quadratis & rectangulis. [3] Idem in cubis et paralle[le]pipidis ordi-

torū, cuborum &c vel rectangulorum [aut] parallelipipedorum ex ipsis applicata ad datam, ut & exemplū in data ratione contentorum sub segmentis inclinatione datis. Et in diametris Parabolicis et Hyperbolicis ubi ordinatæ tendunt ad plagam infinitorum crurum. Et addatur (tanquam consectariū) hæc proprietas quod Hyperbolæ plenæ eadem est Asymptotorum & Hyperbolæ diameter, et probetur per coincidentiam asymptotorum et crurum ad infinitā distantiam(29)

Porisma 5. Si lineæ inclinatione datæ intersectio cum curva per mutationem figuræ abit in infinitum & intersectio illa sit cruris Parabolici, supponi potest datam esse rationem infinitæ illius rectæ ad rectam datam. Sin intersectio illa sit cruris Hyperbolici, supponi potest datam esse rationem infinitæ illius ad distantiam ejus ab Asymtoto cruris, [ac] si distantia illa vel data est vel nulla, supponi potest datam esse rationem infinitæ illius ad rectam datam.

Ponantur exempla 1 in Parabola conica et regula generali de Parabolis, 2 in Hyperbola conica et regula duplici generali de Hyperbolis, 3 in utrisq.

Porisma 6. Si data rectæ inclinatio mutetur donec duæ intersectiones evadunt absurdæ, et amplius donec recta illa versus punctum conjugatū infinitè distans tendit, concipiendū est partes rectæ ad duas illas intersectiones terminatas converti in rectas infinitas quæ ad datas sunt in data ratione.

Porisma 7. Si linea ad datum punctum convergit et punctum illud abit in infinitum concipiendū datā esse rationem rectæ illius infinitæ ad datam.

Porisma 6. Si punctū conjugatum infinitè distat, concipiendū est ordinatas rectas ad eam desinentes datam habere rationem ad rectas datas.(30)

---

[6](31) Quæ conveniunt lineis duarum(32) specierum convenire solent toti generi

---

natarum &c. 4 Idem in ratione contenti sub omnibus segmentis ad contentum sub omnibus alijs. 5 In quæstione Veterum latissime sumpta, ubi incognitæ sunt unius dimensionis ad rectam[,] duarum ad lineas 2di ordinis, trium ad eas tertij'. The last would seem to be the generalization of Descartes' reduction of the Greek 3/4-line locus to a quadratic relationship connecting two coordinate 'unknown' line-lengths (see IV: 219–20) in which 'geometrical' curves of *n*-th algebraic degree are represented by an analogous Cartesian equation of *n*-th degree connecting these coordinate unknowns.

(29) See page 442 above. Newton had stated the cubic instance of this theorem long before (see II: 102), and its present generalized enunciation is an immediate extension.

(30) This preliminary scheme—far from all of whose roughnesses, false starts and tentative redraftings we have here indicated—was at once amplified by Newton in the preliminary version reproduced in [6], and then further broadened and deepened by him in the extended revise which we give in [7].

(31) Add. 3963.3: 14v/3963.10: 107r. In this initial enlargement upon the scheme set out in [5] preceding, Newton presents in improved form his basic criterion for determining the general properties of algebraic curves—those, namely, which hold true *mutatis mutandis* for all their orders and in all the component species of these—and then elaborates two such 'kinds'

& quæ conveniunt lineis duorum(32) generum convenire solent ordini & quæ conveniunt lineis duorum(32) ordinum observato progressionis tenore convenire solent ordinibus universis. Hic autem(33) duæ lineæ primi ordinis(34) positione datæ vicem supplent generis linearum secundi ordinis, tres lineæ primi ordinis(34) positione datæ vel una primi et una secundi vicem supplent generis linearum tertij ordinis. Quatuor lineæ primi ordinis(34) vel duæ primi et una secundi vel duæ secundi vel deniq; una primi et una tertij supplent locum generis linearum quarti ordinis. Et sic in infinitum.(35)

Considero igitur quænam sint generales proprietates linearum primi ordinis combinatarum in infinitum(36) & eas quæ conveniunt curvæ alicui & soli et cum rectis combinatæ retineo penitius examinandas. Fieri enim vix potest quin hæ conveniant curvis omnibus.

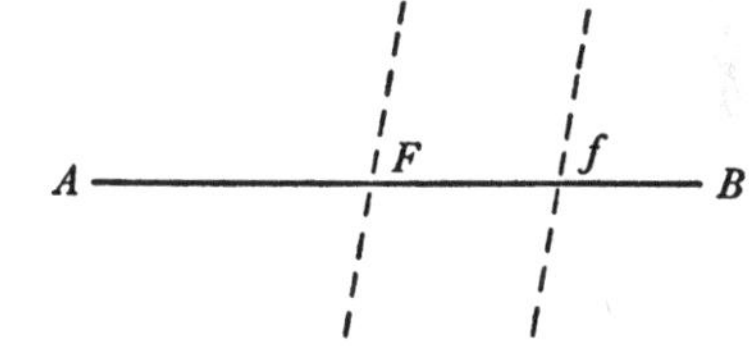

Proprietatum genus primū. Si rectæ quotcunq; parallelæ secuerint rectas quotcunq; positione datas et agatur recta quæ parallelarum duarū segmenta hinc inde æqualia fecerit, ea faciet parallelarum omnium segmenta hinc inde æqualia. Si una est recta positione data AB, segmenta(37) parallelarum hinc inde nulla erunt. Si duæ rectæ positione datæ AB, AC a parallelis quotcunq; FG, fg, KL secentur, recta Rr quæ parallelas duas $FG\,fg$ bisecans fecerit segmenta RF & RG, rf & rg hinc inde æquales etiam bisecando re-

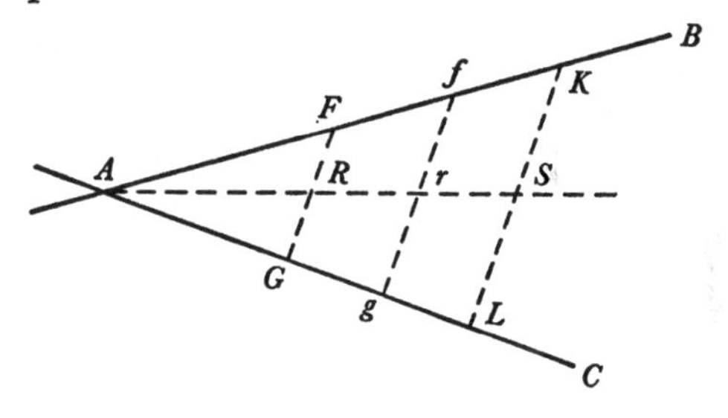

of *proprietas generalis*, sketching the application of the latter genus to construct the tangent at an arbitrary point on a given conic.

(32) Newton has at these places deleted 'diversarum' and 'diversorum' respectively, doubtless on second thought regarding it as superfluous to stress the evident distinction of the two *species* or *genera* here specified.

(33) Originally 'Et in hoc Poristico'.

(34) The equivalent 'rectæ' is in each case replaced.

(35) In sequel Newton provisionally recast this codicil to read 'Transire [autem] potest linea secundi ordinis in duas primi, linea tertij in unam secundi et unam primi vel etiam in tres primi, linea quarti in unam tertij et unam primi vel in duas secundi vel in unam secundi et duas primi vel deniq; in quatuor primi[,] et sic in infinitum. Et ubi hoc fit illarum proprietates generales transeunt in proprietates harum conjunctim sumptarum'. Quickly, however, he discarded this refashioned statement to return to his first version of it.

(36) In a cancelled preceding sentence Newton asserted more narrowly that 'Considero igitur quænam proprietates circuli (siquidem hæc sit curva notissima) applicari possint ad duas rectas, deinde an hæ applicari possint ad seriem rectarum in infinitū'. He here in redraft initially went on to add '& si hæ proprietates conveniant curvæ alicui, summè probabile est quod convenient omnibus'.

(37) How a single line can cut off such segments terminated at their either end Newton does not bother to explain.

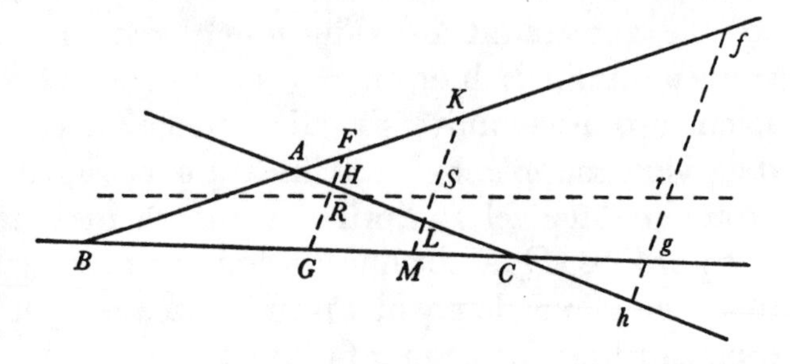

liquas omnes parallelas *KL kl* faciet earum segmenta *SK* & *SL*, *sk* & *sl* hinc inde æquales. Si tres rectæ positione datæ *AB*, *AC*, *BC* a parallelis *FHG*, *fhg* secentur in *F*, *H*, *G* & *f*, *h*, *g* rectâ quæ parallelas duas ita secat ut earum segmenta duo ex una secantis parte & segmentum tertiũ ex altera æquentur, nimirum segmenta *RF* et *RH* segm[en]to *RG* & segmenta *rg* et *rh* segmento *rf*, hæc recta parallelas omnes *KLM* ita secabit [ut] earum segmenta [*SK*, *SL* et *SM*] hinc inde æqualia [sint].(38)

Proprietatum genus secundum.(39) Si rectæ duæ inclinatione datæ secuerint rectas quotcunq͡ȝ positione datas,(40) contentum sub partibus unius ad alteram terminatis erit ad contentum sub partibus alterius ad priorem terminatis in ratione data.

Secent rectæ duæ inclinatione datæ alias quotcunq͡ȝ rectas positione datas, & considerentur partes secantium ad communem intersectionem terminatæ. Si secuerint unam rectã positione datam, pars unius secantis est ad partem alterius in ratione data, si duas positione datas secuerint, contentũ planum sub duabus partibus unius secantis erit ad contentum planum sub duabus partibus alterius in ratione data, si tres positione datas secuerint contentum solidum sub tribus partibus unius secantis erit ad contentum solidum sub tribus partibus alterius in ratione data. Et sic in infinitum. Nam ratio contenti ad contentũ componitur ex rationibus datis partium ad partes. Considero igitur utrum pro rectis duabus positione datis substitui possit circulus et ex successu colligo propositionem in curvis lineis substitutis universaliter valere.(41) Id est si rectæ duæ positione datæ secuerint curvam in pleno numero punctorũ contentum sub partibus unius secantis ad communem secantiũ intersectionem termi-

(38) In the two cases, of course, the line *RS* will be the Newtonian diameter of the line-pair *AB*/*AC* and of the line-triple *AB*/*AC*/*BC* respectively. By his preceding criterion, therefore, all conics and cubics have analogously determinable rectilinear diameters *RS* with respect to a given direction of the defining parallels *FG*, *KL* as Newton had indeed long known to be true. (See II: 96–8, and compare our restoration of his proof of the property in the cubic instance—readily extendable to the general case—in *ibid.*: 98—9, note (36).) Doubtless intending there to insert some remark to this effect, he has in sequel left the remainder of his manuscript page (f. 14v) blank.

(39) Afterwards altered to be '*tertium*' by writing this replacement in overhead; correspondingly, a lone, abandoned half-title '*Proprietatum genus tertium*' at the end of the manuscript text—here omitted—has been converted to be '...*secundum*' (see note (43) below).

(40) '& considerentur omnia duarũ segmenta ad earum communem intersectionem terminata' is deleted in sequel.

(41) As Newton had again long known; see IV: 358, note (18).

natis est ad contentū sub partibus alterius secantis ad eandem intersectionē terminatis in ratione data. Ut si &c.(42)

Et hinc ubi basis secat curvam in pleno numero punctorum et numerus Ordinatarum plenus est, datur contentum sub omnibus Ordinatis, eo quod datur ratio contenti illius ad contentum sub segmentis basis.

Hinc etiam patet quomodo ad datum curvæ cujusvis descriptæ punctum tangens duci potest.(43)

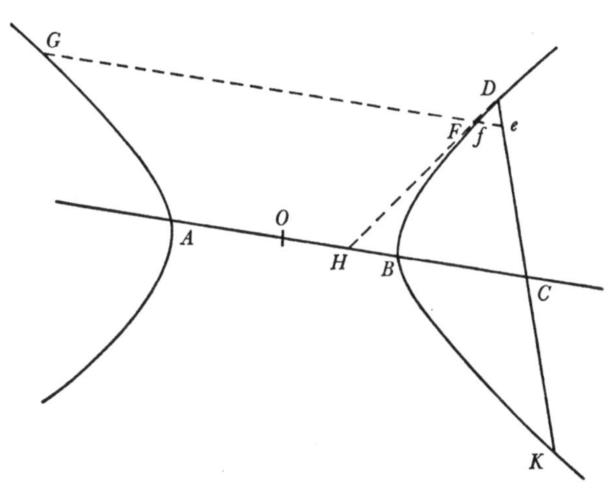

Hinc deniq; patet Regula expedita ducendi tangentes in Conicis Sectionibus. Quoniam est DCK ([hoc est] $-DC^q$) ad ACB ut DeK ad $[f]eG$ seu ut DC,eK ad HC,eG, et punctis e D F coincidentibus ut DC,DK ad HC,DG id est ut DC^q ad $HCO_{[,]}$ erit $HCO = ACB = OC^q - OB^q$. His ablatis de OC^q erit $COH = OB^q$, id est $OC.OB::OB.OH$.(44)

---

(42) Though he here gives no example of this general property of algebraic curves, Newton has set the accompanying figure with his manuscript text, manifestly to illustrate the cubic instance in which, for all parallels to CBD and FGH meeting each other in A and the three fixed lines BF, CH and DG in the respective triples of points B, C, D and F, H, G, the ratio of the 'content' $AB \times AC \times AD$ of the parts cut off in the former to the similar 'content' $AF \times AH \times AG$ of those cut off in the latter is constant.

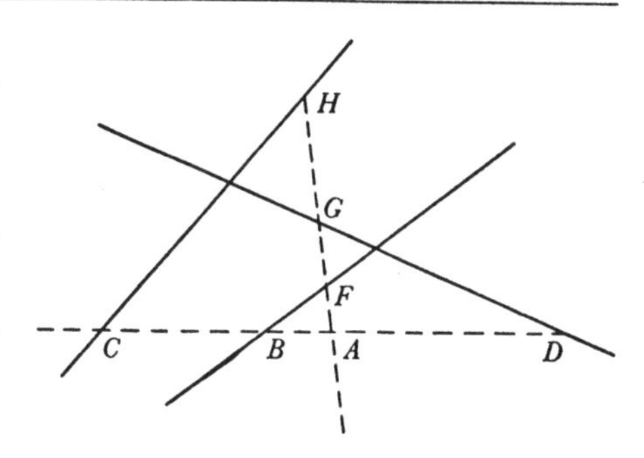

(43) See page 442 above for the full construction. In his present manuscript draft Newton leaves a short gap for its future insertion and passes to illustrate its method in the case of a central conic.

(44) Here understand, of course, that O is the conic's centre and DCK a general ordinate conjugate to the diameter AB, with DH the tangent at D and Ge a parallel to AB, meeting DH in F and the conic arc $\widehat{BD}$ in f, which is assumed to pass ultimately through D 'punctis *e, f, F* coincidentibus'. In this case—that of a hyperbola as Newton draws it in his figure—his preceding general property reduces to that enunciated by Apollonius in his *Conics* III, 17/18 (compare IV: 284, note (8)), whereby it is specified that for all parallels to DK and fG meeting in e the ratio of $eD \times eK$ to $ef \times eG$ is constant, and is here accordingly equal to that of $CD \times CK$ to $CB \times CA$. By considering the limit-ratios of the simultaneously vanishing 'moments' De, $\widehat{Df}$ and DF of DC, $\widehat{DB}$ and DH Newton attains an elegant derivation of the pole-polar property demonstrated (by a limit-avoiding *reductio* proof) in *Conics* I, 34/36/37.

In the manuscript there follows immediately afterwards the heading '*Proprietatum genus*

[7][45] *Propr. genus primū.* Si a basi aliqua rectilinea positione data in dato angulo erigantur ab eodem puncto incedentes[46] ad rectas quotcunꝗ positione datas terminatæ[,] incedens his omnibus ordinatis conjunctim sumptis æqualis terminabitur ad rectam, incedens quæ fit applicando harum omnium quadrata vel rectangula omnia sub singulis binis contenta ad datam rectam, terminatur ad parabolam quadraticam Veterum seu Conicam tenditꝗ ad plagam ejus crurum infinitorum.[47] Incedens[46] quæ fit applicando harum omnium cubos vel parallelipipeda omnia sub singulis ternis contenta ad datū planum, terminatur ad Parabolam simplicem cubicam seu Wallisianam tenditꝗ ad plagam crurum infinitorum ejus. Et sic in infinitum[48] Cum igitur hæ rectæ positione datæ, possint esse vel una vel duæ vel tres vel quatuor et ita deinceps, serie in infinitū progrediente; considero utrum in aliquo casu pro his rectis substitui possit curva[49] linea correspondentis ordinis, id est curva linea secundi ordinis pro rectis duabus; una tertij pro tribus & sic deinceps ita ut pro rectarum illarum incedentibus habeantur totidem incedentes curvæ substitutæ, et facto experimento in circulo substituendo hunc pro duabus rectis, ex successu colligo Propositionem valere in curvis omnibus si modo signa ordinatarum probe observentur. Id est duæ incedentes *BC*, *BD* ad lineam quamvis secundi ordinis *CPD*[50] terminatæ, tres [*BC*, *BD*, *BE*] ad quamvis tertij, quatuor ad quamvis

tertium'—subsequently altered by Newton to be '...*secundum*' (see note (39) above)—devoid of any following text. Which other general property of curves he intended to go on to enunciate is impossible to specify, but it was presumably one or other of the additional ones listed at the end of the revised version which here follows in [7].

(45) Add. 3963.3: 21r–26r. An augmented revise of Newton's discussion of the general properties of curves reproduced in [6] preceding.

(46) Originally 'ordinatæ' and 'Ordinata' respectively.

(47) This replaces 'ejus[ꝗ] diametris parallela est'.

(48) In a first version of the sequel at this point Newton initially continued: 'Considero igitur utrū hæ proprietates conveniant lineæ alicui curvæ, puta circulo, et ex successu colligo eas convenire curvis omnibus. Id est si a basi aliqua *AB* & ab eodem ejus puncto *B* agātur in eodem dato angulo *ABC* ad curvam aliquam incedentes sintꝗ tot incedentes quot curva illius ordinis hēre [= habere] potest, id est duæ *BC*, *B*[*D*] si curva sit secundi ordinis, tres *BC*, *BD*, *BE* si tertij & sic deinceps: aggregatum incedentium omnium sub signis suis ($BC+BD$ vel $BC+BD-B[E]$ &c) componet incedentem (*B*[*F*]) quæ terminabitur ad datam lineam rectam [*FG*]; aggregatum verò quadratorum ex incedentibus applicatorum ad datam *N*, fac[i]et incedentem $\frac{BC^q}{N}+\frac{BD^q}{N}$ vel $\frac{BC^q}{N}+\frac{BD^q}{N}+\frac{BE^q}{N}=B[F]$ quæ terminatur ad Parabolam Conicam, et idem faciet aggregatū rectangulorum sub si[n]gulis binis incedentibus applicatorum ad datam $\frac{BC,BD}{N}$ vel $\frac{BC,BD}{N}-\frac{BC,BE}{N}-\frac{BD,BE}{N}=B[F]$'.

(49) 'aliqua' is deleted.

(50) In his manuscript Newton confusingly here refers to the relevant portion of the figure illustrating 'Prob. 1' below, one which he afterwards much altered without changing his present designations. For consistency and clarity we here reproduce the sketch (slightly simplified and altered in its marking of points) which he had a little before set with the cancelled passage quoted in note (48) above.

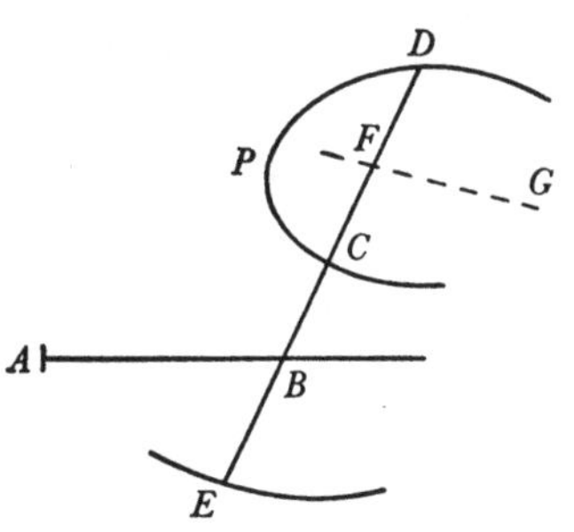

quarti &c, si sub signis suis conjungantur, faciunt longitudinem incedentis *BF* quæ terminatur ad lineam rectam *FG*, & earundem quadrata conjuncta et applicata ad datam rectam efficiunt longitudinem incedentis quæ terminatur ad Parabolam Conicā et ad plagam crurum infinitorum Parabolæ tendit. Quod idem fiet si rectangula sub singulis binis incedentibus conjungantur sub signis suis & applicentur ad rectam positione datam. Et ita de cubis et parallelipipedis & contentis altioribus in infinitum observato progressionis tenore.(51)

Et hinc sequentia Problemata conficiuntur.

Prob. 1. Invenire summam incedentium super quavis basi ad Curvam quamvis propositam.

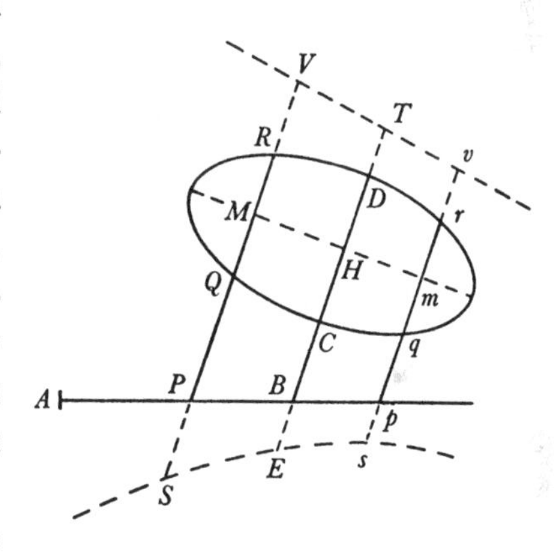

A basis punctis duobus *P* et *p* agantur ad curvam incedentes(52) *PQ*, *P*[*R*], *PS* &c & *pq*, *pr*, *ps*. Debent autem tot esse incedentes ad utrumꝗ punctum quot curva illius ordinis admittit[,] id est duæ ut *PQ*, *PR*, & *pq pr* si curva sit linea secundi ordinis, tres ut *PQ*, *PR*, *PS* & *pq*, *pr*, *ps* si tertij, et ita deinceps. In his incedentibus capiantur novæ incedentes *PV* & *pv* quæ sint harum sub signis suis conjunctarū aggregata ut $PQ+PR$ & $pq+pr$ si curva sit secundi ordinis linea vel $PQ+PR-PS$ & $pq+pr-ps$ si tertij. Et per harum novarum terminos agatur recta *Vv*. Hæc erit locus quem tanget aggregatum incedentium in omni casu.(53) Quare si incedentium aliarum quarumvis et $BC+BD$ vel $BC+BD-BE$

(51) If, in standard Cartesian equivalent, we denote the common abscissa $AB = x$ and take the corresponding ordinates $BC = y_1$, $BD = y_2$, $BE = y_3$, ... to be the roots of some given equation $y^n-(ax+b)\,y^{n-1}+(cx^2+dx+e)\,y^{n-2}-(fx^3+gx^2+hx+l)\,y^{n-3}+\ldots = 0$, it results that $\sum_{1\leqslant i\leqslant n} y_i = ax+b$, $\sum_{i\neq j} y_iy_j = cx^2+dx+e$, $\sum_{i\neq j\neq k} y_iy_jy_k = fx^3+gx^2+hx+k$, and so forth. If, then, the *incedens* $BF = Y$ satisfies $Y = (1/n).\Sigma y_i$, the locus of F is manifestly the straight line $nY = ax+b$; but if there is $Y = (1/n).\Sigma y_iy_j$ or $Y = (1/n).\Sigma y_i^2 = (1/n)\,((\Sigma y_i)^2-2\Sigma y_iy_j)$, then F lies respectively on the conic parabolas $nY = cx^2+dx+e$ or $nY = (ax+b)^2-2(cx^2+dx+e)$; while if there is $Y = (1/n).\Sigma y_iy_jy_k$ or $Y = (1/n).\Sigma y_i^3 = (1/n)\,((\Sigma y_i)^3-3\Sigma y_i.\Sigma y_iy_j-3\Sigma y_iy_jy_k)$, then F lies on the cubic parabolas defined by $nY = fx^3+gx^2+hx+l$ or

$$nY = (ax+b^3)-3(ax+b)\,(cx^2+dx+e)-3(fx^3+gx^2+hx+l)$$

respectively; and so on. The expansions here invoked of the sum of the powers Σy_i^p, $p = 2, 3, \ldots$, of the roots y_i in terms of their symmetric functions $\Sigma y_iy_jy_k \ldots$ had, of course, been one of Newton's earliest discoveries in the theory of equations; see I: 519.

(52) In a cancelled preliminary draft immediately preceding, Newton had specified the proviso 'in pleno numero punctorum secantes' which he here defers to his next sentence.

aggregatum desideratur, tangant hæ incedentes productæ lineam Vv in T et erit BT aggregatum quæsitum.

Et hinc patet curvas omnes lineas habere diametros[54] et simul quomodo hæ diametri inveniri possint. Nimirum si Curva $QqrR$ sit linea secundi ordinis, biseca parallelas duas PV pv in M et m, vel quod perinde est biseca parallelarum partes ad curvam utrinq; terminatas QR et qr et acta Mm bisecabit alias omnes parallelas BT, atq; adeo omnium partes CD ad curvam utrinq; terminatas. Est igitur $M[m]$ basis habens incedentes hinc inde æquales $HD = HC$. Et talem basem nominamus *diam[etrum]*.[55] Quod si curva sit linea tertij ordinis cape PM pm tertias partes ipsarum PV, pv et acta Mm basis erit habens incedentes hinc inde æquales[,] id est duas ex una parte æquales tertiæ ex altera ut

$$HE+HC = HD.$$

Et talem basem nominamus *diametrum* linearum hujus[56] ordinis. Atq; ita in lineis quarti ordinis diameter invenietur capiendo PM pm quartam partem linearum PV pv, in ijs quinti capiendo PM pm quintam partem et sic in infinitum. Incedentes verò super his diametris ut MR HD mr nominamus *O[r]dinatim applicatas*, et si ducantur incedentes parallelæ diametro alicui et quæratur alia diameter ad quam incedentiū illarum partes ordinatim applicantur, hanc in curvis universis dicimus *diametrum secundam* et utramq; alterius *conjugatam*. Et concursum diametrorum *centrum Ordinationis* et absolutè *centrum* ubi diametri omnes in eodem puncto concurrunt. Et quoniam si pro linea secundi ordinis[57] substituantur duæ lineæ primi ordinis, earum diametri omnes quæ ad invicem inclinantur concurrunt in eodem puncto atq; id etiam verum est in circulo, colligo id verum esse in omnibus lineis secundi ordinis quorū diametri ad invicem inclinantur[,] id est has omnes diametros in communi centro concurrere. Et præterea cum omnis diameter a conjugata diametro bisecetur, patet has omnes diametros in centro[58] illo bisecari.

Præterea ex præcedentibus facilè colligitur quod si parallelæ duæ AC, BD secuerint curvam quamvis in pleno numero punctorum id est in duobus si curva sit secundi ordinis[,] in tribus si tertij &c, et arcus intercepti AB, CD, EF &c subtendantur rectis; Dein tertia parallela GK ducatur secans arcus et sub-

(53) For, in the terms of note (51) preceding, with respect to the base-line

$$nY = ax+b = \sum_{1\leqslant i\leqslant n} y_i$$

there is at once $\sum_{1\leqslant i\leqslant n} (y_i - Y) = 0$.

(54) That is, 'rectilineas'.

(55) To accord with what follows understand '*secundi ordinis*'.

(56) Namely 'tertij'.

(57) This replaces an equivalent 'pro conicis' in a preceding cancelled opening to the sentence.

(58) Originally 'in communi concursus puncto'.

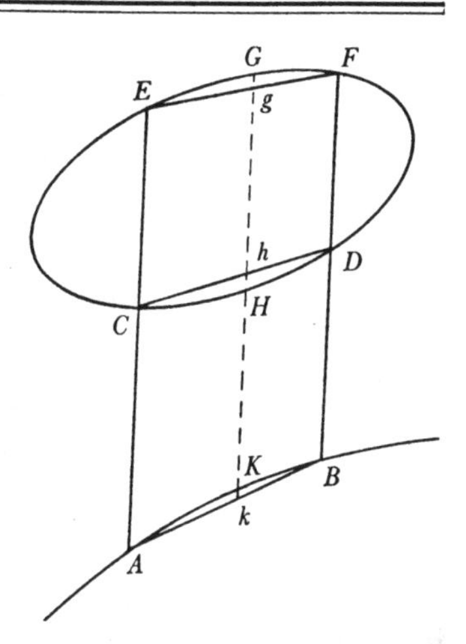

tensas omnes: partes hujus parallelæ inter subtensas & arcus quæ ex una parte subtensarum jacent æquales erint ijs quæ jacent ex altera parte [ut] $Kk = Hh$ vel $Kk + Gg = Hh$ [&c, nec non] areæ arcubus et subtensis comprehensæ quæ ex una parte subtensarum jacent æquales erunt areis ex altera parte jacentibus, nimirum $ABK = CDH$ si linea secundi est ordinis, vel $ABK + EFG = CDH$ si tertij et sic in infinitum.(59)

Ex præcedentibus deniq; colligitur quod si curva habet plenum numerū Asymptotorū id est duas si secundi sit ordinis linea, tres si tertij &c et agatur recta asymptotos omnes totidemq; curvæ crura secans, partes hujus rectæ ab aliquibus asymptotis ad totidem crura Hyperbolica versus unam plagam ductæ(60) æquales erunt ijs quæ a reliquis Asymptotis ad reliqua crura versus plagam oppositam ducuntur. Nam quia Asymptoti & crura conveniunt ad infinitam distantiam[,] ideo communis est eorum diameter adeoq; ordinatæ hinc inde æquales de ordinatis hinc inde æqualibus deductæ relinquunt differentias hinc inde æquales.(61)

Prob. 2. Invenire summam assignatorū dignitatum(62) linearum Incedentium super quavis basi ad curvam quamvis propositam.

A basis punctis tribus si dignitates sunt gradus secundi,(63) a quatuor si tertij,

(59) Newton initially went on to explain: 'Nam rectarum AB, CD vel AB, CD, EF et arcuum AKB, CHD vel AKB, CHD, EGF eadem est diameter adeoq; ordinatarum hinc inde æqualium æquales sunt differentiæ $Kk = Hh$ vel $Kk+Gg = Hh$ et æqualibus differentijs æquales describuntur areæ'. Evidently, since the parallels ACE and BDF meet both the curve $AKB/CHDFGE$ and the line-multiple $AkB/ChD/EgF$ in the same groups of points $A, C, E, \ldots$ and $B, D, F, \ldots$, the rectilinear Newtonian diameters of these (corresponding to the ordinate direction) meet ACE and BDF in identical points—α and β, say, where

$$\alpha A + \alpha C + \alpha E + \ldots = \beta B + \beta D + \beta F + \ldots = 0$$

—and are hence coincident. At once, if the arbitrary parallel KHG meets this diameter in γ, there is $\gamma K + \gamma H + \gamma G + \ldots = 0 = \gamma k + \gamma h + \gamma g + \ldots$ and in consequence

$$(\gamma k - \gamma K) + (\gamma h - \gamma H) + (\gamma g - \gamma G) + \ldots = 0,$$

so that universally (if due account be taken of the sense of the individual segments) there is $Kk + Hh + Gg + \ldots = 0$.

(60) 'jacentes' is replaced.

(61) For, as an immediate corollary to the above, when the parallel BDF comes to coincide with ACE—that is, where Ak, Ch, Eg, ... are tangents to the curve at $A, C, E, \ldots$—it holds true that the arbitrary parallel ordinate KHG meets these such that (see note (59) preceding) $Kk + Hh + Gg + \ldots = 0$. Newton merely enunciates the particular instance of this when ACE is the line at infinity and the tangents Ak, Ch, Eg, ... become asymptotes.

(62) Originally specified by Newton as '*quadratorum, cuborum, quadrato-quadratorū aut aliorum graduum sub signis suis conjunctorū*'.

a quinq; si quarti &c agantur ad curvam Incedentes sitq; numerus incedentium plenus ad omne punctum, id est sint incedentes duæ ad omne punctum si curva est secundi ordinis, tres si tertij et ita deinceps. Dein sum̄a dignitatum incedentium ad unum punctorum capiatur a puncto illo in linea incedentium pro incedente nova. Et idem fiat in incedentibus ad reliqua puncta et per terminos novarum incedentium describatur Parabola Conica si incedentes novæ sunt tres, aut Parabola Wallisiana si quatuor sunt aut Parabola quædam quarti ordinis si quinq; sunt et sic in infinitū observata Parabolarum serie; tendant autem infinita crura Parabolæ ad plagam incedentium et incedens ad Parabolam illam in eodem angulo super eadem basi semper erit summa dignitatum incedentium ad curvam propositam. Problema parum differt a superiore nisi quod loco rectæ lineæ *Vv* ibi per inventionem duorum punctorum *V* et *v* determinatæ, hic habetur linea parabolica per inventionem trium pluriumve punctorū determinanda.(64) Et puncta quidem eodem fere modo inveniuntur in utroq; Problemate, at Parabolam invenire quæ per puncta tria vel plura inventa transibit difficilius est in hoc Problemate quam lineam rectam per inventa puncta ducere in illo. Tota igitur difficultas redit ad Problema sequens.

Prob 3. Parabolam invenire quæ per data quotcunq; puncta transibit et cujus crura infinita tendent ad plagam imperatam.(65)

Possibile est tamen curvam in paucioribus punctis(66) a recta secari quam pro genere suo. Sic enim recta quæ axi Parabolæ vel Asymptoto Hyperbolæ parallela est, curvas illas in unico tandem puncto secat et tamen curvæ illæ sunt secundi generis. Atq; ita recta quæ tendit ad plagam ultimarum partium cruris infiniti figuræ cujuscunq;, secat figuram illam in pūctis uno paucioribus quam pro genere figuræ. Cujus rei ratio est quod rectæ si gyret(67) et cruris infiniti inter-

(63) Initially '...si summa cuborum quæritur', after which he went on to add analogously 'a quatuor si summa □^{to}□^{torū}, a quinq; si summa ^{to}□-cuborum' before likewise, replacing these sums of 'quadrato-quadrates' and 'quadrato-cubes' (fourth and fifth powers).

(64) Compare note (51) above. Where the *incedens* Y is determined, corresponding to the common abscissa $AB = x$, by the condition $Y = (1/n).\Sigma y_i^p$, $p = 2, 3, 4, \ldots$, the locus of its end-point will evidently lie on the p-th degree parabola $nY = rx^p+sx^{p-1}+tx^{p-2}+\ldots+v$. In the manuscript Newton has cancelled at this point a figure (which we see no point in reproducing) where this parabolic locus is illustrated as a broken-line arc curving down towards the base-line AB on either hand.

(65) Newton has left a long space blank in immediate sequel (on f. 22r) for the future elaboration of this problem—doubtless, as on page 492 below (see note (92)), by an appropriate central-differences interpolation in line with his earlier schemes on IV: 60–8, or maybe by an equivalent advancing-differences formula as in his *Principia* (see IV: 70–3). However resolved, the problem at once constructs the p-th order diameter of a given curve through the single points, each readily determinable, in which it meets $p+1$ parallel ordinates AE, BF, KG ....

(66) Understand these to be 'finitis et distinctis'.

(67) Initially 'rectæ per curvæ punctū infinitè distans gyrantis' and thereafter 'si gyrasse concipiatur'.

sectio in infinitū abiens evanescit quam primum recta ad plagam dirigitur ad quam crus pergit in infinitum.(68) Ut igitur numerum intersectionum impleamus, intersectionem quæ in infinitum abijt atq; adeo intersectio non est amplius, vocabimus tamen *intersectionem infinite distantem*. Et siquando intersectiones vel ad eandem plagam vel ad plagas oppositas in infinitum abeunt, ita ut rectæ & curvæ desint duæ intersectiones, vocabimus evanescentes illas *duplicem intersectionem infinite distantem*; & sic *de triplici*, *quadruplici* cæterisq;.

Quinetiam curva in gyrum flexa seipsam nonnunquam decussat & tunc punctum decussationis pro duplici habendum est. Et si curva Ovalem habet conjugatam ut sæpe fit, potest Ovalis illa diminui donec in punctum conjugatum evanescat et quoniam recta secare potest ovalem in duobus punctis, transitus ejus per punctum conjugatum rationem habebit intersectionis duplicis. Sic et ubi curva cuspidata est ad modum Cissoidis, punctum extremum cuspidis pro duplici(69) habendum est. Possunt et intersectiones tres vel plures coincidere: quo in casu punctum ubi coincidunt pro triplici vel multiplici habendum erit. Hæc autem puncta sive duplicia sunt sive multiplicia, sive ad finitā distantiam posita sive ad infinitam uno nomine *puncta composita* vocabo.

Et his punctis distinguuntur genera(70) curvarum in gradus subordinatos. Nam (71) simpliciores sunt curvæ ejusdem generis quæ puncta talia habent quam quæ non habent. Curvæ secundi generis, nimirum Conicæ sectiones talia puncta non habent. Tertij generis aliquæ punctum duplex habent, aliæ non habent. Priores igitur sunt primi gradus, posteriores secundi. In quarto genere aliquæ habent vel punctum triplex vel tria puncta duplicia et has statuo primi gradus, aliæ duo puncta duplicia, aliæ unum duplex, aliæ nullum punctum compositum et hæ sunto gradus secundi tertij et quarti. Rursus in quinto genere aliquæ quæ sunt primi gradus habent vel punctū quadruplex, vel tria duplicia et unum triplex, vel sex duplicia. Et hi gradus rursus in ordines distingui possunt ut in quarti generis gradu primo ordinis sint primi quæ habent punctum triplex[,] secundi quæ tria puncta duplicia. In quinti generis gradu primo ordinis sunt primi quæ habent punctum quadruplex, secundi quæ habent punctum triplex et tria puncta duplicia, tertij quæ habent sex duplicia.

Et rursus figuræ generis gradus & ordinis cujusq; sunt multarum specierum, prout crura habeant vel Hyperbolica vel Parabolica vel utraq; vel ex his

(68) A first cancelled continuation at this point reads: '*Plagam* verò cruris infiniti voco ad quam tendit Asymptoto[s] si crus hyperbolici est generis, vel ad quam tangens cruris quod de genere Parabolico est, ultimo dirigitur ubi recta ipsū secans incipit ad eandem plagam cum cruris partibus ultimis in infinitum pergentibus dirigi'.

(69) Or of course 'pro triplici' if the line intersecting at the cusp is there tangent to the curve.

(70) Originally 'Et horum punctorum consideratione oritur distinctio generum'.

(71) A too forceful 'longe' is here deleted.

cruribus aliqua imaginaria sunt[72] vel etiam Elliptica. *Crus Hyperbolicum* voco quod excurrit in infinitum ea lege ut si punctū aliquo[d] in eo movens abeat in infinitum et recta linea semper tangat hoc crus ad punctum illud; hæc recta dum punctum contactus abit in infinitum, magis et magis accedat ad aliam aliquam positione datam rectam donec tandem ubi punctum contactus in infinitum abijt coincidat cum recta illa quæ positione datur. Recta vero cum qua tangens jam coincidit *Asymtotos* cruris dicitur. Unde Asymptotos est recta linea quæ curvam tangit ad infinitam distantiam, vel ut magis proprie loquar, est tangentium finitarum limes. *Crus Parabolicum* est cujus tangens si punctum contactus in infinitum abit, etiam ipsa tota in infinitum abit et evanescit, per nullum spatium finite distans jam amplius transiens. Et si curvæ alicujus conditiones ita mutentur ut crura duo quæ infinita erant, jam non amplius in infinitum excurrentes sed ad invicem convergentes jungi[73] incipiant ad finitam distantiā & in se mutuo propagari, hæc curva jam *deficere* dici potest vel *Elliptica* esse idq; totupliciter[74] quot sunt hujusmodi defectus crurum.

Jam vero ut quot modis curva sit Hyperbolica Parabolica ac Deficiens innotescat, sciendum est lineam primi generis (quæ sola recta est) habere unicam tantum intersectionem in infinitum abeuntem, lin[e]ā secundi duas habere posse intersectiones Hyperbolicas (seu crurum Hyperbolicorū) [in] infinitum abeuntes & non plures; lineam tertij tres & non plures, lineam quarti quatuor & sic in infinitum.[75] Tot vero numero intersectiones in infinitum abeuntes [esse] quot abibunt in infinitum dum recta lineam propositam secans circa punctum aliquod volvitur, quarum unaquæq; quamprimum in infinitum versus plagam aliquam abijt, incipit a plaga opposita ab infinito redire[: lineam primi generis habere] duo tantum crura infinita, lineam secundi habere posse duo paria crurum infinitorum et non plura, tertij tria tantum, quarti quatuor et sic in infinitum. *Par crurum* vel *crura conjugata* voco quæ ad oppositas plagas vel forte ad easdē tendunt in infinitum. Ubi vero Curva habet omnia crura quæ pro genere suo habere potest[,] crura illa solent esse omnia Hyperbolica: unde Curva dici potest *Hyperbola pura*;[76] aut si mavis pro numero parium crurum *Hyperbola duplex*, *triplex*, *quadruplex* [&c]; vel *curva bis*, *ter*, *quater Hyperbolica.* Pro omni pari crurum Parabolicorum[77] vel deficientium deerunt paria duo crurum Hyperbolicorum[,] ut si [linea est] tertij generis et par Parabolicū

(72) Originally—and less precisely—'aliqua deficiunt'.

(73) 'coalescere' was first written.

(74) Newton initially began to pen an equivalent 'tot mo[dis]'.

(75) An unfinished first sequel at this point begins: 'Una vero hujusmodi intersectio Parabolica [? esse potest]'.

(76) Originally 'plena'.

(77) Newton has deleted a following parenthesis where he initially explained 'talia voco quæ tendunt ad eandem plagam & versus spatium interjectum con[cav]a sunt'.

habet, non habebit nisi unum par Hyperbolicum. Excipiendum est tamen ubi par crurum Parabolicorum divergunt a[d] plagas oppositas, co[n]vexitatem ad easdem partes habentia. Nam hujusmodi par crurum Parabolicorum indicat unum tantum par crurum Hyperbolicorum deesse. Rursus si crura duo parabolica vel duo Hyperbolica & non plura tendunt ad punctum duplex infinite distans vel si duo paria crurum quorumvis et non plura tendunt ad plagam puncti triplicis infinitè distantis, vel tria paria ad plagam puncti quadruplicis, indicio est unum adhuc par hyperbolicum deesse, vel si unum par crurū tendit ad plagam puncti quadruplicis, indicio est duo adhuc paria crurum Hyperbolicorum deesse. Atq; ita si ad plagam aliquā ad quam crura nulla infinite tendunt, sit punctum duplex vel triplex vel quadruplex infinite distans, indicio est paria duo, tria, quatuor deesse. Nam puncta infinite distantia quæ plura sunt quam sunt paria crurum ad punctorum illorum plagas vergentium, indicant tot paria crurum figuræ in infinitum abijsse, & crurum illorum quasi ad infinitam distantiam positorum vicem supplent; ita ut *puncta conjugata infinite distantia* haud male denominari possint. Quo nomine tamen significamus puncta illa vel figuras non revera existere ad infinitam distantiam, sed ad plagas illas in infinitum recedendo evanuisse. Colligendus est itaq;(78) numerus talium punctorū conjugatorū & numerus parium crurum Hyperbolicorū et parium crurum Parabolicorum divergentium & duplus numerus parium crurum reliquorum Parabolicorum[,] & totus numerus subductus de maximo numero parium crurum quæ figura pro genere suo habere possit relinquet numerum parium crurum deficientium.

Et rursus si Figuræ conditiones ita mutentur ut ejus pars aliqua conjugata in infinitum recedens evanescat, ea pars cum cruribus suis jam *imaginaria* vel *conjugatum imaginarium* dici possit. Conjugatum Imaginarium sic dignoscitur. Omne crus infinitum habet crus conjugatū. Crura *conjugata* voco quorum communis est tangens ad infinitam distantiam. Unde si crus aliquod non habet conjugatum reale habebit imaginarium. Rursus si punctum compositū infinite distat ad plagam aliquam ad quam figuræ crus nullū tendit, illud punctum si duplex est indicat paria duo crurum imaginariorū; si quadruplex, quatuor paria; si sextuplex, sex paria. Quod si crurum par unum respicit plagam puncti compositi imaginarij, punctum illud si duplex est indicat unum par crurum imaginariorum; si triplex, duo paria; si quadruplex, tria paria. Et [ubi] crurum paria duo respiciunt plagam illam, punctum illud si triplex est indicat unum par crurum imaginariorum, si quadruplex duo paria, et sic in infinitum. Et hinc ex tangentium & punctorum imaginariorum speculatione (de quibus posthac agetur) facile erit crurum Hyperbolicorum Parabolicorum et Imaginariorum paria omnia numerare. Quibus subductis de numero maximo parium

(78) 'universaliter' was added in an equivalent cancelled preceding draft at this point.

crurum quæ figura pro genere suo potest habere, relinquetur numerus parium crurum deficientium.(79) Numerus vero maximus parium crurum qu[æ] figura cujusqȝ generis habere potest indicatur denominatione generis. Figura primi generis (quæ sola recta linea est) non nisi unum par crurum habet. Figura secundi generis duo paria et non plura, tertij tria &c. Et hinc Figuræ propositæ species assignari potest, dicendo quod sit semel, bis, ter vel sæpius Hyperbolica, Parabolica, Deficiens.(80)

Sunt et aliæ divisiones figurarum in species subordinatas prout crura Hyperbolica conjugata tendant ad contrarias plagas vel ad easdem jaceantve ad diversas partes Asymptoti suæ vel ad easdem, aut secent Asymptoton vel non secent, habeant[ve] Asymptotos aut parallelas, aut in uno puncto concurrentes aut in punctis dive[r]sis; et prout crura Parabolica tendunt ad eandem plagam vel ad plagas oppositas & convexitate sua eandem plagam respiciant vel diversas. Et rursus ex varijs formis quas crura induunt ubi junguntur, ut si se decussant vel tangunt vel cuspidem cissoidalem constituu[nt] vel figuram ovalem habent conjugatam. Sed horum cognitio ad compositionem Loci non est necessaria.(81) Describemus potius generales proprietates curvarum.

Imprimis si curva aliqua duabus rectis parallelis secetur in tot punctis quot potest ejus generis curva a rectis secari, et aliâ recta transversim ducta dividantur parallelæ ita ut in utraqȝ seorsim sumpta summa partium inter transversam illam & singulas curvæ intersectiones ex uno latere jacentium si plures sunt æquetur sum̄æ partium consimilium ex altero latere: dividet illa transversa alias omnes duabus illis parallelas curvamqȝ in totidem punctis secantes rectas, in partes ad singula intersectionum puncta extensas quarum summa ex uno latere æqualis erit summæ ex altero.(82) Nimirū in Figuris secundi generis quæ Conicas sectiones sunt, si parallelæ duæ a transversa bisecantur cæteræ omnes bisecabuntur. In illis tertij si parallelæ duæ ita secantur a transversa ut duæ partes ex uno latere simul sumptæ æquētur tertiæ ex altero, parallelæ omnes eadem

(79) Newton afterwards began to alter this to 'Elliptic[orum]' but quickly changed his mind. In sequel he initially adjoined: 'Quæ quidem paria vel duo erunt vel sex vel octo. Nam defectus quatuor crurum efficit figuram semel Ellipticam'.

(80) Originally 'Elliptica'; compare the previous note. In illustration Newton started in sequel to cite the example of the 'Parabola per quam Cartesius [in *Geometria* sua construxit æquationem sex dimensionum]' (see I: 485, note (15))—which 'Parabola tridens', as he will soon name it in his final cubic 'Eunumeratio' (see **3**, 1, §1: note (31) below), is in his present terms a 'figura semel Hyperbolica, semel Parabolica'—but almost at once broke off to insert in its place a note to himself to 'Here ad an instance de figura tertij generis ter Hyperbolici changed [by 'optical' projection?] first in figuram semel Hyperbolicam semel Parabolicam[,] then in figuram semel Hyperbolicam semel deficientem [Newton began to write 'Elliptic[am]' yet again] then into ye Cissoides'.

(81) Newton initially excused himself from a more careful investigation of the variety of ways in which the infinite branches of curves may be joined and augmented by cusps, nodes and conjugate ovals with a more honest 'Sed omnia determinare non suscepi'.

ratione secabuntur &c. Dicitur vero linea illa transversa *Diameter* figuræ. Et partes parallelarum quæ hinc inde conjunctim æquales sunt dicuntur *ordinatim applicatæ*. Punctum vero ubi curva secatur a diametro Figuræ *vertex* dicitur, & ubi diametri omnes per idem punctū transeunt illud *centrum* est Figuræ. Deniq; si ordinatim applicatæ cum diametro rectos angulos faciant, sintq; singulæ ex una parte diametri æquales singulis ex altera, Diameter dicitur *Axis* Figuræ. Et hinc Figurae cujuscunq; inveniri potest Diameter, angulus applicationis, & vertex figuræ ut et centrum si modo centrum habeat. Et si ordinatæ omnes ad punctum aliquod diametri præter unam dantur dabitur et illa.

Secundo si ducatur recta quævis secans curvam in tot punctis quot potest secare curvam ejus generis et per singulas intersectiones ducantur aliæ rectæ quæ curvam tangant in intersectionibus illis: diameter tangentium rectæ secanti parallela erit diameter figuræ curvilineæ. Et hinc datis diametro et tangentibus curvæ ad terminos omnium ordinatarum præter unam dabitur tangens curvæ ad terminum o[r]dinatæ reliquæ.

Tertio si curva secetur a rectis parallelis in tot punctis uno dempto quot potest illius generis curva a recta secari, nec in pluribus producendo utcunq; curvam et parallelas secari potest, parallelæ dirigentur ad plagam cruris infiniti Figuræ. Et linea transversim acta quæ secet omnes ejusmodi parallelas ad modum diametri[,] id est ita ut omnes partes earum ad curvam et transversam illam terminatæ ex uno latere transversæ jacentes æquales sint omnibus ex altero latere: transversa illa erit aut linea recta aut Conica sectio, Parabola nempe vel Hyperbola, cujus crura infinita tendent ad plagam parallelarum. Et hinc in figuris altioribus descripta transversa illa ex datis omnibus ordinatis præter unā dabitur et illa.

Quarto si Figura sit omninò Hyperbolica tot habens paria crurū infinitorū quot curva illius generis habere potest, et recta secet curvam in tot punctis quot sunt paria crurum, et sumantur partes rectæ illius inter singulas intersectiones & totidem Asymptotos figuræ: summa partium tendentium ab Asymptotis versus unam plagam(83) æquabitur summæ partium tendentium ab Asymptotis versus plagam oppositam(83) et diameter Asymptotorum erit, diameter figuræ curvilineæ. Ut si curva sit Hyperbola scdi(84) generis &c.(85) Et hinc datis omnibus talium figurarum Asymptotis dantur diametri ad ordinatas positione datas.

Quinto si datis positione duabus rectis parallelæ duæ ducantur & sese & curvam datam secantes idq; in tot punctis quot potest curva ejus generis a recta

(82) Or in other words, as he here more pithily first concluded, 'diameter Figuræ erit illa transversa'.

(83) Initially 'jacentium ex uno latere Asymptotôn' and 'jacentium ex altero' respectively.

(84) 'secundi', that is, cubic in degree.

(85) Immediately following, Newton has left a small space—enough to take a couple of lines of his writing—clear to receive the full version of this illustration.

secari & sumantur partes utriusq; lineæ extendentes a communi earum intersectione ad singulas intersectiones: Contentum sub partibus unius rectæ est ad contentum sub partibus alterius in data ratione.(86) Et hinc per imperatum curvæ descriptæ punctum potest linea recta duci quæ curvam illam ibidem tanget.(87)

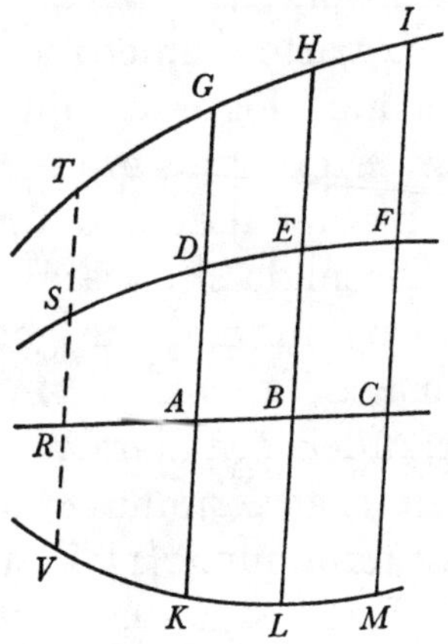

[Sexto](88) datis summis quadratorum ordinatarum ad tria diametri puncta æqualiter distantia, datur summa quadratorum ad quartum quodvis. (89)Si tres ducantur [ordinatæ](90) GK, HL, IM æquali ab invicem distantia secantes & [diametrum](91) AC in punctis A, B, C & curvam datam in tot punctis GDK HEL IFM quot linea illius generis a recta secari potest, dein ducatur [quarta ordinata](90) TV secans itidem [diametrum](91) in R et curvam tot punctis STV quot priores secabant: colligatur

$$AD^q + AG^q + AK^q = M, \quad BE^q + BH^q + BL^q = N$$

et $CF^q + CI^q + CM^q = O$. [tum] Fac $BA.BR :: \dfrac{M-O}{2} . P :: \dfrac{M-2N+O}{2} . Q :: Q.R$ et erit $N+P+R$ summa [quadratorum partium $RS^q + RT^q + RV^q$].(92)

(86) A short gap is again left empty in sequel; in it Newton no doubt intended to give an example of this (now) familiar theorem of his on the constant ratio of the products of the segments cut out on a given curve by two transversals in fixed direction. Compare note (42) above.

(87) For the general method see page 442 above. It is applied to construct the pole-polar property of a central conic at the end of [6] preceding; see note (44).

(88) In the manuscript (f. 25v) this is set as 'Quarto vel potius septimo', but we assign it its correct ordinal numbering in our edited scheme. Newton has cancelled an immediately preceding first version of the following *proprietas generalis* without more than sketchily subsuming its specifications into his revise: in our version we have (with a few necessary alterations) interpolated much of the prior draft in the interest of good sense.

(89) Except for the minimal changes detailed in notes (90) and (91), the ensuing sentence—down to 'Fac...'—is inserted from Newton's cancelled prior draft (see the previous note).

(90) Newton here wrote more vaguely 'parallelæ rectæ' and 'quarta quævis linea...prioribus parallela' respectively.

(91) We amend Newton's 'quartam positione datam rectam' and 'rectam quartam' respectively.

(92) That is, as Newton initially wrote in his draft,

$$\text{'}RS^q + RT^q + RV^q = N + \frac{M-O}{2BA} \times BR + \frac{M-2N+O}{2BA^q} \times BR^q\text{'}.$$

In the terms of note (51) above, if (measured from some origin A' in ABC) we set the common abscissa of $BE = y_1$, $BH = y_2$, $BL = y_3$, ... to be $A'B = x$, where the y_i, $i = 1, 2, 3, \ldots, n$, are the roots of some given equation $y^n - (ax+b)\,y^{n-1} + (cx^2+dx+e)\,y^{n-2} - \ldots = 0$ defining the n-th degree curve $DEF/GHI/KLM$, then

$$\sum_{1 \leqslant i \leqslant n} y_i^2 = (ax+b)^2 - 2(cx^2+dx+e) \equiv \alpha x^2 + \beta x + \gamma.$$

Et eodem modo summa cuborum, [quadratoquadratorum,] quadrato-cuborum &c determinari posset.(93)

---

[8](94) Si rectis duabus positione datis et angulum continentibus parallelæ duæ Curvam quamvis lineam in pleno punctorum numero secuerint, contentum sub omnibus rectæ unius segmentis ad commun[em] rectarum intersectionem & curvam hinc inde terminatis(95) erit ad simile contentum sub omnibus alterius rectæ segmentis in ratione data.

Linea secundi generis duas assympto[to]s ea tertij tres ea quarti quatuor et non plures [habere potest]. Et si curva generis cujuscunq; plenum habeat Asymptotorum numerum et a recta aliqua in pleno punctorum numero secetur et sumantur partes rectæ inter singula intersectionum pūcta et asymptotos totidem, summa partium ab asymptotis in unam plagam ductarum æqualis erit summæ partium ductarum in plagam contrariam.

Linea quæ nulla habet crura in infinitum procurrentia vel secundi generis est (nempe Ellipsis aut circulus) vel quarti vel sexti. Quæ duo tantum habet crura infinita eaq; in plagas oppositas jacentia vel primi generis est (nempe recta) vel tertij vel quinti. Quæ duo tantum habet crura infinita eaq; in plagam eandem jacentia vel secundi est generis (nempe Parabola) vel quarti vel sexti. Quæ quatuor et non plura habent crura infinita,(96) quorum duo in plagas oppositas protenduntur, generis sunt vel secundi (nempe Hyperbola) vel quarti vel sexti.

In general, put

$$f_r \equiv \alpha(x+re)^2+\beta(x+re)+\gamma = (\alpha x^2+\beta x+\gamma)+(2\alpha x+\beta)\,re+\alpha r^2e^2,$$

and there will then, corresponding to $AB = BC = e$, be $M = f_{-1}$, $N = f_0$, $O = f_1$ and hence $\frac{1}{2}(M-O) = -(2\alpha x+\beta)$, $\frac{1}{2}(M-2N+O) = \alpha$; accordingly, corresponding to $BR = -re$, there will be $f_{-r} = N+\frac{1}{2}(M-O)\,r+\frac{1}{2}(M-2N+O)\,r^2$, as Newton decrees. This is, of course, merely the 'Newton–Stirling' expansion $f_{-r} = f_0+\binom{-r}{1}\mu\Delta^1_{-\frac{1}{2}}-\frac{1}{2}r\binom{-r}{1}\Delta^2_{-1}$ of the quadratic function defined by 'summa quadratorum partium ordinatarum' here constructed by Newton. (Compare IV: 58–9, note (19).)

(93) That is, by analogous central-difference expansions of the functions of third, fourth, fifth, ...degree which (see note (51) above) likewise ensue therefrom. These would excellently illustrate the 'Prob. 3' whose enunciation alone is set down by Newton in the above (see page 486).

(94) Add. 3963.15: 181r. Yet one more enunciation of two of Newton's preceding *proprietates generales curvarum*, followed by some remarks on how consideration of a curve's infinite branches allow a ready identification of its genus, and lastly an outline of his organic construction—'the best method for determining lines passing through given points, and also there to touch given straight lines'. The manuscript is, though penned with some care, merely an 'ideas sheet' which was discarded when Newton penetrated over far into the detail of his earlier constructions of a conic to pass through five points or touch an equivalent number of lines.

(95) 'ductis' is replaced.

(96) Originally 'in infinitum protensa'.

Igitur si punctum indeterminatum in Curva aliqua locatur, considerandum est quot modis punctum illud in infinitum abire possit et in quas plagas[,] et inde genus curvæ ut et species ejus si secundi est generis absq; computo innotescet.

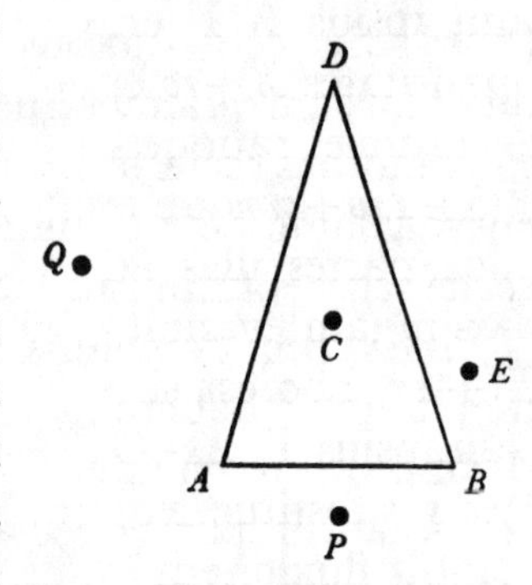

Modus optimus determinandi lineas est per inventionē punctorum per quæ transibunt et rectarum quas tangent in ijsdem punctis. Si linea est conica sectio determinabitur vel per quinq; puncta vel per quatuor puncta et tangentem vel per tria et tangentes duas. Dentur puncta quinq; *A*, *B*, *C*, *D*, *E*. Et junctis tribus quibusvis *AB*, *AD*, *BD* rotetur angulus *DAB* circa punctum *A* et angulus *DBA* circa punctū *B* et ubi eorum crura *AD BD* concurrunt ad puncta reliqua duo *C* et *E* concurrant altera duo crura *AB* et *BA* ad puncta *P* et *Q*[,] et ubi hæc crura *AB* et *BA* concurrunt ad rectam infinitam per puncta *P* et *Q* transientem altera duo crura intersectione sua concurrent ad sectionem conicam quam describere oportuit.

Si quatuor sunt puncta et tangens vel tria puncta et tangentes duæ concipiend[(97)]

---

[9][(98)] Proponantur datæ aliquot quantitates *A*, *B*, *C*, *D* et fluentes *X*, *Y*, *Z* quarum fluxiones respectivè designentur *r*, *s*, *t*[,] et requirātur fluxiones linearum quæ ex his generantur ut[(99)] lineæ $X'Y$, $X'Y'Z$, X^2, Y^3 et similiū. Fluant *X* et *Y* et

---

(97) Newton obviously meant to assert in continuation that 'concipiendum est quod duo puncta ibi coincidunt', but here broke off, evidently realizing how far these particular organic constructions of conics (on which see II: 120 ff. and IV: 298–302) digress from his main theme of the general properties of curves.

(98) Add. 4004: 135r, a first version of the proof of the algorithm for deriving the fluxion of a product set out on page 448 above (see §1: note (114)). Rather than make, as he afterwards preferred to do, a loose appeal to the reader's ill-formed notion of the 'instantaneous' fluxional 'speed' of growth (or diminishing) of a fluent quantity, Newton here specifies the increment *o* of 'time' in which such a 'flow' takes place and then explicitly considers the 'last ratios' of the related increments of the fluent quantities as the interval in which they are contemporaneously generated becomes a vanishingly small 'instant'.

(99) As in his earlier equivalent expositions on III: 72–4 and IV: 566, Newton here initially took a more complicated particular example: 'Proponantur datæ aliquot quantitates *a*, *b*, *c*, *d* et fluentes *x*, *y*, *z* sitq; ea fluentiū relatio inter se quæ in hac æquatione exprimitur

$$ax - \frac{b}{a}xx = zz - yz.$$

Fluant *x y* et *z* et in ea fluxione sit *ro* augmentum ipsius *x*, *so* ipsius *y* & *to* ipsius *z* ita ut *x y* et *z* jam fiant $x+ro$, $y+so$, $z+to$ et perinde has pro *x y* et *z* respectivè scribendo æquatio superior fiet

$$ax + aro - \frac{b}{a}xx - \frac{2b}{a}xro - \frac{b}{a}rroo = zz + 2zto + ttoo - yz - yto - zso - stoo.$$

in ea fluxione sit ro augmentum ipsius X et so augmentum ipsius Y ita[q̃] X et Y jam evaserint $X+ro$ & $Y+so$ et $X'Y$ fiet $X'Y+Xso+Yro+sroo$ adeóq̃ augmentum ipsius $X'Y$ erit $Xso+Yro+sroo$. Defluant jam quantitates $X+ro$ et $Y+so$ donec ad X et Y redierint et ultimæ rationes partium evanescentium ro, so, $Xso+Yro+sroo$ eæ erunt quæ sunt fluxionum.(100) Applica partes illas ad communem factorem o, et quia pars ro jam evanuit dele sro et fluxiones erunt ut r, s, $Xs+Yr$, hoc est si r denotet fluxionem ipsius X et s eam ipsius Y, $Xs+Yr$ denotabit eam ipsius XY. Pro r et s substitue æquipollentes x et y et $Xy+Yx$ denotabit fluxionem ipsius XY.(101)

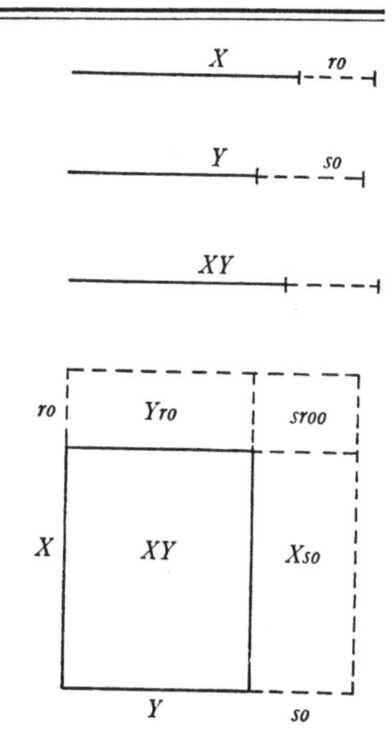

[10](102) In figuris hæc est methodus. In triangulo ABC dentur angulus C et latus CB, fluant cætera. Et ubi CA fluendo evasit Ca simul BA evadat Ba. In Ba

Quantitates vi primæ æquationis æquales dele et restabunt

$$aro-\frac{2b}{a}xro-\frac{b}{a}rroo = 2zto+ttoo-yto-zso-stoo.$$

Applica omnia ad o et fiet $ar-\frac{2b}{a}xr-\frac{b}{a}rro = 2zt+tto-yt-zs-sto$'. Had he not there broken off he would evidently have concluded: 'Defluant jam quantitates $x+ro$, $y+so$, $z+to$ donec evaserint iterum x, y, z et momento o in infinitum diminuto evadit æquatio

$$ar-\frac{2b}{a}xr-\frac{b}{a}rr = 2zt-yt-zs$$

exprimens relationem fluxionum r, s, t quantitatum fluentium x, y, z'.

(100) Understand 'quantitatum fluentium X, Y et XY' as Newton initially added, continuing thereafter: 'Sunt igitur fluxiones illæ ut r, s, $Xs+Yr+sro$'—that is, 'ro jam evanescente', as r, s and $Xs+Yr$ in the limit. And so, no more rigorously, he derives in his revised sequel by way of an equally abrupt 'quia pars ro jam evanuit, dele sro'.

(101) Compare Newton's analogous demonstration of this basic fluxional algorithm some twenty years earlier in the cancelled portion of his addendum to his 1671 tract reproduced on III: 330–2.

(102) Add. 4004: 136r[+135v]/137r. In this cancelled initial opening to the section of the 'Geometriæ Liber primus' whose content we epitomized in the preceding main text (see §1: note (118)) as 'Fluxiones linearum et angulorum' Newton gives a new proof of Proposition 22 of his earlier 'Geometria Curvilinea' (see IV: 458), successively specifying the fluxions of the individual sides and a cited angle $\hat{C}$ of the freely fluent triangle ABC in terms of their respective contemporaneous increments *initio fluxionis*, and then setting the aggregate of these component instantaneous 'speeds' of increase to be the total fluxion. He was in revise (see §1: note (126) above) to replace this step-by-step argument from first principles by a shorter, more sophisticated demonstration making direct appeal to the fluxional derivatives of sines and cosines which here appear in heavily disguised geometrical form.

cape BD æqualem BA, et partes genitæ erunt Aa et Da. Defluant jam lineæ Ca Ba donec ad priorem magnitudinem et positionem CA et BA redierint. (103)Ad AC demitte normalem BG[,] et quia $BG^2+GA^2=AB^2$ et BG datur erit

$$2GA\text{'}ga = 2AB\text{'}ba.$$

Sed fluxiones ga et ca eædem sunt.(104) Ergo

$$2GA\text{'}ca = 2AB\text{'}ba \quad \text{et} \quad \frac{GA\text{'}ca}{AB} = ba.$$

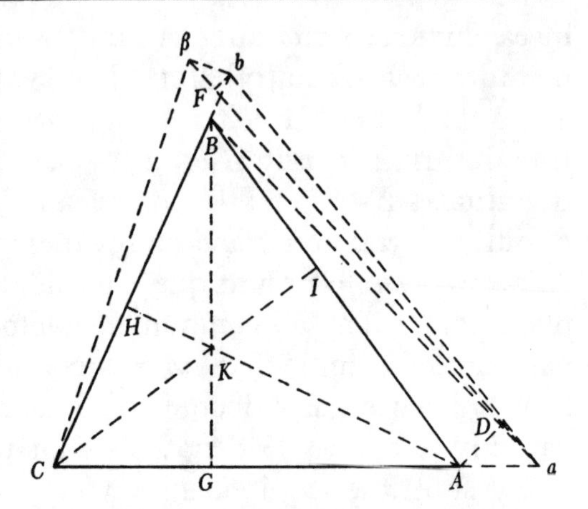

(105)Fluat jam CA donec evaserit Ca. Dein manente Ca fluat CB donec evaserit Cb et demisso ad CB normali AH linea AB fluxione sua $\frac{GA\text{'}ca}{AB}$ convertitur in lineam aB et similiter linea aB deinceps fluxione sua $\frac{HB\text{'}cb}{aB}$ convertitur in ab. Ergo utraqȝ fluxione conjunctim convertitur AB in ab. Id est si CA et CB simul fluunt erit $\frac{GA\text{'}ca}{AB}+\frac{HB\text{'}cb}{aB}$ fluxio ipsius AB, et quia AB et aB initio æquales sunt, erit ea(106) $\frac{GA\text{'}ca+HB\text{'}cb}{AB}$(107).

Ubi CA et CB fluendo evaserint Ca et Cb fluat insuper angulus ACB et fluendo fiat $AC\beta$. In $a\beta$ cape $aF=a[b]$ et junge bF, $b\beta$. Et ipso fluxionis initio bF per-

(103) Newton initially went on to the same effect: 'et si ad CA demittatur perpendiculum BG, ultima ratio partium evanescentium Aa, Da erit ea quæ est linearum AB AG eo quod in eo casu AD perpendiculare fit ad Ba adeoqȝ triangulū aAD triangulo aBG id est triangulo ABG evadit simile. Quare cum fluxiones sint in ultima ratione partium evanescentium erit ut AB ad AG ita fluxio ca ad fluxionem ba seu $\frac{AG\text{'}ca}{AB}=ba$.'

(104) 'quia CG non fluit' and hence its fluxion cg is nil.

(105) A cancelled first version (on ff. 136r/137r) of the following paragraph reads little differently: 'Fluat jam rursus CA et BA donec evaserint Ca et Ba et dein fluat etiam CB donec ipsa evaserit Cb et Ba evaserit ba et demisso ad CB normali AH fluxio ipsius aB hac secunda vice ex jam inventis erit $\frac{BH,\ cb}{aB}$. AB fluxione sua priori convertitur in aB et illud aB fluxione posteriori convertitur in ab. Ergo utraqȝ fluxione conjunctim convertetur AB in ab: id est si CA et CB simul fluant ita ut AB evadat ab, fluxio ipsius AB erit $\frac{AG\text{'}ca}{AB}+\frac{BH\text{'}cb}{aB}$ seu quia initio fluxionis AB et aB æquales sunt, erit ea $\frac{AG\text{'}ca+BH\text{'}cb}{AB}$.'

(106) That is, 'fluxio ipsius AB'.

(107) Where understand that the triangle ACB 'flows' to be the triangle aCb 'manente angulo $\hat{C}$', as in Corollary 3 to Proposition 22 on IV: 458.

pendiculare erit ad ab vel AB adeoq parallelum KI[,] et $b\beta$ perpendiculare erit ad Cb adeoq parallelū AK. Unde triangula βFb & AIK æquales habentia angulos ad b et K et rectos ad F et I similia erunt[,] et $b\beta$ erit ad $F\beta$ id est augmentum lineæ ab ipso fluxionis initio ad augmentum arcus subtendentis angulum ACB radioq Cb descripti vel quod perinde est fluxio lineæ ab ad fluxionem arcus illius(108) ut AI ad AK, hoc est ut CI ad CB.(109)

Cæterum ad fluxionem arcus [circularis] significandam pono angulum literis minusculis nominando illud ejus latus ultimo loco quod radius est. Aut si neutrum ejus latus radius est ubi angulum literis minusculis posui subjungo radium literis minusculis cum litera r ijsdem suffixa. Est itaq lineæ ab fluxio ad arcus præfati fluxionem acb ut CI ad BC. Adeoq $\frac{CI'acb}{BC}$ vel quod perinde $\frac{CI'acb\,ABr}{AB}$(110) fluxio est lineæ ab. Hanc fluxionem qua ab convertitur in $a\beta$ adde fluxioni qua AB conversa fuit in ab et habebis totam fluxionem nempe

$$\frac{AG'ca+BH'cb+CI'acb\,ABr}{AB}$$

qua AB per fluxionē laterū CA, CB et anguli C convertitur in $a\beta$.(111) Est itaq $AG'ca+BH'cb+CI'c\,ABr = AB'ab$ relatio(112) inter fluxiones trium laterum et anguli trianguli cujuscunq, cujus beneficio si tres ex his fluxionibus cognoscantur possumus quartam invenire.

---

[11](113) Porro in triangul[o] ACE ubi fluxio anguli recti E nulla est, per

---

(108) Namely $R.\widehat{ACB}$ where the radius $R = CB$.

(109) That is, $\text{fl}(ab):\text{fl}(R\hat{C}) = \text{Sin}\,B:R$. Here the triangle aCb 'flows' to be the incremented triangle $aC\beta$ 'datis lateribus CA, CB angulum $\hat{C}$ comprehendentibus', as in Corollary 1 to Proposition 22 on IV: 458.

(110) By 'ABr' Newton evidently means that AB is to be taken as the trigonometrical radius (previously understood to be BC), so that '$acb\ ABr$' is to be read 'fluxio arcus angulum aCb subtendentis posito AB pro radio'. We may be grateful that he did not elsewhere persevere with so unwieldy a contracted notation: it is here confusing enough that, because of his double use of lower-case letters, acb has to stand both for 'fluxio ipsius ACB' and 'fluxio ipsius aCb'.

(111) That is, since $AG/AB = \cos\hat{A}$ and $BH/AB = \cos\hat{B}$,

$$\text{'Cos}\,A\times\text{fl}(CA)+\text{Cos}\,B\times\text{fl}(CB)+CG\times\text{fl}(\text{arc}\,C) = \text{rad}\times\text{fl}(AB)$$

posito AB pro radio', much as in the full Proposition 22 of Newton's earlier 'Geometria Curvilinea' (see IV: 458).

(112) Initially—and a little more precisely—'expressio relationis'.

(113) Add. 4004: 138r–144r, a cancelled first draft of the concluding section of the preceding 'Geometriæ Liber primus'—one closely founded, it will be clear, on the equivalent portions of Newton's earlier 'Geometria Curvilinea' (see IV: 420–504). In a cancelled paragraph on f. 138r he had, much as in its revise on page 454 above (on which see § 1: note (127)), immediately before observed—in sequel to his previous propositions adducing the various

Theorema ex tribus novissimis(114) primum mutatis mutandis fit

$$sA'ac^{(115)}+CE'c = R'ae.$$

Est et in triangulo ABC, per Theorematum secundum,

$$sA'ac = sB'bc - AB'a - BG'c.$$

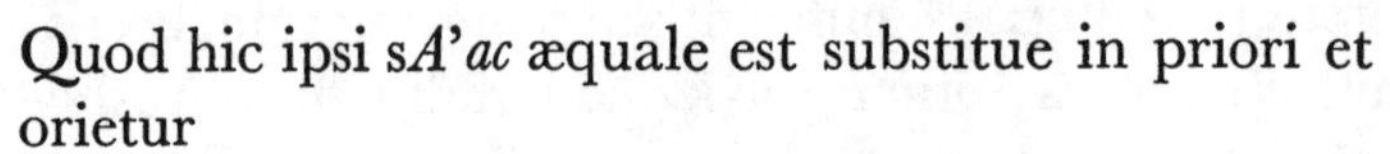

Quod hic ipsi $sA'ac$ æquale est substitue in priori et orietur

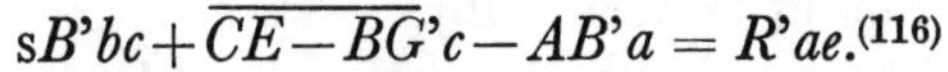

$$sB'bc+\overline{CE-BG}'c-AB'a = R'ae.^{(116)}$$

Ubi si pro $-a$ substituas æquipollentem $b+c$, fit

$$sB'bc+\overline{CE+AG}'c+AB'b = R'ae.$$

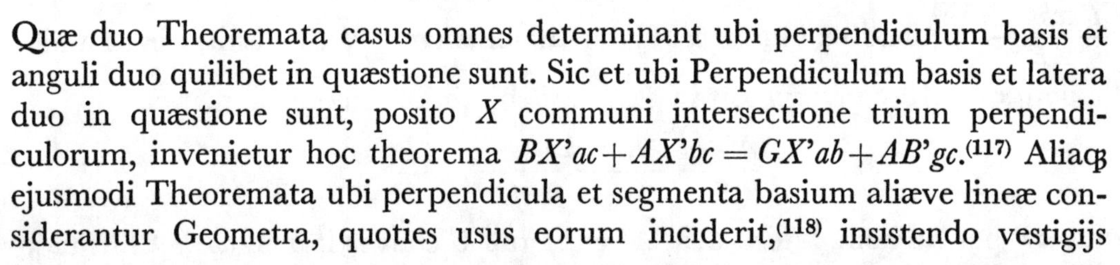

Quæ duo Theoremata casus omnes determinant ubi perpendiculum basis et anguli duo quilibet in quæstione sunt. Sic et ubi Perpendiculum basis et latera duo in quæstione sunt, posito X communi intersectione trium perpendiculorum, invenietur hoc theorema $BX'ac+AX'bc = GX'ab+AB'gc$.(117) Aliaqȝ ejusmodi Theoremata ubi perpendicula et segmenta basium aliæve lineæ considerantur Geometra, quoties usus eorum inciderit,(118) insistendo vestigijs

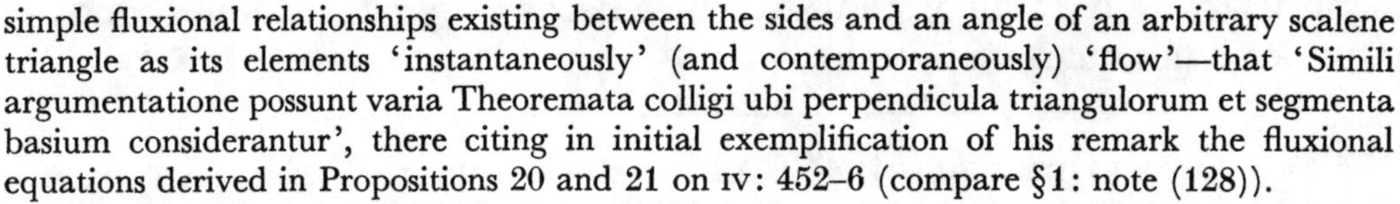

simple fluxional relationships existing between the sides and an angle of an arbitrary scalene triangle as its elements 'instantaneously' (and contemporaneously) 'flow'—that 'Simili argumentatione possunt varia Theoremata colligi ubi perpendicula triangulorum et segmenta basium considerantur', there citing in initial exemplification of his remark the fluxional equations derived in Propositions 20 and 21 on IV: 452–6 (compare §1: note (128)).

(114) Those here reproduced on pages 452–4 above. The 'primum Theorema' invoked in immediate sequel, however, is rather the unstated corollary to the first which deduces the consectary $sA'ac+$ ($sa'AC$ or) $AG'a = R'cg$ more straightforwardly derivable from the starting equation Sin $A\times AC = R\times CG$.

(115) This should be $sC'ac$. In interchanging A with C and E with G (and correspondingly replacing the fluxions a and ac and gc) in the original form $sA'ac+AG'a = R'cg$ of his *primum Theorema* (see the previous note) as he here first wrote it down without fault, Newton has carelessly omitted to convert his first 'A' into a 'C'. The two theorems which he proceeds to derive from this mistaken equation are, we need scarcely say, erroneous. For the correct deductions see the next note.

(116) Newton has here deleted 'Theorema ubi Perpendiculum, angulus verticalis, basis et angulus alteruter ad basem in quæstione sunt'. On making correct substitution of

$$sA'ac = R'cg - AG'a$$

(see the previous note) in his *Theorema secundum* above there results $R'cg = sB'bc - BG'\overline{a+c}$, wherein, on accurately mending the following phrase, 'si pro $-\overline{a+c}$ substituas æquipollentem $+b$ fit $sB'bc+BG'b = R'cg$'. This is a mere variant upon his *Theorema primum*.

(117) This unaltered borrowing from Proposition 21 of the earlier 'Geometria Curvilinea' (see IV: 456) was to pass unchanged into the revised text reproduced on page 454 preceding. (Compare §1: note (128) and see also note (113) above.) On Newton's presumption of the unique existence of the orthocentre X see §1: note (129) preceding.

(118) This replaces the less precise stock phrase 'pro re nata'.

methodi hic patefactæ, facilè inveniet; adeò ut rem plenius prosequi supervacaneum ducam.

Cæterum Theorematum quæ frequentius usui fuerint et ex quibus cætera, siquando opus erit, licebit derivare, seriem convenit subjungere ut promptius citari possint.[119]

De Proportionalibus[120]

Th. 1. Si quotcunq continue proportionalium S, T, A, V, X, Y unum A datur, fluxiones eorum erunt ut ipsa multiplicata per numerum locorum quibus distant a dato termino. $s.t.v.x.y::-2S.-T.V.2X.3Y.$[121]

Th. 2. Positis tribus continuè proportionalibus $V.X.Y$ si summa extremorum datur, erit ut excessus unius extremi supra alterum ad duplum medij ita fluxio medij ad fluxionem extremi illius alterius vel ad defluxionem prioris.

$$V-Y.2X::x.y::x.-v.$$[122]

Et si differentia extremorum datur, erit ut summa extremorū ad duplum medij, ita fluxio medij ad fluxionem alterutrius extremi. $V+Y.2X::x.y::x.v.$[122]

Th. 3. Ijsdem positis si summa medij et extremi datur, a duplo medio aufer alterum extremum & residuum erit ad extremum prius ut fluxio alterius extremi ad fluxionem medij vel ad defluxionem extremi prioris. Detur $V+X$, erit $2X-Y.V::y.x::y.-v.$[123] Et si differentia medij et extremi datur, ad duplum medium adde alterum extremum, et summa erit ad extremum prius ut fluxio alterius extremi ad fluxionem tam medij quam extremi prioris. Detur $V-X$ vel $X-V$, erit $2X+Y.V::y.x::y.v.$[123]

Th. 2.[124] Positis quatuor proportionalibus $V.X::Y.Z$ fluxiones extremorum

(119) As we have already stated (see note (113) above), the ensuing sets of fluxional theorems are remoulds, somewhat extended and superficially reshaped, of equivalent propositions in Newton's 'Geometria Curvilinea'. We give more detailed correlations in following footnotes.

(120) The two following theorems encapsulate the content of Propositions 1–10 of the earlier 'Geometria' (see IV: 428–36) and their more lavish revise under the present title on IV: 486 ff.

(121) A minimal recasting of the enunciation of Proposition 8 on IV: 434 (= Proposition 14 on IV: 490).

(122) These complementary enunciations are essentially Propositions 4 and 5 respectively on IV: 432, here more neatly (and logically) combined into a single theorem. Like 'Th. 3' which immediately follows, it was subsequently discarded by Newton as being, no doubt, but a simple corollary of the uncancelled 'Th. 2' below.

(123) These complementary assertions are evidently intended to combine the enunciations of Propositions 6 and 7 on IV: 432–4, but are carelessly intermixed. They should, correctly, read 'si summa medij et extremi datur, ad duplum medium adde alterum extremum & summa erit...prioris. Detur $V+X$, erit $2X+Y.V::y.x::y.-v$' and 'si differentia medij et extremi datur, a duplo medio aufer alterum extremum, et residuum erit...prioris. Detur $V-X$ vel $X-V$, erit $2X-Y.V::y.x::y.v$'.

(124) Originally—before the preceding 'Th. 2' and 'Th. 3' were (see note (122)) discarded —this was '4'. The following Theorems 3–6 were likewise initially numbered 5–8 respectively.

mutuò ductæ in extremos æquantur fluxionibus mediorum mutuò ductis in medias. $Vz+Zv = Xy+Yx$.(125)

De Contentis.(126)

Th. 3. Contenti fluxio est quæ fit ducendo sigillatim lateris cujusqȝ fluxionem in contentum applicatum ad latus illud et summam productorum capiendo.(127) Sic fluxio XY est $Xy+Yx$ et fluxio AY est Ay. Unde conjunctim f[l]uxio $AY+XY$ est $Ay+Xy+Yx$.

De Triangulis.(128)

Th. 4. In triangulo quovis ad basem acutangulo, si perpendiculū ad basem demittatur, et fluxio lateris utriusqȝ ducta in sinum anguli sibi contermini ad basem, seorsim addatur fluxioni arcus subtendentis angulum illum ductæ in conterminum segmentum basis; æquales erunt summæ.(129) In triangulo ABC, demisso ad basem AB perpendiculo CG, est

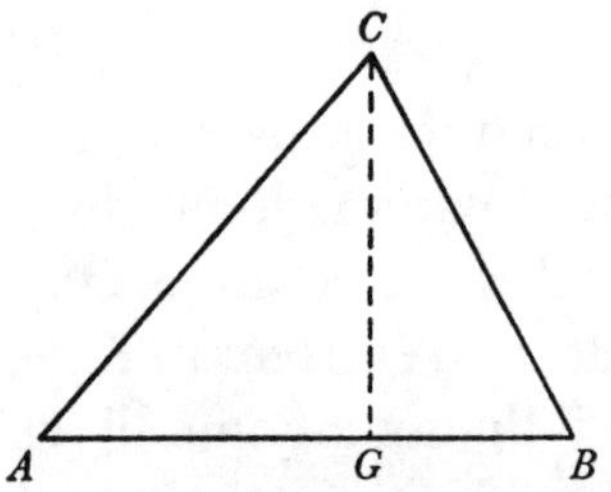

$$sB'bc+BG'b = sA'ac+AG'a.$$

Ad quod Theorema recurrendum est quoties anguli duo et latera duo opposita in quæstione sunt.

Th. 5. In triangulo quovis ad basem acutangulo, si perpendiculum demittatur,(130) fluxio lateris alterutrius ducta in sinum contermini anguli ad basem æqualis est summæ fluxionis lateris alterius ductæ in sinum contermini anguli ad basem & fluxionis arcus subtendentis angulum illum ductæ in basem et fluxionis arcus subtendentis angulum ad verticem ductæ in segmentum basis lateri primo conterminum.(131) In triangulo ABC demisso perpendiculo CG est

$$sB'bc = sA'ac+AB'a+BG'c.$$

(125) Proposition 10 on IV: 436 (= Proposition 13 on IV: 490).

(126) Newton initially set down the title 'De Triangulis' of the next section. The following newly introduced theorem merely states that the operation of addition (and subtraction) is preserved when fluxions are taken of an aggregate 'content'.

(127) In sequel Newton initially further specified 'Et fluxio aggregati ex contentis componitur ex fluxionibus partium' before recognizing that this assertion is contained, if not quite so clearly, in the main statement of his present theorem.

(128) This section embraces Propositions 22–25 of the 'Geometria Curvilinea' (see IV: 458–64).

(129 Proposition 23 on IV: 460. Newton here initially announced: 'In triangulo quovis acutangulo, perpendiculo ad basem demisso, differentia fluxionum laterum ductarū in sinus conterminorum angulorum ad basem, æqualis est differentiæ fluxionum arcuum subtendentium angulos ad basem ductarum in contermina segmenta'.

(130) A superfluous 'ad basem' is deleted here.

(131) Proposition 24 on IV: 462; it may well be that portions of Proposition 25 following (*ibid.*: 462–4) would here be additionally subsumed.

Ad quod Theorema recurrendum est ubi anguli duo et latera totidem interjectum et oppositum in quæstione sunt.

Th. 6. In triangulo quovis ad basem acutangulo, fluxio basis ducta in radium æqualis est summæ fluxionum laterum seorsim ductarum in sinus complementorum conterminorum angulorum ad basem & fluxionis arcus angulum verticalem subtendentis ductæ in perpendiculum ab angulo isto ad basem demissum.(132) In triangulo ABC demisso perpendiculo CG est

$$R'ab = s'A'ac + s'B'bc + CG'c.$$

Ad quod Theorema recurrendum est ubi angulus et latera tria in quæstione sunt.

Scholium. Positis his cas[ib]us tantum triangulorum quæ acutos habent angulos ad basem, Theoremata extenduntur ad omnes casus. Nempe si angulus alteruter ad basem obtusus sit, sinus complementi ejus et conterminum segmentum basis pro retrorsis haberi debent et perinde signa eorum de + in − mutari, positivis Theorematum terminis in quibus reperiuntur transformatis in subductitios. Et quamvis hæc Theoremata de quatuor fluentibus proponuntur, continent tamen omnes casus triangulorū ubi tria vel duo tantum fluentia sunt. Numerus enim quaternarius ex datis terminis trianguli implendus [est] et in Theoremate delendæ fluxiones datorum. Ut si in triangulo ABC daretur angulus A et latus oppositū et ex cognita fluxione alterius anguli B invenire vellem fluxionem lateris huic oppositi AC: quatuor termini quæstionem ingredientes erunt anguli duo datus A et fluens B et latera duo opposita datum BC et fluens AC. Consulo igitur Theorema quartum(133) quod hunc casum includit, et ibi deleta fluxione datorum terminorum A et BC invenio $BG'b = sA'ac$; id est fluxionem lateris AC esse ad fluxionem arcus subtendentis angulum B ut BG ad sinum anguli A. Et eodem modo ubi trianguli alicujus dantur duo termini quilibet, Theoremata se semper resolvunt in proportiones. Unde et proportiones illas nomine Theorematum tunc licebit citare perinde ac si eædem in Theorematis expressæ fuissent. In triangulis autem rectangulis angulus rectus pro dato termino semper habendus est, et numerum quatuor terminorum implet. Hoc modo semper incides in quatuor terminos. Quibus cognitis in promptu est per annotata in calce Theorematum legitimum Theorema consulere. Nec minori promptitudine recurretur ad Triangulorum casus sequentes.(134)

(132) Proposition 22 on IV: 458.

(133) Originally 'sextum'; compare note (122) above.

(134) This last sentence is a late insertion by Newton, who initially here passed straightaway to pen the next section-head 'DE MOTUUM PLAGA' (thus *tout court*) and the preliminary opening to 'Th. 7' which we quote in note (136) below. In sequel he began correspondingly to list '*Cas. 1.* De Perpendiculo basi et lateribus' but abruptly broke off without proceeding further. In his revised scheme of propositions to be expounded as the elements of 'Geometria Curvilinea' (see IV: 486–504) he had earlier listed ten theorems 'De Triangulis rectangulis' (*ibid.*: 494–6), eleven 'De triangulis obliquangulis datum angulum...habentibus' (*ibid.*:

DE MOTUUM PLAGA ET CELERITATE.(135)

Th. 7. (136)Ubi puncti alicujus motus pendet a diversis linearum et angulorum fluxionibus[,] colligenda sunt sigillatim loca in quæ fluxiones singulæ seorsim vel etiam plures earum conjunctim temporibus æqualibus punctum illud transferrent si modo plagas et celeritates servarent immutatas quas habent ipso fluendi initio. Et locorum ultimus vel ultimorum intersectio ipse erit locus ad quem omnes fluxiones eodem temporis spatio punctum illud rectà transferent: adeoq; si a loco primo ad hunc locum recta linea ducatur, hæc et Curvam motu puncti descriptam tanget et puncti describentis celeritatem seu fluxionem curvæ exponet.(137)

Exempli gratia(138) si linea mobilis BC super immobili AB feratur et cognoscantur fluxiones lineæ illius mobilis BC, arcûs angulum ABC ad datam distantiam subtendentis et lineæ immobilis AB: exponantur fluxiones illæ tres per totidem lineas S, T, V. In BC producta cape CD æqualem S, erige normalem $[C]E$ quæ sit ad T ut BC ad Radium, et age CF parallelam AB et æqualem V. Et eodem tempore quo fluxio lineæ BC si sola esset faceret punctum C transferri ad D, fluxio anguli B si sola esset ipsum transferri faceret ad E, et fluxio lineæ AB si sola esset, ad F. Duco igitur DG parallelam et æqualem CE et GH parallelam et æqualem CF, et concipiendo quod æqualibus temporibus punctum C

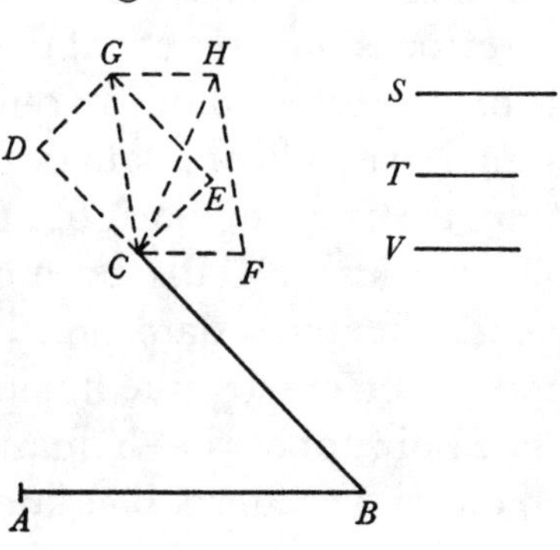

496–8) and ten further ones—other than Propositions 20–25, that is—'De Triangulis obliquangulis nullum datum angulum habentibus' (*ibid.*: 498–500).

(135) Originally '...ET FLUXIONE CURVARUM' in a cancelled first opening to this section on f. 141r. Newton proceeds to elaborate and extend the theorem of his 'Geometria Curvilinea'—namely Proposition 27 (see IV: 468–70)—where he had before given concise (but there unexplained) construction of the instantaneous, tangential direction of motion at an arbitrary point of a curve defined in a generalized Cartesian system of coordinates where the angle between abscissa and ordinate is also allowed to vary.

(136) In his preliminary draft of this theorem on f. 141r (see the previous note) Newton passed immediately, again without any introductory explanation of the way in which the speeds and directions of the various limit-motions are there compounded, merely to repeat the fluxional construction of the *plaga motus* in the following generalized Cartesian instance, propounding (virtually as on IV: 468) before breaking off: 'Si recta mobilis BD rectæ positione datæ et ad datum punctum A terminatæ AB tanquam basi insistat et habeantur fluxiones basis illius AB[,] rectæ insistentis BD et arcus subtendentis angulum quem hæ rectæ comprehendunt, exponantur fluxiones illæ tres per totidem rectas S, T, V respectivè [...].'

(137) Originally 'longitudine sua exhibebit'.

(138) As we have said, the following prime instance of the preceding general rule is but lightly recast—by way of the abandoned preliminary revise cited in note (136) above—from Proposition 27 of Newton's 'Geometria Curvilinea' (IV: 468–70).

sola fluxione prima transferretur de C in D, dein sola secunda de D in G, postea sola tertia de G in H, concludo quod omnibus conjunctis rectà perget eodem temporis spatio de C ad H.(139) Duo motus CD, CE componunt motum CG in diagonali parallelogrami $ECDG$, et motus ille cum tertio motu CF componit motum CH in diagonali parallelogrammi $FCGH$. Erit igitur hæc linea CH tam tangens Curvæ motu puncti C [descriptæ] quàm exponens fluxionis ejus.

Rursus si a rectis duabus positione datis AB, AI ad idem Curvæ alicujus punctum C conveniant lineæ duæ mobiles BC, IC, et ex cognitis fluxionibus anguli B et linearum BC, AI, IC(140) determinandus esset motus puncti C, imprimis motum quem punctum C haberet si linea BC solummodo flueret expono quoad plagam et celeritatem per lineam CD: Dein motum quem idem punctum haberet si angulus B solummodo flueret expono etiam quoad plagam et celeritatem(141) per lineam DG:(142) Deniq; quia fluxio lineæ AB ignoratur, motum puncti C ab ea oriundum expono quoad plagam tantum per lineam indefinitam GH ductam parallelam AB. Et considero lineam GH ut locum ultimum indefinitum puncti C. Tum pergens ad alteram lineam mobilem IC motum puncti C a fluxione lineæ IC oriundum expono per lineam CK(143) et motum ejus a fluxione lineæ AI oriundum per lineam KL,(144) et ultimò motum ejus ab ignota fluxione anguli I expono quoad plagam per indefinitam lineam LM.(144) Cum igitur hi motus faciant ut punctum

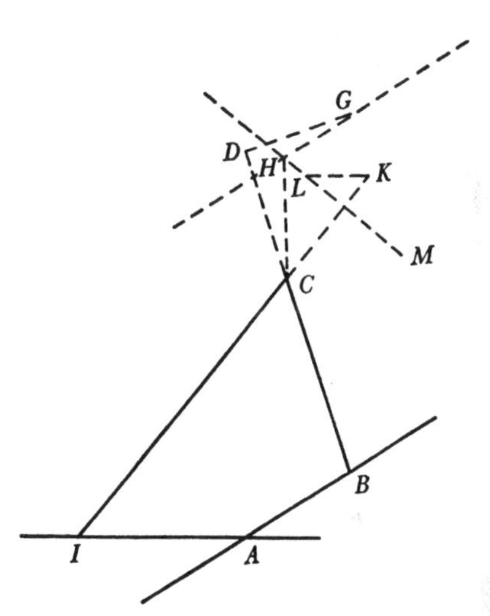

(139) Newton first went on: 'Comple parallelogramma $ECDG$ & $CGFH$ [et] si unus motus transferre valet punctum C de linea CE ad lineam DG et alius eodem tempore de linea CD ad lineam EG[,] uterq; conjunctim transferent in diagonali CG utriusq; lineæ concursum G. Et ex motu de C ad G et de C ad F rursus componetur motus in diagonali CH. Quare linea CH curvam puncto C descriptam tanget & puncti illius celeritatem seu fluxionem Curvæ exponet.'

(140) The system of coordinates here implicitly posited has, we may notice, the grave disadvantage of not uniquely determining the locus-point C corresponding to each given value of the angle $\hat{B}$ and the lengths BC, AI and IC; for fixing those of $\hat{B}$ and BC determines C to lie in a straight line parallel to AB, while given values of AI and IC set it to be in a circle of centre I, radius IC which meets the preceding parallel in two distinct points, so that any given relationship $f(\hat{B}, BC, AI, IC) = 0$ defines a pair of loci (C). We will not insist on the ambiguities which ensue.

(141) Newton first wrote just 'similiter' but here on second thought chooses again to specify his meaning.

(142) Perpendicular to BCD, of course.

(143) In line with its progenitor IC, namely.

(144) That is, 'quoad plagam motuum' parallel to AI and perpendicular to IC respectively.

C ultimo locetur alicubi in linea *LM*, et priores motus ut ultimo locetur in linea *GH*, necesse est ut ultimo locetur in harum linearum communi intersectione *H*, adeoq; recta *CH* motum ejus exponet et Curvam ab eo descriptam tanget.

Neq; res difficilior est si detur fluxionum duarum summa differentia vel proportio. Detur fluxio anguli *B* et summa fluxionum linearum *AB*, *BC*. Motum puncti *C* a fluxione anguli illius oriundum expone per lineam *CG*,(145) dein positionem rectæ *OQ* ea lege quære ut si summa fluxionum linearum *AB* et *BC* dividatur utcunq; in duas partes et motus duo puncti *C* qui ab his partibus seorsim orirentur exponantur(146) per *GP* et *PQ*, punctum *Q* semper incidat in hanc rectam.(147) Et erit hæc recta ille ultimus locus indefinitus puncti *C* cujus intersectione cum alio ejusdem ultimo loco ex alijs quibusvis datis inveniendo determinabitur tangens. Determinabitur autem positio rectæ *OQ* producendo *GP* ad *O* ut sit *PO* æqualis *PQ* et jungendo *OQ*.

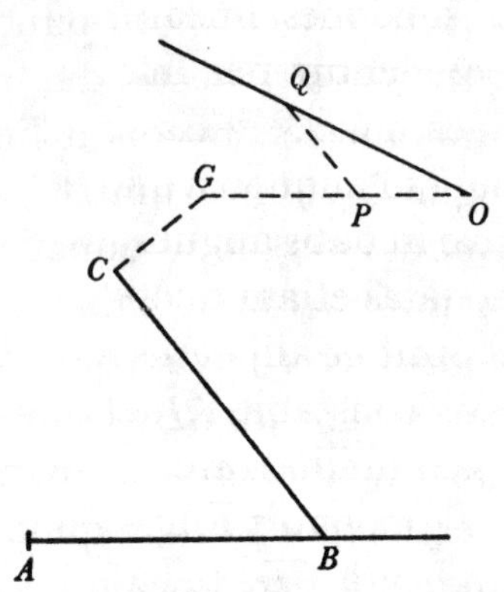

Quinetiam si plurium linearum fluxiones sigillatim ducerentur in cognitas quantitates et productorum aliquorum summa poneretur æqualis summæ reliquorum, problema nihil minùs solvi posset. A mobili puncto *C* ducantur tres lineæ, *CA* ad datum punctum *A*, *CB* et *CD* in datis angulis ad rectas positione datas *BI* ac *DK*, et fluxionibus earum respectivè ductis in tres datas lineas *L*, *M*, *N* æquentur producta duo priora producto tertio, nempe

$$L'ac + M'bc = N'dc.^{(148)}$$

Et Curvæ a puncto *C* descriptæ tangens ita ducetur. Ad *AC* erigo normalem *CE* quæ plagam motus puncti *C* circa *A* gyrantis exhibeat, et rectis *BI DK* parallelas

(145) Perpendicular to *BC*, it is understood.

(146) 'quoad plagam et quantitatem' is here deleted. Newton meant to specify, of course, that *GP* and *PQ* shall represent the fluxions of *AB* and *BC* in direction and *relative* magnitude (and hence be drawn parallel to these).

(147) The locus (*Q*), namely, such that $GP+PQ = GO$, where *GO* is fixed and *PQ* is free to move parallel to itself.

(148) Whence the defining fluent equation of the locus (*C*) is $L.AC+M.BC-N.DC = k$, constant.

duco CF, CG quæ plagas motuum puncti C secundum rectas BI DK exhibeat. Dein quæro punctum aliquod H ita ut si ad CE ducatur HE parallela AC, ad CF autem HF parallela BC et ad CG HG parallela CD, fiat

$$L\text{'}HE+M\text{'}HF = N\text{'}GH$$

et fluxiones linearum AC, BC, DC motusꝗ relativi puncti C inde oriundi exponentur per lineas HE, HF, HG et motus absolutus puncti C per actam lineam CH.(149) Hæc igitur erit tangens quæsita. Quomodo vero inveniri potest punctū aliquod a quo si ad rectas quotcunꝗ positione datas CE, CF, CG totidem aliæ in datis angulis agantur earum aliquæ in datas lineas ductæ æquentur alijs in datas etiam ductis patebit e sequentibus.(150)

Sunt et alij casus difficiliores,(151) sed ex his credo sensus et vis Theorematis satis constabit. Quapropter pergo jam exemplis aliquot methodum hic propositam illustrare.

Proponatur Ellipsis(152) ADB cujus centrum sit G, diameter AB,(153) ordinatim applicata CD, et latus rectum N[,] et ex natura figuræ erit ut AB ad N ita ACB ad CD^2 et (ob datam hanc rationem) ita prioris fluxio $AC\text{'}cb+CB\text{'}ac$, id est $AC\text{'}cb-CB\text{'}cb$ seu $2GC\text{'}cb$, ad posterioris fluxionem $2CD\text{'}cd$. Exponantur (juxta Theor 7) fluxiones cb et cd per FC et CD ita ut FD tangens fiat Ellipseos ad D, & convertentur $2GC\text{'}cb$ et $2CD\text{'}cd$ in $2FCG$ et $2CD^2$: quorum itaꝗ dimidia sunt ut ACB ad CD^2, adeoꝗ FCG et ACB æqualia sunt. Cape igitur FC ad AC ut CB ad CG et acta FD figuram tanget in D.

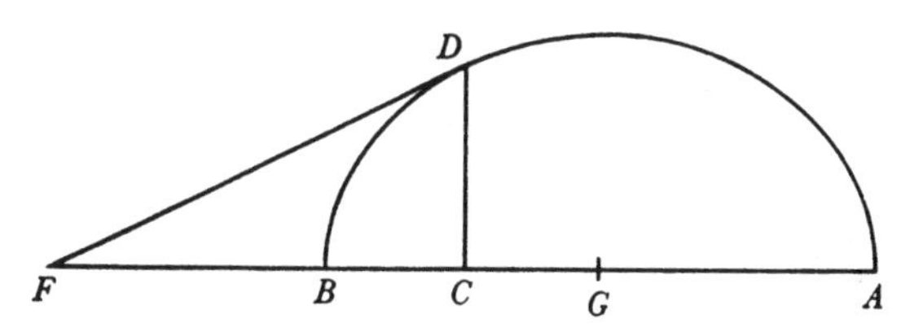

Propo[natur](154)

(149) This will, it is clear, be the locus of points H satisfying $L.EH+M.FH-N.GH = 0$; compare the previous note.

(150) Newton never reached so far, abandoning his present draft after penning but two more paragraphs of its text.

(151) An unwanted 'ad quas pergere liceret' is deleted in sequel.

(152) An alternative 'vel Hyperbola' is cancelled. In his accompanying figure, correspondingly, Newton initially extended GA to B and made BD the ordinate of a hyperbolic arc AD going upwards and rightwards from its vertex A.

(153) Originally Newton equivalently specified 'vertices A et B'.

(154) Here, in the middle of the first word setting out his proposal of a second worked example instancing his general construction of the tangential *plaga motus* at a point of a curve defined in some multi-coordinate system of the type discussed above, Newton abruptly broke off to cancel the whole of his present intended conclusion to his 'Geometriæ Liber primus', proceeding at once to pen (on the following ff. 146r–150r) the less discursive and more tightly organized revised termination reproduced on our previous pages 448–68.

[12][155] *Super recta KL positione data et ad datum punctum A terminata incedit recta AL in dato angulo ALK et termino suo A curvam lineam KA*[156] *describit. Datur relatio linearū KL et AL*[157] *ad invicem et recta ducenda est quæ curvam hanc tangat in A.*

Quoniam punctum A ex assumptione fluentium KL et LA determinatur, postquam earum fluxiones exposui per LM et AN motum quem punctum A haberet si LA assumeretur et fluxio solius KL maneret expono quoad plagam et quantitatem ducendo AO parallelam et æqualem LM, eo quod punctum illud A moveret per hanc AO eodem tempore quo KL evaderet KM. Deinde quoniam plaga motus puncti A foret AN si vice versa flueret LA et altera KL assumeretur in hac plaga[,] per exponentis terminum O duco indefinitam lineam OF et concludo hanc OF esse unum locum metæ. Et simili argumento quoniam AN exponens est tam motus puncti A ubi KL assumitur quam fluxionis lineæ LA[,] per terminum ejus N in plaga motus quem punctum A ex sola fluxione ipsius KL haberet duco lineam indefinitam NF pro altero loco metæ. Et locorum concursus F meta erit ad quam tangens quæsita AF ducenda est. Quam conclusionem ut concinniorem reddas, produc AF donec occurrat LK etiam productæ in Q,[158] et ob similitudinem triangulorum QLA, AOF erit QL ad LA ut AO ad OF seu LM ad AN, adeoq; vice exponentium LM et AN adhiberi possunt QL et LA, eo ut inveni[a]tur punctum Q. Comple parallelogrammum $KLAH$ cujus latus HK tangentem secet in I[158] & ob proportionales AH, HI et LM AN adhibere licet AH, HI eo ut inveniatur punctum I. Utrumvis punctum Q vel I prout commodum videbitur quære.

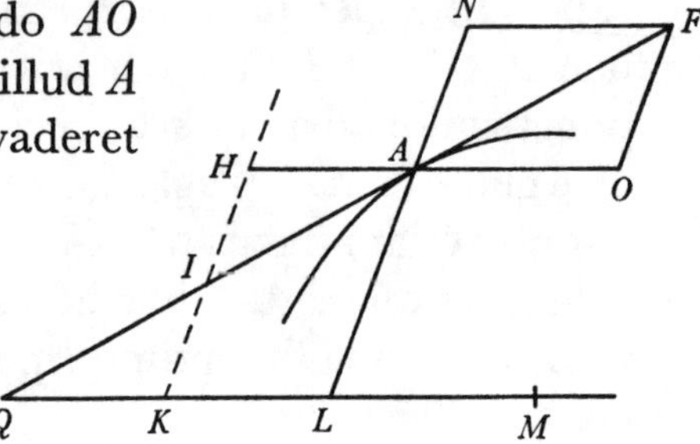

(155) Add. 4004: 146r, a first version of the trivially generalized 'Exemplum 1' on page 460 above. Little differently from the manner of its solution which he had first set down nearly thirty years before in earlier pages of his Waste Book—in May 1665 as a 'universall theorem for drawing tangents to crooked lines when [the coordinates] intersect at any determined angle' (see I: 279–80) and in full fluxional equivalent the following November (*ibid.*: 386–7)—Newton elaborates the construction of the subtangent at an arbitrary point of a curve defined in oblique Cartesian coordinates which straightforwardly ensues from determining the 'goal-point' of the instantaneous fluxions of the coordinate lines whose relationship defines the locus.

(156) In his figure Newton does not in fact draw the curve (A) through K.

(157) Or, equivalently, of the (modern) coordinate lines AH and AL drawn parallel to the axes KL and KH. In his reworked 'Exemplum' Newton afterwards permits these lines —without any significant gain in generality (compare §1: note (138)) if it is an added elegance—to be drawn at any fixed angles to the axes. Here, in a discarded opening which in the manuscript immediately precedes his present enunciation, he initially specified the standard oblique Cartesian coordinates of his following problem in the words 'A mobili puncto A qua linea curva describitur ad rectam KL positione datam in dato angulo ALK [ducatur recta LA...]'.

(158) Where $QL = LA.\text{fl}(KL)/\text{fl}(LA)$ and $KI = LA - KL.\text{fl}(LA)/\text{fl}(KL)$.

§2. 'BOOK 2': THE RATIONAL QUADRATURE OF CURVES.[1]

From the autograph[2] in the University Library, Cambridge

(1) In this latter book of the final text of his 'Geometria' as it now survives Newton presents a severely abridged version of the unfinished treatise 'De quadratura Curvarum' (**1**, §2 preceding) which he had left off only a few months earlier. Abandoning the latter's original *raison d'être*, he here foregoes making explicit claim to the quadrature series which he had communicated to Leibniz in 1676—evidently trusting to Wallis' publication of it and its extension (Proposition V below) in his Latin *Algebra* in 1693 (see **1**, §2, Appendix 3) to safeguard his priority there—and lightly revises and amplifies its middle Propositions IV, V and VII–X to be Propositions I–VI of the present 'Liber Secundus', further elaborating their technique of rational quadrature in four additional Propositions VII–X. In introduction he now gives a brief conspectus of the notations for general indices, fluxions and fluents—still novel to his contemporaries at large—which he employs in expounding and demonstrating their detail, and adjoins a *scholium generale* wherein he sets out afresh the twin tables of algebraic integrals (of those exactly quadrable, and of ones evaluable in terms of conic areas respectively) which he had initially compiled more than twenty years before for his yet unpublished 1671 tract (see III: 236–54), appending thereto a short final Proposition XI in which he tersely describes, without any justification of their skeletal structure, the bare bones of the geometrically recast reduction of the simplest types of multiple integrals to canonical 'areæ curvarum' which he had earlier contrived in more generally pondering the fluents of 'complex' algebraic quantities (see **1**, §2, Appendix 2 above). Unlike that of the preceding 'Liber primus', the text of this second book of his 'Geometria' was not allowed to pass unnoticed into the limbo of Newton's forgotten private papers, to be resurrected only by some modern editor of that dusty heritage. As we shall see in our concluding volume, Newton himself separated this renamed 'Tractatus De Quadratura Curvarum' from its geometrical surround to become once more an independent treatise on algebraic quadrature, adding an introductory sketch of how fluxional analysis may be applied geometrically to determine limit curvilinear motions, and also a final scholium—one to have unforeseen repercussions during the ensuing squabble over calculus priority when the Continentals sought thereby to delimit Newton's expertise in handling higher-order derivatives—in which the successive terms in the 'Taylor' series expansion of the general power of some quantity and its increment were loosely related to its corresponding fluxions. And in this augmented form the 'Tractatus' was in 1704 given to the world by Newton in the second of the two treatises 'Of the Species and Magnitude of Curvilinear Figures' which he appended to the *editio princeps* of his *Opticks*.

(2) Add. 3962.1: 5^r–19^r, a set of folded quarto sheets on whose recto sides Newton has carefully penned out his present text, numbering the alternate pages (f. 5^r, 7^r, 9^r, ..., 17^r, 19^r) at their top right corner as 1, 3, 5, ..., 13, 17. His *paginæ* 14/15 (which would here find their place between ff. 18 and 19) and all those after 17 (which would continue after f. 19) have not survived: the former, containing the text of Proposition X and its corollaries, is restored by us from the autograph draft now Add. 3962.3: 50^v and—in respect of the corollaries which are there only outlined—from the equivalent passage in the *editio princeps* (*Opticks*, 1704: ${}_2196$–7); the latter, embracing all but the first entry of the 'Ordo primus curvarum' in the second tabulation of integrals together with the terminal Proposition XI, we have not chosen to reconstitute in their detail, believing it enough to refer for the missing lists of integrals to their equivalents in the 1671 tract's *Catalogus posterior* which the present tabulation (it will be evident) narrowly repeats, and to adjoin the text of the preliminary version of Proposition XI which is found on Add. 3962.3: 64^r. In the extant portion of his main manuscript Newton has, in preparation for its 1704 printing, made a number of changes and insertions in what is

GEOMETRIÆ
LIBER SECUNDUS(3)

Quo tempore(4) incidi in methodum serierum interminatarum convergentium, necesse habui mutare notationem quæ tunc in usu erat et pro $\sqrt{x}$ $\sqrt{x^3}$, $\sqrt{3:x}$, $\sqrt{3:x^2}$, $\frac{1}{x}$, $\frac{1}{xx}$, $\frac{1}{\sqrt{x}}$, $\sqrt{aa-xx}$, $\frac{1}{\sqrt{aa-xx}}$ &c scribere $x^{\frac{1}{2}}$, $x^{\frac{3}{2}}$, $x^{\frac{1}{3}}$, $x^{\frac{2}{3}}$, x^{-1}, x^{-2}, $x^{-\frac{1}{2}}$, $\overline{aa-xx}^{\frac{1}{2}}$, $\overline{aa-xx}^{-\frac{1}{2}}$. Hac enim ratione computationes magis uniformes & expeditæ & theoremata magis generalia evaserunt. Qua de causa etiam exponentes dignitatum indefinite designavi cum Slusio(5) in hunc modum x^{μ}, x^{ν}, $\overline{aa-xx}^{\nu}$ ponendo dignitatum exponentes μ et ν & similes pro numeris quibusvis, integris an fractis, affirmativis an negativis. Quas quidem notarum formulas, cùm jam in usu esse cœperint, non opus est ut fusius exponam.

Quantitates indeterminatas ut motu perpetuo crescentes vel decrescentes, id

---

clearly a later hand; these we specify at appropriate points in footnote to our reproduction of what we adjudge to be the original text of the 'Liber Secundus'. (There are also a very few extraneous printer's marks in the manuscript indicating divisions in the 1704 pages—notably a '*Zz* pr[ima]/169' at the opening of this signature—which establish that this was the actual copy from which the initial edition was set in type; compare D. F. McKenzie, *The Cambridge University Press: 1696—1712*, **1** (Cambridge, 1966): 118.) We further cite in sequel a number of additional corrections and variants introduced in the early printed editions—by Newton in that of 1704 and the minimally differing text analogously appended to the Latin *Optice* in 1706, and by William Jones in the somewhat reset version which appeared on pages 49–66 of his *Analysis Per Quantitatum Series, Fluxiones, ac Differentias* (London, 1711)—or entered by him in hand in the margins of his private library copy (now ULC. Adv. b. 39.4) of the 1706 *Optice*. During his stay with Newton at Cambridge in early May 1694, we should add, David Gregory was shown the present manuscript (or at least its opening folios)—already bearing the revised title which more accurately specified its content (see the next note)—and was permitted to make an incomplete transcript of it, down to midway through its Proposition V (see note (39) below), which survives in private possession. Though Gregory's copy has little intrinsic value, it allows us to be firm in editing out many of Newton's subsequent additions to the pristine text of the 'Geometriæ Liber Secundus' here given. Shortly after his return to Oxford, in one of the many memoranda in which he endeavoured to order the jumbled impressions of his Cambridge stay when the crowded mass of what he had learnt from Newton was still fresh in his mind, Gregory observed of this '[de] Quadraturis Curvarum tractatu[s] (quem vidi) 3 aut 4 foliorum si ab utraꝗ parte scriberetur' that it 'istam rem [*sc.* quadraturam curvarum] mire et ultra quam credi facile poterat promovet', accurately noticing that 'per series abrumpentes procedit [et non quadrabilia spatia] cum simplicissimis comparat' and further reporting more dubiously that Newton '[illum] post *Principia* editurus est'. (See Royal Society. Gregory Volume/C44, printed in *The Correspondence of Isaac Newton*, **3**, 1961: 334–6, especially 336. A virtually identical passage from Gregory's contemporary memorandum C42 has been cited on pages 196–7 above.) While we ourselves—with all the advantage of hindsight and direct access (which Gregory did not have) to the text of Newton's far fuller and more wide-reaching 1691 treatise 'De quadratura Curvarum' (**1**, §2 preceding)—may be more guarded and lukewarm in our estimate of the 1693 tract which it sired, we may readily pardon him for giving enthusiastic welcome to so concise and elegant a generalized exposition

GEOMETRY
BOOK TWO.[3]

At the time[4] I fell upon the method of unterminated converging series I deemed it necessary to change the notation which was then in use, and in place of $\surd x$, $\surd x^3$, $\sqrt[3]{x}$, $\sqrt[3]{x^2}$, $1/x$, $1/x^2$, $1/\surd x$, $\surd(a^2-x^2)$, $1/\surd(a^2-x^2)$ and so on to write $x^{\frac{1}{2}}$, $x^{\frac{3}{2}}$, $x^{\frac{1}{3}}$, $x^{\frac{2}{3}}$, x^{-1}, x^{-2}, $x^{-\frac{1}{2}}$, $(a^2-x^2)^{\frac{1}{2}}$, $(a^2-x^2)^{-\frac{1}{2}}$. For by this means computation turned out to be more uniform and speedy, and theorems more general. And for this reason also I, with Sluse,[5] indefinitely denoted the exponents of powers in this manner: x^μ, x^ν, $(a^2-x^2)^\nu$, putting the exponents μ, ν and the like of the powers for any numbers, integers or fractions, positive or negative. Since these forms of notation have now, indeed, begun to be in use, there is no need for me elaborately to explain them.

In the sequel I consider indeterminate quantities as increasing or decreasing

---

of a technique of series quadrature whose binomial instance it had half a dozen years before (see pages 4–6) cost him a deal of effort (and very nearly Newton's friendly support of his academic advancement) 'independently' to discover.

(3) Afterwards, in a reversion to the title of the parent 1691 treatise, altered by Newton to be 'TRACTATUS/DE QUADRATURA CURVARUM' (A TRACT/ON THE QUADRATURE OF CURVES): in which form it was in May 1694 copied by David Gregory (see the previous note).

(4) According to a memorandum which—prompted no doubt by Fatio de Duillier's recent declaration recognizing him to be 'primum, ac pluribus Annis vetustissimum...Calculi Inventorem, ipsa rerum evidentia coactus' (see VI: 467, note (25))—Newton was soon, in July 1699, to attach to his earliest notes on the Wallisian interpolation of finite binomial expansions, it was 'in winter between the years 1664 & 1665' that he 'found the method of Infinite series' (ULC. Add. 4000: 14v, quoted in full on I: 7–8; compare *ibid.*: 96–134). As we have seen (I: 343 ff., 369–89) he did not begin to cast his earliest researches into problems of tangents and curvature in fluxional form till the next autumn, and compiled no systematic tract on his new 'calculus' till October 1666 (*ibid.*: 400–48). In introduction to the printed *Tractatus de Quadratura Curvarum* he afterwards accurately specified that 'motuum vel incrementorum velocitates nominando *Fluxiones* & quantitates genitas nominando *Fluentes*, incidi paulatim Annis 1665 & 1666 in Methodum Fluxionum qua hic usus sum in Quadratura Curvarum' (*Opticks*, 1704: $_2$166); in his preliminary autograph on Add. 3962.1: 1r the time-interval of discovery had been less warrantably shortened to be 'in anno 1665' only.

(5) Newton should have written 'Wallisio' (Wallis). Though in communicating his method of algebraic tangents to Oldenburg in 1673 (in letters of 7 January and 23 April printed with minimal delay in *Philosophical Transactions*, 7, No. 90 [for 20 January 1672/3]: 5143–7, and No. 95 [for 23 June 1673]: 6059 respectively; see also A. R. and M. B. Hall, *The Correspondence of Henry Oldenburg*, 9, 1973: 386–92, 617–18) Sluse had referred his method without restriction to general 'potestates' and 'dignitates' of the component variables in the defining Cartesian equations of his curves, he had developed no literal notation for the concept of an arbitrary power-index. That advance had been earlier made in crude form by Wallis in his *Mathesis Universalis* in 1657 (compare F. Cajori, *A History of Mathematical Notations*, 1 (La Salle, 1928): 355), but it was Newton himself who in his early researches into the binomial expansion (see I: 131–4, 318–21)—and in his 'public' enunciation of the general series which he afterwards communicated to Leibniz in his *epistola prior* in 1676 (see IV: 667)—first established this most fertile convention in standard Cartesian form.

est ut fluentes vel defluentes in sequentibus considero(6) designoqꝫ literis z, y, x, v et earum fluxiones seu celeritates(7) crescendi noto ijsdem literis punctatis $\dot{z}, \dot{y}, \dot{x}, \dot{v}$. Sunt et harum fluxionum fluxiones seu mutationes magis aut minus celeres quas ipsarum z, y, x, v fluxiones secundas nominare licet & sic designare $\ddot{z}, \ddot{y}, \ddot{x}, \ddot{v}$, et harum fluxiones primas seu ipsarum z, y, x, v fluxiones tertias sic $\dddot{z}, \dddot{y}, \dddot{x}, \dddot{v}$, et quartas sic $\ddddot{z}, \ddddot{y}, \ddddot{x}, \ddddot{v}$. Et quemadmodum $\dddot{z}, \dddot{y}, \dddot{x}, \dddot{v}$ sunt fluxiones quantitatum $\ddot{z}, \ddot{y}, \ddot{x}, \ddot{v}$ et hæ sunt fluxiones quantitatum $\dot{z}, \dot{y}, \dot{x}, \dot{v}$ & hæ sunt fluxiones quantitatum primarum z, y, x, v: sic hæ quantitates considerari possunt ut fluxiones aliarum quas sic designabo(8) $\acute{z}, \acute{y}, \acute{x}, \acute{v}$ et hæ ut fluxiones aliarum $\acute{\acute{z}}, \acute{\acute{y}}, \acute{\acute{x}}, \acute{\acute{v}}$, et hæ ut fluxiones aliarum $\acute{\acute{\acute{z}}}, \acute{\acute{\acute{y}}}, \acute{\acute{\acute{x}}}, \acute{\acute{\acute{v}}}$. Designant igitur $\acute{\acute{z}}, \acute{z}, z, \dot{z}, \ddot{z}, \dddot{z}, \ddddot{z}$, $\overset{\cdot}{\ddddot{z}}$ &c seriem quantitatum quarum quælibet postcrior est fluxio præcedentis et quælibet prior fluens quantitas fluxionem habens subsequentem. Similis est series $\overset{\prime\prime\prime}{\sqrt{az-zz}}$. $\overset{\prime\prime}{\sqrt{az-zz}}$. $\sqrt{az-zz}$. $\overset{\cdot}{\sqrt{az-zz}}$. $\overset{\cdot\cdot}{\sqrt{az-zz}}$. $\overset{\cdot\cdot\cdot}{\sqrt{az-zz}}$, ut et series $\overset{\prime\prime\prime}{\frac{az+z^2}{a-z}}$. $\overset{\prime\prime}{\frac{az+z^2}{a-z}}$. $\frac{az+z^2}{a-z}$. $\overset{\cdot}{\frac{az+z^2}{a-z}}$. $\overset{\cdot\cdot}{\frac{az+z^2}{a-z}}$. $\overset{\cdot\cdot\cdot}{\frac{az+z^2}{a-z}}$.(9) Et notandum est quod quantitas quælibet prior in his seriebus est area figuræ curvilineæ cujus ordinatim applicata rectangula est quantitas posterior et abscissa z: uti $\overset{\prime\prime}{\sqrt{az-zz}}$(10) area curvæ cujus ordinata est $\sqrt{az-zz}$ et abscissa z. Hanc aream sic etiam designo $[\sqrt{az-zz}]$ vel sic $\boxed{\sqrt{az-zz}}$.(11) Quo autem spectant hæc omnia patebit in Propositionibus quæ sequuntur.

---

(6) A superfluous 'et per earum fluxionem intelligo celeritatem crescendi' (and by their 'fluxion' I understand the speed of their increase) is cancelled in sequel. Newton initially continued thereafter: 'Hujusmodi quantitates designo per literas...' (Quantities of this sort I denote by means of the letters...).

(7) Literally 'swiftnesses'; 'velocitates' (velocities) was first written.

(8) In his *editio princeps* at this point (*Opticks*, 1704: $_2$171, l. 1) Newton interpolated '$\acute{z}, \acute{y}, \acute{x}, \acute{v}$, & hæ ut fluxiones aliarum' ($\acute{z}, \acute{y}, \acute{x}, \acute{v}$, and these as fluxions of others), correspondingly deleting the terminal instance 'et hæ ut fluxiones aliarum $\acute{\acute{\acute{z}}}, \acute{\acute{\acute{y}}}, \acute{\acute{\acute{x}}}, \acute{\acute{\acute{v}}}$' (and these as fluxions of...others $\acute{\acute{\acute{z}}}, \acute{\acute{\acute{y}}}, \acute{\acute{\acute{x}}}, \acute{\acute{\acute{v}}}$)—improvements designed, we may be sure, solely to render the numerical correspondence between the ranks of superscript 'points' and 'accents' which denote analogous orders of fluxions and fluents more exact, rather than with an eye to greater visual elegance or any thought of lightening the printer's labour in setting these complicated symbols in type. He had already made some use of this novel designation of fluents in the preliminary scheme of revision of Proposition XI of the 1691 'De quadratura Curvarum' which is reproduced in **1**, §2, Appendix 2 preceding (see page 165 above).

(9) Newton originally set the denominators of these fractions to be '$aa-bz$'; the alteration converts the basic quantity $(az+z^2)/(a-z)$ to be of the same unit dimension as those in earlier lines. In the *editio princeps* (*Opticks*, 1704: $_2$171)—where (compare the previous note) the first

by a perpetual motion, that is, as onwards or backwards fluent.(6) I denote them by the letters z, y, x, v and mark their fluxions, that is, speeds(7) of increase, by the same letters with points on: $\dot z$, $\dot y$, $\dot x$, $\dot v$. There are also fluxions, or changes of varying speed, of these fluxions which it is permissible to name second fluxions of z, y, x, v and denote thus: $\ddot z$, $\ddot y$, $\ddot x$, $\ddot v$; and first fluxions of these, that is, third fluxions of z, y, x, v, which may be denoted thus: $\dddot z$, $\dddot y$, $\dddot x$, $\dddot v$; and fourth ones, thus: $\ddddot z$, $\ddddot y$, $\ddddot x$, $\ddddot v$. And just as $\dddot z$, $\dddot y$, $\dddot x$, $\dddot v$ are fluxions of the quantities $\ddot z$, $\ddot y$, $\ddot x$, $\ddot v$, and these are fluxions of the quantities $\dot z$, $\dot y$, $\dot x$, $\dot v$, while these are fluxions of the primary quantities z, y, x, v; so too these quantities can be considered as fluxions of others which I shall denote thus: (8) $\overset{\prime}{z}$, $\overset{\prime}{y}$, $\overset{\prime}{x}$, $\overset{\prime}{v}$, and these as fluxions of other ones $\overset{\prime\prime}{z}$, $\overset{\prime\prime}{y}$, $\overset{\prime\prime}{x}$, $\overset{\prime\prime}{v}$, and these as fluxions of yet others $\overset{\prime\prime\prime}{z}$, $\overset{\prime\prime\prime}{y}$, $\overset{\prime\prime\prime}{x}$, $\overset{\prime\prime\prime}{v}$. So that $\overset{\prime\prime\prime}{z}$, $\overset{\prime\prime}{z}$, $\overset{\prime}{z}$, z, $\dot z$, $\ddot z$, $\dddot z$, $\ddddot z$, ... denote a series of quantities of which any latter one is the fluxion of that preceding it, and any former the fluent quantity having the subsequent one as its fluxion. The series $\surd(az\overset{\prime\prime\prime}{-}z^2)$, $\surd(az\overset{\prime\prime}{-}z^2)$, $\surd(az\overset{\prime}{-}z^2)$, $\surd(az-z^2)$, $\surd(az\overset{\cdot}{-}z^2)$, $\surd(az\overset{\cdot\cdot}{-}z^2)$, $\surd(az\overset{\cdot\cdot\cdot}{-}z^2)$ is similar, as also is the series $(az+z^2)\overset{\prime\prime\prime}{/}(a-z)$, $(az+z^2)\overset{\prime\prime}{/}(a-z)$, $(az+z^2)\overset{\prime}{/}(a-z)$, $(az+z^2)/(a-z)$, $(az+z^2)\overset{\cdot}{/}(a-z)$, $(az+z^2)\overset{\cdot\cdot}{/}(a-z)$, $(az+z^2)\overset{\cdot\cdot\cdot}{/}(a-z)$.(9) And note that any prior quantity whatever in these series is the area of a curvilinear figure whose right-angled ordinate is the latter quantity and its abscissa z: for instance, $\surd(az\overset{\prime\prime}{-}z^2)$ is the area of the curve whose ordinate is $\surd(az-z^2)$ and abscissa z. This area I also designate thus: $[\surd(az-z^2)]$ or thus: $\boxed{\surd(az-z^2)}$.(11) What, however, is the regard of all these things will become evident in the propositions which follow.

---

term $\frac{az+z^2}{a\overset{\prime\prime\prime}{-}z}$ is suppressed and a new one '$\frac{az+z^2}{a\overset{\prime}{-}z}$' introduced in its stead (in sequence after the second)—Newton was to replace the cumbersome 'doubly dotted' fluxional fractions by singly pointed equivalents '$\frac{az+z^2}{a-z}\overset{\cdot}{}.\frac{az+z^2}{a-z}\overset{\cdot\cdot}{}.\frac{az+z^2}{a-z}\overset{\cdot\cdot\cdot}{}$'; correspondingly, in the preceding line—where in 1704 the initial term $\sqrt{az\overset{\prime\prime\prime}{-}zz}$ gave way to a singly accented '$\sqrt{az\overset{\prime}{-}zz}$' inserted in sequel to the second—the fluxional radicals were lightly refashioned to become respectively '$\sqrt{az\overset{\cdot}{-}zz}.\sqrt{az\overset{\cdot\cdot}{-}zz}.\sqrt{az\overset{\cdot\cdot\cdot}{-}zz}$'. In the spirit of these emendations we have dared in our English version to 'translate' Newton's present symbols merely by inserting the appropriate number of superscript fluent 'accents' and fluxional points over our slightly modernized renderings, $\surd(az-z^2)$ and $(az+z^2)/(a-z)$, of the respective base quantities.

(10) Changed to be '$\sqrt{az\overset{\prime}{-}zz}$' in the 1704 printed version; compare the previous note.

(11) This sentence was to be omitted by Newton from the 1704 printed version. To be sure, although he had long ago introduced the latter 'box' notation for the operation of squaring an area into his 'De Analysi' in 1669 (see II: 226 ff.) and more recently made extensive employment of it in the later propositions of his 1691 'De quadratura Curvarum' (1, §2 preceding), in his present text—or at least the portion of it which he went on to write—he finds no use for any such pictographic symbols for quadrature, everywhere preferring the equivalent verbal expression 'area curvæ cujus ordinata est...' or simple variations thereupon.

Prop. I. Prob I.(12)

Data æquatione quotcunq; fluentes quantitates involvente, invenire fluxiones.

Solutio.

Multiplicetur omnis æquationis terminus per indicem dignitatis quantitatis cujusq; fluentis quam involvit, et in singulis multiplicationibus mutetur dignitatis latus in suam fluxionem et aggregatum factorum omnium sub proprijs signis erit æquatio nova.

Explicatio. (13)

Sunto a, b, c, d &c quantitates determinatæ & immutabiles, et proponatur æquatio quævis quantitates fluentes z, y, x &c involvens uti $x^3 - xyy + aaz - b^3 = 0$. Multiplicentur termini primo per indices dignitatum x & in singulis multiplicationibus pro dignitatis latere seu x unius dimensionis, scribatur $\dot{x}$ & summa factorum erit $3\dot{x}x^2 - \dot{x}yy$. Idem fiat in y et prodibit $-2xy\dot{y}$. Idem fiat in z et prodibit $aa\dot{z}$. Ponatur summa factorum æqualis nihilo et habebitur

$$3\dot{x}x^2 - \dot{x}yy - 2xy\dot{y} + aa\dot{z} = 0.$$

Dico quod hac æquatione definitur relatio fluxionum.

Demonstratio.(13)

Nam sit o quantitas infinite(14) parva et sunto $o\dot{z}$, $o\dot{y}$, $o\dot{x}$ quantitatum z, y, x momenta id est incrementa momentanea synchrona. Et si quantitates fluentes jam sunt z, y et x, hæ post momentum temporis incrementis suis infinite parvis(15)

(12) An augmented and considerably extended refashioning of Proposition IV of the 1691 'De quadratura Curvarum' on pages 62–6 above. A much less drastically recast preliminary version (on Add. 3962.3: 62r/63r) is reproduced in section [1] of the Appendix to show the line of succession.

(13) These jointly replace 'Cas. 1' of the proposition in its earlier versions (on which see the previous note).

(14) In the 1704 *editio princeps*, where he made (see the next note) a concerted effort to appeal to an infinitesimal—as distinct from indefinitely small—'moment' of the base variable of 'time' only at a subsequent stage as in the limit the *momenta synchrona* of the related fluent quantities become likewise evanescent, Newton amended this crucial adverb to be 'admodum' (extremely).

(15) These specifications of the contemporaneous increments as being from the outset 'infinitesimal' and as 'absolutely infinitesimal' in the limit as the moment of their flow diminishes 'infinitely' were discarded by Newton in 1704, the phrases 'infinite parvis' and 'ut momenta fiant infinitissime parva' being simply omitted from the text of the *editio princeps*. Where, however, the interval o of 'time' is not regarded from the first as infinitely small, the

Proposition I, Problem I.(12)

Given an equation involving any number of fluent quantities, to find the fluxions.

Solution.

Multiply every term in the equation by the index of each fluent quantity which it involves, and in the separate multiplications change a root of the power into its fluxion, and the aggregate of all the products under their proper signs will be a fresh equation.

Explanation. (13)

Let a, b, c, d, ... be determinate and unchangeable quantities and let any equation involving the fluent quantities z, y, x, ... be proposed, for instance:

$$x^3-xy^2+a^2z-b^3=0.$$

Multiply the terms first by the indices of the powers of x and in the separate multiplications in place of the power's root, that is, x of one dimension, write $\dot{x}$, and the sum of the products will be $3\dot{x}x^2-\dot{x}y^2$. Do the same in the case of y and there will result $-2xy\dot{y}$. Do the same in that of z and there will ensue $a^2\dot{z}$. Set the sum of the products equal to nothing and there will be had

$$3\dot{x}x^2-\dot{x}y^2-2xy\dot{y}+a^2\dot{z}=0.$$

I state that by this equation the relationship of the fluxions is defined.

Demonstration.(13)

For let o be an indefinitely(14) small quantity and $o\dot{z}$, $o\dot{y}$, $o\dot{x}$ moments of the quantities z, y, x, that is, their contemporaneous momentary increments. And if the fluent quantities are now z, y and x, these when increased after a moment of time by their infinitely small increments(15) $o\dot{z}$, $o\dot{y}$, $o\dot{x}$ will come to be $z+o\dot{z}$,

---

complication arises that the several *momenta* $o\dot{z}$, $o\dot{y}$ and $o\dot{x}$ of the related fluents z, y and x no longer accurately represent their *incrementa synchrona* since 'quantitates fluentes... post momentum o temporis... evadent $z+o\dot{z}+\frac{1}{2}o^2\ddot{z}+\frac{1}{6}o^3\dddot{z}$ &c, $y+o\dot{y}+\frac{1}{2}o^2\ddot{y}+\frac{1}{6}o^3\dddot{y}$ &c, $x+o\dot{x}+\frac{1}{2}o^2\ddot{x}+\frac{1}{6}o^3\dddot{x}$ &c'. While Newton was well aware of the fluxional nature of the successive coefficients in the general 'Taylor' expansion of an incremented quantity—in Corollary 4 to Case 3 of Proposition XII of his 1691 'De quadratura', to be sure, he had (compare **1**, §2: note (107) preceding) listed their progression in exactly the form we here prescribe—he seems never, either in the text of the 1704 *editio princeps* or in any subsequent comment upon it, to have appreciated the necessity of here positing an analogous expansion of the respective *incrementa momentanea* of the fluents z, y and x when the *momentum temporis* is no longer taken initially to be infinitely small. The revised 'Demonstratio' of this basic Newtonian construction of the fluxion of a complex algebraic quantity wears, in consequence, a spuriously neat logical dress.

$o\dot{z}$, $o\dot{y}$, $o\dot{x}$ auctæ, evadent $z+o\dot{z}$, $y+o\dot{y}$, $x+o\dot{x}$, quæ in æquatione prima pro z, y et x scriptæ dant æquationem

$$x^3+3xxo\dot{x}+3xoo\dot{x}\dot{x}+o^3\dot{x}^3-xyy-o\dot{x}yy-2xo\dot{y}y-2\dot{x}oo\dot{y}y+xoo\dot{y}\dot{y}$$
$$+\dot{x}o^3\dot{y}\dot{y}+aaz+aao\dot{z}-b^3=0.$$

Subducatur æquatio prior et residuum divisum per o erit

$$3\dot{x}x^2+3\dot{x}\dot{x}ox+\dot{x}^3oo-\dot{x}yy-2x\dot{y}y-2\dot{x}o\dot{y}y+xoo\dot{y}\dot{y}+aa\dot{z}=0.$$

Minuatur quantitas o in infinitum ut momenta fiant infinitissime parva(15) & neglectis terminis evanescentibus restabit $3\dot{x}x^2-\dot{x}yy-2x\dot{y}y+aa\dot{z}=0$. Q.E.D.

Explicatio plenior.(16)

Ad eundem modum si æquatio esset $x^3-xyy+aa\sqrt{ax-yy}-b^3=0$, produceretur $3x^2\dot{x}-\dot{x}yy-2x\dot{y}y+aa\dot{\sqrt{ax-yy}}=0$. Ubi si fluxionem $\dot{\sqrt{ax-yy}}$ tollere velis pone $\sqrt{ax-yy}=z$ et erit $ax-yy=z^2$. et (per hanc Propositionem) $a\dot{x}-2y\dot{y}=2z\dot{z}$ seu

$$\frac{a\dot{x}-2\dot{y}y}{2z}=\dot{z},$$

hoc est $\dfrac{a\dot{x}-2\dot{y}y}{2\sqrt{ax-yy}}=\dot{\sqrt{ax-yy}}$. Et inde $3x^2\dot{x}-\dot{x}yy-2x\dot{y}y+\dfrac{a^3\dot{x}-2aa\dot{y}y}{2\sqrt{ax-yy}}=0$.

Et per operationem repetitam pergitur ad fluxiones secundas tertias et sequentes. Sit æquatio $zy^3-z^4+a^4=0$, et fiet per operationem primam $\dot{z}y^3+3z\dot{y}y^2-4\dot{z}z^3=0$, per secundam

$$\ddot{z}y^3+6\dot{z}\dot{y}y^2+3z\ddot{y}y^2+6z\dot{y}^2y-4\ddot{z}z^3-12\dot{z}^2z^2=0,$$

per tertiam

$$\dddot{z}y^3+9\ddot{z}\dot{y}y^2+9\dot{z}\ddot{y}y^2+18\dot{z}\dot{y}^2y+3z\dddot{y}y^2+18z\ddot{y}\dot{y}y+6z\dot{y}^3-4\dddot{z}z^3$$
$$-36\ddot{z}\dot{z}z^2-[24]\,\dot{z}^3z^{(17)}=0.$$

Ubi vero sic pergitur ad fluxiones secundas tertias et sequentes, convenit quantitatem aliquam ut uniformiter fluentem considerare & pro ejus fluxione prima unitatem scribere, pro secunda verò et sequentibus nihil. Sit æquatio $zy^3-z^4+a^4=0$ ut supra & fluat z uniformiter sitꝙ ejus fluxio unitas: et fiet per operationem primam $y^3+3z\dot{y}y^2-4z^3=0$, per secundam

$$6\dot{y}y^2+3z\ddot{y}y^2+6z\dot{y}^2y-12z^2=0,$$

per tertiam $$9\ddot{y}y^2+18\dot{y}^2y+3z\dddot{y}y^2+18z\ddot{y}\dot{y}y+6z\dot{y}^3-[24]\,z=0.$$

(16) The first paragraph following is a lightly refashioned repeat of 'Cas. 2' in the preliminary versions of this proposition (on which see note (12) above). The ensuing extension successively to construct second, third and higher-order fluxions and the concluding observa-

$y+o\dot{y}$, $x+o\dot{x}$; and when these are written in the first equation in place of z, y and x they yield the equation

$$x^3+3x^2o\dot{x}+3xo^2\dot{x}^2+o^3\dot{x}^3-xy^2-o\dot{x}y^2-2xoy\dot{y}-2\dot{x}o^2y\dot{y}+xo^2\dot{y}^2$$
$$+\dot{x}o^3\dot{y}^2+a^2z+a^2o\dot{z}-b^3=0.$$

Subtract the former equation, and when the residue is divided by o there will be

$$3\dot{x}x^2+3\dot{x}^2ox+\dot{x}^3o^2-\dot{x}y^2-2x\dot{y}y-2\dot{x}o\dot{y}y+xo^2\dot{y}^2+a^2\dot{z}=0.$$

Let the quantity o diminish infinitely so that the moments come to be utterly infinitesimal[15] and with the vanishing terms discarded there will remain $3\dot{x}x^2-\dot{x}y^2-2x\dot{y}y+a^2\dot{z}=0$. As was to be proved.

Fuller explanation.[16]

In the same manner, were the equation $x^3-xy^2+a^2\sqrt{(ax-y^2)}-b^3=0$, there would be produced $3x^2\dot{x}-\dot{x}y^2-2xy\dot{y}+a^2\dot{\sqrt{(ax-y^2)}}=0$. Should you here wish to remove the fluxion $\dot{\sqrt{(ax-y^2)}}$, set $\sqrt{(ax-y^2)}=z$ and there will be $ax-y^2=z^2$, and so (by this present Proposition) $a\dot{x}-2y\dot{y}=2z\dot{z}$ or $(a\dot{x}-2\dot{y}y)/2z=\dot{z}$, that is, $(a\dot{x}-2\dot{y}y)/2\sqrt{(ax-y^2)}=\dot{\sqrt{(ax-y^2)}}$; and thence

$$3x^2\dot{x}-\dot{x}y^2-2x\dot{y}y+(a^3\dot{x}-2a^2\dot{y}y)/2\sqrt{(ax-y^2)}=0.$$

And by repetition of this operation you may proceed to second, third and following fluxions. Let the equation be $zy^3-z^4+a^4=0$ and there shall by a first operation come to be $\dot{z}y^3+3z\dot{y}y^2-4\dot{z}z^3=0$, by a second one

$$\ddot{z}y^3+6\dot{z}\dot{y}y^2+3z\ddot{y}y^2+6z\dot{y}^2y-4\ddot{z}z^3-12\dot{z}^2z^2=0,$$

by a third

$$\dddot{z}y^3+9\ddot{z}\dot{y}y^2+9\dot{z}\ddot{y}y^2+18\dot{z}\dot{y}^2y+3z\dddot{y}y^2+18z\ddot{y}\dot{y}y+6z\dot{y}^3-4\dddot{z}z^3$$
$$-36\ddot{z}\dot{z}z^2-24\dot{z}^3z^{(17)}=0.$$

When, however, you proceed to second, third and following fluxions in this way, it is convenient to consider some quantity as uniformly fluent and in place of its first fluxion to write unity, but for the second and following ones nil. Let the equation be $zy^3-z^4+a^4=0$, as above, and let z flow uniformly and its fluxion be unity: by a first operation there will come to be $y^3+3z\dot{y}y^2-4z^3=0$, by a second $6\dot{y}y^2+3z\ddot{y}y^2+6z\dot{y}^2y-12z^2=0$, by a third

$$9\ddot{y}y^2+18\dot{y}^2y+3z\dddot{y}y^2+18z\ddot{y}\dot{y}y+6z\dot{y}^3-24z=0.$$

tions on 'brother' terms in a given fluxional equation are here freshly added (though not in themselves novel).

(17) By a slip of his pen Newton in the manuscript set the coefficient of this term to be '48', a numerical error of near-trivial significance which he himself afterwards caught in the proof of the 1704 *editio princeps*. We similarly follow the printed text in making parallel adjustment of the mistaken coefficient '48' at corresponding places in ensuing equations.

In hujus autem generis æquationibus concipiendum est quod fluxiones in singulis terminis sint ejusdem ordinis, id est vel omnes primi ordinis $\dot{y}$, $\dot{z}$, vel omnes secundi $\ddot{y}$, $\dot{y}^2$, $\dot{y}\dot{z}$, $\dot{z}^2$, vel omnes tertij $\dddot{y}$, $\ddot{y}\dot{y}$, $\ddot{y}\dot{z}$, $\dot{y}^3$, $\dot{y}^2\dot{z}$, $\dot{y}\dot{z}^2$, $\dot{z}^3$ &c. Et ubi res aliter se habet complendus est ordo per subintellectas fluxiones quantitatis uniformiter fluentis. Sic æquatio novissima complendo ordinem tertium fit

$$9\dot{z}\ddot{y}y^2+18\dot{z}\dot{y}^2y+3z\dddot{y}y^2+18z\ddot{y}\dot{y}y+6z\dot{y}^3-[24]\,z\dot{z}^3=0.$$

Termini æquationis primæ ut genitores terminorum æquationis cujusvis quæ per hanc Propositionem ex prima prodit, et termini ex eodem genitore nati ut fratres vel socij[18] considerari possunt. Innotescunt fratres et ex fratribus genitor mutando fluxiones in fluentes quantitates.[19] Sic in æquatione novissima termini omnes præter ultimum mutando fluxiones in fluentes quantitates[19] evadunt zy^3 ideoq; fratres sunt, et genitorem habent mzy^3, et terminus ultimus qui solitarius est eadem ratione migrat in [z^4 & habet] genitorem nz^4. Hic m et n pro coefficientibus determinatis[20] indefinitè ponuntur, et inveniri possunt quærendo fratres ex his genitoribus & comparando cum fratribus datis.[21] Sic ex genitore mzy^3 prodeunt fratres $9m\dot{z}\ddot{y}y^2$, $18m\dot{z}\dot{y}^2y$, $3mz\dddot{y}y^2$, $18mz\ddot{y}\dot{y}y$, $6mz\dot{y}^3$ & [ex nz^4 fit $24n$]$z\dot{z}^3$, qui cum fratribus datis[21] collati dant $m=[n=]1$.[22]

Prop. II. Prob. II.[23]

Invenire Curvas quæ quadrari possunt.

Sit ABC figura invenienda, BC Ordinatim applicata rectangula, et AB abscissa. Producatur CB ad E ut sit $BE=1$,[24] et compleatur parallelogrammum $ABED$: et arearum ABC, $ABED$ fluxiones erunt ut BC et BE. Assumatur igitur æquatio quævis qua relatio arearum definiatur, et inde dabitur relatio ordinatarum BC, et BE per Prop. I. Q.E.I.

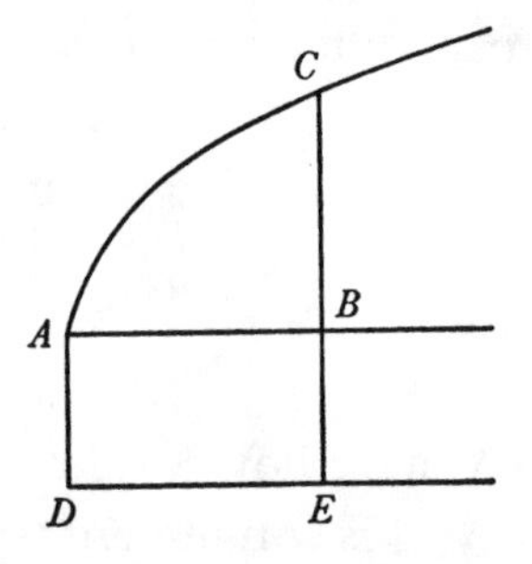

Hujus rei exempla habentur in Propositionibus duabus sequentibus.

(18) In Case 3 of Proposition XI of his 1691 'De quadratura' Newton had earlier dubbed such terms 'relativi' (relatives); see **1**, §2: note (58).

(19) 'et negligendo coefficientes datas' (and neglecting given coefficients) was afterwards adjoined by Newton in each instance.

(20) Originally 'quibusvis datis' (any given).

(21) The 'datis' (given) was subsequently deleted by Newton, who replaced it in the first case by 'in æquatione' (in the equation) and in the latter by '$9\dot{z}\ddot{y}y^2+18\dot{z}\dot{y}^2y$ &c'.

(22) Newton later added in sequel that 'Si quantitas m vel n tali collatione non prodit, fratres non sunt germani' (If the quantity m or n is not forthcoming from such a collation, the brothers are not german [*sc.* fully related in the present sense]). Whether or not it was his intention to go on to include further instances of the exact resolution of fluxional equations, no others of the motley collection of such methods which he had assembled in Proposition XI

In equations of this kind we are to conceive that the fluxions in the separate terms shall be of the same order, that is, all of first order ($\dot{y}, \dot{z}$), or all of second ($\ddot{y}, \dot{y}^2, \dot{y}\dot{z}, \dot{z}^2$), or all of third ($\dddot{y}, \ddot{y}\dot{y}, \ddot{y}\dot{z}, \dot{y}^3, \dot{y}^2\dot{z}, \dot{y}\dot{z}^2, \dot{z}^3$), and so on. And when the situation is different, their order needs to be completed by means of fluxions of the uniformly fluent quantity taken as understood. So by completing its third order the most recent equation becomes

$$9\dot{z}\ddot{y}y^2 + 18\dot{z}\dot{y}^2y + 3z\dddot{y}y^2 + 18z\ddot{y}\dot{y}y + 6z\dot{y}^3 - 24z\dot{z}^3 = 0.$$

Terms in the first equation can be considered as generators of the terms of any equation which results by this Proposition from the first, and terms begotten from the same generator regarded as brothers or relatives.(18) Brothers—and from brothers their generator—are ascertained by changing fluxions into fluent quantities.(19) Thus in the most recent equation all terms except the last come, on changing their fluxions into fluent quantities,(19) to be zy^3 and are consequently brothers, having mzy^3 as generator, while the last term which is on its own passes over by the same reason into z^4, having nz^4 as its generator. Here m and n are put indefinitely for determinate(20) coefficients, and can be found out by seeking the brothers from these generators and comparing them with the brothers given.(21) Thus from the generator mzy^3 there result the brothers $9m\dot{z}\ddot{y}y^2$, $18m\dot{z}\dot{y}^2y$, $3mz\dddot{y}y^2$, $18mz\ddot{y}\dot{y}y$, $6mz\dot{y}^3$, while from nz^4 there comes $24nz\dot{z}^3$, and when these are collated with the brothers given(21) they yield $m = n = 1$.(22)

Proposition II, Problem II.(23)

To find curves which can be squared.

Let ABC be the figure to be found, BC a right-angled ordinate and AB the abscissa. Produce CB to E so that $BE = 1$(24) and complete the rectangle $ABED$: the fluxions of the areas (ABC) and ($ABED$) will then be as BC and BE. Assume, therefore, any equation whereby the relationship of the areas shall be defined, and therefrom will be given the relationship of the ordinates BC and BE by Proposition I. As was to be found.

Examples of this procedure are had in the two following propositions.

---

of the 1691 'De quadratura' are here found, and this lone hit-or-miss technique appears in awkward isolation in the present context. Evidently realizing the incongruity, Newton afterwards deleted this concluding paragraph *in toto* and it is consequently omitted from the 1704 printed version. We have expanded Newton's text to fill an evident lacuna.

(23) This repeats Proposition V of the 1691 'De quadratura Curvarum' (**1**, §2) without essential variation in its detail. An intervening draft on Add. 3962.3: 63r has one small novelty which is recorded in the next note.

(24) In a preliminary version (see the previous note) Newton minimally elaborated this to read 'ut sit *BE* longitudo quævis data' (so that *BE* shall be any given length). Any such linear length may, of course, be taken as the geometrical unit.

Prop. III. Theor. I.(25)

Si pro abscissa AB et area AE seu $AB\times 1$ promiscuè scribatur z, et si pro $e+fz^{\eta}+gz^{2\eta}+hz^{3\eta}+$ &c scribatur R: sit autem area Curvæ $z^{\theta}R^{\lambda}$, erit ordinatim applicata $BC=\theta e\begin{smallmatrix}+\theta\\+\lambda\eta\end{smallmatrix}fz^{\eta}\begin{smallmatrix}+\theta\\+2\lambda\eta\end{smallmatrix}gz^{2n}\begin{smallmatrix}+\theta\\+3\lambda\eta\end{smallmatrix}hz^{3\eta}+$ &c in $z^{\theta-1}R^{\lambda-1}$.

Demonstratio.

Nam si sit $z^{\theta}R^{\lambda}=v$, erit per Prop. I, $\theta\dot{z}z^{\theta-1}R^{\lambda}+\lambda z^{\theta}\dot{R}R^{\lambda-1}=\dot{v}$. Pro R^{λ} in primo æquationis termino et z^{θ} in secundo scribe $RR^{\lambda-1}$ et $zz^{\theta-1}$, et fiet $\theta\dot{z}R+\lambda z\dot{R}$ in $z^{\theta-1}R^{\lambda-1}=\dot{v}$. Erat autem $R=e+fz^{\eta}+gz^{2\eta}+hz^{3\eta}$ &c et inde per Prop I fit $\dot{R}=\eta f\dot{z}z^{\eta-1}+2\eta g\dot{z}z^{2\eta-1}+3\eta h\dot{z}z^{3\eta-1}+$ &c. Huic ductæ in λz adde $\theta\dot{z}R$ et summa ducta in $z^{\theta-1}R^{\lambda-1}$ si modo pro $\dot{z}$ scribatur BE seu 1(26) fiet

$$\theta e\begin{smallmatrix}+\theta\\+\lambda\eta\end{smallmatrix}fz^{\eta}\begin{smallmatrix}+\theta\\+2\lambda\eta\end{smallmatrix}gz^{2\eta}\begin{smallmatrix}+\theta\\+3\lambda\eta\end{smallmatrix}hz^{3\eta}+\text{&c}\quad\text{in}\quad z\theta^{-1}R^{\lambda-1}=\dot{v}=BC.\quad\text{Q.E.D.}$$

Prop. IV. Theor. II.(27)

Si Curvæ abscissa AB sit z, et si pro $e+fz^{\eta}+gz^{2\eta}+$ &c scribatur R et pro $k+lz^{\eta}+mz^{2\eta}+$ &c scribatur S; sit autem area Curvæ $z^{\theta}R^{\lambda}S^{\mu}$: erit ordinatim applicata $BC=$

$$\left.\begin{matrix}\theta ek\begin{smallmatrix}+\theta\\+\lambda\eta\end{smallmatrix}fkz^{\eta} & \begin{smallmatrix}+\theta\\+2\lambda\eta\end{smallmatrix}gkz^{2\eta} & * \ \ldots & * \ \ldots\\ \begin{smallmatrix}+\theta\\+\mu\eta\end{smallmatrix}elz^{\eta} & \begin{smallmatrix}+\theta\\+\lambda\eta\\+\mu\eta\end{smallmatrix}flz^{2\eta} & \begin{smallmatrix}+\theta\\+2\lambda\eta\\+\mu\eta\end{smallmatrix}glz^{3\eta} & * \ \ldots\\ & \begin{smallmatrix}+\theta\\+2\mu\eta\end{smallmatrix}emz^{2\eta} & \begin{smallmatrix}+\theta\\+\lambda\eta\\+2\mu\eta\end{smallmatrix}fmz^{3\eta} & \begin{smallmatrix}+\theta\\+2\lambda\eta\\+2\mu\eta\end{smallmatrix}gmz^{4\eta}\,[\text{&c}]\end{matrix}\right\}\text{in } z^{\theta-1}R^{\lambda-1}S^{\mu-1}.$$

Demonstratur ad modum Propositionis superioris.(28)

(25) Proposition VII of the 1691 'De quadratura', here lightly refashioned in its enunciation and now given explicit proof. Rough preliminary drafts of the latter 'Demonstratio' (too trivially different from that here reproduced to be worth quoting) are to be found on Add. 3962.3: $49^{r}/46^{v}$.

(26) Afterwards altered to read more succinctly 'Quibus substitutis et scripta BE seu 1 pro $\dot{z}$' (Once these are substituted and with BE, that is, 1, written in place of $\dot{z}$).

(27) Proposition VIII of the 1691 'De quadratura', now supplied with the remark that its

Proposition III, Theorem I.[25]

If for the abscissa AB and area (AE), that is, $AB \times 1$, there be indiscriminately written z, and for $e+fz^{\eta}+gz^{2\eta}+hz^{3\eta}+\ldots$ there be written R, let, however, the area of the curve be $z^{\theta}R^{\lambda}$, then the ordinate BC will be equal to

$$(\theta e+(\theta+\lambda\eta)fz^{\eta}+(\theta+2\lambda\eta)\,gz^{2\eta}+(\theta+3\lambda\eta)\,hz^{3\eta}+\ldots)\times z^{\theta-1}R^{\lambda-1}.$$

Demonstration.

For if there be $z^{\theta}R^{\lambda}=v$, then by Proposition I $\theta\dot{z}z^{\theta-1}R^{\lambda}+\lambda z^{\theta}\dot{R}R^{\lambda-1}=\dot{v}$. For R^{λ} in the equation's first term and z^{θ} in the second write $RR^{\lambda-1}$ and $zz^{\theta-1}$, and there will come to be $(\theta\dot{z}R+\lambda z\dot{R})\times z^{\theta-1}R^{\lambda-1}=\dot{v}$. There was, however, $R=e+fz^{\eta}+gz^{2\eta}+hz^{3\eta}\ldots$ and thence by Proposition I there comes

$$\dot{R}=\eta f\dot{z}z^{\eta-1}+2\eta g\dot{z}z^{2\eta-1}+3\eta h\dot{z}z^{3\eta-1}+\ldots.$$

To this multiplied by λz add $\theta\dot{z}R$, and when the sum is multiplied by $z^{\theta-1}R^{\lambda-1}$ there will, provided BE or 1 be written in place of $\dot{z}$,[26] come to be

$$\begin{pmatrix}\theta e+(\theta+\lambda\eta)fz^{\eta}+(\theta+2\lambda\eta)\,gz^{2\eta}\\ \qquad +(\theta+3\lambda\eta)\,hz^{3\eta}+\ldots\end{pmatrix}\times z^{\theta-1}R^{\lambda-1}=\dot{v}=BC.$$ As was to be proved.

Proposition IV, Theorem II.[27]

If the curve's abscissa AB be z, and if for $e+fz^{\eta}+gz^{2\eta}+\ldots$ there be written R and for $k+lz^{\eta}+mz^{2\eta}+\ldots$ be written S, let the area of the curve be $z^{\theta}R^{\lambda}S^{\mu}$, then the ordinate BC will be equal to

$$\begin{pmatrix}\theta ek+((\theta+\lambda\eta)fk\ +(\theta+\mu\eta)\,el)\,z^{\eta}\\ +((\theta+2\lambda\eta)\,gk+(\theta+\lambda\eta+\mu\eta)fl\ +(\theta+2\mu\eta)\,em)\,z^{2\eta}\\ +(\quad\ldots\quad +(\theta+2\lambda\eta+\mu\eta)\,gl+(\theta+\lambda\eta+2\mu\eta)fm)\,z^{3\eta}\\ +(\quad\ldots\quad\ldots\quad +(\theta+2\lambda\eta+2\mu\eta)\,gm)\,z^{4\eta}\\ \ldots\quad\ldots\quad\ldots\end{pmatrix}\times z^{\theta-1}R^{\lambda-1}S^{\mu-1}.$$

This is proved in the manner of the previous proposition.[28]

---

proof follows the style of the demonstration of the preceding theorem which is its particular case (on which see the next note).

(28) Namely by here setting $z^{\theta}R^{\lambda}S^{\mu}=v$ and computing

$$BC=\dot{v}/\dot{z}=(\theta RS+\lambda z(\dot{R}/\dot{z})\,S+\mu zR(\dot{S}/\dot{z}))\,z^{\theta-1}R^{\lambda-1}S^{\mu-1},$$

where $z(\dot{R}/\dot{z})=(fz^{\eta}+2gz^{2\eta}+3hz^{3\eta}+\ldots)\,\eta$ and $z(\dot{S}/\dot{z})=(lz^{\eta}+2mz^{2\eta}+3nz^{3\eta}+\ldots)\,\eta$. As with the previous proposition, duplicate rough computations to this effect are to be found on Add. 3962.3: $49^{r}/46^{v}$.

Prop. V. Theor. III.(29)

Si Curvæ abscissa AB sit z, et pro $e+fz^{\eta}+gz^{2\eta}+hz^{3\eta}+$ &c scribatur R: sit autem ordinatim applicata $z^{\theta-1}R^{\lambda-1}$ in $a+bz^{\eta}+cz^{2\eta}+dz^{3\eta}[+\epsilon z^{4\eta}]+$&c(30) et ponatur $\frac{\theta}{\eta}=r$. $r+\lambda=s$. $s+\lambda=t$. $t+\lambda=v$. &c[,] erit area

$$z^{\theta}R^{\lambda} \text{ in } \frac{\frac{1}{\eta}a}{re}+\frac{\frac{1}{\eta}b-sfA}{\overline{r+1,e}}z^{\eta}+\frac{\frac{1}{\eta}c\frac{-s}{-1}fB-tgA}{\overline{r+2,e}}z^{2\eta}$$

$$+\frac{\frac{1}{\eta}d\frac{-s}{-2}fC\frac{-t}{-1}gB-vhA}{\overline{r+3,e}}z^{3\eta}+\frac{\frac{1}{\eta}[\epsilon]\frac{-s}{-3}fD\frac{-t}{-2}gC\frac{-v}{-1}hB}{\overline{r+4,e}}z^{4\eta}+\&\text{c.}^{(31)}$$

Ubi A, B, C, D,&c denotant totas coefficientes datas terminorum singulorum in serie cum signis suis $+$ et $-$, nempe A primi termini coefficientem $\frac{\frac{1}{\eta}a}{re}$, B secundi coefficientem $\frac{\frac{1}{\eta}b-sfA}{\overline{r+1,e}}$, C tertij coefficientem $\frac{\frac{1}{\eta}c\frac{-s}{-1}fB-tgA}{\overline{r+2,e}}$, et sic deinceps.

Demonstratio.(32)

Sunto juxta Propositionem tertiam

	Curvarum Ordinatæ		*& earundem areæ*
1.	$\theta eA\ {}^{+\theta}_{+\lambda\eta}fAz^{\eta}\ {}^{+\theta}_{+2\lambda\eta}gAz^{2\eta}\ {}^{+\theta}_{+3\lambda\eta}hAz^{3\eta}$ &c		$Az^{\theta}R^{\lambda}$.
2.	... $\overline{\theta+\eta},eBz^{\eta}\ {}^{+\theta+\eta}_{+\lambda\eta}fBz^{2n}\ {}^{+\theta+\eta}_{+2\lambda\eta}gBz^{3\eta}$ &c	$z^{\theta-1}R^{\lambda-1}$.	$Bz^{\theta+\eta}R^{\lambda}$.
3.	.... $+\theta+2\eta eCz^{2\eta}\ {}^{+\theta+2\eta}_{+\lambda\eta}fCz^{3\eta}$ &c		$Cz^{\theta+2\eta}R^{\lambda}$.
4.	.... $+\overline{\theta+3\eta},eDz^{3\eta}$ &c		$Dz^{\theta+3\eta}R^{\lambda}$.

(29) Proposition IX of the 1691 'De quadratura', now supplied with a concise demonstration outlining the pattern of a general proof and also with an elaborate set of explanatory notes and worked instances. Rough preparatory computations and a preliminary verbal draft of the complete theorem exist on Add. 3962.3: 45r–46v; the former are somewhat shaky in their detail and we see no gain in reproducing their structure and pattern (which differ only minimally from those of their finished equivalents), but we cite a couple of variants in the latter in following footnotes.

Proposition V, Theorem III.(28)

If the curve's abscissa AB be z and for $e+fz^{\eta}+gz^{2\eta}+hz^{3\eta}+\ldots$ there be written R, let, however, the ordinate be $z^{\theta-1}R^{\lambda-1}(a+bz^{\eta}+cz^{2\eta}+dz^{3\eta}+\epsilon z^{4\eta}+\ldots)$(30) and put $\theta/\eta=r$, $r+\lambda=s$, $s+\lambda=t$, $t+\lambda=v$, and so on, and the area will then be

$$z^{\theta}R^{\lambda}\times\left(\begin{array}{l}\dfrac{a/\eta}{re}+\dfrac{b/\eta-sfA}{(r+1)\,e}z^{\eta}+\dfrac{c/\eta-(s+1)fB-tgA}{(r+2)\,e}z^{2\eta}\\ \quad+\dfrac{d/\eta-(s+2)fC-(t+1)\,gB-vhA}{(r+3)\,e}z^{3\eta}\\ \quad+\dfrac{\epsilon/\eta-(s+3)fD-(t+2)\,gC-(v+1)\,hB\ldots}{(r+4)\,e}z^{4\eta}\ldots\end{array}\right)^{(31)}.$$

Here $A, B, C, D, \ldots$ denote the totality of the given coefficients of the separate terms in the series with their signs $+$ and $-$; namely, A the coefficient $(a/\eta)/re$ of the first term, B the coefficient $(b/\eta-sfA)/(r+1)\,e$ of the second, C the coefficient $(c/\eta-(s+1)fB-tgA)/(r+2)\,e$ of the third, and so on in turn.

Demonstration.(32)

In line with the third proposition let there be:

$$\begin{array}{lll} & \textit{Ordinates of curves} & \textit{their areas}\\ 1. & \left.\begin{array}{l}\theta eA+(\theta+\lambda\eta)\,fAz^{\eta}+(\theta+2\lambda\eta)\,gAz^{2\eta}+(\theta+3\lambda\eta)\,hAz^{3\eta}\ldots\\ \ldots(\theta+\eta)\,eBz^{\eta}+(\theta+\eta+\lambda\eta)fBz^{2\eta}+(\theta+\eta+2\lambda\eta)gBz^{3\eta}\ldots\\ \ldots\quad\ldots\quad+(\theta+2\eta)\,eCz^{2\eta}+(\theta+2\eta+\lambda\eta)fCz^{3\eta}\ldots\\ \ldots\quad\ldots\quad\ldots\quad+(\theta+3\eta)eDz^{3\eta}\ldots\end{array}\right\}\times z^{\theta-1}R^{\lambda-1}. & \begin{array}{l}Az^{\theta}R^{\lambda}.\\ Bz^{\theta+\eta}R^{\lambda}.\\ Cz^{\theta+2\eta}R^{\lambda}.\\ Dz^{\theta+3\eta}R^{\lambda}.\end{array}\end{array}$$

(rows numbered 1., 2., 3., 4.)

(30) We insert an extra term in this sequence so as fully to prepare the way for Newton's ensuing series expansion of the *area*; see the next note.

(31) The term in $z^{4\eta}$ is not listed by Newton in his preliminary draft on Add. 3962.3: 45r (on which see note (29) above). In here introducing it he entered the first term in the numerator of its coefficient as '$\dfrac{1}{\eta}e$', subsequently, when he came to realize that e is already pre-empted as the initial term of R, choosing to delete it *in toto* rather than take the small trouble of replacing the duplicate letter—and so it went into print (*Opticks*, 1704: $_2$177) without even a surrogate '&c' in its stead. In our reproduction of the manuscript text we prefer to follow his original intention, preserving the full coefficient intact merely by a minimal editorial amendment here and (see the previous note) correspondingly above.

(32) As in his preliminary draft on Add. 3962.3: 45v/46r, Newton initially adjoined this outline proof at the very end of his following notes on the theorem he here enunciates with (see note (49) below) a curt 'Sed demonstranda est Propositio'. Our earlier justification of the equivalent Proposition IX of the 1691 'De quadratura' (see **1**, §2: note (49) preceding) is, if more generally couched, not essentially different.

Et si summa ordinatarum ponatur æqualis ordinatæ $a+bz^{\eta}+cz^{2\eta}+dz^{3\eta}+$ &c in $z^{\theta-1}R^{\lambda-1}$, summa areарū $z^{\theta}R^{\lambda}$ in $A+Bz^{\eta}+Cz^{2\eta}+Dz^{3\eta}$ &c æqualis erit areæ Curvæ cujus est ordinata. Æquentur igitur Ordinatarum termini correspondentes, et fiet $a=\theta eA$, $b=\begin{matrix}\theta\\+\lambda\eta\end{matrix}fA\begin{matrix}+\theta\\+\eta\end{matrix}eB$, $c=\begin{matrix}\theta\\+2\lambda\eta\end{matrix}gA\begin{matrix}+\theta+\eta\\+\lambda\eta\end{matrix}fB+\overline{\theta+2\eta},eC$ &c et inde $\dfrac{a}{\theta e}=A$. $\dfrac{b-\overline{\theta+\lambda\eta},fA}{\overline{\theta+\eta},e}=B$. $\dfrac{c-\overline{\theta+2\lambda\eta},gA-\overline{\theta+\eta+\lambda\eta},fB}{\overline{\theta+2\eta},e}=C$. Et sic deinceps in infinitum. Pone jam $\dfrac{\theta}{\eta}=r$. $r+\lambda=s$. $s+\lambda=t$ &c(33) et in area $z^{\theta}R^{\lambda}\times\overline{A+Bz^{\eta}+Cz^{2\eta}+Dz^{3\eta}}$ &c scribe ipsorum A, B, C, &c valores inventos et prodibit series proposita. Q.E.D.

Et nota(34) quod Ordinata omnis duplici modo in seriem resolvitur. Nam index η vel affirmativus esse potest vel negativus. Proponatur Ordinata $\dfrac{3k-lzz}{zz\sqrt{kz-lz^3+mz^4}}$. Hæc vel sic scribi potest $z^{-\frac{5}{2}}\times\overline{3k-lzz}\times\overline{k-lzz+m[z]^3}\big|^{-\frac{1}{2}}$, vel sic $z\times\overline{-l+3kz^{-2}}\times\overline{m-lz^{-1}+kz^{-3}}\big|^{-\frac{1}{2}}$.(35) In casu priore est $a=3k$. $b=0$. $c=-l$. $e=k$. $f=0$. $g=-l$. $h=m$. $\lambda=-\frac{1}{2}$. $\eta=1$. $\theta-1=-\frac{5}{2}$. $\theta=-\frac{3}{2}=r$. $s=-1$. $t=-\frac{1}{2}$. $v=0$. In posteriore est $a=-l$. $b=0$. $c=3k$. $e=m$. $f=-l$. $g=0$. $h=1$. $\lambda=-\frac{1}{2}$.(36) $\eta=-1$. $\theta-1=1$. $\theta=2$. $r=-2$. $s=-1\frac{1}{2}$. $t=-1$. $v=-\frac{1}{2}$. Tentandus est casus uterq3. Et si serierum alterutra ob terminos tandem deficientes abrumpitur ac terminatur,(37) habebitur area Curvæ in terminis finitis. Sic in exempli hujus priore casu scribendo in serie valores ipsorum $a, b, c, e, f, g, h, \lambda$, $[\eta,]$ θ, r, s, t, v, termini omnes post primum evanescunt in infinitum et area Curvæ prodit $-2\sqrt{\dfrac{k-lzz+mz^3}{z^3}}$. Et hæc area ob signum negativum adjacet abscissæ ultra ordinatam productæ. Nam area omnis affirmativa adjacet tam abscissæ quàm ordinatæ, negativa verò cadit ad contrarias partes ordinatæ et adjacet abscissæ productæ. Hoc modo series alterutra et nonnunquam utraq3 semper terminatur et finita evadit, si Curva geometricè quadrari potest. At si Curva talem quadraturam non admittit, series

(33) Whence there successively ensues

$$a/\eta = reA,\quad b/\eta = sfA+(r+1)\,eB,\quad c/\eta = tgA+(s+1)\,fB+(r+2)\,eC,\ \ldots.$$

(34) This replaces a slightly more forceful 'notandum est' (it must be noted), a variant to which Newton for some reason reverted in the 1704 printed text.

(35) The index of the initial power of z should be entered as '-2', so determining in sequel ($\theta-1=-2$ and hence) $\theta=-1$, $r=1$, $s=1\frac{1}{2}$, $t=2$, $v=2\frac{1}{2}$, ... And so we have adjusted our English version.

(36) Read '$h=k$. $\lambda=\frac{1}{2}$' and so we again mend our English version to be. Though Gregory caught both these slips in his 1694 transcript of the manuscript (on which see note (2) above), they endured—together with the ensuing erroneous evaluations of θ, r, s and t

Then, if the sum of the ordinates be set equal to the ordinate

$$(a+bz^{\eta}+cz^{2\eta}+dz^{3\eta}+\dots)\,z^{\theta-1}R^{\lambda-1},$$

the sum of the areas $z^{\theta}R^{\lambda}(A+Bz^{\eta}+Cz^{2\eta}+Dz^{3\eta}\dots)$ will be equal to the area of the curve whose ordinate it is. Equate corresponding terms in the ordinates, therefore, and there will ensue $a=\theta eA$, $b=(\theta+\lambda\eta)fA+(\theta+\eta)\,eB$,

$$c=(\theta+2\lambda\eta)\,gA+(\theta+\eta+\lambda\eta)fB+(\theta+2\eta)\,eC,$$

and so on, and thence $a/\theta e=A$, $(b-(\theta+\lambda\eta)fA)/(\theta+\eta)\,e=B$, $(c-(\theta+2\lambda\eta)\,gA-(\theta+\eta+\lambda\eta)fB)/(\theta+2\eta)\,e=C$, and so forth indefinitely. Now set $\theta/\eta=r, r+\lambda=s, s+\lambda=t, \dots$ (33) and in the area $z^{\theta}R^{\lambda}(A+Bz^{\eta}+Cz^{2\eta}+Dz^{3\eta}\dots)$ write the values of $A, B, C, \dots$ found, and the series propounded will result. As was to be proved.

And note(34) that every ordinate is resolved in a double way into a series. For the index η can be either positive or negative. Let the ordinate

$$(3k-lz^2)/z^2\sqrt{(kz-lz^3+mz^4)}$$

be proposed. This can be written either in this fashion

$$z^{-\frac{5}{2}}(3k-lz^2)\,(k-lz^2+mz^3)^{-\frac{1}{2}}$$

or in this $z^{[-2]}(-l+3kz^{-2})\,(m-lz^{-1}+kz^{-3})^{-\frac{1}{2}}$. In the former case there is $a=3k$, $b=0$, $c=-l$, $e=k$, $f=0$, $g=-l$, $h=m$, $\lambda=-\frac{1}{2}$, $\eta=1$, $\theta-1=-\frac{5}{2}$, $\theta=-\frac{3}{2}=r$, $s=-1$, $t=-\frac{1}{2}$, $v=0$; in the latter there is $a=-l$, $b=0$, $c=3k$, $e=m$, $f=-l$, $g=0$, $h=[k]$, $\lambda=[\frac{1}{2}]$, $\eta=-1$, $\theta-1=[-2]$, $\theta=[-1]$, $r=[1]$, $s=[1\frac{1}{2}]$, $t=[2]$, $v=[2\frac{1}{2}]$. Each case needs to be tested. And if either one of the series, because its terms at length run out, breaks off and terminates,(37) the area of the curve will be had in finite terms. Thus in the former case of the present example, on entering in the series the values of $a, b, c, e, f, g, h, \lambda, \eta, \theta, r, s, t, v$ all terms after the first vanish without end and the area of the curve proves to be $-2\sqrt{[(k-lz^2+mz^3)/z^3]}$. This area because of its negative sign lies along the extension of the abscissa beyond the ordinate. For every positive area adjoins both the abscissa and ordinate, but a negative one falls on the opposite side of the ordinate and lies along the abscissa's extension. In this way one or other series—and sometimes both—ever terminates and proves to be finite if the curve can be geometrically squared. But if a curve does not admit of such a quadrature,

---

consequent (see the previous note) on the omission of the index $(\theta-1\ =)-2$ in the alternative preceding expression for the ordinate—into the 1704 *editio princeps*, to be caught by Newton only in his handwritten *errata* to its 1706 reprinting in his library copy of the Latin *Optice* (ULC. Adv. b. 39.4) and set right publicly in Jones' re-edition of the text in 1711 (on which again see note (2)).

(37) Where, that is, 'curva quadrari potest' (the curve can be squared) as Newton here first wrote.

utraq continuabitur in infinitum, et earum altera converget et aream dabit approximando præterquam ubi r (propter aream infinitam) vel nihil est vel numerus integer et negativus, vel ubi $\frac{z}{e}$ æqualis est unitati. Si $\frac{z}{e}$ minor est unitate, converget series in qua index η affirmativus est: sin $\frac{z}{e}$ unitate major est converget series altera. In uno casu area adjacet abscissæ,(38) in altero adjacet abscissæ ultra ordinatam productæ.

Nota insuper quod si Ordinata contentum est sub factore rationali Q et factore surdo irreducibili R^{π}, et factoris surdi latus R non dividit factorem rationalem Q; erit $\lambda-1=\pi$ et $R^{\lambda-1}=R^{\pi}$. Sin factoris surdi latus R dividit factorem rationalem semel, erit $\lambda-1=\pi+1$ et $R^{\lambda-1}=R^{\pi+1}$; si dividit bis, erit $\lambda-1=\pi+2$ et $R^{\lambda-1}=R^{\pi+2}$: si ter, erit $\lambda-1=\pi+3$ et $R^{\lambda-1}=R^{\pi+3}$: & sic deinceps.(39)

Si Ordinata est fractio rationalis irreducibilis cum Denominatore ex duobus vel pluribus terminis composito:(40) resolvendus est denominator in divisores suos omnes primos. Et si divisor sit aliquis cui nullus alius est æqualis, Curva(41) quadrari nequit. Sin duo vel plures sint divisores æquales rejiciendus est eorum unus, et si adhuc alij duo vel plures sint sibi mutuò æquales et prioribus inæquales, rejiciendus est etiam eorum unus, et sic in alijs omnibus æqualibus si adhuc plures sint: dein divisor qui relinquitur vel contentum sub divisoribus omnibus qui reliquuntur si plures sunt ponendum est pro R & ejus quadrati reciprocum R^{-2} pro $R^{\lambda-1}$ præterquam ubi contentum illud est quadratum vel cubus vel quadratoquadratum &c(42) quo in casu ejus latus ponendum est pro R et potestatis index 2 vel 3 vel 4 negative sumptus pro λ, et Ordinata ad denominatorem R^2 vel R^3 vel R^4 vel R^5 &c reducenda.

(43)Ut si ordinata sit $\frac{z^5+z^4-8z^3}{z^5+z^4-5z^3-z^2+8z-4}$; quoniam hæc fractio irreduci-

(38) That is, 'ad usq ordinatam ductæ' (drawn up to the ordinate) as Newton subsequently specified in an inserted phrase.

(39) This mode of prior reduction to a simpler form is set out rather more clearly and concisely in Newton's preliminary draft on Add. 3962.3: 45r, where he urged equivalently that 'ex radicali...extrahi debet quicquid est rationale et reliquum poni pro $R^{\lambda-1}$ et si coefficiens hujus $R^{\lambda-1}$ dividi potest per R vel per R^2 vel R^3 &c pro facto ex hoc $R^{\lambda-1}$ et R vel R^2 vel R^3 hoc est pro R^{λ} vel $R^{\lambda+1}$ vel $R^{\lambda+2}$ &c scribi debet [de novo] $R^{\lambda-1}$' (from the radical you ought to extract anything which is rational and put the remainder for $R^{\lambda-1}$, and if the coefficient of this $R^{\lambda-1}$ can be divided by R or R^2 or R^3..., for the product of this $R^{\lambda-1}$ and R or R^2 or R^3..., that is, in place of R^{λ} or $R^{\lambda+1}$ or $R^{\lambda+2}$..., you ought to write $R^{\lambda-1}$ [afresh]).

David Gregory's 1694 transcript (see note (2) above) terminates at this point. Why he did not go on to copy the rest of Newton's text we can only conjecture. It may not be mere coincidence that the present Proposition V is the multinomial generalization of the quadrature series whose binomial instance—that communicated by Newton to Leibniz in his *epistola posterior* in 1676—he had himself, primed by Craige, 'independently' discovered eight years previously (see our pages 4–6 above); one may conceive that he stuck over the present extension and found no time during his brief Cambridge visit to transcribe the sequel to it.

each series will continue to infinity, and one or other of them will converge and yield the area by approximating it, except when r (because the area is infinite) is either nil or a negative integer, or when z/e is equal to unity. If z/e is less than unity, the series in which the index η is positive will converge; but if z/e is greater than unity, the other series will converge. In one case the area lies along the abscissa,(38) in the other it adjoins the extension of the abscissa beyond the ordinate.

Note in addition that if the ordinate is the product of a rational factor Q and an irreducible surd factor R^{π}, and the root R of the surd factor does not divide the rational factor Q, then $\lambda-1=\pi$ and $R^{\lambda-1}=R^{\pi}$. But if the root R of the surd factor divides the rational factor once, there will be $\lambda-1=\pi+1$ and $R^{\lambda-1}=R^{\pi+1}$; if it divides it twice, then $\lambda-1=\pi+2$ and $R^{\lambda-1}=R^{\pi+2}$; if three times, then $\lambda-1=\pi+3$ and $R^{\lambda-1}=R^{\pi+3}$; and so on.(39)

If the ordinate is an irreducible rational fraction with a denominator composed of two or more terms,(40) you need to resolve the denominator into all its prime divisors. Then if there be some divisor to which no other is equal, the curve(41) cannot be squared. But should there be two or more equal divisors you must reject one, and if there still be two or more others equal to one another but unequal to the previous ones you must reject one of those also, and the like in all other equal ones should there be still more; then you must put the divisor which is left, or the product of all the divisors which are left if there are several, for R and the reciprocal R^{-2} of its square for $R^{\lambda-1}$, except when that product is a square or a cube or a fourth power and so on(42)—in which case its root must be put for R and the index 2 or 3 or 4... of its power, taken negative, for λ and its ordinate then be reduced to the denominator R^2 or R^3 or R^4 and so on.

(43)For instance, if the ordinate be $(z^5+z^4-8z^3)/(z^5+z^4-5z^3-z^2+8z-4)$,

The manuscript itself is evenly written at this stage and bears no trace that Newton even rested his pen between paragraphs.

(40) Originally just 'cum divisore composito' (with a compound divisor), as in the preliminary draft on Add. 3962.3: 45r.

(41) Newton has here cancelled a (not quite completed) superfluous insertion 'ex earum numero est [qui]' (is of the number of those which).

(42) Initially this proviso read 'præterquam ubi quadrati illius R^2 radix cubica vel quadrato-quadratica vel quadrato-cubica extrahi potest' (except when the cube or fourth or fifth root of that square R^2 can be extracted), after which Newton went on: 'quo in casu radix illa ponenda est pro R et potestatum reciproca R^{-3}, R^{-4}, R^{-5} &c pro $R^{\lambda-1}$' (in which case that root is to be put for R and the reciprocal powers R^{-3}, R^{-4}, R^{-5}, ... in place of $R^{\lambda-1}$). Since in passing from *area* to *ordinata* any factor R^{λ} of the former is reduced to a corresponding $R^{\lambda-1}$ in the latter's expression, the underlying basis of this working rule will be clear.

(43) Newton here first began to copy from his draft on Add. 3962.3: 45v a first worked example whose full version there reads: 'Ut si ordinata esset $\frac{z^4}{z^3-4z^2+5z-2}$, quoniam fractio irreducibilis est et denominator divisores habet $z-1$, $z-1$, $z-2$ quarum $z-2$ est sine pari, concludo quod curva geometrice quadrari nequit. Obtinebitur tamen area [in] serie infinita

bilis est et denominatoris divisores sunt pares, nempe $z-1, z-1, z-1$ & $z+2$, $z+2$, rejicio magnitudinis[44] utriusq; divisorem unam et reliquorum $z-1, z-1$, $z+2$ contentum z^3-3z+2 pono pro R et ejus quadrati reciprocum $\frac{1}{R^2}$ seu R^{-2} pro $R^{\lambda-1}$. Dein Ordinatam ad denominatorem R^2 seu $R^{1-\lambda}$ reduco et fit $\frac{z^6-9z^4+8z^3}{z^3-3z+2^{\text{quad}}}$. id est $z^3\times\overline{8-9z+z^3}\times\overline{2-3z+z^3}|^{-2}$. Et inde est $a=8$, $b=-9$, $c=0$, $d=1$.[45] $e=2$. $f=-3$. $g=0$. $h=1$. $\lambda-1=-2$. $\lambda=-1$. $\eta=1$. $\theta-1=3$. $\theta=4=r$. $s=3$. $t=2$. $v=1$. Et his in serie scriptis prodit area $\frac{z^4}{z^3-3z+2}$, terminis omnibus in tota seric post primum evanescentibus.

Si deniq; Ordinata est fractio irreducibilis et ejus denominator contentum est sub factore rationali Q et factore surdo irreducibili R^{π}, inveniendi sunt lateris R divisores omnes primi, et rejiciendus divisor unus magnitudinis cujusq; et per[46] divisores qui restant, siqui sint, multiplicandus est factor rationalis Q: et si factum æquale est lateri R vel lateris illius potestati alicui[47] cujus index est numerus integer, esto index ille m, et erit $\lambda-1=-\pi-m$, & $R^{\lambda-1}=R^{-\pi-m}$. Ut si Ordinata sit $\frac{3q^5-q^4x+9q^3xx-qqx^3-6qx^4}{qq-xx\sqrt[3]{q^3+qqx-qxx-x^3}}$[48], quoniam factoris surdi latus R seu $q^3+qqx-qxx-x^3$ divisores habet $q+x$, $q+x$, $q-x$ qui duarum sunt magnitudinum[,] rejicio divisorem unum magnitudinis utriusq; et per divisorem $q+x$ qui relinquitur multiplico factorem rationalem $qq-xx$. Et quoniam factum $q^3+qqx-qxx-x^3$ æquale est lateri R pono $m=1$ et inde, cum π sit $\frac{1}{3}$, fit $\lambda-1=-\frac{4}{3}$. Ordinatam igitur reduco ad denominatorem $R^{-\frac{4}{3}}$ et fit

$$z^0\times\overline{3q^6+2q^5x+8q^4xx+8q^3x^3-7qqx^4-6qx^5}\times\overline{q^3+qqx-qxx-x^3}|^{-\frac{4}{3}}.$$

Unde est $a=3q^6$. $b=2q^5$. &c. $e=q^3$. $f=qq$. &c. $\theta-1=0$. $\theta=1=\eta$. $\lambda=-\frac{1}{3}$. $r=1$. $s=\frac{2}{3}$. $t=\frac{1}{3}$. $v=0$. Et his in serie scriptis prodit area

$$\frac{3qqx+3x^3}{\sqrt[3]{q^3+qqx-qxx-x^3}},$$

terminis omnibus in serie tota post tertium evanescentibus.[49]

ponendo denominatorem $= R$' (Were, for instance, the ordinate $z^4/(z^3-4z^2+5z-2)$, because the fraction is irreducible and its denominator has divisors $z-1$, $z-1$ and $z-2$, of which $z-2$ is without pair, I conclude that the curve cannot geometrically be squared. The area will, however, be obtained in an infinite series by putting the denominator equal to R). The indefinite integral of $z^4/(z-1)^2(z-2) = z+4-1/(z-1)^2-5/(z-1)+16/(z-2)$ is in fact $\frac{1}{2}z^2+4z+1/(z-1)-5\log(z-1)+16\log(z-2)$.

(44) 'generis' (kind) was originally written.

(45) Correctly so specified by Newton in the catchword at the end of f. 11^r in the manuscript, but in a momentary lapse converted to be '$d=-1$' in the first line of his next page. The slip passed unnoticed into the 1704 *editio princeps*, to be at length corrected by Jones in his

because this fraction is irreducible and its denominator has matching divisors, namely $z-1$, $z-1$, $z-1$ and $z+2$, $z+2$, I reject one divisor of each size[44] and put the product z^3-3z+2 of the remainder $z-1$, $z-1$ and $z+2$ for R and the reciprocal $1/R^2$ of its square, that is, R^{-2} for $R^{\lambda-1}$. Then I reduce the ordinate to the denominator R^2 or $R^{1-\lambda}$ and there comes $(z^6-9z^4+8z^3)/(z^3-3z+2)^2$, that is, $z^3(8-9z+z^3)\,(2-3z+z^3)^{-2}$. And thence there is $a=8$, $b=-9$, $c=0$, $d=1$,[45] $e=2$, $f=-3$, $g=0$, $h=1$, $\lambda-1=-2$, $\lambda=-1$, $\eta=1$, $\theta-1=3$, $\theta=4=r$, $s=3$, $t=2$, $v=1$. And when these are entered in the series the area $z^4/(z^3-3z+2)$ ensues, with all the terms in the total series vanishing after the first one.

If, finally, the ordinate is an irreducible fraction and its denominator is the product of a rational factor Q and an irreducible surd factor R^π, you need to find all prime divisors of its root R and reject one divisor of each size, and then by the [46]divisors which remain, if there be any, multiply the rational factor Q: if the product is equal to the root R or to some power[47] of that root whose index is an integer, let that index be m and there will then be $\lambda-1=-\pi-m$ and $R^{\lambda-1}=R^{-\pi-m}$. Should, for instance, the ordinate be

$$\frac{3q^5-q^4x+9q^3x^2-q^2x^3-6qx^4}{(q^2-x^2)\sqrt[3]{(q^3+q^2x-qx^2-x^3)}},\ \text{[48]}$$

because the root R, namely $q^3+qx^2-qx^2+x^3$, has the divisors $q+x$, $q+x$ and $q-x$ which are of two sizes, I reject one divisor of each size and by the divisor $q+x$ which is left I multiply the rational factor q^2-x^2. Then because the product $q^3+q^2x-qx^2-x^3$ is equal to the root R I put $m=1$ and thence, since π is $\frac{1}{3}$, there comes $\lambda-1=-\frac{4}{3}$. I therefore reduce the ordinate to the denominator $R^{-\frac{4}{3}}$ and it comes to be

$$z^0.(3q^6+2q^5x+8q^4x^2+8q^3x^3-7q^2x^4-6qx^5)\,(q^3+q^2x-qx^2-x^3)^{-\frac{4}{3}}.$$

Whence there is $a=3q^6$, $b=2q^5, \ldots,$ $e=q^3$, $f=q^2, \ldots,$ $\theta-1=0$, $\theta=1=\eta$, $\lambda=-\frac{1}{3}$, $r=1$, $s=\frac{2}{3}$, $t=\frac{1}{3}$, $v=0$. And when these are entered in the series the area $(3q^2x+3x^3)/\sqrt[3]{(q^3+q^2x-qx^2-x^3)}$ ensues, with all the terms in the total series vanishing after the third one.[49]

1711 re-edition. We here omit a superfluous '&c' which comes immediately afterwards in the manuscript (and is consequently to be found in all the printed versions).

(46) 'divisorem vel' (divisor or) is deleted.

(47) Initially instanced by Newton as 'R^2 vel R^3 vel R^4 &c' (R^2 or R^3 or R^4, and so on).

(48) Newton's printer and editor permitted themselves a deal of liberty in transposing this to the published page. In the 1704 *editio princeps* the denominator is rendered in the more readily typeset form '$\overline{qq-xx}\sqrt{\overline{\text{cub}.\,q^3+qqx-qxx-x^3}}$', while in Jones' 1711 re-edition the whole fraction was (on repairing a couple of trivial slips in its expression) standardized to be

$$\text{'}\frac{3q^5-q^4x+9q^{[3]}x^2-q^2x^3-6[q]x^{[4]}}{\overline{q^2-x^2}\times\overline{q^3+q^2x-qx^2-x^3}|^{\frac{1}{3}}}\text{'}.$$

(49) As in his preliminary draft, Newton here initially went on not a little awkwardly to

Prop. VI. Theor. IV.(50)

Si Curvæ abscissa AB sit z et scribantur R pro $e+fz^{\eta}+gz^{2\eta}+hz^{3\eta}+$&c & S pro $k+lz^{\eta}+mz^{2\eta}+nz^{3\eta}$ &c: sit autem ordinatim applicata $z^{\theta-1}R^{\lambda-1}S^{\mu-1}$ in $a+bz^{\eta}+cz^{2\eta}+dz^{3\eta}$ &c et si terminorum e, f, g, h &c et $k.\ l, m, n$ &c rectangula sint

$$\begin{matrix} ek & fk & gk & hk & \&c \\ el & fl & gl & hl & \&c \\ em & fm & gm & hm & \&c \\ en & fn & gn & hn & \&c \end{matrix}$$

et si rectangulorum illorum coefficientes numerales sint respectivè

$$\begin{matrix} \frac{1}{\eta}\theta=r. & r+\lambda=s. & s+\lambda=t. & t+\lambda=v. & \&c \\ r+\mu=s'. & s+\mu=t'. & t+\mu=v'. & v+\mu=w'. & \&c \\ s'+\mu=t''. & t'+\mu=v''. & v'+\mu=w''. & w'+\mu=x''. & \&c \\ t''+\mu=v'''. & v''+\mu=w'''. & w''+\mu=x'''. & x''+\mu=y'''. & \&c \end{matrix}$$

area curvæ erit hæc

$$z^{\theta}R^{\lambda}S^{\mu} \text{ in } \frac{\frac{1}{\eta}a}{rek}+\frac{\frac{1}{\eta}b \begin{matrix}-sfk \\ -s'el\end{matrix} A}{\overline{r+1}, ek}z^{\eta}+\frac{\frac{1}{\eta}c \begin{matrix}-\overline{s+1}, fk \\ -\overline{s'+1}, el\end{matrix} B \begin{matrix}-tgk \\ -t'fl \\ -t''em\end{matrix} A}{\overline{r+2}, ek}z^{2\eta}$$

$$+\frac{\frac{1}{\eta}d \begin{matrix}-\overline{s+2}, fk \\ -\overline{s'+2}, el\end{matrix} C \begin{matrix}-\overline{t+1}, gk \\ -\overline{t'+1}, fl \\ -\overline{t''+1}, em\end{matrix} B \begin{matrix}-vhk \\ -v'gl \\ -v''fm \\ -v'''en\end{matrix} A}{\overline{r+3}, ek}z^{3\eta}+\&c.$$

Ubi A denotat termini primi coefficientem datam $\frac{\frac{1}{\eta}a}{rek}$ cum signo suo $+$ vel $-$, B coefficientem datam secundi, C coefficientem datam tertij, & sic deinceps. Terminorum verò a, b, c, &c. e, f, g &c. k, l, m &c unus vel plures deesse possunt. Demonstratur Propositio ad modum præcedentis, et quæ ibi notantur hic obtinent.(51) Progressio seriei manifesta est et facile continuatur in infinitum.(52)

---

add 'Sed demonstranda est Propositio' (But the proposition needs to be proved). Having done no more, however, than enter a following subhead, '*Demonstratio*', he broke off to cancel this intended *addendum*, advancing his proof to the more logical site (f. 9v of the manuscript) which it occupies in the finished text.

(50) A slightly refashioned statement of Proposition X of the 1691 'De quadratura' as we restored it on page 70 above. A fuller preliminary outline of the present theorem which survives on Add. 3962.3: 47v is reproduced in section [2] of the Appendix.

Proposition VI, Theorem IV.[50]

If the curve's abscissa AB be z, and R be written for $e+fz^{\eta}+gz^{2\eta}+hz^{3\eta}+\ldots$ and S for $k+lz^{\eta}+mz^{2\eta}+nz^{3\eta}\ldots$, let the ordinate, however, be

$$z^{\theta-1}R^{\lambda-1}S^{\mu-1}(a+bz^{\eta}+cz^{2\eta}+dz^{3\eta}\ldots)$$

and then, if the products of the quantities $e, f, g, h, \ldots$ and $k, l, m, n, \ldots$ be

$$\begin{matrix} ek, & fk, & gk, & hk, & \ldots \\ el, & fl, & gl, & hl, & \ldots \\ em, & fm, & gm, & hm, & \ldots \\ en, & fn, & gn, & hn, & \ldots \end{matrix}$$

and the numerical coefficients of those products be respectively

$$\begin{matrix} (1/\eta)\,\theta = r, & r+\lambda = s, & s+\lambda = t, & t+\lambda = v, & \ldots \\ r+\mu = s', & s+\mu = t', & t+\mu = v', & v+\mu = w', & \ldots \\ s'+\mu = t'', & t'+\mu = v'', & v'+\mu = w'', & w'+\mu = x'', & \ldots \\ t''+\mu = v''', & v''+\mu = w''', & w''+\mu = x''', & x''+\mu = y''', & \ldots, \end{matrix}$$

the area of the curve will be this:

$$z^{\theta}R^{\lambda}S^{\mu}\times$$

$$\left(\begin{aligned} &\frac{a/\eta}{rek}+\frac{b/\eta-(sfk+s'el)\,A}{(r+1)\,ek}z^{\eta} \\ &\quad+\frac{c/\eta-((s+1)fk+(s'+1)\,el)\,B-(tgk+t'fl+t''em)\,A}{(r+2)\,ek}z^{2\eta} \\ &+\frac{\begin{matrix} d/\eta-((s+2)fk+(s'+2)\,el)\,C \\ -((t+1)\,gk+(t'+1)fl+(t''+1)\,em)\,B-(vhk+v'gl+v''fm+v'''en)A \end{matrix}}{(r+3)\,ek}z^{3\eta} \\ &\quad+\quad\ldots\quad\ldots\quad\ldots \end{aligned}\right)$$

where A denotes the first term's given coefficient $(a/\eta)/rek$ with its sign $+$ or $-$, B the given coefficient of the second, C the given coefficient of the third, and so on. Of the terms $a, b, c, \ldots; e, f, g, \ldots; k, l, m, \ldots$, however, one or more can be lacking. The proposition is demonstrated in the manner of the preceding one, and what is there noted holds true here.[51] The progression of the series is manifest and it is easily to be continued indefinitely.[52]

(51) Compare our modernized proof of the earlier equivalent theorem of the 1691 'De quadratura' (see **1**, §2: note (50) preceding). As we have previously stated, Newton's own preliminary 'Demonstratio' is given in section [2] of the Appendix.

(52) The two halves of this concluding sentence were afterwards inverted by Newton to read a deal less awkwardly 'Pergit autem series talium Propositionum in infinitum et progressio seriei manifesta est' (The series of such propositions, however, proceeds indefinitely and the progression of their sequence is manifest).

In his preliminary outline of the sequel which survives on Add. 3962.3: 51r–52r Newton

Prop. VII. Theor. V.(53)

Si pro $e+fz^\eta+gz^{2n}+$ &c scribatur R ut supra et in Curvæ alicujus Ordinata $z^{\theta\pm\eta\sigma}R^{\lambda\pm\tau}$ maneant quantitates datæ θ, η, λ, e, f, g &c et pro σ ac τ scribantur successivè numeri quicunꝙ integri: et si detur area unius ex Curvis quæ per Ordinatas innumeras sic prodeun[t]es designantur si Ordinatæ sunt duorum nominum in vinculo radicis, vel si dentur areæ duarum ex curvis si Ordinatæ sunt trium nominum,(54) vel areæ trium ex Curvis si Ordinatæ sunt quatuor nominum,(54) & sic deinceps in infinitum: dico quod dabuntur areæ curvarum omnium. Pro nominibus hic habeo terminos omnes in vinculo radicis (55) tam deficientes quàm plenos quorum indices dignitatum sunt in progressione arithmetica. Sic ordinata $\sqrt{a^4-ax^3+x^4}$ ob terminos duos inter a^4 et $-ax^3$ deficientes pro quinquinomio haberi debet. At $\sqrt{a^4+x^4}$ binomium est et $\sqrt{a^4+x^4-\frac{x^8}{a^4}}$ trinomium cùm progressio jam per majores differentias(56) procedat. Propositio verò sic demonstratur.

Cas. 1.(57) Sunto Curvarum duarum Ordinatæ $pz^{\theta-1}R^{\lambda-1}$ et $qz^{\theta+\eta-1}R^{\lambda-1}$, et areæ pA et qB existente R quantitate trium nominum $e+fz^\eta+gz^{2\eta}$. Et cùm per Prop. III sit $z^\theta R^\lambda$ area curvæ cujus Ordinata est $\theta e \begin{smallmatrix}+\theta\\+\lambda\eta\end{smallmatrix} fz^\eta \begin{smallmatrix}+\theta\\+2\lambda\eta\end{smallmatrix} gz^{2\eta}$ in $z^{\theta-1}R^{\lambda-1}$, subduc Ordinatas et areas priores de area et Ordinata posteriori, et manebit $\begin{smallmatrix}\theta e\\-p\end{smallmatrix} \begin{smallmatrix}+\theta\\+\lambda\eta\\-q\end{smallmatrix} fz^\eta \begin{smallmatrix}+\theta\\+2\lambda\eta\end{smallmatrix} gz^{2\eta}$ in $z^{\theta-1}R^{\lambda-1}$ Ordinata nova Curvæ, et $z^\theta R^\lambda-pA-qB$ ejusdem area. Pone $\theta e=p$ & $\theta f+\lambda\eta f=q$ et Ordinata evadet $\begin{smallmatrix}\theta\\+2\lambda\eta\end{smallmatrix} gz^{2\eta}$ in $z^{\theta-1}R^{\lambda-1}$, et area $z^\theta R^\lambda-\theta eA-\theta fB-\lambda\eta fB$. Divide utramꝙ per $\theta g+2\lambda\eta g$, et aream prodeuntem dic C, et assumpta utcunꝙ r erit rC area Curvæ cujus Ordinata est $rz^{\theta+2\eta-1}R^{\lambda-1}$. Et qua ratione ex areis pA et qB aream rC Ordinatæ $rz^{\theta+2\eta-1}R^{\lambda-1}$ congruentem invenimus, licebit ex areis qB et rC aream quartam puta sD ordinatæ $sz^{\theta+3\eta-1}R^{\lambda-1}$ congruentem invenire, et sic deinceps in

introduced his next two propositions with the editorial aside: 'Hac ratione quadrantur Curvæ quarum ordinatæ non constant ex partibus pluribus per signa + vel − connexis. Quæ ordinatas habent ex partibus duabus constantes ['quarum neutra seorsim quadrari potest' is deleted] quadrantur per Theoremata [*sc.* VII/VIII] in progressione sequenti' (By this means are squared curves whose ordinates are not made up of several parts connected by the signs + and −. Those having ordinates consisting of two parts [neither of which is individually quadrable] are squared by the theorems in the following progression).

(53) A light revamping of Theorem VII in the first version (**1**, §1 preceding) of the 1691 'De quadratura Curvarum'; compare pages 40–4 above. A prior draft of the last portion of the opening paragraph exists (see note (55) following) on Add. 3962.3: 49r, while a recast

Proposition VII, Theorem V.[53]

If for $e+fz^{\eta}+gz^{2\eta}+\ldots$ there be written R, as above, and in the ordinate $z^{\theta\pm\eta\sigma}R^{\lambda\pm\tau}$ the quantities θ, η, λ, e, f, g, ... stay given, while in place of σ and τ any integers whatever be successively written; and if there be given the area of one of the curves denoted by the innumerable ordinates thus ensuing if the ordinates are binomial within the root's vinculum, or the areas of two of the curves if the ordinates are trinomial,[54] or the areas of three if the ordinates are quadrinomial,[54] and so forth indefinitely: then I assert that the areas of all the curves will be given. I here regard as 'nomials' all terms within the root's vinculum,[55] both ones lacking and those complete, the indices of whose powers are in an arithmetical progression. Thus the ordinate $\sqrt{(a^4-ax^3+x^4)}$, because of the two terms lacking between a^4 and $-ax^3$, ought to be regarded as a quinquinomial. But $\sqrt{(a^4+x^4)}$ is a binomial and $\sqrt{(a^4+x^4-x^8/a^4)}$ a trinomial since the progression now proceeds by greater differences.[56] The proposition, however, is thus proved.

Case 1.[57] Let the ordinates of two curves be $pz^{\theta-1}R^{\lambda-1}$ and $qz^{\theta+\eta-1}R^{\lambda-1}$, and their areas pA and qB, R being a trinomial quantity $e+fz^{\eta}+gz^{2\eta}$. Then, since by Proposition III $z^{\theta}R^{\lambda}$ is the area of the curve whose ordinate is

$$(\theta e+(\theta+\lambda\eta)fz^{\eta}+(\theta+2\lambda\eta)\,gz^{2\eta})\,z^{\theta-1}R^{\lambda-1},$$

take away the former ordinates and areas from the latter ordinate and area, and there will remain $(\theta e-p+((\theta+\lambda\eta)f-q)\,z^{\eta}+(\theta+2\lambda\eta)\,gz^{2\eta})\,z^{\theta-1}R^{\lambda-1}$ for the new ordinate of a curve and $z^{\theta}R^{\lambda}-pA-qB$ as its area. Put $\theta e=p$ and $(\theta+\lambda\eta)f=q$, and the ordinate will prove to be $(\theta+2\lambda\eta)\,gz^{2\eta}.z^{\theta-1}R^{\lambda-1}$, while the area is $z^{\theta}R^{\lambda}-\theta eA-(\theta+\lambda\eta)fB$. Divide both by $(\theta+2\lambda\eta)\,g$ and call the resulting area C, and then, where r is arbitrarily assumed, rC will be the area of the curve whose ordinate is $rz^{\theta+2\eta-1}R^{\lambda-1}$. And with the method whereby from the areas pA and qB we attained the area rC corresponding to the ordinate $rz^{\theta+2\eta-1}R^{\lambda-1}$ we will be free to find from the areas qB and rC a fourth area, sD say, corresponding to the ordinate $sz^{\theta+3\eta-1}R^{\lambda-1}$, and so on indefinitely. The mode

opening to the ensuing 'Cas. 1' and two brief calculations checking details in 'Cas. 2' (see notes (57) and (60) below) are found at *ibid.*: 51^r.

(54) Newton subsequently specified 'in vinculo radicis' (within the root's vinculum) at these places also.

(55) The prior draft of the rest of this sentence and of the two following which is roughed out on Add. 3962.3: 49^r does not differ in its final text, though it originally ended in a slightly different manner (see the next note).

(56) By 4's, namely, instead of units as previously. Newton in his draft (see the previous note) initially here concluded with the vaguer explanation 'cum progressio loca vacua jam non occupet' (since the progression does not now fill the empty places).

(57) Here the index θ is successively reduced each time by η. As we have said (note (53)), the slightly recast opening words which follow are drafted on Add. 3962.3: 51^r.

infinitum. Et par est ratio progressionis in partem contrariam, præterquam ubi terminorum θ et $2\lambda\eta$ alteruter deficit et seriem abrumpit: quo casu si assumatur area pA in principio progressionis unius et area qB in principio alterius, ex his duabus dabuntur areæ omnes in progressione utraq3. Et contra ex alijs[58] duabus areis assumptis regreditur per analysin ad areas A et B, adeo ut ex duabus datis cæteræ omnes dentur. Q.E.O.[59] Hic est casus Curvarum ubi ipsius z index θ augetur vel diminuitur perpetua additione vel subductione quantitatis η. Casus alter est Curvarum ubi index λ augetur vel diminuitur unitatibus.

Cas. 2. Ordinatæ $pz^{\theta-1}R^{\lambda}$ & $qz^{\theta+\eta-1}R^{\lambda}$, quibus areæ pA et qB jam respondeant, si in R seu $e+fz^{\eta}+gz^{2\eta}$ ducantur ac deinde ad R vicissim applicentur, evadunt $\overline{pe+pfz^{\eta}+pgz^{2\eta}}\times z^{\theta-1}R^{\lambda-1}$ et $\overline{qez^{\eta}+qfz^{2\eta}+qgz^{3\eta}}\times z^{\theta-1}R^{\lambda-1}$. Et per Prop. III est $az^{\theta}R^{\lambda}$ area Curvæ cujus Ordinata est $\theta ae \begin{matrix}+\theta\\+\lambda\eta\end{matrix} afz^{\eta} \begin{matrix}+\theta\\+2\lambda\eta\end{matrix} agz^{2\eta}$ in $z^{\theta-1}R^{\lambda-1}$, et $bz^{\theta+\eta}R^{\lambda}$ area Curvæ cujus ordinata est $\begin{matrix}\theta\\+\eta\end{matrix} bez^{\eta} \begin{matrix}+\theta\\+\eta\\+\lambda\eta\end{matrix} bfz^{2\eta} \begin{matrix}+\theta\\+\eta\\+2\lambda\eta\end{matrix} bgz^{3\eta}$ in $z^{\theta-1}R^{\lambda-1}$.

Et harum quatuor arearum summa est $pA+qB+az^{\theta}R^{\lambda}+bz^{\theta+\eta}R^{\lambda}$ et summa respondentium ordinatarum

$$\begin{matrix}\theta ae\\+pe\end{matrix} \;\; \begin{matrix}\begin{matrix}+\theta\\+\lambda\eta\end{matrix} afz^{\eta}\\ \begin{matrix}+\theta\\+\eta\end{matrix} be\\ +pf\\ +qe\end{matrix} \;\; \begin{matrix}\left.\begin{matrix}+\theta\\+2\lambda\eta\end{matrix}\right\} agz^{2\eta}\\ \left.\begin{matrix}+\theta\\+\eta\\+\lambda\eta\end{matrix}\right\} bf\\ +pg\\ +qf\end{matrix} \;\; \begin{matrix}\left.\begin{matrix}+\theta\\+\eta\\+2\lambda\eta\end{matrix}\right\} bgz^{3\eta}\\ +qg\end{matrix} \text{ in } z^{\theta-1}R^{\lambda-1}.$$

Si terminus primus tertius et quartus ponatur seorsim æqualis nihilo, per primum fiet $\theta ae+pe=0$ seu $-\theta a=p$, per quartum $-\theta b-\eta b-2\lambda\eta b=q$, et per tertium (eliminando p et q)[60] $\frac{2ag}{f}=b$. Unde secundus fit[60] $\frac{\lambda\eta aff-4\lambda\eta age}{f}$, adeoq3 summa quatuor Ordinatarum est $\frac{\lambda\eta aff-4\lambda\eta age}{f}z^{\theta+\eta-1}R^{\lambda-1}$ et summa totidem respondentium arearum $az^{\theta}R^{\lambda}+\frac{2ag}{f}z^{\theta+\eta}R^{\lambda}-\theta aA-\frac{2\theta+2\eta+4\lambda\eta}{f}agB$. Dividantur hæ summæ per $\frac{\lambda\eta aff-4\lambda\eta age}{f}$, et si Quotum posterius dicatur D, erit D area curvæ cujus ordinata est Quotum prius $z^{\theta+\eta-1}R^{\lambda-1}$. Et eadem ratione ponendo

(58) 'quibuscunq3' (any) is deleted.

(59) More generally, if $R=\sum_{0\leqslant i\leqslant n}a_i z^{i\eta}$ where $a_0=e$, $a_1=f$, $a_2=g$, ..., since on setting $I_{r,s}\equiv\int z^{\theta+r\eta-1}R^{\lambda+s-1}.dz$ there is $z^{\theta+r\eta}R^{\lambda+s}=\sum_{0\leqslant i\leqslant n}(\theta+r\eta+i(\lambda+s)\eta)\,a_i I_{r+i,s}$ (by Proposition III), any unknown integral $I_{n,0}$ is at once evaluated in terms of the n given integrals

of progression in the opposite direction is no different, except when one or other of the terms θ and $2\lambda\eta$ is lacking and so breaks off the series: in which case, if the area pA be assumed at the start of one progression and the area qB at the start of the other, from these two all the areas in both progressions will be given. And conversely, when [58]two other areas are assumed, from these there is a regress by analysis to the areas A and B, so that when any two are given all the rest are given therefrom. As was to be shown.[59] This is the case of curves in which the index θ of z is increased or diminished by the perpetual addition or subtraction of the quantity η.[59] The other case is one of curves in which the index λ is increased or diminished by units.

Case 2. If the ordinates $pz^{\theta-1}R^{\lambda}$ and $qz^{\theta+\eta-1}R^{\lambda}$—to which let the areas pA and qB now relate—be multiplied by R, that is, $e+fz^{\eta}+gz^{2\eta}$, and then in turn divided by R, they come to be

$$(pe+pfz^{\eta}+pgz^{2\eta})\,z^{\theta-1}R^{\lambda-1} \quad\text{and}\quad (qez^{\eta}+qfz^{2\eta}+qgz^{3\eta})\,z^{\theta-1}R^{\lambda-1}.$$

And, by Proposition III, $az^{\theta}R^{\lambda}$ is the area of the curve whose ordinate is

$$(\theta ae+(\theta+\lambda\eta)\,afz^{\eta}+(\theta+2\lambda\eta)\,agz^{2\eta})\,z^{\theta-1}R^{\lambda-1},$$

and $bz^{\theta+\eta}R^{\lambda}$ the area of the curve whose ordinate is

$$((\theta+\eta)\,bez^{\eta}+(\theta+\eta+\lambda\eta)\,bfz^{2\eta}+(\theta+\eta+2\lambda\eta)\,bgz^{3\eta})\,z^{\theta-1}R^{\lambda-1}.$$

The sum of these four areas is then $pA+qB+az^{\theta}R^{\lambda}+bz^{\theta+\eta}R^{\lambda}$, and the sum of their corresponding ordinates

$$\begin{pmatrix}(\theta a+p)\,e+((\theta+\lambda\eta)\,af+(\theta+\eta)\,be+pf+qe)\,z^{\eta}\\ +((\theta+2\lambda\eta)\,ag+(\theta+\eta+\lambda\eta)\,bf+pg+qf)z^{2\eta}\\ +((\theta+\eta+2\lambda\eta)\,bg+qg)\,z^{3\eta}\end{pmatrix}\times z^{\theta-1}R^{\lambda-1}.$$

If the first, third and fourth terms be set separately equal to nothing, by the first there will come $(\theta a+p)\,e=0$ or $-\theta a=p$, by the fourth $-(\theta+\eta+2\lambda\eta)\,b=q$, and by the third (on eliminating p and q)[60] $2ag/f=b$. Consequently, the second comes to be[60] $\lambda\eta a(f^2-4ge)/f$, and hence the sum of the four ordinates is $(\lambda\eta a(f^2-4ge)/f)\,z^{\theta+\eta-1}R^{\lambda-1}$, while the sum of the equal number of their corresponding areas is

$$az^{\theta}R^{\lambda}+2(ag/f)\,z^{\theta+\eta}R^{\lambda}-\theta aA-2(\theta+\eta+2\lambda\eta)\,(ag/f)\,B.$$

Divide these sums by $\lambda\eta a(f^2-4ge)/f$ and, if the latter quotient be called D, then D will be the area of the curve whose ordinate is the former quotient

$(A=)\,I_{0,0}$, $(B=)\,I_{1,0}$, ..., $I_{n-1,0}$ as $a^{\theta}R^{\lambda}-\sum_{0\leqslant i\leqslant n-1}(\theta+i\lambda\eta)\,a_iI_{i,0}$. And the like *mutatis mutandis* as θ passes successively to be $\theta+\eta$, $\theta+2\eta$, $\theta+3\eta$, ...

(60) Since these coefficients, when substitution of $-\theta a$ and $-(\theta+\eta+2\lambda\eta)\,b$ is made for p and q, reduce (as Newton noted down on Add. 3962.3: 51^r) to be '$2\lambda\eta ag-\lambda\eta bf[=0]$' and '$\lambda\eta af-2\lambda\eta be$' respectively.

omnes Ordinatæ terminos præter primum æquales nihilo potest area Curvæ inveniri cujus Ordinata est $z^{\theta-1}R^{\lambda-1}$.(61) Dicatur area ista C, et qua ratione ex areis A et B inventæ sunt areæ C ac D, ex his areis C ac D inveniri possunt aliæ duæ E et F ordinatis $z^{\theta-1}R^{\lambda-2}$ et $z^{\theta+\eta-1}R^{\lambda-2}$ congruentes, et sic deinceps in infinitum. Et per analysin contrariam regredi potest ab areis E et F ad areas C ac D, et inde ad areas A et B, aliasq; quæ in progressione sequuntur. Igitur si index λ perpetua unitatum additione vel subductione augeatur vel minuatur, et ex areis quæ Ordinatis sic prodeuntibus respondent duæ simplicissimæ habentur; dantur aliæ omnes in infinitum. Q.E.O.(62)

Cas. 3. Et per casus hosce duos conjunctos, si tam index θ perpetua additione vel subductione ipsius η, quam index λ perpetua additione vel subductione unitatis utcunq; augeatur vel minuatur[,] dabuntur areæ singulis prodeuntibus Ordinatis respondentes. Q.E.O.

Cas. 4. Et simili argumento si ordinata constat ex quatuor nominibus in vinculo radicali et dantur tres arearū, vel si constat ex quinq; nominibus & dantur quatuor arearū, et sic deinceps: dabuntur areæ omnes quæ addendo vel subducendo numerum η indici θ vel unitatem indici λ generari possunt.(63) Et par est ratio Curvarum ubi ordinatæ ex binomijs conflantur et area una earum quæ non sunt geometricè quadrabiles datur. Q.E.O.

Prop. VIII. Theor. VI.(64)

Si pro $e+fz^{\eta}+gz^{2\eta}+$&c et $k+lz^{\eta}+mz^{2\eta}+$&c scribantur R et S ut supra et in Curvæ alicujus Ordinata $z^{\theta\pm\eta\sigma}R^{\lambda\pm\tau}S^{\mu\pm\nu}$ maneant quantitates datæ θ, η, λ, μ, e, f,

(61) Equation of the second, third and fourth coefficients to zero yields respectively

$$(\theta+\lambda\eta)\ af+(\theta+\eta)\ be+pf=-qe, \quad (\theta+2\lambda\eta)\ ag+(\theta+\eta+\lambda\eta)\ bf+pg=-qf$$

and

$$(\theta+\eta+2\lambda\eta)\ b=-q,$$

whence on eliminating q there is $((\theta+\lambda\eta)\ a+p)\ f = 2\lambda\eta be$ and $((\theta+2\lambda\eta)\ a+p)\ g = \lambda\eta bf$, and upon eliminating θ and b in turn in these there comes

$$b/a = fg/(f^2-2eg) \quad \text{and} \quad p/a = -(\theta+\lambda\eta)+2\lambda\eta eg/(f^2-2eg),$$

and thence $q/a = -(\theta+\eta+2\lambda\eta)\ fg/(f^2-2eg)$. For what it is worth Joseph Raphson went badly wrong in computing these values for p and q in his *History of Fluxions* (London, 1715): 84–5, their correct expressions being first published by John Stewart in his commentary on *Sir Isaac Newton's Two Treatises of the Quadrature of Curves* (London, 1745): 168–9.

(62) In general, if we again (as in note (59)) put

$$I_{r,s} \equiv \int z^{\theta+r\eta-1}R^{\lambda+s-1}.dz \quad \text{where} \quad R = \sum_{0\leqslant i\leqslant n} a_i z^{i\eta},$$

it is Newton's present purpose to develop a systematic procedure for passing from the group of given integrals $I_{k,1}$, $k = 0, 1, 2, ..., n-1$, to that of the analogous set $I_{k,0}$ in whose ordinates the coefficient of R is reduced by unity. Because at once $I_{k,1} = \sum_{0\leqslant i\leqslant n} a_i I_{k+1,0}$ while (by Proposition III) there is $z^{\theta+k\eta}R^{\lambda} = \sum_{0\leqslant j\leqslant n} (\theta+(k+j\lambda)\ \eta)\ a_j I_{k+j,0}$, the arbitrary aggregate $\sum_{0\leqslant k\leqslant n-1} (b_k I_{k,1}+c_k z^{\theta+k\eta}R^{\lambda})$ in which $b_0 = p$, $b_1 = q$, ...; $c_0 = a$, $c_1 = b$, ... may straight-

$z^{\theta+\eta-1}R^{\lambda-1}$. And by the same method, putting all terms in the ordinate except the first equal to nothing, the area can be found of the curve whose ordinate is $z^{\theta-1}R^{\lambda-1}$.(61) Call that area C, and by the procedure whereby from the areas A and B the areas C and D were attained, from these areas C and D there can be found two others, E and F, corresponding to the ordinates $z^{\theta-1}R^{\lambda-2}$ and $z^{\theta+\eta-1}R^{\lambda-2}$, and so forth indefinitely. And by the contrary analysis regression can be made from the areas E and F to the areas C and D, and thence to the areas A and B and to the others which follow in the progression. Therefore if the index λ be increased or diminished by the perpetual addition or subtraction of units, and from the areas which correspond to the ordinates thus resulting the two simplest are had, all the others are given infinitely. As was to be shown.(62)

Case 3. And by these two cases jointly, if both the index θ be increased or diminished in any manner whatever by the perpetual addition or subtraction of η, and the index λ by the perpetual addition or subtraction of unity, the areas corresponding to individual ordinates resulting will be given. As was to be shown.

Case 4. And by a similar argument, if the ordinate is quadrinomial within the radical bar and three of the areas are given, or if it is quinquinomial and four of the areas are given, and so on, all the areas generatable by adding or subtracting the number η to or from the index θ or unity to or from the index λ will be given.(63) And the procedure is equivalent for curves where the ordinates are a combination of binomials and one of the areas not geometrically quadrable is given. As was to be shown.

Proposition VIII, Theorem VI.(64)

If for $e+fz^{\eta}+gz^{2\eta}+\ldots$ and $k+lz^{\eta}+mz^{2\eta}+\ldots$ there be written R and S, as above, and in the ordinate $z^{\theta\pm\eta\sigma}R^{\lambda\pm\tau}S^{\mu\pm\upsilon}$ of some curve the quantities $\theta, \eta, \lambda, \mu, e,$

---

forwardly be identified with the sum $\sum_{0\leqslant l\leqslant 2n-1} d_l I_{l,0}$ in which $d_0 = a_0(b_0+\theta c_0)$, and so on; the $2n$ linear equations evaluating in this way the coefficients $d_0, d_1, d_2, \ldots, d_{2n-1}$ in terms of the $b_0, b_1, b_2, \ldots, b_{n-1}, c_0, c_1, c_2, \ldots, c_{n-1}$ can then—directly from first principles, as Newton here does it, or by some more sophisticated means—straightforwardly be inverted to yield the several b_k and c_k individually in terms of linear combinations of the various d_l. In particular, when each of the first $d_0, d_1, \ldots, d_{n-1}$ is successively taken equal to unity and all the remaining $2n-1$ coefficients are put to be zero, there will be given, as Newton requires, the linear combinations of the given integrals $I_{k,1}, k = 0, 1, 2, \ldots, n-1$ which singly generate the corresponding integrals $I_{l,0}, l = 0, 1, 2, \ldots, n-1$. Lastly, by transposing the index λ to be in turn $\lambda-1, \lambda-2, \lambda-3, \ldots$ and repeating the above procedure *mutatis mutandis* we may similarly reduce the power of R in the ordinate by any number of units it is desired.

(63) Namely, by appropriately combining the generalized procedures which we have set out—in modern notation but without essential anachronism—in notes (59) and (62) preceding.

(64) A remould of Proposition XIII of the first version (**1**, §1 preceding) of the 1691 'De quadratura Curvarum'; see pages 46–8 above. Newton's preliminary draft of this on Add. 3962.3: 51ʳ differs only in having the cancelled variant cited in the next note.

g, k, l, m &c et pro σ, τ, et υ scribantur successivè numeri quicunꝗ integri: et si dentur areæ duarum ex curvis quæ per ordinatas sic prodeuntes designantur si quantitates R et S sunt binomia, vel si dentur areæ trium ex curvis si R et S conjunctim ex sex nominibus constant(65) & sic deinceps in infinitum: dico quod dabuntur areæ curvarum omnium.

Demonstratur ad modum Propositionis superioris.(66)

Prop. IX. Theor. VII.(67)

Æquantur Curvarum areæ inter se quarum Ordinatæ sunt reciprocè ut fluxiones Abscissarum.

Nam contenta sub Ordinatis et fluxionibus Abscissarum erunt æqualia, et fluxiones arearum sunt ut hæc contenta.

Corol. 1. Si assumatur relatio quævis inter Abscissas duarum Curvarum, et inde per Prop. I quæratur relatio fluxionum Abscissarum et ponantur Ordinatæ reciprocè proportionales fluxionibus: inveniri possunt innumeræ Curvæ quarum areæ sibi mutuò æquales erunt.(68)

Corol. 2. Sic enim Curva omnis cujus hæc est Ordinata

$$z^{\theta-1}\times\overline{e+fz^{\eta}+gz^{2\eta}+\&c}|^{\lambda}$$

(65) In draft (see the previous note) Newton here originally specified that R and S 'sunt nominum sex' (are (possessed) of six nomials), initially passing thereafter to add 'vel areæ ex quinꝗ curvis si R et S conjunctim sunt nominum septem' (or the areas of five curves if R and S are jointly of seven nomials).

(66) By straightforward extensions of the generalized procedures outlined in notes (59) and (62) above, whereby respectively the index of the initial power of z alone is at each step reduced by η, and those of R and S are successively lowered by a unit each time. These were afterwards worked out at great length by Roger Cotes in his 'De Continuatione Formarum et Fluentium ad Infinitum' (*Harmonia Mensurarum, sive Analysis & Synthesis Per Rationum & Angulorum Mensuras* (Cambridge, 1722): 66–76/248–9). In his *Sir Isaac Newton's Two Treatises of the Quadrature of Curves* (see note (61)): 170–4 John Stewart in 1745 outlined an analogous division of the demonstration of this theorem into four component 'Casus' exactly after the style of Proposition VII, departing from the basic expansion of the *area Curvæ* in terms of instances of the sequence of integrals $I_r \equiv \int z^{\theta+r\eta-1}R^{\lambda-1}S^{\mu-1}.dz$, $r = 0, 1, 2, \ldots$, as outlined in Proposition IV above, namely:

$$z^{\theta}R^{\lambda}S^{\mu} = \theta ekI_0+((\theta+\lambda\eta)fk+(\theta+\mu\eta)\,el)\,I_1 \\ +((\theta+2\lambda\eta)\,gk+(\theta+\lambda\eta+\mu\eta)fl+(\theta+2\mu\eta)\,em)\,I_2+\ldots.$$

An even more elaborate analogous 'Demonstratio', now distinguished into six cases, was subsequently set out by Samuel Horsley in footnote (dd) to his re-edition of the printed *Tractatus De Quadratura Curvarum* in his *Isaaci Newtoni Opera quæ exstant Omnia*, **1** (London, 1779): 333–86, especially 365–9. All very prettily done, but—as Newton no doubt himself thought in here foregoing the detailed particulars of specification and proof—to no great theoretical purpose or practical gain.

(67) A much extended and generalized version of Theorem III of the 1691 'De quadratura' (see pages 56–62 above), which is now—after being initially (see the next note) reshaped to be an intended prime corollary—itself drastically shortened and subsumed into Corollary 7

$f, g, k, l, m, \ldots$ remain given while in place of σ, τ and v any integers whatever be successively entered; then, if there be given the areas of two of the curves denoted by the ordinates thus resulting if the quantities R and S are binomials, or of three of the curves if R and S jointly consist of six nomials,(65) and so on indefinitely, I assert that the areas of all the curves will be given.

This is proved in the manner of the previous proposition.(66)

Proposition IX, Theorem VII.(67)

The areas of curves whose ordinates are reciprocally as the fluxions of their abscissas are equal to one another.

For the rectangles contained by the ordinates and the fluxions of the abscissas will be equal, and the fluxions of the areas are as these contents.

Corollary 1. If there be assumed any relationship whatever between the abscissas of two curves, and therefrom by Proposition I the relationship of the fluxions of the abscissas be sought and the ordinates then set reciprocally proportional to the fluxions, innumerable curves can be found whose areas will be mutually equal one to the other.(68)

Corollary 2. In this way, of course, every curve having this ordinate

$$z^{\theta-1}(e+fz^{\eta}+gz^{2\eta}+\ldots)^{\lambda}$$

---

below. A full-scale preliminary draft of this theorem (under the head of 'Lemma vel Prop.') exists on Add. 3962.3: 51ʳ–52ʳ. Significant variations in the detail of this are, together with similar novelties in a jumble of rough preliminary computations on *ibid.*: 41ʳ–44ᵛ, cited in following footnotes or reproduced at fuller length in sections [3], [4] and [5] of the Appendix.

(68) At this point in his prior draft (ff. 51ʳ/51ᵛ) Newton proceeded in a following paragraph to illustrate this prime corollary by outlining its manner of application—much as earlier on III: 383–5 and in the equivalent portion of Theorem III of the 1691 'De quadratura' (see page 56 above)—to the area-preserving transmutation of a curve defined by a polynomial Cartesian equation, viz: 'Ut si æquatio qua r[elatio inter unius] curvæ [a]bscissam z et ordinatã y definitur sit $py^{\alpha}z^{\beta}+qy^{\gamma}z^{\delta}+[ry^{\epsilon}z^{\zeta}+\&c]=0$ et assumatur relatio inter abscissas $x^{\pi}=z$. Erit $\pi\dot{x}x^{\pi-1}=\dot{z}$ per Prop. 1 ideoq $\dot{z}.\dot{x}::\pi x^{\pi-1}.1$. Sunto ordinatæ v et y in ea ratione et erit $y=\frac{1}{\pi}vx^{1-\pi}$. In æquatione prima pro y et z substituantur earum valores et producetur æquatio $\frac{p}{\pi\alpha}v^{\alpha}x^{\alpha+\pi\beta-\pi\alpha}+\frac{q}{\pi\gamma}v^{\gamma}x^{\gamma+\pi\delta-\pi\gamma}+\frac{r}{\pi\epsilon}v^{\epsilon}x^{\epsilon+\pi\zeta-\pi\epsilon}+\&c=0$, qua relatio inter Curvæ alterius ordinatam v et abscissam x definitur' (If, for instance, the equation whereby the relationship between the abscissa z and ordinate y of one curve is defined by $py^{\alpha}z^{\beta}+qy^{\gamma}z^{\delta}+ry^{\epsilon}z^{\zeta}+\ldots=0$ and the relationship between the abscissas be assumed to be $x^{\pi}=z$, then by Proposition 1 will there be $\pi\dot{x}x^{\pi-1}=\dot{z}$ and consequently $\dot{z}:\dot{x}=\pi x^{\pi-1}:1$. Let the ordinates v and y be in that ratio, and there will be $y=(1/\pi)\,vx^{1-\pi}$. In the first equation in place of y and z substitute their values and there will be produced the equation

$$(p/\pi\alpha)\,v^{\alpha}x^{\alpha+\pi\beta-\pi\alpha}+(q/\pi\gamma)\,v^{\gamma}x^{\gamma+\pi\delta-\pi\gamma}+(r/\pi\epsilon)\,v^{\epsilon}x^{\epsilon+\pi\zeta-\pi\epsilon}+\ldots=0$$

by which the relationship between the ordinate v and abscissa x of the second curve is defined).

assumendo quantitatem quamvis pro ν et ponendo $\frac{\eta}{\nu} = s$ et $z^s = x$(69) migrat in aliam sibi æqualem(70) cujus ordinata est

$$\frac{\nu}{\eta}x^{\frac{\nu\theta-\eta}{\eta}} \text{ in } \overline{e+fx^{\nu}+gx^{2\nu}+\&c}|^{\lambda}.$$

Corol. 3. Et Curva omnis cujus ordinata est

$$z^{\theta-1}\times\overline{a+bz^{\eta}+cz^{2\eta}+\&c}\times\overline{e+fz^{\eta}+gz^{2\eta}+\&c}|^{\lambda}$$

assumendo quantitatem quamvis pro ν et ponendo $\frac{\eta}{\nu} = s$ et $z^s = x$(69) migrat in aliam sibi æqualem cujus ordinata est

$$\frac{\nu}{\eta}x^{\frac{\nu\theta-\eta}{\eta}}\times\overline{a+bx^{\nu}+cx^{2\nu}+\&c}\times\overline{e+fx^{\nu}+gx^{2\nu}+\&c}|^{\lambda}.$$

Corol. 4. Et Curva omnis cujus Ordinata est

$$z^{\theta-1}\times\overline{a+bz^{\eta}+cz^{2\eta}+\&c}\times\overline{e+fz^{\eta}+gz^{2\eta}+\&c}|^{\lambda}\times\overline{k+lz^{\eta}+mz^{2\eta}+\&c}|^{\mu}$$

assumendo quantitatem quamvis pro ν et ponendo $\frac{\eta}{\nu} = s$ et $z^s = x$(69) migrat in aliam sibi æqualem cujus ordinata est

$$\frac{\nu}{\eta}x^{\frac{\nu\theta-\eta}{\eta}}\times\overline{a+bx^{\nu}+cx^{2\nu}+\&c}\times\overline{e+fx^{\nu}+gx^{2\nu}+\&c}|^{\lambda}\times\overline{k+lx^{\nu}+mx^{2\nu}+\&c}|^{\mu}.$$

Corol. 5. Et Curva omnis cujus Ordinata est $z^{\theta-1}\times\overline{e+fz^{\eta}+gz^{2\eta}+\&c}|^{\lambda}$ ponendo $\frac{1}{z} = x$(71) migrat in aliam sibi æqualem cujus ordinata est

$$\frac{1}{x^{\theta+1}}\times\overline{e+fx^{-\eta}+gx^{-2\eta}+\&c}|^{\lambda}$$

id est $\frac{1}{x^{\theta+1+\eta\lambda}}\times\overline{f+ex^{\eta}}|^{\lambda}$ si duo sunt nomina in vinculo radicis, vel

$$\frac{1}{x^{\theta+1+2\eta\lambda}}\times\overline{g+fx^{\eta}+ex^{2\eta}}|^{\lambda}$$

si tria sunt nomina; et sic deinceps.

Corol. 6. Et Curva omnis cujus Ordinata est

$$z^{\theta-1}\times\overline{e+fz^{\eta}+gz^{2\eta}+\&c}|^{\lambda}\times\overline{k+lz^{\eta}+mz^{2\eta}+\&c}|^{\mu}$$

ponendo $\frac{1}{z} = x$(71) migrat in aliam sibi æqualem cujus ordinata est

$$\frac{1}{x^{\theta+1}}\times\overline{e+fx^{-\eta}+gx^{-2\eta}+\&c}|^{\lambda}\times\overline{k+lx^{-\eta}+mx^{-2\eta}+\&c}|^{\mu}$$

id est $\frac{1}{x^{\theta+1+\eta\lambda+\eta\mu}}\times\overline{f+ex^{\eta}}|^{\lambda}\times\overline{l+kx^{\eta}}|^{\mu}$ si bina sunt nomina in vinculis radicum, vel $\frac{1}{x^{\theta+1+2\eta\lambda+\eta\mu}}\times\overline{g+fx^{\eta}+ex^{2\eta}}|^{\lambda}\times\overline{l+kx^{\eta}}|^{\mu}$ si tria sunt nomina in vinculo radicis

on assuming any quantity you wish for ν and setting $\eta/\nu = s$ and $z^s = x$[69] passes into another one equal[70] to it whose ordinate is

$$(\nu/\eta)\, x^{(\nu\theta-\eta)/\eta}(e+fx^{\nu}+gx^{2\nu}+\ldots)^{\lambda}.$$

Corollary 3. And every curve whose ordinate is

$$z^{\theta-1}(a+bz^{\eta}+cz^{2\eta}+\ldots)\,(e+fz^{\eta}+gz^{2\eta}+\ldots)^{\lambda}$$

on assuming any quantity for ν and setting $\eta/\nu = s$ and $z^s = x$[69] passes into another equal to it whose ordinate is

$$(\nu/\eta)\, x^{(\nu\theta-\eta)/\eta}(a+bx^{\nu}+cx^{2\nu}+\ldots)\,(e+fx^{\nu}+gx^{2\nu}+\ldots)^{\lambda}.$$

Corollary 4. And every curve whose ordinate is

$$z^{\theta-1}(a+bz^{\eta}+cz^{2\eta}+\ldots)\,(e+fz^{\eta}+gz^{2\eta}+\ldots)^{\lambda}\,(k+lz^{\eta}+mz^{2\eta}+\ldots)^{\mu}$$

on assuming any quantity for ν and setting $\eta/\nu = s$ and $z^s = x$[69] passes into another equal to it whose ordinate is

$$(\nu/\eta)\, x^{(\nu\theta-\eta)/\eta}\,(a+bx^{\nu}+cx^{2\nu}+\ldots)\,(e+fx^{\nu}+gx^{2\nu}+\ldots)^{\lambda}\,(k+lx^{\nu}+mx^{2\nu}+\ldots)^{\mu}.$$

Corollary 5. While every curve whose ordinate is $z^{\theta-1}(e+fz^{\eta}+gz^{2\eta}+\ldots)^{\lambda}$ on setting $1/z = x$[71] passes into another equal to itself whose ordinate is

$$x^{-(\theta+1)}(e+fx^{-\eta}+gx^{-2\eta}+\ldots)^{\lambda},$$

that is, $x^{-(\theta+1+\eta\lambda)}(f+ex^{\eta})^{\lambda}$ if there is a binomial within the root's vinculum, or $x^{-(\theta+1+2\eta\lambda)}(g+fx^{\eta}+ex^{2\eta})^{\lambda}$ if there is a trinomial, and so on.

Corollary 6. And every curve whose ordinate is

$$z^{\theta-1}(e+fz^{\eta}+gz^{2\eta}+\ldots)^{\lambda}\,(k+lz^{\eta}+mz^{2\eta}+\ldots)^{\mu}$$

on setting $1/z = x$[71] passes into another equal to itself whose ordinate is

$$x^{-(\theta+1)}(e+fx^{-\eta}+gx^{-2\eta}+\ldots)^{\lambda}\,(k+lx^{-\eta}+mx^{-2\eta}+\ldots)^{\mu},$$

that is, $$x^{-(\theta+1+\eta\lambda+\eta\mu)}(f+ex^{\eta})^{\lambda}\,(l+kx^{\eta})^{\mu}$$

if there are binomials in the vincula of the roots, or

$$x^{-(\theta+1+2\eta\lambda+\eta\mu)}(g+fx^{\eta}+ex^{2\eta})^{\lambda}\,(l+kx^{\eta})^{\mu}$$

---

(69) That is, $z^{\eta} = x^{\nu}$ or $z = x^{\nu/\eta}$, whence $\dot{z}/\dot{x}$ (or dz/dx) $= (\nu/\eta)\, x^{\nu/\eta-1}$.

(70) Here and correspondingly in Newton's following corollaries understand 'quoad aream' (in regard to their area).

(71) That is, $z = x^{-1}$ and so $\dot{z}/\dot{x} = -x^{-2}$. Newton will in sequel absorb the negative sign into his basic equality of areas by positing this now to be $\int y.dz = -\int v.dx$ (see the next note); whence $v/y = -\dot{z}/\dot{x} = x^{-2}$.

prioris ac duo in vinculo posterioris: et sic in alijs. Et nota quod areæ duæ æquales in novissimis hisce duobus Corollarijs jacent ad contrarias partes ordinatarum.(72) Si area in alterutra curva adjacet abscissæ, area huic æqualis in altera curva adjacet abscissæ productæ.(73)

Corol. 7. Si relatio inter Curvæ alicujus Ordinatam y et Abscissam z definiatur per æquationem quamvis affectam hujus formæ

$$y^{\alpha} \text{ in } \overline{e+fy^{\eta}z^{\delta}+gy^{2\eta}z^{2\delta}+hy^{3\eta}z^{3\delta}+\&c} = z^{\beta} \text{ in } \overline{k+ly^{\eta}z^{\delta}+my^{2\eta}z^{2\delta}+\&c}:$$

hæc figura assumendo $s=\dfrac{\eta-\delta}{\eta}$, $x=\dfrac{1}{s}z^{s}$ & $\lambda=\dfrac{\eta-\delta}{\alpha\delta+\beta\eta}$, migrat in aliam sibi æqualem cujus Abscissa x ex data Ordinata v determinatur per æquationem non affectam $\dfrac{1}{s}v^{\alpha\lambda}\times\overline{e+fv^{\eta}+gv^{2\eta}+hv^{3\eta}+\&c}|^{\lambda}\times\overline{k+lv^{\eta}+mv^{2\eta}+\&c}|^{-\lambda}=x.$(74)

Corol. 8. Si relatio inter Curvæ alicujus Ordinatam y et Abscissam z definitur per æquationem quamvis affectam hujus formæ

$$y^{\alpha} \text{ in } \overline{e+fy^{\eta}z^{\delta}+gy^{2\eta}z^{2\delta}+\&c}$$
$$= z^{\beta} \text{ in } \overline{k+ly^{\eta}z^{\delta}+my^{2\eta}z^{2\delta}+\&c}+z^{\gamma} \text{ in } \overline{p+qy^{\eta}z^{\delta}+ry^{2\eta}z^{2\delta}+\&c}:$$

hæc figura assumendo $s=\dfrac{\eta-\delta}{\eta}$, $x=\dfrac{1}{s}z^{s}$, $\mu=\dfrac{\alpha\delta+\beta\eta}{\eta-\delta}$ & $\nu=\dfrac{\alpha\delta+\gamma\eta}{\eta-\delta}$ migrat in aliam sibi æqualem cujus Abscissa x ex data Ordinata v determinatur per æquationem minus affectam

$$v^{\alpha} \text{ in } \overline{e+fv^{\eta}+gv^{2\eta}+\&c}$$
$$= s^{\mu}x^{\mu} \text{ in } \overline{k+lv^{\eta}+mv^{2\eta}+\&c}+s^{\nu}x^{\nu} \text{ in } \overline{p+qv^{\eta}+rv^{2\eta}+\&c}.$$

Corol. 9. Curva omnis cujus ordinata [y] est $\dfrac{fz^{\eta-1}+2gz^{2\eta-1}+3hz^{3\eta-1}+\&c}{e+fz^{\eta}+gz^{2\eta}+hz^{3\eta}+\&c}$ assumendo $fz^{\eta}+gz^{2\eta}+hz^{3\eta}+\&c=x$ migrat in Hyperbolam sibi æqualem cujus ordinata [v] est $\dfrac{1}{\eta e+\eta x}$.(75)

(72) In other words Newton here posits the equality $\int y.dz = -\int v.dx$ where y and v are the ordinates of the given and derived curves. He might equally well have set the ordinate v to be negative, thence producing the equivalent transmutation $-v/y = \dot{z}/\dot{x} = -x^{-2}$ (compare the previous note).

(73) The two preceding Corollaries 5 and 6 are a late addition (on Add. 3962.3: 41r) to the main preliminary draft (*ibid.*: 51r–52r), where their place was initially taken (f. 51v) by the particular case and preliminary general version of Corollary 7 next following which we reproduce in section [3] of the Appendix.

(74) There survive on Add. 3962.3: 52r Newton's rough initial computations (reproduced in section [4] of the Appendix) for the analogous reduction in which the relationship between the abscissas is set more simply to be $x = z^{s}$ at the cost of a slightly more complicated ensuing

if there is a trinomial within the vinculum of the former root and a binomial in the latter's, and the like in other cases. And note that the two equal areas in these two most recent corollaries lie on opposite sides of their ordinates.(72) If the area in either curve adjoins the abscissa, the area equal to this in the other curve lies along the extension of the abscissa.(73)

Corollary 7. If the relationship between the ordinate y and abscissa z of some curve be defined by any affected equation of this form:

$$y^{\alpha}(e+fy^{\eta}z^{\delta}+gy^{2\eta}z^{2\delta}+hy^{3\eta}z^{3\delta}+\ldots) = z^{\beta}(k+ly^{\eta}z^{\delta}+my^{2\eta}z^{2\delta}+\ldots),$$

this curve, on assuming $s = (\eta-\delta)/\eta$, $x = (1/s)\,z^{s}$ and $\lambda = (\eta-\delta)/(\alpha\delta+\beta\eta)$, passes into another equal to itself whose abscissa x is determined from the given ordinate v by the non-affected equation

$$(1/s)\,v^{\alpha\lambda}(e+fv^{\eta}+gv^{2\eta}+hv^{3\eta}+\ldots)^{\lambda}\,(k+lv^{\eta}+mv^{2\eta}+\ldots)^{-\lambda} = x.^{(74)}$$

Corollary 8. If the relationship between the ordinate y and abscissa z of some curve is defined by any affected equation of this form:

$$y^{\alpha}(e+fy^{\eta}z^{\delta}+gy^{2\eta}z^{2\delta}+\ldots)$$
$$= z^{\beta}(k+ly^{\eta}z^{\delta}+my^{2\eta}z^{2\delta}+\ldots)+z^{\gamma}(p+qy^{\eta}z^{\delta}+ry^{2\eta}z^{2\delta}+\ldots),$$

this figure, on assuming $s = (\eta-\delta)/\eta$, $x = (1/s)\,z^{s}$, $\mu = (\alpha\delta+\beta\eta)/(\eta-\delta)$ and $\nu = (\alpha\delta+\gamma\eta)/(\eta-\delta)$, passes into another equal to it whose abscissa x is determined from the given ordinate v by the less affected equation

$$v^{\alpha}(e+fv^{\eta}+gv^{2\eta}+\ldots) = s^{\mu}x^{\mu}(k+lv^{\eta}+mv^{2\eta}+\ldots)+s^{\nu}x^{\nu}(p+qv^{\eta}+rv^{2\eta}+\ldots).$$

Corollary 9. Every curve whose ordinate y is

$$(fz^{\eta-1}+2gz^{2\eta-1}+3hz^{3\eta-1}+\ldots)/(e+fz^{\eta}+gz^{2\eta}+hz^{3\eta}+\ldots)$$

by assuming $fz^{\eta}+gz^{2\eta}+hz^{3\eta}+\ldots = x$ passes into a hyperbola equal to it whose ordinate v is $1/\eta(e+x)$.(75)

---

expression for the abscissa x in terms of the ordinate v. Here of course the amended relationship $x = (1/s)\,z^{s}$, that is, $z = (sx)^{1/s}$ determines correspondingly that $y = (v\dot{x}/\dot{z}$ or) $v(sx)^{1-1/s}$ and consequently, on making $(1-1/s)\,\eta = -\delta/s$ or $s = 1-\delta/\eta$, that $y^{\eta}z^{\delta} = v^{\eta}$, while also

$$y^{-\alpha}z^{\beta} = v^{-\alpha}(sx)^{\mu}$$

where $\mu = 1/\lambda = -\alpha+(\alpha+\beta)/s = (\alpha\delta+\beta\eta)/(\eta-\delta)$. A like reduction holds *mutatis mutandis* in the following corollary, where similarly $y^{-\alpha}z^{\gamma} = v^{-\alpha}(sx)^{\nu}$ on taking $\nu = (\alpha\delta+\gamma\eta)/(\eta-\delta)$.

(75) For the given ordinate is evidently $y = \eta^{-1}(\dot{x}/\dot{z})/(e+x)$ and so at once

$$v(\text{or } y\dot{z}/\dot{x}) = \eta^{-1}/(e+x).$$

This discarded corollary states, we need scarcely point out, merely the particular case $\pi = \theta = \tau = 1$, $\lambda = \nu = \omega = 0$, $x \to x-e$ of the generalized one which follows, and so is here strictly superfluous; when exactly it was that Newton decided to suppress it—by crossing out its text in the manuscript and renumbering the two ensuing corollaries to correspond (see notes (76) and (78) below)—is not easy to specify, but insofar as we may date such non-verbal textual adjustments with any accuracy they were perhaps made at a later date in preparation for the 1704 printing.

Corol. 9.(76) Curva omnis cujus ordinata est

$$\pi z^{\theta-1}\times\overline{ve \begin{smallmatrix}+\nu\\+\eta\end{smallmatrix} fz^{\eta} \begin{smallmatrix}+\nu\\+2\eta\end{smallmatrix} gz^{2\eta}+\&c}\times\overline{e+fz^{\eta}+gz^{2\eta}+\&c}|^{\lambda-1}$$
$$\times\overline{a+b\overline{ez^{\nu}+fz^{\nu+\eta}+gz^{\nu+2\eta}+\&c}|^{\tau}}|^{\omega}$$

si sit $\theta=\nu\lambda$ et assumantur $x=\overline{ez^{\nu}+fz^{\nu+\eta}+gz^{\nu+2\eta}+\&c}|^{\pi}$, $\sigma=\frac{\tau}{\pi}$ et $\vartheta=\frac{\lambda-\pi}{\pi}$ migrat in aliam sibi æqualem cujus ordinata est $x^{\vartheta}\overline{a+bx^{\sigma}}|^{\omega}$.(77) Et nota quod ordinata prior in hoc Corollario evadit simplicior ponendo $\lambda=1$, vel ponendo $\tau=1$ et efficiendo ut radix dignitatis extrahi possit cujus index est ω, vel etiam ponendo $\omega=-1$ et $\lambda=1=\tau=\sigma=\pi$, ut alios casus prætereeam.

Corol. 10.(78) Pro $ez^{\nu}+fz^{\nu+\eta}+gz^{\nu+2\eta}+\&c$, $vez^{\nu-1}\begin{smallmatrix}+\nu\\+\eta\end{smallmatrix}fz^{\nu+\eta-1}\begin{smallmatrix}+\nu\\+2\eta\end{smallmatrix}gz^{\nu+2\eta-1}+\&c$, $k+lz^{\eta}+mz^{2\eta}+\&c$ et $\eta lz^{\eta-1}+2\eta mz^{2\eta-1}+\&c$ scribantur R, $\dot{R}$, S et $\dot{S}$ respectivè, et Curva omnis cujus ordinata est $\overline{\pi\dot{R}S+\phi R\dot{S}}$(79) in $R^{\lambda-1}S^{\mu-1}\times\overline{aS^{v}+bR^{\tau}}|^{\omega}$, si sit $\frac{\mu-v\omega}{\lambda}$(79) $=\frac{v}{\tau}=\frac{\phi}{\pi}$, $\frac{\tau}{\pi}=\sigma$, $\frac{\lambda-\pi}{\pi}=\vartheta$, et $\frac{R^{\pi}}{S^{\phi}}=x$, migrat in aliam sibi æqualem cujus ordinata est $x^{\vartheta}\times\overline{a+bx^{\sigma}}|^{\omega}$.(80) Et nota quod Ordinata prior evadit simplicior ponendo unitates pro τ, v et λ vel μ et faciendo ut radix dignitatis extrahi possit cujus index est ω, vel ponendo $\omega=-1$ vel $\mu=0$.

---

(76) Initially '10' (see the previous note). In the manuscript Newton immediately beforehand began to transcribe the revised preliminary draft of this corollary as he had set it down on Add. 3962.3: 44ᵛ (and as it is reproduced therefrom in section [5] of the Appendix; compare *ibid.*: note (15)), then broke abruptly off in mid-equation to pen a lightly condensed equivalent '*Corol. 10.* Pro $e+fz^{\eta}+gz^{2\eta}+\&c$ scribatur R, et curva omnis cujus ordinata est

$$z^{\theta-1}R^{\lambda-1}\times\overline{ve \begin{smallmatrix}+\nu\\+\eta\end{smallmatrix} fz^{\eta} \begin{smallmatrix}+\nu\\+2\eta\end{smallmatrix} gz^{2\eta}+\&c}\times\overline{a+bz^{\nu\tau}R^{\tau}}|^{\mu},$$

si sit $\theta=\nu\lambda$, et assumatur $x=z^{\nu\pi}R^{\pi}$, $\sigma=\frac{\tau}{\pi}$ et $\vartheta=\frac{\lambda-\pi}{\pi}$ migrat in aliam sibi æqualem cujus ordinata est $\frac{1}{\pi}x^{\vartheta}\overline{a+bx^{\sigma}}|^{\mu}$' (*Corollary 10.* In place of $e+fz^{\eta}+gz^{2\eta}+\ldots$ let there be written R and then every curve whose ordinate is

$$z^{\theta-1}R^{\lambda-1}(ve+(\nu+\eta)fz^{\eta}+(\nu+2\eta)gz^{2\eta}+\ldots)(a+b(z^{\nu}R)^{\tau})^{\mu},$$

if there be $\theta=\nu\lambda$ and we assume $x=(z^{\nu}R)^{\pi}$, $\sigma=\tau/\pi$ and $\vartheta=(\lambda-\pi)/\pi$, passes into another equal [in area] to it whose ordinate is $\pi^{-1}x_{\vartheta}(a+bx^{\sigma})^{\mu})$. Except that in it Newton has again spelled out the polynomial expansion of R, made trivial substitution of ω for μ, and multiplied each ordinate by π, his final version which follows is unchanged from this.

(77) Newton's computations for the analogous case in which $\theta\rightarrow\theta+1$ and $\lambda\rightarrow\lambda+1$ (with,

Corollary 9.[76] Every curve whose ordinate is

$$\pi z^{\theta-1}(\nu e+(\nu+\eta)fz^{\eta}+(\nu+2\eta)gz^{2\eta}+\ldots)\,(e+fz^{\eta}+gz^{2\eta}+\ldots)^{\lambda-1}$$
$$\times\,(a+b(ez^{\nu}+fz^{\nu+\eta}+gz^{\nu+2\eta}+\ldots)^{\tau})^{\omega},$$

if there be $\theta=\nu\lambda$ and we assume $x=(ez^{\nu}+fz^{\nu+\eta}+gz^{\nu+2\eta}+\ldots)^{\pi}$, $\sigma=\tau/\pi$ and $\vartheta=(\lambda-\pi)/\pi$, passes into another equal to it whose ordinate is $x^{\vartheta}(a+bx^{\sigma})^{\omega}$.[77] And note that the former ordinate in this corollary comes to be simpler on setting $\lambda=1$, or setting $\tau=1$ and making extractable the root of the power whose index is ω, or again by setting $\omega=-1$ and $\lambda=1=\tau=\sigma=\pi$, to pass other cases by.

Corollary 10.[78] For $ez^{\nu}+fz^{\nu+\eta}+gz^{\nu+2\eta}+\ldots$,

$$\nu ez^{\nu-1}+(\nu+\eta)fz^{\nu+\eta-1}+(\nu+2\eta)\,gz^{\nu+2\eta-1}+\ldots,\quad k+lz^{\eta}+mz^{2\eta}+\ldots$$

and $\eta lz^{\eta-1}+2\eta mz^{2\eta-1}+\ldots$ let there be written R, $\dot{R}$, S and $\dot{S}$ respectively, then every curve whose ordinate is $(\pi\dot{R}S[-]\phi R\dot{S})\,R^{\lambda-1}S^{\mu-1}(aS^{\upsilon}+bR^{\tau})^{\omega}$ passes, if there be $(\mu-\upsilon\omega)/[-]\lambda=\upsilon/\tau=\phi/\pi$, $\tau/\pi=\sigma$, $(\lambda-\pi)/\pi=\vartheta$ and $R^{\pi}/S^{\phi}=x$, into another equal to it whose ordinate is $x^{\vartheta}(a+bx^{\sigma})^{\omega}$.[80] And note that the former ordinate comes to be simpler by putting units for τ, υ and λ or μ and making extractable the root of the power whose index is ω, or by setting $\omega=-1$ or $\mu=0$.

---

again, μ trivially replacing ω) are reproduced in section [5] of the Appendix. In gist, where $e+fz^{\eta}+gz^{2\eta}+\ldots=R$, his primary ordinate is $(y=)\ \pi z^{\theta-1}(\nu R+z\dot{R}/\dot{z})\,R^{\lambda-1}(a+b(z^{\nu}R)^{\tau})^{\omega}$, that is, $\pi S^{\lambda-1}\,\dot{S}(a+bS^{\tau})^{\omega}$ on putting $z^{\nu}R=S$; hence, where $x=S^{\pi}$ is the abscissa of a curve of equal area, the latter's ordinate $\upsilon=y\dot{z}/\dot{x}$, that is, $y/\dot{x}$ (since $\dot{z}=1$ here) will come to be

$$\pi(x^{1/\pi})^{\lambda-1}.(1/\pi)\;x^{1/\pi-1}(a+bx^{\tau/\pi})^{\omega}=x^{\vartheta}(a+bx^{\sigma})^{\omega}$$

on taking $\vartheta=\lambda/\pi-1$ and $\sigma=\tau/\pi$. The preceding reverse argument by which Newton initially proceeded from the ordinate υ to the ordinate y is readily restored.

(78) Originally '11'; compare note (75) above. Newton's preparatory calculations and preliminary verbal drafts of this corollary, too, are reproduced in section [5] of the Appendix.

(79) These should, to take account of Newton's following equation $x=R^{\pi}/S^{\phi}$, that is, $R^{\pi}S^{-\phi}$, read '$\overline{\pi\dot{R}S-\phi R\dot{S}}$' and '$\dfrac{\mu-\upsilon\omega}{-\lambda}$' respectively. And so we have amended our English version. The 1704 printed text (*Opticks:* ${}_2$193) makes equivalent adjustment by setting '$R^{\phi}S^{\mu}=x$' in the sequel, but this further necessitates the replacement of υ by $-\upsilon$.

(80) Much as with the previous Corollary 9 (see note (77) above), in this generalization Newton's primary ordinate is $y=(\pi\dot{R}S-\phi R\dot{S})\,R^{\lambda-1}S^{\mu-1}(aS^{\phi\tau/\pi}+bR^{\tau})^{\omega}$, that is,

$$\dot{x}R^{\lambda-\pi}S^{\phi+\mu-\upsilon\omega}(a+bx^{\tau/\pi})^{\omega}$$

on setting $R^{\pi}S^{-\phi}=x$ and $\phi\tau/\pi=\upsilon$; hence, where x is the abscissa of a curve of equal area, the ordinate $\upsilon=(y\dot{z}/\dot{x}$ or) $y/\dot{x}$ of this will be $x^{\lambda/\pi-1}S^{\mu-\upsilon\omega+\lambda\phi/\pi}(a+bx^{\tau/\pi})^{\omega}$, which on making $\mu-\upsilon\omega=-\lambda\phi/\pi$ reduces to be $x^{\vartheta}(a+bx^{\sigma})^{\omega}$, where $\vartheta=\lambda/\pi-1$ and $\sigma=\tau/\pi$.

Prop. X. Prob. III.[81]

Invenire figuras simplicissimas cum quibus Curva quævis geometricè comparari potest cujus ordinata per æquationem non affectam ex data abscissa determinatur.

Cas. 1. Sit ordinata abscissæ dignitas quælibet in datam quantitatem ducta, puta[82] $az^{\theta-1}$, et area erit $\frac{1}{\theta}az^{\theta}$, ut ex Prop. [V] (ponendo

$$b=0=c=d=f=g[=h] \ \&\text{c}$$

et $e=1$) facile colligetur.

Cas. 2. Sit ordinata[83] $[a]z^{\theta-1}\times\overline{e+fz^{\eta}+gz^{2\eta}+\&\text{c}}|^{\lambda-1}$ et si curva Geometrice comparari potest cum figuris rectilineis, quadrabitur per Prop. V [ponendo $b=0=c=d$]. Sin minus[,] convertetur in aliam curvam sibi æqualem cujus ordinata est $\frac{[a]}{\eta}x^{\frac{\theta-\eta}{\eta}}\times\overline{e+fx+gxx+\&\text{c}}|^{\lambda-1}$ per Corol. 2 Prop. IX. Deinde si de dignitatum indicibus $\frac{\theta-\eta}{\eta}$ et $\lambda-1$ per Prop. VII rejiciantur unitates [donec dignitates illæ fiant quam minimæ], devenietur tandem ad figuras simplicissimas quæ hac ratione colligi possunt. Dein harum unaquæq; per Corol. 5 Prop. IX dat aliam quæ nonnunquam simplicior est. [Et ex his per Prop. III et Corol. 9 & 10 Prop. IX inter se collatis figuræ adhuc simpliciores quandoq; prodeunt.] Deniq; ex figuris simplicissimis facto regressu [84]determinabitur (per Prop. VII et Corollaria 2, 5, 9 et 10 Prop. IX) figura cujus ordinata est

$$[a]z^{\theta-1}\times\overline{e+fz^{\eta}+gz^{2\eta}+\&\text{c}}|^{\lambda-1}.$$

Cas. 3. Sit ordinata

$$z^{\theta-1}\times\overline{a+bz^{\eta}+cz^{2\eta}+\&\text{c}}\times\overline{e+fz^{\eta}+gz^{2\eta}+\&\text{c}}|^{\lambda-1}$$

et hæc [figura] si quadrari potest quadrabitur per Prop. V. Sin minus[,] distinguenda est ordinata in partes $z^{\theta-1}\times a\times\overline{e+fz^{\eta}+gz^{2\eta}+\&\text{c}}|^{\lambda-1}$,

$$z^{\theta-1}\times bz^{\eta}\times\overline{e+fz^{\eta}+gz^{2\eta}+\&\text{c}}|^{\lambda-1} \ \&\text{c}$$

(81) A much augmented elaboration of Proposition VIII of the prior version (**1**, §1 preceding) of Newton's 1691 treatise 'De quadratura Curvarum', which is itself restyled to be Case 2 of the present extended problem. As we have explained in note (2) above, the leaf on which Newton penned this proposition has failed to survive with the other sheets of the final manuscript text. In lieu, we reproduce its main portion from the initial draft which is now Add. 3962.3: 52v[+43v], rounding out this roughly written and partially lacunary preliminary sketch—notably in 'Cas. 5' where a hole in the manscript has obliterated many of Newton's words—with appropriate excerpts from the 1704 printed text, whence also (see note (86) below) we borrow the text of the three corollaries whose content is but shadowily epitomized in the preliminary autograph.

(82) In the 1704 printed text (*Opticks:* $_2$194) this was curtailed to read simply 'Sit ordinata'

Proposition X, Problem III.[81]

To find the simplest figures with which any curve can geometrically be compared, the ordinate of which is determined from the given abscissa by an equation which is not affected.

Case 1. Let the ordinate be any power of the abscissa multiplied by a given quantity, say[82] $az^{\theta-1}$, and the area will be $(1/\theta)\,az^{\theta}$, as will easily be gathered from Proposition V on setting $b=0=c=d=f=g=h=\ldots$ and $e=1$ (therein).

Case 2. Let the ordinate be[83] $az^{\theta-1}(e+fz^{\eta}+gz^{2\eta}+\ldots)^{\lambda-1}$ and, if the curve can be geometrically compared with rectilinear figures, it will be squared by means of Proposition V on setting $b=0=c=d$. If not, it will be converted into another curve equal to it whose ordinate, by Corollary 2 to Proposition IX, is

$$(a/\eta)\,x^{(\theta-\eta)/\eta}(e+fx+gx^2+\ldots)^{\lambda-1}.$$

Next, if by Proposition VII units be cast out from the indices $(\theta-\eta)/\eta$ and $\lambda-1$ of the powers till they become the smallest possible, you will at length arrive at the simplest figures which are gatherable by this procedure. Then each one of these yields, by Corollary 5 to Proposition IX, another which is sometimes simpler. And when these are, by Proposition III and Corollaries 9 and 10 to Proposition IX, compared one with another still simpler figures on occasion ensue from them. Finally, when regression is made, in terms of these simplest figures [84]there will (by Proposition VII and Corollaries 2, 5, 9 and 10 of Proposition IX) be determined the figure whose ordinate is

$$az^{\theta-1}(e+fz^{\eta}+gz^{2\eta}+\ldots)^{\lambda-1}.$$

Case 3. Let the ordinate be

$$z^{\theta-1}(a+bz^{\eta}+cz^{2\eta}+\ldots)\,(e+fz^{\eta}+gz^{2\eta}+\ldots)^{\lambda-1}$$

and this figure, if it can be squared, will be squared by means of Proposition V. But if not, the ordinate is to be separated into the parts

$$z^{\theta-1}.a(e+fz^{\eta}+gz^{2\eta}+\ldots)^{\lambda-1},\quad z^{\theta-1}.bz^{\eta}(e+fz^{\eta}+gz^{2\eta}+\ldots)^{\lambda-1},$$

(Let the ordinate be). In sequel we have corrected a small inconsistency in the manuscript draft (f. 43v) where Newton went on to write '$bz^{\theta-1}$' for the ordinate and '$\frac{1}{\theta}bz^{\theta}$' for the corresponding area.

(83) Here, similarly, we standardize the manuscript draft (f. 52v) to the norm of the printed *Tractatus De Quadratura Curvarum* by introducing an initial coefficient *a* in reproducing the former's expression for the ordinate. Other phrases here taken over from the 1704 printed text—and doubtless faithfully repeating the amended sentences of Newton's lost earlier revise which we aim here to restore—are set within square brackets in the Latin text, following our usual editorial habit.

(84) In his 1704 text Newton went on to conclude much more briefly that 'computabitur area quæsita' (there will be computed the required area).

et per Cas. 2 inveniendæ [sunt] figuræ simplicissimæ [(85)]cum quibus pars quælibet comparari potest. Nam partes omnes cum ijsdem figuris comparabuntur.

Cas. 4. Sit ordinata

$$z^{\theta-1}\times\overline{a+bz^{\eta}+cz^{2\eta}[+\&c]}\ \text{in}\ \overline{e+fz^{\eta}+gz^{2\eta}+\&c}|^{\lambda-1}\times\overline{k+lz^{\eta}+mz^{2\eta}+\&c}|^{\mu-1}$$

et si curva quadrari potest quadrabitur per Prop. VI. Sin minus[,] convertetur in simpliciorem per Corol. 4 Prop. IX ac deinde comparabitur cum figuris simplicissimis per Prop. VIII & Corol. 6, 9 et 10 Prop. IX ut fit in Casu 2 et 3.

Cas. 5. Si ordinat[a e]x varijs partibus constat, partes singulæ pro ordi[natis c]urvarum totidem habendæ sunt et curvæ quotquot quadrari possunt sigillatim quadrand[æ sunt earumq] ordinatæ et areæ de Ordinata et area tota demendæ. Dei[n curva quam] ordinatæ pars residua designat seorsim (ut in Casu 2, [3] et 4) cum figuris simplicissimis comparanda est cum quibus comparari potest. Et [summ]a arearum omnium pro area Curvæ propositæ habenda est.

[(86)]*Corol. 1.* [Hinc etiam Curva omnis cujus Ordinata est radix quadratica affecta æquationis suæ, cum figuris simplicissimis seu rectilineis seu curvilineis comparari potest. Nam radix illa ex duabus partibus constat quæ seorsim spectatæ non sunt æquationum radices affectæ. Proponatur æquatio

$$aayy+zzyy=2a^3y+2z^3y-z^4,^{(87)}$$

& extracta radice erit $y=\dfrac{a^3+z^3+a\sqrt{a^4+2az^3-z^4}}{aa+zz}$, cujus pars rationalis $\dfrac{a^3+z^3}{aa+zz}$ & pars irrationalis $\dfrac{a\sqrt{a^4+2az^3-z^4}}{aa+zz}$ sunt ordinatæ curvarum quæ per hanc Propositionem vel quadrari possunt vel cum figuris simplicissimis comparari cum quibus collationem geometricam admittunt.]

Corol 2. [Et curva omnis cujus Ordinata per æquationem quamvis affectam definitur quæ per Corol. 7 Prop. IX in æquationem non affectam migrat, vel quadratur per hanc Propositionem si quadrari potest vel comparatur cum figuris simplicissimis cum quibus comparari potest. Et hac ratione Curva omnis quadratur cujus æquatio est trium terminorum. Nam æquatio illa si affecta sit

(85) The 1704 text here goes on: 'cum quibus figuræ partibus illis respondentes comparari possunt' (with which the figures corresponding to those parts can be compared), to which Newton there adjoined in confirmation 'Nam areæ figurarum partibus illis respondentium signis suis + & − conjunctæ component aream totam quæsitam' (For, of course, the areas of the figures corresponding to those parts will, when conjoined with their signs + and −, make the total area sought).

(86) The manuscript draft (f. 52ᵛ) here vaguely foreshadows the inclusion of 'Corol. 1. 2. 3. De æqu[ationibus] affec[tis]' (...On affected equations), and otherwise adds only a first

and so on, and then by Case 2 there are to be found the simplest figures [85] with which any individual part can be compared; for all the parts will be compared with the same figures.

Case 4. Let the ordinate be

$$z^{\theta-1}(a+bz^{\eta}+cz^{2\eta}+\ldots)\,(e+fz^{\eta}+gz^{2\eta}+\ldots)^{\lambda-1}\,(k+lz^{\eta}+mz^{2\eta}+\ldots)^{\mu-1}$$

and, if the curve can be squared, it will be squared by Proposition VI. But if not, it will be converted into a simpler one by Corollary 4 to Proposition IX, and then compared with the simplest figures by Proposition VIII and Corollaries 6, 9 and 10 to Proposition IX, as is done in Cases 2 and 3.

Case 5. If the ordinate consists of various parts, the separate parts are to be regarded as the ordinates of an equal number of curves, and as many of those curves as can be squared are individually to be squared and their ordinates and areas taken from the total ordinate and area. Then the curve which the residual portion of the ordinate denotes is separately (as in Cases 2, 3 and 4) to be compared with the simplest figures with which it is comparable. And the sum of all the areas is to be regarded as the area of the curve proposed.

[86] *Corollary 1.* Hence also every curve whose ordinate is the affected square root of its equation can be compared with the simplest figures, be they rectilinear or curvilinear. For that root consists of two parts which, regarded separately, are not affected roots of equations. Let the equation be

$$a^2y^2+z^2y^2 = 2a^3y+2z^3y-z^4,\text{[87]}$$

and when the root is extracted it will be

$$y = (a^3+z^3+a\sqrt{[a^4+2az^3-z^4]})/(a^2+z^2);$$

of this the rational part $(a^3+z^3)/(a^2+z^2)$ and irrational part

$$a\sqrt{[a^4+2az^3-z^4]}/(a^2+z^2)$$

are ordinates of curves which, by the present proposition, can either be squared or compared with the simplest figures with which they admit geometrical comparison.

Corollary 2. And every curve whose ordinate is defined by any affected equation which by means of Corollary 7 to Proposition IX passes into a non-affected equation is either squared by the present proposition if it can be squared, or compared with the simplest figures with which it is comparable. By this procedure is squared every curve whose equation is of three terms. For, if that equation be affected, it is transformed into a non-affected one by

---

version (see the next note) of the quadratic in y set, in the 1704 text which we now reproduce in sequel within editorial brackets, to illustrate the opening corollary.

(87) In Newton's draft (f. 52^r) this was framed to be '$[z]zyy+aayy = 2a^3y-2z^3y-z^4$', thence yielding the extracted root '$[y=]\dfrac{[a]^3-z^3\pm\sqrt{a^6-2a^3z^3-a^2z^4}}{[aa+]zz}$'.

transmutatur in non affectam per Corol. 7 Prop. IX ac deinde per Corol. 2 & 5 Prop. IX in simplicissimam migrando, dat vel quadraturam figuræ si quadrari potest, vel curvam simplicissimam quacum comparatur.]

Corol. 3. [Et Curva omnis cujus Ordinata per æquationem quamvis affectam definitur quæ per Corol. 8 Prop. IX in æquationem quadraticam affectam migrat, vel quadratur per hanc Propositionem & hujus Corol. 1 si quadrari potest, vel comparatur cum figuris simplicissimis cum quibus collationem geometricam admittit.]

Scholium.

Ubi quadrandæ obvenerint figuræ, ad Regulas hasce generales semper recurrere nimis molestum esset: Præstat figuras quæ simpliciores sunt et magis usui esse possunt semel quadrare et quadraturas in Tabulam referre, deinde Tabulam consulere quoties ejusmodi Curvam aliquam quadrare oportet. Hujus autem generis sunt Tabulæ duæ sequentes, in quib[u]s z denotat Abscissam, y Ordinatam & t Aream Curvæ quadrandæ, & d, e, f, g, h, η sunt quantitates datæ.

CURVÆ SIMPLICIORES QUÆ QUADRARI POSSUNT.(88)

Curvarum ordines.	*Curvarum areæ.*
Ordo primus.	
$dz^{\eta-1} = y.$	$\frac{d}{\eta} z^{\eta} = t.$
Ordo secundus.	
$\frac{dz^{\eta-1}}{ee+2efz^{\eta}+ffz^{2\eta}} = y.$	$\frac{dz^{\eta}}{\eta ee+\eta efz^{\eta}} = t,$ vel $\frac{-d}{\eta ef+\eta ffz^{\eta}} = t.$
Ordo tertius	
1. $dz^{-1}\sqrt{e+fz^{\eta}} = y.$	$\frac{2d}{\eta f}R^3 = t,$ existente $R = \sqrt{e+fz^{\eta}}.$
...	... ...(89)
Ordo quartus.	
1. $\frac{dz^{\eta-1}}{\sqrt{e+fz^{\eta}}} = y.$	$\frac{2d}{\eta f}R = t.$
...	... ...(89)

(88) Under this new title Newton proceeds to repeat without essential variant the parallel columns of the first four orders of the 'Catalogus Curvarum aliquot ad rectilineas figuras

Corollary 7 to Proposition IX, and on then, by Corollaries 2 and 5 to Proposition IX, passing into the simplest it yields either the quadrature of the figure if it can be squared, or the simplest figure with which it is comparable.

Corollary 3. While every curve whose ordinate is defined by any affected equation which, by way of Corollary 8 to Proposition IX, passes into an affected quadratic equation is either squared by this proposition and Corollary 1 of the present one if it can be squared, or compared with the simplest figures with which it admits geometrical comparison.

Scholium.

Where figures present themselves to be squared, it would be excessively troublesome always to recur to these general rules. It is better once for all to square the figures which are simpler and can be of more use, and to list their quadratures in a table: we can thereafter consult the table each time we need to square some curve of the kind. Of this type are the two following tables, wherein z denotes the abscissa, y the ordinate and t the area of the curve to be squared, while d, e, f, g, h, η are given quantities.

THE SIMPLER CURVES WHICH CAN BE SQUARED.(88)

The curves' orders.		*Their areas.*
First order.		
$dz^{\eta-1} = y.$		$(d/\eta)\ z^{\eta} = t.$
Second order.		
$dz^{\eta-1}/(e^2+2efz^{\eta}+f^2z^{2\eta}) = y.$		$dz^{\eta}/\eta e(e+fz^{\eta}) = t$, or $-d/\eta f(e+fz^{\eta}) = t.$
Third order.		
1. $dz^{\eta-1}\sqrt{(e+fz^{\eta})} = y.$		$(2d/\eta f)\ R^3 = t$, where $R = \sqrt{(e+fz^{\eta})}.$
...	...	...(89)
Fourth order.		
1. $dz^{\eta-1}/\sqrt{(e+fz^{\eta})} = y.$		$(2d/\eta f)\ R = t.$
...	...	...(89)

relatarum' which he had originally set in annexe to Problem 9 of his 1671 fluxional tract (see III: 236–8).

(89) We omit to reproduce the analogous tabulations of the areas $\int y.dz$ corresponding to the ordinates $y = dz^{i\eta-1}\sqrt{(e+fz^{\eta})}$ and $y = dz^{i\eta-1}/\sqrt{(e+fz^{\eta})}$, $i = 2, 3, 4$, respectively which Newton takes over unaltered from his 1671 tract (except for carelessly excluding a necessary factor 'dR^3' in his entry for the third area in the former 'Ordo tertius').

CURVÆ SIMPLICIORES QUÆ PER QUADRATURAM SECTIONUM CONICARUM QUADRARI POSSUNT,(90) existente x Abscissa, v Ordinata et s Area Sectionis Conicæ.

Curvarum [*ordines.*]	*Sectionis Conicæ*		*Curvarum*
Ordo primus.	*Abscissa.*	*Ordinata.*	*areæ.*
1. $\frac{dz^{\eta-1}}{e+fz^{\eta}}=y.$	$z^{\eta}=x.$	$\frac{d}{e+fx}=v.$	$\frac{1}{\eta}s=t=\frac{\alpha GDB}{\eta}.$(91)
...	...	...	...(92)

Prop. XI. Theor. VIII.(93)

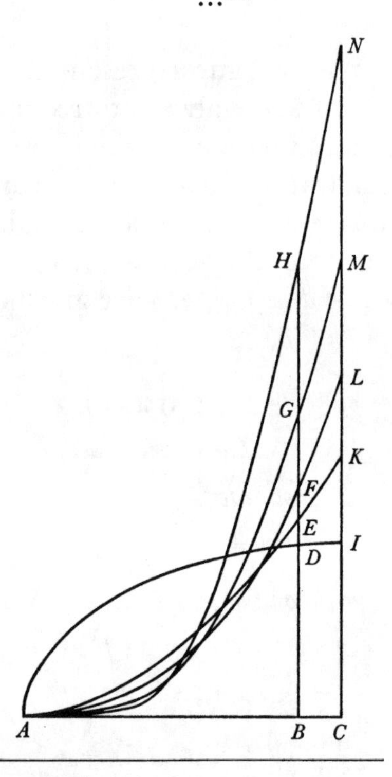

Sit $ADIC$ curva quævis abscissam habens $AB=z$ et ordinatam $BD=y$, et sit $AEKC$ curva alia cujus ordinata BE æqualis est prioris areæ ADB ad unitatem applicatæ, et $AFLC$ curva tertia cujus ordinata BF æqualis est secundæ areæ AEB ad unitatem applicatæ, et $AGMC$ Curva quarta cujus ordinata BG æqualis est tertiæ areæ AFB ad unitatem applicatæ, et $AHNC$ curva quinta cujus ordinata BH æqualis est quartæ areæ AGB ad unitatem applicatæ & sic deinceps in infinitum.(94)

Sunto A, B, C, D, E &c areæ curvarum ordinatas habentiū y, zy, z^2y, z^3y, z^4y [&c] & abscissam communem z.

Detur abscissa quævis $AC=t$, sitq

$$BC=t-z=x,$$

(90) Without any prior explanation of the conic figures to which this tabulation, here (see the next note) as earlier (see III: 240–2), is visually keyed, Newton begins straightaway to transcribe with minor verbal variations the complementary 'Catalogus Curvarum aliquot ad Conicas Sectiones relatarum' which he had likewise adjoined to Problem 9 of his 1671 tract (see III: 244 ff.).

(91) Understand the first figure to the left on III: 242, as we earlier specified at the corresponding point of Newton's 1671 tabulation (see our note (551) to III: 244).

(92) The folios (beginning '18' in Newton's pagination; see note (2) above) on which the remainder of this 'Ordo primus' of the 1671 catalogue was transcribed, along with, we may presume, its ten ensuing orders of algebraic integrals evaluable 'through the quadrature of conics' (see III: 244—54), do not now survive with the preceding sheets of the manuscript. They were doubtless discarded by Newton himself when he came, ten years after he penned it, to revise the present scholium for the press, at which time he made a number of small changes in the prior list of integrals—retitling it to be a 'Tabula Curvarum simpliciorum quæ quadrari possunt' and relabelling its 'Ordines curvarum' to be 'Formæ'—and replaced

THE SIMPLER CURVES WHICH CAN BE SQUARED BY MEANS OF THE QUADRATURES OF CONICS,(90) x being the conic's abscissa, v its ordinate and s its area.

The curves' orders.	*The conics'*		*The curves'*
First order.	*Abscissa*	*Ordinate.*	*areas.*
1. $dz^{\eta-1}/(e+fz^{\eta}) = y$.	$z^{\eta} = x$.	$d/(e+fx) = v$.	$(1/\eta)\, s = t = (1/\eta)\, (\alpha GDB)$.(91)
...	...	...	...(92)

Proposition XI, Theorem VIII.(93)

Let $ADIC$ be any curve having abscissa $AB = z$ and ordinate $BD = y$, and let $AEKC$ be another curve whose ordinate BE is equal to the preceding one's area (ABD) divided by unity, and $AFLC$ a third curve whose ordinate BF is equal to the second's area (AEB) divided by unity, and $AGMC$ a fourth curve whose ordinate BG is equal to the third one's area (AFB) divided by unity, and $AHNC$ a fifth curve whose ordinate BH is equal to the fourth's area (AGB) divided by unity, and so on indefinitely.(94)

Take $A, B, C, D, E, \ldots$ to be the areas of curves having ordinates y, zy, z^2y, z^3y, z^4y, ... and the common abscissa z. Let there be given any abscissa $AC = t$ and let $BC = t-z = x$, and take P, Q, R, S, T to be the areas of curves having ordinates y, xy, x^2y, x^3y, x^4y and the common abscissa z.(95)

this initial opening to the latter list of conic integrals by a reworked 'Tabula Curvarum simpliciorum quæ cum Ellipsi et Hyperbola comparari possunt' (Add. 3962.1: 20r–22r), likewise now subdivided into eleven 'Formæ'. The revised scholium will be reproduced in full at its due place in our final volume and we may here defer further comment upon it.

(93) In this last proposition of his 'Geometriæ Liber secundus' (as we here restore its concluding part) Newton presents a geometrical model which elegantly and concisely encapsulates the technique for reducing simple multiple integrals which he had earlier expounded in Case 1 of his determination of the fluents of 'complex' fluxional quantities, as we have reproduced it on page 165 above in appendix to the text of his 1691 'De quadratura Curvarum'. The bare statement which Newton gives of his construction—which cost his contemporaries not a little effort generally to justify (see **1**, §2, Appendix 2: notes (9) and (21) preceding)—is amplified and explained in following footnotes in our usual manner.

(94) In modern terms Newton here successively constructs the sequence of multiple integrals

$$(ADB) = \int_0^z y.dz,\ (AEB) = \int_0^z (ADB).dz = \int_0^z\int_0^z y.dz^2,\ (AFB) = \int_0^z (AEB).dz,$$
$$(AGB) = \int_0^z (AFB).dz,\ (AHB) = \int_0^z (AGB).dz, \ldots,$$

and thence, corresponding to some given abscissa $AC = t$ (as he will set it to be in sequel), defining the areas

$$(ADIC) = \int_0^t y.dz,\ (AEKC) = \int_0^t (ADB).dz = \int_0^t\int_0^z y.dz^2,\ (AFLC) = \int_0^t (AEB).dz,$$
$$(AGMC) = \int_0^t (AFB).dz,\ (AHNC) = \int_0^t (AGB).dz, \ldots.$$

et sunto P, Q, R, S, T areæ curvarum ordinatas habentium y, xy, xxy, x^3y, x^4y et abscissam communem z.(95)

Terminentur autem hæ omnes areæ ad abscissam totam datam AC, nec non ad ordinatam positione datam & infinite productam CI: et erit arearum sub initio positarum prima $ADIC = A = P$, Secunda $AEKC = tA - B = Q$. Tertia $AFLC = \dfrac{ttA - 2tB + C}{2} = \frac{1}{2}R$. Quarta $AGMC = \dfrac{t^3A - 3ttB + 3tC - D}{6} = \frac{1}{6}S$. Quinta $AHNC = \dfrac{t^4A - 4t^3B + 6ttC - 4tD + E}{24} = \frac{1}{24}T$. [&c].(96)

Ut si fuerit $y = z^2$ adeoq(97)

---

(95) Whence in turn

$$P = \int_0^t y.dz = A, \quad Q = \int_0^t (t-z)\,y.dz = \left(t\int_0^t y.dz - \int_0^t zy.dz \text{ or}\right) tA - B,$$

$$R = \int_0^t (t-z)^2\,y.dz = t^2A - 2tB + C, \quad S = \int_0^t (t-z)^3\,y.dz = t^3A - 3t^2B + 3tC - D,$$

$$T = \int_0^t (t-z)^4\,y.dz = t^4A - 4t^3B + 6t^2C - 4tD + E,$$

and so on.

(96) Briefly to demonstrate the *raison d'être* of these successive evaluations we may set

$$A_z \equiv \int_0^z y.dz, \quad B_z \equiv \int_0^z zy.dz, \quad C_z \equiv \int_0^z z^2y.dz, \quad D_z \equiv \int_0^z z^3y.dz, \ \ldots$$

(so that $A_t = A$, $B_t = B$, $C_t = C$, $D_t = D$, ... and also there is

$$dA_z/dz = z^{-1}.dB_z/dz = z^{-2}.dC_z/dz = z^{-3}.dD_z/dz = \ldots)$$

and then in turn compute:

$$(ADIC) = \int_0^t y.dz = A = P;$$

$$(AEKC) = \int_0^t A_z.dz = \Big[zA_z - B_z\Big]_{z=0}^{z=t} = tA - B = Q;$$

$$(AFLC) = \int_0^t (zA_z - B_z).dz = \Big[\tfrac{1}{2}(z^2A_z - C_z) - (zB_z - C_z)\Big]_{z=0}^{z=t},$$

$$\text{or } \Big[\tfrac{1}{2}(z^2A_z - 2zB_z + C_z)\Big]_{z=0}^{z=t} = \tfrac{1}{2}(t^2A - 2tB + C) = \tfrac{1}{2}R;$$

$$(AGMC) = \int_0^t \tfrac{1}{2}(z^2A_z - 2zB_z + C_z).dz = \tfrac{1}{2}\Big[\tfrac{1}{3}(z^3A_z - D_z) - (z^2B_z - D_z) + (zC_z - D_z)\Big]_{z=0}^{z=t},$$

$$\text{or } \Big[\tfrac{1}{6}(z^3A_z - 3z^2B_z + 3zC_z - D_z)\Big]_{z=0}^{z=t} = \tfrac{1}{6}(t^3A - 3t^2B + 3tC - D) = \tfrac{1}{6}S;$$

and so on. The general pattern of the development is readily proved much as before (see **1**, §2, Appendix 2: notes (9) and (21)) by showing that $I_n \equiv \sum_{0 \leqslant i \leqslant n} (-1)^n \binom{n}{i} \int_0^z z^i y.dz$ satisfies the recursion: $I_0 = \int_0^z y.dz$ and also, for $n = 1, 2, 3, \ldots$,

Let all these areas terminate, however, at the total given abscissa AC and also at the ordinate CI given in position and infinitely extended: then of the areas posited at the start the first one $(ADIC) = A = P$, the second one $(AEKC) =$

$$tA-B = Q,$$

the third $(AFLC) = \frac{1}{2}(t^2A-2tB+C) = \frac{1}{2}R$, the fourth $(AGMC) =$

$$\tfrac{1}{6}(t^3A-3t^2B+3tC-D) = \tfrac{1}{6}S,$$

the fifth $(AHNC) = \frac{1}{24}(t^4A-4t^3B+6t^2C-4tD+E) = \frac{1}{24}T$, and so on.(96)

If, for instance, there were $y = z^2$ and hence(97)

$$\int_0^z I_{n-1}.dz = \sum_{0\leqslant i\leqslant n-1}(-1)^{n-1}\binom{n-1}{i}.\frac{1}{n-i}\left(z^{n-i}\int_0^z z^iy.dz-\int_0^z z^ny.dz\right) = \frac{1}{n}I_n,$$

whence there is $$\int_0^t\int_0^z\int_0^z\dots\int_0^z y.dz^n = \int_0^t\frac{1}{(n-1)!}I_{n-1}.dz,$$

that is, $$\frac{1}{n!}.\sum_{0\leqslant i\leqslant n}(-1)^n\binom{n}{i}t^{n-i}\int_0^t z^iy.dz = \frac{1}{n!}\int_0^t(t-z)^n\,y.dz.$$

(97) The manuscript draft which we reproduce here breaks off, but it is here readily computed that $A = \int_0^t y.dz = \frac{1}{3}t^3$, $B = \int_0^t zy.dz = \frac{1}{4}t^4$, $C = \int_0^t z^2y.dz = \frac{1}{5}t^5$, ..., so that

$$(ADIC) \equiv \tfrac{1}{3}t^3 = \int_0^t(t-z)^0\,y.dz,\ (AEKC) = \tfrac{1}{12}t^4 = \int_0^t(t-z)\,y.dz,$$

$$(AFLC) = \tfrac{1}{30}t^5 = \int_0^t(t-z)^2\,y.dz,$$

and so on. We have already displayed Newton's working of the instance in which $y = z$ in **1**, §2, Appendix 2.2 preceding. In the rewritten version of the present proposition on Add. 3962.1:23r no example of any kind is given, but in compensation Newton appended (*ibid.*: 25r) a short '*Corol.*' where he further remarked: 'Unde si Curvæ quarum ordinatæ sunt y, zy, z^2y, z^3y &c vel y, xy, x^2y, x^3y &c quadrari possunt, quadrabuntur etiam Curvæ $ADIC$, $AKEC$, $AFLC$, $AGMC$ &c et habebuntur ordinatæ BE, BF, BG, BH areis Curvarum illarum proportionales' (Consequently, if the curves whose ordinates are y, zy, z^2y, z^3y, ... or y, xy, x^2y, x^3y, ... are quadrable, the curves $ADIC$, $AKEC$, $AFLC$, $AGMC$, ... will also be squared, and there will be obtained the ordinates BE, BF, BG, BH, ... proportional to the areas of those curves).

While it must surely have been tempting for Newton here in sequel to present a digest of the highlights of the final Propositions XI–XIII of his 1691 treatise 'De quadratura Curvarum' —indeed, as we have seen (note (22) above), he had already in his Proposition 1 taken over from the first of these a *résumé* of the method of grouping together in a given fluxional equation 'brother' terms which derive from a common fluent, if only subsequently to discard it—in the mass of his surviving preparatory drafts for the 'Geometriæ Liber secundus' the only evidence that he ever thought seriously of so doing is a lone, roughly scrawled and unfinished sketch of a cumbrously phrased 'Lemma' where he outlined his technique for determining the 'altissimus terminus' in the series development of the 'root' of a simple first-order fluxional equation. Merely to permit the reader to perceive the jejune quality of this crude, undeveloped tentative we reproduce its text (Add. 3962.3: 43v) in section [6] of the Appendix.

APPENDIX. VARIANT PRELIMINARY DRAFTS FOR THE FINAL 'LIBER SECUNDUS'.(1)

From the originals in the University Library, Cambridge

[1](2) *Prop. I.*

Data æquatione quotcunq; fluentes quantitates involvente, invenire fluxiones.

Cas. 1. Si æquatio quantitates irrationales ex fluentibus compositas non continet, multiplicetur omnis ejus terminus per indicem dignitatis quantitatis cujusq; fluentis quam involvit, et in singulis multiplicationibus mutetur dignitatis latus(3) in suam fluxionem et aggregatum factorum omnium sub proprijs signis erit æquatio nova.

Sunto a, b, c, d &c quantitates determinatæ et immutabiles; z, y, x, v &c quantitates fluentes, id est indeterminatæ ac perpetuo motu crescentes vel decrescentes; $\dot{z}$, $\dot{y}$, $\dot{x}$, $\dot{v}$ &c earum fluxiones seu velocitates crescendi vel decrescendi; o quantitas infinite parva et $o\dot{z}$, $o\dot{y}$, $o\dot{x}$, $o\dot{v}$ &c quantitatum momenta id est incrementa vel decrementa momentanea synchrona. Et si proponatur æquatio quævis rationalis quantitates fluentes involvens uti

$$x^3 - xyy + aav - b^3 = 0,$$

multiplicentur termini primò per indices dignitatum x et in singulis multiplicationibus pro dignitatis latere, seu x unius dimensionis, scribatur $\dot{x}$, et summa factorum erit $3\dot{x}x^2 - \dot{x}yy$. Idem fiat in y et prodibit $-2x\dot{y}y$. Idem fiat in v et prodibit $+aa\dot{v}$. Ponatur summa factorum $=0$ et habebitur æquatio $3\dot{x}x^2 - \dot{x}yy - 2x\dot{y}y + aa\dot{v} = 0$. Dico quod hac æquatione definitur relatio fluxionum.

Nam si quantitates fluentes jam sunt x, y, et v, hæ post momentum temporis incrementis infinitè parvis $o\dot{x}$, $o\dot{y}$, $o\dot{v}$ auctæ, evadent $x+o\dot{x}$, $y+o\dot{y}$, $v+o\dot{v}$ quæ in æquatione prima pro x y et v scriptæ dant æquationem

$$x^3 + 3x^2o\dot{x} + 3xoo\dot{x}^2 + o^3\dot{x}^3 - xyy - o\dot{x}yy - 2xo\dot{y}y - 2\dot{x}oo\dot{y}y$$
$$[-]xoo\dot{y}\dot{y}[-]\dot{x}o^3\dot{y}\dot{y} + aav + aao\dot{v} - b^3 = 0.$$

(1) A selection from the much greater number of initial castings and rough scrap-sheet computations which are preserved both in Cambridge University Library and in private possession. Most of these are too fragmentary and their content too repetitive and jejune to be worthy of more than a summary note which cites their location and specifies their minimal place in the grand scheme of composition. From this heterogeneous mass of preliminary jottings we here extract certain more connected passages which, if of essentially transitory interest, afford a glimpse of the several more significant variations of argument discarded by Newton *en route* to the revised tract 'De Quadratura Curvarum' which he came (in late 1692?) to set in the preceding finale to his reworked treatise on geometry.

(2) Add. 3962.3: 62r/63r, a minimal reshaping of Proposition IV of the 1691 'De quadratura Curvarum' (1, §2 preceding) which is further restyled and considerably extended in Newton's final version on pages 512–16 above.

(3) Originally 'quantitas illa unius dimensionis'.

Subducatur æquatio prior et residuum divisum per o erit

$$3\dot{x}x^2+3\dot{x}\dot{x}ox+\dot{x}^3oo-\dot{x}yy-2x\dot{y}y-2\dot{x}o\dot{y}y[-]x[o\dot{y}\dot{y}-\dot{x}]oo\dot{y}\dot{y}+aa\dot{v}=0.$$

Minuatur quantitas o in infinitum ut momenta fiant infinitissimè parva & neglectis terminis evanescentibus restabit $3\dot{x}x^2-\dot{x}yy-2x\dot{y}y+aa\dot{v}=0$. Q.E.D.

Cas. 2. Si æquatio quantitates surdas(4) ex fluentibus compositas involvit, pro quantitatibus istis scribendæ sunt novæ quantitates fluentes [,] dein operandum ut in casu priore. Ut si æquatio sit $x^3-xyy+aa\sqrt{ax-yy}-b^3=0$, pone $\sqrt{ax-yy}=v$ et habebuntur duæ æquationes

$$x^3-xyy+aav-b^3=0 \quad \text{et} \quad ax-yy=vv,$$

et inde ut in casu priore prodeunt fluxionum relationes

$$3\dot{x}x^2-\dot{x}yy-2xy\dot{y}+aa\dot{v}=0,$$

et $a\dot{x}-2y\dot{y}=2v\dot{v}$. Dein si per æquationem posteriorem quæratur valor fluxionis $\dot{v}$ nempe $\dfrac{a\dot{x}-2y\dot{y}}{2v}$ seu $\dfrac{a\dot{x}-2y\dot{y}}{2\sqrt{ax-yy}}$ et scribatur idem pro $\dot{v}$ in priore: prodibit æquatio quæsita $3\dot{x}x^2-\dot{x}yy-2xy\dot{y}+\dfrac{a^3\dot{x}-2aay\dot{y}}{2\sqrt{ax-yy}}=0.$(5)

---

[2](6) [Si pro $e+fz^\eta+gz^{2\eta}+hz^{3\eta}$ &c et $k+lz^\eta+mz^{2\eta}+nz^{3\eta}$ &c scribantur R et S respectivè, sint autem z abscissa et $z^{\theta-1}R^{\lambda-1}S^{\mu-1}$ in $\overline{a+bz^\eta+cz^{2\eta}+dz^{3\eta}\ \&c}$ ordinata curvæ cujus area est $z^\theta R^\lambda S^\mu$ in $\overline{A+Bz^\eta+Cz^{2\eta}+Dz^{3\eta}\ \&c}$: erit(7)]

a	$+bz^\eta$	$+cz^{2\eta}$	$+dz^{3\eta}$	[&c]
θekA	$\begin{matrix}+\theta\\+\lambda\eta\end{matrix} fkAz^\eta$	$\begin{matrix}+\theta\\+2\lambda\eta\end{matrix} gkAz^{2\eta}$	$\begin{matrix}+\theta\\+3\lambda\eta\end{matrix} hkAz^{3\eta}$	[&c]
	$\begin{matrix}+\theta\\+\mu\eta\end{matrix} elA$	$\begin{matrix}+\theta\\+\lambda\eta\end{matrix} flA$	$\begin{matrix}+\theta\\+2\lambda\eta\end{matrix} glA$	
	$+\overline{\theta+\eta}, ekBz^\eta$	$+\mu\eta$	$+\mu\eta$	

(4) Newton here replaces an initial equivalent specification 'irrationales' probably more for stylistic effect than to insist that the quantities are—as, of course, they need not be—quadratic irrationals. In Proposition IV of his 1691 'De quadratura Curvarum' he had more summarily proposed an 'æquatio irrationalis'.

(5) In immediate extension of his present construction of the first fluxions of given 'compound fluents', in his revised Proposition I on pages 512–16 above Newton here passes analogously to derive their second, third and higher fluxions, 'ubi vero...convenit quantitatem aliquam ut uniformiter fluentem considerare & pro ejus fluxione prima unitatem scribere, pro secunda et sequentibus nihil'.

(6) Add. 3962.3: 47v a fuller preliminary outline of Proposition VI on page 528 above in which the basis for Newton's allocation of the coefficients in his series expansion of the quadrature of a curve whose ordinate is $z^\theta R^\lambda S^\mu$, $R=e+fz^\eta+gz^{2\eta}\ldots$ and $S=k+lz^\eta+mz^{2\eta}\ldots$, more clearly appears; compare note (8) following.

(7) By Proposition IV of the preceding 'Liber secundus' (page 518 above).

$$\substack{+\theta\\+2\mu\eta}emA\,z^{2\eta}\quad \substack{+\theta+\eta\\+\lambda\eta}fkB\quad \substack{+\theta[+\eta]\\+\mu\eta}elB\quad \overline{+\theta+2\eta},\,ekC$$

$$\substack{+\theta\\+\lambda\eta\\+2\mu\eta}fmA[z^{3\eta}]\quad \substack{+\theta+\eta\\+2\lambda\eta}gkB\quad \substack{+\theta+\eta\\+\lambda\eta\\+\mu\eta}flB\quad \substack{+\theta+\eta\\+\mu\eta}emB\quad \substack{+\theta+2\eta\\+\lambda\eta}fkC\quad \substack{+\theta[+2\eta]\\+\mu\eta}elC\quad \overline{+\theta+3\eta},\,ekD$$

[Fit] $\dfrac{a}{\theta ek}=A.\qquad \dfrac{b\ \substack{-\theta\\-\lambda\eta}fkA\ \substack{-\theta\\-\mu\eta}elA}{\overline{\theta+\eta},\,ek}=B.$

$$\frac{c\ \substack{-\theta\\-2\lambda\eta}gkA\ \substack{-\theta\\-\lambda\eta\\-\mu\eta}flA\ \substack{-\theta\\-2\mu\eta}emA\ \dfrac{\overline{-\theta-\eta-\lambda\eta},fkB}{\overline{-\theta[-\eta]-\mu\eta},elB}}{\overline{\theta+2\eta},\,ek}=C.\ [\&c.]$$

[Vel concinnius si]

ex coefficientibus conflentur rectangula				et ex dignitatum indicibus hi rectangulorum coefficientes numerales			
ek	*fk*	*gk*	*hk* [&c]	$\frac{1}{\eta}\theta=r.$	$r+\lambda=s.$	$s+\lambda=t.$	$t+\lambda=v.$ [&c]
el	*fl*	*gl*	*hl*	$r+\mu=s'.$	$s+\mu=t'.$	$t+\mu=v'.$	$v+\mu=w'.$
em	*fm*	*gm*	*hm*	$s'+\mu=t''.$	$t'+\mu=v''.$	$v'+\mu=w''.$	$w''+\mu=x''.$
en	*fn*	*gn*	*hn*	$t''+\mu=v'''.$	$v''+\mu=w'''.$	$w''+\mu=x'''.$	$x''+\mu=y'''.$(8)

(8) Whence there ensues in succession

$$\begin{aligned}
a/\eta &= rekA,\\
b/\eta &= (sfk+s'el)\,A+(r+1)\,ekB,\\
c/\eta &= (tgk+t'fl+t''em)\,A+((s+1)\,fk+(s'+1)\,el)\,B+(r+2)\,ekC,\\
d/\eta &= (vhk+v'gl+v''fm+v'''en)\,A+((t+1)\,gk+(t'+1)\,fl+(t''+1)\,em)\,B\\
&\qquad +((s+2)\,fk+(s'+2)\,el)\,C+(r+3)\,ekD,
\end{aligned}$$

and so on. The sequel is immediate.

[erit area curvæ $= z^{\theta}R^{\lambda}S^{\mu}$ in]

$$\frac{\frac{1}{\eta}a}{rek} + \frac{\frac{1}{\eta}b \begin{matrix}-sfk\\-s'el\end{matrix}A}{r[+1], ek}z^{\eta} + \frac{\frac{1}{\eta}c \begin{matrix}-tgk\\-t'fl\\-t''em\end{matrix}A \begin{matrix}-\overline{s+1}, fk\\-\overline{s'+1}, el\end{matrix}B}{\overline{r+2}, ek}z^{2\eta}$$

$$-\frac{\frac{1}{\eta}d \begin{matrix}-vhk\\-v'gl\\-v''fm\\-v'''en\end{matrix}A \begin{matrix}-\overline{t+1}, gk\\-\overline{t'+1}, fl\\\overline{t''+1}, em\end{matrix}B \begin{matrix}-\overline{s+2}, fk\\-\overline{s'+2}, el\end{matrix}C}{\overline{r+3}, ek}z^{3\eta} \text{ [\&c].}$$

[3](9) *Corol. 5.* Et curva omnis cujus ordinata y definitur [ex ordinata z] per æquationem quamvis affectā triū terminorum $z^{\alpha} = ez^{\beta}y^{\gamma} + fy^{\delta}$ assumendo $\pi = \dfrac{\delta-\gamma}{\beta-\gamma+\delta}$ et ponendo $x^{\pi} = z$ & $\lambda = \dfrac{\gamma-\beta-\delta}{\alpha\gamma+\beta\delta-\alpha\delta}$ migrat in aliam sibi æqualem cujus abscissa x est $\overline{e\pi^{-\gamma}v^{\gamma}+f\pi^{-\delta}v^{\delta}}|^{\lambda}$ existente ordinata v.(10)

Corol. 6. Et si relatio inter curvæ alicujus ordinatam [y] et abscissam [z] definiatur per æquationem quamvis affectam

$$e+fy^{\alpha}z^{\beta}+gy^{2\alpha}z^{2\beta}+\&c = [y^{\gamma}z^{\delta}] \text{ in } k+ly^{\alpha}z^{\beta}+my^{2\alpha}z^{2\beta}+\&c$$

quæ in duas partes distingui potest in quibus seorsim spectatis indices dignitatum ordinatæ et abscissæ secundum eundem terminorum ordinem sunt in progressionibus arithmeticis, ac progressiones in una æquationis parte sunt eædem cum progressionibus in altera; hæc figura curvilinea assumendo $\eta = \dfrac{\alpha}{\alpha-\beta}$ et ponendo $x^{\eta} = z$ et $\lambda = \dfrac{1}{\gamma+\eta\delta-\eta\gamma}$ migrat in aliam sibi æqualem cujus abscissa [x] ex data ordinata [y] determinatur per æquationem non affectam

$$\eta^{\gamma\lambda}v^{-\gamma\lambda} \times \overline{e+f\eta^{-\alpha}v^{\alpha}+g\eta^{-2\alpha}v^{2\alpha}[+\&c]}|^{\lambda} \times \overline{k+l\eta^{-\alpha}v^{\alpha}+m\eta^{-2\alpha}v^{2\alpha}+\&c}|^{-\lambda} = x.\text{(11)}$$

(9) Add. 3962.3: 51v, two superseded corollaries enunciating a particular case and an initial general version of Corollary 7 to Proposition IX on page 540 above (see §2: note (73) preceding).

(10) The given relationship $z = x^{\pi}$ and the posited equality of areas $\int y.dz = \int v.dx$ (whence $y\dot{z} = v\dot{x}$) together determine $y = (v\dot{x}/\dot{z}$ or) $\pi^{-1}vx^{1-\pi}$, so that the derived defining equation is $x^{\alpha\pi} = e\pi^{-\gamma}x^{\beta\pi+\gamma(1-\pi)}v^{\delta} + f\pi^{-\delta}x^{\delta(1-\pi)}v^{\delta}$, which on setting $\beta\pi+\gamma(1-\pi) = \delta(1-\pi)$ or $\pi = (-\gamma+\delta)/(\beta-\gamma+\delta)$ and dividing through by $x^{\delta(1-\pi)} \equiv x^{\beta\delta/(\beta-\gamma+\delta)}$ reduces to be

$$x^{1/\lambda} = e\pi^{-\gamma}v^{\gamma} + f\pi^{-\delta}v^{\delta}$$

where $1/\lambda = (-\alpha\gamma+\alpha\delta-\beta\delta)/(\beta-\gamma+\delta)$.

(11) Much as before, the given relationship $z = x^{\eta}$ between the abscissas determines there to be correspondingly $y = (v\dot{x}/\dot{z}$ or) $\eta^{-1}vx^{1-\eta}$, whence on substituting these and putting $\alpha(1-\eta)+\beta\eta = 0$ or $\eta = \alpha/(\alpha-\beta)$ there ensues $y^{\alpha}z^{\beta} = \eta^{-\alpha}v^{\alpha}$ and also $y^{\gamma}z^{\delta} = \eta^{-\gamma}v^{\gamma}x^{1/\lambda}$ where $1/\lambda = \gamma(1-\eta)+\delta\eta$.

[4][12] [Sit] $y^{\alpha}\times\overline{e+fy^{\eta}z^{\delta}+\&c}=z^{\beta}\times\overline{k+ly^{\eta}z^{\delta}+\&c}$. [Pone] $x=z^{s}$.[13] [erit] $\dot{x}=s\dot{z}z^{s-1}$. [adeoq3] $\frac{\dot{x}}{\dot{z}}=\frac{y}{v}=sz^{s-1}$. [hoc est] $z=x^{\frac{1}{s}}$. [et] $y=svx^{\frac{s-1}{s}}$.

[Prodit ergo] $s^{\alpha}v^{\alpha}x^{\frac{s\alpha-\alpha}{s}}\times\overline{e+fs^{\eta}v^{\eta}x^{\frac{s\eta-\eta+\delta}{s}}+\&c}=x^{\frac{\beta}{s}}\times\overline{k+ls^{\eta}v^{\eta}+\&c}$.

[Fac $s\eta-\eta+\delta=0$ sive] $s=\frac{\eta-\delta}{\eta}$. [fitq3]

$$s^{\alpha}v^{\alpha}\times\overline{e+fs^{\eta}v^{\eta}+\&c}=x^{\frac{\alpha+\beta-s\alpha}{s}}\times\overline{k+ls^{\eta}v^{\eta}+\&c}.$$

[unde capiendo $\frac{\alpha+\beta-s\alpha}{s}=\Big]\frac{\beta\eta+\delta\alpha}{\eta-\delta}=\frac{1}{\lambda}$. [evadit æquatio]

$$s^{\alpha\lambda}v^{\alpha\lambda}\times\overline{e+fs^{\eta}v^{\eta}+\&c}\big|^{\lambda}\times\overline{k+ls^{\eta}v^{\eta}+\&c}\big|^{-\lambda}=x.]$$

---

[5][14] [Sit] $x^{\theta}\times\overline{a+bx^{\sigma}}\big|^{\mu}=v$. $\overline{ez^{\nu}+fz^{\nu+\eta}+gz^{\nu+2\eta}}\big|^{\pi}=x$. [erit]

$$\overline{vez^{\nu-1}\begin{smallmatrix}+\nu\\+\eta\end{smallmatrix}fz^{\nu+\eta-1}\begin{smallmatrix}+\nu\\+2\eta\end{smallmatrix}gz^{\nu+2\eta-1}}\times\dot{z}=\frac{1}{\pi}x^{\frac{1-\pi}{\pi}}\times\dot{x}.$$

[seu] $\frac{\dot{x}}{\dot{z}}=\frac{y}{v}=\pi\times\overline{ez^{\nu}+fz^{\nu+\eta}+gz^{\nu+2\eta}}\big|^{\pi-1}\times\overline{vez^{\nu-1}\begin{smallmatrix}+\nu\\+\eta\end{smallmatrix}fz^{\nu+\eta-1}\begin{smallmatrix}+\nu\\+2\eta\end{smallmatrix}gz^{\nu+2\eta-1}}$.

[Unde]

$$\pi z^{\nu\pi\theta+\nu\pi-1}\times\overline{ve\begin{smallmatrix}+\nu\\+\eta\end{smallmatrix}fz^{\eta}\begin{smallmatrix}+\nu\\+2\eta\end{smallmatrix}gz^{2\eta}}\times\overline{e+fz^{\eta}+gz^{2\eta}}\big|^{\pi\theta+\pi-1}$$
$$\times\overline{a+b\times\overline{ez^{\nu}+fz^{\nu+\pi}+gz^{\nu+2\pi}}\big|^{\pi\sigma}}\big|^{\mu}=y.$$

---

(12) Add. 3962.3:52^r, Newton's rough preparatory calculations for Corollary 7 to Proposition IX on page 540 above (compare §2: note (74) preceding). Their result was slightly simplified in his revised version by the assumption of an amended relationship between the abscissas (see the next note) but the effect of this is merely to alter the form of certain of the constants in the final equation. We have, following our usual editorial practice, introduced a few verbal phrases and connectives to make clear the sequence of Newton's argument.

(13) In his revised corollary Newton here set there to be $x=(1/s)\,z^{s}$, whence $z=(sx)^{1/s}$ and consequently $y=(v\dot{x}/\dot{z}$ or$)\ v(sx)^{1-1/s}$, so that on again making $s=1-\delta/\eta$ there resulted $y^{\eta}z^{\delta}=v^{\eta}$ simply; see page 540 above.

(14) Add. 3962.3: 44^v/42^r/44^r; edited extracts from Newton's preliminary computations for Corollaries 9 and 10 to Proposition IX on page 542 above, followed by his working up of these into two initial verbal drafts of them, here (see §2: note (75) preceding) numbered Corollaries 10 and 11 in sequel to a suppressed previous Corollary 9 which is but a special case of the former. Since the preparatory texts here reproduced— unlike their terse and over-polished final version!—clearly exhibit the steps of his argument, we shall in commentary pick up only a couple of minor lapses in their mathematical detail, otherwise allowing Newton to speak wholly for himself.

[Pone] $\nu\pi\theta+\nu\pi-1=\vartheta$. $\pi\theta+\pi-1=\lambda$. [hoc est] $\vartheta-\nu\lambda=\nu-1$. [sive] $\nu=\dfrac{\vartheta+1}{\lambda+1}$. [ut et] $\pi\sigma=\tau$. [unde] $\nu\lambda+\nu-1=\vartheta$. $\dfrac{\lambda+1}{\theta+1}=\pi=\dfrac{\vartheta+1}{\nu\theta+\nu}$. $\dfrac{\tau}{\sigma}=\pi$. $\dfrac{\lambda+1-\pi}{\pi}=\theta$. [Erit]

$$\pi\times z^{\vartheta}\times\overline{\nu e\begin{matrix}+\nu\\+\eta\end{matrix}fz^{\eta}\begin{matrix}+\nu\\+2\eta\end{matrix}gz^{2\eta}}\times\overline{e+fz^{\eta}+gz^{2\eta}}|^{\lambda}\times\overline{a+b\times\overline{ez^{\nu}+fz^{\nu+\eta}+gz^{\nu+2\eta}}|^{\tau}}|^{\mu}=y.$$

[vel aliter]

$$\pi\times z^{\vartheta[+\pi\sigma\mu\nu]}\times\overline{\nu e\begin{matrix}+\nu\\+\eta\end{matrix}fz^{\eta}\begin{matrix}+\nu\\+2\eta\end{matrix}gz^{2\eta}}\times\overline{e+fz^{\eta}+gz^{2\eta}}|^{\lambda+\pi\sigma\mu}$$
$$\times\overline{b+a\times\overline{ez^{\nu}+fz^{\nu+\eta}+gz^{\nu+2\eta}}|^{-\pi\sigma}}|^{\mu}=y.$$

[Pone] $ez^{\nu}+fz^{\nu+\eta}+gz^{\nu+2\eta}=R$. $k+lz^{\eta}+mz^{2\eta}=S$. [et fac] $R^{\pi}S^{\rho}=x$. [unde] $\overline{\pi S\dot{R}+\rho R\dot{S}}\times R^{\pi-1}S^{\rho-1}=\dot{x}$. [Sit] $x^{\theta}\,\overline{a+bx^{\sigma}}|^{\lambda}=v$. [Quia $v\dot{x}=y\dot{z}$, erit]

$$R^{\pi\theta+\pi-1}S^{\rho\theta+\rho-1}\,\overline{a+bR^{\pi\sigma}S^{\rho\sigma}}|^{\mu}\times\overline{\pi S\dot{R}+\rho R\dot{S}}\ [=y].$$

[hoc est] $R^{\pi\theta+\pi-1}S^{\rho\theta+\rho-1-\rho\sigma\mu}\,\overline{aS^{-\rho\sigma}+bR^{\pi\sigma}}|^{\mu}\times\overline{\pi S\dot{R}+\rho R\dot{S}}\,[=y]$.

[Fac] $\lambda=\pi\theta+\pi$. [sive] $\dfrac{1}{\lambda-\pi}=\theta$. [ut et] $\rho=\dfrac{1}{\theta-\sigma\mu+1}$. $\lambda+1=\mu$. [Erit]

$$\overline{\pi\dot{R}S+\rho R\dot{S}}\times R^{\lambda-1}\times\overline{aS^{[-\rho\sigma]}+bR^{[\pi\sigma]}}|^{\lambda+1}\,[=y].$$

Corol. 10. Et curva omnis cujus ordinata [y] est

$$z^{\nu\lambda+\nu-1}\times\overline{\nu e\begin{matrix}+\nu\\+\eta\end{matrix}fz^{\eta}\begin{matrix}+\nu\\+2\eta\end{matrix}gz^{2\eta}+\&c}\times\overline{e+fz^{\eta}+gz^{2\eta}+\&c}|^{\lambda}$$
$$\times\overline{a+bez^{\nu}+bfz^{\nu+\eta}+bgz^{\nu+2\eta}+\&c}|^{\mu}$$

ponendo $\dfrac{1}{\lambda+1}=\sigma$ et $\overline{ez^{\nu}+fz^{\nu+\eta}+gz^{\nu+2\eta}+\&c}|^{\lambda+1}=x$ migrat in curvam aliam sibi æqualem cujus ordinata [v] est $\sigma\times\overline{a+bx^{\sigma}}|^{\mu}$.

Corol. 11. Et curva omnis cujus ordinata est

$$z^{\nu\lambda+\nu-1}\times\overline{\nu e\begin{matrix}+\nu\\+\eta\end{matrix}fz^{\eta}\begin{matrix}+\nu\\+2\eta\end{matrix}gz^{2\eta}+\&c}\times\overline{e+fz^{\eta}+gz^{2\eta}+\&c}|^{\lambda}$$
$$\times\overline{b+aez^{\nu}+afz^{\nu+\eta}+agz^{\nu+2\eta}+\&c}|^{\mu}$$

ponendo $\dfrac{1}{\lambda+\mu+1}=\sigma$ et $\overline{ez^{\nu}+fz^{\nu+\eta}+gz^{\nu+2\eta}+\&c}|^{\lambda+\mu+1}=x$ migrat in curvam aliam sibi æqualem cujus ordinata est $\sigma\times\overline{a+bx^{-\sigma}}|^{\mu}$.

Et nota quod ordinatæ in Corollario decimo et undecimo migrant in simpliciores efficiendo vel ut radix dignitatis extrahi possit cujus index est μ vel ut indices λ et μ ijdem sint & per multiplicationem dignitatum coalescant.

Corol. 10. Et curva omnis cujus ordinata est

$$z^{\theta}\times\overline{ve\begin{smallmatrix}+\nu\\+\eta\end{smallmatrix}fz^{\eta}\begin{smallmatrix}+\nu\\+2\eta\end{smallmatrix}gz^{2\eta}+\&c}\times\overline{|e+fz^{\eta}+gz^{2\eta}+\&c|}^{\lambda-1}$$
$$\times\overline{|a+b\overline{|ez^{\nu}+fz^{\nu+\eta}+gz^{\nu+2\eta}+\&c|}^{\tau}|}^{\mu},$$

si sint $\theta=\nu\lambda-1$, $x=\overline{ez^{\nu}+fz^{\nu+\eta}+gz^{\nu+2\eta}+\&c|}^{\pi}$, $\dfrac{\tau}{\pi}=\sigma$, $\dfrac{\lambda-\pi}{\pi}=\vartheta$ migrat in aliam sibi æqualem cujus ordinata est $\dfrac{1}{\pi}x^{\vartheta}\overline{|a+bx^{\sigma}|}^{\mu}$(15).

Et nota quod ordinata prior in hoc Corollario proposita fit simplicior ponendo $\lambda=0$, vel $\tau=1$, vel insuper etiam efficiendo ut radix dignitatis extrahi possit cujus index est μ. Ut si sit $\nu=0$, $\tau=1$ et $a+be=\dfrac{4bff}{g}$(16), vel si sit $\nu=-\eta$, $\tau=1$ et $a+bf=\pm 2beg$(16) vel de[n]iqꝫ [s]i sit $\nu=-2\eta$, $\tau=1$ et $a+bg=\dfrac{4bff}{e}$(16), extrahetur radix quadratica.

Corol. 11. Pro $ez^{\nu}+fz^{\nu+\eta}+gz^{\nu+2\eta}+\&c$, $vez^{\nu-1}\begin{smallmatrix}+\nu\\+\eta\end{smallmatrix}fz^{\nu+\eta-1}\begin{smallmatrix}+\nu\\+2\eta\end{smallmatrix}gz^{\nu+2\eta-1}+\&c$, $k+lz^{\eta}+mz^{2\eta}+\&c$, et $\eta lz^{\eta-1}+2\eta mz^{2\eta-1}+\&c$ scribantur R, $\dot{R}$, S & $\dot{S}$ respective: et Curva omnis cujus ordinata est $\overline{\pi\dot{R}S+\phi R\dot{S}}$(17) in $R^{\lambda-1}S^{\mu-1}\times\overline{aS^{v}+bR^{\tau}|}^{\xi}$, si sit $\dfrac{\mu-v\xi}{\lambda}$(17) $=\dfrac{v}{\tau}$ & assumatur σ et ponatur $\dfrac{v}{\sigma}=\phi$. $\dfrac{\tau}{\sigma}=\pi$. $\dfrac{\lambda-\pi}{\pi}=\vartheta$ et $\dfrac{R^{\pi}}{S^{\phi}}=x$[,]

(15) In this form Newton began to copy the corollary without further variation into his main text but broke off, only halfway through doing so, first to remodel it into the equivalent enunciation which we have quoted in §2: note (76) preceding, and then straightaway further to convert this into the minimally refashioned 'Corol. 9' as it appears on page 542 above.

(16) Read '$\dfrac{bff}{4g}$', '$\pm 2b\sqrt{eg}$' and '$\dfrac{bff}{4e}$' respectively. In each case Newton not only sets $\tau=1$ but implicitly assumes that the last factor in the expression for the ordinate is rounded off to be the μ-th power of the quadrinomial $a+bez^{\nu}+bfz^{\nu+\eta}+bgz^{\nu+2\eta}$, further specifying the three values of ν for which this reduces to be a trinomial in powers of z^{η} and then stating the corresponding condition between the coefficients which determines this trinomial to be the exact square of a binomial.

(17) On correctly replacing ϕ by $-\phi$ in the preceding calculation these will be '$\overline{\pi\dot{R}S-\phi R\dot{S}}$' and '$\dfrac{\mu-v\xi}{-\lambda}$' respectively;

migrat in aliā sibi æqu[alem] cujus ord[inata] est $x^{\vartheta}\overline{|a+bx^{\sigma}|}^{\xi}$. Et nota quod Ordinata prior evadit simplicior ponendo unitates pro v, τ, et λ vel μ et faciendo ut radix dignitatis extrahi possit cujus index est ξ.

---

[6][(18)] *Lemma.* Quantitas V ducitur in factum[(19)] sub data quanti[tat]e a & quantitatis R fluxione $\dot{r}$, et quantitatis V fluxio $\dot{v}$ ducitur in quantitatem R, et contenti utriusq; habetur summa vel differentia P[(20)] una cum quantitate R: requiritur quantitas V.

Finge primum seu altissimum terminū quantitatis quæsitæ V esse $cx^m y^n$, et ejus fluxio erit $mc\dot{x}x^{m-1}+ncx^m\dot{y}y^{n-1}$. Duc hanc flux[ionem] in R et terminum $[a]cx^m y^n$ in fluxionem $\dot{r}$ et fac ut contentorum summæ vel differentiæ termini altissimi vel congruant cum terminis altissimis quantitatis P, vel se mutuo destruant[,] et si se mutuo destruunt fac ut termini altissimi eorum qui relinquuntur congruant cum terminis altissimis quantitatis P. Hoc modo invenietur terminus $cx^m y^n$.[(21)]

---

(18) Add. 3962.3: 43ᵛ, a stray pair of paragraphs (entered beneath a preliminary drafting of Case 1 of Proposition X of the 'Geometriæ Liber secundus') in which Newton, vaguely in the spirit of Proposition XII of his earlier 'De quadratura Curvarum' (1, §2 preceding; see especially pages 92–4 above), roughs out a crude *ad hoc* technique for identifying the 'altissimus terminus' in a required descending series-expansion of the root of a given first-order fluxional equation. It was doubtless (as we have supposed in §2: note (97) preceding) briefly his intention—one equally swiftly abandoned, if so—to adjoin to his main text as we now have it some account of the novelties of method and technique which he had presented in the later propositions of his previous tract on the quadrature of curves. We know no more finished version of this lone 'Lemma', and its content would seem never to have been communicated by Newton to his contemporaries. In here reproducing its text we ignore a couple of wholly minor, cancelled first turns of phrase which it will add nothing here to repeat.

(19) Newton here trivially replaces an equivalent 'contentum'.

(20) In other words it is assumed that $P = R\dot{v} \pm a\dot{r}V$, where $\dot{r}$ and $\dot{v}$ are the fluxions of R and V. Newton in sequel supposes that $V = cx^m y^n + dx^p y^q + \ldots$, $m \geqslant p$, $n \geqslant q$, and presumably takes the quantities P and R to have similar (given) polynomial expansions in terms of the variables x and y.

(21) Analogously, of course, a similar application of the procedure will determine the further, successively 'less higher' terms in the expansion of V. One may even hope that in given particular cases the process will terminate after a finite number of steps or yield an infinite polynomial series in x and y which suitably converges! But the method is at once extremely cumbersome and highly laborious to perform in all but a few well-chosen particular examples, and Newton, having reached this point in his unwieldy explanation, was doubtless happy to forego any further exploration of its subtleties.

PART 3

CARTESIAN ANALYSIS OF HIGHER ALGEBRAIC CURVES AND FINITE-DIFFERENCE APPROXIMATIONS

(*c.* summer 1695)

INTRODUCTION

In this miscellaneous gathering of what are otherwise terse snippets and unexplained calculations recording the train of Newton's mathematical thoughts in the quiet of his last year at Cambridge pride of place is rightly taken by the only considerable treatise which he then wrote, even though this gave only a final hone and polish to the many-faceted researches into the classification and enumeration of cubic curves and his accompanying analytical explorations of the general properties of those of higher algebraic degree which he had conducted on and off over the previous thirty years.(1)

The text of this definitive 'Enumeratio Linearum Tertii Ordinis'(2) has been available in a variety of published forms ever since Newton himself, little more than eight years after he wrote it, sent it to be printed as the first of the 'Two Treatises of the Species and Magnitude of Curvilinear Figures' appended by him in 1704 to the *editio princeps* of his *Opticks*.(3) Here we need give no elaborate breakdown of its technical content. Not merely does he fulfil his title by presenting (in its central sections VIII–XXVIII) a more straightforwardly geometrical refashioning of his earlier analytical subdivisions of the general cubic curve into his preferred scheme of nine principal 'cases', sixteen component *genera* and—here(4)—72 individual species which ensue on considering every

---

(1) See I: 155–212; II: 10–88, 498–508; IV: 346–405.

(2) Newton arrived at this final title, fixed irrevocably in print in 1704, only after much hesitation and resuming of previously superseded alternatives; see 1, §2: note (3) below.

(3) *Opticks: Or, a Treatise of the Reflexions, Refractions, Inflexions and Colours of Light* (London, $_1$1704): $_2$138–62. The surviving autograph fair copy (now ULC. Add. 3961.2: 1^r–14^r) which Newton sent to the press in late 1703—to be subsequently returned to him, shorn of its figures, when it had fulfilled its primary purpose—is the text of the 'Enumeratio' which, with some slight adjustment of its detail, we reproduce in 1, §2 following; our version also, in accompanying footnotes and appendices, takes account of a number of variants in an immediately preliminary draft (ULC. Add. 3962.1: 38^r–50^r) which *inter alia* does not (see 1, §2: note (2)) distinguish between the final species 14/16, 15/17 and 28/29 of cubic curve, and we further cite four important, hitherto unpublished emendations (see 1, §2: notes (25), (28), (49) and (101)) which Newton afterwards entered in his library copy (now ULC. Adv. b. 39.4) of the ensuing Latin *Optice: sive de Reflexionibus, Refractionibus, Inflexionibus & Coloribus Lucis Libri Tres* to which the unchanged texts of the 'Two Treatises' were (newly reset and with separate paginations) again appended. His introductory tabulation (ULC. Add. 3961.4: 15^r) of the 69[+3] species of cubic which he there enumerates, and which should afford a useful guide to their broad plan, is reproduced in 1, §1. These manuscript originals are, we should add, described—not entirely adequately—by W. W. Rouse Ball in his pioneering essay 'On Newton's Classification of Cubic Curves' (*Proceedings of the London Mathematical Society*, **22**, 1891: 104–43): 128–9.

(4) The six further species which escape Newton's present net—namely (see 1, §2: notes (67), (72) and (84) below) the two monodiametral and two tridiametral redundant hyperbolas

possible (non-degenerate) variation in the four reduced canonical forms of its Cartesian defining equation. This detailed tabulation he now prefaces with an introductory paragraph defining his novel notion of the dimensional 'order' of an algebraic curve, followed by (sections II–VII) a condensed account[5] of the principal characteristics and some common properties of all such 'geometrical' curves; and in sequel to his enumeration of cubics passes first (section XXIX) succinctly to assert the possibility of generating each of his listed species as a plane perspective 'shadow' of one or other of the five (projectively distinct) divergent parabolas,[6] thereafter (sections XXXI–XXXIII) more fulsomely[7] to expound his favourite 'organic' construction—by fixed angles rotating round their vertices as poles—of the general conic from a straight line and of cubics possessed of a double point from a conic through the pole at which this is located, and lastly (section XXXIV) to sketch how the Wallisian cubic parabola and the diametral hyperbolic parabolism can each be employed geometrically to construct by their intersections with lines, conics and other cubics the real roots of algebraic equations (reduced as these need be) of up to ninth degree.[8] A surfeit of richnesses indeed. When faced for the first time with this boiled-down purée of Newton's lifetime of discovery in the pure and analytical geometry of curves, one is filled with admiration for the range and deftness of its techniques and minimally demonstrated arguments, and also less than confident in one's capacity to digest its drily bottled intricacies. And so it proved: rather than risk the griping pains which might ensue from trying to assimilate its rich essence too rapidly, Newton's contemporaries chose almost to a man to lay his tract aside after briefly tasting its supra-sweetness, and its insights were not to be absorbed into the common store of mathematical knowledge for many years.

However little they could have understood of its subtleties at so brief a glance, the favoured few among Newton's acquaintances who in the years before its publication were allowed a private preview of his manuscript tract were, we

---

and two parabolic cubics, each possessing either an oval or conjugate point, whose discovery is customarily attributed to James Stirling and the quartet of François Nicole, Niklaus Bernoulli, de Gua de Malves and Edmund Stone who later independently found them and, in 1717 and during 1731–40 respectively, publicized their existence—had, we have seen (2, 3, §1: notes (54) and (65) preceding), earlier been tabled by him in his equivalent projectively ordered listing of cubics.

(5) Digested from the more elaborate previous treatments reproduced on II: 90–104 and IV: 354–8.

(6) But without hinting that he himself had already contrived the detailed enumeration along these lines which is set out on pages 418–32 above.

(7) In summary, we now know, of his lavish exposition of the wide varieties of this 'organica descriptio curvarum' in the late 1660's; see II: 106–50.

(8) Compare I: 496–502 and II: 498–506.

need scarcely say, much impressed and eager to have it appear quickly in print.[9] The most insistent of these was perhaps David Gregory, who on being shown the text of the 'Enumeratio', in London towards the middle of July 1698, tentatively arranged with Newton to be allowed to edit it.[10] But its author, grown increasingly suspicious—not without good reason[11]—of the shallowness of Gregory's mathematical insight and judgement, was clearly reluctant to go along with this plan, and almost three years later, on 21 May 1701, the latter set down among his provisional *agenda* for a forthcoming meeting with Newton a much less optimistic (and, it would seem, in the event equally fruitless) reminder 'To see to gett at least his book *De Curvis secundi generis*'.[12] That Gregory was concerned largely to have the honour of presenting Newton's tract to the world, and in no way to sound its technical depths, is suggested by his failure anywhere in his bulky surviving memoranda of this period to pass comment, however minimal, upon its content after the 'Enumeratio' was published in early April 1704.[13] Nor is any more positive English reaction to the tract recorded during the next half dozen years; in particular, like its companion treatise 'De Quadratura Curvarum', it received no review in the contemporary pages of the then ailing *Philosophical Transactions*.[14]

(9) In issuing it to the public in 1704 Newton remarked in his prefatory 'Advertisement' to the parent *Opticks* (on its preliminary leaf without page-signature) that his appended 'small Tract concerning the Curvilinear Figures of the Second Kind...was...written many Years ago, and made known to some Friends, who have solicited the making it publick'.

(10) In a memorandum of this date Gregory noted that 'Sunt 16 genera Curvarum secundi generis, et 76[!] Curvæ. Newtonus conscripsit tractatum de illis quem mihi impertietur ut eum edam' (ULE. Gregory C62, printed in *The Correspondence of Isaac Newton*, **4**, 1967: 277).

(11) Compare v: 522, note (7).

(12) Item 9 in Royal Society MS 247.74, first printed by S. P. Rigaud in his *Historical Essay on the First Publication of Sir Isaac Newton's 'Principia'* (Oxford, 1838): Appendix: $_2$79 [= *Correspondence of Newton*, **4**: 354–5]. Since in his surrounding paragraphs Gregory similarly reminded himself 'To endeavour to gett his book of Light & Colours, & to have it transcrib'd if possible' and also 'To see if he has any design of reprinting his *Principia Mathematica* or any other thing', he retained some hope—fulfilled?—that Newton would at least show him what he had by him ready for the press.

(13) All that is found in his subsequent memoranda is a brief and, in the event, none too well-informed note on 25 March 1708 (just a few months before Gregory's death) that 'Sr Isaac...has begun to reprint his *Principia Philosophiæ* at Cambridge. He is to add his Quadrature ['De Quadratura Curvarum'] & Curves of the second genre to [it]' (Christ Church MS 346: end-sheet, printed in W. G. Hiscock, *David Gregory, Isaac Newton and their Circle. Extracts from David Gregory's Memoranda, 1677–1708* (Oxford, 1937): 41).

(14) The first reference in the Royal Society's *Transactions* to the 'Enumeratio' would seem to be in 'A ready Description and Quadrature of a Curve of the Third Order, resembling that commonly call'd the Foliate [*sc.* of Descartes]...' (*Philosophical Transactions* **29**, No. 345 [for September/October 1715]: IV: 329–31) where Abraham de Moivre, having specified the (now) familiar construction of the strophoid by points, went on to observe (*ibid.*: 331) that both this cubic and the Cartesian *folium* to which it is akin are 'comprehended under the

By and large, Continental mathematicians were equally unforthcoming in their appreciation, if a little more active in acknowledging the tract's published existence. In thanking de Moivre in mid-November 1704 for sending him Newton's gift copy of the *editio princeps* of the *Opticks*, Johann Bernoulli added dully that he would take the 'first' opportunity to read through its two appended Latin treatises and framed the conventional hope of finding 'fine things' in them.(15) A fortnight later Bernoulli wrote to Leibniz that he had 'still had no leisure to study them thoroughly'(16) and he never again mentioned the matter in his subsequent letters to London. Leibniz was then himself busy preparing his own recension of the appended 'Isaaci Newtoni Tractatus Duo, De Speciebus & Magnitudine Figurarum Curvilinearum', but when this appeared the next month—anonymously as usual—in the Leipzig *Acta* it proved to contain, as far as the 'Enumeratio' was concerned, but a colourless (if lengthy) summary of Newton's opening paragraphs, followed by a quick sketch of the main individual species of cubic which lays stress principally upon the divergent parabolas—among whose number Leibniz places the 'vulgar' Wallisian parabola!—and then, without mention of Newton's organic description of curves or his application of cubics geometrically to construct the roots of equations, abruptly concluded (as is ever the way of reviewers who seek to cloak their deep-seated ignorances beneath a superficial show of cleverness) with an irrelevant pointer

---

fortieth Kind of the Curves of the third Order, as they stand enumerated by Sir *Isaac Newton*, in his incomparable Treatise on that Subject'. De Moivre is, we may add, surely to be forgiven for not knowing that his curve had earlier been discovered by Roberval and—from the soaring shape of its hyperbolic 'wings'—christened by him 'pteroid' (see **2**, 3, §1: note (29) preceding), but he ought perhaps to have been aware that a construction of the strophoid identical to his had been published by Isaac Barrow in his *Lectiones Geometricæ* (London, ${}_1$1670): Lectio VIII, §XIX: 69 as a prelude (compare III: 249, note (51)) to determining the tangent at an arbitrary point of this (by him un-named) cubic *curva*.

(15) 'Je m'attacherai au premier loisir à lire ses deux autres traités écrits en latin sur le dénombrement des lignes du 3e ordre, et sur la quadrature des courbes; j'espere d'y trouver de belles choses...'; see K. Wollenschläger, 'Der mathematische Briefwechsel zwischen Johann I Bernoulli und Abraham de Moivre' (*Verhandlungen der Naturforschenden Gesellschaft in Basel*, **43**, 1933: 151–317): 187. In his previous sentence Bernoulli had written that 'Je suis fâché de n'entendre pas assez la langue angloise pour entendre bien son traité d'Optique'; to which de Moivre responded on 13 March 1705 that 'M. Newton...m'a chargé...de vous dire qu'on imprime à présent son traité d'Optique en latin et qu'il vous l'enverra aussitôt qu'il sera imprimé: ce qui sera dans peu de temps' (*ibid.*: 198), but without eliciting further reply from Bernoulli.

(16) 'Accepi...Newtoni opus recens editum, continens...Opticam de coloribus, Anglice scriptam...Enumerationem linearum tertii ordinis, Latine [et] Quadraturas Curvarum Geometricarum, etiam Latine; nondum mihi vacavit eas perlegere' (letter of 6 December 1704 (N.S.), first printed in [ed. G. Cramer] *G. G. Leibnitij et Johan. Bernoullij Commercium Philosophicum et Mathematicum*, **2** (Lausanne/Geneva, 1745): 123 [= C. I. Gerhardt, *Leibnizens Mathematische Schriften*, **3** (Halle, 1856): 759]).

to Newton's omission to 'touch upon' the foci of cubics and higher curves.[17] That woefully myopic vision of the 'Enumeratio' was to cloud the reactions of Leibniz' fellow mathematicians to it for many years to come. In his private response to Bernoulli shortly afterwards he was only marginally less reticent in passing any positive opinion on the technical merit of Newton's tract, though he did express his especial pleasure at its author's notion of the general rectilinear diameter of a higher curve (corresponding to any given ordinate direction).[18] That Leibniz nearly three years afterwards was, however, still no nearer to verifying the correctness and completeness of Newton's exhibited enumeration of the cubic's sub-species, nor any closer to demonstrating the ubiquitous existence of these linear diameters in curves whose unproven Newtonian rule of construction had previously taken his fancy is clear from yet another letter of his to Johann Bernoulli in which, freshly stimulated to the topic by the recent posthumous publication of L'Hospital's lavish treatise on the analytical geometry of conic sections,[19] he requested the latter's views on the accuracy of this generalized rule for diameters and on the propriety of similarly extending other fundamental properties of conics to hold *mutatis*

(17) *Acta Eruditorum* (January 1705): 30–6, especially 31–4. In illustration of Newton's 72 species of cubic Leibniz reproduces simplified drawings of the cissoid, Cartesian trident, all five divergent parabolas and the 'parabola cubica vulgaris' which is—as he would purblindly have it be, at least—'ex numero parabolarum divergentium' (see *ibid.*: 33–4); and closes with the words: 'Reliquæ figuræ, quæ cum his omnes septuaginta duas curvas secundi generis repræsentant, apud Autorem videri possunt. Cæterum Autor non attingit *focos* vel *umbilicos* curvarum secundi generis, & multo minus generum altiorum. Cum ergo ea res abstrusioris sit indaginis, & maximi tamen in hoc genere usus, tum ad descriptiones tum ad alias proprietates curvarum, & doctrina hæc focorum ab Ill[ustri] D[omi]n[o] *D[e] T[schirnhausio]* [*viz.* in Tschirnhaus' *Medicina Mentis* ([Amsterdam, $_1$1687 →] Leipzig, $_2$1695): 91–102] profundius sit versata; supplementum ejus pro his curvis ab [i]psius ingenio expectamus' (*ibid.*: 34). How a cubic—or indeed any higher algebraic curve of odd degree—can possibly have a Tschirnhausian focus he has evidently not stopped to ponder. John Keill was to single out this irrelevant concluding remark of Leibniz' when he brought the *Acta* review to Newton's notice (principally, as we shall see in our final volume, for its infinitely more wounding ensuing comments on the originality of the accompanying 'De Quadratura Curvarum') in a lately published letter of 3 April 1711 (ULC. Add. 3985.1) where he desired that the latter should 'read from pag 34 at these words. *Cæterum autor non attingit focos vel umbilicos curvarum* &c to the end'. (See also *Correspondence of Isaac Newton*, **5**, 1975: 115.)

(18) Leibniz to Bernoulli, 25 January 1705 (N.S.): 'Determinatio numeri linearum tertij gradus &, ut credo, recta est, & habenda est pro incremento non contemnendo Geometriæ; & placet imprimis quod de diametro observat ordinatam curvæ altioris secante in plures quam duas partes, servata utrimque æqualitate' (*Commercium Philosophicum*, **2**: 124 [= *Mathematische Schriften*, **3**: 761]). In preceding sentences he persevered in his published opinion that, while 'Opus Newtoni de coloribus profundum videtur', yet 'In Tractatu de Quadraturis Curvarum ordinariarum, nihil puto esse quod nobis sit valde novum aut arduum'.

(19) *Traité Analytique des Sections Coniques, et de leur Usage pour la Resolution des Equations dans les Problêmes tant déterminez qu'indéterminez* (Paris, $_1$1707).

mutandis for curves of higher degree, as Newton had done:[20] to which Bernoulli cautiously answered that the rule was 'elegant, if true' but that he 'feared' it to be 'but a mere conjecture' on Newton's part.[21] So much for the leading proponents of the new wave in European mathematics. The one person on the Continent who might fruitfully have responded to Newton's condensed insights into the theory of higher plane curves—Johann's elder brother Jakob, who had himself in the early 1690's drafted a provisional, still unpublished subdivision of the cubic 'hypersolid locus' into 33 individual types[22]—was to be dead within scarcely a year of the 'Enumeratio' appearing in public, and his reactions to Newton's tract, whatever they were (and if indeed he ever studied it), died with him.

With the passing years interest in (and concern for) the 'Enumeratio' slowly grew, though understanding of its detail lagged sadly far behind. In the summer of 1710 John Harris set his somewhat muddled and opaque English rendering of its Latin text in print in the second volume of his widely circulated *Lexicon Technicum* as its article on 'Curves'.[23] More importantly, William Jones

(20) Leibniz to Bernoulli, 12 October 1707: 'Nescio an D^{nus} Marchio Hospitalius protulerit cogitationem ad altiores gradus. Velim nosse, an Tibi aliquando vacaverit examinare, quæ de ijs D^{nus} Newtonus dedit. Si regula ejus generalis de Diametris vera est, oportet ut non difficili calculo deprehendi possit' (*Commercium Philosophicum*, **2**: 180 = *Mathematische Schriften*, **3**: 818–19).

(21) Bernoulli to Leibniz, 19 November 1707: 'Librum Hospitalii de Sectionibus Conicis percurri, sed...[non] protulit cogitationem ad curvas altiorum generum, uti Newtonus fecit in sua diatribe..., cujus regulam generalem de diametris, elegantem certe si vera est, nondum vacavit examinare; mallem autem ab ipso Newtono demonstrationem videre, vereor enim ne quod pro regula venditat, mera tantum sit conjectura' (*Mathematische Schriften*, **3**: 820).

(22) Jakob Hermann's transcript of this 'Jacobi Bernoulli Typus Locorum Hypersolidorum', with additions in the hand of Niklaus I Bernoulli, was located only in 1948 by Otto Spiess (see the 'Vorwort' to his edition of *Der Briefwechsel von Johann Bernoulli*, **1** (Basel, 1955): 23, 71) and is now in the University Library at Basel. Gabriel Cramer no doubt omitted it from his collected edition of Jakob's works two centuries before because of its numerous weaknesses and deficiencies in comparison with Newton's listing of cubics, there (*Jacobi Bernoulli Opera*, **1** (Geneva, 1744): 539 note (h), 540 note (i)) preferring to key even the two divergent parabolas —one with a node and one *cum ellipsi*—which Jakob had identified and sketched in ΕΠΙΜΕΤΡΑ II and III of his *Positiones Arithmeticæ de Seriebus Infinitis, Earumque Summa Finita* (Basel, 1692) to Newton's species 68 and 67 respectively, rather than to their author's corresponding 'types'. We have already pointed out (II: 469, note (120)) that Jakob had, independently of Newton, four years previously employed the Wallisian 'vulgare paraboloides cubicale' to construct the real roots of the general reduced cubic equation.

(23) *Lexicon Technicum. Or, An Universal Dictionary of Arts and Sciences*, **2** (London, $_1$1710): signatures M4^r–P1^r, reprinted (with its columns recut to suit a smaller page-area) in *The Mathematical Works of Isaac Newton*, **2** (New York, 1967): 137–61. Though Harris indicates in an introductory paragraph that 'The incomparable Sir Isaac Newton gives this following Enumeration of Geometrical Lines of the Third or Cubick Order...', he omits to specify that his article is an exact word-for-word English rendering of Newton's Latin text.

included a light remoulding of its *editio princeps* in the attractively produced collection of Newton's minor mathematical works which he brought out the next year.(24) This elegant version was to be the basis of all subsequent eighteenth-century editions of the tract, those in the *Opera* edited by Castiglione (1744)(25) and Samuel Horsley (1779)(26) and the separate republication by J. B. M. Duprat in 1797:(27) Horsley, indeed, was to use the very plates of figures which Jones had had engraved half a century before.(28) In his anonymous review of Jones' *Analysis* in the Leipzig *Acta* in 1712 Leibniz could, as we would expect, again find words only weakly to complain that Newton had not yet seen fit to adjoin full supporting documentation of his breakdown of the cubic into its component species,(29) and some months afterwards in a private letter to him Pierre Varignon

(24) *Analysis Per Quantitatum Series, Fluxiones, ac Differentias: Cum Enumeratione Linearum Tertii Ordinis* (London, 1711): 67–92. The whole work was reprinted at Amsterdam in 1723, being then issued both separately and in appendix to a pirated 'Editio Ultima' of the *Principia*'s second edition (Cambridge, 1713); compare I. B. Cohen, *Introduction to Newton's 'Principia'* (Cambridge, 1971): 257. A few minimal textual ameliorations apart, Jones' sole improvement upon the structure of the 1704 *editio princeps* was to rearrange two groups of Newton's numberings of his figures into correct ordinal sequence; see 1, §2: notes (58) and (95) below.

(25) *Isaaci Newtoni Opuscula Mathematica, Philosophica et Philologica*, 1 (Lausanne/Geneva, 1744): 245–70. We have not seen the annotated Russian translation by D. D. Mordukay-Boltovskoy of this first volume of Castiglione's collected edition which appeared as *Matematicheskie Raboty* (Moscow/Leningrad, 1937).

(26) *Isaaci Newtoni Opera quæ exstant Omnia*, 1 (London, 1779): 529–60. The division of Newton's text into eight 'Capita' which Horsley there makes has no precedent.

(27) *Isaaci Newtoni Enumeratio Linearum Tertii Ordinis* (Paris, 1797).

(28) See his printed 'Proposals For publishing [Newton's *Opera*] by Subscription' of 'January 1, 1776' (we have used the copy in Christ Church, Oxford. Z. 286^5), where Horsley announces (page 7) that 'For the enumeration of the lines of the third order, we are furnished with the original plates, engraved for the elegant edition of that work by the late *William Jones*, esq.; a favour which we owe to the son of that eminent mathematician...'.

(29) *Acta Eruditorum* (February, 1712): 74–7. In the sentences on page 76 which we omitted in previously reprinting this review (on II: 259–62) Leibniz merely added: 'Recudi una fecit Cl. Editor duos Tractatus, *Opticæ* Newtonianæ ad calcem adjectos, quorum alter de Quadratura Curvarum agit, alter lineas tertii ordinis enumerat. De utroque diximus [see note (17) above] in Actis A[nni] 1705 p. 30 & seqq. Sed egregie de Geometris meritus fuisset Cl. Editor, si demonstrationem numeri linearum tertii ordinis, quam petenti non denegaturus erat *Newtonus*, una exhibuisset: immo adhuc bene mereri poterit, si per modum appendicis aut alia occasione eandem edat.' In the second of his draft counter-'Observations' on Leibniz' 'synopsis' whose text (ULC. Add. 3968.32: 460^r–461^v) was correspondingly in large part reproduced on II: 265–70 following, we also there discarded Newton's rejoinder (f. 461^v; compare II: 270, note (30)) that he had in his tract by no means failed to give adequate demonstration of the validity of his enumeration of the cubic's species, since 'In the seventh section [Mr Newton] shews how to find the Asymptotes of the *crura Hyp*[*erb*]*olica* of these Curves & the *plagæ crurum infinitorum Parabolicorum*. In the eighth ni[n]th tenth & eleventh sections he teaches how by the Asymptotes & *plagæ infinitæ* to find the position of the Abscissa & its Angle wth ye Ordinate by wch the species of ye Curve is to be known. And supposing you know how from the nature of the curve by vulgar Analysis to find the equation expressing the relation between that Abscissa &

made similar criticism of Newton's 'undemonstrated' and 'over dry' demonstration.(30) Two years later, we still find Leibniz writing to Johann Bernoulli to query, out of his own ignorance, whether Newton might not 'rightly have determined the number of curves of third degree'(31)—to receive once more a negative reply and the coupled rebuff that 'I have not yet [!] been able to bring myself to examine this matter, since I do not willingly embroil myself with intricacies of this sort, utterly useless as they are indeed'.(32) Nor, in Newton's own university, was the response of the editor of the forthcoming 'Editio Secunda Auctior et Emendatior' of his *Principia* much more acute in its stumbling understanding: in commenting upon its author's intention in the spring of 1712 of appending thereto 'a small Treatise of Infinite Series & ye Method of Fluxions'—Newton's yet unprinted 'Analysis per Quantitates Fluentes et earum Momenta', which will be reproduced in our final volume—Roger Cotes wrote to him from Cambridge on 26 April 1712 to 'beg leave to make another Proposal to You', that of editing a separate collection of Newton's mathematical works which 'together will make a Volume nearly of a Size with your *Principia* & may be printed in the same Character';(33) and then sought to bolster his case by adding:

that Ordinate, he enumerates all the cases of these equations & in the following sections shews how many forms of curves there are in every case'. A Latin revision on Add. 3968.3: 17^r adds the clincher that 'Plures autem non sunt curvarum species quam sunt æquationum casus et formæ'.

(30) Varignon to Leibniz, 19 November 1712 (N.S.): '...Je souhaiterois aussi une demonstration *numeri linearum tertii ordinis* que M. Newton donne trop sechement' (C. I. Gerhardt, *Leibnizens Mathematische Schriften*, **4** (Halle, 1859): 189).

(31) Leibniz to Bernoulli, 10 January 1714: 'Paralogismi quos in scriptis ejus [*viz.* the 1687 *editio princeps* of Newton's *Principia*; compare our earlier footnotes on VI: 71, 349–50, 359] notasti, faciunt ut dubitare cogamur utrum recte numerum Curvarum tertii gradus determinaverit' (*Commercium Philosophicum*, **2**, 334 = *Mathematische Schriften*, **3**: 929).

(32) Bernoulli to Leibniz, 28 February 1714: 'An Newtonus non erraverit in Enumeratione Linearum tertii ordinis, omittendo forte quasdam [species], quasdam alias bis sumendo pro diversis quæ eædem sunt habendæ, ...asseverare non ausim: ejus enim Tractatum hac de materia ut examinarem, nondum a me impetrare potui, quia non libenter hisce tricis, utpote haud valde utilibus me immisceo. Frater meus [Jakob] aliquando hoc vadum tentavit; quo vero successu non memini, fortasse in Scriptis ejus aliquid invenire est' (*Commercium Philosophicum*, **2**: 334 = *Mathematische Schriften*, **3**: 929). To atone for his uncle's show of petulance Niklaus Bernoulli wrote to Leibniz the same day that 'Newtonum in enumeratione...errasse, mihi non videtur verisimile; examinabo tamen, quia ita jubes, ejus hac de re scriptum, quam primum per otium licuerit' (*Mathematische Schriften*, **3**: 992). Other than that Niklaus made a few notes on Hermann's surviving transcript of Jakob Bernoulli's 'Typus Locorum Hypersolidorum' (see note (22) above) to which Johann had obliquely referred in his letter, and also afterwards discovered one of the six species of cubic omitted by Newton from his published enumeration, nothing is known of his subsequent efforts to keep this promise. (Compare Eneström's addition to Cantor's *Vorlesungen über die Geschichte der Mathematik* [**3**: 426] on this point in *Bibliotheca Mathematica*, $_3$**14**, 1913–14: 267–8.)

(33) ULC. Add. 3983.28; the minimally variant draft of this letter (now Trinity College,

'Your Treatise of y[e] Cubick Curves should be reprinted, for I think y[e] Enumeration is imperfect, there being as I reckon five cases of Æquations. vizt

$$\begin{aligned} xyy+ey &= ax^3+bxx+cx+d \\ yy+gxxy &= ax^3+bxx+cx+d \\ xxy+ey &= ax^3+bxx+cx+d \\ xy &= ax^3+bxx+cx+d \\ y &= ax^3+bxx+cx+d. \end{aligned}$$

I should have acquainted You with this before Mr Jones's Book was published if I had known any thing of y[e] Printing of it, for I had observed it two or three Yeares ago. I think there are some other things of lesser moment amiss in y[e] same Treatise'.

While Newton would not entirely have disagreed with the last remark (though he might not have been ready to admit so), this thoroughly misguided(34)

Cambridge. R. 16. 38: 212/213) which Cotes retained has been published by Edleston in his edition of *The Correspondence of Sir Isaac Newton and Professor Cotes* (London, 1850): (100–2+) 119. In his preceding sentence Cotes specified that 'When this Book [*Principia*, $_2$1713] shall be finished I intended to have importun'd You, to review your *Algebra* [*sc. Arithmetica Universalis*; compare v: 7–8] for a better Edition of it, & to have added to it those things [*viz.* Newton's 'De Analysi', 'De Quadratura', 'Enumeratio' and 'Methodus Differentialis'] which are published by M[r] Jones & what others You have by You of the like nature'.

(34) It will be clear that Cotes' first, fourth and fifth equations are exactly Newton's canonical reduced first, second and fourth cases respectively, while when $g = 0$ his second equation is the latter's general third case of the divergent parabolas. When $g \neq 0$, on resolving Cotes' second equation as a quadratic in y there ensues

$$y = \tfrac{1}{2}g(-x^2 \pm [x^2+2(a/g^2)\,x+2(-a^2/g^4+b/g^2)+O(x^{-1})]),$$

so that the cubic defined by it has for its asymptotes the straight line $y = (a/g)\,x-a^2/g^3+b/g$ and the parabola $y = -gx^2-(a/g)\,x+a^2/g^3-b/g$, and is therefore a parabolic hyperbola, readily reducible to the standard Newtonian form $XY^2+EY = BX^2+CX+D$ by the linear transformation of coordinates $\begin{Bmatrix} x = kY-a/g^2 \\ y = X+(a/g)\,kY-2a^2/g^3+b/g \end{Bmatrix}$ for the appropriate value of k. Similarly, since Cotes' third equation may be rearranged to be

$$(y-ax-b)\,(x^2+e) = (c-ae)\,x+d-be,$$

this defines a cubic having the oblique asymptote $y = ax+b$ and the two parallel ones $x = \pm\sqrt{-e}$, and which is therefore reducible by the linear transformation $\begin{Bmatrix} x = X+akY+b \\ y = kY \end{Bmatrix}$ to the standard Newton form $XY^2+EY = AX+B$ of a conic hyperbolism (hyperbolic, parabolic or elliptical according as $e = k^2E$ is less than, equal to or greater than zero). In his note in *Bibliotheca Mathematica*, $_3$**14**, 1913–14: 268 (see note (32) above) Eneström follows Edleston's mistaken surmise (*Correspondence of Newton and Cotes:* 119) that he 'probably never sent' this paragraph of his letter to Newton in charitably conjecturing that 'Vermutlich entdeckte Cotes recht bald, dass seine Bemerkung unzutreffend war'. Whatever the truth of this, Cotes certainly did not here reveal his geometrical talents and expertise to best advantage. Subsequently, his cousin Robert Smith recorded in the 'Editoris Præfatio' to the posthumous collection of Cotes' *Harmonia Mensurarum, sive Analysis & Synthesis Per Rationum & Angulorum Mensuras Promotæ* and *Alia Opuscula Mathematica* which he brought out at Cambridge in 1722, when he came (signature ** 3[v]) to make mention 'de aliis nonnullis quæ...in lucem edere in

attempt to fault the completeness of the four reduced canonical forms of Cartesian defining equation on which he had founded his enumeration of the cubic's species was met by him with a diplomatic silence in his answering letter;(35) nor did he, we may be sure, furnished with such an instance of Cotes' geometrical perspicuity, take up this well-intentioned offer to be his mathematical editor, and no more is heard of it.

The turning-point in contemporary appreciation of the essential accuracy and technical accomplishment of the 'Enumeratio' came five years later in 1717 when the young James Stirling, still *in statu pupillari* at Balliol in Oxford, produced the first proficient exposition and percipient critique of Newton's tract,(36) elaborating its presuppositions, proving most of its undemonstrated theorems and arguments, and adding four (of the six) minor species of cubic which Newton himself had there overlooked, further exemplifying the power and effectiveness of its mode of reduction by instancing how readily one could thereby determine the species of any given cubic from its defining equation and

---

animo habeo': 'Scripsit etiam [Cotesius] Tractatum *de Natura Curvarum*; quarum genera & species distinguit figurasque describit & primarias proprietates eruit ex æquationibus algebraicis. Et quanquam agat præcipue de curvis primi Generis (scilicet de Conicis Sectionibus...); insigniores tamen earum proprietates ad superiorum generum curvas pertinere similibus argumentis ostendit: quibus alia subnectit, cum de Locorum & Æquationum Constructione tum de Problematum inde pendentium Resolutione'. Of Cotes' treatise there has survived a single theorem, the 'pulcherrima Linearum geometricarum proprietas' defining the general rectilinear harmonic polar of a point with respect to any algebraic curve which Smith communicated to Colin Maclaurin at some unknown after-date (in the late 1720's we presume) and from which the latter afterwards drew many of the 'Theoremata generalia...quæ...ad arduam hanc Geometriæ partem augendam & illustrandam conducere viderentur' which he set forth in the monograph 'De Linearum Geometricarum Proprietatibus generalibus' published after his death in appendix (pages $_2$1–65, see especially $_2$2) to his popular *Treatise of Algebra* (London, $_1$1748).

(35) Newton to Cotes, 10 May 1712 (Edleston, *Correspondence:* 113–14).

(36) *Lineæ Tertii Ordinis Neutonianæ, sive Illustratio Tractatus D. Neutoni De Enumeratione Linearum Tertii Ordinis* (Oxford, 1717); reprinted by Duprat in appendix to his edition of the 'Enumeratio' in 1797 (see note (27) above). Newton showed his goodwill towards this youthful, able commentator by pledging himself (as its introductory list of subscribers shows) 'for two [copies]'. Only pages 83–120 of the book are given over in a narrow sense to the 'Enumeratio Linearum tertii Ordinis' in which (see 1, §2: notes (67) and (72) below) Stirling adds four new species to the 72 of Newton's tract; the preceding pages 1–83 expound the various algebraic techniques of solving equations, exactly and by expansion into series, and the ways in which these may be applied to derive the general properties of curves enunciated by Newton, and to determine the number and direction of their infinite branches; while the concluding pages 120–8 broach the 'Determinatio Locorum Geometricorum' from their given Cartesian equations (see the next note). A fuller breakdown of the contents is given in C. Tweedie, *James Stirling: A Sketch of his Life and Works, along with his Scientific Correspondence* (Oxford, 1922): 23–9; see also H. Wieleitner, 'Zwei Bemerkungen zu Stirlings *Lineæ tertii ordinis Neutonianæ*', *Bibliotheca Mathematica*, $_3$**14**, 1913–14: 55–62.

hence trace its shape.[37] This 'illustration' of Newton's tract set the style for later eighteenth-century accounts of the analytical geometry of cubic and higher plane curves—notably those by Leonhard Euler,[38] Gabriel Cramer[39] and Edward Waring[40]—which were in turn to lead directly into the yet more complex ramifications of the topic developed by Plücker, Cayley and a host of others in the nineteenth century.[41] To go any way into these later antecedents of the modern field of algebraic geometry would take us too far from our present limited purpose of sketching the immediate impact of Newton's pioneering essay, but the reader who finds our running commentary (in the ensuing foot-notes to its text) to be deficient will consult with profit the fulsome technical

---

(37) Or so Stirling set his aim to be in his concluding 'Determinatio Locorum Geometricorum', but we have previously observed (IV: 228, note (48)) that in drawing the figures of the two cubics, $y^3 = x^3+a^3$ and $y(y-x)^2 = a^3$, which he added in worked example he chose in each case to simplify their equation by a preliminary change of coordinate axes.

(38) *Introductio in Analysin Infinitorum* (Lausanne, 1748): *Liber Secundus, Continens Theoriam Linearum Curvarum*... [= *Leonhardi Euleri Opera Omnia* (1) **9** (Zurich, 1951)]; see especially Caput IX. 'De Linearum tertii ordinis subdivisione in species': 114–26, and Caput X. 'De præcipuis Linearum tertii ordinis proprietatibus': 127–39.

(39) *Introduction à l'Analyse des Lignes Courbes Algébriques* (Geneva, 1750).

(40) *Miscellanea Analytica, de Æquationibus Algebraicis, et Curvarum Proprietatibus* (Cambridge, 1762): Pars Secunda: 66–162, whose first four chapters (*ibid.*: 66–147) were reissued in a much revised and augmented separate version as *Proprietates Algebraicarum Curvarum* (Cambridge, 1772). It is customary to point out that in Caput V of the former's second book (*Miscellanea:* 148–62), where it is shown in preliminary that 'Omnes curvæ tertii generis [= quarti ordinis] reduci possunt ad duodecim...æquationum casus', Waring departs from these twelve canonical reduced forms of defining equation to prove that by a similar enumeration of the totality of the 'Newtonian' species of the general quartic 'non plures sunt quam 84551 curvæ quarti ordinis' (*ibid.*: 161); but little known that in his latter work ten years afterwards he excused his omission of the derivation of this loose upper bound to the quartic's species with the pessimistic words: 'Non hic enumerantur lineæ quarti ordinis, quoniam talis enumeratio nec in detegendis curvarum proprietatibus, nec in quâvis aliâ re ulli usui inservire potest; etiamque in enumeratione linearum superiorum ordinum secundum methodos a *Newtono* vel *Eulero* usitatas occurrunt aliæ curvæ, quæ quadraturam recipiunt; aliæ vero curvæ ejusdem speciei, quæ minime quadrari possunt, ob has & alias rationes omnes enumerationes hujusce generis hic rejiciuntur.' No one has ever attempted to emulate Waring's feat by setting a bound to the number of species of a quintic or any higher order of curve, but we may notice that Colin Maclaurin had earlier made several observations on the problem of classifying quartics in his *Geometria Organica* (London, 1720): Pars Prima, Propositio XIX, Scholium; compare Charles Tweedie, 'The "Geometria Organica" of Colin Maclaurin: A Historical and Critical Survey' (*Proceedings of the Royal Society of Edinburgh*, **36**, 1915–16: 87–150): 118.

(41) See George Salmon, *A Treatise on the Higher Plane Curves* (London, $_3$1879): *passim* for a systematic account of the technical theory of algebraic curves as it evolved during the century after Euler; and consult C. B. Boyer's *History of Analytical Geometry* (New York, 1956): 174–268 for a broader historical view of Newton's impact, through Euler and his fellow mid-18th century geometers, upon the course of this development. Of the latter Rouse Ball gives a useful summary in his essay 'On Newton's Classification of Cubic Curves' (see note (3) above): 131–2.

notes appended by C. R. M. Talbot to his standard English edition of the 'Enumeratio' in 1860.(42)

Of the remaining, disconnected fragments which are reproduced in sequel we neither need nor can say very much since, other than that their date of composition can fairly narrowly be established from their manuscript context, the lack of any independent documentation of their background and circumstances allows only unsupported circumstantial conjecture, plausible as it may be, regarding their destined rôle in the grand scheme of Newton's researches in exact science.

Whatever their ultimate purpose, Newton's two essentially equivalent instrumental constructions(43) of the class of conchoidal cubics describable as the loci of points fixed in the plane of a moving jointed ruler, one 'leg' of which is constrained ever to pass through a stationary pole while the other at its end-points travels continuously along on a given straight line, are straightforward extensions of his earlier 'organic' tracing(44) of the 'cissoid of the ancients' which is here, in each case,(45) specified as their primary instance: as such they afford only a minor complement to what has gone before. A deal more original are the unfinished computations(46) in which Newton sought initially to define the most general Cartesian equations of bifoliate and trifoliate quartic curves—the first by simple considerations of bilateral symmetry round the axes, the latter by a more sophisticated appeal to the subtle insight that the linear transformation from one vertex of a trefoil to a second which also rotates the axes through 60° merely passes the defining equation into itself—as a preliminary to determining their

(42) *Sir Isaac Newton's Enumeration of Lines of the Third Order, Generation of Curves by Shadows, Organic Description of Curves, and Construction of Equations by Curves. Translated from the Latin with Notes and Examples* (London, 1860); the English version itself is on pages 7–30, followed immediately by Talbot's 'Notes to the foregoing text' (pages 33–88) and then by less immediately relevant observations 'On the Analytical Parallelogram' (pages 88–112), 'Examples of Propositions producing Loci of the Third Order' (pages 112–25) and, finally, 'Miscellaneous Problems having reference to Lines of the Third Order' (pages 125–40). We may also cite the more recent (but considerably less substantial) surveys by H. Hutton ('Newton on Plane Cubic Curves', in (ed. W. J. Greenstreet) *Isaac Newton, 1642–1727*, London, 1927: 115–16) and Maximilian Miller ('Newton, Aufzählung der Linien dritter Ordnung', *Wissenschaftliche Zeitschrift der Hochschule für Verkehrswesen Dresden*, **1**, 1953: 5–32).

(43) Reproduced in 2, §1.1/2 below; in note (2) thereto we suggest that one or other of these may have been drafted to fill a lacuna in Newton's preliminary version of the 'Enumeratio', in whose section XXVII he had (see page 566 above) originally set himself to present just such a 'Facilis descriptio curvarum per polos'.

(44) In Problem 20 of his Lucasian lectures on algebra (see v: 218–20) and in Problem 7 of his subsequent 'Geometria Curvilinea' (see iv: 480).

(45) See 2, §1: notes (5) and (11).

(46) Printed, with appropriate editorial interlardings to bring out their sense, in 2, §2.1/2 following.

instantaneous tangential direction at any point and thence their curvature at a vertex. Leaving aside again what may have stimulated Newton to attack such a brace of problems,(47) we may admire for itself the expert and adroit manner in which he conducts his geometrical analyses and be grateful that these now, after the passage of nearly three centuries, at last see the public light.

In the concluding rag-bag of miscellaneous items and scraps on functional interpolation and area approximation by adapting an appropriate curve to a given set of values or ordinates, Newton has for the most part a more practical end in view. While he had in the middle 1670's developed(48) a variety of modes of intercalating median terms in given sets of suitably parametrized quantities by fitting a parabolic polynomial to these and thereafter determining its coefficients from the successive first, second, third and higher-order differences of the given terms, for all the subtlety of his earlier essay expounding his derived 'Regula Differentiarum'(49) he had there—ignorant, it would seem, of Henry Briggs' earlier canons(50) for such subtriplication and subquintuplication—failed to frame a workable scheme by which a given array of functional values listed at equal intervals of the argument might be densely subtabulated; nor, though much show had been made by him in the third book of his *Principia* in 1687 (without his there adducing any worked example) of setting his 'master' adjusted-differences formula to round out the observed apparent path of a comet from a few timed terrestrial sightings of its orbit,(51) had he evinced the power of his general interpolation theorems in any other astronomical or physical context, where indeed he preferred, as we have seen,(52) to employ such simple approximating curves and functions as he thought fit to the occasion. Here we rob his unpublished scientific worksheets of the middle 1690's to give an indication of how occasionally in his computational practice he did appeal to Briggsian notions in subtabulating his empirical and calculated data,(53) and also to present an instance(54) (unique we believe) where he accurately applied his *Principia* lemma to interpose a parabolic path between five of Flamsteed's sightings of the 'great' comet of 1680–1, so determining the apparent slope of

(47) In 2, §2: note (3) we present our highly tentative conjecture that Newton might here initially have had it in mind to explore some generalized Schootenian locus of the kind set forth in our note (17) to I: 30.

(48) See IV: 36–68.

(49) Reproduced on IV: 36–50.

(50) Delivered by him in his *Arithmetica Logarithmica* (London, 1624): Caput XIII: 27–32 and equivalently in his posthumous *Trigonometria Britannica* (Gouda, 1633): Liber I, Caput XII: 35–41; see IV: 45, note (25).

(51) *Principia*, $_1$1687: Liber 3, Lemma VI: 483; compare IV: 73, note (5).

(52) VI: 434, note (32); 450, note (22).

(53) See 3, §1.1–4 below.

(54) In 3, §2.1 following.

the comet's orbit to the meridian on December 30, and thereafter[55] ingeniously —but not without a couple of numerical slips—checking this by determining the limiting angle at which it is there met by a great circle of the celestial sphere. In the final, unfinished paper 'Of Quadrature by Ordinates'[56] he seeks to develop a viable alternative to the straightforward (though computationally intricate) approximation of an area by squaring the corresponding portion of the general parabola which is adapted to pass through the end-points of a given set of parallel, equidistant ordinates; but his variant approach by refining successive prior 'Cotesian' approximations of this kind proves not to be theoretically exact in the general case[57]—whether Newton himself realized this or no—and the resulting formulas become rapidly unwieldy and all but useless for numerical practice.

But now to the texts themselves, which will reveal far better than any editorial fine phrases the enduring quality and undulled acuity of Newton's mathematical mind in the middle of his fifty-third year.

---

(55) See 3, §2.2. We should add that this sequence is our editorial conjecture, and cannot be confirmed from the internal evidence of the two manuscript fragments.

(56) Reproduced in 3, §3 below.

(57) See 3, §3: notes (12) and (13), where we both identify Newton's errors of presumption and correct them in line with the accurate formulas of approximation afterwards first computed by Roger Cotes (compare *ibid.*: note (1)).

1

IMPROVED ENUMERATION OF THE COMPONENT SPECIES OF THE GENERAL CUBIC CURVE[1]

[Summer 1695]

§1. A PRELIMINARY BREAKDOWN INTO 69[2] SPECIES.

From the original draft[3] in the University Library, Cambridge

---

(1) Whether or not David Gregory had (as we have suggested in our preceding introduction) already urged him to compile such a 'book *de Curvis secundi generis*' during his visit to Cambridge in May of the previous year, it is certain that Newton began, in or soon after the early summer of 1695, to condense the multi-layered bulk of his earlier researches into the theory and construction of cubic curves (see especially II: 10–104, 498–508; IV: 346–405) into a short text which should briefly enunciate their main properties, but above all give an exact enumeration of their component species according to the scheme of such division which he had laid down a quarter of a century before. The *terminus ante quem non* of early June for the composition of the tract whose finished version, along with a preparatory breakdown into heads of species, we here reproduce is fixed by the fact that Newton has entered the text of part of one of its drafts (see Appendix 1.1: note (2) below) on the blank portion of a letter addressed to him from London on 'ye 6 June '95'. The uniformity of the hand and ink with which the autographs are penned correspondingly guarantees beyond reasonable doubt that they were composed over a fairly narrow interval of time not long after that—but in any event, surely, before April 1696 when their author finally renounced the quiet of academic life to take up in the metropolis a public career whose insistent administrative demands and responsibilities long left him too little private leisure to pursue such unworldly matters?

(2) As in the ensuing detailed preliminary 'Enumeratio Curvarum secundi Ordinis' (ULC. Add. 3961.1: 38^r–50^v; see §2: note (2) below), though the denominations of the various component *genera* and individual species are here somewhat different; in the final enumeration (Add. 3961.2: 1^r–14^r) which we reproduce in sequel Newton at a late stage further subdivided his species 14/16, 15/17 and 28/29 (here paired) to make altogether 72—to which we should add yet six more, here wholly overlooked, which he had earlier distinguished in his projective listing (see **2**, 3, §1: notes (54) and (65) above). It will be evident that the layout of his present catalogue of cubic curves narrowly follows the pattern of his

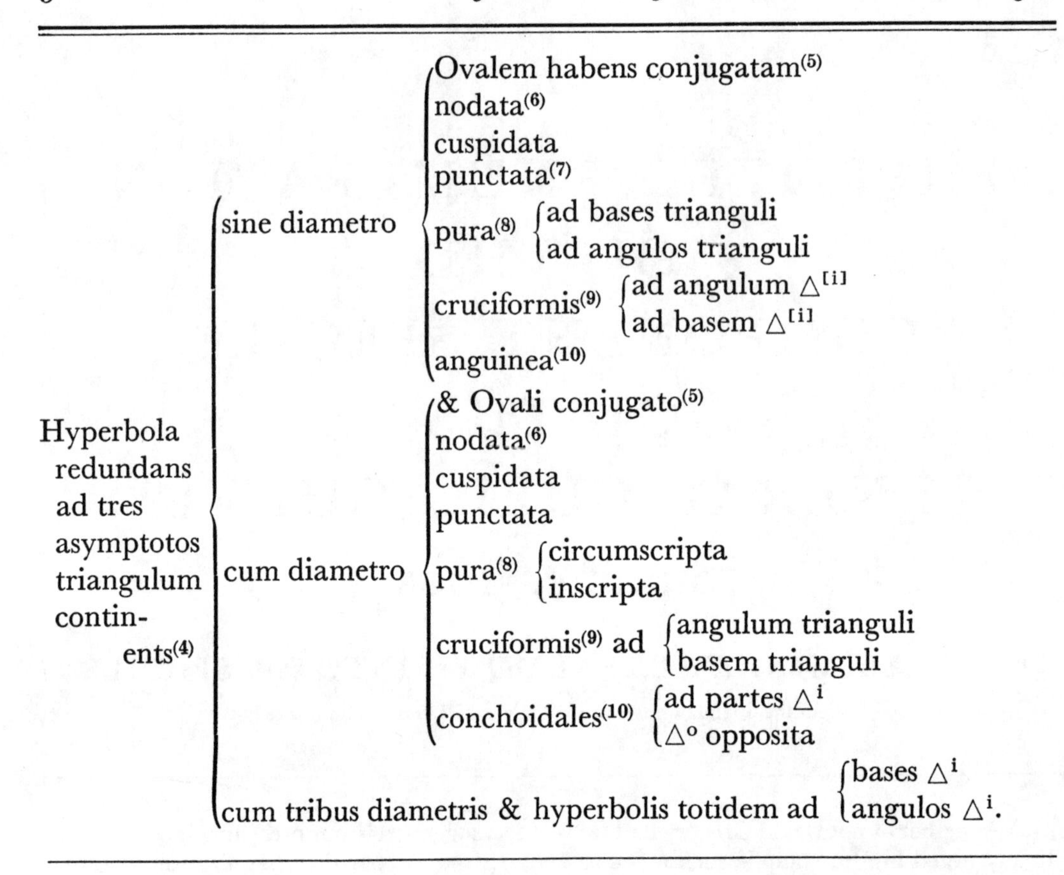

previous *tabula* on IV: 370–4, even though Newton now drastically changes his nomenclature. In regard to the last we may add that on Add. 3961.2: 16r Newton has roughed out a long column of possible revised appellations descriptive of the various configurations which the cubic 'Hyperbola' may in particular circumstances assume, namely: 'convergens' (convergent), 'divergens' (divergent), 'circumscripta' (circumscribed), 'inscripta' (inscribed), 'ambigena' (double-kinded), 'conchoidalis' (conchoidal, shell-shaped), 'cissoidalis' (cissoidal, prickly pointed), 'anguinea, anguis' (snakish, snaky), 'parabolica' (parabolic), 'redundans' (redundant, fully stocked [*sc.* with asymptotes]), 'defectiva' (defective, lacking [*sc.* a full quota of asymptotes]), 'reflexa' (back-bent), 'asy[m]ptotis || in eandē/contrar. plagā' (with parallel asymptotes [tending] to the same/contrary direction(s)), 'cuspidata' (cusped), 'nodosa' (knotty, *sc.* nodate), 'cruribus contrarijs' (with opposing infinite branches), 'punctata' (punctate), 'pura' (pure [*sc.* without node, conjugate oval or conjugate point]) and lastly 'conjugatam decussans, cruciformis' (crossing over a conjugate branch, cross-shaped). Most of these designations are taken up again—many on numerous occasions—in the sequel.

(3) Add. 3961.4: 15r. A preliminary listing on *ibid.*: 16r omits to take any account of the particular case of the redundant hyperbola where its three asymptotes are concurrent at a point; other variants in its denominations of individual *genera* and species are cited at the appropriate place in following footnotes. In a first state of the present tabulation no terminal bifurcation is made by Newton, according to the site of the hyperbolic branch(es) in relation to the asymptotes, for the six twin species (as here distinguished) of the general redundant hyperbola *sine diametro, pura/cruciformis; cum diametro, pura/cruciformis/conchoidalis;* and *cum tribus diametris*, or for those of the defective hyperbola *sine diametro, pura;* and *cum diametro, punctata.* The remaining species he has consecutively numbered 1–61 at their end, but this superseded

Translation

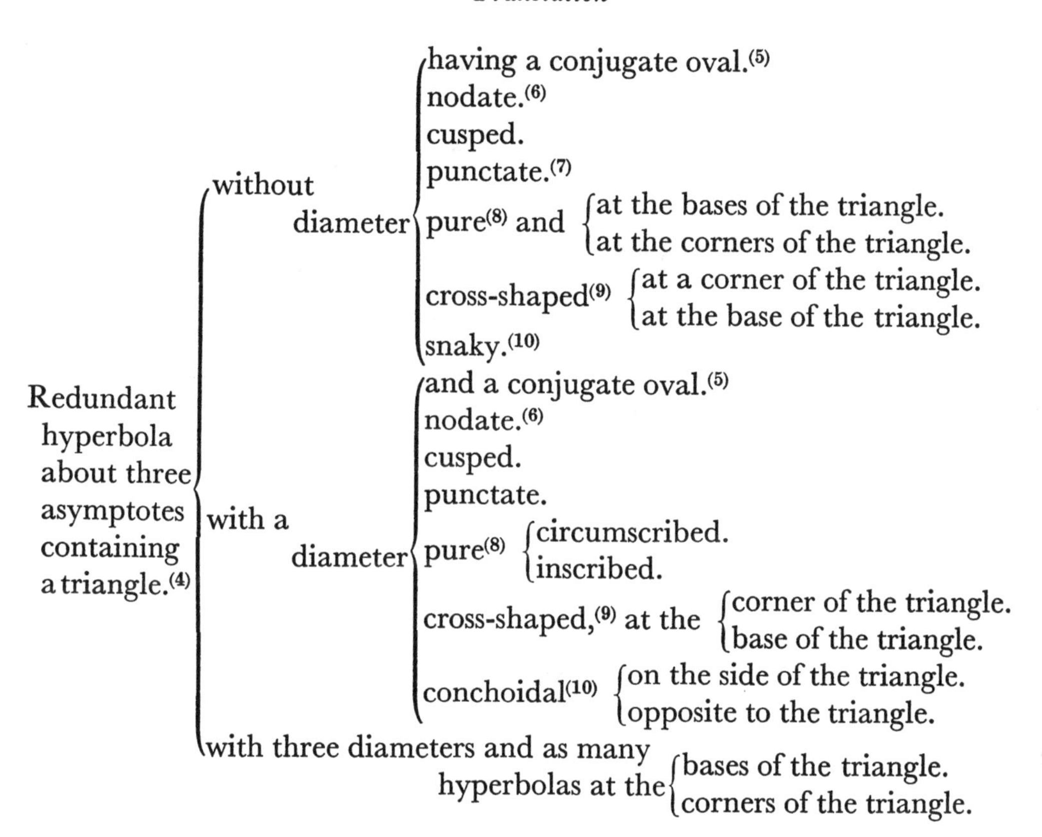

listing we here neglect to reproduce so as to avoid any confusion with the revised enumeration into 69 species which is grafted on to it. On f. 15ᵛ, we may add, Newton tentatively set out an alternative tabulation of the species of the cubic curve in which the general *Hyperbola redundans* is divided into its component *genera* according as it is *sine diametro*, *cum diametro* or *cum tribus diametris*, and the first two divisions are only then further distinguished into the cases of *tres habens asymptotos triangulum continentes* and *tres habens asymptotos convergentes* [*sc. ad punctum commune*], before passing to a detailed breakdown into species in his usual manner; but he did not further proceed with this somewhat more logical subclassification.

(4) Originally 'Hyperbola triplex ad tres Asymptotos inclinatas ipsam secantes' (Threefold hyperbola about three inclined asymptotes cutting it) on f. 16ʳ; 'triplex' was initially 'triformis' (three-limbed; see II: 88 and IV: 368, note (48)) and then 'plena' (full). The 9+10+2 species here distinguished by Newton in his three ensuing *genera* are respectively species 1–9; 10–13, 14/16, 15/17, 18–21; and 22, 23 in his final enumeration in § 2 below.

(5) Just 'cum Ovali' (with an oval) on f. 16ʳ.

(6) Earlier 'se decussans' (crossing itself) on f. 16ʳ, and then 'nodosa' (knotty) on f. 15ᵛ; compare IV: 370–2.

(7) Originally 'cum puncto conjugato' (with a conjugate point) on f. 16ʳ, afterwards minimally altered on f. 15ᵛ to be 'punctum [habens] conjugatum' (having a conjugate point).

(8) Initially 'deficiens' (deficient) as on f. 16ʳ, then 'mere hyperbolica' (merely hyperbolic) as on f. 15ᵛ, and yet again 'simplex' (simple).

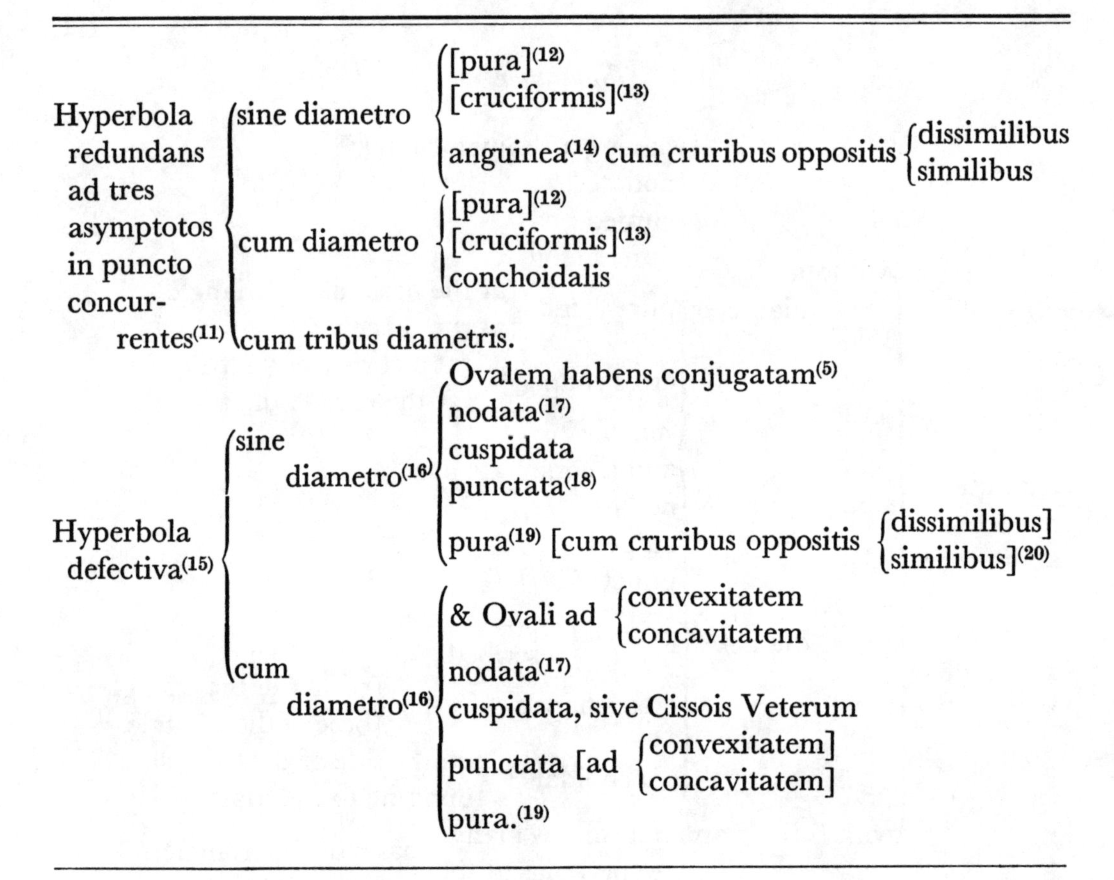

(9) First 'conjugatam decussans' (crossing a conjugate) as on ff. $16^{r}/15^{v}$ and in his previous cubic enumerations (see IV: 370–2).

(10) Originally 'circumflexa' (around-bent) on ff. 16^{r} and 15^{v}; compare II: 86–8.

(11) Newton has here replaced 'convergentes' (converging). As we have already remarked (note (3) above), the corresponding case is lacking in the otherwise similar draft tabulation on f. 16^{r}, but no doubt, had Newton chosen to repair his oversight *in situ*, he would there have made the head to read 'Hyperbola triplex ad tres Asymptotos convergentes ipsam secantes' or some near-equivalent (compare his title for the preceding case, quoted in note (4)). The $4+3+1$ types of cubic severally distinguished in the three component *genera* are respectively species 24–7; 28/29, 30, 31; and 32 in the final enumeration in §2 following.

(12) For consistency's sake we here convert an unaltered 'simplex' (simple), which itself replaces 'mere hyperbolica' (merely hyperbolic); compare note (8) preceding.

(13) Here, likewise, we substitute Newton's standardized equivalent for an unreplaced 'conjugatam decussans' (crossing a conjugate); compare note (9) above.

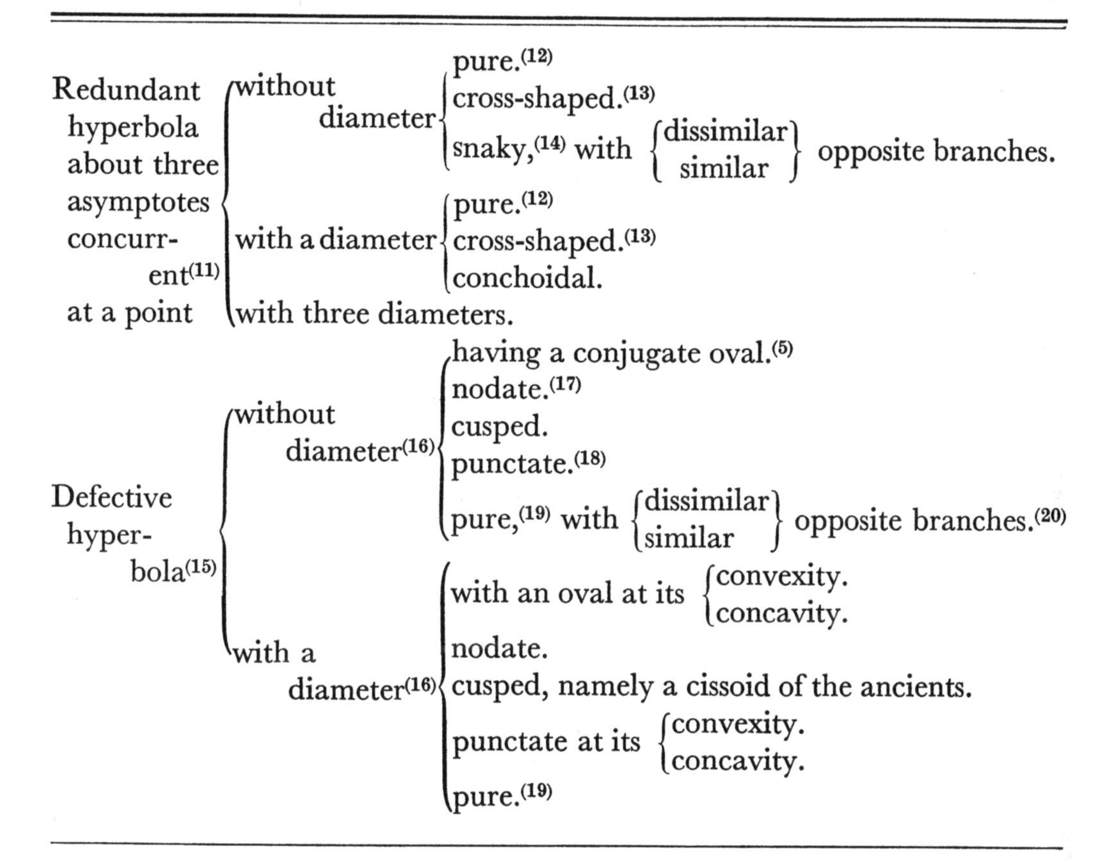

(14) Originally 'anguillaris' (eel-like).

(15) Named 'Hyperbola simplex et ad unam Asymptoton' (Simple hyperbola, about a single asymptote) in the draft on f. 16^{r}. The $6+7$ cubics into which the two ensuing *genera* are further subdivided are respectively species 33–8 and 39–45 of the final enumeration in §2 below.

(16) On f. 16^{r} these *genera* are further qualified as being 'circumflexa' (around-bent) and 'conchoidalis' (shell-shaped) respectively.

(17) Originally (compare note (6) above) 'se decussans' (crossing itself) both here and in the draft on f. 16^{r}.

(18) Trivially altered from 'cum puncto conjugato' (with a conjugate point) in the draft on f. 16^{r}; compare note (7).

(19) Initially 'deficiens' (deficient) on f. 16^{r}, then here changed first to be 'mere hyperbolica' (merely hyperbolic) and then to be 'simplex' (simple); compare note (8).

(20) Originally 'extra asymptotos {non amplexa' / amplexa} (not entwined/entwined about the asymptotes).

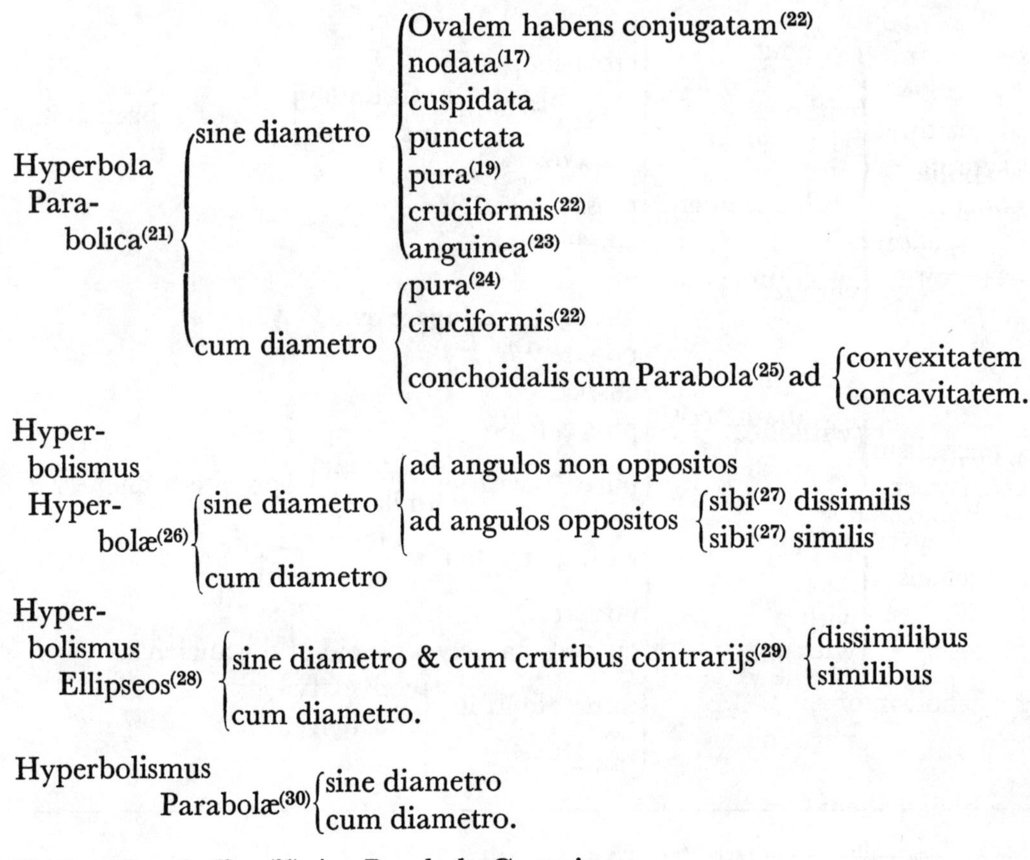

Tridens Parabolica,[31] sive Parabola Cartesiana.

(21) 'Parabola Hyperbolica' (Hyperbolic parabola) in the draft tabulation on f. 16r, there momentarily inverting—maybe without conscious intent to do so?—the earlier appellation (see IV: 372) to which he now returns. The 7+4 types of cubic severally distinguished by Newton in the two *genera* into which this class subdivides are species 46–52 and 53–6 respectively in his final enumeration in §2 below.

(22) Originally 'conjugatam decussans' (crossing a conjugate); compare note (9) above.

(23) Initially 'circumflexa' (around-bent), as twice before; compare notes (10) and (16) preceding.

(24) 'Parabolam aperiens' (disclosing [but] a parabola) in the draft on f. 16r.

(25) Initially '& versus Parabolam' (facing...). On f. 16r Newton's preliminary denomination is 'clausa et verticibus ad {easdem partes / contrarias partes} concavis' (closed in and with its vertexes concave the same/opposite way(s)).

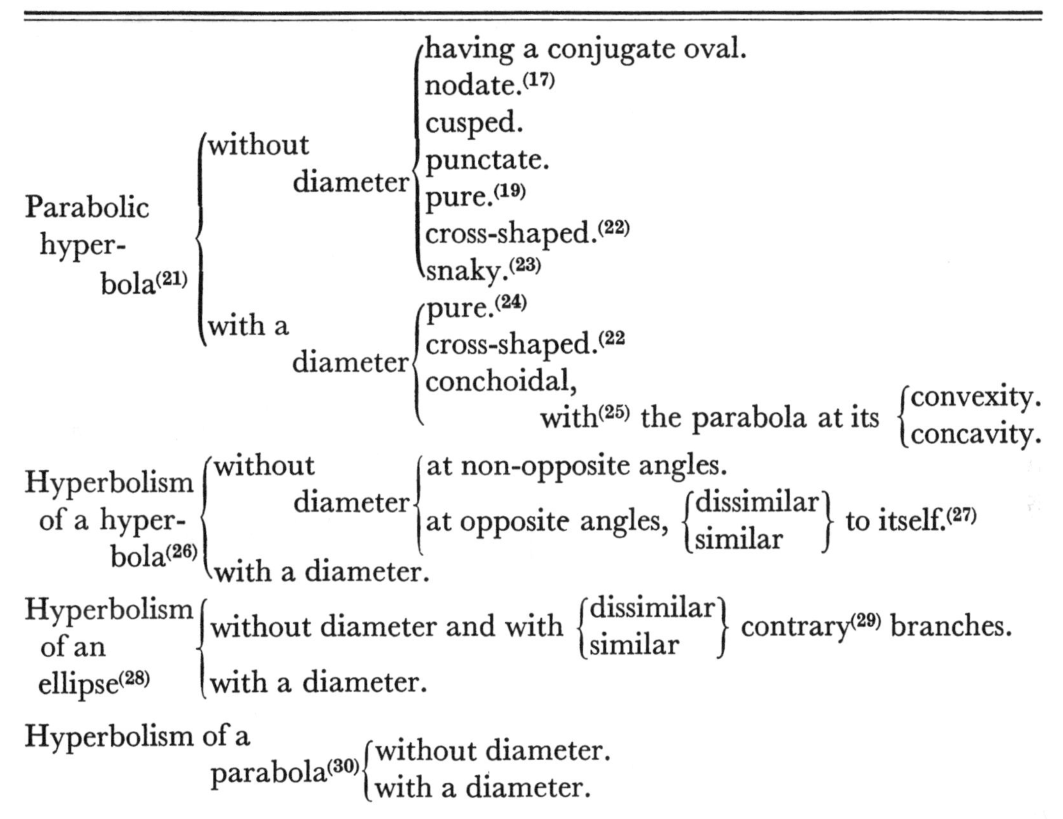

(26) The 3+1 cubics into which the two *genera* of this case subdivide are respectively species 57–9, and 60 in the final enumeration which follows.

(27) A superfluous 'ex opposito' (oppositely) in the draft on f. 16^{r} is here eliminated.

(28) The 2+1 cubics in the two *genera* of this case are respectively species 61, 62; and 63 in Newton's ensuing 'Enumeratio' (§ 2 below).

(29) Originally 'oppositis' (opposite).

(30) The two component *genera* of this case are species 64 and 65 in the final enumeration following.

(31) The first appearance of this now universal epithet for Descartes' 'parabole', though Newton had been familiar with the shape and Cartesian generation of the curve itself, of course, since his first reading of the *Geometrie* while still a Cambridge undergraduate (see I: 495, and compare II: 70, note (92) and 363, note (166); IV: 478, note (155); and **2**, 3, §1: note (48) above). As in his draft tabulation on f. 16^{r}—there merely continuing his preferred earlier designation (on which see IV: 374, note (73))—Newton's initial choice of name here, too, was 'Parabolismus Hyperbolæ' (Parabolism of a hyperbola). The unique *genus* of this case of cubic is species 66 in his final 'Enumeratio' below.

Parabola divergens(32) { cum Ovali
punctata
cruciformis(33)
cuspidata, semicubica,(34) Neiliana
pura.(35)

Parabola cum cruribus contrarijs, cubica, Wallisiana.(36)

---

(32) Originally 'Parabola semicubica [!] sive cum diametro' (Semicubic or diametral parabola) in the draft on f. 16r. Newton here, we may add, returns to his earlier—and now standard—designation for this class of cubic curve (see IV: 374). The five subdivisions here distinguished of its single *genus* are respectively species 67–71 in his final enumeration following.

(33) Initially 'se decussans' (crossing itself), as in the draft on f. 16r. Newton's revised choice of appellation is surprising; since the parabola's nose is 'closed' one would rather expect him to have settled for 'nodata' (nodate), as indeed he does in his final enumeration in §2 following. (Compare notes (6) and (17) above.)

(34) Newton first inserted 'sive Parabola' (that is, parabola) as in his draft on f. 16r, where the further qualification 'cissoidalis' (prickly pointed) was also initially introduced.

(35) Originally 'deficiens' (deficient), both here and on f. 16r. (Compare notes (8) and (19) above.)

(36) Initially just simply 'Parabola cubica' (Cubic parabola) on f. 16r, onto which Newton thereafter tacked—for the very first time (see §2: note (100) following)—the alternative appellation 'sive Wallisiana' (that is, Wallis'). The unique cubic of this last *genus* is species 72 in the detailed enumeration to which we now pass.

Divergent parabola[32]	with an oval.
	punctate.
	cross-shaped.[33]
	cusped, semicubic[34] or Neilian.
	pure.[35]

Parabola with contrary branches, cubic or Wallisian.[36]

§2. NEWTON'S FINAL ENUMERATION INTO 72 SPECIES.[1]

From the autograph[2] in the University Library, Cambridge

Enumeratio Linearum tertij Ordinis.[3]

I Linearum[4] Ordines.

Lineæ Geometricæ secundum numerum dimensionum æquationis qua relatio inter Ordinatas et Abscissas definitur, vel (quod perinde est) secundum numerum punctorum in quibus a linea recta secari possunt, optimè[5] distinguuntur in Ordines. Qua ratione linea primi Ordinis erit recta sola, eæ secundi sive quadratici Ordinis erunt Sectiones conicæ & circulus,[6] et eæ tertij sive cubici Ordinis Parabola Cubica, Parabola Neiliana, Cissois Veterum & reliquæ quas hic enumerare suscepimus. Curva autem primi generis (siquidem recta inter curvas non est numeranda,) eadem est cum linea secundi Ordinis, et curva secundi Generis eadem cum linea Ordinis tertij.[7] Et linea Ordinis infinitesimi[8] ea est quam recta in punctis infinitis secare potest, qualis est Spiralis,[9]

---

(1) Now, in a minimal amplification of the preceding breakdown (see §1: notes (4) and (11)), distinguishing between three pairs of monodiametral hyperbolas (species 14/16, 15/17 and 28/29 following) according to the differing lie of their hyperbolic branches about the triangle formed by the asymptotes, but continuing here to neglect six further species—namely, a couple of variant monodiametral redundant hyperbolas (see note (67) below), a pair of tridiametral redundant hyperbolas (see note (72)) and two monodiametral parabolic hyperbolas (see note (84)), each respectively 'cum Ovali' and 'punctata'—which Newton had embraced a little earlier in his equivalent projective classification of the individual varieties of cubic as the optical 'shadows' variously cast by the five divergent parabolas (compare **2**, 3, §1: notes (54) and (65) preceding). In this familiar definitive text of his 'Enumeratio Linearum tertij Ordinis', which in all but trivial verbal and syntactical details is that which he subsequently gave out to the world as the first of the 'Two Treatises of the Species and Magnitude of Curvilinear Figures' appended by him to his *Opticks* in 1704 (see the next note), Newton deftly blends the gist of his previous many-forked explorations of the nature, species, properties and applications of the general curve of third degree. While we may accept the justice of Rouse Ball's criticism that this essay is but 'a catalogue of results' ('On Newton's classification of Cubic Curves', *Proceedings of the London Mathematical Society*, **22**, 1891: 104–43, especially 105), it is yet a marvellous epitome of results whose subtleties were only just becoming to be understood by mathematicians in the last decade of Newton's life, half a century after their initial discovery.

(2) Add. 3961.2: 1^r–14^r; except for the accompanying figures, which were copied by Newton on separate sheets which have not been preserved (and are here reproduced, with some small but necessary corrections, from the *editio princeps*), this is the very text sent to the printer in 1704 and has been marked up by him for composition with marginal symbols here omitted. (Compare D. F. McKenzie, *The Cambridge University Press, 1696–1712*, **1** (Cambridge, 1966): 118.) Three long, greatly variant passages in Newton's preparatory draft on Add. 3961.1: 38^r–50^r(+3961.2: 15^v)—where (see §1: note (2)) the cubic is subdivided into the 69 species tabulated in the scheme printed in our previous section—are reproduced in Appendix 1 below; other minor differences in this preliminary version are reported in footnotes here following.

Translation

ENUMERATION OF LINES OF THE THIRD ORDER.(3)

I
Orders of lines.(4)

Geometrical lines are best(5) distinguished into orders according to the number of dimensions of the equation by which the relationship between their ordinates and abscissas is defined, or (what is the very same) according to the number of points in which they can be cut by a straight line. On this basis a line of first order will be the straight line alone, those of second or quadratic order will be conics and the circle,(6) and those of third or cubic order the cubic parabola, Neilian parabola, cissoid of the ancients and the rest which we have here undertaken to enumerate. A curve of first kind however—seeing that the straight line is not to be counted among curves—is the same as a line of second order, and a curve of second kind the same as a line of third order;(7) while a line of infinite order is one which a straight line can cut in an infinity of points, such as the (9)spiral,

---

We may add that among other *adversaria* in his library copy (now ULC. Adv. b. 39. 4) of the first Latin edition (*Optice*, London, 1706)—to which its reset but unaltered text was again appended as the first of two alien 'Tractatus duo...de Speciebus & Magnitudine Figurarum Curvilinearum'—Newton about the year 1710 (see note (25) below) entered in the printed version of his 'Enumeratio' a number of marginal emendations and longer inserts on loose paper slips which have not hitherto been published; these, too, we introduce at the pertinent points in our ensuing commentary. Other than that the accompanying figures are there (see note (58)) more logically renumbered into the sequence in which Newton cites them, the text printed by William Jones in his *Analysis Per Quantitatum Series, Fluxiones, ac Differentias: cum Enumeratione Linearum Tertii Ordinis* (London, 1711): 69–92 is essentially unchanged from those of its 1704 and 1706 issues; but see note (58) below.

(3) Originally '...CURVARUM CUBICI GENERIS' (...CURVES OF CUBIC KIND), with 'CUBICI' initially amended to be 'SECUNDI' (...SECOND). In his draft on Add. 3961.1: 38r Newton's first title proclaimed '...CURVARUM TERTIJ ORDINIS' (... CURVES OF THIRD ORDER), the 'CURVARUM' being initially replaced—as in his final version—by 'LINEARUM' (...LINES), but the whole thereafter converted into the further alternative '...CURVARUM SECUNDI ORDINIS' (...CURVES OF SECOND ORDER).

(4) Newton originally wrote 'Curvarum' (curves) in his preliminary draft on f. 38r; compare the previous note. We will not systematically take notice of such replacements in the sequel.

(5) Just 'commode' (conveniently) in Newton's first version on f. 38r.

(6) In his draft on f. 38r Newton initially added as a sop to his reader 'qui utiqꝫ species est Ellipseos' (which is, of course, a species of ellipse).

(7) After the many vicissitudes of framing an appropriate title to his piece (see note (3) above) Newton now settles for what will henceforth be his standard terminology.

(8) Better 'infiniti'—since the order of such a 'mechanical line' is of necessity infinitely great (and not infinitely small)—and so we render its sense in our English version.

(9) In present context understand 'Archimedea' (Archimedean). The trio of the cycloid, quadratrix and Archimedean spiral ever remained Newton's favourite examples of such 'lines of infinite order' from his earliest mathematical days (compare I: 378–9 and p. 396 above).

Cyclois, Quadratrix et linea omnis quæ per radij vel rotæ revolutiones infinitas generatur.

II Proprietates Sectionum Conicarū competunt curvis superiorum generum.[11]

[10]Sectionum Conicarum proprietates præcipuæ[12] a Geometris passim traduntur. Et consimiles sunt etiam proprietates Curvarum secundi generis & reliquarum, ut ex sequenti proprietatum præcipuarum enumeratione constabit.

III Curvarum secundi generis[14] Ordinatæ, Diametri, Vertices, Centra, Axes.

Nam si rectæ plures parallelæ & ad conicam sectionem utrinq; terminatæ ducantur, recta duas earū bisecans bisecabit alias omnes,[13] ideoq; dicitur *Diameter* figuræ, et rectæ bisectæ dicuntur *Ordinatim applicatæ* ad diametrum, et intersectio curvæ et diametri *Vertex* nominatur, et diameter illa *Axis* est cui ordinatim applicatæ insistunt ad angulos rectos. Et ad eundem modum in Curvis secundi generis, si rectæ duæ quævis parallelæ ducantur occurrentes Curvæ in tribus punctis: recta quæ ita secat has parallelas ut summa duarum partium ex uno secantis latere ad curvam terminatarum æquetur parti tertiæ ex altero latere ad Curvam terminatæ, eodem modo secabit omnes alias his parallelas curvæq; in tribus punctis occurrentes rectas, hoc est ita ut summa partium duarum ex uno ipsius latere semper æquetur parti tertiæ ex altero latere. Has itaq; tres partes quæ hinc inde æquantur, *Ordinatim applicatas*, et rectam secantem cui ordinatim applicantur *Diametrum*, et intersectionem diametri et curvæ *Verticem*, et concursum duarum diametrorum *centrum* nominare licet.[15] Diameter autem ad Ordinatas rectangula si modò aliqua sit, etiam *Axis* dici potest; et ubi omnes diametri in eodem puncto concurrunt istud erit *centrum generale*.

IV Asymptoti et earum proprietates.

Hyperbola primi generis[16] duas *Asymototos*, ea secundi tres, ea tertij quatuor et non plures habere potest et sic in reliquis. Et quemadmodum partes lineæ cujusvis rectæ inter Hyperbolam Conicam et duas ejus Asymptotos sunt hinc inde æquales: sic in Hyperbolis secundi generis, si ducatur recta quævis secans tam Curvam quàm tres ejus Asymptotos in tribus punctis, summa duarum partium istius rectæ quæ a duobus quibusvis Asymptotis in eandem plagam ad duo puncta Curvæ extenduntur æqualis erit parti tertiæ quæ a tertia Asymptoto in plagam contrariam ad tertium Curvæ punctum extenditur.[17]

(10) The next five paragraphs are taken over by Newton with little essential change from the opening to his earlier cubic enumeration a decade and a half before (see IV: 354–8). The marginal titles are, of course, here new.

(11) In the equivalent postil in his draft (f. 38ʳ) Newton's epitome was 'Curvarum proprietates principales quæ singulis ordinibus competunt' (The principal properties of curves which hold for the individual orders singly).

(12) In the draft on f. 38ʳ Newton first wrote 'generales' (general) and then—as in his accompanying marginal head (see the previous note)—'principales' (principal).

(13) Originally specified in draft as 'alias omnes parallelas rectas ad curvam utrinq; terminatas' (all other parallel straight lines terminated at the curve on their either hand), much as previously on IV: 354.

cycloid, quadratrix and every line which is generated by an infinity of revolutions of a radius or wheel.

II The properties of conics fit curves of higher kinds.(11)

(10)The chief(12) properties of conics are recounted *passim* by geometers. Closely similar also are the properties of curves of the second kind and the rest, as will be agreed from the following enumeration of their principal properties.

III Ordinates, diameters, vertices, centres and axes of curves of second kind.(14)

Specifically, if several parallel straight lines terminating at a conic at their either end be drawn, the straight line bisecting two of them will bisect all the others(13) and is consequently called a *diameter* of the figure, and the straight lines bisected are called *ordinates* to the diameter, while the intersection of the curve and a diameter is named the *vertex*, and the diameter to which the ordinates stand at right angles is an *axis*. And in much the same way if in curves of second kind any two parallel straight lines be drawn meeting the curve in three points, the straight line which cuts these parallels so that the sum of the two parts terminating at the curve on one side of the cutting line shall equal the third part terminating at the curve on its other side will then cut all other straight lines parallel to these and meeting the curve in three points in the same manner, that is, so that the sum of the two parts on one of its sides shall always equal the third part on the other. It is accordingly permissible to name these three parts which are equal on one and the other side *ordinates*, and the secant straight to which they are ordinate a *diameter*, the intersection of a diameter and the curve a *vertex*, and the meet of two diameters a *centre*.(15) A diameter, however, pertaining to right-angled ordinates—provided there be one—can also be called an *axis*; and, when all diameters are concurrent at the same point, that will be a *general centre*.

IV Asymptotes and their properties.

A hyperbola of first kind(16) can have two *asymptotes* and no more, one of second three, one of third four, and so forth in the rest. And, just as the parts of any straight line between a conic hyperbola and its two asymptotes are equal on one side and the other, so in hyperbolas of second kind, if any straight line be drawn cutting both the curve and its three asymptotes in three points, the sum of the two parts of that line extending from any two asymptotes in the same direction to two points of the curve will be equal to the third part extending from the third asymptote in the opposite direction to the third point of the curve.(17)

---

(14) Initially 'Linearum Ordinis tertij' (...lines of third order) in the draft on f. 38r; compare note (3) above. An equivalent replacement was made by Newton in his text alongside.

(15) In his draft (f. 38v) Newton here inserted: 'Hujusmodi diametros vertices et centra nominabimus *ordinis secundi*' (Diameters, vertices and centres of this type we shall name *of second order*).

(16) Initially 'Linea secundi ordinis' (A line of second order) in the draft on f. 38v; in sequel, correspondingly, Newton there referred to 'ea tertij..., ea quarti...' (one of third..., one of fourth...).

(17) Theorem 2 on II: 102, here repeated from IV: 356.

V
Latera recta et transversa.[18]

Et quemadmodum in Conicis sectionibus non Parabolicis quadratum ordinatim applicatæ, hoc est rectangulum ordinatarum quæ ad contrarias partes diametri ducuntur est ad rectangulum partium diametri quæ ad vertices Ellipseos vel Hyperbolæ terminantur ut data quædam linea quæ dicitur *Latus rectum* ad partem diametri quæ inter Vertices jacet et dicitur *Latus transversum*: sic in Curvis non Parabolicis secundi generis Parallelipipedum ex tribus ordinatim applicatis est ad parallelipipedum ex partibus diametri ad ordinatas & tres vertices figuræ abscissis[19] in ratione quadam data: in qua ratione si sumantur tres rectæ ad tres partes diametri inter vertices figuræ sitas, singulæ ad singulas, tunc illæ tres rectæ dici possunt *Latera recta* figuræ, et illæ partes diametri inter vertices *Latera transversa*. Et sicut in Parabola Conica, quæ ad unam et eandem diametrum unicum tantum[20] habet verticem, rectangulum sub ordinatis æquatur rectangulo sub parte diametri quæ ad Ordinatas et verticem abscinditur et recta quadam data quæ *Latus rectum* dicitur: sic in curvis secundi generis quæ non nisi duos habent vertices ad eandem diametrum, parallelipipedum sub ordinatis tribus æquatur parallelipipedo sub duabus partibus diametri ad ordinatas et vertices illos duos abscissis & recta quadam data, quæ proinde *Latus rectum* dici potest.

VI
Ratio contentorum[21] sub parallelarum segmentis.

Deniqȝ sicut in Conicis sectionibus ubi duæ parallelæ ad curvam utrinqȝ terminatæ secantur a duabus parallelis ad curvam utrinqȝ terminatis, prima a tertia et secunda a quarta, rectangulum partium primæ est ad rectangulum partium tertiæ ut rectangulum partium secundæ ad rectangulum partium quartæ: sic ubi quatuor tales rectæ occurrunt Curvæ secundi generis singulæ in tribus punctis, parallelipipedum partium primæ rectæ erit ad parallelipipedum partium tertiæ ut parallelipipedum partium secundæ ad parallelipipedum partium quartæ.[22]

VII
Crura Hyperbolica et Parabolica & eorũ plagæ.

Curvarum secundi et superiorum generum æque atqȝ primi crura omnia in infinitum progredientia vel *Hyperbolici* sunt generis vel *Parabolici*. Crus *Hyperbolicum* voco quod ad Asymptoton aliquam in infinitum[23] appropinquat, *Parabolicum* quod Asymptoto destituitur. Hæc crura ex tangentibus optime dignoscuntur. Nam si punctum contactus in infinitum abeat tangens cruris Hyperbolici cum Asymptoto coincidet et tangens cruris Parabolici in infinitum recedet evanescet et nullibi[24] reperietur. Invenitur igitur Asymptotos cruris cujusvis quærendo tangentem cruris ad punctum infinite distans. Plaga autem cruris infiniti invenitur quærendo positionem rectæ cujusvis quæ tangenti parallela est

(18) Newton's original marginal head in his draft (ff. $38^v/39^r$) stated that 'Data ratio contenti sub Ordinatis ad contentum sub partibus diametri' ([There is] given the ratio of the product of the ordinates to the product of the parts of a diameter).

(19) In the draft (f. 39^r) originally 'hinc inde terminatis' (terminated on either hand).

(20) Newton wrote 'non nisi unic[u]m' to precisely the same effect in his draft.

(21) Initially 'parallelipipedorum' (solid product) on f. 39^r, as in Newton's text alongside.

And just as in non-parabolic conics the square of the ordinate, that is, the rectangle of the ordinates which are drawn to opposite sides of their diameter, is to the rectangle of the parts of the diameter which terminate at the ellipse's or hyperbola's vertices as a certain line which is called the *latus rectum* to the part of the diameter lying between the vertices which is called the *latus transversum* (*transverse diameter*); so in non-parabolic curves of second kind the 'parallelepiped' (solid product) of the three ordinates is to that of the parts of the diameter cut off[19] at the ordinates and the three vertices of the figure in a given ratio, while if three straight lines be taken in this ratio to the three parts of the diameter situated between the vertices of the figure, each to a separate one, then those three lines can be called *latera recta* of the figure, and the parts of the diameter between the vertices *transverse diameters*. And as in the conic parabola, which has pertaining to one and the same diameter but a single vertex, the rectangle contained by the ordinates is equal to the rectangle beneath the part of the diameter cut off at the ordinates and the vertex and under a given straight line which is called the *latus rectum*; so in curves of second kind which have but two vertices pertaining to the same diameter the 'parallelepiped' contained by the three ordinates is equal to that contained beneath the two parts of the diameter cut off at the ordinates and those two vertices and under a given straight line, which can in consequence be called the *latus rectum*.

V
Latera recta and transverse diameters.[18]

Finally, just as in conics, when two parallels terminating on either side at the curve are cut by two parallels terminating on either side at the curve, the first by the third and the second by the fourth, the rectangle of the parts of the first is to that of the parts of the third as the rectangle of the parts of the second to that of the parts of the fourth; so when four such straight lines meet a curve of second kind, each individually in three points, the 'parallelepiped' (solid product) of the parts of the first line will be to that of the parts of the third as the 'parallelepiped' of the parts of the second to that of the parts of the fourth.[22]

VI
The ratio of the [products] contained[21] under segments of parallels.

In curves of the second and higher kinds equally as in those of the first all branches progressing to infinity are of either *hyperbolic* or *parabolic* kind. I call a branch 'hyperbolic' which[23] infinitely approaches some asymptote, and 'parabolic' one which is devoid of an asymptote. These branches are best ascertained from their tangents; for, if the point of contact should go off to infinity, the tangent of a hyperbolic will come to coincide with its asymptote, while the tangent of a parabolic branch will recede to infinity, so vanishing and being found nowhere.[24]. The asymptote of any branch is therefore discovered by seeking the tangent to the branch at an infinitely distant point. The direction of an infinite branch, however, is found out by seeking the position of any

VII
Hyperbolic and parabolic branches and their directions.

(22) See IV: 358, note (18).
(23) Newton initially added a superfluous 'semper' (always) at this point in his draft (f. 39^v).
(24) At a finite distance, that is.

ubi punctum contactûs in infinitum abit. Nam hæc recta in eandem plagam cum crure infinito dirigitur.(25)

VIII Reductio Curvarum omnium generis secundi ad æquationum casus quatuor.

(26)Lineæ omnes ordinis primi, tertij, quinti, septimi et imparis cujusꝗ duo habent ad minimum crura in infinitum versus plagas oppositas progredientia. Et lineæ omnes tertij ordinis duo habent ejusmodi crura in plagas oppositas progredientia in quas nulla alia earum crura infinita (præterquam in Parabola Cartesiana(27)) tendunt.(28)

Casus primus. Si crura illa sint hyperbolici generis, sit AS eorum Asymptotos et huic parallela agatur recta quævis CBc ad Curvam utrinꝗ (si fieri potest) terminata, eademꝗ bisecetur in puncto X et locus puncti illius X erit Hyperbola Conica (puta $X\Phi$) cujus una asymptotos est AS. Sit ejus altera asymptotos AB.(29) Et æquatio qua relatio inter ordinatam BC et abscissam AB definitur,

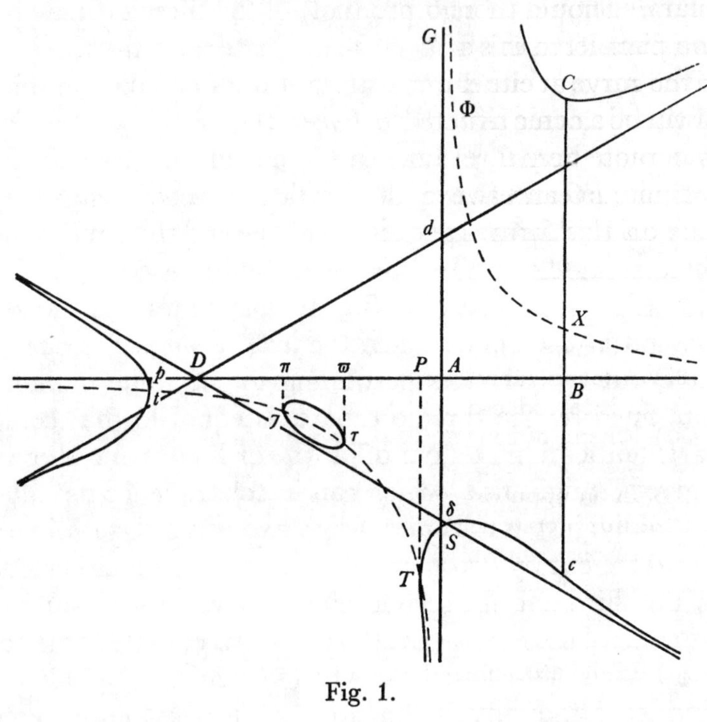

Fig. 1.

si AB dicatur x et BC y, semper induet hanc formam

$$xyy+ey = ax^3+bxx+cx+d.^{(30)}$$

(25) In his annotated copy of the first Latin edition of his *Optice* (see note (2) above) Newton at this point in the appended text of his 'Enumeratio' here inserted two extra sentences: 'Ex Asymptotis vero et Crurum infinitorum plagis dabuntur positiones Abscissæ et Ordinatæ, quarum relatio per æquationem definiri debet, ut inde species Curvæ innotescat. Æquatio autem ex datis Curvæ conditionibus colligenda est: et sunt æquationum sic prodeuntium casus quatuor sequentes' (From the asymptotes, however, and the directions of the infinite branches there will be given the positions of the abscissa and ordinate, the relationship of which ought to be defined by an equation so that the species of the curve may be established therefrom. The equation, of course, must be gathered from the given circumstances of the curve: and of the equations resulting in this way there are the four following cases). Since a first draft of this unpublished addition—where the latter sentence is rendered more shortly as 'Provenient autem æquationum casus quatuor '(Four cases of the equations will ensue)—was entered by him on the verso of a letter-cover (ULC. Add .3964.8: 16) addressed 'To the R:t Hon:ble S:r Isaac Newton in S:t Martins Street near Leister fields', the insertion must have been made

straight line which is parallel to the tangent when the point of contact goes off to infinity; for this line is directed to the same region as the infinite branch.(25)

VIII
Reduction of all curves of second kind to four cases of equations.
First case.

(26)All lines of first, third, fifth, seventh and each odd order have at least two branches advancing to infinity in opposite directions. And all lines of third order have two branches of the sort progressing in opposite directions to which no other of their infinite branches—except in the Cartesian parabola(27)—tend.(28) Should those branches be of hyperbolic kind, let AS be their asymptote and parallel to this draw any straight line CBc terminating (if it can so be done) at the curve at either end, then bisect it in the point X and the locus of that point X will be a conic hyperbola ($X\Phi$ say), one asymptote of which is AS. Let its other asymptote be AB,(29) and the equation by which the relationship between the ordinate BC and abscissa AB is defined will, if AB be called x and BC y, always take on this form: $xy^2+ey = ax^3+bx^2+cx+d$,(30) where the terms e, a, b, c, d

some time in or after 1710 when he took up residence at the house at 35 St Martin's Street, off Leicester Square, where he dwelt almost the whole of his remaining life.

(26) A considerably different first version (on Add. 3961.2: 16v) of the next four paragraphs VIII–XI is reproduced in Appendix 1.1 below.

(27) Where, of course, the corresponding point at infinity is a node (in which its *crura infinita hyperbolici generis* are intersected by the curve's second, parabolic branch).

(28) In an unpublished manuscript addition at this point in the 1706 printed text of the 'Enumeratio' (on which see note (2) above) Newton afterwards interpolated: 'Crura autem vel Hyperbolica sunt vel Parabolica; et recta infinita cruribus parallela Curvam vel in duobus punctis vel in unico tantum secat. Casus non sunt plures' (Such branches, however, are either hyperbolic or parabolic; and an infinite straight line parallel to the branches cuts the curve [*sc.* respectively] either in two points or in but a single one. There are no further cases).

(29) In the accompanying figure as in those which follow—all, for want (see note (2)) of their holograph originals, here taken over from the equivalents which illustrate the 1704 *editio princeps* of the 'Enumeratio'—the asymptotes GAS and DAB are for convenience (not of course out of necessity) drawn to be at right angles: a 'convention' which is obeyed in all the printed editions. In his preliminary version (f. 40r), exactly as in the initial draft reproduced in Appendix 1.1 below, Newton originally here went on to add: 'Et siquando Hyperbola illa conica cum asymptotis suis coincidendo in rectas duas migraverit, sunto rectæ illæ AS et AB' (And whenever that conic hyperbola, by coinciding with its asymptotes, passes into two straight lines, let those lines be AS and AB).

(30) In Newton's preliminary version (f. 40r) these four equations—defining the general hyperbolic cubic, the Cartesian trident, the general divergent parabola and the Wallisian cubic parabola respectively—were, as in his initial draft (see Appendix 1.1) originally set to be

$$`xyy = dx^3+gxx+hy+kx+l', \quad `xy = dx^3+gxx+kx+l', \quad `yy = dx^3+gxx+kx+l'$$
$$\text{and } `y = dx^3+gxx+kx+l';$$

and thereafter

$$`xyy = \pm ax^3 \pm bxx \pm cx \pm d \pm ey', \quad `xy = \pm ax^3 \pm bxx \pm cx \pm d', \quad `yy = \pm ax^3 \pm bxx \pm cx \pm d'$$
$$\text{and } `y = \pm ax^3 \pm bxx \pm cx \pm d'$$

in a standardization no longer revealing their origin. His present semi-geometrical derivation of these four reduced canonical forms of the cubic's Cartesian equation, valid though it is (if its assumptions are clarified), wears something of the air of a mathematical sleight-of-hand, and

Ubi termini e, a, b, c, d designant quantitates datas cum signis suis $+$ et $-$ affectas quorum quilibet[31] deesse possunt, modo ex eorum defectu figura in sectionem Conicam non vertatur. Potest autem Hyperbola illa Conica cum asymptotis suis coincidere, id est punctum X in recta AB locari: et tunc terminus $+ey$ deest.[32]

IX Casus secundus.

At si recta illa CBc non potest utrinꝗ ad Curvam terminari sed Curvæ in unico tantum puncto occurrit: age quamvis positione datam AB asymptoto AS occurrentem in A ut et aliam quamvis BC asymptoto illi parallelam et curvæ occurrentem in puncto C, et æquatio qua relatio inter ordinatam BC et abscissam AB definitur semper induet hanc formam $xy = ax^3 + bxx + cx + d$.[30]

X Casus tertius.

Quod si crura illa duo opposita Parabolici sint generis, recta CBc ad Curvam utrinꝗ (si fieri potest) terminata in plagam crurum ducatur & bisecetur in B, & locus puncti B erit linea recta. Sit ista AB terminata ad datum quodvis punctum A, et æquatio qua relatio inter ordinatam BC et abscissam AB definitur semper induet hanc formam, $yy = ax^3 + bxx + cx + d$.[30]

XI Casus quartus.

At verò si recta illa CBc in unico tantum puncto occurrat Curvæ,[33] ideoꝗ ad Curvam utrinꝗ terminari non possit: sit punctum illud C, et incidat recta illa ad punctum B in rectam quamvis aliam positione datam et ad datum quodvis punctum A terminatam AB: et æquatio qua relatio inter ordinatam BC et abscissam AB definitur, semper induet hanc formam, $y = ax^3 + bxx + cx + d$.[30]

XII Nomina formarum.[34]

Enumerando curvas horum casuum [35] Hyperbolam vocabimus *inscriptam* quæ tota jacet in Asymptotôn angulo ad instar Hyperbolæ conicæ, *circumscriptam* quæ Asymptotos secat & partes abscissas in sinu suo amplectitur, *ambigenam* quæ uno crure infinito inscribitur & altero circumscribitur, *con-*

---

it is not surprising that his contemporaries greeted it with some suspicion upon the appearance of the *editio princeps* in 1704. (Compare the preceding introduction.) Newton himself had in earlier years twice (see II: 10–16; IV: 362–8) been privately at considerable pains to produce this quartet of equations by appropriate linear transformations of the coordinates. Here too, more simply, he might have taken as basis the defining equation

$$bxy^2 + cx^2y + dx^3 + ey^2 + fxy + gx^2 + hy + kx + l = 0$$

of a general cubic which has a (real) asymptote parallel to the y-axis ($x = 0$): at once, since the equation may be rewritten as $(bx+e)\,y^2 + (cx^2+fx+h)\,y + dx^3 + gx^2 + kx + l = 0$, the locus of the mid-points of its chords parallel to the asymptote is $2(bx+e)\,y + cx^2 + fx + h = 0$, and if the asymptotes $x = -e/b$ and $2by + cx = ce/b - f$ of this diametral conic hyperbola are taken as the new axes there is produced Newton's primary reduced canonical equation; but if $b = 0$, whence also $c = 0$ (or else the cubic has a single real asymptote parallel to $cy + dx = 0$ *contra hypothesin*), the locus of the mid-points of the parallel chords degenerates to be the straight line $2ey + fx + h = 0$, and Newton's third reduced form ensues on taking this to be the new x-axis; while if also $e = 0$, then we need only put $x = -h/f$ for the new y-axis to have his second form; and if $e = f = 0$ (so that the diametral line is at infinity), his final form results (on trivially putting $h = -1$ at least).

(31) In the draft (f. 40^r) initially just 'Ubi termini signis ambiguis notati' (where the terms

denote given quantities furnished with their signs + and −, any of which(31) can be wanting provided that in consequence of their lack the figure does not turn into a conic. That conic hyperbola can, moreover, coincide with its asymptotes, that is, the point X be located in the straight line AB; and in that case the term $+ey$ is wanting.(32).

IX
Second case.

But if that line CBc cannot terminate on either side at the curve, but meets the curve in merely a single point, draw any AB given in position meeting the asymptote AS in A, and also any other, BC, parallel to that asymptote and meeting the curve in the point C, and the equation by which the relationship between ordinate BC and abscissa AB is defined will then always assume this form: $xy = ax^3+bx^2+cx+d$.(30)

X
Third case.

While if the two opposite branches be of parabolic kind, draw the straight line CBc terminating at the curve (if it can so be) at its either end and bisect it in B, and the locus of the point B will then be a straight line. Let that be AB terminating at any given point A, and the equation whereby the relationship between the ordinate BC and abscissa AB is defined will always take on this form: $y^2 = ax^3+bx^2+cx+d$.(30)

XI
Fourth case.

But if, however, the straight line CBc should meet the curve in a single point only,(33) and consequently cannot terminate at the curve at both its ends, let that point be C and let that line fall at the point B upon any other straight line AB given in position and terminating at any given point A; the equation by which the relationship between the ordinate BC and abscissa AB is defined will always assume this form: $y = ax^3+bx^2+cx+d$.(30)

XII
Names of the forms.(34)

In enumerating the curves of these cases(35) we shall call a hyperbola *inscribed* when it lies wholly in the angle of its asymptotes in the fashion of a conic hyperbola, *circumscribed* when it cuts the asymptotes and embraces the parts cut off in its fold, *double-kinded* when it is inscribed in one of its infinite branches and

marked with ambiguous signs) when the preceding equation was (see the previous note), first set equivalently to be '$xyy = \pm ax^3 \pm bxx \pm cx \pm d \pm ey$'.

(32) Since in this particular case the two values of y determined by the equation must be equal but opposite in sign. Newton originally in his draft wrote '$\pm ey$' in line with his prior denotation of the cubic (compare the two preceding notes).

(33) In his draft Newton first went on to specify 'puta in C' (say in C).

(34) In the draft (f. 50r; see the next note) 'hujus Hyperbolæ [*sc.* redundantis]' (of this [redundant] hyperbola) was initially specified when the text alongside occupied its original site after the two paragraphs which next follow.

(35) This connecting phrase was in the preliminary version—where the present assignation of descriptive names for the various forms and configurations of cubic was, in afterthought, appended (on ff. 50r/50v) to follow the next paragraph but one below—first made to be a separate sentence: 'Cæterum ut Hyperbolas hujus ordinis commode enumeremus varijs ejus formis imponenda sunt sequentia nomina' (In order conveniently to enumerate the hyperbolas of this order, however, we need to set the following names upon their various forms). A first draft of the ensuing text on Add. 3961.2: 15v is reproduced in Appendix 1.2.

vergentem cujus crura[36] concavitate sua se invicem respiciunt et in plagam eandem diriguntur,[37] *divergentem* cujus crura convexitate se invicem respiciunt et in plagas contrarias diriguntur, *cruribus contrarijs præditam*[38] cujus crura[36] in partes contrarias convexa sunt et in plagas contrarias infinita, *Conchoidalem* quæ vertice concavo et cruribus divergentibus ad Asymptoton applicatur, *anguineam* quæ flexibus contrarijs asymptoton secat et utrinꝗ in crura contraria producitur, *cruciformem* quæ conjugatam decussat, *nodatam* quæ seipsam decussat in orbem redeundo, *cuspidatam* cujus partes duæ in angulo contactus concurrunt et ibi terminantur, *punctatam* quæ conjugatam habet Ovalem infinite parvam id est punctum, et *puram* quæ per impossibilitatem duarum radicum[39] Ovali, Nodo, Cuspide et Puncto conjugato privatur. Eodem sensu Parabolam quoꝗ *convergentem, divergentem, cruribus contrarijs præditam,*[38] *cruciformem, nodatam, cuspidatam, punctatam* et *puram* nominabimus.

XIII
De Hyperbola redundante et ejus tribus Asymptotis.[41]

[40]In casu primo si terminus ax^3 affirmativus est Figura erit Hyperbola triplex cum sex cruribus hyperbolicis[42] quæ juxta tres Asymptotos (quarum nullæ sunt parallelæ) in infinitum progrediuntur binæ juxta unamquamꝗ in plagas contrarias. Et hæ Asymptoti si terminus bxx non deest se mutuò secabunt[43] in tribus punctis triangulum ($Dd\delta$) inter se continentes, sin terminus bxx deest convergent omnes ad idem punctum. In priori casu cape

$$AD = \frac{-b}{2a} \quad \text{et} \quad Ad = A\delta = \frac{b}{2\sqrt{[a]}}\,{}^{(44)}$$

ac junge Dd, $D\delta$ & erunt $A[d]$,[45] Dd, $D\delta$ tres asymptoti.[46] In posteriori duc

(36) In his draft (f. 50^r) Newton has here cancelled 'infinita' (infinite) and 'duo continua' (two continuous) respectively.

(37) A superfluous 'ad modum Hyperbolæ conicæ' (in the style of the conic hyperbola) is deleted in sequel in the preliminary draft.

(38) Initially 'sibi contrariam' (contrary to itself) in each case.

(39) In his draft (f. 50^v) Newton originally started to specify in sequel 'æquationis qua curva d[efinitur]' (of the equation whereby the curve is defined)—that is, of

$$ax^3+bx^2+cx+d = 0$$

by which the cubic's intersection with the abscissa $y = 0$ is in each of the four preceding canonical forms of Cartesian equation determined.

(40) The next sixteen paragraphs (XIII—XXVIII) in which Newton sets out his final algebraic enumeration of cubics in sixteen *genera* (adiametral/monodiametral/tridiametral general redundant hyperbolas; adiametral/monodiametral/tridiametral redundant hyperbolas with concurrent asymptotes; adiametral/monodiametral defective hyperbolas; adiametral/monodiametral parabolic hyperbolas; adiametral/monodiametral hyperbolic hyperbolisms; adiametral/monodiametral elliptical hyperbolisms; adiametral/monodiametral parabolic hyperbolisms; the Cartesian trident; the divergent parabolas; and the Wallisian parabola) are, it will be evident, closely styled on the equivalent tabulation (see II: 40–70) which he had originally framed nearly thirty years before, though now of course the individual appellations of curves are greatly altered, and the 58 types of cubic initially distinguished (see II: 37, note (1))

circumscribed in the other, *convergent* when its [36]branches face one another in their concavity and are directed the same way,[37] *divergent* when its branches face one another in their convexity and are directed in opposing ways, *endowed with opposite branches*[38] when its [36]branches are convex in opposing ways and infinite in opposite directions, *conchoidal* (*shell-shaped*) when it is applied along the asmyptote with a concave vertex and divergent branches, *snaky* when it cuts its asymptote with an inflection and is extended either way in opposing branches, *cross-shaped* when it crosses over its conjugate curve, *nodate* when in circling back it crosses itself, *cusped* when its two parts concur at an angle of contact and there terminate, *punctate* when it has an infinitely small conjugate oval, that is, a conjugate point, and *pure* when through the impossibility of two roots[39] it is deprived of oval, node, cusp and conjugate point. With the same sense we shall name a parabola, too, *convergent, divergent, endowed with opposite branches,*[38] *cross-shaped, nodate, cusped, punctate* and *pure*.

XIII The redundant hyperbola and its three asymptotes.[41]

[40] In the first case, if the term ax^3 is positive the figure will be a triple hyperbola with six hyperbolic branches[42] which proceed to infinity in line with three asymptotes (none of which are parallel), a pair along one each in opposite directions. And if the term bx^2 is not lacking, these asymptotes will mutually intersect[43] one another in three points, so containing between themselves a triangle ($Dd\delta$); but if the term bx^2 is wanting, they will all converge on the same point. In the former case take $AD = -\frac{1}{2}b/a$ and $Ad = A\delta = \frac{1}{2}b/\sqrt{a}$[44] and join Dd, $D\delta$, and then Ad,[45] Dd, $D\delta$ will be the three asymptotes.[46] In the latter,

are now increased by minor additions to be altogether

$$(9+12+2;\ +4+4+1;\ +6+7;\ +7+4;\ +3+1;\ +2+1;\ +1+1;\ +1;\ +5;\ +1\ =\ )\ 72$$

separate species. While, as before, we continue to register in our accompanying footnotes the significant textual variations in Newton's preliminary draft (on Add. 3961.1: 40ᵛ–45ᵛ) of the revised text here reproduced, let us add that we do not systematically record the changes in the numbering of the single species consequent upon his decision in this final version to distinguish species 14/16, 15/17 and 28/29 which were not previously (compare note (1) above) so divided.

(41) In the draft at this point (f. 40ᵛ) Newton has set down a rather more appropriate variant marginal head 'XIII/Formarum hujus/Hyperbolæ descrip-/tio generalis' (XIII. A general description of the forms of this hyperbola) which was not carried over into his final text.

(42) An unnecessary 'ad invicem inclinatis' (inclined to each other) is deleted in sequel in Newton's draft at this point.

(43) Initially 'decussabunt' (...cross) in the draft.

(44) Here forgetting to allow for his earlier switch in the coefficient of x^3 in his primary equation (see note (30)), Newton set the quantity in the radical in the denominator of this fraction to be 'd'. The slip lingered on into the 1704 *editio princeps*, to be corrected in the 1706 reprint.

(45) In both draft (f. 40ᵛ) and revise Newton here carelessly wrote 'AD', and again the slip was perpetuated in print in 1704, to be caught in 1706.

(46) Understand Figure 1 following. In the cubic's defining equation

$$xy^2+ey\ =\ ax^3+bx^2+cx+d$$

the highest-order terms $xy^2-ax^3 \equiv x(y+\sqrt{a})\ (y-\sqrt{a})\ =\ 0$ determine the directions of the

ordinatam quamvis BC & in ea utrinq; producta cape hinc inde BF et Bf sibi mutuò æquales et in ea ratione ad AB quam habet $\sqrt{d}$ ad a jungeq; AF, Af et erunt $A[G]$, AF, Af tres Asymptoti.(47) Hanc autem Hyperbolam vocamus redundantem quia numero(48) crurum Hyperbolicorum sectiones Conicas superat.

XIV De hujus Hyperbolæ diametris & situ crurum infinitorum.

In Hyperbola omni redundante si neq; terminus ey desit neq; sit $bb-4ac$ æquale $\pm ae\sqrt{a}$,(49) curva nullam habebit diametrum, sin eorum alterutrum accidat curva habebit unicam, et tres si utrumq;. Diameter autem semper transit per intersectionem duarum Asymptotôn et bisecat rectas omnes quæ ad asymptotos illas utrinq; terminantur & parallelæ sunt Asymptoto tertiæ.(50) Estq; abscissa AB diameter Figuræ quoties terminus ey deest. Diametrum verò absolutè dictam hic et in sequentibus in vulgari significatu usurpo, nempe(51) pro abscissa quæ passim habet ordinatas binas æquales ad idem punctum hinc inde insistentes.

XV Hyperbolæ novem redundantes quæ diametro destituuntur & tres habent Asymptotos triangulum capientes.

Si Hyperbola redundans nullam habet diametrum quærantur Æquationis hujus

$$ax^4+bx^3+cx^2+dx+\tfrac{1}{4}ee^{(52)}=0$$

radices quatuor seu valores ipsius x. Eæ sunto AP, $A\varpi$, $A\pi$, Ap. Erigantur ordinatæ PT, $\varpi\tau$, $\pi\jmath$, pt & hæ tangent Curvam in punctis totidem T, τ, $\jmath$, t, et tangendo dabunt limites Curvæ per quos species ejus innotescet.

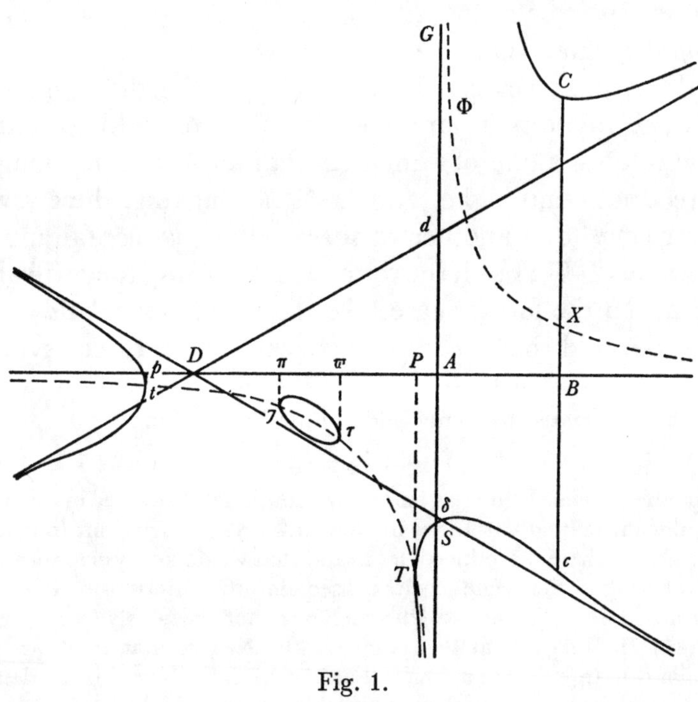

Fig. 1.

Fig. 1, 2. Nam si radices omnes AP, $A\varpi$, $A\pi$, Ap sunt reales, ejusdem signi et inæquales, Curva constat ex tribus Hyperbolis, (inscripta circumscripta et ambigena) *cum*

asymptotes. Of these Ad $(x=0)$ is manifestly one, while the two others Dd, $D\delta$ are evidently defined by the equations $y=\pm\sqrt{a}.(x-X)$. To determine X we need only substitute to find $2aX+b$ as the coefficient of x^2, of necessity zero if the given pair of lines are asymptotes; while $Ad=-A\delta=\frac{1}{2}b/\sqrt{a}$ and $AD=-\frac{1}{2}b/a$ are readily fixed by setting respectively $x=0$ and $y=0$ in their equation.

(47) Here understand Figure 30 on page 614 below.

(48) Originally 'multitudine' (multitude) in Newton's draft.

(49) Newton here perpetuates a numerical error—corrected in our English version—into

draw any ordinate BC and in its extension either way take on opposite sides BF Bf equal to each other and in the ratio to AB which $\surd d$ has to a, then join AF, Af and AG, AF, Af will be the three asymptotes.(47) We call this hyperbola 'redundant', however, because in the number(48) of its hyperbolic branches it surpasses conic ones.

XIV The diameters of this hyperbola and the position of its infinite branches.

In every redundant hyperbola, if neither the term ey should be wanting nor b^2-4ac equal to $\pm[4]ae\surd a$,(49) the curve will have no diameter; but if one or other of these circumstances should occur the curve will have a single one, and three if both happen. A diameter, however, always passes through the intersection of two asymptotes and bisects all straight lines which terminate at those asymptotes at their either end and are parallel to the third asymptote.(50) And the abscissa AB is a diameter of the figure each time the term ey is wanting. 'Diameter' absolutely so-called I here, of course, and in the sequel employ in its common meaning, namely(51) to signify an abscissa which everywhere has a pair of equal ordinates standing on the same point to its either side.

XV The nine redundant hyperbolas which are devoid of diameter and have three asymptotes containing a triangle.

If a redundant hyperbola has no diameter, seek the four roots, or values of x, in this equation $ax^4+bx^3+cx^2+dx+\frac{1}{4}e^{2}$(52) $=0$. Let them be AP, $A\varpi$, $A\pi$ and Ap; erect the ordinates PT, $\varpi\tau$, $\pi 7$, pt and these will touch the curve in as many points T, τ, 7, t and, in so touching, yield limits to the curve by which its species will come to be known.

To be sure, if all the roots AP, $A\varpi$, $A\pi$ and Ap are real, of the same sign and unequal, the curve consists (Figures 1, 2) of three hyperbolas—inscribed,

which he had equivalently fallen in his youthful enumerations at this point (see II: 42, note (19) and 54, note (40)). Though he himself caught the slip in about 1710 in a marginal correction to his annotated copy of the 1706 printing (now ULC. Adv. b. 39.4; see note (2) above), it was adjusted publicly in none of the editions of the 'Enumeratio' printed in Newton's lifetime, nor —a silent comment upon the capacity of his subsequent editors to detect such mistakes in mathematical detail!—was it afterwards trapped either by Castiglione in 1744 or Horsley in 1779. In general, of course, the two oblique asymptotes $y=\pm\surd a.(x-\frac{1}{2}b/a)$ each meet the cubic in a finite point fixed by $x=(4ad\mp 2be\surd a)/(b^2-4ac\pm 4ae\surd a)$; but when the cubic is tridiametral this also must lie at infinity.

(50) Originally in the draft 'Diametri autem semper transeunt per intersectionem duarum Asymptotorum et bisecant latera opposita trianguli quod Asymptoti continent' (Diameters, however, always pass through the intersection of two asymptotes and bisect the opposite sides of the triangle which the asymptotes contain).

(51) In his draft Newton here inserted 'pro Diametro primi ordinis id est' (as a diameter of first order, that is), after which he initially went on with 'pro abscissa cui Ordinatæ binæ ad idem punctum hinc inde insistunt quæ sibi mutuò semper æquantur' (to signify an abscissa upon which at the same point and to its either side stand pairs of ordinates which are ever equal one to the other).

(52) Omitting there to carry through his change in the coefficient of y in his primary equation (see note (30) above), Newton wrote '$+\frac{1}{4}hh$' in his preliminary version (f. 40v). His quartic resolvent determines, it will be clear, the abscissas of the four points in which the cubic $xy^2+ey=ax^3+bx^2+cx+d$ meets its diametral hyperbola $xy+\frac{1}{2}e=0$ (and at which the corresponding ordinates are evidently tangent to the cubic).

Ovali.(53) Hyperbolarum una jacet versus D, altera versus d, tertia versus δ, & Ovalis semper jacet intra triangulum $Dd\delta$(54) atq; etiam inter medios limites 7 et τ in quibus utiq; tangitur ab ordinatis π7 et $\varpi\tau$. Et hæc est species prima.

Fig. 3, 4. Si e radicibus duæ maximæ $A\pi$, Ap, vel duæ minimæ AP, $A\varpi$ æquentur inter se, et ejusdem sunt signi cum alteris duabus, Ovalis et Hyperbola circumscripta sibi invicem junguntur, coeuntibus earum punctis contactûs 7 et t vel T et τ, et crura Hyperbolæ sese decussando in Ovalem continuantur, figuram *nodatam* efficientia.(55) Quæ species est secunda.

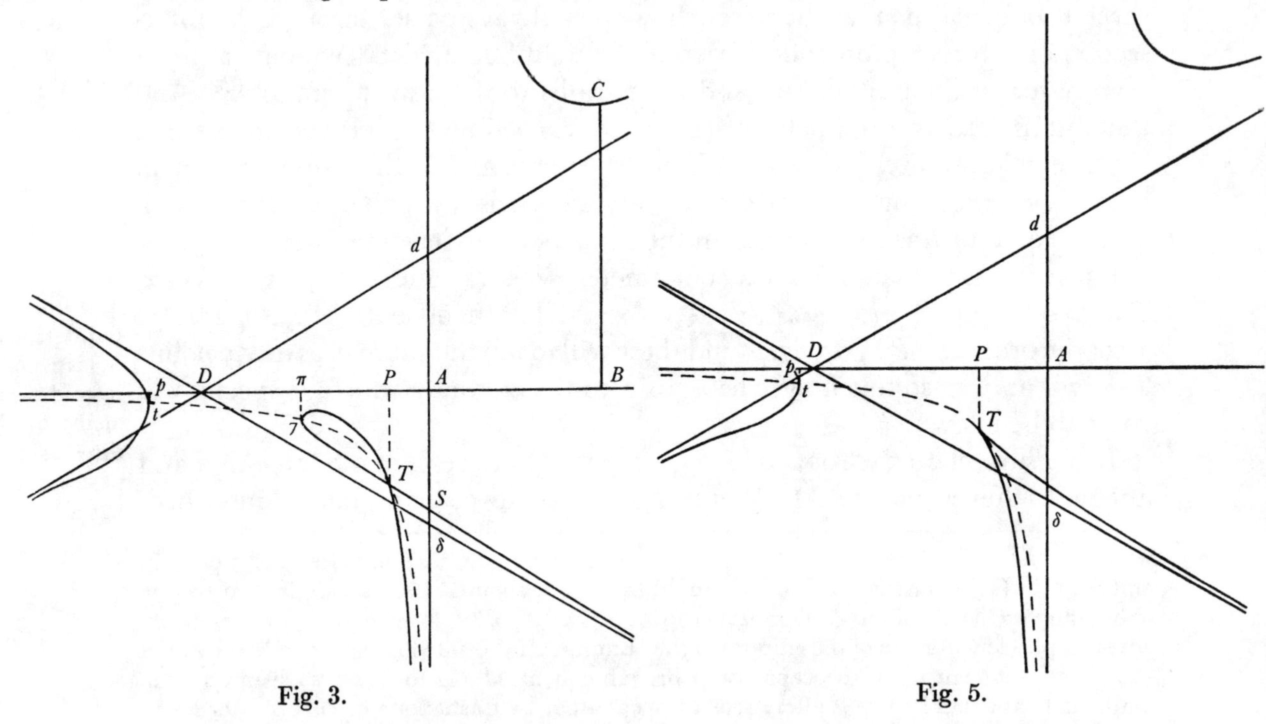

Fig. 3. Fig. 5.

Fig. 5, 6. Si e radicibus tres maximæ Ap, $A\pi$, $A\varpi$, vel tres minimæ $A\pi$, $A\varpi$, AP æquentur inter se, (56)nodus in *cuspidem* acutissimum convertetur. Nam crura duo Hyper-

(53) Initially '*Ellipsi conjuga*[*ta*]' (*conjugate oval*) in the draft (f. 42r). Newton has there made a number of analogous replacements of 'Ellipsis' by 'Ovalis' in sequel.

(54) The oval cannot, of course, touch or intersect any asymptote; for then the latter would meet the cubic in three finite points, which is impossible. Nicole neglected to heed this basic restriction when in his 'Traité des lignes du troiséme order' (*Mémoires de l'Académie Royale des Sciences. Année* MDCCXXIX (Paris, 1731): 194–224) he asserted (*ibid.*: 217, §31) the existence of a phantom tenth species of this primary *genus* of adiametral redundant hyperbolas in which the oval was continued outside the triangle of the asymptotes to connect with an opposing ambigenal hyperbolic branch.

(55) 'formam *nodi* constituentes' (forming the shape of a *node*) in Newton's draft version.

(56) 'Ovalis illa in punctum decussationis evanuit et' (the oval has vanished into the crossing point and) is deleted in the draft.

circumscribed and double-kinded—*with an oval*.(53) Of the hyperbolas one lies towards D, the second towards d, the third towards δ, while the oval lies always within the triangle $Dd\delta$(54) and also between the middle limits γ and τ in which, of course, it is touched by the ordinates $\pi\gamma$ and $\varpi\tau$. This is the first species.

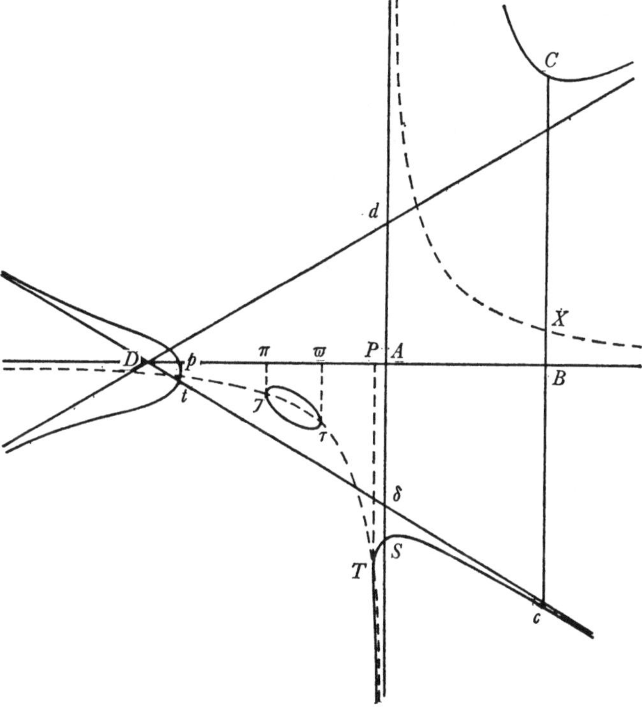

Fig. 2.

If among the roots the two greatest $A\pi$ and Ap, or the two least, AP and $A\varpi$, be equal to one another and of same sign as the two others, the oval and circumscribed hyperbola are (Figures 3, 4) joined to each other, their points of contact γ and t or T and τ coalescing, and the branches of the hyperbola are, on crossing themselves, continued into the oval, rendering the figure *nodate*.(55) And this is the second species.

If of the roots the three greatest, Ap, $A\pi$ and $A\varpi$, or three least, $A\pi$, $A\varpi$ and AP, should be equal to one another,(56) the node will (Figures 5, 6) be converted

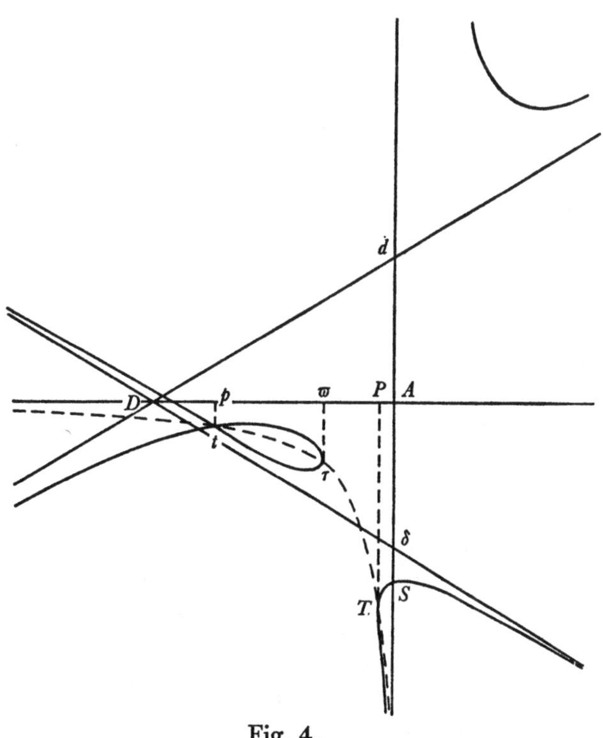

Fig. 4.

bolæ circumscriptæ ibi in angulo contactus concurrent et non ultra producentur. Et hæc est species tertia.

Si e radicibus duæ mediæ $A\varpi$ et $A\pi$ æquentur inter se, puncta contactûs τ et γ coincidunt, et propterea Ovalis interjecta in punctum evanuit, et constat figura ex tribus Hyperbolis, inscripta, circumscripta & ambigena cum *puncto* conjugato. Quæ est species quarta.

Fig. 7.

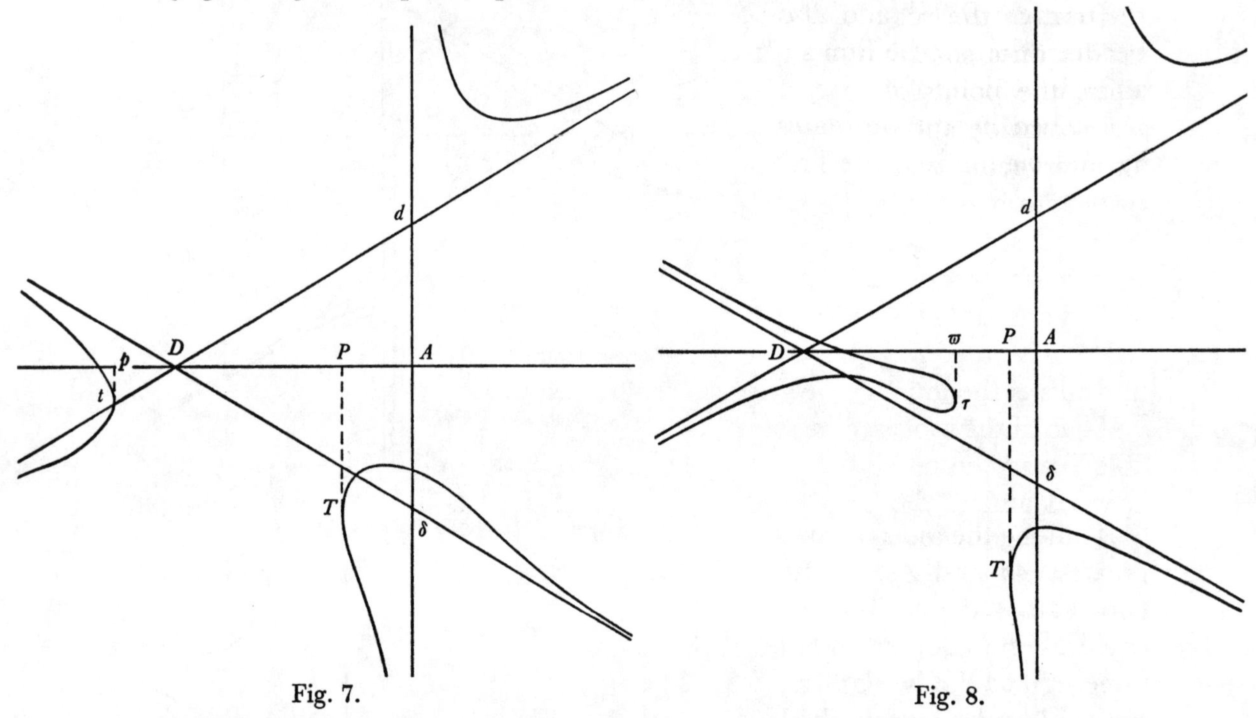

Fig. 7. Fig. 8.

Si duæ ex radicibus sunt impossibiles(57) et reliquæ duæ inæquales et ejusdem signi (nam signa contraria habere nequeunt) *puræ* habebuntur Hyperbolæ tres sine Ovali vel nodo vel cuspide vel puncto conjugato, & hæ Hyperbolæ vel ad latera(59) trianguli ab Asymptotis comprehensi vel ad angulos ejus jacebunt & perinde speciem vel quintam vel sextam constituent.

Fig. 7, 8, 13, 14.(58)

(57) Newton first began to write 'ima[ginariæ]' (imaginary) in his draft.

(58) In the minimally re-edited text of the 'Enumeratio' which William Jones presented in his 1711 collection of Newton's tracts and other miscellaneous fragments on *Analysis* (see note (2) above) he trivially reorganized the numbering of these and the immediately following figures into correct ordinal sequence, permuting 13 and 14 to be 9 and 10; 9 and 10 to become 11 and 12; 15 and 16 to be the new 13 and 14; 11 to be 15; and 12 to be 16. An analogous recycling of the numbers of Newton's figures illustrating the Cartesian trident and the several species of divergent parabola was also there made by him, not wholly consistently; see note (95) below. These changes are, in full awareness of Newton's original numberings or no, maintained in all subsequent printed editions of the 'Enumeratio'.

(59) Newton wrote 'bases' (bases) in his draft version (f. 41v).

into an acutely sharp *cusp*. The two branches of the circumscribed hyperbola there, of course, will concur at an angle of contact and not extend beyond. This is the third species.

If out of the roots the two middle ones shall equal each other, the points of contact τ and 7 coincide, and accordingly the intervening oval has shrunk away into a point, and so the figure consists (Fig. 7) of three hyperbolas, inscribed, circumscribed and double-kinded, together with a conjugate point. And this is the fourth species.

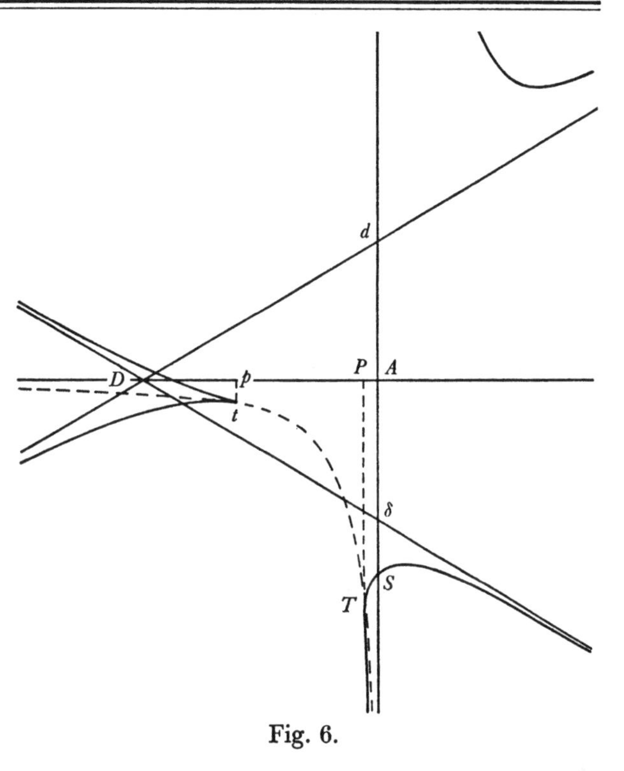

Fig. 6.

If two of the roots are impossible[57] and the remaining two unequal and of the same sign (to be sure they cannot have contrary signs), there will be had (Figures 7, 8, 13, 14[58]) three *pure* hyperbolas without oval, node, cusp or conjugate point, and these hyperbolas will lie either along the sides[59] of the triangle comprised by

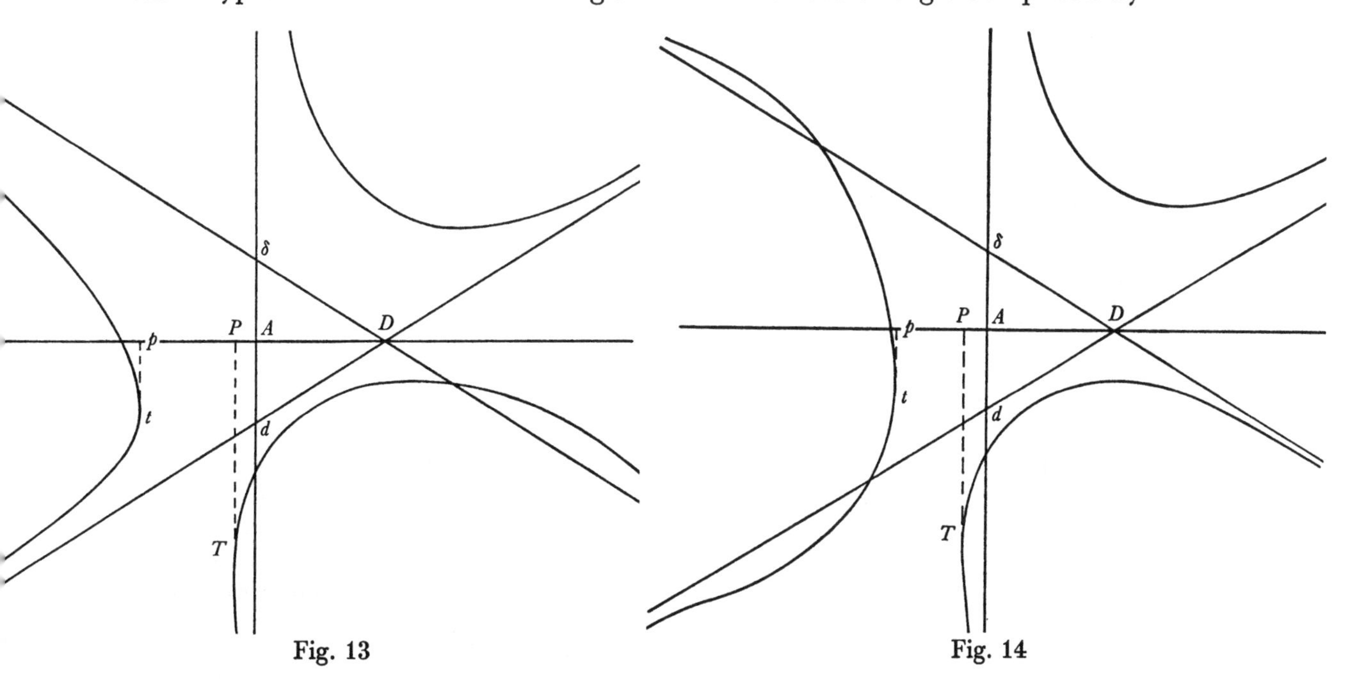

Fig. 13 Fig. 14

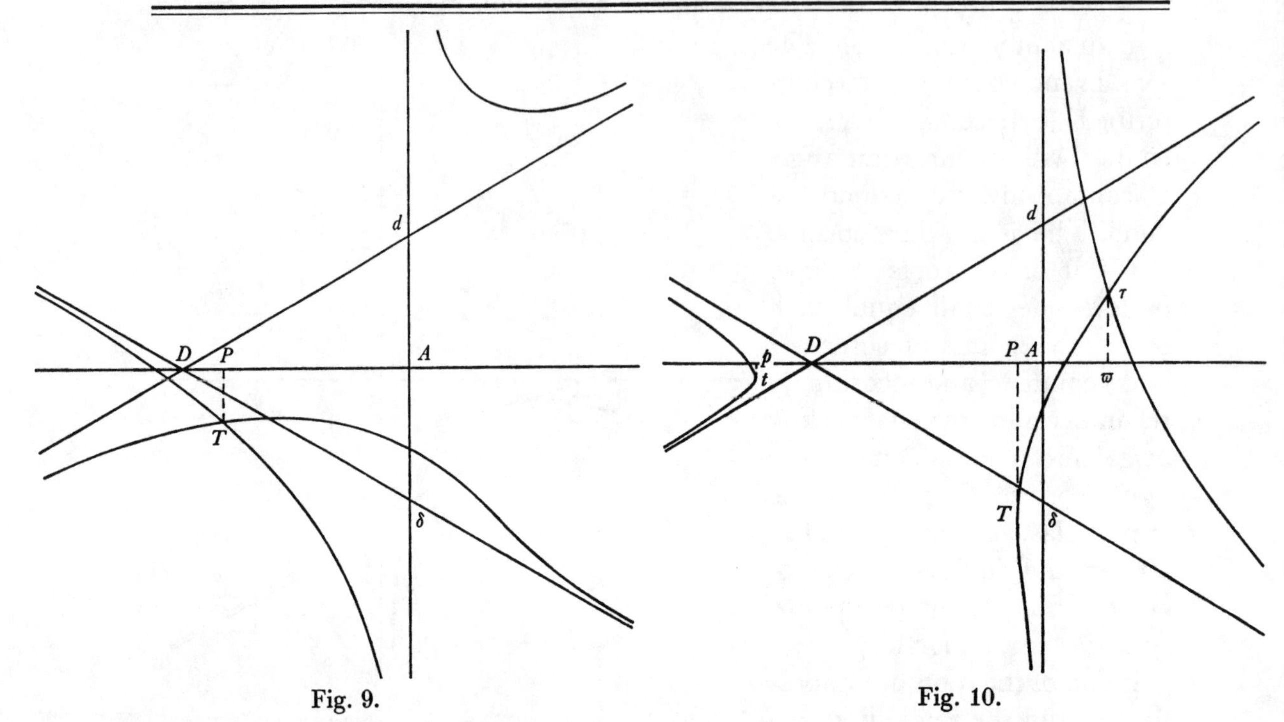

Fig. 9.

Fig. 10.

Si e radicibus duæ sunt æquales & alteræ duæ vel impossibiles sunt vel reales cum signis quæ a signis æqualium radicum diversa sunt: figura *cruciformis* habebitur,(60) nempe duæ ex hyperbolis se invicem decussabunt idq; vel ad verticem trianguli ab Asymptotis comprehensi vel ad ejus basem. Quæ duæ species sunt septima et octava.

Fig. 9, 10, 15, 16.

Si deniq; radices omnes sunt impossibiles vel si omnes sunt reales & inæquales & earum duæ sunt affirmativæ et alteræ duæ negativæ: tunc duæ habebuntur Hyperbolæ ad angulos

Fig. 11, 12.

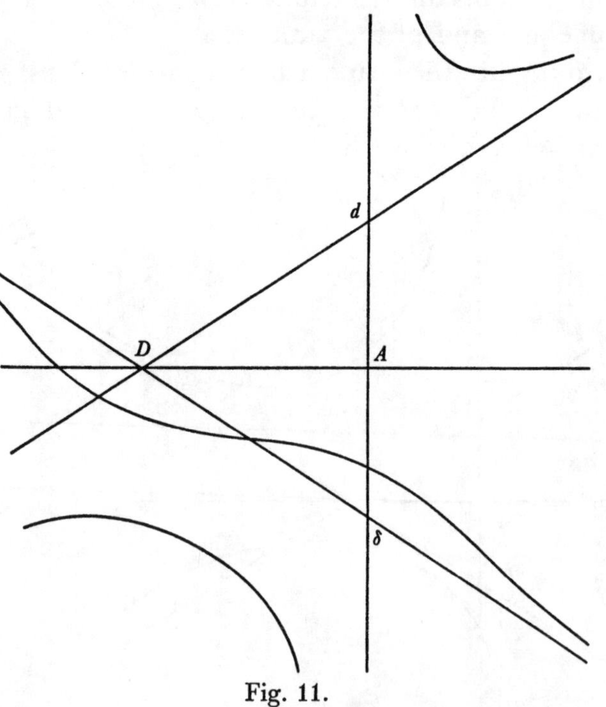

Fig. 11.

(60) This summary encapsulation is not found in the draft—though Newton there makes explicit the connective 'tunc' (then) which is merely understood in the present revise—and the following 'nempe' (specifically) is correspondingly absent. In sequel he initially there

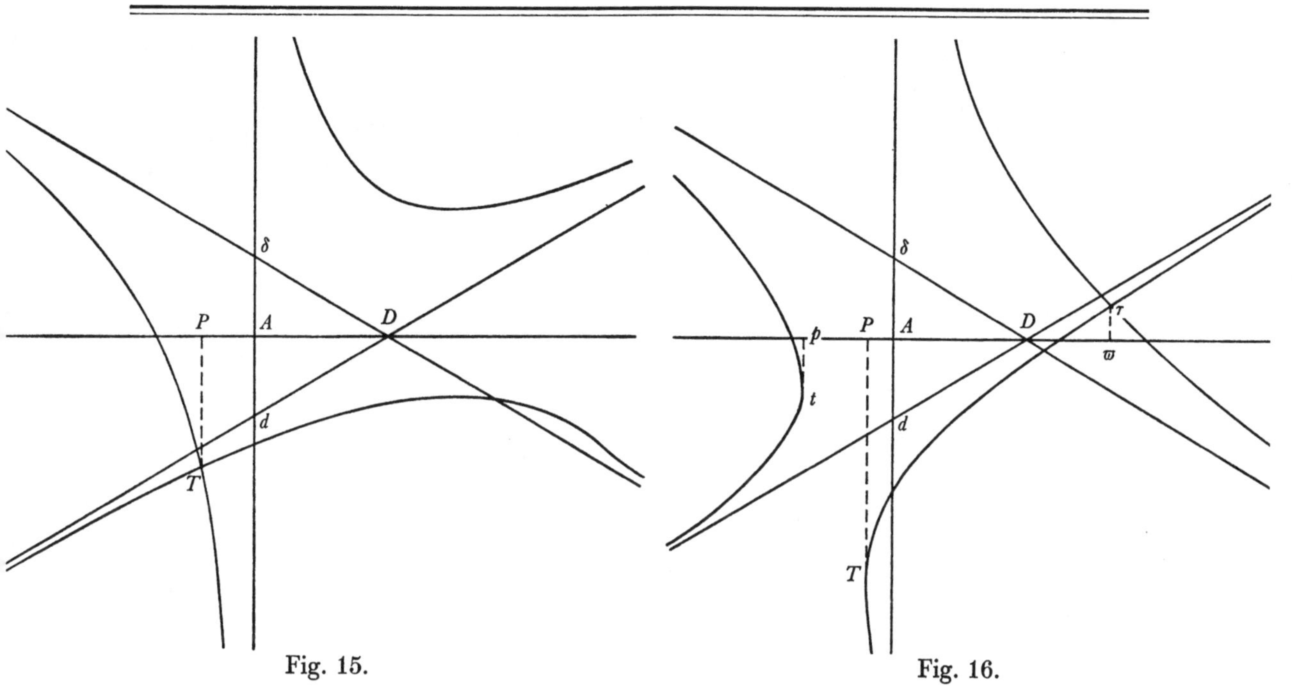

Fig. 15.

Fig. 16.

the asymptotes or at its angles, and will correspondingly constitute either the fifth or the sixth species.

If of the roots two are equal and the other two either impossible or real with signs differing from those of the equal roots, then (Figures 9, 10, 15, 16) a *cross-shaped* figure will be obtained:[(60)] specifically, two of the hyperbolas will cross one another and this either at the vertex of the triangle comprehended by the asymptotes or at its base. These two species are the seventh and eighth.

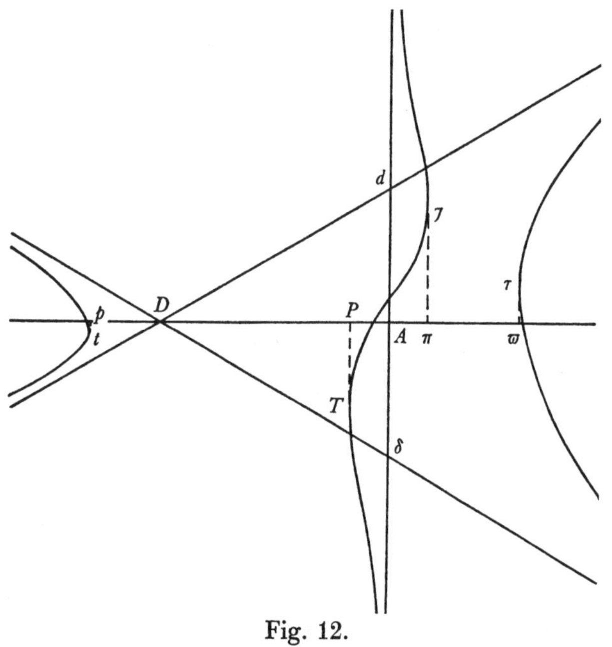

Fig. 12.

If, finally, all the roots are impossible, or if all are real and unequal and two of them are positive and the other two negative, then (Figures 11, 12) there will be had two hyperbolas at

went on to affirm that 'duæ ex hyperbolis in specie quinta jungentur et per contactum suum mutabuntur in alias duas quæ se mutuò in puncto contactus decussabunt' (two of the hyperbolas in the fifth species will be joined and through their contact changed into two others which

oppositos duarum Asymptotôn cum Hyperbola *anguinea* circa Asymptoton tertiam.(61) Quæ species est nona.

Et hi sunt omnes radicum casus possibiles. Nam si duæ sunt æquales inter se et aliæ duæ sunt etiam inter se æquales, Figura evadet Sectio Conica cum linea recta.(62)

XVI Hyperbolæ duodecim redundantes cum unica tantum Diametro.(63)

Si Hyperbola redundans habet unicam tantum Diametrum sit ejus Diameter Abscissa AB,(64) et æquationis hujus

$$ax^3+bxx+cx+d=0$$

quære tres radices seu valores x.

Si radices illæ sunt omnes reales(65) et ejusdem signi, Figura constabit (66)ex *Ovali* intra triangulum $Dd\delta$ jacente et tribus Hyperbolis ad angulos ejus, nempe circumscripta ad angulum D et inscriptis duobus ad angulos d et δ. Et hæc est species decima.(67) [Fig. 17.]

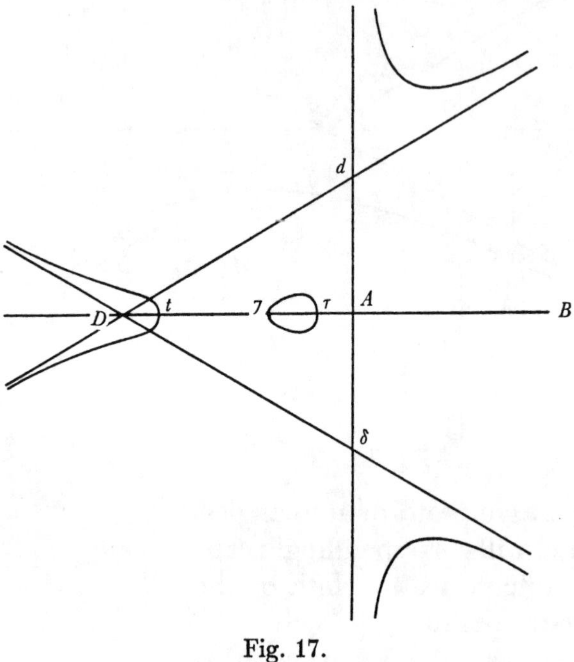

Fig. 17.

Si radices duæ majores sunt æquales et tertia ejusdem signi, crura Hyperbolæ jacentis versus D sese decussabunt in forma *nodi* propter contactum Ovalis. Quæ species est undecima. [Fig. 18.]

will cross one another at the point of contact), and in turn to state that 'tres habebuntur hyperbolæ quarū duæ *se mutuò decussabunt*, tertia jacebit in angulo asymptotωn qui puncto contactus maxime [finitimus est]' (there will be had three hyperbolas, of which two will *cross one another* and the third will lie in the angle of the asymptotes which [is clos]est to the point of contact).

(61) Newton concluded the equivalent passage in his prior draft with the words '...cum linea tertia quæ in longum cum flexibus contrarijs juxta asymptoton tertiam porrigitur, quamq; ideo lineam *anguineam* vocare licet. Et sic duæ habebuntur Hyperbolæ cum *Anguinea*' (...with a third line which is stretched out with contrary flexes along the length of the third asymptote, and which we may in consequence be permitted to call a *snaky* line. And in this way there will be had two hyperbolas with a *snaky* branch).

(62) No straight line can meet a cubic in a pair of double points unless it wholly coincides with it—in the degenerate case, namely, where the cubic is itself the pair of this straight line and some complementary conic (or line-couple).

(63) Initially in the draft Newton set this marginal head to read: 'Species Hyperbolæ redundantis unicam tantum habentis diametrum & tres asymptotos triangulum continentes ejusdem Figuræ' (The species of the redundant hyperbola having but a single diameter, where the three asymptotes of this same figure contain a triangle).

opposite angles of two of the asymptotes with a *snaky* hyperbola round the third asymptote.(61) This species is the ninth.

And these are all the cases of the roots possible. For, if two are equal to one another and the other two are also equal to one another, the figure will prove to be a conic together with a straight line.(62)

XVI
The twelve redundant hyperbolas with but a single diameter.(63)

If a redundant hyperbola has but a single diameter, let its diameter be the abscissa AB(64) and seek the three roots or values of x in this equation:

$$ax^3+bx^2+cx+d=0.$$

If those roots are all real(65) and of the same sign, the figure will consist (Fig. 17)(66) of an *oval* lying within the triangle $Dd\delta$ and three hyperbolas at its corners, namely a circumscribed one at the corner D and two inscribed ones at the corners d and δ. This is the tenth species.(67)

If the two greater roots are equal and the third of the same sign, the branches of the hyperbola lying (Fig. 18) towards D will cross each other in the form of a *node* because the oval now makes contact with them. This is the eleventh species.

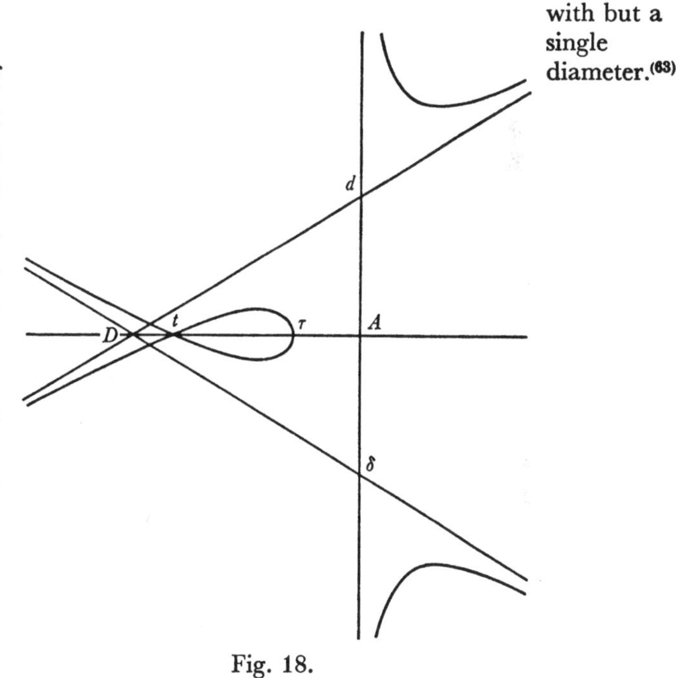

Fig. 18.

(64) In which case the diametral hyperbola $xy = -\frac{1}{2}e$ reduces to be the pair of the axes Ad $(x = 0)$ and AB $(y = 0)$, so that $e = 0$.

(65) In his draft Newton further specified 'inæquales' (unequal).

(66) The draft sentence (f. 41v) initially concluded: '...ex tribus Hyperbolis et *Ovali*; et Hyperbolarum una jacebit extra suas asymptotos ad partes D, alteræ duæ jacebunt intra suas asymptotos, Ovalis autem reperietur intra triangulum $Dd\delta$' (of three hyperbolas and an *oval*; and of the hyperbolas one will lie outside its asymptotes in the direction of D, the other two within theirs, while the oval will be located within the triangle $Dd\delta$).

(67) There are variants on these two species, *cum ovali* and *punctata*, in which the infinite branch at the angle D is a simple hyperbola wholly contained within the asymptotes Dd and $D\delta$, while the branches at the angles d and δ are ambigenal hyperbolas curling (one each) round these same asymptotes. While Newton had included both these in his earlier tabulation of the individual species of cubic as optical 'shadows' of the divergent parabolas—where they are respectively projection 8 of Genus 5 and projection 6 of Genus 3 (see **2**, **3**, §1: notes (54) and (65) preceding)—he here in his complementary algebraic enumeration neglects to notice them. Whence it has passed into the historical record that these additional species were

Si tres radices sunt æquales Hyperbola ista fit *cuspidata* sine Ovali. Quæ species est duodecima. (Fig. 19)

Si radices duæ minores sunt æquales et tertia ejusdem signi, Ovalis in *punctum* evanuit. Quæ species est decima tertia.(67) In speciebus quatuor novissimis Hyperbola quæ jacet versus D Asymptotos in sinu suo amplectitur, reliquæ duæ in sinu Asymptotôn jacent. (Fig. 20)

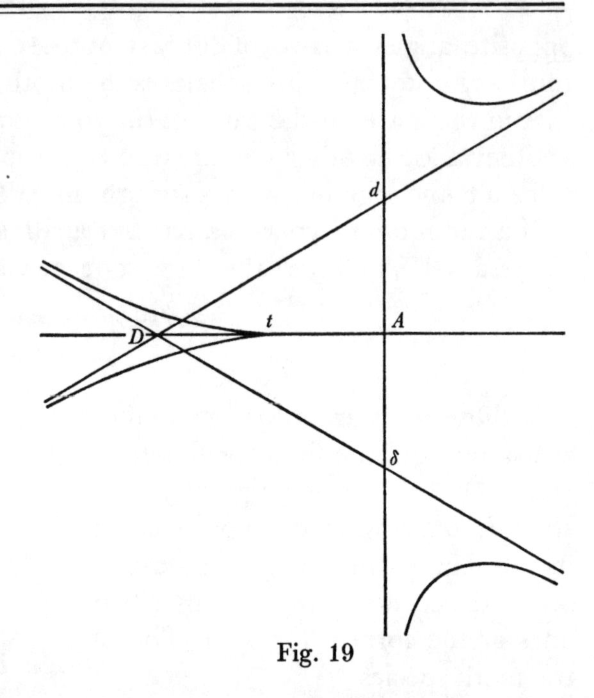

Fig. 19

Si duæ ex radicibus sunt impossibiles, habebuntur tres Hyperbolæ puræ sine(68) Ovali decussatione vel cuspide. Et hujus casus species sunt quatuor, nempe decima quarta si Hyperbola circumscripta jacet versus D et decima quinta si Hyperbola inscripta jacet versus D, decima sexta si Hyperbola circumscripta jacet sub basi $d\delta$ trianguli $Dd\delta$ & decima septima si Hyperbola inscripta jacet sub eadem basi.(69) (Fig. 20, Fig. 21, Fig. 22, Fig. 23)

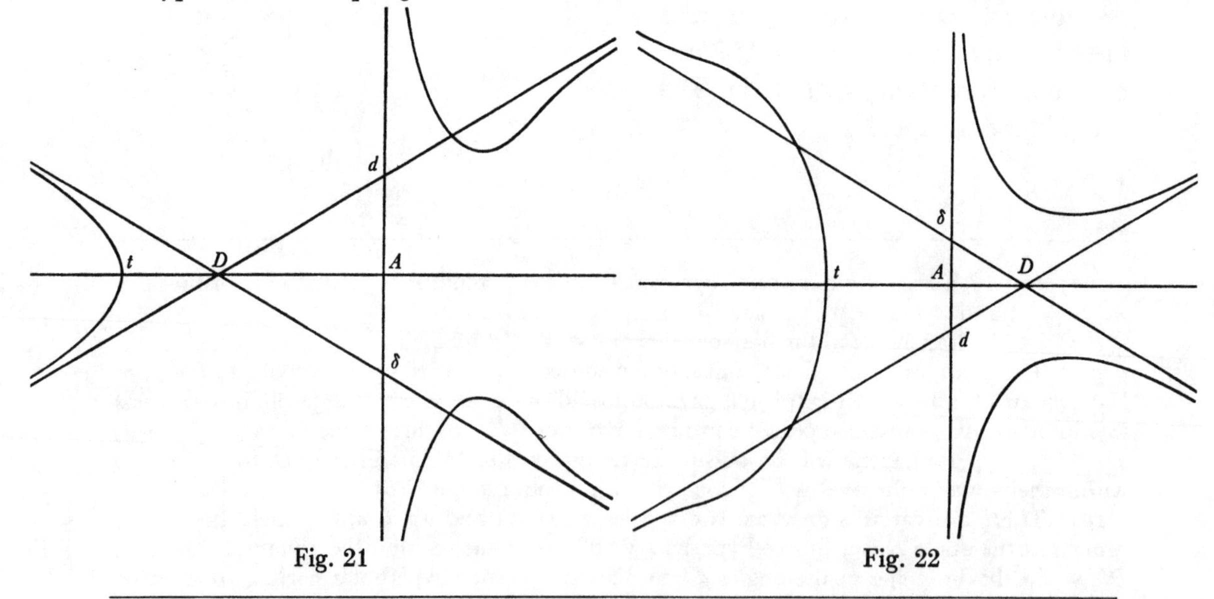

Fig. 21 Fig. 22

'discovered' by James Stirling, who first publicly rectified Newton's present omission in his *Lineæ Tertii Ordinis Neutonianæ, sive Illustratio Tractatus D. Neutoni De Enumeratione Linearum Tertii Ordinis* (Oxford, 1717), there displaying them (*ibid.*: 99/100) as new species 11 and 15.

(68) 'aliquo' (any) is deleted by Newton in his draft (f. 42r).

If the three roots are equal, that hyperbola comes (Fig. 19) to be *cusped* without an oval. This species is the twelfth.

If the two lesser roots are equal and the third of the same sign, the oval has vanished into a point (Fig. 20). This species is the thirteenth.(67) In the four most recent species the hyperbola lying towards D embraces the asymptotes in its fold, while the remaining two lie in the fold of the asymptotes.

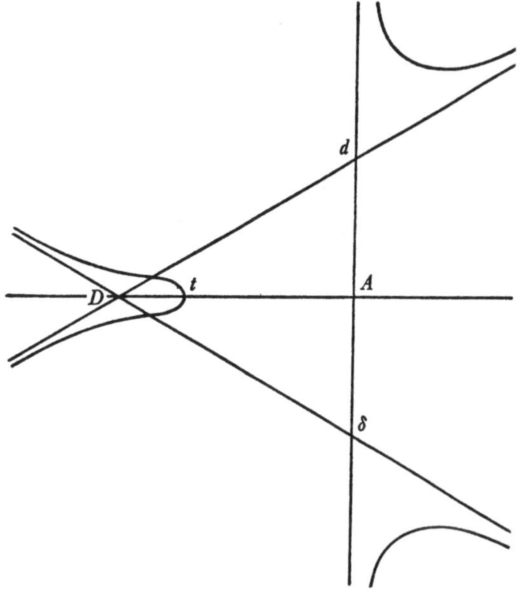

Fig. 20.

If two of the roots are impossible, there will be had three pure hyperbolas without(68) oval, crossing or cusp. Of this case there are four species, namely the fourteenth if (Fig. 20) the circumscribed hyperbola lies towards D, and the fifteenth if (Fig. 21) the inscribed hyperbola lies towards D, the sixteenth if (Fig. 22) the circumscribed hyperbola lies below the base $d\delta$ of the triangle $Dd\delta$ and the seventeenth if (Fig. 23) the inscribed hyperbola lies below the same base.(69)

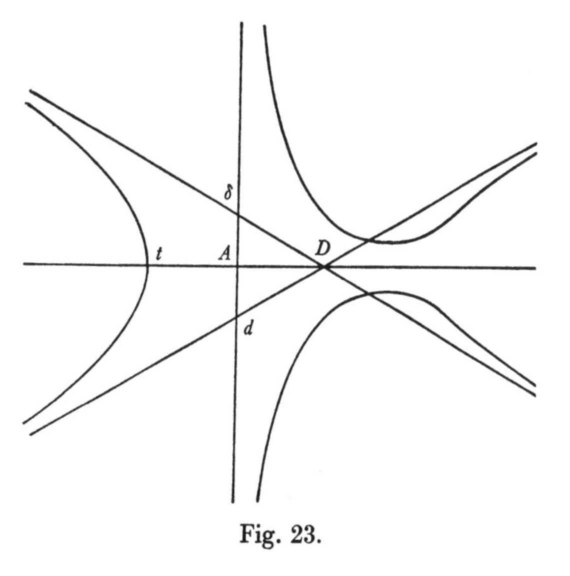

Fig. 23.

(69) In his preliminary draft Newton did not divide species 14/16 and 15/17, there stating that 'hujus casus species sunt duæ, decima quarta quando Hyperbola versus D in sinu suarum asymptotôn comprehenditur, decima quinta quando hyperbola illa asymptotos suas in sinu suo comprehendit' (of this case there are two species: the fourteenth when the hyperbola [lying] towards D is comprehended in the fold of its asymptotes, the fifteenth when that hyperbola embraces its asymptotes in its fold); afterwards shortening this to read '...decima quarta si Hyperbola inscripta jacet versus D & decima quinta si circumscripta est' (the fourteenth if the inscribed hyperbola lies towards D and the fifteenth if it is circumscribed).

Si duæ radices sunt æquales & tertia signi diversi, figura erit *cruciformis*. Nempe duæ ex tribus Hyperbolis se invicem decussabunt idq̧ vel ad verticem trianguli ab asymptotis comprehensi vel ad ejus basem. Quæ duæ species sunt decima octava et decima nona.

Fig. 25 Fig. 24

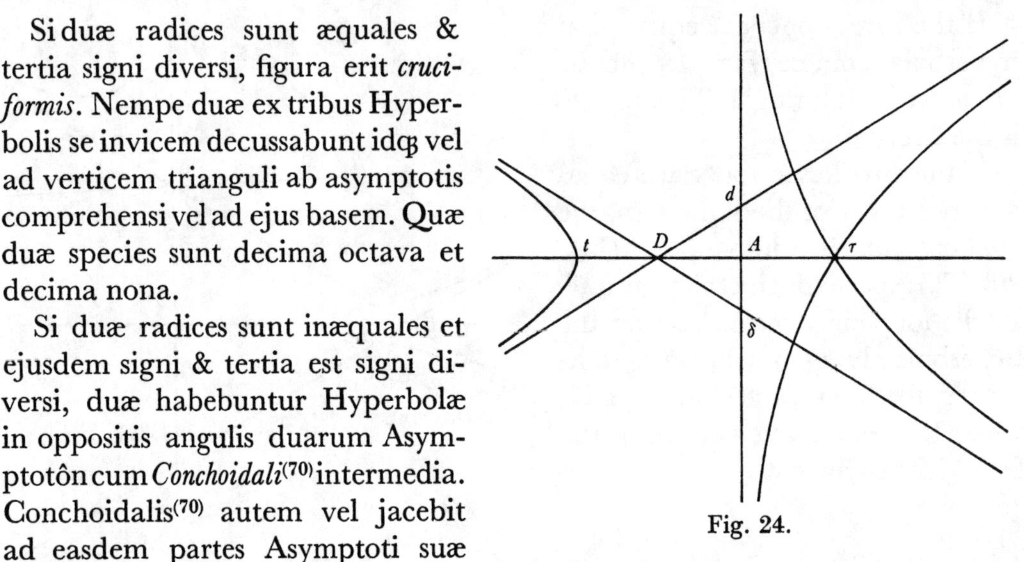

Fig. 24.

Si duæ radices sunt inæquales et ejusdem signi & tertia est signi diversi, duæ habebuntur Hyperbolæ in oppositis angulis duarum Asymptotôn cum *Conchoidali*[70] intermedia. Conchoidalis[70] autem vel jacebit ad easdem partes Asymptoti suæ cum triangulo ab Asymptotis constituto, vel ad partes contrarias; et hi duo casus constituunt speciem vigesimam & vigesimam primam.

Fig. 27 Fig. 26

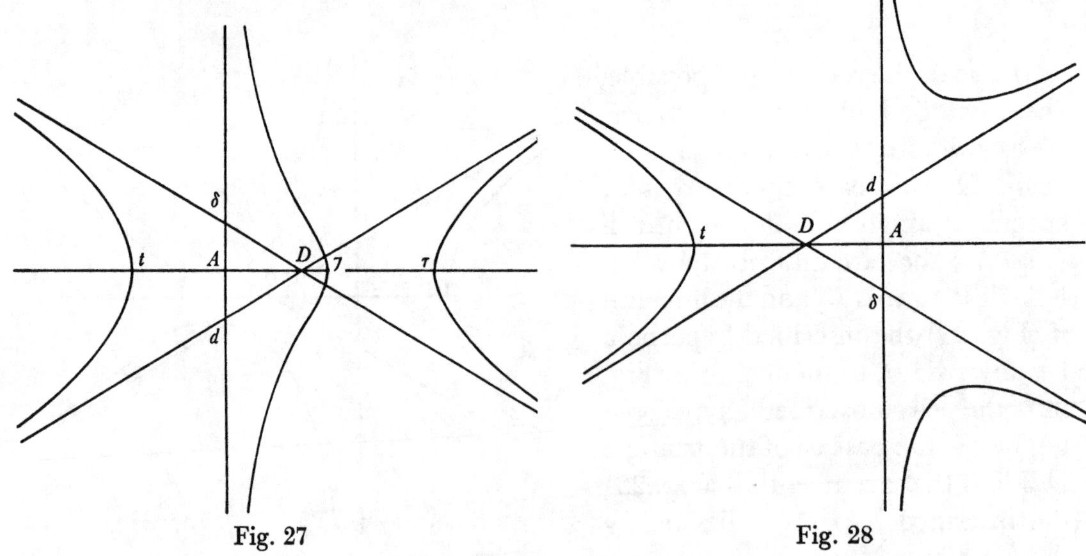

Fig. 27 Fig. 28

XVII Hyperbolæ duæ redundantes[71] cum tribus Diametris. Fig. 28 Fig. 29

Hyperbola redundans quæ habet tres diametros constat ex tribus Hyperbolis in sinubus Asymptotôn jacentibus, idq̧ vel ad angulos trianguli ab Asymptotis comprehensi vel ad ejus latera.[72] Casus prior dat speciem vigesimam secundam & posterior vigesimam tertiam.

(70) 'Concha' (shell) in each instance in the draft (f. 42r).

(71) Initially just 'Species ejusdem figuræ' (Species of the same figure) in Newton's draft marginal head.

(72) Or 'bases' (bases), as was here first written in the draft (compare note (59) above).

If two roots are equal and the third of different sign, the figure will be *cross-shaped*: to be specific, two of the three hyperbolas will cross one another and this either (Fig. 25) at the vertex of the triangle comprehended by the asymptotes or (Fig. 24) at its base. These two species are the eighteenth and nineteenth.

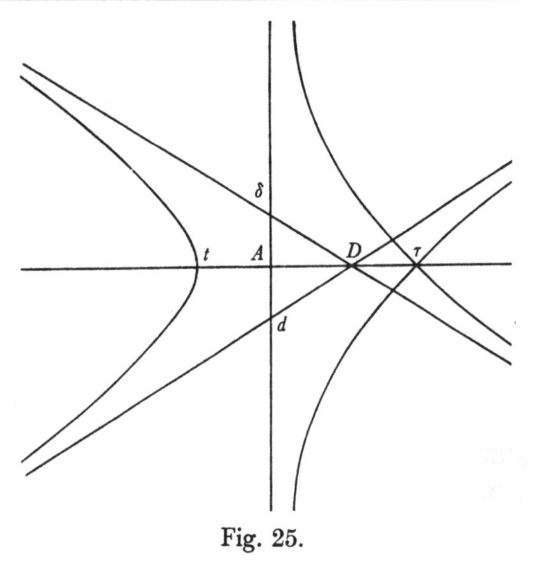

Fig. 25.

If two roots are unequal and of the same sign while the third is of different sign, there will be had two hyperbolas in opposite angles of the two asymptotes with an intermediate *conchoidal* branch.[70] The conchoidal branch,[70] however, will lie either (Fig. 27) on the same side of its asymptote as the triangle constituted by the asymptotes, or (Fig. 26) on the opposite side; and these two cases constitute the twentieth and the twenty-first species.

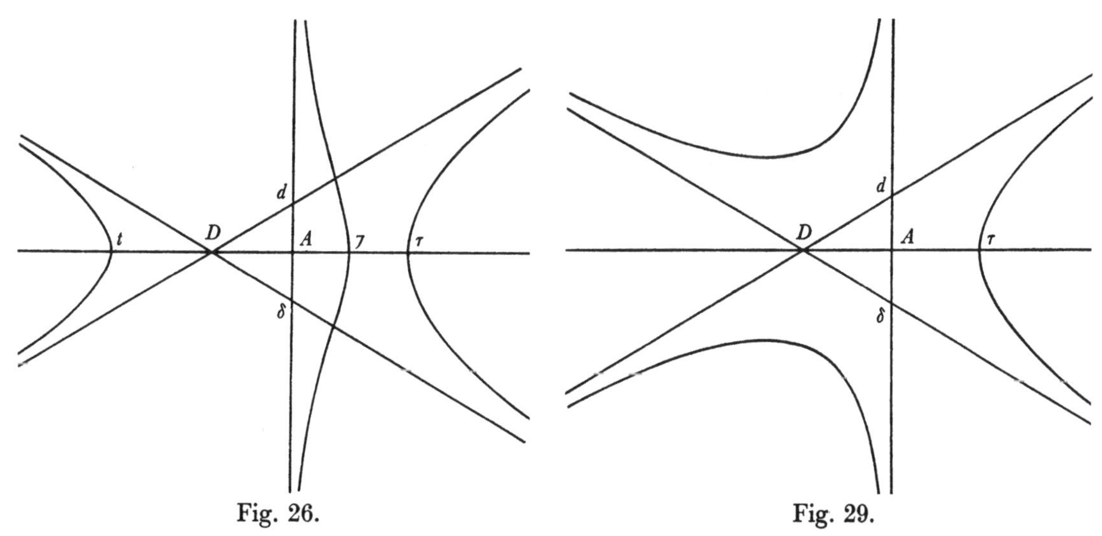

Fig. 26. Fig. 29.

XVII
The two redundant hyperbolas[71] with three diameters.

The redundant hyperbola which has three diameters consists of three hyperbolas lying in the folds of the asymptotes, and this either (Fig. 28) at the corners of the triangle comprehended by the asymptotes or (Fig. 29) at its sides.[72] The former case yields the twenty-second species and the latter the twenty-third.

In the former case where the hyperbolic branches lie at the corners of the triangle of the asymptotes this type of cubic may also have an oval or conjugate point lying within that triangle. Though Newton here fails to record these additional species *cum ovali* and *punctata* of the tridiametral redundant hyperbolas, we have seen (**2**, 3, §1: notes (54) and (65) preceding)

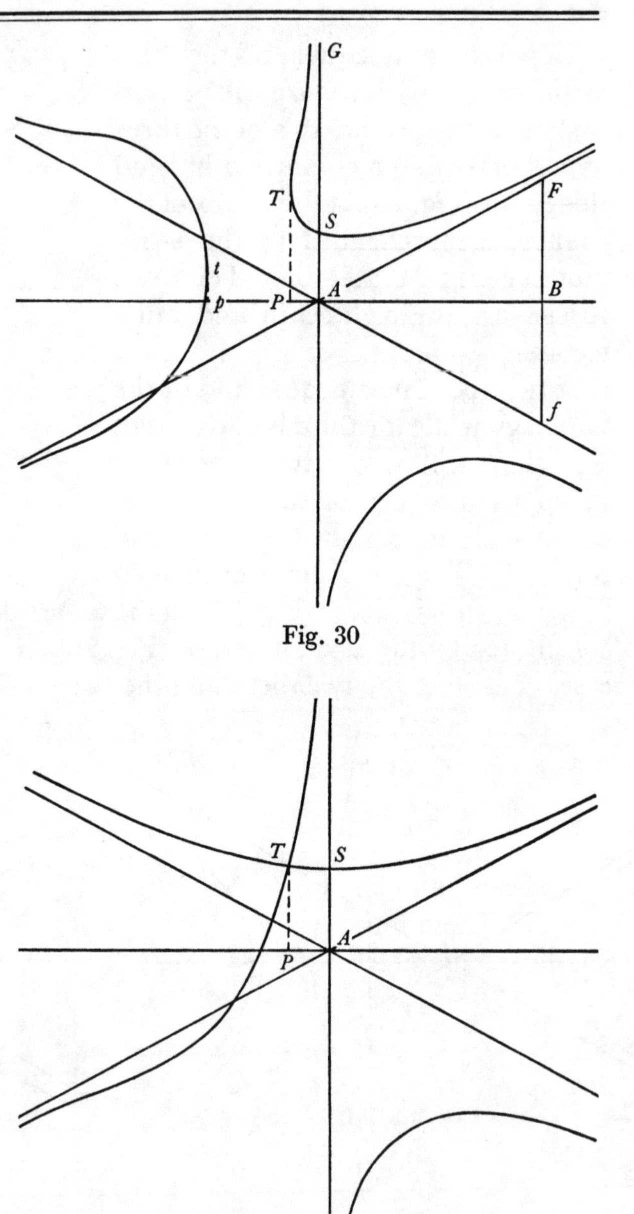

Fig. 30

Fig. 31

XVIII
Hyperbolæ novem redundantes cum Asymptotis tribus ad commune punctum convergentibus.

Fig. 30 Fig. 31
Fig. 32 Fig. 33

Si tres Asymptoti in puncto communi se mutuò decussant, vertuntur species quinta et sexta in vigesimam quartam, septima et octava in vigesimam quintam & nona in vigesimam sextam ubi Anguinea non transit per concursum Asymptotôn, & in vigesimam septimam ubi transit per concursum illum, quo casu termini *b* ac *d* desunt et concursus Asymptotôn est centrum figuræ ab omnibus ejus partibus oppositis æqualiter distans.(73) Et hæ quatuor species Diametrum non habent.

that he had earlier embraced them in his complementary tabulation of the varieties as they derive in optical 'shadow' from the divergent parabolas—namely, as projection 9 of Genus 5 and projection 7 of Genus 3 respectively. Their existence was afterwards publicly exhibited by James Stirling in his *Lineæ Tertii Ordinis Neutonianæ* (see note (67) above): 102, where they are listed as species 24 and 25 in Stirling's emendation of Newton's present algebraic enumeration.

XVIII
The nine redundant hyperbolas with three asymptotes converging to a common point.

If the three asymptotes cross each other at a common point, the fifth and sixth species turn into (Fig. 30) the twenty-fourth, the seventh and eighth into (Fig. 31) the twenty-fifth, and the ninth into (Fig. 32) the twenty-sixth when the snaky branch does not pass through the common meet of the asymptotes, and into the twenty-seventh when (Fig. 33) it passes through that concourse—in which case the terms in b and d are wanting, and the common meet of the asymptotes is the centre of the figure equally distant from all its opposing parts.[73] These four species do not have a diameter.

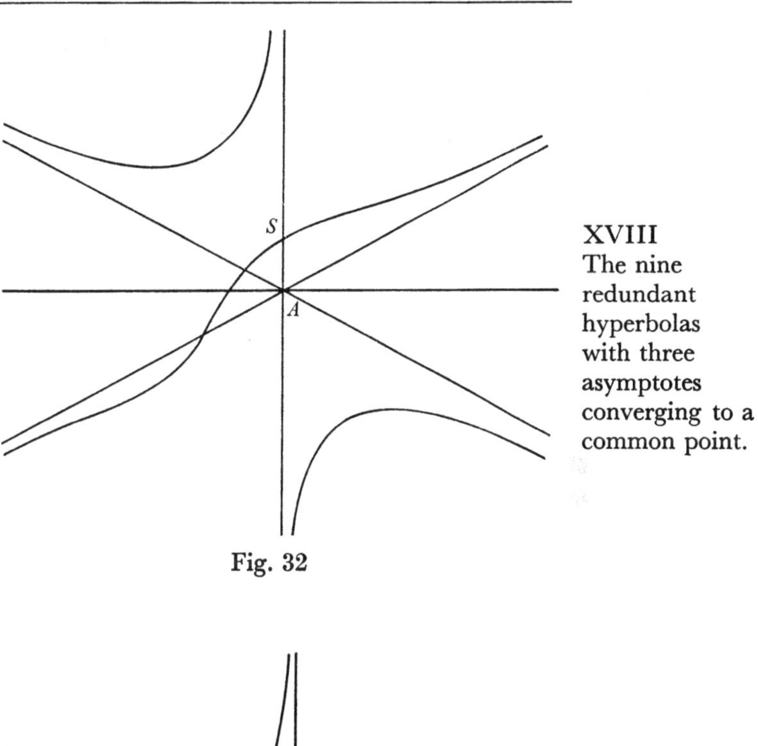

Fig. 32

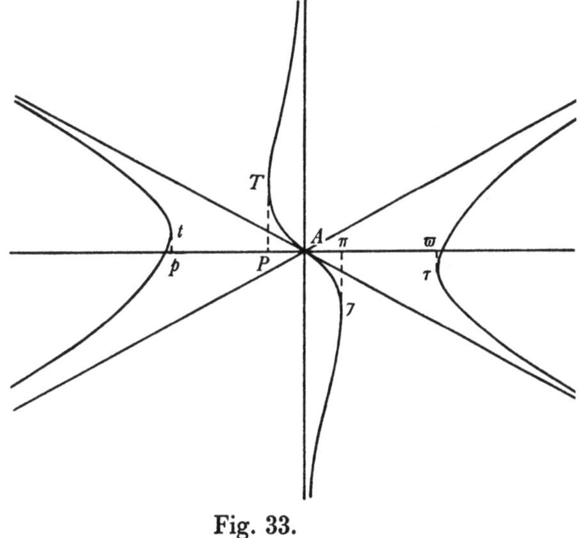

Fig. 33.

(73) This last clause was absent from Newton's original version (on f. 42v), where 'centrum' (centre) first filled the place of 'concursus' (concourse) in the preceding line. In initial revise, having made the change-over to the latter, he here wrote: 'Quo in casu concursus ille est centrum figuræ in quo rectæ omnes ad figuram utrinq; productæ bisecantur' (in which case that common meet is the figure's centre wherein all straight lines extended either way to the figure are bisected).

Fig. 34 Fig. 35 Fig. 36 Fig. 37 Vertuntur etiam species decima quarta ac decima sexta in vigesimam octavam, decima quinta ac decima septima in vigesimam nonam,(74) decima octava & decima nona in tricesimam & vigesima cum vigesima prima in tricesimam primam. Et hæ species unicam habent diametrum.

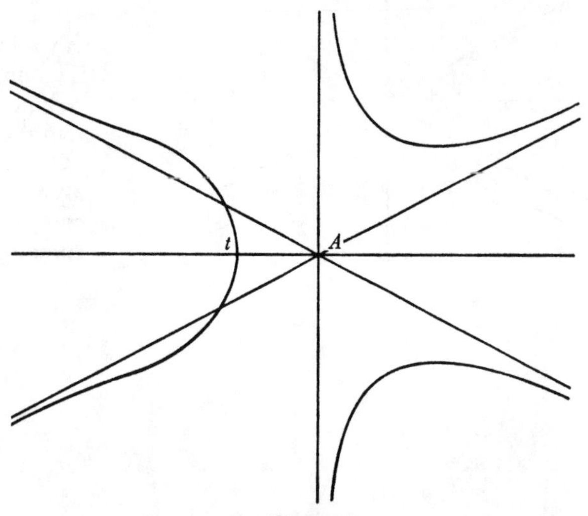

Fig. 34

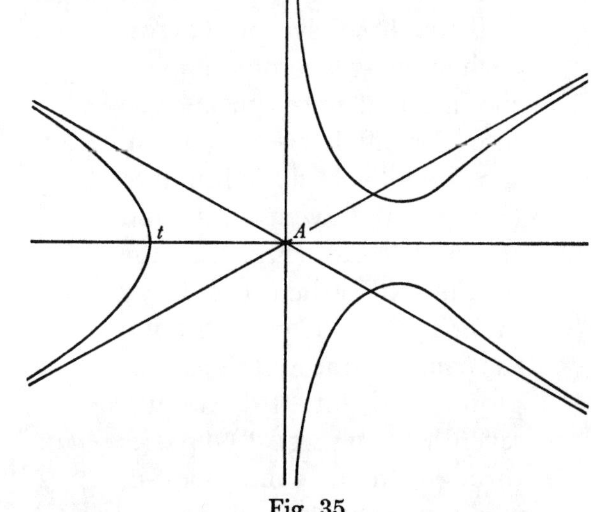

Fig. 35

Ac deniq; species vigesima secunda et vigesima tertia vertuntur in speciem tricesimam secundam cujus tres sunt Diametri per concursum Asymptotôn transeuntes. Quæ omnes conversiones facillimè intelliguntur faciendo ut triangulum ab Asymptotis comprehensum diminuatur donec in punctum evanescat. (Fig. 38)

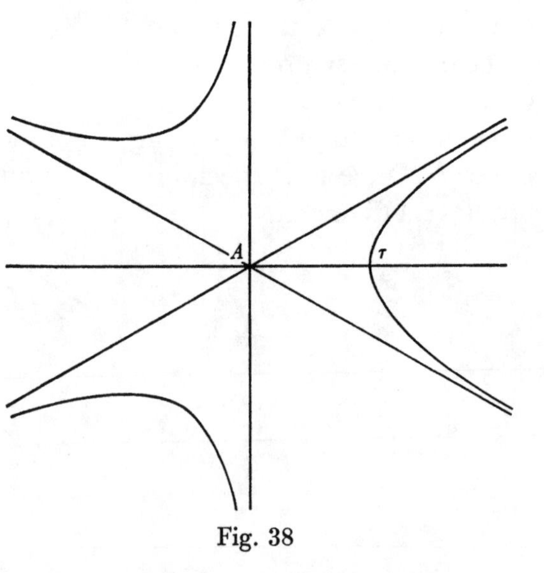

Fig. 38

XIX Hyperbolæ sex defectivæ diametrum non habentes.

Si in primo æquationum casu terminus ax^3 negativus est,(75) Figura erit Hyperbola defectiva unicam habens Asymptoton & duo tantum crura Hyperbolica juxta Asymptoton illam in plagas contrarias infinite progredientia. Et Asymptotos illa est ordinata prima & principalis *AG*. Si terminus *ey* non deest figura nullam habebit Diametrum, si deest habebit unicam.

(74) Following on his omission (see note (69)) there to separate the prior pairs of species 14/16 and 15/17, Newton in his preliminary draft here likewise failed to distinguish between species 28 and 29.

Also the fourteenth and sixteenth species turn into (Fig. 34) the twenty-eighth, the fifteenth and seventeenth into (Fig. 35) the twenty-ninth,(74) the eighteenth and nineteenth into (Fig. 36) the thirtieth, and the twentieth together with the twenty-first into (Fig. 37) the thirty-first. And these species have single a diameter.

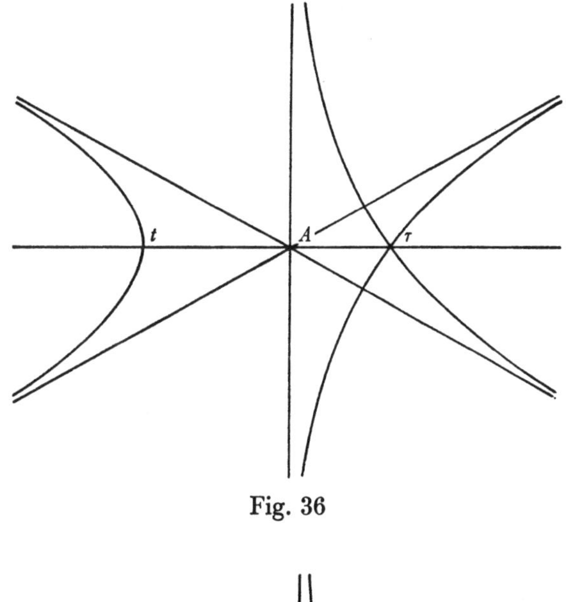

Fig. 36

Finally, the twenty-second and twenty-third species turn into (Fig. 38) the thirty-second species, of which there are three diameters passing through the common meet of the asymptotes. All these conversions are very easily understood by making the triangle comprehended by the asymptotes diminish till it vanishes into a point.

XIX
The six defective hyperbolas not having a diameter.

If in the first case of the equations the term ax^3 is negative,(75) the figure will be a defective hyperbola having a single asymptote and merely two hyperbolic branches progressing to infinity along that asymptote in opposite directions. And that asymptote is the prime and principal ordinate AG. If the term ey is not lacking, the figure will have no diameter; if so, it will have a single one.

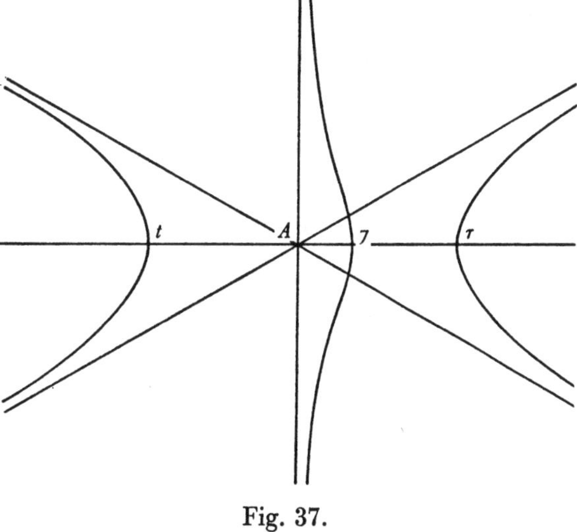

Fig. 37.

(75) A superfluous following specification

'id est si æquatio sit hujus formæ $xyy+ey = -ax^3+bxx+cx+d$'

(that is, if the equation be of this form $xy^2+ey = -ax^3+bx^2+cx+d$) is deleted at this point in the draft.

Si æquationis hujus

$$ax^4 = bx^3 + cxx + dx + \tfrac{1}{4}ee^{(76)}$$

radices omnes $A\pi$, AP, Ap, $A\varpi$ sunt reales & inæquales, Figura erit Hyperbola anguinea Asymptoton flexu contrario amplexa cum *Ovali* conjugata. Quæ species est tricesima tertia. (Fig. 39)

Si radices duæ mediæ AP & Ap æquentur inter se, Ovalis et Anguinca junguntur sese decussantes in forma *nodi*. Quæ est species tricesima quarta. (Fig. 40)

Si tres radices sunt æquales, nodus vertetur in *cuspidem* acutissimum in vertice Anguineæ. Et hæc est species tricesima quinta. (Fig. 41)

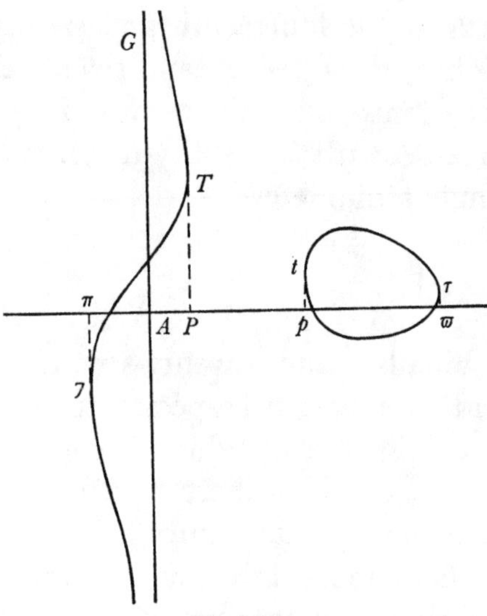

Fig. 39

Si e tribus radicibus ejusdem signi duæ maximæ Ap et $A\varpi$ sibi mutuò æquantur, Ovalis in *punctum* evanuit. Quæ species est tricesima sexta. (Fig. 42)

Si radices duæ quævis imaginariæ sunt, sola manebit Anguinea *pura* sine Ovali, decussatione, cuspide vel puncto conjugato. (Fig. 43) Si Anguinea illa non transit per punctum A species est tricesima septima, sin transit per punctum illud A (Fig. 43(′)) (id quod contingit ubi termini b ac d desunt) punctum illud A erit centrum

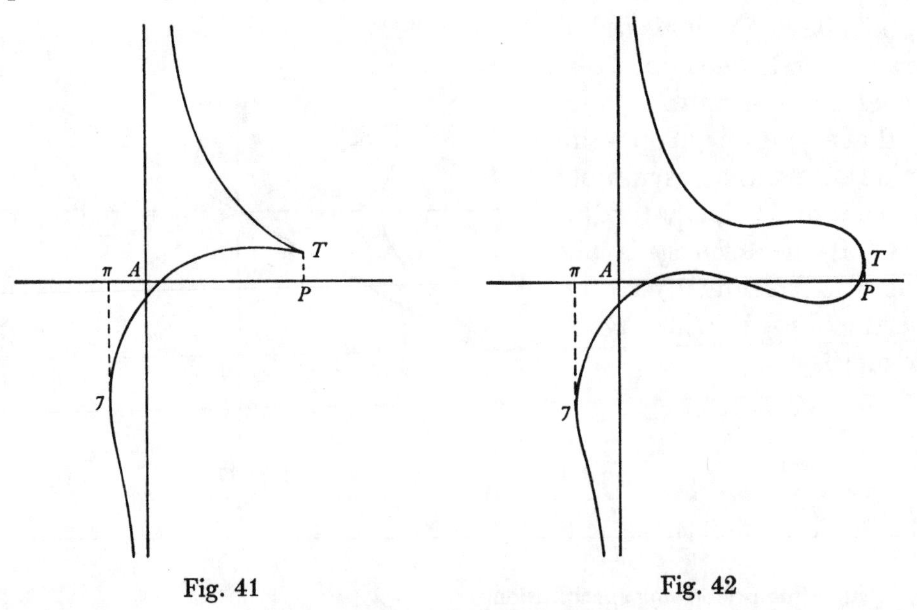

Fig. 41 Fig. 42

(76) Much as before (compare note (52) preceding) this quartic equation determines the abscissas of the meets of the given cubic with its diametral hyperbola $xy + \frac{1}{2}e = 0$.

If all the roots $A\pi$, AP, Ap, $A\varpi$ of this equation $ax^4 = bx^3 + cx^2 + dx + \frac{1}{4}e^2$(76) are real and unequal, the figure will be a snaky hyperbola (Fig. 39) coiled with an inflection round the asymptote and with a conjugate *oval*. This species is the thirty-third.

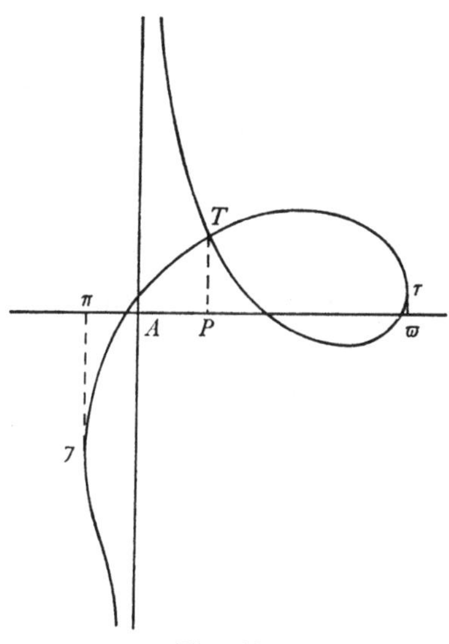

Fig. 40.

If the two middle roots AP and Ap are equal to one another, the oval and snaky branch are (Fig. 40) joined, crossing each other in the form of a *node*. This is the thirty-fourth species.

If the three roots are equal, the node will turn into (Fig. 41) an acutely sharp *cusp* at the vertex of the snaky branch. This is the thirty-fifth species.

If of the three roots the two greatest of same sign, Ap and $A\varpi$, are equal to one another, the oval has (Fig. 42) vanished into a *point*. This species is the thirty-sixth.

If any two roots are imaginary, there will remain a *pure* snaky curve without oval, crossing, cusp or conjugate point. If (Fig. 43) that snaky curve does not pass through the point A, the species is the thirty-seventh; but if (Fig. 43[']) it does pass through the point A—which happens when the terms in b and d are missing—that point A will be the centre of the figure, bisecting all straight lines

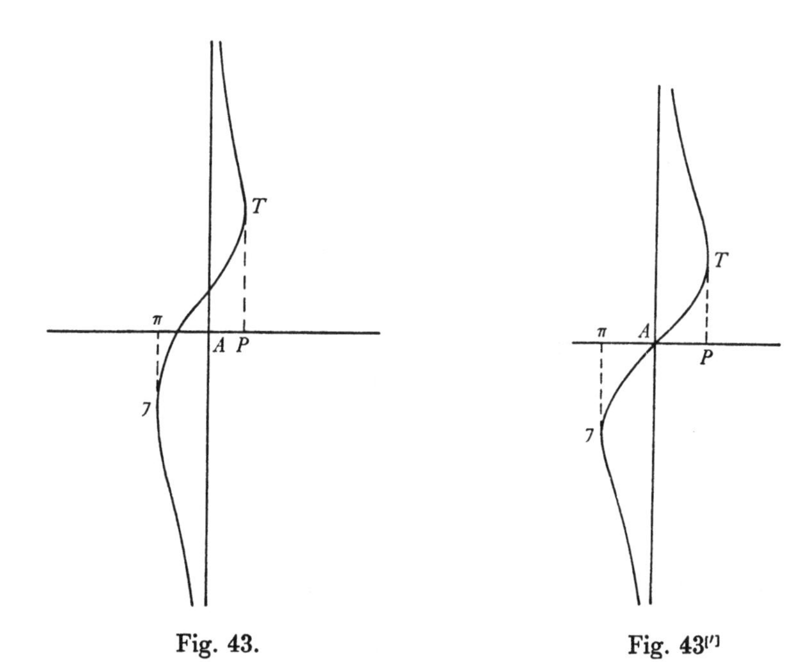

Fig. 43. Fig. 43[']

figuræ rectas omnes per ipsum ductas et ad curvam utrinq; terminatas bisecans. Et hæc est species tricesima octava.

XX Hyperbolæ septem defectivæ diametrum habentes.

In altero casu ubi terminus ey deest et propterea figura diametrum habet, si æquationis hujus $ax^3 = bxx + cx + d$[(77)] radices omnes AT, At, $A\tau$ sunt reales inæquales et ejusdem signi, figura erit Hyperbola Conchoidalis cum *Ovali* ad convexitatem. Quæ est species tricesima nona. (Fig. 45)

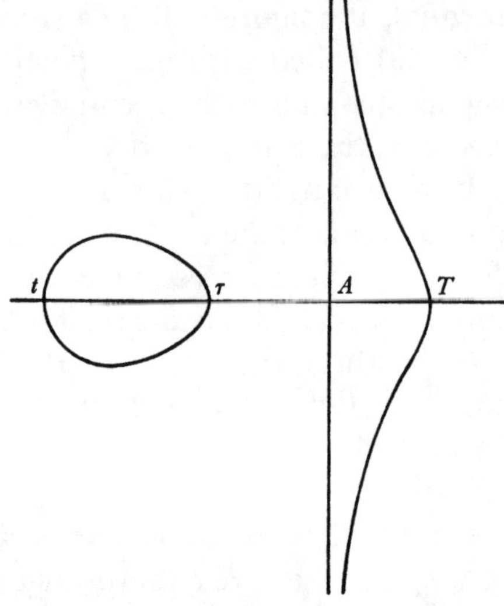

Fig. 44

Si duæ radices sunt inæquales et ejusdem signi et tertia est signi contrarij,[(78)] Ovalis jacebit ad concavitatem Conchoidalis.[(79)] Estq; species quadragesima. (Fig. 44)

Si radices duæ minores AT, At sunt æquales & tertia $A\tau$ est ejusdem signi, Ovalis & Conchoidalis[(80)] jungentur sese decussando in modum *nodi*. Quæ species est quadragesima prima. (Fig. 46)

Si tres radices sunt æquales, nodus mutabitur in *cuspidem* et figura erit *Cissois Veterum*. Et hæc est species quadragesima secunda. (Fig. 47)

Si radices duæ majores sunt æquales et tertia est ejusdem signi, Conchoidalis[(80)] habebit *punctum* conjugatum ad convexitatem suam, estq; species quadragesima tertia. (Fig. 4[8])

Si radices duæ sunt æquales et tertia est signi contrarij[,] Conchoidalis[(80)] habebit *punctum* conjugatum ad concavitatem suam estq; species quadragesima quarta. (Fig. 49)

Si radices duæ sunt impossibles habebitur Conchoidalis[(80)] *pura* sine Ovali, nodo, cuspide vel puncto conjugato. Quæ species est quadragesima quinta. (Fig. 48, 49)

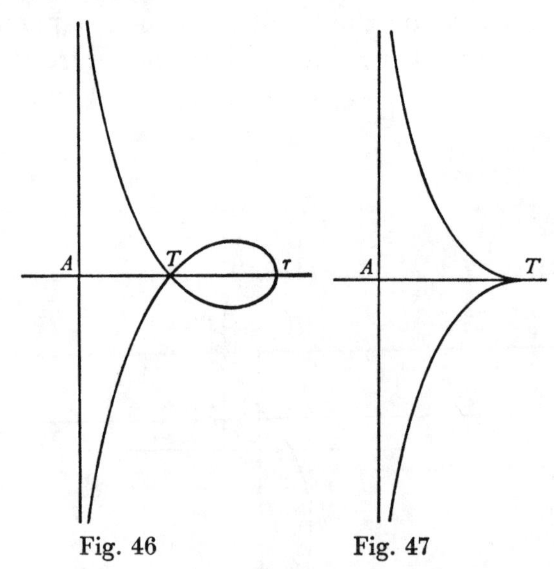

Fig. 46 Fig. 47

(77) Here, of course, the related diametral hyperbola collapses into the pair $xy = 0$ of the abscissa and the primary asymptote; compare note (64) above.

(78) In his draft (f. 43ʳ) Newton wrote to the same effect 'Sin aliqua radicum est signi a reliquis duabus contrarij' (But if some one of the roots is contrary in sign to the remaining two).

(79) 'Conchæ' (shell) in the draft; compare note (70).

drawn through it and terminating on the curve at their either end. This is the thirty-eighth species.

XX
The seven defective hyperbolas having a diameter.

In the other case where the term ey is wanting and the figure accordingly has a diameter, if all the roots AT, At and $A\tau$ of this equation

$$ax^3 = bx^2 + cx + d^{(77)}$$

are real, unequal and of the same sign, the figure will be (Fig. 45) a conchoidal hyperbola with an *oval* at its convexity. This the thirty-ninth species.

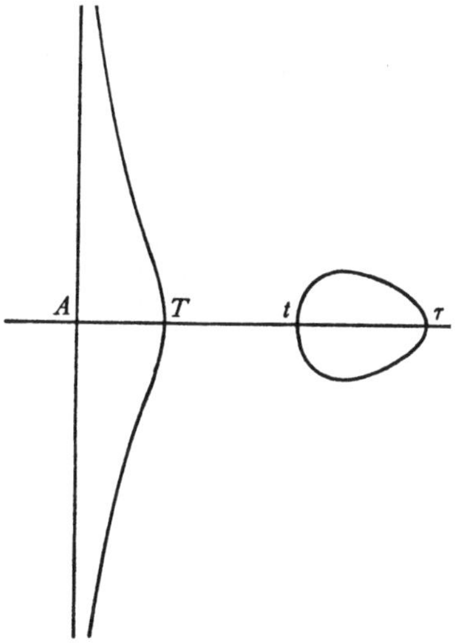

Fig. 45

If two roots are unequal and of the same sign while the third is of the contrary sign,(78) the oval will (Fig. 44) lie at the concavity of the conchoidal branch.(79) This is the fortieth species.

If the two lesser roots AT, At are equal and the third $A\tau$ is of the same sign, the oval and conchoidal branch(80) will (Fig. 46) be joined, crossing each other in the manner of a *node*. This species is the forty-first.

If the three roots are equal, the node will change into (Fig. 47) a cusp and the curve will be the *cissoid of the ancients*. This is the forty-second species.

If the two greater roots are equal and the third is of the same sign, the conchoidal branch(80) will (Fig. 48) have a conjugate *point* at its convexity, and the species is the forty-third.

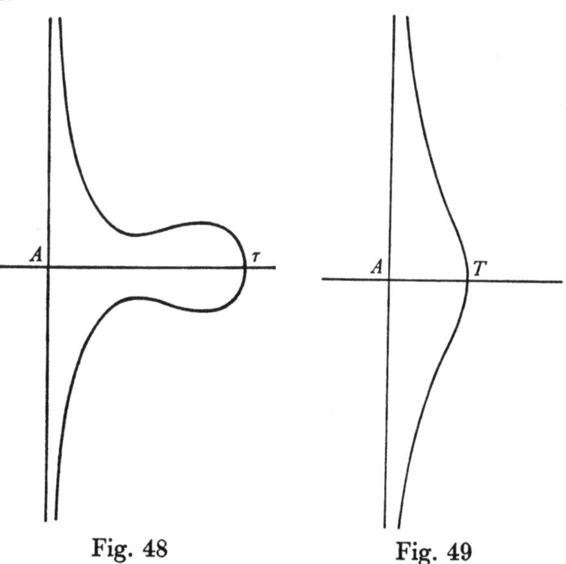

Fig. 48 Fig. 49

If two roots are equal and the third is of converse sign, the conchoidal curve(80) will (Fig. 49) have a conjugate *point* at its concavity, and the species is the forty-fourth.

If two roots are impossible, there will be had (Figures 48, 49) a *pure* conchoidal curve(80) without oval, node, cusp or conjugate point. This species is the forty-fifth.

(80) In his draft (ff. $43^r/43^v$) Newton again wrote 'Concha' (shell) in each instance, there qualifying the last (pure) conchoidal curve as 'simplex' (simple).

XXI
Hyperbolæ septem Parabolicæ Diametrum non habentes.

Siquando in primo æquationum casu terminus ax^3 deest & terminus bxx non deest, Figura erit Hyperbola Parabolica duo habens crura Hyperbolica ad unam Asymptoton *SAG* & duo Parabolica in plagam unam & eandem(81) convergentia. Si terminus *ey* non deest figura nullam habebit diametrum, sin deest habebit unicam. In priori casu species sunt hæ.

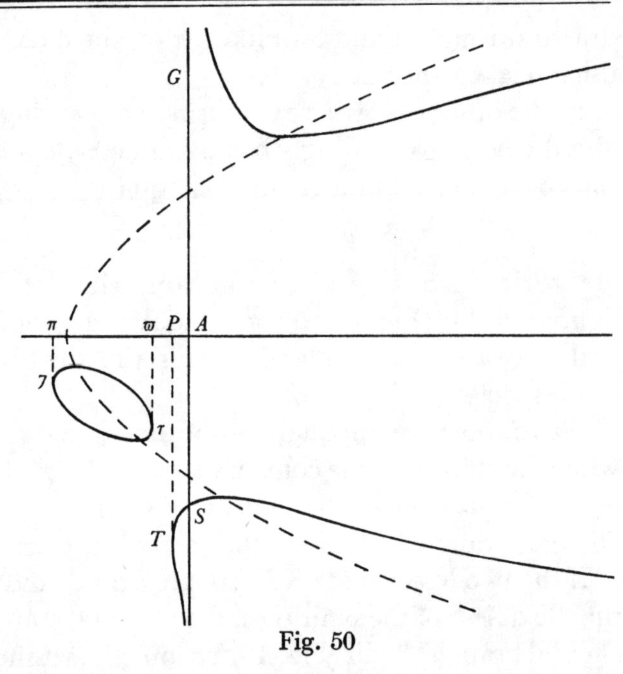

Fig. 50

Si tres radices *AP*, *Aϖ*, *Aπ* æquationis hujus

$$bx^3+cx^2+dx+\tfrac{1}{4}ee = 0^{(82)}$$

sunt inæquales & ejusdem signi, figura constabit ex *Ovali* et alijs duabus curvis quæ partim Hyperbolicæ sunt et partim parabolicæ. Nempe crura Parabolica continuo ductu junguntur cruribus Hyperbolicis sibi proximis. Et hæc est species quadragesima sexta. [Fig. 50]

Si radices duæ minores sunt æquales et tertia est ejusdem signi, Ovalis et una curvarum Hyperbolo-Parabolicorum junguntur & se decussant in forma *nodi*. Quæ species est quadragesima septima. [Fig. 51]

Si tres radices sunt æquales, nodus ille in *cuspidem* vertitur. Estq; [species] quadragesima octava. [Fig. 52]

Si radices duæ majores sunt æquales et tertia est ejusdem signi, Ovalis in *punctum* conjugatum evanuit. Quæ species est quadragesima nona. [Fig. 53]

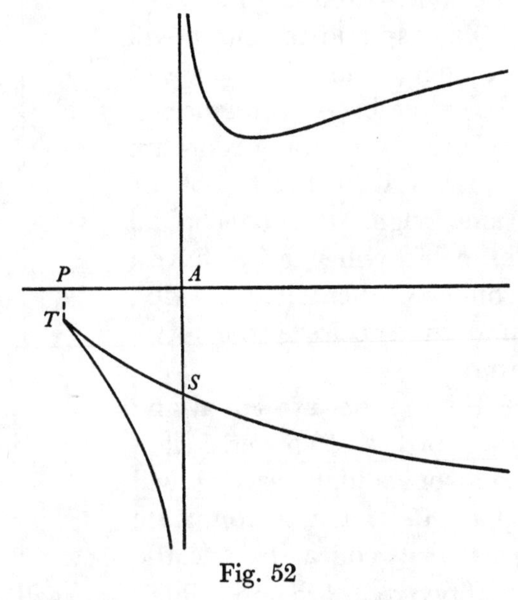

Fig. 52

(81) This specification is in the draft written in over 'communem' (a common), so cancelling it. The broken-line curve in the accompanying figure is the parabola $y^2=bx+c$ to which these branches tend asymtotically.

XXI
The seven parabolic hyperbolas not having a diameter.

Whenever in the first case of the equations the term ax^3 is lacking and the term bx^2 is not, the figure will be a parabolic hyperbola having two hyperbolic branches along one asymptote SAG and two parabolic ones converging in one and the same[81] direction. If the term ey is not wanting, the figure will have no diameter, but if so, the figure will have a single one. In the former case the species are these.

If the three roots AP, $A\varpi$ and $A\pi$ of this equation

$$bx^3+cx^2+dx+\tfrac{1}{4}e^2 = 0^{(82)}$$

are unequal and of the same sign, the figure will consist (Fig. 50) of an *oval* and two other curves which are partly hyperbolic and partly parabolic—specifically, the parabolic branches are joined without break of drawing to the hyperbolic branches nearest them. This is the forty-sixth species.

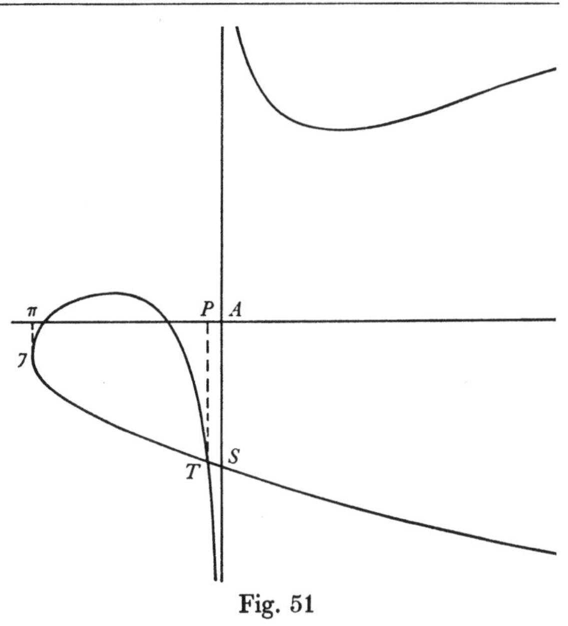

Fig. 51

If the two lesser roots are equal and the third is of the same sign, then (Fig. 51) the oval and one of the hyperbolo-parabolic curves are joined and cross each other in the form of a *node*. This species is the forty-seventh.

If the three roots are equal, that node turns into (Fig. 52) a *cusp*. The species is the forty-eighth.

If the two greater roots are equal and the third is of the same sign, the oval has (Fig. 53) vanished into a conjugate *point*. This species is the forty-ninth.

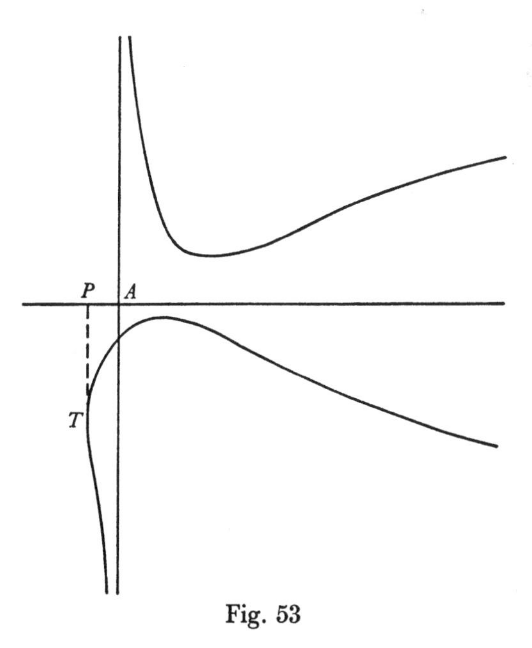

Fig. 53

(82) Once more (compare notes (52) and (76) preceding) this quartic determines the abscissas of the meets of the given cubic with its diametral hyperbola $xy+\frac{1}{2}e = 0$.

Fig. 53, 54 Si duæ radices sunt impossibiles, manebunt *puræ* illæ duæ curvæ Hyperbolo-Parabolicæ sine Ovali, decussatione, cuspide vel puncto conjugato, et speciem quinquagesimam constituent.

Fig. 55 Si radices duæ sunt æquales et tertia est signi contrarij, curvæ illæ Hyperbolo-Parabolicæ junguntur sese decussando in morem *crucis*. Estq; species quinquagesima prima.

Fig. 56 Si radices duæ sunt inæquales & ejusdem signi et tertia est signi contrarij, figura evadet Hyperbola *anguinea* circa Asymptoton AG, cum Parabola conjugata. Et hæc est species quinquagesima secunda.

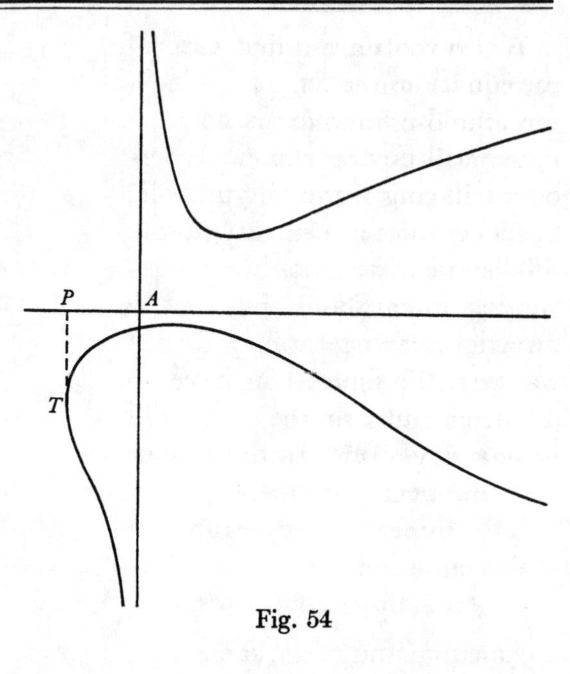

Fig. 54

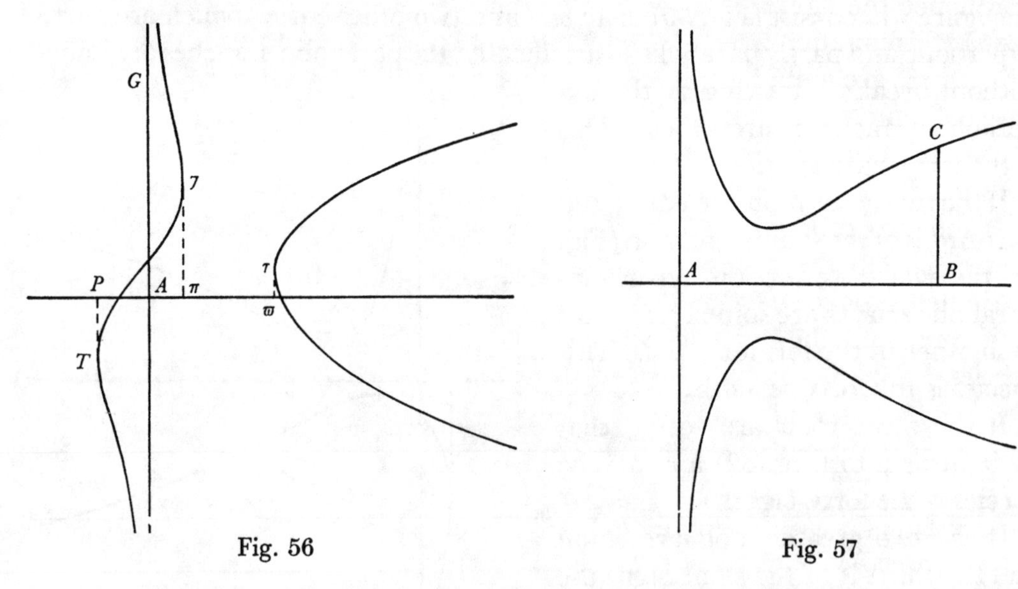

Fig. 56 Fig. 57

XXII Hyperbolæ quatuor Parabolicæ Diametrum habentes. Fig. 57 Fig. 58

In altero casu ubi terminus ey deest et figura Diametrum habet, si duæ radices æquationis hujus $bxx+cx+d=0$ sunt [83]impossibiles, duæ habentur figuræ[84] hyperbolo-parabolicæ a Diametro AB hinc inde æqualiter distantes. Quæ species est quinquagesima tertia.

Si æquationis illius radices duæ sunt æquales, Figuræ hyperbolo-parabolicæ junguntur sese decussantes in morem *crucis*, et speciem quinquagesimam quartam constituunt.

If two roots are impossible, there will remain (Figures 53, 54) those two *pure* hyperbolo-parabolic curves without oval, crossing, cusp or conjugate point, and they will constitute the fiftieth species.

If two roots are equal and the third is of contrary sign, then (Fig. 55) those hyperbolo-parabolic curves are joined, crooked over each other in the fashion of a *cross*. The species is the fifty-first.

If two roots are unequal and of the same sign while the third is of the opposite sign, the figure will prove to be (Fig. 56) a *snaky* hyperbola round the asymptote *AG* together with a conjugate parabola. This is the fifty-second species.

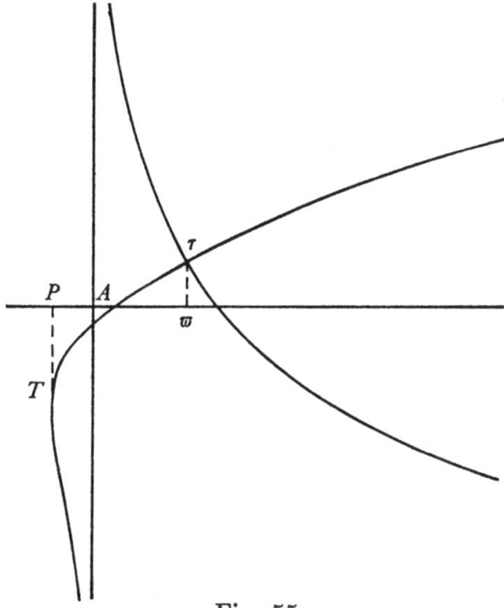

Fig. 55

XXII
The four parabolic hyperbolas having a diameter.

In the other case where the term *ey* is lacking and hence the figure has a diameter if the two roots of this equation $bx^2+cx+d=0$ are(83) impossible, there are had (Fig. 57) two(84) hyperbolo-parabolic figures equally distant from the diameter *AB* on its either side. This species is the fifty-third.

If the two roots of that equation are equal, then (Fig. 58) the hyperbolo-parabolic figures are joined, crooked over each other in the fashion of a *cross*, and they constitute the fifty-fourth species.

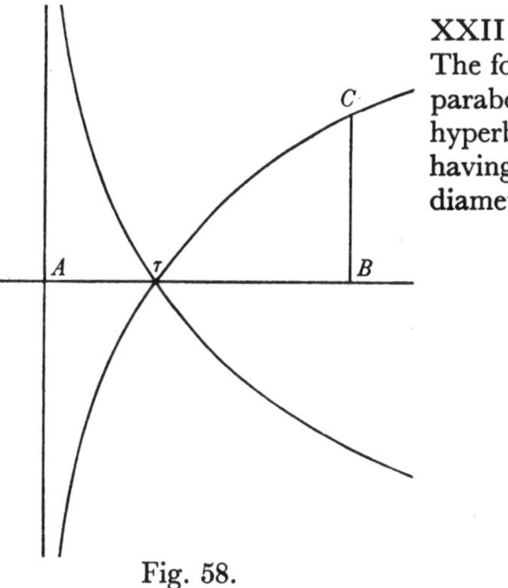

Fig. 58.

(83) An unwanted—and indeed illogical—'inæquales &' (unequal and) is deleted at this point (f. 44r) in the draft.

(84) Understand 'puræ' (pure) as Newton has drawn his accompanying figure. There can also be, as he omits here to notice, other *figuræ hyperbolo-parabolicæ* possessed of an oval or conjugate point lying to the left of *A*—those, namely, which he had listed in his earlier equivalent tabulation of the cubic 'shadows' of the divergent parabolas as, respectively, projection 7 of Genus 5 and projection 5 of Genus 3 (see **2**, 3, §1: notes (54) and (65) preceding). Briefly to recapitulate our previous editorial observations thereon, the parabolic hyperbola *cum ovali* was first mentioned in print by Nicole in the *Mémoires de l'Académie Royale des Sciences. Année MDCCXXXI* (Paris, 1733): 501: §xii, while that with a conjugate point, though known to Niklaus I Bernoulli in 1733, was simultaneously described by Edmund Stone in *Philosophical Transactions*, **41**, 1740: 320 and by De Gua de Malves in his *Usages de l'Analyse de Descartes* (Paris, 1740): 367–8.

Fig. 59 Si radices illæ sunt inæquales et ejusdem signi, habetur Hyperbola *Conchoidalis* cum Parabola ex *eodem latere* Asymptoti. Estq; species quinquagesima quinta.

Fig. 60 Si radices illæ sunt signi contrarij, habetur *Conchoidalis* cum Parabola ad *alteras partes* Asymptoti. Quæ species est quinquagesima sexta.

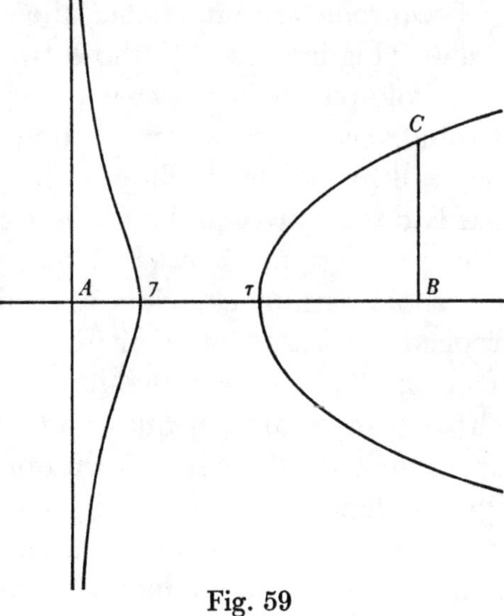

Fig. 59

XXIII Quatuor Hyperbolismi Hyperbolæ. Siquando in primo æquationum casu terminus uterq; ax^3 et bxx deest, figura erit Hyperbolismus Sectionis alicujus Conicæ. Hyperbolismum figuræ[(85)] voco cujus ordinata prodit applicando contentum sub ordinata figuræ illius[(85)] & recta data ad abscissam communem.[(86)] Hac ratione linea recta vertitur in Hyperbolam conicam, et Sectio omnis Conica vertitur in aliquam figurarum quas hic Hyperbolismos sectionum Conicarum voco. Nam æquatio ad figuras de quibus agimus, nempe $xyy+ey = cx+d$, seu $y = \frac{[-]\,e \pm \sqrt{ee+4dx+4cxx}}{2x}$, generatur[(87)] applicando contentum sub ordinata Sectionis Conicæ $\frac{[-]\,e \pm \sqrt{ee+4dx+4cxx}}{2m}$ et recta data m ad curvarum abscissam communem x.[(88)] Unde liquet quod figura genita Hyperbolismus erit Hyperbolæ, Ellipseos vel Parabolæ perinde ut terminus cx[(89)] vel affirmativus est vel negativus vel nullus.

Hyperbolismus Hyperbolæ tres habet Asymptotos quarum una est ordinata prima et principalis $A[G]$, alteræ duæ sunt parallelæ Abscissæ AB et ab eadem hinc inde æqualiter distant. In ordinata principali $A[G]$ cape Ad, $A\delta$ hinc inde

(85) 'notæ' (known) was specified in the draft in each case.

(86) Originally just 'applicando ordinatam figuræ ad ejus abscissam' (on dividing the ordinate of the figure by its abscissa). In other words, if $y = f(x)$ is the Cartesian equation of a given curve, then $y = f(x)/x$ is the equation defining its 'hyperbolism'—more generally, if $F(x, y) = 0$ is the given Cartesian equation, then $F(x, xy) = 0$ is the hyperbolism (where Newton understands that we may split off any extraneous factors $x = 0$ that there may be). In the present instance (the pair of $x = 0$ and) $xy^2+ey = cx+d$, that is

$$(xy)^2+e(xy) = cx^2+dx,$$

is the hyperbolism of $y^2+ey = cx^2+dx$, the equation of a conic (a hyperbola when $c > 0$). Newton had, we have seen (IV: 372, note (69)), introduced this notion of hyperbolism—as elegant as it is strictly here irrelevant—in his previous tabulation of the *genera* of cubic curves a decade and a half earlier, but without there justifying the name. (The operation, of course, transmutes the general straight line $y = mx+n$ into the hyperbola $xy = mx+n$.)

If those roots are unequal and of the same sign, there is had (Fig. 59) a *conchoidal* hyperbola together with a parabola on the *same side* of the asymptote. The species is the fifty-fifth.

If those roots are opposite in sign, there is had (Fig. 60) a *conchoidal* branch together with a parabola on the *other side* of the asymptote. This species is the fifty-sixth.

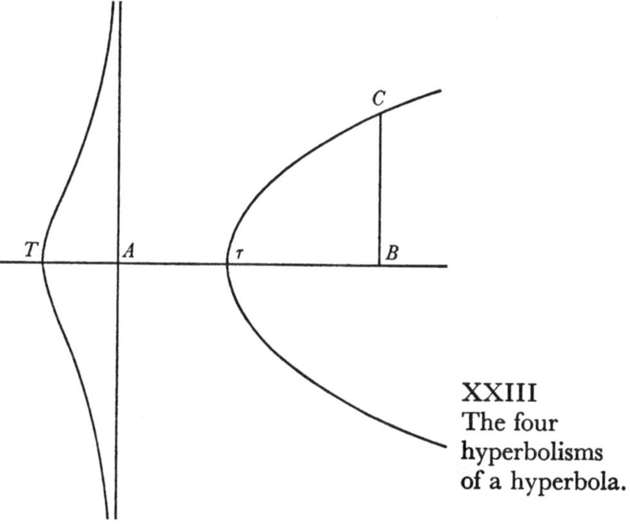

Fig. 60.

XXIII
The four hyperbolisms of a hyperbola.

Whenever in the first case of the equations both of the terms ax^3 and bx^2 are wanting, the figure will be a hyperbolism of some conic. I call a 'hyperbolism' of a(85) figure one whose ordinate ensues on dividing the product of the ordinate of that(85) figure and a given straight line by the common abscissa.(86) By this means a straight line is turned into a conic hyperbola, while every conic is converted into some one of the figures which I here call hyperbolisms of conics. To be precise, the equation to the figures with which we are concerned, namely $xy^2+ey=cx+d$ or $y=(-e\pm\sqrt{[e^2+4dx+4cx^2]})/2x$ is generated(87) by dividing the product of the conic's ordinate $(-e\pm\sqrt{[e^2+4dx+4cx^2]})/2m$ and the given line m by the common abscissa x of the curves.(88) Whence it is clear that the figure so gotten will be the hyperbolism of a hyperbola, ellipse or parabola correspondingly as the term cx(89) is either positive, negative or nil.

The hyperbolism of a hyperbola has three asymptotes, one of which is the prime and principal ordinate AG, while the other two are parallel to the abscissa AB and equally distant from it on its either side. In the principal ordinate AG

(87) In his draft (f. 44v) Newton first wrote 'producitur' (is produced). This somewhat unwieldy phrase passed unaltered into print in 1704 in the *editio princeps* of the 'Enumeratio', but was with considerable gain amplified in its *errata* to read '...nempe $xy^2+ey=cx+d$, dat ordinatam $y=\ldots$ quæ generatur' (...namely $xy^2+ey=cx+d$, yields the ordinate $y=\ldots$ which is generated)—and so it appears in all subsequent editions.

(88) In the draft this is (compare note (86) preceding) just

$$\text{'applicando ordinatam sectionis conicæ } \frac{[-]e\pm\sqrt{ee+4dx+4cxx}}{2} \text{ ad ejus abscissam } x\text{'}$$

(by dividing the conic's ordinate $\frac{1}{2}(-e\pm\sqrt{[e^2+4dx+4cx^2]})$ by its abscissa x).

(89) Corresponding to the term cx^2 in the defining equation $xy+ey=cx^2+dx$ of the conic (see note (86) above).

æquales quantitati $\sqrt{c}$, et per puncta d ac δ age dg, $\delta\gamma$ Asymptotos abscissæ AB parallelas.[90]

Ubi terminus ey non deest figura nullam habet diametrum. In hoc casu si æquationis hujus $cxx + dx + \frac{1}{4}ee = 0$ radices duæ AP, Ap sunt reales & inæquales (nam æquales esse nequeunt nisi figura sit Conica Sectio[91]) figura constabit ex tribus Hyperbolis sibi oppositis quarum una jacet inter asymptotos parallelas et alteræ duæ jacent extra. Et hæc est species quinquagesima septima. Fig. 61

Si radices illæ duæ sunt impossibiles habentur Hyperbolæ duæ oppositæ extra Asymptotos parallelas et Anguinea Hyperbolica intra easdem. Hæc figura duarum est specierum. Nam centrum non habet ubi terminus d non deest; sed si terminus ille deest punctum A est ejus centrum. Prior species est quinquagesima octava, posterior quinquagesima nona. Fig. 62 Fig. 63

Quod si terminus ey deest figura constabit ex tribus Hyperbolis oppositis quarum una jacet inter Asymptotos parallelas et alteræ duæ jacent extra ut in specie quinquagesima [septima][92] & præterea Diametrum habet quæ est Abscissa AB. Et hæc est species sexagesima. Fig. 64

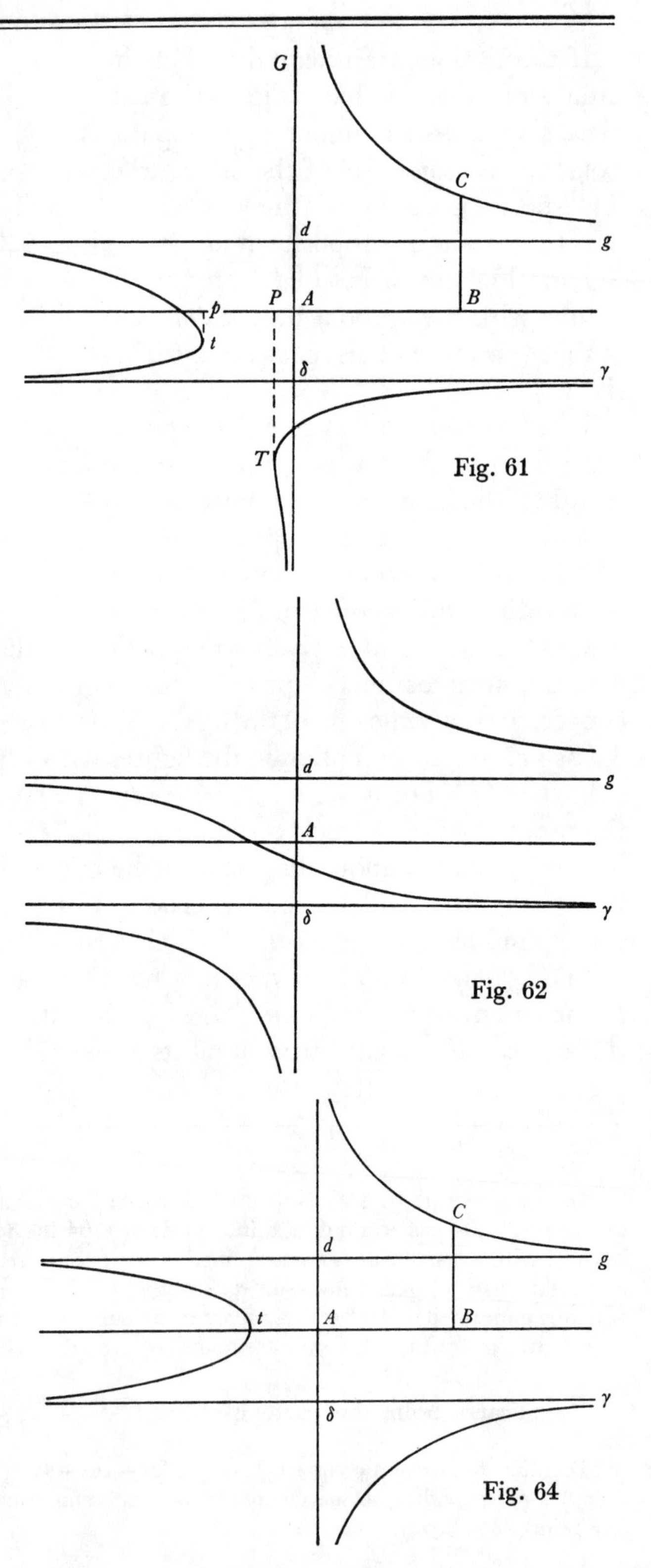

Fig. 61

Fig. 62

Fig. 64

take Ad, $A\delta$ to either side equal to the quantity $\sqrt{c}$, and through the points d and δ draw the asymptotes dg, $\delta\gamma$ parallel to the abscissa AB.(90)

When the term ey is not lacking, the figure has no diameter. In this case, if the two roots AP, Ap of this equation

$$cx^2+dx+\tfrac{1}{4}e^2 = 0$$

are real and unequal (they cannot, of course, be equal unless the figure should be a conic(91)), the figure will consist (Fig. 61) of three hyperbolas opposite to one another, one of which lies between the parallel asymptotes while the other two lie outside. This is the fifty-seventh species.

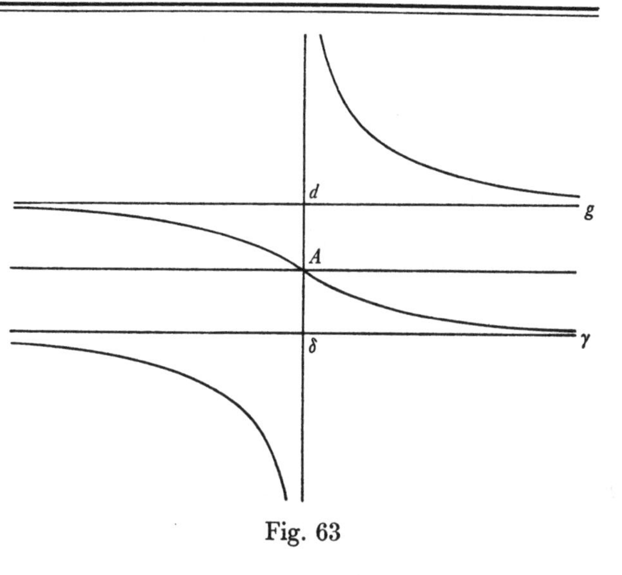

Fig. 63

If those two roots are impossible, there are had two opposite hyperbolas outside the parallel asymptotes and a snaky hyperbolic branch between them. This figure comprehends two species. For (Fig. 62) it does not have a centre when the term d is not wanting; but if that term is lacking, then (Fig. 63) the point A is its centre. The former species is the fifty-eighth, the latter the fifty-ninth.

If, however, the term ey is wanting, the figure will consist (Fig. 64) of three opposite hyperbolas, one of which lies between the parallel asymptotes while the other two lie outside, as in the fifty-seventh(92) species, and moreover it has a diameter which is the abscissa AB. This is the sixtieth species.

---

(90) The asymptotes to the conic hyperbolism $xy^2+ey = cx+d$ are evidently defined by $xy^2 = cx$, and hence are the pair of the primary ordinate $x = 0$ and the parallels $y = \pm\sqrt{c}$. The draft here adds the superfluous sentence 'Nam hæ erunt asymptoti duæ alteræ' (For these will be the two other asymptotes).

(91) When the roots of $cx^2+dx+\frac{1}{4}e^2 \equiv c(x+\frac{1}{2}d/c)^2-\frac{1}{4}(d^2/c-e^2) = 0$ are equal there is evidently $e = \pm d/\sqrt{c}$; in either of which cases the cubic $xy^2+ey = cx+d$ degenerates to be the pair of one of the parallels dg or $\delta\gamma$ ($y = \pm\sqrt{c}$)—it will be $\delta\gamma$ in Newton's accompanying figure—and the conic hyperbola $x(y\pm\sqrt{c}) = \pm d/\sqrt{c}$ which has the other for an asymptote (along with the primary ordinate $x = 0$).

(92) Here forgetting to adjust for the additional three which ensue by his distinction in this final revise of the tract between the preceding pairs of species 14/16, 15/17 and 28/29 (see notes (69) and (74) above), Newton has unthinkingly copied 'quinquagesima quarta' (fifty-fourth) from his draft (f. 45r). The slip passed into the 1704 *editio princeps* and was unseeingly parroted in all subsequent eighteenth-century reissues of the 'Enumeratio'.

XXIV
Tres Hyperbolismi Ellipseos.
Fig. 65

Hyperbolismus Ellipseos[93] per hanc æquationem definitur $xyy+ey=-cx+d$, et unicam habet Asymptoton quæ est ordinata principalis $A[G]$. Si terminus ey non deest Figura est Hyperbola anguinea sine diametro atq; etiam sine centro si terminus d non deest. Quæ species est sexagesima prima.

Fig. 66

At si terminus d deest figura habet centrum sine diametro et centrum ejus est punctum A. Species verò est sexagesima secunda.

Fig. 67

Et si terminus ey deest et terminus d non deest, figura est Conchoidalis ad Asymptoton AG habetq; diametrum sine centro, et Diameter ejus est Abscissa AB. Quæ species est sexagesima tertia.

XXV
Duo Hyperbolismi Parabolæ.
Fig. 68
Fig. 69

Hyperbolismus Parabolæ[94] per hanc æquationem definitur $xyy+ey=d$, et duas habet Asymptotos Abscissam AB et Ordinatam primam & principalem AG. Hyperbolæ verò in hac figura sunt duæ, non in Asymptotôn angulis oppositis sed in angulis qui sunt deinceps jacentes, idq; ad utrumq; latus abscissæ AB, vel sine Diametro si terminus ey habetur, vel cum Diametro si terminus ille deest. Quæ duæ species sunt sexagesima quarta et sexagesima quinta.

XXVI
Tridens Parabolica.
Fig. 76[95]

In secundo Æquationum casu habebatur Æquatio

$$xy=ax^3+bxx+cx+d.$$

Et figura in hoc casu habet quatuor crura infinita quorum duo sunt Hyperbolica circa Asymptoton AG in contrarias partes tendentia et duo Parabolica convergentia & cum prioribus speciem *Tridentis* fere efformantia. Estq; hæc Figura Parabola illa per quam Cartesius æquationes sex dimensionum construxit.[96] Hæc est igitur species sexagesima sexta.

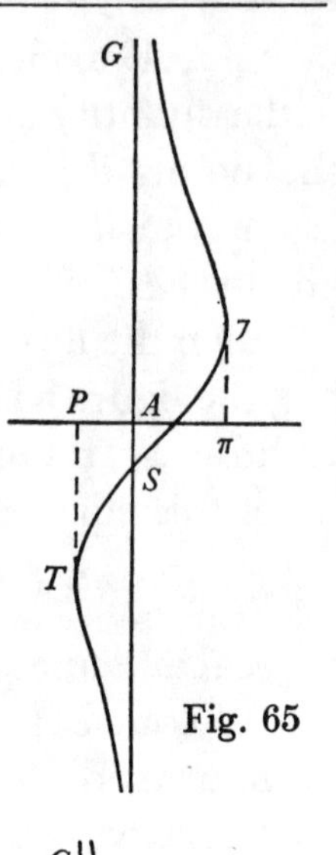

Fig. 65

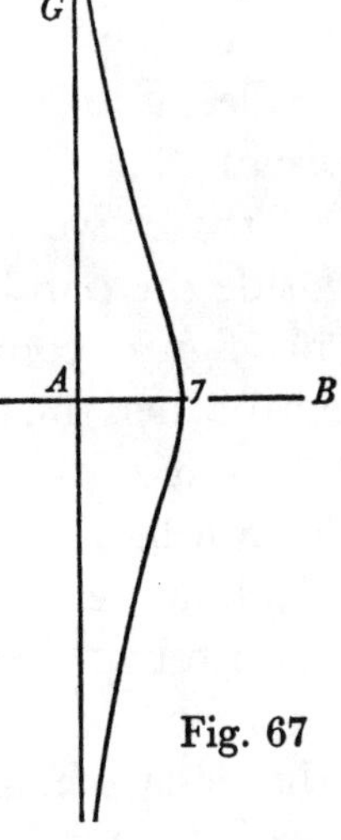

Fig. 67

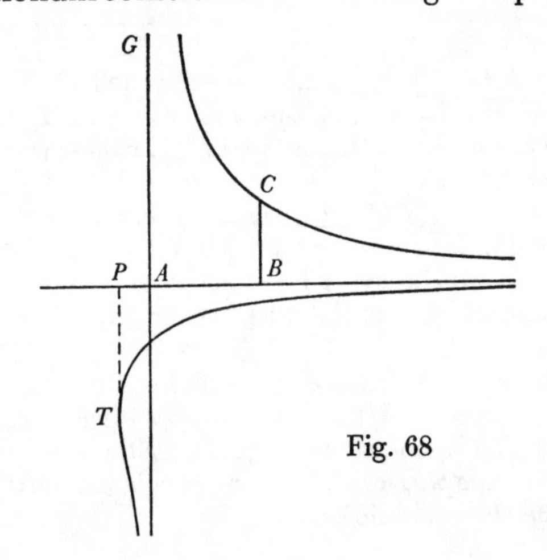

Fig. 68

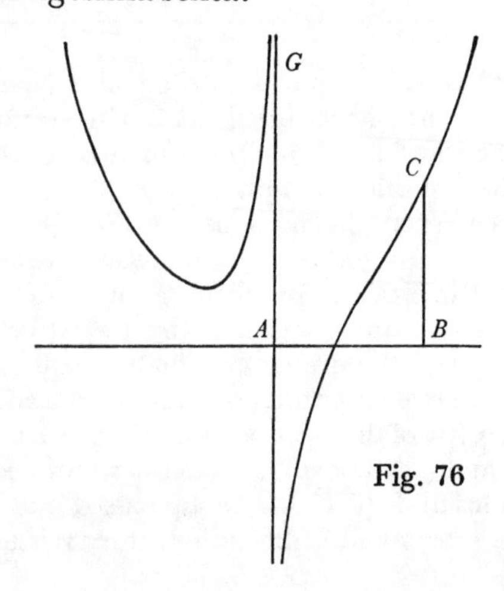

Fig. 76

The hyperbolism of an ellipse(93) is defined by this equation $xy^2+ey=-cx+d$, and has a single asymptote, namely the principal ordinate AG. If the term ey is not lacking, the figure is (Fig. 65) a snaky hyperbola without a diameter, and also without a centre if the term d is not wanting. This species is the sixty-first.

XXIV The three hyperbolisms of an ellipse.

But if the term d is lacking, the figure has (Fig. 66) a centre without a diameter, and its centre is the point A. The species, to be sure, is the sixty-second.

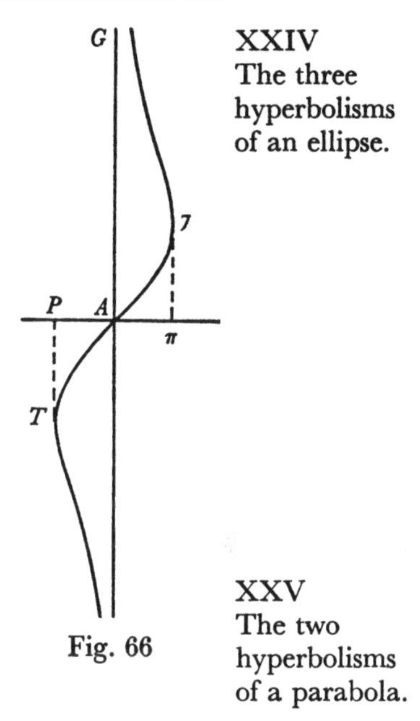

Fig. 66

While if the term ey is lacking and the term d not wanting, the figure is (Fig. 67) conchoidal along the asymptote AG, having a diameter but not a centre, and its diameter is the abscissa AB. This species is the sixty-third.

XXV The two hyperbolisms of a parabola.

The hyperbolism of a parabola(94) is defined by this equation $xy^2+ey=d$ and has two asymptotes, the abscissa AB and the prime and principal ordinate AG. The hyperbolas in this figure, however, are two, lying not in opposite angles of the asymptotes but in immediately adjacent ones, and this on each side of the abscissa AB, being either (Fig. 68) without a diameter if the term ey is had, or (Fig. 69) with a diameter if that term is lacking. These two species are the sixty-fourth and sixty-fifth.

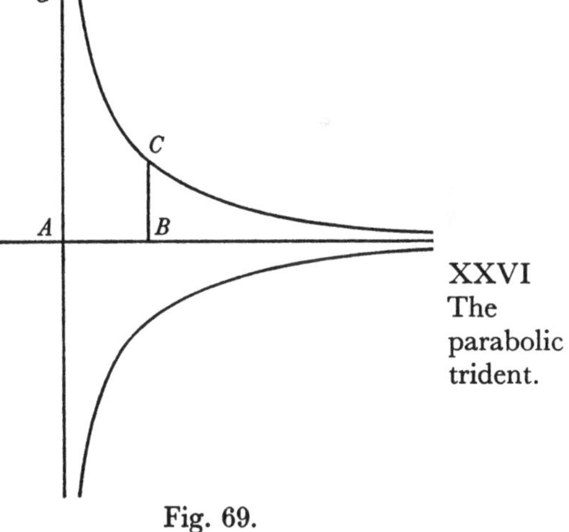

Fig. 69.

XXVI The parabolic trident.

In the second case of equations there was had the equation $xy=ax^3+bx^2+cx+d$. The figure in this instance has (Fig. 76(95)) four infinite branches, of which two are hyperbolic ones tending in opposite directions round the asymptote AG and two converging parabolic ones all but forming with the previous ones a species of *trident*. And this figure is the parabola by means of which Descartes constructed equations of six dimensions.(96) This is therefore the sixty-sixth species.

(93) Namely $y^2+ey=-cx^2+dx$; compare note (86).

(94) That is, $y^2+ey=dx$ or $(y+\frac{1}{2}e)^2=d(x+\frac{1}{4}e^2/d)$.

(95) In his 1711 edition of the 'Enumeratio' William Jones, oblivious to the inconsistency of still having this precede the unchanged 'Fig. 70, 71' which follow, renumbered this in improved ordinal sequence as '72', correspondingly increasing the denominations of Figures 72, 73, 74 and 75 below by unity; compare his analogous cyclical adjustment of Figures 9–16 above (on which see note (58)).

(96) In the third book of his *Geometrie*; see I: 495, note (15). On Newton's name—now standard, but here freshly minted—for this Cartesian 'parabole' see §1: note (31) preceding.

XXVII Parabolæ quinq; divergentes. In tertio casu Æquatio erat $yy = ax^3 + bxx + cx + d$ et Parabolam designat cujus crura divergunt ab invicem & in contrarias partes infinitè progrediuntur. Abscissa AB(97) est ejus Diameter, et species ejus sunt quinq; sequentes.

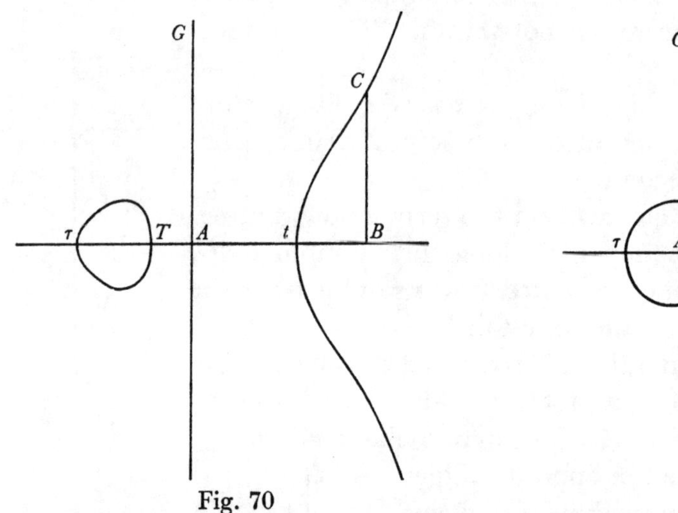

Fig. 70

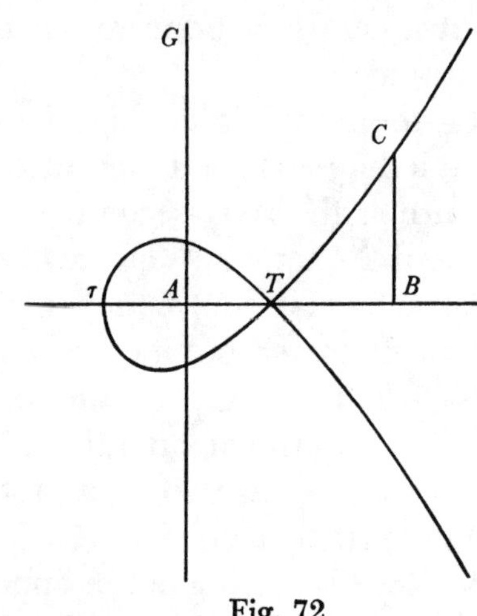

Fig. 72

Fig. 70, 71 Si æquationis $ax^3 + bxx + cx + d = 0$ radices omnes $A\tau$, AT, At sunt reales et inæquales, figura est Parabola divergens campaniformis cum *Ovali* ad verticem. Et species est sexagesima septima.

Fig. 72 Fig. 73 Si radices duæ sunt æquales Parabola prodit vel *nodata* contingendo Ovalem, vel *punctata* ob Ovalem infinitè parvam. Quæ duæ species sunt sexagesima octava et sexagesima nona.

Fig. 75 Si tres radices sunt æquales Parabola erit *cuspidata* in vertice. Et hæc est Parabola *Neiliana*(98) quæ vulgo *semicubica* dicitur. [Et species est septuagesima.](99)

Fig. 73, 74 Si radices duæ sunt impossibiles habetur Parabola *pura* campaniformis speciem septuagesimam primam constituens.

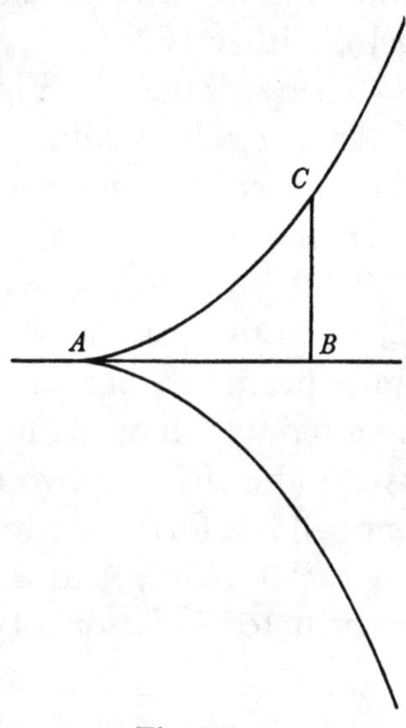

Fig. 75

(97) That is, $y = 0$.

(98) See II: 68, note (88) and IV: 374, note (72) on William Neil's connection with this species of divergent parabola.

(99) Both manuscript versions of the 'Enumeratio'—and, correspondingly, its 1704 and 1706 printings—lack any specification at this point of the number of the place occupied by the

XXVII The five divergent parabolas.

In the third case the equation was $y^2 = ax^3+bx^2+cx+d$ and denotes a parabola whose branches diverge from one other and progress to infinity in contrary directions. The abscissa AB[97] is its diameter, and its species are the five following.

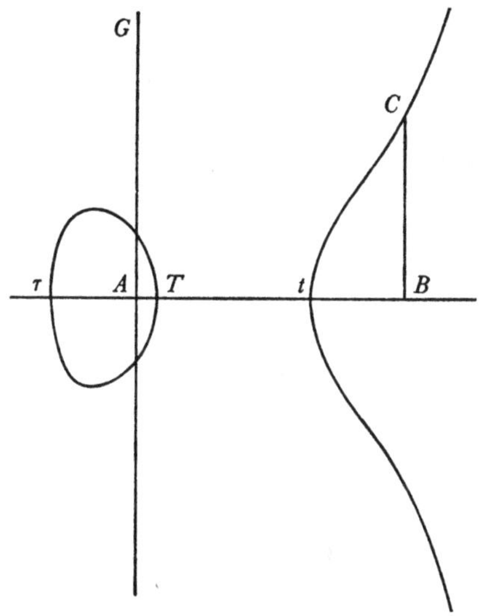

Fig. 71

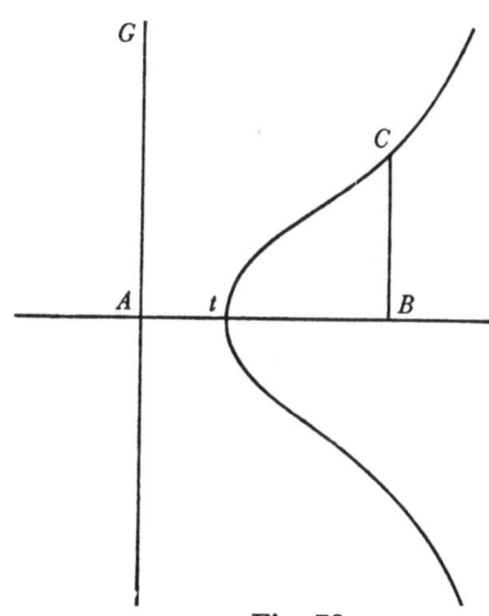

Fig. 73

If all the roots $A\tau$, AT, At of the equation

$$ax^3+bx^2+cx+d = 0$$

are real and unequal, the figure is (Figures 70, 71) a bell-shaped divergent parabola with an *oval* at its vertex. The species is the sixty-seventh.

If two roots are equal, the parabola proves to be either (Fig. 72) *nodate* by touching the oval, or (Fig. 73) *punctate* because the oval is infinitely small. These two species are the sixty-eighth and sixty-ninth.

If the three roots are equal, the parabola will (Fig. 75) be *cusped* at the vertex. And this is the *Neilian*[98] parabola, which is commonly called the *semicubic*. The species is the seventieth.[99]

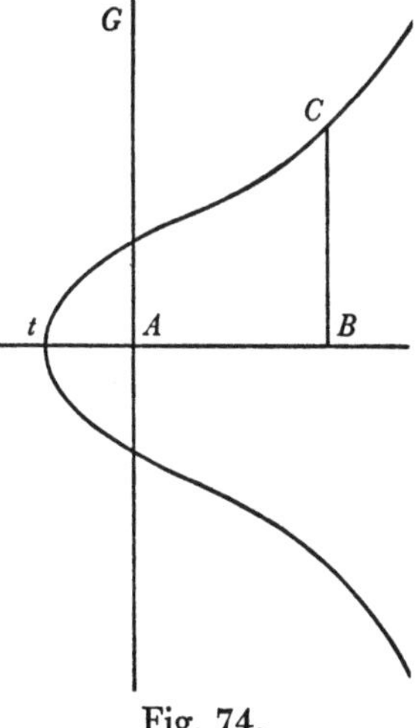

Fig. 74.

If two roots are impossible, there is had (Figures 73, 74) a *pure* bell-shaped parabola which constitutes the seventy-first species.

Neilian parabola in Newton's present tabulation of the species of cubics. Our interpolation, styled on the pattern of what has gone before, differs only by a trivial inversion of word order from an equivalent insertion made by Jones in his 1711 edition.

XXVIII
Parabola cum cruribus contrarijs.

In quarto casu æquatio erat

$$y = ax^3 + bxx + cx + d,$$

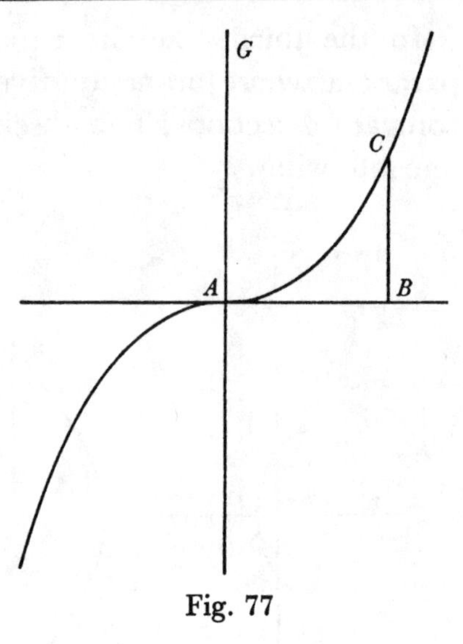

Fig. 77

et hæc æquatio Parabolam illam *Wallisianam*(100)

Fig. 77 designat quæ crura habet contraria & *cubica* dici solet. Et sic species omninò sunt septuaginta duæ.(101)

XXIX
Genesis Curvarū per Umbras.

(102)Si in planum infinitum(103) a puncto lucido illuminatum umbræ figurarum projiciantur, umbræ Sectionum Conicarum semper erunt Sectiones Conicæ, eæ Curvarum secundi generis semper erunt Curvæ secundi generis, eæ curvarum tertij generis semper erunt Curvæ tertij generis & sic deinceps in infinitum. Et quemadmodum Circulus umbram projiciendo generat sectiones omnes conicas, sic Parabolæ quinq; divergentes umbris suis generant et exhibent alias omnes secundi generis curvas, et sic Curvæ quædam simpliciores aliorum generum inveniri possunt quæ alias omnes eorundem generum curvas umbris suis a puncto lucido in planum projectis delineabunt.(104)

XXX
Curvarum puncta duplicia.

(105)Diximus curvas secundi generis a linea recta in punctis tribus secari posse. Harum duo nonnunquam coincidunt. Ut cùm recta per Ovalem infinitè parvam(106) transit vel(107) per concursum duarum partium curvæ se mutuò

(100) This now familiar name for the 'parabola of ye second kind' which John Wallis had employed forty years earlier (see II: 292–3, note (62)) geometrically to construct the roots of the general reduced cubic is here (compare §1: note (36) preceding) Newton's fresh invention. He will himself use it to the same purpose in his last section. Notice that in strict truth the figure alongside depicts the reduced cubic $y = ax^3$ (on which see IV: 368, note (47)).

(101) Understanding the last species to be 'septuagesima secunda' (the seventy-second), that is. Newton concluded his preliminary draft (f. 45^v)—where (see notes (69) and (74) above) species 14/16, 15/17 and 28/29 are not separately distinguished—with an explicit statement that 'Hæc est species sexagesima nona' (This is the sixty-ninth species). In a manuscript addition at this point on an interleaf in his library copy of the 1706 *Optice*, whereto a reprint of the 'Enumeratio' was added (see note (2) above), Newton repeated for the benefit of his future readers the prescription: 'Proposita igitur Curva aliqua secundi generis, ut innotescat ejus species, invenienda est ejus tangens, et ex puncto contactus in infinitum abeunte habebuntur plagæ crurum infinitorum ejus una cum Asymptotis crurum Hyperbolicorum. Et inde innotescet positio Abscissæ *AB* & ordinationis angulus *ABC*, ut & Æquatio qua relatio inter Abscissam *AB* et Ordinatam *BC* definitur. Et ex Æquationis forma et conditionibus habebitur species Curvæ ut supra' (Where, therefore, there is proposed [the Cartesian equation of] some curve of second kind, to get to know its species you need to ascertain its [general] tangent, and then from the point of contact passing off to infinity there will be had the directions of its infinite branches together with the asymptotes of its hyperbolic branches. And therefrom you will come to know the position of the abscissa *AB* and the ordination angle

XXVIII
The parabola with contrary branches.

In the fourth case the equation was $y = ax^3+bx^2+cx+d$: this equation denotes (Fig. 77) the *Wallisian*(100) parabola which has contrary branches and is usually called *cubic*. And thus there are altogether seventy-two species.(101)

XXIX
The genesis of curves by shadows.

(102)If onto an infinite plane(103) lit by a point-source of light there should be projected the shadows of figures, the shadows of conics will always be conics, those of curves of second kind will always be curves of second kind, those of curves of third kind always curves of third kind, and so on without end. And just as the circle by projecting its shadow generates all conics, so the five divergent parabolas by their shadows generate and exhibit all other curves of second kind; while in this manner certain simpler curves of other kinds can be found which by their shadows cast by a point-source of light onto a plane shall delineate all other curves of the same kinds.(104)

XXX
Double points of curves.

(105)We have stated that curves of second kind can be cut by a straight line in three points. Of these, two on occasion coincide—for instance, when the line passes through an infinitely small oval(106) or is drawn(107) through the meeting-

---

$\widehat{ABC}$, as also the equation whereby the relationship between the abscissa AB and the ordinate BC is defined. From the form and circumstances of this equation there will then be had the species of the curve as above). In addition to the 72 species now listed by Newton there are, as is well known, three further pairs of hyperbolas *cum ovali* and *punctata* whose existence he here (see notes (67), (72) and (84)) fails to record.

(102) The following paragraph is not found in Newton's initial draft, which at this point straightaway continues (ff. 45[v] *et seq.*) with the preliminary version of the remainder of the text which is reproduced in Appendix 1.3 below.

(103) 'positione datum' (given in position) is deleted.

(104) In this paragraph Newton has made a number of later changes of 'Ordinis'/'Ordinum' into 'generis'/'generum'; compare note (3) above. We will not be surprised that this brief announcement of the possibility of projectively generating all the species of cubic above enumerated as the (plane) optical 'shadows' of the five divergent parabolas all but surpassed the capacity of his contemporaries—ignorant of the 'Geometriæ Liber primus' where (see pages 416–32 above) Newton had earlier elaborated it—merely to comprehend, let alone demonstrate. Trust him not to mention the key insight that on projecting the tangent at an inflection point to infinity every cubic passes into its corresponding divergent parabola! But then again he himself may not fully have appreciated the generality of this property of all cubic curves (compare **2**, 3, §1: notes (55) and (60) preceding).

(105) Of the remaining text Newton's manuscript draft contains both (ff. 45[v]–47[v] [+3961.2: 15[v], 16[r]]) a widely variant, partially cancelled initial version which for its several points of special interest we reproduce *in toto* in Appendix 1.3 below, and also (ff. 48[r]–50[r]) a preliminary recasting of this which differs little in its essentials from the final 'Enumeratio', as it is here printed, other than in retaining the original numbering (XXVI–XXX; see note (19) to Appendix 1.3) of the accompanying marginal heads and in omitting (see note (134) below) the latter's terminal paragraph. A few not entirely trivial verbal variations in the preliminary revise are cited at pertinent places in following footnotes.

(106) Where, that is, the oval has vanished to become a conjugate double point of the cubic.

(107) In his draft (f. 48[r]) Newton initially continued by writing 'per intersectionum punctum in quo Curva seipsam vel conjugatam suam decussat vel in cuspidem terminatur'

secantium vel in cuspidem coeuntium ducitur. Et siquando rectæ omnes in plagam cruris alicujus infiniti tendentes curvam in unico tantum puncto secant (ut fit in ordinatis Parabolæ Cartesianæ[108] et Parabolæ cubicæ nec non in rectis abscissæ Hyperbolismorum Hyperbolæ et Parabolæ parallelis) concipiendum est quod rectæ illæ per alia duo curvæ puncta ad infinitam distantiam sita (ut ita dicam) transeunt. Hujusmodi intersectiones duas coincidentes sive ad finitam sint distantiam sive ad infinitam vocabimus punctum duplex. Curvæ autem quæ habent punctum duplex describi possunt per sequentia Theoremata.[109]

XXXI Theoremata de Curvarum descriptione organica.[110]

1. Si anguli duo dati PAD, PBD circa polos positione datos A, B rotentur ut eorum crura AP, BP concursu suo P percurrant lineam rectam; crura duo reliqua AD, BD concursu suo D describent Sectionem Conicam per polos A, B transeuntem: præterquam ubi linea illa recta transit per polorum alterutrum A vel B, vel anguli BAD, ABD simul evanescunt,[111] quibus in casibus punctum D describet lineam rectam.[112]

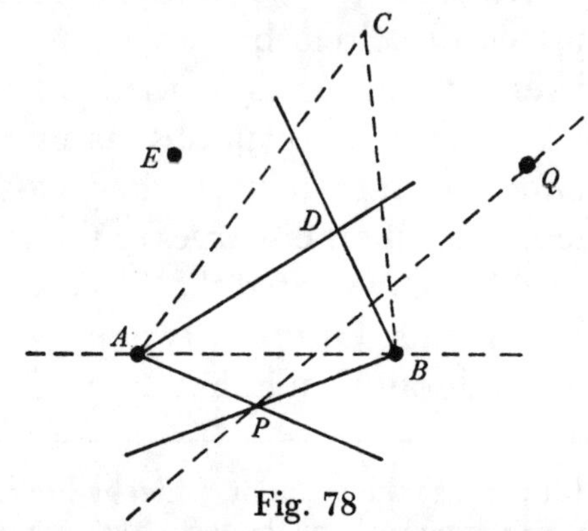

Fig. 78

2. Si crura prima AP, BP concursu suo P percurrant Sectionem Conicam per polum alterutrum A transeuntem, crura duo reliqua AD, BD concursu suo D describent Curvam secundi generis per polum alterum B transeuntem et punctum duplex habentem in polo primo A per quem Sectio Conica transit: præterquam ubi anguli BAD, ABD simul evanescunt,[113] quo casu punctum D describet aliam Sectionem Conicam per polum A transeuntem.[114]

3. At si Sectio Conica quam punctum P percurrit transeat per neutrum

(through the intersection point in which a curve crosses itself or its conjugate branch or comes to an end at a cusp).

(108) Compare note (27) above.

(109) It will be evident that these three theorems on the description by means of angular 'legs' pivoting round fixed poles of conics and of cubics and quartics possessed of a double point are abstracted from the yet more general lemmatical rules (see II: 106–16) which Newton had nearly thirty years before—and in more than one version—listed in preliminary, as here, to his unified organic 'Modus describendi Conicas sectiones et curvas trium Dimensionum [punctum duplex habentium] &c' (*ibid.*: 118–32, further recast and broadened in its scope on *ibid.*: 132–50).

(110) That is, 'per organa facta' (instrumental); compare page 392 above. This appellation—intended by Newton to place this present mode of construction in the class of all such 'mechanical' descriptions of curves, but now universally regarded (however illogically) as its distinguishing name—does not appear in the draft head. We have previously (see II: 8) pointed to Schooten's earlier fruitless efforts to find just such a unified 'organic' description of all the species of conic as the stimulus which led Newton initially to contrive his own highly successful method.

point of two parts of a curve cutting one another or coalescing at a cusp. And whenever all straight lines tending in the direction of some infinite branch cut the curve in but a single point (as happens in the case of the ordinates of the Cartesian parabola(108) and the cubic parabola, and also in lines parallel to the abscissa of hyperbolisms of a hyperbola and a parabola), we need to conceive that those straight lines pass through another two points of the curve situated (as I might say) at an infinite distance. Two coincident intersections of this sort, whether they be at a finite distance or at an infinite one, we shall call a 'double point'. Curves, however, which have a double point can be described by means of the following theorems.(109)

XXXI Theorems on the organic(110) description of curves.

1. If the two given angles $P\widehat{A}D$, $P\widehat{B}D$ rotate round the poles A, B given in position so that their legs AP, BP at their meet P traverse a straight line, the two remaining legs AD, BD will at their meeting point D describe a conic passing through the poles A and B; except when that straight line passes through one or other of the poles A or B, or the angles $B\widehat{A}D$, $A\widehat{B}D$ simultaneously vanish,(111) in which cases the point D will describe a straight line.(112)

2. If the first pair of legs AP, BP at their meet P traverse a conic passing through one or other pole A, the two remaining legs AD, BD will at their meet D describe a curve of second kind passing through the other pole B and having a double point at the first pole A through which the conic passes; except when the angles $B\widehat{A}D$, $A\widehat{B}D$ simultaneously vanish, (113) in which case the point D will describe another conic passing through the pole A.(114)

3. But if the conic which the point P traverses should pass through neither of

(111) In his draft Newton initially wrote at far greater length: 'vel per punctum tertium in quod concursus crurum $A[P]$, $B[P]$ incidit ubi crura altera $A[D]$, $B[D]$ applicantur ad rectam AB et cum ea coincidunt' (or through the third point at which the meet of the legs AP, BP falls when the other legs AD, BD are applied to the straight line AB and coincide with it)—we emend the text to agree with his accompanying figure, where the points P and D were originally designated as R and S—and then shortened this to be 'vel crura illa reliqua AD, BD simul coincidunt cum recta AB polos jungente' (or the remaining legs AD, BD simultaneously coincide with the straight line AB joining the poles), concluding thereafter: 'quibus in casibus crura illa reliqua AD, BD describent lineam rectam' (in which cases those remaining legs AD, BD will describe a straight line). A number of similar replacements of phrases in the draft citing 'crura' (legs) by ones referring to 'anguli' (angles) are noticed in sequel.

(112) A minimal elaboration of Theorem 1 of section [2] on II: 110; compare the equivalent Theorem 1 on II: 106 preceding.

(113) Much as in the previous paragraph (see note (111)) Newton's draft (f. 48v) here reads 'præterquam ubi crura AD, BD simul coincidunt cum recta AB' (except when the legs AD, BD simultaneously coincide with the straight line AB), proceeding thereafter to affirm 'quo in casu earum intersectio D describet...' (in which case their intersection D will describe...).

(114) An amalgam of Theorems 2–4 on II: 110.

polorum *A*, *B*, punctum *D* describet curvam secundi vel tertij generis punctum duplex habentem. Et punctum illud duplex in concursu crurum describentium *AD*, *BD* invenietur ubi anguli *BAP*, *ABP* simul evanescunt.(115) Curva autem descripta secundi erit generis si anguli *BAD*, *ABD* simul evanescunt,(116) alias erit tertij generis & alia duo habebit puncta duplicia in polis *A* et *B*.(117)

XXXII Sectionum Conicarum descriptio per data quinqȝ puncta.

Jam Sectio Conica determinatur ex datis ejus punctis quinqȝ et per eadem sic describi potest. Dentur ejus puncta quinqȝ *A*, *B*, *C*, *D*, *E*. Jungantur eorum tria quævis *A*, *B*, *C* et trianguli *ABC* rotentur anguli duo quivis *CAB*, *CBA* circa vertices suos *A* et *B* et ubi crurum *AC BC* intersectio *C* successive applicatur ad puncta duo reliqua *D*, *E*, incidat intersectio crurum reliquorum *AB* et *BA* in puncta *P* et *Q*. Agatur et infinite producatur recta *PQ* et anguli mobiles ita rotentur ut intersectio crurum *AB*, *BA* percurrat rectam *PQ*, et crurum reliquorum intersectio *C* describet propositam Sectionem Conicam per Theorema primum.(118)

XXXIII Curvarum secundi generis punctum duplex habentium descriptio per data septem puncta.

Curvæ omnes secundi generis punctum duplex habentes determinantur ex datis eorum punctis septem quorum unum est punctum illud duplex, et per eadem puncta sic describi possunt. Dentur Curvæ describendæ puncta quælibet septem *A*, *B*, *C*, *D*, *E*, *F*, *G* quorum *A* est punctum duplex. Jungantur punctum *A* et alia duo quævis e punctis puta *B* et *C*; et trianguli *ABC* rotetur tum angulus *CAB* circa verticem suum *A* tum angulorum reliquorum alteruter *ABC* circa verticem suum *B*. Et ubi crurum *AC*, *BC* concursus *C* successivè applicatur ad puncta quatuor reliqua *D*, *E*, *F*, *G* incidat concursus crurum reliquorum *AB* et *BA* in puncta quatuor *P*, *Q*, *R*, *S*. Per puncta illa quatuor et quintum *A* describatur Sectio Conica, et anguli præfati *CAB*, *CBA* ita rotentur ut crurum *AB*, *BA* concursus percurrat Sectionem illam Conicam, et concursus reliquorum crurum *AC*, *BC* describet Curvam propositam per Theorema secundum.(119)

---

(115) In his draft Newton first wrote: '...punctum *D* describet curvam tertij Ordinis[!] tria habentem puncta duplicia, unum in polo *A*, alterum in polo *B* ac tertium in concursu crurum describentium *AD*, *BD* ubi crura altera *AP*, *BP* applicantur ad rectam *AB* et cum ea coincidunt' (the point *D* will describe a curve of third [kind] having three double points, one at the pole *A*, a second at the pole *B*, and the third at the meet of the describing legs *AD*, *BD* when the other legs *AP*, *BP* are applied to the straight line *AB*, coinciding with it); and only then began to make the proviso 'præterquam ubi crura describentia simul coincidunt cum recta *AB*, quo in casu [curva secundi erit generis, nullum punctum duplex habens ad *A* vel *B*]' (except when the describing legs simultaneously coincide with the line *AB*, in which case [the curve will be of second kind, having no double point at *A* or *B*]) before breaking off. In preliminary revise he then recast his sentence to affirm more broadly, much as in his final version, that 'punctum *D* describet curvam secundi vel tertij generis punctum duplex habentem in concursu crurum describentium *AD*, *BD* ubi crura altera *AP*, *BP* applicantur ad rectam *AB* et cum ea coincidunt' (the point *D* will describe a curve of second or third kind having a double point at the meet of the describing legs *AD*, *BD* when the other legs *AP*, *BP* are applied to the straight lines *AB* and coincide with it).

the poles A or B, the point D will describe a curve of second or third kind having a double point. And that double point will be found at the meet of the describing legs AD, BD when the angles $B\widehat{A}P$, $A\widehat{B}P$ simultaneously vanish.(115) The curve described will be of second kind, however, if the angles $B\widehat{A}D$, $A\widehat{B}D$ simultaneously vanish,(116) otherwise it will be of third kind and have two other double points at the poles A and B.(117)

XXXII
The description of conics through five given points.

Now a conic is determined from five of its points given and can be described through them in this way. Let there be given five of its points A, B, C, D, E. Join any three of them A, B, C and rotate any two angles $C\widehat{A}B$, $C\widehat{B}A$ of the triangle ABC round their vertices A and B, and when the intersection C of the legs AC, BC is successively applied to the two remaining points D and E, let the intersection of the remaining legs AB and BA fall at the points P and Q. Draw and indefinitely extend the straight line PQ, and let the mobile angles so rotate that the intersection of the legs AB, BA traverses the line PQ: the intersection C of the remaining legs will then describe the proposed conic by the first Theorem.(118)

XXXIII
The description of curves of second kind having a double point through seven given points.

All curves of second kind having a double point are determined from seven of their points given, one of which is that double point, and can be described through these same points in this way. In the curve to be described let there be given any seven points A, B, C, D, E, F, G, of which A is the double point. Join the point A and any two other of the points, say B and C, and rotate both the angle $C\widehat{A}B$ of the triangle ABC round its vertex A and either one, $A\widehat{B}C$, of the remaining angles round its vertex, B. And when the meeting point C of the legs AC, BC is successively applied to the four remaining points D, E, F, G, let the meet of the remaining legs AB and BA fall at the four points P, Q, R, S. Through those four points and the fifth one A describe a conic, and then so rotate the before-mentioned angles $C\widehat{A}B$, $C\widehat{B}A$ that the meet of the legs AB, BA traverses that conic, and the meet of the remaining legs AC, BC will by the second Theorem describe the curve proposed.(119)

(116) In continuation of what we have quoted in the previous note, the draft here reads: 'Et hæc curva secundi erit generis ubi crura *AD*, *BD* simul coincidunt cum recta *AB*' (And this curve will be of second kind when the legs *AD*, *BD* come simultaneously to coincide with the line *AB*).

(117) This amplifies Theorem 6 on II:110 in making exception for the particular case in which the described quartic curve degenerates to be the pair of the line *AB* and a cubic locus (*D*).

(118) A minimal reshaping of Problem 1 on II: 118–20; see also the equivalent converse (IV: 300) to Proposition 7 of Newton's subsequent 'Solutio Problematis Veterum de Loco solido' (*ibid.*: 282–320).

(119) Exactly as in the equivalent Problem 9 on II: 128–30 on making trivial changes in the designation of points.

Si vice puncti C datur positione recta BC quæ Curvam describendam tangit in B, lineæ AD, AP coincident, et vice anguli DAP habebitur linea recta circa polum A rotanda.(120)

Si punctum duplex A infinite distat, debebit Recta ad plagam puncti illius perpetuò dirigi & motu parallelo ferri interea dum angulus ABC circa polum B rotatur.

Describi etiam possunt hæ curvæ paulo aliter per Theorema tertium, sed descriptionem simpliciorem posuisse sufficiat.

Eadem methodo Curvas tertij, quarti et superiorum generum describere licet, non omnes quidem sed quotquot ratione aliqua commoda per motum localem(121) describi possunt. Nam curvam aliquam secundi vel superioris generis punctum duplex non habentem commode describere Problema est inter difficiliora numerandum.(122)

XXXIV Constructio æquationum per descriptionem Curvarum.(123)

Curvarum usus(124) in Geometria est ut per earum intersectiones Problemata solvantur. Proponatur æquatio construenda dimensionum novem

$$x^9 * +bx^7+cx^6+dx^5+ex^4 \begin{matrix}+f\\+m\end{matrix} x^3+gxx+hx+k = 0$$

Ubi b, c, d &c significant quantitates quasvis datas cum signis suis + et − affectas. Assumatur æquatio ad Parabolam cubicam $x^3 = y$,(125) et æquatio prior scribendo y pro x^3 evadet

$$y^3+bxyy+cyy+dxxy+exy+my+fx^3+gxx+hx+k = 0,$$

æquatio ad Curvam aliam secundi generis. Ubi m vel f deesse potest vel pro

(120) The equivalent paragraph in Newton's draft (f. 49r) here repeats that of his initial version (see Appendix 1.3 below) to affirm: 'Si punctum C accedat ad punctum B et cum eo coincidat, evanescet angulus BAC, et recta BC tanget curvam describendam in puncto B. Unde si vice puncti C, detur positione recta BC quæ curvam describendam tangat in B, describetur Curva per Regulam rectilineam circa polum A & angulum ABC circa polum B rotantes. Nam si Regula illa concurrat cum anguli latere BA ad Sectionem conicam, concurret cum altero ejus latere BC ad Curvam describendam' (If the point C should approach the point B and coincide with it, the angle $\widehat{BAC}$ will vanish, and the straight line BC will touch the curve to be described at the point B. Whence if, instead of the point C, there be given in position the straight line BC which is to touch the curve to be described at B, that curve will be described by a straight-line ruler rotating round the pole A and an angle $\widehat{ABC}$ rotating round the pole B; for, should that ruler meet the side BA of the angle at [a point of] the conic, it will [ever] meet its other side BC at [a point of] the curve to be described). The individual organic constructions in this way of conic and cubic describends are set out in Problems 2 and 10 on II: 122–4/130, while the general observation is made in Lemma 6 'De Angulis Mobilibus' on *ibid.*: 138.

(121) These last three words are not found in Newton's draft (f. 49v). He understands, of course, that this motion shall be smoothly continuous in its progress from place to place.

(122) In his draft Newton initially wrote 'minime contemnendum' (not in the least to be

If instead of the point C there is given in position the straight line BC which the curve to be described touches at B, the lines AD, AP will coincide and instead of angle $D\widehat{A}P$ will be had a straight line to be rotated round the pole A.(120)

If the double point A is infinitely distant, a straight line will need to be perpetually directed in the direction of that point and be carried along in a parallel motion all the while that the angle $A\widehat{B}C$ rotates round the pole B.

These curves can also be described a little differently by means of the third Theorem, but let it be enough to have set out the simpler description.

By the same method we are free to describe curves of third, fourth and higher kinds—not all, indeed, but as many as can in some convenient manner be described by local motion:(121) to be sure, conveniently to describe some curve of second or higher kind not having a double point is a problem to be counted among the more difficult.(122)

XXXIV
The construction of equations by the description of curves.(123)

The(124) use of curves in geometry is that by means of their intersections problems shall be solved. Let there be proposed for construction an equation of nine dimensions $x^9 * +bx^7+cx^6+dx^5+ex^4+(f+m)\,x^3+gx^2+hx+k=0$, where $b, c, d, \ldots$ signify any given quantities affected with their signs $+$ and $-$. Let the equation $x^3 = y$ to a cubic parabola(125) be assumed, and on writing y in place of x^3 the former equation will turn out to be

$$y^3+bxy^2+cy^2+dx^2y+exy+my+fx^3+gx^2+hx+k=0,$$

one to another curve of second kind. Here m or f can be lacking or assumed at

despised) and then changed this to read 'solutione dignum' (worthy of solution) before thus accentuating the problem's difficulty in his final version. It had originally been his intention—in section XXVI of the preparatory text of this concluding portion of the 'Enumeratio' which is reproduced in Appendix 1.3 below—to present an analogous 'organic' construction of a particular class of cissoidal cubics on the pattern of that which he had set out in Problem 20 of his Lucasian lectures on algebra (see v: 218–20, and compare *ibid.*: 462–4) but, carelessly carrying over an earlier error (on which see v: 466–8, note (686)), he botched his exposition of it and, rather than make the necessary amendations rectifying his account (see notes (27), (29) and (30) to Appendix 1.3), deleted it from his draft.

(123) In this last section of his tract Newton resumes, after more than a quarter of a century, his youthful researches (see I: 496–7, 500–1) into the application of the Wallisian cubic parabola $n^2y = x^3$ (of which he subsequently gave an elegant 'organic' contruction in his 'Problems for construing æquations'; see II: 498–500, 502–4)—for convenience he again sets the coefficient n to be unity—geometrically to construct, by its intersection with another curve of at most third degree, the real roots of equations (lacking their second term as and when necessary) of up to nine 'dimensions'. He also now for the first time—but almost certainly unaware of Viviani's prior employment of this 'iperbole mesolabica' to duplicate the cube (see note (129) below)—outlines how the parabolic hyperbolism $y = x^{-2}$ may be analogously employed to the same end.

(124) Newton here took over the exaggeration 'principalis' (principal) from his draft (f. 49v) before rightly deleting the adjective.

(125) The 'Parabola of ye 2d kind' as he had before (see I: 500) dubbed its equal $x=y^3$.

lubitu assumi. Et per harum Curvarum descriptiones & intersectiones dabuntur radices æquationis construendæ. Parabolam cubicam semel describere sufficit.(126)

Si æquatio construenda per defectum duorum terminorum ultimorum hx & k reducatur ad septem dimensiones Curva altera delendo m habebit punctum duplex in principio Abscissæ(127) et inde facilè describi potest ut supra.

Si æquatio construenda per defectum terminorum trium ultimorum $gxx+hx+k$ reducatur ad sex dimensiones, Curva altera delendo f evadet Sectio Conica.

Et si per defectum sex ultimorum terminorum æquatio construenda reducatur ad tres dimensiones incidetur in constructionem Wallisianam per Parabolam cubicam et lineam rectam.(128)

Construi etiam possunt æquationes per Hyperbolismum Parabolæ cum diametro.(129) Ut si construenda sit hæc æquatio dimensionum novem termino penultimo(130) carens,

$$a+cxx+dx^3+ex^4+fx^5 \begin{matrix}+m\\+g\end{matrix} x^6+hx^7+kx^8+lx^9=0;$$

assumatur æquatio ad Hyperbolismum illum $xxy=1$, et scribendo y pro $\frac{1}{xx}$, æquatio construenda vertetur in hanc(131)

$$ay^3+cyy+dxyy+ey+fxy+mxxy+g+hx+kxx+lx^3=0$$

(126) In his prior account Newton specified 'uppon a [tem]plate' (I: 496).

(127) For, since its equation then lacks both a constant term and ones in x and y, the curve's tangential direction at the origin is indeterminate. In his draft (f. 45^v, where this whole paragraph is a later insertion—sideways in the margin in the case of its revised version here repeated word for word, or perhaps there afterwards copied as a record of the revision) Newton initially began: 'Si æquatio septem dimensionum construenda sit deleantur termini tres $m[x^3]$, hx et k et Curva altera habebit...' (If an equation of seven dimensions should need to be constructed, delete the terms mx^3, hx and k, and the other curve will have...).

(128) While John Wallis' method of constructing the reduced cubic equation geometrically as the intersection of a cubic parabola and a straight line—expounded by him in his *Adversus Marci Meibomii De Proportionibus Dialogum, Tractatus Elencticus* (*Opera Mathematica*, **1** (Oxford, 1657): Dedicatio: 43–6), even as Newton wrote freshly reissuing 'E Theatro Sheldoniano' in Wallis' *Opera Mathematica*, **1** (Oxford, 1695): 229–90 (see especially 232–4)—is indeed in its essentials no different from Newton's, it is in its detail (as we have already pointed out in II: 292–3, note (62)) somewhat more complicated, in effect determining the real root(s) of the cubic $x^3 = mx+n$ through the meet of the cubical parabola $k^2(x+n/m) = y^3$ with the straight line $y = x\sqrt[3]{(k^2/m)}$, drawn parallel to the parabola's tangent at $x = \sqrt{(\frac{1}{27}m)}-n/m$.

In his draft at this point (ff. 49^v/50^r), we may add, Newton has cancelled a following paragraph which affirms: 'Quinetiam assumendo quantitatem quamcunque determinatam m, et pro fy scribendo $fy-my+mx^3$; curva altera ubi secundi est ordinis, modis infinitis variari potest, et casus facillimus eligi. Sic ad septusectionem anguli, si ponatur $m=f$ & ut æquatio construenda ad dimensiones tantum septem ascendat deleantur termini duo ultimi hx et k: curva altera habebit punctum duplex in principio abscissæ et inde facile describi potest ut supra' (To be sure, by assuming any determinate quantity m whatever and in place of fy

will. And through the descriptions of these curves and their intersections the roots of the equation to be constructed will be given. It is sufficient to describe the cubic parabola just once.(126)

If the equation to be constructed should through the lack of its two last terms hx and k be reduced to seven dimensions, the other curve will, on deleting m, have a double point at the origin of the abscissa(127) and can thence easily be described as above.

If the equation to be constructed should through lack of its three last terms gx^2+hx+k be reduced to six dimensions, the other curve will, on deleting f, prove to be a conic.

And if through wanting its six last terms the equation to be constructed should be reduced to three dimensions, you will fall on the Wallisian construction by a cubic parabola and a straight line.(128)

Equations can also be constructed by a hyperbolism of a parabola without diameter.(129) Should you need to construct, for instance, this equation of nine dimensions lacking its penultimate term(130):

$$a+cx^2+dx^3+ex^4+fx^5+(m+g)\,x^6+hx^7+kx^8+lx^9=0,$$

assume for the hyperbolism the equation $x^2y=1$, and on writing y in place of $1/x^2$ the equation to be constructed will turn into this:(131)

$$ay^3+cy^2+dxy^2+ey+fxy+mx^2y+g+hx+kx^2+lx^3=0$$

---

writing $(f-m)\,y+mx^3$, the other curve can, when it is of second order, be varied in an infinity of ways and its easiest case be chosen. Thus if, to effect the septusection of an angle, there be put $m=f$ and, in order that the equation to be constructed may rise but to seven dimensions, the two final terms hx and k be deleted, the other curve will have a double point at the origin of the abscissa and can thence easily be described as above). He presumes it as understood, of course, that the septusection of the angle $\theta=\sin^{-1}(q/2r)$ is, in some way or other, to be achieved by a construction which determines the roots $x=2r\sin\frac{1}{7}(\theta+2i\pi)$, $i=0,1,2,\ldots,6$, of the equation $x^7-7r^2x^5+14r^4x^3-7r^6x=r^7q$ (compare I: 476).

(129) Of general Cartesian equation $x^2y=cx+d$, though Newton here employs only the particular form in which $c=0$ and $d=1$. In his draft (f. 50r) he first added 'semel descriptū' (described [just] once)—on a template we may again understand, as with his preceding cubical parabola (see note (126) above)—but afterwards deleted this, presumably as evident. While it seems to us utterly unlikely that Newton would previously even have heard of this obscure Italian work, we should add that at the end of his *Quinto Libro di Euclide spiegato con la Dottrina del Galileo* (Florence, 1674) Vincenzo Viviani had made an equivalent use of this 'iperbole mesolabica' (as he named it) to effect the problem of duplicating the cube: in a minimal generalization of Viviani's approach, the first of two mean proportionals, $x=\sqrt[3]{(a^2b)}$, between two given magnitudes a and b is readily constructible geometrically as the meet of the parabolic hyperbola $x^2y=a^3$ and the straight line $ax=by$. (Compare G. Loria, *Storia delle Matematiche dall'Alba della Civiltà al Tramonto del Secolo XIX* (Milan, $_2$1950): 435.)

(130) That in x, namely, understanding that the terms of the following equation are ranged in descending powers of x, *viz.* as $lx^9+kx^8+hx^7+\ldots+cx^2*+a=0$.

(131) Notice how, by adroit substitution of $y=x^{-2}$, the term $(m+g)\,x^6$ in the previous equation is set to be $(mx^2y+g)\,x^6$.

quæ curvam[(132)] secundi generis designat, cujus descriptione Problema solvetur. Et quantitatum m ac g alterutra hic deesse potest, vel pro lubitu assumi.

Per Parabolam cubicam et curvas tertij generis construuntur etiam æquationes omnes dimensionum non plusquam duodecim, et per eandem Parabolam et curvas quarti generis construuntur omnes dimensionum non plusquam quindecim, et sic deinceps in infinitum. Et curvæ illæ tertij quarti et superiorum generum describi semper possunt inveniendo eorum puncta per Geometriam planam.[(133)] Ut si construenda sit æquatio

$$x^{12} * +ax^{10}+bx^9+cx^8+dx^7+ex^6+fx^5+gx^4+hx^3+ixx+kx+l=0,$$

& descripta habeatur Parabola Cubica; sit æquatio ad Parabolam illam cubicam $x^3=y$, et scribendo y pro x^3 æquatio construenda vertetur in hanc

$$y^4 \begin{matrix}+ax\\+b\end{matrix} y^3 \begin{matrix}+cxx\\+dx\\+e\end{matrix} yy \begin{matrix}+fxx\\+gx\\+h\end{matrix} y \begin{matrix}+ixx\\+kx\\+l\end{matrix} =0,$$

quæ est æquatio ad Curvam tertij generis cujus descriptione Problema solvetur. Describi autem potest hæc Curva inveniendo ejus puncta per Geometriam planam propterea quod indeterminata quantitas x non nisi ad duas dimensiones ascendit.[(134)]

---

(132) 'aliam' (further) is here deleted in Newton's draft.

(133) That is, by the intersection of straight lines and circles in the classical manner, and hence (in modern algebraic equivalent) by constructing the roots of quadratic equations. Newton elaborates upon this bare observation in his final sentence below.

(134) Whence, corresponding to any given y, the two values of x ensuing will be determinable by a 'plane' (that is, quadratic) construction; compare the previous note. This last paragraph—apparently added (along with the last four words of the preceding one) just as the 'Enumeratio' was put to press in late 1703—is not found in the preliminary text, which in lieu abruptly terminates with a 'Finis'. Its rough draft, entered by Newton at the head of a separate sheet (ULC. Add. 3963.14: 156^r) which otherwise contains a revamping of the essay on 'Analysis Geometrica' reproduced in **2**, 1, §1 above of no present importance (to this we shall return at the appropriate place in the final volume), differs from it only in the most minor verbal details except in having an unfinished accompanying comment, wholly omitted in the revise, where Newton began to specify that 'Constructiones Problematum hic traditæ componuntur ad mod[um præcedentium]' (The constructions of the problems here delivered are composed in the style [of those preceding]).

which denotes a[132] curve of second kind by whose description the problem will be solved. And either one of the quantities m and g can here be wanting or assumed at will.

Through the cubic parabola and curves of third kind there are constructed also all equations of not more than twelve dimensions, while through the same parabola and curves of fourth kind are constructed all of not more than fifteen dimensions, and so on indefinitely. And those curves of third, fourth and higher kinds can always be described by finding their points by plane geometry.[133] If, for instance, you should need to construct the equation

$$x^{12}+ax^{10}+bx^9+cx^8+dx^7+ex^6+fx^5+gx^4+hx^3+ix^2+kx+l=0$$

and have the cubic parabola described, let the equation to that cubic parabola be $x^3=y$, and on writing y in place of x^3 the equation to be constructed will turn into this:

$$y^4+(ax+b)\,y^3+(cx^2+dx+e)\,y^2+(fx^2+gx+h)\,y+ix^2+kx+l=0,$$

which is the equation to a curve of third kind by whose description the problem will be solved. This curve can, moreover, be described by finding its points through plane geometry seeing that the indeterminate quantity rises to no more than two dimensions in it.[134]

APPENDIX 1. VARIANT PARTIAL DRAFTS OF THE 'ENUMERATIO'.(1)

From originals in the University Library, Cambridge

[1](2) Lineæ omnes ordinis primi, tertij, quinti & imparis cujusq(3) duo ad minimum crura in infinitum versus plagas oppositas progredientia habent. Et lineæ tertij(4) ordinis de quibus agimus duo habent ejusmodi crura in plagas oppositas progredientia in quas nulla alia earum crura infinita tendunt. Si crura illa sint hyperbolici generis sit FAS eorum Asymptotos communis, et(5) huic parallela agatur recta quævis Cc ad curvam utrinq terminata si fieri potest eademq bisecetur in puncto $X_{[,]}$ et locus puncti illius X erit(6) Hyperbola conica (puta $X\Phi$) cujus una Asymptotos est AS. Sit ejus altera Asymptotos AB. Et siquando Hyperbola illa Conica cum asymptotis suis coin-

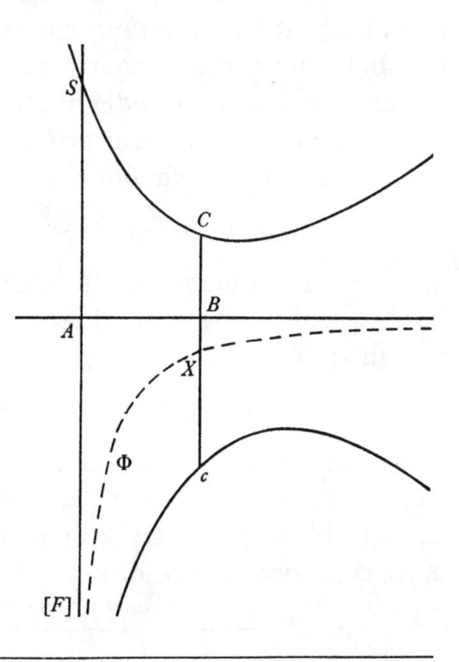

(1) Three considerably different preliminary versions of portions of the preceding main work (compare §2: note (2) above). While we reproduce their texts largely for completeness' sake, they each have their own small interest for the specialist student of Newton's geometry.

(2) Add. 3961.2: 16ᵛ, a first draft of paragraphs VIII–XI in the preceding 'Enumeratio'; see §2: note (26). Newton has entered his text in the blank spaces of a letter addressed to him from 'London yᵉ 6 June 95' by a Richard Chrisloe 'thatt is youre mostt humble searvantt to command', announcing that 'I have sent you Mʳ Tutt receatt of 108ˡⁱ witch I hope you will recuife safe from him' (see *The Correspondence of Isaac Newton*, **4**, 1967: 131).

(3) Initially '&c' *tout court*.

(4) Newton originally added 'seu curvæ secundi'; compare §2: notes (3) and (14) above.

(5) The text here first continued in a cancelled passage: '...huic parallelæ agantur tres rectæ quævis CD, ${}^2C^2D$, ${}^3C^3D$ ad curvam utrinq terminatæ eædemq bisecentur in E, ${}^2E$ et ${}^3E$ et in ipsis capiantur ${}^2EG$, ${}^3E^2G$ Asymptotis occurrentes in [F, ${}^2F$ et G, ${}^2G$].' In a rough accompanying sketch of a general redundant hyperbola (which we see no need here to reproduce) Newton has correspondingly inserted three ordinates BCD, ${}^2B^2C^2D$ and ${}^3B^3C^3D$ through points B, ${}^2B$ and ${}^3B$ in the abscissa AB, parallel to the cubic's asymptote AF and meeting it in the pairs of points C, D; ${}^2C$, ${}^2D$; ${}^3C$, ${}^3D$; then having drawn the (conic) diametral hyperbola through their mid-points E, ${}^2E$, ${}^3E$, along with its second asymptote, he has extended the chords E^2E and E^3E to meet the primary asymptote AF and this latter one respectively in the points F, ${}^2F$ and G, ${}^2G$. His intention is mysterious. That he abandoned his construction before here completely stating it would suggest that he had himself no clear notion of how to proceed.

(6) Newton initially went on to write 'vel linea recta vel Hyperbola conica...', further specifying in sequel that 'Si locus ille est linea recta sit ista $AB_{[,]}$ sin Hyperbola Conica sit ejus altera Asymptotos AB'.

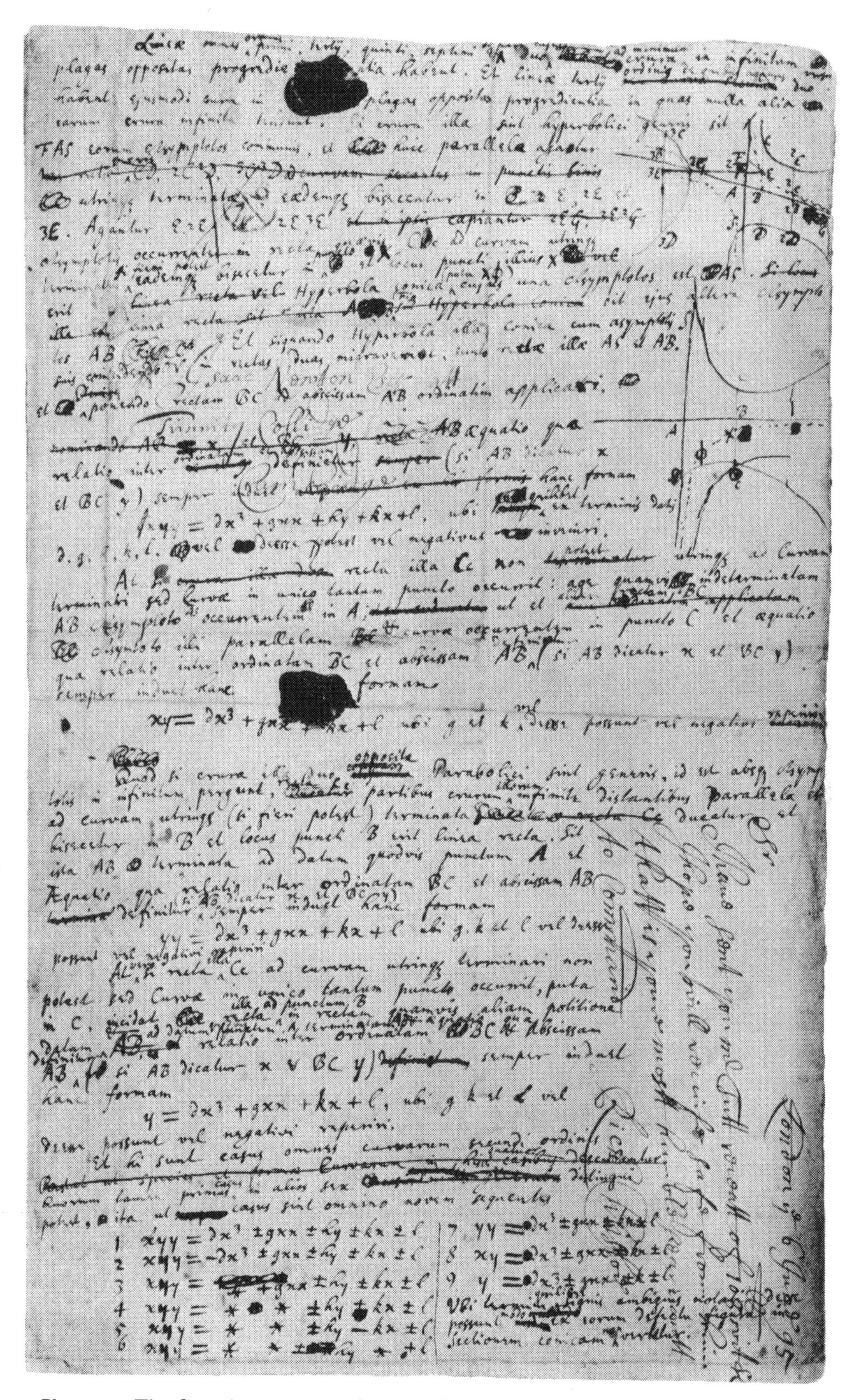

Plate III. The Cartesian equation of a general cubic reduced to its four canonical forms and nine principal cases (**3**, 1, Appendix 1.1).

cidendo in rectas duas migraverit, sunto rectæ illæ *AS* et *AB*. Et ponendo rectam *BC* ad abscissam *AB* ordinatim applicari, æquatio qua relatio inter ordinatam et abscissam definiatur (si *AB* dicatur x et *BC* y[(7)]) semper induet hanc formam[(8)] xyy[(9)] $= dx^3+gxx+hy+kx+l$, ubi quilibet ex terminis datis d, g, h, k, l vel deesse potest vel negativus inveniri.

At si recta illa *Cc* non potest utrinꝗ ad Curvam terminari sed Curvæ in unico tantum puncto occurrit: age quamvis indeterminatam *AB* Asymptoto *AS* occurrentem in *A*, ut et aliam rectam *BC* Asymtoto illi parallelam & curvæ occurrentem in puncto *C* et æquatio qua relatio inter ordinatam *BC* et abscissam *AB* definiatur (si *AB* dicatur x et *BC* y) semper induet hanc formam

$$xy = dx^3+gxx+kx+l$$

ubi g et k vel deessse possunt vel negativi reperiri.

Quod si crura illa[(10)] opposita Parabolici sint generis, id est absꝗ Asymptotis in infinitum pergunt, recta *Cc* partibus crurum illorum infinite distantibus parallela et ad curvam utrinꝗ (si fieri potest) terminata ducatur et bisecetur in *B*[,] et locus puncti *B* erit linea recta. Sit ista *AB* terminata ad datum quodvis punctum *A* et Æquatio qua relatio inter ordinatam *BC* et abscissam *AB* definiatur (si *AB* dicatur x et *BC* y) semper induet hanc formam

$$yy = dx^3+gxx+kx+l$$

ubi g, k et l vel deesse possunt vel negativi reperiri.

At vero si recta illa *Cc* ad curvam utrinꝗ terminari non potest sed Curvæ in unico tantum puncto occurrit, puta in *C*, incidat recta illa ad punctum *B* in rectam quamvis aliam positione datam & ad datum quodvis punctum *A* terminatam *AB*, & æquatio qua relatio inter ordinatam *BC* & abscissam *AB* defini[a]tur (si *AB* dicatur x & *BC* y) semper induet hanc formam

$$y = dx^3+gxx+kx+l,$$

ubi g k et l vel deesse possunt vel negativi reperiri.

Et hi sunt casus omnes curvarum secundi ordinis. Quorum tamen primus casus in alios sex[(11)] distingui potest, ita ut casus sint omnino novem sequentes

(7) Initially 'nominando $AB = x$ et $BC = y$'.

(8) Originally 'aliquam ex his formis', when it was evidently Newton's first impulse to list all four of the following reduced forms of equation in immediate sequel.

(9) '$bxyy$' was first written, as in the source text (see IV: 364 ff.) which Newton here takes over in equivalent geometrical mode.

(10) 'infinita' is deleted; in sequel Newton began to write 'contrar[ia]'.

(11) 'sequentes minus generales' is cancelled. Newton here first wrote less precisely 'Restat ut species et formæ Curvarum in hisce quatuor casibus describantur'.

$$
\begin{array}{lll}
1 & xyy = & dx^3 \pm gxx \pm hy \pm kx \pm l \\
2 & xyy = & -dx^3 \pm gxx \pm hy \pm kx \pm l \\
3 & xyy = & * [\pm] gxx \pm hy \pm kx \pm l \\
4 & xyy = & * \quad * \quad \pm hy + kx \pm l \\
5 & xyy = & * \quad * \quad \pm hy - kx \pm l \\
6 & xyy = & * \quad * \quad \pm hy \ * \ \pm l \\
7 & yy = & dx^{3(12)} \pm gxx \pm kx \pm l \\
8 & xy = & dx^{3(12)} \pm gxx \pm kx \pm l \\
9 & y = & dx^{3(12)} \pm gxx \pm kx \pm l.
\end{array}
$$

Ubi termini quilibet signis ambiguis notati deesse possunt modo ex eorum defectu figura in Sectionem conicam non vertatur.

---

[2][13] Hyperbolam vocamus *inscriptam* quæ tota jacet in asymptotôn angulo, *circumscriptam* quæ angulum asymptotôn in sinu suo amplectitur, *ambigenam* quæ uno crure inscribitur altero circumscribitur; *convergentem* cujus crura concavitate sua se invicem respiciunt, *divergentem* cujus crura convexitate sua se invicem respiciunt, *cum cruribus contrarijs*[14] *productam* cujus crura in partes contrarias convexa sunt & in plagas contrarias infinita; *cruciformem* quæ conjugatam decussat, *nodatam*[15] quæ seipsam decussat in orbem redeundo, *cu[s]pidatam* cujus partes duæ in angulo contactus concurrunt & ibi terminantur, *punctatam* quæ loco Ovalis conjugatæ habet punctum aliquod solitarium conjugatum in quod Ovalis diminuta et in nihilum contracta evanuit, *puram* quæ per impossibilitatem duarum radicum Ovali, nodo, cuspide & puncto conjugato privatur; [16] *conchoidalem* quæ cruribus divergentibus asymptoton appropinquat, *undulatam* vel *anguineam* quæ verticibus in contrarium flexis et cruribus contrarijs ad ipsam[17] applicatur. Eodem sensu Parabolam quoqꝫ vocamus *convergentem*, *divergentem*, *cum cruribus contrarijs productam*, *cruciformem*, *cuspidatam*, *nodatam*, *punctatam* & *puram*.

---

(12) Understand '$\pm dx^3$' in each case, of course.

(13) Add. 3961.2: 15^v, a first draft of paragraph XII in the final text of the 'Enumeratio'; see §2: note (35) above. Compare the preliminary list of appellations for the various possible configurations of cubic, roughed out by Newton on *ibid.*: 16^r, which has been quoted earlier in §1: note (2).

(14) Afterwards changed to be 'in crura contraria'.

(15) Initially 'nodosam'.

(16) In sequel Newton has deleted '*redundantem* quæ plura habet crura hyperbolica quam quatuor, *defectivam* quæ duo tantum habet crura infinita eaqꝫ hyperbolica, *parabolicam* quæ præter duo crura hyperbolica habet etiam duo parabolica'.

(17) Understand 'asymptoton', as indeed Newton here first wrote.

[3][18] ... ...

XXVI[19] Curvarum Poli et puncta duplicia.

Curvarum quas hic enumeravimus[20] aliquæ polum habent aliæ non habent. Polum voco punctum illud[21] a quo si recta ducatur et in orbem rotetur, hæc curvam alicubi extra polum in uno puncto secabit & nunquam in pluribus. Nam Polus pro duobus Curvæ punctis habendus est et recta curvam[22] non nisi in tribus punctis secare potest. Est autem Polus vel Cuspis Curvæ vel punctum decussationis ejus vel Ovalis infinite parva quam punctum conjugatum vocamus.[23] Et siquando hujusmodi polus vel punctum duplex in infinitum abeat, rectæ omnes in plagam puncti illius tendentes Curvam in unico tantum puncto secabunt: et contra si rectæ omnes in plagam cruris alicujus infiniti tendentes curvam in unico tantum puncto secant (ut fit in Ordinatis Parabolæ Cartesianæ et Parabolæ cubicæ nec non in rectis Abscissæ Parabolismorū Hyperbolæ & Parabolæ parallelis) concipiendum est quod rectæ illæ per alia duo curvæ puncta ad infinitam distantiam sita transeunt.

XXVII[24] Facilis descriptio curvarum per polos.

Curvæ quæ vel polum habent vel punctum duplex infinite distans, faciliùs describi solent; eæ quæ non habent, difficiliùs. Et descriptio prioris generis optime fit[25] per rectam quæ vel circa polum rotatur vel tendit ad punctum duplex infinite distans. Sic Conchois Veterum facillime describitur per Radium qui[26] circa polum ejus rotatur et polus ille est Ovalis infinite parva. Sic etiam Cissois Veterum per Regulam circa cuspidem[27] ejus rotatam optimè describi

(18) Add. 3961.1: 45ᵛ–47ᵛ/ 3961.2: 16ʳ, a widely different initial version of the concluding sections XXX–XXXIV of the final 'Enumeratio', where it was set to follow section (XXVI =) XXIX; see §2: notes (102) and (105) preceding.

(19) In the numbering of the sections in the draft this should be 'XXVII'. (Newton there initially set both of the two preceding sections to be 'XXV' and, in later adjusting the latter one to be 'XXVI', omitted likewise to augment the numbers of the ensuing sections by unity.) The accompanying paragraph, it will be clear, corresponds *grosso modo* to section XXX in the revised 'Enumeratio': the three preceding sections not comprehended within the draft's main scheme are I (an introductory paragraph—of no great intrinsic import—which is a late insertion in the final text on Add. 3961.2: 1ᵛ), XII (already added as an afterthought to the draft on its ff. 50ʳ/50ᵛ; see §2: note (35) above) and XXIX (wholly new in the revised version; see §2: note (102)).

(20) Originally 'descripsimus'.

(21) 'Figuræ' was initially added.

(22) That is, 'cubicam'.

(23) Initially just 'seu punctum conjugatum'.

(24) Rather than mend its faults (on which see note (30) below) Newton afterwards preferred merely to cancel this paragraph *in toto* and it therefore does not appear in the final text of the 'Enumeratio' as we reproduce it in §2 preceding.

(25) Newton first wrote 'fieri debet'.

(26) Originally 'describi solet per Regulam quæ'.

(27) Newton here repeats the error into which he had fallen some dozen years earlier in framing this same mechanical construction of (what he intended to be) the cissoid; see v: 466. For the correct position of the rotatory pole see our emendation of Newton's ensuing directions in note (29) below.

potest.(28) Sit *PQR* Cissois, *P* cuspis vel polus(29) ejus, *AB* asymptotos et *PA* perpendiculū ad asymptoton. Bisecetur *PA* in *G* et agatur *GH* Asymptoto parallela, et ad polum *P*(29) et regulam *GH* applicetur norma mobilis *PDE* ea lege ut ejus crus infinitum *DP* semper transeat per polum *P* interea dum crus alterum finitum *DE* termino suo *E* labatur super Regula *GH*. Et si crus illud alterum *DE* æquale sit perpendiculo *PA* et bisecetur in *Q*, ejus punctum *Q* describet Cissoidem *PQR*. Quinetiam si a puncto *E* ad normæ crus *PD* in angulo quovis dato *PdE* ducatur recta *Ed* et bisecetur eadem in *q*, norma acutangula vel obtusangula *PdE* puncto suo [*q*] describet aliam Cissoidem *Pqr*. Et hac ratione Cissoides innumeræ describi possunt quæ omnes sunt secundi ordinis curvarum & quarum simplicissima est rectangula *PQR*.(30)

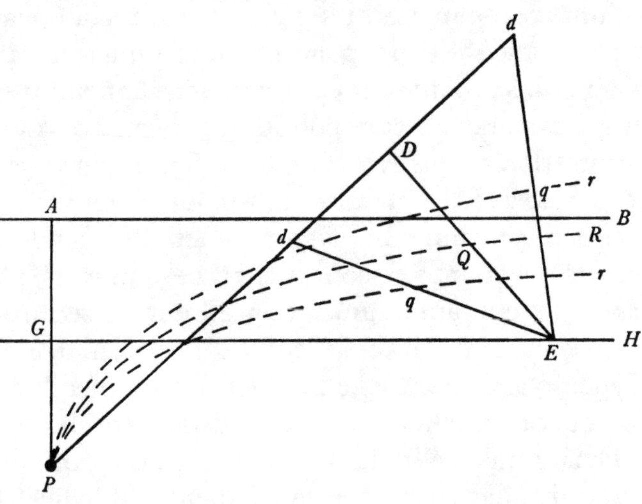

(28) A more emphatic 'describitur' is replaced.

(29) At these places, respectively, we need in correction (compare v: 466–8, note (686)) to delete 'vel polus' and in sequel read '...ad polum P' ubi punctum P' situm est in recta *AP* producta ad distantiam $PP' = \frac{1}{2}AP = GP'$. (See our amended figure in the next note.)

(30) Even on correctly relocating the pole at P' in the extension of *P* such that $GP = PP'$ (see the previous note) Newton's figure has yet further to be adjusted to have the cusps *p* of these more general cissoidal cubics *pqr* placed at distinct points in the perpendicular at *P* to *AGP* such that $pP = qQ$. Rather than take the trouble to think out this necessary emendation, however, Newton chose to cancel the whole of the preceding paragraph, omitting it from the revised text of the 'Enumeratio'.

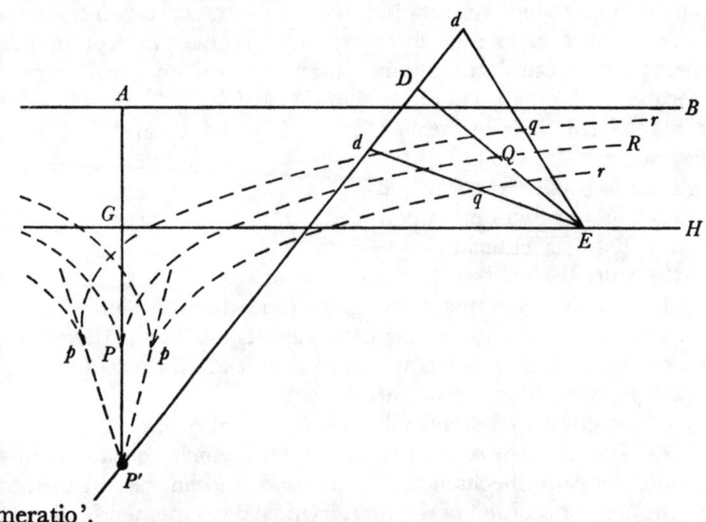

(31) In cancelling the whole of this section (see the previous note) Newton added to XXVI

Sed missis descriptionibus particularibus præstat generalem aliquam subjungere. Quem in finem proponam sequentia Theoremata.(31)

XXVIII Theoremata de curvarum descriptione.(32)

1. Si anguli duo dati *EAD*, *EBD* circa polos positione datos *A*, *B* rotentur et eorum crura *AD*, *BD* concursu suo *D* describant(33) lineam rectam,(34) crura reliqua *AE BE* concursu suo *E* describent(33) Sectionem conicam per polos *A*, *B* transeuntem præterquam ubi linea illa recta transit per polum *A* vel *B* vel per punctum tertium *G* in quod concursus *D* incidit ubi concursus alter *E* incidit in rectam *AB*. Nam concursus ille alter *E* hoc in casu describet lineam rectam.

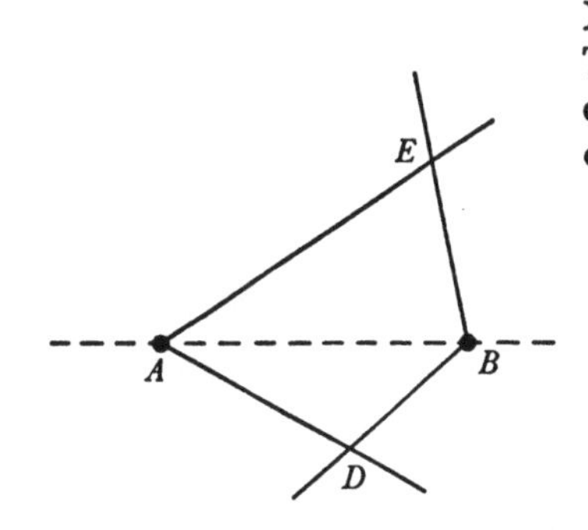

2. Si crura prima *AD*, *BD* concursu suo *D* describunt sectionem conicam per polum alterutrum *A*(35) transeuntem, crura altera concursu suo *E* describent curvam secundi generis per polum alterum *B* transeuntem & punctum duplex habentem in polo *A*.

3. At si sectio conica transit per neutrum polorum *A*, *B*, concursus *E* describet Curvam tertij generis tria habentem puncta duplicia, unum in polo *A*, alterum in polo *B*, ac tertium in concursu *E* ubi concursus alter *D* incidit in rectam *AB*.

XXIX Sectionum conicarū descriptio per data quinque puncta.

Sectio conica ex datis ejus punctis quinqꝫ determinatur et per eadem sic describi potest. Dentur ejus puncta quinqꝫ *A*, *B*, *C*, *D*, *E*. Jungantur eorum tria quævis *A*, *B*, *C* et trianguli *ABC* rotentur anguli duo quivis *CAB*, *CBA* circa vertices suos *A* et *B*[,] et ubi crurum *AC*, *BC* intersectio *C* successive incidit in puncta duo reliqua *D* et *E* incidat intersectio crurum reliquorum *AB* et *BA* in puncta *F* ac *G*. Agatur et infinite producatur recta *FG*, et anguli ita rotentur ut

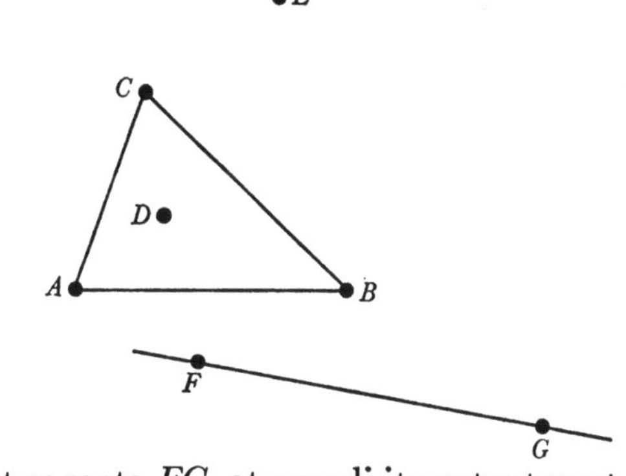

preceding the equivalent bridging sentence (f. 46r): 'Conducit autem punctum duplex ad descriptionem Curvarū per Theoremata sequentia'.

(32) As their equals in sections XXXI–XXXIII of the revised 'Enumeratio' (§2 preceding), these theorems on the organic construction of conics, cubics and quartics summarize the relevant portions of Newton's researches nearly thirty years before (see II: 106–50) into this universal 'Modus describendi Conicas sectiones et curvas trium dimensionum [punctum duplex habentium] &c'.

(33) Originally 'tangant' and 'tangent' respectively. A number of like replacements have been made by Newton in the sequel.

(34) A clause 'quæ non transit per polum alterū *A* vel *B*' here initially following has been cancelled by Newton to permit the exception thereto which he makes in his next sentence.

(35) A superfluous 'vel *B*' is deleted in sequel.

crurum AB et BA intersectio percurrat rectam FG et crurum reliquorum AC BC intersectio C describet Sectionem Conicam[36] quæsitam per Theorema primum.

XXX

Curvarum secundi ordinis punctum duplex habentium descriptio per data septem puncta.[37]

Curvæ omnes secundi generis punctum duplex habentes determinantur ex datis earum septem punctis quorum unum sit punctum duplex, et per eadem sic describi possunt. Inveniantur Curvæ propositæ puncta quælibet septem A, B, C, D, E, F, G, quorum A sit punctum duplex. Jungantur punctum A et alia duo quævis e punctis puta B et C. Et trianguli ABC rotetur angulus CAB circa verticem suum A, ut et angulorum reliquorum alteruter ABC circa verticem suum B. Et ubi crurum AC, BC concursus C successive applicatur ad puncta quatuor reliqua D, E, F, G incidat concursus crurum reliquorū AB et BA in puncta quatuor F, G, H, I.[38] Per puncta illa quatuor et quintum A describatur sectio Conica, et anguli CAB, CBA ita rotentur ut crurum AB, BA concursus semper incidat in sectionem illam Conicam, & concursus reliquorum crurum AC et BC describet Curvam propositam, per Theorema secundum.

Si punctum C accedat ad punctum B et cum eo coincidat, angulus BAC evanescet et recta BC tanget curvam describendam in puncto B. Unde si vice puncti C detur positione recta BC quæ curvam describendam tangat in B, describetur curva per regulam rectilineam circa polum A et angulum ABC circa polum B rotantes. Nam si Regula illa concurrat cum anguli latere BA ad sectionem conicam, concurret cum altero ejus latere BC ad Curvam describendam.[39]

Si punctum duplex A infinite distat, debebit Regula ad plagam puncti illius perpetuò dirigi & motu parallelo ferri, interea dum angulus ABC circa polum B rotatur.

Eadem ratione Curvas tertij, quarti et superiorū ordinum describere licet, non omnes quidem sed quotquot ratione[40] aliqua commoda describi possunt.

XXXI Curvarum usus principalis in Geometria est ut per earum intersectiones

(36) Originally just 'curvam'.

(37) Newton's initial marginal head (on f. 47^r) proclaimed more precisely his 'Descriptio curvarum secundi generis [= ordinis] per data sex puncta simplicia & punctum duplex'.

(38) Newton here intends, of course, four wholly new points 'H, I, K, L' (to follow the alphabetical sequence) or 'P, Q, R, S' as—yet more closely following his earlier 'Prob 9' (see II: 128–30)—he will afterwards denote them in the equivalent section XXXIII of his revised 'Enumeratio' (see page 638 above).

(39) A preliminary augmentation on Add. 3961.2: 15^v here continues by specifying 'punctum duplex habentem in polo A', and thereafter adds two new paragraphs, affirming respectively that 'Si crurum primorum concursus D tangit sectionem conicam transeuntem per neutrum polorum A et B; crurum reliquorum concursus C tanget curvam secundi generis quæ transibit per polos A et B et habebit punctum duplex in concursu C ubi concursus D incidit in rectam AB', while 'Si concursus D tangit curvam secundi generis[,] concursus C tanget curvam tertij et nonnunquam secundi vel primi'.

(40) 'lege' was first written.

Problemata solvantur. Proponatur æquatio construenda dimensionum novem

Constructio æquationum.(41)

$$x^9 * +bx^7+cx^6+dx^5+ex^4+fx^3+gx^2+hx+k=0$$

ubi b, c, d &c significant quantitates datas cum signis suis. Assumatur æquatio ad Parabolam cubicam $x^3=y$ & æquatio prior scribendo y pro x^3 evadet

$$y^3+bxyy+cyy+dxxy+exy+fy+gx^2+hx+k=0,$$

æquatio(42) ad Curvam aliam secundi generis. Et per harum curvarum intersectiones determinantur radices æquationis construendæ. Parabolam Cubicam semel describere sufficit. Curva altera si terminis h et k deficientibus, æquatio construenda sit septem dimensionum & pro fy scribatur fx^3, habebit punctum duplex in principio abscissæ, per quod et alia sex puncta utcunq; inventa facile describi potest ut supra. Si terminus etiam g deficit, & propterea æquatio construenda sit sex dimensionum & restituatur fy, Curva altera erit sectio Conica, ut et vel si per defectum terminorum f et e æquatio sit quinq; vel quatuor dimensionum. Et si per defectum etiam termini d æquatio sit dimensionum trium incidetur in Constructionem Wallisianam per Parabolam cubicam & lineam rectam.(43)

Construi etiam possunt æquationes per Hyperbolismum Parabolæ cum diametro semel descript[u]m. Construenda sit hæc æquatio dimensionum novem termino penultimo carens

$$a+cxx+dx^3+ex^4+fx^5+gx^6+hx^7+kx^8+lx^9[=0]$$

et assumatur $x^2y=1$, & æquatio construenda, si scribatur $\frac{1}{y}$ pro x^2, evadet

$$ay^3+cyy+dxyy+ey+fxy+g^{(44)}+hx+kxx+lx^3=0.^{(45)}$$

---

(41) Namely 'per descriptionem Curvarum' as Newton specified in the corresponding marginal head of the equivalent section XXXIV of his revised 'Enumeratio'.

(42) The text at this point (the bottom of f. 47v) continues, none too evidently so, in a sequel written longways across the sheet to the right of Newton's tentative listing of appellations for the various possible configurations of cubic (see §1: note (2) preceding) and preliminary tabulation of its component species (see *ibid.*: note (3)) on Add. 3961.2: 16r, where also at the page-bottom are entered—further to confuse!—the wholly alien jotted recomputations which he made at this time of the approximate construction of a central conic's general arc-length communicated by him in his *epistola prior* to Leibniz in 1676, and which we here reproduce in Appendix 2 following. Before inserting the connecting 'æquatio' he there initially went on: 'et habentur æquationes duæ novæ quibus Curvæ duæ secundi generis [= ordinis] communem habentes interse[ctionem definiuntur]'.

(43) Again (compare §2: note (128) preceding) see II: 292–3, note (62).

(44) Afterwards augmented to be the equivalent '$+gxxy$'.

(45) Where understand, as in the penultimate paragraph of the revised 'Enumeratio' (see page 644 above), 'quæ curvam secundi generis designat, cujus descriptione Problema solvetur'. At this point Newton left off to begin his preliminary revision of the preceding text which now follows (see §2: note (105) above) on Add. 3961.1: 48r–50r.

APPENDIX 2. NEWTON'S 'EXAMINATION' IN JULY 1695 OF THE APPROXIMATE RECTIFICATION OF A CENTRAL CONIC COMMUNICATED BY HIM TO LEIBNIZ IN 1676 IN HIS 'EPISTOLA PRIOR'.(1)

From the original jotting(2) in the University Library, Cambridge

[Positis $AB = x$. $BD = y$. ut et æquatio] $yy = rx - \frac{r}{q}xx$.(3) [Fit](4)

(1) In answer to an anxious letter from John Wallis on 3 July 1695 (ULC. Add. 3977.17, printed in *The Correspondence of Isaac Newton*, **4**, 1967: 139–40) inquiring about 'a Transcript of your two letters [Newton's letters to Leibniz of 13 June and 24 October 1676] which I wished might be printed [in Wallis' *Opera Mathematica*]' and which Wallis had sent to Cambridge 'about a month or five weeks since'—this secretary copy (now ULC. Add. 3977.2) he had earlier rather surreptitiously had made from primary ones by Collins and Oldenburg not in Newton's possession—Newton drafted a reply (Add. 3977.3) to say how 'very much obliged' he was to Wallis 'for your pains in transcribing my two Letters of 1676 & much more for your kind concern of right being done me by publishing them', and that he had 'perused your transcripts of them & examined ye calculations' (see *Correspondence*, **4**: 140). The present rough jotting, checking the Huygenian approximation to the arc-length of a conic (measured from a principal vertex) which Newton had initially set down nineteen years earlier in his *epistola prior* to Leibniz (see *Correspondence*, **2**, 1960: 31), is evidently one such re-examination of a point of technical detail in this first of his two celebrated 1676 letters. While the computation itself has nothing to do with the preceding 'Enumeratio'—and is contingently related to it only by the mere chance that Newton found a convenient blank space to rough it out at the bottom of a page of a scrap sheet which was otherwise (see the next note) crowded with a miscellany of recent notes and stray drafts for his essay on cubic curves—we may appropriately reproduce it here for its own intrinsic interest. That we can independently refer its genesis to early July 1695 will, of course, strongly support our assignation of the date of Newton's composition of the 'Enumeratio' itself as the summer of that year (see §1: note (1) above).

(2) At the bottom of Add. 3961.2: 16^r, where (see Appendix 1.3: note (42) preceding) it fills a small space previously left blank in a page otherwise covered from top to toe (and partially also from right to left lengthways) by draft tabulations and the end of a continuous verbal passage preparing the way for the final text of the 'Enumeratio' as we give it in §2 above. Pursuing our usual editorial practice in such cases, we have rounded out Newton's curt lines of mathematical computation with insertions which (it is our hope) will better convey their sense, silently adjusted a few minor numerical slips which are left uncorrected in the manuscript, and also adapted the whole to an accompanying figure in which we have inserted the designations D and G which best agree with Newton's corresponding text.

(3) So defining the curve AD to be an ellipse of vertex A and main axis AB, but a parallel argument manifestily holds for the complementary case of the hyperbola $y^2 = rx + (r/q)\,x^2$ on replacing q by $-q$ throughout the sequel. It is understood that the ordination angle $A\widehat{B}D$ is right. Given that the segment AG of the tangent at A is equal in length to the arc $\widehat{AD}$, the problem is approximately to construct the point C in which GD meets AB; whence, conversely to rectify the arc $\widehat{AD}$ we need only to determine C and therefrom draw CD cutting off the equal length AG in the tangent at A. (See our brief notice of this Huygenian construction on IV: 669—70.)

(4) On computing (in an auxiliary calculation not here recorded) that

$$\dot{x}/\dot{y} \text{ (or } dx/dy) = y/(\tfrac{1}{2}r - (r/q)\,x)$$

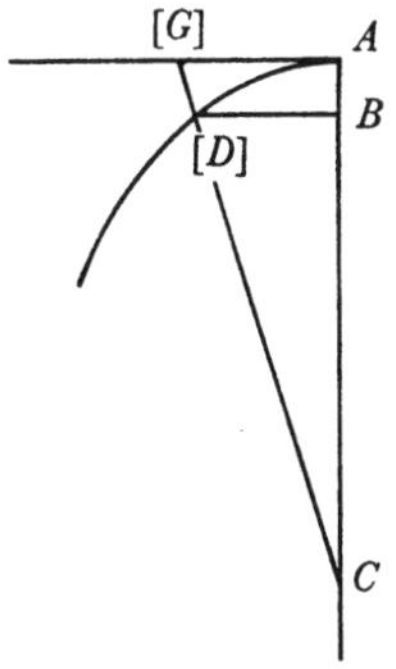

$$[A]G = y + \frac{2}{3rr}y^3 + \frac{8}{5r^3q}y^5 \&c : [BD =] y$$

$$-\frac{2}{5r^4}$$

$$:: 1 + \frac{2yy}{3rr} + \frac{8y^4}{5qr^3} - \frac{2y^4}{5r^4} [\&c] : 1 :: AC : BC.$$

[G] A B [D] C

[unde] $\frac{2yy}{3rr} + \overline{\frac{8r-2qy^4}{5qr^4}} . 1 :: AB . BC.$ [adeoq̧][(5)]

$$\frac{2}{3}x - \frac{2}{3q}xx + \frac{8r-2q}{5qr^3}\overline{rrxx - \frac{2rr}{q}x^3 + \frac{rr}{qq}x^4} . r :: x . BC. \text{ [et facta divisione]}$$

$$10x + \frac{14r-6q}{qr}xx\Big) \quad 15rx \qquad \Big(\frac{3}{2}r \frac{+9q-21r}{10q}x [= BC].$$

$$-15rx \frac{-21r+9q}{q}xx$$

[itaq̧] $\frac{3}{2}r + \frac{19q-21r}{10q}x$ [$= AC$, proxime].[(6)]

---

and therefore $(\dot{x}/\dot{y})^2 = y^2/(\frac{1}{4}r^2 - (r/q)\,y^2) = 4(y^2/r^2)\,(1 - 4y^2/qr)^{-1}$, whence

$$AG = \widehat{AD} = \int_0^y \surd[1 + (\dot{x}/\dot{y})^2] . dy$$

$$= \int_0^y \surd[1 + (4/r^2)\,y^2 + (16/qr^3)\,y^4 + \ldots] . dy$$

$$= \int_0^y (1 + (2/r^2)\,y^2 + (8/qr^3 - 2/r^4)\,y^4 + \ldots) . dy.$$

(5) On substituting $rx - (r/q)\,x^2$ for y^2 and $(rx - (r/q)\,x^2)^2$ for y^4 in the left-hand member of the preceding ratio.

(6) Which is of course, *mutatis mutandis*, as Newton had set it to be in his letter to Leibniz of 13 June 1676.

2

MISCELLANEOUS INVESTIGATIONS IN THE ANALYTICAL GEOMETRY OF CUBIC AND QUARTIC CURVES(1)

[c. 1695?]

§1. CONCHOIDAL CUBICS CONSTRUCTIBLE BY THE MOVEMENT OF A JOINTED RULER.(2)

From stray drafts in private possession and in the University Library, Cambridge

[1] Eodem modo quo sectiones Conicæ derivantur a circulo possunt figuræ tertij generis(3) derivari a notioribus in eo genere figuris. Illarum vero notissima et celeberrima est Cissois, & omnium facillimè describitur. Norma *PGF* ita

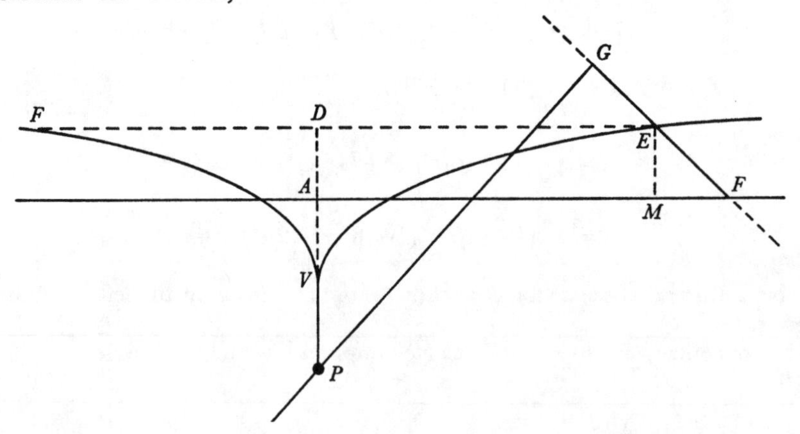

moveatur ut ejus crus infinitum *GP* semper transeat per polum *P* dum alterū crus *GF* datæ longitudinis termino suo *F* habeatur super regulâ positione data *AF*.(4) Et si Crus illud *GF* æquale sit distantiæ *PA* quæ est poli *P* a regula positione data

(1) These stray pieces, respectively outlining the construction of a range of conchoidal cubics by a generalization of Newton's intrumental construction of the cissoid by a moving jointed ruler, and also, conversely, seeking the simplest defining equation of the class of symmetric quartic trefoils, are here gathered together more for editorial convenience than because of their underlying unity of theme, though they each in their way, of course, furnish

Translation

[1] In the same way as conics are derived from the circle, figures of third kind[(3)] can be derived from the better known figures in that class. Of these, indeed, the best known and most celebrated is the cissoid, and it is the most easily described of all. Let the square joint *PGF* move so that its infinite leg *GP* shall ever pass through the pole *P* while the other leg *GF* of given length shall at its end-point *F* be held on the ruler *AF* given in position.[(4)] Then, if that leg *GF*

---

excellent illustration of the Cartesian techniques which he had at his disposal to apply in resolving such problems of higher-order algebraic curves. Newton's handwriting in the manuscripts is, both in size and quality, all of a one, and the following date conjectured for their composition cannot be far out, since the first two fragments on cubics bear a close relationship to the preceding 'Enumeratio' (see the next note) and the ensuing jottings on quartics are set down on sheets where Newton has also penned—no later than the early 1690's—preliminary verbal revisions and checking computations relating to the text of the 1687 *editio princeps* of his *Principia* (for details of which see §2: note (2) below).

(2) These extensions of the elegant 'mechanical' description of the classical cissoid—here (see notes (5) and (11) following) set to be but their particular instance—which Newton had first presented more than a decade and a half previously as Problem 20 of his *soi-disant* Cambridge lectures on algebra (v: 218–20) had already, in the case where the locus-point remains sited in the sliding leg of the jointed ruler but no longer bisects it, been foreshadowed in his ensuing construction there of a cubic equation 'simili normæ applicatione' (see *ibid.*: 466), while he had adumbrated the yet broader generalization where the locus-point is fixed in arbitrary position in the moving plane of the ruler in Problem 7 of his contemporary 'Geometria Curvilinea' (see IV: 480). His slip in the former extension in accurately positioning the pole *P* (on which see v: 466–9, note (686)) similarly spoils, as we have noticed above (1, §2: Appendix 1.3: notes (27), (29) and (30)), the equivalent 'Facilis descriptio curvarum [tertij ordinis]' which Newton initially planned to add in the summer of 1695 to his enumeration of the component species of these (conchoidal) and other cubics. For want of any recorded clue to their progeny and purpose, indeed, we might hazard the guess that one or other of the accurate descriptions of generalized 'Cissoides Veterum' which here follow was composed with an eye to its filling the place of that faulty section XXVII in Newton's preliminary conclusion to his 'Enumeratio'; but, if so, it was discarded without trace from the latter's final version (1, § 2 preceding).

(3) Understand either 'lineæ tertii ordinis' (lines of third order) or 'curvæ secundi generis' (curves of second kind), that is, cubics. Newton, however, standardized his terminology in this regard—including the straight line in his set of *ordines linearum*, but banishing it from the parallel grouping (numerically one less in each corresponding instance) of *genera curvarum*—only at a late stage in the composition of his 'Enumeratio' (compare the many attempts he made correctly to fashion its title which we have cited in 1, §2: note (3) above) and even thereafter was far from consistent in maintaining this distinction, often (as here) referring interchangeably to the *ordo/genus* of a *linea/figura curva*.

(4) It is understood that the angles $\widehat{PAF}$ and $\widehat{PGF}$ are each right. Newton first began his sentence with the less precise instruction: 'Circa Polum *P* rotetur regula *PG* quæ cum *GF* longitudinis datæ angulũ *PGF* continet...' (Round the pole *P* let there rotate the ruler *PG* which with *GF* of given length contains the angle $\widehat{PGF}$...).

et bisecetur in E[,] punctum E hoc motu describet Cissoidem Veterum EVF.(5) Et alia normæ puncta imò et omnia puncta quæ sunt in mobili normæ plano PFG describent curvas tertij generis.(6)

---

[2](7) DESCRIPTIO CUJUSDAM GENERIS CURVARUM SECUNDI ORDINIS.(8)

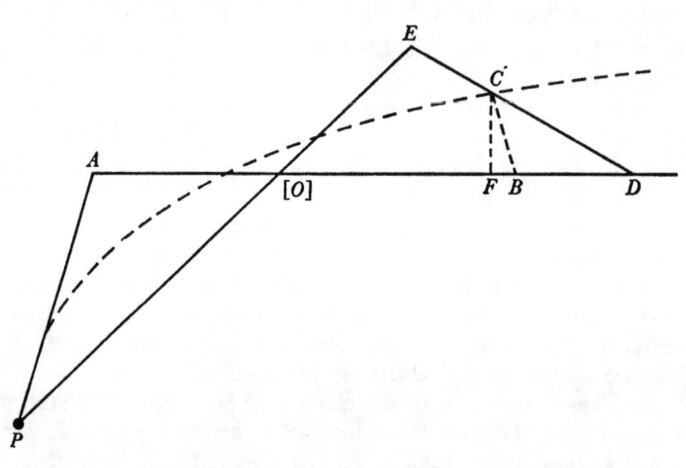

Concipe lineas(9) PED datum angulum PED continentes ita moveri ut una earum EP perpetuo transeat per polum P positione datum, et altera ED datæ longitudinis existens perpetuò tangat rectam AB positione datam. Age PA constituentem angulum PAD æqualem angulo PED sitqꝫ ED æqualis AP et quodvis punctum C in recta ED datum describet curvam secundi ordinis.

Age CB constituentem angulum CBD æqualem angulo PED, et ad AD demitte normalem CF, et dictis $DC = c$. $CE = b$. $AB = x$. $BC = y$. et posito $1.e::BC.BF$, erit

$$y^3 \begin{array}{l} +2b \\ +c \\ -4eeb \\ -4eec \\ -2ex \end{array} yy \begin{array}{l} +bb \\ +2bc \\ +2b[e]x \\ +4c[e]x \\ +xx \end{array} y \begin{array}{l} +cbb \\ -cxx \end{array} = 0.$$ (10)

---

(5) See the next note. The reader will have no great trouble in mentally distinguishing between the two points F in Newton's accompanying figure.

(6) Not all cubics of course, but only those which are 'conchoidales' (conchoidal/possessed of but a single asymptote) and having no conjugate oval or double point. In general, if C is an arbitrary point fixed in the plane of the moving *norma PGF* and perpendiculars CB, CH, CL are let fall therefrom to PA, PG and AF respectively (meeting these in B, H and L) and also, where HC is extended to meet AF in K, the normal FN is dropped to HK, then on calling

$$PA = GF = a, \quad GH = FN = b, \quad HK = c, \quad HC(= a-c) = d$$

and the variable lengths $AB = x$, $BC = y$, $FK = z$ there is $NK = \sqrt{[z^2-b^2]}$ and also, where PG and AF meet in O,

$$AO = OG = GF \times FN/NK = ab/\sqrt{[z^2-b^2]} \quad \text{and} \quad OF = GF \times FK/NK = az/\sqrt{[z^2-b^2]};$$

whence, because $CL/CK = FN/FK$ and $AF = BC + (LK \text{ or}) \ CK \times NK/LK - FK$, there ensues respectively $x/(d+\sqrt{[z^2-b^2]}) = b/z$ and

$$(c+d)\ (b+z)/\sqrt{[z^2-b^2]} = y + (d+\sqrt{[z^2-b^2]})\ \sqrt{[z^2-b^2]}/z - z = y + d\sqrt{[z^2-b^2]}/z;$$

be equal to the distance PA of the pole P from the ruler given in position and be bisected in E, the point E will by this motion describe the cissoid of the ancients EVF.[5] While other points of the jointed ruler—and, indeed, all points which are in its mobile plane PFG—will describe curves of third kind.[6]

---

[2][7] THE DESCRIPTION OF A CERTAIN CLASS OF CURVES OF SECOND ORDER.[8]

Conceive the [9]lines PE/ED containing the given angle $P\widehat{E}D$ to move so that one of them, EP, shall perpetually pass through the pole P given in position while the other, ED of given length shall perpetually touch the straight line AB given in position. Draw PA constituting the angle $P\widehat{A}D$ equal to $P\widehat{E}D$ and let ED be equal to AP, and then any point C given in the line ED will describe a curve of second order.

Draw CB constituting the angle $C\widehat{B}D$ equal to $P\widehat{E}D$ and to AD let fall the normal CF, and then, on calling $DC = c$, $CE = b$, $AB = x$, $BC = y$ and setting $1:e = BC:BF$, there will be

$$y^3+(2b+c-4e^2(b+c)-2ex)\,y^2+(b^2+2bc+(2b+4c)\,ex+x^2)\,y +cb^2-cx^2 = 0.^{(10)}$$

---

and when z is eliminated from these the defining equation of the locus $C(x, y)$ in the (stationary) plane of the Cartesian axes PA ($y = 0$) and AF ($x = 0$) reduces with some little trouble to be $(x+d)\,((x+c)^2\,(x+d)-2b(x+c)\,(y+d)+(x-d)\,(y+b)^2) = 0$, that is, the pair of the straight line $x = c-a$ traced parallel to the *regula* AF (in the case where the *norma* PGF is placed below this, with its leg GF travelling ever parallel to PA) and the conchoidal cubic $X^2(X+a-2c)-2bXY+(X-a)\,Y^2 = 0$ referred to new coordinates $X = x+c$ and $Y = y+b$, parallel to the old, and having a double point at its origin. Where in particular (compare v: 468, note (686)) the point C is stationed in the leg itself (so that $b = 0$), the latter equation becomes $X^2(X+a-2c)+(X-a)\,y^2 = 0$; and if (when also $c = \frac{1}{2}a$) it is the mid-point E of GF, then the locus (E) described by the rotating/sliding *norma* is the 'cissois Veterum'

$$y^2 = X^3/(a-X)$$

whose cusp V is the mid-point of PA.

(7) ULC. Add. 4004: 78ʳ, an isolated but (like the preceding) fairly carefully written note at the head of a page in Newton's Waste Book which is otherwise blank.

(8) In strict Newtonian parlance this should read 'GENERIS' (KIND). Further to confuse, 'TERTIJ ORDINIS' (OF THIRD ORDER) was first written.

(9) Understand 'junctas' (coupled).

(10) Since, on letting fall the perpendicular DG from D to PE (see our figure on the next page), the locus-point C is fixed in the plane of the rectangular *norma* PGD whose leg PG passes through the pole P while the end-point D of its other leg GD slides along on the *regula* AF, this description is only a slight variation on that outlined in [1] preceding, and the present equation is readily derived from the cubic one obtained in note (6) above, *mutatis mutandis*. From first principles, equivalently, since in Newton's terms there is $FB = ey$, $CF = y\sqrt{[1-e^2]}$ and thence $FD = \sqrt{[c^2-y^2(1-e^2)]}$, from the equality

$$AB+FD-FB = AD = (AO \text{ or}) \; OE+OD = ED(BC+BD)/BD$$

Et nota quod ubi angulus *PED* rectus est, et recta *ED* bisecatur in *C*, curva erit Cissoides Veterum.(11)

---

there comes $(\sqrt{[c^2-y^2(1-e^2)]}-ey)\,(x+\sqrt{[c^2-y^2(1-e^2)]}-ey) = (b+c)\,(y+c)$, whence on multiplying out and squaring there ensues a quartic which splits into the pair of $y = -c$ (the straight line $C'N$ traced out by the locus-point C' parallel to the *regula* AF when the *norma*

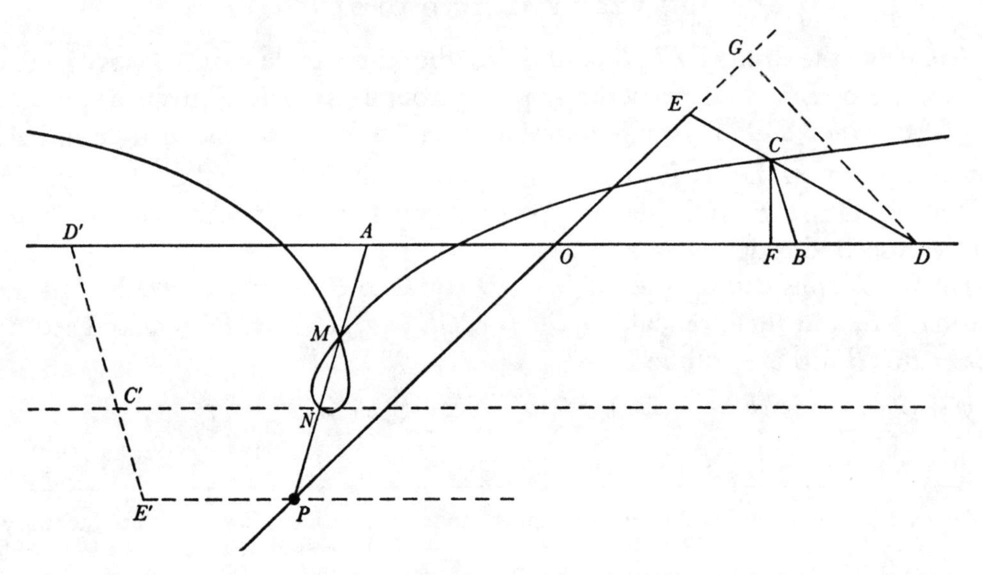

initially occupies such a position as $PE'D'$ in which its leg $E'D'$ can move only parallel to itself) and—when (as is here done) we interpolate two *e*'s trivially omitted by him in setting it out—Newton's present irreducible equation. The resulting cubic locus (C) so defined, a unicursal, conchoidal curve having a double point at its first intersection $M(-2eb, -b)$ with the line PA ($x = 2ey$) and meeting it again in the second point $N(-2ec, -c)$, is none too carefully depicted in his manuscript sketch (here reproduced with its proportions preserved), which would seem roughly to have been drawn by 'eye' without taking account of the singularities of its deduced Cartesian equation; compare our own more accurate figure.

And note that when the angle $P\widehat{E}D$ is right and the line ED is bisected in C, the curve will be a cissoid of the ancients.[11]

---

(11) In this particular case there is $e = 0$ and $b = c$, so that Newton's preceding equation reduces to be $(y+c)^3+x^2(y-c) = 0$ or $x^2 = Y^3/(2c-Y)$ where the coordinate line-length $Y = y+c$ is referred to its cusp $(0, -c)$ as origin; compare note (6) above.

§2. THE DETERMINATION OF BIFOLIATE AND TRIFOLIATE QUARTICS FROM THEIR SIMPLEST CARTESIAN EQUATIONS.(1)

From original worksheets(2) in the University Library, Cambridge

[1](3) [Dic] $S[g]=x$. $[gh]=y$. [et pone æquationem esse]

$$y^4 \begin{array}{l} -2ayy \quad +bb \\ +2cxxyy -2dxx \\ \qquad\qquad\quad +eex^4 \end{array} = 0.^{(4)}$$

(1) Of the background to these rough, verbally unexplained computations—here liberally filled out with inserted Latin connectives to elucidate their sense—we know nothing, and it must remain anyone's guess as to what it was that stimulated Newton to seek the defining equations of the symmetrical quartic *bifolia* and trefoils which he now sets himself to determine. Our own highly tentative conjecture that his analysis of the bifoliate curve in [1] may have had something to do with his youthful annotations on Schooten's *Exercitationes* is conveyed in note (3) below.

(2) Add. 3963.6: $44^r/42^r-42^v$; ancillary calculations on *ibid.*: 44^r, $45^r/44^v$ [inverted] and 45^v are digested in following footnotes. The folded sheet ff. 44^r—45^v was initially employed by Newton in the mid-1680's first to rough out (in vertical line on f. 44^v) three virtually identical versions of the left-hand figure reproduced (from ULC. Dd. 9. 68: 218) on v: 432, and thereafter to pen (on its ff. 44^v–45^v) a number of draft phrases and calculations in preparation for the 1687 *editio princeps* of his *Principia*. Of the latter (all which remain unpublished) we may single out two items for special notice. The first (on f. 44^v) is a preliminary casting of Definition 13 on VI: 191, in which Newton stated that 'Corporis vis exercita est [qua id reluctatur et repugnat mutationi →] conatus ejus ne status suus movendi vel quiescendi mutetur estq͡ mutationi status illius singulis momentis [effectæ →] illatæ proportionalis, et non immerito reluctatio vel resistentia corporis dici potest' (compare *ibid.*: note (18)). The latter (f. 45^r) is an ingenious variation on the 'Moon test' in Proposition IV of Book 3 of the *Principia*, $_1$1687: 406–7 wherein, positing that a body falls $(\frac{1}{2}g=)$ $15.082\approx 15\frac{1}{12}$ Paris feet in the first second of free fall at the surface of the Earth and that the latter's circumference is $(2\pi R=)$ 123249600 Paris feet, Newton in the first place computes that the period of revolution of a satellite just grazing the Earth's surface in circular orbit (at a speed $v=\sqrt{[gR]}$) would be

$$(2\pi R/v=\sqrt{[\pi.2\pi R/\tfrac{1}{2}g]}=)\sqrt{[\pi.123249600/15.082]}=5066.912'';$$

since, departing from the Moon's observed period of 27 days 7 hours 43′ 9″, that is, 2360589″ and the assumption that its mean radius of orbit is '60^{semidiam}' of the Earth, there is likewise $2360589''/60^{\frac{3}{2}}\approx 5067''$, it follows at once that Kepler's third planetary law—and hence the inverse-square hypothesis of (here terrestrial) gravity from which it depends—is confirmed to fit in the case of the Moon. (In his calculation, we may note, Newton initially took for his value of π the simple Archimedean bound $\frac{22}{7}$ and subsequently, in a more refined computation, Metius' highly accurate approximation $\frac{355}{113}=3.141593$.) These accompanying jottings, both assignable to 1684, would suggest a like date of composition for the present piece—certainly its terminus *ante quem non*—in the middle 1680's, but we would not care to attach such rigid temporal limits to an essentially independent algebraic computation which is a subsequent addition in the manuscript and all but lacks any text by which a more definite identification might be made. At the risk of being a decade out in our chronology we prefer to reproduce it at a place in our edition where it appropriately rounds off Newton's other researches into the Cartesian geometry of cubics and quartics.

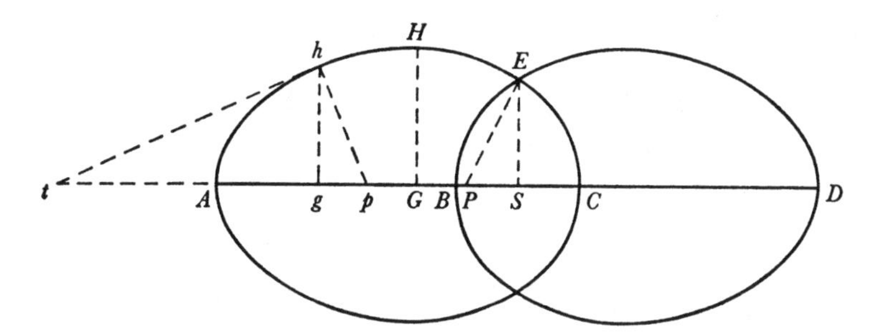

[Ubi] $x = 0$. [est] $y^4 - 2ayy + bb = 0$. [Cape igitur] $a = b$.(5) [fit]

$$y^4 \begin{array}{l} -2byy \quad +bb \\ +2cxxyy \; -2dxx \\ \qquad\qquad +eex^4 \end{array} = 0.$$

[Unde capiendo fluxiones erit] $\begin{array}{c} y^3Y - byY + cxxyY \\ + cxXyy[-]\, dxX + eex^3X \end{array} = 0.$(6) [adeoq͡;]

$gh.gp :: tg.gh :: X. -Y :: y^3 - by + cxxy . cxyy - dx + eex^3 :: y.gp = \dfrac{cxyy - dx + eex^3}{yy - b + cxx}$.

[Igitur si] $y = 0$. erit gp(7) $= \dfrac{-dx + eex^3}{+b - cxx}$. [ubi] $eex^4 - 2dxx + bb = 0$. [sive]

$xx = \dfrac{d}{ee} \pm \sqrt{\dfrac{dd - eebb}{e^4}}$. [Hoc est] $gp = \dfrac{\pm eex\sqrt{dd - eebb}}{eeb - cd \mp c\sqrt{dd - eebb}}$.(8)

[Jam sume] $AS.SC :: Cp = gp$.(9) $Ap = gp$.(9) [aut etiam]

---

(3) Add. 3963.6: 44r. Newton sets himself the problem of finding the Cartesian equation of the two-looped quartic which he sketches in his figure, supposing that it is symmetrical round its axis $ABCD$ and the vertical ES (which meets it in the double points E and its mirror twin in respect to BC). The only comparable instance of such a quartic curve to be found in our preceding pages occurs in I: 30, note (17) where we corrected a youthful generalization by Newton of a construction for the conic ellipse given by Schooten in his *Exercitationes Geometricæ* in 1657 (see *ibid.*: 31, note (19)). It might be that Newton himself had some such Schootenian locus in mind here, but this is pure conjecture.

(4) The symmetry of the quartic round the axes ASD and SE (understood to be mutually perpendicular) requires that the Cartesian equation defining (x, y) shall have no terms involving odd powers of $Sg = x$ and $gh = y$.

(5) This because $E(0, \sqrt{[a \pm \sqrt{(a^2 - b^2)}]})$ is posited to be a double point on the quartic. When also there is $e = c$, we may note, the quartic reduces to be the pair of the ellipses

$$y^2 + cx^2 \pm \sqrt{[2(d - bc)]}\, x = b.$$

(6) Where X and Y are taken proportional to the fluxions $\dot{x}$ and $\dot{y}$ of Sg and gh.

(7) Strictly '$-gp$'. This limiting value of the subnormal at h when this comes to coincide with the vertices A, B, C, D in which the quartic meets its axis $y = 0$ will evidently be the radius of curvature at these points.

(8) Where understand that $x = \sqrt{[d \pm \sqrt{(d^2 - e^2b^2)}]}/e$, of course.

(9) Understand that Cp and Ap are the limiting values of the subnormal gp as h comes to coincide with C and A respectively. To set their ratio to be that of AS to SC or, alternatively,

$AS^q.CS^q::Cp = gp.$[9] $Ap = gp.$[9] [Hoc est fac]

$$\frac{ee,AS\sqrt{dd-eebb}}{eeb-cd-c\sqrt{dd-eebb}}\cdot\frac{ee,CS\sqrt{dd-eebb}}{eeb-cd+c\sqrt{dd-eebb}}::\frac{AS}{eeb-cd-c\sqrt{\ })}\cdot\frac{CS}{eeb-cd+c\sqrt{\ })}$$

$::CS.AS.$[aut]
$CS^q.AS^q.$

[In primo casu erit] $AS^q\times\overline{eeb-cd+c\sqrt{\ })} = CS^q\times\overline{eeb-cd-c\sqrt{\ })}$. [unde]

$$\frac{d+\sqrt{\ })}{ee}\times\overline{eeb-cd+c\sqrt{\ })} = \frac{d-\sqrt{\ })}{ee}\times\overline{eeb-cd-c\sqrt{\ })}.$$

[id est] $2cd\sqrt{\ })+\overline{2eeb-2cd\sqrt{\ })} = 0.$ [sive] $cd+eeb-cd = 0.$ [itaq3] $eeb = 0.$[10]

[Pone] $x = 0.$ [fit] $y^2 = +b.$[11] [At generalius est]

$$yy = b-cxx\pm\sqrt{-2bcxx+ccx^4+2dxx-eex^4}.$$

[et inde] $gp = \dfrac{cyy-d+eexx}{\sqrt{2d-2bc+ccxx-eexx}}$. [Ergo] $\dfrac{cb-d}{-\sqrt{2d-2cb}} = \sqrt{\dfrac{d-cb}{2}} = PS.$[12]

[Extrahendo radicem quadraticam prodit successive]

$$yy = b-cxx\pm x\sqrt{2d-2bc}\pm\frac{cc-ee}{2\sqrt{2d-[2bc]}}x^3 \text{ \&c. [tum]}$$

$$y = b^{\frac{1}{2}}\pm\frac{x}{2}\sqrt{\frac{2d-2bc}{b}}\frac{-dxx-bcxx}{4b^{\frac{3}{2}}} \text{ [\&c].}$$[13]

---

of AS^2 to SC^2 is an independent assumption on Newton's part for which we can see no reason other than his mild hope of attaining some interesting particular instance of his quartic *bifolium.* In sequel only the first premise is examined—the latter is probably but an even more tentative proposal for its subsequent revision which he did not bother further to pursue—and, when (see the next note) it led nowhere very far, it was quickly abandoned.

(10) Whence, not very rewardingly, there is either $b = 0$ or $e = 0$.

(11) Originally '$\pm b^2$' by a slip of Newton's pen. This once more (compare note (5) above) determines $E(0, b^{\frac{1}{2}})$ and its mirror-image $(0, -b^{\frac{1}{2}})$ in the axis AD to be double points of the quartic.

(12) The subnormal at the double points $(0, \pm b^{\frac{1}{2}})$. This cannot, of course, be computed by substituting $x = 0$ and $y^2 = b$ in the preceding formula $gp = (cy^2-d+ex^2)\,x/(y^2-b+cx^2)$.

(13) Which again does not get very far. At this point Newton abandons his quartic *bifolium* to discuss the analogous three-looped quartic to which he now passes without break in his manuscript (f. 44r).

[2][14] [Dic $Dp = x$. $pq = y$. et pone æquationem esse]

$$y^4 \begin{array}{l} +ayy+dx^2 \\ +bxyy+ex^3 \\ +cxxyy+fx^4 \end{array} = 0.^{(15)}$$

[Ubi] $DE = x$. $EF = y$. [erit] $3yy = xx$. [sive] $\pm\sqrt{3}y[=x]$.[16] [Evadet][17]

$$\tfrac{1}{9}xx+\tfrac{1}{3}a+\tfrac{1}{3}bx+\tfrac{1}{3}cxx+d$$
$$+ex+fxx = 0.$$

[seu]

$$\overline{\tfrac{1}{9}+\tfrac{1}{3}c+f}\times xx+\overline{\tfrac{1}{3}b+e}\times x$$
$$+\tfrac{1}{3}a+d[=0].$$

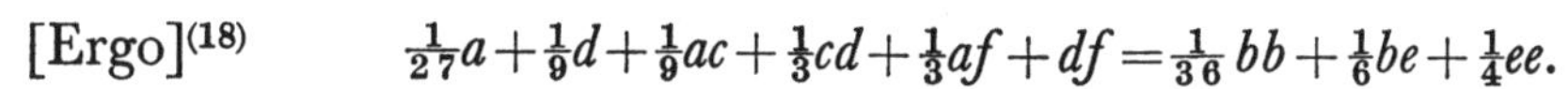

[Ergo][18] $\frac{1}{27}a+\frac{1}{9}d+\frac{1}{9}ac+\frac{1}{3}cd+\frac{1}{3}af+df = \frac{1}{36}bb+\frac{1}{6}be+\frac{1}{4}ee.$

[quo casu] $2x = \dfrac{\frac{1}{3}b+e}{\frac{1}{9}+\frac{1}{3}c+f} = 2n$. [posito $ED = n$, quo casu erit] $CD = \frac{2}{3}n$.[19] [adeoq] $Cp = \frac{2}{3}n - x$. [Insuper demissâ qsr perpendiculari in CF erit] $qs = 2y$.

(14) Add. 3963.6: 44r/42r–42v. In the pattern of his preceding investigation of the bifoliate quartic, Newton now seeks to determine the Cartesian equation of the trefoil shown in the accompanying figure, assuming it to be symmetrical round each of the axes *ACD*, *CF* and *CG* (understood to be inclined to each other at 60°). This problem was manifestly here contrived by him on the spur of the moment, for in the manuscript he drew a first rough, unlettered diagram in which the quartic, after looping clockwise through *A*, almost circled *C* before again attaining a maximum distance from it in the direction *CF*.

(15) By choosing its double point *D* as origin and the axis of symmetry *AD* as abscissa ($y = 0$) of his (mutually perpendicular) Cartesian coordinates, Newton ensures that the equation defining the trefoil's general point $q(x, y)$ shall lack both a constant term and ones in x and y alone, and also all those involving odd powers of y.

(16) For the double points *D*, *F* and *G* evidently lie at the corners of an equilateral triangle, and so $FE = \frac{1}{2}FD = \frac{1}{2}\sqrt{[FE^2+ED^2]}$. The equations $x = \pm y\sqrt{3}$ define, we need scarcely say, the two straight lines through the origin *D* which intersect the quartic in its two other double points *F* and *G*.

(17) On substituting $y^2 = \frac{1}{3}x^2$ in the previous equation of the trefoil to determine the abscissas of its meets with the pair of lines *DF* and *DG*, and omitting the ensuing double root $x^2 = 0$ which marks the double point at *D*.

(18) Because the intersections *F* and *G* are each double, and so the preceding quadratic must be the perfect square $(\frac{1}{9}+\frac{1}{3}c+f)\ (x+\frac{1}{2}(\frac{1}{3}b+e)/(\frac{1}{9}+\frac{1}{3}c+f))^2 = 0$, whence

$$(\tfrac{1}{9}+\tfrac{1}{3}c+f)\ (\tfrac{1}{3}a+d) = \tfrac{1}{4}(\tfrac{1}{3}b+e)^2.$$

(19) The trefoil's centre *C* is the centroid of the equilateral triangle *DFG* in which *DE* is the altitude (and so median) from the vertex *D* to the opposite side *FG*.

$ps = y\sqrt{3}$.(20) [unde fit] $Cs = \frac{2}{3}n - x - y\sqrt{3}$. [id est](21) $Cr = \frac{1}{3}n - \frac{1}{2}x - y\sqrt{\frac{3}{4}}$. [et inde] $Fr = n - \frac{1}{2}x - y\sqrt{\frac{3}{4}} = z$. [Item](21) $rs = \frac{n}{\sqrt{3}} - x\sqrt{\frac{3}{4}} - \frac{3}{2}y$. [et inde]

$rq^{(22)} = \frac{n}{\sqrt{3}} - x\sqrt{\frac{3}{4}} + \frac{1}{2}y = v$. [seu] $\frac{n}{3} - \frac{1}{2}x + \frac{y}{2\sqrt{3}} = \frac{v}{\sqrt{3}}$. [Prodit igitur]

$$2n - 2x = z + v\sqrt{3}. \text{ [necnon] } \tfrac{2}{3}n - \frac{2y}{\sqrt{3}} = z - \frac{v}{\sqrt{3}}.$$

[unde tandem]

$$x = n - \frac{z}{2} - \frac{v\sqrt{3}}{2}. \text{ [et] } y = \frac{n}{\sqrt{3}} - \frac{z\sqrt{3}}{2} + \tfrac{1}{2}v.^{(23)}$$

[Quibus inventis valoribus pro x et y respectivè scriptis in æquatione, orietur]

... ...(24)

[Hic oportet sumere] $c = 2. f = 1.$(25) [unde erit]

$$\tfrac{1}{27}a + \tfrac{1}{9}d + \tfrac{2}{9}a + \tfrac{2}{3}d + \tfrac{1}{3}a + d = \tfrac{16}{27}a + \tfrac{16}{9}d = \frac{bb}{36} + \frac{be}{6} + \frac{ee}{4}.^{(26)}$$

(20) Since $\widehat{pqs} = \widehat{rCs} = 60°$, there is $qs:qp:ps = 2:1:\sqrt{3}$.

(21) Here similarly there is $Cs:Cr:rs = 2:1:\sqrt{3}$.

(22) That is, $rs + (sq \text{ or}) \, 2pq$.

(23) These equations represent, it will be evident, the product of a rotation of the coordinates anticlockwise through $\frac{2}{3}\pi$ radians round D and of a translation of these to the new origin F. The point of this linear transformation will be clear when it is seen that, by the axial symmetry of its figure, the trefoil's general point $q(x, y) \equiv (z, v)$ should be defined by exactly the same equation with respect to the coordinates $Dp = x$, $pq = y$ as with regard to the new ones $Fr = z$, $rs = v$ (since in both cases the origin, D and F respectively, is a double point of the curve and the corresponding abscissa the related axis, DC and FC)—in other words, the Cartesian equation of $q(z, v)$ will be $v^4 + (a + bz + cz^2)\,v^2 + dz^2 + ez^3 + fz^4 = 0$. Newton now proceeds to substitute for x and y in the original defining equation to see what further restrictions on the coefficients a, b, c, d, e, f obtain.

(24) In these omitted calculations (which we see no point in here reproducing) Newton, having—to no great purpose—recast the trefoil's equation to be

$$\tfrac{16}{27}(y^4 + cx^2y^2 + fx^4 \ldots) \equiv (\tfrac{2}{3}y)^4 + c(\tfrac{2}{3}y)^2\,(\tfrac{2}{3}x)^2 + f(\tfrac{2}{3}x)^4 \ldots = 0,$$

began therein to substitute

$$(\tfrac{2}{3}y)^2 = \tfrac{1}{9}v^2 - \frac{2}{3\sqrt{3}}zv + \tfrac{1}{3}z^2 + \frac{4}{9\sqrt{3}}nv - \tfrac{4}{9}nz + \tfrac{4}{27}n^2 \quad \text{and} \quad (\tfrac{2}{3}x)^2 = \tfrac{1}{3}v^2 + \frac{2}{3\sqrt{3}}zv \ldots - \frac{4}{3\sqrt{3}}nv \ldots;$$

but, having all but fully expanded $((\frac{2}{3}y)^2)^2$ and a few opening terms of $c((\frac{2}{3}y)^2(\frac{2}{3}x)^2)$ and $f((\frac{2}{3}x)^2)^2$, broke off to enter the following improved reduction.

(25) No supporting computations for this assertion, entered halfway down f. 44^r in a prominent rectangle in a space immediately to the right of his initially posited equation for the trefoil, occur in the manuscript, but it is manifestly entailed by the condition that the coefficients, $\frac{1}{4}\sqrt{3}.(-1-c+3f)$ and $\frac{1}{4}\sqrt{3}.(-3+c+f)$, of the terms in zv^3 and z^3v in the transformed defining equation of $q(z, v)$ shall (see note (23) above) each be zero.

(26) That is, $a + 3d = \frac{3}{64}(b + 3e)^2$. This is the revised condition (compare note (18) preceding) for F and G to be double points of the curve.

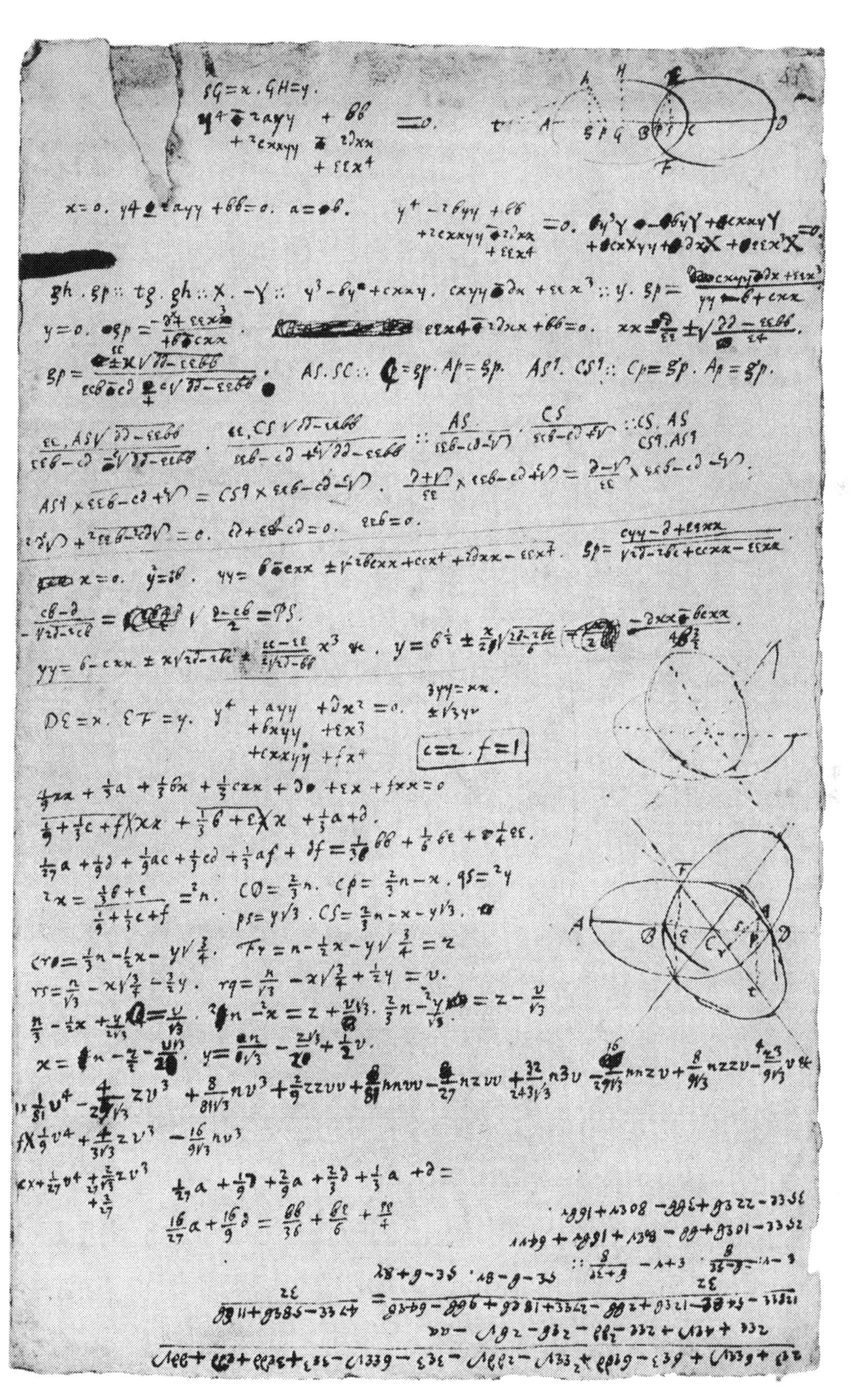

Plate IV. Computing the defining Cartesian equation of a quartic trifolium (**3**, 2, §2).

[atq; in æquatione prodeunte] $y^4 \begin{matrix} +2xx \\ +bx \\ +a \end{matrix} yy \begin{matrix} +x^4 \\ +ex^3 \\ +dxx \end{matrix} = 0$. [substituendo valores jam inventos] $x = n - \frac{1}{2}z - \frac{1}{2}v\sqrt{3}$. [et] $y = \frac{n}{\sqrt{3}} - \frac{z\sqrt{3}}{2} + \frac{1}{2}v$. [habebitur]

$$n^4 - 2n^3z + \tfrac{3}{2}nnzz - \tfrac{1}{2}nz^3 + \tfrac{1}{16}z^4 - 2\sqrt{3} \times n^3v + 3\sqrt{3} : nnzv - \frac{3\sqrt{3}}{2}nzzv + \frac{\sqrt{3}}{4}z^3v$$

$$+\tfrac{9}{2}nnvv - \tfrac{9}{2}nzvv + \tfrac{9}{8}zzvv - \frac{3\sqrt{3}}{2}nv^3 + \frac{3\sqrt{3}}{4}zv^3 + \tfrac{9}{16}v^4 : + \frac{n^4}{9} - \tfrac{2}{3}n^3z + \frac{3nnzz}{2}$$

$$-\frac{3nz^3}{2} + \tfrac{9}{16}z^4 + \frac{2}{3\sqrt{3}}n^3v - \sqrt{3} \times nnzv + \frac{3\sqrt{3}}{2}nzzv - \frac{3\sqrt{3}}{4}z^3v + \frac{nn}{2}v^2 - \tfrac{3}{2}nzvv$$

$$+\tfrac{9}{8}zzvv + \frac{n}{2\sqrt{3}}v^3 - \frac{\sqrt{3}}{4}zv^3 + \frac{v^4}{16} : + \frac{2n^4}{3} - \frac{2n^3z}{3} - \frac{2n^3v}{\sqrt{3}} + \frac{nn}{6}zz + \frac{1}{\sqrt{3}}nnzv$$

$$+\tfrac{1}{2}nnvv;\ -2n^3z + 2nnzz + 2nnvz\sqrt{3} - \tfrac{1}{2}nz^3 - \sqrt{3}nzzv - \tfrac{3}{2}nzvv; + \tfrac{3}{2}nnzz - \tfrac{3}{2}nz^3$$

$$-\frac{3\sqrt{3}}{2}nvzz + \tfrac{3}{8}z^4 + \frac{3\sqrt{3}}{4}vz^3 + \tfrac{9}{8}zzvv + \frac{2}{\sqrt{3}}n^3v - \frac{2}{\sqrt{3}}nnvz - 2nnvv + \frac{1}{2\sqrt{3}}nvzz$$

$$+nvvz + \frac{\sqrt{3}}{2}nv^3 - \sqrt{3}nnzv + \sqrt{3}nzzv + 3nvvz - \frac{\sqrt{3}}{4}vz^3 - \tfrac{3}{2}zzvv - \frac{3\sqrt{3}}{4}zv^3,$$

$$+\tfrac{1}{2}nnvv - \tfrac{1}{2}nvvz - \tfrac{1}{2}\sqrt{3}nv^3 + \tfrac{1}{8}vvzz + \frac{\sqrt{3}}{4}zv^3 + \tfrac{3}{8}v^4 :\ + \frac{bn^3}{3} - bnnz + \frac{3bn}{4}zz$$

$$+\frac{1}{\sqrt{3}}bnnv - \frac{\sqrt{3}}{2}bnzv + \tfrac{1}{4}bnvv,\ -\tfrac{1}{6}bnnz + \tfrac{1}{2}bnzz - \tfrac{3}{8}bz^3 - \frac{bn}{2\sqrt{3}}vz + \frac{\sqrt{3}}{4}bzzv$$

$$-\tfrac{1}{8}bzvv,\ -\frac{1}{2\sqrt{3}}bnnv + \frac{\sqrt{3}}{2}bnzv - \frac{3\sqrt{3}}{8}bvzz - \frac{bn}{2}vv + \tfrac{3}{4}bzvv - \frac{\sqrt{3}}{8}bv^3 :\ + en^3$$

$$-\frac{3e}{2}znn + \frac{3ne}{4}zz - \frac{e}{8}z^3 - \frac{3\sqrt{3}}{2}nnve + \frac{3\sqrt{3}}{2}nzve - \frac{3\sqrt{3}}{8}zzve + \tfrac{9}{4}envv - \tfrac{9}{8}evvz$$

$$-\frac{3\sqrt{3}}{8}ev^3 : + \frac{ann}{3} - anz + \frac{3a}{4}z^2 + \frac{anv}{\sqrt{3}} - \frac{\sqrt{3}}{2}azv + \frac{a}{4}vv :\ + dnn - dnz - dnv\sqrt{3}$$

$$+\frac{d}{4}zz + \frac{\sqrt{3}}{2}dzv + \frac{3d}{4}vv = 0.\text{(27)}$$

---

(27) That is, on collecting up the 106 terms of this equation,

$$v^4 + 2z^2v^2 + z^4 + Av^3 + Bzv^2 + Cz^2v + Dz^3 + Ev^2 + Fzv + Gz^2 + Hnv + Inz + Kn^2 = 0$$

where
$$A = C = -\frac{\sqrt{3}}{8}(\tfrac{32}{3}n + b + 3e), \quad B = -\tfrac{1}{8}(32n - 5b + 9e),$$

$$D = -\tfrac{1}{8}(32n + 3b + e), \quad E = \tfrac{1}{4}(16n^2 + (-b + 9e)\,n + a + 3d),$$

$$F = \frac{1}{2\sqrt{3}}(16n^2 + (-b + 9e)\,n - 3a + 3d), \quad G = \tfrac{1}{4}(\tfrac{80}{3}n^2 + (5b + 3e)\,n + 3a + d),$$

$$H = \frac{1}{\sqrt{3}}(-\tfrac{16}{3}n^2 + \tfrac{1}{2}(b - 9e)\,n + a - 3d), \quad I = -\tfrac{16}{3}n^2 - \tfrac{1}{6}(7b + 9e)\,n - a - d$$

and $K = \tfrac{1}{3}(\tfrac{16}{3}n^2 + (b + 3e)\,n + a + 3d)$.

[Hoc est] $v^4+2zzvv+z^4+bzvv+ez^3+avv+dzz=0$[28] posito quod

$$\left[-\frac{3b+9e}{32}=n.\right]\quad \frac{9bb+18eb-27ee}{128}=a.\quad \frac{27ee+6eb-bb}{128}=d.\quad \text{et [inde]}$$

$$\frac{3eb+bb}{16}=a+d.\quad \frac{5bb+6eb-27ee}{64}=a-d.^{(29)}$$

[Vel sic melius. Dic] $AD=x$. $DE=y$. [et curvæ AGE æquatio sit]

$$y^4 \begin{array}{l}+2xx\\ -bx\\ +a\end{array} yy \begin{array}{l}+x^4\\ -ex^3\\ +dxx\end{array} = 0.^{(30)}$$

[posito quod] $\dfrac{9bb+18eb-27ee}{128}=a.$

[et] $\dfrac{27ee+6eb-bb}{128}=d.$ [unde fit]

$$\frac{3eb+bb}{16}=a+d. \text{ [et]}$$

$$\frac{5bb+6eb-27ee}{64}=a-d.$$

[Erit] $\dfrac{3b+9e}{32}=AF=r.$[31]

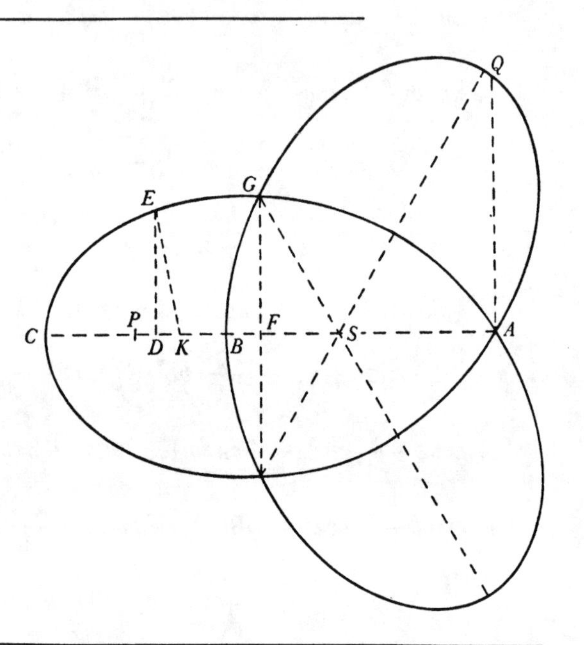

(28) See note (23) above (with now $c=2$ and $f=1$). In the terms of the previous note, it follows that $B=b$, $D=e$, $E=a$, $G=d$ and $A=C=F=H=I=K=0$, a supernumerary set of equations which are simultaneously satisfied by taking b and e to be arbitrary and then giving n, a and d the values which Newton proceeds to assign. Although in his ensuing computation of these on f. 45v he listed the six derived equations

$$4(E-a)=2F\sqrt{3}=4(G-d)=H\sqrt{3}=I=3K=0,$$

he at once realized that the last three alone suffice to attain them.

(29) That is,

$$n=-\tfrac{3}{32}(b+3e),\quad a=\tfrac{9}{128}(b+3e)\,(b-e),\quad d=\tfrac{1}{128}(b+3e)\,(-b+9e)$$

and so $$a+d=\tfrac{1}{16}b(b+3e),\quad a-d=\tfrac{1}{64}(b+3e)\,(5b-9e);$$

whence more simply $a=-\frac{3}{4}(b-e)\,n$, $d=-\frac{1}{12}(-b+9e)\,n$ and correspondingly $a+d=-\frac{2}{3}bn$, $a-d=-\frac{1}{6}(5b-9e)\,n$. In his prior computation on f. 45v Newton also confirmed that '$a+3d=\dfrac{3bb+18be+27ee}{64}$' (compare note (26) above). Having thus obtained his initial end of obtaining a general Cartesian equation for his quartic trefoil, he now proceeds to refine his choice of its coefficients and to connect them with the geometrical elements of his figure.

(30) To avoid troublesome negatives, in a first amelioration of the preceding he replaces the coefficients b and e in his previous equation by $-b$ and $-e$ respectively, henceforward relating his argument to a relettered figure.

(31) This was previously called n, of course.

$$\frac{b+3e}{16}=AS.\ \text{[hoc est]}\ \frac{3br-3er}{4}=a.\ \text{[et]}\ \frac{2br-3a}{3}=d.\ \text{[seu]}\ \frac{2br}{3}=a+d.$$

$$\text{Si } y=0,\ \text{fit}^{(32)}\ x=\tfrac{1}{2}e\pm\sqrt{\tfrac{1}{4}ee-d}=\tfrac{1}{2}e\pm\sqrt{\frac{5ee-6eb+bb}{128}}=\begin{matrix}AC.^{(33)}\\ AB.\end{matrix}$$

[Vel adhuc concinnius, cape] $AP=e$. $AQ=a$. [et] $BP=PC=\sqrt{ee-dd}$. [hoc est] $AP^q-PC^q=dd$. [ut et] $\frac{b+3e}{8}=AS$. [id est] $8AS-3AP=b$. [Curvæ æquatio erit] $y^4\begin{matrix}+2xx\\-2bx\\-aa\end{matrix}yy\begin{matrix}+x^4\\-2ex^3\\+ddx^2\end{matrix}=0.$[34] [positis ut ante] $AD=x$. $DE=y$.

[Fit] $\frac{9e-9b}{4}AS=aa=AQ^q$. [ut et] $2b\times AS+aa=dd=AP^q-PC^q$. [seu] $\frac{9e-b}{4}AS=dd=\frac{27ee+6eb-bb}{32}$. [Insuper capiendo fluxiones erit][35]

$$4y^3Y+4xxyY+4xXy^2+4x^3X-4bxyY-2aayY-2bXyy-6exxX+2ddxX=0.$$

[adeoq; demissa EK perpendiculari ad curvam]

$$X.Y::2y^3+2xxy-2bxy-aay.-2xyy-2x^3+byy+3exx-ddx::y.DK.$$

[Ergo] $\frac{3exx-2x^3-ddx}{2xx-2bx-aa}=DK$. [ubi] $y=0$. [hoc est ubi] $x=e\pm\sqrt{ee-dd}$.

[Sic optime. Dic] $AD=x$. $DE=y$. [et pone æquationem esse]

$$y^4\begin{matrix}+2xx\\-2bx\\-aa\end{matrix}yy\begin{matrix}+x^4\\-2ex^3\\+ddx^2\end{matrix}=0.^{(36)}$$

---

(32) On excluding the two roots $x=0$ (marking the double point A) in the quartic $x^2(x^2-ex+d)=0$ and solving the quadratic which remains.

(33) But, correspondingly, 'Si $x=0$, fit [$y^4+ayy=0$, adeoq;] $y=\sqrt{-a}=AQ$', so that the coefficient a must always be negative. Newton tries again!

(34) In further refinement Newton replaces his previous coefficient a, which must be negative (see the previous note), by $-a^2$ and likewise for homogeneity's sake transposes $d\rightarrow d^2$, while to avoid messy fractions in the values of

$$AP=\tfrac{1}{2}(AC+AB)\quad\text{and}\quad BP=PC=\tfrac{1}{2}(AC-AB)$$

he also doubles his earlier coefficients b and e.

(35) Again understand (compare note (6) above) that X and Y in the sequel are proportional to the fluxions of x and y.

(36) Exactly as before, that is.

[Erit] $AS = \frac{b+3e}{8} = c.$ [ut et] $\frac{9ec-9bc}{4} = \frac{27ee-18eb-9bb}{32} = aa = AQ^q.$ [nec non] $\frac{9ee-bc}{4} = dd = \frac{27ee+6eb-bb}{32}.$ [adeoq; ponendo] $AP = e.$ [erit]

$$PC = PB = \sqrt{ee-dd} = \sqrt{\frac{5ee-6eb+bb}{32}} = r.^{(37)}$$

[Insuper capiendo fluxiones orietur][35]

$$2y^3Y+2xxyY-2bxyY-aayY+2xyyX-byyX+2x^3X-3exxX+ddxX = 0.$$

[sive demissâ EK perpendiculari ad curvam]

$$ED = y.DK::X.Y^{(38)}::2y^3+2xxy-2bxy-aay.2xyy-byy+2x^3-3exx+ddx.$$

[adeo ut sit] $\frac{2xyy-byy+2x^3-3exx+ddx}{2yy+2xx-2bx-aa} = DK.$ [Unde] $DK = \frac{2x^3-3exx+ddx}{2xx-2bx-aa}$

[ubi] $y = 0.$ [adeoq;] $x = e \pm \sqrt{ee-dd}.$ [hoc est si] $x = e+\sqrt{ee-dd}\,[=AC].$ [erit]

$$\frac{2e^3-2edd+2ee[r]-dd[r]}{4ee-2dd-2eb-aa+\overline{4e-2b}[r]} = DK = \frac{64err+64eer-32ddr}{47ee-58eb+11bb+128er-64br}.^{(39)}$$

[at si] $x = e-\sqrt{ee-dd}\,[=AB].$ [fit]

$$dk = \frac{64err-64eer+32ddr}{47ee-58eb+11bb-128er+64br} = \frac{64err-37eer+6ebr-bbr}{47ee-58eb+11bb-128er+64br}.^{(40)}$$

[Quare]

$$DK.dk::\frac{64er+37ee-6eb+bb}{47ee-58eb+11bb+128er-64br}.\frac{-64er+37ee-6eb+bb}{47ee-58eb+11bb-128er+64br}$$

(37) This 'r[adix]' $\sqrt{[\frac{1}{32}(5e-b)(e-b)]}$ should not be confused with Newton's earlier denomination of AF. In his preliminary computations on f. 44r (at the bottom) he employed $\sqrt{)}$ to the same purpose; we have adjusted two places in the ensuing argument where this notation is forgetfully carried over.

(38) Read '$X.-Y$' strictly, but Newton is here interested only in the absolute length of the subnormal $DK = yY/X$ at $E(x, y)$.

(39) On putting e^2-r^2 for d^2 in the numerator of the preceding fraction and equivalently replacing it by $\frac{1}{32}(27e^2+6eb-b^2)$ in the numerator. A like substitution of the latter in the present numerator further reduces it to be $64er^2+(37e^2-6eb+b^2)\,r$.

(40) As before (see also the previous note) but with $-r$ in place of r. These limiting values of the length of the subnormal when the corresponding point E of the trefoil comes to coincide with the vertices C and B in which it intersects its axis AS are (as in the analogous limit-subnormals to the quartic *bifolium* in [1]; compare note (7) above) the respective radii of curvature at these latter points. In the following final portion of his calculation Newton (again much as in [1]) attempts in some manageable way to equate their ratio to some (inverse) power of that of the distances of C and B from the curve's centre S, but quickly breaks off when only purposeless complication results.

[puta]$::BS^q.CS^q::35ee-22be+3bb-80er+16br.35ee-22be+3bb+80er-16br$

$::7e-3b-16r.7e-3b+16r.$[41]

[vel]$::BS^{\frac{3}{2}}.C[S]^{\frac{3}{2}}::$[42]

---

(41) Since (as Newton set down in his preliminary computation here on f. 44r)

$$\text{‘}SC = e+[r]\frac{-b-3e}{8} = \frac{5e-b}{8}+[r]\text{’ and ‘}SB = e-[r]\frac{-b-3e}{8}\left[=\frac{5e-b}{8}-r\right]\text{’}$$

where (see note (37) above) $r = \sqrt{[\frac{1}{32}(5e-b)\ (e-b)]}$, so that SC^2 and SB^2 share the common factor $5e-b$. The equation of this ratio to that of the preceding radii of curvature at C and B respectively evidently leads nowhere.

(42) Newton abandoned this still less forthcoming alternative without bothering here to enter ‘$\overline{7e-3b-16r}|^{\frac{3}{2}}.\overline{7e-3b+16r}|^{\frac{3}{2}}$’ or its equivalent. The remaining half page in the manuscript (f. 42v) is left wholly blank, so it is just possible that he intended to return further to pursue his investigation into the quartic trefoil at a future time.

3

INTERPOLATION AND APPROXIMATE QUADRATURE BY CURVE-FITTING[1]

[c. autumn 1695]

Extracted from original worksheets and drafts in the University Library, Cambridge

§1. SIMPLE SCHEMES OF SUBTABULATION BY FINITE DIFFERENCES.[2]

---

(1) This miscellany of contemporary notes and drafts on interpolation—adumbrating *ad hoc* rules for the subtabulation of listed numerical values of a function by variously adapting a parabolic polynomial to these much as he had done nearly twenty years before (see IV: 28–68), and also, in more particular application, outlining inter-related schemes for intercalating the apparent celestial path of a comet between individual timed sightings of its orbit by similarly fitting both a general parabola and a hyperbola to its elements, and then passing to elaborate in more finished form an ingenious technique (not entirely free from intrinsic error as it is developed) for successively refining parabolic approaches to the area under a given curvilinear arc—contains no penetrating new insight on Newton's part into the topic, but jointly presents a revealing illustration of the continuing subtlety and dexterity of his mathematical mind on the eve of his final departure from Cambridge for the Wardenship of the Mint. The manuscripts of all these pieces occur, contingently or no, in the context of his revised computations of cometary orbits in the early 1690's, and the more precise date we adjoin is founded on the fact that both Newton's jottings on intercalating a median term set out here in §1.4 and his preparatory casting of a passage in the concluding essay 'Of Quadrature by Ordinates' (§3 below) are entered on a sheet (Add. 3965.14: 611ʳ) which also (see note (20) following) contains his response to a letter from Halley in early October 1695.

(2) From first principles, that is, using an approximating parabolic polynomial both in the Briggsian advancing-differences form $f_x = a+b\binom{x}{1}+c\binom{x}{2}+\ldots$ $(a = f_0,\ b = \delta_0^1,\ c = \delta_0^2, \ldots)$ which he had employed in 1665 in intercalating the binomial series (see I: 130–3, and compare IV: 4, note (4)) and also in the equivalent direct form $f_x = a+b'x+c'x^2+\ldots$ whereby he had derived his more general theorems on finite-difference interpolation ten years later (compare IV: 37, note (3)). Throughout, the values f_i of the function to be subtabulated are understood to be given at unit intervals of the argument i.

[1][3] $[a+bx+\frac{1}{2}c, \overline{x^2-x}]$[4]

a					
	b				
$a\ +b$		c			
	$b+c$				
$a+2b\quad +c$		c			
	$b+2c$				
$a+3b\ +3c$		c			
	$b+3c$		a		
$a+4b\ +6c$		c		$[4b]\ +6c$	
	$b+4c$		$a+4b\ [+6c]$		$16c$[5]
$a+5b+10c$		$[c]$		$[4b]+22c$	
	$b+5c$		$a+8b[+28c]$		
$a+6b+15c$		$[c]$			
	$b+6c$				
$a+7b+21c$		$[c]$			
	$b+7c$				
$a+8b+28c$		$[c]$			
	$b+8c$				
$a+9b+36c$					

(3) Add. 3965.18: 671r/671v. Newton's first use of this manuscript sheet was to draft (on f. 671v) a letter to 'Mr Martin' specifying the terms of an annuity to be paid to his recently widowed half-sister Hannah Barton, the gist of which we quoted in IV: 205, note (5), but whose penning we there too rigidly assigned to the period (late 1693) immediately following Robert Barton's death—which furnishes, of course, only a *terminus ante quem non* to its composition. Since the other, subsequent jottings on the letter's paper include (f. 671r) Newton's computation of the maximum (semi-monthly) angular departure of the lunar apsides from its mean motion—in the Horroxian theory of the moon's motion brought freshly to his notice by Flamsteed in October 1694 (see *The Correspondence of Isaac Newton*, **4**, 1967: 26–7)—to be '12deg⌊295' (afterwards rounded off to be '12gr.18'' in *Principia*, ₂1713: 432), it would be preferable to set a date upon all of them nearer to our above one of '*c*. autumn 1695'.

(4) We insert this general Briggsian interpolating quadratic, whose instances $x = 0, 1, 2, \ldots 9$, Newton proceeds to list, in parallel to his explicit statement of the straightforward cubic approximating polynomial which he adduces in sequel.

(5) Following, consciously or no, in the footsteps of Gabriel Mouton in 1670 in Liber III, Caput III of his *Observationes Diametrorum Solis et Lunæ Apparentium* (on which see IV: 4–5, note (5)), Newton evidently here has it in mind to quadrisect the given functional values f_0, f_4, f_8 in terms of their first- and second-order differences

$$\Delta_0^1 = f_4 - f_0, \quad (\Delta_1^1 = f_8 - f_4 \text{ and}) \quad \Delta_0^2 = (\Delta_1^1 - \Delta_0^1 =) f_8 - 2f_4 + f_0$$

by assuming that the f_i are approximated by the quadratic $a+b\binom{i}{1}+c\binom{i}{2}$, where $a = f_0$, $b = f_1 - f_0 = \delta_0^1$, $(f_2 - f_1 = \delta_1^1$ and) $c = \delta_1^1 - \delta_0^1 = \delta_0^2$. At once, as Newton readily computes, there is $\Delta_0^1 = 4b + 6c$ and $\Delta_0^2 = 16c$; from which inversely there comes $c = \frac{1}{16}\Delta_0^2$ and

$$a+bx+cxx+dx^3$$

$a+2b+4c+8d=p.$	$a+c=\dfrac{q+s}{2}.\quad a+4c=\dfrac{p+t}{2}.$
$a\ +b\ +c\ +d=q.$	
$a\qquad\qquad =r.$	[unde] $c=\dfrac{p+t-q-s}{6}.$
$a\ -b\ +c\ -d=s.$	
$a-2b+4c-8d=t.$	$\Big[$adeoq; $r=a=\dfrac{4q+4s-p-t}{6}.\Big]^{(6)}$

$a+5b+25c+125d=p.$	
$a+3b\ +9c\ +27d=q.$	$a+c=\dfrac{r+s}{2}.\quad b+d=\dfrac{r-s}{2}.$
$a\ +b\ \ +c\ \ \ +d=r.$	
a	$a+9c=\dfrac{q+t}{2}.\quad b+9d=\dfrac{q-t}{6}.$
$a\ -b\ \ +c\ \ \ -d=s.$	
$a-3b\ +9c\ -27d=t.$	[unde] $8a=\dfrac{9r+9s-q-t}{2}.$ (8)
$a-5b+25c-125d=v.$ (7)	

$b=\frac{1}{4}\Delta_0^1-(\frac{6}{4}c \text{ or})\frac{3}{32}\Delta_0^2$, whence $f_i=f_0+\frac{1}{32}\binom{i}{1}(8\Delta_0^1-3\Delta_0^2)+\frac{1}{16}\binom{i}{2}\Delta_0^2$, though he does not here make this final deduction. As we have seen (IV: 5, note (6)), Henry Briggs had in Chapter 12 of his *Arithmetica Logarithmica* (London, 1624): 24–7 set down a formula, elaborating a like subtabulation into tenths, which equivalently adduces such an 'advancing-differences' approximating quadratic, but there is (compare IV: 3, note (1)) nothing to show that Newton ever read this work.

(6) We adjoin what is manifestly the goal of Newton's computation, one attainable yet more rapidly on noticing that, since the approximating polynomial f_x is cubic, the fourth difference $p-4q+6r-4s+t$ of $p=f_2$, $q=f_1$, $r=f_0$, $s=f_{-1}$ and $t=f_{-2}$ is zero. But the given values p, q, s and t are not placed at equal intervals of the argument x, and Newton breaks off to remedy this; see the next note.

(7) Here $p=f_5$, $q=f_3$, $r=f_1$, $s=f_{-1}$, $t=f_{-3}$ and $v=f_{-5}$ are now spaced more conveniently at equal (2-unit) intervals of the argument x. Since the approximating polynomial f_x is taken to be a cubic, the 'outer' terms p and v are evidently superfluous and Newton makes no use of them in his sequel.

(8) That is, $(a+9c)-(a+c)$; whence the intercalated median term f_0 is

$$a=\tfrac{1}{16}(9(r+s)-(q+t)).$$

This is the fourth of the rules for subtabulation which Newton had originally intended to communicate to Leibniz in his *epistola posterior* in October 1676 (see IV: 30); the refined 'Reg. 6' which he there appended (*ibid.*: 32) here similarly follows by assuming an approximating quartic $f_x=a+bx+cx^2+dx^3+ex^4$ and deriving therefrom

$$\tfrac{1}{2}(r+s)=a+c+e,\quad \tfrac{1}{2}(q+t)=a+9c+81e\quad\text{and}\quad \tfrac{1}{2}(p+v)=a+25c+625e,$$

so that $a=\frac{1}{128}(150.\frac{1}{2}(r+s)-25.\frac{1}{2}(q+t)+3.\frac{1}{2}(p+v))$. In [4] below he elegantly deduces both these approaches to the 'terminus intermedius' f_0 by setting $x=0$ in his earlier general 'Newton–Bessel' central-differences formula (see IV: 60) for interpolating the equi-separate

[2][9] $[A+bx+cxx+dx^3]$

$$
\begin{array}{l|l|l|l}
A-10b+100c-1000d & & & \\
 & 5b-75c+875d\,(D) & & \\
A\;-5b\;+25c\;-125d & & 50c-750d\;(K) & \\
 & 5b-25c+125d\,(E) & & 750d\,(P) \\
A & & 50c\qquad (L) & \\
 & 5b+25c+125d\,(F) & & 750d\,(Q). \\
A\;+5b\;+25c+\;125d & & 50c+750d\;(M) & \\
 & 5b+75c+875d\,(G) & & \\
A+10b+100c+1000d & & &
\end{array}
$$

$$
\begin{array}{l|l|l|l}
A & & & \\
 & b\;+c\;\;+d\,(T) & & \\
A\;+b\;\;\;+c\;\;\;\;\;+d & & 2c+6d\,(S) & \\
 & b+3c\;+7d & & 6d\,(R) \\
A+2b\;+4c\;\;\;+8d & & 2c+12d & \\
 & b+5c+19d & & 6d \\
A+3b\;+9c\;+27d & & 2c+18d & \\
 & b+7c+37d & & 6d \\
A+4b+16c\;+64d & & 2c+24d & \\
 & b+9c+61d & & \\
A+5b+25c+125d & & &
\end{array}
$$

[Hic est] $\dfrac{P}{125}=6d=R.\quad \dfrac{L}{25}+R=2c+6d=S.\quad \dfrac{F}{5}=b+5c+25d.$

[seu] $\dfrac{F}{5}-2S-2R=T=b+c+d.$[10]

---

array p, q, r, s, t, v and considering the first two and three terms respectively of the resulting expansion $f_0 = \mu f_0 - \frac{1}{8}\mu\Delta^2_{-1} + \frac{3}{128}\mu\Delta^4_{-2}\ldots$, in which there is

$$\mu f_0 = \tfrac{1}{2}(r+s),\quad \mu\Delta^2_{-1} = \tfrac{1}{2}((q-2r+s)+(r-2s+t)) = \tfrac{1}{2}(q+t)-\tfrac{1}{2}(r+s)$$

and

$$\mu\Delta^4_{-2} = \tfrac{1}{2}((p-4q+6r-4s+t)+(q-4r+6s-4t+v)) = \tfrac{1}{2}(p+v)-3.\tfrac{1}{2}(q+t)+2.\tfrac{1}{2}(r+s).$$

(9) Add. 3965.18: 697r. Newton seeks *ab initio* to formulate a general rule for quinquisecting an array of values f_x ($x = \ldots, -10, -5, 0, 5, 10, \ldots$) given at equal (5-unit) intervals of the argument, supposing that their third-order differences are effectively constant, and hence that they may adequately be approximated by the cubic $A+bx+cx^2+dx^3$.

(10) That is, $R = P/5^3$, $S = L/5^2+P/5^3$ and $T = F/5-2L/5^2+P/5^3$. In modern terms, where the given values $f_{5i} = a+b(5i)+c(5i)^2+d(5i)^3$, are to be subtabulated into f_i differing only by unit-intervals of the argument, Newton successively computes the differences

$$\Delta^1_k = f_{5(k+1)}-f_{5k},\quad \Delta^2_k = \Delta^1_{k+1}-\Delta^1_k \quad\text{and}\quad \Delta^3_k = \Delta^2_{k+1}-\Delta^2_k,$$

setting $\Delta^1_0 = 5(b+5c+25d) = F$, $\Delta^2_{-1} = 25(2c) = L$ and $\Delta^3_{-2} = 125(6d) = P$; and then, where

$$[A+bx-cxx-dx^3]$$

$A-10b-100c+1000d$			
	$5b+75c-875d$		
$A-\ 5b-\ 25c\ +125d$		$50c+750d$	
	$5b+25c-125d$		$750d$
A		$50c\quad (G)$	
	$5b-25c-125d\,(F)$		$750d(H)$
$A\ +5b\ -25c\ -125d$		$50c-750d$	
	$5b-75c-875d$		
$A+10b-100c-1000d$			

$A-2b-4c\ +8d$			
	$b+3c\ -7d$		
$A\ -b\ -c\ \ +d$		$2c\ -6d$	
	$b\ +c\ \ -d$		$6d$
A		$2c$	
	$b\ -c\ \ -d$		$6d$
$A\ +b\ -c\ \ -d$		$2c\ +6d$	
	$b-3c\ -7d$		$6d$
$A+2b-4c\ -8d$		$2c+12d$	
	$b-5c-19d$		
$A+3b-9c-27d$			

[Hic ponendo] $\dfrac{F}{5}=P.\ \dfrac{G}{25}=Q.\ \dfrac{H}{125}=R.$ [erit] $6d=R.\ 2c=Q.$ [et]

$P=b-5c-25d.$ [vel] $P+2Q+4R=b-c-d.$(11)

[Hoc est ubi] $\frac{8}{1000}H=R.$ [et] $\frac{4}{100}G=Q.$ [evadit] $\frac{2}{10}F+2Q+4R$

$=[b-c-d].$(12)

$\delta^1_k=f_{k+1}-f_k$, $\delta^2_k=\delta^1_{k+1}-\delta^1_k$ and $\delta^3_k=\delta^2_{k+1}-\delta^2_k$, likewise determines the quinquisected differences $\delta^3_0=R=6d=\underline{\underline{\Delta}}^3_{-2}$, $\delta^2_0=S=2c+6d=\underline{\underline{\Delta}}^2_{-1}+\underline{\underline{\Delta}}^3_{-2}$ and lastly

$\delta^1_0=T=b+c+d=\underline{\underline{\Delta}}^1_0-2\underline{\underline{\Delta}}^2_{-1}-4\underline{\underline{\Delta}}^3_{-2}$ on putting $\underline{\underline{\Delta}}^1_k=\frac{1}{5}\Delta^1_k$, $\underline{\underline{\Delta}}^2_k=(\frac{1}{5})^2\,\Delta^2_k$ and $\underline{\underline{\Delta}}^3_k=(\frac{1}{5})^3\Delta^3_k$. Though we again insist that there is nothing to show that Newton ever looked at his predecessor's work, this rule evidently agrees with the scheme of quinquisection earlier devised by Henry Briggs (see IV: 45, note (25)), according to which there is

$$F=\underline{\underline{\Delta}}^1_0=\delta^1_2+\delta^3_1+\tfrac{1}{5}\delta^5_0=\delta^1_0+2\delta^2_0+2\delta^3_0\ldots,$$

and thence $L=\underline{\underline{\Delta}}^2_{-1}=\delta^2_{-5}+4\delta^3_{-5}\ldots=\delta^2_0-\delta^3_0\ldots$ and $P=\underline{\underline{\Delta}}^3_{-2}=\delta^3_{-10}\ldots=\delta^3_0\ldots.$

(11) This follows exactly as before on replacing c and d by $-c$ and $-d$ respectively, since $Q=2c=S-R$.

(12) For convenience of calculation Newton takes the trivial further step of converting the preceding fractions to have denominators which are powers of 10 by multiplying them top and bottom by the corresponding power of 2. In sequel in the manuscript he adapted his two equivalent rules to quinquisecting a numerical array—evidently astronomical in character,

[3][13]

$$[A+bx+cxx+dx^3+ex^4+fx^5]^{(14)}$$

$A-3b \quad +9c \quad -27d \quad +81e \quad -243f$					
	$b-5c \quad +19d \quad -65c \quad +211f$				
$A-2b \quad +4c \quad -8d \quad +16e \quad -32f$		$2c-12d \quad +50e \quad -180f$			
	$b-3c \quad +7d \quad -15e \quad +31f$		$6d-36e+150f$		
$A \quad -b \quad +c \quad -d \quad +e \quad -f$		$2c-\ 6d \quad +14e \quad -30f$		$24e-120f$	
	$b \quad -c \quad +d \quad -e \quad +f$		$6d-12e \quad +30f$		$120f$
A		$2c \quad \pm 0 \quad +2e \quad \pm 0 = Q$		$24e \quad \pm 0 = S$	
	$b \quad +c \quad +d \quad +e \quad +f = P$		$6d+12e \quad +30f = R$		$120f = T.$
$A \quad +b \quad +c \quad +d \quad +e \quad +f$		$2c \quad +6d \quad +14e \quad +30f$		$24e+120f$	
	$b+3c \quad +7d \quad +15e \quad +31f$		$6d+36e+150f$		$120f$
$A+2b \quad +4c \quad +8d \quad +16e \quad +32f$		$2c+12d \quad +50e \quad +180f$		$24e+240f$	
	$b+5c \quad +19d \quad +65e \quad +211f$		$6d+60e+390f$		
$A+3b \quad +9c \quad +27d \quad +81e \quad +243f$		$2c+18d \quad +110e \quad +570f$			
	$b+7c \quad +37d \quad +175e \quad +781f$				
$A+4b \quad +16c \quad +64d \quad +256e \quad +1024f$					

$$[A+gx+hxx+kx^3+lx^4+mx^5]^{(14)}$$

A		$2h \quad +0 \quad +2l \quad [+0]$		$[24l \quad +0]$	
	$g \quad +h \quad +k \quad +l \quad [+m]$		$6k+12l \quad [+30m]$		$[120m].$
$A \quad +g \quad +h \quad +k \quad +l \quad [+m]$		$2h \quad +6k \quad +14l \quad [+30m]$		$[24l+120m]$	
	$g+3h \quad +7k \quad +15l \quad [+31m]$		$[6k+36l \quad +150m]$		$[120m].$
$A+2g \quad +4h \quad +8k \quad +16l \quad [+32m]$		$2h+12k \quad [+50l \quad +180m]$		$[24l+240m]$	
	$g+5h \quad +19k \quad +65l \quad [+211m]$		$[6k+60l \quad +390m]$		
$A+3g \quad +9h \quad +27k \quad +81l \quad [+243m]$		$2h+18k \quad [+110l \quad +570m]$			
	$g+7h \quad +37k \quad [+175l \quad +781m]$				
$A+4g \quad +16h \quad +64k \quad [+256l \quad +1024m]$		$2h+24k$			
	$g+9h \quad +61k$				
$A+5g \quad +25h \quad +125k$					

[The notes for this Table appear on p. 678]

[Quibus comparatis fit] $\frac{T}{120}=f=3125m.$ $\frac{S}{24}=e=625l.$ $\frac{R}{6}=d+2e+5f.$ $\frac{Q}{2}=c+e.$ $P=b+c+d+e+f=5y+25h+125k+625l+3125m.$ [unde]

$$\frac{R}{6}-\frac{S}{12}-\frac{T}{24}=d=125k. \quad \frac{Q}{2}-\frac{S}{24}=c=25h.$$

$$\text{[itaq}\text{ȝ]}\ g+h+k+l+m=\frac{P}{5}-\frac{2Q}{25}+\frac{S}{150}-\frac{4R}{125}+\frac{2S}{125}+\frac{T}{125}-\frac{24}{625}e-\frac{624}{3125}f$$

$$=\frac{P}{5}-\frac{2Q}{25}-\frac{4R}{125}+\frac{8\frac{1}{6}S}{625}+\frac{T}{125}.^{(15)}$$

---

though we have not been able exactly to specify its context—whose third-order differences proved, unfortunately, not to be constant. Aware that a more refined approach was here called for, he turned his page over and began—straightaway, we presume—to derive the refined scheme of subtabulation, keyed to the supposition that the fifth-order differences are constant, which we reproduce in [3] following.

(13) Add. 3965.18: 695ᵛ. Newton now (on the verso of the sheet on which he penned his simpler rule in [2] preceding) proceeds to derive a more accurate rule for quinquisection which appeals to an approximating quintic polynomial.

(14) We again specify the polynomials which Newton here fits to the given functional values F_x $(x=-3,-2,-1,0,1,\ldots,4)$ and the quinquisections f_x $(x=1,2,3,4)$ to be intercalated between $F_0=f_0=A$ and $F_1=f_5$, understanding generally that $F_x\equiv f_{5x}$, whence $b=5g$, $c=25h$, $d=125k$, $e=625l$ and $f=3125m$. From the differences

$$\Delta_k^1=F_{k+1}-F_k,\quad \Delta_k^2=\Delta_{k+1}^1-\Delta_k^1,\quad \Delta_k^3=\Delta_{k+1}^2-\Delta_k^2,\ \ldots$$

which he computes in the first array he singles out $\Delta_0^1=P$, $\Delta_{-1}^2=Q$, $\Delta_{-1}^3=R$, $\Delta_{-2}^4=S$ and $\Delta_{-2}^5=T$, and out of the analogous latter array of the quinquisection differences $\delta_k^1=f_{k+1}-f_k$, $\delta_k^2=\delta_{k+1}^1-\delta_k^1$, $\delta_k^3=\delta_{k+1}^2-\delta_k^2$, ... he will in sequel aim to determine $\delta_0^1=g+h+k+l+m$, $\delta_{-1}^2=2h+2l$ and—though he does not in fact go so far—also $\delta_{-1}^3=6k+12l+30m$, $\delta_{-2}^3=24l$ and $\delta_{-2}^5=120m$ in terms of the adjusted given differences

$$\underline{\Delta}_0^1=\tfrac{1}{5}P=g+5h+25k+125l+625m,\quad \underline{\Delta}_{-1}^2=\tfrac{1}{25}Q=2h+50l,$$

$$\underline{\Delta}_{-1}^3=\tfrac{1}{125}R=6k+60l+750m,\quad \underline{\Delta}_{-2}^4=\tfrac{1}{625}S=24l$$

$$\text{and } \underline{\Delta}_{-2}^5=\tfrac{1}{3125}T=120m.$$

In the second array we have rounded out Newton's manuscript entries—initially keyed but to an approximating cubic $A+gx+hx^2+kx^3$, as in [2] preceding—to the extent that his following text requires: the final entry (f_5) on the left might similarly have been completed by inserting the further terms '$+625l+3125m$', so necessitating the rounding out of the differences δ_4^1 and δ_3^2 by '$+369l+2101m$' and '$+191l+1320m$' respectively, together with the addition of $\delta_2^3=$ '$6k+84l+750m$', $\delta_1^4=$ '$24l+360m$' and $\delta_0^5=$ '$120m$'.

(15) When correctly computed from the previous line, the last two terms are '$+\frac{9S}{625}+\frac{99T}{15625}$', the latter of which (since T itself is small in comparison with P, Q, R and S) may without loss be rounded off to be '$\ldots+\frac{4T}{625}$'.

[et] $2h+2l = \frac{Q}{25} - \frac{S}{12,25} + \frac{S}{12,625} = \frac{Q}{25} - \frac{S}{3,625}$.(16)

[Habebitur ergo]

$$1 \quad \text{Diff} = \frac{2P}{10} - \frac{4,4Q}{100} - \frac{4,8R}{1000} + \frac{131S}{10000} + \frac{8T}{1000}.\text{(17)}$$

$$2 \quad \text{Diff} = \frac{4Q}{100} - \frac{5\frac{1}{3}S}{10000}.\text{(18)}$$

$$3 \quad \text{Diff} = \text{(19)}$$

---

(16) A second slip: this latter term should be $-\frac{2S}{625}$. *recte*. In the same fashion the third difference is readily computed to be $6k+12l+30m = \frac{R}{125} - \frac{2S}{625} - \frac{6T}{3125}$; the corresponding fourth and fifth differences are at once, of course, $24l = \frac{S}{625}$ and $120m = \frac{T}{3125}$.

(17) The last two terms should (see note (15) above) be '$+\frac{144S}{10000} + \frac{6336T}{1000000}$', the second o which may again safely be rounded off as $+\frac{64T}{10000}$.

(18) This latter term should (see note (16)) read '$-\frac{32S}{10000}$'.

(19) Newton breaks off before entering his computed value—whatever it might have been —for this subtabulated third difference, which should correctly (again see note (16)) have been $\frac{8R}{1000} - \frac{32S}{10000} - \frac{192T}{100000}$ on converting the denominators of the fractions to be powers of 10. In terms of note (14) preceding, this mode of quinquisection successively inverts the adjusted differences

$$\underline{\Delta}^5_{-2} = \delta^5_{-2}, \quad \underline{\Delta}^4_{-2} = \delta^4_{-2}, \quad \underline{\Delta}^3_{-1} = \delta^3_{-1} + 2\delta^4_{-2} + 6\delta^5_{-2}, \quad \underline{\Delta}^2_{-1} = \delta^2_{-1} + 2\delta^4_{-2}$$
$$\text{and } \underline{\Delta}^1_0 = \delta^1_0 + 2\delta^2_{-1} + 4\delta^3_{-1} + 3\delta^4_{-2} + 4\tfrac{1}{5}\delta^5_{-2}$$

thence to derive in turn

$$\delta^5_{-2} = \underline{\Delta}^5_{-2}, \quad \delta^4_{-2} = \underline{\Delta}^4_{-2}, \quad \delta^3_{-1} = \underline{\Delta}^3_{-1} - 2\underline{\Delta}^4_{-2} - 6\underline{\Delta}^5_{-2}, \quad \delta^2_{-1} = \underline{\Delta}^2_{-1} - 2\underline{\Delta}^4_{-2}$$
$$\text{and lastly } \delta^1_0 = \underline{\Delta}^1_0 - 2\underline{\Delta}^2_{-1} - 4\underline{\Delta}^3_{-1} + 9\underline{\Delta}^4_{-2} + 19\tfrac{4}{5}\underline{\Delta}^5_{-2}.$$

(But how much simpler is the parallel scheme of quinquisection set out by Henry Briggs in Chapter 13 of his *Arithmetica Logarithmica* in 1624! One has here but to fix on

$$\delta^1_2 = g+5h+19k+65l+211m, \quad \delta^3_1 = 6k+60l+390m \quad \text{and} \quad \delta^5_0 = 120m,$$

and then (compare IV: 45, note (25)) invert $\underline{\Delta}^0_1 = g+5h+25k+125l+625m = \delta^1_2 + \delta^3_1 + \frac{1}{5}\delta^5_0$, $\underline{\Delta}^3_{-1} = 6k+60l+750m = \delta^3_1 + 3\delta^5_0 \ldots$ and $\underline{\Delta}^5_{-2} = 120m = \delta^5_0 \ldots$ to deduce the more concise equivalent subtabulation proceeding from

$$\delta^5_0 = \underline{\Delta}^5_{-2} \ldots, \quad \delta^3_1 = \underline{\Delta}^3_{-1} - 3\underline{\Delta}^5_{-2} \ldots \quad \text{and} \quad \delta^1_2 = \underline{\Delta}^1_0 - \underline{\Delta}^3_{-1} + \tfrac{14}{5}\underline{\Delta}^5_{-2} \ldots,$$

correct to $O(\underline{\Delta}_7)$.) We may add that the generalization of Newton's present method which identifies $F_x \equiv f_{nx}$, $n = 2, 3, \ldots$, to derive therefrom similar rules for *n*-section by evaluating the subtabulated differences δ^p_j in terms of the given adjusted differences $\underline{\Delta}^q_k = n^{-q}\Delta^q_k$ was independently attained some dozen years later by Roger Cotes and developed by him in the 'Canonotechnia sive Constructio Tabularum per Differentias' which was published only posthumously in 1722 along with the other *Opera Miscellanea* which Robert Smith appended to his edition of Cotes' *Harmonia Mensurarum*; compare IV: 43–4, notes (23) and (25).

[4][20] [Sint C, B, A, a, b, c sex continui termini in medio interpolandi. Evadent differentiæ]

$$\begin{array}{llllll}
C & & & & & \\
 & B-C & & & & \\
B & & A-2B+C & & & \\
 & A-B & & a-3A+3B-C & & \\
A & & a-2A+B & & b-4a+6A-4B+C & \\
 & a-A & & b-3a+3A-B & & c-5b+10a-10A+5B-C. \\
a & & b-2a+A & & c-4b+6a-4A+B & \\
 & b-a & & c-3b+3a-A & & \\
b & & c-2b+a & & & \\
 & c-b & & & & \\
c & & & & &
\end{array}$$

[Terminus intermedius erit] $k-\frac{1}{8}m+\frac{9}{384}o$.[21] [ubi] $\dfrac{A+a}{2}=k$. $\dfrac{B-A-a+b}{2}=m$.

[et] $\dfrac{C-3B+2A+2a-3b+c}{2}=o$. [Fit igitur]

$$\frac{A+a}{2}+\frac{-B+A+a-b}{16}+\frac{9C-27B+18A+18a-27b+9c}{768}.$$

---

(20) Add. 3965.14: 611^r, where these rough jottings on determining the median of 4 or 6 (or more) equally spaced values by direct appeal to his general 'Newton–Bessel' interpolation theorem (see the next note) are flanked by his preliminary casting of a paragraph in the concluding paper 'Of Quadrature by Ordinates' (§3 below) and by a verbal draft revising his computed elements of the comet of 1680–1 (as published in his *Principia*, ${}_1$1687: 490–4; compare VI: 504–5, note (26)) which responds to Edmond Halley's letter to him thereon of 7 October 1695 (see *Correspondence*, **4**, 1967: 173–4). Without dwelling too much on the latter draft *hors de contexte*, we may notice that this begins: 'Attamen orbis hujus Cometæ nondum satis accurate determinavimus. Si nodus ascendens locetur in ♑ 2.2′, & angulus quem planum orbitæ continet cum plano Eclipticæ sit 60^{gr} 56′, & angulus in plano orbitæ inter nodum et aphelium Cometæ sit 9^{gr} 22′ 48″ et latus rectum Parabolæ sit partium 245 [originally '243'; compare Newton's answering letter on 17 October (*Correspondence*, **4**: 180)] qualium radius Orbis magni est 10000 et maneat tempus Perihelij Decem. 8. 0^h. 4′, P.M. Tabula sequens exhibebit loca Cometæ in hoc orbe computata...'. There then follows a preliminary, incomplete (but unified) version of the tables in *Principia*, ${}_2$1713: 459, 463.

(21) The value of the 'Newton–Bessel' central-differences formula on IV: 60 when $x=0$ (and so the point P in the figure on *ibid.*: 54 coincides with the median point O). In thus adapting his present array of terms and their differences to his earlier theorem Newton understands, of course, that $C=A^2B^2$, $B=A^3B^3$, $A=A^4B^4$, $a=A^5B^5$, $b=A^6B^6$, $c=A^7B^7$, whence

$$B-C=b^2,\quad A-B=b^3,\quad a-A=b^4,\ldots;\quad A-2B+C=c^2,\quad a-2A+B=c^3,\ldots;$$
$$a-3A+3B-C=d^2,\ldots;\quad b-4a+6A-4B+C=e^2,\quad c-4b+6a-4A+B=e^3,\ldots;$$

and so on.

[hoc est] $\dfrac{9A+9a-B-b}{16}$ [proxime, atꝙ accuratius]

$$\frac{9C+9c-75B[\overline{+b}]+450A[\overline{+a}]}{768}=\frac{3C[\overline{+c}]-25B[\overline{+b}]+150A[\overline{+a}]}{256}.^{(22)}$$

---

(22) See our comment (note (8) above) on the equivalent approximation to the 'terminus intermedius' derived by Newton from first principles in [1] preceding.

Let us finally remark that among Newton's contemporary astronomical worksheets there are found yet other variants on the preceding schemes of subtabulation by finite differences. Among his preserved computations on lunar theory, notably, he appealed at one place (ULC. Add. 3966.13: 135^{v}) to a cubic approximating polynomial $f_x \equiv A+bx+cx^2+dx^3$ in quinquisecting the four given values $E=f_{15}$, $F=f_5$, $G=f_{-5}$ and $H=f_{-15}$ by means of their first, second and third central differences $L=\Delta^1_{-5}=10b+250d$, ($P=\Delta^2_{-5}=200c+3000d$, $Q=\Delta^2_{-15}=200c-3000d$ and hence) $\frac{1}{2}(P+Q)=\mu\Delta^2_{-10}=200c$ and $R=\Delta^3_{-15}=6000d$: therefrom, by considering the array of corresponding differences δ^i_j of the values $F=f_5, f_3, f_1, f_{-1}, f_{-3}$ and $G=f_{-5}$, he was at once able inversely to calculate that $\delta^3_{-3}=48d=\frac{1}{125}R$, $\mu\delta^2_{-2}=8c=\frac{1}{50}(P+Q)$, $\delta^1_{-1}=2b+2d=\frac{1}{5}L-\frac{1}{125}R$ and $\mu f_0=A+c=\frac{1}{2}(F+G)-\frac{3}{50}(P+Q)$, but yet again betrayed no hint that he was aware of the underlying Briggsian pattern of these results, namely:
$$\delta^i_{-i}=(\delta^1_{-1})^i=(\tfrac{1}{5}\Delta^1_{-1}-(\tfrac{1}{5})^3\Delta^3_{-3}+\tfrac{14}{5}(\tfrac{1}{5})^5\Delta^5_{-5}\ldots)^i$$
$$=(\tfrac{1}{5})^i\Delta^i_{-i}-\binom{i}{1}(\tfrac{1}{5})^{i+2}\Delta^{i+2}_{-(i+2)}\ldots.$$
It was here evidently enough for him that the particular rule for quinquisection thus derived *ad hoc* should work in empirical practice.

§2. DETERMINING THE SLOPE OF THE APPARENT PATH OF THE COMET OF 1680–1 BY FITTING BOTH A QUINTIC PARABOLA AND A CONIC HYPERBOLA TO IT[(1)]

[1][(2)] [Deductâ tabula]

	Dec 21	26	30	Jan 5	13
Temp[us]	[−]359⌞4108.	[−]240⌞6817.	[−]141⌞8533.	0⌞0000.	193⌞1214.
Long[itudo]	[−]63⌞6922.	[−]40⌞4178.	[−]21⌞1912.[(3)]	0⌞0000.	17⌞1733.
Lat[itudo]	21⌞75833.	27⌞01583.	28⌞18666.	26⌞25722.	22⌞29333.

[cape $AA^4 = 63,6922.$
$A^2A^4 = 40,4178.$
$A^3A^4 = 21,1912.$
$A^4A^5 = 17,1733.$

ut et $AP = 21,75833 = a.$
$A^2P^2 = 27,01583 = {}^2a.$
$A^3P^3 = 28,18666 = {}^3a.$
$A^4P^4 = 26,25722 = {}^4a.$
$A^5P^5 = 22,29333 = {}^5a.$

et evadent][(4)]

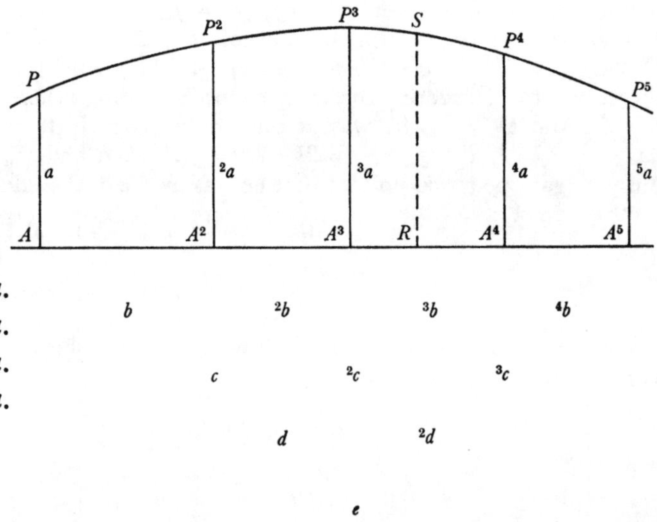

(1) When in our fourth volume we reprinted Lemma V of Book 3 of the *Principia* from its *editio princeps* ($_1$1687: 481–3) we remarked in regard to the general theorem for interpolating by divided differences which Newton presented in its 'Cas. 2' (see IV: 72–3, note (5)) that he never, in that or any subsequent edition, published a worked instance of its application to round out the observed apparent path on the celestial sphere of some past solar comet, given 3, 4 or more accurately timed sightings of its course from the Earth, in the way he went on to outline in his accompanying Lemma VI—understandably so, since (in a Newtonian scheme of things) only the true orbit of a comet round the Sun is of substantial astronomical significance. However, among the many (still all but wholly unstudied) autograph worksheets recording how deeply Halley stimulated him in the autum of 1695 to renew the cometary researches and computations which Newton had left off ten years before when he sent the manuscript of his *Principia* to press, there is preserved a calculation in which he does indeed appeal to this general theorem in fitting a quintic polynomial to the apparent course in the heavens of the 'great' comet of 1680–1, and thereafter differentiates to obtain the slope of this path at a given point. This we reproduce in [1], fleshing out its bare skeleton (and in some small part daring to rearticulate its members) in our now familiar editorial manner so as fully to bring out the subtleties of its sequence of argument. In [2] we adjoin our similarly amplified

Translation

[1][2] [Having deduced the table]

	December 21	–26	–30	January 5	–13
Time	–359·4108	–240·6817	–141·8533	0·0000	193·1214
Longitude	–63·6922	–40·4178	–21·1912[3]	0·0000	17·1733
Latitude	21·75833	27·01583	28·18666	26·25722	22·29333

[take $A_1A_4 = 63{\cdot}6922$ and also $A_1P_1 = 21{\cdot}75833 = a_1$,
$A_2A_4 = 40{\cdot}4178$ $A_2P_2 = 27{\cdot}01583 = a_2$
$A_3A_4 = 21{\cdot}1912$ $A_3P_3 = 28{\cdot}18666 = a_3$
$A_4A_5 = 17{\cdot}1733$ $A_4P_4 = 26{\cdot}25722 = a_4$
$A_5P_5 = 22{\cdot}29333 = a_5$

and there will prove to be][4]

---

rendition of what is basically a checking computation, but one in which Newton makes ingenious use of an approximating conic hyperbola to attain the same end of determining the angle in which the comet's apparent path intersects the meridian at any observed point.

(2) Add. 3965.14: 586r. Except that by implication the longitude of the sighting on December 30 is now corrected to have been in '♓ 17.37.42' (see the next note), Newton founds this calculation on the pertinent portion of the improved 'tablet' of the observed motion of the comet of 1680–1 which John Flamsteed had privately sent him on 26 September 1685 (see *Correspondence*, **2**, 1960: 422; compare also v: 525, note (3)) and which he had subsequently himself published in the first edition of his *Principia* ($_1$1687: 490), namely, in more modern style:

	True time		Longitude	Latitude
1680	December	21. 6h. 36′. 59″	305°. 7′. 38″	21°. 45′. 30″
	—	26. 5 . 20 . 44	328 . 24 . 6	27 . 0 . 57
	—	30. 8 . 10 . 26	347 . 37 . 42	28 . 11 . 12
1681	January	5. 6 . 1 . 38	8 . 49 . 10	26 . 15 . 26
	—	13. 7 . 8 . 55	25 . 59 . 34	22 . 17 . 36.

For computational convenience Newton proceeds to norm the *tempora vera* and *longitudines* of these sightings to the observation on January 5, decimalizing the ensuing entries to be parts of a degree.

(3) This is Newton's value 'correctè' (so amended to make the following divided-differences scheme accurately apply?), corresponding to an adjustment of the equivalent longitude in Flamsteed's 1685 'tablet' (see the previous note) to be '♓ 17.37.42'; a second value 'obs. 21,2014' below this in the manuscript—not there afterwards used—relates exactly to Flamsteed's observed longitude '♓ 17.37.5' as published (*Principia*, $_1$1687: 490).

(4) Following the usual contemporary convention in which the ecliptic and ordinate meridians on the celestial sphere are transposed onto a plane rectangular plot, Newton adapts the preceding table to his divided-differences scheme in the *Principia* (see iv: 72–3), making A^4R to be the longitude and RS the related latitude in each of the observed sightings of the comet—namely, on December 21 (corresponding to P), December 26 (P^2), December 30 (P^3), January 5 (P^4) and January 13 (P^5). Notice, however, that in so doing he has inverted the sense of the comet's motion which (as viewed from the Earth) was from right to left.

Diff[erentiæ] long[itudinum]	Diff[erentiæ] 1[mæ] lat[itudinum]	Quoti b(5)
$[AA^2 =]$ 23,2744	$[a-{}^2a =]$ −5,2575	$[b =]$ −2,2589197
$[A^2A^3 =]$ 19,2266	$[{}^2a-{}^3a =]$ −1,17083	$[{}^2b =]$ − .6089636
$[A^3A^4 =]$ 21,1912	$[{}^3a-{}^4a =]$ 1,92944	$[{}^3b =]$ .9104911
$[A^4A^5 =]$ 17$_{\llcorner}$1733	$[{}^4a-{}^5a =]$ 3,96389	$[{}^4b =]$ +2,3081700
	Diff 2	Quoti c(5)
$[AA^3 =]$ 42.5010	$[b-{}^2b =]$ −1.6499501	$[c =]$ −.388216
$[A^2A^4 =]$ 40.4178	$[{}^2b-{}^3b =]$ −1.5194547	$[{}^2c =]$ −.375937
$[A^3A^5 =]$ 38.3645	$[{}^3b-{}^4b =]$ −1.3976789	$[{}^3c =]$ −.3643157
	Diff 3	Quoti d(5)
$[AA^4 =]$ 63.6922	$[c-{}^2c =]$ −0.012279	$[d =]$ −0,001927866
$[A^2A^5 =]$ 57.5911	$[{}^2c-{}^3c =]$ −0.0116213	$[{}^2d =]$ −0,00201790
	Diff 4	Quotus e(5)
$[AA^5 =]$ 80,8655	$[d-{}^2d =]$ +0,000090034	+0.000011134.

[Datâ igitur longitudine quavis $A^4R = x$ erit latitudo respondens

$$RS = a+mb+nc+od+pe$$

ubi $m = x-AA^4$, $n = m\times\overline{x-A^2A^4}$, $o = n\times\overline{x-A^3A^4}$ et $p = ox$. Ut computus comprobetur, pone $x = -A^4A^5 = -17.1733$ et evadet]

$$-8.08655 = m. \quad m\times -5.75911 = n = 46.571331.$$

$$n\times -3.83645 = o = -178.6686. \quad o\times -1.71733 = p = +306,833.$$

$$[\text{adeoq}] \; mb = 18.266867. \quad nc = -18.079735.$$

$$od = +0.3444912. \quad pe = +0.0034163.$$

$$[\text{Quare}] \left.\begin{array}{rl} a = & 21.75833 \\ +mb = & 18.266867 \\ [+]nc = & [-]18.079735 \\ +od = & 0.344491 \\ +pe = & 0.003416 \end{array}\right\} = 22.2933[= {}^5a \text{ rectissime}].^{(6)}$$

[Unde(7) motus in latitudinem in S est ad motum respondentem in longitudinem ut $b+n'c+o'd+p'e$ ad 1, positis $n' = m+x-A^2A^4$, $o' = n+n'\times\overline{x-A^3A^4}$

(5) These 'quotients' are, in fact, the successively decupled divided differences

$${}^ib(=b_i) = 10({}^ia-{}^{i+1}a)/A^iA^{i+1} = 10(a_i, a_{i+1}),\ i = 1, 2, 3, 4;$$
$${}^ic(=c_i) = 10({}^ib-{}^{i+1}b)/A^iA^{i+2} = 100(a_i, a_{i+1}, a_{i+2}),\ i = 1, 2, 3;$$
$${}^id(=d_i) = 10({}^ic-{}^{i+1}c)/A^iA^{i+3} = 1000(a_i, a_{i+1}, a_{i+2}, a_{i+3}),\ i = 1, 2;$$

and $$e(=e_1) = 10(d-{}^2d)/AA^5 = 1000(a_1, a_2, a_3, a_4, a_5).$$

In sequel Newton adjusts for these multiplying factors—evidently introduced in the dubious hope of reducing the complexity of the ensuing calculation—by conversely taking $m = \frac{1}{10}AA^5$, $n/m = \frac{1}{10}A^2A^5$, $o/n = \frac{1}{10}A^3A^5$ and $p/o = \frac{1}{10}A^4A^5$. The last figures in ${}^4b$ and c are out by 2.

(6) As it should be, since in the terms of the previous note there is readily seen to be

Differences in longitude	Differences in latitude	
	1st differences	Quotients b[5]
$(A_1A_2 =)$ 23·2744	$(a_1-a_2 =)$ −5·2575	$(b_1 =)$ −2·2589197
$(A_2A_3 =)$ 19·2266	$(a_2-a_3 =)$ −1·17083	$(b_2 =)$ −0·6089636
$(A_3A_4 =)$ 21·1912	$(a_3-a_4 =)$ 1·92944	$(b_3 =)$ 0·9104911
$(A_4A_5 =)$ 17·1733	$(a_4-a_5 =)$ 3·96389	$(b_4 =)$ 2·3081700
	2nd differences	Quotients c[5]
$(A_1A_3 =)$ 42·5010	$(b_1-b_2 =)$ −1·6499501	$(c_1 =)$ −0·388216
$(A_2A_4 =)$ 40·4178	$(b_2-b_3 =)$ −1·5194547	$(c_2 =)$ −0·375937
$(A_3A_5 =)$ 38·3645	$(b_3-b_4 =)$ −1·3976789	$(c_3 =)$ −0·3643157
	3rd differences	Quotients d[5]
$(A_1A_4 =)$ 63·6922	$(c_1-c_2 =)$ −0·012279	$(d_1 =)$ −0·001927866
$(A_2A_5 =)$ 57·5911	$(c_2-c_3 =)$ −0·0116213	$(d_2 =)$ −0·00201790
	4th difference	Quotient e_1[5]
$(A_1A_5 =)$ 80·8655	$(d_1-d_2 =)$ 0·000090034	0·000011134.

[Correspondent, therefore, to any given longitude $A_4R = x$ will be the latitude $RS = a+mb_1+nc_1+od_1+pe_1$, where $m = x-A_1A_4$, $n = m(x-A_2A_4)$, $o = n(x-A_3A_4)$ and $p = ox$. To test these figures put $x = -A_4A_5 = -17{\cdot}1733$ and there will come]

$$m = -8{\cdot}08655, \quad n = m.(-5{\cdot}75911) = 46{\cdot}571331,$$

$$o = n.(-3{\cdot}83645) = -178{\cdot}6686, \quad p = o.(-1{\cdot}71733) = 306{\cdot}833$$

and hence

$$mb_1 = 18{\cdot}266867,\ nc_1 = -18{\cdot}079735,\ od_1 = 0{\cdot}3444912 \text{ and } pe_1 = 0{\cdot}0034163.$$

$$\text{Consequently}\ \left.\begin{aligned} a_1 &= 21{\cdot}75833 \\ +mb_1 &= 18{\cdot}266867 \\ +nc_1 &= -18{\cdot}079735 \\ +od_1 &= 0{\cdot}344491 \\ +pe_1 &= 0{\cdot}003416 \end{aligned}\right\} = 22{\cdot}2933 = a_5 \text{ exactly right.}^{(6)}$$

[It follows from this[7] that the motion in latitude at S is to the corresponding motion in longitude as $b_1+n'c_1+o'd_1+p'e_1$ to 1 on setting $n' = m+x-A_2A_4$,

---

$$mb = AA^5.(a_1, a_2),\quad nc = AA^5\times A^2A^5.(a_1, a_2, a_3),\quad od = AA^5\times A^2A^5\times A^3A^5.(a_1, a_2, a_3, a_4)$$
and $pe = AA^5\times A^2A^5\times A^3A^5\times A^4A^5.(a_1, a_2, a_3, a_4, a_5)$,

while

$$a_1+AA^5\{(a_1, a_2)+A^2A^5[(a_1, a_2, a_3)+A^3A^5((a_1, a_2, a_3, a_4)+A^4A^5.(a_1, a_2, a_3, a_4, a_5))]\} = a_4.$$

Notice, however, that the product od_1 is in fact 0.34444912.

(7) By differentiating the preceding expression for RS with respect to the abscissa $A^4R = x$, whence m' (or dm/dx) $= 1$, n' (or dn/dx) $= m+m'(x-A^2A^4)$, o' (or do/dx) $= n+n'(x-A^3A^4)$ and p' (or dp/dx) $= o+o'x$.

et $p' = o+o'x$. Ubi igitur $x = A^3A^4$, fit $m = -AA^3$, $n = m \times -A^2A^3$ et $o = 0$, adeoq; $n' = m - A^2A^3$, $o' = n$ et $p' = o' \times A^3A^4$; unde] Ut est 10^{gr}(8) ad

$$-b+c\times\overline{AA^3+A^2A^3}-d\times AA^3\times A^2A^3+e\times AA^3\times A^2A^3\times A^4A^3$$

ita motus in longitudinem [in P^3] ad motum in latitudinem. [Hic est]

$$\overline{AA^3+A^2A^3} \text{ in } c = 6_{\llcorner}17276\times 0_{\llcorner}388216 = 2_{\llcorner}3963642.$$

$$AA^3\times A^2A^3 = 4_{\llcorner}2501\times 1_{\llcorner}92266 = 8_{\llcorner}0714973.$$

[atq; adeo] $$AA^3\times A^2A^3\times d = 0_{\llcorner}015560765.$$

[et] $AA^3\times A^2A^3\times A^4A^3 = 17_{\llcorner}10447$. hoc in $e = 0_{\llcorner}0001904$. [itaq; evadet]

$$-b+c\times\overline{AA^3+A^2A^3}-d\times AA^3\times A^2A^3+e\times AA^3\times A^2A^3\times A^4A^3$$

$$= +2_{\llcorner}2589197-2_{\llcorner}396364+0_{\llcorner}0155608+0_{\llcorner}0001904 = -0_{\llcorner}1216933.$$

Ergo motui $10^{[gr]}$ in longit[udinem] congruit motus $-0^{[gr]}{}_{\llcorner}1216933$ [in latitudinem].(9)

---

[2](10) Circulus(11) jungens locum cometæ Dec 30 cum locis Dec 21, 26, Jan 5 [et] 13 continet angulos cum meridiano eclipticæ per locum cometæ Dec 30 transeuntis hosce(12) $90^{[gr]}{}_{\llcorner}3012$. $90^{[gr]}{}_{\llcorner}59429$. $90^{[gr]}{}_{\llcorner}88793$ [et] $90^{[gr]}{}_{\llcorner}8291$ [respective] orientem et Aquilonem versus.(13) Ergo constituantur

$0_{\llcorner}0000$.	$23_{\llcorner}2744$.	$42_{\llcorner}501$.	$63_{\llcorner}6922$.	$80_{\llcorner}8655$	Bases.
$0_{\llcorner}0000$.	$0_{\llcorner}2931$.	*.	$0_{\llcorner}5867$.	$0_{\llcorner}5279$	Ordinatæ

---

(8) Newton has to compensate for his earlier decupling (see note (5) above) of the various 'quotients' b, c, d, e. An error in entering the product $AA^3\times A^2A^3$ botches the sequel.

(9) It would follow that the slope of the (quintic) parabola $\widehat{PP^2\ldots P^5}$ at P^3 is

$$\cot^{-1}(-0{\cdot}012169\ldots) \approx 90{\cdot}6967°.$$

Here, however, there lurks a subtlety which Newton fails to notice: in the apparent path of the comet as actually observed on the celestial sphere the longitudinal motion at any point is only '10^{gr} in cosinum latitudinis' ($10°\times$cosine of the latitude) and hence the true slope of the comet's course, as viewed, to the meridian through P^3 is

$$\cot^{-1}(-0{\cdot}012169\ldots/\cos 28{\cdot}18666\ldots) \approx 90{\cdot}79°.$$

(10) Add. 3965.11: 164r. While this variant computation of the inclination of the apparent path of the comet of 1680–1 to the meridian at its sighting on December 30 is set (by itself) on a separate manuscript sheet, its close affinity with the preceding will be evident; in particular, it departs from the same observational data—those furnished by Flamsteed in 1685, but with the longitude on December 30 again corrected to be '♓ 17.37.42' (see note (2) above)—as before.

(11) That (in the celestial sphere) whose plane passes through the eye of the terrestrial observer.

$o' = n+n'(x-A_3A_4)$ and $p' = o+o'x$. When $x = A_3A_4$, therefore, there comes to be $m = -A_1A_3$, $n = m.(-A_2A_3)$ and $o = 0$, and thus $n' = m-A_2A_3$, $o' = n$ and $p' = o'.A_3A_4$; in consequence] as $10°$(8) is to

$$-b_1+c_1(A_1A_3+A_2A_3)-d_1.A_1A_3\times A_2A_3+e_1.A_1A_3\times A_2A_3\times A_4A_3,$$

so is the motion in longitude [at P_3] to the motion in latitude. [Here]

$$c_1(A_1A_3+A_2A_3) = 0{\cdot}388216\times 6{\cdot}17276 = 2{\cdot}3963642,$$

$$A_1A_3\times A_2A_3 = 4{\cdot}2501\times 1{\cdot}92266 = 8{\cdot}0714973$$

[and therefore] $d_1.A_1A_3\times A_2A_3 = 0{\cdot}015560765$, [while]

$$A_1A_3\times A_2A_3\times A_4A_3 = 17{\cdot}10447$$

and e_1 times this $= 0{\cdot}0001904$; [accordingly there will prove to be]

$$-b_1+c_1(A_1A_3+A_2A_3)-d_1.A_1A_3\times A_2A_3+e_1.A_1A_3\times A_2A_3\times A_4A_3$$

$$= +2{\cdot}2589197-2{\cdot}396364+0{\cdot}0155608+0{\cdot}0001904 = -0{\cdot}1216933.$$

Therefore to a motion of $10°$ in longitude there is correspondent a motion of $-0{\cdot}1216933°$ [in latitude].(9)

---

[2](10) The (great) circle(11) joining the place of the comet on December 30 with its places on December 21 and 26 and on January 5 and 13 contains the following angles(12) with the meridian to the ecliptic passing through the place of the comet on December 30: 90·3012°, 90·59429°, 90·88793° and 90·8291° respectively (going) north-eastwards.(13) Therefore set up

$$\begin{cases} 0{\cdot}0000, & 23{\cdot}2744, & 42{\cdot}501, & 63{\cdot}6922, & 80{\cdot}8655 & \text{as bases} \\ 0{\cdot}0000, & 0{\cdot}2931, & [?], & 0{\cdot}5867, & 0{\cdot}5279 & \text{as ordinates} \end{cases}$$

---

(12) In the terms of Theorem 3 in the 'Speciall resolution' of 'sphæricall triangles' on I: 469 these are the angles

$$d = \tan^{-1}[\cot\tfrac{1}{2}b\times\sin\tfrac{1}{2}(e-g)/\sin\tfrac{1}{2}(e+g)]+\tan^{-1}[\cot\tfrac{1}{2}b\times\cos\tfrac{1}{2}(e-g)/\cos\tfrac{1}{2}(e+g)],$$

where g (= 61.81334°) is the colatitude—the complement of the latitude measuring the angular distance from the North celestial pole—of the comet's observed position on December 30, the second sides e (= 68·24167°, 62·98417°, 63·74278° and 67·70667° respectively) are the colatitudes of the sightings on December 21, 26 and January 5, 13, while the included (polar) angles b (= 42·5010°, 19·2266°, 21·1912° and 38·3645° respectively) are the related differences in the comet's observed longitudes on these dates from that —again 'corrected' from Flamsteed's original value (see note (2) above)—of the sighting on December 30. Our computations of the angles of slope to the meridian at these points do not exactly agree with the values for these which Newton proceeds to list, but we will not insist upon the discrepancies since he will in sequel commit far more egregious blunders in his calculation (see notes (17) and (18) below).

(13) Since the comet of 1680–1 pursued a direct course which also circled the moving Earth, reaching its ascending node soon after attaining perihelion at noon on 8 December, and thereafter (as seen by an observer in the northern hemisphere) proceeding ever upwards and to the east from the ecliptic; see Newton's own informative depiction of the comet's true path in Book 3 of his *Principia* (at page 496 of the 1687 *editio princeps*) and also VI: 505, note (26).

respectivè.(14) [Ubi basis est x, pone ordinatam respondentem esse] $= ax + \dfrac{bx \text{ in } x - 23{\llcorner}2744}{c+x}$. [eo quod] Ut basis $23{\llcorner}2744$ ad ordin[atam] $0{\llcorner}2931$ ita bases [$0{\llcorner}0000$: $23{\llcorner}2744$:] $42{\llcorner}5010$: $63{\llcorner}6922$: $80{\llcorner}8655$ ad ordinatas [$0{\llcorner}0000$: $0{\llcorner}2931$:] $0{\llcorner}53522$: $0{\llcorner}80209$: $1{\llcorner}018356$. [et] His subductis de prioribus ordinatis relinquuntur ordinatæ $0{\llcorner}0$. $0{\llcorner}0$. $*-0{\llcorner}53522$. $-0{\llcorner}21539$. $-0{\llcorner}49046$.(15) [Hic selecta est coefficiens] $a = \dfrac{0{\llcorner}2931}{23{\llcorner}2744} = 0{\llcorner}012593$. [Ad coefficientes b et c insuper determinandas cape](16)

Cas. 1. $x = 63{\llcorner}6922$. [Prodit] $x \times x - 23{\llcorner}274[4] = 2574{\llcorner}2986$. [adeoq$\mathfrak{z}$]

$$\frac{2574{\llcorner}3b}{c+63{\llcorner}6922} = -0{\llcorner}21539.$$

[unde] $$\frac{b}{c+63{\llcorner}6922} = \frac{-0{\llcorner}21539}{2574{\llcorner}3} = -0{\llcorner}000083669.$$

Cas. 2. $x = 80{\llcorner}8655$. [Fit] $x \times x - 23{\llcorner}274[4] = 4657{\llcorner}133$. [indeq$\mathfrak{z}$]

$$\frac{4657{\llcorner}133b}{c+80{\llcorner}8655} = -0{\llcorner}49046$$

[sive] $$\frac{b}{c+80{\llcorner}8655} = \frac{-0{\llcorner}49046}{4657{\llcorner}133} = -0{\llcorner}000105313.$$

[Quare] $c+63{\llcorner}6922 . c+80{\llcorner}8655 :: -0{\llcorner}83669 . -1{\llcorner}05313$.(17) [adeoq$\mathfrak{z}$]

$c+63{\llcorner}6922 . 17{\llcorner}1733 :: -0{\llcorner}83669 . +0{\llcorner}21644$.(18)

[unde] $c+63{\llcorner}6922 = -66{\llcorner}3867$. [itaq$\mathfrak{z}$] $c = -130{\llcorner}0789$. [Quo substituto in Cas. 1 evadit] $-0{\llcorner}000083669$ in $-66{\llcorner}3867 = b = +0{\llcorner}00555451$.

Jam sit $x = 42{\llcorner}501$. et erit

$$\frac{x \text{ in } x - 23{\llcorner}2744 \text{ in } b}{c+x} = \frac{42{\llcorner}501 \times 19{\llcorner}2266 \times 0{\llcorner}00555451}{-87{\llcorner}5779} = -0{\llcorner}05183.$$

[Igitur] $90{\llcorner}3012 + 0{\llcorner}53522 - 0{\llcorner}05183 = 90^{[gr]}{\llcorner}78459$ = angulo quem via cometæ continet cum meridiano in loco cometæ Dec 30.(19)

---

(14) Newton lists the differences in the comet's longitude at its sightings on December 26, 30 and January 5, 13 from that observed on December 21, along with the related differences between the angles contained with the meridian at the comet's place on December 30 by the great circles intersecting there and also passing through its sighted positions at these other times, indicating that the resulting pairs (23·2744, 0·2931), (63·6922, 0·5867) and (80·8655, 0.5279) are, along with (0, 0), as points (x, y) on a Cartesian grid in which the 'bases' x are the abscissas on whose end-points $(x, 0)$ stand the ordinates y. It will in sequel be his aim to fit an appropriate approximating curve $y = f_x$ through these four plotted points and so thereby compute the angle $(f_{42.501} + 90.3012)°$ contained with the meridian there by the great circle touching the comet's path at its observed place on December 30—which is, of course, the inclination of the path itself to the meridian at that point.

(15) Since the approximating curve $y = f_x$ has to pass both through the origin (0, 0) and the point (23·2744, 0·2931), it is clearly simplest to set $f_x \equiv x . g_x$ where $g_{23\cdot 2744} = a = 0\cdot 2931/23\cdot 2744$

respectively.[14] [Where the base is x, put the corresponding ordinate] equal to $ax+bx(x-23\cdot2744)/(c+x)$ [inasmuch] as the base 23·2744 is to the ordinate 0·2931 as the bases 0·0000, 23·2744, 42·5010, 63·6922, 80·8655 to the ordinates 0·0000, 0·2931, 0·53522, 0·80209, 1·018356, and when these are taken away from the previous ordinates there remain 0·0000, 0·0000, [?]−0·53522, −0·21539, −0·49046.[15] [Here we have chosen the coefficient]

$$a = 0\cdot2931/23\cdot2744 = 0\cdot012593.$$

[To determine in addition the coefficients b and c take][16]

Case 1. $x = 63\cdot6922$. [There ensues] $x(x-23\cdot2744) = 2574\cdot2986$ [and therefore] $2574\cdot3b/(c+63\cdot6922) = -0\cdot21539$, [whence]

$$b/(c+63\cdot6922) = -0\cdot21539/2574\cdot3 = -0\cdot000083669.$$

Case 2. $x = 80\cdot8655$. [There comes] $x(x-23\cdot2744) = 4657\cdot133$ [and hence] $4657\cdot133b/(c+80\cdot8655) = -0\cdot49046$, [that is]

$$b/(c+80\cdot8655) = -0\cdot49046/4657\cdot133 = -0\cdot000105313.$$

[In consequence] $(c+63\cdot6922):(c+80\cdot8655) = -0\cdot83669:-1\cdot05313$[17] [and therefore] $(c+63\cdot6922):17\cdot1733 = -0\cdot83669:+0\cdot21644$,[18] [whence] $c+63\cdot6922 = -66.3867$ [and so] $c = -130\cdot0789$. [And when this is substituted in Case 1 there comes to be]

$$b = -0\cdot000083669 \times -66\cdot3867 = +0\cdot00555451.$$

Now let $x = 42\cdot501$ and there will be

$$\frac{x(x-23\cdot2744)\,b}{c+x} = \frac{42\cdot501 \times 19\cdot2266 \times 0\cdot00555451}{-87\cdot5779} = -0\cdot05183.$$

[Consequently] (90·3012+0·53522−0·05183 or) 90·78459° is equal to the angle which the path of the comet contains with the meridian at the comet's place on December 30.[19]

---

and hence in general $g_x \equiv a+(x-23\cdot2744).h_x$. Because the other two given points (63·6922, 0·5867) and (80·8655, 0·5279) can determine only two more constants, b and c, in the expression for f_x and since $f_x - ax \equiv x(x-23\cdot2744).h_x$ is roughly quadratic in x (as Newton's calculation shows), we do best to choose $h_x \equiv b/(c+x)$, so defining the approximating curve to be the conic hyperbola which Newton specifies.

(16) Taking the defining equation of the fitted hyperbola in its trivially re-ordered form $y-ax = x(x-23\cdot2744)\,b/(c+x)$, Newton passes to evaluate its two cases $x = 63\cdot6922$ and $x = 80\cdot8655$.

(17) A reduction of the preceding equations which is unorthodox, to say the least! The blunder is perpetuated in the remaining calculation and further compounded by a mistake in sign in the next line in carrying through the ensuing simplification *dividendo*.

(18) Another careless slip. Read '$-0_{\llcorner}21644$' correctly, whence in sequel there should follow—somewhat differently!—the evaluation '$c+63_{\llcorner}6922 = +66_{\llcorner}3867$' and thence '$c = 2_{\llcorner}6945$'.

(19) Undeterred by his double miscalculation *en route* (on which see the two previous notes), Newton contrives with his customary skill to attain a final result which is (compare note (9) above) all but exactly true.

§3. THE TRACT 'OF QUADRATURE BY ORDINATES.'(1)

From the incomplete original(2) in the University Library, Cambridge

Of Quadrature by Ordinates

If upon the base A at equal distances be erected ordinates(3) to any Curve AK, BL, CM &c the Curve may by ye Ordinates be squared *quamproxime*(4) as follows.

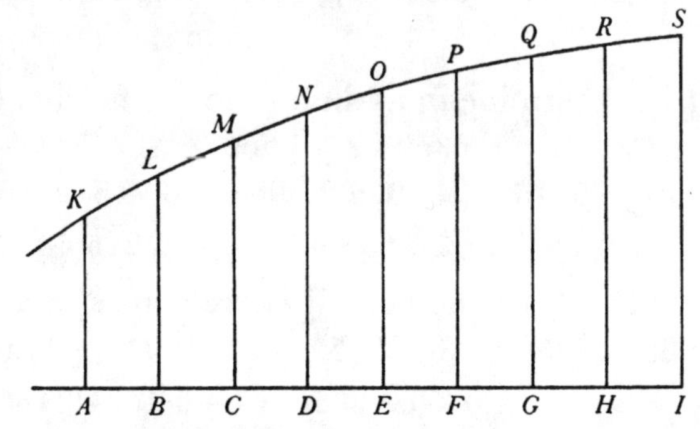

Cas. 1. Si dentur duæ tantũ ordinatæ AK, BL fac aream

$$AKLB = \frac{AK+BL}{2}AB^{(5)}$$

Cas. 2. Si dentur tres AK, BL, CM, dic $\frac{AK+CM}{2}AC = \square AM$, et rursus $\frac{AK+BL}{4} + \frac{BL+CM}{4}$ in $AC = AK+2BL+CM$ in $\frac{1}{4}AC = \square AM$ (per Cas. 1) et errorem solutionis prioris esse ad errorem solutionis posterioris ut AC^q ad AB^q

(1) Though the period when Newton penned its text can (see note (2) of the Appendix) be narrowed with confidence to early October 1695, we are unable, yet once more, to furnish any sure insight into what stimulated him to compose this (our final) piece on approximately squaring the areas beneath given curves in terms of appropriate sets of their equally spaced ordinates, presupposing that their arcs are adequately approached by the parabola which is drawn through the latter's end-points. Manifestly, however, the simpler instances of such precepts for approximate area-quadrature have some practical value, and in note (24) below we cite a hitherto unpublished proposition from his contemporary, amplified lunar theory where Newton applies his first three Cases following—the second in its uncorrected, non-Simpsonian form—to 'find the annual variation of the moon's mean motion', given two, three or four successive values of the 'hourly' area traversed at the octants. The ingenious short cut which Newton here adopts, bypassing the rigours of a full computation of the approximating parabolic areas *ab initio* in each case by proportionately adjusting the 'errors' of the rules deduced in earlier ones, is, it will be evident from the sequel, neither wholly exact nor in working his Case 5 (as we shall specify in note (12) below) properly pursued by him. The correct precepts for Cases 5 and 6 derived by direct computation from the pertinent approximating parabola were, along with those of Cases 2 and 4 and also in the higher-order instances where 6, 8, 10 or 11 equally distant ordinates are given, subsequently (see IV: 73–4, note (6)) deduced by Roger Cotes after reading the concluding Scholium to Newton's 1711 'Methodus Differentialis', where the 'pulcherrima & utilissima Regula' (as it appeared to Cotes) which is the present Case 3 was first publicly enunciated, devoid of any demonstration. (We will take this point up again in our final, eighth volume when we come to analyse the 'Methodus'.) These more lavish Cotesian formulas will be found listed in the *Postscriptum* (page 33) to his posthumously edited commentary 'De Methodo Differentiali Newtoniana' (*Opera Miscellanea* [appended to his *Harmonia Mensurarum* (Cambridge, 1722)]: 23–33).

(2) Add. 3964.4: 21r; two minor preliminary partial drafts on Add. 3965.14: 611r/612v

Translation

OF QUADRATURE BY ORDINATES.

If upon the base A at equal distances be erected ordinates(3) to any Curve AK, BL, CM &c the Curve may by y^e Ordinates be squared *quamproxime*(4) as follows.

Case 1. If there be given but two ordinates AK and BL, make the area $(AKLB) = \frac{1}{2}(AK+BL)\,AB$.(5)

Case 2. If there be given three AK, BL and CM, say that

$$\tfrac{1}{2}(AK+CM)\,AC = \square(AM)$$

and again, by Case 1,

$$\tfrac{1}{2}(\tfrac{1}{2}(AK+BL)+\tfrac{1}{2}(BL+CM))\,AC = \tfrac{1}{4}(AK+2BL+CM)\,AC = \square(AM),$$

and that the error in the former solution is to the error in the latter as AC^2 to

---

are reproduced in the Appendix which follows. The concluding lines, evidently entered on an accompanying sheet which has now disappeared, are here conjecturally restored. As his original cataloguing slip (now loose in ULC. Add. 4005) establishes, Samuel Horsley in 1777 found this 'fragment in English [*sic*] concerning quadratures by equidistant ordinates' in 'N° 4 of [Thomas Pellet's] selected papers' (compare I: xxi), along with the 'fragment in Latin concerning Interpolations' (now Add. 3964.5: 3^r–4^v) which is printed on IV: 22–34 and 'Two Problems' which are too vaguely titled to be positively identifiable. His myopic accompanying judgement that 'None of this [is] to be publishd unless the entire pieces should be found. S. Horsley. Octr 23^d 1777' has already been quoted in IV: 22, note (2).

(3) Understand that these stand at right angles to the base. In the manuscript figure (as in that to which the preliminary version of 'Cas. 4' reproduced in section [2] of the Appendix is keyed) the ordinate end-points K, L, M, N, ... were originally marked I, K, L, M, ... respectively. Only at a late stage did Newton, evidently appreciating the confusion which ensues from the duplicate designation 'I' (both of the terminus of the base and of the end-point of the primary ordinate at A), come to make the transition, and in doing so he omitted correspondingly to amend his verbal text to suit; in our edited version of this, however, we silently introduce the permutations $I \to K \to L \to M \to N \to O \to P \to Q \to R \to S$ which are thereby necessitated.

(4) More precisely, it is—as he will later specify (see note (14) below)—Newton's intention that the approximation shall in each of the ensuing cases be that which, for the stated number of equidistant ordinates, exactly measures the area under the parabola drawn through their end-points; indeed, he has rejected from this revised text an initial version of 'Cas. 4' (reproduced in section [2] of the Appendix) where this is not so. As we have already observed, however, his preferred procedure of refining successive 'errors' by an appropriate factor will not—even when correctly applied—in general attain this perfection.

(5) The simple trapezoidal rule. In the pattern of what follows, on taking the mid-point of the base $AB = 1$ to be the origin of (mutually perpendicular) Cartesian coordinates of its general point (x, y), Newton's assumption here is that $\widehat{KL}$ is adequately approximated by the straight line $y = f_x \equiv a+bx$ through the end-points of its ordinates $AK = f_{-\frac{1}{2}}$ and $BL = f_{\frac{1}{2}}$, whence $(AL) \approx \int_{-\frac{1}{2}}^{\frac{1}{2}} f_x . dx = a = \frac{1}{2}(f_{-\frac{1}{2}}+f_{\frac{1}{2}})$.

seu 4 ad 1[(6)] adeoq; solutionum differentiam $\frac{AK-2BL+CM}{4}AC$ esse ad errorem posterioris ut 3 ad 1, et error posterioris erit $\frac{AK-2BL+CM}{12}AC$. Aufer hunc errorem et solutio posterior evadet

$$\frac{AK+4BL+CM}{6}AC = \Box AM^{(7)}.\text{ Solutio quæsita.}^{(8)}$$

Cas. 3. Si dentur 4 Ordinatæ *AK*, *BL*, *CM*, *DN*: dic $\frac{AK+DN}{2}AD = \Box AN$.

Item $$\frac{AK+BL}{6}+\frac{BL+CM}{6}+\frac{CM+DN}{6}\text{ in } AD$$

(id est $\frac{AK+2BL+2CM+DN}{6}AD$) $= \Box AN$. Et solutionū errores erunt ut AD^q ad AB^q seu 9 ad 1 adeoq; errorum differentia (quæ est solutionū differentia $\frac{2AK-2BL-2CM+2DN}{6}AD$) erit ad errorem posterioris ut 8 ad 1. Aufer hunc errorem et posterior manebit

$$\frac{AK+3BL+3CM+DN}{8}AD = \Box AN.^{(9)}$$

(6) A superficially plausible—and here correct—premiss which needs more concrete justification (see the next note). Newton in sequel seems to adopt the argument *per analogiam* that, where $n+1$ equidistant ordinates are given, the corresponding ratio of the errors is $n^2:1$, but this will fail him in 'Cas. 5' (see note (12) below).

(7) This approach is equivalent to fitting a parabola $y = f_x \equiv a+bx+cx^2$ (where the mid-point *B* of the abscissa *ABC* is the origin) to pass through the end-points of the ordinates $AK = f_{-1} = a-b+c$, $BL = f_0 = a$ and $CM = f_1 = a+b+c$; for then the errors

$$\epsilon_1 = \Box_1 AM-(AM) \quad\text{and}\quad \epsilon_2 = \Box_2 AM-(AM)$$

in the deviation of

$$\Box_1 AM = \tfrac{1}{2}.2(AK+CM) = 2a+2c \quad\text{and}\quad \Box_2 AM = \tfrac{1}{2}(AK+BL)+\tfrac{1}{2}(BL+CM) = 2a+c$$

from the true parabolic area $(AM) = \int_{-1}^{1} f_x.dx = 2a+\frac{2}{3}c$ are indeed in the ratio of $(\frac{4}{3}c/\frac{1}{3}c =)$ 4 to 1.

(8) Though it is here, in this hitherto unpublished paper, presented for the first time in its explicit form—one which we may more generally set to be $\int_0^h f_x.dx \approx \frac{1}{6}h(f_0+4f_{\frac{1}{2}h}+f_h)$, this eminently practical formula of approximation to an area cut off between parallel ordinates (and *mutatis mutandis* to a volume cut off between parallel plane surfaces) is rightly named in our modern textbooks after Thomas Simpson, who publicized it in his *Mathematical Dissertations* (London, 1743): 109–13: 'Of the Areas of Curves &c. by Approximation'. It is less well known that particular cases of 'Simpson's rule' had earlier been given by Bonaventura Cavalieri in Problem 80 of his *Una Centuria di varii Problemi nella Prattica Astrologica* (Bologna, 1639): 446 in measuring the volume of a symmetrical wine-cask whose general circular cross-

AB^2 or 4 to 1,(6) and hence the difference $\frac{1}{4}(AK-2BL+CM)\,AC$ of the solutions is to the error in the latter as 3 to 1, and the error in the latter will be

$$\tfrac{1}{12}(AK-2BL+CM)\,AC.$$

Take away this error and the latter solution will come to be

$$\tfrac{1}{6}(AK+4BL+CM)\,AC = \square(AM),^{(7)}$$

the solution required.(8)

Case 3. If there be given four ordinates AK, BL, CM and DN, say that $\frac{1}{2}(AK+DN)\,AD = \square(AN)$; likewise, that

$$\tfrac{1}{3}(\tfrac{1}{2}(AK+BL)+\tfrac{1}{2}(BL+CM)+\tfrac{1}{2}(CM+DN))\,AD,$$

that is, $\frac{1}{6}(AK+2BL+2CM+DN)\,AD = \square(AN)$. The errors in the solutions will be as AD^2 to AB^2 or 9 to 1, and hence the difference in the errors—which is the difference $\frac{1}{6}(2AK-2BL-2CM+2DN)\,AD$ in the solutions—will be to the error in the latter as 8 to 1. Take away this error and the latter will remain as $\frac{1}{8}(AK+3BL+3CM+DN)\,AD = \square(AN)$.(9)

---

section is f_x and for which $f_0 = f_h$; and also by James Gregory in his 'Methodus Componendi Tabulas Tangentium & Secantium Artificialium Ex Tabulis Tangentium & Secantium Naturalium Exactissimè & minimo cum Labore' [= *Exercitationes Geometricæ* (London, 1668): 25–6] in evaluating $\log \sec h = \int_0^h f_x . dx$ where $f_x = \tan x$ and so $f_0 = 0$. (Gregory also went on to give a much more complicated cubic approximation to $\int_0^h f_x . dx$ in which, it may be shown, he departed from $f_x \approx cx^2 + dx^3$, but we will not here go into its detail.) With Gregory's *Exercitationes* Newton was well familiar: his (unmarked) library copy of it is now in Trinity College, Cambridge. NQ. 9.48. See also Georg Heinrich, 'Notiz zur Geschichte der Simpsonschen Regel', *Bibliotheca Mathematica*, $_3$**1**, 1900–1: 90–2; and D. T. Whiteside, 'Patterns of Mathematical Thought in the later Seventeenth Century' (*Archive for History of Exact Sciences*, **1**, 1961: 179–388): 248–9.

(9) Where, much as before but now taking the mid-point of the abscissa AD as origin, we equivalently set the cubic parabola $y = f_x \equiv a+bx+cx^2+dx^3$ to pass through the end-points $AK = f_{-\frac{3}{2}}$, $BL = f_{-\frac{1}{2}}$, $CM = f_{\frac{1}{2}}$ and $DN = f_{\frac{3}{2}}$, it is readily verified that the errors

$$\epsilon_1 = \square_1 AN - (AN) \quad \text{and} \quad \epsilon_2 = \square_2 AN - (AN)$$

in the differences of $\square_1 AN = \frac{1}{2}.3(AK+DN) = 3a+\frac{27}{4}c$

and $\square_2 AN = \frac{1}{2}(AK+BL)+\frac{1}{2}(BL+CM)+\frac{1}{2}(CM+DN) = 3a+\frac{11}{4}c$

from the area $(AN) = \int_{-3/2}^{3/2} f_x . dx = 3a+\frac{9}{4}c$ are indeed in Newton's asserted ratio of $(\frac{9}{2}c/\frac{1}{2}c =)$ 9 to 1, so confirming the accuracy of the resulting approximation, namely $\frac{1}{8}(9\square_2 AN - \square_1 AN)$. We have already indicated in IV: 73, note (6) that this formula, as elegant as it is easy to apply, for approaching the area under a curve in terms of four equidistant ordinates was afterwards set by Newton in the terminal Scholium of his 'Methodus Differentialis', but without any hint as to how he derived it.

Cas. 4.(10) Si dentur 5 Ordinatæ, dic (per Cas. 2)

$$\frac{AK+4CM+EO}{6}AE = \square AO.$$

Item $\frac{AK+4BL+CM}{12}+\frac{CM+4DN+EO}{12}$ in $AE = \square AO$ et errores esse ut AE^q ad AB^q seu 16 ad 1[,] et cum errorum differentia sit

$$\frac{AK-4BL+6CM-4DN+EO}{12}AE.$$

error minoris erit $\frac{AK-4BL+6CM-4DN+EO}{180}AE$ quem aufer et manebit

$$\frac{7AK+32BL+12CM+32DN+7EO}{90}AE = \square AO.^{(11)}$$

Cas. 5. Eodem modo si dentur 7 Ordinatæ, fiet

$$\frac{17AK+54BL+51CM+36DN+51EO+54FP+17GQ}{280}AG = \square AQ.^{(12)}$$

Cas. 6. Et si dentur 9, fiet

$$\frac{217AK+1024BL+352CM+1024DN+436EO+1024FP+352GQ+1024HR+217IS}{5670}AI = \square AS.^{(13)}$$

(10) Newton initially went on in the manuscript to enter the preliminary version of this case which is reproduced (from Add. 3965.14: 612^v) in section [2] of the following Appendix, namely—on making (see note (3) above) appropriate transpositions in its designation of the points of the curve—'Si dentur 5 Ordinatæ, dic $\frac{AK+EO}{2}AE = \square AO$. Item

$$\frac{AK+2BL+2CM+2DN+EO}{8}[AE] = \square AO$$

et errores esse ut AE^q ad AB^q seu 16 ad 1 adeoq; cum errorum differentia [sit...]'. As he must have appreciated in here replacing it, the result '$\frac{3AK+8BL+8CM+8DN+3EO}{30}AE = \square AO$' does not (compare note (8) of the Appendix) exactly measure the area under the quartic parabola through the ordinate end-points K, L, M, N and O.

(11) Here again, on taking the mid-point C of the base $AE = 5$ to be origin of coordinates, since the quartic parabola $y = f_x \equiv a+bx+cx^2+dx^3+ex^4$ set to pass through the end-points of the ordinates $AK = f_{-2}$, $BL = f_{-1}$, $CM = f_0$, $DN = f_1$ and $EO = f_2$ fixes the errors

$$\epsilon_1 = \square_1 AO-(AO) \quad \text{and} \quad \epsilon_2 = \square_2 AO-(AO)$$

in the deviations of $\square_1 AO = \frac{1}{6}.4(AK+4CM+EO) = 4a+\frac{16}{3}c+\frac{64}{3}e$

and $\square_2 AO = \frac{1}{6}(AK+4BL+CM)+\frac{1}{6}(CM+4DN+EO) = 4a+\frac{16}{3}c+\frac{40}{3}e$

from the area $(AO) = \int_{-2}^{2} f_x.dx = 4a+\frac{16}{3}c+\frac{64}{5}e$ to be exactly in Newton's prescribed ratio of $(\frac{128}{15}e/\frac{8}{15}e =)$ 16 to 1, this refined approximation $\frac{1}{15}(16\square_2 AO-\square_1 AO)$ is exact.

Case 4.(10) If there be given five ordinates, say (by Case 2) that

$$\tfrac{1}{6}(AK+4CM+EO)\,AE = \square(AO),$$

likewise that $\tfrac{1}{2}(\tfrac{1}{6}(AK+4BL+CM)+\tfrac{1}{6}(CM+4DN+EO))\,AE = \square(AO)$, and that the errors are as AE^2 to AB^2 or 16 to 1; then, since the difference in the errors is $\tfrac{1}{12}(AK-4BL+6CM-4DN+EO)\,AE$, the error in the lesser will be $\tfrac{1}{180}(AK-4BL+6CM-4DN+EO)\,AE$, and when this is taken away there will remain

$$\tfrac{1}{90}(7AK+32BL+12CM+32DN+7EO)\,AE = \square(AO).^{(11)}$$

Case 5. In the same way, if there be seven ordinates there will come to be

$$\tfrac{1}{280}(17AK+54BL+51CM+36DN+51EO+54FP+17GQ)\,AG = \square(AQ).^{(12)}$$

Case 6. While if there be given nine there will come

$$\tfrac{1}{5670}(217AK+1024BL+352CM+1024DN+436EO$$
$$+1024FP+352GQ+1024HR+217IS)\,AI = \square(AS).^{(13)}$$

---

(12) Because this is $\tfrac{1}{35}(36\square_2 AQ-\square_1 AQ)$ where

$$\square_1 AQ = \tfrac{1}{8}(AK+3CM+3EO+GQ)\,AG$$

and $$\square_2 AQ = \tfrac{1}{8}(AK+3BL+3CM+DN)\,AD+\tfrac{1}{8}(DN+3EO+3FP+GQ)\,DG,$$

Newton has here chosen the errors $\epsilon_1 = \square_1 AQ-(AQ)$ and $\epsilon_2 = \square_2 AQ-(AQ)$ to be—by analogy with the preceding? (compare note (6) above)—in the ratio of $(6^2 =)$ 36 to 1. Since, however, on fitting the parabola $y = f_x \equiv a+bx+cx^2+dx^3+ex^4+fx^5+gx^6$ through the end-points of the ordinates $AK = f_{-3}$, $BL = f_{-2}$, $CM = f_{-1}$, $DN = f_0$, $EO = f_1$, $FP = f_2$ and $GQ = f_3$ there is $\square_1 AQ = 6a+18c+126e+1098g$ and $\square_2 AQ = 6a+18c+99e+693g$, so that, because $(AQ) = 6a+18c+\tfrac{486}{5}e+\tfrac{4374}{7}g$, there comes

$$\epsilon_1 = 16(\tfrac{9}{5}e+\tfrac{207}{7}g) \quad\text{and}\quad \epsilon_2 = \tfrac{9}{5}e+\tfrac{477}{7}g = \tfrac{1}{16}\epsilon_1+\tfrac{270}{7}g,$$

the 'true' factor is $(\epsilon_1/\epsilon_2 \approx)$ 16, yielding the refined approximation

$$\frac{49AK+168BL+147CM+112DN+147EO+168FP+49GQ}{840}\,AG = \square AQ,$$

larger than (AQ) by $\tfrac{288}{7}g$. But, as listed by Roger Cotes nearly twenty years afterwards in the *Postscriptum* to his 'De Methodo Differentiali Newtoniana' (see note (1) above), the aggregate of the seven ordinates AK, BL, ..., GQ which exactly measures the area (AQ) is

$$\tfrac{1}{840}(41(AK+GQ)+216(BL+FP)+27(CM+EO)+272DN)\,AG.$$

Some preliminary jottings by Newton on Add. 3965.14: 611r (here printed, with suitable editorial interlardings, in section [1] of the Appendix) can (see *ibid.*: note (5)) be manipulated to yield yet a further refined 'quadrature by seven ordinates' in this case.

(13) That is, $\tfrac{1}{63}(64\square_2 AS-\square_1 AS)$, where there is taken

$$\square_1 AS = \tfrac{1}{90}(7(AK+IS)+32(CM+GQ)+12EO)\,AI$$

and similarly

$$\square_2 AS = \tfrac{1}{90}(7(AK+EO)+32(BL+DN)+12CM)\,AE$$
$$+\tfrac{1}{90}(7(EO+IS)+32(FP+HR)+12GQ)\,EI,$$

corresponding to Newton's prior assumption—again by analogy? (compare note (6) above and that here preceding)—that the errors, ϵ_1 and ϵ_2 respectively, by which $\square_1 AS$ and $\square_2 AS$

Hæ sunt quadraturæ Parabolæ quæ per terminos Ordinatarum omniū transit.(14)

Investigantur etiam hæ quadraturæ in hunc modum. Sit *a* summa Ordinatæ primæ et ultimæ, *b* summa secundæ et penultimæ[,] *c* tertiæ & antepenultimæ[,] *d* summa 4^{tae} a principio et quartæ a fine[,] [*e*] summa quintæ a principio et quintæ a fine &c[,] *m* Ordinata media(15) et Abscissa *A* et sit

$$\frac{za+yb+xc+vd+te+sm}{2z+2y+2x+2v+2t+s}A = \square^{æ\,(16)} \text{ quæsitæ,}$$

et pro ordinatis primo scribantur termini totidem primi hujus seriei numerorū quadratorum 0.1.4, 9.16.25 &c[,] dein termini totidem primi hujus cubicorum 0.1.8.27.64.125 &c[,] dein totidem hujus quadrato-quadraticorū 0.1.16. 81.256. &c pergendo (si opus est) ad tot series una dempta quot sunt incognitæ quantitates *z*, *y*, *x*, *v* &c, et pro quadratura quæsita scribe quadraturam Parabolæ cui hæ ordinatæ congruunt. Et provenient æquationes ex quibus collatis determinabuntur *z*, *y*, *x* &c.(17)

Ut si Ordinatæ sint quatuor, pono $a = 0+9$ & $b = 1+4$. $A = 3$ et quadraturam $= 9$.(18) Sic fit $\frac{9z+5y}{2z+2y}\times 3 = 9$. et inde $y = 3z$.(19) Igitur pro *y* scribo $3z$ et æquatio $\frac{za+yb}{2z+2y}A = \square$ fit $\frac{a+3b}{8}A = \square$. ut in casu tertio.

---

surpass (AS) are in the ratio of $(8^2 =)$ 64 to 1. On setting, however, the parabola

$$y = f_x \equiv a+bx+cx^2+dx^3+ex^4+fx^5+gx^6+hx^7+ix^8$$

to pass through the end-points of the ordinates $AK = f_{-4}$, $BL = f_{-3}$, $CM = f_{-2}$, $CN = f_{-1}$, $EO = f_0$, $FP = f_1$, $GQ = f_2$, $HR = f_3$ and $IS = f_4$, it follows that

$$\square_1 AS = 8(a+\tfrac{16}{3}c+\tfrac{256}{5}e+\tfrac{2048}{3}g+\tfrac{155648}{15}i)$$

and likewise $\square_2 AS = 8(a+\frac{16}{3}c+\frac{256}{5}e+\frac{1760}{3}g+\frac{111968}{15}i)$, while the parabola's area

$$(AS) = \int_{-4}^{4} f_x.dx = 8(a+\tfrac{16}{3}c+\tfrac{256}{5}e+\tfrac{4026}{7}g+\tfrac{65536}{9}i),$$

so that $\epsilon_1 = 8(\frac{2048}{21}g+\frac{139264}{45}i)$ and $\epsilon_2 = 8(\frac{32}{21}g+\frac{8324}{45}i) = \frac{1}{64}\epsilon_1+\frac{49184}{45}i$; it may be verified that Newton's refined approximation is in consequence a little larger than (AS), transcending it by a little less than $2428i$. As computed by Cotes (see note (1)) the exact 9-ordinate aggregate for (AS) is, we may add,

$$\tfrac{1}{28350}(989(AK+IS)+5888(BL+HR)-928(CM+GQ)+10496(DN+FP)-4540EO)\ AI.$$

(14) Or such is Newton's ideal. This notwithstanding, the intention is fulfilled in neither 'Cas. 5' nor 'Cas. 6' above; see the two previous notes.

(15) When, that is, the equidistant ordinates are odd in number; there can, of course, be no middle one where their number is even.

(16) Read 'quadraturæ' (quadrature)—in other words 'square'.

(17) Going by the pattern of his successive approximations in Cases 1–6 preceding, Newton here supposes—on taking, for simplicity, the ordinates to be spaced at unit intervals—that,

These are quadratures of the parabola which passes through the end-points of all the ordinates.[14]

These quadratures may also be ascertained in this manner. Let a be the sum of the first and last ordinates, b the sum of the second and next to last ones, c that of the third and next but one to last, d the sum of the fourth from the beginning and the fourth from the end, e the sum of the fifth from the beginning and the fifth from the end, ..., m the middle ordinate[15] and A the abscissa, and let there be $\dfrac{za+yb+xc+vd+te+sm}{2z+2y+2x+2v+2t+s}A = \square^{\text{re}}$ required. Then in place of the ordinates write first an equal number of the opening terms of this series of square numbers $0, 1, 4, 9, 16, 25, \ldots$; next, as many first terms of this one of cubes $0, 1, 8, 27, 64, 125, \ldots$; thereafter the same number of this one of fourth powers $0, 1, 16, 81, 256, \ldots$, so proceeding (if need be) to as many series, less one, as there are unknown quantities $z, y, x, v, \ldots$, and in place of the quadrature required write the quadrature of the parabola to which these ordinates accord. There will ensue equations from which, when they are collated, $z, y, x \ldots$ will be determined.[17]

For instance, if there be four ordinates, I set $a = 0+9$, $b = 1+4$, $A = 3$ and the quadrature $= 9$.[18] In this way there comes to be $\dfrac{9z+5y}{2z+2y}\times 3 = 9$ and thence $y = 3z$.[19] Therefore in place of y I write $3z$ and the equation $\dfrac{za+yb}{2z+2y}A = \square$ becomes $\frac{1}{8}(a+3b)\,A = \square$, as in the third case.

---

where $n+1$ ordinates $f_0, f_1, f_2, \ldots, f_{n-1}, f_n$ stand perpendicularly to the base $A = n$ of the curve $y = f_x \equiv \sum_{0\leqslant i\leqslant n} a_i x^i$, the area $\square = \int_0^h f_x . dx$ contained by the latter can be represented by $k^{-1}(za+yb+xc+\ldots+te+sm)\,n$, in which $a = f_0+f_n$, $b = f_1+f_{n-1}$, $c = f_2+f_{n-2}, \ldots$ and (when n is even) $e = f_{\frac{1}{2}n-1}+f_{\frac{1}{2}n+1}$, $m = f_{\frac{1}{2}n}$ or (when n is odd) $e = f_{\frac{1}{2}(n-1)}+f_{\frac{1}{2}(n+1)}$, $m = 0$. Since this general formula must hold both in the simplest instance in which $f_x \equiv 1$ for all x, yielding $a = b = c = \ldots = e = 2$, $m = 1$, $\square = n$ and thence $k = 2z+2y+2x+\ldots+2t+s$, and also in each case $f_x \equiv x^i$, $i = 1, 2, 3, \ldots, n$, for which $a = 0+n^i$, $b = 1+(n-1)^i$, $c = 2^i+(n-2)^i, \ldots$ and correspondingly $\square = \int_0^n x^i . dx = \dfrac{1}{i+1}n^{i+1}$, the *raison d'être* of this method of determining the values of the coefficients $z, y, x, \ldots t$ and s will be obvious. As Newton himself is in sequel quick to realize (compare note (19) following), the odd values of i yield exactly the same equations as the next lowest even ones, and we need therefore compute only the equations ensuing from the latter.

(18) That is, $\int_0^3 x^2 . dx = \frac{1}{3}.3^3$.

(19) Newton initially went on to add

'Rursus pono $a = 0+27$, $b = 1+8$. et $\square = \dfrac{27z+9y}{2z+2y}\times 3 = [\frac{1}{4}.3^4 \text{ or}]\ 20\frac{1}{4}$

Rursus si ordinatæ sint quinq; pono $a = 0+16$, $b = 1+9$, $c = 4$. $A = 4$, $\square = \frac{64}{3}$.(20) Et sic fit $\frac{16z+10y+4s}{2z+2y+s} \times 4 = \frac{64}{3}$ seu $8z = y+2s$. Rursus pono

$$a = 0+64,\ b = 1+27,\ c = 8,\ A = 4 \text{ et } \square = 64.^{(21)}$$

et sic fit $\frac{64z+28y+8s}{2z+2y+s} \times 4 = 64$. seu $8z = y+2s$ ut supra. Pono igitur(22)

[$a = 0+256$, $b = 1+81$, $c = 16$, $A = 4$ et $\square = \frac{1024}{5}$.(23) Et sic fit

$$\frac{256z+82y+16s}{2z+2y+s} \times 4 = \frac{1024}{5}$$

seu $384z-88s = 51y = 408z-102s$, hoc est $14s = 24z$ sive $7s = 12z$ adeoq; $7y = 32z$. Igitur pro s et y scribo $\frac{12}{7}z$ et $\frac{32}{7}z$ respective et æquatio

$$\frac{za+yb+sm}{2z+2y+s} A = \square \quad \text{fit} \quad \frac{7a+32b+12m}{90} = \square$$

ut in casu quarto.](24)

---

seu $3z+y = \frac{3z+3y}{2}$ [adeoq;] $3z = y$' (Again, I put

$$a = 0+27,\ b = 1+8 \text{ and } \square = \frac{27z+9y}{2z+2y} \times 3 = \frac{81}{4},$$

that is $3z+y = \frac{1}{2}(3z+3y)$ and hence $3z = y$) before suppressing this second derivation of the preceding equation as superfluous.

(20) Namely $\int_0^4 x^2.dx = \frac{1}{3}.4^3$.

(21) That is, $\int_0^4 x^3.dx = \frac{1}{4}.4^4$.

(22) The manuscript text, as it survives, here finishes at the very end of the page (f. 21r). It is a natural surmise—one which we make in note (2) above—that Newton entered its continuation on a further sheet which is now lost; however, an identical draft of this present paragraph on Add. 3965.14: 611r (was it written earlier or later than it?) terminates at this very same point and it might therefore be the case that Newton, though he intended to complete this worked example verifying the approximation in 'Cas. 4' above, never in fact wrote any more. Whatever be the historical truth of the matter, we here in sequel restore its conclusion in the mould of what has gone before.

(23) That is, $\int_0^4 x^4 = \frac{1}{5}.4^5$.

(24) If Newton had gone on similarly to check the accuracy of the area-approximation in his preceding 'Cas. 5', he would (see note (12) above) have had a shock! Or maybe he did so and broke off drafting his essay, nonplussed that the formula

$$\tfrac{1}{840}(41a+216b+27c+272m)\ A = \square$$

thence (correctly) ensuing did not at all agree with his earlier expression? We shall probably never know.

Again, if there be five ordinates, I set $a = 0+16$, $b = 1+9$, $c = 4$, $A = 4$ and $\square = \frac{64}{3}$,[20] and thus there comes to be $\frac{16z+10y+4s}{2z+2y+s} \times 4 = \frac{64}{3}$, that is, $8z = y+2s$. Again, I set $a = 0+64$, $b = 1+27$, $c = 8$, $A = 4$ and $\square = 64$,[21] and in this way there comes $\frac{64z+28y+8s}{2z+2y+s} \times 4 = 64$, or $8z = y+2s$ as above. I therefore put[22] [$a = 0+256$, $b = 1+81$, $c = 16$, $A = 4$ and the quadrature $= \frac{1024}{5}$.[23] And in this way there comes to be $\frac{256z+82y+16s}{2z+2y+s} \times 4 = \frac{1024}{5}$ or $384z - 88s = 51y = 408z - 102s$, that is, $14s = 24z$ or $7s = 12z$ and hence $7y = 32z$. Accordingly, in place of s and y I write $\frac{12}{7}z$ and $\frac{32}{7}z$ respectively, and the equation $\frac{za+yb+sm}{2z+2y+s}A = \square$ becomes $\frac{1}{90}(7a+32b+12m) = \square$, as in the fourth case.][24]

Let us remark, in finale, that Newton did make some limited practical use of the simpler of these approximations in his contemporary scientific computations. Most notably, in a tentative appendix at about this time to a problem in his yet unpublished researches (in extension of Propositions XXV–XXXV of Book 3 of his 1687 *Principia*) into the moon's motion—'Prop. XXVIII' on ULC. Add. 3966.1: 5^r, namely, where he set himself 'Invenire variationem annuam motus medij Lunaris'—he successively embraced the present Case 1, (the unrefined form of) Case 2 and Case 3 in noting: 'Sin magnitudo illa accuratius determinanda sit, id sic [cons]tabitur. Sunto A, B, C, D, &c areæ horariæ in Octantibus quibusvis quatuor vel pluribus continuis, et area mediocr[is] erit $\frac{A+B}{2}$ vel accuratius $\frac{A+2B+C}{4}$ vel adhuc accuratius $\frac{A+[3B]+3C+D}{8}$ &c.' Elsewhere he was most times content—and indeed it was for him almost always enough—to employ merely the trapezoidal rule of Case 1.

APPENDIX. PRELIMINARY CASTINGS FOR THE 'OF QUADRATURE BY ORDINATES'.[1]

From a stray worksheet[2] in the University Library, Cambridge

[1][3] [Si curvæ IQ dentur septem ordinatæ $AI=$] a. [$BK=$] b. [$CL=$] c. [$DM=$] d. [$EN=$] e. [$FO=$] f. [$GP=$] g. [ab invicem distantes æqualibus intervallis

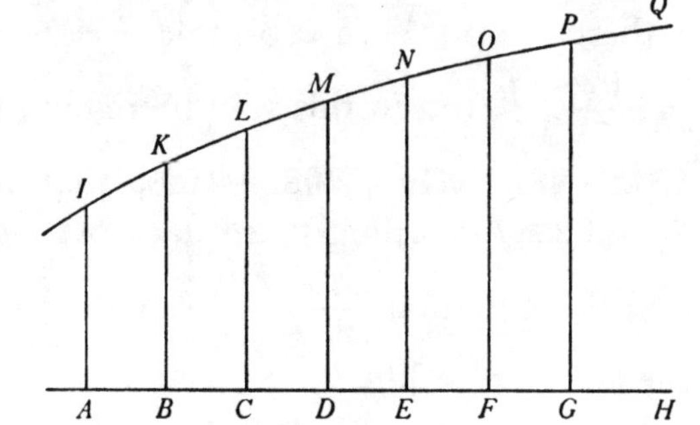

$AB=BC=CD=DE=EF=FG=1$,

quia proxime[4]

$$\square AL=\frac{a+4b+c}{6}\times 2.$$

necnon $\square CN=\dfrac{c+4d+e}{6}\times 2.$ et $\square EP=\dfrac{e+4f+g}{6}\times 2.$ erit addendo

$$\square AP=]\ \frac{a+4b+2c+4d+2e+4f+g}{18}\times 6.$$

[Simili modo, quia proxime[4] $\square AK=\dfrac{a+b}{2}.\ \square BL=\dfrac{b+c}{2}.$ &c. erit

$$\square AP=]\ \frac{a+2b+2c+2d+2e+2f+g}{12}\times 6.$$

[Et quia proxime[4] $\square AM=\dfrac{a+3b+3c+d}{8}\times 3.\ \square DP=\dfrac{d+3e+3f+g}{8}\times 3.$

(1) Some initial jottings on approximate area-quadrature in terms of seven equidistant ordinates—evidently in prelude to the more sophisticated formula enunciated in Case 5 of the mature essay 'Of Quadrature' which precedes—together with an analogously derived (but inadequately accurate) first version of its Case 4.

(2) Add. 3965.14: 611–12. We have earlier remarked (see §2.4: note (20) above) that Newton subsequently penned on this scrap-sheet several slightly different records of the improved elements of the 1680–1 comet which Halley stimulated him to compute early in October 1695, unintentionally thus providing later ages with a firm *terminus post quem non* for his composition of these preparatories to the preceding main tract.

(3) These sparse entries on f. 611ʳ are here, as ever, reproduced with editorial intercalations to bring out their sense. The accompanying figure is restored in accord with that which Newton afterwards initially set to illustrate his main essay (and on which see §3: note (3) preceding).

(4) The following triplicated, sextuplicated and duplicated approximations to the pertinent segments under the arc $\widehat{IP}$ are those obtained respectively in Cases 2, 1 and 3 of the preceding 'Of Quadrature'. Newton at once discards the first two to concentrate on the third.

erit $$\square AP =]\,\frac{a+3b+3c+2d+3e+3f+g}{16}\times 6.$$

[atq; etiam $\square AP =]\,\dfrac{a+3c+3e+g}{8}\times 6.$ [Differentia existente]

$$\frac{-a+3b-3c+[2]\,d-3e+3f-g}{16}\times 6.^{(5)}$$

[Aliter est $\square AP =]\,\dfrac{a+g}{2}\times 6.$ [tum] $\dfrac{a+2d+g}{4}\times 6.$ [tum] $\dfrac{a+2c+2e+g}{6}\times 6.$

[ut et] $\dfrac{a+2b+2c+2d+2e+2f+g}{12}\times 6$ [vero successive propius].(6)

---

[2](7) *Cas. 4.* Si dentur quinq; Ordinatæ dic $\dfrac{AI+EN}{2}AE = \square AN.$ Item

$$\frac{AI+BK}{8}+\frac{BK+CL}{8}+\frac{CL+DM}{8}+\frac{DM+EN}{8}\text{ in }AE = \square AN$$

---

(5) The difference, that is, between the two immediately previous approximations

$$\square_2 AP = \tfrac{1}{8}.[3(a+3b+3c+d)+3(d+3e+3f+g)] \quad\text{and}\quad \square_1 AP = \tfrac{1}{8}.6(a+3c+3e+g)$$

to the required area $(AGPI)$. If we take the curve $\widehat{IP}$, with respect to D as origin and DH the x-axis, to be approximated by the sextic parabola

$$y = F_x \equiv A+Bx+Cx^2+Dx^3+Ex^4+Fx^5+Gx^6$$

drawn through the end-points of the ordinates $AI = a = F_{-3}$, $BK = b = F_{-2}$, $CL = c = F_{-1}$, $DM = d = F_0$, $EN = e = F_1$, $FO = f = F_2$ and $GP = g = F_3$; it follows that

$$(AP) = \int_{-3}^{3} F_x.dx = 6(A+3C+\tfrac{81}{5}E+\tfrac{729}{7}G)$$

is approached by

$$\square_1 AP = 6(A+3C+21E+183G) \quad\text{and}\quad \square_2 AP = 6(A+3C+\tfrac{33}{2}E+\tfrac{231}{2}G)$$

to within the respective errors

$$\epsilon_1 = \square_1 AP-(AP) = 6(\tfrac{24}{5}E+\tfrac{552}{7}G)$$

$$\text{and } \epsilon_2 = \square_2 AP-(AP) = 6(\tfrac{3}{10}E+\tfrac{159}{14}G) = \tfrac{1}{16}\epsilon_2+\tfrac{45}{7}G.$$

Whence on taking $\epsilon_2 \approx 16\epsilon_1$ there ensues the near (if not quite best possible) approximation $\frac{1}{15}(16\square_2 AP-\square_1 AP)$ to the area $(AGPI)$.

(6) Newton—rather pointlessly after what has gone before—successively adduces the increasingly refined approximations to the area $(AGPI)$ which ensue by applying the trapezoidal rule to its halves $(ADMI)+(DGPM)$, thirds $(ACLI)+(CENL)+(EGPN)$, and sixths $(AK)+(BL)+(CM)+(DN)+(EO)+(FP)$.

(7) f. 612v. In this first version of Case 4 of the preceding 'Of Quadrature', evidently rejected from the final text because of its relative inefficiency (on which see the next note), Newton elaborates his mode of refining the errors in successive approximations to an area in an instance where but five equally distant ordinates are given.

et errores esse ut AE^q ad AB^q seu 16 ad 1. Et cum errorum differentia sit $\frac{-3AI+2BK+2CL-3AN}{8}AE$, error minor erit $\frac{-3AI+2BK+2CL-3AN}{120}AE$ quem aufer, et manebit

$$\frac{12AI+32BK+32CL+32DM+12EN}{120}AE = \square AN.$$

seu

$$\frac{3AI+8BK+8CL+8DM+3EN}{30}AE = \square AN.^{(8)}$$

---

(8) In refining the trapezoidal approximations

$$\square_1 AN = \tfrac{1}{2}(AI+EN)\ AE$$

and

$$\square_2 AN = \tfrac{1}{2}(AI+BK)\ AB+\tfrac{1}{2}(BK+CL)\ BC+\tfrac{1}{2}(CL+DM)\ CD+\tfrac{1}{2}(DM+EN)\ DE$$

to the area $(AENI) \equiv (AN)$, Newton's supposition that the errors $\epsilon_1 = \square_1 AN-(AN)$ and $\epsilon_2 = \square_2 AN-(AN)$ are in the ratio $(AE/AB)^2 = (4^2$ or) 16 to 1 implies that the curve $\widehat{IN}$ is accurately fitted by a parabola of only third degree, whereas its improved replacement set as 'Cas. 4' in the finished essay 'Of Quadrature' (see §3: note (11) above) presupposes an exact adaptation of the curve to a quartic one. For if, on again setting $AB = BC = CD = DE = 1$ for simplicity and taking C as the origin of Cartesian coordinates, the parabola

$$y = F_x \equiv A+Bx+Cx^2+Dx^3+Ex^4$$

be drawn to pass through the end-points of the ordinates $AI = F_{-2}$, $BK = F_{-1}$, $CL = F_0$, $DM = F_1$ and $EN = F_2$, the respective errors in the approaches

$$\square_1 AN = 4A+16C+64E \quad\text{and}\quad \square_2 AN = 4A+6C+18E$$

from $(AN) = \int_{-2}^{2} F_x.dx = 4A+\frac{16}{3}C+\frac{64}{5}E$ prove to be

$$\epsilon_1 = \tfrac{32}{3}C+\tfrac{256}{5}E \text{ and } \epsilon_2 = \tfrac{2}{3}C+\tfrac{26}{5}E = \tfrac{1}{16}\epsilon_1+2E;$$

whence the approximation $\frac{1}{15}(16\square_2 AN-\square_1 AN)$ which Newton here derives from positing that $\epsilon_2 = \frac{1}{16}\epsilon_1$, exactly, is correspondingly too large by $(\frac{16}{15}\times 2E$ or) $\frac{32}{15}E$.

INDEX OF NAMES